16th European Microelectronics and Packaging Conference and Exhibition 2007

(EMPC 2007)

AA002403

Oulu, Finland
17-20 June 2007

ISBN: 978-1-62276-466-2

Printed from e-media with permission by:

Curran Associates, Inc.
57 Morehouse Lane
Red Hook, NY 12571

Some format issues inherent in the e-media version may also appear in this print version.

Copyright© (2007) by International Microelectronics and Packaging Society, Europe
All rights reserved.

Printed by Curran Associates, Inc. (2012)

For permission requests, please contact International Microelectronics and Packaging Society, Nordic
at the address below.

International Microelectronics and Packaging Society, Nordic
PO Box 277
SE-431 24 Molndal, Sweden

info@imapsnordic.org

Additional copies of this publication are available from:

Curran Associates, Inc.
57 Morehouse Lane
Red Hook, NY 12571 USA
Phone: 845-758-0400
Fax: 845-758-2634
Email: curran@proceedings.com
Web: www.proceedings.com

TABLE OF CONTENTS

001 - INVITED PAPERS

Printed Intelligence – Opportunity for Innovations and New Business 1
    Harri Kopola, Jani-Mikael Kuusisto, Tomi Erho, Jukka Hast, Antti Kemppainen, Terho Kololuoma, Markku Kansakoski, Ari Alastalo, Eero Hurme, Pia Qvintus-Leino, Caj Sodergard, Arto Maaninen

Advanced Packaging, Friend or Foe 3
    Ken Ball, Reiner Schuetz, Wolfgang Mackrodt, Georg Meyer-Berg, Klaus Pressel, Alun Jones, Jean Roggen

Technical Trends on Flat-Panel-Displays (FPDs) in Japan 7
    Fumio Miyashiro

002 - SIP & EMBEDDED COMPONENTS 1

Design Tools for System-in-Package Applications 13
    Anna Fontanelli

Embedded Passives in Multi-dielectric Layer Printed Wiring Board for IEEE 802.11a/b/g Tri-Mode Dual-band Wireless Card Bus Adapter 19
    Cheng-Hua Tsai, Chang-Sheng Chen, Chang-Lin Wei, Kuo-Chiang Chin, Wei-Ting Chen, Chin-Sun Shyu

Embedding of Chips in Flex: A Global Optimization from Thermal, Mechanical and Electrical RF Perspective 25
    B. Vandevelde, S. Brebels, C. Okoro, H. Oprins, L.C. Chen, W. Christiaens, J. Vanfleteren

Wafer Level Packaging/Wirefree Die-On-Die Applications to FLASH Memories and Smart Cards 31
    Christian Val, Pascal Couderc, Christophe Serrano

003 - LTCC 1

LTCC Materials Selection for the Realization of 10GHz Interdigital Filters 35
    B. Pierce, P. Barnwell, M. Ehlert, T. Vincent

Designing SMD Lowpass Filters in Multilayer LTCC Technology 40
    R. Kulke, G. Mollenbeck, P. Uhlig, K. Maulwurf, M. Rittweger

High Frequency Characterization of VIAs up to 100 GHz in LTCC and Thick Film Structures 45
    M. Henry, Benito Sanz Iszquierdo, Charles Free, John Batchelor, Paul Young

005 - POSTER 1

3D Approaches of Optical Waveguide Fabrication on Flexible Substrate and its Application 50
    Tze Yang Hin, Changqing Liu, Paul P. Conway

Modeling the Effect of Assembly Parameters on Warpage and Stresses of Molded Package for Inkjet Printing 56
    Kimmo Kaija, Jani Miettinen, Matti Mantysalo, Pauliina Mansikkamaki, Mikiharu Kuchiki, Mikihiko Tsubouchi, Risto Ronkka

Printed Antenna Designs 62
    Roland Reitbauer, Manuela Midl, Wolfgang Stocksreiter

Inter-plane Coupling Structures for PCB-integrated Multilayer Optical Interconnections 68
    N. Hendrickx, J. Van Erps, H. Thienpont, P. Van Daele

Prototyping of Pluggable Out-of-plane Coupling Components for Multilayer Board-level Optical Interconnections 73
    J. Van Erps, N. Hendrickx, Christof Debaes, P. Van Daele, H. Thienpont

Opto-mechanical Tolerance Modelling of a Free-space Intra-MCM Optical Interconnect System 79
    Michael Vervaeke, Christof Debaes, H. Thienpont

Comparison of Different Silicon Solar Cell Structures 85
    Ondrej Hegr, Jaroslav Bousek, Ales Poruba, R. Barinka, Jaroslav Sobota

A Microcontroller-Based Neurostimulator in Thick-film Technology 90
    Krzysztof Zaraska, Barbara Groger

Wire Bonding Power Interconnection 95
    M. Novotny, T. Dvorak, I. Szendiuch

Thermal Capabilities of Integrated Resistors in Organic Substrates 98
    Ciprian Ionescu, Norocel Dragos Codreanu, Virgil Golumbeanu, P. Svasta

Impact of Package Ground Ball and Plated Through Hole (PTH) Count Reduction to Differential Interfaces Signal Integrity.................103
Jiun Kai Beh

006 - ADVANCED PACKAGING

UBM Structures for Lead Free Solder Bumping using C4NP.................107
Klaus Ruhmer, Eric Laine, Karin Hauck, D. Manessis, A. Ostmann, Michael Töpper
ECD Wafer Bumping and Packaging for Pixel Detector Applications.................113
T. Fritzsch, Rafael Jordan, O. Ehrmann, H. Reichl
Interface Reactions during Au-Ball/Wedge and AlSi1-Wedge/Wedge Bonding at Room Temperature.................119
M. Schneider-Ramelow, K.-D. Lang, U. Geißler, W. Scheel, H. Reichl
Surface Acoustic Wave Component Packaging.................125
G. Feiertag, H. Kruger, C. Bauer

007 - OPTOELECTRONICS 1

Low Cost Hybrid Integration of Laser Diode on Silicon PLC for Optoelectronic Applications.................130
Luca Maggi, Arturo Canali, Danilo Caccioli, Gianni Preve, Stefano Lorenzotti
Packaging of Miniaturized Optical Encoder.................136
Y. Jourlin, O. Parriaux, K. Keranen, J.T. Makinen, M. Karppinen, P. Karioja, M. Johnson
High Resolution Optical Component Technology for Advanced Novel Encoder (Hi-OCTANE).................140
John Carr, Marc P.Y. Desmulliez, Eitan Abraham, Nick Weston, David McKendrick, Matt Kidd, Geoff McFarland, Wyn Meredith, Andrew McKee, Conrad Langton
Multimode Optical Interconnections Embedded in Flexible Electronics.................146
E. Bosman, G. Van Steenberge, N. Hendrickx, P. Geerinck, W. Christiaens, J. Vanfleteren, P. Van Daele

008 - NAMIS NETWORK-NANO & MICROSYSTEMS 2

Integration of Fluidic and Detection Functions in Sensing Systems.................N/A
C. Khan Malek, S. Ballandras
Low Temperature Direct Bonding for Wafer-scale Packaging of MEMS.................N/A
H. Kattelus, T. Suni, K. Henttinen
Reliability Issues of Optical MEMS Related to Device Packaging.................N/A
H. Toshiyoshi, K. Isamoto, A. Morosawa, C. Chong, H. Fujita

009 - POSTER 2

Smart Packaging of Microlenses over a UV-LED Array.................153
Markus Luetzelschwab, Dominik Weiland, Marc P.Y. Desmulliez
LTCC Gas Flow detector.................159
Dominik Jurkow, Leszek J. Golonka, Henryk Roguszczak
Film Assisted Molding Technologies and Applications for High Volume Mems and Sensor Packages.................163
Frank Boschman, Ton van Weelden, Arnold Bos
Transmission Line Pulsing Behavior of Thin Film Resistors.................165
D. Bonfert, H. Wolf, H. Gieser, G. Klink, P. Svasta
The Application and Characteristics of the Wire-penetration Films in the Use of Stack-die CSP (Chip Scale Package).................171
C.L. Chung, S.L. Fu
Inertial Bridge with Electronic Compass for Auto Pilots.................175
P. Svasta, Iaroslav-Andrei Hapenciuc
High Performance Low-firing Temperature Thick-film Pressure Sensors on Steel.................179
C. Jacq, T. Maeder, N. Johner, G. Corradini, P. Ryser
Nonferroelectric Lead-Free Ceramics and Thick Films with High Dielectric Permittivity - Synthesis, Sintering and Properties.................183
J. Kulawik, D. Szwagierczak
Development of Micro-Integrated Sensors and Actuators Based on PZT Thick Films.................189
S. Gebhardt, Thomas Rodig, U. Partsch, Andreas J. Schonecker
High Volume MEMS Packaging.................194
Mark Shaw, Federico Ziglioli, Anne Marie Grech, Mario Cortese
Screen-printed Ruthenium Dioxide pH-electrodes.................200
Kathrin Reinhardt, Christel Kretzschmar, Claudia Feller

RF – Membrane Filter Production and Packaging Challenges 205
M. Chatras, N. Onda, P. Nigg, Wolfgang Tschanun

New Type of Micro Sensor with Active Nanoparticle Surface 207
Radim Hrdy, Jaromir Hubalek, Katerina Klosova

Creation of Nanostructured Metal Surface by Template-Based Electrodeposition Method and Its Employment in Sensor Technology 211
Katerina Klosova, Jaromir Hubalek

010 - MANUFACTURING TECHNOLOGIES 1

Capillary Flow Kinetics in the Presence of Uneven Gaps 216
Horatio Quinones

Additional Stresses of ECA Joints due to Moisture Induced Swelling 221
Richard C. Low, Ralf Miessner, Jurgen Wilde

Novel Packaging Technology for Combo Memory Package 227
Ville Pekkanen, Matti Mantysalo, Jani Miettinen, Pauliina Mansikkamaki

Cost Efficient Quality and Production Strategies for Electronics Production 233
Martin Oppermann, Wilfried Sauer, Klaus-Jurgen Wolter, Thomas Zerna

011 - OPTOELECTRONICS 2

Towards Low Cost Coupling Structures for Short-Distance Optical Interconnections 239
N. Hendrickx, J. Van Erps, T. Alajoki, N. Destouches, D. Blanc, J. Franc, P. Karioja, H. Thienpont, P. Van Daele

Surface Mounted Coupling Elements for PCB Embedded Optical Interconnects 245
Thomas Kuhner, Marc Schneider

Optical Coupling and Optoelectronics Integration Studied on Demonstrator for Optical Interconnects on Board 250
T. Alajoki, N. Hendrickx, J. Van Erps, Samuel Obi, M. Karppinen, H. Thienpont, P. Van Daele, P. Karioja

Study of Thermal Behavior in a Multi-Chip-Composed Optoelectronic Package 254
J. Tian, S. Sinaga, M. Bartek

012 – GBC

The Market Situation of Ceramic Micro Circuits 260
Erwin Effenberger

Electronics Manufacturing in Europe – Competing in a Global Market 264
Indro Mukerjee

NXP System-in-Package Vision and Latest 3D Technology Developments 265
J.-M. Yannou, F. Neuilly, J.-O. Moreno, M. Pommier, S. Bellenger, P. Biermans

013 – MATERIALS

Conductive Adhesives for Blue LEDs 270
Noritsuka Mizumura, Sen-ichi Ikarashi, Michinori Komagata, Yukio Shirai

Short and Long Term Reliability of In-mould Sealed Bare and Globtop Shielded Led Devices 275
K. Keranen, M. Silvennoinen, A. Lehto, J. Ollila, T. Salmi, J.T. Makinen, A. Ojapalo, M. Schorpp, P. Hoskio, P. Karioja

Embedding a Thin Polymer Voltage Esd Suppressing Core in a Chip Package - Alternative to on Chip ESD Protection 280
Karen Shrier, Paul Collander

Tunable Dielectric Material Embedded in LTCC for GHz-Frequency-Range Applications 285
S. Rentsch, Tao Hu, J. Muller, R. Stephan, M. Hein, H. Jantunen

014 - SENSORS & MEMS PACKAING

Vacuum Package design for a MEMS based IR Detector Array 289
J. Ollila, M.F. Toy, O. Ferhanoglu, P. Karioja, H. Urey

Capacitive Micro Force Sensors Manufactured with Mineral Sacrificial Layers 293
Y. Fournier, S. Wiedmer, T. Maeder, P. Ryser

Cavity Formation in a Silicon Cap Wafer Using Aluminum Etch-Stop Layer 299
S. Sosin, J. Tian, M. Bartek

Structural and Electrical Investigation of PZT Films on Different Substrates 304
Darko Belavic, Marko Hrovat, Hana Ursic, Silvo Drnovsek, Mitja Jerlah, Jena Cilensek, Janez Holc, Marina Santo Zarnik, Marija Kosec

015 - EUROPEAN PROJECTS ON EMBEDDING ACTIVE AND PASSIVES

Reliability Aspects of Embedded Chips 309
A. Ostmann, D. Manessis, M. Cauwe, Johann-Peter Sommer
Manufacturing of Demonstrator Printed Circuit Boards with Embedded Active Components 315
A. Kriechbaum
Embedded Actives Technology, from Functional Densification to Fan-out Redistribution 320
A. van der Lugt, W. Peels
Lamination Process Studies for Realisation of Chip Embedding Technologies-Current Applications and Technical Challenges 324
D. Manessis, A. Ostmann, S-F. Yen, R. Aschenbrenner, H. Reichl
High-frequency Modeling and Measurements of Tracks Running on Top of Active Components Embedded in Printed Circuit Boards 330
M. Cauwe, J. De Baets
Cost Modelling for Embedded Component Technology 336
J. De Baets, G. Willems, A. Ostmann, A. Kriechbaum, H. Kostner

016 - SIP & EMBEDDED COMPONENTS 2

Double and Triple Stacked Solder Joint Technology for Further Miniaturization of 3D-SIP 340
Serguei Stoukatch, Christophe Winters, Tom Torfs, Walter De Raedt, Eric Beyne, Chris Van Hoof
Challenges Of 3D/ Stack Die Integration for Thin Large Die 346
Gaurav Mehta, Tan H. Hong, Wilson Ong, Y.C. Koh, John D. Beleran, Ravi Kolan
Through Wafer Interconnect Technologies for 3D System-in-Package 352
E. van Grunsven, F. Roozeboom, F. Sanders, F. van den Heuvel, M. Burghoorn, T. Grob
Thermal Performance of Embedded IC Structure 354
Tanja Karila

017 - LTCC 2

Application of Metallo-organic Pastes on LTCC Substrates 359
Jaroslaw Kita, Ralf Moos
Fabrication of High Performance RF-MEMS Structures on Surface Planarised LTCC Substrates 364
Massimiliano Dispenza, Roberta Buttiglione, Anna Maria Fiorello, Jarkko Tuominen, K. Kautio, J. Ollila, Pentti Korhonen, Manu Lahdes, Kari Ronka, Simone Catoni, Daniele Pochesci, Romolo Marcelli, Vittorio Foglietti, Elena Cianci, Andrea Coppa
LTCC Ultra High Isostatic Pressure Sensors 370
T. Maeder, B. Afra, Y. Fournier, N. Johner, P. Ryser
LTCC-Based Sensors for Mechanical Quantities 376
U. Partsch, S. Gebhardt, D. Arndt, H. Georgi, H. Neubert, D. Fleischer, M. Gruchow

018 - POSTER 3

Epoxy-based Polymer for Capacitive Chemical Sensors 384
Marijan Macek, Marta Klanjsek Gunde, Nina Hauptman
Leakage Current, Noise and Reliability of NbO and Ta Capacitors 389
V. Sedlakova, J. Sikula, H. Navarova, J. Pavelka, J. Hlavka, Z. Sita
Analysis Methods for Characterization of Unleaded Solderable Surfaces 395
Thomas Hetschel, Klaus-Jurgen Wolter
Influence of Environmental Conditions on Pb-free Solder Joints Quality 401
A. Skwarek, K. Witek
Forward Compatibility Assessment for Aeronautical and Military Communication Systems (GEAMCOS project) 406
O. Maire, A. Chaillot, C. Munier, I. Lombaert-Valot, S. Bousquet, C. Chastanet, D. Plouseau, E. Munier, D. Maron, P. Raynal, S. Villard, R. Dumonteil
Eco-design Workflow Process 412
Cyril Vasko, I. Szendiuch, Karsten Schischke
Observations on Particle Loaded Silver Inks 416
Ulrike Currle, Klaus Krueger

Preheating in Solderability Testing 422
F. Steiner, P. Harant

Reliability Qualification of Flexible Printed Circuits 426
Markus Detert, Thomas Zerna, Klaus-Jurgen Wolter

Studies of Selected Inspection and Failure Analysis Techniques for LTCC Micromodules 432
Kari Remes, Leena Palmu, Petri Ronkainen

Characterization of Failure Modes and Analysis of Joint Strength Using Various Conditions for High Speed Solder Ball Shear and Cold Ball Pull Tests 435
Fubin Song, S.W. Ricky Lee, Stephen Clark, Bob Sykes, Keith Newman

Comparison of Accelerated Life-time Test Methods of Pb-free Solder Joints 441
Zsolt Illyefalvi-Vitez, Pal Nemeth, Oliver Krammer, Janos Pinkola

On the Simulation of Flexible Circuit Boards 447
Luciano Arruda

Reliability of Flexible Circuits with Different Lead-Free Technologies 452
Balint Balogh, Peter Gordon, Zsolt Illyefalvi-Vitez, Graham Farmer, Anna Girulska, Tom Harvey, Gyorgy Kotora, Damien Kirkpatrick

019 - INTERCONNECTION TECHNOLOGIES

Prospects and Yield of Electrochemical Wafer Plating for Bumping and Signal Routing 458
L. Dietrich, M. Töpper, T. Fritzsch, O. Ehrmann, H. Reichl

Development of Chip to Antenna Interconnections for Contact-less Smart Card Applications 467
Jaakko Lenkkeri, Sari Kivela, Eveliina Juntunen, Tuomo Jaakola, Kaj Nummila, Mark Allen, Toni Kaskiala, Gerhard Hillmann, Alan Mathewson

Reliability of ACA Bonded Flip Chip Joints on LCP and FR-4 Substrates 473
Laura Frisk, Anne Cumini

Second-Level Interconnect – "Package to PCB" – Future Challenges and Solutions 479
Ashok N. Kabadi

020 – RF

Design and Technology Considerations on LTCC Microwave Modules for Fixed Radio Link Equipment 490
M. Piloni, A.G. Milani

LTCC Multilayer Technology Enables Very Compact 20 GHz Switch Unit for Space Applications 495
K.-H. Drue, M. Hein, J. Muller, R. Perrone, S. Rentsch, R. Stephan, J. Trabert

A New Integrated Waveguide Antenna using Multi-Layer Photoimageable Thick-Film Technology 500
M. Henry, Benito Sanz Iszquierdo, Charles Free, John Batchelor, Paul Young

Microstrip and Wave-guide SMT Package up to 60 GHz (MWgSP) 505
Carlo Buoli, Paolo Bonato, Luigi Negri, Fabio Morgia

021 - POSTER 4

Influences of the Layout on the Lifetime of Direct Copper Bonding Substrates (DCB) 511
Michael Günther, Klaus-Jurgen Wolter

Solvent Resistance of Silicones when Used in Electronic Chemical Cleaning Environment 517
Bill Riegler, Michelle Velderrain, Scott Duffer

Silicone Polymer Coating for Piezo Actuator Protection 523
Marko Pudas, Markus Polet, Jouko Vahakangas

Selected Perovskite Type LSFO Thin Films for the Infrared Detectors 526
Andrzej Lozinski, Pawe- Wierzba

Immersion Tin Wetting Behaviour with Lead-free Soldering 530
Mustafa Oezkoek, Nigel White

Advanced Thin Film Substrates in Cu-AlN Technology 536
E. Feurer, B .Holl, J. Vanselow, K. Ruess, A. Kaiser

Laser Soldering of LTCC Hermetic Packages with Minimal Thermal Impact 540
F. Seigneur, Y. Fournier, T. Maeder, J. Jacot

The Deposition of Thick Film Paste by Direct Writing 545
J. Hladik, J. Vanek, I. Szendiuch

Direct Write Technique Used for Solar Cell Fabrication 548
J.Hladik, R. Barinka, I. Szendiuch

Study of the Impact of High-voltage Trimming on Several Characteristics of Model TFRs and Their Stability............551
N. Johner, T. Maeder, C. Jacq, P. Ryser

Analysis of Fine-pitch BGA Placement Accuracy............556
Johannes Hurtig, Timo Liukkonen

Fine Line Technology And Panel Plating - Opposing Directions, One Solution............560
Stephen Kenny, Bert Reents

Electro-ultrasonic Spectroscopy of Polymer Based and Cermet Thick Film Resistors............564
V. Sedlakova

022 - MANUFACTURING TECHNOLOGIES 2

Experimental Study of Solder Joint Reliability on Injection Moulded Substrates............570
Minna Arra, Ilkka Harkonen, Esko J. Paakkonen

ASPACT® Additive Circuit Transfer Technology............576
Juha Hagberg, Teija Kekonen, Terho Kutilainen

Advanced Electronics Packaging via M³D Direct Writing............580
Martin Hedges, Bruce King, Mike Renn

NanoCT: Visualizing of internal 3D-Structures with Submicrometer Resolution............585
Andre Egbert

023 - LAMINATES & QUALITY ISSUE

Performance and Reliability of Flexible Substrates when subjected to Lead-Free Processing............589
M.J. Rizvi, C.Y. Yin, C. Bailey, H. Lu

Multilayer Flexible Wiring Board based on Screen Printed Conductors............595
J. Petaja, K. Kautio, H. Funck, M. Karppinen, P. Karioja, R. Vatanparast

Investigation on Printed Wiring Board Failures During Reliability Assessment for Telecommunication Products............601
Yujie Dong, Markku Tammenmaa, Visa Ruuhonen, Virpi Pennanen

024 - EU & INTERNATIONAL PROGRAMMES

Global Joint Effort to Solve Microelectronics Supply Chain Technology Issues............606
Paul Collander, Marshall Andrew, Ruben Bergman

Integration of Thin Flexible RF Structures into Flexible PCB............611
W. Christiaens, H. Burkard, J. Link, J. Vanfleteren

3D Chip Size Packaging for Highly Integrated Memory Cards for Consumer-products............617
Reiner Gotzen, Andrea Reinhardt, Helge Bohlmann

Realization of Large Area Stretchable Electronic Systems Using Lamination Processes............620
Thomas Loeher, D. Manessis, A. Ostmann, H. Reichl

3DµTune: High-Q Micromachined Cavities for Millimetre-wave Filters and Oscillators............625
J.B. Mills, B. Giesbers, M. Matters-Kammerer, I. Ocket, B. Nauwelaers, A. Jourdain, W. Gautier, B. Schonlinner

DAVID - A Strategic Research Project for Chip-Scale MEMS / ASIC Co-integration............630
N. Marenco, W. Reinert, S. Warnat

025 - SIP & EMBEDDED COMPONENTS 3

3D Package-on-Package Solution for Next Generation Cameras............633
Vern Solberg, Giles Humpston

Competitive Environment for 3D Semiconductor Assembly, Applications, Strategies & Cost............639
C. Bauer, H.J. Neuhaus

Effects of Underfill and Molding Compounds on Reliability of System in Package............644
Shan Gao, Jupyo Hong, Jinsu Kim, Seogmoon Choi, Sung Yi

Via Fabrication Techniques for High Density Vertical Interconnections............650
Zsolt Illyefalvi-Vitez, Laszlo Gal, Oliver Krammer, Janos Pinkola

026 - THERMAL MANAGEMENT

3D-Fluidic Cooling Structures in LTCC............656
M. Mach, J. Muller

High-brightness RGB LED Modules Based on Alumina Substrate 662
Veli Heikkinen, Eveliina Juntunen, K. Kautio, Antti Kemppainen, Pentti Korhonen, J. Ollila, Aila Sitomaniemi, Timo Kemppainen, Heikki Korkala, Terho Kutilainen, Hannu Sahavirta

Performance of Thin and Thick Film Resistors Exposed to High Temperature and High Pressure (200^{0}C @ 1000 Bar) 668
Rolf Johannessen, Froydis Oldervoll, Frode Strisland, Per Ohlckers

Novel Diamond Al and Diamond Cu Composites 673
Renaud de Langlade, Maxim Seleznev

027 - MEDICAL ELECTRONICS

Packaging Concepts for Neuroprosthetic Implants 677
M. Topper, M. Klein, M. Wilke, H. Oppermann, S. Kim, P. Tathireddy, F. Solzbacher, H. Reichl

Low Cost, Biocompatible Elastic and Conformable Electronic Circuits and Assemblies Using MID in Stretchable Polymer 681
F. Axisa, D. Brosteaux, E. De Leersnyder, F. Bossuyt, M. Gonzalez, M. Vanden Bulcke, N. DeSmet, J. Vanfleteren

Packaging of an Implantable Accelerometer for Measurements of Heart Motion 687
Kristin Imenes, Knut Aasmundtveit, Ellen M. Husa, Jan Olav Hogetveit, Steinar Halvorsen, Ole Jakob Elle, Erik Fosse, Lars Hoff

Development of a Reliable LTCC-BGA Module Platform for RF/Microwave Telecommunication Applications 693
Tero Kangasvieri, Olli Nousiainen, Jouko Vahakangas, K. Kautio, Markku Lahti

028 - NANO TECHNOLOGIES

2D and 3D X-ray Inspection for Nano-technology 699
Keith Bryant, David Bernard

Author Index

Printed Intelligence – Opportunity for Innovations and New Business

Harri Kopola, Jani-Mikael Kuusisto, Tomi Erho, Jukka Hast, Antti Kemppainen, Terho Kololuoma, Markku Känsäkoski, Ari Alastalo, Eero Hurme, Pia Qvintus-Leino, Caj Södergård and Arto Maaninen

VTT –Technical Research Centre of Finland, Center for Printed Intelligence

Kaitovayla 1, 90570 Oulu, Finland, email: Harri.Kopola@vtt.fi, tel. +358-20-7222369

Abstract

Emerging printed intelligence markets present fascinating opportunities for new multidisciplinary product innovations and businesses arising from different technology interfaces. Disposable sensors for home and point-of-care diagnostics, large area user interfaces, flexible displays and interactive packaging are some examples.

Key words: roll-to-roll, flexible, multidisciplinary, OLED, disposable

Introduction

The primary function of printing has been and continues to be the delivery of data and information for visual inspection and further interpretation by humans or machines. Printing-like manufacturing methods have awaken much interest as a potential cost-efficient mass-manufacturing method of electronics and functionalities on large areas. The introduction of new printable functional materials, print production processes and reading mechanisms is now paving the way for expanding the role and function of printing toward printed intelligence.

With printed intelligence, printed items will become more active, self-sensing, and self-controlling. Printed products are already becoming more integrated to the smart environments and information systems around us. They are helping us access and use digital content and services. With new intelligent features the consumption experiences with printed products will become increasingly engaging, entertaining, informative and easy to use. Printed intelligence will also bring added safety as they sense the environment around us, contents within packages, or just simply verify authenticity of the product. Printed items like disposable sensors will also promote health through point of care diagnostics and various other applications for health care.

Despite several years of research and developments printed intelligence as an industry is still in its infancy. However, the research and development activity in this area is ever increasing and market researchers are forecasting rapid market growth for various printed devices. New companies in this field are emerging to offer printed intelligence components and or solutions, and established companies in various fields are evaluating the impacts and possibilities of these new technologies on their businesses.

VTT Technical Research Centre of Finland has investigated and developed technologies for printed intelligence, electronics and optics and their applications since 1999. In August 2006, VTT established the "Center for Printed Intelligence" (CPI) as a strategic initiative to contribute to the more effective exploitation of research results for the generation of applications and profitable business.

Generic Enabling Technologies

Technologically printed intelligence is based on the application of innovative fluid processable materials onto various printable matters with printing like mass-manufacturing methods.

New advanced materials in liquid phase – 'printable inks' – are a corner stone in this business. The variety of materials and their numerous application possibilities introduce opportunities far beyond those offered by traditional 'silicon' based electronics for ubiquitous computing. Examples of these materials are conductive polymers, organic semiconductors, nanoparticulate materials and bioactive materials.

Another generic corner stone is the high-throughput, cost-efficient manufacturing methods like continuously running roll-to-roll printing, hot-embossing, coating, laser processing and their combinations. The third important generic capability is the integration of those multifunctional components, circuits and systems on web, sheet or foil. The capabilities of these processes and new advanced manufacturing equipment and automated production lines are in a key position when developing and commercializing products where new intelligent functionalities are embedded to high through put large area surfaces.

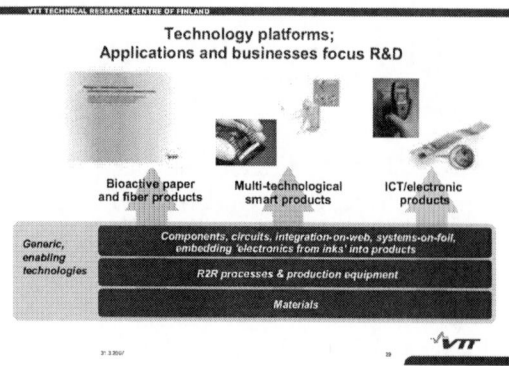

Figure 1: Technology platforms of the Center for Printed Intelligence.

Intelligent printed components and modules are created through the combination of materials and processes. On the top of these generic technology developments, VTT concentrates on three application and customer oriented technology platforms:

1. Multitechnological smart products - manufacturing of electronic, optic and optoelectronic components and circuits, simple displays, disposable sensors, indicators and their integration into applications like smart packages, printed products, home diagnostics,

2. Bioactive paper and fibre products - large area manufacturing of products, which provide added value based on biomolecules or other active materials, and

3. ICT/electronic products - integration and communication with ICT/electronics products and services, large area electronic systems on foil or flexible substrates.

From components to printed modules and systems

Electronic, bio, chemical, optic, optoelectronic etc. functionalities are created using variety of different devices. Printed resistors, capacitors, inductors, antennas, optical gratings, light guides, optical read-only memory are examples of passive electrical and optical components. OLEDs (organic light emitting devices), organic solar cells, organic transistors and diodes are examples of active electronic and optoelectronic components. Printed indicators and sensors for hydrogen sulphide and oxygen are examples of food packaging quality measures.

The ongoing R&D activities aim to more challenging goals like R2R manufactured OLED simple displays and signage, printed organic transistor circuits and solar cell modules, memory devices, miniaturized fuel cells and multilayer electronic circuits. For example VTT is coordinating the European Comission FP6 ROLLED-project 2004-2008 – Roll-to-roll

manufacturing of arbitrary size and shape OLEDs on web. Printable bioactive materials and disposable biosensors for diagnostic and healthcare applications, and electric components for applications like RFID are examples of modules for intelligent systems. Different code activated ICT/hybrid media systems and applications have been demonstrated and some are now in the commercialisation phase.

Applications and business arenas

The market for organic and printable electronics is expected to be a $35 billion industry by 2015 and reach over $300 billion in 2025, that is, almost twice the size of the silicon industry today (Frost & Sullivan).

This is the opportunity gap between traditional paper, packaging and printing industry products and ICT/electronics industry products and can be realised for example as disposable sensors, simple 'electronic' components and circuits, large area functional paper-like intelligent products, smart packages, tag and code technology based ICT and hybrid media services etc. According to technology roadmaps the first generation of roll-to-roll and inkjet printed flexible electronic components, disposable sensors and simple displays will be commercialized around 2008-2010. Passive 'functionalities' like diffractive optical effects for product packages and printed products and non-electronic disposable biosensors are in the early stages of making the commercial transition to large area applications beyond their traditional labelling and lamination applications. Active R2R printed electronic and optoelectronic components and simple circuits need 2-4 years to reach initial commercialisation phase. Autonomous printed electronic systems will need 5 + years to market entries.

Acknowledgements

All co-workers at VTT and our collaborating partners are greatly acknowledged for their contribution in these development efforts. Tekes, European Comission and industrial partners are acknowledged for their continued financial support.

Advanced Packaging, friend or foe

Ken Ball[1], Reiner Schuetz[2], Wolfgang Mackrodt[2], Georg Meyer-Berg[3], Klaus Pressel[3], Alun Jones[4], Jean Roggen[5],

[1]KBTeC and Technical Manager for ENCASIT, www.kbtec.co.uk, www.encasit.org
[2]Bosch
[3]Infineon
[4]Austin Semiconductors Europe
[5]IMEC

Tel: +44 1548 531414 email: ken.ball@kbtec.co.uk

Abstract

Advanced packaging has become an important feature of new electronic components which are being developed to serve the requirements of mobile phone and handheld electronics. The main driver for these developments is reduced space, more functionality and lower cost. The development of this advanced packaging has led to completely new ways of component assembly from the chip up. Whereas older technologies, such as wire bonding, are mature and well understood, new packaging techniques are focused less and less on mature processes and more and more on innovation. The constant drive for cost reduction means that anything which can be changed or removed, for the sake of cost, is considered. It is also said that all materials used in advanced packaging will change over the next 10 years. This paper will take up this theme and present a view of Technology Roadmaps for Advanced Packaging that have be gathered by the ENCASIT consortium. Following on from this, the paper will examine the effect of these developments on industries such as automotive, medical and aerospace, that have different application and product life-cycle requirements to the mobile phone and handheld electronics. In particular, the paper will highlight what must be done to avoid problems while taking advantage of new developments that will come through Advanced Packaging.

Key words: Advanced Packaging, System-in-Package, System Integration, ENCASIT

Advanced Packaging, friend or foe

In the past, the IC was designed separately from package – an IC could be assembled into a different package and would still function, albeit maybe with different AC performance. Military and aerospace applications drove the quest for higher performance in smaller, lighter package.

Today, high performance ICs require the package to be designed alongside the IC, preferably in a co-design environment, in order to get the required overall performance. The high bandwidth requirement of next generation processors require memory to be stacked above or below the processor and linked using a true 3D interconnect method. Arguably, some of the silicon now being designed could not be re-packaged using an alternative packaging method without a complete redesign of the chip itself. For example, IC designers can exploit the advantages of reduced interconnection inductance and capacitance offered by flip-chip attach to reduce the drive current of the I/O transistors and hence their size and over all size of the silicon. This has the added advantage of reducing cost and improving yield.

Mobile electronics, especially the mobile phone has taken over as the primary driver for more functionality in smaller and lighter package. But with the growth of consumer electronics driving advanced mobile technology came an even more important driver – that of cost reduction. Now, most of the advanced packaging development is driven by cost reduction, not improved technology. Yes, increased performance is required, but this must be delivered at the same or less cost. East Asia has now begun to take over development of consumer electronics items, and their number one consideration is cost.

This had led to a re-evaluation of what materials and processes are actually required for advanced packages. Anything that can be dispensed with or removed from a IC or package is considered for the sake of cost reduction so long as the basic functionality and quality requirements are met. However, these quality requirements are not based on Military specifications and standards, as in the past with MIL883 etc., but with ad-hoc on-the-fly rules to assess quality and reliability for a particular application. For example, the most important test for a mobile phone, or any hand-held electronics, is the drop test. The drop-test is carefully specified as a standard test, but everything about the electronics is designed to be able to pass this test, and at the Lowest Cost. In this case, the use of, for instance, underfill, would help CSP and uBGA packages pass a drop test. But underfill is a slow and expensive process, so there is much effort put into designing package solutions that will pass the drop test for the assembly, but without having to use underfill. Wouldn't it be better just to use undefill as a belt-and-braces approach and be done with? – no, that would result in excessive cost.

So, does it really matter? We all get what we want – more functionality at a cheaper price? For consumer items, this may be acceptable, but note woes of Apple with the reliability problems of the early iPods. 'Iprod, Ipoke, my iPod is broke' [1]. Here a consumer product lasted for the duration of the warranty – 12 months – but there were many failures just outside the warranty period. Is this acceptable quality?

Behind all the consumer products we love to own, is a whole raft of industrial electronics, including for instance, base-stations for mobile phones, that require a different level of intrinsic quality and reliability. It's no good having the smallest, lightest, cheapest, most functions, mobile phone, if you can't get a signal because the nearest base station has a fault. And what of all the other critical elements of infrastructure – the public utilities of electricity generation and distribution, water, gas etc.

Unfortunately, there is little we can do to change the way the electronics industry is currently set on cutting cost while maintaining the ever insatiable thirst for high performance, smaller form factor etc.

So, is advanced packaging a 'friend' or 'foe'?

Sometimes, people can become our enemies because we really don't know them or understand them – and end up with a misconception of what they really are like. Once we get to know someone, to find out what they are really like, our perception of them can fundamentally change. Likewise, with advanced packaging, we really have to know in detail what developments are going on to really know how to handle it.

So where is advanced packaging now, where is it going and what specific developments are on the horizon? How will any of these technologies affect hi-rel applications?

Industry Roadmaps and Future Packaging Trends

The two main industry roadmaps covering advanced packaging are ITRS and iNEMI. These two organisations are coming from different directions, ITRS from the IC and iNEMI from the system, but both acknowledge overlap in the area of IC packaging, including SiP and stacked die. In addition, the ENCASIT Consortium in Europe has been attending advanced packaging conferences around the world to build up a picture of what types of developments are important for the future.

A number of major trend have been noticed, some of which are:

Die Stacking – currently most manufacture uses wire-bonding, with some flip-chip staring to be used and future stacking methods will use TSV technology. The average number of die being stacked is increasing from 2 to 3, but some manufacturers e.g.. Samsung, have developed two stacks of 8 die in one package. Overall package height is limited to 1.2mm

Stacking & 3D thermal issues – the majority of die currently stacked are memory with low thermal dissipation. As other types of die are stacked, thermal dissipation is becoming a huge design and simulation issue. If a memory is stacked on top of a processor, then it will run at the temperature of the processor which requires high temperature memory die and many thermal vias.. If the memory is placed below the processor, then there will need to be some way of getting up to 100A up through the memory die to power the processor. Either way thermal performance will be an issue.

3D design – die are designed to be stacked, without having to use redistribution of bond pads, and I/O is optimized to take advantage of 3D interconnect. One of the first applications to take advantage of this type of design will be imaging, where the pixel processor will be directly under each pixel.

IC/package co-design – where the IC is designed concurrently with the package, and changes can be made in either with the effect simulated for the IC/package combination. Most work currently is being done in thermally-aware routing, to minimize power dissipation on a chip, but the co-design environment will extend to take in specific material properties and characteristics to be integral in the IC design. Changes to the package construction, including selecting alternative materials, without assessment using design simulation, will lead to differences in performance of the IC or even non-functionality.

Embedding of passives and some active die e.g. HIDING DIES project and others. Here, space is saved by embedding the passives, and even active die, inside the pwb laminate. Currently, there is a lack of suitable materials to make this efficient for inductive components, but there is much research into new materials. Most current use of embedded components is limited to embedded decoupling capacitors.

Direct Bonding of Die to wafer and wafer to wafer, using anodic, plasma etch or other methods. These constructions are potentially extremely robust with the bond between the die to wafer or wafer to wafer being stronger than the silicon. Some methods propose capacitive or inductive interconnection for

I/O signal paths. Plasma etch bonding is based on BCB polymers typically either with a Cu-Cu interconnection bond or Cu-Sn-Cu. Seal rings are required to ensure that the copper posts are protected from moisture ingress and consequential corrosion. There is plenty of work being done in this area and future applications will use this technology, but not for the next 5 to 10 years.

Optical interconnect using light guides in the laminate substrate or motherboard. The technology exists today to make this type of device, but the extra cost involved is not likely to make it worthwhile till at least 2017, according to the iNEMI roadmap, even though data bandwidths are reaching a bottleneck.

PoP v SiP – PoP is continuing to be a preferred method of assembly for some applications, rather than a true SiP package, due to business issues and costs. PoP is preferred by companies such as Amkor, due to 'margin stacking'. Business issues like this can determine the technology used, not because the intrinsic technology is better, but because the overall solution is lower cost.

Plastic packages for Power, based on Liquid Crystal Polymers (LCP) are being developed for power devices. These will replace ceramic and metal packages for power products since the packaging is cheaper. However, an LCP package may not truly hermetic since the main material is a polymer.

Plastic packages in general – new packages being developed generally have poor Moisture Sensitivity Limits and perform poorly in traditional high humidity/temperature tests. This requirement is not important for most consumer electronic applications

Legislation and regulation

The introduction of RoHS and WEEE has contributed to the changes in materials used for all packaging types, in this case not for cost reduction, but to satisfy a European Law.

The largest change of all has been in the eradication of lead from all electronics. Total eradication is impossible since any metals used may contain traces of lead and so a 'lead-free percentage level of lead was specified. Unfortunately, SnPb solder was an ideal interconnect material and all alternatives have drawbacks due to reduced ductility and increased process temperatures. Instead of one replacement technology being developed, a plethora of different materials and processes have emerged, many based on SAC (Tin/Silver/Copper) alloys. Tin whiskers, once eradicated in electronics, are a major concern and applications outside of commercial environmental limits should do material analysis to check that there are no potential reliability issues.

China has introduced their own version of legislation which is different to European which further compounds the problem. As the USA moves further towards embracing environmental issues, further, different legislation may emerge.

ELFNET (www.europeanleadfree.net) is a useful resource for this type of information. ENCASIT have been monitoring the situation mainly to advise on obsolescence issues that result from changes in materials in microelectronics packaging.

WEEE is designed to control hazardous substances which are harmful to the environment. However, the removal of some substances for example Brominated flame retardants in the plastic, leads to other problems. The addition of alternative materials, such as rubbers, can cause potential contamination and corrosion of the silicon metallization due to Sulpher ions.

Tomorrow's engineer

Tomorrow's engineer, assessing whether a component type is suitable for hi-rel applications, will have to know about the mechanical construction and materials used in the IC packaging. In may be possible to get all the required information from the manufacturer, but it may also be required to do some form of constructional analysis to determine all that is required. Even where the manufacturer has told you about the construction, it may be necessary to do constructional analysis to confirm the information and to check on process tolerances. There are a number of companies who already offer this type of service and this technique is used by IC manufacturers to help analyze failure modes in rejects. What may be required, however, is for this type of analysis to be done on a more routine basis to assess whether a certain component type can be used in an application. Strict change control would also be required to ensure that further analysis is carried out when the construction is changed. However, this may be a problem to component manufacturers where such a system is just not commercially worth their while.

Summary and Conclusion

Developments in advanced packaging are to improve performance, reduce size etc., but the primary driver is cost reduction. This has resulted in an unprecedented rate of change of packaging methods, processes and materials which will continue if not accelerate. The market for electronic components is driven by mobile phones and handheld electronics.

To ensure that current and future advanced electronic components are suitable for other applications, especially where higher reliability is required, detailed constructional analysis may be required on a component by component and even batch by batch basis.

In order to be aware of what to look out for, designers for these types of applications should be aware of the advanced packaging techniques being developed and to make some assessment of their suitability to the application environment.

Supplementary information about ENCASIT

The European project that has sponsored the work done in producing this paper, is ENCASIT (formally also known as Gooddie and ENCAST). This is a Networking project to gather and disseminate information on systems integration and, in particular, advanced packaging of microelectronics. Members of the consortium, listed as joint authors to this paper, include IC manufacturers and users, each with an expertise in different technology areas including IC/package design, test, standards, whole-life support,and handling, and in application area of automotive and harsh environments. Part of the work that KBTeC contributes is to attend conferences on advanced packaging and to report them in the ENCASIT Newsletter and via the Die Products User Club (see www.encasit.org and www.DPUC.net).

References

[1] Cover story for The Guardian Money supplement, Saturday 27th May 2006.

Technical Trends on Flat-Panel-Displays (FPDs) in Japan

Fumio Miyashiro

Japan Institute of Electronics Packaging (JIEP), IMAPS Fellow

e-mail : miyashirof@pop06.odn.ne.jp

Abstract

The packaging technology for TV-set had not been thought as an important technology item in CRT-age. But, in case of FPD, "A Flat Panel TV-set" must be composed of several films, substrates, display materials assembled by high density packaging technologies. Moreover, various FPDs, LCD, PDP, Rear Projection Display, OELD, and FED (SED) have been produced or developed in Japan. Each FPD is operated by different principle and has different packaging problems. In Japan, ordinary analog TV-set must be replaced to digital TV-set by 2011. So, by 2010, all kind of FPDs must be on market. And severe survival competition will be occurred. In this paper, various technical items to be solved will be reported and discussed from view point of packaging technologies and materials.

Key words : Flat Panel Display, TV-set, LCD, PDP, OELD, FED, Packaging technology

Introduction

There are several FPDs which operates by different principles and has different mechanisms and technologies. Now, these FPDs are still competitive and not focused to one or two dominants yet. Never the less, the leading role will be easily changed by a new emerging technology like new system, new material or new manufacturing method. My lecture time will be finished soon, if I want to explain all FPD's principle, mechanism and technologies in detail. Application field of FPDs are also wide like handy phone, DSC,PC and TV. Today, I would like to focus TV application.

By JEITA's forecast, world-wide needs of color TV will exceed two hundred millions per year on 2011 by growth of FPDs. Color TV market of 2006 was 1.7071 hundred millions (+ 7.5% in comparison with 2005) and this is the maximum value in history (Fig.1).

In Japan, the flat TV(LCD-TV plus Plasma-TV) ratio was more than that of CRT in 2005. In 2007 that of USA and Europe will occur and in 2009 that of ww will be over 50%. The number of LCD TV over 10" will be 1.0617 hundred millions and Plasma-TV will be 22.77 million in 2011. These FPD will be classified as follows;

    A. FPDs onmarket: **LCD, Plasma, Projection**

    B. FPDs under development : **SED, OLED**

    C. FPDs under research: **Electronic Paper**

I will explain about principle, market, merit & demerit and status of A & B category briefly.

Present position of each FPD

LCD

Once, it is said that LCD-TV covers less 40" and PDP does over 40". But, now, there is no difference in size repertory. For instance, Panasonic announced 103" Plasma-TV in 2006 and Sharp exhibited 108" LCD-TV at CES show, Jan. 2007 This means LCD-TV manufacturers had solved many tasks like contrast, motion picture correspondence, larger size manufacturing problems, so on. Fig.2 shows the share of LCD-TVs.

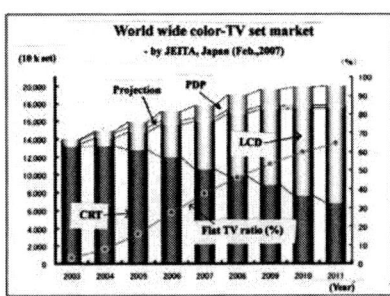

Figure 1 World wide Color TV market

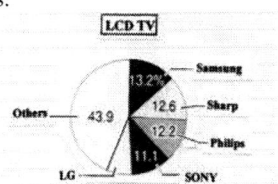

Figure 2 LCD-TV share (3Q,2006, DisplaySearch)

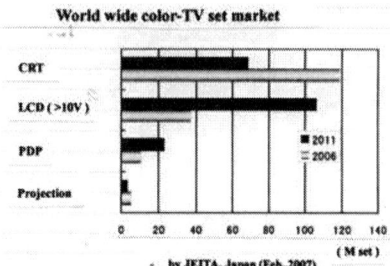

Figure 3 FPD-TV market forecast

Fig.3 shows 2011 future forecast on main FPD-TV including CRT-TV comparing with 2006 by JEITA.

The growing rate of LCD-TV looks bigger than that of Plasma-TV and Projection TV's growth will be low. The reason why LCD-TV's share is so high is as follows;
(1) As for the market of LCD-TV, Korea and Japan share market almost equal ratio competitively. (2) From companies of Korea and Taiwan supply over 75% of world needs LCD-panel also competitively. (3) Price down of LCD-TV is very drastic like 30% per year. (4) Key factor to establish lower cost of LCD-TV is mainly due to parts & material cost. Fifty to sixty percent cost of LCD-panel is occupied by parts & materials cost. (5) Demerits of LCD-TV like narrow view angle, low contrast were solved by technical development. (6) Development of larger size glass substrate has promoted. 46"panel size x 8 pieces are obtained from 2200 x 2500mm (G8). Now a new factory using G10 glass (3000 x 3000mm) is planned in Japan. But, in case of G10, the glass cannot be carried on public road by law. So, the glass factory must be built within LCD-TV site. (7) On the other hand, the flexible LCD-TV composed of all organic materials is under development in Japan. (8) Big size LCD-TV over 65" was tentatively on market. This trial is successful because bigger display's needs are in industrial, logistics, railway, airport and so on besides home use. (9) LCD-TV group is in good circulation in R&D, parts supply and manufacturing. But this state is not proved to maintain in future.

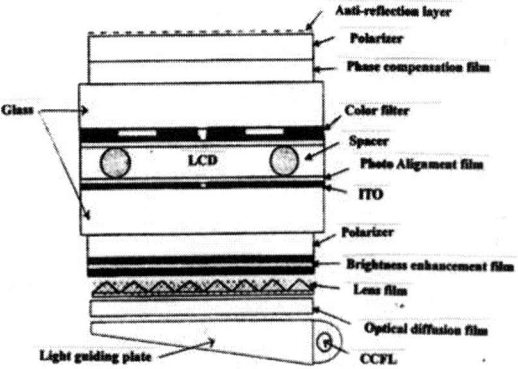

Figure 4 Cross section of LCD module

Fig.4 shows cross section of LCD-TV. LCD operates by supplying light from backlight. So, many parts are necessary to treat light optically.

PDP-TV
Fig 5 shows PDP-TV products and share in 2006.

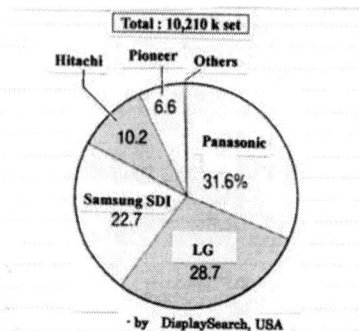

Figure 5   PDP-TV's share (2006)

The five manufacturers of Korea and Japan obtained 100% share in the world.

The principle of PDP is shown in Fig. 6. The feature of PDP-TV is as follows;
(1)   PDP-TV is self-emitting light system.
(2)   No viewing angle correspondence is exist.
(3)   Response to moving picture is excellent.
(4)   The number of parts & materials of Plasma display is less.

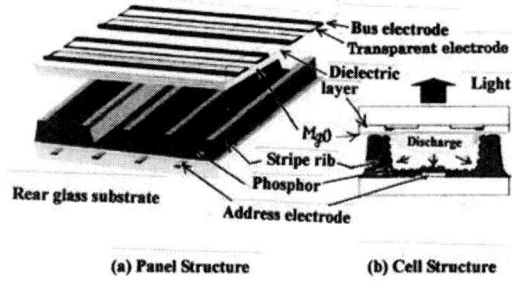

Figure 6   Panel & Cell structure of PDP

Fig. 7 shows comparison between LCD-TV vs PDP-TV.

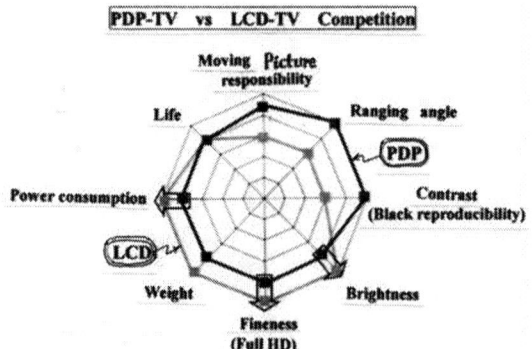

Figure 7 LCD-TV vs PDP-TV

Development trends from the radar chart indicates to improve power consumption, weight, fineness and brightness. Manufacturing of PDP is rather simple because no active devices like TFT in the system. Electrodes and phosphor layers are screen printed and baked. Dielectric layers and ribs are formed by coating and baking of glass powder paste which have low melting point. MgO protect layer is made by vacuum deposition. But procedures are basically simple.

Projection TV
The Projection-TV has rather long history. There was products in CRT age. Image obtained small devices like Micro Display is projected to a Large screen using an optical system and projection lamp (ex: high pressure mercury lamp). In place of CRT, now LCD, LCOS (Liquid Crystal On Silicon), DLP (Digital Light Processing) are used (Fig.8).

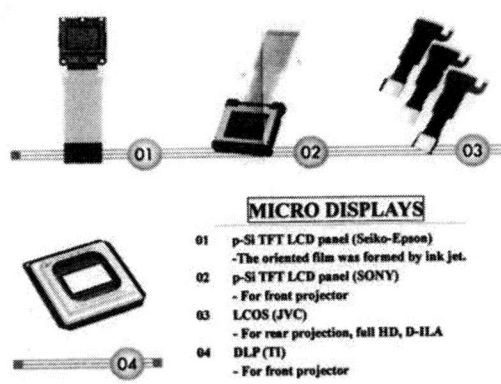

Figure 8   Micro displays for Projection TV

The Projection-TV is very popular in USA and China. But in other country, it is not so popular because life of the lamp is rather short, good screen in low cost is difficult. So, few forecasts say promoting of it in future.

But, recently, the new Projection-TV system using semiconductor laser light source in stead of mercury amp has been developed. Fig. 9 shows the new system.   This system is expected to be new star in Projection-TV.

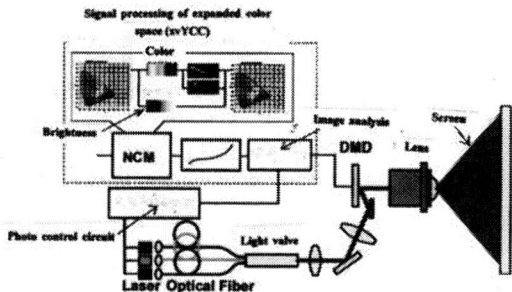

Figure 9 A new Projection TV using laser light

OLED-TV
OLED (Organic light Emitting Diode) is the device which has the structure that organic phosphor materials are sandwiched by electrodes (Fig.10).

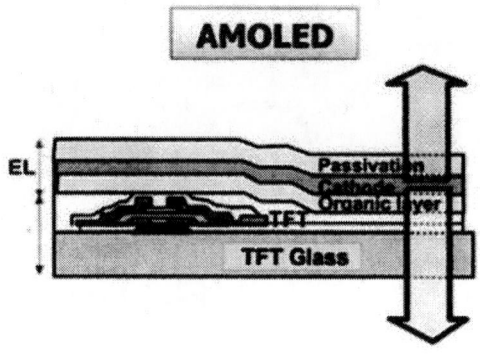

Figure 10   Active matrix OLED

OLED is also self-emitting light device and image quality is very good.   But, life of phosphor materials has not been guaranteed yet and larger size manufacturing seems to be rather difficult. Recently, however, longer life new materials were developed. So, such smaller size instruments as handy phone, music player have adopted OLED.

Mr. Izumiya, a journalist in Japan said "OLED market will be enlarged to 800 B¥ (this correspond to 10% of LCD-TV market)."

SONY exhibited 27" and 11" OLED-TV at CES held in USA in January 2007.(Fig.10) . The panel thickness of these TV were 10mm, 3mm respectively. Number of pixels of !1" OLED was 1024 x 600 , contrast was beyond 100M:1, brightness was 200 cd/m^2 in white picture, and 600 cd/m^2 at peak.

FED (SED)
SED (Surface–conduction Electron-emitter-Display) is one of FED (Field Emission-Display). SED is also self-emitting light system in which electrons emitted from electron source is accelerated and collide to phosphor.

Like conventional CRT, SED utilize the collision of electrons with a phosphor-coated screen to the electron gun in a CRT, are distributed in equal number to the pixels on the display. The electron emitters, at the heart of the SED, are characterized by a "nanogap", an extremely narrow gap measuring only a few nanometers in width, formed between two electrodes on the electron-emitting layer. When a voltage of approximately 10 volts is applied, electrons are e3mitted from one side of the nanogap. Some of these electrons are scattered at the opposite electrode and accelerated by the roughly 10 kV applied between the front and back glass substrates causing light to be emitted when they collide with the phosphor-coated glass substrate (Fig. 11).

SED-TV is expected as a monitor TV of broadcasting because of CRT-like characteristics.

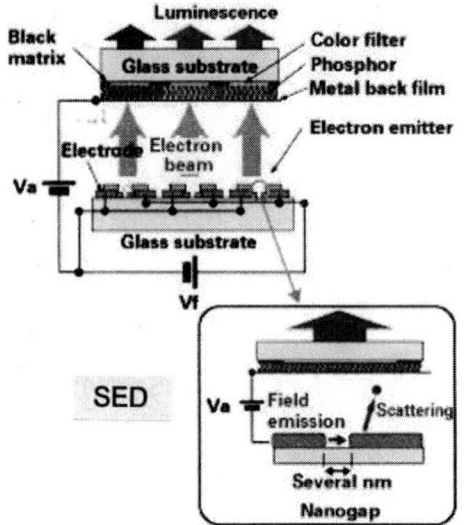

Fig. 11   The principle of SED-TV

Technical trend and tasks of LCD-TV

In this chapter, technology topics of the most popular LCD-TV will be introduced and discussed.

Needs for bigger size TV

G10 (3m x 3m) will be prepared for cost saving of LCD-TV. On the other hand, such new bigger size TV market plans as PC-TV or Internet-TV as a information center of home, built in wall TV, monitor display in public space are proposed.

Liquid Crystal filling

The liquid crystal filling between glass substrates method will be changed from absorption method by vacuum to dropping method. As for the latter method, liquid crystal is dropped from dispenser nozzle to a glass substrate, then another glass is covered on it and pressed. The case of 40" set, it took about 2 days by the former method and only 45 seconds by the latter method.

Color filter manufacturing by inkjet method:

Inkjet technology is one of very important, simple and economy technologies and will be widely used in future in place of photo lithography and wet processes. Seiko-Epson developed practical inkjet systems through "Future Vision Project". In 2005, this method was applied to first product, photo alignment layer of poly-Si TFT LCD panel (Fig. 12).

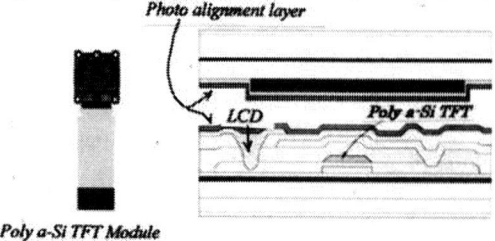

Fig. 12   Photo alignment layer by Inkjet

And in 2006, Sharp has adopted this method in color filter manufacturing line (G8) in Kameyama works. Merits obtained using inkjet method are (1) Film surface flatness was very good because of non-contact method and picture quality was improved. The used quantity of polyimide was reduced from 580 μg / mm^2 to only 2μg /mm^2. (2) Circumstance load reduced because much solvent and detergent were used to wash in case of former printing method but almost no solvent and detergent is necessary because of dry process. They could reduce 76% resources in color filter manufacturing process. And they could also reduce 68% energy consumption.

Cost structure of LCD-module

LCD-module cost consist of material fee 66% and manufacturing fee 34%. Swo, material cost is dominant. Among materials & components, back light, color filter, alignment film, driver IC & circuit and glass are worst 5. To solve the program, (1) Revolution in law material, (2) Study of functional structure of LCD, (3) New manufacturing method, (4) Global supply system must be achieved.

Multifunctionalization and integration of optics films

Several optical films are used in LCD-TV panel. Recently, multi-functional films or functionally integrated films have been developed. For example, Toppan has developed a color filter with spacers for liquid crystal. Ordinarily, beads are mixed as spacers in the liquid crystal. But beads don't spread always uniformly over screen area. So, Toppan attached periodical spacers to inner surface of color filter in place of bead spacer. As for polarizer, polarizer with phase compensation film or polarizer with phase compensation film or polarizer with anti-reflection function was developed.

Compensation methods for viewing angle problem:

This problem is peculiar to TN-LCD. So, various mode structures have been proposed and used as shown in Fig, 13

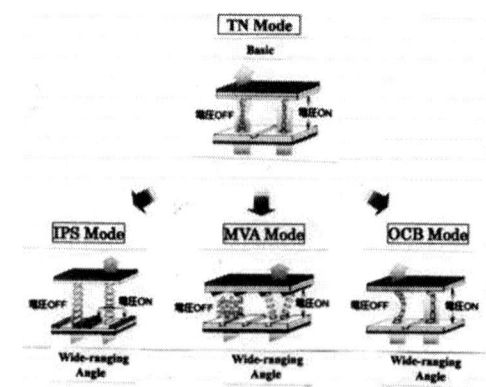

Fig. 13   Solution for viewing angle problem

LED Back Light :

The cost of LED-BL will be 1/10 of CCFL in future although now, LED-BL is more expensive than CCFL-BL. LED chip cost will be expected to ¥10/@ and high emission efficiency like 200lm/W will be achieved in near future. Then, low power consumption also will be realized to introduce area control technology.

The flexible LCD module by TRADIM

In Europe, Roll to Roll method is very popular. In Japan, a national project "TRADIM" has studied all plastic LCD module. Fig.14 shows the mission and target. The flexible LCD-panel will be very useful in coming Ubiquitus age.

The Concept of TRADIM

Fig. 14   Flexible LCD panel (TRADIM)

Recent FPD works in Japan

(1) Perfect "Lead-free" was achieved in Plasma-TV works .

Panasonic is manufacturing PDP-TV in Amagasaki-Plant in Japan and they obtained 33% share of world wide in 2006 summer. And they achieved perfect leade-free manufacturing process in Nov. over 140 types of PDP-TV. They could replace all lead containing materials like electrodes, dielectrics and, sealing materials into new materials. They investigated this program systematically and

solved.

- Amagasaki Plant (Panasonic) -

(2) Sharp : Kameyama Plant (Japan)

Kameyama plant is Sharp's main works manufacturing LCD-TV(G8). Kameyama Plant is also famous for concept to global environment and energy.

    1) Solar power generation system : 5,150kw
        by window glass see through type : 112kw
    2) Co-generation system
    3) Fuel Cell System : 1,000kw
    4) Perfect water recycle system
    5) Power storage by superconductor : 10,000 kw
    6) Damper system for earthquake

図1　亀山工場の全景

- Kameyama Plant (Sharp) -

Future trend and Roadmap

Nikkei Microdevice announced "40" LCD-TV price will be ¥100,000 ($833, €645) in 2010."

If the price is realized, world wide market will be drastically enlarged. So, cost-war will continue over design, devices, components, materials and manufacturing. Now there is an opinion "Cost reduction is going 0.1%/day. If this is correct, 100day development leads to 10% cost-up and 40day's logistics by ship results 4% price down.

Table 1 shows "A roadmap of display

Table 1: A Roadmap for Display

Category	Technical Element	Development Item	2005	2010	2015	2020
Current	*Social current	*Constraction of Infra-		Ubiquitas Network Age/ Green & Low Energy Consumption Age		
	*Broadband Infra-	structure enable to	Digital BS/CS			
	technology of	communicate everyone		Digital Bradcasting System		
	Broadcasting	from everywhere	Next Stage DVD			
	& Communication	*Low energy consump.		FTTH, PLC		
		& Eco-Society	ADSL			
	*High speed & high	*Realising revolutional	3.0–3.5	4.0 Next Stage Celluer Phone		
	volume treatment	targets by emerging	Wireless LAN	UWS/Next Stage Wireless LAN		
	technology for media	technologies	Polymer Li Battery			
	& communication			Fuel Cell		
Products Devices	*TV, Ultrafine HD,	*Key displays for	PDP 720p	High Efficiency PDP 1080p	*Wall paper like Large area TV	
	Thin, High color	Ubiquitas age are TV,	LCD Wide V	High Speed & Fine RPTV=QHDTV	*Large area 3D TV	
	Reproduction, Low	Mobile and Electronic	Slim RPTV	Ultra Slim & Fine RPTV=QHDTV	*High presence TV	
	power consumption	papar		FED-TV, OLED-TV	*Roll screen TV	
	*Mobile displays	*High function & chara-	TFT-LCD QVGA	*Handy phone with TV		
		cteristics must be	OLED QVGA	QVGA –		
	*Paper-like displays	realised by Emerging	Electronic Book B/W	Electronic Book color	*Personal Yubiquitas	
		technologies	Flexible B/W	Flexible Color		
Emerging techno- logies (Nano techno- logies)	*Active elments technologies	*Nano-tech new mat- erials which can make	Electronic Tag B/W	Electronic Poster Color		
			a-Si/LTPS,TFT array	Few masks LTPS/Single grain-Si TFT array + High integrate passiv		
		ultra-low power driven	MEMS array	Ultra fine MEMS/Photonic array		
		, devices, ultra fine wir-	HTPS/LCOS-LV –	QXGA – QHDTV		
		ing, low cost devices	Org-TFT	Org-TFT array + Integrated passives		
	*Process technologies	*Low cost ultra fine		Self organised Organic Semiconducter thin film – Large area		
		patterning using nano		Ultrafine Ink Jet patterning + High speed & Multilayer		
		processes		Nano Inprint patterning – Large area		
				High gas barrier Nano coating – Highly flexible & tough		
	*Nano material technologies	*High efficiency, high		High efficiency, high reliability Nano Crystal Fluorescent material		
		reliability and ultra-fine		CNT for High current cathode + Self healing		
		structure by nano		High dispersion Nano particle + High mobility		
		technologies		Transparent plastic film with high Tg + Multi-Optical function		
				High tenacity & strength Nano glass substrate + Ultra thin & flexible		

Design Tools for System-in-Package Applications

Anna Fontanelli

Mentor Graphics Corporation – Systems Design Division – Milan, Italy

Phone: +39 02 249894 230, Fax: +39 02 249894 200

E-mail: Anna_Fontanelli@Mentor.com

Abstract

In 2005, consumer electronics became the largest slice of the electronic industry pie. Consumer applications, though, demand for an ever increasing integration of very different silicon process technologies to meet their performance (power, clock), size and cost targets. Advanced packaging techniques are needed, such as Multi-Chip-Module (MCM) and System-in-Package (SiP), in order to develop each component of the application in its native process technology. It is then a matter of integrating them all, combined with the required active and passives components, into a single package. Packaging technology is radically moving from horizontal, "one-story" integration to vertical "skyscrapers-like" integration, driven by the increasing constraints of "real-estate" availability, cost, and reliability. In addition to this, no system is possible today without passive components. Mobile phones typically contain more than 50 passive components for each IC, overall there are more than 200 passive components and only a handful of ICs. This has had a dramatic impact on the EDA industry, so far too IC-centric. Schematic and Layout editors must work in a 3D world while extraction and simulation tools must handle ICs, their packages and many diverse passive components simultaneously, taking into account also the Board effects. To solve this Challenge, a brand new IC/package/board co-design and co-verification methodology, based on open standards, is required to address the intricacies of SiP. IC, Package, and Board have become a single whole, and must be treated as such to increase total system performance and promote right first time.

Key words: System-in-Package (SiP), System-on-Chip (SoC), Wire Bonding, Flip-Chip, and Embedded Passives.

Introduction

Today consumer's demand is driven by personal, mobile communications and entertainment: more than 700 million mobile phones, approximately 80 million digital cameras, 60 million laptops computers (which have surpassed the desktop ones for the first time in history) and more than 30 million iPods.

In September 2005, the two billionth worldwide mobile phone subscriptions were celebrated and the forecast was that the three billionth subscriptions would have been achieved by 2010. In December 2005 the forecast was corrected: the new threshold will be achieved by 2008.

Mobile phones have become the highest volume electronics product ever conceived, and are now integrating mega-pixels camera, MPEG player (both audio and video), and internet access. Complexity and cost considerations rule and all the players – from OEMs such as Nokia, Motorola and Samsung, to IDMs such as TI, Freescale, Qualcomm and ST – are striving to reduce their cost in order to address the broadest possible customer base. The design and the manufacturing of a sub-$20 mobile phone is hardly compatible with the System-on-Chip (SoC) approach[1].

System-on-Chip Concept

For decades, the least expensive way to add functions to an electronic system was to integrate more functions on a single chip.

IC manufacturing, though, is entering new territory, where delays for longer on-chip interconnects are starting to have a significant impact on the IC's performance. Putting circuit components closer together by putting them on the same die will not necessarily improve signal propagation time between them. Also, once a die's size reaches a certain point, its yield starts to drop significantly. Going beyond that point makes little economic sense.

So, although technologically feasible[2], a SoC design at 90, 65 or 45 nm is becoming a blessing and a curse. A true blessing, since the number of possible gates in one of these ICs will approach 30M by 2008, which offers the opportunity to implement large and varied functions in a single chip, when the target application requires the highest performance, high production volumes, longer time-to-market and not mixed technology.

The curse is that to effectively design and verify a chip with this many gates require many designers for several months if not years.

Even with advances in IC design techniques to higher levels of abstraction, re-use of cores and new verification functions, short time to market is not achievable and the NRE investment is significant.

New alternatives to System-on-Chip

Moreover, the poor integration of passive devices makes the term SoC itself an overstatement as no system is possible today without passive components. Mobile phones typically contain more than 50 passive components for each IC, overall there are more than 200 passive components and only a handful of ICs.

Driven by the handheld industry, System-in-Package production is increasing at a 14% CAGR (source: Semico Research Corp.) in several different segments (source Prismark).

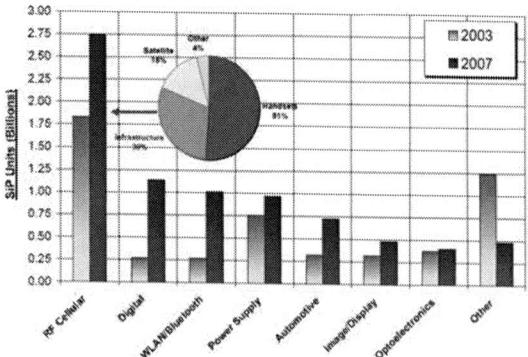

Figure 1: SiP Market growth in most market segments

These consumer applications demand new packaging techniques, such as Multi-Chip Modules (MCM)[3] and SiP, in order to reduce the form factor while increasing functionality. Implementing multiple ICs in a single package, the complexity of the PCB is reduced, thus further reducing the form factor while reducing the product cost: two primary goals of the highly competitive handheld market.

These challenges are driving the evolution of SoC into SiP. This trend is further accelerated by the technical difficulties of keeping the pace in the rush towards the atomic limits of Moore's Law.

Actually, SoC and SiP do not necessarily compete[4] with each other. "More Moore" will continue to drive the design and manufacturing of SoC-based computing and storage functions, while "More than Moore" will be required to integrate them all into a SiP, which is likely to be very heterogeneous, and to incorporate a broad range of both CMOS- and non-CMOS-based technologies, such as RF[2], sensors, actuators, passives and MEMS.

IC Packaging

IC Packaging is the final step in the complex process of turning silicon into a useful semiconductor device.

In recent years, major innovations have occurred in IC packaging technology, which have led to the industrialization of several kinds of new packages – more powerful and more flexible – in the attempt to cope with the challenges posed by multi-million gates and multi-GHz SoCs.

These designs demand higher integration, resulting in high packaging pin counts and density.

High frequency designs require complicated flip-chip technologies to avoid the high inductance of wire bonding.

According to the latest update of the International Technology Roadmap for Semiconductors (ITRS 2005)[5], at 65 nanometers a 500 million gate ASIC, with 4,400 pads and running at 10GHz, must be packaged in a giant, and yet very expensive ($1.61 c/pin) flip-chip BGA[6], with 4,000 pins.

In contrast to the costly SoC, packaging technologies are evolving into SiP to keep up with the need for "smaller, faster, cheaper" products. Within one single package, multiple wire bonded or flip-chip ICs are integrated on a common substrate, including several passives – such as IPAD (Integration of Active and Passive Devices), SMD, embedded passives – not achievable in IC technology.

The ability to drastically simplify board-level design complexity by integrating large numbers of discrete elements into the interconnection structure as embedded components is one of the strongest points of the SiP approach.

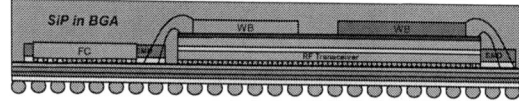

Figure 2: BGA Module for System-in-Package design

Mixed signal communication products, such as mobile phones, require the digital, analog and RF portions of the design to work reliably and in close proximity to each other. This means a number of passives in both the analog and RF design sections, and integration leads to a substantial size reduction and performance improvements, solving issues such as signal noise, crosstalk, ground bounce etc.

The higher integration capacity of SiP reduces the number of components in the system, thus reducing both the size and the routing complexity of the printed circuit board (PCB).

IC Packaging: from SoC to MCM

Multi Chip Module has existed for many years in the computer and telecom industries, even if it was limited to expensive high end applications.

Today, after the extreme reduction of package size to Chip Scale Package (CSP) it is used for very high volume applications, where cost and real estate are the key drivers.

Partitioning of the SoC has led to a new design architecture, where all the modules can be designed in the most appropriate technology, and then placed and connected in the same package substrate

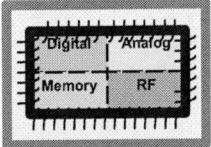 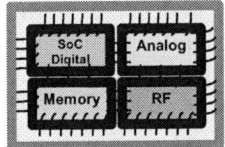

Figure 3: Evolution from SoC into MCM architecture

In addition, the designer selects IP components – logic, memory, analog, optimized cost ICs – from multiple sources that are optimized to minimize manufacturing costs and also to help shorten development time.

Next Step: 3D Integration

Another level of integration is a 3-dimensional approach, offered by stacking dice[7].

IC Packaging technology is radically moving from horizontal, "one-story" integration to vertical "skyscrapers-like" integration, driven by the increasing constraints of "real-estate" availability, cost, and reliability. Combining several levels of memory in a stack[8], a multi-layers Memory Combos can be realized, as shown in the picture below.

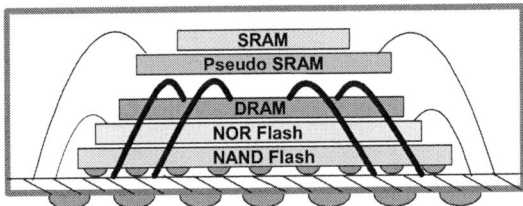

Figure 4: Memory Combo in CSP

To address the many needs for system integration[9], combining ASIC, Flash, SRAM, Integrated passives, sensors … there are many stacking solutions for integration, such as Pyramidal Stack, Twin Stack, Mixed Wire-Bonding and Flip-Chip with or without die-to-die connections. All the solutions can be combined, multiplied (N-layers) or mixed (Wire Bonding/Flip-Chip).

Figure 5: Pyramidal and Twin Stacking

Figure 6: Die-to-Die connections and mixed Wire-Bond / Flip-Chip stacking

An additional level of complexity is introduced by the possibility of placing side-by-side two ICs on the package substrate, and then stacking a bigger component on top of the two.

Figure 7: Bridge Stacked CSP

Interconnection can be realized with flip chip as well as wire bonding technologies[10], providing all the necessary die-to-substrate and die-to-die connections.

Wire Bonding is still dominating for chip & package connections also for rather high pin count devices: it's really cheap, it's relatively simple and, above all, a very well proven technology. It's a typical first choice due to cost and yield.

Flip-Chip is exactly the opposite: it's rather costly and is subject to several yield issues. So, it is used only when there is a technical reason:

- Extremely high pin count: number of peripheral I/O cells exceeding 1500-2000, not enough room to place them around the core block perimeter.
- Ultra high-speed: Flip chip connectivity has shorter interconnection distance with lower parasitic inductance which result in better signal propagation than you may have in wire-bond.

An Interposer could also be needed as a separation layer between two stacked components. It is a rectangular block of metal or dielectric material which is required and must be added every time the pads of the bottom die are overlapped. Its dimensions and thickness must be carefully and automatically designed.

System-in-Package Architecture

A SiP electronic product[11] will be characterized by increasing design heterogeneity:

- Technology: digital, analog, RF, optoelectronics, MEMS, passives
- Frequency: from MHz range in digital and RF to GHz in microwave component Energy: extremely low power consumption for handheld.
- Thermal issues
- Form factor and weight constraints.
- Architecture: low and high power analog functions together with event-driven, data-driven, and time-driven digital structures.

This requires an extremely flexible architecture: a 'real' SiP is not only a MCM or a stacked die CSP but is a combination of the two with fewer added components.

After an intelligent partitioning of the system, the diverse functions can be placed in the package substrate side-by-side or in a stacked fashion with the appropriate interposers. They are then properly connected with wire bonding and/or direct chip attachment.

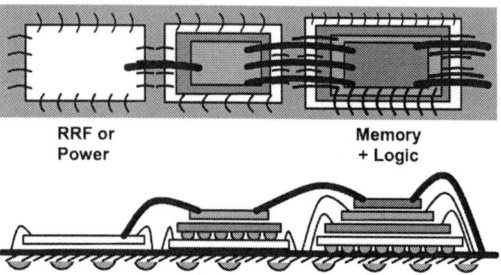

Figure 8: System-in-Package is one single module integrating multiple system or sub-system functions, with Wire Bonding and/or direct chip attachment

Wire bonding can be used to connect die-to-die and/or die-to-substrate, linear and/or staggered pads using a standard or a reverse bonding technology, even with tri-tier package configurations.

SiP Design Tools

While offering a great deal of opportunities, complex SiP substrate design and verification requires an unprecedented level of integration between ICs and the package[12].

The package is not only just a package anymore. Multiple ICs and a large number of passive components, such as resistors, capacitors and inductors, active devices such as ESD protection diodes, all into a single substrate, make the package substrate itself 'a system' that must be effectively co-designed and co-verified.

Unfortunately, today's design methodologies result in a segregated relationship between IC and package design, making coordinated planning a difficult and time-consuming task.

The serial nature of the traditional silicon to package design flow limits the effectiveness of existing tools for concurrent planning. Both IC and package design tools lack the needed visibility into their respective environments. This serial approach may lead to a poor IC to package connectivity, resulting in longer cycle time and sometimes preventing the co-verification of the entire system.

Increasing functionality of System-in-Package designs and shortening design and product life cycles have increased the pressure for first-time right design solutions. This means an increasing emphasis on co-design, simulation and on-the-fly verification EDA tools.

SiP Co-Design Challenges

SiP designers, a multidisciplinary team, will have to cooperate to solve several optimization problems[13], including system I/O requirements, thermal and signal integrity constraints, die placement and orientation, stacking configurations, package substrate and interposer design, interconnect design, at IC plus package level, taking into account also the customers' constraints on the PCB.

Starting step, to achieve optimum results, is the availability of a SiP Connectivity Environment. It must be able to read and write a top-level hierarchical netlist of all the heterogeneous pieces (ICs -digital and analog-, package, board) and to merge them all, even if coming from different design environments. It must also be able to compose schematics and generate the complete models.

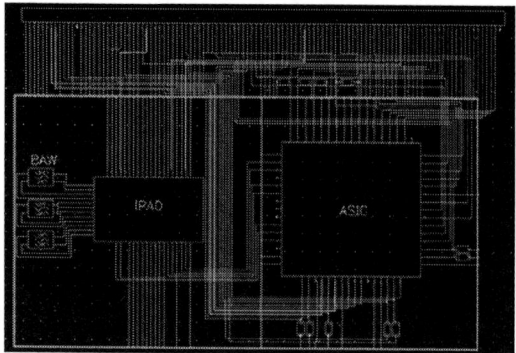

Figure 9: Schematic view of a simple SiP Design

Then, the complete floor-planning of all the heterogeneous SiP elements must be realized.

A new class of algorithms, with the ability to work in a 3D world, must be available to properly stack the required components, to place them along with the rest of the SiP components and properly plan the interface between all the elements including package bond pads.

A smart I/O Planning strategy has to be applied to all the ASICs involved in order to concurrently validate initial I/O periphery configuration and/or optimize it, respecting different sets of rules in order to satisfy silicon process, electrical and assembly rules.

Prerequisites to start the IO Planning process are the availability of all the different IO Libraries involved, even if of different nature in term of silicon process, pad pitch, P&G levels etc.., the SiP overall connectivity, and the complete set of Rules to be respected.

16

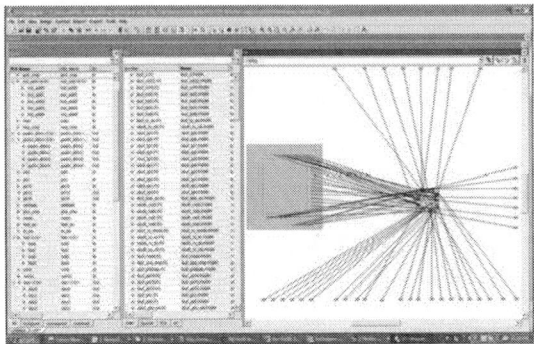

Figure 10: Rule's based IO Planning and optimization co-design flow

The design of the SiP substrate also presents many challenges to the designer and design process. It requires the support of advanced technologies such as:

- High Density Interconnect (HDI)
- Micro Vias
- Embedded Passives (EP)
- RF shapes
- 3D Wire Bonding
- Bond Pattern Generation
- Substrate Routing
-

3D Wire-bonding poses a number of challenges for IC Packaging design, especially if you stack several dies and pin counts go from several hundreds up to 2000 I/Os.

A new wire bonding technology, that utilizes isolated bond-wires, has emerged: X-Wire technology.

This technology allows bond-wires to cross in any direction making a very dense bond pattern possible. If the bond pattern synthesizer can follow the specific X-Wire rule sets, there is an opportunity to realize a higher pin counts wire bonding. Of course, there is a physical limit also to the density of X-Wires!

Figure 11: Very dense X-wire bonding technology

Design tools must support automatic bond pattern generation of entire die stacks, while respecting all the 3D aspect of IC packaging and obey numerous 3D related rules and constraints, in order to successfully synthesize die patterns. A manual bond pattern generation is no longer realistic, and there is a lot of time to gain on an advanced automatic solution.

Once a bond pattern has been established, the challenge of connecting the bond pads to the package becomes obvious. Often, the only means to obtain a routable solution is to let the router assign package pins while routing, and this may require updates to the IC's I/O planning.

SiP Co-Verification requirements

In addition to physical implementation, system verification is a critical part of IC Packaging design.

Accurately simulating ultra-fast signaling technologies requires a system-level analysis, involving a range of physical structures and modeling technologies. Simulation of a complete signal path (or "channel") requires accurate modeling of I/O buffers, all chip-to-chip interconnect (including IC packages, PCB traces, connectors, and vias with associated bypass capacitors), and associated power-distribution structures (planes in the package and PCB, decoupling capacitors, and stitching vias).

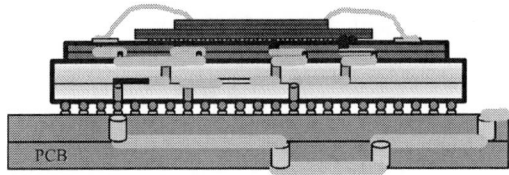

Figure 12: IC & Package model for System Verification

Power integrity concerns itself with proper delivery of power to IC buffers; and proper quality of the launched signal as it interacts with plane layers in the PCB or package. To fully understand how to distribute power to an IC requires simulation in the context of the package; and possibly even the context of the PCB.

"Classic" signal-integrity concerns (like delay) are becoming an issue in package design, as timing margins shrink and packages become larger. For SERDES signals, very high frequencies (that require careful simulation) are involved. In addition to simulating, package designers typically follow very tight physical rules, as they attempt to maintain differential impedances, eliminate reflections and discontinuities, etc.

Some structures in typical packages can be extracted/modeled using well-proven, mature techniques from the signal-integrity world (e.g., trace routes, vias, and sometimes, bond wires).

Others require more-tedious 3-D extraction (less-controlled bond wires, and possibly balls). Accurate wideband models of such structures are usually captured in the form of S parameters.

Although many package designers are tempted to rely heavily on 3-D simulation, such electromagnetic tools tend to be difficult to use and very slow.

Large-scale 3-D extraction involving thousands (or even just hundreds) of wire bonds and other interconnect elements is completely impractical. Thus, whenever possible, the problem should be decomposed into sub-elements, each one of which is modeled with the best available engine: one that provides sufficient accuracy but also reasonable speed and ease-of-use.

A promising solution is a suite of tools, where decomposition of the problem is automatic and all appropriate engines are bundled and available.

Conclusions

A revolution is taking place in the marketplace, driven by high volume applications with fast innovation and a large variety of new options.

The integration roadmap defined by Moore's Law and ruling SoC is no longer sufficient. Packaging technology can integrate a number of heterogeneous functions in the same device. SiP is a modular design approach offering unprecedented flexibility in the development of systems.

An intelligent partitioning of all the components of the electronic system is key to achieving greater functionality in a smaller area, combining dissimilar device types and high-yielding memory devices with similar size and wiring requirements.

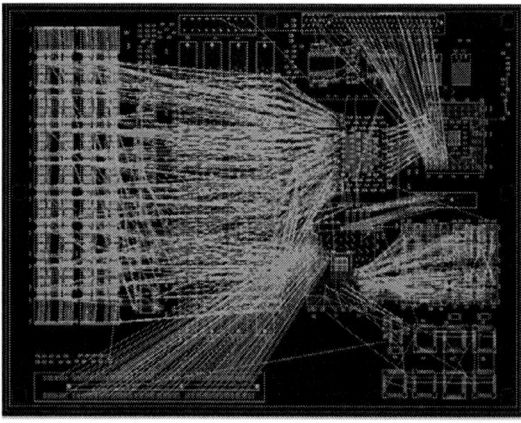

Figure 13: System-in-Package Floor planning

A multi-disciplinary team has to execute the SiP design and verification, in order to solve the overall system optimization problems thus achieving all the requirements.

New methodologies and new tools are becoming a must in order to facilitate coordinated planning and sharing of data across the domains of silicon, package and Printed Circuit Board (PCB).

Having coordinated planning during the early stages of silicon floor-planning, before elements within the chip are fixed, can result in an optimized silicon/package interface to reduce cycle-time while enhancing overall device performance and reducing costs.

References

[1] K. Lim et alt, "RF-system-on-package (SOP) for wireless communications", IEEE Microwave Magazine, pp. 88-99, 2002.

[2] H. Bernhard Pogge – IBM Fellow, "Realizing Effective SOCs with Viable 3D Interconnect", First International Workshop on SoP, SiP, SoC Electronics Technologies 2005.

[3] Rao Tummala, "Multichip packaging – a Tutorial", Proceedings of the IEEE 1992, pp. 1924-1941.

[4] Rao Tummala and Vijay Madisetti, "System on Chip or System on Package?", IEEE Design & Test of Computers, pp. 48-56, 1999.

[5] "The International Technology Roadmap for Semiconductors, 2005"; www.itrs.com.

[6] John H. Lau, "BGA Technology", McGraw-Hill, Inc., 1995.

[7] Y. Fukui et al., "Triple-Chip Stacked CSP", 50th Electronic Components & Technology Conference, 2000.

[8] Takahiro Oka, "System-In-a-Package Stacked Flash Memory and Logic LSI", Semicon Japan, 2000.

[9] Y. Yano et al., "Three-dimentional Very Thin Stacked Packaging Technology for SiP", 52nd Electronic Components & Technology Conference, 2002.

[10] George Harmon, "Wire Bonding in Microelectronics", Second Edition, McGraw Hill, 1997.

[11] Joe Adam, Vice President SkyWorks, "System-in-Package Roadmap", First International Workshop on SoP, SiP, SoC Electronics Technologies 2005.

[12] Rao Tummala, "System-on-Package (SOP)", First International Workshop on SoP, SiP, SoC Electronics Technologies 2005.

[13] K. Tai, "System-in-Package (SiP): challenges and opportunities", Asia and South Pacific Design Automation Conference, pp. 191-196, 2000.

Embedded Passives in Multi-dielectric Layer Printed Wiring Board for IEEE 802.11a/b/g Tri-Mode Dual-band Wireless CardBus Adapter

Cheng-Hua Tsai, Chang-Sheng Chen, Chang-Lin Wei, Kuo-Chiang Chin,

Wei-Ting Chen, and Chin-Sun Shyu

Electronics and Optoelectronics Research Laboratories, Industrial Technology Research Institute

Room166, Bldg. 14, 195, Sec. 4, Chung Hsing Rd. Chutung, Hsinchu, Taiwan 310, R.O.C.

Phone: 886-3-591-3324, Fax: 886-3-582-0374, E-mail: CHTsai@itri.org.tw

Abstract

This study investigates the application of special high dielectric constant (Hi-DK) material, which is suitable for Multi-Layer Printed Wiring Board lamination process, and is used to fulfill embedded capacitors. In the past years, Bluetooth™ module and other RF circuits, such as Power Amplifier had been successfully designed by utilizing this technology. Today, this novel technology also applied to the Wireless LAN (WLAN) module. Embedded passives technology is more economy than any other system in package (SiP) processes, such as LTCC substrate or silicon substrate; because the additional material can fully compatible with the conventional HDI-PWB process. In this paper, a wireless cardbus adapter was successfully designed by built many passive components into its board. And not only capacitors are embedded, but also some functional devices, such as filters, diplexer, dualband coupler and antenna...etc, are been designed in the Hi-DK organic substrate. There are 51 passive components built into the substrate. Compared with the original 115 SMD passive components on this wireless cardbus adapter, the maximum replacing rate by embedded passives is up to 44.3 %. And the performance of the adaptor in different mode is also as good as original SMD version. Thus, this kind of organic substrate can reduce the total module cost and suitable for high frequency products. The core of the technology that making this design successful is a systematic design library which integrate all 3D model for embedded structures with considering the variation of material character and the tolerance of process. In the future, embedded resistors will also be applied into the cardbus adaptor to increase the replacing rate.

Key words: high dielectric constant (Hi-DK), Wireless LAN (WLAN), system in package (SiP), embedded passives.

Introduction

The main considerations of system in package (SiP) technology for consuming electronic products are cost-effective and small size. Therefore, there are two kinds of suitable integral substrates: silicon substrate, which has the best combination property with chip; and organic substrate, which is the conventional multi-layer printed wiring board (PWB). Further, SiP with PWB process on organic substrate is considered to be more cost-effective than SiP with other process. If modules were fabricated by another SiP processes such as LTCC, still factories have to mount these modules onto the system board. Packaging the whole module on PCB substrate directly would save steps and cost. In the past years, many RF circuits with high Hi-DK in 6-layer multi-material organic substrates laminated by normal HDI process had implemented. By using advanced Hi-DK and low-loss material, size reduction of functional components are accomplished with concerned electrical and critical effect.

This paper will introduce and explain how to design the functional devices of embedded passive elements, which include capacitor, low pass filter, band pass filter, balun, diplexer, and coupler on organic substrate. And, the main applied frequency is 2.4GHz and 5.2GHz to verify that SiP technology can be applied on more wide band and high frequency. In addition, we use all these RF circuits to form a IEEE 802.11a/b/g [1]-[3] Tri-Mode Dual-band Wireless CardBus Adapter on organic substrate.

Substrate and Material

Base on low-cost target for wire-less communication, EOL/ITRI devoted to develop the circuit design technology that applies embedded passives to RF modules for six years. The verified modules, including the 2.4GHz VCO, BPF, balun, power amplifier and the Bluetooth module, are successful results in the past years [4]-[6]. Now, six layers substrate structure with high dielectric constant material were also applied in those RF circuits and Dual-band Wireless CardBus Adapter. The six layers substrate structure is illustrated in Figure 1. The PWB structure is three symmetrical copper layers. The dielectric layer 3 is a material carrier that supports the mechanical strength for upper and lower RCC (resin coated copper) materials. Dielectric layer 2 and 4 are built up with high dielectric material (Hi-DK40) to reduce the size of the embedded capacitors. And dielectric layer 1 and 5 are built up with lower dielectric material mainly for the RF signal processing. Figure 2 shows the continuous coating process for mass production of the dielectric material. The electrical properties of each dielectric layer were characterized in Table 1.

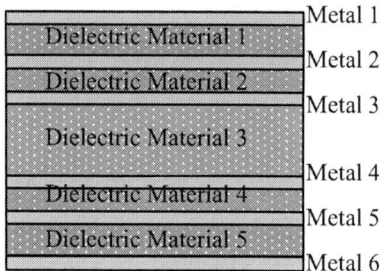

Figure 1: PCB Substrate Structure

Figure 2: Coating process for Hi-DK RCC Type Substrate Material

Table 1: Material Characters of Substrate.

Layer	Material	Thickness
Metal 1~6	Copper foil	0.7~1.4mil
Dielectric material 1, 5	DK=3.9; DF=0.017	4 mil
Dielectric material 2,4	DK=38; DF=0.04	2 mil
Dielectric material 3	DK=4.3; DF=0.014	12mil

Capacitor

The basic equation of capacitor is shown as equation (1). Where, d is the thickness of the dielectric layer and A is the area of the parallel capacitor plates. Therefore, high dielectric constant materials can save valuable circuit area. Ideally, the capacitance is directly proportional to the electrode area. However, the fringing effect increases the effective dielectric constant obviously by the special material, Hi-DK40. A simplified SPICE model for 1-port capacitor is shown in Figure 3(a). The resistors represent the loss factor for the capacitor and Ls represent parasitic inductance. If the capacitor has less Ls then the Self-Resonate-Frequency (SRF) will get higher. In general, for a simple capacitor, Rs and Ls are increased with capacitance. Cp is the parasitic capacitor of grounding effect.

$$C = \varepsilon_0 \varepsilon_r \frac{A}{d} \qquad (1)$$

For high frequency circuit design, the coupling effect of embedded passive element in the Hi-DK40 material is quite severe. Hence, some parasitic effect of critical elements may decrease the system performance, and the electromagnetic simulation software is useful to get more accurate information. Figure 3(b) shows the electromagnetic (EM) structure of the embedded capacitor. In fact, the EM model should combine with the high frequency material model.

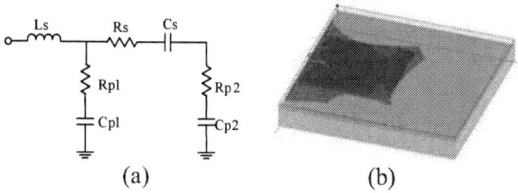

(a)   (b)

Figure 3: One Port Capacitor (a) SPICE Model of (b) 3D EM Model

Inductor

Inductors are fundamental components in high frequency circuits. There are various kinds of embedded inductors, and sometimes we use a trace as small inductors. Figure 4 shows the spiral inductor, and equation (2) is the approximate inductance formula. The symbol n denotes for the number of turns and the symbols d_o and d_i are explained in Figure 4. In Figure 5(a), the basic equivalent circuit model is illustrated. The scaling model of mathematics of inductor can be obtained from the experimental data.

$$L \ (nH) = \frac{39.39 n^2 a^2}{8a + 11c}$$
$$a = \frac{1}{4}(d_o + d_i) \ ; \ c = \frac{1}{2}(d_o - d_i) \qquad (2)$$

The structure of embedded inductor is more complicated than that of embedded capacitor. We also use different structures of embedded inductor such as solenoid and meander [7] for specific purpose. Usually, the equivalent circuit model of inductor is not easy to predict. Using the electromagnetic simulation software can help us to get scaling models faster. Figure 5(b) shows the electromagnetic (EM) structure of the embedded inductor. After verify the models by experimental data, we get a whole library of embedded inductors.

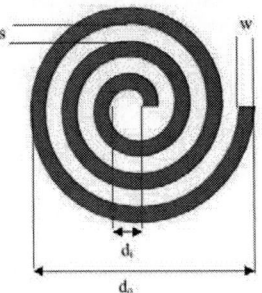

Figure 4: Spiral Inductor

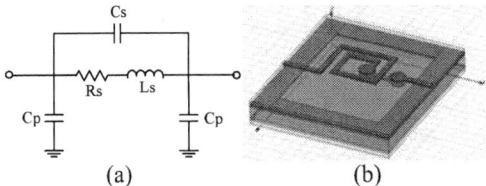

(a)   (b)

Figure 5: Two Ports Spiral Inductor (a) SPICE Model (b) 3D EM Model

Low Pass Filter

One stepped-impedance hairpin resonator structure [8] is used to design low pass filter. Figure 6 shows the geometry and equivalent circuit of the low pass filter. As show in figure 6(b), L_s is the equivalent inductance of the single transmission line of the filter. C_g is the equivalent capacitance of the coupled lines and C_{ps} is sum of the capacitances of the transmission line l_1 and the coupled lines.

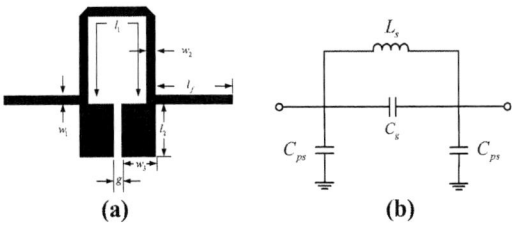

(a)   (b)

Figure 6: Low Pass Filter Using One Hairpin Resonator (a) Layout Model (b) Equivalent Circuit

The exact dimensions are adjusted to optimize the performance of the LPF using EM simulation tool. Figure 7(a) shows the 3D EM model of the LPF and the photograph of realized circuit is shown on Figure 7(b). Observing measured results

in Figure 8, the LPF has a 3-dB passband from dc to 9.3 GHz. The insertion loss is less than 0.24 dB, and the return loss is better than 26 dB from 2.41 to 2.485 GHz. Furthermore, from 5.15 to 5.35 GHz, the insertion loss is less than 0.51 dB, and the return loss is better than 27 dB. The rejection is greater than 20dB within 12.3~20 GHz.

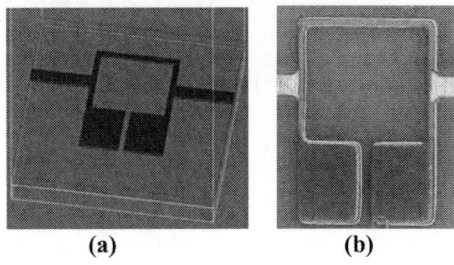

(a)   (b)

Figure 7: Low Pass Filter (a) 3D EM Model (b) Photograph

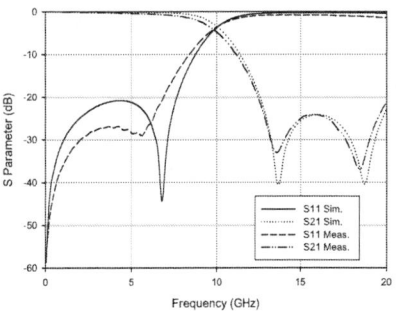

Figure 8: Simulated and Measured Response of Low Pass Filter

Band Pass Filter

A tapped combline structure [9] is used to design 2.4 GHz and 5.2 GHz band pass filters individually. Figure 9 shows the geometry of band pass filter. The tapped combline filter is comprised of an N array of coupled resonators which are short-circuited at one end, with a lump capacitance C_{Li} loaded between the other end of each resonator line element and ground. The filter used the tapped lines as the input and output. With the lumped capacitors present, the resonator lines will be less than $\lambda_{g0}/4$ long at resonance, where λ_{g0} is the guided wavelength in the medium of propagation at the mid-band frequency of filter. Therefore, the embedded capacitors are taken the place of lump capacitors.

The exact dimensions are adjusted to optimize the performance of the BPF using EM simulation tool. Figure 10(a) shows the 3D EM model of the 2.4 GHz BPF and the photograph of realized circuit is shown on Figure 10(b). Observing measured results in Figure 11(a), the insertion loss is less than 2.4 dB, and the return loss is better than 11 dB from 2.41 to 2.485 GHz. The insertion loss is less than 3.2dB, and the return loss is better than 8.7 dB from 5.15 to 5.35 GHz, as show in Figure 11(b).

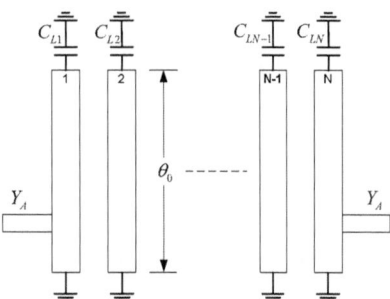

Figure 9: A Tapped Combline Filter Model

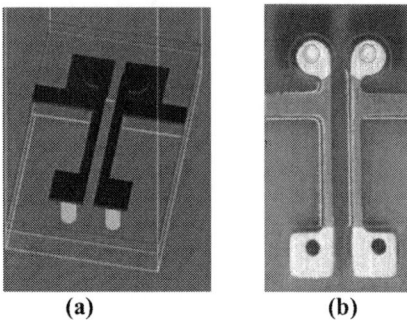

(a)	(b)

Figure 10: 2.4 GHz Band Pass Filter (a) 3D EM Model (b) Photograph

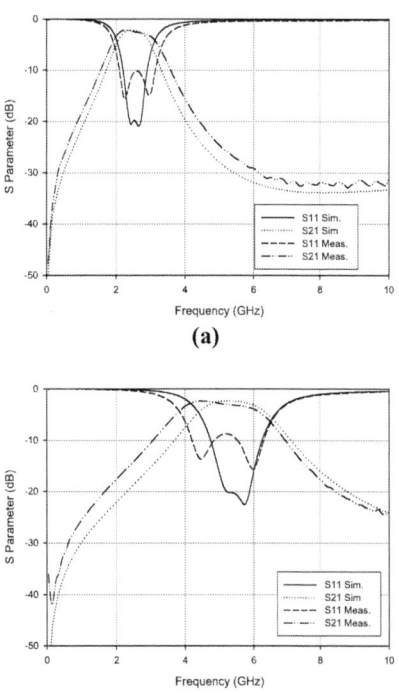

Figure 11: Simulated and Measured Response of Band Pass Filter (a) 2.4 GHz (b) 5.2GHz

Balun

The quarter-wave transformer is easy to design a balun. Figure 12 shows $\lambda/4$ and $3\lambda/4$ transformers are taken place of lumped capacitors and inductors by using π-type equivalent circuit [10]. The exact dimensions are also adjusted to optimize the performance of the balun using EM simulation tool. Figure 13(a) shows the 3D EM model of the 5.2 GHz balun and the photograph of realized circuit is shown on Figure 13(b). Observing measured results in Figure 14, the amplitude imbalance is -0.45~-1.4 dB and the phase imbalance is $177.3^\circ \sim 180.5^\circ$ from 5.15to 5.35 GHz.

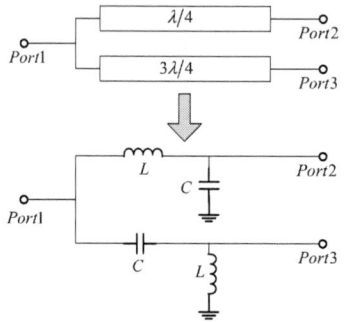

Figure 12: Equivalent Circuit of Balun

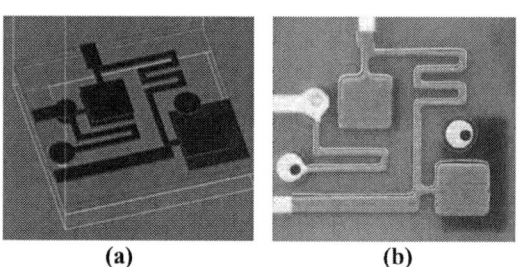

(a)	(b)

Figure 13: 5.2 GHz Balun (a) 3D EM Model (b) Photograph

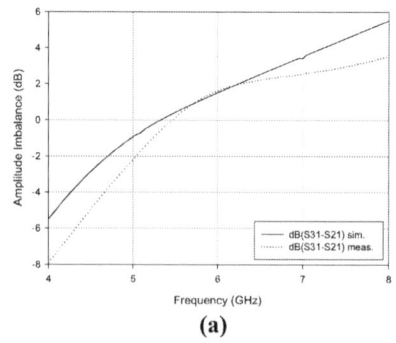

(a)

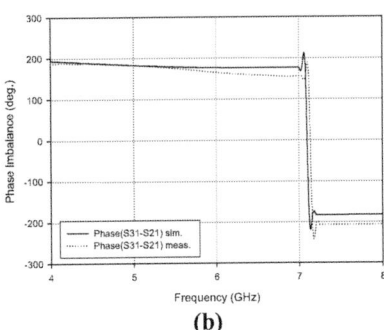

(b)

Figure 14: Simulated and Measured Response of 5.2 GHz Balun

Diplexer

The simple high-pass and low-pass, two ports networks, are used to design diplexer [11]. Figure 15 shows equivalent circuit of diplexer. The 3D EM simulation tool is used to adjust the dimensions and performance of the diplexer. Figure 16(a) shows the 3D EM model of the diplexer and the photograph of realized circuit is shown on Figure 16(b). Observing measured results in Figure 17, the insertion loss are less than 5.4 dB for high-pass and less than 1.4 dB for low-pass. This measured results is not good, therefore the scattering parameters at port 2 is observed by using the Smith chart for tuning this mismatch. Additional parallel capacitor is connected to port 2 of this diplexer. The capacitance is 1.1pF. After tuning the diplexer, the insertion loss is less than 2.3 dB for high-pass and less than 1.2 dB for low-pass in Figure19.

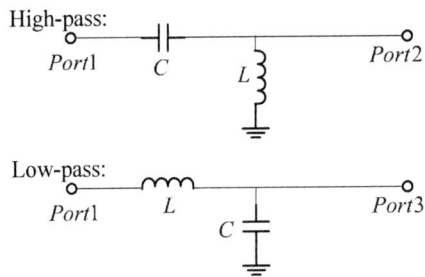

Figure 15: Equivalent Circuit of Diplexer

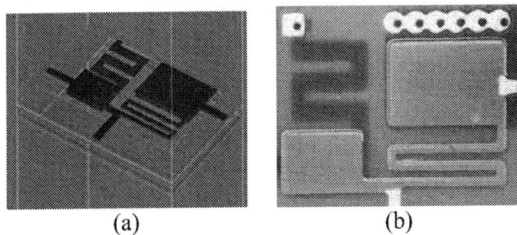

(a) **(b)**

Figure 16: Diplexer (a) 3D EM Model (b) Photograph

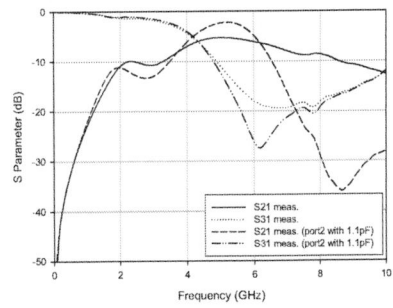

Figure 17: Measured Response of Diplexer

Coupler

The coupler is used the space between the substrate of Metal 1 and Metal 2 to bring the coupling. Further, the length of Metal 2 line is adjusted the magnitude of the coupling by using 3D EM simulation tool. Figure 18(a) shows the 3D EM model of the coupler and the photograph of realized circuit is shown on Figure 18(b). Observing simulated and measured results in Figure 19, the simulated and measured results are similar from 0.045~6 GHz. The coupling is -29~-28.6 dB from 2.41to 2.485 GHz and -23.5~-22.8dB from 5.150 to 5.350 GHz.

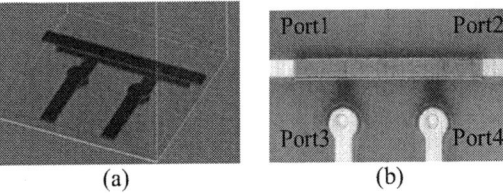

(a) **(b)**

Figure 18: Coupler (a) 3D EM Model (b) Photograph

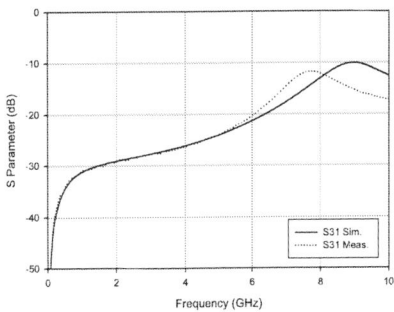

Figure 19: Measured Response of Coupler

Conclusion

Using all kinds of above-mentioned embedded elements, an IEEE 802.11a/b/g tri-mode dual-band wireless cardbus adapter was implemented in figure 20, which performance is as good as original module. The wireless cardbus adapter is an evaluation circuit to verify the electrical characteristics of these embedded passives in a system package. There were 51 elements replaced by using functional embedded passives, so the 6-layer organic built up substrate consists of almost 44.3 % embedded passives. Table 2 lists the comparison of the original passive components and the embedded version with some SMD components left outside of the substrate. The embedded passive technology of organic substrate leads advantageous position of SiP. Not only the organic substrate serves a variety of low cost process, but also can produce large size board for mass production. Since the resistors are not embedded to the board, and there are still some capacitors and inductors left outside the boards. In the future, we will develop special material for embedded resistor, and use higher dielectric constant material to achieve higher capacitance, and further reduce the size of embedded capacitor. We also plan to develop high permeability material to rise to the inductance. Of course, these new materials also needs to compatible with common PCB process.

Our next challenges on the way toward SiP integration are the MIMO and WiMAX RF system design. Moreover, we will combine embedded resistor material into organic substrate with accurate and stable property. Consequently, the variety of design can not only reduce the circuit area, but also produce cost-effective SiP technology. Besides, the proposed SiP process on organic substrate is the most compatible technology with the traditional PCB fabrication, although we have also developed many embedded passives on silicon substrate, which can be used for high frequency circuit. Moreover, we will use this cost-effective technology to design analog and digital circuits on the same board, so that the technology of SiP and SoC will be co-designed.

Figure 20: Photograph of Cardbus Adaptor

Table 2: Passive Components of Cardbus Adaptor

Item	Quantities of Components	Cardbus	
		SMD	Embedded Passives
Antenna	2	0	2
BPF	3	0	3
LPF	1	0	1
Coupler	1	0	1
Diplexer	2	0	2
Balun	1	0	1
Capacitor	65	32	33
Inductor	10	2	8
Resister	30	30	0

References

[1] IEEE Std 802.11a/D7.0, "Part11: Wireless LAN Medium Access Control (MAC) and Physical Layer (PHY) Specifications: High-speed Physical Layer in the 5 GHz Band," 1999.

[2] IEEE Std 802.11b, "Part11: Wireless LAN Medium Access Control (MAC) and Physical Layer (PHY) Specifications: Higher-speed Physical Layer Extension in the 2.4 GHz Band," 1999.

[3] IEEE P802.11g/D8.2, "Part11: Wireless LAN Medium Access Control (MAC) and Physical Layer (PHY) Specifications: Further Higher Data Rate Extension in the 2.4 GHz Band," Apr. 2003.

[4] L. K. Wu, B. C. Tseng, L. C. Liao, "Integral Passives in Multi-dielectric Substrate," International Symposium on Microelectronics, Boston'2000.

[5] C. S. Chen, S. F. Liu, C. K. Wu, etc, "Embedded Capacitors Technology in 2.4GHz Power Amplifier with Multi-Layer Printed Wiring Board (PWB) Process", Electronic Materials and Packaging, 2002. Proceedings of the 4th International Symposium on Electronic Materials and Packaging, 4-6 Dec. 2002, pp. 348-355.

[6] C. L. Weng, P. S. Wei, etc, " Embedded Passives Technology for Bluetooth Application In Multi-layer Printed Wiring Board (PWB)", Proc. 54th Electronic Components and Technology Conf., Las Vegas, Nevada, May. 2004.

[7] U. M. Jow, S. C. Liau, Ying-Jiunn Lai, etc, "Cost Effective SiP Solution for the RF Wireless Applications", IMAPS'2004.

[8] L. H. Hsieh, Kai Chang, "Compact Elliptic-Function Low-Pass Filters Using Microstrip Stepped-Impedance Hairpin Resonators", IEEE Transactions on Microwave Theory and Techniques, Vol. 51, pp. 193 – 199, Jan. 2003.

[9] J. S. Hong, M. J. Lancaster, "Microstrip Filters for RF/Microwave Applications", Wiley-Interscience Publication, New York, Chapter 5, pp. 142-151, 2001.

[10] D. W. Lew, J. S. Park, D. Ahn, N. K. Kang, C. S. Yoo, and J. B. Lim, "A Design of the Ceramic Chip Balun Using the Multilayer Configuration", IEEE Transactions on Microwave Theory and Techniques, Vol. 49, No 1, pp.220-224, Jan. 2001.

[11] M. Fritz and W. Wiesbeck, "A Diplexer Based on Transmission Lines, Implemented in LTCC", IEEE Transactions on Advanced Packaging, Vol. 29, No. 3, pp. 427-432, August 2006.

[12] H. C. Tung, W. S. Chen, and K. L. Wong, "Integrated Rectangular Spiral Monopole Antenna for 2.4/5.2 GHz Dual-Band Operation", in 2002 IEEE Antennas Propagat. Soc. Int. Symp. Dig., Vol. 3, pp. 446-449.

Embedding of chips in flex: a global optimization from thermal, mechanical and electrical RF perspective

B. Vandevelde[1], S. Brebels[1], C. Okoro[1], H. Oprins[1], L. C. Chen[1], W. Christiaens[2], J. Vanfleteren[2]

[1]IMEC, Leuven, Belgium

[2]IMEC, Gent, Belgium

Tel.: +32 16 281 513, Fax: +32 16 28 85 00; Bart.Vandevelde@imec.be

Abstract

More and more applications require flexible packaging. The highest integration is reached when also chips can be embedded in the flex. However, an optimum has to be found in between thermal resistance, mechanical bendability/flexibility and RF performance. To have the lowest thermal resistance, thick copper layers are needed to spread out the heat from the chip. From mechanical bending point of view, all layers should be as thin as possible and the chip should be lying in the neutral fibre. For good RF performance, microstrip lines are used for the introduction on flex and RF losses decrease by increasing the layer thicknesses. Between these conflicting trends, an optimum has to be found to have acceptable thermal, mechanical and RF performance. Using simulation techniques - Finite Element Modelling for thermal and mechanical analysis and 3D electromagnetic simulator for RF simulation -the influence of parameters such as copper thickness, chip thickness, location of the chip and polyimide thickness and number of layers is calculated. The simulation results give as function of these parameters the thermal (spreading) resistance, the maximum bendability (before chip fracture or copper fatigue) and the RF losses. The results are implemented in a spreadsheet which will be used by flex designers.

Key words: High density flex, chip embedding, finite element modelling, thermal resistance, mechanical bendability, electrical RF characterisation

Introduction

There is a continuous need for denser packaging of chips in order to cope with everlasting thirst for miniaturization and increased performance. Additionally, more and more applications require also a certain mechanical flexibility of the packaged solution in order to fit them better in the system or to adjust them to the body. As a consequence, the use of flexible substrates based on PI substrates and copper metallisation increased interest during the past years. Assembly of chips and components on top of the flex became already established, but this automatically involves that the assembled areas on the flex are rigid (mechanical bending is not allowed as it would destroy the components and/or interconnections to the flex). It becomes a challenge to also embed the passive (RF) structures and active components/chips into the flex to finally reach full flexibility over the whole area. There will still be differences in size of flexibility between the different regions, but at least the substrate could be bent to for example fit the human arm (Figure 1).

Accordingly, reliable technology for embedding chips inside flex would generate a lot of interest for many advanced applications such as portable mobile communication systems, smart textiles, robotics, implantable medical applications, etc (Figure 1).

As reliability and performance are important, several issues will show up. First, the thermal resistance, expressed as the temperature increase per unit watt dissipation, increases due to thinner copper tracks (compared to FR4) [4]. Second, the allowed mechanical bending of the components will be limited by its induced mechanical stress which may not exceed the ultimate strength values. Third, electrical RF losses will be larger with thinner copper tracks and thinner dielectric layers.

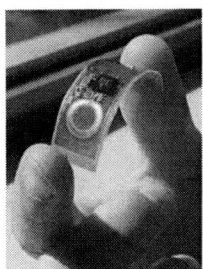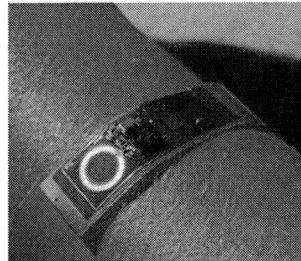

Figure 1: Application of flexible EMG/EEG/ECG circuit for personal health/wellness application (a chip embedded version is developed in the SHIFT project [2]).

Concept of global optimisation

Optimising the thermal, mechanical and electrical properties will result in conflicting trends (Figure 2).

From thermal point of view, the amount of copper should be optimized in order to decrease the thermal spreading resistance. This includes thick copper layers (preferably at the outside) and thermal vias.

From mechanical point of view, all stiff layers such as the silicon chip and the copper layers should be localized in the neutral fibre area, which is in the centre of the flex.

From electrical point of view, the distance between signal conductor and return path (=ground) is a compromise between low interference/radiation (small cross section) and low transmission line loss (large cross section). The width of the conductors is fixed to reach a good impedance matching.

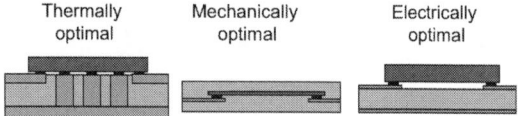

Figure 2: Optimal designs from thermal (left), mechanical (middle) and RF electrical (right) point of view.

Optimization approach

For the parametric study, a generic concept of chip embedded in flex is selected (Figure 3). A chip is glued to a first PI laminate. A second PI layer is applied to planarise the structure that allows applying an almost full first copper layer (for electrical RF reasons). A third PI layer is applied to cover the chip and first copper layer. A second copper layer is applied and again covered by a final PI layer. The vias contacting the copper layers with the chip are made using laser technologies. The concept is based on the embedding technology developed at IMEC Gent, but the concept represents a broad scope of chip-in-flex embedding principles [3].

Figure 3: generic chip embedded in flex structure (red colour = chip, green colour = polyimide, orange colour = copper).

For the parametric study of the above chip embedding concept, following parameters have been kept fixed:
- Silicon chip:
  - dimensions: 5 x 5 mm^2
  - 169 GPa, 2.3 ppm/°C, 150 W/mK,
  - epsr = 11.9, tanδ = 0.001 (= values for high-resistivity silicon)

- Polyimide
  - thickness = parameter P2; size = 40 by 40 mm^2
  - 3.4 GPa, 27 ppm/°C, 0.15 W/mK
  - epsr = 3.2, tanδ = 0.001
- BCB glue
  - Thickness = 5 μm
  - 3 GPa, 50 ppm/°C, 0.2 W/mK
  - epsr = 2.65, tanδ = 0.0008
- Copper layers:
  - 120 GPa, 17 ppm/°C, 390 W/mK,
  - sigma=58 10^6

The only parameters which are varied are the chip thickness, the thickness of the polyimide layers and the thickness of the copper layers. The variations are shown in the Table 1:

Table 1: Parameter variations in this study

Parameter	Nominal	Min	Max
Si thickness (μm)	25	10	100
PI thickness (μm)	15	5	50
Cu thickness (μm)	12	5	25

Thermal parametric study

The parametrically defined Finite Element Model is shown in Figure 4. As boundary conditions, natural convection is taken with a uniform heat transfer coefficient of 10 W/m^{2}K.

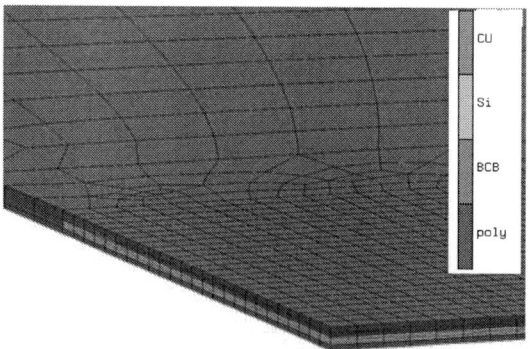

Figure 4: 3D finite element model to simulate the temperature distribution in an chip embedded in flex structure (software code = Msc.Marc [4]).

The sensitivity study showed that from the three studied parameters, the dominating parameter is the copper thickness. In the natural environment conditions, it is important to have an optimal spreading over the complete area in order to increase the area in contact with the surrounding air. As the thermal conductivity of copper is 2600 times higher than for polyimide, the heat will mainly distributed over the copper tracks/planes. Similar results based on both simulations and measurement has been already published in previous work [4].

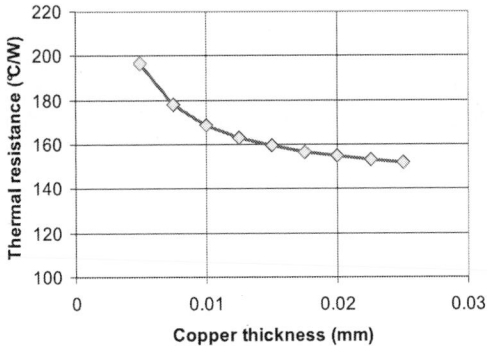

Figure 5: Impact of the copper thickness on the thermal resistance of chip embedded in flex.

The thermal resistance are defined as a function of these parameters. Therefore, the software code Noesis.Optimus has been used to create a response surface model. The formula is given by

$$R_{th} = 145.3$$
$$- 179.0 * chip\_thickness$$
$$- 6.87 * polymer\_thickness$$
$$+ 0.28 * \frac{1}{Copper\_thickness}$$

with chip thickness, polymer thickness and copper thickness given in **mm**. As expected, the thermal resistance is inversely proportional to the copper thickness.

Mechanical bending strength of silicon chips

For the mechanical part of the optimisation study, it is important to get a value for the ultimate stress that the chip can withstand before cracking. A dummy Si wafer has been cut into pieces of 5 x 50 mm² (standard cutting procedure with a diamond saw). The samples have been taken from wafers with two thicknesses: 0.700 mm and thinned 0.300 mm. Testing of thinner structures was not possible as the samples becomes so flexible that would not break anymore. The idea of using two thicknesses is to check if it would be possible to extrapolate to thicknesses lower than 300 μm.

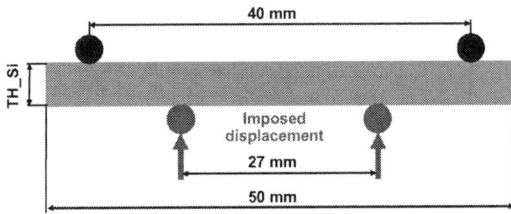

Figure 6: Four point bending test of diced wafers (schematic drawing and test equipment).

Figure 7 shows the results of the test for 30 samples at each thickness. For both thickness, almost the same average ultimate strength was found equal to around 300 MPa. This is about one order lower than the intrinsic strength of silicon (~ 2 GPa). The main reason is that due to dicing, small cracks are induced which lowers the ultimate strength. It is also envisaged that the distribution of the ultimate force over the 30 samples is also large. For example for the 300 μm thick silicon test samples, the maximum force varies between 5 and 10 N, which is a factor of two. The reason is that the dicing can cause cracks, which are random in size, number and location. Larger micro cracks result in lower force to fracture.

As a conclusion of this work, a safe ultimate stress of **150 MPa** is taken as the maximum allowed stress in the package.

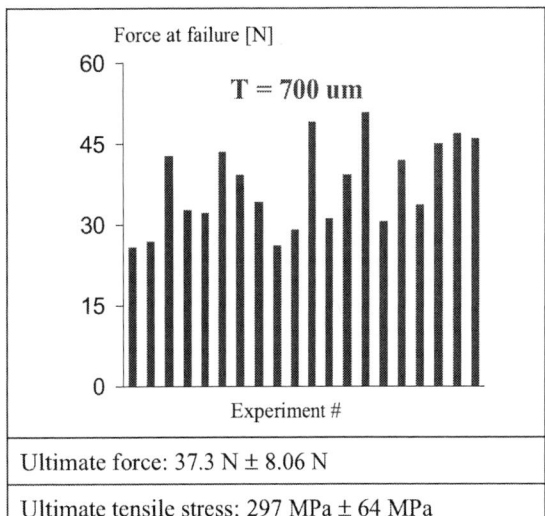

Ultimate force: 37.3 N ± 8.06 N

Ultimate tensile stress: 297 MPa ± 64 MPa

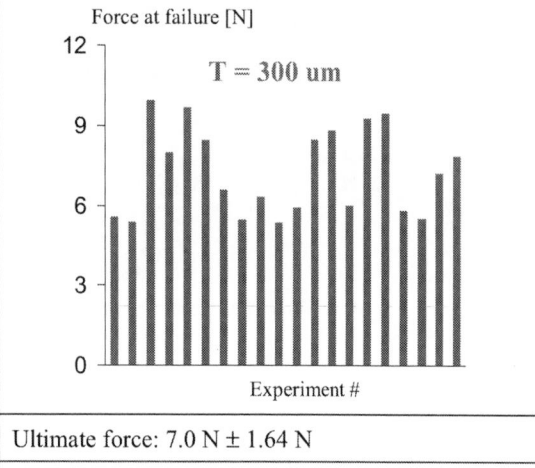

Ultimate force: 7.0 N ± 1.64 N

Ultimate tensile stress: 303 MPa ± 71 MPa

Figure 7: Results for 4 point ending tests on 700µm (top) and 300 µm (bottom) wafers.

Mechanical bending parametric study

In a first simplification, we can assume that the chip is located near the neutral fibre of the structure. The minimum curvature allowed to mechanically bend a multilayer flex is basically determined by the location of the critical layer. The stress in a layer is linearly dependent with the distance to the neutral fibre and inversely proportional to the curvature.

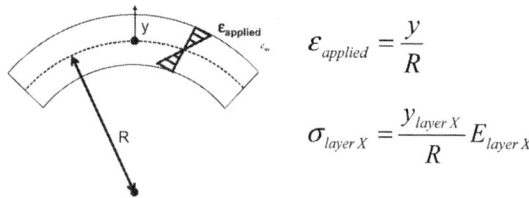

$$\varepsilon_{applied} = \frac{y}{R}$$

$$\sigma_{layer\,X} = \frac{y_{layer\,X}}{R} E_{layer\,X}$$

Assuming that the ultimate stress is 150 MPa, the minimum radius to avoid chip cracking should be equal to:

$$R_{min} = \frac{0.5 * Chip\_Thickness}{150 MPa} 169000 MPa$$

$$= 563.3 * Chip\_Thickness$$

with R_{min} and chip_thickness defined in mm.

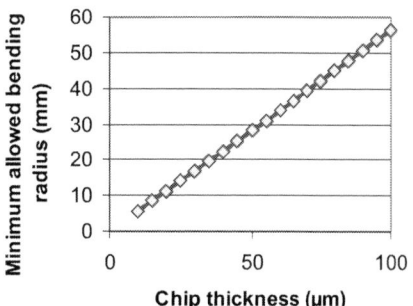

Figure 8: Minimum allowed bending radius versus chip thickness

In order to get a more detailed view, a 2D plains strain model is constructed for performing the bending experiment (Figure 9). Figure 10 shows the warpage for different chip thicknesses under the same applied force of 0.5 N. Figure 11 shows the maximum principle stress and curvature as a function of the silicon thickness (for two applied forces). At very low chip thicknesses, stress in silicon decreases again as the PI surrounding takes over the bending rigidity.

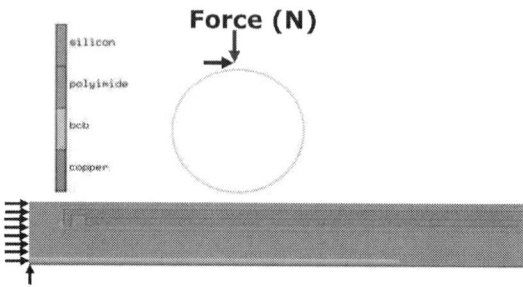

Figure 9: 2D plain strain model for simulating the mechanical bending of embedded chips

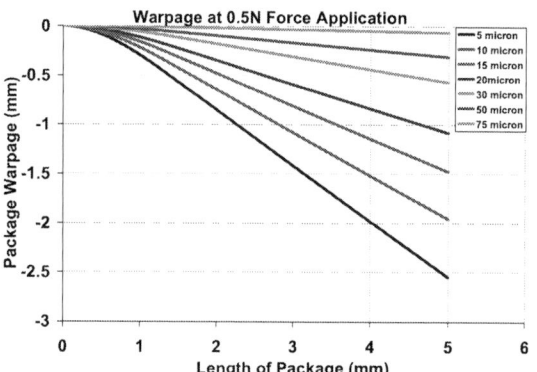

Figure 10: Warpage at 0.5 N force application for different chip thicknesses

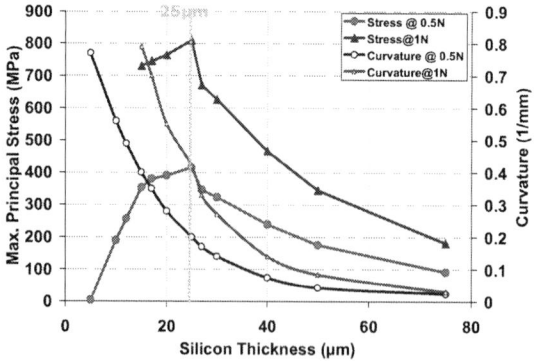

Figure 11: Results for 4 point ending tests on 700µm (top) and 300 µm (bottom) wafers.

Electrical RF study

The electrical RF performance is simulated of a microstrip transmission line on flex connected with vias to three 100 µm wide coplanar bond pads on an embedded thinned silicon chip (central bond pad for signal line, two outer for ground). Two high-frequency structures are compared in this study: a first one shown in Figure 12 with 25 µm thick silicon chip and polyimide layers and a second one shown in Figure 14 with 100 um thick layers. The microstrip line width is optimized to realize a 50 Ω transmission line (50 µm for 25 µm layers, 200 µm for 100 µm layers).

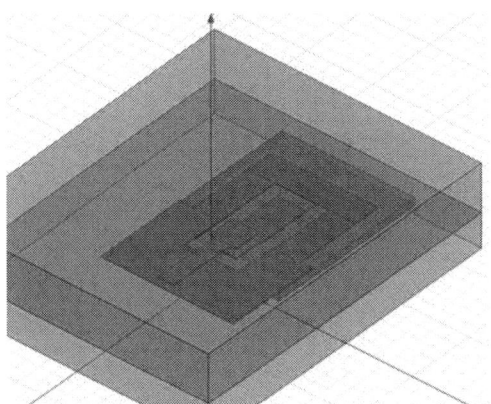

Figure 12: 3D view of microstrip to embedded chip connection with 25 µm thick layers.

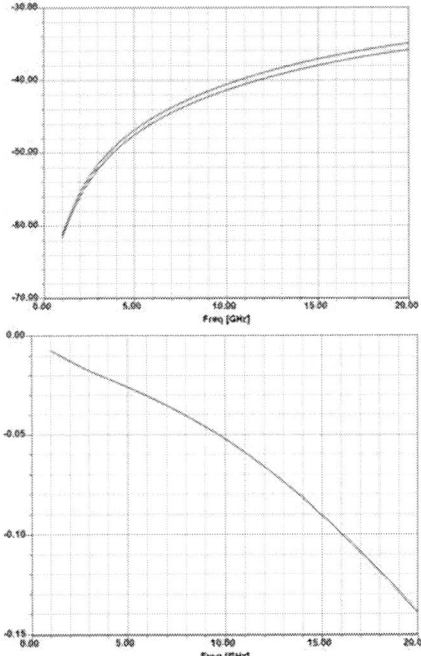

Figure 13: Electrical performance of simulated structure with 25 µm layers (see Figure 12). In the top graph the return loss in dB is given (blue for the flex and red for the chip side), the bottom graph shows the insertion/transmission loss in dB of the flex-to-chip connection.

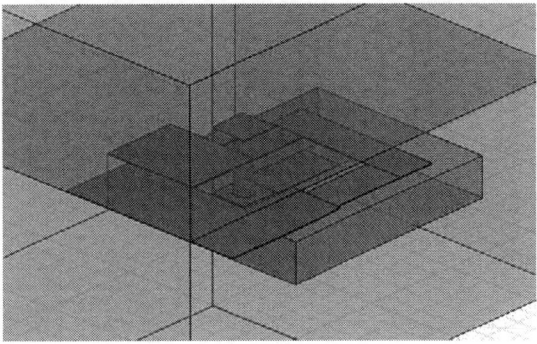

Figure 14: 3D view of microstrip to embedded chip connection with 100 µm thick layers.

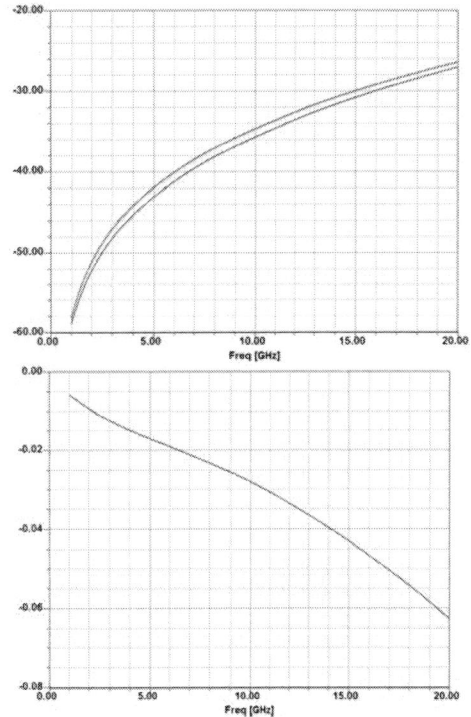

Figure 15: Electrical performance of simulated structure with 100 µm layers (see Figure 12). In the top graph the return loss in dB is given (blue for the flex and red for the chip side), the bottom graph shows the insertion/transmission loss in dB of the flex-to-chip connection.

The reflection (return loss) increases with the layer thickness due to a larger inductance of the interconnect via through a thicker substrate. The losses however decrease when the layer thickness increases because most of the loss originates in the microstrip transmission line. The wider line needed for a 50 Ω microstrip line on a thicker substrate leads to a lower loss. A good high-frequency

29

interconnection can be realized between flex and chip using 25-100 μm layers.

Conclusions

A parametric study has been carried out for a chip embedded structure, investigating the thermal, mechanical bending and RF performance. The results are implemented in a spreadsheet which will be used by flex designers.

A good high-frequency interconnection can be realized between flex and chip using 25-100 μm layers. The electrical choice of thicker or thinner layers depends on the application: thicker layers are selected for lower losses at lower frequencies while thinner layers are preferred for the higher frequencies.

Acknowledgements

The work was carried out under the EC IST funded project IP-SHIFT (contract number 507745). This work is a joint publication by the partners from the EU-SHIFT project. The authors therefore would like to thank the support from the EU commission and our colleagues who give support to the research in both projects. The authors would also like to thank Dominiek Degryse for the chip bending strength measurements he performed during his PhD work at IMEC.

References

Articles from conference proceedings

[1] J. Fjelstad, Flexible Circuit technology, http://www.flexiblecircuittechnology.com.

[2] www.shift-project.org

[3] Christiaens W., Bosman E., Vanfleteren J., Ultra-thin chip package using embedding in spin-on polyimides, 4th European Microelectronics and Packaging Symposium, Terme Catez, Slovenia, May 22-24, 2006.

[4] http://www.marc.com

[5] Chen L.C., Lehtiniemi R., Vandevelde B., Arslan A., Steady State and Transient Thermal Characterization for Flip Chip Interconnection on Flexible Substrate, 7th Eurosime conference, Como, Italy, 23-26 April, 2006.

[6] Yu, F.; Sang, W.; Pang, E., Liu, D. and Teng, J., 'Stress Analysis in Silicon Die under Different Types of Mechanical Loading by Finite Element Method (FEM)', IEEE Transactions on Advanced Packaging, vol. 25, no. 4, 2002, pp. 522-527

[7] http://www.noesissolutions.com

Wafer Level Packaging/Wirefree Die-On-Die
Applications to FLASH Memories and Smart Cards

Christian Val, Pascal Couderc, Christophe Serrano

3D PLUS

641, rue Hélène-Boucher, 78532 Buc, France

Tel : 00 33 1 30832650, Fax : 00 33 1 39562589, E-mail : sales@3d-plus.com

Abstract

Due to the consumers needs, the Ultra Dense Stacking of dice and mainly of memories is currently undergoing a heavy competition. Manufacturers of FLASH media for mobile phones and smart cards with Mega SIM, are demanding very small thickness and areas as close as possible to the die. The important developments, which have been carried out in Europe with the EUREKA/WALPACK project, allowed 3D PLUS to appraise this technology compared to the competition (as Chip-on-Chip technique) and, as a consequence, to improve it with regard to the cost. The use of industrial equipments at the same time Front-End and Back-End, thanks to the collaboration with Philips Semiconductors/NXP, allowed to simplify the WALPACK process by making a parallel process from A to Z. This Wirefree Die-on-Die process ("WDoD") follows the criteria which we established in the years 1990 when the standard 3-D was started. Commercial applications will be presented. As a conclusion, a comparison Wafer-to-Wafer versus Die-to-Wafer will be made and with perspectives of ultra miniaturized applications (of around 1 mm3).

Key words: Wirefree, Die-on-Die, Stacking, 3-D, Thru-Polymer via

Introduction

The needs for packaging for the commercial applications are well known:

- Need for thinning which definitively leads to the use of bare die.
- Need for multi functionality: the SiP notion is now totally admitted since it allows to quickly build electronics systems from available heterogeneous components.
- Need for low cost: this is the main need and it leads to following criteria :
  - Very good yield after stacking
  - Wafer level process /parallel processing from A to Z
  - Very low added value
  - Use of Front-End and Back-End standard equipments (this leads to contemplate a new category of manufacturing line: the "Mid-End". This WDoD technique does not demand the same cleanness level as with the Front-End
  - No proprietary packaging solution, which limits the number of supply sources (as PiP and PoP).

To satisfy these criteria, an important European programme, WALPACK, from 2001 to 2005, funded up to 20 M€ with ST Microelectronics, CEA/LETI, THALES, 3D PLUS etc, has allowed to establish the feasibility of a stacking technique totally "Wafer Level Process".

The prototyping has been designed thanks to a agreement between 3D Plus and NXP/Philips.

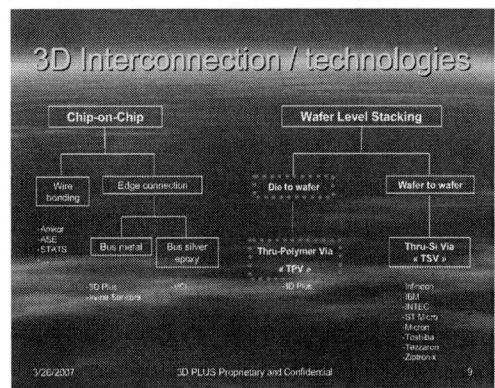

Figure 1

State of the Art

Regarding the stacking of bare die, 2 main families can be considered (see figure 1):

Chip-on-Chip and Wafer Level Stacking

The quasi totality of the 3-D modules are built in Chip-on-Chip by very large companies such as AMKOR, ASAT, ASE, STATS, ChipPack).

The limitations of this technique are, on the one hand the yield, above 3 or 4 stacked die, and on the other hand the size.

Another category of Chip-on-Chip concerns the "Edge Connection".

These techniques, started by Irvine Sensors and Thomson-CSF, then by 3D Plus in the beginning of the 1990's, allow to stack the die and to interconnect them on the sides of the module thanks to the Bus Metal (Figure 2).

Figure 2

As far as Wafer Level Stacking is concerned, the Wafer-to-Wafer is studied by most semiconductors largest companies worldwide and is based on the Thru-Silicon-Via (TSV).

These are ambitious studies which present following weaknesses :
- Use of non standard wafers which must be designed to have Thru-Silicon-via
- Chips generally not having the same size (except for memories) it is nearly impossible to stack them as they are.
- Impossibility to have a 100 % good wafer, which after stacking leads to a very low global yield.

Of course, particularly cases exist if an important memories manufacturer decides to rebuild a set of masks, in order to place the TSV inside the die. They can reach a very high density, at least as important as the one obtained with the WDoD, but it will be a mono-source proprietary product (Elpida/NEC/OKI - NADO programme – and Samsung Electronics ; presented at the 3-D architecture for Semiconductor NIV 2006, San Francisco).

Remembering the failure of the Wafer Scale Integration (WSI) some 20 years ago mainly for yields problems as well as the failure of the multi-chips modules (MCM), the criteria already retained in 1990 during the launch of the stacking of components have been almost reconducted for the Wafer Level package:
- Use of standard components, i.e. wafer without holes (price of mask set: around 500 to 800 000 USD)
- Use of standard components not always thinned

- Stacking of die of any size
- Stacking of Known Good Rebuilt Wafers (KGRW)
- Wirefree stacking with totally mature Bus Metal technology (applied to the Wafer Level package, leads to the Thru-Polymer- Kerf (TPK) or Thru-Polymer-Via (PPV) .
- Significant miniaturization: 100 µm/level, 50 µm in a near future
- Surface: XY: + 100 µm around the largest die.

Process

The flowchart is presented on Figure 3. (see annexe). It comprises 2 parts:

a) **Building of Known Good Rebuilt Wafers (KGRW) based on the Pick, Flip and Place of known good die (step 2) from wafers**

Use of the technique named "compression moulding" which allows moulding without moving material. Two Japanese manufacturers are currently supplying this type of equipment (step 3). See Figure 4.

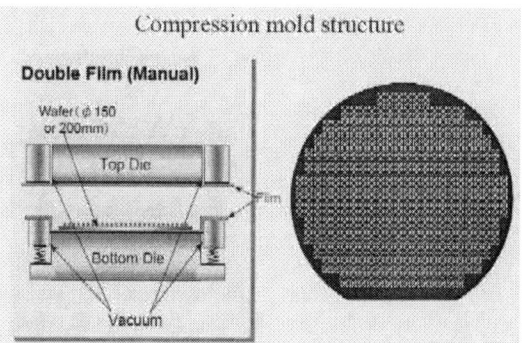

Figure 4

Step 4 concerns the grinding and it allows to thin the whole of the die whatever their original thickness (wafer of standard thickness). Step 6 consists in a redistribution layer (RDL) which is perfectly known in Front-End. These 2-D 1 to 6 steps are fortunately more and more studied and developed by several semiconductors manufacturers (Infineon, Freescale, NEC, etc) in order to simplify the traditional manufacturing of micro packages (µBGA). Steps 7 to 12 are optional and exceptional; they concern the die which must be burned-in (SDRAM, DDR2, DDR3, etc. In step 7, an operation Electroless Ni/Au (UBM) has been performed in order to obtain a Land Grid-array, which will allow placing such miniature CSP inside standard sockets. This is Step 10. Steps 11 and 12 are the same as steps 2 and 3, i.e. Pick, Flip and Place (11) and Compression Moulding (12)

b) Stacking of KGRW

On each KGRW, an adhesive 2-side sheet is pressed. Step 14 serves only to set up XY accurate locations Step 15 consists in placing the KGRW one upon the others with their adhesive sheets. Step 16 consists in doing a dicing of the stacked wafers in the same way as if it were an only wafer. After dicing, we get a network of kerf streets and step 17 allows building kerf streets edges plating process (electroless Ni + Au). This technique was set up in the beginning of the 1990's during the starting-up of the 3-D process at 3D Plus and is totally mature. Step 18, direct laser patterning, is a parallel process (by 200 modules) to etch 1, 2, 3 or 4 sides of the modules. This operation is totally mature, as it has been used since 1990 for the building of 3-D modules.

One alternative solution, allowing lowering the interconnection pitch consists in using Thru-Polymer -Vias (TPV). The use of 1, 2 or several staggered rows allows reaching pitches of 20 to 30 µm.

The Thru-Polymer-Via (TVP) is technically very close to the Thru-Polymer-Kerf (TPK).

Electrical tests, marking, etc operations follow.

Applications

For the small volumes 3D Plus carries out the design and the manufacturing.

For large volumes of the commercial type (mobile phones, etc) a Joint Development Agreement was signed in 2007 with NXP/Philips.

Micro implants

For a medical application and for very small volumes we have integrated 5 die on a 0.6 mm thickness and a total volume of 3 mm^3.

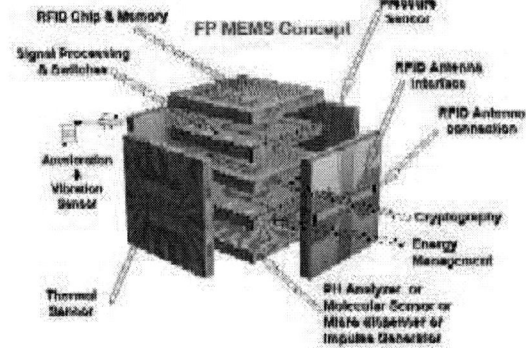

Figure 5

Extremely dense micro-systems for medical applications are contemplated thanks to a future European project (7e PCRD) – see figure 5

SiP with memories

Memories and memory-based SiP naturally represent a considerable market.
On Figure 6 we present a stacking of 8 SRAM memories on a 0.8 mm thickness.

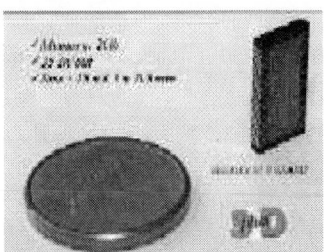

Figure 6

The FLASH memories represent an important part of the applications for the nomad products.
We can distinguish:

The FLASH media with a stacking of FLASH memories + the micro controller; discussions are in progress with the most important manufacturers of Nomad products.

The smart cards: for applications such as Mega SIM, it will be necessary to stack FLASH memories. This System in Package will be used inside the SIM card, i.e. within the 800 µm thickness of this card. The 5 levels (4 FLASH memories + secure controller + oscillator + passive components) must absolutely have a thickness lower than 550 µm in order to be embedded in the SIM card.
We are currently developing functional prototypes for the largest Smart Cards manufacturer.
With regards the mobile phones, base band applications are being defined by NXP.

Conclusions

Miniaturisation for commercial applications demands very high interconnection densities and low costs.

Wiser for former experiences, multi-chip modules, wafer scale integration, 3-D modules for Space, Defense and professional (3D Plus) applications, we learned that the yield constituted an important part of the production costs.

The WDoD process allows stacking Known Good Rebuilt Wafer (KGRW) only.

The main large-volume applications, which are, contemplated concern the SiP for smart cards, the SiP for mobile phones and stacked FLASH memories with micro controller for nomad applications.

This extremely important densification of 10 to 20 levels per mm with only 100 µm around the largest die allows launching extremely ambitious applications:

- Wearable microsystem for medical applications (Figure 5)

- Beyond the nano satellite application, the Satellite in a Cube concept becomes perfectly conceivable

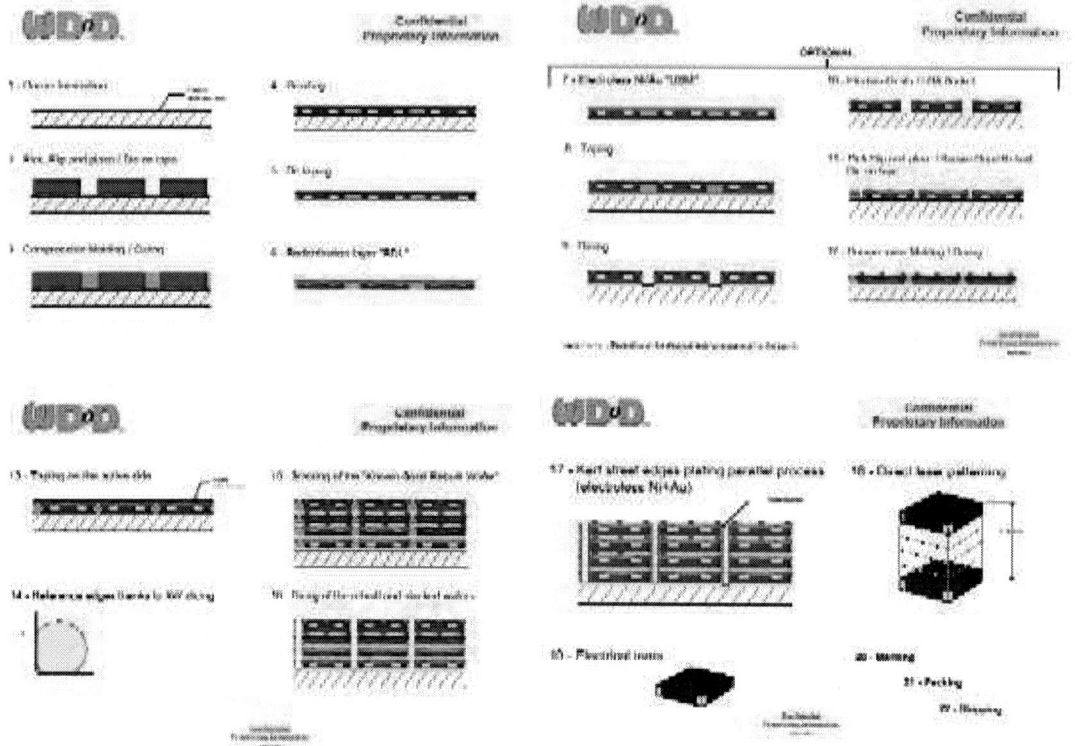

Figure 3

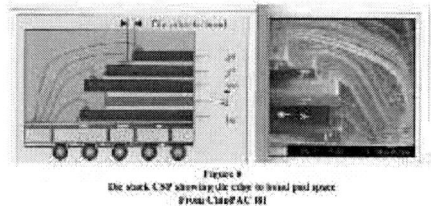

LTCC Materials Selection for the Realization of 10GHz Interdigital Filters

B Pierce, P Barnwell, M Ehlert, and T Vincent

Barry Industries, 60 Walton Street, Attleboro, MA 02703, USA

Tel. 508-226-3350, Fax 508-226-3317, Brian.pierce@barryind.com, Peter.barnwell@barryind.com

Abstract

Low Temperature Cofired Ceramic (LTCC) materials are a promising and ubiquitous technology for cost effective, reduced size packaging in microelectronics for radio frequency products. The selection of the optimum materials system is the first of several key variables in the successful creation of an LTCC product. Of the several commercially available materials, DuPont offers two systems, namely "DuPont 951 Green Tape®" and "DuPont 943 low loss Green Tape®". Conventional wisdom is that on the balance of design and cost trade-offs, 951 is the material of choice for applications up to 4GHz, while 943 is deemed the necessary material for higher frequencies designs. This paper describes the design, fabrication and characterization of a set of stripline interdigital 10GHz filters. The filters are constructed in both 951 and 943 with comparable design philosophies. Particular attention is paid to the fabrication variables and resulting insertion losses. These results will give the reader additional guidance in the selection of the proper material for the application.

Key words: LTCC, filter, DuPont, materials

Introduction

The objective of this study is to decide, on the balance of material and design trade-offs, which LTCC DuPont material is the best choice for the fabrication of an interdigital filter at 10GHz.

It is well understood that the dielectric loss factor is lower in DuPont 943 than DuPont 951 at this frequency. However, the dielectric loss factor cannot be considered in isolation; DuPont 951 is a mature and well-understood tape system that has fewer process and material variables. It is of benefit to work in a materials system where the variables can be tightly controlled for the fabrication of this filter topology.

Previous studies have demonstrated the dielectric loss comparison of 951 and 943 tape systems up to 100GHz [1]. Permittivity and dielectric loss tangents of these tape systems have been compared using split-cylinder technique over temperature [2]. There is published work on interdigital filter design within LTCC [3]. LTCC can offer a low cost, low power interdigital filter with a relatively small size at high frequencies. The design trade-offs including LTCC processing has not been studied for RF response comparing the two material systems.

In this work a 10GHz interdigital filter is fabricated on both 951 and 943 tape systems; the RF performance is examined by measurement and simulation.

Method

A 5 section, stripline construction, 25% bandwidth interdigital filter 0.2"x0.2" was designed; see Figure 1.

The filter response was simulated using SONNET©, CST®, and Ansoft Designer® software for initial conceptual design. A bulk conductivity of 2.8 x 10^7 S/m was used for the simulations.

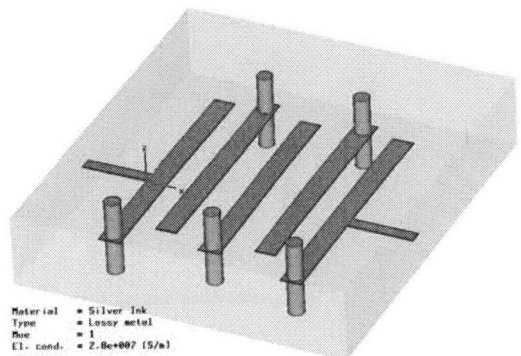

Figure 1: Stripline, 25% Interdigital Bandwidth Filter

The filters were measured using a vector network analyzer (VNA), Anritsu 37397A, to obtain s-parameters. GGB Industries Ground-signal-ground (GSG) probe tips were used to capture the data.

After the filters were fabricated the simulation results and measured results were compared. The dielectric constant, or relative permittivity ε_r, was checked with a ring-resonator technique [4].

Transmission line equations were used to calculate the reflected loss; this was taken away from the total loss to calculate the attenuated signal. Breakdown of the attenuated loss factors into dielectric and conductor loss was carried out using SONNET© simulation software.

Results

Figure 2 shows the simulated return loss of the filter for LTCC tape system DuPont 951. The simulated software response SONNET©, CST®, and Ansoft Designer® are compared.

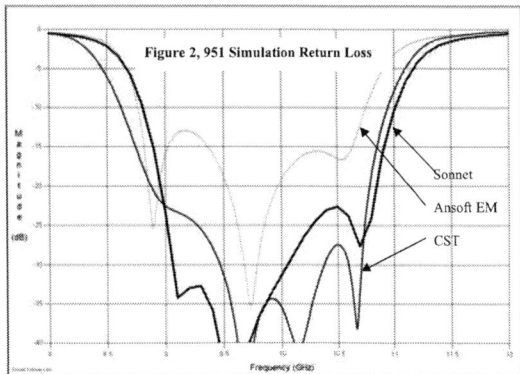

Figure 2: Filter return loss 951 simulation results

Figure 3 shows the simulated insertion loss of the filter for LTCC tape system DuPont 951. The simulated software response SONNET©, CST®, and Ansoft Designer® are compared.

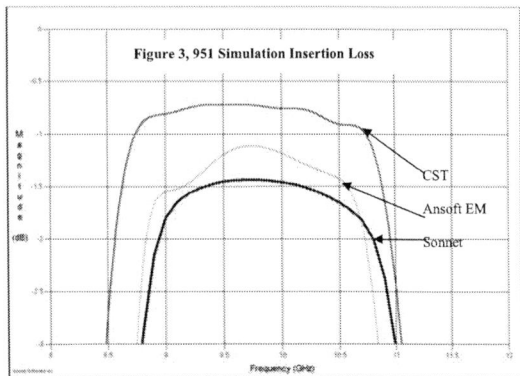

Figure 3: 951 Filter insertion loss simulation results

Figure 4 shows the simulated return loss of the filter for LTCC tape system DuPont 943. The simulated software response SONNET©, CST®, and Ansoft Designer® are compared.

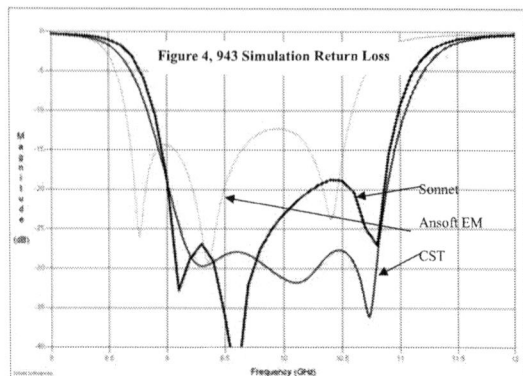

Figure 4: Filter return loss 943 simulation results

Figure 5 shows the simulated insertion loss of the filter for LTCC tape system DuPont 943. The simulated software response SONNET©, CST®, and Ansoft Designer® are compared.

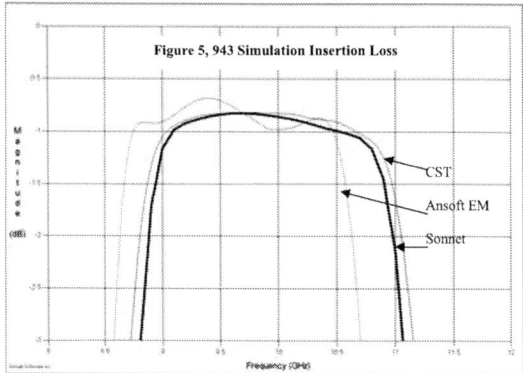

Figure 5: Filter insertion loss 943 simulation results

Figure 6 shows the simulated return loss of 951 as Figure 2 of the filter with the measured result of return loss included.

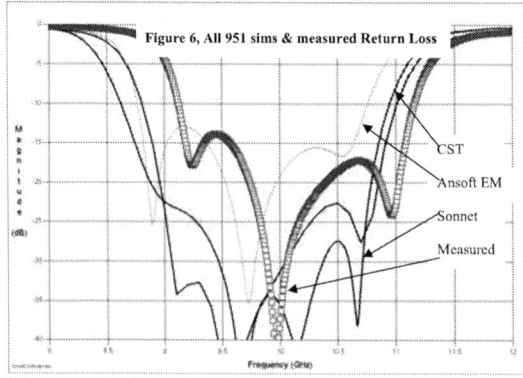

Figures 6: Simulated and measured return loss results for 951

Figure 7 shows the simulated return loss of 951 as per Figure 3 with the measured result of insertion loss included.

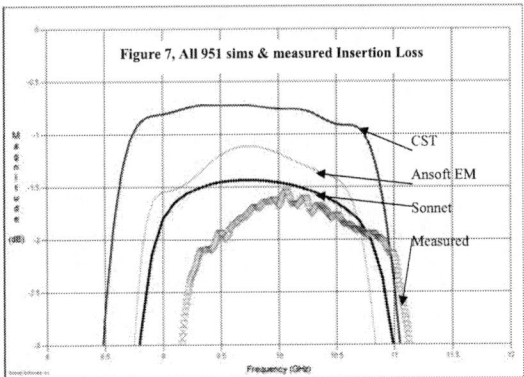

Figure 7: Simulated and measured insertion loss results for 951

Figure 8 shows the simulated return loss of the filter for 943 as per Figure 4 with measured result included.

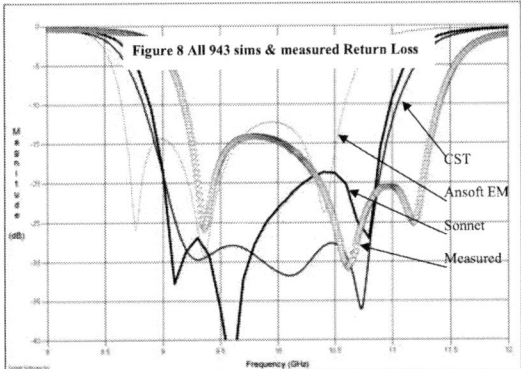

Figures 8: Simulated and measured return loss results for 943

Figure 9 shows the simulated insertion loss of the filter for 943 as per Figure 5 with measured result included.

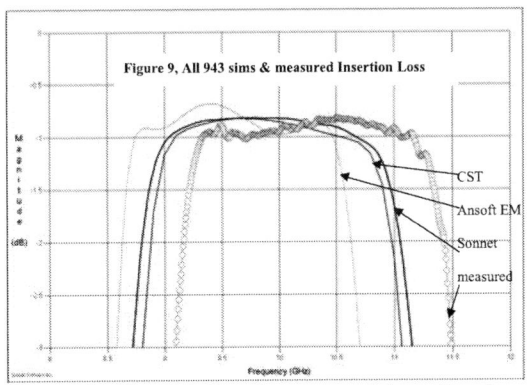

Figure 9: Simulated and measured insertion loss 943 results

Figure 10 compares the 951 and 943 results of return loss using SONNET© simulation package.

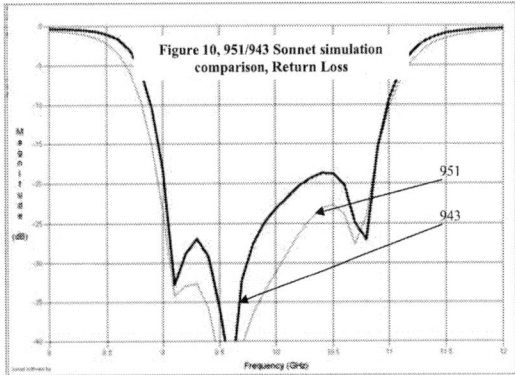

Figure 10: 951 and 943 simulation return loss comparison using sonnet©

Figure 11 compares the 951 and 943 results of insertion loss using SONNET© simulation package.

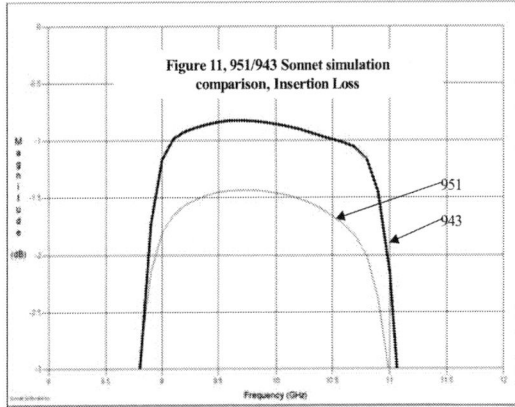

Figure 11: 951 and 943 simulation insertion loss comparison using sonnet©

Figure 12 compares the 951 and 943 measured return loss results.

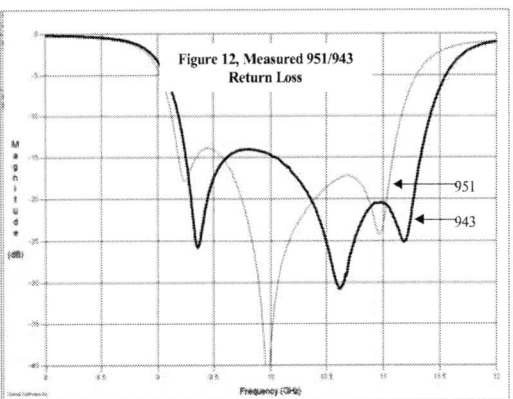

Figure 12: Measured return loss results for 943 and 951

37

Figure 13 compares the 951 and 943 measured insertion loss results.

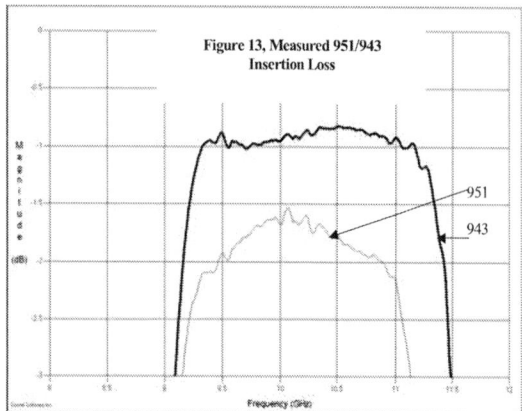

Figure 13: Measured insertion loss results for 943 and 951

Figure 14 compares the 951 and 943 simulated insertion loss results. SONNET© has been used to separate the conductor loss from the dielectric loss factor.

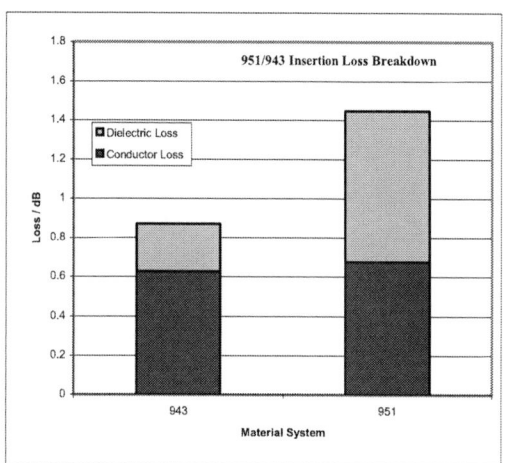

Figure 14: 951 and 943 bar graph attenuation loss factors, dielectric and conductor loss.

Table 1 is a summary of measured and simulated using SONEET© (for initial design) key design drivers.

Table 1: DuPont 951 and 943 measured and simulated insertion loss and center frequency

Table 1		
Insertion Loss	**Measured**	**Simulated**
943	0.9dB	0.9dB
951	1.6dB	1.5dB
Center Frequency	**Measured**	**Simulated**
943	10.29GHz	9.95GHz
951	10.15GHz	9.92GHz

Table 2 is a summary of key process drivers.

Table 2: DuPont 951 and 943 shrinkage and dielectric constant process variables

Table 2		
X,Y Shrinkage	**Measured**	**Specification**
951	12.5%	12.7+/-0.3%
943	8.8%	9.5+/-0.3%
Z Shrinkage	**Measured**	**Specification**
951	16.3%	15.0+/-0.5%
943	11.6%	10.3+/-0.5%
Dielectric Constant	**Measured**	**Specification**
951	7.85@10GHz	7.8 +/-0.1 @3GHz
943	7.05 @10GHz	7.4 @ 40GHz

Discussion

The same thick film printing screens were used for both 943 and 951 tape systems. This resulted in the difference bandwidths, 943 is 25.4% and 951 is 22.2%. The X, Y shrinkage specification difference is 9.5% (943) and 12.7% (951). The shrinkage factor will alter the resonator spacing that are especially sensitive in terms of bandwidth.

Center frequency shift for 943 is largely dependant on dielectric constant variability. Public domain data was used for the simulation using a dielectric constant of 7.4. The resultant 943 filter had a measured dielectric constant of 7.05, this was found using a ring resonator measurement method [4]. When the dielectric constant of 7.05 is inserted into the simulation, the center frequency shifts to 10.2GHz; 90MHz away from measured result.

Other process variables such as shrinkage tolerance and thick film conductor print also have to be considered. The shrinkage specifications given in Table 2 [5-8] are for bare ceramic. The measured shrinkage of 951 and 943 is affected by the metal print loading especially for the formation of the ground planes in this stripline structure. The 943-tape system seems to be more sensitive to this phenomenon. Shrinkage due to metal loading is difficult to determine a priori.

The SONNET© simulation results were the closest to measured results especially for insertion loss measurement where CST® and Ansoft® were optimistic, as seen by Figure 7. The sonnet software package [9-10] was used to compare the return and insertion loss differences between 951 and 943, and to separate the dielectric and conductor loss factors after the reflected losses were subtracted. Figure 14 demonstrated that the increase in insertion loss for 951 is due to dielectric loss factor.

Conclusion

Simulated and measured results for a 10GHz interdigital filter have been determined and compared for 943 and 951 DuPont LTCC material systems. Examinations of the results indicate that process variables, especially dielectric constant and shrinkage have a significant impact on RF performance.

While 943 yields better insertion loss performance, due to the decrease in dielectric loss, it is subject to greater process variability.

References

[1] M. Miller, D Zimmerman, D Nair, M Walsh, B Butler, "Comparison of LTCC Material Systems through Dielectric Transmission Characterization for Microwave/Millimeter Wave Applications" Feature Article, Advancing Microelectronics January/February 2007pp30-34

[2] M. D. Janezic, T. Mobley, D Amey "Temperature-Dependant Complex Permittivity Measurements of Low-Loss Dielectric Substrates with a Split-Cylinder Resonator" CICMT Denver 2006.

[3] Wang, F Barlow, A Elshabini "An Interdigital Bandpass Filter Embedded in LTCC for 5-GHz Wireless LAN Applications", IEEE Microwave Wirelss Components Letters, VOL,. 15, NO. 5, May 2005.

[4] K. Chang, L-H Hsieh "Microwave Ring Circuits and Related Structures" Wiley-Interscience 2nd edition pp139-152 2004.

[5] D Amey, R. Draudt, S. Horowitz "Designing for High Frequency with DuPont Green Tape" DuPont Microcircuit Materials RTP.

[6] 943 Green Tape Design and Process Guideline Addendum, DuPont Microcircuit Materials

[7] 951 Green Tape, Thick Film Composition, DuPont Microcircuit Materials.

[8] 943 Low Loss Green Tape, Thick Film Composition, DuPont Microcircuit Materials.

[9] J. C. Rautio and V. Demir, "Microstrip Conductor Loss Models for Electromagnetic Analysis" IEEE Transactions on Microwave Theory and Techniques, March 2003

[10] J. C. Rautio "An Investigation of Microstrip Conductor Loss" issue of IEEE Microwave Magazine, December 2000 pp.60-67,

Designing SMD Lowpass Filters in Multilayer LTCC Technology

R. Kulke, G. Möllenbeck, P. Uhlig, K. Maulwurf, M. Rittweger

IMST GmbH, Carl-Friedrich-Gauss-Str.2, 47475 Kamp-Lintfort, Germany

Tel.: +49 (2842) 981-214, Fax: +49 (2842) 981-499, E-Mail: kulke@imst.de

Abstract

The design flow for a lowpass filter in SMD technology demonstrating the advantages of multilayer LTCC technology will be presented. State-of-the-art simulation software was utilized to design and optimize the filter response. True 3-dimensional EM analysis, based on Finite Differences Time Domain (FDTD) method was applied to achieve accurate modeling of the filter's components and their interaction. An 800 MHz lowpass filter was selected for realization. The development started with the application's requirements and the derived filter specifications. The first design step consisted of ideal capacitors and inductors. The capacitances and inductances can be derived from filter theory, synthesis tools or from optimization with standard circuit simulation software. The challenge was to define real multilayer L and C components. This was accomplished with an add-on library for ADS^{TM} called MultiLib^{TM}, which consists of parameterized multilayer components. The specified capacitors and inductors were found by optimizing their parameters like electrodes' and spiral turns' geometries in a given LTCC stack. Parasitic parameters from these real components were also taken into account. The entire multilayer module was generated from those Ls and Cs with real via connections, ground layers and terminations of a SMD component in 1206 size (here: 3.2 x 1.6 x 1.2 mm³). Finally a 3D full wave analysis and optimization with the FDTD software Empire^{TM} was carried out including all coupling effects as well as the filter properties in a realistic test environment. The design work was finished with a tolerance analysis considering all tolerances of materials and process steps. The result is a filter in a 13 layer LTCC stack ready for production. DuPont 951 with different tape thicknesses was selected as material system.

Key words: Lowpass Filter, Multilayer LTCC, Surface Mount Technology

Introduction

Arbitrary filter characteristics are frequently required for RF and microwave circuit design. At high GHz frequencies it is useful to build filters utilizing transmission line segments. The length of a line stub is related to the wavelength, thus filters are smaller in size for higher frequencies. In case of low GHz or hundreds of MHz frequencies it is useful to apply capacitors and inductors to achieve acceptable filter dimensions. With today's multilayer RF substrates it is possible to integrate C and L elements in standard sizes for surface mount technology, which are called Surface Mount Devices (SMD). For standard applications filters are available from worldwide suppliers in nearly any dimension and performance. However, in some cases it is still necessary to develop custom tailored filters with specific properties, which are not offered as catalogue components. Another motivation for a specific design might be to integrate a filter into multilayer substrates of an entire module to reduce the number of components, which have to be assembled for that module. Whatever the motivation for a specific design is, the designer needs software tools and a procedure to realize the desired filter. The following chapters are intended to illuminate the mystery of multilayer filter design and provide a systematic approach with less trial and error.

Filter Specifications

The development work starts with the specifications of a lowpass filter. The assumption was made that a pass-band ($|S_{21}| > $-1dB) from DC to 800 MHz and stop-band ($|S_{21}| < $-20dB) starting from 1 GHz applies. An ideal LC-filter with a suitable topology was derived from these parameters as shown in figure 1. The filter coefficients can be obtained from standard literature like [1] or from any filter synthesis software. The multilayer substrate environment has to be chosen in the next development step. In this case the LTCC material system 951 from DuPont was selected, since it is a well known material with suitable properties.

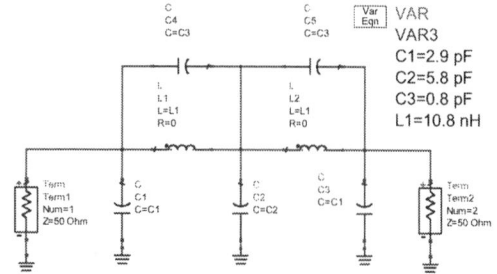

Figure 1: Filter Schematic

Determination of Multilayer Elements

The LTCC material system has a dielectric constant of 7.8 and provides 4 different tapes with unfired thickness from 50 to 254µm. The bottom layers will be used to form capacitors C1 and C2, the center layers for the inductors L1 and the top ones for C3. A first estimation for a capacitor's dimension can be made with the formula for parallel plate capacitors $C = \varepsilon_0 \varepsilon_r A/d$, where ε_r is the dielectric constant, A the area and d the thickness of a parallel plate capacitor. This lowpass filter was to be designed as 1206 SMD component with the total dimensions $3.2 \times 1.6 \times 1.2\text{mm}^3$. This allows for a multilayer stack of 13 ceramic substrate layers from different tapes. The stack is illustrated in figure 2 with the dimensions of fired tape thickness.

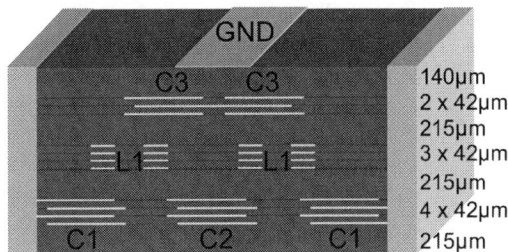

Figure 2: Multilayer LTCC Stack

The 3-dimensional multilayer inductors and capacitors were simulated and optimized with MultiLib [2]. This is a library with multilayer LTCC elements, which is an add-on to the design software ADS[TM] from Agilent. MultiLib allows an accurate 3D modeling of complex predefined and parameterized structures based on the Finite Differences Time Domain (FDTD) method. Figures 3 and 5 show the schematic symbols of a multilayer inductor and capacitor. The desired inductance of the spiral inductor (L1 = 10.8nH) was found with a simple circuit optimization in ADS. The optimization goal was to achieve $L = \text{imag}\{Z_{IN}/(2\pi f)\}$. This was obtained with the following geometry of the inductor: $4 \times 3/4$ turns across 4 conductor layers and 150µm turn width. The layout of the inductor is shown in figure 6. It is obvious, that the turns will not only result in an inductivity but also in parasitic capacitances. It was expected, that these capacitances will have a significant influence on the capacitor dimensions of the lowpass filter. Figure 4 shows the equivalent circuit

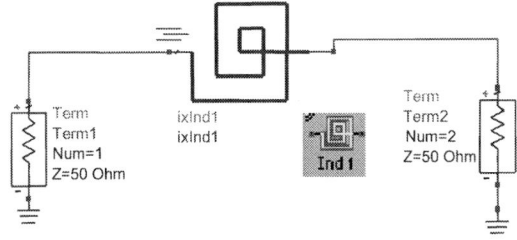

Figure 3: Multilayer Spiral Inductor (Schematic)

of the spiral inductor with parasitic capacitances. The values for C1' and C3' have been found by fitting the S-parameter curves of the equivalent circuit with the results of the multilayer circuit. This results in C1' = 0.35pF and C3' = 0.2pF. The final capacitances of the lowpass filter have been determined by C1-C1', C2-C1' and C3-C3'. C2 can be split up into 2 parallel capacitors, so that all bottom capacitors are equal 2.55pF.

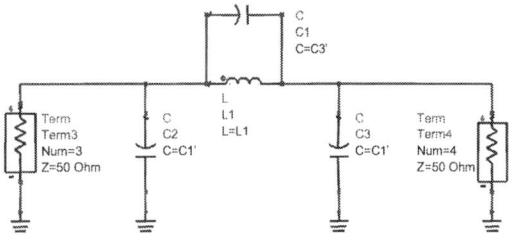

Figure 4: Spiral Inductor (Equivalent Circuit)

The multilayer capacitors have been found in a similar way as the inductor by optimizing the circuit from figure 5 to the desired capacitance value $C = -1/\text{imag}\{Z_{IN}\ 2\pi f\}$. This results in an area of 0.13mm^2 for C3 and 0.34mm^2 for C1 and C2 exhibiting three conductor layers (see layout in figure 6). C1 and C2 have been simulated with additional top and bottom ground plane (see figure 8).

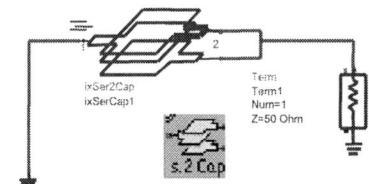

Figure 5: Multilayer Capacitor (Schematic)

Figure 6: Multilayer Inductor and Capacitor (Layout)

The lowpass filter now consists of 2 inductors with an inductivity of 10.8nH, 2 capacitors with 0.6pF and 4 capacitors with 2.55pF. The schematic circuit is shown in figure 7. An ADS simulation and comparison of the ideal and the multilayer filter showed that a transformation of L and C elements results in a suitable circuit. This is illustrated in figure 8, where the insertion losses of both filter configurations are plotted as a function of the frequency. It is obvious that the multilayer filter is still not completely representing a real SMD component, because inner line and via connections as well as the signal and ground terminations are missing. It is not possible yet to perform simulations of an entire SMD filter in ADS.

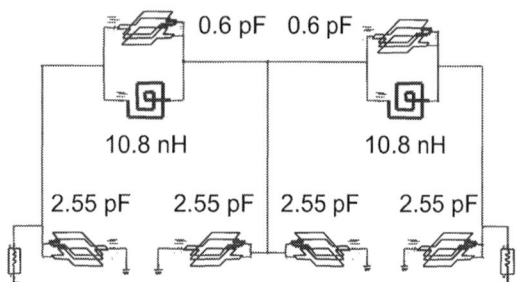

Figure 7: Multilayer Lowpass Filter (Equivalent Circuit)

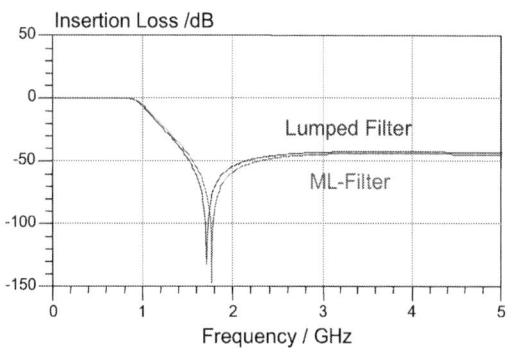

Figure 8: Comparison of Ideal Filter with Lumped Elements vs. Multilayer Filter

Full-Wave Simulation of 3D Filter Structure

In the next design step the multilayer L and C elements were transferred to a real 3D simulation software. Empire™ XCel [3] has been selected since this tool based on the FDTD method has proven to be very accurate. All conductors and via connections were implemented. Additionally two inner ground planes above and below the C1 and C2 capacitors were inserted for decoupling. Figure 9 delivers an insight into the filter structure. Conductors with real thickness and ohmic losses were analyzed as well as dielectric layers with tangent delta . The signal and ground terminations have been added in a final implementation step. This is shown in figure 10 with grey terminations and transparent ceramic layers.

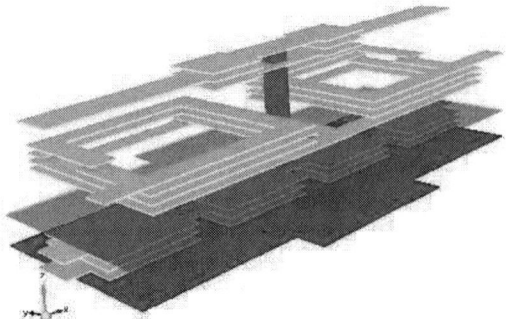

Figure 9: Multilayer Lowpass Filter

Figure 10: SMD Filter with Terminations

The overall simulation was conducted for the SMD filter mounted on a PCB with a microstrip line environment. This ensures that simulation and measurement are comparable at the end. Figure 11 shows return and insertion losses of the entire SMD filter. In comparison with the filter response in figure 8 it is evident, that in the final characteristic the slope between stop and pass-band is even steeper due to the parasitic capacitances of the terminations.

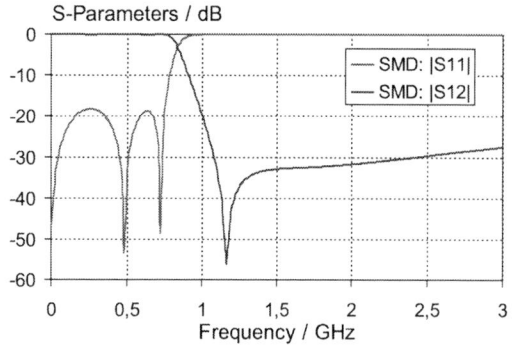

Figure 11: Simulation of 3D Lowpass Filter

Tolerance Analysis

A number of material and processing parameters will influence the filter characteristic. It was investigated how these tolerances will affect the capacitor values. The following parameters were considered:

1. Tape Thickness (C2): 50µm ±3µm
2. Z-Shrinkage: 15% ±0.5%
3. Dielectric Constant: 7.8 ±0.1

X/Y-shrinkage and conductor over- and undersize were neglected. Table 1 summarizes the parameters of the unfired LTCC tapes, which were utilized for the filter design. Table 2 continues these parameters. It shows the resulting fired thickness of each tape from its minimum to its maximum value. The second part of table 2 adds up all tapes in the stack. The nominal height of the filter results in 1.17mm with a tolerance of ±72µm. This is more than the thickness of the thinnest tape C2. The influence of the material tolerances on the capacitor values C1 and C3 of the lowpass filter is summarized in table 3. Capacitances vary from –8% for the thickest C2 tape (45,32μm) with lowest dielectric constant (ε_r =7.7) to +8% with minimum 39,72μm tape thickness and maximum ε_r = 7.9.

Table 1: Unfired Thickness and Tolerance Parameters of LTCC Tapes 951 from DuPont

DuPont 951 Tape System	unfired Thickness μm	Thickness Tolerance μm	Shrinkage %	Shrinkage Tolerance %
C2	50	3	15	0,5
AT	114	8	15	0,5
A2	165	11	15	0,5
AX	254	13	15	0,5

Table 2: Fired Thickness and Tolerance Parameters of LTCC Tapes 951 from DuPont

DuPont 951 Tapes	fired Min. μm	fired Norm. μm	fired Max. μm	Max.-Min. μm
C2	39,72	42,50	45,32	5,60
AT	89,57	96,90	104,31	14,74
A2	130,13	140,25	150,48	20,35
AX	203,65	215,90	228,29	24,64
13 Layers Stack:				
A2 +	130,13	140,25	150,48	20,35
4 x C2 +	288,99	310,25	331,74	42,75
AX +	492,64	526,15	560,03	67,39
3 x C2 +	611,78	653,65	695,97	84,19
AX +	815,43	869,55	924,26	108,83
2 x C2 +	894,86	954,55	1014,89	120,03
AX +	1098,50	**1170,45**	1243,17	144,67

Table 3: Capacitances for C1 and C3 taking Material Tolerances into Account

Capacitances	Min. pF	Nom. pF	Max. pF	Tolerance %
C3	0.56	0.6	0.65	± 8 %
C1	2.36	2.55	2.76	

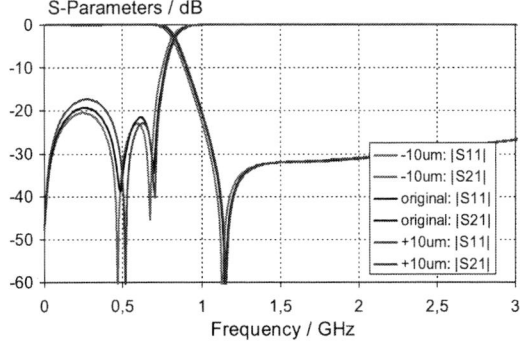

Figure 13: Tolerance Analysis with 10μm Under- and Oversize of Conductors

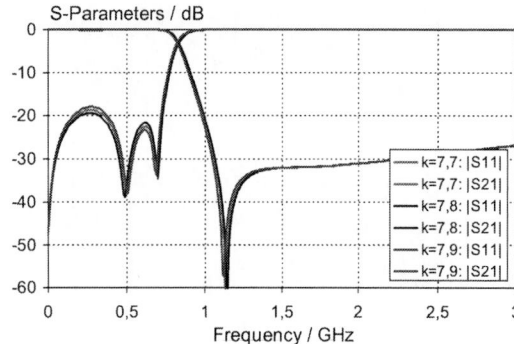

Figure 14: Tolerance Analysis with varying Dielectric Constant 7.8 ±0.1

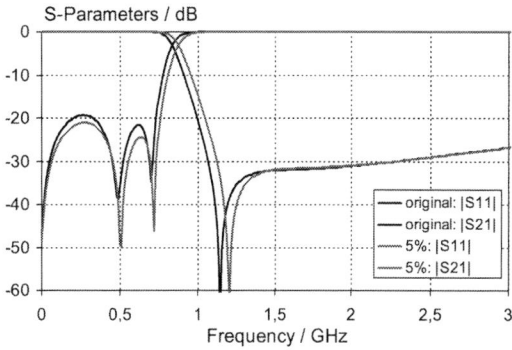

Figure 15: Tolerance Analysis with 5% Increase of X- and Y-Shrinkage

Further tolerance analysis was carried out for filters with varying parameters. First of all a 10μm under- and oversize of all conductors was implemented. The simulation results are plotted in figure 13. The filter response is shifted about -15MHz for 10μm undersize and +10MHz for 10μm oversize. Figure 14 shows the simulations for variation in dielectric constant. The influence is marginal. The highest deviation appears for a 5% increase of the X/Y-shrinkage. The filter response shifts about 40 MHz. However, shrinkage tolerance for DP951 is ±0.5%, so that 5% should not appear in a well controlled manufacturing process. These analyses showed that the lowpass filter is rather insensitive to material and process tolerances.

Manufacturing

The filter was manufactured at IMST's LTCC prototyping line. About 600 modules on four 2 x 2 inch² tiles were processed in 5 different variations. The first layout implemented the original dimensions derived from design optimization. The following variations in geometry were introduced deliberately:

1. 25μm oversize of all conductor
2. 25μm undersize of all conductor
3. 10% increase of shrinkage
4. 20% increase of shrinkage

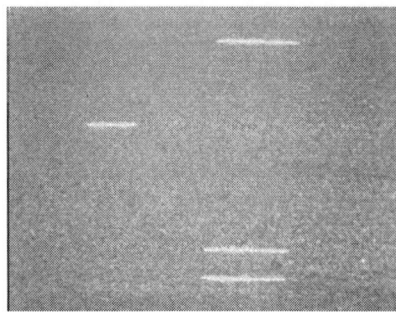

Figure 16: Diced Front Side of Multilayer Filter

The LTCC tiles were diced after processing. A photo of a front side cross section is shown in figure 16. Four conductors are visible: the center electrode of the top capacitor C3, the contact of an inductor and 2 electrodes of the bottom capacitor C1. The diced chips were sent to Syfer Technology Limited, UK. Syfer made the termination and plating process with their FlexiCap[TM] polymer terminations [4]. This termination is a silver loaded epoxy polymer that is flexible and absorbs some of the strain between the PCB and the ceramic of the LTCC chip. The termination is applied using conventional termination techniques, but instead of being sintered at approximately 800°C, the polymer is cured at 180°C. The polymer material has excellent adhesive and conductive properties after curing. Following the termination process stage the LTCC chips are plated with Nickel and Tin using the same methods as for the industry standard sintered Silver terminated capacitors. Figure 17 shows a photograph of the final SMD filters with terminations.

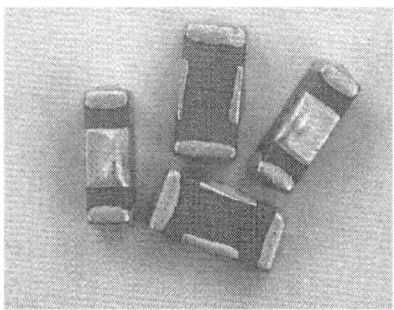

Figure 17: Terminated and Plated LTCC Chips

Characterization of Lowpass Filters

After all processing steps the filters were mounted on test boards and characterized. As test environment a 50Ω microstrip line (w=720µm) on FR4 substrate (h=410µm) with left- and right-hand ground areas was designed. The photograph in figure 18 shows the substrate mounted in a test fixture for characterization. The microstrip ports were contacted with SMA connectors, while the center pin is pressed on the microstrip line. TRL calibration circuits for 100 MHz to 12 GHz were prepared with 2 different line lengths (L$_1$=84.5mm, L$_2$=29.1mm), an

open and a short. A few filters were mounted on test boards and were measured shortly before finishing this paper. The first results of 4 filters are plotted in figure 19. Passband and stopband insertion loss as well as the slope between passband and stopband fit very well with the simulated circuit. Passband return loss is with –15dB slightly worse compared to the aimed performance. Passband insertion loss is <1dB for DC to 800MHz.

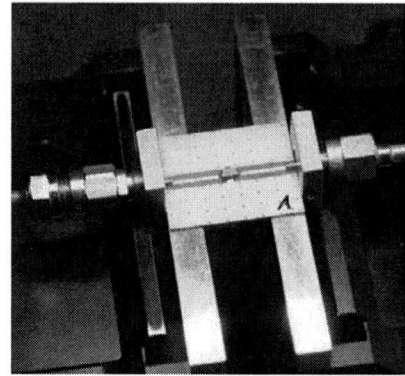

Figure 18: Test Environment for SMD Filter

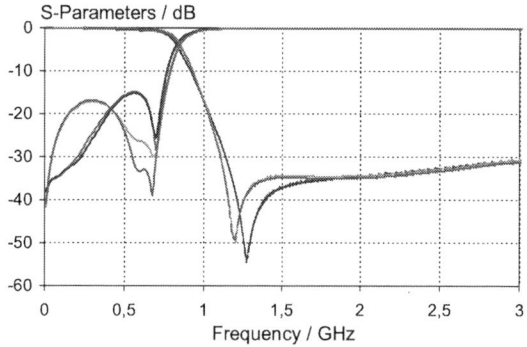

Figure 19: First Measurement Results

Conclusion

The authors demonstrated all design and evaluation steps for a SMD filter development. Accurate real 3D simulation tools were applied to obtain the aimed filter response within one development cycle.

Acknowledgements

The authors wish to thank Syfer Technology Limited, particularly John Shreeve for the extensive support of these activities.

References

[1] R. Saal, "Handbook of Filter Design", AEG-Fachbücher, Hüthig Verlag, 2. Auflage 1988.

[2] MultiLib[TM]: http://www.multilib.de/

[3] Empire[TM]: http://www.empire.de/

[4] FlexiCap[TM] polymer terminations: http://www.syfer.com/

High Frequency Characterization of VIAs up to 100 GHz in LTCC and Thick film Structures

M. Henry[1], Benito Sanz Iszquierdo[2], Charles Free[1] John Batchelor[2] and Paul Young[2]

(1) Advanced Technology Institute, University of Surrey, Guildford, United Kingdom,

(m.henry@surrey.ac.uk)

(2) Department of Electronics, University of Kent, Canterbury, United Kingdom

Abstract

Multi-layer packaging of microwave active and passive devices requires VIA interconnections that are to carry RF signals. Also VIAs can be used to increase the isolation between components and so reduce undesired coupling which leads to reduced circuit performance. The VIA effects will be more pronounced in the upper part of the mm-wave frequency band. In this paper a detailed characterization of VIAs in LTCC and thick film interconnections up to 100 GHz is presented.

Key words: material characterization, Via, LTCC, photo-imageable.

Introduction

Multilayer structures provide the microwave designer with the opportunity to develop highly integrated, high performance components. However, in many situations the microwave signals have to pass through VIAs connecting conductors on different layers of the structure. These VIAs will necessarily be small, and at low frequencies have little effect on the transmission of the signal. At microwave, and particularly millimeter-wave, frequencies the VIAs can introduce significant loss and an excess transmission phase. The excess transmission phase being defined as the phase change beyond that expected from the simple geometry of the structure. The loss is important not only in terms of the attenuation of the signal, but also because of the noise that it introduces. This can be a serious issue at millimetre-wave frequencies, where VIAs are use at the front end of a SOP (system-on-package) to connect an integrated antenna to an embedded filter.

So far there is no information in the literature that enables the circuit designer to predict the loss from a VIA of particular size and geometry. In this paper we have provided some general guidance on the effect of VIAs at different frequencies, and indicated when it is prudent to use a double-VIA structure to take the signal through particularly thick layers of substrate.

Design Structure and Methodology

The test structure consisted of a microstrip line of two embedded VIAs and a reference line as shown in Figure 1. The VIAs connected the microstrip lines on the upper surface, with the microstrip line embedded in the dielectric. In the practical circuit, two microstrip-to-coplanar transitions were incorporated so that the structure could be tested on a coplanar wafer-probing station. The distance separating the VIAs was 12mm; this was chosen to ensure there was no coupling between the evanescent modes surrounding the each VIA.

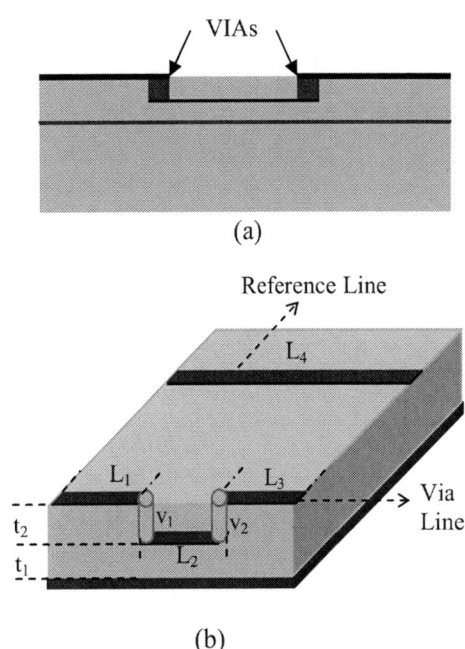

Figure1: (a) Schematic and (b) Cross sectional view of the test structure

By subtracting the measured transmission data from the two lines we were able to deduce the transmission characteristics of the VIAs.

$$Loss(VIA) \cong \frac{S21(via\ line) - S21(straight\ line)}{2}$$

From the figure 1(b) we can calculate the losses in VIAs using the following equation,

$$Loss(v1) + Loss(v2) = Loss(L4) - Loss(L1) - Loss(L2) - Loss(corrected - Loss(L3)$$

The Loss (corrected) represents the correction in loss in line L_2 due to the superstrate loading of thickness t_2 on the microstrip. When t2 is very small the correction due to the effect of superstrate loading can be neglected. The advantage of performing the difference measurement is that by calculating the difference in the transmission characteristics the errors due to mismatches at the feeding probes and the lines can be cancelled out. The VIA characteristics were studied for both rectangular and circular cross sections.

Loss in the VIA

Two rectangular VIAs of cross sectional dimension (80um*80um) were constructed and connected by lines of lengths L_1=12 mm, L_2=12 mm and L_3=12 mm. The reference line constructed was of length L_4=36mm. For the loss calculations the thickness t_1 and t_2 were selected such that the effect of superstrate loading in L_2 can be neglected, and also multiple reflection effects (explained in the following sections) due to the VIA discontinuities are negligible. The thicknesses selected were t_1=70 microns and t_2= 10 microns. The results for a rectangular VIA are plotted in dB/ μm in Figure 2. The experiment was repeated for a circular VIA and the loss characteristic is plotted in Figure 3.

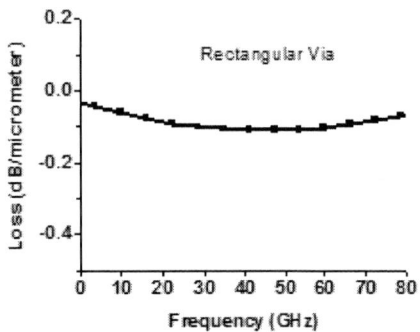

Figure 2: Loss in a rectangular VIA plotted in dB/μm for frequencies up to 80 GHz

The loss data curves shows circular VIAs have improved performance compared to the rectangular VIAs through out the frequency range.

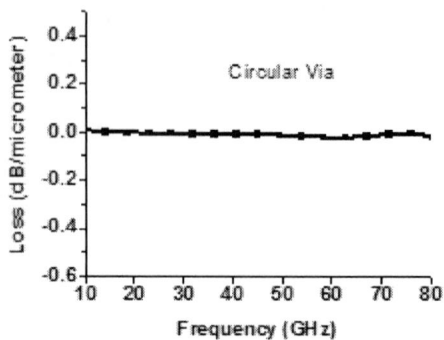

Figure 3: Loss in a circular VIA plotted in dB/μm up to 80 GHz

The loss effect on VIA depth

The structure was simulated using *ADS-Momentum®* , using an adaptive mesh technique to obtain an optimum result. We looked at circular and rectangular VIAs having different diameters, *d,* and thickness, *t*. The results from simulation are shown in figure 4 and figure 5. The effect of the losses are studied varying the depth of the VIA constructed. For a rectangular VIA graphs are plotted in figure 4 (a-e) by varying the thickness t_1 and t_2 keeping the total thickness constant (t_1+ t_2=80microns). In the figures 4 (a) 4 (b) and figure 4 (c) the loss in the reference line and the VIA line show a linear increase with frequency. But when the VIA depth exceeds 20 microns as shown in Figure 4 (d) and figure 4 (e) the loss in the VIA shows periodical ripples of different amplitudes.

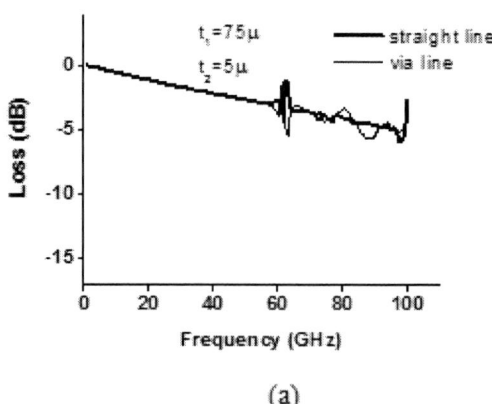

(a)

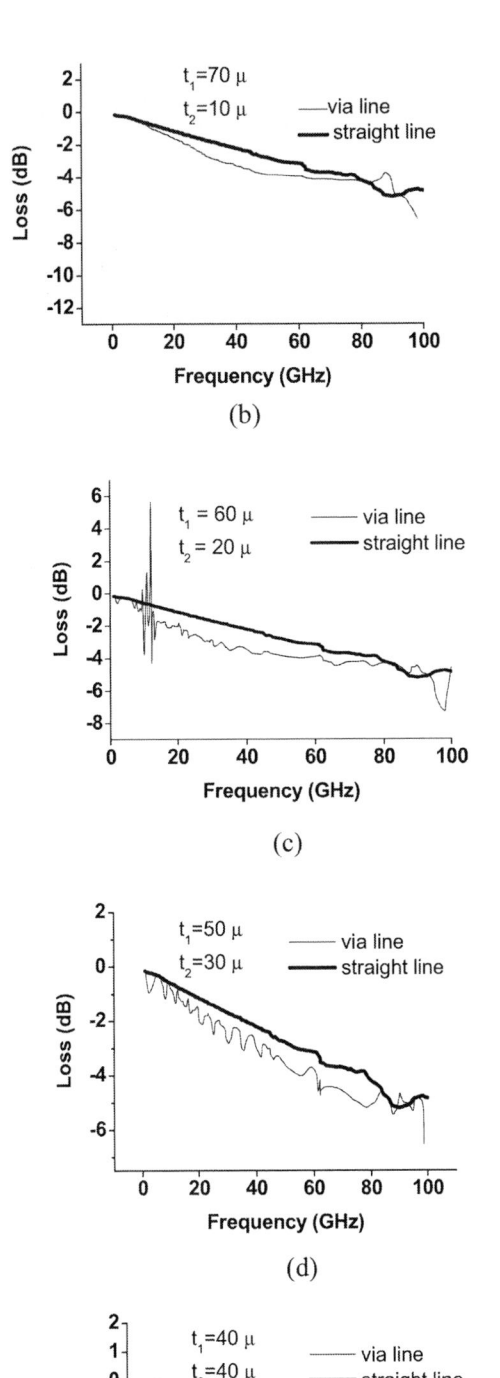

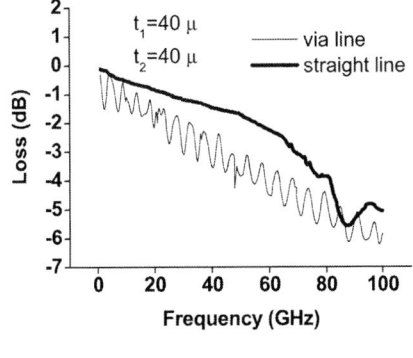

(b)

(c)

(d)

(e)

Figure 4 Loss effect in rectangular VIA of varying depths.

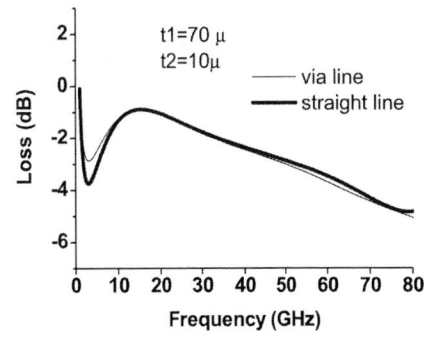

(a)

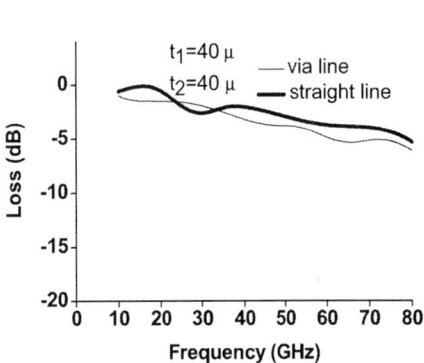

(b)

Figure 5 Loss effect in circular VIA of varying depths.

It can be noted that the VIA depth in circular cross section has less influence than in the rectangular cross sections. It is logical as the discontinuity in circular geometries will be less compared to the rectangular geometries. The significance of the result is that the non-uniformity in the VIA filling due to shrinkage of the paste after firing will have a significant effect on the losses due to the increased discontinuity.

Multiple Reflections

The unusual behavior of the VIAs for depths greater than 20 microns can be due to the following reasons:

(a) The resonance effect in the VIA
(b) Coupling effect of the parallel VIAs
(c) Multiple reflections at the line and VIA interfaces

The resonance effect can only be expected when the VIA dimension is comparable to the wavelength. In all the above cases illustrated the VIA dimension is less than $1/30^{th}$ of the minimum wavelength at 100 GHz. Also, the resonance effect if it exists will be observed only at certain frequencies. Hence the possibility of the resonance effect is ruled out. The VIAs were separated by a distance sufficient to ensure there was no coupling between the evanescent modes around each VIA. Also, the effect was not observed in the parallel VIAs of smaller depths. These rules out the possibility of parallel coupling effect. This leads to the conclusion that the periodical ripples arises in the loss plots are due to the multiple reflections at the VIA and line interface. The effect of multiple reflections at the two interfaces is conveniently illustrated in Figure 6. A theoretical analysis of the multiple reflections will give related information on the magnitude of the first order and second order reflections due to VIAs in the circuit.

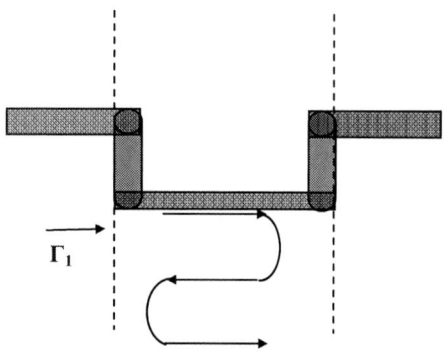

Figure 6: Multiple Reflections at the line and the VIA interfaces

It is these multiple reflections that probably account for the periodic ripples that are observed in the transmission plots.

Analysis on the structure without the VIAs

The results we obtained in the earlier sections showed significant errors associated with the use of long vertical VIAs as interconnections. We have done some simulation measurements on an alternative structure, shown in Figure 7.

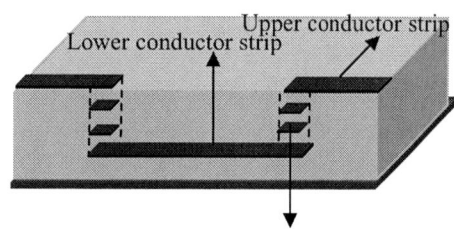

Intermediate overlapping strips forming horizontal VIAs

Figure 7: Proposed VIA-less Structure

The structure shown in figure 7 relies on electromagnetic coupling between the layers. This is an attractive technique at high millimeter-wave frequencies, where there will be very strong coupling across thin dielectric layers. The use of this technique thus overcomes the effect of the VIA discontinuities. Moreover, it is a relatively simple technique to implement, since it does not require the precision alignment associated with the alignment of small VIAs. We have simulated a typical structure, and the results are shown in figure 8. As expected,

there is significant loss through the structure at lower frequencies, where there is less coupling, but at frequencies greater than around 30GHz, the loss is comparable to that in a line containing two vertical VIAs. Thus there has been no significant degradation in electrical performance, but a major benefit has been achieved in terms of manufacturing simplicity. In particular, there is a significant saving in fabrication time since there are no VIA-fill steps, and also no consequent shrinkage effects. In addition, it can be seen that there are no significant ripples in the response, showing that the device can be designed to be well matched.

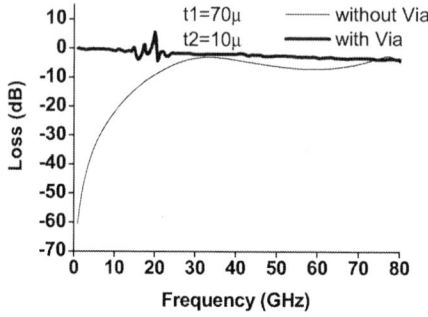

Figure 8: The simulated response of VIA-less structure.

VIAs and trenches in LTCC and photoimageable circuits

LTCC substrates of 90-micrometer thickness with supporting plastic sheet were patterned by pulsed laser ablation using a shadow mask. The layers were covered with the mask containing 100-micron diameter circular patterns prior to laser irradiation. A 3×3 mm^2 25 ns duration uniform KrF excimer laser pulse obtained from a Lambda Physik LPX 210i laser operating at 248 nm was used for laser patterning. The laser pulse with an energy density of 1 Jcm^{-2} was imparted on to each feature for 10 seconds in air at 80 Hz. A reasonable quality of VIAs without burning can be achieved even using high power excimer lasers by properly controlling the power with suitable metal masks.

For the purpose of simulation it was quite important that a realistic shape of VIA was chosen. Figure 9 shows examples of circular and square VIAs fabricated using photoimageable technology. As a further guide, figure 10 also shows a view of a narrow trench used to make vertical walls within a multilayer structure.

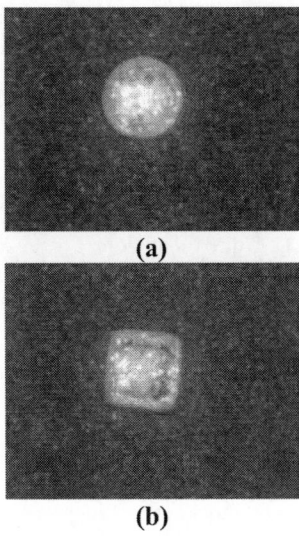

(a)

(b)

Figure 9: (a) Circular filled VIA (diameter = 40µm) (b) Square filled VIA (side = 40µm)

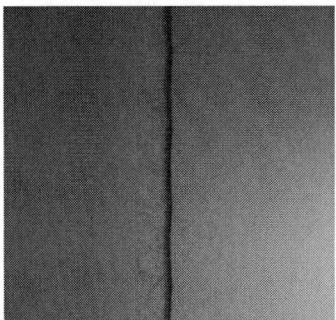

Rectangular trench, width = 50µm

Figure 10: VIAs and trench in LTCC structure

It can be seen that the fabrication process maintains the intended shapes of the VIAs. It can also be seen that the metal fills the VIA quite well leaving a small meniscus on the top. When the next layer of metal is applied it is reasonable to assume that the VIA will be completely filled with metal. This is a rather better situation that that experienced with VIAs in silicon [1, 2] where the VIAs tend to be wedge-shaped, and only partially filled with metal, as shown in figure 11.

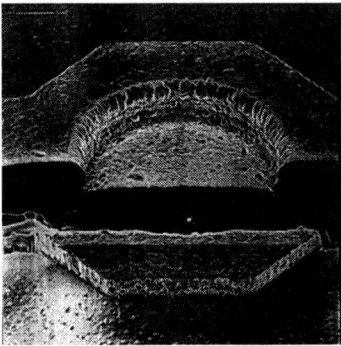

Figure 11: SEM view of a VIA in silicon

In LTCC the situation is much better in that the shape of the VIA is determined by the quality of the mechanical stamping process, or by the laser-drilling. In our case we used laser drilling, and some examples of laser-drilled VIA holes are shown in figure 12.

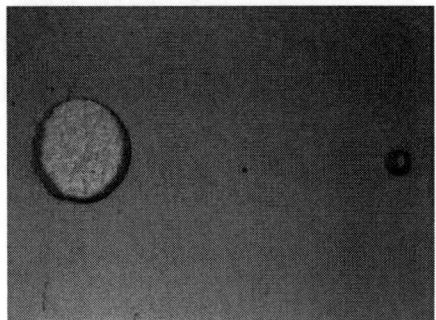

Figure 12: Circular unfilled VIAs in LTCC (diameters, 200 µm & 40µm)

For comparison two different sized VIAs are shown in figure 12. The larger hole had a diameter of 200µm and the smaller a diameter of 40µm. In both case the quality of the VIA shape was maintained.

Conclusion

The work has shown that VIA characterization is an important issue for high millimeter-wave circuits, and that significant loss can occur in VIA structures. A VIA-less structure has been investigated, and this shows good performance for millimeter-wave multilayer structures.

Acknowledgement

The authors acknowledge the financial support given by EPSRC (Engineering and Physical Sciences Research Council, UK) under Grant No. 107724.

References

[1] Free C. E, Hydes A, Grantham A, Aitchison C.S., Chen L, "Multilayer Interconnections in Silicon MCMs", Proceedings 31st International Microelectronics Symposium San Diego, USA, November 1998.

[2] Free C. E, Hydes A, Grantham A, Aitchison C.S., "Silicon MCM-D Interconnections for high speed applications", Proceedings 12th European Microelectronics and Packaging Conference, pp. 428-434, Harrogate, UK, June 1999.

3D Approaches of Optical Waveguide Fabrication on Flexible Substrate and its Application

Tze Yang Hin, Changqing Liu and Paul P. Conway

Loughborough University, Leicestershire, LE11 3TU, United Kingdom

T.Y.Hin@lboro.ac.uk

Abstract

Polymer optical waveguide devices will play a key role in several rapidly developing areas such as optoelectronics, optical networking, micro-opto-electrical-mechanical systems (MOEMS), computer and communications due to their easier processability and integration approach into a system. The first section of this paper reviews the commercially available polymeric waveguide materials and key optical properties. This is followed by a report on the preliminary trial in printing the polymeric material on flexible substrate. The initial trial has successfully demonstrated 70μm thickness epoxy layer printed on both polyimides Kapton and liquid crystal polymer (LCP) Vectra substrates. Both polyester screen and stainless steel stencil approaches were evaluated using DEK Horizon printer. In the last section of this paper, the authors review the 3D approach with the flexible polymeric waveguide application in spectroscopy. The approaches examined in this section are the conventional Waveguide Raman Spectroscopy (WRS) in spectroscopy application. Several key optical properties and design parameters such as refractive index, device's aperture, birefringence, fluorescence and intensity are discussed.

Key words: *optical polymeric waveguide, screen and stencil printing, flexible substrate, waveguide spectroscopy.*

Introduction

With optical fiber communication systems evolving from long haul networks into metropolitan and access networks, optical waveguide technology is increasingly being considered as a solution to the bottleneck of optical component design. Polymeric waveguide material has been recognized for their high thermo-optic coefficient, ease of fabrication, broad range of property adjustability, cost effective and good compatibility with other materials. Many commercial grade polymeric waveguide materials have been made available in the industry. But depending on specific applications, some trade-offs among materials properties are needed. In most cases, optical designers are required to work closely with the polymer scientists in order to tailor-make the required materials. The first section of this paper reviews the list of commercial grade waveguide material and the key optical and material properties.

Both polyimide and liquid crystal polymer (LCP) materials are evaluated as the substrate candidate for foil-based waveguiding material. The additive screen printing approach was investigated as the waveguide material deposition and patterning process where flexible substrate was attached to a backing plate for rigidity during the printing. In the last section of this paper, waveguide application in spectroscopy is reviewed and a novel waveguide coupling approach is proposed. Critical optical properties for this application are discussed.

Material

The report of Ma et al. [1] provides a good overview of available material systems, their properties and processing. The report covers conventional optical polymers such as poly-methylmethacrylate (PMMA), polystyrene, polycarbonate, polyurethane, epoxy materials; novel optical polymers such as dendrimer systems in fluorinated state, polyacrylates, fluorinated polyimides, PFCB aryl ether polymers, BCB, Cytop, Teflon AF, silicone; and non-linear optical polymers were evaluated. As briefly discussed in previous work [2], three waveguide forming process namely buried waveguide inside printed circuit board (PCB), optical layer on top of PCB and flex-foil based optical interconnects, shall be regarded as an add-on step to established optoelectronic production which implies temperature stability for the materials at least for short times over $T \geq 200°C$.

Refractive index is the key optical property in designing a waveguide. Ma [1] has described the refractive index's relationship with material's packing density, polarizability and the used wavelength. Chuang et al. have demonstrated the process for tuning the refractive index of UV epoxy via low electric-field-induced from 0 to 70V [3, 4].

The change in the refractive index may be due to the electric field that induced the change of the polymer molecular orientation. Fresnel diffraction is lowered with lower difference in the refractive index. Masahiro et al. [5] has provided a new method for measuring temperature dependence (dn/dT) of waveguide's material refractive index while Guenthner et al. [6] has proposed a validated Bicerano's method to predict the refractive index of isotropic polymers at visible wavelength.

Initial printing trial of epoxy was performed on substrate materials made of polyimide (Kapton HN) and LCP (Vectra). The epoxy used was ESL's grade 243-S which is the common grade for paste printing. The printer used was DEK Horizon printer.

Table 1 Material Property of commercial grade Kapton and LCP

	Kapton HN	LCP Vectra
Melting temperature	>400°C	280°C
Dielectric constant	2.9	3.5
Moisture absorption	2.8%	<0.02%
Coefficient of thermal expansion	20ppm/°C	0-30ppm/°C
Tensile strength	34 Kpsi	30Kpsi
Tensile modulus	370 Kpsi	1.3Mpsi
Specific gravity	1.42 kg/m^3	1.4kg/m^3

Polyimide is the dominant material in the flexible circuits industry because of its numerous advantages including excellent flexibility at all temperatures, good electrical properties, excellent chemical resistance, very good tear resistance and a high glass transition temperature. Many efforts have been done to incorporate polyimide foil with optical layers but failed due to the difference in material's coefficient of thermal expansion (CTE) resulting high internal stress under temperature difference induced deflection. The liquid crystal polymer is a thermoplastic contains rigid and flexible monomers that link to each other. When flowing in the liquid crystal state, rigid segments of the molecules align next to one another in the direction of shear flow. Once the orientation is formed, their direction and structure persist, even when is cooled below the melting temperature. This is different from most thermoplastic polymers whose molecules are randomly oriented in the solid state. LCP films can bond to other surfaces directly by thermal lamination to another LCP, glass, copper, gold or silicon surface at 260°C-270°C as described by Wang [7] thus providing a good opportunity for multi-layers fabrication.

For optical waveguide application, the conventional polyimides are not the best candidates. Polyimides are polymers that usually consist of aromatic rings coupled by imide linkages that are linkages in which two carbonyl (CO) groups are attached to the same nitrogen atom. They are anisotropic and also tend to be highly colored due to intramolecular charge transfer complexes that form between the electron rich diamine and electron deficient dianhydride. These properties result in high optical absorption and scattering losses. Recent investigations to produce colorless polyimides by interrupting the planarity of the polyimide chain, which the hexafluoro group has been incorporated to reduce the optical anisotropy. One example of the "colorless" polyimide is provided by Amoco Chemicals and was used for the waveguide fabrication [8].

According to Zhou [9], one major issue related to polyimides is the high birefringence which originates from the orientation of phenyl and imide rings during the imidization conversion of polyamic acid on the substrates. Additional birefringence can be induced by the stress if the thermal mismatch exists between polyimides and the substrates. The birefringence of materials and related polarization dependent properties are critical to device performance. Example of low birefringence (n_{TE}-n_{TM} < $1x10^{-6}$) polymers is the polyarcylate system as well as the PMMA (poly methyl methacrylate).

Generally, polymeric waveguide material can be classified into organic, inorganic and hybrid waveguide materials that are currently being used for a wide variety of optical interconnect applications. The conventional transparent polymer materials have high optical losses at the near infrared (NIR) range due to the overtone absorption of C-H vibration. It is well reported that the substitution of hydrogen atoms by halogen atoms, usually fluorine shifts the absorption overtone to longer wavelengths and reduces optical losses at 1.55μm. Introducing fluorine in the aromatic systems provides an efficient approach without sacrificing the thermal properties. However, perfluorinated polymer has poor adhesion to many substrates due to low surface energies and inert nature. Table 2 provides a revised list from previous work [2] on the commercially available waveguide materials.

Table 2 Revised Commercial Waveguide Material from [2]

Manufacturer	Polymer Family	Deposition
Microresist ORMOCER [12, 22]	Fluorinated sol-gel inorganic network silanes crosslinked with organic side chain	Spin coat, UV lithography
Wacker Chemie [23]	Siloxane	Casting, doctor blading
Norland NOA61	Liquid photopolymer	Dispense UV light curing
Rohm & Haas Lightlink [24]	Organic-inorganic hybrid	Coat, photo-lithographic
Terahertz	Fluorinated	Lithographic

Truemode[10]	acrylate	
Corning (Allied Signal)	Halogenated acrylate	Lithographic, RIE, laser
Dow Chemical	Benzocyclobutene (BCB)	RIE
Dupont Polyguide	Acrylate	Lithographic
RPO Pte Ltd	UV curable silicon resin	Wet etching
Luvantix	UV curable epoxy	Soft lithography, UV embossing
Zen Photonics	Fluorinated acrylate	UV imprint, dry/ wet etching
Siemens/ IZM	Polycarbonate	Hot-embossing
Amoco Chemical [8]	Polyimide	Spin-coating
Sigma-Aldrich	PMMA	Spin-coat
Asahi Glass Cytop	Amorphous perfluorinated polymer	Spin-coating, fiber processing
Dupont Teflon AF [25]	Amorphous fluoropolymers	Fiber processing

Experiment

The material deposition process in this experiment is to evaluate the printing approach. Printing method is flexible in customization and additive deposition compared with conventional planar spin-coating and photolithographic patterning. Printing processes available in the industry are screen and stencil printing, inkjet printing and doctor blading. In this preliminary trial, both screen and stencil printing were evaluated. Table 1 shows the parameters and materials used in the printing trial.

Table 3 List of Printing Trials

Substrate	Screen/ Stencil	Paste Material
Polyimide Kapton	polyester screen	Epoxy ESL423
Polyimide Kapton	SS stencil	Epoxy ESL423
LCP Vectra	polyester screen	Epoxy ESL423
LCP Vectra	SS stencil	Epoxy ESL423

The paste was printed with stencil printing and screen printing. A circular track pattern with 1mm track width was selected in the printing trial. Stencil used was made of stainless steel (SS) while the screen was made of polyester. Both squeegees with attack-angles of 45° and 60° were used and the squeegees were made of polyurethane or stainless steel. 60° attack-angle and polyurethane squeegee is used when lower pressure is sufficient for the paste printing. Potential damage to the polyester screen is expected if stainless steel squeegee is used. 45° attack-angle squeegee is used to exert higher pressure for more viscous paste.

The printing speed was set at 20mm/s. Print pressure was at 1.6kg. Squeegee length used was 170mm. The separation speed was 0.5mm/s while the separation distance is a variable to intended print thickness. Printed epoxy on the flexible substrate was subjected to post treatment with curing temperature at 150°C and bake time of 15 minutes. The printed circular track pattern is shown in Figure 1. The track width measured is 1mm while the thickness post-cured was measured at ~70μm. Typical spin-coated commercial waveguide material is reported at 10μm thick [10]. The thickness of the printed track is a controllable parameter of the screen or stencil design and the process variables. Figure 2 showed the printed top epoxy layer at ~70μm on the 50μm thick polyimide Kapton sheet.

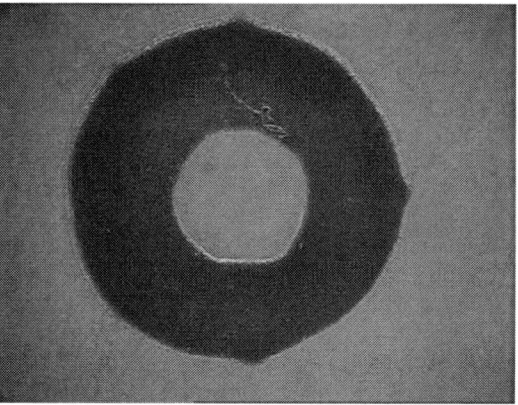

Figure 1 Screen printed epoxy circular track pattern on polyimide

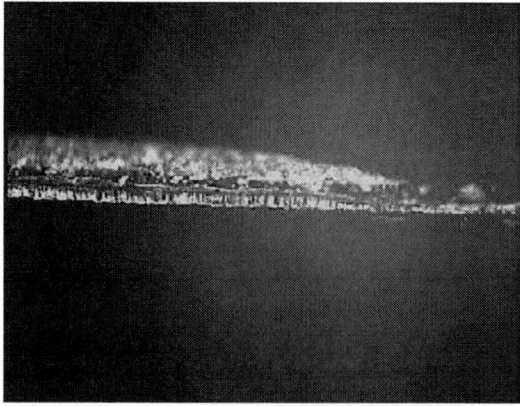

Figure 2 Cross section thickness of the printed epoxy circular track on 75um thick Kapton

Result and Discussion

A simple scratch test was performed to evaluate the adhesion between the printed epoxy and the flexible substrate as shown in Figure 3 and 4. The printed track was found to be better adhered on polyimide compared to the LCP Vectra in this test. The cured epoxy on the LCP was scratched away within the first five scratches, and left a very smooth fracture surface. It is suggested to evaluate the optimization of curing parameters and the use of adhesion promoter to enhance the adhesion of the printed layer with the substrate top surface. When the film has significant adhesion loss after stresses, the failure mode would change most likely from cohesive at initial to adhesive after stress. Scanning electron microscopy (SEM) and X-ray photoelectron spectroscopy (XPS) analysis of fracture surfaces can be used to detect surface contamination and both adhesive-cohesive failure mechanisms.

Apart from the printing process parameters, the thickness of the resulting layer is dependent of the curing process. However, the printed epoxy on the substrate does exhibit low bending radius. A large volume of data has been collected, particularly to understand the rheology of paste printing and the behaviour of the fluid under the squeegee printing process [11]. Other studies include the flow motion of paste material when close to the stencil surface, understanding the pressures involved to overcome aperture generated surface tension, studies of the forces generated by a squeegee passing at different speeds, and etc.

Figure 3 Simple scratch test performed on the printed circular track

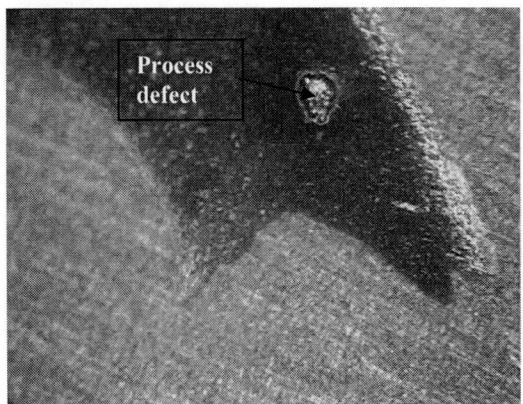

Figure 4 Poor adhesion between printed epoxy on LCP material

In paste printing process, it is important to look at the wetting behaviour of the paste and the underlying surface. One method to test the wetting behaviour is to look at contact angle measurement as described by Uhlig [12]. The printed pattern was well spread as observed in Figure 2 and has high wetting tendency.

As well as the devices being organic, much of the need is for them to be formed on flexible polymeric substrate. The low surface energy of polymers makes the wetting more difficult and the flexing of the substrate implies a need for toughness in the device materials. When a drop of liquid is brought into contact with a solid surface, the wetting tendency of the droplets can be modified with several methods. The common approaches are modifying the surface properties through applying a surface activation method, such as silane coupling agent [13], plasma activation methods [14] and changing the surface free energy of the liquid. To improve the adhesion for Vectra product, it is recommended to apply light sanding or grit blasting followed by a solvent wash on the average lap shear strength. The strength is expected to increase from 4.8 N/mm^2 to 9 N/mm^2 at 23°C with surface treatment.

It is generally accepted that in order to keep optical losses to a minimum, polymer optical films should exhibit variations in thickness of no more than a small fraction (e.g. 1/20th) the wavelength of the guided light. In many cases, deposited polymeric films can become severely wrinkled during drying due to the formation of a polymer "skin" on top surface of the film as reported in Andrew J. Guenther, et al. [15] resulted in high scattering losses.

Application

As the last section in this paper, the waveguide application in spectroscopy is reviewed. In the conventional Waveguide Raman Spectroscopy (WRS), theoretically a combination of a thin, high refractive index layer on top of a low index substrate results in an increase in Raman signal. As first reported by Levy [16] shown in Figure 5; and followed by Swalen [17], due to the small dimensions of the waveguide, large electric field strengths can be obtained in the waveguide over centimeter distances. The combination of high intensities and large sampling volume gives rise to the Raman signal detection of 3-4 orders of magnitude compared to conventional Raman scattering geometries. Kanger et al. [18] continued to utilize the waveguide's intense evanescent field with small film's thickness to enhance Raman signal intensity for thin layers of specimen. Fontaine et al. [19] have developed Variable-angle Internal-reflection Raman Spectroscopy (VAIRRS) technique that is capable to provide molecular signatures as a function of depth in buried interfaces. Another well known technique in Raman enhancement and photon capturing technology is the research project in Surface Enhanced Raman Spectroscopy (SERS) which is looking at the interaction between complex molecules and surfaces [20].

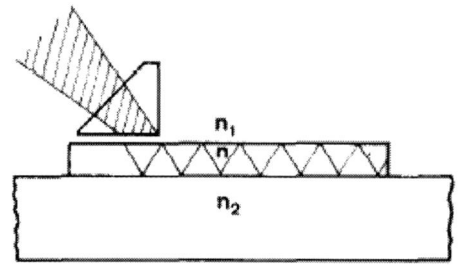

Figure 5 Levy's illustration of Waveguide Raman Spectroscopy [21]

The main challenges in photon-capture are to improve spectrum utilization, avoid the re-combination losses and increase the photon-detection count. As the earlier challenges are more relevant towards material limitation, the approach proposed in this section is to enhance the signal detection.

Chen [8] in his work has reviewed two types of thin film waveguide couplers namely tilted grating couplers and 45° waveguide coupling mirrors. Both designs can be fabricated via reactive ion etching (RIE) process. The output profile from the microcoupler can be determined using near – field diffraction theory of Fresnel approximation.

Fluorescence and photoluminescence are two behaviours in the polymer which can couple noise signal to the waveguide spectroscopy. Fluorescence is usually associated with UV excitation while luminescence is associated with excitation by visible light. Jenekhe et al. [21] have suggested that the fluorescence of conjugated polymer thin films originates from excimer emission and that the generally low quantum yield is the result of self-quenching. Fluorescence intensity is also reported to

be varied in different solvents used. Good solvent can unfold the chain with same polarity well. Future work is recommended to investigate the fluorescence behaviour of various waveguide materials at its application spectrum. Another optical properties worth looked into when selecting the right waveguiding materials is the material's degeneration upon intense illumination like other optical polymers. In order to investigate this behaviour, the degeneration of the chemical structure i.e. life time of the material was investigated by continuous intense illumination until the transmission of the exposure beam begins to decrease.

Altkorn has reported in [24] a method of estimating and optimizing the Raman intensity delivered by a Teflon-AF liquid core optical fiber (LCOF) to a spectroscopic instrument. He has concluded that Raman intensity is directly proportional to this effective fiber length and is a function of (1) the loss coefficient of the waveguiding materials at the laser and Raman wavelength; (2) physical length of the waveguide and (3) the sampling geometry.

For future work, investigation needs to be carried out to evaluate the waveguide material's emission spectra to understand the vibrational modes and hence choosing the right waveguiding material to improve the spectrum utilization. There is limited data available on this area. For structural investigation, Dynamic mechanical analyzer (DMA) measurement is required to evaluate the performance temperature dependent structural change upon processing.

Conclusion

In summary, a complete review on the commercial waveguide materials has been presented and the key optical properties are compared and discussed. Initial printing trial result on flexible substrate with epoxy paste has been reported on both polyimide and LCP materials. In the application of spectroscopy, thin film flexible waveguide has been put forward as a device to enhance signal counts in spectroscopy application. Both design and optical properties are surveyed and discussed briefly to understand the potential design challenges.

Acknowledgements

The authors gratefully acknowledge inputs and the evaluation samples supported by Alan Hobby (DEK), John Whitmarsh (ESL Europe), Jonathan Barclay (Ticona UK) and Roger Jamieson (Dupont UK).

References

[1] Hong Ma, et al., "Polymer-based Optical Waveguides: Materials, Processing, and Devices", Advanced Materials, vol. 14, 2002

[2] Tze Yang Hin, et al., "A Review on 3D Integrated Approaches in Multimode Optical Polymeric Waveguide Fabrication",

Proceedings of the 2007 Electronic Components and Technology Conference (ECTC), 2007

[3] Wei-Ching Chuang, "Fabrication of Polymer Waveguides using a Molding Process", Optical Engineering, vol. 45, 103401, 2006

[4] W. C. Chuang, et al., "A New Method to Fabricate Polymer Waveguides," in Proc. Progress in Electromagnetics Research Symp., Hangzhou, China, 2005

[5] Masahiro Tomiki, et al, "A New Method for Accurately Measuring Temperature Dependence of Refractive Index", Optical Review, vol. 12, 2005

[6] Andrew J. Guenthner, et a., "Methods for Estimating the Refractive Index Profile at Near Infrared Wavelengths of polymers for optical waveguides", Proceedings of SPIE vol. 4798, 2002

[7] Xuefeng Wang, et al., "Liquid Crystal Polymer (LCP) for MEMS: Processes and Applications", J. Micromech. Microeng. 13, 2003

[8] Ray T. Chen, et al., "Fully Embedded Board-level Guided-wave Optoelectronic Interconnects", Proceedings of the IEEE, vol. 88, 2000

[9] Ming Zhou, "Low-loss polymeric materials for passive waveguide components in fibre optical telecommunication", Proceeding of SPIE vol. 4580, 2001

[10] Erwin Bosman, et al., "Optical Connections on Flexible Substrates", Proceeding of SPIE, vol. 6185, Micro-Optics, VCSEL, and Photonic Interconnects II: Fabrication, Packaging and Integration, 2006

[11] Marc Desmulliez and David Clements, "Applications and Opportunities for stencil printing technology", 3D-Mintegration Consortium, private document

[12] Steffen Uhlig, "ORMOCER Materials Characterization, LAP & Micro-processing – Applied to Optical Interconnects and High Frequency Packaging", Linkoping University, Dissertation no. 1011, 2006

[13] Edwin P. Plueddemann, "Silane Coupling Agents", Plenum Press, New York, 2nd edition, 1991

[14] N. Inagaki, "Plasma surface modification and plasma polymerization", Lancaster, Technomic Corp., 1996

[15] Andrew J. Guenthner, et al., "Effect of Processing Conditions on the Properties of Polyimide Films in Optical Waveguides", Proceedings of SPIE, vol. 5212, Linear and Nonlinear Optics of Organic Materials III, 2003

[16] Y. Levy, et al., "Raman Scattering of Thin Films as a Waveguide", Optics Communications, vol. 11, 1974

[17] J. D. Swalen, et al., "Properties of Polymeric Thin Films by Integrated Optical Techniques", IBM J. Res. Develop, 1977

[18] Johannes Sake Kanger, et al., "Waveguide Raman Spectroscopy of Thin Polymer Layers and Monolayers of Biomolecules Using High Refractive Index Waveguides", J. Phys. Chem., 3288-3292, 1996

[19] N. H. Fontaine, et al., "Variable-angle Internal-reflection Raman spectroscopy for Depth-resolved Vibrational Characterization of Polymer Thin Films", The American Physical Society, vol. 57, no. 7, 1998

[20] http://www.npl.co.uk/smd/npl_research/cur_res_sers.html

[21] Samson A. Jenekhe, et al., "Excimers and Exciplexes of Conjugated Polymers" Science 5, vol. 265, no. 5173, 765-768, 1994

[22] R. Houbertz, et al., "Inorganic-organic Hybrid Materials for Application in Optical Devices", Thin Solid Films, 442, 194-200, 2003

[23] A. Neyer, et al., "Electrical-optical Circuit Board Using Polysiloxanes Optical Waveguide Layer",

[24] Nick Pugliano, et al., "Progress Toward the Development of Manufacturable Integrated Optical Data Buses", http://electronicmaterials.rohmhaas.com

[25] Robert Altkorn, et al., "Intensity Considerations in Liquid Core Optical Fibre Raman Spectroscopy", Applied Spectroscopy, vol. 55, no. 4, 2001

Modeling the Effect of Assembly Parameters on Warpage and Stresses of Molded Package for Inkjet Printing

Kimmo Kaija[1], Jani Miettinen[1], Matti Mäntysalo[1], Pauliina Mansikkamäki[1],

Mikiharu Kuchiki[2], Mikihiko Tsubouchi[2], Risto Rönkkä[3]

[1] Tampere University of Technology, Finland

P.O. Box 692

FI-33101 Tampere, Finland

Phone: +358 3 3115 5324, E-mail: kimmo.kaija@tut.fi

[2] Kyoritsu Chemical & CO, Japan

[3] Nokia Research Center

Abstract

Finite element method can be used effectively and fast to test different assembly structures and to find the best alternatives from the desired viewpoint. When inkjet material deposition is used, the target surface should not be excessively warped, if accurate drop placement is required. The components of an example system were encapsulated with mold material and different assembly options were studied with computational modeling to determine their effect on the residual profile of the package. Thermal and cure shrinkage create mechanical stresses that warp the package after the mold resin is cured at elevated temperature. The molding process was modeled with different assembly alternatives and the resulting simulated profiles of the package are compared. Experimental tests were also made to determine film-mold resin combinations for packaging material's performance at the elevated temperatures it has to undergo during inkjet manufacturing of wiring layers.

Key words: FE-modeling, encapsulation, electronics packaging, printable electronics, inkjet

Introduction

The increasing demands in electronics packaging and manufacturing methods are driving the development of current production methods and encourage seeking for alternative manufacturing processes. Printing methods, such as inkjet printing, have gained a lot of attention on manufacturing of electronic circuits. Inkjet manufacturing allows a maskless, fully additive and a non-contact material deposition method [1,2]. Without the need for masking and etching steps, circuits can be processed rapidly, with low cost, and without the material waste that is required with subtractive processes, such as photolithography [3]. The reduction of processing steps reduces also the energy consumption. The development around printable electronics has been fast, and it is likely that inkjetted circuits will soon increase popularity at low-cost commercial products [2].

Multilayer structures are formed by depositing in turns conductive and insulative materials directly on the desired areas. With inkjet manufacturing a multilayer printed circuit board (PCB) with thin profile can be formed. One example

is demonstrated in [4], where a 20 layer PCB with thickness of 200 µm was created by utilizing inkjet process.

Inkjet manufacturing allows also a new approach for system assembly. Instead of making first the PCB and then attaching the components onto it with soldering, the electrical connections to components' pads can be directly formed with conductive ink. With this approach there is no need for a separate component attachment phase.

For inkjet printing it is important that the printable surface has proper conditions for inkjet patterning [2]. Material compatibility and planarity are important aspects [2]. This paper is focused on the effect of assembly parameters on the profile of the patterned substrate. The substrate is formed during the same phase when components are encapsulated. Several discrete components are attached in mold material and the cured mold material on the component side is used as the substrate for inkjet patterning of wiring layers.

During the molding phase the materials of the assembly undergo various amount of dimensional change, which induces stresses and can warp the package. The warpage was seen with bare eye when experimental test modules were made. The acceptable amount of substrate's curvature during inkjet patterning depends on the printing conditions (equipments, deposited ink, surface energy, accuracy requirements). Also the attachment method of the module with e.g. solder ball matrix to the next level substrate has to be taken into account when estimating the tolerable amount of curvature.

The dimensional change of the package occurs due to coefficient of thermal expansion (CTE) and cure shrinkage of mold resin. Thermo-mechanical stresses can be minimized by using material combinations with similar CTE values or by reducing the temperature change that the structure undergoes. In the studied molding process a large temperature change occurs after mold material is cured at an elevated temperature and the temperature is brought down to room temperature.

Finite element method was used to model the molding process and to see the effect of various assembly parameters on the resulting profile of the package. An example package layout was modeled and used to simulate various package options. Experimental tests were also made to verify the computational model and to see the characteristics of different material combinations after high temperature treatment.

Example SiP Assembly

The studied structure is based on System-in-Package (SiP) approach, where several components are integrated inside a single package that forms one functional unit. In the studied manufacturing approach the components are first encapsulated, and then the electrical connections are formed by inkjet material depositing on one side of the module. The discrete components of the system have to be attached before the signal connections can be formed. In this process, the components are attached with epoxy-based molding compound that was found suitable for the required heat resistance (e.g. sintering temperature) during inkjet manufacturing. The molding process is shown in the Figure 1. First the components are assembled at desired locations on a carrier adhesive tape. Next, the tape is placed in a cast and molding compound is poured on top of the components. Finally, the mold material is cured at an elevated temperature. After cure, the carrier tape is removed and the component side can be prepared for inkjet deposition of signal traces.

Figure 1. The molding process of example SiP.

The molding process is used to create a package that is suitable for inkjet deposition of material. After the carrier tape is removed the electrical contacts of the components are accessible and signal traces can be formed with inkjet deposition as long as the surface of the package is suitable for material deposition. Therefore, the component side must be cleaned from dirt and the surface energy level must be adjusted for good drop formation of deposited ink [2]. Also, the mold must provide sufficient mechanical stiffness and have good adhesion to the components. If the components dissipate lot of heat, the thermal characteristics must be also taken into account.

One problem with the molding process is the thermo-mechanical stresses induced from the temperature change after the mold is cured at an elevated temperature. The top side of the package is structurally different from the bottom side. This makes the top and bottom sides to experience different amount of thermal shrinkage, due to coefficient of thermal expansion (CTE). When the shrinkage of the sides is uneven, the package warps, e.g. if the component side shrinks less that the top side, the warpage is convex from the component side and vice versa. A straighter surface allows more accurate alignment of the inkjetted drop over the module area.

Experimental

There are several ways to control the residual warpage. The most common method is to chooce materials with small CTE differences, which minimizes thermo-mechanical stresses and warping. Also, the change of the package structure can affect the warpage; better alternatives for component placement or package dimensions can be found through experimental tests or computational modeling. Additional structures, such as stiffening structures or structures that make the shrinkage of the top and bottom side more even, can be used. For example, a suitable film can be applied on the top of mold material during molding process that stiffens the structure and changes the warpage of the package. A film can be attached such that it is placed on top of the liquid mold material. Capillary forces attach the film to the mold surface and pull the film straight. The resulting profile was experimentally tested with different film-mold resin combinations. The mold cast had dimensions of 17 mm * 17 mm * 1 mm which was filled with mold resin and a film was applied on top of the mold. Curing was done at 230 degrees in order to see the result after high temperature treatment, because the materials have to tolerate this level of temperature during e.g. sintering of inkjetted conductors. The examined mold resins were two types of acrylate, two types of polyurethane, polyimide silicone, and two types of epoxies. The film choices were PET, PEN, PI, PPS, and glass. Figure 2a shows the warping with 100 μm thick PET film with different mold resins. In comparison, Figure 2b shows the case with 220μm thick PI film on different mold resins.

Figure 2 a) Warping of mold samples with directly attached PET film. b) Warping with PI film. Mold resins: Acrylate (1-2), polyurethane (3-4), polyimide silicone (5), and Epoxy (6-7).

FE modeling

The molding process was computationally modeled with finite element method (FEM) to determine the effect of different assembly parameters on residual warpage of the package. In addition to material choices, placements of components and mold dimensions have also an effect on the results. The layout of the simulated structure is shown in the Figure 3. The structure consists of four silicon chips with dimensions of 4.65 mm x 3.82 mm x 160 μm. The chips are placed symmetrically with respect to the symmetry lines shown in Figure 3. The mold material has x/y

dimension of 17 mm. Due to the symmetry, only quarter of the model was needed to be simulated. The location of the chip in quarter symmetry model is determined by the distance of the chips bottom left corner from symmetry axis.

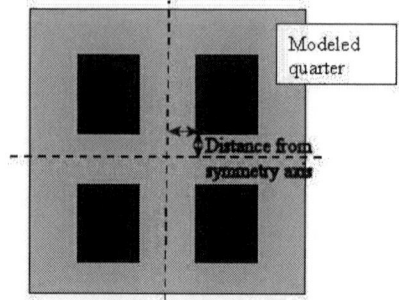

Figure 3. The layout of the simulated model.

Computational modeling is a mathematical description of some phenomena from the viewpoint of interest and with certain accuracy. Practically always, simplifications and assumptions have to be made. However, the assumptions need to be justified either based on expertise and previous knowledge or with experimental tests.

In the finite element analysis (FEA) of molding process it was estimated that the structure is at stress-free state at the cure temperature of 120 degrees. After the mold resin is cured, the material obtains mechanical strength and is able to bear loads. When the assembly is cooled down to room temperature, the dissimilar shrinkage of different materials creates stresses to the package. The shrinkage of the assembly is induced from the temperature change and from the cure shrinkage of the mold material. The volumetric cure shrinkage was estimated as 0.4% and it was set to the model by increasing the stress-free temperature of mold material. The required temperature increase ΔT was computed from equation

$$\Delta T = \frac{Vol.Shrinkage / 3}{CTE_{mold}}, \qquad (1)$$

where CTE_{mold} is the CTE value of the mold material.

Linear elastic material models were used and the values are given in Table 1. The usage of linear material model causes inaccuracy compared to the experimental case, because the material parameters are more or less temperature dependant. In addition, polymeric materials can have significant amount of nonlinear strain due to viscoelastic behavior at elevated temperatures. In this case, the used epoxy resin was specially designed for high temperature applications (up to 330 degrees).

Table 1. Material parameters of the model

Material	Young's Modulus (GPa)	Poisson's ratio	CTE(ppm)
Mold resin	5	0.38	92
Silicon	169	0.28	2.49
PEN	6	0.39	13
PET	5.4	0.39	15
PI	3.63	0.34	20

To estimate the validity of the modeling assumptions, a test mold sample was measured with optical profilometer and the shape was compared with simulation results. Figure 4 shows the measured profile of mold sample from the component side. The shape of the package is convex on the component side, because the backside of the package shrinks more than the combination of components and mold on the component side. Figure 5 shows the simulated displacement contour of the measured sample. The simulated shape for the displacement is similar but the measured maximum difference along the horizontal path differs by 100 μm compared to the experimental results. However, in this study the main interest is about comparative study of different assemblies instead of the absolute values, and therefore the accuracy of the computational model was considered sufficient.

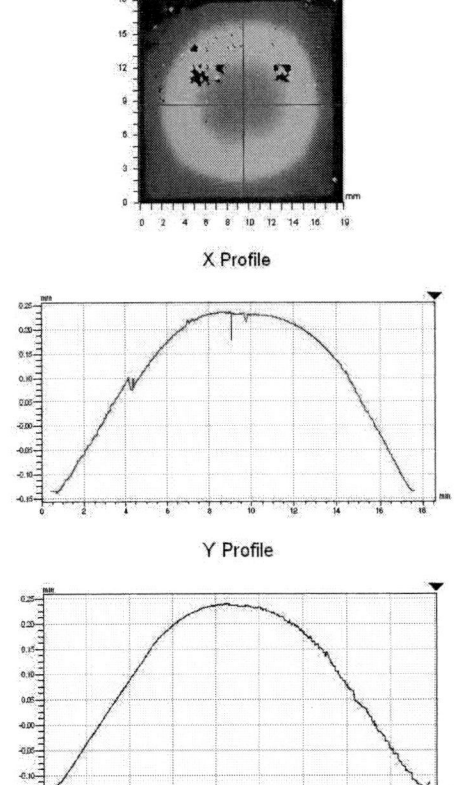

Figure 4. Measured profile of the mold sample.

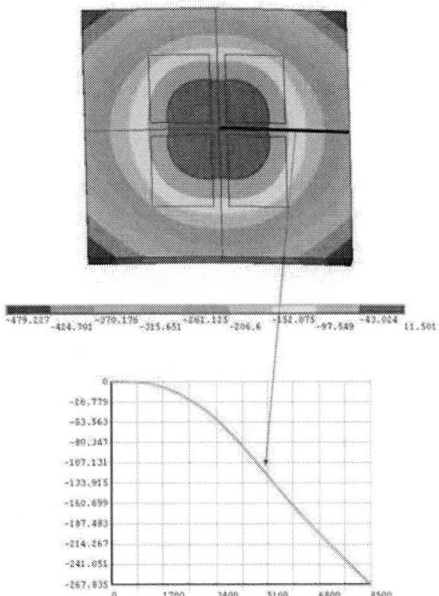

Figure 5. Simulated displacement contour (die location 1).

The studied assembly choices included the effect of component placement, thickness of mold material, and the effect of additional film placed on top of the mold material.

Results and Discussion

Effect of component placement

The effect of component placement was simulated. In the first case the four chips are placed in the center of the package and in the second and third cases those are placed towards the edges of the package. The offset (c.f. Figure 3) was set as 0.5 mm, 2 mm, and 3.5 mm respectively. Mold thickness was set as 1 mm. The simulated displacements in z-direction are shown in the Figures 5-7 (units μm). As can be seen from the results, the placement of the components has a significant effect on the warpage. The displacements were mapped onto a path (c.f. figure 5) and the graphs are plotted in Figure 8.

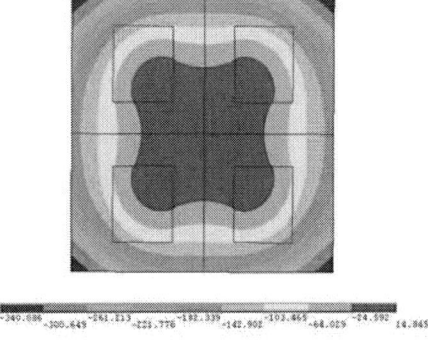

Figure 6. Displacements with die location 2.

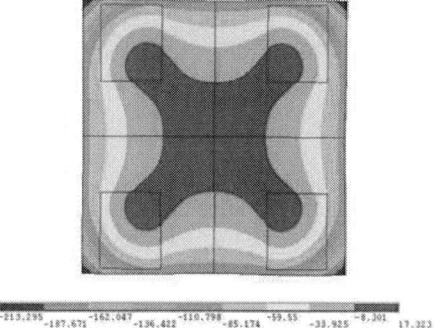

Figure 7. Displacements with die location 3.

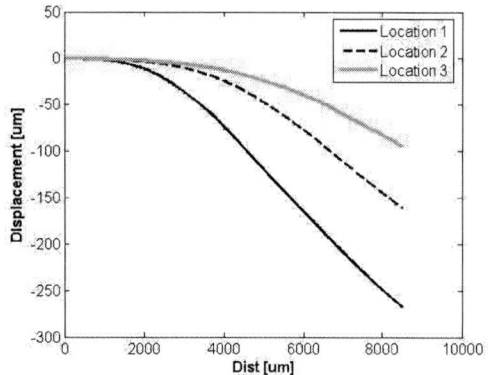

Figure 8. Displacements from the path with different die placements.

Effect of mold thickness

The effect of mold thickness was simulated with three different thicknesses: 1 mm, 1.5 mm, and 2 mm. The component placement was the die location 1 and the displacements were gathered from the previously defined path. Figure 9 shows the displacement of the package with three different mold thicknesses. As can be seen, the thicker mold decreases the warpage of the package because the structure becomes stiffer.

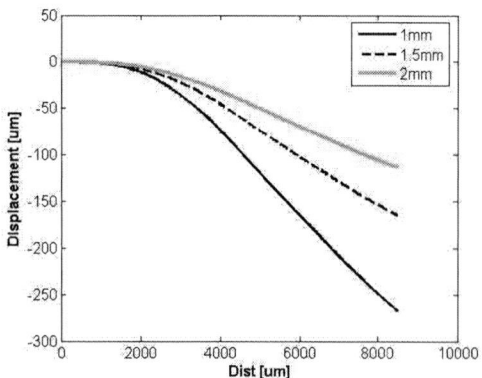

Figure 9. The effect of mold thickness on package warpage.

Effect of film on top of mold

An additional film can be attached on top of the mold material before it is cured. The film stiffens the structure but also adds thermo-mechanical stresses when film and mold material have different CTE values. This effect can be used to compensate the total warpage of the package.

The effect of different films was simulated. The film alternatives were PEN (75 μm), PET (100 μm), and PI (50 μm, 125 μm, 220 μm). The die location 1 and 1 mm mold thickness was used in the simulations. Figure 10 shows the effect of different film materials on the warpage. For comparison the displacements without film is also shown in Figure 10. As can be seen, the addition of film decreases the warpage. From the simulated cases, the 220 μm PI film resulted in the smoothest profile of the package, because the thickest film provides the most stiffness to the package.

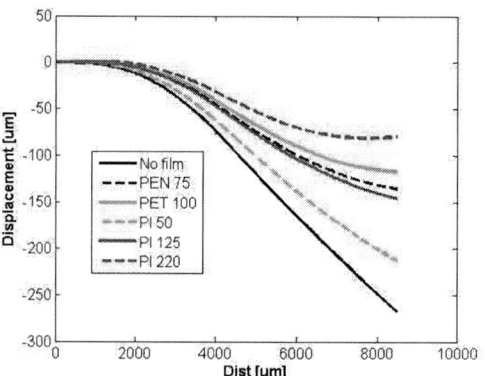

Figure 10. The effect of film on top of the mold.

Conclusions

The inkjet manufacturing of the studied multichip package requires attaching of components before the wiring layers can be processed. In this paper, we studied a structure where the components are encapsulated in mold material such that the mold surface serves as the substrate for inkjet deposition of material. There are several aspects that are required for good pattern formation on the substrate. This study concentrated on how different assembly parameters affect the planarity of the example package after molding process. It was seen that the change of component placement, package's thickness and additional film attached on top of the mold material have a considerable effect on the resulting profile.

Acknowledgements

The authors would like to acknowledge the Finnish Funding Agency for Technology and Innovation (TEKES) for the support.

K. Kaija would also like to thank the Graduate School in Electronics, Telecommunications and Automation (GETA),

Finnish Cultural Foundation and Nokia Foundation for support.

References

[1] Radivojevic, Z., Andersson, K., Hashizume, K., Heino, M., Mäntysalo, M., Mansikkamäki, P., Matsuba, Y., Terada, N., "Optimized Curing of Silver Inkjet Based Printed Traces" In *Proc. THERMINIC 2006,* Nice, France, 2006.

[2] Mäntysalo, M., Mansikkamäki, P., Miettinen, J., Kaija, K., Pienimaa, S., Rönkkä, R., Hashizume, K., Kamigori, A., Matsuba, Y., Oyama, K., Terada, N., Saito, H., Kuchiki, M., Tsubouchi, M., "Evaluation of Inkjet Technology for Electronic Packaging and System Integration", In *Proc. 57th Electronic Components and Technology Conference*, Nevada, USA, 2007.

[3] Saito, H., Matsuba, Y., "Liquid Wiring Technology by Ink-jet Printing Using NanoPaste®, in *Proc 39th International Symposium on Microelectronics*, San Diego, USA, 2006.

[4] Imai, H., Mizuno, S., Makabe, A., Sakurada, K., Wada, K., "Application of Inkjet Printing Technology to Electro Packaging", in *Proc 39th International Symposium on Microelectronics*, San Diego, USA, 2006.

Printed Antenna Designs

Roland Reitbauer, Manuela Midl, Wolfgang Stocksreiter

FH Joanneum University of Applied Sciences, Industrial Electronics,
Werk-VI-Straße 46, 8605 Kapfenberg, Austria
roland.reitbauer.iel04@fh-joanneum.at, manuela.midl.iel04@fh-joanneum.at,
wolfgang.stocksreiter@fh-joanneum.at

Abstract

As printed antennas can easily be produced in any thinkable design during the manufacturing process of printed circuit boards, they become more and more interesting for various applications. Due to their low profile and low manufacturing costs, they are being integrated for example in handheld devices for wireless communication with Bluetooth, or in sensor networks, like ZigBee. Therefore, our research concentrates on the use of printed antennas in the ISM (Industrial, Scientific, and Medical) band, at 2.4 up to 2.4835 GHz. We focused our attention on certain parameters like the geometry of the antenna structure and its effect on the antenna's performance. Field simulation software based on infinitesimal modelling was used to derive parametric influences on the antenna's characteristics. By varying the parameters, we carefully evaluated matching, efficiency and radiation patterns.

Key Words: **Antenna Parameters, Dipole Antenna, Inverted F Antenna, Simulation.**

Description of antenna parameters

What follows now is a brief introduction to the mathematical description of antenna characteristics.

Radiation Patterns

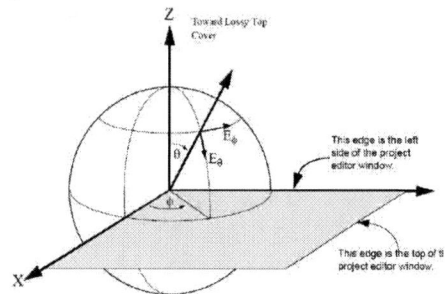

Figure 1: Definitions of angles for the Far Field Plot

The objective of radiation patterns is to show the directivity of the antennas far field at one frequency. Note that the antenna radiation pattern is reciprocal so it receives and transmits signals in the same direction.

Gain

The gain expresses how much greater the peak radiated power density is for an antenna, compared to an isotropic radiator.

Matching

Bad matching is caused by discontinuous impedance transitions. This leads to reflections at the joint and, accordingly, a loss of power available for radiation.

The complex reflection factor $\rho 0$ describes the reflection at such a joint. It is defined as the ratio of the reflected to the incoming wave, given in voltages. The S11 parameter corresponds to the reflection factor of the input port. 10dB explain that a tenth of the input power is being reflected. A near perfect match has an extremely large negative value.

General Information about the software

The software that was used for the simulations is called SONNET. This is a field simulator, which is able to perform an electromagnetic analysis of any microstrip circuits, planar structures or antennas. For the calculation the structure is divided into many subsections. Each of them gets evaluated by calculating its current and therefore the electromagnetic fields can be analysed. The most important point, which has to be considered when analysing radiating elements, is that SONNET simulates them in an enclosing shielding box. The problem is that the surrounding, perfectly conducting box walls affect the radiation characteristics. The big challenge is to choose the correct size of the box, so that the results are accurate.

The box may consist of as many layers as you need, which means that the box can be made up of different materials. In our case, we have three layers. The first layer, beginning from the bottom, is filled with air. This is followed by the substrate, which consists of FR4 material and above the metallization is an air layer again.

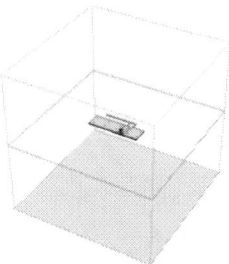

Figure 2: SONNET Box

In order to choose the right box size, we investigated the influence of the sidewalls on the antenna performance. By varying the distances between the antenna structure and the box walls, we analysed the reflection coefficient at the input of the antenna. In order to understand our evaluations, we show the parameterizations of the geometries and several settings in the software concerning an Inverted F antenna in the next chapter. One important condition is that the top cover is placed roughly a half wavelength away from the antenna layer. It is also necessary to set the top cover to free space. This enables the radiating electromagnetic waves to leave the box. The same setting was made with the bottom cover.

Inverted F Antenna

An inverted F antenna is a folded monopole so that it is parallel to the ground plane. This reduces the height of the antenna, but introduces a capacitance to the input impedance, which gets compensated by a short circuit stub. If the ground plane is much larger than the antenna, an image of it is created below the ground plane, so that it acts as an energy reflector.

Table 1: Origin Settings and Parameter Values

x [mm]	y [mm]	
35.7	10	
l [mm]	h [mm]	w [mm]
23	3.5	3
d [mm]	ε_r - FR4	tanδ
1.5	4.7	0.027

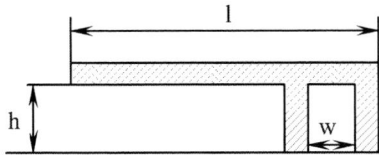

Figure 3: Parameterization on the antenna layer

The metallization, through which the antenna is provided with RF power, was set up in a way that it corresponds with our measuring prototype. The Parameter d corresponds to the thickness of the substrate.

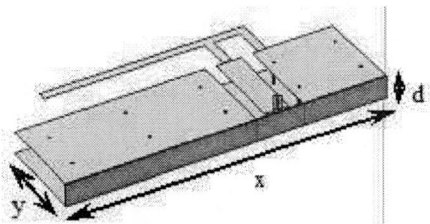

Figure 4: Structure of the Reference Metallization with Antenna

Investigation of Box Size

Before we start our investigations, it should be mentioned that when one parameter is evaluated, all others keep their origin values. Keep in mind that just the varied parameters are declared at the simulation results. To check the others you have to refer to Table 1.

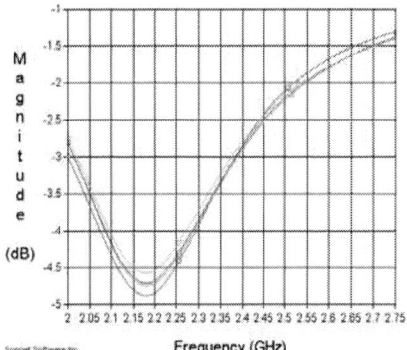

Figure 5: Results of S11 at different box sizes

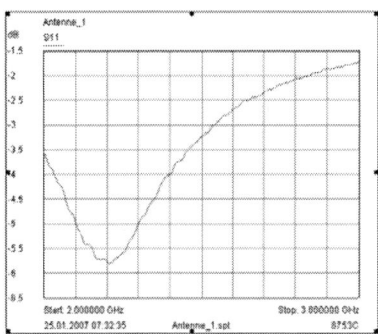

Figure 6: Measured Results of S11

In our case, when both the top metal and the bottom metal are set to 'free space', the entire box shouldn't be greater than a half wavelength at the highest frequency. When this criterion is satisfied, you can minimize the sensitivity of the results towards the box size, otherwise you receive wrong values. In Figure 5 you can see, that a right box size was found.

Investigation of Ground Size

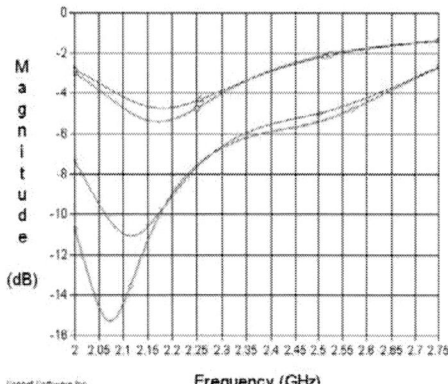

Sonnet Software Inc.

Frequency (GHz)

Figure 7: Results of S11 at different ground widths

In Figure 7 the parameter x, corresponding to the ground plane width, was changed.

The blue graphs represent the results, where the ground plane consists of metallization on the antenna layer and metallization beneath the substrate as shown in Figure 4. The red graphs on the other hand stand for the results that were received with a single ground metallization only below the substrate, just with the feedline on the antenna layer.

As the results of both reference plane formations show similar characteristics, all further simulations are executed with the configuration of a single plane. Another positve side effect of this setup is, that it is simulated much faster due to less metallization being calculated.

Analysis of relative permitivity, thickness and loss factor of substrate

If you make the substrate layer thinner, the resonant frequency moves upwards. You must care that you have to adapt your feeding line to the substrate thickness in order to retain correct feeding impedance.

The resonant frequency can also be varied by the relative permittivity. The lower its value is, the higher the resonant frequency gets.

The loss factor doesn't have any influence on the resonant frequency.

Investigation of antenna geometry

Now we are interested in the optimization of the antenna geometry by varying the appropriate parameters.

The graph with the lowest resonant frequency in Figure 8 is related to a parameter value of w = 2,4mm and those with the highest one corresponds with a value of 5mm.

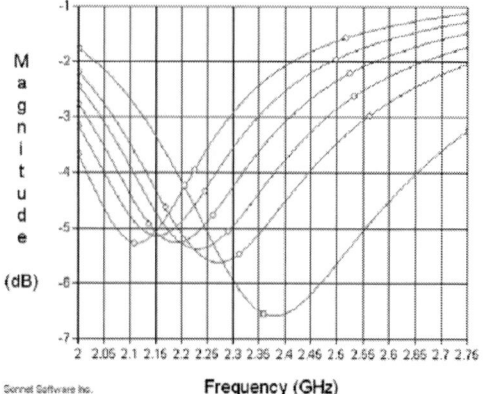

Sonnet Software Inc.

Frequency (GHz)

Figure 8: Different values for parameter w

The more the antenna length decreases, the higher gets the resonant frequency, as shown in Figure 9. At the highest resonant frequency the antenna is 20mm long.

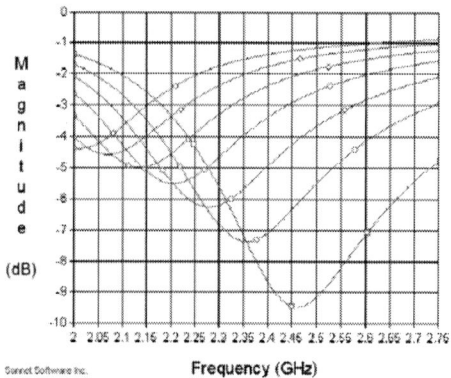

Sonnet Software Inc.

Frequency (GHz)

Figure 9: Different values for parameter l

The antenna height was changed from h = 2,5mm to 7,5mm. By raising this parameter, the resonant position moves to the left.

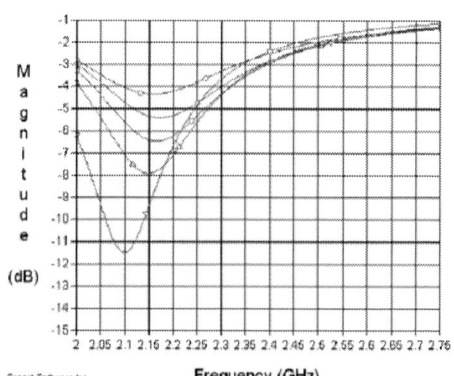

Sonnet Software Inc.

Frequency (GHz)

Figure 10: Different values for parameter h

The next step is to optimize our antenna that way, that it shows a resonant frequency at 2,45GHz, which is the operating frequency of applications in the ISM band. You have to take into consideration that the parameters show different extents of influence when diverse values of them are combined. This means that it isn't necessarily a good solution to combine the individual best results from the previous simulations. Figure 11 shows a good impedance match at the right frequency.

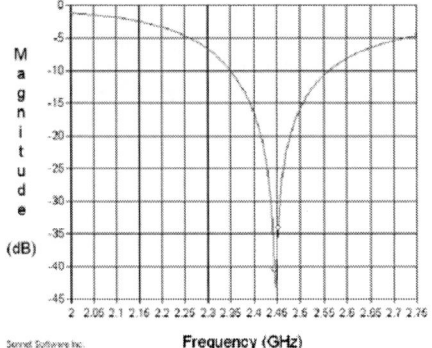

Figure 11: Results of S11 of antenna with optimized geometry parameters: h = 7.5mm, l = 18mm, w = 2mm, ε_r = 4.1

The radiation patterns of the antenna with this geometry are shown in Figure 12 and Figure 13. At the resonant frequency the gain has values in the range of 0dB with all metal and reflection losses included.

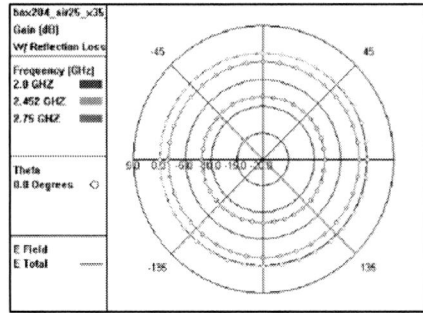

Figure 12: Elevation Radiation Plot: Angle φ is plotted from -180° to +180° at a constant angle θ of 0°

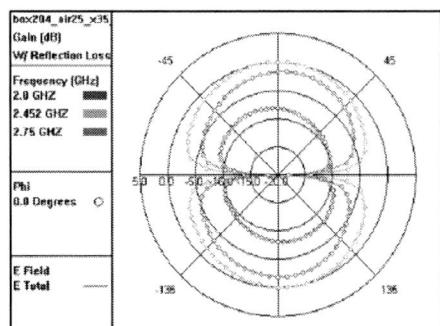

Figure13: Azimuth Radiation Plot: Angle θ is plotted from -180° to +180° at a constant angle φ of 0°

Conclusion

You have to try a lot of parameter variations concerning the geometry before you find the best impedance match for a given environment. Generally you can say that the resonant frequency moves up with a shorter length of the antenna whereas it moves down with greater values for the parameter h and vice versa. But it is impossible to predict the tendency of the magnitude of S11 by modifying these parameters because this depends on the whole constellation. A different behaviour shows the parameter w. The resonant frequency is less sensitive to it whereas the magnitude of S11 can be changed in great dimensions. Therefore an effective method is to set the resonant frequency with the parameters l or h and afterwards the parameter w can be used for tuning the magnitude of S11. A fine tuning for the exact frequency can be achieved by varying the relative permittivity of the substrate.

Concerning the environment, I would advice to avoid metallization planes on the antenna layer and just use a ground underneath the substrate. In this case we could observe that at bigger dimensions for the ground plane, the parameter w has to be increased as well.

Half-Wave Dipole

What follows is a discussion of the half-wave dipole antenna, half-because its physical length has to be about a half of the free-space wavelength to achieve resonance.

The dipole antenna is a basic antenna, consisting of a straight wire cut in the centre, where the feeding point is. The total length of a dipole is a half wavelength of the operating frequency, so that the antenna shows resonance there. The dipole antenna should be placed at some distance from the ground plane to have enough efficiency.

The current and voltage distribution on the antenna is such that the lowest impedance is in the centre. The supply is realized symmetrically, shown in Figure 14. According to Equation (1) the nominal values were estimated to start the simulation.

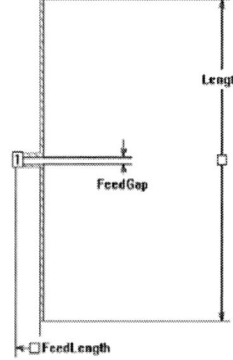

Figure 14: Geometry of the half-wave dipole

$$Length \leq \frac{\lambda}{2} \qquad (1)$$

Variation of FeedGap

To derive the influences of FeedGap a parameter sweep from 1mm to 10mm was done. By evaluation of the S_{11} parameter we discovered that FeedGap has absolutely no effect on the matching of the antenna. Therefore, to reduce the needed amount of space, it should be kept quite narrow.

Determination of the resonant frequency through variation of Length and FeedLength

By varying the two parameters separately we do continually change the sum of Length/2 and FeedLength. In both cases the resonant frequency varies greatly, as well as the magnitude of S_{11}. Assuming that FeedGap is narrow enough this sum corresponds to one arm of the dipole. By leaving this sum constant it's much easier to determine frequency and matching.

$$\frac{Length}{2} + FeedLength = const \qquad (2)$$

We found two major points to find the optimal geometry:

[1] The shorter the arm of the dipole the higher is the resonant frequency.

[2] For each combination of Length/2 and FeedLength there is (unfortunately) a special relation between FeedLenght and Length/2 to receive optimal performance.

$$\frac{FeedLength}{Length/2} \qquad (3)$$

For Length/2 and FeedLength equals 30 we get an optimal relation of 0.25 (6/24). As a rule of thumb we may say, that 1mm reduction in Equation(2) approximately leads to minus 0.02 in the optimal relation (Equation (3)).

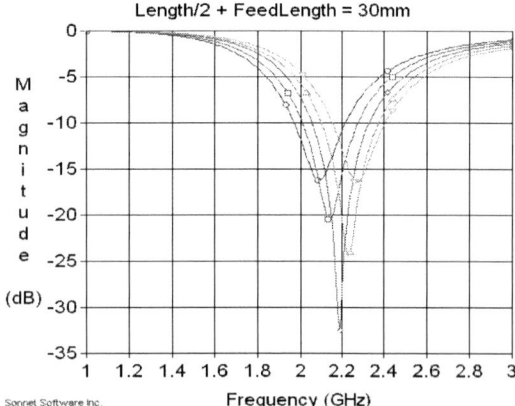

Figure 15: Resonant Frequency at 2.2GHz;
Blue: Length = 52mm, FeedLength = 4mm;
Lavender: Length =50mm, FeedLength = 5mm;
Red: Length = 48mm, FeedLength = 6mm;
Green: Length = 46, FeedLength = 7mm;
Turquoise: Length = 44, FeedLength = 8mm;

With this knowledge the dipole can easily be matched for various boundary conditions.

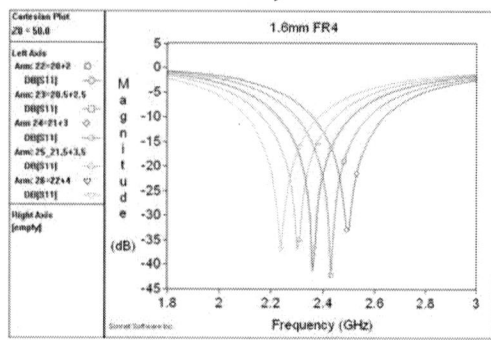

Figure 16: Matching for different dipole lengths

Variation of the substrate

The increase of the substrates thickness or the use of a different material with other ε_r simply push the resonant frequency downwards. If this effect is not desired, the dipole should be a little shorter, according to [1], followed by [2] for optimal matching.

Variation of Conductor width

Figure 17 shows, that the wider the conductor is, the higher is the resonant frequency. If this effect is not desired, the dipole should be a little bit longer to compensate.

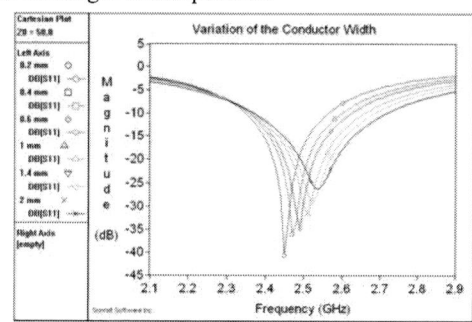

Figure 17: Variation of conductor width

Table 2: optimal values for simple dipole for the 2.4GHz ISM Band

Length [mm]	FeedGap[mm]	FeedLength [mm]
41	0.5	2.5
Substrat thikness [mm]	ε_r - FR4	tanδ
1.6	4.7	0.027

Increase the bandwidth to be less sensitive to external factors

Environmental factors may affect the antenna characteristic. As they are unknown we try to make the antenna less sensitive to these factors by increasing its bandwidth. Therefore stubs, shown in Figure 18, are added.

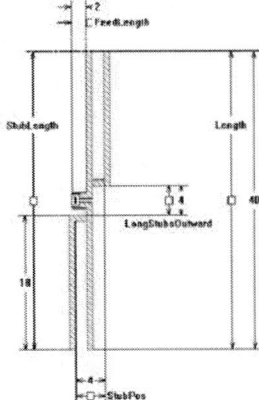

Figure 18: Stubs are added to increase bandwidth

Our research showed that the length of the stubs should approximately be a little shorter than the length of the dipole. In this case the resonant frequency is not being affected.

The Length of the stubs does have a positive effect on the bandwidth. To optimize the magnitude of S_{11} as well there are two ways:

[3] Moving the stubs horizontally outwards or

[4] Moving them vertically to the outer end of the dipole by leaving their length constant.

When reducing the length of the Stubs we do not get a positive effect on the bandwidth.

The two ways of positioning the stubs fortunately have in common, that the farer the stubs are away from the centre of the dipole, the less is the magnitude of S_{11}. If they are too near the centre the bandwidth decreases.

Accordingly there is one optimal solution, to be found by following [3] as well as by following [4], showed in Figure 19.

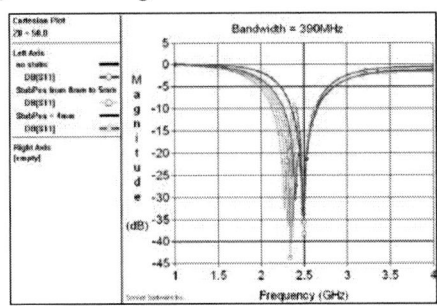

Figure 19: Optimal solution, bandwidth measured at S_{11}= -10dB

Table 3: Dipole for 2.4GHz ISM Band

Length [mm]	Feed-Gap [mm]	Feed-Length [mm]
40	0.5	2
Stub-Length [mm]	Stub-Pos [mm]	LongStubs-Outward [mm]
40	4	4

The radiation patterns of the dipole antenna with the optimal parameters of Table 3 are shown in Figure 20 and Figure 21. At the resonant frequency the gain has values in the range of 0dB with all metal and reflection losses included.

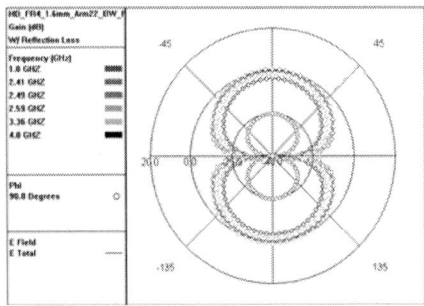

Figure 20 : Elevation Radiation Plot: Angle θ is plotted from -180° to +180° at a constant angle φ of 90°

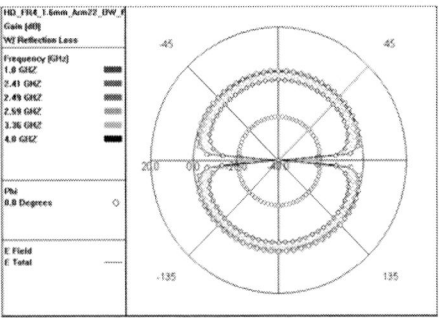

Figure 21: Elevation Radiation Plot: Angle θ is plotted from -180° to +180° at a constant angle φ of 0°

Conclusion

The half-wave dipole is a very good radiator and simple to match through the variation of the total length of one arm. The variation of a single parameter is independent from other parameter constellations, which makes it easy to find an optimal solution.

Acknowledgements

The authors want to thank Dr.Volker Mühlhaus and Dr. Mühlhaus Consulting & Software GmbH for the fast and efficient support in the work with Sonnet.

References

[1] Sonnet Application Notes
[2] Warrwn L. Stutzmann, and Gary A. Thiele, "Antenna Theory and Design," John Wiley & Sons 1998
[3] Hiroyuki Arai, "Measurement of Mobile Antenna Systems," Artech House 2001
[4] Gregor Gronau, "Höchstfrequenztechnik," Springer 2001
[5] Klaus Kark, "Antennen und Strahlungsfelder," Vieweg Studium Technik 2004

Inter-plane Coupling Structures for PCB-integrated Multilayer Optical Interconnections

N. Hendrickx[1], J. Van Erps[2], H. Thienpont[2], P. Van Daele[1]

[1]Ghent University, Technologiepark 914A, B-9052 Ghent, Belgium

Phone: +3292645370, Fax: +3292645374, Email: nina.hendrickx@intec.ugent.be

[2]Vrije Universiteit Brussel, Pleinlaan 2, B-1050 Brussels, Belgium

Abstract

Optical interconnections gradually find their way towards shorter distances as optoelectronic components become available at affordable costs. To fully exploit the two-dimensional character of the available VCSEL- and detector arrays, multilayer optical interconnects have gained interest over the last years. The multi-layer approach allows increased integration density and flexible routing schemes. In these multilayer structures light can be coupled both between and out of the plane of the optical layers. We present two different approaches to achieve this goal: the first configuration contains laser ablated micro-mirrors, the other one consists of a pluggable inter-plane coupler that can be inserted into laser ablated cavities in a two layer optical structure, which consists of two stacked optical layers, integrated on a printed circuit board (PCB) to couple the light between two subsequent layers. Non-sequential ray-tracing is used to perform a tolerance analysis on the two layer optical structure. The fabrication process of the optical board, the ablated mirrors and the pluggable component are described.

Key words: Coupling Structures, Multilayer Optical Structures, Non-Sequential Ray-Tracing, Optical Interconnections

Introduction

Optical interconnections gradually find their way towards shorter distances as optoelectronic components become available at affordable costs. To fully exploit the two-dimensional character of the available VCSEL- and detector arrays, multilayer optical interconnects have gained interest over the last years. The multi-layer approach allows increased integration density and flexible routing schemes. In these multilayer structures light can be coupled both between and out of the plane of the optical layers. The interconnection scheme can in this way be simplified, with less need for passive optical components such as cross-overs.

A two layer optical structure, which consists of two stacked optical layers, integrated on a printed circuit board (PCB), is studied. The optical layer contains multimode waveguides, to guide the light in the plane of the optical layer. Two different approaches are presented to couple the light beam between the two layers. The first configuration consists of a two layer optical structure with integrated laser ablated micro-mirrors. The second configuration makes use of a pluggable inter-plane coupler which can be inserted into a laser ablated micro-cavity.

In the next sections, we will first discuss the fabrication process of the optical board. The fabrication process of the two coupling structures is

then presented. Numerical simulations have been carried out to determine the alignment tolerances in the two layer optical structure. The simulation results will also be briefly discussed.

Two layer optical structure

The integration of the optical interconnection on the PCB is done with the use of a polymer optical layer. The optical material has to be compatible with existing PCB manufacturing and soldering processes and should in addition show very good optical properties such as low propagation loss at the targeted wavelength. We study the use of Truemode Backplane™ Polymer, which is a highly cross-liked acrylate based polymer material with excellent optical, thermal and environmental properties. The main properties are given in Table 1.

Table 1: main properties of the optical material used for the optical layer

Propagation loss @ 850nm	0.04dB/cm
Refractive index core material	1.5562
Refractive index cladding material	1.5422
Coefficient of thermal expansion	60ppm/K
Decomposition temperature	> 350°C

The optical material is spincoated onto the FR4 substrate and in a next step UV-exposed and cured. One optical layer consists of a cladding-core-cladding stack, where the core material has a slightly higher refractive index than the cladding material. The light can in this way be trapped inside the core layer through total internal reflection.

The optical layer contains multimode waveguides which are patterned into the core layer using laser ablation [1]. The excimer laser beam is send through a rectangular aperture and projected onto the sample with an optical projection system. The photon energy of the laser beam is used to remove the core material on both sides of the resulting waveguides core. The principle is given in Fig. 1.

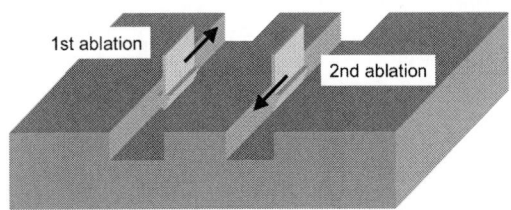

Figure 1: schematic showing the principle of ablating a waveguide core.

The waveguides have a cross-section of 50μm×50μm and are on a pitch of 125μm. The propagation loss at a wavelength of 850nm is 0.12dB/cm which is acceptable for the aimed applications. The waveguide loss is higher than the propagation loss given in Table 1. This is primarily caused by scattering losses that result from the surface roughness of the ablated structures. The average RMS surface roughness of an ablated waveguide is 35nm on a scan area of 49μm×159μm, which is higher than the values obtained for photo lithographical waveguides. Main advantage of using laser ablation is the fact that it is a mask-less technology that can be used for rapid prototyping. This allows us to change the design according to the given specifications in a very flexible way; whereas for photolithography a new mask has to be designed and fabricated each time the design changes. For these reasons, laser ablation is used for the fabrication of the demo optical board.

The two layer optical structure consists of a stack of two optical layers. Each layer contains arrays of ablated multimode waveguides, which are aligned with respect to each other with the use of alignment marks placed onto the substrate. The alignment accuracy that can be obtained in this way is better than 5μm, which is in accordance with the results of the alignment tolerance study that has been done with non-sequential ray tracing. The cross-section of a two layer optical waveguide structure is shown in Fig. 2; the results from the numerical study are given in Fig. 3 for the case of a multimode optical fiber (MMF) with core diameter 100μm and

NA 0.29 as detector and for the case of a flat detector.

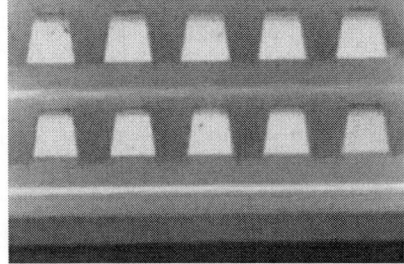

Figure 2: cross-section of a two layer optical waveguide structure.

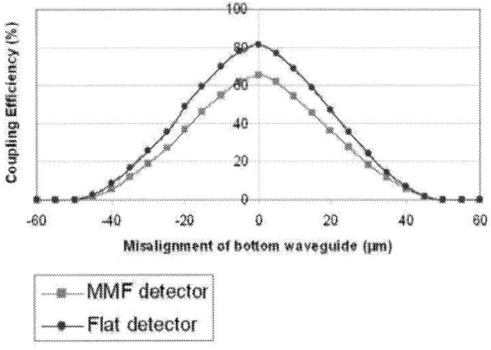

Figure 3: alignment tolerance of the bottom waveguide in two layer optical structure.

Inter-plane coupling structures

Metallized 45° micro-mirrors can be used to deflect the light beam over 90°. The 45° mirror facet is ablated into the optical layer and in a next step Au-coated. The ablated cavity is finally filled with cladding material. The entire process flow is shown in Fig. 4.

A two layer optical structure, which contains a stack of two optical layers, has been studied both experimentally and numerically. Both layers contain multimode optical waveguides and metallized micro-mirrors. The alignment between the different elements in the top and bottom optical layer has to be accurate enough to guarantee a high coupling efficiency. Numerical simulations have been carried out to determine the alignment tolerance curves for the two layer waveguide and mirror structure. The alignment tolerance ranges that are required to obtain an excess loss -1dB are drawn from the tolerance curves. The achievable alignment accuracy of the experimental results can then be compared to the results from this numerical analysis.

ASAP 2005 V2R2 (Breault Research) is used to carry out the simulations. In view of the highly multimodal character of the considered structures, non-sequential ray tracing is used. Light with a wavelength of 850nm is coupled into the ablated Truemode multimode waveguide, which has a numerical aperture (NA) of 0.3, with a MMF with core diameter 50μm and NA 0.2. The light is deflected over 90° using metallized 45° micro-mirrors and is coupled from one layer to the other, with maintenance of the propagation direction. A MMF with core diameter 100μm and NA 0.29 is used to detect the outcoupled power.

An optical microscope picture of a cross section of the studied two layer optical structure is given in Fig. 5. The tolerance curve for a misalignment of the top mirror with respect to the bottom mirror is given in Fig. 6. The alignment tolerance range from the ideal position for an excess loss ≤1dB along the propagation direction is [-10μm, 10μm].

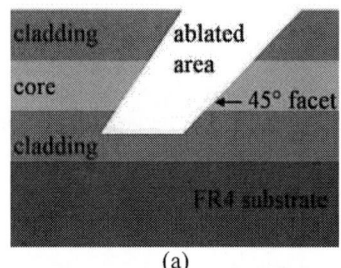

(a)

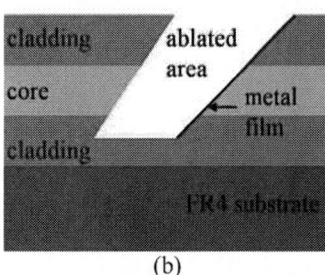

(b)

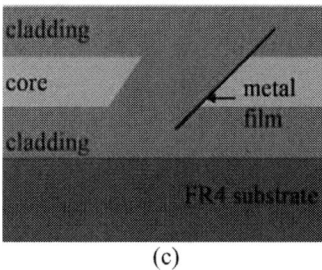

(c)

Figure 4: process flow used to integrate the metallized mirror into the optical layer.

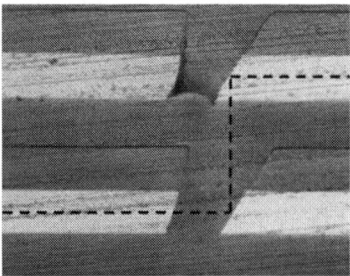

Figure 5: cross-section of a two layer optical structure with metallized mirrors.

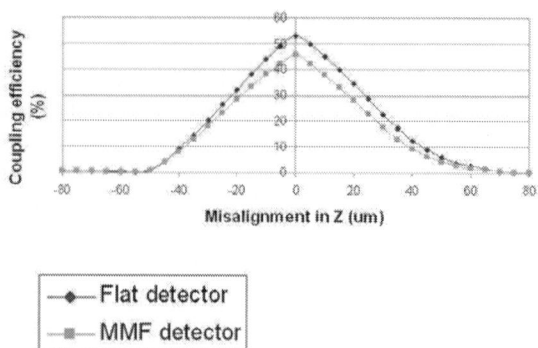

Figure 6: The alignment tolerance of the top and bottom mirror for a misalignment along the waveguide direction for the case of a flat and MMF detector.

Pluggable inter-plane coupling component

The optical layers contain arrays of multimode waveguides which are used to guide light in the plane of the optical layer. The coupling of light from a waveguide in one layer towards the corresponding waveguide in the other layer can be done with a pluggable inter-plane coupler. Two designs can be used: the first one redirects the light beam from one layer to the other with preservation of the propagation direction; the other one in addition also changes the propagation direction. Both principles are given in Fig. 7.

The pluggable coupler contains two 45° micro-mirrors and is patterned in PMMA with Deep Proton Writing [2]. It consists of the following processing steps. First a collimated 8.3MeV proton beam is used to irradiate an optical grade PMMA sample according to a predefined pattern by translating the PMMA sample, changing the physical and chemical properties of the material in the irradiated zones. As a next step, a selective etching solvent is applied for the development of the irradiated regions. This allows for the fabrication of (2D arrays of) micro-holes, optically flat micro-mirrors and micro-prisms, as well as alignment features and mechanical support structures. On the other hand, an organic monomer vapor can be used to expand the volume of the bombarded zones through an in-diffusion process. This enables the fabrication of spherical (or cylindrical) micro-lenses with well-defined heights. If necessary, both processes can be applied to different regions of the same sample, yielding micro-optical structures combined with monolithically integrated micro-lenses. We use high molecular weight PMMA with a thickness of 500μm, which allows the 8.3MeV protons to completely traverse the sample. During the irradiation step, the PMMA sample is semi-continuously translated perpendicularly to the beam in steps of 500nm using high-precision translation stages with an accuracy of 50nm. Optimal surface roughness results are obtained by using a proton dose of 50pC per step of 500nm, with a proton current of 160pA. The fabricated components are shown in Fig. 8 and Fig. 9. Au coating can consequently be evaporated on the micro-mirror facet to increase the coupling efficiency.

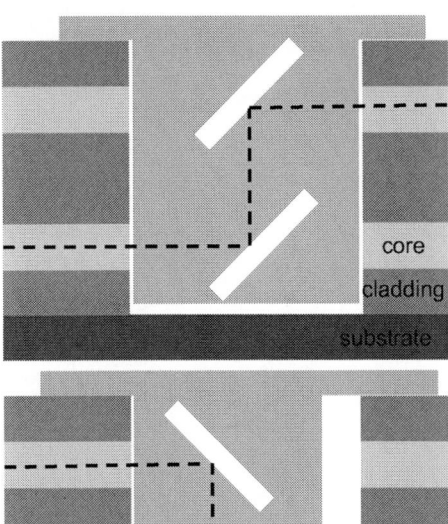

Figure 7: Schematic principle of pluggable inter-plane coupling structures preserving the propagation direction (top) or inverting the propagation direction (bottom)

The thickness of cladding and core layer has to be identical in top and bottom optical layer, and has to correspond to the values used for the design of the coupler in order to guarantee a high coupling efficiency.

The coupling component is inserted into a micro-cavity that is ablated into the two layer optical structure. There is always a certain degree of tapering during the ablation, which also causes the slightly trapezoidal form of the waveguides, which can be measured experimentally and corrected for. The ablated cavity thus has one vertical wall and one wall with the double tapering angle. The insert is placed against the vertical wall, which is in fact the output facet of the waveguides. When the discrete coupler is plugged into the micro-cavity, it has to be aligned with respect to the waveguides. Active alignment is used for this purpose.

For the characterization of the critical optical surfaces of the component, namely the entrance and exit facet, we use a WYKO NT-2000 non-contact optical surface profiler (Veeco). Since the micro-

mirrors themselves are not accessible with the microscope objective, these surface were not measured, but their surface roughness will be analogous to the two others. The surface roughness analysis reveals that the flat top part has an average local RMS surface roughness R_q of 14.1nm ± 2.7nm measured over an area of 60µm by 46µm. We averaged at least 5 measurements of randomly chosen positions. The flatness R_t or peak-to-valley difference along the depth of 500µm of the component is measured to be smaller than 2.5µm. This is due to the scattering of the protons during the interaction with the PMMA. As a conclusion, we can say that our developed DPW surfaces have a very good and reproducible optical quality: almost flat and a very low RMS roughness.

Loss measurements will be performed in the near future to determine the coupling loss of the inter-plane coupling structures. Experimental results will be compared with the results obtained from a numerical study.

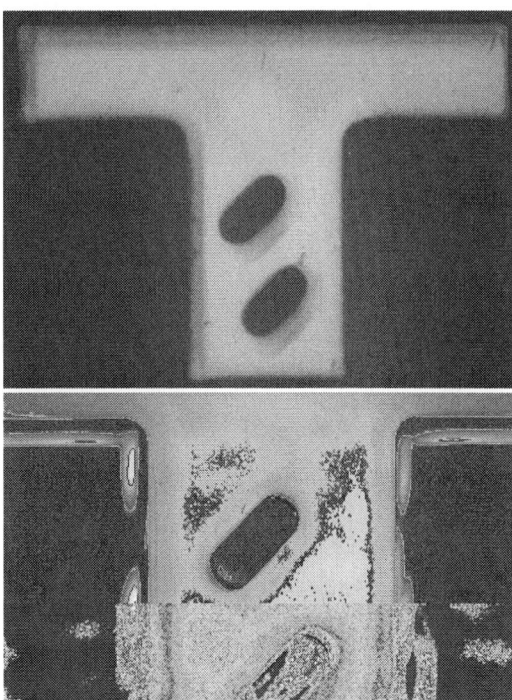

Figure 8: Fabricated pluggable inter-plane coupler with preservation of propagation direction: entire component (top) and zoom on mirrors (bottom)

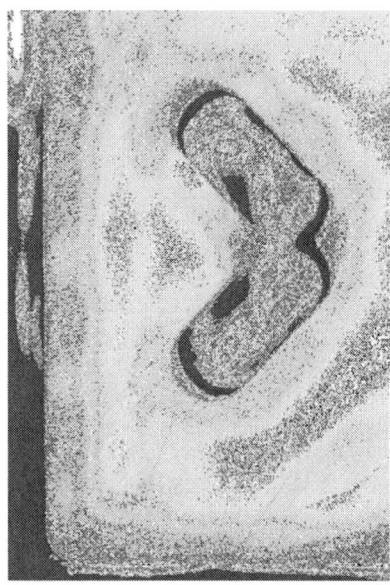

Figure 9: Fabricated pluggable inter-plane coupler with inversion of propagation direction

Conclusions

We have presented two different fabrication technologies allowing the manufacturing of inter-pane coupling structures for multilayer optical waveguides integrated on a Printed Circuit Board. Laser ablation can be used to create monolithically integrated micro-mirrors in the waveguides. A numerical study has been performed to show us the required alignment tolerance in this case. In a second approach, we make use of Deep Proton Writing for the fabrication of dedicated, pluggable inter-plane coupling components which can be readily inserted in cavities formed in the PCB-integrated multilayer waveguides.

Acknowledgements

Nina Hendrickx would like to acknowledge the Institute for the Promotion of Innovation by Science and Technology in Flanders (IWT Flanders) for financial support. Jürgen Van Erps acknowledges the Fund for Scientific Research-Flanders (FWO) for financial support. Part of this work was carried out within the framework of the network of excellence on micro-optics (NEMO), supported by the European Commission through FP6 program.

References

[1] G. Van Steenberge, N. Hendrickx, E. Bosman, J. Van Erps, H. Thienpont, and P. Van Daele, "Laser Ablation of Parallel Optical Interconnect Waveguides", Photonics Technology Letters, Vol. 18, no. 9, May, 2006.

[2] C. Debaes et al., "Deep proton writing: a rapid prototyping polymer micro-fabrication tool for micro-optical modules", New Journal of Physics, vol. 8, 270, 2006.

Prototyping of pluggable out-of-plane coupling components for multilayer board-level optical interconnections

Jürgen Van Erps[1], Nina Hendrickx[2], Christof Debaes[1], Peter Van Daele[2], Hugo Thienpont[1]

[1]Vrije Universiteit Brussel, Dept. of Applied Physics and Photonics (TONA-FirW),

Pleinlaan 2, B-1050 Brussel, Belgium

[2]Ghent University, TFCG Microsystems, Dept. of Information Technology (INTEC),

Technologiepark 914A, B-9052 Zwijnaarde, Belgium

Contact: Jurgen.Van.Erps@vub.ac.be, Tel. : +32 2 477 48 71, Fax: +32 2 629 34 50

Abstract

Board-level optical interconnects offer a possible solution to the bandwidth problems that electrical interconnects are facing in the near future. The integration of the optical interconnection to the board level is done by integrating one or more optical layers on a printed circuit board (PCB). We present Deep Proton Writing (DPW) as a generic rapid prototyping technology for the fabrication of a micro-optical coupling component incorporating a 45° micro-mirror that can be readily inserted into a multilayer optical waveguiding structure integrated on a PCB. Micro-cavities are ablated into the optical layers to accommodate the discrete out-of-plane coupler. The advantage of using a discrete component is that micro-lenses can be incorporated to increase the coupling efficiency with a guaranteed perfect alignment of the lens and the micro-mirror. In case lenses are integrated in the coupling component, the layer thickness of top and bottom optical layer has to be in accordance with the designed value and the alignment of the component with respect to the waveguide is critical. In the case the lenses are not used and a metallized mirror facet is used for out-of-plane coupling, there is quite a large tolerance on the thickness of the layers and the alignment accuracy of the component. The surface quality of the fabricated components was characterized and the coupling efficiency of the out-of-plane coupling components was be measured in a fiber-to-fiber coupling scheme. The coupling component is prototyped in PMMA material, which is not compatible with standard PCB manufacturing. This should however not be considered as a limiting factor since the DPW process is compatible with mass replication technologies such as hot embossing or micro-injection moulding and the master as such can be replicated in a variety of high-tech plastics.

Key words: coupling structures, deep proton writing, micro-optics, optical interconnects, polymers, waveguides

Introduction

In the future, the communication bandwidth inside data processing systems will be severely limited by the properties of galvanic interconnections. These limitations stem from physical constraints imposed by RC time constants, ohmic losses and cross-talk between the conductances of these galvanic interconnections. Optics is a potential alternative route to circumvent the underlying problems of galvanic interconnects and is also said to have the potential to continue to scale with future generations of silicon integrated circuits. Optical interconnects based on low-loss integrated waveguides are a promising solution to overcome the interconnect bottlenecks at board and module level [1]. However, one of the most critical problems is coupling the light in and out of the optical plane. A common approach is the use of 45° micro-mirrors. Various techniques are being applied

for the fabrication of these micro-mirrors. Micro-machining techniques using a 90° V-shaped diamond blade [2],[3] can provide an excellent cut surface, but it is difficult to cut individual waveguides on the same substrate due to the physical size of the machining tool. Reactive ion etching RIE [4] where the slope of the mirror is formed by 45° oblique etching is limited by directional freedom. Temperature controlled RIE [5] is not limited by directional freedom but this method has the disadvantage of being material dependent. Other techniques are tilted X-ray exposure [6] and laser ablation, where generally a KrF Excimer laser is used, depending on the material in which the waveguides are defined [7].

All the above technologies are used to write the micro-mirrors directly in the waveguides. In this paper, we present a completely different approach, where we propose the use of a pluggable out-of-plane coupling component with an integrated 45°

micro-mirror that can be readily inserted in into cavities fabricated into printed circuit boards (PCB) containing multilayer optical waveguides.

Multilayer optical waveguides

Truemode Backplane™ Polymer is used as for the optical layer. It is an acrylate-based highly cross-linked polymer material that shows excellent optical and thermal properties. Each optical layer consists of a cladding-core-cladding stack, where the cladding material has a slightly lower refractive index than the core material. The light can in this way be trapped inside the core layer through total internal reflection. The numerical aperture (NA) of the waveguides is 0.3. Multimode waveguides are patterned into the core layer with either photolithography or laser ablation. In case photolithography is used, the waveguide core features are transferred to the core layer by UV-exposure through a suitable mask which has to be used in proximity mode because of the sticky character of the material. In case laser ablation is used, material is removed on both sides of the resulting waveguide core with a KrF excimer laser beam. The waveguides have a cross-section of 50μm×50μm and have a pitch of 125μm or 250μm.

Figure 1: Cross-section of a two-layer optical waveguiding structure

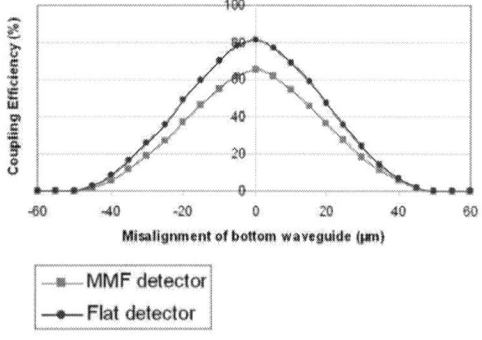

Figure 2: Tolerance for mechanical misalignment of the upper waveguide layer with respect to the lower layer, results from optical simulations.

A multilayer optical structure consists of a stack of optical layers, as shown in Figure 1. Each layer contains multimode optical waveguides which have to be aligned with respect to each other accurately. The alignment is done with the help of Au alignment marks which are evaporated on the substrate through a suitable mask. The achievable alignment accuracy between waveguides in top and bottom layer in a two layer optical structure is better than 5μm. This value is in accordance with the required accuracy obtained from a numerical study, as can be seen in Figure 2.

The coupling component described in the next section is to be plugged into a laser ablated micro-cavity. There is always a certain degree of tapering during the ablation, which also causes the slightly trapezoidal form of the waveguide cores. The KrF excimer laser beam is therefore tilted for the ablation of the micro-cavity to compensate for the tapering. This ensures that the ablated micro-cavity has one vertical wall and one with the double tapering angle. The vertical wall is used as output facet; the coupler is subsequently inserted into the cavity.

Out-of-plane coupling components

As mentioned in the introduction, we are investigating the use of pluggable micro-optical components that can be used to couple the light to or from PCB-integrated waveguides by inserting it into cavities fabricated in the board. We have previously shown that this type of out-of-plane coupling components is capable of achieving high coupling efficiencies for single layer waveguide structures [8]. In this paper, we extend this out-of-plane coupler towards multilayer structures. This can be easily done by increasing the size of the micro-mirror. A schematic working principle is shown in Figure 3. Two cylindrical micro-lenses ensure the collimation (in one direction) of the beam emitted by the PCB-integrated optical waveguides, increasing the coupling efficiency.

Another type of extension of the component, allowing routing of signals between the different layers of optical waveguides, is also possible thanks to the versatility of the Deep Proton Writing fabrication technology. We refer to [9] for more details.

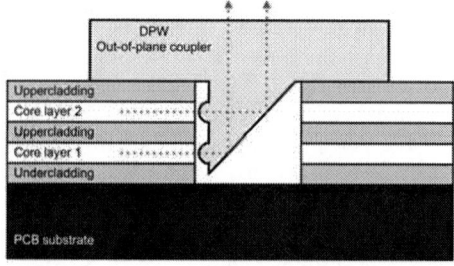

Figure 3: Multilayer out-of-plane coupling component with integrated cylindrical micro-lenses inserted in a cavity formed in the PCB-integrated waveguides. Light paths are indicated by the dotted arrows.

In the following sections, we describe in detail the design, fabrication and characterization of the pluggable multilayer out-of-plane coupling components.

Fabrication using Deep Proton Writing

For the fabrication of the pluggable multilayer out-of-plane coupling component, we make use of our in-house generic rapid prototyping technology of Deep Proton Writing (DPW) [8]. It consists of the following processing steps.

First a collimated 8.3MeV proton beam is used to irradiate an optical grade PMMA sample according to a predefined pattern by translating the PMMA sample, changing the physical and chemical properties of the material in the irradiated zones (see Figure 4). As a next step, a selective etching solvent is applied for the development of the irradiated regions. This allows for the fabrication of (2D arrays of) micro-holes, optically flat micro-mirrors and micro-prisms, as well as alignment features and mechanical support structures. On the other hand, an organic monomer vapor can be used to expand the volume of the bombarded zones through an in-diffusion process. This enables the fabrication of spherical (or cylindrical) micro-lenses with well-defined heights. These processes are illustrated in Figure 5. If necessary, both processes can be applied to different regions of the same sample, yielding micro-mechanical structures combined with monolithically integrated micro-lenses.

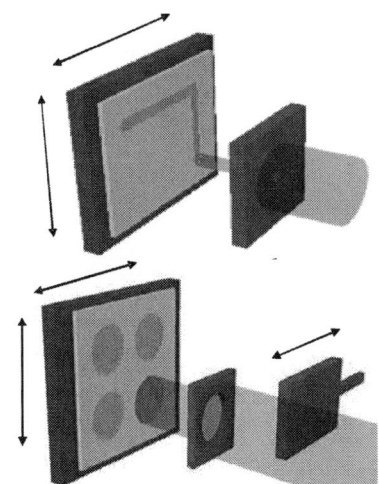

Figure 4: Irradiation step of DPW: Semi-continuous (top) or pointwise (bottom) irradiation of a PMMA sample

We use high molecular weight PMMA with a thickness of 500μm, which allows the 8.3MeV protons to completely traverse the sample. During the irradiation step, the PMMA sample is semi-continuously translated perpendicularly to the beam in steps of 500nm using high-precision translation stages with an accuracy of 50nm. Optimal surface roughness results are obtained by using a proton

dose of 50pC per step of 500nm, with a proton current of 160pA. This current is monitored by measuring the charge that the protons induce on a target located directly behind the sample. The deposited dose at each position can then be determined by integrating this proton current during the irradiation using a precision-switched integrator trans-impedance amplifier that aims at compensating any current fluctuations of the proton source.

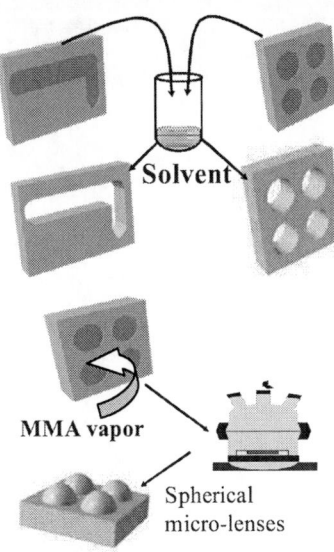

Figure 5: Chemical process steps of DPW: Selective etching process (top) and swelling process (bottom)

After the exposure step, the irradiated zones can be selectively etched in a so-called GG developer (diethylene glycol monobutyl ether 60%, morpholine 20%, 2-aminoethanol 5%, and DI water 15%) during 1h at 38°C, resulting in micro-components with high-quality optical surfaces, as will be discussed in section 3.3. An ultrasonic stirrer is used to enhance the etching process. The resulting component is shown in Figure 6. The total dimensions of the pluggable out-of-plane coupler are 4.5mm x 0.5mm x 0.5mm.

It is obvious that DPW is not a mass fabrication technique as such. However, one of its assets is that, once the master component has been prototyped with DPW, a metal mould can be generated from the master by applying electroplating. After removal of the plastic master, this metal mould can be used as a shim in a final micro-injection molding or hot embossing step [11]. This way, the master components can be mass produced at low cost in a wide variety of high-tech plastics. It is especially important to ensure compatibility of the polymer used with standard PCB fabrication processes – the lamination and solder reflow processes in particular. The PMMA material used for the prototypes has a glass transition temperature T_g around 100°C, which is too low to withstand the temperatures reached during these processes. By replication of the prototype in e.g. cyclo-olefin copolymers (COC), the T_g -and thus standard PCB process compatibility-can be greatly improved.

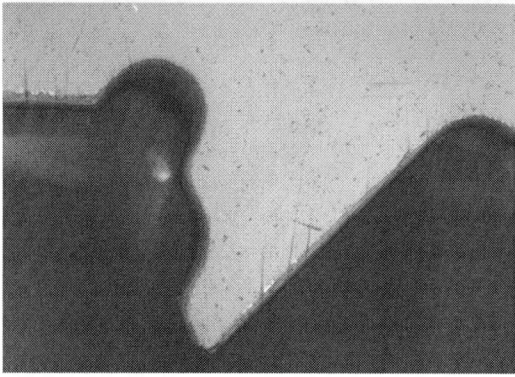

Figure 6: Fabricated multilayer our-of-plane coupling component: overview (top) and zoom on the micro-mirror with integrated cylindrical micro-lenses at the input facet (middle, bottom)

Total Internal Reflection

The propagation of light in the PCB-integrated waveguides as well as the reflection at the micro-mirror facet is based on the phenomenon of Total Internal Reflection (TIR), which can confine light in a material with refractive index n_1 surrounded by another material (or air), with a lower index of refraction n_2. PMMA has a refractive index of 1.4834 at the targeted datacom wavelength of 850nm. To investigate the reflection of light on a PMMA-air interface, we use the Fresnel equations to calculate the reflectance for various incidence angles θ_i, measured from the surface normal [12]. The result is shown in Figure 7 (top), where R_s is the reflectance component perpendicular to the plane of incidence and R_p the component parallel to that plane. We see that we satisfy the TIR condition for incidence angles larger than the critical angle θ_c, which equals 42.39° for a PMMA-air interface. However, for light reflection on a 45° micro-mirror, we are very close to this critical angle, especially

taking into account the NA of 0.3 of the PCB-integrated waveguides. This means that we have a high risk of having light rays inciding on the mirror with an angle θ_i smaller than θ_c and thus not satisfying the TIR condition. To avoid this, we investigate the use of a metal reflection coating on the PMMA mirror facet. If we use gold (Au) for this purpose, having a complex index of refraction of $0.188 + 5.39i$ at a wavelength of 827nm [13], the reflectance at a PMMA-Au interface shown in Figure 7 (bottom) is obtained. We now have a high reflectance, albeit polarization dependent, regardless of the angle of incidence. The use of gold instead of other metals is preferred, since it has the lowest absorption.

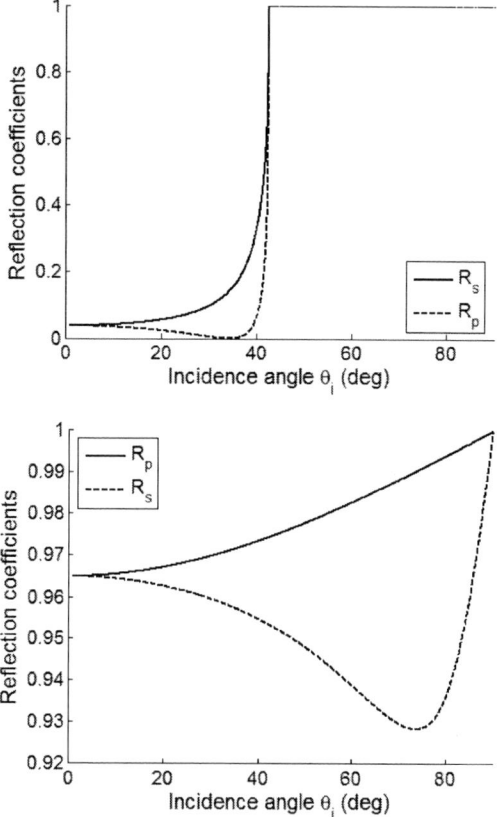

Figure 7: Fresnel reflection coefficients for various angles of incidence in the case of a PMMA-air (top) interface and a PMMA-Au interface (bottom)

Experimental characterization

For the characterization of the critical optical surfaces of the component, namely the flat top exit facet and the 45° mirror facet, we use a WYKO NT-2000 non-contact optical surface profiler (Veeco). Since the entrance facet with the cylindrical microlenses is not accessible with the microscope objective, this surface was not measured, but its surface roughness will be analogous to the two others. The surface roughness analysis reveals that the flat top part has an average local RMS surface roughness R_q of 14.1nm ± 2.7nm measured over an area of 60μm by 46μm. We averaged at least 5 measurements of randomly chosen positions. Applying the same measurement method to the 45° angled facet reveals an RMS roughness R_q of 17.1nm ± 4.2nm. The global mirror surface as well as an example of a locally measured surface profile of the micro-mirror can be seen in Figure 8.

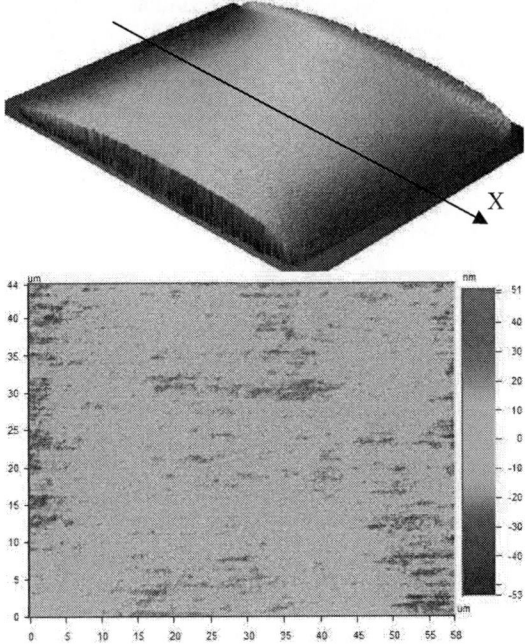

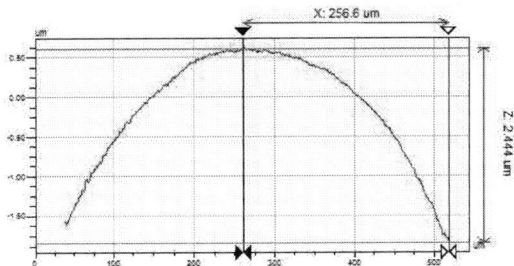

Figure 8: Non-contact optical surface profile measurement of the micro-mirror: overview image (400μm x 500μm, top) and local image (60μm x 48μm, bottom) with resulting RMS roughness R_q=14.98nm.

Figure 9: Cross-sectional profile along X showing the flatness of our optical surfaces. R_t=2.444μm over the total depth of the component (500μm).

The top part shows an overview image of the entire micro-mirror surface, whereas the bottom part shows a local measurement for RMS roughness determination, after removing the sample tilt from the measurement. The overview image shows that the surfaces created by DPW are not completely flat (vertical), due to the scattering of the protons during the interaction with the PMMA. It can be clearly seen by taking a cross-sectional profile along X through this surface, as shown in Figure 9. The

flatness R_t or peak-to-valley difference along the depth of 500μm of the component is measured to be smaller than 2.5μm. As a conclusion, we can say that our developed DPW surfaces have a very good and reproducible optical quality: almost flat with a very low RMS roughness.

We first test our DPW multilayer out-of-plane coupling component in a fiber-to-fiber coupling scheme, as illustrated in Figure 10. For the input, we use a multimode fiber (MMF) with a core diameter of 50μm and a numerical aperture (NA) of 0.2. The detector MMF has a core size of 100μm and a NA of 0.29 and is mounted on a PI F-206 six-axis parallel motion kinematics Hexapod system. This allows us not only to position the detector with an accuracy of 300nm, but also to perform a two-axis scan to check the tolerance for mechanical misalignments of our detector fiber. The resulting 2D scan of the output fiber is shown in Figure 11.

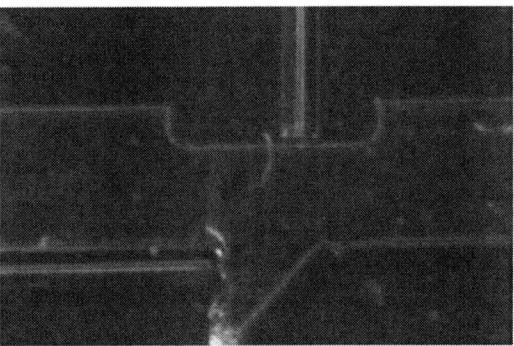

Figure 10: Fiber-to-fiber coupling efficiency measurements (shown in upper channel position)

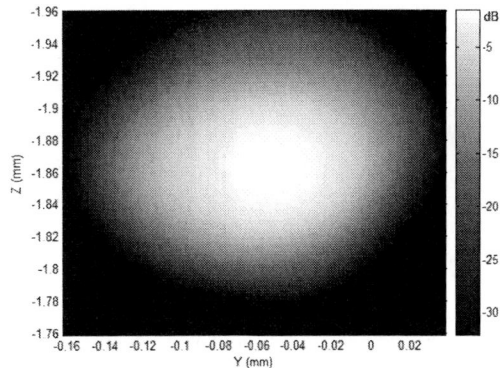

Figure 11: 2D tolerance scan of the spot coupled out by scanning the output MMF

The maximum coupling efficiency measured when using a source wavelength of 850nm was 70% and 75% for respectively the upper and the lower channel position. The reference measurement consisted of a in-line butt coupling of the input and output fiber. In Figure 12, we show a normalized cross-section of the 2D tolerance scan of Figure 11. This plot shows that the -1dB tolerance range for mechanical misalignment of the output MMF is ±26μm. Coupling efficiency measurements when the component is plugged into a multilayer printed circuit board will be performed in the near future.

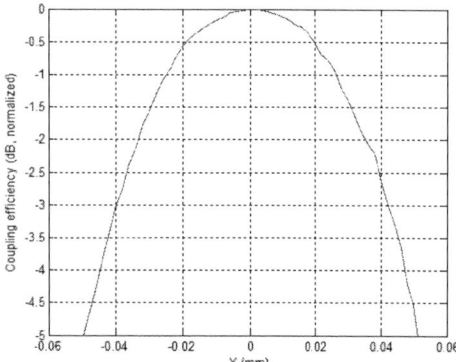

Figure 12: Normalized cross-section of the 2D scan of Figure 11. The -1dB tolerance range for detector multimode fiber misalignments is shown to be ±26μm.

Conclusions

We have shown that Deep Proton Writing is a versatile fabrication technology allowing the rapid prototyping of dedicated pluggable out-of-plane coupling components for multilayer optical waveguides integrated on PCBs. The quality of the fabricated optical surfaces is very high, with average RMS roughness below 20nm and flatness of 2.5μm over a total length of 500μm. Cylindrical micro-lenses can be monolithically integrated into the component to increase the overall coupling efficiency. Coupling efficiencies up to 75% have been measured in a fiber-to-fiber coupling scheme, which can be further increased by applying a metal reflection coating on the micro-mirror.

Acknowledgements

This work was supported in part by DWTC-IAP, FWO, GOA, IWT-GBOU, the European Network of Excellence on Micro-Optics NEMO, and by the OZR of the Vrije Universiteit Brussel. The work of J. Van Erps and C. Debaes was supported by the Fund for Scientific Research-Flanders (FWO) under a research fellowship. N. Hendrickx was financially supported by the Flemish IWT (Institute for the Promotion of Innovation by Science and Technology).

References

[1] R.T. Chen *et al.*, "Fully embedded board-level guided-wave optoelectronic interconnects", Proceedings of the IEEE, vol. 88, no. 6, pp. 780-793, 2000.

[2] R. Yoshimura *et al.*, "Polymeric Optical Waveguide Films with 45° Mirrors Formed with a 90° V-Shaped Diamond Blade", Electronics Letters, vol. 33, no. 15, pp. 1311-1312, 1997.

[3] A. Glebov, J. Roman, M.G. Lee and K. Yokouchi, "Optical Interconnect Modules with Fully Intgrated Reflector Mirrors", IEEE Photonics Technology Letters, vol. 17, no. 7, pp. 1540-1542, 2005.

[4] Y. Liu, L. Lin, C. Choi, B. Bihari and R.T. Chen, "Optoelectronic Integration of Polymer Waveguide Array and Metal-Semiconductor-Metal Photodetector Through Micromirror Couplers", IEEE Photonics Technology Letters, vol. 13, no. 4, pp. 355-257, 2001.

[5] M. Kagami, A. Kawasaki and H. Ito, "A Polymer Optical Waveguide with Out-of-Plane Branching Mirrors for Surface-Normal Optical Interconnections", Journal of Lightwave Technology, vol. 19, no. 12, pp. 1949-1955, 2001.

[6] J.-S. Kim and J.-J. Kim, "Fabrication of Multimode Polymeric Waveguides and Micro-mirrors using Deep X-ray Lithography", IEEE Photonics Technology Letters, vol. 16, no. 3, pp. 798-800, 2004.

[7] G. Van Steenberge *et al.*, "MT-Compatible Laser-Ablated Interconnections for Optical Printed Circuit Boards", Journal of Lightwave Technology, vol. 22, no. 9, pp. 2083-2090, 2004.

[8] J. Van Erps *et al.*, "Prototyping micro-optical components with integrated out-of-plane coupling structures using deep lithography with protons", Proc. SPIE, Micro-Optics, VCSELs, and Photonic Interconnects II: Fabrication, Packaging, and Integration, vol. 6185, 618504, April 2006

[9] N. Hendrickx, J. Van Erps, H. Thienpont and P. Van Daele, "Intra-plane coupler for PCB-integrated multilayer optical interconnects", Poster session 1, EMPC 2007.

[10] C. Debaes *et al.*, "Deep proton writing: a rapid prototyping polymer micro-fabrication tool for micro-optical modules", New Journal of Physics, vol. 8, 270, 2006.

[11] M. Heckele and W.K. Schomburg, "Review on micro molding of thermoplastic polymers", Journal of Micromechics and Microengineering, vol.14, pp. R1-R14, 2004.

[12] M. Born and E. Wolf, Principles of optics, Chapter 1, Cambridge University Press, Cambridge, 1999.

[13] E.D. Palik, Handbook of optical constants of solids, p. 294, Academic Press, London, 1985.

Opto-mechanical tolerance modelling of a free-space intra-MCM optical interconnect system

Michael Vervaeke, Christof Debaes and Hugo Thienpont

Vrije Universiteit Brussel, Dep. Applied Physics and Photonics, Pleinlaan 2, B-1050 Brussels, Belgium

Tel. +32 (0)2 477 48 69, Fax. +32 (0)2 477 48 50, E-mail: michael.vervaeke@vub.ac.be

Abstract

We report on the opto-mechanical design and performances of an intra-multi-chip-module free-space optical interconnect, integrating microlenses and a deflection prism above a dense opto-electronic chip, under various fabrication and assembly errors. To this aim we have built an opto-mechanical model of the complete interconnect. The model allows us to examine the optical performance of the interconnect system under various fabrication and assembly errors using a Monte Carlo simulation. Our design methodology starts with an analysis of the sensitivity of each parameter on the optical efficiency and cross-talk of the module. Sensitivity limits are set depending on the minimum performance specifications of the module. We subsequently simulate the random combination of errors on all parameters in a Monte Carlo simulation in which the variances of the random errors are set to to a fraction of the sensitivity limits. Scaling these variances allows us to asses the effect of a technology accuracy enhancement on the number of fabricated systems that comply with the performance specifications. We will detail an approach to optimize the fabrication and assembly yield. We have used the opto-mechanical simulation framework to predict the process yield of our in-house micro-opto-mechanical fabrication technology, which we call Deep Proton Writing (DPW). DPW is able to define sidewall surfaces with optical quality in 500 µm tick Polymethyl Metacrylate (PMMA) samples. In addition the technique can create microlenses with various lens diameters and sag on the top surface of the substrate. We have been able to integrate our module in a demonstrator with a large opto-electronic device containing Vertical Surface Emitting Lasers and Resonant Cavity Photodetectors. We determined that controlling the adhesive bonding of the opto-mechanical structures and the opto-electronic device to minimize tilt errors will be of utmost importance in a real-world assembly process..

Keywords: Optical interconnects, Tolerancing, Monte Carlo simulation

Introduction

The definition of an adequate set of tolerances for dimensional and geometrical errors is directly related to all the aspects of manufacturing processes. Too tight tolerances will lead to costly production methods, difficult assembly and high fall-out of finalized devices, adversely affecting the total cost and time-to-market. Quality and reliability on the other hand are harmed by too loosely toleranced systems.

The optics community has been advertising optics as an interconnect technology for many decades now. They are regarded as a viable alternative for the electrical wiring at different levels of the computer interconnect hierarchy, in extremis even at the off-chip and on-chip interconnect level[1,2]. However, these approaches can only prove their viability in practical and economical real world systems if they can be made compatible with low-cost mass fabrication technologies and standard semiconductor packaging techniques.

The challenge taken up in this research is to define a set of tolerances for a free-space micro-optical Interconnection Module (OIM) using a combination of optical simulation and geometry calculations based on the Monte Carlo method[3].

The Monte Carlo method is a fairly simple method for solving a wide range of problems, including tolerance calculations. It is based on the ability to generate a large number of random designs which are subsequently checked for their fitness against prescribed specifications. Using statistical analysis tools we are thus able to predict and set the tolerances needed for the free-space optical module.

Earlier research used Monte Carlo-analysis to investigate the tolerance stack-up of micro-optical interconnect schemes[4,5,6], but in general, none of these efforts have dealt with the specific tolerance problems coming with a realistic packaging scheme for free-space intra-chip optical interconnects.

Although some optical simulation tools already provide an interface for tolerancing, this is usually limited to the core of the optical structure, and not the underlying mechanical alignment features. Root-sum-square or worst-case-analysis will not adequately estimate the tolerance stack-up effects[7,8]. For mechanical systems without optics, tolerance tools have been researched heavily. None of currently aviable tools are able to integrate optical simulator data with mechanical tolerancing data.

The article is organized as follows. First, we introduce the free-space concept, along with the optical design of the OIM for maximum performance. The next section of this article highlights the technological requirements for the interconnect demonstrator. We use Monte Carlo simulation for a first-order definition of the technological needs of this demonstrator. In the next section we concentrate on the performances of our in-house prototyping technology currently installed at our facilities. In last section we describe the full-scale Monte Carlo model we built to simulate the fully packaged system. This model will include data from the technology specifications and the real geometry of the demonstrator. The aim is to establish whether it will be possible to manufacture the proposed packaging scheme with an acceptable yield.

The Free-Space Optical Link

For the very short optical interconnects or the Multi Chip Module (MCM) level interconnects, we opt for a free-space approach due its potential for high interconnect densities as was shown in previous studies[9]. The optical interconnect module (OIM) contains micro-optical beamshaping and beamdelivering structures as depicted in Fig. 1.

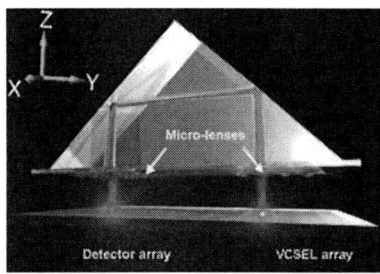

Figure 1: The concept of the OIM.

A microlens array collects and collimates the light coming from a VCSEL array. As the light travels through the structure, it encounters a micromirror which reflects the light such that it is deflected over 90°. A second micromirror reflects the light back to the substrate The OIM containing the necessary features to implement the mentioned functionalities was designed and fabricated in our photonics labs using Deep Proton Writing (DPW)[10]. DPW is a very accurate prototyping technique but is however not suited for low-cost mass-production. We have nevertheless shown the potentialities for mass-production by using vacuum casting[11]. An early component proved the viability of this concept. It was able to create a 4-channel intra-chip OE-VLSI interconnect[9].

Our next generation demonstrator is built around a large interleaved Vertical Cavity Surface Emitting Laser / Resonant Cavity Photo-Detector (VCSEL/RCPD)[12] array on a pitch of 55 µm, arranged in a matrix-addressable way (see Fig. 2 for a prototype assembly of the chip in its package with the OIM sitting above it). The VCSELs emit a 17.5° Full Width Half Maximum (FWHM) beam at 850nm wavelength and the detectors have a 20µmx20µm square aperture. The purpose of it is to demonstrate DPW's capabilities to package optical interconnects.

Figure 2: The matrix addressable VCSEL/RCPD array and the OIM mounted in a pin-grid-array package. The chip resides in a DPW spacer structure onto which 4 LTCC wire carriers are mounted to provide the electrical connections to the chip. Also 4 microspheres are mounted on the spacer to allow the XYZ alignment of the OIM.

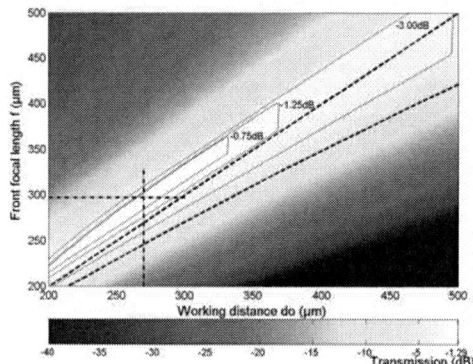

Figure 3: Optical simulation of the efficiency of the OIM versus focal length and working distance. Dashed lines: roots of the analytic solution.

We implemented the optical model using Zemax[13]. The modelling is based on non-sequential tracing of rays emitted in a Gaussian manner with a given angular and spatial distribution. These rays are traced through 3D geometry of the components and integrated on the detector plane. The data returned from the optical simulation tool contains the light transmission in the data channel, and the stray light impinging on 4 neighbouring detectors causing cross-talk. The lens pitch was set to 220µm, enabling the usage of lens diameters up to 125µm. Each channel has an equal optical pathway length of 8 mm, resulting in a flat latency. The throw on the chip varies from 385µm up to 4400µm.

80

The remaining parameters of the OIM optical design, the distance between the chip and the lenses and the focal length of the lenses, are determined using analytical Gaussian beam propagation[9] and optical simulation. Fig. 3 depicts the transmission losses (expressed in dB) of the OIM in the front focal length and working distance design space. The analytical solution does not take into account aberrations which explains why the systems with low losses are at longer focal distances than given by the theory. The nominal design was set to a focal length of 297μm at a working distance of 270μm for 125μm diameter lenses, resulting in a -1.3dB transmission loss.

Monte Carlo analysis of the free-space optical interconnect

We describe the implications of 11 error sources on the fabrication and assembly yield of our technology, together with the influence of different performance specifications. We built a simulation model for the optical interconnect using MatLab[14] to define the perturbed geometry in a Monte Carlo manner and to analyze the data, and Zemax to simulate the optical behavior of the system. This enables us to exploit the versatility of MatLab to perform matrixmanipulations to describe the component geometry and its deformation in conjunction with the raytracing capabilities of Zemax.

The error sources are both related to the assembly and fabrication of the OIM: 3 translations along X, Y and Z (designated dX, dY and dZ), 2 tilts about the X- and Y-axis (dα and dβ) and a rotational error dγ about the vertical Z-axis of the complete optical interconnect with respect to the chip. The other perturbations describe the position of the prism on the baseplate using 1 translation error along Y (dYpB), 2 tilts about X and Y and a rotation error about Z. The last error source defined in this model is the focal length of the lenses (designated df). Other influences were left out of the parameter list at this time due to their complexity or the fact that commercially available components such as high-precision prisms have ample accuracy to meet the constraints established in earlier research.

First, we performed a sensitivity analysis. In such a study we vary one parameter around the nominal value while leaving the other perturbations fixed to zero. We set 4 performance specifications in order to assess the influence of specification settings: cross-talk should not exceed -20dB and the relative loss with respect to the maximum transmission should not exceed -3.00, -1.25 or -0.75dB. We observed that the translation of the complete OIM along the interconnect channel (dY), and the rotation of the complete module about the vertical Z-axis (dγ) are the most sensitive parameters (see Table I).

Table I: Sensitivity analysis of the OIM for 3 loss specification settings. The last column is the result of a Monte Carlo Method (MCM) iteration to update the tolerances of the -0.75dB design for 75% process yield.

Sensitivity setting		-3.00dB	-1.25dB	-0.75dB	MCM
Rotation OIM around X	d α	±0.510°	±0.270°	±0.220°	±0.003°
Rotational OIM around Y	d β	±0.580°	±0.300°	±0.200°	±0.053°
Rotation OIM around Z	d γ	±0.090°	±0.050°	±0.030°	±0.009°
Chip position along X	d X	±2.9μm	±1.5μm	±1.0μm	±0.3μm
Chip position along Y	d Y	±2.6μm	±1.3μm	±0.9μm	±0.2μm
Working distance OIM	d Z	±28.8μm	±14.6μm	±9.6μm	±1.4μm
Rotation prism around X	d α pB	±0.450°	±0.240°	±0.160°	±0.039°
Rotation prism around Y	d β pB	±0.390°	±0.200°	±0.140°	±0.037°
Rotation prism around Z	d γ pB	±0.900°	±0.480°	±0.330°	±0.073°
Prism position along Y	d Y pB	±27.3μm	±14.1μm	±9.6μm	±2.5μm
Lens focal length	d f	±25.3μm	±14.6μm	±10.6μm	±1.5μm

To gain insight into the effect of combined errors in the OIM, we used the earlier obtained sensitivity limits as a driver for the misalignment errors. The probability density function for the random generator in the Monte Carlo simulation was chosen to be Gaussian with a standard deviation σ_i that equals a fraction $1/n$ of the sensitivity limit Δ_i of the corresponding parameter i. Enlarging n for the standard deviation implies a tightening of the misalignments and thus an overall scaling of the technology accuracy.

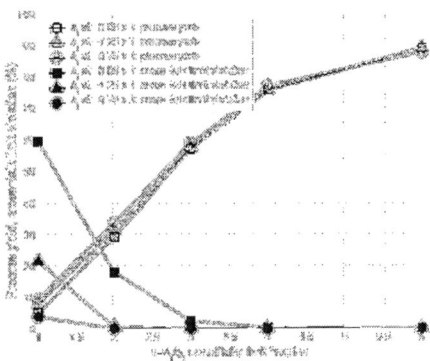

Figure 4: The evolution of the fabrication and assembly yield as a function of the sensitivity limit fraction n used to set the PDFs (with $\sigma_i = \Delta_i/n$).

We observe in Fig. 4 clearly an increase of the manufacturing yield when a more accurate fabrication and assembly is imposed. Furthermore: setting tighter performance specifications increases the initial yield, but at very tight technology specifications the total yield flattens out. If we only look at the failure to comply with the -20dB cross-talk limit in we observe that the failure rate is larger for designs with the most relaxed performance specifications. At $3 \times \sigma_i$ the failure rate goes down to 4% for the -3.00dB performance setting.

The Monte Carlo method also allows us to update the settings of each paramater to obtain a given process yield. We hereby collect the statistical data of each parameter after each successful simulated system. We are thus capable of determining the new nominals and standard deviations of each single parameter and optimize the tolerances to reflect the specified process yield. Table I shows in the last column the Monte Carlo iteration results for the 11 parameters of our simplified OIM model starting from the sensitivity limits at –0.75dB to obtain 75% process yield. Not every parameter contributes in a similar manner to the updating of the process yield. Typically the angular deviations have to be restrained more since they have a severe impact on the cross-talk and efficiency figures of the OIM.

Technology specifications

This section deals with the basic technology specifications of DPW and the packaging challenges that come with the earlier described OIM demonstrator. The data will be a basis to predict manufacturing yields for the OIM design.

Our in-house technology DPW is able to produce high-aspect ratio components in PMMA. The process is a writing process using a single circular mask to deposit a given proton dose on a preprogrammed trajectory onto the PMMA sample[10]. We hereby rely on the accuracy of an XY-translation stage system (50 nm closed loop) and the dose measurement system. We observe a surface roughness Ra of 32 nm over an area of 50μmx50μm or better and a flatness of 1μm over 3.75 mm or better. Microhole diameters have a standard deviation of 1.8μm. We are able to passively mount commercial glass prism with an accuracy of less than ±10μm (1xσ) using a DPW-fabricated mechanical holder to lock the prism on its designated position on a baseplate during adhesive bonding.

We also produce microlenses on the same substrate in one irradiation step. Positional accuracies and circularity of the lens footprints are thus similar to those found with microholes. The relationship between the focal length of the microlenses and the proton dose has been documented in extenso[15] and results in lenses showing a focal length standard deviation of 0.31μm.

The specific layout of the electrical bond pads of the demonstrator chip, being two staggered rows of 256 pads along the 4 sides of the chip, inhibits the usage of easy aligneable packages for the OIM such as kinematic designs. We opted to rely on the dicing edge accuracy of the devices by mating the chips closely to two edges in the cavity of a DPW spacer structure. The spacer structure contains microholes to insert accurate microspheres (508μm±0.254μm at 3xσ)[16] for lateral and longitudinal alignment of the baseplate.

Besides the lateral errors occurring due to positional, assembly and errors on the hole diameter, we have to take into account defects of the hole rim, elastic and plastic deformation of the PMMA by the microspheres and the loads on them, and the influence of the adhesive bonding of the microspheres. We determined that the the most significant errors stem from the static load imposed on the micro-spheres. We performed an analysis of variance on a particular experiment whereby we loaded microspheres with 12 to 24x the nominal weight (50mN) of expect load per microsphere. The microspheres were mounted in microdrilled holes which we believe are more prone to cracking and deformation due to the imperfect hole rim. Even in these worst-case conditions we did not observe more than 7μm impression of the microspheres at the highest load. We expect to obtain a 3.0μm error on the nominal 270μm working distance between the chip surface and the lenses (dZ), which is well beyond the sensitivity limit of the severest design condition (–0.75dB).

The last error sources that can be accounted for, the tilt of the chip and spacer with respect to each other, are more difficult to assess in detail since statistical analysis requires a considerable amount of systems to be integrated in a controllable way. Worst-case measurements set the tilt error to maximum ±0.250°. A careful control of the adhesive dispensing and deposition of the spacer with respect to the chip on the adhesive layer is crucial since the tilt value comes very close to and exceeds the -1.25dB sensitivity limit of the allowable tilt.

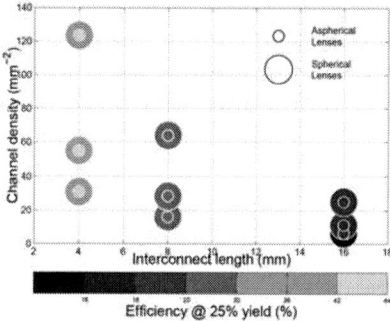

Figure 5: Process yield of DPW as predicted for different pathway lengths and channel densities for both spherical and aspherical lenses. The efficiency limit is set at 25%.

We can give a first estimate of the process yield of DPW using the data from above. When setting a design goal at the –3.00dB loss limit, our technology is performing at most 2.6x better than the sensitivity limit dY . The tilt errors are also 2x better than required while the technology value of the rotational error dγ is yet unknown. The working distance is 10x better specified than set by its sensitivity limit. This leaves us with a total process yield estimation of 30% ($2x\sigma_i$) till 45% ($2.5x\sigma_i$) for the –3.00dB specification. This seems unfavorable but the estimations have to be confirmed yet in the next section of this paper, where a realistic geometrical model is described for extensive calculations.

We were also able to determine the influence of the pathway length, channel density of the interconnect and the usage of aspherical lenses on the process yield. Our analysis yields that aspherical lenses designed to eliminate spherical aberration are able to enhance the efficiency of the OIM, as well as the cross-talk figure. But from a tolerancing standpoint we observed no significant process yield gain using the complex aspherical lenses for large pathway lengths (L>4mm). We also observed that longer pathway lengths lower the total process yield when the technology specifications are remaining constant (see Fig. 5). The same goes for higher channel densities versus the cross-talk. The efficiency values at 25% process yield are close to the values we estimated earlier.

3D geometrical model of the OIM

The sensitivity limits and Monte Carlo analysis of previous section allowed us to draw conclusions on the most critical parameters and the effect of combined misalignments. The technique however suffers from two drawbacks: real world technologies accuracies do not scale uniformly and it does not follow the complete assembly chain. We have studied in this section following error sources: the diameter of the microspheres and their position errors due to microhole deformations, the diameter of the microholes and the tilt of the chip with respect to the OIM. This corresponds closely with a real-life integration step of a replicated OIM onto the spacer mating with the VCSEL/RCPD array.

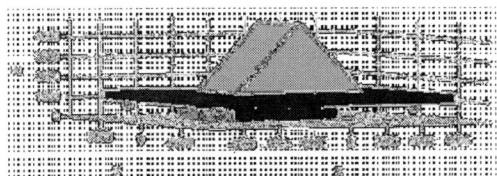

Figure 6: 3D Polygon layout of the OIM.

All elements were implemented in a generic polygondriven geometrical model (see Fig. 6). Collisions between structures when setting up a simulation are detected by calculating the possible intersection points of each polygon with faces of another object, and are resolved by calculating the translation vectors at each collision point[17]. An iterative procedure removes the collision in a random manner ensuring two similar simulation setups to end up in two different systems.

We apply the technology data onto the polygon model in a Monte Carlo way, and build up the OIM as described above. Each Monte Carlo defined geometry is then passed to the optical simulator to determine its transmission loss and cross-talk, as we have described earlier.

A first set of simulations deal with the microsphere alignment scheme. The variation of the distance between optical baseplate and chip was estimated to be ±3.0μm (1xσ$_i$). In reality, this spacing error will be averaged and will induce a tilt error about X and Y. Results from the mechanical simulation using the polygon-drive model with independent perturbations on the 8 microholes and 4 microspheres reveal that an error of ±0.6μm can be expected, with optical baseplate tilt standard deviations of ±6.4 10^{-3}° and a rotational error of ±2.6 10^{-3}°. This can be confirmed using non-contact profilometer measurements of the baseplate tilt. We measured tilts close to ±10 10^{-3}° for a particular system. The optical part of the simulation returned process yields of 91% for the –0.75dB specification setting.

A major point of concern are the tilt errors induced by the bonding of the spacer and chip on the substrate. We estimate them to be worst-case ±0.250°. Setting the worst-case tilt value as standard deviation in the Monte Carlo simulator reveals a poor 2% of the systems complying with the strongest specification, and only 25% are doing better than –3.00dB. A ten-fold decrease of the tilt error yields 77% of the systems complying with –0.75dB. We refer to Fig. 7 for the process yields with transmission below the indicated penalty.

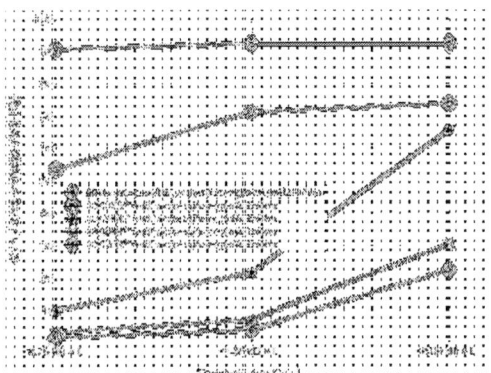

Figure 7: Cumulative yield obtained with the opto-mechanical Monte Carlo simulations for different tilts of the complete OIM combined with the deviation of the micro-spheres.

Conclusions

We reported on the design choices for a large matrixaddressable VCSEL/RCPD array OIM. Using our in-house fabrication technology DLP, we predict to be able to obtain link efficiencies of at most –1.3dB using 125μm diameter lenses with a front focal length of 297μm.

Defining a reasonable set of tolerances for the OIM is imperative for its manufacturability. We have investigated the behavior of the OIM when scaling the technology constraints for 3 performance specifications.

A sensitivity analysis showed the OIM to be very vulnerable to rotation errors along the vertical axis of the module. Monte Carlo simulation of 11 perturbations of the OIM, together with technology data, enabled us to estimate the process yield in first order.

A fully coupled mechanical Monte Carlo simulation and optical simulation tool was developed. We show that our DPW technology in itself is capable of handling the strict requirements set for the OIM (91% process yield at –0.75dB), but the tilt and rotation errors caused by the adhesive bonding of the OIM spacer structure and the VCSEL/RCPD array will deteriorate the yield of the packaging process considerably to 25% at –3.00dB maximum loss.

Acknowledgements

The authors express their gratitude to Mr. K. Geib of Sandia National Laboratories (USA) for kindly providing us samples of the matrix-addressable VCSEL/RCPD arrays, and thank the group of Mr. P. Karioja of NEMO partner VTT Electronics (Oulu, FIN) for taking care of the LTCC wire-carrier fabrication and wire-bonding. This research has been made possible through the financial support of the Fund for Scientifc Research-Flanders (FWO-Vlaanderen), the IAP Photon Network, the Institute for Promotion of Innovation by Science and Technology in Flanders (IWT), the IWT GBOU project "Generic Technologies for Plastic Photonics" and the Research Council (OZR) of the Vrije Universiteit Brussel. This project is involved in the EC Network of Excellence NEMO.

References

[1] D. A. B. Miller, "Rationale and challenges for optical interconnects to electronic chips," in Proceedings of the IEEE, vol. 88, no. 6. IEEE, 2000, pp. 728–749.

[2] G. I. Yayla, P. J. Marchand, and S. C. Esener, "Speed and energy analysis of digital interconnections: comparison of on-chip, off-chip and free-space technologies," Applied Optics, vol. 37, no. 2, pp. 205–227, 1998.

[3] G. S. Fishman, Monte Carlo: Concepts, Algorithms and Applications. Springer–Verlag, 1996.

[4] Kirk, D. V. Plant, M. H. Ayliffe, M. Châteauneuf, and F. Lacroix, "Design rules for highly parallel free-space optical interconnects," IEEE Journal of Selected Topics in Quantum Electronics, vol. 9, no. 2, pp. 531–547, March–April 2003.

[5] S. P. Levitan, T. P. Kurzweg, P. J. Marchand, M. A. Rempel, D. M. Chiarulli, J. A. Martinez, J. M. Bridgen, C. Fan, and F. B. McCormick, "Chatoyant: a computer-aided-design tool for free-space optoelectronic systems," Applied Optics, 1998.

[6] V. N. Morozov, Y.-C. Lee, J. A. Neff, D. O`Brien, T. S. McLaren, and H. Zhou, "Tolerance analysis for three-dimensional optoelectronic systems packaging," Optical Engineering, vol. 35, no. 7, pp. 2034–2044, 1996.

[7] F. and A. G. Kirk, "Tolerance stackup effects in free-space optical interconnects," Applied Optics, vol. 40, no. 29, pp. 5240–5247, October 2001.

[8] S. D. Nigam and J. U. Turner, "Review of statistical approaches to tolerance analysis," Computer-Aided Design, vol. 27, no. 1, pp. 6–15, 1995.

[9] C. Debaes, M. Vervaeke, V. Baukens, H. Ottevaere, P. Vynck, P. Tuteleers, B. Volckaerts, W. Meeus, M. Brunfaut, J. V. Campenhout, A. Hermanne, and H. Thienpont, "Low-cost micro-optical modules for MCM level optical interconnections," IEEE Journal of Selected Topics in Quantum Electronics, vol. 9, no. 2, pp. 518–530, March–April 2003.

[10] B. Volckaerts, H. Ottevaere, P. Vynck, C. Debaes, P. Tuteleers, A. Hermanne, I. Veretennicoff, and H. Thienpont, "Deep Lithography with Protons: a generic fabrication technology for refractive micro-optical components and modules," Asian Journal of Physics, vol. 10, no. 2, pp. 195–214, 2001.

[11] L. Desmet, S. Van Overmeire, J. Van Erps, H. Ottevaere, C. Debaes, and H. Thienpont. "Elastomeric inverse moulding and vacuum casting process characterization for the fabrication of arrays of concave refractive microlenses". Journal of Micromechanics and MicroEngineering, 17(1):81–88, 2007.

[12] K. M. Geib, K. Choquette, D. K. Serkland, A. A.Allerman, and T. W. Hargett, "Fabrication and performance of large (64x64) arrays of integrated VCSELs and detectors," in Proceedings of the SPIE, vol. 4649. SPIE, 2002.

[13] "http://www.zemax.com," Website Zemax Development Corporation.

[14] "http://www.mathworks.com/," Website The MathWorks.

[15] H. Ottevaere, B. Volckaerts, M. Vervaeke, P. Vynck, A. Hermanne, and H. Thienpont, "Plastic microlens arrays by Deep Lithography with Protons: fabrication and characterization," Jap. J. Appl. Phys., Special Issue on MicroOptics, 2004.

[16] "http://www.nemb.com," Website New England Miniature Balls.

[17] P. Jiménez, F. Thomas, and C. Torras, "3D Collision Detection: A Survey," Computers and Graphics, vol. 25, no. 2, pp. 269–285, April 2001.

Comparison of different silicon solar cell structures

Ondřej Hégr[1], Jaroslav Boušek[1], Aleš Poruba[2], Radim Bařinka[2] and Jaroslav Sobota[3]

[1]Faculty of Electrical Engineering and Communication, Brno University of Technology,
Údolní 53, 602 00 Brno, Czech Republic;
Tel/Fax: +420 541 146 392/298; e-mail: xhegro00@stud.feec.vutbr.cz

[2]Solartec s.r.o., Televizní 2618, 756 61 Rožnov pod Radhoštěm, Czech Republic
[3] Institute of Scientific Instruments ASCR, Brno, Czech Republic

Abstract

In our work different types of mono-crystalline solar cell structures based on mentioned technologies are presented. For passivation coatings on both front and rear sides we use reactively sputtered layers of different compositions (SiN_X, AlN, SiO_2 and SiC.) Sputtering technology gives the possibility to replace high-temperature processes which results in decrease of surface and bulk defect density. Consequently this can lead to higher solar cell efficiency. Completed solar cells were tested using LBIC (Light Beam Induced Current) method and V-I characteristics under illumination. Substitute scheme of the cell was measured using dynamic testing method. Surface and bulk recombination was measured by means of microwave photoconductivity decay (μ-PCD).

Key words: Solar cells, Screen-printed contacts, Surface passivation, μ-PCD

Introduction

A standard production process for crystalline silicon solar cells comprises many steps which influence resulting solar cell efficiency. The junction can be made by back-to-back diffusion, by screen-printing on the front side or by gas phase diffusion comprising both front and rear sides. Each process brings some drawbacks so there is no unambiguous selection. In any case consecutive steps depend on technology which is used to PN junction preparation. To decrease the substrate reflectance different substrate textures (pyramids or nano-structures) are shaped on the cell surface. For high-efficiency, a key technology is surface passivation. Silicon nitride films, deposited by thermal activated CVD or low pressure PECVD, are to time standard coatings. Contacting is the last step of solar cells fabrication. To time mostly used are screen-printed contacts on both front and rear sides. For thin wafers a new technology of rear contacting is emerging.

Cell structure requirements

To minimize Auger recombination and free-carrier absorption, the optimum cell should use intrinsic material. It could attain nearly 29% efficiency at one sun. To achieve such ideal cell design, following conditions must be fulfilled:

1. No reflection losses and maximum absorption as achieved by ideal light-trapping techniques.
2. Minimum recombination: SRH and surface recombination are assumed avoidable and only Auger recombination remains.
3. The contacts are ideal: neither shading nor series resistant looses.

No transport losses in the substrate: the carrier profiles in the substrate are flat so that recombination is the minimum possible for a given voltage.[1]

Contacting structures

Screen-printing contacts

In a conventional silicon solar cell, the front contact is formed by screen printing of a silver-loaded paste onto the top surface. These "screen printed" solar cells are reliable and effective, but are able to achieve energy conversion efficiencies of only 12-15%. (Fig. 1)

In most cells, the contacts are placed on both front and rear sides which are technologically simpler. Minority carriers in the substrate are usually collected at the front contact. The diffusion length describes the maximum distance from where they can be collected.

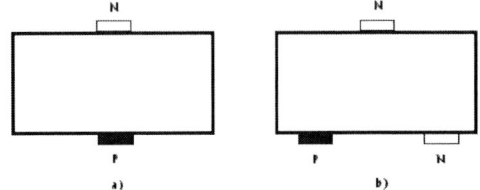

Figure 1: Contacting structure: a) both faces contacted, b) one carrier extracted at both faces. The structures with interchanged *n*- and *p*-types are also possible.

Screen printing drastically affects the design of the emitter: it must be very highly doped to decrease the high-contact resistance and not very shallow to not allow the perforation during paste firing, which can short-circuit the junction. Besides, the wide metal lines must be placed well apart; and in order to keep shading losses moderate, the emitter lateral conductance must be high, which also advises deep and highly doped regions. These characteristics are good to decrease recombination at the contacts, but are far from optimum at the exposed surface.[1]

Buried contacts

The buried contact solar cell is a high efficiency commercial solar cell technology based on a plated metal contact inside a laser-formed groove. The buried contact technology overcomes many of the disadvantages associated with screen-printed contacts technology and this way it allows buried contact solar cell to have performance up to 25% better than commercial screen-printed solar cells.

A key high efficiency feature of the buried contact solar cell is that the metal is buried in a laser-formed groove inside the silicon solar cell. This allows for a large metal height-to-width aspect ratio. A large metal contact aspect ratio in turn allows a large volume of metal to be used in the contact finger, without having a wide strip of metal on the top surface. Therefore, a high metal aspect ratio allows a large number of closely spaced metal fingers, while still retaining a high transparency. For example, on a large area device, a screen printed solar cell may have shading losses as high as 10 to 15%, while in a buried contact structure, the shading losses will only be 2 to 3%. These lower shading losses allow low reflection and therefore higher short-circuit currents.[4]

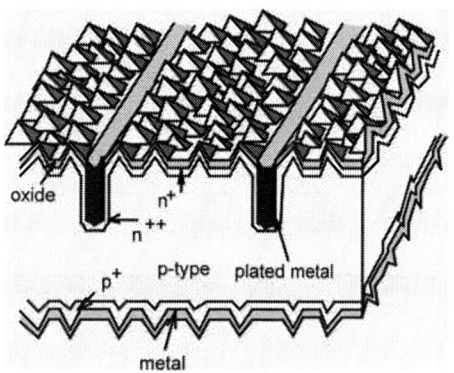

Figure 2: Buried contacts solar cells

Back contact solar cells – high efficiency device

In conventional solar cells, the metal coverage on the front side is a compromise between shadowing and series resistance losses. High efficiency back contact solar cells allow the decoupling of these two factors: the front surface is optimised for optical performance as well as for low surface recombination whereas back surface is optimised for low series resistance losses. Therefore back contact solar cells are especially suitable for concentrator applications which require a low series resistance due to a high I^2R power loss.

In general, high efficiency back contact solar cells have a collecting junction only on the rear surface whereas the front surface is well passivated (Fig. 3). The minority charge carriers mainly generated in the front surface region have to diffuse a long way to the rear junctions. Hence, the devices require a high ratio of bulk diffusion length to cell thickness.

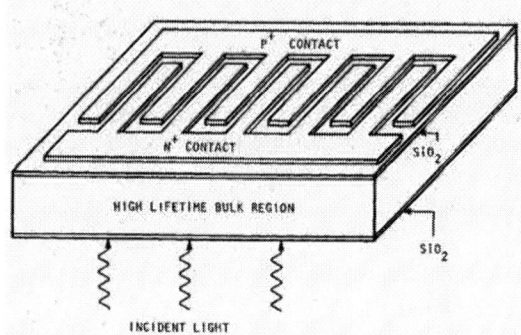

Figure 3: Schematic drawing of point contact solar cell

Antireflection coating and passivation layers

Antireflection coating (ARC) means an optically thin dielectric layer designed to suppress reflection by interface effects. Reflection is at a minimum when the layer thickness is $n_{ARC} \lambda_0/4$, with λ_0 the free space wavelenght, since in this case reflected components interfere destructively. The ARC is usually designed to present the minimum reflectance at around 600 nm, where the flux of photons is a maximum in the solar spectrum.

The most important material for silicon solar cells surface passivation is at present time the silicon nitride (SiNx). It gives optimal combination of optical and passivation material properties. Mostly the layer with thickness of approximately 80 nm is used.

Deposition of SiN_x layers by means magnetron sputtering

The SiN_x layers were deposited in an industrial LEYBOLD-HERAEUS Z550 sputtering device with 150 mm silicon cathode. All films were deposited in the radio-frequency (RF) mode. Sputtering system consists of a stainless-steel vacuum chamber and turbomolecular pump which gives an ultimate pressure of 4.10^{-5} Pa about. The cleaned substrates (Si) are mounted on stainless-steel substrate carriers and were deposited in rotating mode.

The substrates were preheated to 50°C. The RF power for magnetron sputtering was from 400 W up to 600W, deposition rate were from 2 nm/min up to 5 nm/min, depending on the process parameters. The gas mixture (Ar-N$_2$) was controlled in steady pressure condition by mass-flow controllers. In presented SiN$_x$ deposition runs hydrogen was not used in gas mixture. The deposition rates were evaluated from the film thickness using a Talystep profilometers.

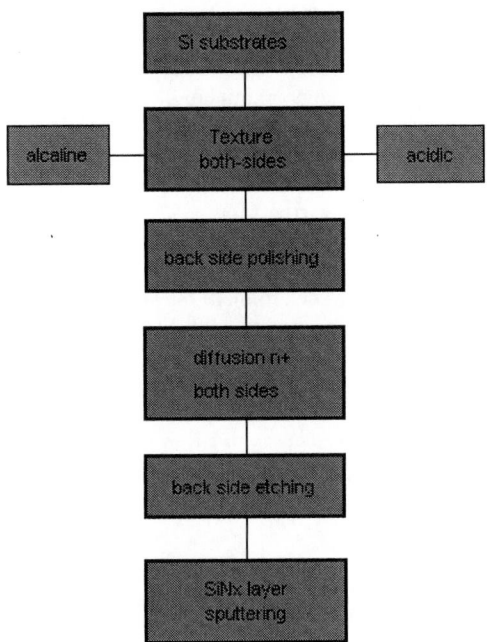

Figure 4: Scheme of silicon substrate adjustment

The first step for crystalline silicon substrates adjustment is etching of damaged layer. Silicon etching in the NaOH solution is most often used. The damaged layer removing by means of cathode sputtering is also possible. (In this case the magnetron power supply is connected to substrate holder and the holder is used as cathode.) Here the requirements are low-energy species that are used to surface "bombardment". High energy species bombardment could damage the surface layers of silicon substrate. Consequently, the surface recombination would be increased. After removal of the damaged layer the standard process for texturing of silicon surface follows. On the surface the pyramidal structure is created by means of etching. This structure reduced the reflectance of light from the silicon surface. Block scheme of silicon substrate adjustment is on the Fig. 4.

Optical properties of sputtered layers

In the Fig. 5 shown is a comparison between passivation layers on different substrates. The

thickness of sputtered SiN$_x$ layers on non-textured substrate are measured.

The left case is silicon nitride layer of 76 nm deposited by means of magnetron sputtering on the alkaline texture. The middle case show the sputtered layer SiN$_x$ of 76 nm on the acidic texture and the right case there is standard Si$_3$N$_4$ deposited by means of LPCVD on the sample with alkaline texture.

The thickness of sputtered layer depends on top angles of texture "pyramids". In case of the alkaline texture the angle is large and the thickness of sputtered silicon nitride is about 40% less with comparison to the smooth surface.

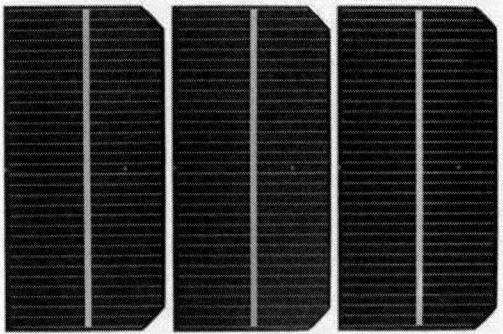

Figure 5: Comparison of different textures with passivation SiN$_x$ layers

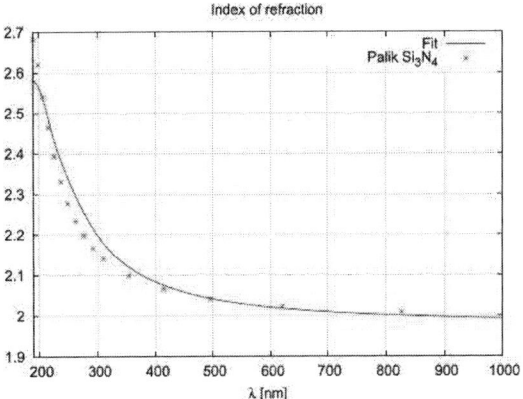

Figure 6: Refractive index of the stochiometric silicon nitride and sputtered silicon nitride

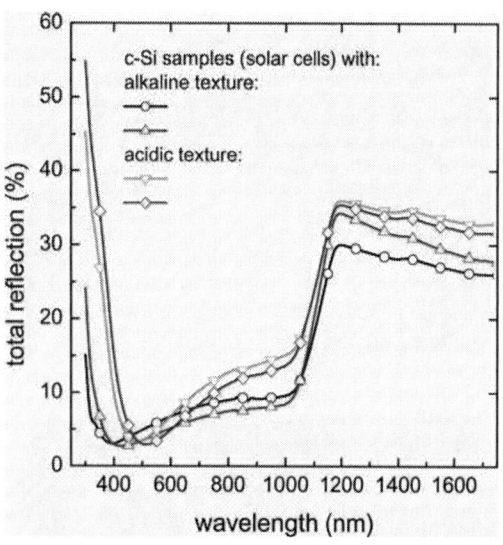

Figure 7: Total reflection of SiN_x on the alkaline and acidic texture

On the sputtered layers the refractive index and reflection was measured. The refractive index was measured by ellipsometry using different elevation. The "Fit" curve in Fig. 6 presents the refractive index of sputtered SiN_x and the "Palik" curve is calculated value Si_3N_4 by Palik model.

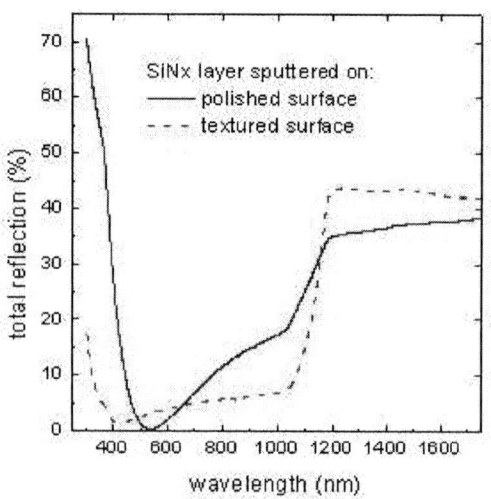

Figure 8: Total reflection of sputtered silicon nitride on polished and textured surface

MW-PCD measurement

In the method called MW-PCD (Microwave Photoconductive Decay), within the wafer under test a short laser pulse (duration 100 ns) generates excess carriers. After termination of the laser pulse, the excess carriers recombine in the bulk and at the surfaces of the wafer. The decay of the excess carrier concentration within the wafer is measured via the reflected microwave power. As the microwaves penetrate deep into the wafer, the recorded signal represents a spattialy averaged excess carrier concentration. The asymptotic part of the resulting transient curve is fitted with an exponential decay function, where the time constant of the exponential function is the effective carrier lifetime τ_{eff} of the wafer under test. The maximum intensity of the bias light is ~ 250 mW/cm^2 using an additional halogen lamp. Hence, MW-PCD system is limited to $\Delta n \leq 2\times10^{15}$ cm^{-3}, depending on the lifetime of the measured material. Standard MW-PCD method gives no information about the excess carrier density at a particular bias light intensity. The bulk injection level is given from the measured bias light intensity and the measured effective lifetime using the semiconductor simulation program.

The bias light illuminates a relatively large section of the sample and the laser beam has a small diameter (a few mm) and hence the measured quantity is a local carrier lifetime.

The quality of silicon material is very important in case the mono-crystalline substrate. The purity of silicon determined the lifetime on the silicon surface and within bulk. The top value of lifetime is 13,3 μs for non-passivate substrate and 10,2 μs for substrate with sputtered passivation layer of SiN_x. (Figure 9)

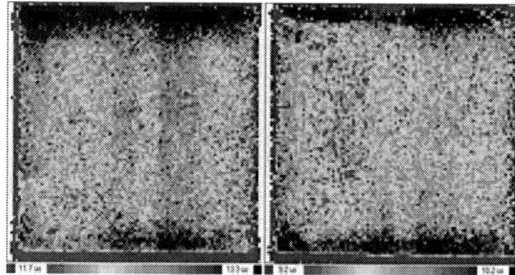

Figure 9: Measurement of mono-crystalline silicon substrates; before passivation (on left) and next to sputtered of SiN_x (on right)

Conclusions

For high efficiency of silicon solar cells, the back contact studies are responsible. In general, high efficiency back contact solar cells have a collecting junction only on the rear surface whereas the front surface must be well passivated. Present, for surface passivation the most popular technique is PECVD for SiN_x:H deposition. In this paper the application by RF reactive magnetron sputtering for SiN_x layers deposition is described. In presented, SiN_x deposition runs hydrogen was not used in the gas mixture.

Silicon substrates with different types of texture were used. The optical parametres of passivation layer are strongly depended on the thickness of sputtered SiN_x, therefore thickness of the layer must be controlled precisely

To carry on the investigation a set of deposition rate experiments is to be done in next future to ensure the optimal thickness of passivation layer on both front and back solar cell sides. In next future also AlN and SiC magnetron sputtered passivation layers will be investigated throughly, especially for the passivation of the back side of the cell.

For determining of minority carrier lifetime were used very know microwave-detected photoconductance decay (MW-PCD) method. By the help of this technique we measured monocrystalline substrates before passivation layer deposition and under sputtered of SiN_x passivation layer. The passivation effect by monocrystalline substrates is decreased or same. The account can be strongly n^+ diffusion layer within.

In the next research, the sputtering condition by SiN_x layers will be optimised. As antireflection coating and passivation layers of silicon solar cells the next materials (AlN, SiO_2 or SiC) will deposited.

Acknowledgements

This research has been supported by the Czech Ministry of Education in the frame of Research Plan MSM 0021630503 MIKROSYN "New trends in Microelectronic System and Nanotechnologies", by the Ministry of Environment of the Czech republic in the frame of R&D project VaV SN/3/172/05 and by project of Ministry of Education, Youth and Sports in Research Plan 1M06031

References

[1] Tobías I., Canizo C., Alonso J.; Crystalline Silicon Solar Cells and Modules; Handbook of Photovoltaic, 2003 John Wiley & Sons, ISBN: 0-471-49196-9

[2] Hégr O., Boušek J., Sobota J., Bařinka R., Poruba A.; Reactive magnetron sputtering silicon nitride layer for passivation of crystalline silicon solar cells, 21st European Photovoltaic solar Energy Conference, 4-8 September 2006

[3] Cuevas A.; Recombination and trapping in multicrystalline silicon, IEEE Transaction on Electron Devices, submitted 15 December 1998

[4] http://www.udel.edu/igert/pvcdrom/MANUFACT/BCSC.HTM

[5] Luque A.; The Requirements of High Efficiency Solar Cells", in Luque A, Araújo G., Physical Limitations to Photovoltaic Energy Conversion, 1-42, Adam Hilger Ltd, Bristol (1990)

[6] Turton R, "Band Structure of Si: Overview", in Hull R (ed), Properties of Crystalline Silicon, INSPEC, Stevenage, UK (1999)

[7] Hegr O.; Surface recombination analysis and quality of special passivation layers, Diploma project, FEKT VUT Brno 2005, Czech Republic

[8] Cuevas A.; Recombination and trapping in multicrystalline silicon, IEEE Transaction on Electron Devices, submitted 15 December 1998

A Microcontroller-Based Neurostimulator in Thick-film Technology

Krzysztof Zaraska and Barbara Gröger

Institute of Electron Technology, ul. Zabłocie 39, 30-701 Kraków, Poland

Phone: +48 12 6565183, fax: +48 12 6563626, e-mail: zrkzaras@cyf-kr.edu.pl

Abstract

This paper describes an experimental realization of an animal implantable neurostimulator using hybrid technology. The device is based on a low-power microprocessor, with the passive components and interconnections made using thick-film technology on a ceramic substrate. A novel aspect of the design is the fact that customization of output signal parameters is performed only in the manner of programming the microcontroller and not by changing the values of the passive components, as it was the case in earlier designs. Furthermore, such approach allowed to introduce automatic mechanisms for adaptively controlling the output signal, depending on the characteristics of the individual test subject. Such capabilities are of great interest, because of inherent significant physiological differences between individual test animals.

Introduction

Stimulation of vegetative neural system by electrical signals is viewed nowadays as a promising therapeutic technique. In order to better understand its mechanism, many experiments on animals (particularly small animals, such as rats) are performed. This paper deals with the issues regarding development and manufacturing of small animal neurostimulators for research purpose. Development of such devices poses numerous (and very interesting) technical problems. The first one is the issue of run size: for a typical experiment, about 30 devices are needed. At the same time, the device unit cost must be low enough to be acceptable in such quantities, with respect to the overall experimental budget. Next, there is the problem of power supply and power consumption. Furthermore, the device must be customizable, in particular with respect to the output signal parameters, allowing a wide range of experiments to be performed. At the same time, a strict device size limit exists, associated with the body size of the test animals.

In this paper, we describe a novel solution of a cheap, implantable neurostimulator circuit, manufactured in a thick-film technology on a ceramic substrate, and encapsulated in a biocompatible Dow Corning 3140 RTV silicon resin. The device is intended for experimental use on animals. It is based on a low power CMOS Atmel AVR ATtiny 13 microcontroller. We discuss the requirements, present our solution along with simulation results and results of the practical evaluation of the implemented device, along with technological constraints.

The present work is based on earlier experiences in the development of fixed-parameter neurostimulators started at the AGH University of Science and Technology in Kraków by M. Lipiński (0), and later continued by ITE; for an overview of the development history, see 0. A detailed treatment of the design challenges and solutions is given in 0.

Design requirements

The device design should meet a number of criteria, namely:

- *Low unit cost.* The device should be characterized by an affordable unit cost at a low run size (typically 30 units).
- *Flexibility.* As the device is intended for research applications, the design should allow for it to be easily customized in order to fit the needs of a particular experiment.
- *Low power consumption.* Our experience suggests that the limiting factor in the design of implantable devices is power source. In our case, we are limited to stock Li-Ion watch batteries. This, along with requirements concerning the operational lifetime of the device, imposes requirements on the power consumption of the device.

In order to fulfill the first two (somewhat contrary) requirements, we have decided to use a following approach: construct a programmable microprocessor-controlled current source, with a wide range of operational parameters. The parameters should be preferably set in software (i.e. by programming the microprocessor) or by selecting the values of the passive (R, C) SMD components. This allows us to achieve a low unit cost by manufacturing a larger series of universal devices and then customizing them according to the client specification at a later stage. The concept of using a low-power CMOS microcontroller for generating a stimulation signal is well established in the literature (0,0,0).

Problem statement

As the natural nerve signal is composed of series of spikes (so-called "spike trains") 0, a typical solution is to construct a device generating rectangular pulses of preset amplitude, duration, and repetition time; the pulses are then introduced into the neuronal axon using a suitable electrode. In some solutions (0, 0) a voltage source is used; whereas research suggest that better results can be achieved by using a current source. This is caused by the variability of the electrical impedance of the nerve-electrode contact; it is known to increase over time due to the living organism's reaction to the implanted foreign body 0 . Also, the contact impedance can vary greatly between the test animals, both as a result of individual physiological differences, and inherent non-repeatability of manual process of implanting and connecting the device. It is therefore desirable, to implement a device generating rectangular current pulses of regulated amplitude and duration.

Description of the device

Consider a typical solution of a regulated current source as shown in Fig. 1 (as shown in 0, although the general concept of such source has been well-established before; see for example 0). Stimulated nerve is connected to the drain of an output transistor. The drain current of the transistor is controlled using the source resistance and gate voltage. An operational amplifier is used to linearize the source over the whole operating region of the output transistor's I-U characteristics. The control voltage is usually generated by a D/A converter. As a source resistance, a fixed resistor, or a digital potentiometer can be used.

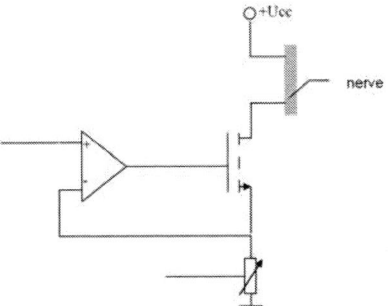

Figure 1. A typical adjustable monolithic current source (based on 0).

Such solution, while correct, has certain disadvantages. Foremost, it requires the use of an operational amplifier to compensate for the non-linearity of the output transistor. This leads to a big diversity of the required elements, which means that the concept, while suitable for a monolithic (ASIC) realization, is not suitable for a space-constrained hybrid realization, where the element count has to be minimized.

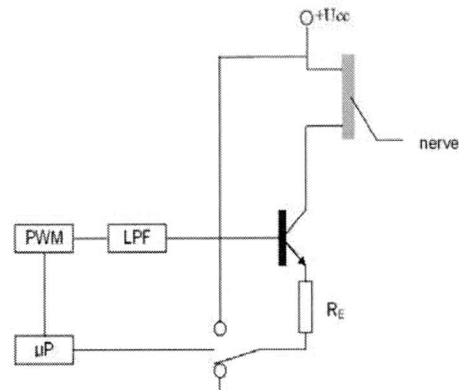

Figure 2. Our concept of the neurostimulator. μP – microprocessor core, PWM – pulse width modulator, LPF – low pass filter.

Our solution of the problem is shown in the Fig. 2. The stimulated nerve is connected, similarly as before, to the collector of the output transistor (the fact that we use a bipolar transistor instead of a MOSFET is of no significant importance for the concept of the circuit). Because of the fixed emitter resistance, R_E, the output current of the device is linearly dependant on the base voltage (U_B) of the output transistor. A switch connecting R_E to the ground or U_{CC} is used to switch the output circuit respectively on and off (the reason for using this switch will become apparent later on).

In order to generate control voltage U_B, and control the output current of the device, we employ a DAC. Similar solutions are well established in the patent literature with respect to monolithic circuits 00. We observe, that instead of using a monolithic converter, we can easily construct a sigma-delta DAC, using a programmable PWM (*Pulse-Width Modulation*) generator and a low-pass filter. A PWM generator can be implemented easily using a microcontroller; either in software, or using a build-in PWM generator, already available on die of numerous integrated microcontrollers. In the simplest realization, we thus connect the base of the output transistor to the output pin of the microcontroller, via a passive low-pass filter.

The output current can be measured by the microcontroller using an on-die A/D converter measuring the voltage drop across R_E; the result can then be used to adjust the duty factor of the PWM modulator to correct the actual output current value. This way, we can compensate for non-linearities of the output stage, without using an operational amplifier: we implement a corresponding feedback mechanism in software. Alternatively, by measuring the voltage on the collector, we can measure the voltage drop across the nerve. This can be used to determine its input impedance, or adjust the source current to achieve present output voltage amplitude. Such possibility opens the door to both adaptively controlling the stimulation parameters (i.e. responding to physiological changes) and using the same device as current or voltage source, depending on the needs.

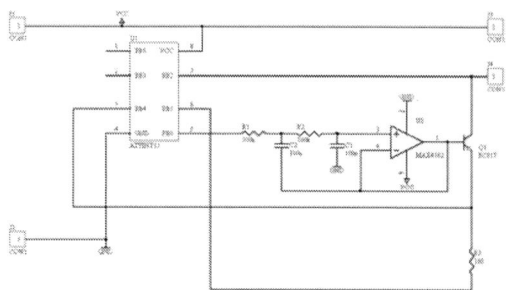

Figure 3. Practical implementation of the device: J1,J2 – battery connectors, J3,J4 – electrode connectors, U1 – microcontroller, PB0 – PWM output, PB1 – switch output, PB2, PB4 – ADC inputs

Practical experiments have shown that in order to minimize power consumption (and thus increase the lifetime) the device has to operate at a low clock rate, in our case, 128 kHz. Since we use an 8-bit PWM modulator, that means that the PWM carrier frequency is about 0.5kHz. As the cut-off frequency of the filter should be several times lower, this rules out application of a passive filter, as it would result both in very large RC element values and an unacceptable ripple in the output signal. Instead, we use an active filter based on a low-power MAX4132 operational amplifier (Fig. 3). According to the simulation results, attenuation of the carrier frequency is on the order of 60 dB.

Thanks to the use of an active filter topology, the device is resilient to changes in the values of the RC components of the filter. In Fig. 4, we show a result of a Monte Carlo simulation of the device, that is histogram of values of 3dB frequency of the filter, assuming deviations of filter elements from their nominal values of 20% for capacitors and 10% for resitors. As can be seen, the whole range of possible 3dB frequency of the filter (9 – 13 Hz) is significantly lower than the carrier frequency (500 Hz). This allows

us to avoid individual correction and selection of filter elements, thus reducing the cost.

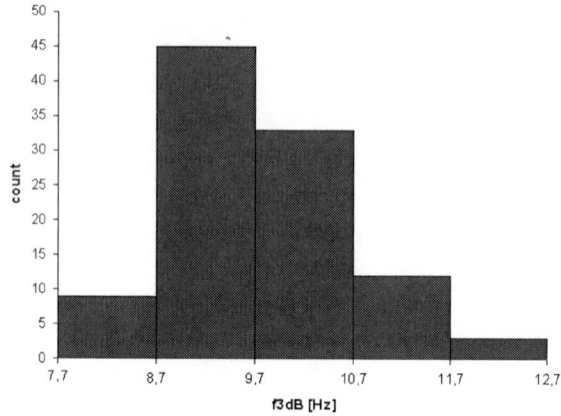

Figure 4. Results of the Monte Carlo simulation.

However, because of the low cutoff frequency of the low-pass filter, we cannot use the DAC for shaping the output pulse; it would result in an very long slope (Fig. 5), which would be physiologically unacceptable. Instead, we use the DAC only to regulate the amplitude of the rectangular output pulse, while the pulse is formed independently by keying the output transistor on and off. This is easily achieved with a switch shorting R_E to ground or U_{CC}, forcing the transistor to enter the active or cut-off state respectively. The role of the switch can be effectively performed by a digital output pin of the microcontroller.

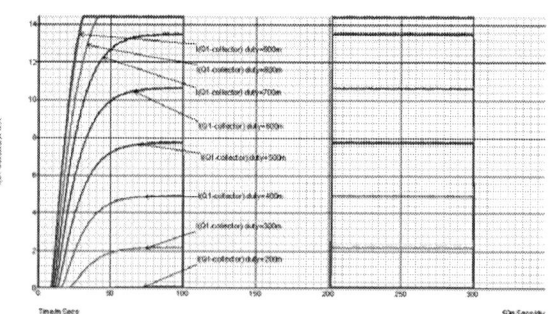

Figure 5. Switch on/off transients for various PWM duty factors (simulation). The first rising slope is achieved using the change of PWM duty factor, from 0 to the desired value whereas the second one is achieved using emiter switching of the output transistor, with a stable base voltage. Note slope lengths in both cases and effective output current regulation range (duty factor 0.2 … 0.8).

While our design initially appears to be complicated, in a practical terms it is not; as seen in Fig. 3, it can be effectively implemented using only 5 passive and 3 semiconductor elements (in discrete or die form): the output transistor, the operational amplifier for the active low-pass filter, and the microcontroller, integrating on die the control logic, PWM generator, and an A/D converter with the input multiplexer.

Practical evaluation

In our example implementation, the device has been manufactured in a standard screen-printing thick-film process with 200μm line width and a single layer of DuPont QS171 conductive paste on alumina substrate. The printed and fired structure can be seen in Fig. 6, whereas the fully assembled structure is shown in Fig. 7. Since we have opted to use the semiconductors in a packaged form, the area below the packages has been used for placing the resistors; such solution has contributed to decreasing the device size.

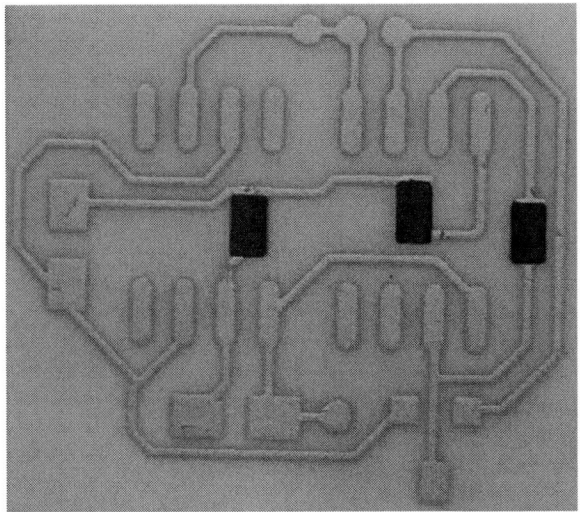

Figure 6. The ceramic structure (pre-assemly).

Figure 7. The assembled device (before encapsulation)

All the resistors in the circuit have been screen-printed with a DuPont BIROX 1651 resistive paste.

For other elements, we have used SMD versions. The capacitors are in 0805 SMD package. Both integrated circuits used (Atmel AVR ATtiny13V microcontroller and MAX4132 operational amplifier) are in SO-8 SMD packages. The output transistor is in the SOT-23 SMD package. The device size is 16X18 mm.

Fig. 8 shows the measured output signal of the device, demonstrating the capability of generating pulse trains with amplitude regulation.

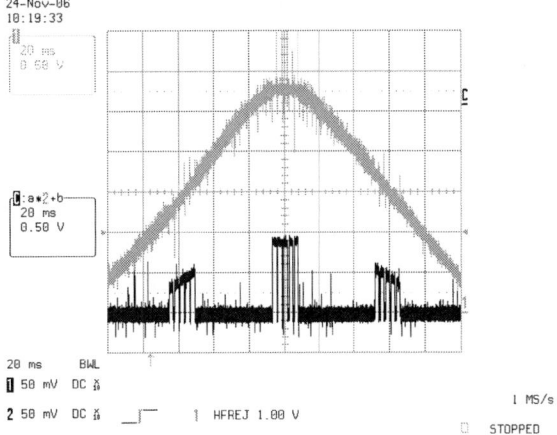

Figure 8. Keyed output current signal (bottom) and DAC output voltage (top)

Conclusions

We have demonstrated the practicality of implementing an experimental research neurostimulator in a low-cost thick film process. The flexibility of the design allows achieving a large range of output parameters, only by reprogramming the microcontroller . The use of thick-film process allows for significant space savings in the device layout, due to stacking of elements.

Acknowledgements

The authors wish to thank Mr. Rafał Gara for his help in preparation of this paper.

The simulations have been performed using SIMetrix Intro 5.2 software.

References

[1] Archer S.T., Pless D.B., "Stimulation Signal Generator for an Implantable Device", U.S. Patent 6,690,974, 2004.

[2] Meadows P. et. al., "Rechargeable spinal cord stimulator system", U.S. Patent 6,516,227, 2003

[3] Królczyk G., Żurowski D., Sobocki J., Laskiewicz J., Thor P.J., "Encoding meal in integrated vagal afferent discharge", J. Physiol. Pharmacol. 2004 Mar;55(1 Pt 1):99-106.

[4] Lipiński M., "Implantable neurostimulator of muscles and nerves" Unpublished manuscript, AGH, 2001

[5] Sobocki J., Królczyk G., Herman R.M., Matyja A., Thor P.J., "Influence of vagal nerve stimulation on food intake and body weight - results of experimental studies", J Physiol Pharmacol. 2005 Dec;56 Suppl 6:27-33.

[6] Tietze U., Schenk Ch., "Advanced electronic circuits", Springer-Verlag, Berlin, 1978

[7] T hil M.-A., Gerard B., Jarvis J.C., Debelke J., "Two way communication for programming and measurement in a miniature implantable neurostimulators", Medical & Biomedical Engineering & Computing, 43, 528-534 (2005)

[8] Turner J.N., Shain W., Szarowski D.H., Andersen M., Martins S., Isaacson M., Craighead H., "Cerebral Astrocyte Response to Micromachined Silicon Implants", Experimental Neurology 156, 33-49 (1999).

[9] Zaraska K., "Evolution of neurostimulators implants", Proc. 43rd IMAPS Nordic Annual Conference, Gothenburg, Sep. 17-19, 2006

[10] Zaraska, W., Thor, P., Lipiński, M., Cież, M., Grzesiak, W., Początek, J., Zaraska, K., "Design and fabrication of neurostimulator implants – selected problems", Microelectronics Reliability, 45 (2005), 1930-1934

Wire Bonding Power Interconnection

M. Novotny, T. Dvorak, I. Szendiuch

Department of Microelectronics, Brno University of Technology, Udolni 53, 602 00 Brno, Czech Republic

Phone:+420 541 146 158, Fax: +420 541 146 259, E-mail: xnovot39@stud.feec.vutbr.cz

Abstract

This paper describes recent developments made to the finite element modeling of wire bonding interconnection, extending its capability to handle viscoplastic behavior. In this project a test system to be used in investigation and research on reliability of interconnections in integrated circuits and printed circuit boards will be developed and realised. Especially the reliability of interconnections of semiconductor chips under high current regime until 10 A, or more, is emphasised. This study discusses the analysis methodologies as implemented in the ANSYS finite element simulation software tool. The aim of this paper is to improve reliability of wire bonding interconnection and to increase durability of these structures.

Key words: Stress, ANSYS, wire bonding, BGA

Introduction

Finite Element Analysis (FEA) is a computer-based numerical technique for calculating the strength and behavior of engineering structures. It can be used to calculate deflection, stress, vibration, buckling behavior and many other phenomena. It can be used to analyze either small or large-scale deflection under loading or applied displacement. It can analyze elastic deformation, or "permanently bent out of shape" plastic deformation. The computer is required because of the astronomical number of calculations needed to analyze a large structure. The power and low cost of modern computers has made Finite Element Analysis available to many disciplines and companies.

Finite Element Analysis makes it possible to evaluate a detailed and complex structure, in a computer, during the planning of the structure. The demonstration in the computer of the adequate strength of the structure and the possibility of improving the design during planning can justify the cost of this analysis work. FEA has also been known to increase the rating of structures that were significantly over designed and built many decades ago.

This work use finite element analysis for calculating stress values rising in the solder joints and wire bonding connection, depending on different substrates, solder pastes and two types of wires (Au, Al).

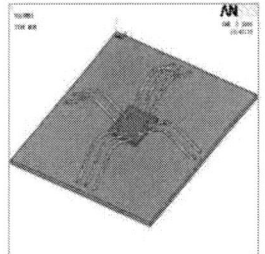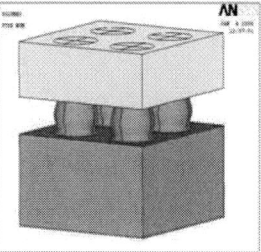

Fig. 1: Models used for simulation.

Wire bonding simulation results

The surveyed sample is the connection of ceramics (Al_2O_3) to substrates (FR4, Al_2O_3) by the help of wire bonding. The gold and aluminium wires are used for the simulation. The most stressed place on the wires is on the bottom side, where is wire connected on the substrate or chip as shown figure 2.

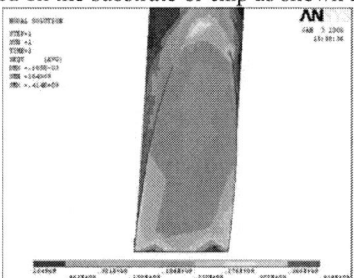

Fig. 2: Bottom side of wire.

Figures 3,4 show stress values along wire. Stress distribution for Al2O3 – Al2O3 structure is similar for both ends of wire (fig.9). However the

Al2O3 – FR4 structure have higher stress values in connection between wire and chip.

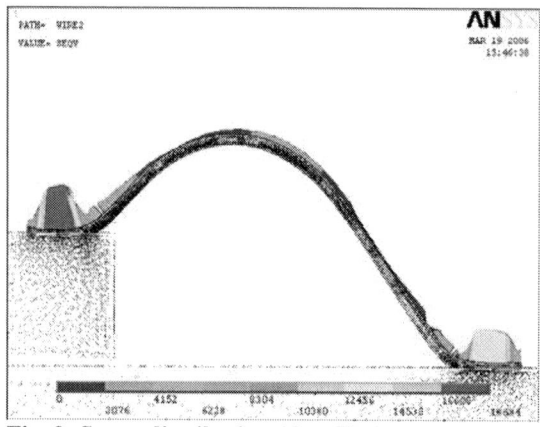

Fig. 3: Stress distribution Alu – FR4 structure.

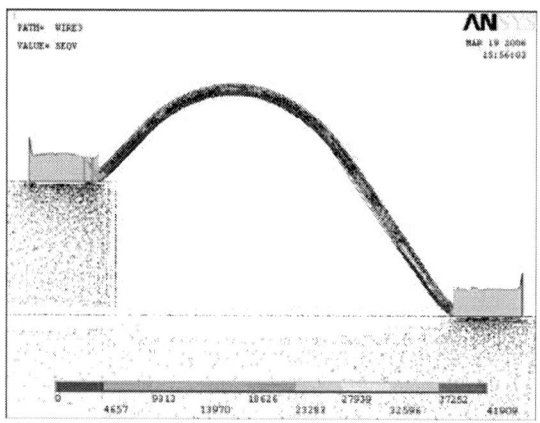

Fig. 4: Stress distribution Alu – Alu structure.

Figure 5 shows stress values for all combinations of substrates and wires. The values with the gold wire are practically identical. The highest stress raised for combination Al2O3 substrate and aluminium wire.

Table 1: Stress values.

Temp. [°C]	Stress [MPa]			
	Alu_Al	Alu_Au	FR4_Al	FR4_Au
20	26,5	26,2	60,3	20,1
40	59	58,4	117	22,3
60	98,1	97,2	171	28,5
80	144	142	220	50,9
100	196	194	267	69,3
120	235	233	320	95,6
140	274	272	373	122
160	313	311	427	148

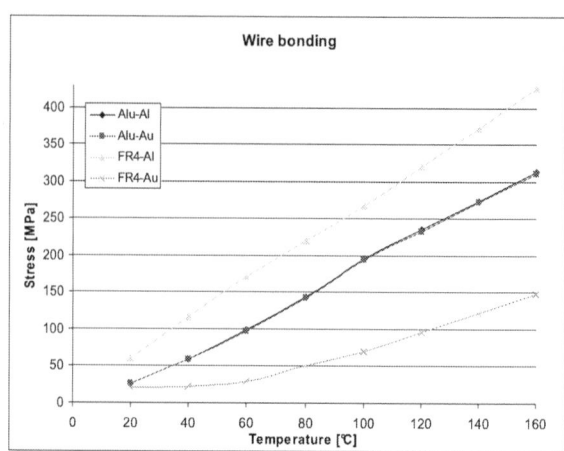

Fig. 5: Thermal dependence of stress in wires.

Alternative way of interconnection

The surveyed sample is the connection of ceramics (Al$_2$O$_3$) to substrates (FR4, Al$_2$O$_3$) by the help of solder bumps. The bump size is 1 mm. The solder paste is lead-free solder 95,5Sn3,9Ag0,6Cu. The program ANSYS creates geometrical and finite element model and will carry out analysis surveyed stress. Simulation is executed for some different solders and substrates. The reason is to the find out materials where the smallest final stress will be. The interpretations of results are divided into two parts. The first part of evaluation discusses the stress distribution in solder joints. Determining a place in the solder with the maximal stress values and determining the stress is distributed for all materials are the results of this investigation. The possible danger solder joint crack is in the place with the maximal stress value. The second part of evaluation discusses the maximal stress value in solder joints. The configuration with the smallest stress value which one will be the most reliable regarding stress has been found.

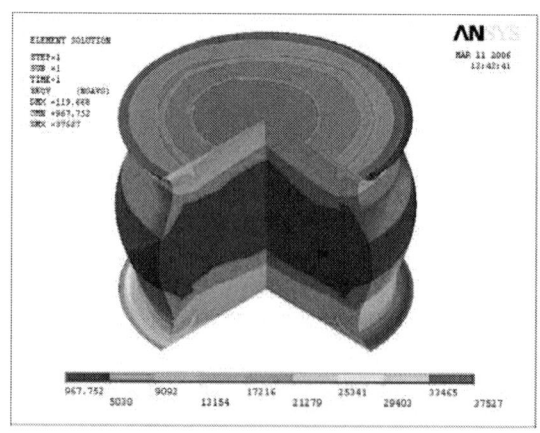

Fig. 6: Part of solder joint.

The maximal stress rises on the top of the bumps (fig.6), where is solder connected on the chip and the stress distribution is similar for all temperatures and substrates. There is stress distribution for all bumps on the figure 7. The bumps on the edge have different stress distribution than bumps in the middle of bump filed as shown figure 7.

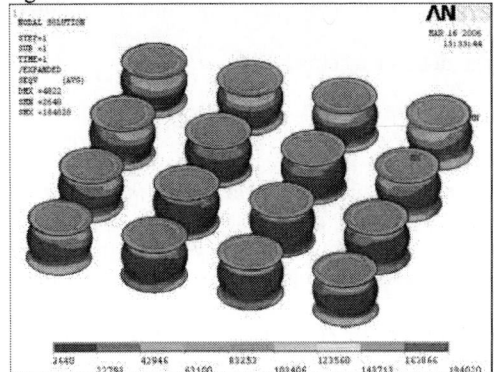

Fig. 7: Stress distribution for all bumps.

Figures 8, 9 shows stress values for one bump. The stress values are different for each substrate. Figure 8 shows Al2O3 – SnAgCu – Al2O3 structure. The maximal stress is on the top of bump. This is the same as Al2O3 – SnAgCu – FR4 structure, but stress value on the bottom of bump is lower for Al2O3 – SnAgCu – FR4 structure.

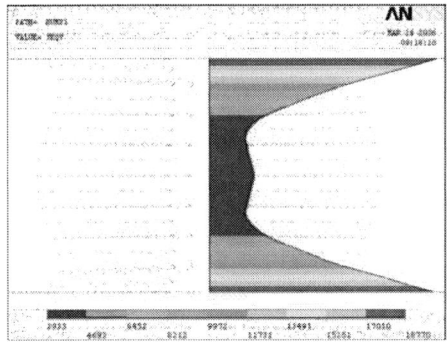

Fig. 8: Stress values in the solder joint. (Alu-Alu)

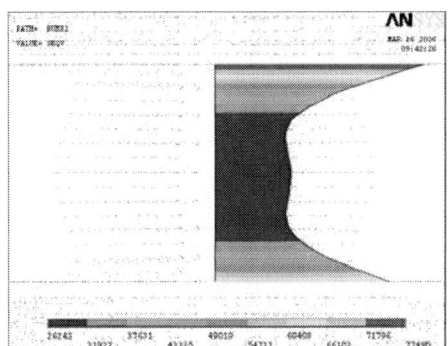

Fig. 9: Stress values in the solder joint. (Alu-FR4)

There are stress values in dependence on temperature on figure 10. The lowest solder joints stress values are in the FR4 – FR4 structure. This experiment simulates cooling down of solder. The stress value climbs with the opprobrious temperature.

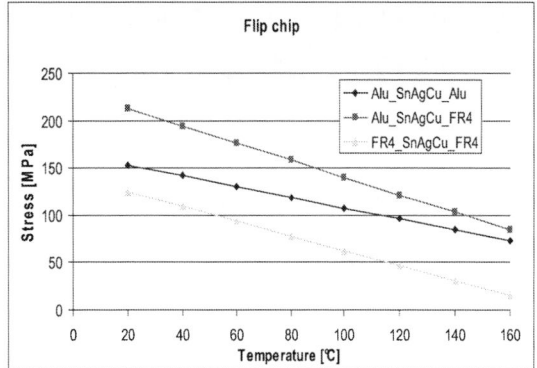

Fig. 10: Thermal dependence of stress in solder joint.

Conclusions

The aim of this work is investigation in the thermo mechanical stress of chip interconnection.

The results will be used for next investigations in this area (e.g. life time of these structures).

Acknowledgment

The paper has been prepared with the support of the Czech Ministry of Education in the frame of Research Plan MSM0021630503 "MIKROSYN New Trends in Microelectronic Systems and Nanotechnologies".

References

[1] Bret A. Zahn , "Impact of ball via configurations on solder joint reliability in taped based chip scale packages",Director, Design & Characterization - ChipPAC Inc.

[2] Matt Warner, John Parry, "Solder life prediction in a thermal analysis software enviroment", Flomerics Limited, Bridge Road, Hampton Court.

[3] R. Darveaux, "Solder Joint Fatigue Life Model", Design and Reliability of Solders and Solder Interconnections, 1997, pp. 213-218, Pub. TMS.

Thermal Capabilities of Integrated Resistors in Organic Substrates

Ciprian Ionescu, Norocel Dragoş Codreanu, Virgil Golumbeanu and Paul Svasta

POLITEHNICA University of Bucharest, Center of Technological Electronics and Interconnection Techniques

Phone: +40 21 3169633 E-mail: ciprian.ionescu@cetti.ro

Abstract

In the present stage of development of electronic technology there is an increased interest in integration of passive components directly into the interconnection substrate. The main advantage of this integration is the fact that, by replacing discrete SMDs, more free space is available on the surface of the printed circuit board. Supplemental, the integrated (embedded) solution could offer an economic advantage for mass production by reducing the placement costs. Many efforts are oriented in direction of improving the parameters of integrated passives, where the material researches are of determinant importance. This paper will be oriented on integrated resistors, as the most suitable to integration among passive components. In the process of design and implementation of integrated resistors there are some rules of thumb derived from practical experience of the manufacturer. Unfortunately, there are no reliable data related to thermal capabilities of these embedded resistors. This paper will present the results of finite element modeling and simulation of planar resistors realized on organic rigid substrate. We propose us to compute the thermal resistance of typical resistor configuration used in embedded structures. The main difference between buried resistors and resistors at board surface is the absence of convection cooling. We will investigate the influence of the position of the resistive layer in the laminate stack. The analysis will be done at various power density levels in order to determine the heat spreading path and the temperature distribution at board surface. The data derived from this analysis will be very useful in the design process of integrated resistors.

Key words: integrated resistors, finite element analysis, modeling and simulation, design

Introduction

The objective in implementing integrated passives is to find these configurations and technologies that are cost effective and also offer improved electrical mechanical, thermal characteristics of these components. The development of modern communications and portable or wireless applications requires the miniaturization of passive components in a higher degree as the actual SMD components can offer. On the other hand, there were made studies regarding the economics of assembly industry that revealed a possibility to decrease the costs by reducing the placement costs of SMD components. The actual idea, that is not quite new (remember "hybrids") is to include the passive components in the interconnection structure, that is the printed circuit board, in organic or inorganic substrate.

The objective in implementing integrated passives is to find these configurations and technologies that are cost effective and also offer improved electrical mechanical, thermal characteristics of these components.

The reason for integrating passives can be summarized in the followings: a) Reduced system mass, volume and footprint (individual packages are eliminated and more space is available on the surface of the board for active ICs. b) Improved electrical performance (integrated passives have in principle lower parasitics). c) Improved reliability because solder joints are eliminated. d) Increased design flexibility (the component's values, resistance, capacitance, or inductance can be tailored to any desired value). e) Reduced unit cost; integrated passives can be realized simultaneously low cost. Another up-to-date issue is that they are inherently lead-free.

There are, of course, some difficulties in using integrated passive components on a large scale. Most of the problems are related to the technology. The technology for passive components must be at the PCB fabricator. Also there are some uncertainties regarding which materials to use. Costs, fabrication yield or tolerances issues are also important. On the other side, the evolution of surface-mount technology toward components smaller and smaller improves the packing densities at a level where integrated passives are not immediate and absolute necessary.

Configurations of Integrated Passive Resistors

The configurations of integrated resistors mainly used in today applications are well known from the classical technologies that can realize planar structures, thick and thin-film technologies.

The main difference in the integrated passive components approach, compared to "classical" components is their presence as embedded in substrate volume.

The most used configurations for integrated resistors are presented in figure 1. The form factor k_f represents the number of squares used to compute resistor value.

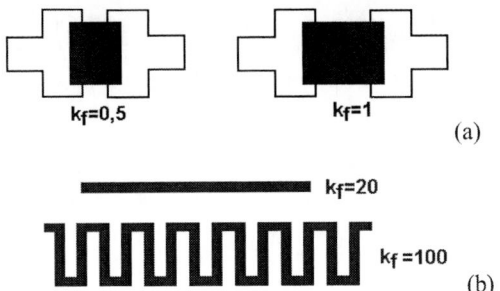

Figure1: Configuration of integrated resistors (a) rectangular resistor b) meander resistor.

Referring to the materials and technologies there are many thin and thick film processes that are commercially available today. The materials used to prepare resisitive layers can be grouped in single component metals, metal alloys, metal-nonmetal compounds, cermets and polymer thick films.

Thermal Model of Integrated Passive Resistors

The maximum allowable heat flux is, by convention the heat flux emerging from only one side of the planar resistor. Actually, the maximal heat flux that can be produced by an integrated resistor depends on the proximity of resistor regarded to the surface of the substrate. The amount of copper in the outer or inner layers is also a very important factor in thermal distribution. The presence of other heat generating devices influences, of course the temperature and hence the thermal capability of integrated resistors

The heat flux itself is not the limiting factor in thermal design of the resistors. The limiting factor is the temperature of the resistive material, the so called "hot spot" that must not exceed the limits declared by the manufacturer. These limits are determined by the degradation phenomena for general purpose resistors or are established on stability issues basis ($\Delta R/R$) for precision resistors. It is also well known the fact that the maximal temperature reached during function by the resistor influences the life time of the circuit, or other saying the reliability,

It is given [1] as a rule of thumb that for resistors on the surface of substrate a thermal flux of about 10-30 W/cm2 is possible. This relative high level of thermal flux can lead to significant temperature rise. For planar resistors on the surface of substrate with free or forced convection there was realized many studies to compute the temperature based on the heat flux. The situation is more complex for integrated resistors. The integrated resistors are realized beneath the surface of the substrate and the substrate is electrically insulated and hence relatively poor thermal conductive. In this case the actual distribution of heat from the two sides may change somehow depending on the distance of each side to the margin of the substrate, the effect of interconnection metal tracks or other devices that are heat generating. The same heat flux that produces a reasonable temperature rise on the surface is possible to overheat the same component if it is buried, depending on the distance to surface, the arrangement of conductive layers in the board, the presence of other integrated resistors nearby or other heat producing devices on top of the board. In this situation the rules of thumb taken from the resistor realized on the surface must be applied with circumspection. In order to have a precise image of the temperature, the solution is to apply finite element modeling of temperature field in the board.

Even without a thermal simulation it is possible to compute the thermal resistance and observe that the temperature gradient is very high in the substrate, due to poor thermal conductivity of insulated materials.

Metal layers, especially power and ground planes will make the temperature distribution to have a lateral profile leading to a more relaxed thermal stress.

For an integrated resistor we can apply the thermal model based on thermal resistances Rth.

$$R_{th} = \Delta T / P_d \qquad (1)$$

where ΔT is the temperature difference between the surfaces where the heat flows and Pd the resistor dissipated power.

In the case of integrated resistors about a half of the thermal flux is oriented to one side of the substrate, the rest to the other side. The thermal model of the integrated resistor is changed compared to resistors on only one side of a board, as in figure 2.

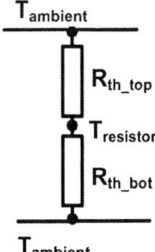

Figure 2: Thermal model of integrated resistor

Finally, the effective thermal resistance *Rth* is obtained by connecting in parallel the two resistances R_{th_top} in parallel with R_{th_bot}.

Results of Modeling

We have the intention to prove the major effect of copper layers on thermal performance of integrated resistors.

In order to have a better understanding of the numerical values, some structures having the same dimension of the resistive element as in case of discrete thick film chip resistors. Based on the other results from literature we will assume a free convection from the upper and bottom sides of the board. We will assume that the lateral dimensions of the board are large enough in order not to influence the temperature distribution. A board of 15 mm × 15 mm was taken into account.

The schematic representation of the model is presented in figure 3.

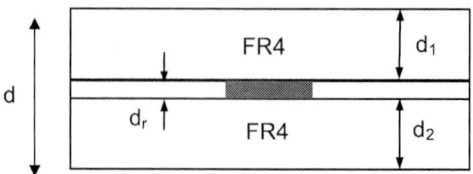

Figure 3: Schematic representation of the model

The inner layer contains a resistor with dimensions RL × RW. The overall thickness of the FR4 laminate with be kept at 0.8 mm, the position of the resistor can be changed by modifying the values of d_1 and d_2. We will investigate the situations with or without copper cladding on the opposite side of the resistor. The thickness of the copper was taken 35 μm. The dimensions of resistors and other data are presented in Table 1.

Table 1 Dimensions of chip resistors; resistive elements with the same dimensions were analyzed

Size	Length L [mm]	Width W [mm]	Resistive layer length RL [mm]	Resistive layer width RW [mm]	Rated power P [W]	Thickness t [mm]
0201	0.6	0.3	0.4	0.2	0.05	0.23
0402	1.0	0.5	0.6	0.4	0.063	0.35
0603	1.6	0.8	1.0	0.6	0.1	0.45
0805	2.0	1.25	1.2	1.0	0.125	0.5
1206	3.2	1.6	2.2	1.2	0.25	0.6
1210	3.2	2.6	2.2	2.0	0.5	0.6
2010	5.0	2.5	4.0	2.0	0.75	0.6
2512	6.3	3.1	5.3	2.5	1	0.6

We have used the ANSYS software for which the modeling and simulation flow includes: building the solid model, define and assign material properties and finite elements, mesh the model, apply loads and boundary conditions, solve and postprocess the results. A characteristic of the model is that a 3D structure was modeled. A parametric model was built which permit us to realize a series of runs without to re-create the solid model.

A major problem in modeling planar structures is the large number of elements that can be generated by the very thin layers that model the resistive, conductive or dielectric depositions. We have used a special modeling technique, which implies the building of the solid model by extrusion of areas along "z" direction and in this way the number of finite element can be dramatically reduced. For our models presented here there were up to 15000 elements, number which is convenient as a compromise between accuracy and running time.

Regarding the thermal analysis itself, the heat source is the resistor volume and the heat is applied as power density, computed accordingly. For the convection film coefficients we have chosen to take some result from literature. A method for determining the heat convection coefficients are presented in Table 2.

The characteristic length *l* that has the expression (1)

$$l = \frac{BL}{2(B+L)} \qquad (2)$$

and will have in the case, of a square resistor the value l=B/4.

It was computed a film coefficient of about 13 W/m²K for the top side and one half of this for the bottom side.

The thermal constant of FR4 used in analysis was 0.2 W/mK and for copper it was 380 W/mK.

Table 2: Heat convection film coefficients for plane surfaces.

	Plane surface heated on top side	Plane surface heated on bottom side
	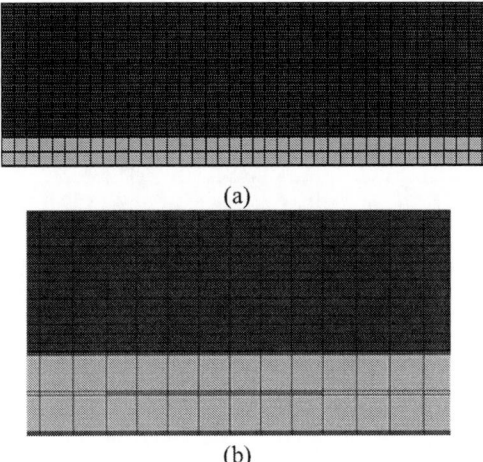	
Small centered surface	$h = 0{,}83 \cdot \left(\dfrac{\Delta T}{l}\right)^{0,33}$	$h = 0{,}415 \cdot \left(\dfrac{\Delta T}{l}\right)^{0,33}$

The model of the laminate with buried resistor is presented in figure 4.

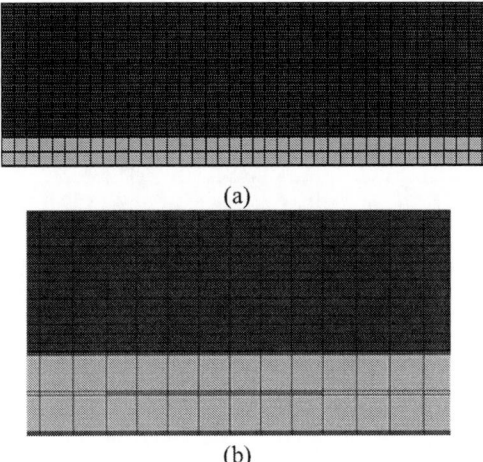

(a)

(b)

Figure 4: Finite element model a) overview b) section detail with a view to resistive element

It can be observed the rectangular mapped meshing that reduces considerably the computation time.

Results of Simulation

Successive thermal analyses were done for all the resistor dimensions presented in Table 1, in two major situations: with and without copper cladding on both substrate surfaces. Of course these are two extreme cases, but this will permit an interpolation of results for configurations with partial copper plane. The typical results are in form of temperature contour plots, as presented in figure 5. It can be seen the major influence of the copper layers on temperature distribution. The temperature of "hot spot" was noted and used to compute thermal resistance. The plots are for the same resistor at the same power level and in the same cooling conditions.

Another issue was to analyze the effect of the position in the layer stack of the resistor in order to find a position that gives better results. Four layer stack configurations were analyzed. The results obtained from simulations are presented in Table 4.

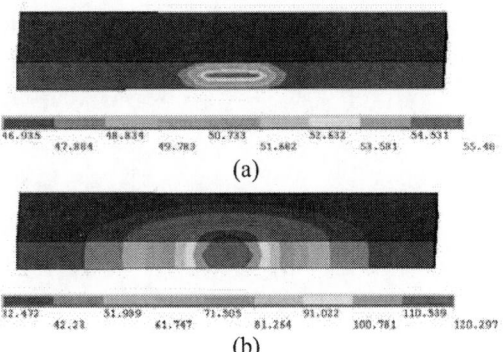

(a)

(b)

Figure 5: Temperature distribution for resistor with 2512 size at 0.1 W (a) with copper cladding, (b) without copper cladding

We present also the graphs for the temperature rise. In these graphs, see figure 6, the maximal temperature, the "hot spot" was plotted,. It can be seen that in case of the presence of copper layers the approaching to the outside of the substrate produces a better cooling of the resistor. In the situation without copper there is no significant difference between the four cases. Contrary to first situation, a little increase in temperature is observed as the resistor approaches the substrate limits.

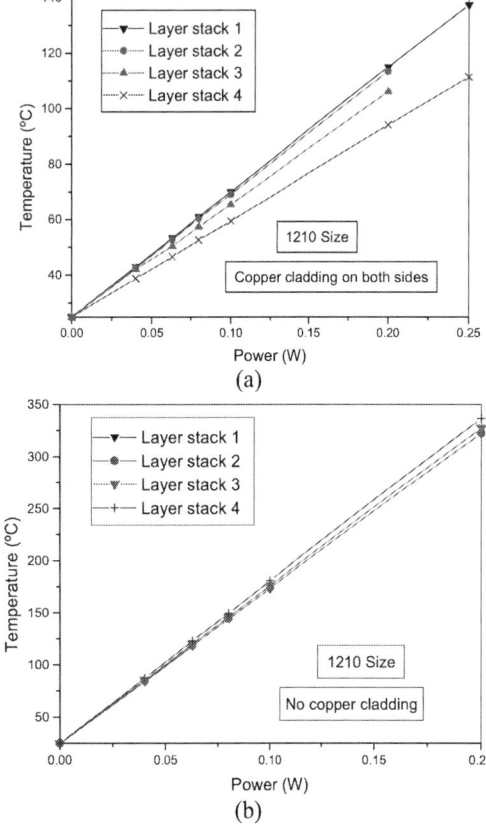

(a)

(b)

Figure 6: Temperature rise in an integrated resistor in 1210 size for four different layer arrangements (a) with copper cladding on both sides, (b) without copper

101

Table 3: Thermal resistance R_{th} from simulation data for 1210 size

Configuration	Layer stack 1 d_1=0.4 mm d_2=0.4 mm	Layer stack 2 d_1=0.5 mm d_2=0.3 mm	Layer stack 3 d_1=0.6 mm d_2=0.2 mm	Layer stack 4 d_1=0.7 mm d_2=0.1 mm
R_{th} (K/W) with copper cladding	449.5	441.8	399.3	345.6
R_{th} (K/W) without copper cladding	1482.3	1486.8	1512.5	1558.1

The effect of copper layer is not quite obvious. It acts as a reservoir for the heat flux generated by the resistor and keeps the heat spreading at relatively low level around the resistor area.

In Table 4 the results of all the simulation were synthesized in form of thermal resistance. This way it is easy to compute the temperature rise and certainly thee real situation will be between the two extreme cases.

Assuming a maximum temperature of the resistive element of 125 °C, for the resistor with the dimensions of 1210, from the graphs in figure 6 it results a maximal power of about 0.06 W without copper and 0.2 W with copper cladding. These values are to be compared with rated power of 0.6 W for the same size of resistive element, if realized on alumina as chip resistor. The integrated resistor presents a serious reduction of dissipation capacity, let's say an integrated resistor with an equivalent size to 1210 will behave like a 0402 discrete.

Table 4: Thermal resistance Rth for all resistive elements used in analysis

Size code	0402	0603	0805	1206	1210	2010	2512
Thermal resistance R_{th} (K/W) with copper cladding	1860	1398.8	913.3	575.6	449.5	354	304
Thermal resistance R_{th} (K/W) without copper cladding	3300	2778.3	2200	1697.2	1482.3	1162.1	953

Conclusions

In this paper a description of a procedure used to model and analyze integrated passive resistors was presented. The procedure implies using of Finite Element Analysis in thermal domain. There were obtained interesting results regarding the major contribution of the copper cladding on the power capabilities of integrated resistors. Some of the results were not obvious at a first look. The copper planes have a major role in heat spreading. Also it minimizes the influences of the convection conditions on top or bottom surface of the board. The differences are caused by the better heat transfer on top of the board, this was supposed from the beginning. The copper layer offers a better cooling for the resistor as the distance to board edge decrease.

Also, a useful estimation of the power capabilities of the integrated resistors was presented. Other computations can be made based on the values of thermal resistance. These will be useful in relation with the manufacturer for an integrated passives board design.

References

[1] R. K. Ulrich and L. W. Schaper, "Integrated Passive Component Technology", IEEE Press, Wiley Interscience, 2003.

[2] ANSYS 6.1, Theory Reference Manual, documentation, 2002.

[3] W. M. Rohsenow, J. P. Hartnett, Young I. Cho (eds.), "Handbook of heat transfer", 3rd edition, McGraw-Hill, 1998.

Impact of Package Ground Ball and Plated Through Hole (PTH) Count Reduction to Differential Interfaces Signal Integrity

Jiun Kai Beh

Intel Microelectronics (M) Sdn. Bhd. FIZ, 11900 Bayan Lepas, Penang, Malaysia

Phone: +604-2534353, Email: jiun.kai.beh@intel.com

Abstract

As the frequency of the high speed interfaces in computer platforms and systems continue to ramp due to architectural advancement and strong market demands, feature sets embedded in a chip has been increasing tremendously. This trend is not helped by the demand for smaller package form factor to be able to fit it into various devices. In order to have a better signal quality for high frequency interfaces, differential signaling is preferred at the expense of doubling the signal count. This has consequently caused complexities in package layout design and extensive electrical modeling while at the same time losing the competitive edge in the market in terms of requiring larger package size and higher cost. Traditionally, ground balls and Plated Through Hole (PTH) are considered to be very critical in providing good return path for the signal and noise reduction. This paper discusses the impact of package ground ball and PTH count reduction with respect to differential interfaces signal integrity. Time and frequency domain analysis has been completed and shows both the analysis results correlates well with each other. This is done by extracting, analyzing and comparing package models with different number of package ground balls and PTH. In order to obtain a 3-Dimensional package electrical model, all the models are extracted and analyzed based on their actual physical dimension from a specific package for accurate results using AnsoftLink™, Ansoft Q3D Extractor® and Ansoft HFSS™. The package models extracted are then connected with other interconnect component models of a motherboard and simulated using HSPICE to obtain its final result in terms of Eye Diagram, Insertion and Return Loss. Ansoft Designer®/Nexxim® is used to simulate Time Domain Reflectometry (TDR) results to understand the impact on package full path impedance matching.

Background

Signal integrity has been a major concern typically in any electronics circuit design and has become even more so with higher speed interfaces. Driven by strong market demand for smaller package form factor and lower cost; feature sets, signal count and density on a package has been rising steadily. At the same time, this is not helped by the trend of faster busses is tracking Moore's Law bus bandwidth doubles every 2 years referring to Figure 1.

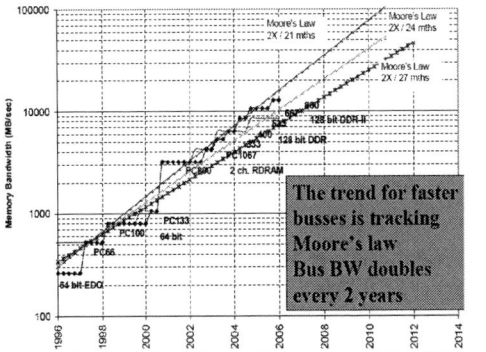

Figure 1: Trend of faster busses tracking Moore's Law [1]

This implies that the challenge is even more intense with more high speed interfaces integrated in a single package as providing good return path for the signals and noise reduction are critical.

Modeling and Analysis

The package PTH and Ball is modeled with actual dimension using AnsoftLink™, Ansoft Q3D Extractor® and Ansoft HFSS™ for accurate results. AnsoftLink™ is a tool used to generate 3D model with actual physical dimension from package layout. The 3D model is then imported, modeled and analyzed using Ansoft Q3D Extractor® to obtain the parasitic resistance, inductance and capacitance. Spice models are exported from Ansoft HFSS™ and connected with other package components to simulate Insertion & Return Loss, Cross Talk, Eye Diagram and TDR using HSpice and Ansoft Designer®/Nexxim®.

The package model used in this study is extracted from a 6 layer Flip Chip Ball Grid Array (FCBGA) package with 6 signals that consist of micro-vias, PTH and balls. The planes and PTH other than the signals in the model are ground. Figure 2 shows the model with 8 ground PTH and Ball together with their arrangement.

103

Figure 2: Full Ground PTH and Ball model

Figure 3 shows the model with only 1 ground PTH and Ball. This model is modified from the initial model by removing 7 ground PTH and Balls from it. The remaining PTH and Ball are purposely separated to different corners in order to increase the current flowing loop and the potential impact to the signal quality.

Figure 3: Reduced Ground PTH and Ball model

In Ansoft Q3D Extractor ®, the current source in the model is assigned at the micro-vias above the PTH and the current sink is assigned at the bottom of the balls. This is assuming the signal current flows from the micro-vias to the balls and the return current flows from the ground balls back to the micro-vias above. This creates a coupling effect between the signals and ground which will reduce the mutual inductance. The model is solved at 1.5GHz as it is the fundamental frequency for Serial ATA II (SATA II) interface, the highest frequency interface currently available in the market.

The signal self parasitic resistance, inductance and capacitance is acquired after performing Reduce Matrix calculation using Ansoft Q3D Extractor®. A random signal is chosen to compare the parasitic values between the two models. Referring to Table 1 and 2, DC-Resistance has an average increase of 44.94% while AC-Inductance has an average increase of 298.06%. The parasitic resistance and inductance however does not have much impact on the signal. Capacitance is an important factor to cause signal delay shows a minor average decrease of 0.95% shown in Table 3. All these results are expected as less ground return path will increase the parasitic resistance and inductance while less ground coupling will decrease the parasitic capacitance.

Table 1: Parasitic resistance of Full vs. Reduced Model

Signal	Full Model DC-R (mΩ)	Reduced Model DC-R (mΩ)	Percentage Increase
1	12.418	18.014	45.06
2	12.508	18.106	44.76
3	12.495	18.101	44.87
4	12.418	18.014	45.06
5	12.428	18.015	44.95
6	12.432	18.021	44.94
Average	12.450	18.045	44.94

Table 2: Parasitic inductance of Full vs. Reduced Model

Signal	Full Model AC-L (nH)	Reduced Model AC-L (nH)	Percentage Increase
1	0.48008	2.0014	317.89
2	0.50601	2.0198	299.16
3	0.51057	1.9969	291.11
4	0.47733	1.9626	311.16
5	0.4975	2.0024	302.49
6	0.55105	2.0485	271.74
Average	0.50376	2.0053	298.06

Table 3: Parasitic capacitance of Full vs. Reduced Model

Signal	Full Model C (pF)	Reduced Model C (pF)	Percentage Decrease
1	0.57329	0.56761	0.99
2	0.66227	0.65568	1.00
3	0.66387	0.65521	1.30
4	0.57503	0.5703	0.82
5	0.57437	0.56943	0.86
6	0.50013	0.49687	0.65
Average	0.59150	0.58585	0.95

A different modeling methodology is used in Ansoft HFSS™ as it is a full wave solver with electromagnetic waves solved in all directions. This requires the signal to reference to other structure or net as a return path. This is done by assigning ports in the model using Lump Port at the micro-vias above the PTH with reference to ground and ports at the bottom of the balls with reference to a virtual ground plane that connects all ground balls together. Cutouts are created on certain regions of the virtual ground plane where the signal ball is in contact with to avoid signals shorted to ground. All ports assigned at the micro-vias and balls must have the same reference terminal line direction, either from ground to signal or vice versa. This is to avoid the output results being inversed. Analyzing complicated models with full wave solver usually requires extremely huge amount of computing resources and longer simulation time. In order to minimize computing resource usage and simulation time, octagonal micro-via, PTH and Ball are used instead of true cylinder and sphere shape. An air box with the dimension slightly larger than the model is also created enclosing the model, acting as a meshing and solving boundary for the tool, any free space outside the air box will not be meshed or solved.

The equivalent Spice circuit model is then exported from Ansoft HFSS™ and connected in HSpice. As this is a study on differential interfaces, the 6 signals are configured and connected to become 3 pairs of differential signals. The four S-parameter results including Insertion Loss, Return Loss, Near End Cross Talk (NEXT) and Far End Cross Talk (FEXT) has been acquired and analyzed after the full path frequency domain simulation is performed. Insertion and Return Losses include effects such as impedance discontinuities and resonance effect, which are not true losses. Insertion Loss is the measure of the power transmitted from source to sink while Return Loss is the measure of the power reflected back to the source. [2]

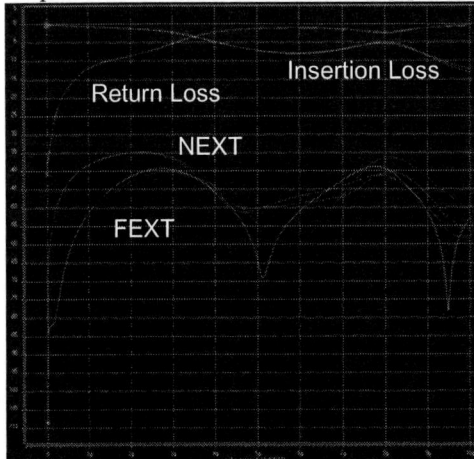

Figure 4: Insertion & Return Loss, NEXT & FEXT

Figure 4 shows a total of eight curves on the same plot which includes the results of both PTH and Ball models after being simulated and compared against each other. From a high level view, all four plots show almost identical results between the two models especially in the lower frequency region. NEXT and FEXT shows some differences in the frequency region above 5GHz, these differences are considered to be insignificant as they at very low magnitude and only affects the higher order harmonics of the signal.

Zooming into the frequency region of 1GHz to 2GHz for each of the curves shown in Figure 5, the small differences between the two models are more visible and measurable. At 1.5GHz, the differences between the two models in Insertion & Return Loss are less than 0.05dB which is insignificant. NEXT & FEXT are less than 1.5dB, but with the results at around -45dB, the differences are considered insignificant as well.

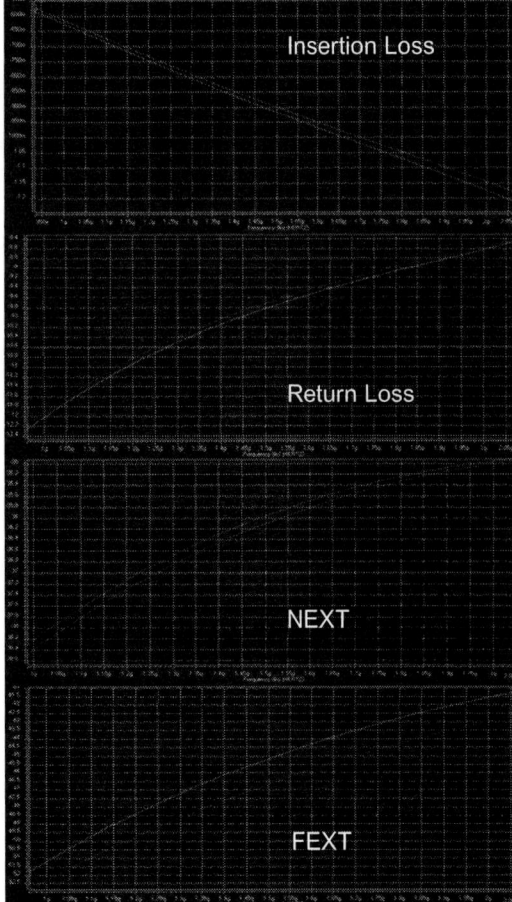

Figure 5: Close Up of Insertion & Return Loss, NEXT & FEXT

To fully understand the signal behavior, time domain analysis is performed. The models are being connected and simulated using an internal tool to obtain the worst case bit pattern eye diagram. Due to confidential issue, details of the internal tool are not disclosed in this paper. In Figure 6, the diamond shape is the SATA II eye height and eye width specification. The red curve is the signal eye without any crosstalk effect applied while the black curve is the signal eye with crosstalk from adjacent signal pairs injected in to it. The eye diagram from both models pass the SATA II eye specification with the reduced ground PTH and Ball model shows better result with a slight 1mV in eye height and 1ps in eye width improvement. The differences however can be considered to be negligible as the improvement is less than 1%.

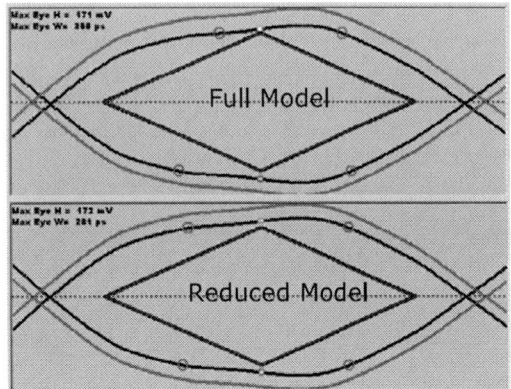

Figure 6: Worst Case Bit Pattern Eye Diagram

To understand the impact on package impedance matching, TDR analysis has been performed to obtain the impedance profile of both the PTH and Ball models connected with other package components using Ansoft Designer®/Nexxim®. The model is probed at the Ball, thus the injected signal propagates through the package from the Ball to the PTH, transmission line and breakout. An open ended package is assumed where there is no silicon die on it, thus an open TDR response is expected from the result. The transmission line length is assumed to be typical 10mm and a delay is introduced before the signal starts to propagate. Referring to the TDR results in Figure 7, there are insignificant differences between the two models. The reduced model has slightly higher impedance at the PTH and Ball region, thus better matching the full path impedance profile. This better full path impedance matching has been suspected to be the main reason for better result in worst case bit pattern eye diagram in certain cases.

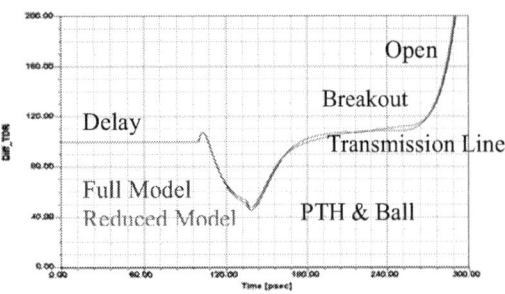

Figure 7: TDR Impedance Profile

Recommendation

For further understanding of the impact on NEXT and FEXT, another set of models need to be created with ground balls in between the signal balls. This is to observe the difference in NEXT and FEXT between models with and without ground balls between the signal balls. As for the impact of better full path impedance matching understanding, more PTH and Ball models with different height and diameter are needed to do an in depth study. A balance between cross talk and impedance matching should be studied to acquire the optimal solution. Manufacturing a sample package with reduced ground PTH and Ball will be important for validation and correlation work. As correlation of analysis data and measurement data using TDR and Vector Network Analyzer (VNA) for impedance and S-Parameters respectively will greatly help to evaluate the validity and performance of the reduced ground design and future implementation in products.

Conclusion

Reduced Ground PTH and Ball model shows insignificant impact to differential interface signal integrity when compared to the Full Ground PTH and Ball model. At package level simulation, Insertion and Return Loss, NEXT and FEXT shows minimum differences. TDR results even shows better impedance matching with slightly higher impedance at the PTH and Ball region. At platform full path level simulation, the eye diagram result shows insignificant as well. This concludes that the reduced ground PTH and Ball design has minimum impact to signal quality. More studies are still required to be done to fully understand the reason behind it before this can be implemented on actual product designs.

Acknowledgements

Thanks to Lim Yun Rou from Intel Corporation ETMS team for conducting training on platform full path time and frequency domain simulation.

References

[1] Falconer, Maynard "Introduction – Signal Integrity" March 2005 pp. 7. [A reference to Intel Corporation internal training material]

[2] Hall, Stephen "Introduction to Frequency Domain Analysis for Digital Engineers" March 2005 pp. 34-45 [A reference to Intel Corporation internal training material]

UBM Structures for Lead Free Solder Bumping using C4NP

Klaus Ruhmer[1], Eric Laine[1], Karin Hauck[2], Dionysios Manessis[2], Andreas Ostmann[2], Michael Toepper[2]

[1] SUSS MicroTec, Inc., Waterbury Center, VT, United States

[2] Fraunhofer IZM, Berlin, Germany

+1 802 244 5181 x 420, eric.laine@suss.com

Abstract

This paper analyzes electroless Ni/immersion Au (ENIG), with and without Pd, as an alternative Under Bump Metallurgy (UBM) structure in combination with lead free solders. Wafers were fabricated with these UBM structures, lead-free solders applied with Controlled Collapse Chip Connection – New Process (C4NP), and chip level stressing performed to determine the robustness of this alternative stack-up. Shear strength of these structures following multiple reflows and thermal cycling is presented. The data in this paper is provided by the analytical laboratory at Fraunhofer Institute, IZM, Berlin, Germany. UBM formation was done at Fraunhofer IZM. Wafer bumping was done using the C4NP process at IBM Hudson Valley Research Park, NY.

Key words: Lead free, solder bumping, UBM, C4NP, ENIG

Introduction

Semiconductor packaging continues to migrate from wire bond to flip chip first level interconnect to meet size, weight and electrical performance requirements. Solder electroplating is commonly employed for wafer bumping, especially for fine pitch applications. Wafer level chip scale packaging (WLCSP) typically uses solder sphere placement to manufacture the bumps. C4NP (Controlled Collapse Chip Connection New Process) has proven to be suitable for a broad range of solder bump pitches, from FCiP to CSP bump dimensions. As the industry migrates to 300mm wafer processing and lead-free flip chip interconnect, C4NP is establishing itself as a viable solder bumping alternative.

C4NP is a novel solder bumping technology developed by IBM which addresses the limitations of existing bumping technologies by enabling low-cost, fine pitch bumping using a variety of lead-free solder alloys. It is a solder transfer technology where molten solder is injected into pre-fabricated and reusable glass molds. The glass mold contains etched cavities which mirror the bump pattern on the wafer. Filled mold and wafer are brought into close proximity/soft contact at reflow temperature and solder bumps are transferred onto the entire 300mm (or smaller) wafer in a single process step without the complexities associated with liquid flux. The simplicity of the process makes it a low cost, high yield and fast cycle time solution for bumping with a variety of high performance lead free alloys.

The under bump metallurgy (UBM) structure is a critical component of any solder interconnect system. The UBM typically serves three functions: adhesion to underlying dielectric and metal, barrier to protect the silicon circuitry, and a solder wettable surface. For lead-free bumps, the barrier layer is key to reliability. A common barrier layer used in the industry is electroplated nickel. This layer provides good protection of the silicon metallurgy from degradation caused by tin rich lead free solders. C4NP provides an opportunity to eliminate electroplating, and its associated costs for plating chemistry, analysis, supply and waste treatment.

UBM Technology Overview

The Under Bump Metal (UBM) is the direct interface between the interconnecting solder and the final chip metallization [1]. The UBM has to provide a low resistance contact between the chip pad and the solder, good adhesion to the chip metallization and the chip passivation and a hermetic seal between the UBM and IC pad. It has to be a reliable diffusion barrier between the IC pad and bump with low film stress and it needs to be sufficiently resistant to stress caused by thermal mismatch during die assembly. In the case of PbSn bumping, common UBM stacks are Cr-Cr:Cu-Cu-Au (original C4 from IBM); Ti-Cu; Ti:W-Cu; Ti-Ni:V; Cr-Cr:Cu-Cu; Al-Ni:V-Cu; Ti:W(N)-Au. Usually, these UBM stacks are sequentially deposited by sputtering or plating.

Intermetallic compounds (IMCs) are formed between Sn and Cu or Ni by the reflow process providing the required adhesion of the bump to the chip pad. IMCs are brittle in nature due to the ordered crystal structure which is in contrast to the solid solutions. The metals which are mostly used in packaging; Cu, Ni, Au and Pd; form binary intermetallics with Sn-based solders of the Hume Rothery type.

In general, the intermetallic growth rate with Sn is much higher for Cu compared to Ni. This is becoming more important for lead-free solders due to their higher Sn content.

As previously described, the barrier integrity provided by the UBM structure is of critical importance for the performance of lead free solders. A common barrier layer used for electroplated lead free solder is electroplated Ni. C4NP provides the opportunity to eliminate electroplating of solder. If electroplating of the UBM can be avoided, the entire infrastructure required for electroplating chemistry procurement, analysis, mixing, pumping and waste treatment can be avoided. A detailed cost model has been developed, which shows that these electroplating infrastructure costs are a significant portion of the overall bumping cost for a wafer.

ENIG is a well known and commonly practiced UBM for lead free solders. It has the advantage of not requiring photolithography to form a "capture pad" for the C4NP solder transfer process. The ENIG metallurgy is applied directly on the wafer metallurgy in the passivation layer via, and then solder is applied with C4NP.

Electroless Nickel UBM

The ENIG process is based on the selective chemical deposition of metal on Al bond pads. Wafers are treated in a sequence of chemical solutions. After each treatment they have to be rinsed carefully in DI water. The principle of the process is shown in Figure 1. First the surface of Al bond pads is cleaned by immersing the wafers into two baths. The first (passivation cleaner) removes possible residues while the second (Al cleaner) removes thick Al oxides and roughens the surface. In a zincate bath a thin Zn layer is deposited on Al by an exchange reaction in order to activate the surface for subsequent Ni plating. The electroless Ni bath contains mainly Ni ions and hypophosphite. A first Ni layer is deposited on the pads by an exchange reaction between Zn and the Ni ions. On this first layer additional Ni is plated by a continuous autocatalytic reaction. The plating rate is 25 um/hr. In the subsequent immersion Au solution, a thin Au film is deposited by an exchange reaction on the surface of the Ni layer. The Au has a thickness of 0.05 – 0.08 um and is required to prevent Ni from oxidation.

The complete process flow is shown in Table 1. In addition to the wet chemical treatments mentioned above, the backsides of the wafers have to be protected in order to prevent Ni deposition on Si. This is done by spin-coating of a protective resist on the backside of the wafer. After UBM plating, the resist is stripped. All chemicals used in this process are commercially available. They are completely cyanide-free and no organic solvents are used.

Figure 1: Principle process of electroless Ni bumping: (a) bond pad in initial state, (b) after zinc deposition, (c) after growth of electroless Ni, (d) after plating of thin immersion Au.

Table 1: Process steps of electroless Ni bumping and their function.

process step	function
1. protective resist coating	protective resist coating on wafer backside
2. passivation cleaning	removes passivation residues from Al pads
3. Al cleaning	removes thick Al oxides and prepares surface for metal deposition
4. zincating	activates Al for Ni deposition
5. electroless Ni	deposition of Ni layer (typ. 5 um)
6. immersion Au	Au finish on Ni (typ. 0.08 um) to prevent Ni from oxidation
7. backside cleaning	removes protective coating from backside

For all wet-chemical treatments, 25 wafers are handled together in one carrier. The process requires tanks with seven different chemical baths and additional rinse tanks. The process times are relatively short. They have a range from 30 s (zincating) to 30 min (immersion Au). By handling the wafer cassettes manually from bath to bath a throughput of 25 wafers per hour can be achieved. In fully automatic systems 100 wafer per hour are possible.

The uniformity of the Ni height is better than +/- 5 % over a 200 mm wafer. The variation within a die is correspondingly lower.

C4NP Technology Overview

The C4NP process starts with a glass mold in which the bump pattern for an entire wafer is replicated as a mirror image of cavities in the glass mold. These cavities are filled with solder as the mold is scanned below a fill head. The fill head contains a reservoir of molten solder and a slot through which the solder is injected into the mold cavities. The cavity geometry determines the volume of the solder bumps that will be subsequently formed on the wafer. The filled mold is inspected automatically and then aligned below a wafer with exposed UBM pads facing the mold. Mold and wafer are heated above the solder melting point and then brought into contact. The solder forms spherical balls which transfer from the mold to the UBM regions on the wafer, where they preferentially wet and solidify. Wafer and mold are separated, and the mold is cleaned for reuse.

Description of Test Cells

The test cells consisted of varying thicknesses of electroless Ni, electroless Pd and immersion Au, as described in Table 2 below.

Table 2: test matrix

Ni	Pd	Au	Solder
3um	No	100nm	Sn/Cu
3um	No	100nm	Sn/Ag
7um	No	No	Sn/Cu
7um	No	No	Sn/Ag
7um	No	100nm	Sn/Cu
7um	No	100nm	Sn/Ag
7um	0.5um	100nm	Sn/Cu
7um	0.5um	100nm	Sn/Ag
15um	No	100nm	Sn/Cu
15um	No	100nm	Sn/Ag

The Sn/Cu alloy used was Sn/0.7%Cu. The Sn/Ag was a proprietary alloy composition. The test chip is based on 3 um thick AlCu pads passivated with 7 um thick PI (HD 4000 series). The chip size is 15 mm x 15 mm square, 150 um pitch. The test chip is shown in Figure 2.

Figure 2: test chip

The AlCu pad size is 126 um and the opening in the polyimide is 54 um. The diameter of the Ni pad depends on the thickness of the e-less Ni.

Experimental Results

Table 3 provides ball shear results for the test matrix. The C4 balls were sheared at a rate of 25um/sec at a height of 25um above the polyimide

passivation surface. The average was calculated from 30 shear measurements from C4 bumps selected randomly across each chip. Data is provided for each cell at time zero and at one, three, five and ten lead free solder reflows. Samples at time zero have undergone only the C4NP process, with one solder reflow cycle ("as-received"). Subsequently, the samples received 1, 3, 5, and 10 additional reflows in a nitrogen convection oven with a lead-free temperature profile. Also provided is data for each cell after 100, 250, 500 and 1000 deep thermal cycles (-50 to + 125C).

Figure 3 shows a cross section of a C4 bump at time zero with 7um Ni and Sn/Ag solder.

Figure 3: 7um Ni with Sn/Ag solder

Several observations can be made from this data:

1. The ball shear forces are roughly proportional to the UBM area. The greater the Ni thickness, the greater the resulting UBM pad diameter and resulting shear strength. It is expected that shear strength will increase with increasing pad area.

2. The shear force of Sn/Ag solder is in general greater than for Sn/Cu solder. This is expected, since Sn/Ag solder has a higher yield strength and hardness than Sn/Cu solder. Also, Ag_3Sn intermetallics strengthen the solder joint.

3. All ball shear failure modes were within the bulk solder. This may have changed if the shear height had been closer to the UBM pad.

4. Shear values are roughly stable through 10X reflows and 1000 thermal cycles. This UBM structure is very robust.

5. Au improves the ball shear strength, all else being equal.

6. Pd does not have a significant effect on the shear strength of the solder joints.

Table 3: Ball shear values

Ni um	Pd um	Au nm	Solder	T0 (g)	1X (g)	3X (g)	5X (g)	10X (g)	100 cy (g)	250 cy (g)	500 cy (g)	1000 cy (g)
3	No	100	Sn/Cu	12.43	13.32	13.97	13.84	14.24	12.74	14.1	12.37	13.87
3	No	100	Sn/Ag	15.51	12.02	14.53	13.67	12.97	13.37	14.82	14.03	15.12
7	No	No	Sn/Cu	12.74	13.78	13.03	13.18	13.49	11.54	11.37	10.97	11.56
7	No	No	Sn/Ag	14.59	15.96	15.76	15.29	15.67	13.86	13.78	14.15	13.58
7	No	100	Sn/Cu	13.22	15.28	14.09	14.44	15	14.6	16.25	13.44	16.58
7	No	100	Sn/Ag	16.64	18.9	18.29	17.34	18.46	17.16	17.76	16.24	16.72
7	0.5	100	Sn/Cu	15.54	15.74	16.24	15.79	16.08	14.74	14.51	14.44	13.96
7	0.5	100	Sn/Ag	17.31	17.77	17.99	17.45	17.17	17.64	17.87	16.86	16.71
15	No	100	Sn/Cu	15.87	17.7	18.27	18.33	18.27	16.06	18.87	14.82	18.36
15	No	100	Sn/Ag	20.04	24.3	23.58	22.68	21.76	20.28	20.5	20.1	19.36

Figures 4 through 13 provide ball shear values for each cell, normalized for pad area.

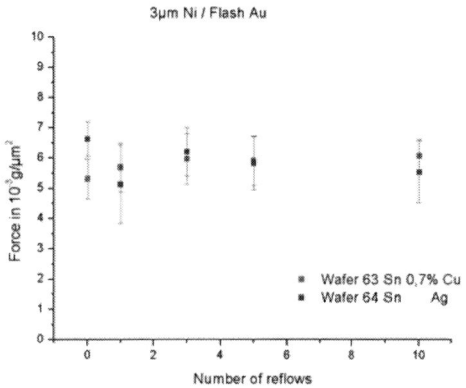

Figure 4: 3um Ni/ Au shear force vs reflows

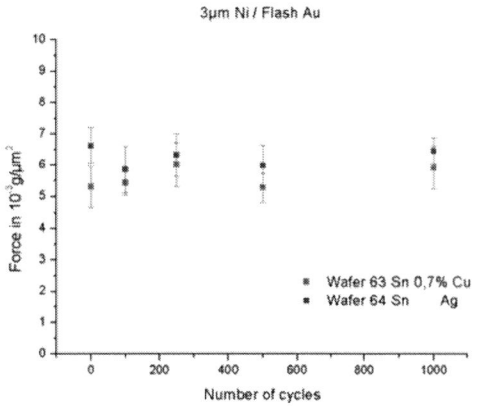

Figure 5: 3um Ni/ Au shear force vs cycles

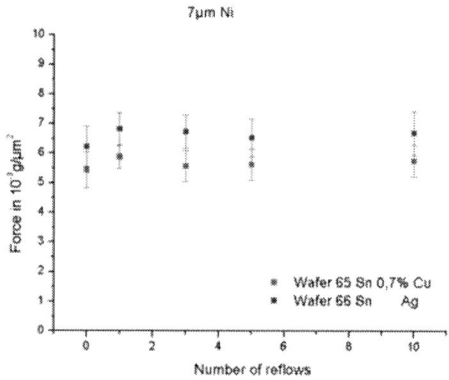

Figure 6: 7um Ni shear force vs reflows

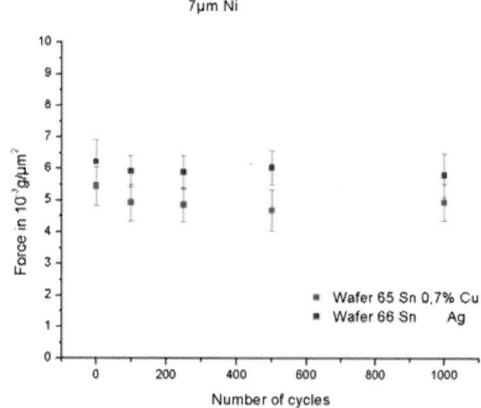

Figure 7: 7um Ni shear force vs cycles

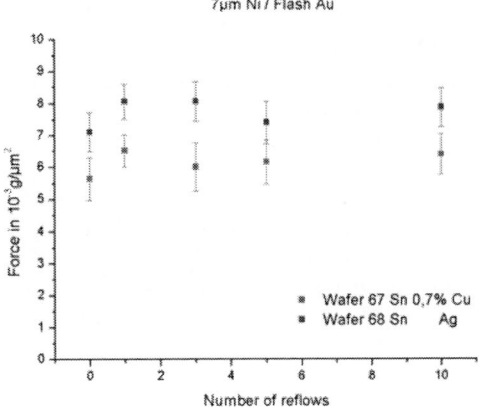

Figure 8: 7um Ni/ Au shear force vs reflows

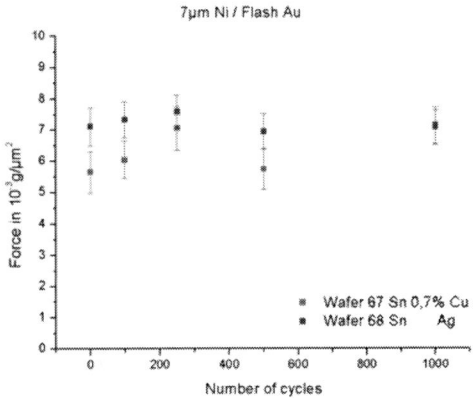

Figure 9: 7um Ni/ Au shear force vs cycles

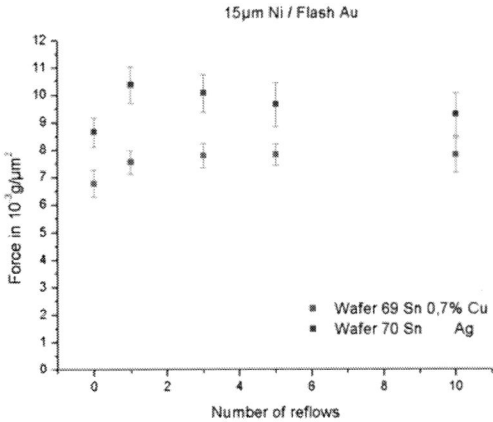

Figure 10: 15um Ni/ Au shear force vs reflows

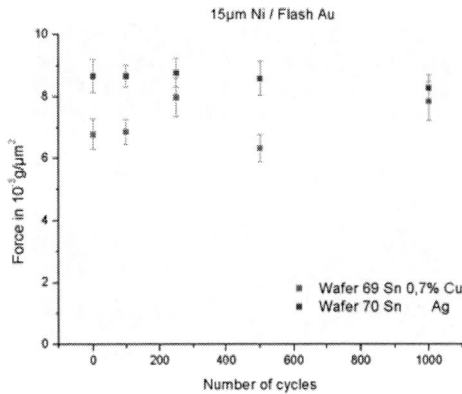

Figure 11: 15um Ni/ Au shear force vs cycles

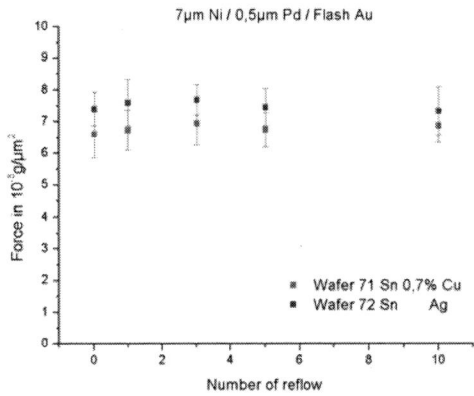

Figure 12: 7um Ni/ Pd/ Au shear force vs reflows

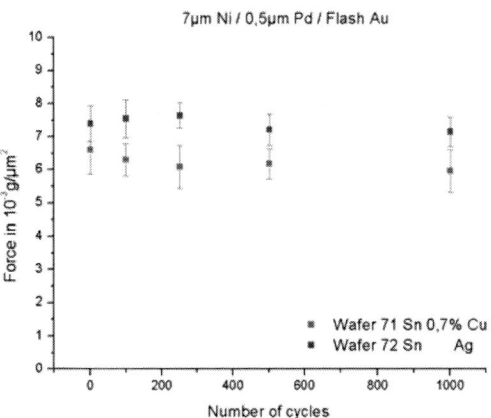

Figure 13: 7um Ni/ Pd/ Au shear force vs cycles

Figure 14 is an SEM image of a 15um Ni/ flash Au cross section at time zero, prior to any additional reflows or thermal cycles. There is a uniform intermetallic layer and continuous Ni coverage of the underlying Al/Cu layer.

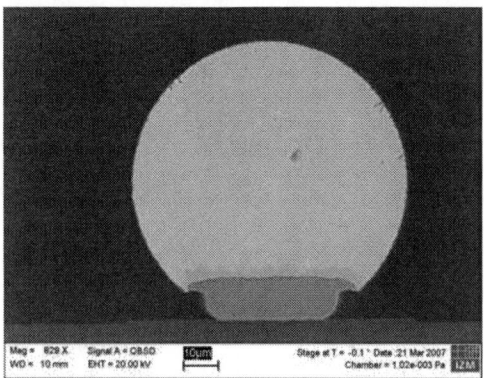

Figure 14: 15um Ni/ Au at time zero

Figure 15 is an SEM image of a 7um Ni cross section, also at time zero. Again, the intermetallic and Ni layers are continuous and uniformly cover the underlying chip last level metallurgy.

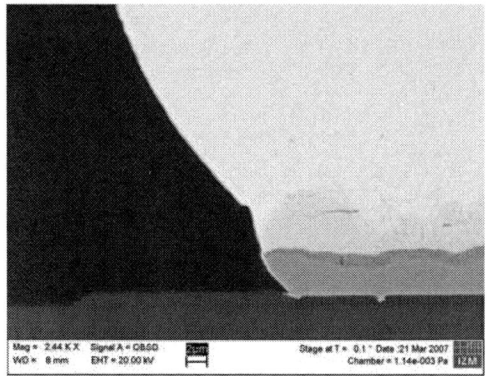

Figure 15: 7um Ni at time zero

Figure 16 is an SEM image of a 15um Ni/ Au C4 pad after 10X reflows and ball shear. The shear failure mode is in the bulk solder, indicating strong intermetallic formation.

Figure 16: 15um Ni/ Au sheared after 10X reflow

Conclusions

Alternative UBM structures have been demonstrated with lead free solders applied with C4NP. The ENIG UBM structures tested have proven to be very robust, with minimal changes in shear strength after 10 reflows and 1000 deep thermal cycles. This paper analyzed chip level data only. A rigorous reliability qualification must be performed, including first and second level interconnection, to determine if a given UBM and bump construction will meet the requirements for a given application.

Acknowledgements

The authors would like to acknowledge the IBM development engineering team for their contributions to this study.

References

[1] "Microelectronic Packaging (Chapter 25)" M. Töpper, D. Tönnies in Semiconductor Manufacturing Handbook (M.h. Geng, Ed.), McGraw-Hill, 2005

ECD Wafer Bumping and Packaging for Pixel Detector Applications

Thomas Fritzsch, Rafael Jordan, Oswin Ehrmann, Herbert Reichl

Fraunhofer IZM, Gustav-Meyer-Allee 25, 13355 Berlin

Tel.: +49-30-46403-681, Fax: -123; Mail: thomas.fritzsch@izm.fraunhofer.de

Abstract

Several hybrid pixel detectors have been manufactured at Fraunhofer IZM. The applications vary from x-ray detection for medical diagnostics to particle detection for fundamental physical research. The different number of readout chips on one sensor substrate and the number of modules for the assembly of the whole detector require different manufacture technologies from single chip processing up to processing on wafer level. The pixel interconnection bumps are produced by SnPb, SnAg or AuSn electroplating. The details of the process flow from sputtering of plating base up to flip chip assembly will be decribed in the first part of the paper. In the second part examples of hybrid pixel detectors manufactured at Fraunhofer IZM are described more in detail, e.g. assembly of modules for the ATLAS and CMS detector at the LHC, the assembly of detector prototypes using CVD diamond as sensor material and the assembly of GaAs x-ray pixel imaging detectors.

Key words: wafer bumping, pixel detector, fine pitch assembly, ATLAS, imaging detectors, particle tracking,

Hybrid Pixel Detectors

After a long period of development hybrid pixel detectors are now state of the art in precise particle tracking in high energy physics applications. Three detectors at the Large Hadron Collider (LHC) at the European Organization for Nuclear Research (CERN) in Geneva, Switzerland, will use so-called vertex detectors closest to the interaction point of colliding particles. The basic elements of hybrid pixel detectors are modules which consist of a particle sensing element, the sensor, and one or more electronic readout chips which are flip chip bonded to the sensor tile. Thus the signal track length between the sensor pixel and the electronic readout cell can be minimized. A schematic cross section of a hybrid detector pixel is shown in figure 1. The sensor pixel metallization, the semiconductor material and the sensor back side electrode form a reverse-biased diode in operation. In case of a particle transition electron-hole pairs are generated and collected. Subsequently the signal will be recorded and amplified in the electronic readout cell and transferred to the data acquisition system outside the detector module.

One advantage of hybrid pixel detectors is the separate development and processing on wafer level of sensor and electronics. The use of leading edge technology for each component is possible in this way. Specifically as components for operation in a harsh radiation environment, such as sensors and electronics inside of high energy particle detectors, which can be optimized to maximize radiation hardness. Further requirements of readout electronic chips are a high readout speed, low power consumption and a low noise and threshold dispersion. The development of pixel sensors is driven by high detection efficiency, fast signal response, and a high spatial resolution. Because of the direct interconnection between pixel and readout cell the high spatial resolution demands small solder joints in a tight pitch for an accurate and fast detection of the particle track location. Electroplating bumping technology is the therefore best suited because it can be used for pitches even below 40 μm.

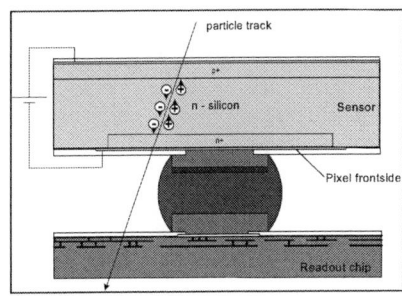

Fig. 1: Schematic view of a hybrid detector pixel

Furthermore electronic readout chips can be flip chip bonded to different pixel detector materials. This has led to a spin-off from particle physics to imaging applications. Compared to the pixel tracking mode, imaging detectors accumulate the incident radiation by counting individual radiation quanta in every pixel cell. Photon counting pixel detectors can be used for x-ray imaging in biomedical applications and protein-crystallography for synchrotron radiation [1].

ECD Bumping of Readout Chip Wafer

As mentioned before the sensor pixels are usually arranged in a very tight pitch to allow a high spatial resolution. If there is no redistribution to a larger pitch possible, the bump interconnections has to be deposited in the same pitch as well. Because solder paste printing and other solder transfer technologies are limited for pitches down to 150 µm, electrochemical deposition of bumping material is the technology of choice for smaller bump sizes and pitches. Beside the low-ohmic electrical interconnection between sensor pixel and readout electronic the bumps have to guarantee the mechanical and thermal contact between chip and substrate. Therefore the bump material has to be adapted to the sensor pixel as well as the electronic chip pad metallization, the selected joining process and the operating conditions of the detector. Different materials, e.g. Au, Ni, In, PbSn, AuSn, SnAg and CuSn can be used for flip chip interconnections. Indium, a very common bumping material for detector applications, can only be deposited in small pitches by evaporation and subsequent wet lift of technology. Depending on the commercial availability of electrolytes the other mentioned materials can be electroplated as stannous alloys or subsequently in a multilayer configuration.

A flip chip technology using miniaturized eutectic Au-Sn bumps was developed at Fraunhofer IZM for fluxless assembly of devices for optical and RF applications [2]. For bump sizes below 50 µm in diameter Au and Sn have to be deposited subsequently and must not be reflowed before assembly in order to achieve a high interconnection yield. The quantity of eutectic AuSn 80/20 can be easily controlled by an appropriate relation of Au and Sn deposition heights. Similar to Au-Sn bumping technology subsequent deposited Cu and Sn layers are used for solid-liquid-interdiffusion (SLID) bonds. After optical alignment, placement onto Cu pads and soldering both parts are electrically and mechanically connected by a very thin layer of thermodynamically stable Cu_3Sn compounds [3]. The other tin based solder materials, eutectic lead tin solder Sn60Pb40 and eutectic tin silver SnAg3.5, are commonly used for flip chip interconnections.

Usually the solder metal can't be deposited directly onto the aluminum or gold readout chip pad. Therefore a so called under bump metallization (UBM) is the basis for a reliable bump interconnection. First a diffusion barrier impedes the interdiffusion between bump and chip pad metallization. Second it is desired to use UBMs showing limited formation of intermetallic compounds at the UBM-solder interface to maintain the plasticity of the bond and the adhesion of the UBM to the diffusion barrier. A number of different UBM-solder configurations are investigated, e.g. Cu-PbSn60, Cu-SnAg3.5, NiV/Cu-SnAg, TiW/Ni/Cu-SnAg3.5, and results of reliability tests are reported. [4-5]

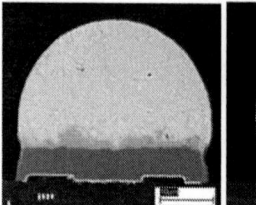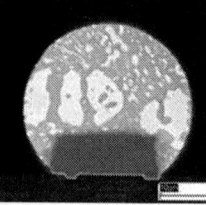

Fig.2: SnAg Bump left and SnPb bump right on readout chips before flip chip assembly

The bumping process on the usually 200 mm electronic readout chip wafers starts with the deposition of a Ti/W - Cu sputter layer on the wafer surface. As mentioned before, the first layer serves as a diffusion barrier to the aluminium I/O-pads. The subsequently sputtered copper layer, which is deposited onto the Ti/W layer without breaking the vacuum, acts as a plating base. It should protect the underlying Ti/W from oxidation and also provides electrical conductivity for the electroplating process. Spin coating is used for deposition of the high-viscous photosensitive resist system onto the wafer surface. The layer thickness depends on the required height of the electroplated UBM-solder structures and can be adjusted by an accurate wafer chuck spin speed. Then the wafer is aligned with a lithographic mask and UV-exposed in the pad regions where later the bump structures will be deposited. The exposed areas dissolve during the development in alkaline fluid and the resist pattern appears. In order to achieve high resist stability within the electrolytic bath a final postbake is made to fix the photoresist. The electroplating process starts with deposition of the under bump metallization directly followed by the deposition of the solder material. After stripping of the photoresist the plating base has to be removed without any residues between the bumps by two differential etching steps. Due to the fact that residues can lead to transverse conductivity between the bumps, the etchant should erode both layers uniformly and completely. On the other hand, the attack of the etch solution on the electroplated structures should be as low as possible. Compared to the small Au-Sn bump structures the electroplated SnPb- or SnAg-solder bump structures have to be transformed into spherical shape by a reflow step. In this case an inert liquid is used because of the homogeneous heat distribution. Next the bumping results as well as probable surface defects caused by particles or residues are optically inspected. Because of the large number of deposited bumps on the whole wafer surface the inspection is done by an automated optical inspection tool. Every die on the wafer is compared with a so called "golden die" image. Particles, plating base residues as well as bumping defects like missing, bridged, small or large bumps can be detected in only a brief time. All defects are stored in a defect file and the relevant chips can be excluded later from module assembly.

For the most particle tracking detector applications a reduced readout chip thickness is required to limit the loss of tracking precision due to multiple scattering. For this reason the readout chip wafer are thinned by backside grinding to a thickness of $150 \, \mu m - 200 \, \mu m$. A special UV curable release tape attached on the wafer front side protects the solder bumps during the grinding process. After removal of this tape without damaging the bumps the thinned wafers are diced by die sawing and sorted into chip trays for transfer to the assembly station.

ECD of Solderable Pixel Pads on Sensor Wafer

Usually aluminium, a nonsolderable pixel metallization, is also used for the senor side of the detector module. Therefore the underbump metallization has to be deposited exactly where the passivation vias are located. A Cu-Ni-UBM, protected from oxidation by a thin gold layer, is a well-proven pad metallization and shows an excellent solderability with eutectic SnPb solder. In case of lead-free SnAg solder bump material a bi-layer Ni-Cu-UBM was designed especially for small solder bump dimensions [4]. If an Au-Sn solder material is used on the readout chip side an Au pad has to be deposited on the sensor pixel. The process flow is similar to the bumping of readout chip wafers. It starts with the deposition of a Ti/W-Cu adhesion layer and plating base. If a Au pad is required on the sensor pixel side a Ti/W(N)-Au plating base has to be sputtered. After patterning of the spin coated photo-resist, the above mentioned solderable under bump metallization is deposited by electroplating to the contact pads of the pixel structures. Subsequently the resist is stripped and the plating base has to be etched without any residues. After the complete UBM deposition process the sensor wafer is cut into individual tiles by a dicing saw followed by an optical inspection step.

Module flip chip assembly

In addition to thermocompression and thermode soldering, reflow soldering is commonly used for detector assembly. Generally the reflow process is divided into two parts. First the die is placed on the substrate and second the die is soldered. This has the advantage that a quick pick and place equipment can be used and a large number of dice can be reflown in one soldering process.

The assembly of detector modules has several challenges. Often the multiple readout chips are larger than the sensor. Because the sensor is used as the substrate, it can not be handled after the placement of the chips. Additionally it must be placed very accurately into the assembly machine to make the automatic alignment of the chips as efficient as possible. To achieve both goals, a special adapter is designed, which will carry the sensor

during the whole pick-and-place and reflow process. Mechanical stop positions make it possible to place the sensor within $\pm 75 \, \mu m$ into the bonder and the thermal conductivity of the carrier is high, to improve a homogenous and quick heating during reflow. The large number of bumps with a very fine pitch down to $50 \, \mu m$ makes a very accurate placement necessary. Often only alignment marks on one edge are available. Therefore the alignment accuracy should be within $3 \, \mu m$ to guarantee no tilt of the die, which will result in undefined interconnections. As long as the overlap area between bump and pad is more than 60%, the surface tension of the liquid solder will self align the chip over the substrate with accuracy up to $2 \, \mu m$. To fix the position of the chips during the transfer to the reflow oven a combination of highly diluted flux or tacking agent and a controlled deformation of the bumps during the placement is used. To avoid flux or at least to minimize the amount a reflow with an active atmosphere is performed. The necessity of flux is highly dependent on the solder used and the pad metallization. After the reflow process an x-ray analysis will give good information whether the position of the chip is correct or not. Also bump shorts can be observed by this method.

Modules for LHC Pixel Detectors

ATLAS (A Toroidal LHC ApparatuS) and CMS (Compact Muon Solenoid) are in two of five particle detectors which are currently under construction at the Large Hadron Collider (LHC) at CERN. The main purpose of the detectors is the analysis of proton-proton collisions at mass center energy of $14 \, TeV$ for the search of the last undiscovered particle in the Standard Model of elementary particles and their interactions, the Higgs Boson. A hybrid pixel vertex system is the innermost part closest to the interaction point of the CMS and ATLAS detector. Part of the raw hybrid pixel modules of both detectors were manufactured at Fraunhofer IZM.

The overall dimensions of the ATLAS pixel detector for instance are 1850 mm in length and 380 mm in diameter. The tracks of particles generated by the proton-proton collision will be detected by three coaxial cylindrical barrels and three respectively forward and backward located end cap wheels. Every barrel consists of several carbon fiber reinforced staves where 13 silicon pixel detector modules are fixed on every stave. The complete ATLAS pixel detector is assembled from about 2000 modules. [6] These hybrid silicon modules with a size of 63.0x18.8 mm² are the basic unit of the ATLAS pixel detector. To achieve a high resolution of the particle tracks every sensor tile consists of an array of 46080 diode pixels with a pixel size of 50x400 μm^2. 16 radiation hard front-end (FE) electronic readout chips, each one is divided into 2880 readout cells, are flip chip bonded to one sensor tile. Electroplated eutectic PbSn solder bumps with a diameter of 25 μm in a pitch of only 50 μm are used for the interconnection.

One part of the hybrid pixel module, the radiation hard ATLAS FE-I readout chips, are designed on 200 mm silicon wafers in deep sub-micron 0.25 μm CMOS technology. These chips fulfill the requirements for electrical performance after an irradiation of 10^{15} n_{eq} cm^{-2} which is the radiation dose after ten years operation time of the ATLAS detector. 288 FE-I chips with a size of 7.4x11 mm² are arranged on one 200 mm wafer. The readout chip wafers were thinned down to a thickness of 180 μm after bumping. The particle sensing element, the ATLAS sensor tiles, are fabricated on double sided processed 100 mm radiation hard oxygen enriched silicon wafers with a thickness of 250 μm. Every sensor has an active area of 60.8x16.4 mm². Three tiles and several test structures are arranged on one sensor wafer.

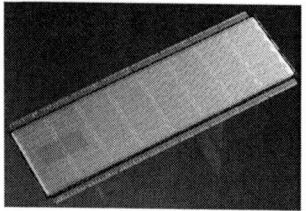

Fig. 3: ATLAS Module after flip chip assembly

After flip chip assembly the bump yield of every raw module was quantified at several laboratories in the ATLAS pixel collaboration. If a module passes the test procedure, it is completed by a Kapton Flex Circuit which is glued to the backside of the sensor module. This flex circuit provides the signal routing between the 16 FE chips and the module controller chip (MCC). Beside the thinned MCC, passive components, and a data bus connection flex are mounted on the flex circuit as well. All chip components are connected to the flex circuit by wire bonds. In order to build up the vertex system 13 of these completed pixel modules are glued on a carbon reinforced stave structure. Finally several staves in turn are assembled to three barrel layers of the detector (Fig. 4).

About 100 readout chip wafers have been bumped during the production phase of the ATLAS Pixel project with approximately 850,000 bumps per wafer. Eutectic PbSn solder bumps with a diameter of 25 μm in a very close pitch of 50 μm were deposited on wafer level using the above described electroplating technology. In order to avoid imprecise particle tracking caused by dead pixels, not more than 30 pixel failures per chip and 150 pixel failures per hybrid module were accepted. Therefore an extensive test and inspection procedure was identified as a key step in this small batch production. Due to the fact that yield can be decreased by other failures too, the number of defects after bumping has to be below this mentioned rate. Chips with more than 5 individual

bump defects were rejected before flip chip assembly. Mechanical damage, shorts by plating base residues and electrical failure after dicing are another reasons for chip rejection before flip chip assembly. Overall 95% of more than 20,000 readout chips passed this inspection and electrical test procedure. Within the two year production phase 1150 modules have been flip chip bonded at Fraunhofer IZM. Due to extensive use of optical inspection and process control and the development of single chip repair technology a total module yield of 98% was achieved.

Fig. 4: complete pixel detector modules mounted on staves, one half shell of barrel layer (courtesy of ATLAS Pixel Collaboration)

A smaller amount of 526 raw silicon modules have been flip chip bonded for the pixel tracker of the CMS detector at the LHC. Unlike the ATLAS project these modules are part of the disk segments of the pixel detector end cap wheels. Therefore these modules have different sizes from single chip tile up to 5x2 multichip tile. But these considerations don't affect the wafer processing. Identical to the ATLAS readout wafers 25 μm SnPb solder bumps were deposited onto the CMS chips and a Cu-Ni-Au UBM as solderable sensor pixel metallisation. The chips were also thinned before assembly down to a thickness of 200 μm.

Failure analysis of ATLAS raw modules

Failure analyses of ATLAS pixel detector modules after flip chip assembly have shown that thinning of the readout chips is a limiting factor. The failure was high ohmic or open connections at the edges of the chips and could be detected with noise measurement of the module (Fig. 5).

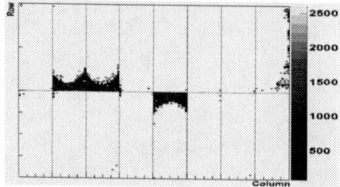

Fig. 5: electrical noise map of an ATLAS raw module

This failure was not observed on previous batches of chips. The only difference identified was a slightly difference in the thickness after thinning. To check the mechanical deformation of the chips during reflow process laser interferometrical measurements at discrete temperatures where done. To avoid errors from moving the gauging head a double interferometer with an angle of 180° is used. The difference distant measurement is done between the die below the interferometer and a temperate polished silicon wafer piece on top. The temperature was applied with an infrared heated thermode and an equilibration time of 5 min is given. To realize this setup a flip chip bonder (FC150, SUSS MicroTec S.A.) is used. Fig. 6 shows the bowing of the chip depending on thickness and temperature. The three measurement points are located at the upper left corner (point 1), center (point 2) and lower right corner (point 3). It is visible that the temperature dependent bowing is significantly correlated with the thickness of the dice. For a chip with the required thickness of 180 µm the center is at soldering temperature about 10 µm lower than the level of the chip edges. This deformation increases for chips with 160 µm and 130 µm thickness to 14 µm and to 17 µm, respectively.

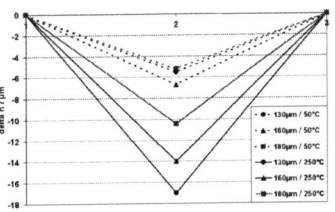

Fig. 6: Chip bending dependence for different chip thicknesses at 50°C and 250°C

As the first of the 25µm bumps in the center of the chip build a 10 µm thick liquid phase when soldering, the outer area of the chip should not be higher then 12 µm to allow a good wetting of the pads on the substrate. Therefore the suggestion is given to use at least 180 µm thick chips, with a thickness of 200 µm being better. The bowing is caused by the four times higher CTE of the six aluminum metal layers used for the wiring intra the bottom of the chip.

Diamond Pixel Detector Module

Due to its properties the chemical vapor deposited (CVD) diamond is a suitable alternative sensor material for detectors that operate in a very high radiation environment. First, CVD diamond material is much more radiation tolerant than silicon. The low dielectric constant and the low leakage current of diamond detectors minimize the noise in the amplifying electronics. Second, the high band gap of about 5.5 eV allows operation at elevated temperatures. The high drift velocity of the electron-hole pairs enables a fast signal collection. But due to the polycrystalline structure of the CVD diamond material (pCVD) the charge collection behavior is not uniformly distributed and the resolution of position is worse compared to silicon detectors. Synthetic diamond material can be grown as a polycrystalline film on a 6 inch substrate to a layer thickness up to 2 mm. Only parts of pCVD wafers

were manufactured as sensor tiles with promising properties. The sensor area size varies from single chip of 1x1 cm² up to a full ATLAS diamond pixel module of 6x2 cm². An aluminium pixel layout similar to the silicon sensors was deposited on the sensor front side and a gold layer on the back side, which serves as the back side electrode. Compared to the production of a large number of silicon sensor modules for the ATLAS and CMS detectors the electroplating process of this prototype requires another technology approach. The embedding technology developed at Fraunhofer IZM plays the key role for the processing of the single diamond substrate. The diamond sensor tile has to be inserted into a ceramic carrier wafer, in which an exactly matched opening was cut by laser before (Fig.7). After the fixing of the diamond by epoxy every process step from the sputtering of the plating base up to the electroplating of the pixel contacts can be performed on wafer level. Finally the sensor was removed out of the ceramic carrier wafer by dicing. 16 radiation hard ATLAS front-end readout chips were flip chip bonded to the diamond sensor likewise the ATLAS silicon detector modules described in the previous chapter. The functionality of the full 16 chip ATLAS diamond pixel module was tested with a 180 GeV pion beam at LHC and with a 6 GeV electron beam at DESY. [7]

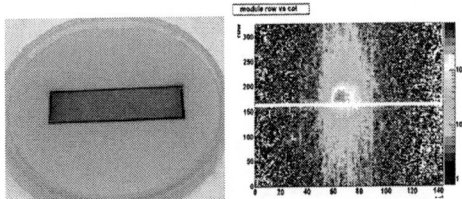

Fig. 7: diamond pixel module embedded into ceramic wafer and recorded beam profile of an 180 GeV pion beam at LHC

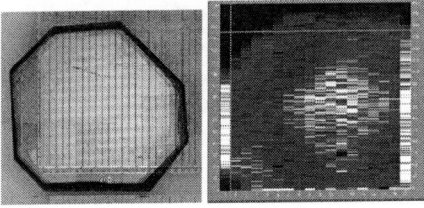

Fig. 8: diamond single crystal detector after UBM processing and test of the hybrid module with an Am241 source (courtesy of Bonn University)

The growth of single crystal CVD (scCVD) diamond has been developed in the last years. This new mono-crystalline material promises to overcome most of the polycrystalline shortcomings, as there is non-uniformity of the charge signal and the poorer resolution of position compared to devices made with Silicon. At the moment only small irregular shaped detectors are available, but these detectors can be processed and assembled similarly. A detector with an octagonal shaped sc-diamond was assembled and successfully tested (Fig.8).

Hybrid Pixel Detectors for X-Ray Imaging

Hybrid pixel detectors have had an impact on imaging applications as detectors that accumulate the incident radiation through the counting of individual radiation quanta in every pixel cell. This technique offers many features which are very attractive for X-ray imaging: full linearity in the response function, in principle an infinite dynamic range, optimal exposure times and a good image contrast compared to conventional film-foil based radiography [1]. In addition to the common silicon material promising high absorption semiconductor materials like GaAs and CdTe are used for x-ray imaging sensors. In order to use single photon counting detector systems for imaging applications, it is necessary to cover a large imaging plane which can be achieved with multichip modules or high-IO readout chips. MEDIPIX and XPAD3 are two examples of photon counting pixel systems assembled at Fraunhofer IZM.

MEDIPIX readout chips are fabricated on 200 mm wafers in deep sub-micron 0.25 μm CMOS technology. A matrix of 256x256 readout pixels (55x55 μm²) is arranged on a chip area of 14x14 mm² in size. Solder bumps were deposited by electroplating on the pads of the readout cells using a SnAg wafer level bumping process. About 6 million bumps with a height of 25 μm in a pitch of only 55 μm are deposited on a 200 mm wafer surface. After dicing the MEDIPIX readout chips were flip chip bonded to Si and GaAs sensor substrates respectively.

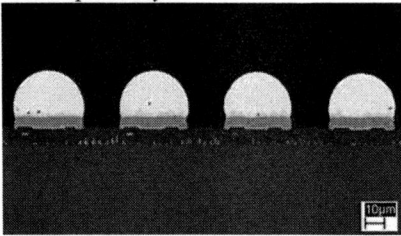

Fig. 9: MEDIPIX readout chip (25 μm SnAg bumps pitch of 55 μm), SEM cross cut

Conclusion

The hybrid pixel technique is an established concept in manufacturing of particle and radiation detectors. Optimized readout chips can be flip chip bonded on different sensor substrates. Several examples of hybrid pixel detectors packaged at Fraunhofer IZM were described in this paper. The combination of a reliable bumping process and process control resulted in high yield MCM technology as shown for the assembly of 1150 ATLAS and more than 500 CMS pixel detector modules. Further pixel detector prototypes were assembled using CVD diamond sensors for beam monitoring and GaAs sensors for x-ray imaging.

The future development of semiconductor detectors may be divided into two groups. First, the material approaches are driven by higher rates and detection

efficiency and radiation hardness. Monocrystalline diamond and hydrogenated amorphous silicon seem promising for high radiation dose applications. High absorption sensor materials like GaAs, CdTe or CdZnTe can be used in low-dose medical imaging applications. The assembly technology has to be adapted to the mechanical properties of these brittle sensor materials. Second, geometric detector parameters will change to smaller pixel sizes, less material per sensor layer but larger sensitive detector area. The development of 3D detectors is an interesting new approach. Electrodes are formed perpendicular to the sensor surface through the silicon wafer in a very fine pitch of 30 μm, which demands a fine pitch of the interconnections too. The advantages are a reduced depletion voltage and charge transition time, improved radiation hardness and an edgeless sensitive detector area. First prototypes of ATLAS 3D single chip sensors were processed and flip chip bonded to ATLAS readout chips at Fraunhofer IZM.

Acknowledgements

The authors would like to thank the staff of department HDI&WLP and department P2SA for thin-film-processing on wafer level and module flip-chip-assembly, respectively.

References

[1] Wermes N. "Pixel Detectors for Tracking and Their Spin-Off in Imaging," Nuclear Instruments & Methods in Physics Research A, No. 541 (2005), pp. 150-165.

[2] Hutter, M. et al: "Assembly and Reliability of Flip Chip Solder Joints Using Miniaturized Au/Sn Bumps", Procceddings 54th Electronic Components and Technology Conference, Las Vegas, June, 2004, p 49-57.

[3] Wieland, R. et al: „3D Integration of CMOS Transistors with ICV-SLID Technology", Proceedings of Materials for Advanced Metallization Conference 2005, in Microelectr. Engineering 82 (2005), p. 529-533

[4] Wolf, et al: "Flip Chip Bumping technology – Status and Update", Nuclear Instruments & Methods in Physics Research A, Vol. 565/1 (2006), p. 290-295

[5] Dietrich, et al: "Conformance of ECD Wafer Bumping to Future Demands on CSP, 3D Integration and MEMS", Proceedings of the 56th Electronic Components and Technology Conference, San Diego, CA, May/June 2006.

[6] Troncon, C.: "The ATLAS Pixel Detector" 32nd International Conference on High Energy Physics, Beijing, August. 2004, p.933-936.

[7] Kagan, et al, "Radiation Hard Diamond Sensors for Future Tracking Applications," Nuclear Instrument & Methods in Physics Research A, Vol 565 (2006), p. 278-283

Interface Reactions during Au-Ball/Wedge and AlSi1-Wedge/Wedge Bonding at room temperature

M. Schneider-Ramelow[1], K.-D. Lang[1], U. Geißler[2], W. Scheel[1], H. Reichl[2]

[1] Fraunhofer Institute for Reliability and Microintegration (IZM), Gustav-Meyer-Allee 25, 13355 Berlin

[2] Technical University Berlin, Research Center for Microperipheric Technologies

++49 03 46403-172, Fax -271, E-mail: martin.schneider-ramelow@izm.fraunhofer.de

Abstract

The presentation addresses results of Au Ball/Wedge Bonding and AlSi1 Wedge/Wedge Bonding from 125°C down to room temperature (RT). In this context, also interfacial reactions on Flash-Au PCB-metallizations as well as microstructural changes in AlSi1 wire material in relationship to bonding conditions will be discussed. This emphasizes the understanding of bonding process remarkable. One topic is the verification of RT bondability of different types of Au wire on specific Al pad metallizations and PCB-substrates with Cu/Ni/Pd/Flash-Au metallization [1]. Investigations include mechanical tests of Au loops and ball contacts as well as microstructural observations of the contacts (FIB, SEM, TEM). A most important result in this context is, that Au/Al intermetallic phases with thicknesses of a few hundred nanometers were found below the Au contacts on Al metallization directly after bonding at room temperature. Secondly new results of the Au-Au interface between wire and Flash-Au finish metallization will be presented. Another related purpose was to investigate wedge microstructures of AlSi1 wire bonds as well as the interface between the bonding wire and a Cu/Ni/Flash-Au metallization layer. Focused Ion Beam (FIB) and Transmission Electron Microscopy (TEM) have been the dominant analytic methods. The interface between the AlSi1 wire and the Cu/Ni/Flash-Au metallization layer of the optimized bonds consists of a closed crystalline Au layer with a thickness of nearly 80 nm. Above this Au layer, a second smaller zone consisting of an intermetallic phase was analyzed and identified by electron diffraction as Au_8Al_3. Further a correlation of wire material structure changes (fiber texture to recrystallized grains) and ultrasonic power during bonding were determined. With this results understanding of room temperature wire bonding could be improved exceedingly.

Key words: AlSi1 and Au wire bonding, wedge structure, interface formation, Au-Al intermetallic phases

Introduction

Nowadays, advanced gold ball auto bonders have the capability of bonding to test pads with 25 µm pitch (wire 12 – 15 µm). High volume industrial production runs in the range of pitches between 65 µm and 35 µm. Aluminum wedge bonders have to fulfill requirements like bonding of very thin metallization layers, temperature sensitive multilayer substrates and highly different pad levels (e.g. chip stacking). That means for both bonding methods varying material properties, very accurate equipment parameters and more complex assemblies have to be combined for high quality bonding results at extremely small bonding areas.

However, necessary infrastructure (e.g. wire and tool development, test technology) have also to be considered in this context. So, there have been for example many developments in very tough ceramic capillaries and in new alloys for very thin wires, but the requirements for bonding with less than 35 µm pitch still remain a challenge.

But, also the bonding procedures (e.g. temperature influence, power dissipation), behavior of materials during bonding (e.g. deformation, hardness), interface reactions between wire and metallization (e.g. contamination influence, phase building) or differences in the phases of interconnection (e.g. length of activation phase) have to be explained more in detail concerning fine pitch bonding.

To give a contribution to meet these demands of future fine pitch wire bonding without or at low temperatures the paper discusses nano scale interfacial reactions between Au and Al as well as microstructural changes in AlSi1 wire material in relationship to bonding conditions. Very new cognitions emphasize in this context the understanding of wire bonding process remarkable.

TS bonding of Au wires down to RT

Standard bonding machine parameters as used for Au or Al metallizations have to be modified to produce high yield and reliable bonding to fine pitch pads. This results also from different material/metallurgical influences such as power and temperature dissipation per volume, smaller bonding areas, surface characteristics and hardness. These influences are in the focus of the described investigations. Materials and parameters of bonding investigations are shown in Table 1.

Table 1: Materials and Properties

material	notation	parameter
bond wire	doped standard bond wire Heraeus HD2	diameter = 32 µm, load ca. 14.5 cN and elongation 2-7%
	Pd alloyed bond wire (HA3)	diameter = 32 µm, load last ca. 24.5 cN and elongation 2-7%
	highly doped bond wire (HA7)	diameter = 32 µm, load ca. 21.4 cN and elongation 2-7%
chip	industrial test chip	metallization: ca. 0.6 µm AlSi0.5Cu0.5
	chip demonstrator 1	metallization: ca. 0.6 µm AlSi0.5Cu1
substrate	PCB with Ni/Pd/Flash-Au	Cu ($\geq$ 50 µm); 7-8 µm Ni; 0.25 µm Pd; 0.06 µm Flash-Au
	flex substrate demonstrator 1	no comments (delivered by a company)

The achievements of the bonding investigations took place with an auto bonder F&K Delvotec Model 6200, 70 kHz and a capillary UTS-46KJ-CM-1/16-16MM at bonding temperatures: 125°C, 75°C and room temperature (RT).

The evaluation of bonding investigations were performed as described in [2] by shear testing (ball bond) and pull testing (loop, wedge bond). The failure codes pull or ball lift off while mechanical testing is valid for material residues on pad metallizations with less than 50%.

For evaluation of ball bonding strength on Al chip metallization shear tests were used. The ball diameter values after bonding have been smaller than 100 µm; a free air ball diameter of 60 – 70 µm was adjusted by adaptation of the flame off conditions. The shear height was fixed to 3 µm over chip passivation. For ball diameters of 100 µm, the average value of the shear force should achieve > 63 cN under laboratory conditions and > 54 cN under manufacturing conditions. Since the deformed ball diameter partially clearly was less 100 µm, in the following one, average values of shear force > 50 cN were regarded as sufficient.

Figure 1 shows the results of the bonding optimization after extensive parameter variation. The HD2 wire as well as the highly doped HA7 wire were bondable with high quality at 125, 75°C and RT under given boundary conditions (ball diameter < 100 µm). Merely the US power had to be increased negligibly to it with decreasing bonding temperature. Shear lift offs occurred with the clearly harder Pd alloyed HA3 wire at all bonding

temperatures (100% at RT, 66% at 75°C and 33% at 125°C). Quality-justly bonding interconnections at RT without shear lift offs could be achieved only with accurately elevated bonding parameters and with interconnected ball diameters > 100 µm (e.g. ball with 124 µm, shear force 99.2 cN). Consequently, also the ball of this wire can become eligible with certain bonding parameters, if greater ball diameters are acceptable.

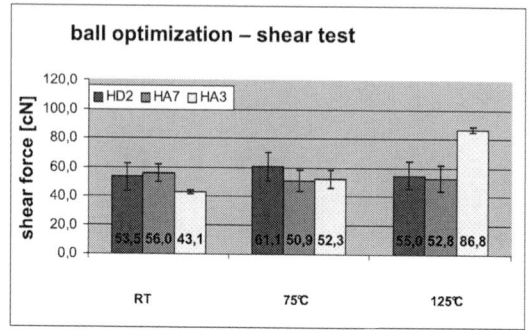

Figure 1: Bonding and shear test results on chip

The phase thicknesses as well as the phase lengths with reference to the ball bond width were intended for the evaluation of the intermetallic phases (IMP, s. Figure 2) by three ball bonds each at five places minimal. The results of these experiments are presented and discussed extensively in [1] in dependence of the bonding frequency and temperature as well as the wire quality.

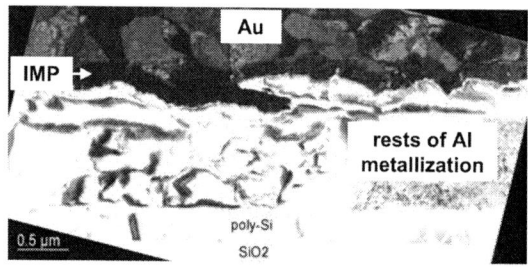

Figure 2: **TEM picture of ball bonding contact area, HD2 without heating**

To the evaluation of the adaptability of the Au wedges on PCB metallizations, bonding were achieved on the PCB material with Ni/Pd/Flash-Au. Thirty (30) loops per parameter set (each bonded in vibration direction) were tested by the destructive pull test in that way, that the angle at both bonds was approximately 30° and no correction factor of the pull force was necessary consequently [2]. The failure codes at pull testing have been: 1 – ball lift off; 2 – neck crack; 3 – wire crack; 4 – wedge crack; 5 – wedge lift off. The optimization of the bonding parameters were running only related to the wedge contact. Following, the quality-like bonding results with optimized parameters are shown (see Figure 3).

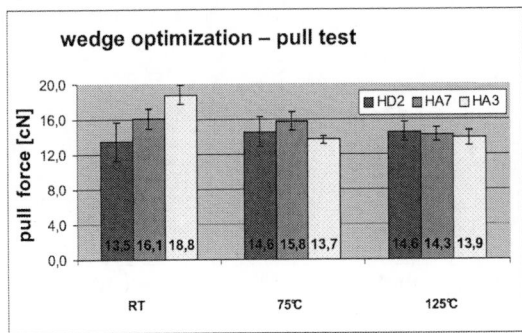

Figure 3: Bonding and pull test results on PCB

For the micro-structural examinations of the wedge contact formation, extensive comparative microstructure analyses about the influence of the bonding temperature, the wire material and bonding machine were achieved (see [1]). Because these results generated no significant difference of wedge contact formation, the shaping of the wedge interface bonded at RT will be discussed exemplarily for the HA7 wire (s. Figure 4).

Figure 4a shows a wedge contact as a general survey, prepared in the cross-section by means of focused ion beam technology and represented with SEM.

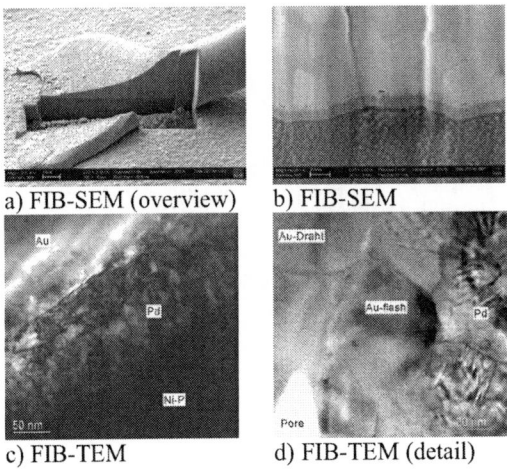

a) FIB-SEM (overview) b) FIB-SEM

c) FIB-TEM d) FIB-TEM (detail)

Figure 4: SEM and TEM pictures of FIB cuts (HA7 on pcb with Ni/Pd/Flash-Au)

The interface region of the wedge is demonstrated as an increased SEM picture in Figure 4b. The contact formation between Au wire and substrate material took place in the uppermost layers of substrate metallization. That means, the thin Au-Flash is not destroyed by the bonding process and can be still recognized clearly. Also the highly dissolving illustrations of the interconnection area in the TEM (Figure 4c-d) confirmed that the direct contact formation occurred directly between the bonding wire and the Flash-Au of the substrate metallization. The Flash-Au between Au wire and

the Pd layer can be recognized clearly in Figure 4d. These results have been observed already at interface formations between bonding wires and PCB metallizations documented in [3-4]. Thickness of Flash-Au metallizations was varying extremely from 30 to 100 nm. Thickness of Pd layers was between 90 and 100 nm (supplier value 250 nm) at all investigated PCBs (Table 1). After checking bondability in principle at RT test vehicles were built up with die and wire bonded chips and structured PCBs. Positive results regarding the processability at RT could be confirmed on that occasion.

Demonstrators were still produced in form of industry-typical smart card constructions afterwards. These were used for the proof of the won realizations at real modules from the industry. Because of the smaller pad dimensions (90 µm) a 25 µm HA7 Au wire was processed for these attempts instead of used 32 µm wire at the principle investigations above. The substrate heating was also given up completely. With the 25 µm HA7 wire (breaking load > 10 cN, elongation 2 – 6%), the following results were achieved (s. Table 2) related to the strength properties of the loops (here exemplary with another half automated bonding machine from 30 measurements each):

Table 2: Results of ball and wedge bond optimization, chip card demonstrator

ball optimization	median value	standard deviation	min.	max.
shear force [cN]	65.2	6.0	54.8	79.2
DVS-Merkblatt	> 42	< 9.8	> 30	-
wedge optimization	median value	standard deviation	min.	max.
pull force [cN]	8.5	1.1	6.2	10.4
DVS-Merkblatt	> 5	< 1.3	> 4	-

After ball and wedge bond optimization at processing the industry-typical demonstrator all quality criterions for TS bonding fixed in the "DVS-Merkblatt 2811" [2] were fulfilled with turned off substrate heating (at RT).

Attendant reliability tests (150°C temperature annealing, 85°C/85%r.H. and temperature cycling -40/RT/+125°C were done on PCB substrates with Ni/Pd/Flash-Au, IZM test chips (see Figure 5) and Au HD2 bonding wire with 14 cN breaking load and 5% elongation. Test samples were built up based on optimized bonding parameters. Pull and shear testing as well as cross sectioning was carried out in reliability test intervals up to 2000 hours resp. cycles. Pull and shear codes (pc/sc) were defined as following: pc 2 neck crack (over the ball), pc 3 wire break, pc 4 heelcrack (near the wedge), sc 2 partial ball shearing (35-75% wire material),), sc 3 shearing through the ball (75-100% wire material), sc 4 metallization lift off, sc 5 cratering and sc 6 break within the metallization. The sc 6 was characterized as to be seen in Figure 6 with metallization residuals on the pad and adhered metallization under the ball. Pull test correction factor was 0.76.

Figure 5: Test vehicle (reliability tests)

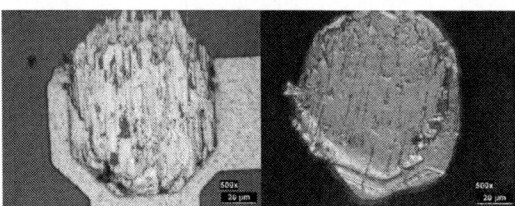

Figure 6: Shear mode 6 with residuals at pad metallization and under the ball

The results of the mechanical test are shown in Figure 7 - Figure 10. No pull or shear lift offs have been observed during all different reliability test.

Slightly increasing pull and shear values were scheduled. Explanation for pull values can be given by changing pull codes. Because of the aging conditions temperature induced interdiffusion and structural changes appeared. Obviously the pull code changes from 4 to 2 resp. 3 dependent on the absolute loading temperature (Figure 9). This effect is smaller at significant lower temperature load at humidity (85°C) and cycling (short periods at 125°C) compared to exposure at 150°C.

The interfacial areas were stressed by ball shear testing. Therefore the constitution of intermetallic phases had a great influence on shear codes and this again is dependent on loading conditions (see Figure 10).

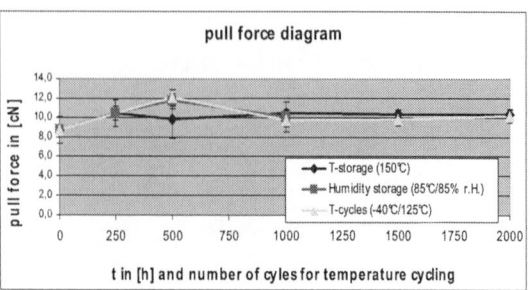

Figure 7: Pull force results during reliability tests

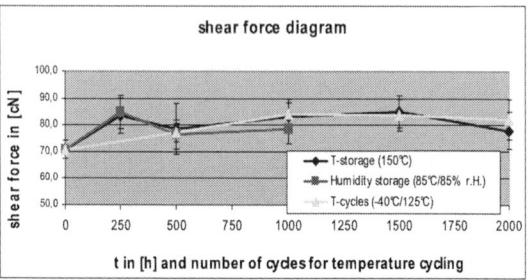

Figure 8: Shear force results during reliability tests

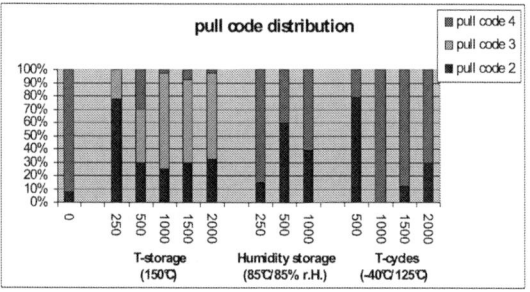

Figure 9: Pull codes

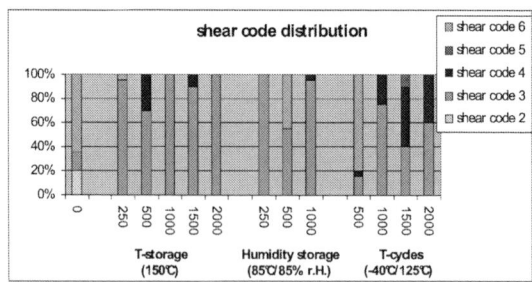

Figure 10: Shear codes

Following figures (cross sections in Figure 11 to Figure 14) present ball and wedge bonds at initial state and after 500 hours resp. cycles. Cross sections confirm the good interconnection quality which were confirmed by pull and shear tests. The growth of intermetallic phases was carried out uniformly without too much pores. The reason for none phase building in the edge areas is that the ball deformation to achieve the best interface formation was so large that the processed ball was bigger than the pad dimension. So the peripheral areas of the balls were located above the chip passivation.

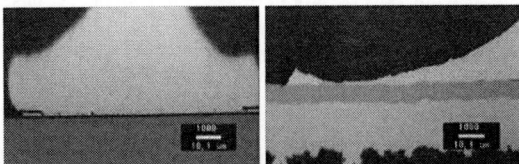

Figure 11: Bonds in initial state

Figure 12: Bonds after 500 h at 150°C

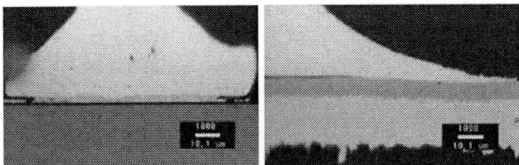

Figure 13: Bonds after 500 h at 85°C/85%r.H.

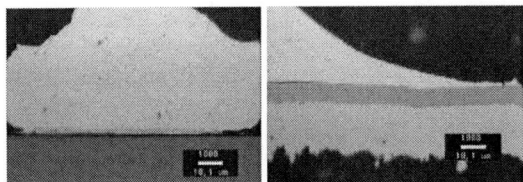

Figure 14: Bonds after 500 temperature cycles

Micro Structural Changes during Ultrasonic AlSi1 Wedge/Wedge Bonding

The wedge microstructure of 25 µm AlSi1 wire bonds as well as the interface between the bonding wire and the Cu/Ni/Au substrate metallization layer has been investigated by Focused Ion Beam (FIB), Transmission Electron Microscopy (TEM) and micro-hardness measurements. The as-received wires were characterized by fiber texture of <111> and <100> orientation and vertical grain diameters of nearly 400 nm. This fiber texture is also characteristic for wires which were only deformed by bonding force (without US power) in the first phase of the wire bonding process. But the wedges deformed only by bonding force are characterized by stronger cold working and hardening than the as-received wires. Following ultrasonic treatment induces softening into the wedges. To investigate the effect of ultrasonic energy on the grain structure and the interface between wire and Cu/Ni/Flash-Au metallization different types of samples were prepared: Optimized bond contacts ([2], Figure 15), overbonded bond contacts (high ultrasonic power, Figure 16) and underbonded wedges (low ultrasonic

power, Figure 17). All contacts were bonded with bonding force of 27 cN and bonding time of 50 ms. All samples processed with US power show small recrystallized grains near the wire/metallization interface (Figure 15 - Figure 17).

With increasing ultrasonic power the results indicate further recrystallization of the grain structure and decreasing micro-hardness inside the bonded wedges (Figure 18).

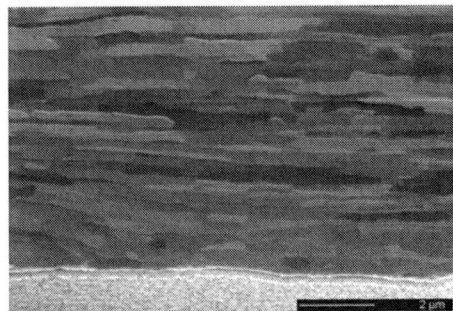

Figure 15: Recrystallized grains of optimized wedges

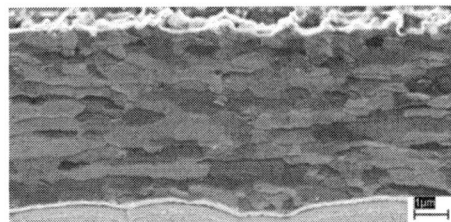

Figure 16: Completely recrystallized grain structure of an overbonded wedge

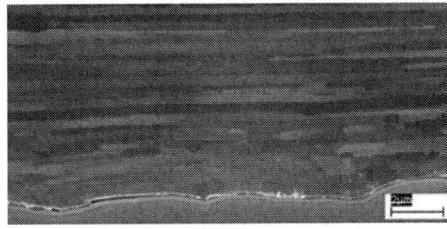

Figure 17: Equiaxed and recrystallized grains of underbonded wedges

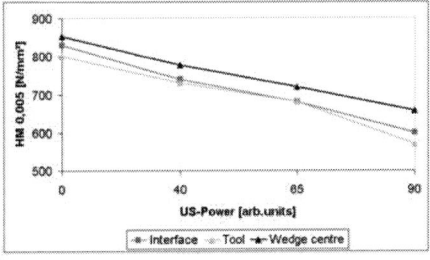

Figure 18: Effect of US-Power on wedge hardness

During US wedge/wedge wire bonding at RT dynamic recrystallization of the AlSi1 wires induced by wire deformation and US energy takes place. Optimized wedges exhibit a fully interconnected interface characterized by very small grains near the interface. Increasing US Power causes a completely recrystallized grain structure inside the wedge. It activates defect movement and diffusion processes in and between the bond partners and results in a closed interface without voids.

Increasing ultrasonic power also changes the interface formation between wire and metallization. It could be shown in TEM images that the thin crystalline Au layer of the metallization system (ca. 50 – 70 nm) is still present after the wedge/wedge wire bonding process (Figure 19). So the interface of the optimized bonds consists of the closed crystalline Au layer and above this Au layer, a second zone consisting of intermetallic phases was analyzed and identified by convergent electron beam diffraction as Au_8Al_3. No other type of intermetallic phases was found. Two zones have been found also in overbonded wedges and partially in the underbonded wedges. By comparing the dimensions of the zone of intermetallic phases detected in TEM images the following thicknesses of the Au_8Al_3 layer were measured: underbonded wedge, 20 nm, optimized wedge, 30 nm and overbonded wedge 85 nm. Increasing US-Power causes an accelerated growth of intermetallic phases during wire bonding.

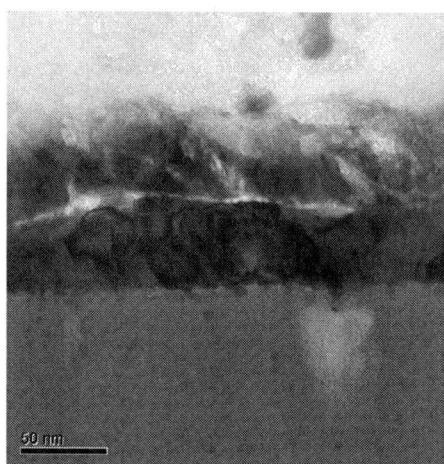

Figure 19: Interface of optimized wedges (TEM)

Conclusion

Currently wire bonding methods have to be conformed more and more to fine pitch structures, varying material properties, extraordinary equipment parameters and complex assemblies.

In this context also particular understanding of the bonding procedures (e.g. temperature influence, power dissipation), the behavior of materials during bonding (e.g. deformation, hardness), the interface reactions between wire and metallization (e.g. contamination influence, phase building) and differences in the phases of interconnection (e.g. length of cleaning phase) have to be improved.

In the realm of RT Au- Ball/Wedge bonding new explanations for interface reactions during bonding of thin Al metallization layers were found. A most important result in this context is, that Au/Al intermetallic phases with thicknesses of a few hundred nanometers have been discovered below the Au contacts directly after bonding at RT.

Another purpose was to investigate wedge microstructures of AlSi1 wire bonds as well as the interface between the wire and very thin Flash-Au metallization layers. Main results are a correlation between wire texture and ultrasonic influence as well as first-time confirmation that the interface between the AlSi1 wire and thin Flash-Au metallization layers consists of two zones, a closed crystalline Au layer (ca. 50 nm) and above a second smaller zone identified as Au_8Al_3 intermetallic phase (optimized bonds 30 nm).

Acknowledgements

Some results of this article were funded by „Bundesministeriums für Wirtschaft über die Arbeitsgemeinschaft industrieller Forschungsver-einigungen "Otto von Guericke" e.V. (AiF-Nr.13309BG) and supported by „Forschungsver-einigung Schweißen und verwandte Verfahren e.V. des DVS". Some other results were funded by "Deutsche Forschungsgemeinschaft" DFG. Authors are deeply grateful for this support.

References

[1] Forschungsbericht: Thermosonic Drahtbonden bei Verfahrenstemperaturen unter 100°C. BMWi/AiF-Abschlussbericht AiF 13.309 B (Laufzeit des Vorhabens: 05.2002 - 04.2004).

[2] American Society for Testing and Materials ASTM Designation F 1269 – 89, Test methods for destructive shear testing of ball bonds, and F 459 – 84, Standard methods for measuring pull strength of microelectronic wire bonds, and Merkblatt DVS 2811, Prüfverfahren für Drahtbondverbindungen, Deutscher Verband für Schweißtechnik e.V., 1996.

[3] Geißler, U.; Schneider-Ramelow, M.; Lang, K.-D.; Reichl, H.: Investigation of Microstructural Processes during Ultrasonic Wedge/Wedge Bonding of AlSi1 Wires. Journal of ELECTRONIC MATERIALS 35/1, 2006.

[4] Geißler, U.; Schneider-Ramelow, M.; Lang, K.-D.; Reichl, H.: Grenzflächenuntersuchungen beim Ultraschall-Wedge-Wedge-Bonden von 25 µm AlSi1-Draht. PLUS 4/2004, S.636.

Surface Acoustic Wave Component Packaging

G. Feiertag, H. Krüger and C. Bauer

EPCOS AG, Anzinger Str. 13, 81671 München, Germany

Phone: +49 89 636 21620, gregor.feiertag@epcos.com

Abstract

Surface Acoustic Wave (SAW) filters are key components of mobile phones and TV sets. In SAW components mechanical waves propagate on the surface of a chip, so the package must provide a cavity. Solid materials on the chip would inhibit the propagation of the surface waves. Reduction of size and cost, improved reliability and electrical performance are the main trends of SAW component evolution. In the past SAW filters were exclusively soldered directly on the PCBs. Now more and more filters are integrated in modules, which are often packaged by a transfer molding process. So the ability to withstand pressures up to 100 bars was added to the list of requirements. In this paper the SAW package evolution from packages using bonded wires to flip chip packages is reviewed. Our new SAW package developments for single SAW filters, 2in1 filters and duplexers using HTCC or LTCC interposers are presented.

Key words: SAW Filter, Surface Acoustic Wave, Cavity Package, MEMS Package, BAW, FBAR

Introduction

Surface Acoustic Wave (SAW) filters have conquered one application after another over the past 30 years, first in TV sets and later in mobile phones [1]. Since SAW filters have a superior filter functionality and a small size, they play an important role in the evolution of mobile phones. The reduction in size of SAW filters has made a major contribution first to the miniaturisation of mobile phones and later to extending their functionality [2]. Several billion SAW filters are now produced per year, more than any other Micro-Electromechanical (MEMS) component.

In a SAW component mechanical waves propagate on the surface of a solid. To convert an electrical signal into a mechanical wave so called Interdigital Transducers (IDT) patterned on a piezoelectric solid are used. A second IDT converts this mechanical wave back into an electrical signal. Piezoelectric materials like Lithium Tantalate (LiTaO$_3$), Lithium Niobate (LiNbO$_3$) or Quartz are used as substrate materials. In Figure 1 the basic layout of a SAW Filter is shown.

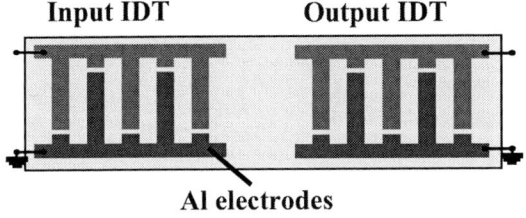

Figure 1: Basic SAW Filter Layout

The electrodes of the IDTs are arranged in a periodic pattern so only the frequencies of the electrical signal which match this pattern are converted into a mechanical wave. The velocity of the SAW wave divided by the frequency results in the SAW wavelength. The wavelength must be twice the pitch of the IDT fingers. For other frequencies the interference of the waves generated between the fingers is not constructive. This leads to the frequency filtering function. With more sophisticated layouts filters with low insertion loss within a wide frequency band and good out of band rejection near the pass band can be designed.

In mobile phones SAW components are applied in the transmit (Tx) and receive (Rx) path near the antenna. The following SAW components are used:

- Single band pass filters. Input and output of the filter can be single ended or balanced.
- 2in1 filters which combine two band pass filters in one package. The filters can share one common chip or two chips are integrated in one package.
- Duplexers which direct the transmit signals from the power amplifier to the antenna and simultaneous direct the incoming signal to the low noise amplifier. Duplexers are necessary for CDMA and W-CDMA (UMTS) telephone systems.

For the propagation of the surface wave the package must provide a cavity above the chip surface. For 2 GHz filters the IDTs are typically made of Aluminium lines with a thickness of 150nm and a width below 500 nm. These structures must be protected against humidity to prevent corrosion. This protection can be either a very thin passivation layer on the chip or a hermetic package. The resistance against corrosion is tested using high temperature at high humidity.

The package must withstand temperature changes. This is tested by temperature cycling or shock tests. For package development it has to be taken into account that the thermal expansion of $LiTaO_3$ and $LiNbO_3$ is anisotropic.

A cavity package is also needed for Bulk Acoustic Wave (BAW) components or Film Bulk Acoustic Resonators (FBAR). Packaging technologies developed for SAW filters are now also applied for BAW packaging [3].

In the past SAW components were exclusively soldered directly on the PCB of the mobile phone. Now more and more SAW filters are integrated into modules [4] [5]. A wide variety of modules are now used in mobile phones. Typical examples are:

- Filter banks integrating more than two filters and impedance matching.
- Frontend modules integrating filters, switching and matching mainly for GSM applications.
- Transceiver modules integrating the transceiver IC, filters and matching.
- PaiD modules integrating power amplifiers and duplexers.

Typically modules use a LTCC or a FR4 substrate. In LTCC substrates many passive components can be integrated. On the substrate SAW filters, other passives and semiconductor ICs are assembled. The module is then closed by a metal cap, a glob top or a molding process. For molding processes pressures up to 100 bar and temperatures up to 180°C are applied. So the cavity of the SAW filter package must be robust enough to withstand up to 10 N/mm² at 180°C. This "moldability" had to be added to the list of requirements for SAW components. This is not an easy task for miniaturized cavity packages.

Ways to integrate bare SAW chips into modules are described in [6] and [7]. As far as we know modules with bare SAW chips have not appeared on the market yet. Up to now packaged SAW components are integrated into modules.

Besides the moldability both small size and height are required when SAW components are used for modules.

The Start with Wire Bonding

First SAW components were packaged into hermetic metal packages. Bonded wires to the terminal pins connected the chip. With the introduction of Surface Mount Devices (SMD) ceramic packages with flat solder pads or lands were used. In Figure 2 such a ceramic cavity package is shown.

Figure 2: 3.8 x 3.8 mm² Ceramic SAW Filter Package. Top Open, Bottom Closed.

The ceramic package is essentially a box with the SAW chip bonded into the cavity using an adhesive. Wire bonding is used to create the electrical connection between the chip and the pads inside the package. These pads are connected by vias in the ceramic or metallized castellations with the SMD pads on the bottom side of the package. The box is closed with a metal cap by seam welding or soldering.

The Move to Flip Chip Packages

It is difficult to achieve package sizes below 3 x 3 mm² and a height below 1 mm with ceramic packages and bonded wires. So around the year 2000 the major manufacturers of SAW components switched over to flip chip packages. Besides the need to reduce the size, another reason for the move to flip chip packaging were performance limits caused by the inductivity of the bonding wires.

Most Japanese SAW component manufacturers continued to use ceramic "box" packages. They flipped the chip and used Au stud bumps instead of bonding wires [8].

At EPCOS a fundamentally new approach was applied to achieve smaller package sizes. This so called "Chip Sized SAW Package" (CSSP) will be presented in the next chapters. Since 2000 three generations of CSSP technologies have been

126

introduced on the market. All CSSP generations use a flat cofired ceramic interposer and solder balls for the electrical connection between the SAW chip and the interposer. To achieve low costs up to several thousand components are processed simultaneously by using large ceramic panels. At the end of the packaging process the components are singulated by a dicing process. The difference between the CSSP generations is mainly the way how the cavity between chip and substrate is closed.

CSSP1

Start of production of the first CSSP generation was in the year 2000. A schematic cross section is shown in Figure 3. The area on the chip where the acoustic surface waves are activated is protected by a polymeric cavity. This cavity is made on the wafer level using the proprietary PROTEC process. A ceramic interposer made of High Temperature Cofired Ceramic (HTCC) or Low Temperature Cofired Ceramic (LTCC) is used. The electrical connection between the chip and the interposer is realised by solder balls having a diameter of 200 µm. The solder is screen printed on the ceramic substrate. In the beginning only SnPb solder was used. Later also lead free SnAgCu solder was introduced. After the soldering of the chip an underfiller is deposited between chip and interposer. The PROTEC cavity keeps the underfiller away from the active area of the chip. Copper and Nickel metal layers then seal the package hermetically.

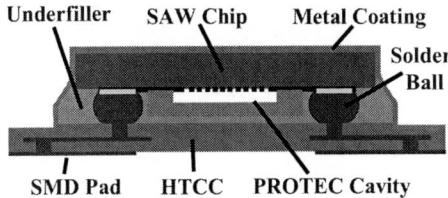

Figure 3: Schematic cross section of CSSP1

With CSSP1 technology the smallest package size is 2.0 x 2.0 mm². As a result this technology is now being phased out for mobile phone applications.

CSSP2

To achieve a smaller package size the second CSSP generation was developed [9]. With CSSP1 technology too much space on the chip was used for the PROTEC cap and the 200µm diameter solder balls.

The production of the second CSSP generation has been running since 2002. This technology is also named CSSPlus.

In Figure 4 a schematic cross section of CSSP2 is shown. The chip is soldered on a HTCC or LTCC substrate using lead free SnAgCu solder balls. The solder balls have a diameter of 100µm.

The Under Bump Metallization (UBM) on the chip side has a diameter of 125µm. Laminating a thin polymeric foil over the chips forms the cavity. A special process was developed to ensure a good adhesion of the foil and to achieve small tolerances of the foil geometry at the gap between chip and substrate. The foil is then removed at the edge of the component using a laser before the metallization of the package top. This is necessary to prevent that humidity can diffuse into the package. The top metallization is made by sputtering a plating base and plating of Cu and Ni layers. The components are separated by sawing the ceramic panel.

SAW chips have anisotropic thermal expansion. Cuts of Lithium Tantalate which are often used for mobile phone filters have a inplane thermal expansion of 8 ppm/K in x and 16 ppm/K in y direction. So the thermal expansion of a substrate with isotropic thermal expansion can never be matched to the chip. The solder bumps, the polymer foil and the top metallization of CSSP2 components are compliant so well above 500 temperature cycles are reached even for SAW duplexers where the length of the chip can be above 2 mm.

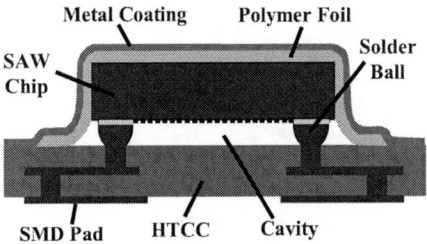

Figure 4: Schematic cross section of CSSP2

In figure 5 and 6 two examples of SAW components in CSSP2 technology are shown.

A 2in1 filter with a package size of 2.0 x 1.6 mm² can be seen in Figure 5. 2in1 filters integrate two band pass filters in one package. Compared with two CSSP2 single filters with a package size of 2.0 x 1.4 mm² significantly less space on the phone PCB is needed for one 2.0 x 1.6 mm² 2in1 filter.

The typical height of CSSP2 components on a 300µm HTCC substrate is 0.7 mm. With a wafer grinding process a maximum height of 0.6 mm can be achieved.

In figure 6 a 3.0 x 2.5 mm² duplexer is shown. For a duplexer two SAW or BAW filters and a matching network are necessary. The matching network can be integrated in a LTCC substrate. At EPCOS duplexers with one SAW chip and duplexers with two BAW chips are in production. Prototypes of a duplexer combining a BAW and a SAW chip for W-CDMA Band II were completed in February 2007 [10]. In addition to the CSSP2 process an adhesive foil is placed on top of the two chips. This foil is necessary if standard nozzles are used for the pick and place process. Standard pick and place nozzles have a vacuum hole in the center so they can not hold components with do not have a flat surface.

Figure 5: 2.0 x 1.6 mm² 2in1 Filter in CSSP2 Technology on HTCC Substrate

Figure 6: 3.0 x 2.5 mm² Duplexer in CSSP2 Technology on LTCC Substrate with one SAW and one BAW chip

CSSP3 non Hermetic

The third CSSP Generation started with a non-hermetic package. In figure 7 a schematic cross section is shown. As for CSSP1 and CSSP2 the chip is flip chip bonded to a HTCC interposer. SnAgCu solder balls are used for the electrical connection between chip and substrate. Compared to CSSP2 the diameter of the solder balls was reduced. The diameter of the UBM on the chip side is 90 μm.

The package is closed by a laminated polymer foil and a glob top. This material is applied without filling the cavity between chip and substrate. Humidity can diffuse trough the foil and the glob top so this package is not hermetic. To avoid corrosion of the Aluminium structures on the chip a very thin anorganic passivation layer is deposited on the chip.

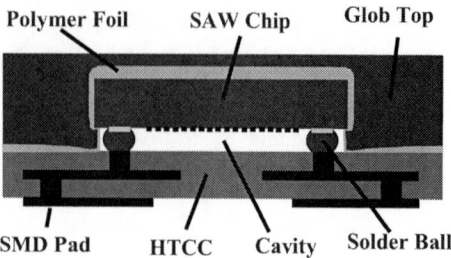

Figure 7: Schematic cross section of CSSP3 using glob top

With CSSP3 technology less space around the chip is necessary for the package. The distance between the chip edge and the component edge is only 200 μm. Due to the smaller UBM diameter a smaller chip size is possible. So a package size of 1.4 x 1.1 mm² could be achieved for single filters. In Figure 8 a filter in CSSP3 technology is shown. By using a chip grinding process and ceramic substrates with a thickness of only 150 μm a typical height of 0.4 mm was achieved.

Figure 8: 1.4 x 1.1 mm² Single Filter in CSSP3 Technology

CSSP3 Hermetic

In addition to the non-hermetic CSSP3 a hermetic third generation CSSP technology has been developed in the last two years.

For module applications thin SAW components which are stable in molding processes are required. Single and 2in1 Filters in CSSP2 and CSSP3 (non hermetic) technology are stable in molding processes up to 100 bar. For SAW duplexers which have larger chip sizes a new package development was necessary to reach stability up to 100 bars.

A schematic cross section of this package technology is shown in Figure 9. First a Copper frame and Copper pillars are patterned on the ceramic substrate. This is done by photolithography and electroplating processes.

The chip is then flip chip bonded to the substrate. After soldering the process is similar to the CSSP2 process sequence. A foil is laminated, the foil is removed on the Copper frame and a Copper/Nickel metallization is used to close the package hermetically.

The Copper frame and the Copper pillars give the necessary stability against the high pressure during a molding process. In addition, the copper frame leads to less tolerance of the foil geometry compared to CSSP2 technology. So a reduced distance from the chip edge to the component edge is possible.

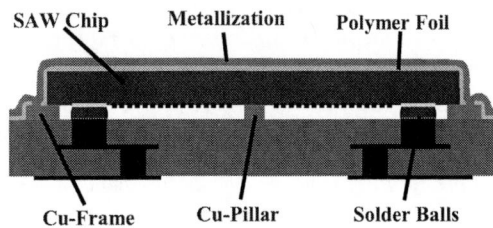

Figure 9: Schematic Cross Section of Hermetic CSSP3

Samples of duplexers with a package size of 2.5 x 2.0 mm² and 2in1 filters with a package size of 1.7 x 1.3 mm² have been made. Another technology for hermetic packaging of single and 2in1 SAW filters with similar size was published by Fujitsu [11] [12].

In Figure 10 a 2.5 x 2.0 mm² duplexer in CSSP3 technology is shown. With a HTCC substrate the maximum height is 0.5 mm. With its small size and low height it is well suited for the integration into modules.

Figure 10: 2.5 x 2.0 mm² Duplexer in CSSP3 Technology with Cu-Frame

Conclusion

Driven by the high demand of the mobile phone industry advanced cavity package technologies have been developed for SAW components. They offer low cost, high reliability and small

size. For the integration into modules moldability was also achieved.

References

[1] C. Ruppel, L. Reindl, R. Weigel, "SAW Devices and their Wireless Communications Applications", IEEE microwave magazine, June 2002, pp 65-71

[2] P. Selmeier, R. Grünwald, A. Przadka, H. Krüger, G. Feiertag and C. Ruppel, "Recent Advances in SAW Packaging", Proc. 2001 IEEE Ultrasonic Symp., pp 283-291

[3] S. Marksteiner, M. Handtmann, H.-J. Timme, R. Aigner, R. Welzer, J. Portmann, U. Bauernschmitt, "A Miniature BAW Duplexer using Flip-Chip on LTCC", Proc. 2003 IEEE Ultrasonic Symp.

[4] P. Hagn, A. Przadka, A. Leidl, S. Seitz, C. Ruppel, "Acoustic Frontend Modules", Frequenz 59 (2005) 1-2, pp 18-23

[5] U. Bauernschmitt, C. Block, P. Hagn, G. Kovacs, E. Leitschak, A. Przadka, C. Ruppel, "RF Front-Ends For Multi-Mode, Multi-Band Cellular Phones", Third International Symposium on Acoustic Wave Devices, Ciba, 2007

[6] M. Ha, J. Lee, Y. Kwon, "Chip Scale Package for SAW Filter on the Oxidized Porous Silicon using Flip-chip Bonding and Cu plated Metal Wall", Proc. IEEE Electr. Components and Technology Conference 2002, pp 372-377

[7] V. Georgel, F. Verjus, E. Grunsven, P. Poulichet, G. Lissorgues, S. Chamaly, "Integration of SAW filter on PICS substrate using polymer sealing", IEEE, 2006, P. D. Research in Microelectronics and Electronics, pp 421-424

[8] H. Yatsuda, T. Horishima, T. Eimura, T. Ooiwa, "Miniaturized SAW Filters Using a Flip-Chip Technique", IEEE Trans. on Ultrasonics Vol. 43, No. 1, January 1996, pp. 125-130

[9] G. Feiertag, H. Krüger, P. Selmeier., "Advanced Packages for Surface Acoustic Wave Components", Proc. of Micro System Technology Conference, Munich, 2003

[10] S. Marksteiner, D. Ritter, E. Schmidhammer, M. Schmiedgen, T. Metzger, "Hybrid SAW/BAW System–in-Package Integration for Mode-Converting Duplexers", Third International Symposium on Acoustic Wave Devices, Ciba, 2007

[11] O. Kawachi, K Sakinada, Y. Kaneda, S. Ono, "Packaging of SAW Devices with Small, Low Profile and Hermetic Performance", Proc. 2006 IEEE Ultrasonic Symp., Vol. 27, pp 281-282

[12] O. Ikata, Y. Kaneda, S. Ono, K. Sakinada, O. Kawachi, Y. Tanimoto, "Miniaturized SAW Package with Hermetic Performance", Third International Symposium on Acoustic Wave Devices, Ciba, 2007

Low Cost Hybrid Integration of Laser Diode on Silicon PLC for Optoelectronic Applications

Luca Maggi, Arturo Canali, Danilo Caccioli, Gianni Preve, Stefano Lorenzotti

Pirelli Labs SpA, Optical Innovation, Viale Sarca 222 – 20126 Milano- Italy

Phone +39 02 6442 9210, Fax +39 02 6442 5522, E-mail luca.maggi@pirelli.com

Abstract

This paper describes cost effective passive alignment of a standard Fabry-Perot laser in front of a high Δn waveguide on SiOB PLC using flip chip technology. The present approach try to differentiate from other studies on the fact that it is based on the use of standard technology (scalable to wafer level), avoiding the need for complicate and expensive process steps. An high index contrast ($\Delta n = 2.5\%$) PLC has been used. The waveguide has been properly tapered and the laser diode is placed in a "hole" realized by etching the substrate. Simple alignment markers have been implemented in the PLC. The performances of the laser diodes have been evaluated, a proper assembly process has been developed, an appropriate set up of measurements has been realized and some optical simulations of the system have been performed. The comparison between experimental results and simulation is presented. The first results indicate that a process within 1-1.5 µm alignment precision and a good coupling loss has been reached. The paper finally describes possible upgrades and developments.

Key words: Low Cost, Passive Alignment, Flip Chip

Introduction

In recent years the demand of access bandwidth is continuously growing. It's generally accepted that The Fiber To The Home (FTTH) or Fiber To The Premise (FTTP) can satisfy all the present and future requirements. However, despite its unique bandwidth capability, the introduction of optical fiber is limited by the cost of optoelectronic components. Most of the costs are due to packaging and mainly to the alignment process [1]. In order to realize a "low cost" optoelectronic component the main driving forces are integration and passive alignment [2-5]. By hybrid integration the active optical devices are mounted on Planar Lightwave Circuit (PLC), which can perform some passive optical operations (filtering, adding, dropping etc). The cost of alignment a laser in front of a waveguide can be drastically reduced using a passive approach, thus avoiding the need of biasing and monitoring the laser performances during the alignment. The passive alignment requires a very precise die placement and proper designed substrate for achieving the desired coupling efficiency.

In the present paper we describe our approach to low cost hybrid integrated optoelectronic component based on passive alignment of a standard laser in front of a waveguide using flip chip technology. Following the effort toward real low cost component, we decide to use only commercially available laser diodes and standard technology for PLC.

The sources are Fabry-Perot laser diodes operating at 1310 nm. They are designed for emitting high output power and can be used in uncooled applications. The laser diodes are suitable for FTTH applications and are compliant with ITU recommendations. [6]

The PLC is based on silica on silicon platform with high index contrast ($\Delta n = 2.5\%$). For a better coupling efficiency the waveguide has been properly tapered [7-8].

The alignment process has been carried out using high precision flip chip equipment.

The present measurements have been performed on different chip sets, each of them was customized according to the laser diode. However the present approach is scalable to wafer level, so reducing furthermore the cost.

We have developed a complete theoretical model, describing the whole process. The optical and thermo-mechanical aspects of the system have been investigated using different software.

We described the basic structure of PLC in section II, and the process in section III. In section IV we discuss the experimental set up, while the results are shown in section V. Finally in section VI, we provide conclusions and recommendations.

Basic Structure

The Test Vehicles are based on silica on silicon platform and are developed using the typical processes of integrated photonics [9]. Basically they implement and optical waveguide, a trench in which the laser diode is allocated and proper markers for alignment. On the bottom of the trench there are some metal pads over which eutectic AuSn 80/20 solder is deposited. The waveguide is properly tailored so that the optical coupling is optimized.

The process flow is the following one:

- Start with lower cladding deposited on Si Bulk
- Core deposit and etch
- Upper cladding deposit
- Dry etch of the trench
- Lift off of the metal
- Evaporation of eutectic solder

In the following picture a draft of the process flow diagram is summarized.

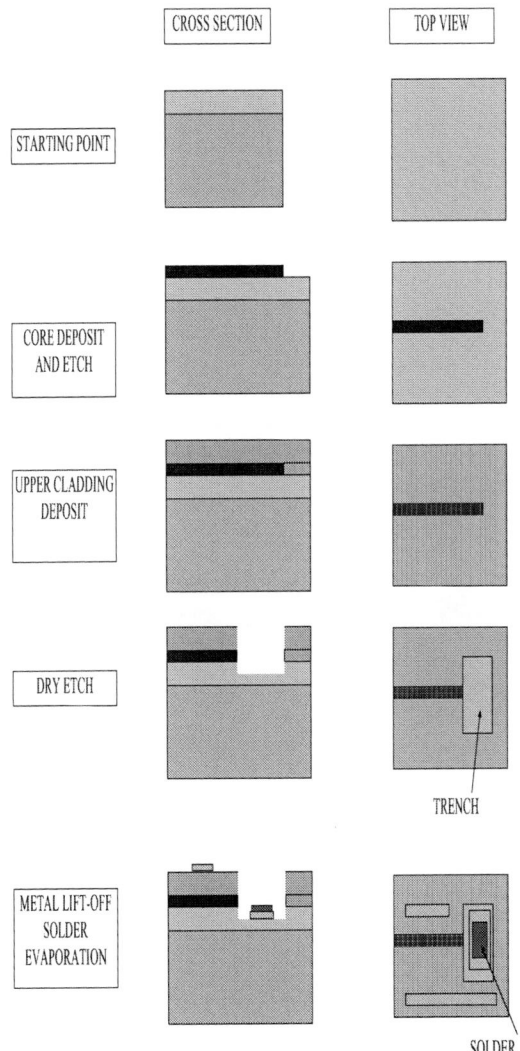

Figure 1: Process Flow Diagram

The depth of the dry etch is calibrated considering the thickness of the evaporated solder so that, as a final results, the laser ridge is at the same level of the core. Thanks to such a tailored process is

possible to minimize or avoid any placement correction along the vertical axis (Z axis). Of course each structure must be customized according to the laser diode that will be placed in. We have developed the proper designs and have selected the right process parameter for achieving such a goal. The metal in the trench is patterned so that it can be used for aligning the laser along horizontal direction (X and Y axis, being the X axis the optical one). According to optical simulations the position along optical axis is less critical. The crucial placement is along the Y axis, where the tolerances are tight.

Furthermore some proper markers are added to simplify the pattern recognition and the die placement.

The laser diodes are firstly flipped chip and then placed in the trench. Flip chip permits to work only with very well determined edges and thickness. In fact the position of the ridge (emitting channel) respect to diodes top surface is well known and it's affected only by epitaxial growth tolerances, which are absolutely negligible. Moreover the position of laser metal patterns respect to emitting channel can be determined within lithographic errors. So using the metal pattern on laser chip as first marker and the metal at the bottom of the trench as the second marker, a precise placement can be reached. A repeatable and well controlled process is mandatory for achieving the required placement tolerances.

Process Description

For placing the laser into the trench we use a precise flip chip bonder by Suss. The Suss Microtec FC150 flip chip bonder is a precision instrument used to align and bond one or more chips onto a substrate using pressure and heat. The current configuration of our system allows the following "flip chip" techniques: solder reflow and z- leveling bonding of different die-size to substrates.

Operation of the FC150 can be divided into four different basic steps: loading, alignment, bonding, and unloading. Loading and unloading are controlled by the pre-programmed information contained in the template libraries.

We developed internally all the alignment procedures and we implemented them in software programs. A routine using a pattern recognition system has been also written. An overall control of the alignment process has been achieved.

In our equipment the motorized alignment stage is upgraded with high resolution X/Y stage which has a resolution of 0.1 micron and a minimum step in rotation of 9 mrad, with a standard range of ± 7°.

The Reflow Arm controls the vertical (z-axis) motion of the upper part, and the rotations about the x and y-axes (pitch and roll). The minimum step in the z direction is 0.5 microns while the minimum step for pitch and roll is 0.05 mrad with a range of ± 0.6°.

The FC150 uses a set of optical systems, shown in Figure 1, to align the parts in all six degrees of freedom.

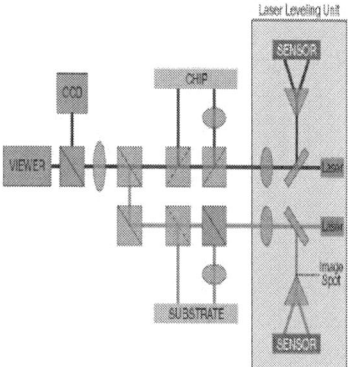

Figure 2: FC150 Alignment System

The primary alignment system on the FC150 is a bi-directional microscope used for x and y alignment. The bi-directional microscope is positioned between the parts before aligning, and withdrawn before bonding. It allows simultaneously viewing both the upper and the lower parts, while positioning them in x and y.

Theta alignment requires aligning two pairs of alignment marks. To maximize theta alignment accuracy, the distance between the two pairs of alignment marks should be as large as possible.

The FC150 offers two alignment systems for leveling (pitch and roll) alignment. The first is an autocollimator, and the second is a laser leveling system. We use both of them in our application, to reach a perfect parallelism between laser diode and substrate.

The coarse alignment of the laser to the waveguide has been performed collimating the gold metallization of the laser diode and one corner of the bump integrated on substrate by using an automatic pattern recognition system.

Patterns need to be trained by specifying a pattern image, a pattern origin and training parameters. Once you have trained them, you can search for them in image.

The system aligns the two parts till 0.1 micron tolerance value is reached. The final alignment of the laser ridge to optical waveguide was performed by using the manual alignment system, controlling the position of the substrate to the laser in X and Y axes.

The designs of the substrate and the process for realizing it have been tailored so that it was possible to precisely determine the position of the waveguide. This reference is used for achieving the fine alignment.

Using high resolution movement of the chuck (0.1 micron) and checking the alignment in the monitor (magnification 400 X) the ridge of the laser diode has been aligned to the optical waveguide on the substrate.

The distance between the edge of the laser and the optical waveguide was set to 10 micron according to optical simulations.

We performed the alignment process heating both chuck and arm at 270 °C in order to minimize the thermal expansion mismatch between the laser and substrate materials. Once the parts are fully aligned they are soldered applying heat and force. During the soldering process we controlled the Z position and can compensate any unwanted misalignment

In our test vehicle the solder was evaporated, but it can be eventually deposited onto components by sputtering, electroplating, or as solder preforms. [10-11]

The solder used for our test vehicle is Gold Tin (80-20 weight by weight) evaporated eutectic alloy.

Our bonding process consists in a rapid thermal process with Z level control. The chuck and the arm are both heated up to 290°C in 5 seconds and the arm place the laser at Z= 0 micron after 20 seconds. After 23 seconds starts the cooling of both arm and chuck in order to freeze the solder. The force is maintained as the temperature reach 260°C at this point it is released and the bonding of the laser is ready. The samples are wire bonded by using standard gold ball bonding process for electrical connection.

After bonding the structures of the test vehicle is as visible in Figure 3.

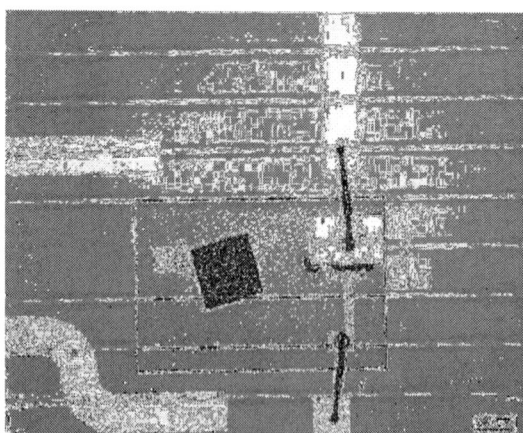

Figure 3: Laser Diode Soldered and Wire Bonded

Experimental Set up

Before evaluating the quality of our alignment process, a characterization of the laser was performed. In this step we measured the out put power of a flip chip mounted laser diode and, for comparison, the output power of a standard attached laser diode. In this way we could optimize the flip chip process regardless the optical coupling. Then we evaluated the precision of passive alignment.

The performed test and set up are the following ones:

Laser diode front facet output power efficiency (LDP$_{out}$) evaluation. The test is performed by using a free space optical bench. The method enable the comparison between several samples mounted using the flip chip technology and the standard technology. The characterization bench acquires the laser diode output power versus laser diode current (LDI) at different temperatures.

This test is carried out using a properly designed Silicon Optical Bench (SiOB) as laser substrate.

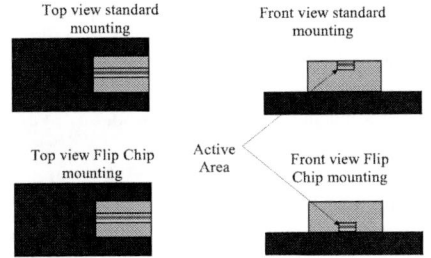

Figure 4: Laser Diode Front Facet Output Power Test Vehicles

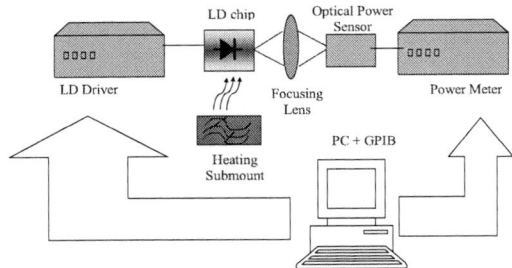

Figure 5: Optical Set up for First Method

Laser diode coupling efficiency. The test is performed by using a pigtailing bench and a fiber array of single mode fiber (SMF). The method enables the coupling efficiency analysis between several laser diodes flip chip mounted and than passively aligned in front an integrated waveguide. The characterization bench acquires the optical power coupled into a standard fiber array with single mode fiber.

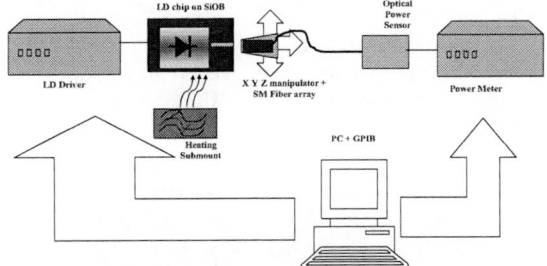

Figure 6: Optical Set up for Second Method

Results

The output power has been selected as the reference parameter in order to evaluate the performances of the assembled modules. A diode laser mounted p-side up on silicon was considered as "gold sample" or better as a reference target. This reference was called "laser pill".

The IL curve of such a device is represented hereafter:

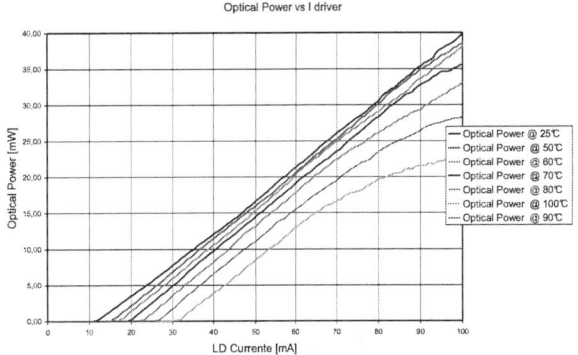

Figure 7: IL Curve for Standard Laser Pill

As it can be seen, the laser is extremely linear and the output power is large also at high temperature.

As described in previous section, a laser diode, mounted p-side down on the same substrate, has been used to verify if any intrinsic limitation is present in the flip chip technology.

The IL curve for flip chip laser pill is the following one:

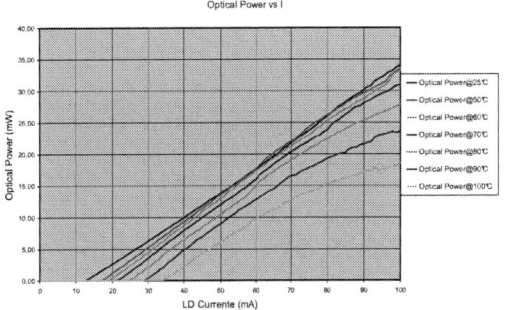

Figure 8: LI Curve for Flip Chip Laser Pill

It seems that the flip chip component works as well as the standard one. This assertion is confirmed by a direct comparison of the performances of the two technologies.

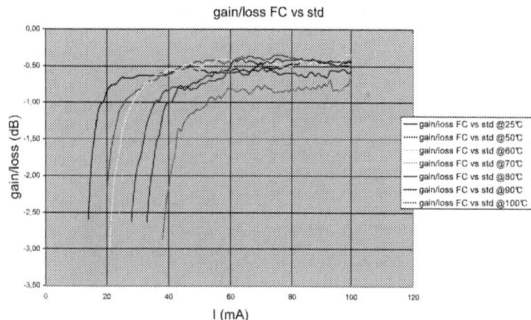

Figure 9: comparison of flip chip and standard die attach

The minimal differences in threshold (4 mA) and in output power (-0.5 dB) are only statistical errors. If an intrinsic technological limitation should be present, the performances would have been affected more dramatically.

After this test, a laser diode was mounted in flip chip configuration on passive substrate with integrated waveguide.

Measuring the output power and normalizing it respect the case without waveguide it's possible to determine the losses due to the misalignment.

The output power is shown in the following figure.

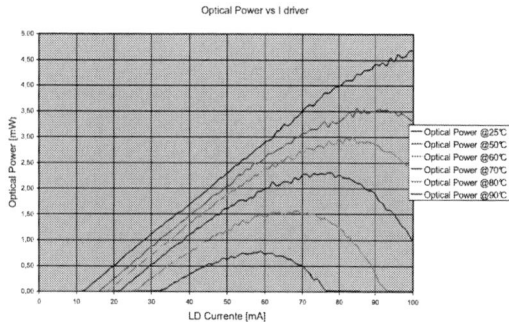

Figure 10: LI Curve for Passive Aligned Flip Chip Laser

The coupling losses are represented hereafter.

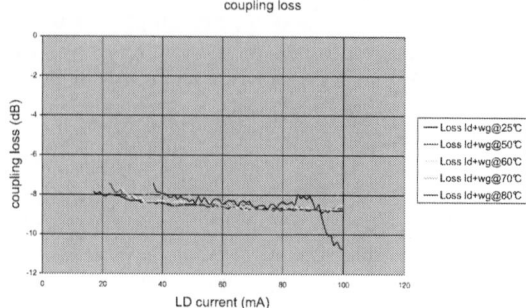

Figure 11: Coupling Loss

Over the threshold the coupling loss is constant over the whole temperature and current range.

This means that the only misalignment which is present is that of the die attachment process.

No further thermal or mechanical misalignment occurs.

The coupling loss is -8.5 dB.

Comparing the experimental value of coupling loss with the predictions of our theoretical model it is possible to evaluate the global misalignment. The model used in optical simulations was developed and optimized fitting the data of active alignment between laser diodes and waveguide.

The results of the optical simulations are shown in the following figure:

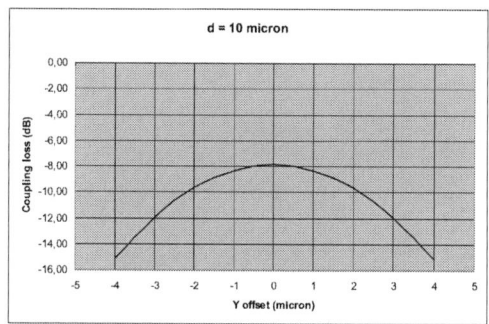

Figure 12: Optical Simulations

The experimental measured coupling loss (-8.5 dB) corresponds to a global misalignment of 1-1.5 µm.

This extremely good placement precision makes passive alignment of a laser diode suitable for telecom applications based on single mode waveguide.

Most of the optical loss is due to the modal mismatch rather than misalignment due to passive process. For improving the optical coupling it's possible to modify the Δn or the shape of the taper or using a more symmetric laser beam. Any changes in Δn or in taper dimension lead to a complicate lithographic technology and front end process. This complexity increases the cost of the substrate and so can cancel the advantages of passive alignment approach. Using a laser diode characterized by symmetric beam reduces the modal mismatch without any modification in realization and alignment process. [12-13]

If a laser diode with Spot Size Converter (SSC) is used the coupling loss can be improved by 2 or 3 dB.

This was verified mounting some engineered samples of a laser with SSC. In that case the coupling losses are about 6 dB. These devices are in their development phase and hopefully become commercially available in short time.

Conclusions

The results of detailed study of passive alignment of a flip chip laser diode in front a waveguide have been presented.

The equivalence of standard attach technique and flip chip technology was demonstrated for a laser chip.

A complete theoretical and predictive model of the passive alignment has been carried out. Using such a model, the layout of test structure has been designed.

A proper passive alignment process has been realized and finally a global tolerance of ± 1- 1.5 μm was reached. This precision makes passive alignment of a laser diode suitable for telecom applications based on single mode waveguide.

Finally some possible improvements using a SSC laser have been analyzed.

References

[1] H. Hentzell, M. Olson, G. Arvidsson, "Optical Technology – The Next Innovations", Proceedings of the 14th European Microelectronics and Package Conference (EMPC), Friedrichshafen, Germany, June 23-25, pp 382-387, 2003

[2] H. Koh, M. Choi, H. Chun, J. Song, J. Bu, "On-wafer Process for Mass Production of Hybridly Integrated Optical Components Using Passive Alignment on Silicon Motherboard", Proceedings of 49th Electronic Components and Technology Conference (ECTC), San Diego, CA, USA, June 1-4, 1999

[3] T. Mitze et al. "Hybrid Integration of III/V Lasers on a Silicon-On-Insulator (SOI) Optical Board", 2nd IEEE Conference on Group IV Photonics, Antwerpen, Belgium, September 21-23, pp 210-212, 2005

[4] H. Okano et al., " Hybrid Integrated Optical WDM Transceiver Module for FTTH Systems", Hitachi Cable Review, No 17, August, 1998

[5] H. Imam, J.P. Rasmussen, M. Pearson, "Integrated Bi-directional Transceiver for Access Applications Based on a Cost Effective PLC Hybridized Platform", Optoelectronic

Integrated Circuits VIII, Proceedings of the SPIE, volume 6124, pp 321-330, 2006

[6] ITU G 983, IEEE 802.3ah and ITU G984

[7] Y. Shani, C.H. Henry, R.C. Kistler, R.F. Kazarinov, K.J. Orlowsky, "Integrated Optic Adiabatic Devices on Silicon", IEEE J. Quantum Electron., vol. 27, pp 556-566, 1991

[8] Y. Shani, C.H. Henry, R.C. Klister, K.J. Orlowsky, D.A. Ackermann, "Efficient Coupling of a Semiconductor Laser to an Optical Fiber by Means of a Tapered Waveguide on Silicon", Appl. Phys. Lett., vol. 55, pp. 2389-2391, 1989

[9] M. Salib et al. "Silicon Photonics", Intel Technology Journal, vol. 87, pp. 143-160, 2004

[10] J. Wei, "A New Approach of Creating AuSn Solder Bumps from Electroplating", Cryst. Res. Technol., vol 41, No 2, pp. 150-153, 2006

[11] J. Yoon, H. Chun, J. Koo, S. Jung, "AuSn Flip Chip Solder Bump for Microelectronic and Optoelectronic Applications", Proceedings of Symposium on Design, test, Integration and Packaging of MEMS/MOEMS (DTIP), Stresa, Italy, April 26-28, 2006

[12] K. Kawano et al. "Design of a Spot Size Converter Integrated Laser Diode" IEEE Journ. of Selected Topics in Quantum Electronics, vol. 2, No 2, pp 348-354, 1996

[13] K. Yokoyama et al. "Design and Analysis of High Coupling Efficiency Spot Size Converter Integrated Laser Diode by Three Dimensional BPM", Electron. Comm. Jpn., vol. 81, No 12, pp. 1-9, 1999

Packaging of miniaturized optical encoder

Y. Jourlin[1], O. Parriaux[1], K. Keränen[2], J.T. Makinen[2], M. Karppinen[2], P. Karioja[2],
M. Johnson[3]

[1]Laboratoire Hubert Curien (formerly TSI) – UMR CNRS 5516
18 rue du Prof. Benoit Lauras, 42000 Saint Etienne, France
yves.jourlin@univ-st-etienne.fr, Tel : +33 477 91 58 21

[2]VTT Technical Research Centre of Finland
Kaitoväylä, Oulu, P.O. Box 1100, FI-90571 Oulu, Finland

[3]Instrumentation Design Ltd.
7 Dorset Road, Altrincham, Cheshire, WA14 4QN, UK

Abstract

A displacement measurement technology of diffractive interferometry has been developed in the objective of miniaturization. The interference principle has been implemented in the form of a miniature read head prototype according to a novel device configuration. The read head developed at TSI can be miniaturized utilizing 3D hybrid integration and packaging techniques. Integration leads to a very small head comprising a light source (LED or VCSEL), a detector array and needed optical components such as lens, gratings and mirrors. A prototype of hybrid integration read head will be implemented after choosing critical components, such as, the source and detectors and designing the read head in detail.

Key words: High resolution optical encoders, optoelectronics packaging

Introduction

Incremental optical encoders consisting of a read head that moves along a periodic diffraction grating scale find more and more applications in metrology and manufacturing automation [1] . Not satisfied with short scales and output pulses every one or two microns, users now demand total scale lengths of many hundreds of millimeters and interpolation of the fundamental scale period by 1000 or even 10000 times. Even if absolute location to 100 pm is not required, resolution at this level is advantageous when the sensor is used in a high bandwidth feedback-controlled positioning system. Further, high speed position control requires small size and low mass of the moving read-head.

We have developed a new, high-resolution encoder configuration which offers many advantages for these applications. We present in this paper the integrated head configuration. A wireless configuration has been developed yet [2].

It is sensitive, generating an electrical output with a spatial period of one quarter of the scale period. It is wavelength-insensitive, allowing the use of multimode lasers or even an LED source, and its simple optical path allows in principle significant miniaturization. However, this is limited by current fabrication techniques because of the disparate combination of sources, multiple detectors, gratings, mirrors, and lenses in a single package. Significant reduction in size will only come from a radical approach to micro-assembly, and 3D hybrid integration and packaging techniques offer a very attractive route to a new palette of devices.

Optical Encoder description

A typical encoder read head consists of a VCSEL chip, a collimation lens, a small area of "swapping" grating and an array of detectors (see figure 1). Light from the source is split at the scale grating into two diffraction orders. These are differently diffracted at the swapping grating and then recombined in further diffraction events at the scale grating. The combined beams interfere, in a configuration resembling a Mach-Zehnder which has been twisted to cross the usually parallel arms. In the perfectly aligned case the output beams' overlap is perfect and largely independent of wavelength. The interferometer is achromatic. With a modest-resolution 4μm scale and a 2μm swapping grating the output intensity cycles from dark to bright and back again for a 1μm relative movement of the two gratings.

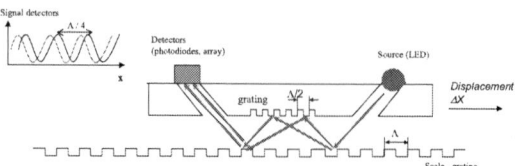

Figure 1: sensor principle

For any real application other than as a unidirectional tachometer, two detector outputs are needed, typically chosen as sin and cos of the mechanical phase angle between the two grating structures. This can be arranged in a number of ways, one of which is described below.

Interpolation of the 1μm base period requires signal processing of a pure sine-wave intensity output from the detectors as show in the figure 2.

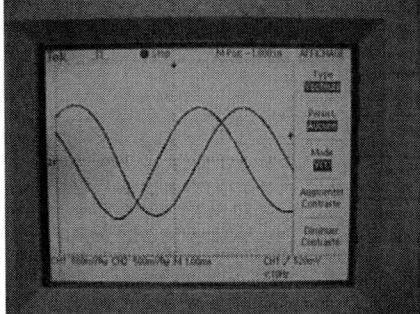

Figure 2: sin/cos signals

For instance, using the arctan function on the ratio of the sin and cos signals gives a continuous measure of phase angle. The degree of useful interpolation is limited essentially by the signal-to-noise ratio of the two analog signals. The linearity of that interpolation (the constancy of true period of the interpolated steps) is further a function of the purity of the sin/cos signals. For interpolation by a factor 1000 the harmonic distortion levels of the signals must also be of the order of 1/1000. With scattered light present in any real sensor, this is a stringent demand. Nevertheless the balanced design of our encoder can deliver this. The key is to deliberately misalign the two gratings to give an output fringe system projected in space, which can then be sampled by a suitable array of detectors. Adjusting the known misalignment is no more, no less critical than aligning for perfect output beam overlap and a uniform light field.

We have fabricated this and similar structures using conventional high-precision machining techniques, and shown excellent performance, albeit in a small but not miniature package. As shown, in figure 3, where all the elements are separated: source swapping grating and detectors.

Figure 3: optical encoder demonstrator

In order to fully miniaturize the optical encoder demonstrator we now have to master, 3D micro-assembly has to be mastered. The first step in the design is to generate the available list of bare-die or low-level packaged components. In the case of laser diodes, there is a further choice of edge- or surface-emitting geometries. Either can be more convenient depending on the overall configuration.

The second step is a detailed analysis of the tolerances allowed for each component and for each of its six degrees of freedom. In a single design tolerances can easily span a 1000:1 range. In the case of a laser collimator with a short focal length lens small transverse errors can lead to drastic pointing errors. In some cases the combination of source and lens should be pre-assembled and tested as a module, for integration in a later step. What alignment accuracy should be provided by the component itself, what by the subsequent handling? A single-mode optical fiber with ±0.5μm tolerances on outer diameter and centration already does much of the difficult work. Insertion into a precision V-groove or ceramic ferrule perhaps reduces adjustment to a single degree of freedom, the axial position. This is orders of magnitude simpler than grasping the fiber and aligning it with a motorized hexapod.

In the present encoder it may not even be sufficient to align the system alone. As the swapping and scale gratings have to be *mutually* aligned, a "gold-standard" scale will probably be included in the assembly jigging. Alignment will hence be performed with active feedback from operation of the full encoder.

Lastly the fixing of every component must be considered. In some cases the adjustability of a heated solder joint will allow enough ability to align some elements. In general however, separation of the functions of adjustment and fixing will be preferable. As component sizes decrease, liquid surface tension plays an ever bigger role in alignment. Surface mount electronics and properly designed printed circuit pads make use of this effect for auto-alignment. Whether this is useful for the next factor ten of size reduction remains to be seen. One of our encoder configurations is exceptional in that it uses optical fibers for source, detector-array or both. While the tolerances demanded are similar to those of opto-electronic components, the presence of meter-length fibers or even bundles add the complication of how to handle the free ends. Again, analysis of the mundane aspects of jigging and fiber/cable handling will greatly ease the job of achieving high precision.

Optoelectronics packaging

For photonic integration, the following issues must be considered [3]:
- Devices alignment with micron accuracy by the use of 3D passive alignment structures.
- Integration of high-speed/low-noise/high-power electronics as close as possible to the critical photonic devices, such as, a diode laser, optical modulator, photo detector or MEMS device.
- Matching of the thermal expansion coefficients between the devices and substrate.
- Provide required internal and external optical, electrical and mechanical interconnects.
- Provide required thermal management for critical power devices, such as, high-power lasers, laser drivers and amplifiers.
- Protect devices against environmental conditions during the use and storage of the modules.
- Provide local hermetic sealing of critical devices, such as, diode lasers.

These requirements must be fulfilled by the most cost-efficient means in high- volume production with good production yield and reliability.

For optical sensor integration, the capabilities of Low Temperature co-fired Ceramics, LTCC, technique to form 3D precision alignment structures for passive device alignment has proven out to be strong enabler. In the LTCC process, open cavities can be made in the LTCC substrate. The precision structures can be used to assemble discrete semiconductor devices into the cavities allowing local hermetic sealing of devices. U-grooves can be processed using punching and photolithographic processes on the LTCC substrate. The groove structure provides precise alignment and firm attachment for the adhesive bonded optical fibers and other optical elements.

The hermetic fiber pigtailed diode laser module shown in Figure 3 is a good example of LTCC based 3D photonic integration [4]. The 3D structure of the substrate was fabricated by the use of conventional via punching technique to create the necessary alignment structures including holes, grooves and cavities. Packaging was carried out in two steps: firstly, device assembly and secondly,

hermetic sealing. Device assembly included the assembly of the diode laser, Kovar frame and fiber onto the LTCC carrier. Hermetic sealing included preparing the hermetic fiber feed-through and laser welding of a sealing the lid on the module.

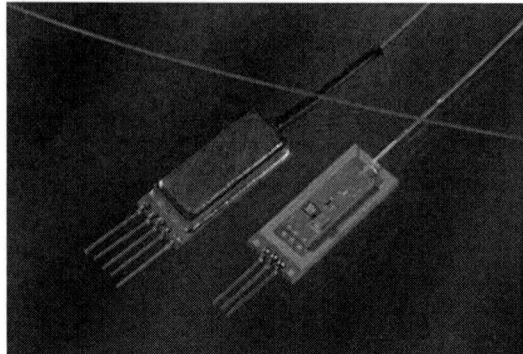

Figure 4: Encapsulated Hermetic Fiber Pigtailed Laser Module and an open Subassembly.

For cost-effective optoelectronics integration, injection molding provides a very powerful tool to add optical functions on active devices on electronic substrates. Injection molding capabilities are depicted in Figure 5 showing a VCSEL based illuminator module fabricated by the use of an inmold integration technique [5]. The module was successfully injection molded using thermoplastic Lexan 123 polyacryle material. The operation of the encapsulated device was tested immediately after the surface temperature was decreased to the room temperature. An immediate observation target in the test runs was to follow the device failure dependency on the used holding pressure in the injection molding. Both bare and glob-top shielded devices were used in the test processing. The most typical cause for module failure was loosening of the bonding wire during the injection molding process.

Figure 5: Inmold Integrated VCSEL Illuminator Module

Injection molding allows the fabrication of precision optical elements. In many cases, however, mold modifications combined with processing parameter iterations are required in order to attain the designed surface profile and the dimensional accuracy of the optical elements. Figure 5 shows an imaging macrolens integrated with illuminating LEDs manufactured by the use of inmold integration technique [6]. Several series of macrolenses were injection molded with Plexiglas PMMA material type 6N. Several series were needed in order to find the optimal injection molding parameters that produced good quality optical structures without any air bubbles and other typical defects associated with injection molded parts. The set of parameters included the temperature of the mold, injection speed and holding pressure.

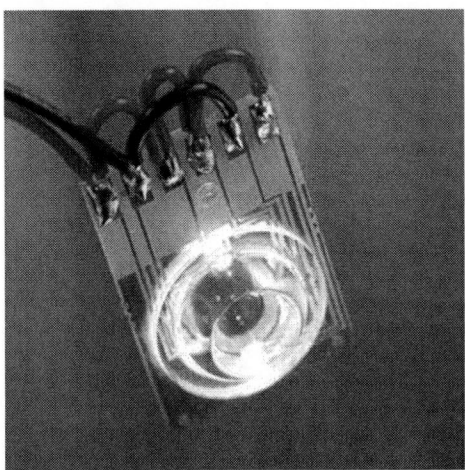

Figure 6: Injection Molded Macrolenses with Inmold Integrated Illuminating LEDs

The use of bare active devices and precision optics integration in injection molding process enable extremely miniature and high performance module implementations. These demonstrators of 3D photonics integration based on LTCC and injection molding technologies provide us a cost-efficient tools for effective optical encoder miniaturization and packaging.

Conclusion

TSI and VTT plan to fabricate a demonstrator in the next months in the framework of NEMO by submitting an application oriented project. The objectives of the demonstrator is an integrated head whose size is below 2 cm^3 in order to offer a miniaturized and high resolution optical encoder (1 nm resolution) which is needed in microelectronics equipment, metrology and high precision mechanics. Industrial partners in the field of optical encoders or electro-mechanical modules manufacturers (translation stages, gauges...) who want to introduce a non-intrusive measurement function in their modules are welcome to pursue and strengthen the project.

Acknowledgements

The authors acknowledge the European Network of Excellence in MicroOptics (NEMO, http://www.micro-optics.org/) for its support in the read head development.

References

[1] Spies A. "Linear and angular encoders for the high-resolution range". Progress in Precision engineering and Nanotechnology. Proceedings of the 9th International Precision Engineering Seminar, Germany, 26-30 May 1997.

[2] Y. Jourlin, O. Parriaux, S. Reynaud, J. C. Pommier, M. Johnson, A. Last, and M. Guttmann "A new wireless and miniaturized high-resolution optical displacement sensor", Proc. SPIE Vol. 6188, 61881B (Apr. 28, 2006).

[3] K. Keränen et al., "Cost-efficient hermetic fibre pigtailed laser module utilizing passive device alignment on an LTCC substrate". Proceedings of SPIE - The International Society for Optical Engineering, Vol. 6185, Strasbourg, France, 3.– 7.4. 2006.

[4] P. Karioja et al., "LTCC toolbox for photonics integration", Proc. of Ceramic Interconnect and Ceramic Microsystems Technologies (CICMT 2006), Denver, CO, USA, April 24–27, 2006.

[5] J-T. Mäkinen et al., "Inmould integration of a microscope add-on system to a 1.3 Mpix camera phone", Proceedings of SPIE, Vol. 6585, 2007.

[6] K. Keränen et al., "Injection moulding integration of a red VCSEL illuminator module for a hologram reader sensor", Proceedings of SPIE, Vol. 6585, 2007.

High Resolution Optical Component Technology for Advanced Novel Encoder (Hi-OCTANE)

John Carr[1,2*], Marc P.Y. Desmulliez[1], Eitan Abraham[1], Nick Weston[2], David McKendrick[2], Matt Kidd[2], Geoff McFarland[2], Wyn Meredith[3], Andrew McKee[3], Conrad Langton[3]

[1]Heriot-Watt University, School of Engineering and Physical Sciences, Microsystems Engineering Centre, Riccarton, Edinburgh, EH14 4AS, UK

[2]Renishaw Plc, Heriot-Watt Research Park, Riccarton, Edinburgh, UK

[3]CST Global Ltd, Block 7, Kelvin Campus, West of Scotland Science Park, Maryhill Road, Glasgow, G20 0TH, UK,

Tel: +44 (0)131 451 3774, E-mail: jpc1@hw.ac.uk

Abstract

Today's optical encoders comprise several discrete components: a photodiode array, an emitting light source as well as reference and analyzer diffraction gratings., The critical alignment requirements between the optical gratings and the photodiode array, the bulky nature of the encoder devices and the subsequent packaging mean that optical encoders can be prohibitively expensive for some applications and unsuitable for others. This paper reports a novel optical encoder integrated design for high resolution systems. The components of the micro-engineered encoder are monolithically integrated onto a single compound semiconductor chip radically reducing the footprint and cost of assembly. The new photodiode configuration not only facilitates increased performance but also improves the alignment tolerances. Fabrication of the optical gratings at the wafer level allows the alignment to the photodiodes in a single process step removing the need for individual alignment thereby dramatically reducing assembly costs. Introducing a back emitting light source allows all electrical contacts to be located on the same side of the encoder chip ensuring no obtrusion in the optical path. The chip is to be flip-chip bonded onto a substrate in order to reduce cabling costs by time multiplexing the output signals to the downstream interface control unit.

Key words: Optical Encoder, Monolithic Integration, Photodiode Array, Assembly and Packaging.

Introduction

The classic configuration of an opto-electronic encoder consists of a patterned disc also called scale and a readhead that can be moved relative to the scale to convert a mechanical position into a representative electrical signal. The readhead includes (1) a light source for producing light rays incident on the scale, (2) an index grating for diffracting readable rays into fringes in at least one order of diffraction, (3) an analyser grating for converting these fringes into light intensity modulation patterns at a rate which is a function of the displacement between the readhead and the scale and (4) arrays of photo-sensitive elements that are divided into sets within sub-cells and are interleaved in a repeating pattern with the outputs of a given set being connected in common [1].

A three grating system is studied here whereby the pitch of the analyser grating is set to extend or curtail the overall grating length relative to the index grating. This generates a Moiré fringe pattern over the photo-detectors positioned behind the analyser grating. The photo detectors are configured to produce four phases shifted (usually 90° in the quadrature method), cyclically varying electrical signals from which the magnitude and displacement of the relative movement between the readhead and the scale is determined [2].

The optical methods used in encoding technology are now well established. The technology is still however impeded by technical challenges. Most notable are the critical alignment required between the optical gratings and also from the gratings to the photo detector arrays. The bulky nature and subsequent packaging mean also that optical encoders can be prohibitively expensive for some applications and unsuitable for others.

Proposed Solution

In order to reduce the device size and costs of assembly and packaging MicroSystems Technology (MST) manufacturing and assembly techniques are employed in order to monolithically integrate the readhead components onto a single semiconductor chip. The die is then subsequently flip-chip bonded to a carrier substrate or package which houses some processing electronics. Figure 1 provides a schematic displaying a partial cross-section of the proposed encoder readhead.

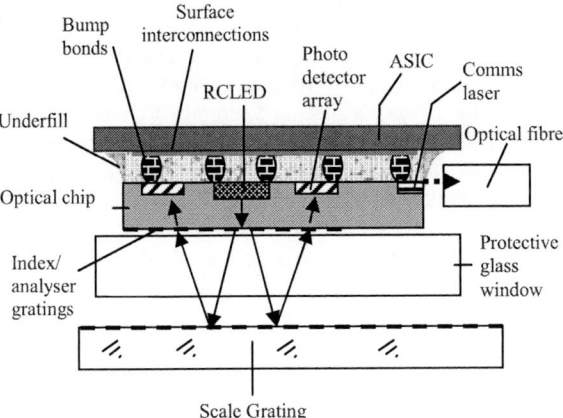

Figure 1: Proposed encoder readhead

The optical chip comprises a back emitting Resonant Cavity Light Emitting Diode (RCLED) and photo-detector arrays made of III-V semiconductor material. Additionally in preparation for the second generation of encoders, an additional light source, an edge-emitting laser, has been fabricated to transmit optically the current signals produced by the photo-detectors to a down stream evaluation unit outside the encoder. In the first generation of encoder, the electrical signals from the photo-detectors will be time multiplexed through a twisted wire pair. The index and analyser gratings are manufactured on the opposite side of the optical chip using standard lithographic techniques. The optical chip is then flip-chip bonded to a suitable carrier substrate that has been patterned with electrically conductive tracks.

A glass protective cap is fixed over the optical chip. The wavelength of the RCLED optical output has been chosen such that little attenuation is experienced as the light is passing through the substrate of optoelectronic chip. The light then interacts with the index grating, before passing through the glass protective cap and onto the reflective scale grating. The rays are then reflected back into the optical chip via the analyser grating such that a Moiré fringe pattern is created over the photo detectors for detection of the resulting intensity pattern. The detectors convert the light into electrical signals, which are passed through the processing electronics mounted onto the carrier substrate via flip-chip bonds. The processed signals are then communicated through a multimode optical fibre or a twisted wire pair as discussed previously.

The flip chip bonding process allows the distribution of the bonds over the entire surface of the chip [3], allowing thereby the required interconnections between photo detectors to be located on both the optical chip and the carrier substrate. This bonding method facilitates the design

of some novel photo-detector array arrangements, as shown in figure 2.

Figure 2a shows a layout whereby the photo-detector arrays are arranged in a four-by-four subcells. The detectors are marked A, B, C or D to correspond to the relevant signal phase shift experienced at that detector as dictated by the analyser grating. The subcell arrays are repeated around a centrally located light source, such that the summed optical paths at each phase are identical. Detectors experiencing similar signals are interconnected with diagonal tracks on the optical chip surface with the remaining connections being made on the carrier substrate. A total of 128 detectors make up 32 repeats of the ABCD arrays. The increased number of diodes reduces the sensitivity against the loss of signal as a result of obstructions in the optical path. At least one of the subcells may be replaced by a series of reference markers and/or additional light source(s) should more power be required.

Figures 2b and 2c present enhanced configurations of the detector array. Here the diagonal tracks and corresponding photo-detectors are combined to form a single entity such that detectors now occupy the whole diagonal area. The geometries of the detectors are such that the overall areas are increased for the same given footprint, generating more photocurrent per usable detector area. The magnitude of the electrical signals generated is therefore increased while maintaining a similar robustness to line of sight defects in the optical path as in figure 2a. Robustness against such defects is further improved by flipping one set of diagonal detectors to create a chevron style arrangement as shown figure 2c. Now defects having the same orientation as the detector will only eliminate a detector on one side rather than both. Again the vacant space to the left and right of the light source may be utilized for reference markers and/or additional light source(s).

Advantages

The proposed device configuration offers many distinct advantages over previous encoders. Firstly, the monolithic integration of the optical components not only offers substantial cost savings from reduced assembly (no need for individual component assembly and alignment) times and labour costs but also significantly reduces the footprint required by these components by up to 90%, therefore increasing the number of devices per wafer. Furthermore patterning of the optical grating using backside alignment techniques is performed at the wafer level. Thus the critical alignment between gratings and photo-detectors can be performed for many devices in one single step.

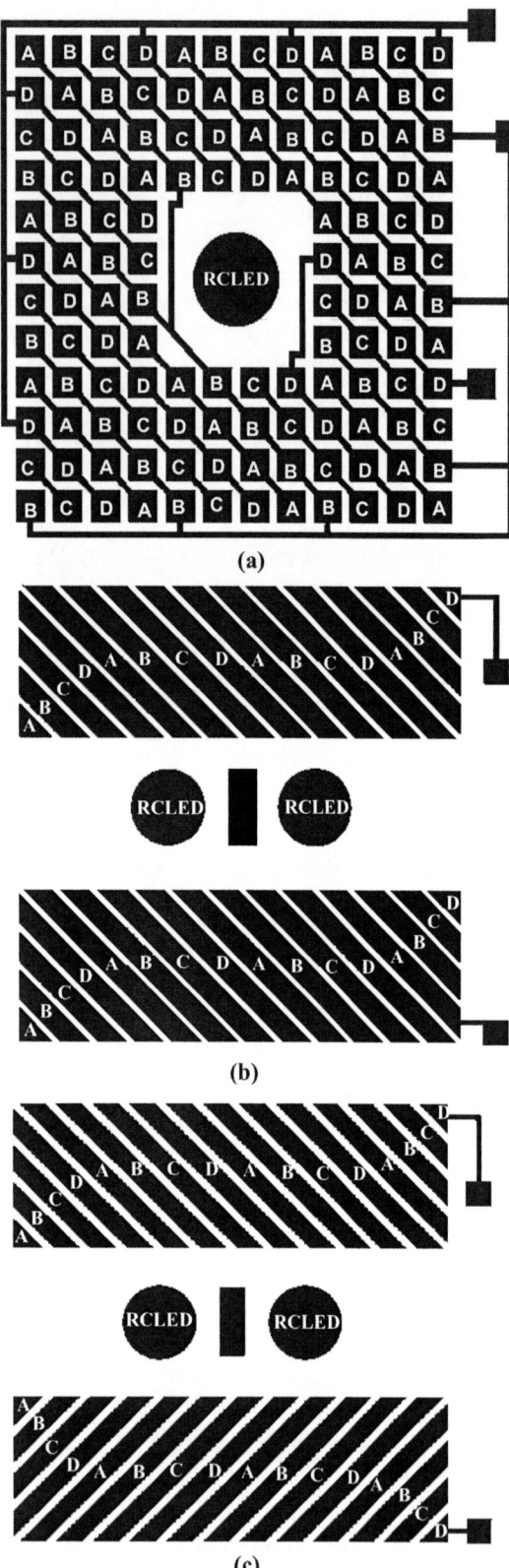

(a)

(b)

(c)

Figure 2: Diagrams showing photo-detector arrangements (a) Square pixels with diagonals interconnecting tracks, (b) Diagonals configuration and (c) Chevron configuration

All the contacts are located on one side of the chip in order not to obscure the optical path. The geometries shown in figure 2(a) and 2(b) aid also in the assembly process as the tolerances for the linear (X-Y) alignment of the gratings becomes a trivial issue with such geometries. The angular tolerances remain the same but these are more easily achieved. Finally the present designs encoders require up to 16 cables for power delivery and signal transfer. By time multiplexing the output signals the number of connections can be reduce to less than 4 depending on whether the twisted wire or optical fibre method is employed. This solution reduces the cost and increases the cable flexibility.

Fabrication

Figure 3 presents a detailed schematic of the construction of the optical chip including the light emitting diode and photo-sensitive elements. The epitaxy structure is deposited on the semiconductor substrate using metal organic phase vapour epitaxy (MOPVE) deposition techniques and comprises two n-type layers and p-type region, to form a junction surface for the light emitting region and photo-sensitive region. The n-type and p-type layers are electrically connected by gold deposited film electrodes. An isolation layer is deposited to ensure appropriate electrical contact to only the required location on the n-type and p-type layers. The emitting regions and absorbing regions are located at different levels within the epitaxy structure. Therefore precision controlled electroforming of gold contact pillars is required on the p-type contact electrode, on top of the emitter, to planarise the contact electrode for subsequent flip-chip bonding. The additional metal over the emitting elements provides the added advantage of acting as a heat sink at these locations.

The bumps are fabricated by controlled electroplating methods onto a glass substrate carrier for initial test and characterization of the optical chips. The glass wafers are prepared for the deposition of a seed layer by Physical Vapour Deposition (PVD) techniques. The surface metal is subsequently patterned, using standard photolithography techniques and acts as a metal mask for fabrication of the interconnecting tracks as well as the surface material to which the bumps are electroplated at specified locations. Once grown the remaining seed layer is selectively etched before the wafer is appropriately diced.

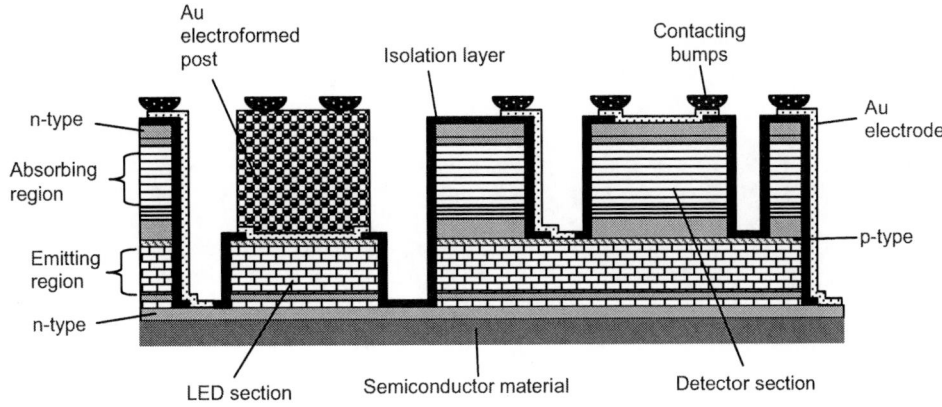

Figure 3: Schematic of the cross-section of the optical encoder

Results and Discussion

First generation prototype opto-electronic chips have been manufactured using indium phosphate, InP, as the semiconductor material onto which the epitaxy structure is deposited. Four detector arrangements are considered in this iteration, the three shown in figure 2, along with a parallel line design (not shown). This fourth arrangement, used for performance comparison, is an exact representation of what presently used by the company Renishaw in some of their encoders. Figure 4 shows the photos of the fabricated chip having the pixel configuration illustrated in figure 2a. On the topside image the grating areas corresponding to the photo detectors and LED can just be made out. Alignment arrows for positioning the optical fibre against the laser can also be seen. Figures 4 c and d highlight the footprint reduction of the integrated encoder chip when bonded to a test glass carrier substrate against the current Renishaw encoder assembly.

Testing and analysis of discrete LED and lasers devices has also been performed. Figure 5(a) shows the output power versus DC drive current for LEDs with diameters of 100um, 200um and 300um. The larger diameter device emits more power for a given drive current which is probably due to less internal heating from a lower current density.

The spectrum of a 300um device is shown in Figure 5(b) for 100mA and 300mA drive currents. The peak emission wavelength red-shifts slightly from 1290nm to 1300nm as the drive current is increased from 100mA to 300mA which is again due to internal heating of the chip causing electronic bandgap shrinkage. The spectra, as expected, are very broad with 3dB bandwidths of around 115nm.

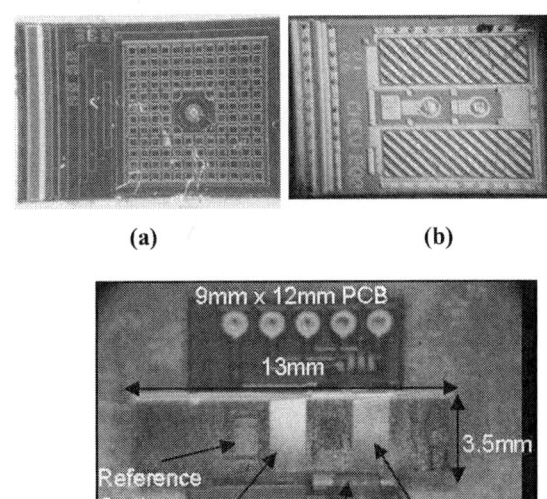

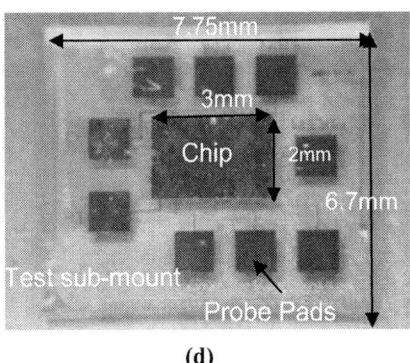

Figure 4: Images of integrated optical encoder chip showing bottom side detector/ emitter for pixel configuration (a) diagonal configuration (b) and footprint comparison between old (c) and new (d)

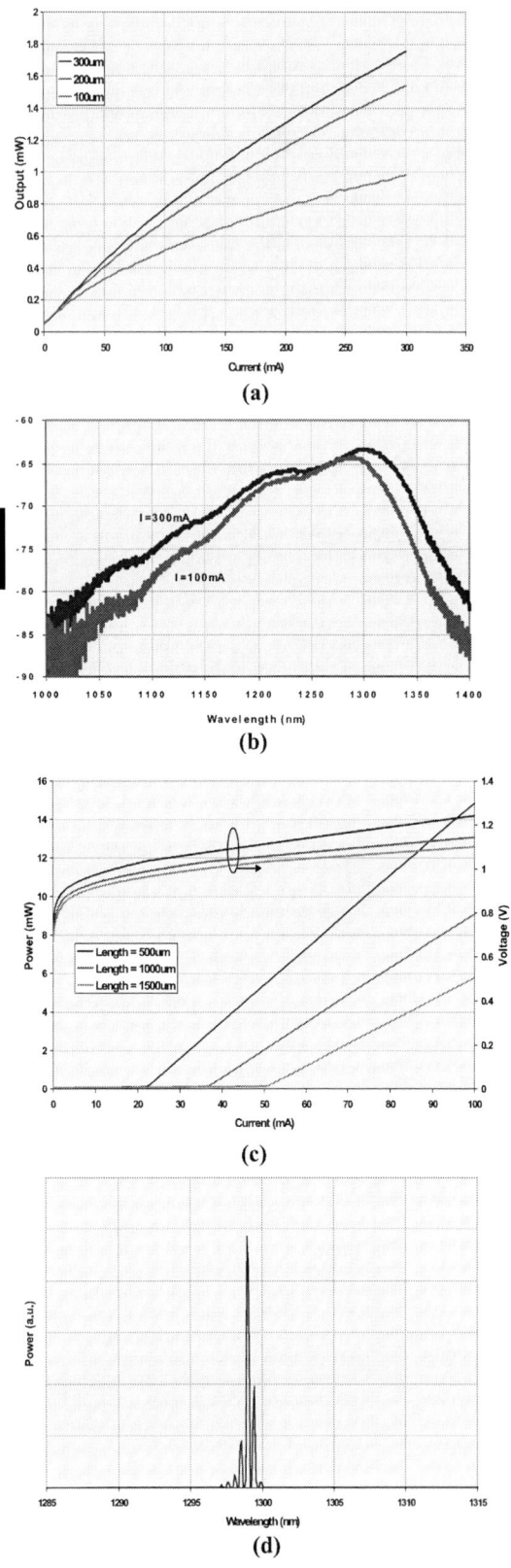

(a)

(b)

(c)

(d)

Figure 5: Discrete analysis showing P vs I for three LED diameters (a) LED spectra (b) P vs I for three lengths of laser (c) and laser output spectra (d)

Figure 5(c) shows the output power versus DC drive current for as-cleaved lasers with cavity lengths of 500um, 1000um and 1500um. As expected, the threshold current increases and the slope efficiency decreases as the cavity length is increased. The output spectra is shown in Figure 5(d) and shows a very typical narrow spectral linewidth of around 1.5nm with clear Fabry-Perot peaks resulting from the cavity resonance. These results indicate that the epi material is of very good quality as the performance is amongst the best reported for InGaAsP based Ridge Waveguide Laser devices operating around 1300nm.

Two electroplated bump structures have been investigated: gold capped copper pillars with a thin nickel intermediate layer to act as a diffusion barrier preventing gold migrating into the copper [4], and pure gold bumps. The optical chip is then flip-chip bonded to the test substrate carrier forming gold to gold bonds.

For the Cu/Ni/Au bumps both ultrasonic and thermo-compression bonding methods have been attempted. Ultrasonic bonding failed to yield any reliable bonds. Bonds were achieved using thermo compression but the yield was extremely low. Analysis revealed that failures occur at the three interfaces glass-seed layer, chip-seed layer and at the Au-Au interface. Failures during thermo-compression also appeared at the three interfaces discussed. However successful bonding has been achieved but yields are too low to consider this method as a viable production scale bonding technique.

Measures taken to improve adhesion include sputtering of seed layers, increased bump cap thickness, tighter tolerance over electroplated cap surface profile, and improved electroplating and thermo-compression parameters. As can be seen from figure 6 measurements have improved the shear strength of successful bonds by over 50%. However yields remained unacceptably low. One reason for this low yield is the use of Cu to form the core of the bump. Cu hardens during the bonding procedure and forms a very rigid bump, capable of penetrating the 250µm contact pad on the glass substrate and reducing the amount of deformation. Therefore further amendments have been made.

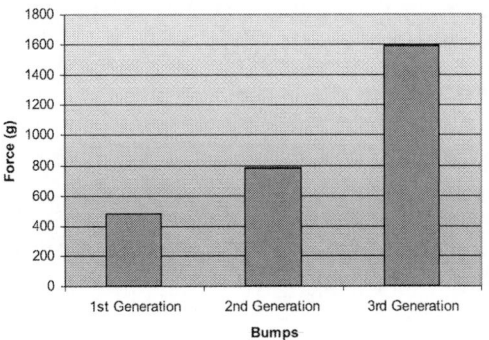

Figure 6: Average shear strength for three generations of bump structure

In the third generation bumps Cu has been replaced by Au. The contact pads are also gold electroplated increasing the overall thickness to 1μm. This improves deformation and further promotes adhesion. As shown in figure 6, near 100% improvement compared to the second generation in average shear strength (1580g) has been achieved which represent over eight times the MIL standard [5] from the second generation bump results. Furthermore yield with this configuration is greater than 90%, so within the required limits for higher volume manufacture. The large shear strength achieved might allow the optical encoder not to necessitate underfill, reducing further the cost of assembly.

Conclusions

A novel optical encoder with monolithically integrated components and unique photo detector arrangements has been proposed. The development is a substantial improvement in encoder chip manufacturing by significantly reducing manufacturing costs through waver scale manufacturing and assembly, and with novel detector arrangements that permit wafer scale alignment of the optical gratings to the detector arrays.

The first prototype chips have been manufactured and the light sources and detectors have to date been tested. With the improved the flip-chip bonding process fully functional testing of the bonded devices can be carried out.

Acknowledgements

The authors would like to acknowledge John Gorman, OptoCap, for his efforts with the ultrasonic flip bonding. One of the authors, Mr. John Carr, acknowledges the financial support of the Company Renishaw. The project was also sponsored by the Engineering and Physical Sciences Research Council (EPSRC) through a CASE award studentship. The authors also acknowledge the financial support of the EPRSC-funded Grand Challenge project 3D-Mintegration, grant number EP/C534212/1.

References

[1] Alfons Ernst, "Digital Linear and Angular Metrology – Position Feedback for Machine Manufacture and the Electronics Industry", Verlag Moderne Industrie, third edition, 1998.

[2] R.M. Pettigrew and L.A. Sayce, "Measurement Apparatus", G.B. Patent 1,504,691, March 22, 1978.

[3] J.H. Lau, "Low Cost Flip Chip Technologies for DCA, WLCSP, and PBGA Assemblies", McGraw-Hill, Chapter 2, pp. 28-33, 2000.

[4] C. Wang and A.S. Holmes, "Laser-assisted bumping for flip chip assembly", IEEE Transactions on Electronics Packaging Manufacturing, Vol. 24, No. 2, pp. 109-114, April 2001.

[5] MIL-STD-883E, Notice 5, Method 2019.7, "Die Shear Strength", March 7, 2003. Last accessed at
http://www.royceinstruments.com/Technology/ Standards/Die%20Shear%20Test%20MIL-STD-883_2019.pdf on March 5, 2007.

Multimode optical interconnections embedded in flexible electronics

E. Bosman, G. Van Steenberge, N. Hendrickx, P. Geerinck, W. Christiaens, J. Vanfleteren
and P. Van Daele

Ghent University, INTEC-TFCG-Microsystems, Technologiepark Building 914A, 9052 Zwijnaarde, Belgium
Phone: +32/92645370, Fax: +32/92645374, E-mail Address: Erwin.Bosman@intec.ugent.be

Abstract

This work aims at developing of a flexible substrate or foil in which optical waveguides, light sources, detectors, and electronic circuitry are embedded or integrated on compact signal processing boards. Our generic technology offers an integrated solution to the increasing demand for flexible optical sensors to be applied to irregular surfaces or to be folded into compact modules. On the other hand, the technology platform allows the establishment of flexible high-speed optical data connections, based on optical multimode wave guiding layers. Truemode Backplane^{TM} Polymer, Ormocer® and Epocore material meets the requirements for this application. These optical materials are sandwiched between two spin-coated Polyimide (PI) layers which absorb all stress and pressure during bending. The resulting stack shows good adhesion and very good flexible behavior (down to bending radii lower than 3mm). Patterning of the core-layer to create multimode waveguides is done with a standard photolithography process and by laser ablation using a 248 nm wavelength Kr-F Excimer laser. The fabricated waveguides have low propagation losses for 850 nm wavelength light and low bending losses. Besides passive optical waveguides and coupling structures, active components as light sources and detectors are embedded, using a photolithographic cavity and underfill adhesives. The chips need to be thinned down to 30 μm by lapping to show flexible behavior. In this stage of the research, mechanical chips were successfully embedded in the optical cladding layers and in the PI layer, resulting in a very thin final substrate of 150 μm thick.

Key words: Waveguide, Flexible, Laserablation, Embedding

Introduction

Over the past 5 years the demand for flexible substrates and the applications in which these flexible printed circuit boards are being used has been constantly growing [1]. Because of their flexible behavior, the use of these substrates can significantly lower the over-all substrate thickness and weight and most of all they can ease the assembly, increase the module compactness and can be applied to a flat and a curved surface. Most common applications are all the portable applications (mobile phone, digital camera, smart cards…).

While the electrical assembly of such flexible substrates is reaching maturity, the optical assembly is only commencing. The demand and interest for flexible optical communication is however growing for some dedicated applications, many of which triggered by the development of optical sensing techniques.

The use of these optical sensors however always implies the use of a light-source, detectors and electronic circuitry to be coupled and integrated with these sensors. The coupling of these fibers with these light sources and detectors is a critical packaging problem and as it is well-known the costs for packaging, especially with optoelectronic components and fiber alignment issues are huge. Due to these problems optical sensing is not yet implemented in many possible and practical applications.

This research is therefore aiming at developing a generic technology that offers an integrated solution by means of developing a flexible substrate or foil in which the sensing elements can be integrated and in which also the light sources, detectors and electronic circuitry are embedded.

On the other hand, this technology platform allows the establishment of flexible high-speed optical data connections, based on optical waveguiding layers. Optical data transmission has become the obvious choice for communication over longer distances, but new trends force designers to use optical interconnections also to bridge short distances [2]. Many research institutes (including ours [3]) have proven the integration concept on a rigid printed circuit board. This paper describes the extension to flexible substrates.

Materials

Optical layers (undercladding, core and uppercladding) have been successfully deposited on rigid substrates in the past by the TFCG Microsystems-group and optical elements as waveguides, out-of-plane 45° micro-mirrors and micro-lenses were integrated [3]. These rigid circuit boards have their main applications in optical backplanes, demanding very low light propagation losses of the optical path. Bulk materials for these optical layers were chosen to have low losses at a typical wavelength of 850 nm for data communication and 1.3 or 1.55 µm for telecommunication applications. They must have the right processibility properties in view of UV-crosslinking ability, spin-coating, temperature- and chemical resistance and mechanical brittleness. Special care must be taken to ensure the compatibility of the production process with the standard PCB production processes. This means the materials should be inert to production solvents and temperatures used during the production steps of the electronic assembly afterwards. The substrate should be physically and chemically stable in temperatures of about 280 degrees during solder reflows and in temperature cycling's from -40 to +85 degrees.

Truemode Backplane™ Polymer [4], Ormocer® [5] and Epocore [5] are materials which meet these requirements and have shown good results when applied on rigid substrates in the past [3]. Further material development is ongoing to create novel cross-linkable polymers with improved properties for the flexible applications.

The existing optical materials are however not flexible and strong enough to be bended without cracking or damaging. Therefore these layers are sandwiched between two spin-coated Polyimide (PI) layers, one at the top and one at the bottom, which absorb all stress and pressure during bending. PI is the dominant material in the flexible circuits industry because of its good electrical, chemical, temperature and mechanical behavior [6]. Stacking of materials with such a different chemical and mechanical behavior demands special measures like CTE (Coefficient of Thermal Expansion) matching adjusted cure-temperatures and good adhesion.

Best results have been obtained by using spin-coated PI [7]. The spin coat version of Polyimide is actually a polyamic acid solution. A cure heating cycle converts the acid to the insoluble imide and drives out remaining solvent. As an alternative way to fabricate the stack, the optical layers are spin-coated on a release layer and are laminated between prefabricated commercially available PI-foil after release. The optical layers can also be deposited immediately on top of a PI wafer which functions as the carrier and the final substrate at the same time [8].

Fabrication of flexible optical foil

Isolating tracks in the core-layer and consequently surrounding the track completely with cladding material, results in the creation of optical waveguides. This is a well proven principle for optical interconnections on rigid boards. Light that will be coupled into these waveguides will be guided in that layer due to total internal reflection, caused by the different refractive indexes of the core- and the cladding layers.

Experiments have shown that the flexibility, strength and reliability of the complete structure is significantly improved when the optical layers are sandwiched between two Polyimide layers. Hereby we have become an almost symmetrical build-up (Fig. 1), releasing the internal stress from the central core layer. Mechanical stress caused by bending will mainly be taken up by the outer Polyimide layers, which own a high tensile strength (35 kg/mm$^{2)}$) and elongation (25 %). PI is often used as a stress buffer in the semiconductor industry.

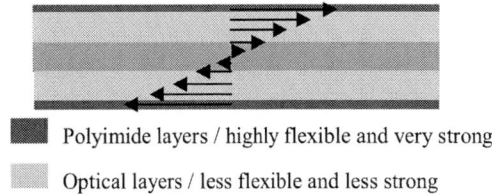

■ Polyimide layers / highly flexible and very strong

▦ Optical layers / less flexible and less strong

Figure 1: Symmetrical layer build-up releases the stress from the inner layers.

PI is a very inert material and basically it adheres quite bad to the proposed optical materials. For this reason a plasma etch is performed on the bottom PI layer. A combination of a mixed O_2/CHF_3 (2 minutes) plasma treatment and a pure O_2 plasma treatment (for 2 minutes) results in an optimized adhesion. Before applying the top PI layer, a very thin adhesion promoter layer is spin coated and no plasma etch is needed.

The most important process parameters are the bake temperatures of the different layers. Each layer must be hardbaked at a temperature that is lower or equal than the hardbake temperature of the underlying layer. For example: to cure the top PI layer, a temperature of minimum 210 °C is needed, so it is highly advisable to hardbake each layer of the structure at least at 210 °C to ensure no more solvents will be driven out of the layers. Experiments have shown that, if done otherwise, the chemical structure of the top Polyimide layer can be drastically changed, loosing its strength and flexibility properties. We make use of a special low cure-temperature PI to avoid reaching the degradation temperatures of the optical materials.

The proposed process layout results in a very light, thin (160 µm total thickness) foil with a high tensile strength due to that of Polyimide. Very high flexibility is achieved and the minimum mechanical bending radius before damaging the structure is set to less then 0.5 cm. The foil has been realized and the adhesion matters have been optimized for the 3 optical materials (Truemode Backplane™ Polymer [4], Ormocer® [5] and Epocore [5]).

Fabrication of optical waveguides

Optical multimode waveguides are fabricated in two different approaches, laserablation and standard lithography. Fig. 2 shows the schematic overview of both the principles. The dimensions are chosen to be 50x50 µm^2 to be compatible with standard optical multimode fibers which have a typical core size of 50 µm.

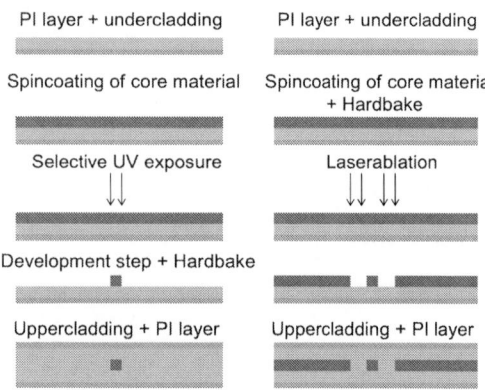

Figure 2: Schematic overview of *left*: standard lithography and *right*: Laser ablation

A scanning electron microscope (SEM) image of a single waveguide fabricated using KrF Excimer laser ablation is shown in Fig. 3; an array of 10 waveguides in a full stack is shown in Fig. 4. The picture shows the structured core layer prior to application of the upper cladding layer. Optimization of the ablation parameters is required in order to achieve the very smooth sidewalls and the absence of debris.

This results in the smoothness of a laser ablated sidewall of 27-nm Ra and 35-nm root mean square (rms) roughness (cylinder and tilt removed), measured on an area of 60x46 µm) for a Truemode Backplane^{TM} Polymer waveguide.

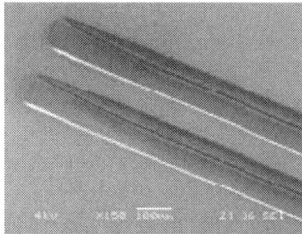

Figure 3: SEM picture of a laser ablated waveguide in Truemode Backplane^{TM} Polymer

Laser ablation of Ormocer® and Epocore generates too much debris (laser ablated particles resettling on the surface) and the roughness of the sidewalls is too high, which will result in high light propagation losses.

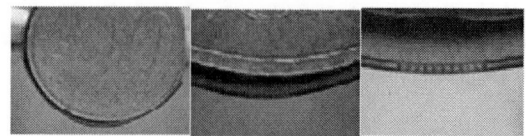

Figure 4: Complete stack with laser ablated waveguides in Truemode BackplaneTM Polymer, over a coin with radius of curvature of 1 cm.

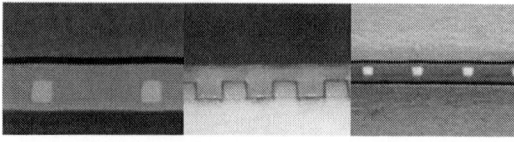

Figure 5: Complete stack with photolithographic waveguides in *left*: Ormocer®, *middle*: Epocore (on FR4-material) and *right*: Truemode^{TM}

For Ormocer® and Epocore [5], well defined waveguides can be fabricated with a standard lithography step. The material acts as a negative resist. Areas of the core layers which are exposed to the UV source will cross-link while the unexposed areas remain soluble in a developer solution. Very sharp edges and smooth sidewalls can be obtained with this technique. Ormocer® material however is still sticky after the prebake step, so any contact with the mask must be avoided to avoid contamination. UV-exposure in proximity mode demands for a highly controllable height between substrate and mask and results in less sharp, but acceptable edges of the waveguide (Fig. 5). Experiments have pointed out that for Ormocer®, the adhesion of the waveguides to the cladding layer depends on the UV exposure time. Best adhesion is obtained at the longest exposure times. After the developing of the non-exposed areas, the waveguides are all that is left from the core layer, so the top cladding layer must have twice the thickness of the waveguides to obtain a symmetrical build-up.

The resulting waveguides have low propagation losses lower than 0.15 dB per cm and bending losses lower than 0.15 dB per cm for a bending radius of 15 mm and lower than 0.25 dB per cm for a 8 mm bending radius for Truemode Backplane^{TM} Polymer. Fig. 6 shows the optical measurement set-up: A controlled Laserdiode emits 850 nm wavelength light into a glass fiber. The other end of this fiber is actively aligned with one end of the waveguides. When the light has propagated through the waveguide, it is coupled out into an actively aligned glass fiber which leads towards a photodiode. This way the propagation and coupling losses can be measured using the cut back method. The waveguides can now be bended as we wish. The extra power loss measured is the bending loss. Fig. 7 shows the propagation-, coupling- and bending losses for 10 waveguides with a length of 8 cm. The waveguides were bended in the set-up that is shown next to the graph.

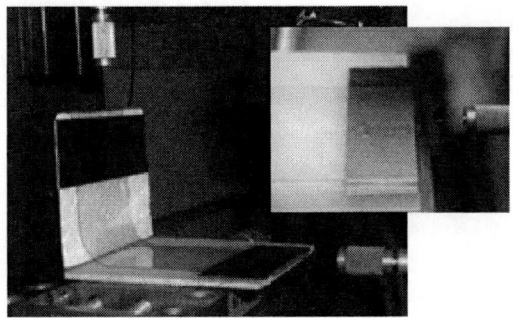

Figure 6: Optical loss measurement set-up

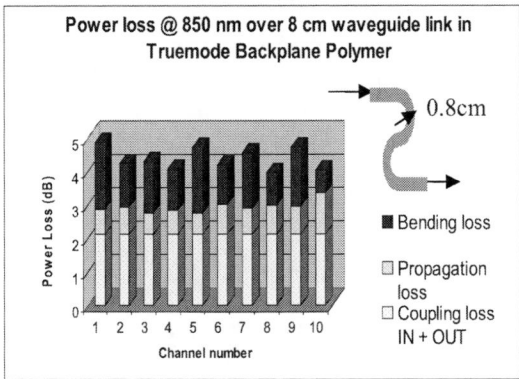

Figure7: Propagation-, coupling- and bending losses

Fabrication of 45 degrees out of plane turning mirrors

The data-carrying light can be vertically coupled in- and out of the waveguides with 45 degrees out of plane deflecting micro-mirrors, terminating the waveguides and connecting them with laser diodes, receivers, optical fibers, open air or optical elements.

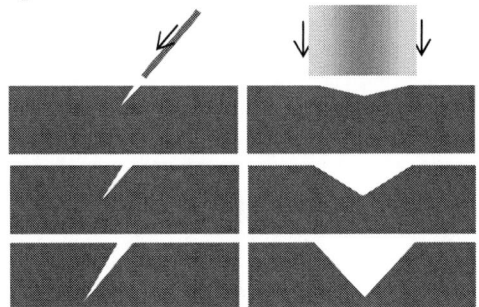

Figure 8: Schematic overview of the fabrication of micro-mirrors: *left* **tilted laserbeam ablation ;** *right* **vertical distributed laser ablation**

Two basic fabrication methods are proposed and shown in Fig. 8.

By tilting the beam delivery optics of the Excimer KrF laser setup the laser beam impacts on the surfaces with a tilt of 45°. This results in a tilted laser ablated cavity (Fig. 9). Deposition of a metallization layer on the sidewall of this cavity creates a 45° reflecting mirror. Later on the cavity is filled with the cladding material while ultrasonic treatment avoids air bubbles captured in the cavity. It can be seen that using the Excimer laser, there is always a certain tapering of the edges. The tapering effect is shown to be highly reproducible; therefore this effect can be compensated for, in order to achieve an angle of 45° at the positive facet.

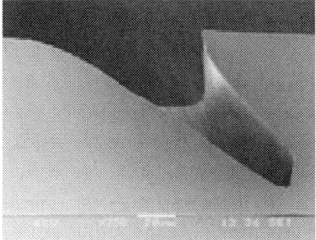

Figure 9: Micro-mirror in Truemode Backplane[TM] Polymer, using the tilted laserbeam

The second fabrication approach consists of a distributed exposure of the surface to the laser beam. The sample is moved horizontally while a triangular shaped laser beam is ablating the surface. This way a smooth V-groove with 45° edges is ablated in the surface. Fig. 10 shows a Scanning Electron Microscope picture of such a mirror structure.

Figure 10: SEM picture of a micro-mirror in Truemode Backplane[TM] Polymer, using the distributed laser ablation technique.

After metallization, this structure acts as a mirror. The advantages of this second method is the absence of a negative facet (no air bubbles can be trapped in the cavity), the ease of alignment (the beam impacts vertically on the substrate) and a better depth control (important when mirror has to be ablated on top of an active component). The depth of the V-groove and thus also the mirror angle can be fine-tuned by controlling the laser power.

The smoothness of a laser ablated sidewall of such a V-groove is 39-nm Ra and 50-nm root mean square (rms) roughness (cylinder and tilt removed), measured on an area of 90x50 μm) for Truemode Backplane[TM] Polymer. Ongoing fine-tuning of the other laser parameters shows promising results to lower down these roughness values.

Embedding of active optoelectronic devices

This research aims for a generic technology that offers an integrated compact solution for optical communication in foil. Integrating active components in the foil is therefore essential to reach fully embedded optical links. VCSEL's and Photodiodes have been embedded in optical layers before, to implement them onto rigid boards [9], but not yet to form an autonomous opto-electrical foil.

The VCSEL's, Photodiodes and electrical IC's will be thinned down by lapping and polishing to reach a thickness of about 40 μm. These thin chips are embedded in a cavity in the first cladding layer.(Fig. 11 shows a 30 μm thick copper dummy chip embedded in the flexible foil). When standard lithography is used to outline the cavities, problems occur during the development step. The adhesion of the optical polymers with the underlying PI layer before hardbake seems to be too bad, so the development solution creates delamination of the two layers in the region of the cavity.

Figure 11: 30 μm thick copper samples, used as dummy chip, embedded in the cladding layer of the optical foil. Waveguides can be seen on top of the embedded sample.

A well proven alternative is laser ablation with the KrF Excimer laser (wavelength 248 nm). To ensure a good depth control, a copper stop is applied to the PI layer. This copper stop consist of a local sputtered layer of TiW (50 nm) for adhesion promotion and a sputtered and electroplated copper (9 μm). During laser ablation, all the polymer material is removed selectively from the cavity while the copper layer is not ablated and becomes the bottom of the cavity (Fig. 12).

Figure 12: Laser ablated cavity for VCSEL array embedding; dimension: 3000 x 280 μm²

After cleaning of the substrate, an adhesive is deposited into the cavity. Underfill adhesives for flip chip bonding with very low viscosity and low thermal cure-temperatures have shown good results for this application. The thin dies are then placed manually into the cavity. Optimised dimensions of the cavity can result in alignment errors < 15 μm (Fig. 13). This is not sufficient for optical coupling, but the real fine alignment happens later. A thermode levels the die in the cavity and cures the underfill adhesive.

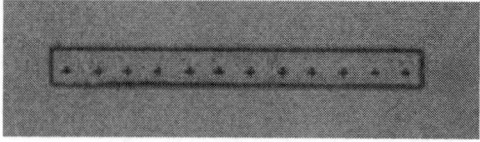

Figure 13: Dummy VCSEL array embedded in a laser ablated cavity using underfill material as adhesive.

In a next step, the core layer is spin coated on top of this structure. The fabrication of waveguides in the core layer is described earlier. As well for the laser ablated waveguides as for the lithography waveguides, alignment of <5 μm in relation to the active areas of the optical active dies, can be achieved after optimization (Fig. 14). Same accuracy is realised for the laser ablated micro-mirrors which are fabricated afterwards. The 5 μm precision is crucial for the feasibility of a low loss optical connection.

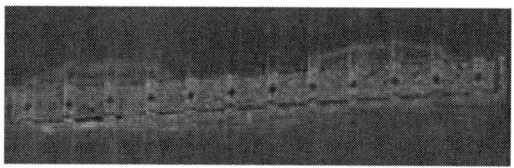

Figure 14: Epocore waveguides on top of an embedded VCSEL dummy. Alignment of the waveguides has been done trough the help of laser ablated marks on the copper dummy

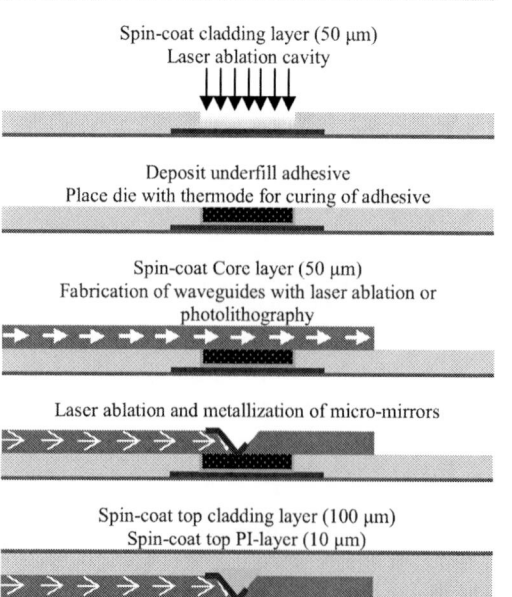

150

Figure 15: Process flow for the production of an optical foil with embedded active opto-electronic components

Alignment and process issues were optimized for mechanical chips. An overview of the process flow can be seen in Fig. 15. Pictures of the resulting optical foil are in Fig. 16. Ongoing research aims at the characterisation of functional optical links with functional VCSEL's and PD's. As an alternative to thinned-down dies, standard components with 150 μm thickness can be embedded by using a 150 μm thick cladding layer or multiple 50 μm layers. Optimisation of the adjusted process flow has been done. Using thicker chips than 50 μm is however a draw-back for the flexible behaviour of the optical foil.

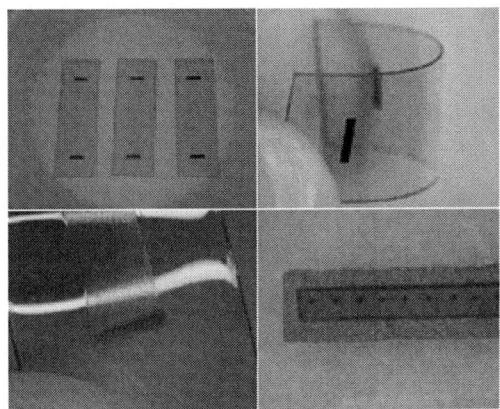

Figure 16: Pictures of the optical foil with embedded dummy VCSEL arrays, waveguides and micro-mirror's.

Electrical assembly of the substrate

All active components should be electrically connected to each other. Micro-via's are ablated using a frequency tripled Nd:YAG laser (355 nm wavelength) and metallized by sputtering and plating to fan out the contacts on the top PI layer where all other electrical assembly can be done with standard flex assembly processes. The optical layers have proven to withstand the temperature cycles during these procedures. The optical components have been chosen to have all electrical pads at the top side to simplify the assembly to one sided flex technology. However double-sided routing of the electrical connections is well known nowadays and could be perfectly done is this application because of the symmetrical build-up.

In ongoing research, a combination of all the above techniques will be united in a proof-of-principle demonstrator.

Conclusion

The increasing need for flexible modules and the integration of photonics on printed circuit board's results in a challenging and competitive research which combines both needs by embedding passive optical interconnections and active optoelectronic devices on flexible Polyimide substrates, resulting in a complete autonomous opto-electrical and flexible module. Dummy electrical and opto-electrical dies were successfully embedded in the foil. Low loss waveguides, micro-mirrors and micro-via's were fabricated using laser ablation and standard lithography and aligned with the dies, meeting the application requirements. The resulting opto-electrical foil has a thickness of about 160μm and shows high flexibility. Materials are chosen to be compatible with standard flex and PCB fabrication technologies.

Acknowledgements

Part of this work was carried out within the framework of the Network of Excellence on Micro-Optics (NEMO), supported by the European Commission through the FP6 Programme and the authors would also like to thank the Flemish IWT (Institute for the Promotion of Innovation by Science and Technology) for financial support through the FAOS (Flexible Artificial Optical Skin) - project.

References

[1] www.bpaconsulting.com
[2] A.L. Glebov, M.G. Lee, K. Yokouchi, "Integration technologies for board-level optical interconnects", Proceedings of SPIE Photonics Europe, April 2006.
[3] G. Van Steenberge, N. Hendrickx, E. Bosman, J. Van Erps, H. Thienpont, P. Van Daele, "Laser Ablation of Parallel Interconnect Waveguides", IEEE Photonics Technology Letters, Vol. 18, No. 9, May 2006.
[4] www.exxelis.com Truemode Backplane™ Polymer data sheets
[5] http://www.microresist.de Ormocer® and Epocore material datasheets
[6] Joseph Fjelstad, "An engineer's guide to flexible circuit technology", Elektromechanical Publications LTD, 1997
[7] www.hdmicrosystems.com/tech/polyimid.html
[8] J. Farah, "Polished Polyimide Substrate", IMAPS Electronic Bulletin, january 2007
[9] C.Choi, L Lin and R.T. Chen, "Performance Analysis of 10-μm-thick VCSEL Array in Fully Embedded Board Level Guided-Wave Optoelectronic Interconnects", Journal of lightwave Technology, Vol. 21, No. 6, June 2003

Smart Packaging of Microlenses over a UV-LED Array

Markus Luetzelschwab, Dominik Weiland and Marc P. Y. Desmulliez

Heriot-Watt University, MIcroSystems Engineering Centre (MISEC), School of Electrical Engineering and Physical Sciences, Earl Mountbatten Building, Edinburgh, EH14 4AS, United Kingdom

Phone:+44 (0)131 451 3942, Fax:+44(0)131 451 4155, E-mail: M.Luetzelschwab@hw.ac.uk

Abstract

This article presents a fully integrated packaging solution that permits the static and active alignment of a microlens array placed on top of a micro-UV-LED array. The manufacturing technology uses a modified UV-LIGA process based on the photoresist SU8. The microlens array rests on four photoresist posts created on top of the micro-UV-LED array, in order to reduce the contact area and hence reduce the probability of vertical misalignment. The holes, matching the posts, are situated on the micro-UV-LED array structure. By electroplating the electrodes with a certain thickness, the distance of the microlens array can be finely adjusted. The proposed concept is, with minimal modification, transferred to a microactuator, which enables the movement of the microlens array in the vertical direction. For that purpose, an electrostatic actuation method is applied. In this case the posts and holes serve as a guidance to limit the lateral movement. A further possibility is the lateral actuation using magnetic force simultaneously with vertical actuation. Measurements of the actuation mechanisms are presented.

Key words: active alignment, electro-magnetic actuation, monolithic integration, optoelectronics packaging

Introduction

The precise positioning of micro-optical components over Surface Active Devices (SAD) has long been a critical issue in optoelectronics packaging [1, 2]. Different alignment methods between SADs and optical components have been utilised [3]. These methods however make predominantly use of external alignment tools and therefore restrict the device to be rigidly aligned with no means for actuation and therefore active alignment [4, 5].

Herein is presented a fully integrated packaging solution, which permits the static and active alignment of a microlens array placed on top of a micro-UV-LED array (µLED). The active area where the 4096 LEDs are situated measures 2 mm x 2 mm. The static alignment process consists of the accurate positioning and locking of the microlens array (µLens) in the vertical and lateral directions. The dynamic approach allows the µLens to be moved in the vertical direction to change the focal point of the optoelectronic system.

Figure 1 depicts the µLens with its four posts, which are inserted into the corresponding cavities on the µLED during the assembly process. Posts have been chosen to reduce the contact area and hence the probability of misalignment due to particulates and process imperfections. Each cavity has two electrodes, which can be electroplated individually to a certain height for adjustment of the vertical distance between the µLED and the µLens. This structure also allows the tilting of the µLens to a certain degree in order to compensate for inaccuracies from earlier processing steps. In addition, the posts and the cavities serve for lateral positioning. After alignment, the assembly can be fixed in place using UV-curable adhesive and sealed afterwards.

The dynamic approach permits the active alignment of the µLens and uses the same configuration with some slight modifications. Four floating electrodes made of titanium are evaporated on the µLens for electrostatic actuation. This layer of titanium serves also as a seed layer for the ferromagnetic posts in the case of magnetic actuation (Figure 1). Each of the four pairs of electrodes together with the floating electrodes acts as an actuator. The floating electrode does not have to be connected to any potential. Hence no wire is attached to the µLens, which would mechanically disturb its movement. Electrostatic forces are only attractive in nature. In order to permit movement in the positive vertical direction and/or restore the µLens to its original position, gel bumps have been used as a test vehicle in preparation for more sophisticated PDMS structures planned to be used in the future. Lateral actuation is also feasible by electroplating both the tracks and pads and the posts with a ferromagnetic material such as nickel. With this method it is possible to simultaneously align the µLens vertically and laterally with an appropriate driving device as explained later on.

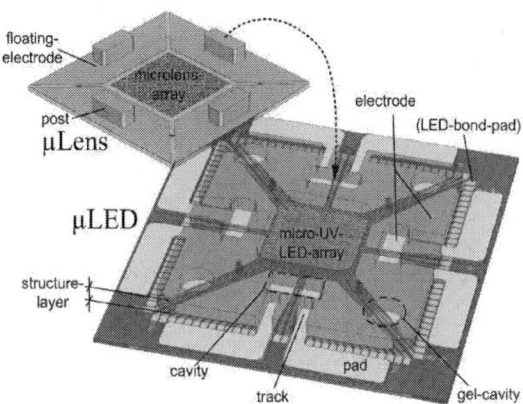

Figure 1: Packaging structure of the μLens and μLED.

Fabrication of chip sized dummies

Both the μLED as well as the μLens are usually supplied as bare dies rather than as wafers. This is usually the situation that packaging specialists are confronted with when design for packaging is omitted at the device design stage. This renders the entire fabrication process of the packaging or alignment structures more difficult due to issues such as stiction and surface tension effects caused by different surface energies of the materials involved. In our case, we are concerned with the reproducible manufacturing of posts onto the μLens. Figure 2 shows a simplified process for the μLens.

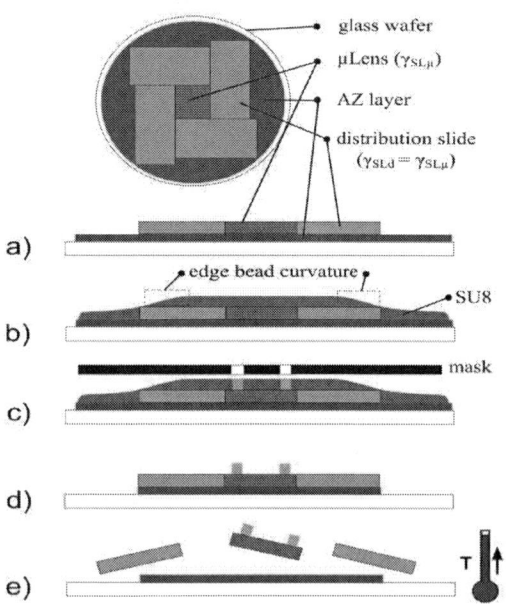

Figure 2: Chip size processing of the μLens.

After having spun a thin layer (≈3 μm) of AZ® 9260 onto the glass wafer, the wafer is baked on a hotplate for 1min at 110° C. The μLens is then placed at the center of the wafer, and distribution slides of the same height are placed around it. For a given liquid considered (here a photoresist) the distribution slides must be of the same material or at least must have a similar surface tension as the lens material ($\gamma_{SLd} = \gamma_{SL\mu}$). Alternatively, OmniCoat® spread over the whole assembly, and baked, could be used to achieve the same effect. The assembly is then shortly placed on a hotplate to temporarily fix the pieces (a). In the following step, SU8 is spun onto the structure. Due to the presence of distribution slides, the edge bead curvature is shifted away from the μLens, leaving a uniform height distribution on it (b). The assembly is then exposed to UV-light in a photolithographic step. It is important to protect the distribution slides from exposure for better removal of the structure (c). EC-solvent is then used for the development of the SU8 which also dissolves the AZ which is not covered by the μLens or the distribution slides (d). Due to the area of the distribution slides and μLens, a solvent would need a long time to penetrate the gap where the AZ is situated. Hence, in order to gently release the μLens, the assembly is placed on a hotplate for approximately 10 s with a temperature of 90° C (e). This process permits the deposition of photoresist at a constant height across the whole structure under consideration by removing the edge bead curvature away from the centre.

Electroplating of individual electrodes in static approach

For the static approach, the posts on the μLens are made of SU8, as shown in Figure 1. The intended height of the posts, t_{post}, can only be approximated by spin coating. It is observed that an accuracy of around ±10 % is normally obtained. This is not satisfactory in most cases where precise alignment of the μLens is needed on top of the SADs. In order to satisfy that requirement, it is necessary to achieve an accurate distance, t_{total}, between the μLens and the μLED. Various dimensions need to be known in order to achieve this total distance as shown in Figure 3. The height of the post, t_{post}, as well as the thickness of the base layer and evaporated electrode, t_{base}, are measured with a white light interferometer. From these dimensions, the thickness, t_{elec}, which needs to be electroplated, can be derived.

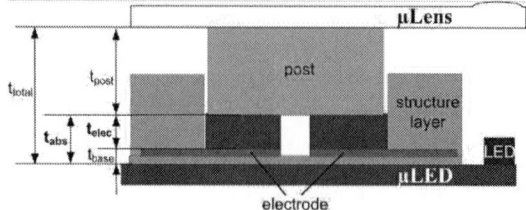

Figure 3: Side view of the cavity and post with the important dimensions for the electroplating process.

Prior to electroplating the electrode is analyzed by using the Zygo® white light interferometer. In one particular example the initial average height, t_{base}, was measured to be 9.7 μm with ±0.3 μm deviation. One prominent peak with a height of around 1 μm was observed. This defect was caused prior to electroplating and can be neglected since only the electroplating itself is of interest in this study. A first electroplating step is carried out in order to determine the growth rate. From that, the remaining electroplating time is calculated to achieve a height, t_{abs}, of 15 μm. Measurements revealed an average height of 15.3 μm with ± 0.3 μm deviation. Each of the electrodes can be electroplated. This offers the possibility to achieve different thicknesses on different electrodes. In case the heights of the μLens posts are not all the same, the tilt can be compensated by the electrodes as shown in Figure 4. For instance, if post C is small then the electrodes at the position C need to be electroplated longer than for example at position A where post A is tall.

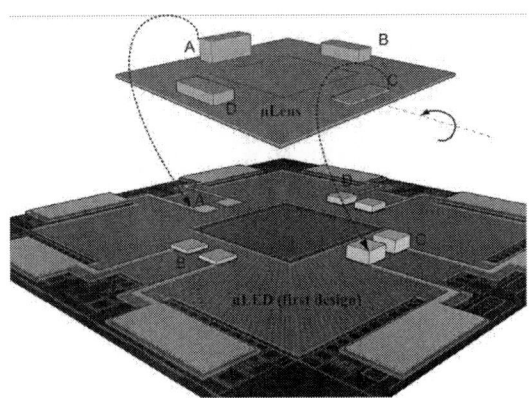

Figure 4: The compensation of a tilt, caused by different post heights, is possible by individual electroplating of electrodes.

Modeling of the electrostatic force

The majority of electrostatic actuators have two electrodes both connected to a potential, where one of the electrodes is moving. In the present approach, another type of electrostatic actuation has been chosen [6]. It uses the pair of electrodes, situated on the μLED to generate an electrostatic field and hence a force. Evaporated onto the μLens is a floating electrode, which "reflects" the electrostatic field from one electrode to the other, as shown in Figure 5. Neglecting the fringing fields, this assembly can be regarded as two capacitors connected in series each having an area of k times b and a gap distance of d. The reduction of area and the series connection of the two capacitors has the disadvantage of producing approximately only a fourth to a fifth of the force that would be achievable

if the floating electrode was connected to a potential and the two electrodes shared the same potential.

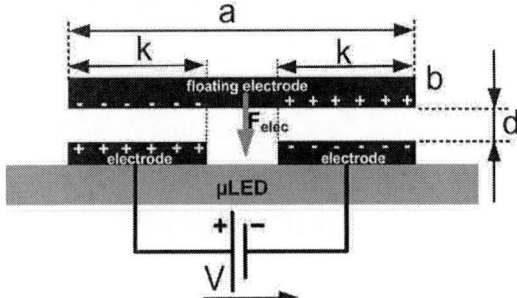

Figure 5 Electrostatic actuation principle using a floating electrode.

The results of the simulation using COMSOL®, fit very well with the analytical equation with error of less than < 1 % between the two methods. The main advantage of this actuation method is firstly the ease of using the same structure as in the static approach where parts of the electrodes have been used for electroplating. No wire needs to be attached to the μLens, since only a conductive surface is required. This simplifies the fabrication and assembly process tremendously. Furthermore, considering the relatively small size of the μLens itself (4.5 mm x 4.5 mm) an attached wire would certainly have a disturbing effect on the movement due to its mass and rigidity. With the four pairs of electrodes, it is possible to tilt the electrodes along the horizontal axis. Due to the tilt of the floating electrode towards the electrode, as shown in Figure 6, the charges tend to shift to the area where the floating electrode is the closest to the electrodes.

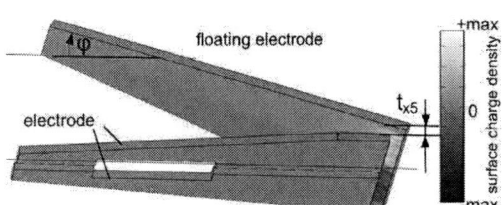

Figure 6: Side view of the tilted floating-electrode with respect to the electrode simulated with COMSOL®. The surface charge density is qualitatively depicted

Comparison of the force generated in the dynamic approach has been carried out using numerical and simulated values. Numerical values for the analytical equation have been calculated using MATLAB® and the results compared to the corresponding COMSOL® simulation. Figure 7 shows this comparison where the force is a function of the closest distance between electrodes and floating electrode, t_{x5}; the tilt has been taken at φ=10° , and the thickness of the dielectric structure layer, t_s,

155

is chosen as a variable parameter. The error between the two results is less than 17%.

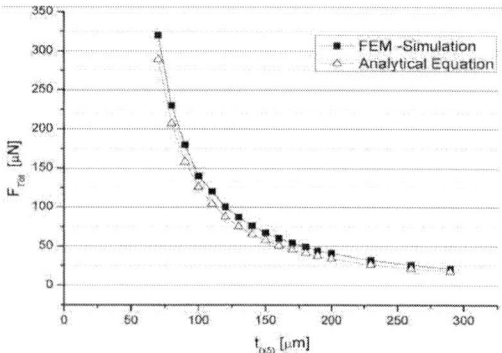

Figure 7: Comparison between simulated values and the analytical equation.

Electrostatic Actuation

In order to restore the μLens back to its original position, the use of the buoyancy effect of a viscous fluid, such as the UV-curable adhesive NOA 61 was considered a possibility. The fluid is filled into the cavities (Figure 1) and the μLens then placed into position. The vertical displacement was measured using optoNCDT 2400, a confocal chromatic displacement sensor from micro-epsilon. Figure 8 shows that the displacement behavior is rather unpredictable and possesses a large hysteresis. Except for the third 0 V slot, the displacement never settles to the previous equilibrium position. It was observed that due to capillary forces, the adhesive went out of the cavities and caused stiction between the μLens and μLED. An improvement was achieved when a large amount of adhesive was taken with the entire μLens floating above a thick adhesive layer. Due to practical considerations, this approach was not pursued any further.

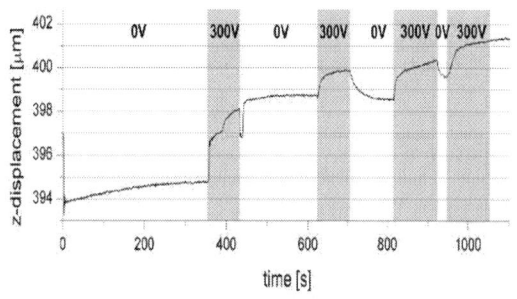

Figure 8: Displacement measurement by using viscous adhesive within the cavities.

Gels have densities similar to liquids but possess the structural coherence of solids. The use of a gel was therefore deemed not to be prone to capillary actions and to have good restoring properties. As shown in Figure 9, gel bumps were placed within the gel cavities

(Figure 1) of the μLED. The μLens was then placed on top of the bumps.

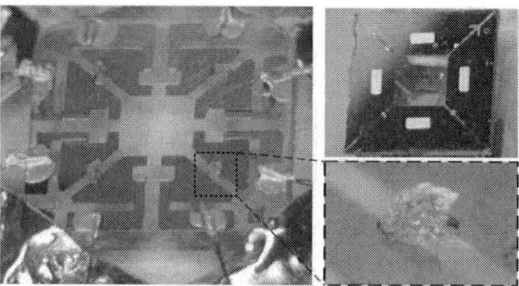

Figure 9: Dummy μLED (left) and μLens (right, top) for actuation measurements. On the right bottom, one of the four gel bumps that are responsible for the restoring force.

Different voltage steps were applied with a duty cycle of 50 % and a voltage-on-time of 60 s. In each period, the voltage was increased by 50 V. After having reached 600 V the voltage was then set back to 0 V, as shown in Figure 10. The high voltages can be drastically reduced in future designs, since the gel bumps were relatively large therefore the distance between the electrodes and the floating electrode was large as well. It can be seen that the voltage-on-time and voltage-off-time were too short to allow the gel to return to a stable equilibrium position. The slopes of the plateau seem to follow a parabolic curve.

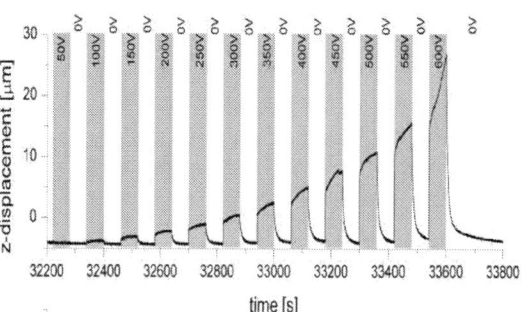

Figure 10: Step response measurement for voltages between 0 and 600 V in 50 V increments.

In a second experiment, the gel bumps of the same device have been reduced alongside the initial distance between electrode and floating electrode. Furthermore, the voltage-on-time and the voltage-off-time were increased from 1 min to 4 min as shown in Figure 11.

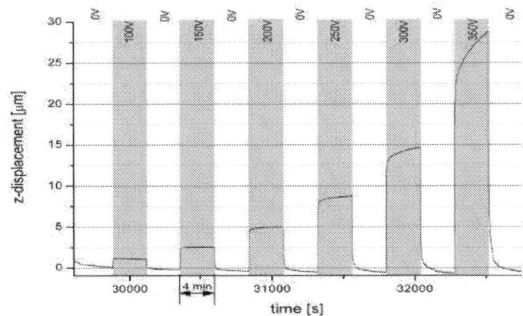

Figure 11: Additional measurement with longer duty cycles and reduced gel bump heights

Up to 200 V the voltage-on-time seems to be long enough to reach an equilibrium position. From 250 V upwards, the gel bump is not able to reach an equilibrium position any more. Due to the reduced initial distance, a displacement of nearly 30 μm can already be obtained at 350 V. The long term behavior can be described as good, as long as the displacement is approximately 20 % of the maximal deflection. Further investigation is however needed to confirm this assertion. In the future, it is envisaged, that instead of using gel bumps, photo-patternable PDMS is to be used [7]. In this way, the geometry of the restoring element can be controlled more accurately and characterized. A further advantage of using gel like restoring means is in the elasticity in the z-direction but also in the x,y-plane due to shear forces. This is an interesting property that can be used for lateral movement and alignment as described in the following section.

Lateral magnetic movement

The structure allows magnetic actuation, by electroplating the tracks and the pads with a ferromagnetic material (Ni) as shown in Figure 1. Also, as seen on the right side of Figure 9, the posts need to be made of a ferromagnetic material. The magnetic flux Φ, leaving and entering the tracks, attract the post (Ni) with a force F_m, to make the distance d_m, as small as possible, hence reducing the magnetic resistance of the whole path. (Figure 12).

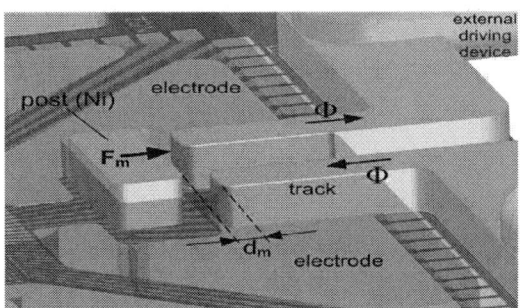

Figure 12: Close-up of the nickel post that gets attracted by the tracks due to the magnetic flux, Φ.

Figure 13 shows the external driving device necessary for the lateral and vertical movement of the μLens. The coil and its current source, generate the magnetomotive force. The magnetic field lines are able to penetrate through the electrical insulator provided that its thickness is as small as possible. The electrical insulator is essential for the electrostatic movement, since the magnetic path, with its electrical properties, would shorten out the voltage source. With one actuator on each side, the lateral movement can be performed in both x- as well as y-direction. Preliminary tests and calculations have been conducted which prove the feasibility of the magnetic movement but further investigations are needed.

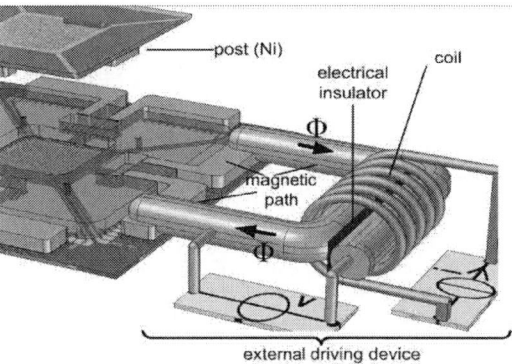

Figure 13: External driving device allows lateral magnetic movement and vertical electrostatic movement.

As a restoring means, a gel-like structure will be used, since it offers not only elastic properties in vertical but also in lateral direction. This combined actuation can also be used for the static alignment, by locking the μLens, once the position is obtained, with UV-curable adhesive. The gel bumps are then expected to counteract the retarding influence of the capillary effect of the liquid adhesive. The external driving device would only be attached during the alignment process.

Conclusion

A versatile structure for aligning a microlens array over a micro-UV-LED array has been demonstrated. Mechanical movement in the vertical direction has been tested by using a simple electrostatic actuator and gel structure for the restoring force. A concept for simultaneous magnetic and electrostatic movement has been proposed and partially tested.

Acknowledgements

The authors would like to acknowledge the financial support from the UK Engineering and Physical Sciences Research Council (EPSRC) through its Basic Technology Programme. The work was carried out under the project entitled "A thousand Micro-emitters per square millimetres" referenced GR/S85764.

References

[1] C. Pusarla and A. Christou, "Solder bonding alignment of microlens in hybrid receiver for free space optical interconnections," 1996.

[2] S. S. Lee, L. Y. Lin, K. S. J. Pister, M. C. Wu, H. C. Lee, and P. Grodzinski, "Passively aligned hybrid integration of 8 × 1 micromachined micro-Fresnel lens arrays and 8 × 1 vertical-cavity surface-emitting laser arrays for free-space optical interconnect," *IEEE Photonics Technology Letters*, vol. 7, pp. 1031-1033, 1995.

[3] G. C. Boisset, B. Robertson, W. S. Hsiao, M. R. Taghizadeh, J. Simmons, K. Song, M. Matin, D. A. Thompson, and D. V. Plant, "On-die diffractive alignment structures for packaging of microlens arrays with 2-D optoelectronic device arrays," vol. 8, pp. 918, 1996.

[4] M. T. Gale, J. Pedersen, H. Schutz, H. Povel, A. Gandorfer, P. Steiner, and P. N. Bernasconi, "Active alignment of replicated microlens arrays on a charge-coupled device imager," *Optical Engineering*, vol. 36, pp. 1510-1517, 1997.

[5] S. Eitel, S. J. Fancey, H. P. Gauggel, K. H. Gulden, W. Bachtold, and M. R. Taghizadeh, "Highly uniform vertical-cavity surface-emitting lasers integrated with microlens arrays," vol. 12, pp. 459, 2000.

[6] J. Jin, T. Higuchi, and M. Kanemoto, "Electrostatic levitator for hard disk media," *IEEE Transactions on Industrial Electronics*, vol. 42, pp. 467-473, 1995.

[7] W. O. J. C. Loetters, P. H. Veltink, P. Bergveld, "The mechanical properties of the rubber elastic polymer polydimethylsiloxane for sensor applications," *Journal of Micromechanic and Microengineering*, vol. 7, pp. 145-147, 1197.

LTCC Gas Flow detector

Dominik Jurków, <u>Leszek J. Golonka</u>, Henryk Roguszczak

Wroclaw University of Technology, Faculty of Microsystem Electronics and Photonics

Wybrzeze Wyspianskiego 27, PL 50-370 Wroclaw, Poland

E-mail: dominik.jurkow@wp.pl

Abstract

A new construction of Low Temperature Cofired Ceramics (LTCC) gas flow detector is presented in this article. The detector consists of gas channel and a cavity with axle and turbine. The turbine and axle are made independently without sacrificial layer. The rotational speed of the moving turbine depends on the gas flow velocity. The speed is measured by the optical method. The gas flow can be calculated on the base of the frequency. The optical components are integrated with the LTCC module. The test device is made with a transparent polymer cover. It allows on the observation of the turbine movement in a stage of a preliminary design. The construction of the detector and technological steps are described in the paper. The experimental results are presented.

Key words: LTCC, gas, detector, flow

Introduction

Standard gas flow detector works on the base of the thermal effects. The first LTCC flow sensor with heater and thermistors was presented by Gongora Rubio [1,2]. There is a great influence of gas/liquid temperature on a detector accuracy of measurements. Moreover, this type of detector can not work at too high temperature. The working temperature can be increased by use a special high temperature thermistor. Problem with different gas/liquid temperatures during measure can be solved by measurement of gas/liquid temperature. This improvement complicates the sensor structure and increases its cost.

The presented LTCC (Low Temperature Cofired Ceramics) gas flow detector works on the other principle – instead of heater and temperature sensor there is a moving turbine. The rotational speed of the turbine depends on the gas flow velocity. The speed is measured by the optical method. There are many advantages of the construction. The detector can work at high temperature with reactive gases because all gas channels and moving elements are made from ceramics. The problems appeared in the flow sensor based on thermal effects are solved in the mechanical device. Silicon microturbines were presented by Dziuban in his book [3]. The first information on moving turbines made in LTCC technology was presented by Peterson [4,5]. He used a special method with sacrificial materials for making such structures.

Presented in this paper gas flow detector is made in LTCC technology. The structure consists of gas channel and a cavity with axle and turbine. The turbine and axle is made independently without sacrificial layer. The test device is made with transparent polymer cover. It allows on the observation of the turbine movement at the stage of a preliminary design. The main parts of the final model are made in the LTCC ceramics. The turbine rotational speed is measured by optical method. Light source and detector measure the light impulses frequency. The gas flow can be calculated on the base of the frequency. The optical components are integrated with the LTCC module. The construction of the detector and technological steps are described in the paper. Moreover, the experimental results are presented.

Moving turbine technology

The turbines, vias, chambers and channels were made inside the green LTCC foils using an Nd-YAG laser (Aurel NAVS 30 laser trimming and cutting system, AUREL, Modigliana, Italy) [6,7].

The design of the turbines is shown in Figure 1. Four layers of DuPont 951 A2 foil were used to make the LTCC turbine. The thickness of each foil was equal to 137 µm after firing.

There were two methods of making the turbines. In the first method the turbines were cut independently in each tape and then stacked together and laminated. The process could be carry out in the other way. In the second method the tapes were laminated together and then the structures were cut by laser. The second solution gave better results. The elements were made more precisely.

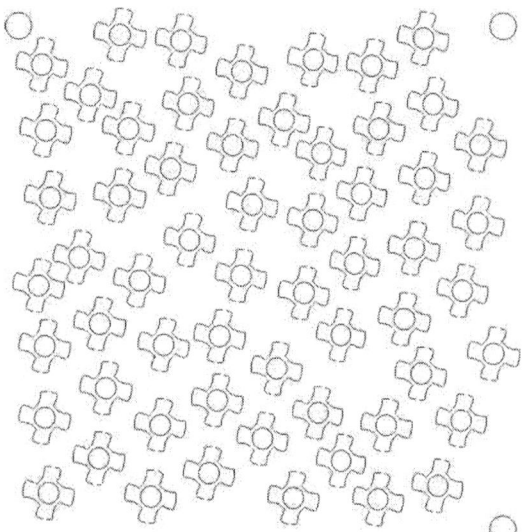

Figure 1: Design of turbines prepared for laser cuting

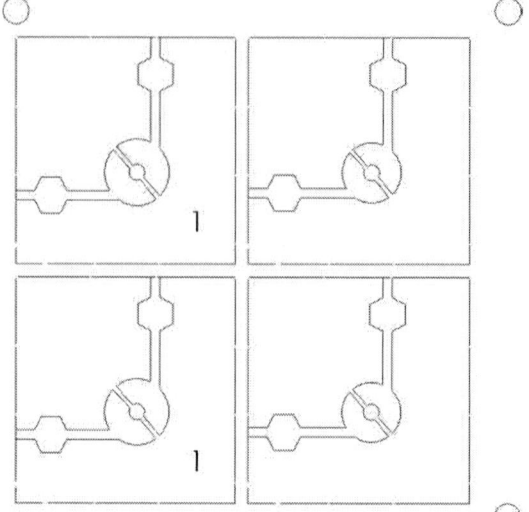

Figure 3: Design of the structures with axles

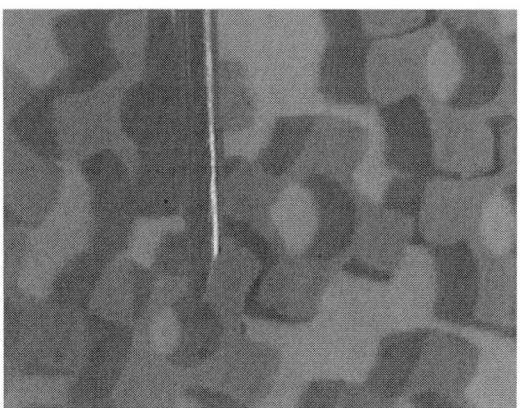

Figure 2: Green turbines before cofiring process

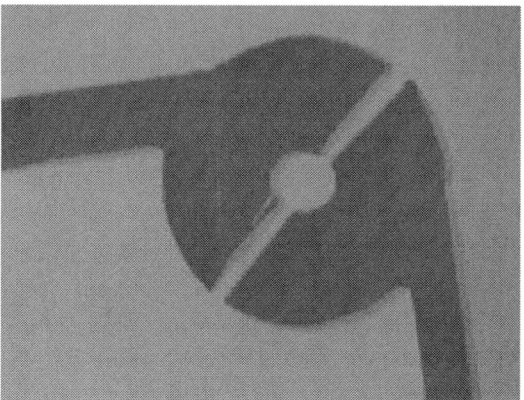

Figure 4: Part of the laminated structure with spokes carry the axle

Picture of green turbines before cofiring process is shown in Figure 2. Additionally a needle is presented in the Figure for a size comparison. Diameter of a single turbine was equal to 4 mm before and about 3.4 mm after cofiring process, respectively. The hole diameter inside the turbine was about 400 µm larger than an axle diameter. It was impossible for turbine to rotate when that different was smaller than 200 µm.

Technology of the structure with axle

The shape of the LTCC structure with channel, chamber, axle and optical windows was cut in green tape using the Nd-YAG laser. The detector part consists of 5 layers of tape. Two different diameters of turbine chambers were made equal to 5 mm or 4.5 mm before cofiring. The design of the LTCC structures with channel and axle is presented in Figure 3. The chamber diameter of structures marked with "1" is equal to 5 mm.

The structure was made in a few lamination steps. The laminated structure with spokes carry the axle is shown in Figure 4. The spokes were removed from the structure before the final lamination step. They could be removed, for example by cutting spokes ends by laser. In each lamination process the proper lamination parameters were very important. The device was leaky if the pressure was too low. The structure was collapsed in case of too high lamination pressure.

Deformation of the test structure is presented in Figure 5. The lamination pressure was too high in this case.

The final chamber structure with the axle and removed spokes is shown in Figure 6. The hole for optical fiber is visible on the lower part of the chamber. Holes for fiber made in the top and bottom part of the structure must be matched. Gas inlets and outlets were glued to the structure after firing. The glue had to bear high temperature and reactive gases.

160

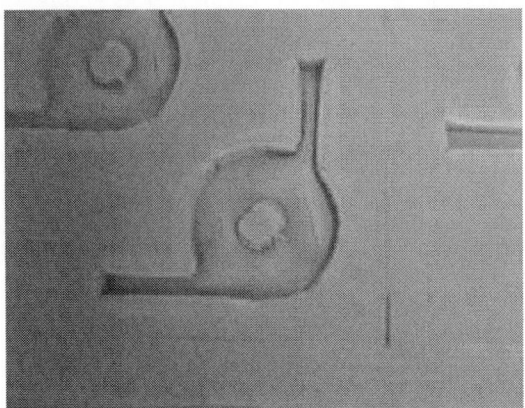

Figure 5: Deformation caused by too high lamination pressure

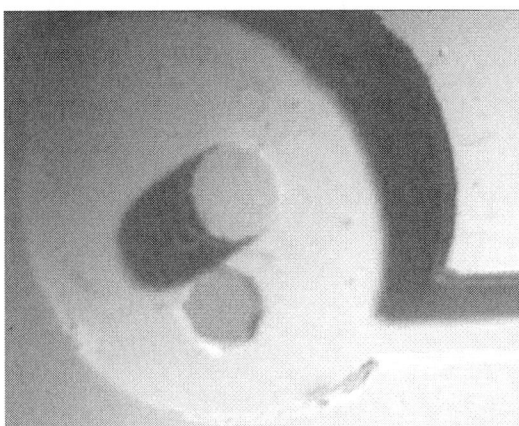

Figure 6: Final chamber structure with the axle

Optical system integration

The optical system was applied for detection of turbine rotation velocity. As a light source SMD EL 12-21UBGC LED was used. It is a high performance electro- luminescence diode with nominal wave-length equal to 502 nm. The light detection circuit consisted of high-sensitivity light to voltage converter TSLB257. This converter combines a photodiode, optical filter and transimpedance amplifier in a single monolithic CMOS integrated circuit. Its output voltage is proportional to the input light intensity.

The light source and detector can be integrated directly to the structure or mounted separately and connected with the structure via optical fiber glued to structure using polymeric, transparent adhesive. It assured good optical coupling and mechanical stability of the optical detection circuit.

Experimental

A light source circuit consisted of blue LED diode, resistor 1 kΩ and voltage supply 2.7 V. The proper supplying voltage was chosen. The light was so strong in case of too high voltage, that the detector was constantly detecting the light, even when the wing of the turbine overcast the light from the LED diode. In this case it was impossible to measure frequency of the movement of the turbine. Too low voltage gave not strong enough light intensity to the detector.

A light detector system consisted of integrated element TSLB 257, a resistor 322Ω in voltage supply circuit and a oscilloscope (Figure 7).

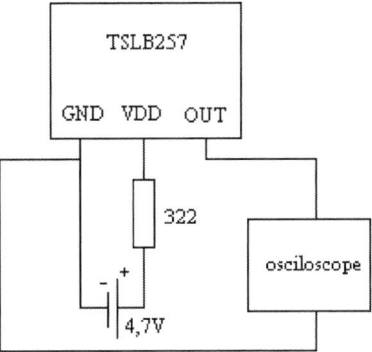

Figure 7: Schematic of the light detector test circuit

During experiment the frequency was measured by a digital oscilloscope. A constant line was observed on the oscilloscope screen when it a gas did not flow. The picture on the oscilloscope screen was changed when gas flow reached a detectable value. The frequency of switching from low signal to high signal and vice versa gave information on a flow velocity. Up to now the device is used as a flow detector.

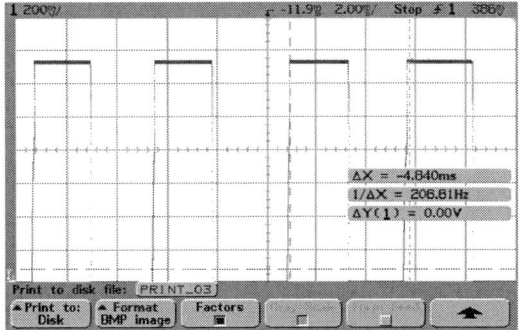

Figure 8: Detected flow visible on the oscilloscope screen

The oscilloscope screen with trace of signals measured at the light detector for gas flow about 7 ml/s is presented in Figure 8. The trace shows two main advantages of the device: possibility of not only detection but also measurement of a gas flow and moreover, high sensitivity. Each impulse will have the same frequency as long as flow does not change. Turbine has four wings. One full rotation detector generates four signals. Signal frequency is changing immediately when gas flow velocity is varied. There is a very small inertia of turbine because of its low mass.

Conclusion

The detector was made successfully. The technology process is simple and inexpensive.

The rotational speed of the moving microturbine depends on the gas flow velocity. The speed is measured by the optical method. Wings of turbine cause generation of the light impulses with various frequency which are detected by detector. The light impulses are then transform to the electric signals. It is possible to calculate a gas flow on the base of the frequency measurement. The device can work at high gas temperature. Rotation velocity of turbine does not depend of the gas temperature and therefore there is no influence on the detector accuracy. The detector is suited for various applications. Detection can be made in environment with a great danger of explosion, because there is no electric signal in the channel with the measured gas. Detector is sensitive enough to be used as a flow sensor. The future work will be devoted to this topic.

Acknowledgements

The authors wish to thank the Polish Ministry of Science and Higher Education for financial support (grant No. R02 017 02).

References

[1] M.R. Gongora-Rubio, P. Espinoza-Vallejos, L. Sola-Laguna, J.J. Santiago-Aviles, "Overview of low temperature co-fired ceramics tape technology for meso-system technology (MsST)", Sensors and Actuators A 89, pp. 222-241, 2001.

[2] M. Gongora-Rubio, L.M. Solá- Laguna, P.J. Moffett, J.J. Santiago-Avilés, "The utilisation of low temperature co-fired ceramic (LTCC-ML) technology for meso-scale EMS, a simple thermistor based flow sensor", Sensors and Actuators A 73, pp. 215-221, 1999.

[3] J.A. Dziuban, „Bonding in microsystem technology", Springer, Chapter 3, pp. 91-104, 2006.

[4] K.A. Peterson, K.D. Patel, C.K. Ho, S.B. Rohde, C.D. Nordquist, C.A. Walker, B.D. Wroblewski,
M. Okandam, "Novel microsystem applications with new techniques in Low Temperature Co-Fired Ceramics", International Journal of Applied Ceramics Technology, vol. 2, pp. 345–363, 2005.

[5] K.A. Peterson, K.D. Patel, C.K. Ho, B.R. Rohrer, C.D. Nordquist, B.D. Wroblewski, K.B. Pfeifer, "LTCC microsystems and microsystem packaging and integration applications", Proc. IMAPS/AcerS 2nd Int. Conf. and Exhib. on Ceramic Interconnect and Cer. Microsystem Technologies (CICMT), Denver (USA), April 2006.

[6] J. Kita, A. Dziedzic, L.J. Golonka, T. Zawada, "Laser treatment of LTCC for 3D structures and elements", Microelectronics International 19 (3), pp. 14-18, 2002.

[7] L.J. Golonka, "Technology and applications of Low Temperature Cofired Ceramic (LTCC) based sensors and microsystems", Bulletin of the Polish Academy of Sciences Technical Sciences, Vol. 54, No. 2, pp. 223-233, 2006.

Film assisted molding technologies and applications for high volume MEMS and Sensor packages

Frank Boschman, Ton van Weelden, Arnold Bos

Boschman Technologies B.V. Stenograaf 3, Duiven, The Netherlands

Phone +31 26 319 4900, E-mail Address: FBoschman@boschman.nl

Abstract

The increase of the application of Film Assisted Molding technology confirms the importance of this new encapsulation technology for the Semiconductor Industry. The introduction of film in the mold has enabled a process technology using standard pellets that addresses the packaging of advanced silicon devices. The new challenge is the encapsulation of MEMS or sensor packages. MEMS packages often need the access to the environment to make its measurements. FAM technology is an enabling material and process technology that addresses the packaging technology in keeping the functional area open during the encapsulation. FAM technology is also the solution for clean room molding. By carefully analyzing the features of the FAM technology it is possible to manufacture economically advanced packages in high volume production, achieving the highest device reliability. The presentation will explore the possibilities and the benefits that FAM can bring to the encapsulation of MEMS packages.

Key words: Film assisted molding, Low cost film solutions,

Sensing new package opportunities for MEMS and Sensors

Many manufacturers in the semiconductor industry have been inspired by the possibilities of micro-electro-mechanical systems (MEMS), particularly in the search for new market opportunities. MEMS offer this hope, but there have been logjams in the flow from good ideas to successful, money-making products. While many interesting structures have been produced in silicon and other materials using traditional front-end processing techniques - lithography, etch, deposition . . . – there has been less success in the back-end, particularly in finding a suitable packaging technology.

As often happens, packaging considerations have only come in as an afterthought at the end of the design process. While electronic devices only need electronic contact with the external world, MEMS often need both protection from and intimate connection with various physical parameters. Further, many of the packaging production processes commonly used in mass production of electronics can be quite brutal. For example, a pressure sensor chip, while requiring protection from mechanical stress, also needs access to the environment to make its measurements. Such components can contain vacuum regions that while able to withstand the normal temperatures and pressures of the atmosphere would break under the temperatures, pressures and mechanical stress of some packaging technologies such as moulding. In such situations, MEMS package designers have resorted to more elaborate techniques such as dispensing, where the package is built up from material dispensed from a needle. However, this technique is expensive, time-consuming and produces less robust structures. For example, some dispense materials can absorb moisture with the potential for failing in its job of protecting the device. Alternatively, plastic hoods can be manufactured separately and glued over the component. Again, the resulting parts are generally subject to stress problems, making for less reliable and expensive products.

Ideally, package designers would like to use a semiconductor mass-production moulding technology that creates packages with encapsulation-free surfaces where the chip needs to contact the environment.

Low cost film solutions

Moulding expert Boschman Technologies of the Netherlands believes it has the answer with the Film

Assisted Molding (FAM) technique developed by its technologists since the early 1990s. FAM tackles two problems with moulding - release from the mould and keeping die or metal surfaces clear of moulding compound. The first problem - release from the mould – is tackled by a "seal film" made of a Teflon-based material. This overcomes many problems with other release techniques such as coating the mould with Teflon or using releasing agents. Teflon surfaces are easily damaged and mould compounds contain abrasive "filler" particles. Releasing agents are designed to stop the encapsulant sticking to the mould metal, yet at the same time the mould compound is required to make a firm attachment to the metal of the component. In any case, these techniques further need mould cleaning after each production step, introducing compound and other particles into the atmosphere - clearly not desirable in clean environments. With sensors, the problem is even more critical since micron-size particles can be sufficient to damage the active surface. In microelectronics, the component can be sealed with a passivation layer in the front-end processing facility - something that is not always possible with sensors. With Boschman's seal film, the encapsulant never comes in contact with the mould metal. The protection for each production shot is provided by a fresh piece of film. Since the release problem has been essentially solved, one can consider different mould compounds with stickier properties or ultra low viscosity. One can even remove the filler from the compound and produce packages from the extremely sticky clear epoxy base. The films also act as "gaskets", providing a good seal and stopping encapsulant material flowing out of the moulding area. Without film, the metal-on-metal contact areas of the two mould halves have to be pressed together at extremely high pressures - creating more potential for damage of delicate components. The gasket effect allows these pressures to be reduced.

In the last two years, Boschman has developed its technique so that film can be used on both sides of the device with sealing film on the top and either sealing or adhesive film on the bottom, depending on application. Further, the sealing film has enabled the creation of encapsulant-free surfaces at less mechanical stress than rubber insert techniques. The technique with seal film on both sides of the device is particularly useful for leadframe mounted chips requiring package structuring on both sides. The adhesive film option is used in array packaging and leadless parts. Ton van Weelden, Boschman vice-president for process research, sees a number of opportunities for double film encapsulation in MEMS and sensor production. "Such optical, pressure, temperature or humidty sensors are expensive to produce at the moment because of the limitations of packaging technology," says van Weelden. The company has worked with customers to produce sensor packages that have encapsulation free areas with dimensions from the order of centimetres down to 400x200microns. Van Weelden sees no essential problem in making even smaller encapsulation structures for biochemical (e.g. for DNA testing) or other sensing applications.

Successful application

Van Weelden and company owner Frank Boschman stress the need for early involvement of packaging considerations in product design and development. The semiconductor industry has often fallen down in this respect, ad-hoc, piecemeal fixes are rushed into place at the last moment. This creates

a preference for well-understood processes that "work". While such caution and conservatism has served the semiconductor industry, it is no longer adequate to the needs of new sensor technologies such as MEMS. "It is vital that MEMS sensor designers involve package specialists in the design flow early on in the development phase to ensure a successful product," says van Weelden. Boschman has built up facilities to enable it to provide cost-effective, low-volume samples to support such development work.

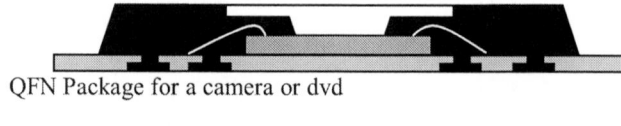

QFN Package for a camera or dvd

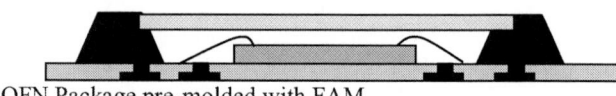

QFN Package pre-molded with FAM

BGA Package for fingerprint scanning

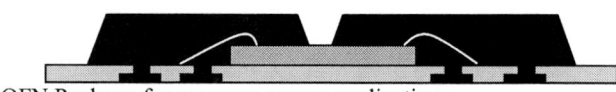

QFN Package for pressure sensor application

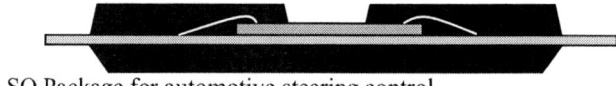

SO Package for automotive steering control

QFP Package for MEMS application

Transmission Line Pulsing Behavior of Thin Film Resistors

D. Bonfert[a], H. Wolf[a], H. Gieser[a], G. Klink[a], P. Svasta[b]

[a] Fraunhofer Institute Reliability and Microintegration (FhG – IZM-M)
Hansastr. 27d, 80686 Munich, Germany
Phone: (+49)089-54759-195; Fax: (+49)089-54759-100, E-mail: detlef.bonfert@izm-m.fraunhofer.de

[b] University Politehnica of Bucharest, Romania
Center for Technological Electronics and Interconnection Techniques, UPB-CETTI
Phone: (+40)21-316 9633; Fax.: (+40)21-316 9634, E-mail: paul.svasta@cetti.ro

Abstract

One of the most important issues of resistor properties is the value stability under different electrical and non electrical influences. Mechanical or thermal stress together with the electrical one represents the main factors that have a major contribution in resistor value stability. With shrinking resistor geometries of discrete, integrated and integral passives, the electrical stress gains more and more importance, especially under short duration voltage pulses. The analytical test technique of Transmission Line Pulsing (TLP) allows, on the basis of square pulses, the in-situ monitoring of the voltages and currents at the DUT and helps to gain fundamental insights into the electrical behavior. The resistance change due to an applied high voltage pulse is a measure of the electrostatic discharge susceptibility of thin film resistors. The influence of the pulse width (1.5 ns, 10 ns, 100 ns and 300 ns) and amplitude on the susceptibility was investigated on thin film resistors. By means of the Transmission Line Pulsing technique (TLP), the ESD behavior of thin film resistors is presented

Key words: Transmission line pulsing; Thin film resistors; Resistance change; ESD behavior

Introduction

Passive components are becoming susceptible to ESD events as their geometries are shrinking. Integrating and embedding makes them even more vulnerable. Thin film is one main technology in producing resistive passive components of high stability, widely used as chip components, having geometries in the mm- and even sub-mm-range.

The ESD behavior of thin film resistors has been reported in several studies, using rectangular pulses [1], [2] or RC-pulses [3], [4] and [5]. They all report a resistance change after pulse application. A thermal failure model for thin film resistors is developed by D.C. Wunsch [2], similar to that for semiconductor devices. Overheating is considered to be the principle failure mechanism.

An efficient technique for ESD measurements is to use square pulses. Rectangular pulses from Transmission Line Pulsers (TLP) are of particular interest for ESD measurements. They have been well established for the analysis of ESD-protection structures [6]. This analytical test technique on the basis of square pulses allows the in-situ monitoring of the voltages and currents at the DUT, during pulsing and helps to gain fundamental insights into the electrical behavior.

The TLP test-method was applied to characterize the ESD-susceptibility of film chip resistors on alumina substrates of different, commercially available values and sizes, as they are used in today's systems [7], [8]. In this work the TLP-behavior of Al thin film resistors on silicon substrates will be presented

Test Setup

The test setup for pulsing thin film chip resistors is shown in Figure 1. It consists of the pulse generator (TLP), the DC measurement and the test fixture with the DUT. The used Transmission Line Pulser (TLP) is a Celestron-I type from Oryx Instruments Corporation. It is a flexible, two terminal bench top system for fast, accurate and reliable characterization of advanced semiconductor structures [9]. The system is capable of delivering up to 10 A into a 50 Ω load, representing a dissipated power per pulse of 5 kW.

For the characterization of thin film chip resistors the Time Domain Reflection (TDR) configuration of the TLP was used, with 1.5 ns, 10 ns, 100 ns, and 300 ns wide pulses. Figure 2 presents a simplified schematic of the TLP system for this configuration.

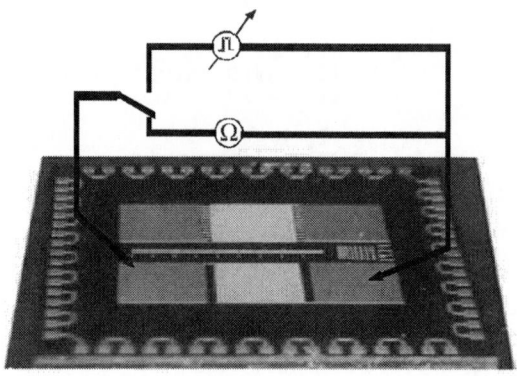

Figure 1: Principle of DUT stressing and measuring

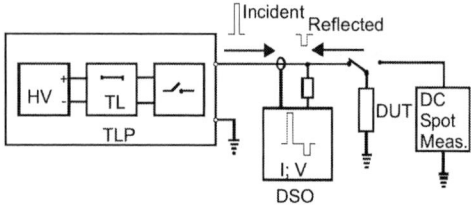

Figure 2: Transmission Line Pulser (TLP) schematic

An incident voltage pulse, defined by the length of charging transmission line (TL) travels from the pulse generator to the DUT and may be reflected at the DUT. The voltage of the incident and reflected pulse is measured with a passive voltage probe close enough to the DUT, so that both pulses can be recorded. The current of the incident and reflected pulse is measured by a current probe. For pulse widths above 30 ns incident and reflected current pulses overlap, so that they can be directly measured (TDR-O configuration). For pulse widths below 10 ns incident and reflected voltage and current pulses are delayed, so they need to be superimposed, according to the following equations (TDR-S configuration) in order to get the voltage at and the current through the device:

$$V_{DUT}(t) = V_{incid}(t-t_{incid}) + V_{refl}(t-t_{refl}) \quad (1)$$

$$I_{DUT}(t) = I_{incid}(t-t_{incid}) + I_{refl}(t-t_{refl}) \quad (2)$$

Typical measurement results of the incident and reflected voltage across and the current into the DUT are presented for 10 ns in Figure 3. Applying (1) and (2) to these curves yields the voltage at and the current through the DUT, represented in Figure 4.

For 100 ns and wider pulses the measured incident and reflected voltage and current overlap, as presented in Figure 5, so no further calculation is needed.

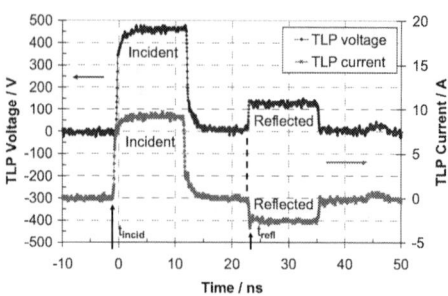

Figure 3: Incident and reflected voltage at and Incident and reflected current through the DUT for 10 ns TLP (TDR-S)

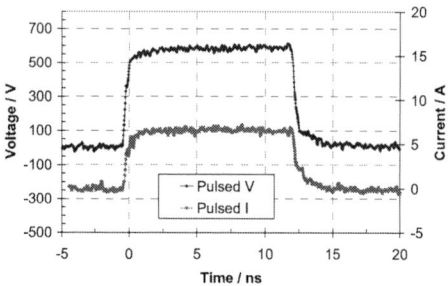

Figure 4: Superimposed pulsed voltage at and pulsed current through the DUT for 10 ns TLP (TDR-S)

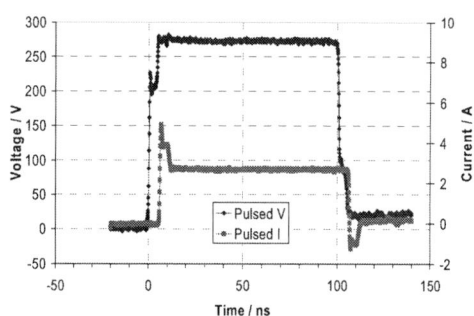

Figure 5: Pulsed voltage across and pulsed current into the DUT for 100 ns TLP (TDR-O)

Measurement

Prior to the pulsing, a DC measurement was performed to measure the initial value of the resistor. During this measurement heating of the resistor was avoided, measuring well below its maximum DC power ratings.

During the TLP measurement the pulse amplitude was stepwise increased, as shown in figure 6. The resistors voltage and current transients were recorded for each pulse, leading to the pulsed I/V- characteristic of the DUT. From this characteristic the R/V-behavior during the pulses is

166

calculated (pulsed R) as well as the pulsed relative resistance change to the starting value (pulsed dR/R_0).

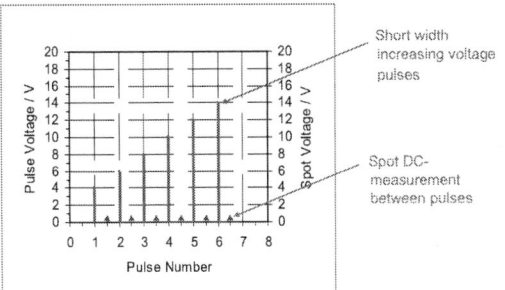

Figure 6: Principle of measurement technique

Additionally, after each pulse, a DC spot measurement was performed in order to detect a possible resistance change. This leads to the DC R/V characteristic, as well as to the relative resistance change to the starting value (DC dR/R_0).

To see whether the resistance shift after the first measurement remains constant, successive pulsed measurements were performed, unless the devices failed.

Applying multiple high voltage pulses with constant height gives information about the robustness of the resistor to the TLP stress.

The tested resistors were thin film aluminum meander lines patterned on a silicon substrate, with a thickness of 1 μm. The meander resistor covers an area of 1000x1000 μm², the width and the spacing of the lines is 20 μm. The resistor is isolated from the Si substrate by a 500 nm SiO_2 layer.

Measurement results

Influence of pulse amplitude on resistor

The TLP measurements were performed increasing the pulse amplitude from 4 V to 1000 V by a 2 V step. Recording the measured voltage and current transients at the DUT during each voltage step, makes possible a 3D representation of the measurement, as depicted in figure 7 and figure 8. As an example the 100 ns pulses are represented, showing, that during the pulse, the pulsed resistance, calculated from each corresponding point of the current- and voltage transients, increases with applied TLP voltage, due to Joule heating.

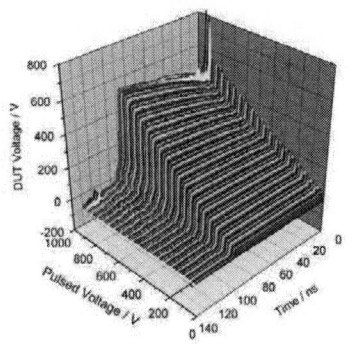

Figure 7: Pulsed voltage at the thin film resistor during 100ns TLP stress

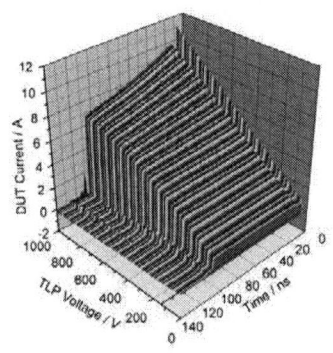

Figure 8: Pulsed current through the film resistor during 100ns TLP stress

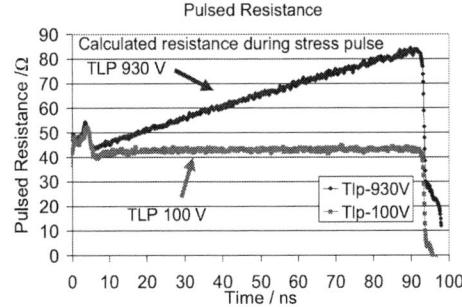

Figure 9: Pulsed resistance during pulsing (100 ns TLP) w/o and with heating, due to applied voltage level

Representing the pulsed resistance versus time (pulse width) for two different applied pulse voltages, reveals the amount of resistance change during the high voltage pulse of about 100%, as shown in figure 9. The initial peak during the first 5 ns is due to the TDR-O measurement method and is of no significance. If the voltage pulse is too high, voltage breakdown followed by resistor melting will occur, as depicted in figure 10. The corresponding failure signature can be seen in figure 11, were the bottom line of the meander is melted away.

167

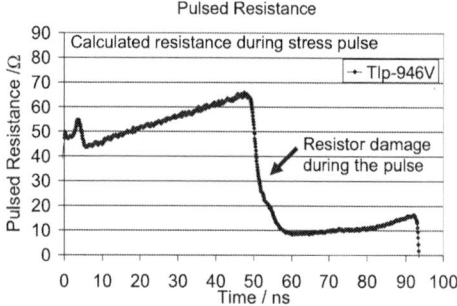

Figure 10: Pulsed resistance during pulsing (100 ns TLP) with breakdown, due to applied over-voltage level

Figure 11: Failure signature after over-voltage pulsing (100 ns TLP) with resistor distruction

Influence of pulse width on resistor breakdown

Making TLP measurements with increasing pulse amplitude and for different pulse widths gives the information of the high voltage behavior of the resistor. Figure 12 shows the current voltage characteristic for the tested pulse widths of 1.5 ns, 10 ns, 100 ns and 300 ns. Figure 13 and figure 14 depict the corresponding relative DC- resistance change and the relative pulsed resistance change, respectively. For 1.5 ns and 10 ns the system limit is about 400 V at the DUT and no damage can be seen, the DC resistance change is low, less than 0.05 %. The pulsed resistance shows some heating.

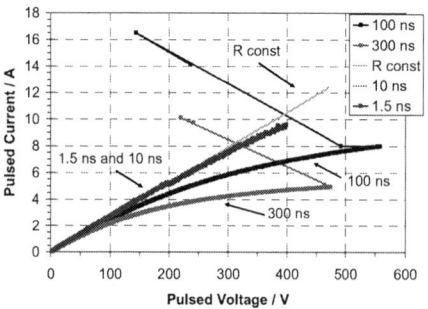

Figure 12: Pulsed high current-voltage characteristic of the resistor for different pulse widths

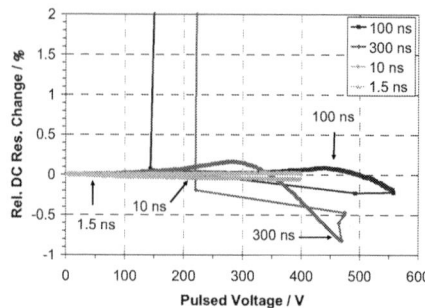

Figure 13: Relative DC resistance-change versus voltage for different pulse widths

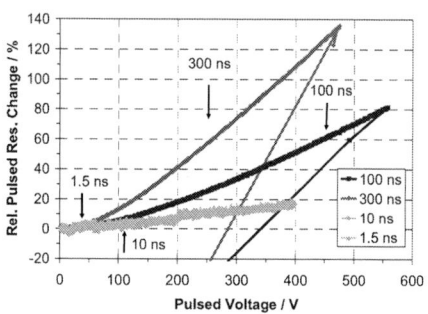

Figure 14: Relative pulsed resistance change versus voltage for different pulse widths

For longer pulses the changes are more dramatically. For 100 ns pulses breakdown occurs at 558 V at the DUT, for 300 ns pulses damage appears at 475 V. The relative DC resistance changes are even at the breakdown voltage relative low (-0.2 % and -0.8 % respectively).

The pulsed resistance shows changes for all pulse widths because of heating during the pulse. Beginning with 20 % of change for 1.5 ns and 10 ns, the changes will rise up to 80 % for 100 ns and 135 % for 300 ns.

Influence of pulse number on resistor

Applying multiple high voltage pulses, below the breakdown value, and with constant height gives information about the robustness of the resistor to the TLP stress. Up to 4000 pulses were applied and at the beginning the resistance decreases up to a minimum value after which the resistance increases. The results are summarized in figure 15, indicating for 100 ns a change of -/+ 0.4 % and for 300 ns - 1.2 % to +1.6 %. During this stress the structure of the resistive material (Al) changes as can be seen from figure 16 and figure 17. Re-crystallization follows the current lines, which are non- uniform at the corners, leading to increased current densities and even to melting of the metallization.

168

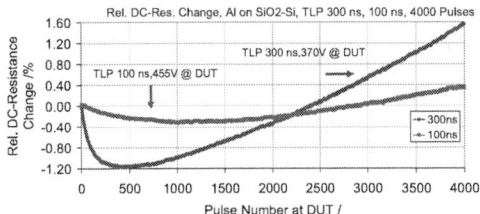

Figure 15: Relative DC resistance-change during repetitive constant pulsing, for 100 ns and 300 ns TLP

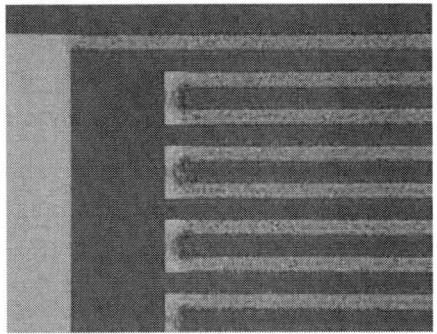

Figure 16: Resistive material changes after repetitive constant pulsing, for 100 ns TLP

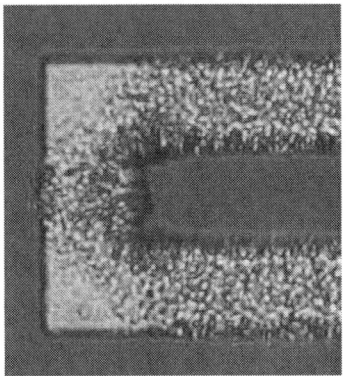

Figure 16: Close view-up of meander edge resistive material changes after repetitive constant pulsing, for 100 ns TLP

Influence of successive pulsing on resistance

Applying successive pulsed measurements with stepwise increasing voltages, up to the same final non destructive voltage, tells about the irreversible changes that take place in the resistor. The highest deviation is obtained with long 300 ns pulses. Figure 17 depicts the DC resistance changes for five successive measurements, M1 – M5, and the corresponding relative resistance changes are shown in figure 18. After the first pulsed measurement the resistor undergoes the highest change of -0.2 %. For the following measurements the changes will reduce settling to a value of +0.05 %, showing that the changes are finished

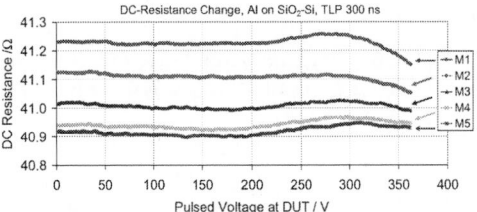

Figure 17: DC resistance change versus applied pulse voltage for successive measurements, for 300 ns TLP

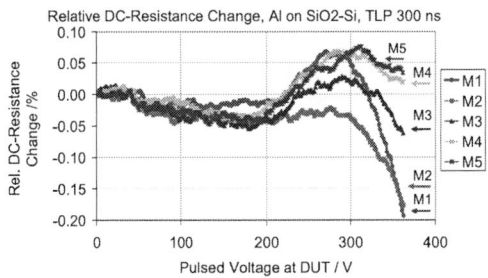

Figure 18: Relative DC resistance change versus applied pulse voltage for successive measurements, for 300 ns TLP

The pulsed resistance changes obtained during the pulses are much higher because of self heating of the meander resistor and reach values of about 100%, as shown in figure 19. But these changes are reversible because the curves do not differ between different successive measurements.

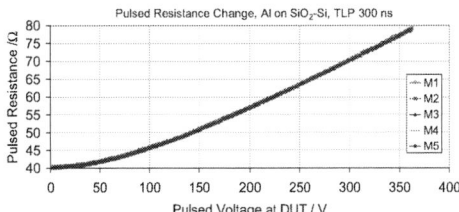

Figure 19: Pulsed resistance change versus applied pulse voltage for successive measurements, for 300 ns TLP

Conclusions

The behavior of thin film resistors due to applied Transmission Line Pulses was presented. It was shown that these resistors are susceptible to high energy pulses. The applied pulses can bring the devices to parametric failure, altering the resistance value and eventually exceeding the accepted tolerance, although the device is still functional. The amount of resistance change depends on pulse width and amplitude. The applied pulses can also cause a catastrophic damage of the resistor if the current capabilities of the resistor are exceeded or if voltage breakdown occurs.

References

[1] Domingos, H. *et al*, "High Pulse Power Failure of Discrete Resistors", *IEEE Trans-PHP*, Vol. PHP-11, No. 3 (1975), pp. 225-229.

[2] Wunsch, D.C, "An Overview of EOS Effects on Passive Components" *3rd EOS/ESD*, 1981, pp 167-173.

[3] Lai, T., "Electrostatic Discharge (ESD) Sensivity of Thin-Film Hybrid Passive Components", *IEEE Trans-CHMT*, Vol. 12, No. 4 (1989), pp. 627-638.

[4] Bos, L., "Performance of Thin-Film Chip Resistors", *IEEE Trans-CPMT*, Vol. 17, No. 3 (1994), pp. 359-365.

[5] Vishay Intertechnology, Inc., Tech Note TN-0027-0702, "Resistor Sensitivity to Electrostatic Discharge (ESD), Rev. 20-Feb-07.

[6] Gieser, H. *et al,* "Very-Fast Transmission Line Pulsing of Integrated Structures and the Charged Device Model", *EOS/ESD Symposium Proceedings,* 1996, pp 85-94.

[7] Bonfert, D., *et al* , "ESD Susceptibility of Thick Film Resistors by Means of Transmission Line Pulsing", *IEEE ESTC, Symposium Proceedings* , Dresden, Sept. 5. – 7., 2006.

[8] Bonfert, D., *et al* , "ESD Susceptibility of Thin Film Resistors by Means of Transmission Line Pulsing", *IEEE SIITME, Symposium Proceedings* , Jassy, Sept. 21. – 24., 2006.

[9] Grund, E. and Gauthier, R., "VF-TLP System Using TDT and TDRT for Kelvin Wafer Measurements and Package Level Testing" *EOS/ESD Symposium Proceedings,* 2004, pp 183-190.

The Application and Characteristics of the Wire-penetration Films in the Use of Stack-die CSP (Chip Scale Package)

C.L. Chung* and S. L. Fu

I-Shou University, Department of Materials Science and Engineer, No.1, section 1, Shiuecheng Rd., Dashu Shiang, Kaohsiung Country, Taiwan, 84008, R.O.C.

Tel: 886-7-6577711 Ext.3121, Fax: 886-7-6578444, E-Mail: markchun@isu.edu.tw

Abstract

For the request of higher device density, the designs of stacked die mount process should be simplified and wire-penetrated. In this paper, we focus on defining the characteristics of the two wire-penetration films (WPF) materials, and then designing the process to stack dies. The cure kinetic and thermal resistances of the WPF material were analysis by Differential Scanning Calorimetry (DSC) and Thermogravimetric Analysis (TGA). The thermo-deformation and pressure-induced flow behaviors of the WPF were evaluated by penetration mode in Thermal Mechanical Analyzers (TMA) and dynamic mode in Rheology test for simulating die mount process, respectively. Commercial software ANSYS® was used to analyze the residual stress induced form the cooling shrinkage of the curing process after die mount. Our results revealed the cure kinetic, thermal stability, Rheology behavior and microstructure in detail for WPF material evaluation and process design reference that obviously would influence the quality of Stack-die CSP. Besides, the stress and warpage simulation of package with WPF were fully discussed.

Key words: Polymer, Materials, wire-penetration, Films, Stack-die CSP (Chip Scale Package).

Introduction

For the current advanced packages, Bismaleimide Triazine (BT) substrates are widely used for the purpose of PBGA, MCM with high-density I/O applications[1,2]. Stacked Chip Scale Package (CSP) are developed under the concept not only for the above advantages but also multifunctional applications[3]. There are some important papers discussed about triple-Chip stacked CSP[4], mold flow simulation[5], failure and fatigue life[6] and stress analysis of spacer paste[7]. Base on the mass product requirements, the die mount process should be simplified and improved. Cross-section of stacked CSP package can be seen in Figure 1. There are 4 dice in the package and some another information shown in this table.

Within the limited space/thickness, to carefully evaluate the WPF materials are vital for the quality controls of the stacked CSP package. Reported here were our updated studies about the roles played by WPF characters and microstructure during stacked die mount processes.

Process Flow of stack CSP.

Figure 2 shows the schematic illustrate of the process flow of stacked CSP package with WPF. Noteworthy, the support film using in wafer saw process also as the dice mount material in the following die bond process8.

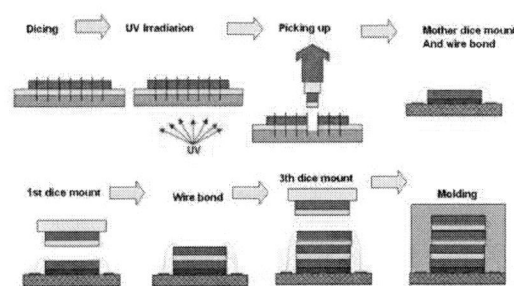

Schematic Process Flow of Stacked CSP

Figure 2: Schematic illustrate of process flow of stacked CSP package with WPF.

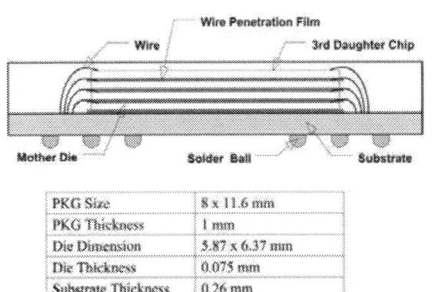

PKG Size	8 x 11.6 mm
PKG Thickness	1 mm
Die Dimension	5.87 x 6.37 mm
Die Thickness	0.075 mm
Substrate Thickness	0.26 mm

Figure 1: Detailed structure and cross-section of 4-layer stacked CSP package.

Wire-Penetration Films (WPF):

Table 1 lists the material data and process parameters of WPF, which were recommended from the engineering experiences specifically for the attached material. The films were a multifunctional tape, which consists of the function for dicing tape and die bonding tape.

Table 1: The materials data and process parameters of two wire-penetration films (WPF) in the stacked CSP package.

	WPF -H	WPF -L
UV Irradiations	100~200 mJ/cm^2	70~200 mJ/cm^2
UV function	Before die mount for release the die from base film	After wafer mount to cured the film for good adhesion on wafer
Resin formula	Thermosetting resin type. Multi-function epoxy resin. High flow epoxy resin. Acrylic rubber , Inorganic filler	UV cured and thermal cured thermosetting resin
filler	0.2-2um silica powder	3-30um silica powder
Thickness	60μm	Same
Lamination temperature	60-80 °C	40-50 °C
Die mount condition	150 °C, 1sec, 200-800g/cm^2	Same
Before molding thermal history	Wire bond 130 °C, mold preheat 150 °C	Same
Die attach curing condition	130°C,20min	120-140°C, 1hr
Mold temperature	175 °C	Same
Tg	153 °C	142 °C

Penetration test

A routinely calibrated Dynamic Mechanical Analysis (Perkin-Elmer DMA7) equipped with an ice or LN$_2$ bath cooling accessories is used. Heating rate of normal run is 5°C/min. Scanning range was from 25 °C to 230 °C for WPF after UV irradiation process. The touching area on film was circular shape about 1mm in diameter.

Scanning Electron Microscope (SEM).

Hitachi$^@$ S-3500N is used to observe the cross-sectional failed sample sealed with epoxy resin under high magnification after the reliability test.

Rheology test

A routinely calibrated ARES strain control type (Rheometric Scientific) equipped with an ice or LN$_2$ bath cooling accessories is used. Heating rate of normal run is 5°C/min. Scanning range was from 25 °C to 230 °C for UV cured WPF. The dynamic test for G' and G" were examined by parallel plate disk tool with 1 Hz frequency.

Stress Simulation

Commercial software ANSYS® was used to analyze the residual stress induced from the cooling shrinkage of the curing process after die mount process. A three-dimensional finite element model contained substrate, WPF, and die were built for the simulation. A temperature drop form 150°C to 25°C was applied as thermal loading for the package structure to investigate the stress distribution due to the thermal mismatch from the thermal expansion coefficient of each material after the WPF material curing. The results of warpage and 1st principal stress were investigated for the die crack probability, and the conditions of different WPF materials and die thickness were compared. Also two-dice and fore-dice stacked conditions were studied in this analysis.

Results and discussion

Microstructrue of WPF

Figure 3 show the SEM and EDS elment analysis on the (a) WPF-H and (b) WPF-L package cross-section. The filler distribution of WPF-H and WPF-L are mono-model (filler diameter ca. 0.5-2.0um) and bi-model (5-15um and 30-50um), respectively. The EDS results show the fillers material are SiO$_2$ that is very harden material. There is a concern about die surface damage during the die mount process even a few special big filler in the WPF-L system.

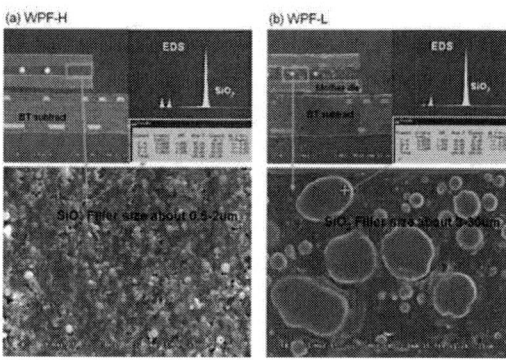

Figure 3: SEM and EDS analysis of (a) WPF-H and (b) WPF-L after die mount process.

Curing Kinetic Analysis

Shown in Figure 4 were the DSC thermograms of (a) WPF-H and (b) WPF-L after UV irradiation. The WPF-H revealed an earlier and less exothermic heat than WPF-L. The initial cure point of WPF-H and WPF-L are at 130°C and 145°C, and the exothermic heats are -72.7 and -95.9 J/g for WPF-H and WPF-L system. It is mean that the WPF-L system is more stable than WPF-H system during the ambit storage and followed assembly process.

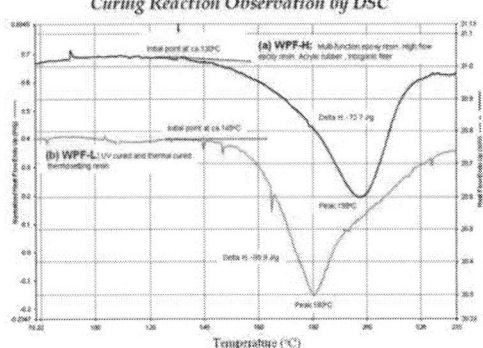

Figure 4: DSC heating thermogram of (a) WPF-H and (b) WPF-L after UV Irradiation

Thermal Resistance Analysis

Given in Figure 5 were the TGA thermograms of (a) WPF-H and (b) WPF-L after UV Irradiation. Both WPF materials revealed very stable and without serious degradation before 350 °C. The thermogram curves show the SiO_2 filler contents of WPH-H and WPF-L are 47.5 wt% and 42.5wt%.

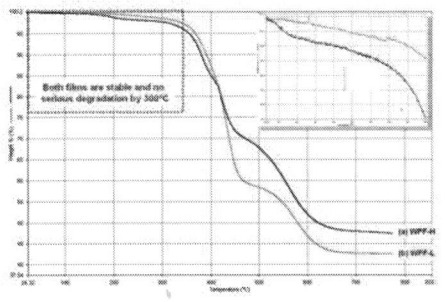

Figure 5: TGA heating thermogram of (a) WPF-H and (b) WPF-L after UV irradiation

Simulating Die Mount Process by Penetration Test

The penetration tests were designed to simulate the practice die mount process. Shown in the Figure 6 were the height% evolution of (a) WPF-H and (b) WPF-L during the heating process. The soften point of WPF-H and WPF-L were 40 and 50 °C, respectively. The remained heights of WPF-H and WPF-L are 0 and 27 %, respectively. The data indicate that WLF-H exhibit better flowability and good for wire penetration process than WLF-L under the same force and temperature conditions.

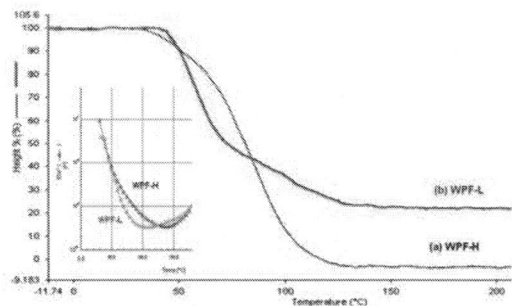

Figure 6: Height% evolution of (a) WPF-H and (b) WPF-L after UV Irradiation during the heating process.

Rheology test

Combination of Rheology behavior and cure kinetic are very important information for die mount condition designs. Shown in Figure 7 were the viscosity curves of (a) WPF-H and (b) WPF-L after UV Irradiation. The viscosity curve increase point identified the initial point of crosslinkage that are almost the start point of crosslinking reaction. The viscosity of WPF-H and WPF-L were increased located ca. 150 and 170°C for, respectively. The WPF-L with wider lowest part of the viscosity curve that revealed wider process window than WPL-H system. Noteworthy, WPF-H exhibit higher the viscosity at ambit temperature will induce die broken issue during die pick up process.

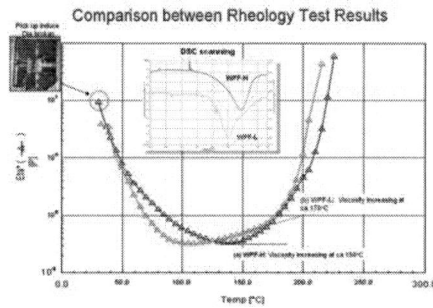

Figure 7: The viscosity curve of (a) WPF-H and (b) WPF-L after UV Irradiation.

Stress Simulation

Figure 8 and 9 show the package warpage in Z direction pictures and the first principal stress of dice with WPF-L after die mount process. Comparing with the first three figures, thinner the die introduces much serious effect to the total package warpage than change the WPF materials same effect could be also found for the warpage of the dice. Besides, more dice stacked will constrain the warpage from the substrate and reduce the total warpage of the package. As the mother die located between the soft substrate and the hard ceramic dices, the maximum stress could be obtained at the mother die from the red regions of the figures.

Compare with the Figure 9, it is found that much dice stacking causes higher stress to the dice, especially for the mother die. Moreover, the maximum and minimum stresses located closely at the corner of the mother dice makes cracks at the corner of the chips much easily.

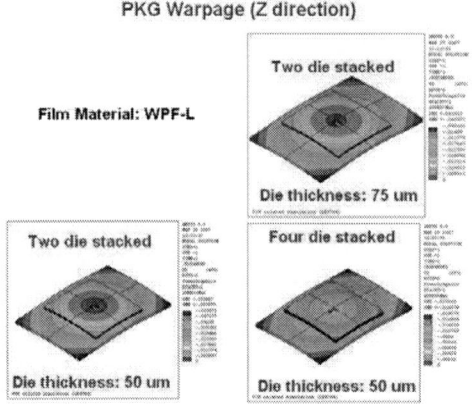

Figure 8: Package warpage in z-direction pictures with WPF-L after die mount process.

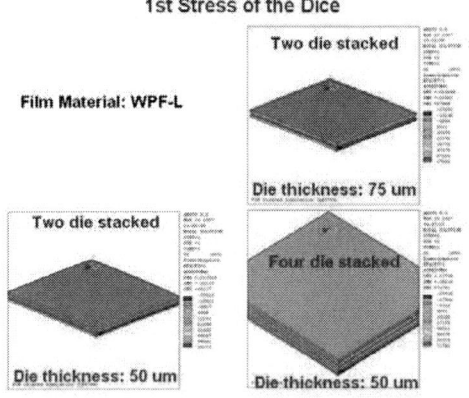

Figure 9: First stress of dice with WPF-L after die mount process.

Conclusions

Base on cure kinetic study, it is clear that the WPF-L system is more stable than WPF-H system for the ambit storage and whole assembly process. Rheology and penetration test results also indicate The WPF-L with relative wider working window in die mount process than WPL-H system. Stress simulation reveal thinner the die introduces much serious effect to the total package warpage than change the WPF materials same effect could be also found for the warpage of the dice. It is found that much dice stacking causes higher stress to the dice, especially for the mother die.

Noteworthy, WPF-H exhibit higher the viscosity at ambit temperature will induce die broken issues during die pick up process. WPF-L system with a few special big filler has a concern about die surface damage during the die mount process.

Acknowledgements

Many thanks are presented to Dr. Lin, HC, Alex, Dr Hsu and professor Liu for his great helps in the study.

References

[1] John H. Lau, 'Chip on board technologies for mulitchip modules', Van Nostrand Reinhold, New York, NY, Chapter 6, pp. 251-274, 1994

[2] Basavanhally, N. R., D. D. Chang, B. H. Craston, and S. S. Seger, Jr., 'Direct chip interconnect with adhesive-connector films', Proceedings 42nd IEEE Electronic Components and Technology Conference, pp. 487-491, 1992

[3] Estes, R. H., 'A practical approach to die attach adhesive selection', Hybrid Circuit Technology, pp. 44-47, June 1991.

[4] Yasuki Fukui, Yuji Yano, Hiroyuki Juso, Yuji Matsune, Koji Miyata, Atsuya Narai, Yoshiki Sota, Yoshikazu Takeda, Kazuya Fujita, Morihiro Kada, 'Triple-Chip Stacked CSP', 2000 Electronic Components and Technology Conference, PP. 385-389.

[5] Min Woo Lee, Jin Young Khim, Min Yoo, JiYoung Chung and Choon Heung Lee. 'Rheological Characterization and Full 3D Mold Flow Simulation in Multi-Die Stack CSP of Chip Array Packaging', 2006 Electronic Components and Technology Conference, pp. 1030-1037

[6] J. D. Wu, S. H. Ho, P. J. Zheng, C.C. Liao, and S. C. Hung, 'An Experimental Study of Failure and Fatigue Life of a Stacked CSP Subjected to Cyclic Bending', 2001 Electronic Components and Technology Conference.

[7] Jack Zhang , James T. Huneke, 'Stress,Analysis Of Spacer Paste Replacing Dummy Die In A Stacked CSP Package', ICEPT2003, pp. 82-85.

[8] C.L. Chung, S. L. Fu, Timmy Lin, Alex Lu, Michael Ho, Debby Kuo and Steven Chou, ' A Study on the Characteristic of UV cured Die-attach Films In Stack CSP (Chip Scale Package)', ICM 2003, Cairo, Egypt. pp.235-368, 2003, Dec. 9.

Inertial Bridge with Electronic Compass for Auto Pilots

Paul Svasta, Iaroslav-Andrei Hapenciuc

CETTI, Polytechnic University of Bucharest, Bucharest, 7000, Romania

Email; paul.svasta@cetti.ro, andrei.hapenciuc@cetti.ro

Abstract

The present electronic module is an inertial bridge that has the possibility to measure the acceleration, rotation movement and earth magnetic field in all 3 axes. It can be used alone to determine the movement of a system. This device can be used alone or together with other navigation equipments like GPS or "dead reckoning" to improve the results. In the developing of this equipment where used sensors done in MEMS technology which are very small in size and reliable in comparison with piezoceramic sensors. Taking advantage of all technological improvements in the sensor and packaging fields was obtained a reliable and precise navigation system that ca be used in wide variety of situations.

Introduction

This inertial bridge was designed to work in a intelligent autonomous vehicle capable of following a predetermined route and able to take decisions in unexpected situations. This equipment will work as an electronic module next many others to reach this desiderate.

Functional Description

The information's that inertial bridge is giving on rotational movement, linear movement and the orientation of earth magnetic field in 3D. The system is presented in figure 1.

The rotational movement is gathered using MEMSIC gyros. The model used ADXRS150 is a high sensitivity one, with a maximum range of 150deg/sec. This is manufactured by Analog Device and it is encapsulated in a BGA32 package.

This type of gyroscope had been chosen because of its reduced noise: $0.05deg/sec/Hz^{-2}$. Since the acquisition chain have bandwidth of 2Hz then is resulting an noise of 0.1 deg/sec. at a sensitivity of 12mV/deg/sec is resulting and electrical noise of 1.2mVp-p. With a 16bit ADC and 5V reference this noise will affect the last 4bits. To improve the accuracy a digital filtering is possible. The most simple technique is to average multiple samples. But this is equivalent with decreasing of bandwidth so this is limited buy the desired response time which is decided by the application. The internal structure of the gyroscope is presented in the Figure 2.

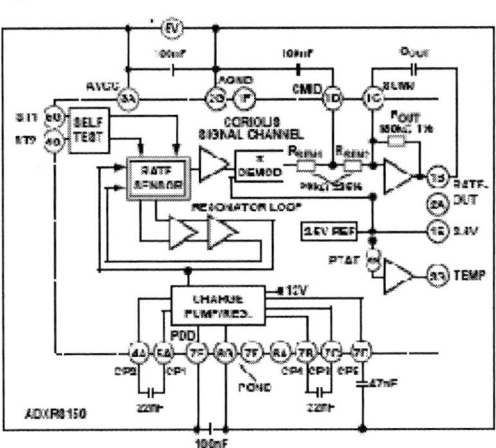

Figure 2 - Gyroscope Diagram

The accelerometer used is also an Analog Device model: ADXL 330. This is a triaxial accelerometer with a range of 3g in LFCSP package.

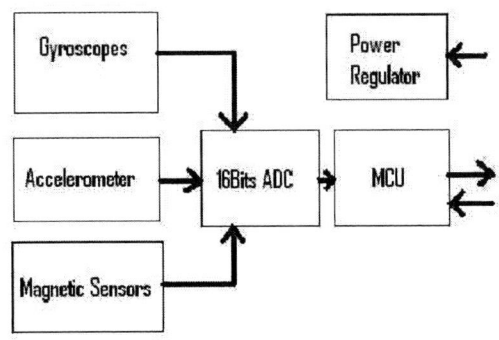

Figure 1 – System Block Diagram.

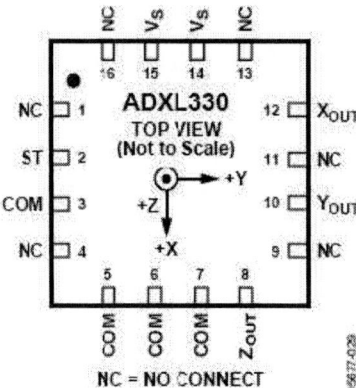

Figure 3 Accelerometer pins name

This sensor had been chosen because is a three axial one and because of its low noise: 300ug/Hz^-2. This sensor is radiometric one. This mean that the output voltage is directly proportional with the alimentation voltage and the dividing factor is dependant on acceleration. This is very different than the gyroscope case when the output was absolute which mean it is not dependant on the power voltage.

The power voltage can be in the 2V to 3.6V range. For this equipment the power for the accelerometer is 2.5V and it is taken from the gyroscope reference voltage. At this voltage the sensitivity is of 250mV/g, so the noise will be 0.125mV which is almost twice of the ADC performance.

The accelerometer is giving us information on tilt, centrifugal force of break efficiency. The triaxial accelerometer pins are given in Figure 3.

To complete the acquisition possibility of this device a triaxial magnetic sensor was needed. To achieve this had been chosen one two axes magnetometer and one single axis magnetometer specially designed for low magnetic fields.

The models are HMC1001 (for Z axis) and HMC1002 (for X and Y) made by Honeywell.

The accuracy of these sensors is impressive. This way was possible to measure the orientation of the earth magnetic field with an accuracy of less than 0.5degree. The sensors are presented in figure 4.

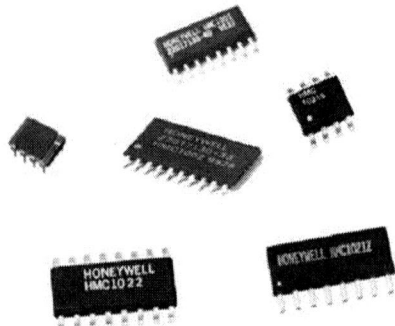

Figure 4 - Magnetic Sensors

All these performances where possible thanks the low noise of the sensors, but also thanks the ADC high accuracy.

The ADC is ADS8344 from Texas Instruments. This is a 16bit ADC with 8 single channels that can also work as 4 differential channels. It is able to reach 100ksps speed.

It is connected to a ATmega8 uC thru SPI bus. The uC also have 8 analog inputs with 10 bit resolution. These had been connected to a 8 bin connector just in case extra analog inputs are necessary.

The schematic of entire system is presented in Figure 5.

Measurement performance

The electronic circuit has been developed to obtain maximum of accuracy from the available parts.

The main part is the 16bit ADC - ADS8344. The main issue that had to be solve is the ADC noise that is affecting the last two bits.

This noise is always added to the noise of the sensors, resulting a degraded signal. That is way in the acquisition process the data is averaged on 64 samples. A longer buffer for averaging would have been possible but the conversion speed it would have decreased too much. For particular applications where very low sampling frequency is allowed the noise can be decreased even more.

The ADC speed is rated at 100ksps and is adjusted from the SPI speed. Because the no of channel is large: 11 and for each channel 4 bytes must be send on SPI between ADC and uC the resulted ADC sampling rate will be:

ADCspeed = SPI speed/ (no of channels*frame length*byte length).

For a SPI speed of 1Mbps the ADC speed is 1Mbps/(11chanells*4 bytes*8bits) = 2840 sps (samples per second). The real value is a bit lower: 2000sps and is given by a internal timer interrupt that is controlling the SPI communication. With the 64values averaging we end up with a 31sps, just enough for most applications.

The equivalent noise of ADC will decrease from around 600uV to 70uV which is close to the ADC accuracy for a 5V reference

At the output of each sensor and analog low pass filter is inserted to eliminate the high frequency noise.

The 3db frequency can be computed with the formula:

$$f = \frac{1}{2 * \pi * R * C} \qquad (1)$$

Where R = 1000k and C=100nF, so the frequency value is: 1.59Hz. The noise at the output of this filter is added to the ADC one.

For the magnetic sensors the noise is 29nV/sqr(Hz) at around 1Hz. The spectral density is decreasing to higher frequencies and is increasing to lower ones according to 1/f noise spectral density. The noise at the output of the analog filter is: 36.6nV. This value is much smaller than the ADC accuracy so the signal can be amplified.

That way a instrumentation amplifier was inserted at the output of the sensors. The gain of the amplifier can't be too high because the signal mast not gets out of the measurement range of the ADC: 0-5V.

Because the only magnetic field measured is the earth magnetic field: +/-0.5Gauss the total voltage span at the output of the sensor is:

Vspan = Field span*Sensitivity*Sensor voltage (2)

$$= 1Gauss*4mV/V/Gauss*5V$$
$$= 20mV$$

From 20mV to 0-5V a gain of 250x is possible but the maximum sensor offset is: -60mV to which the Vspan must be added 20mV, so a total of 80mV from ideal value. The new value of maximum gain is 50x to be certain the signal doesn't get out of the range.

Now the output noise will be 2.4uV which is still a lot under the ADC accuracy.

To evaluate the electronic compass accuracy the new sensitivity must be computed. The ADC accuracy is 100uV; this is decreasing 50X because of the magnetic amplifier to a value of 20uV. At 20mV/Gauss sensitivity the accuracy of magnetic field measurement is 1mGauss. This is equivalent to a sensitivity of 0.5 degree.

For the Gyros was not needed an amplifier because of the high sensitivity 12.5mV/degree/s. The rated noise density is 0.05 degree/s/sqrt(Hz). At a bandwidth of 1.56Hz the rms noise voltage is:

RMS noise = Sensitivity*Noise density*SQRT(LPF Bandwidth) (3)

$$= 12.5*0.05*SQRT(1.56)$$
$$= 788uV$$

This value is larger than the ADC accuracy so the last 3 bits will be unstable. To obtain a RMS value of noise N times lower than the actual one. The average must be done on a time interval N^2 larger than the one use by the LPF (1/1.59Hz).

To obtain an RMS value close to the ADC accuracy the value must improve at least N=8 times so the time interval must be 64 times higher. From here result a value of 41 seconds witch is not acceptable for many applications.

The main issue of the gyroscope is that the value it is giving is not directly indicating the angle but the angle speed. To find the angle the value must be integrated in time. The smallest offset will add in time and give an error that is increasing linear in time.

The minimum offset achievable practically is close to the noise RMS value. The angular noise can be computed as

Angle noise = RMS Noise*Sensitivity (4)

$$= 0.788mV/12.5mV/degree/s;$$
$$= 0.063degree/s$$

This mean 3.78degree/min or 226 degree/hour.

So, the actual setup can be used to determine the direction for relatively short time intervals, shorter than a minute. For larger periods of times the result can be averaged on longer periods of time as previously indicated and the accuracy will improve drastically: For 41seconds averaging the error will became: 0.5 degree/minute and 28 degree/hour.

As comparison, Laser Interferometer Gyroscopes used in space applications have an error of 2degree/hour.

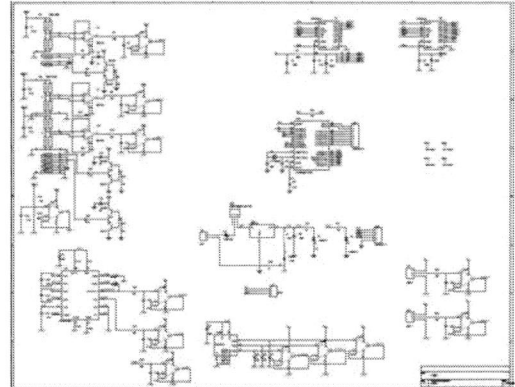

Figure 4 – Electronic Diagram

Figure 5 – Test vehicle of the inertial bridge circuit

Communication Protocol

To make the acquired information available to different application, on the uC was implemented and reliable industrial communication protocol: ModBus.

This is enabling the user to connect to a system more such inertial boards for different applications.

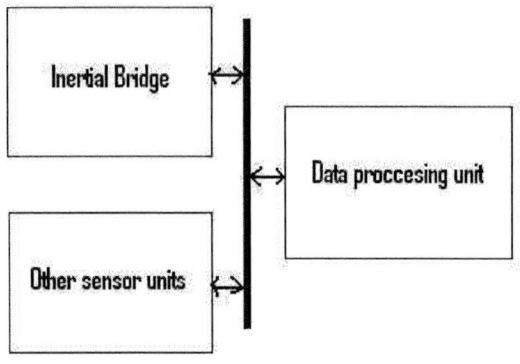

Figure 6 – Communication bus topology

The values from the ADC is averaged on 32 samples, this is for a more reliable conversion.

The information in contained in 22registes that are creating 11words containing information from: 3 gyroscopes, 3 accelerometers, 3 magnetic sensors, 1 temperature sensor and the 2.5V reference.

Conclusions

This paper shows that accurate inertial measurements are possible with affordable sensors thanks to the iMEMS technology.

Also proper design of the PCB can help having reliable equipment to mechanical and thermal stress.

The information from Gyroscopes can be corrected by the electronic compass realized from the triaxial magnetic sensor and the triaxial accelerometer used as inclinometer.

For complex signal processing the results can be send to a complex computation unit with high computation power.

Acknowledgements

A part of this work has been supported by the Romanian Ministry of Education and research in the frame of the Integrated Automotive Engineering Platform project.

References

[1] Col Eng I Musat, Gen. PhD. N. Irimie "Investigation of system performances for parameters computation in missile guiding" Military Publisher, Bucharest 1996.

[2] N. Nita, I Aron "Inertial Navigation ", Bucharest, 1971.

[3] Pierre Faurre, "Optimal Inertial Navigation", Paris 1971.

[4] Paul Svasta, I. Hapenciuc "Inertial Navigation Module for Dead Reckoning Proceedings of SIITME 2006 September 12-15, Cluj, Romania, page 244-247

[5] www.Atmel.com : "ATmega8 Datasheet"

[6] www.ModBus.org: "MODBUS Application Protocol Specification V1.1"

[7] www.FTDI.com : "FT232BM data sheet"

[8] www.NI.com : "LabVIEW user Manual"

[9] Analog Device, iMEMS Accelerometers:

[10] Analog Device, iMEMS Gyroscopes:

High performance low-firing temperature thick-film pressure sensors on steel

C. Jacq, T. Maeder, N. Johner, G. Corradini, P. Ryser

Laboratoire de Production Microtechnique, Ecole Polytechnique Fédérale de Lausanne

BM – Station 17, CH-1015 Lausanne, Switzerland

+41.21.693.53.85, +41.21.693.38.91, caroline.jacq@epfl.ch

Abstract

Performance of piezoresistive thick-film pressure sensors on steel substrates is hindered by the high required firing temperatures of the layers, which cause degradation of the substrate mechanical properties. Special steels or alloys that are relatively unaffected by these temperatures are available, but their cost is very high. A much more promising alternative is the reduction of the thick-film firing temperature, in order to allow application of the standard high-strength steels used in thin-film cells. Therefore, a series of thick-film materials systems (dielectrics, resistors and conductors) firing at low temperatures (<700°C) on steel substrates has been developed and electrically characterised (sheet resistance and thermal coefficient of resistance – TCR). It was found that the main issue in these systems lies in mastering the materials interactions during firing, especially at the silver-based resistor terminations. The interaction of silver, resistor and dielectric tends to give rise to highly resistive zones at the terminations, affecting reliability. This can be circumvented by post-firing the resistor terminations at a moderate temperature.

Key words: Thick-film materials, steel substrates, low-temperature processing, resistors, termination effects

Introduction

Pressure and force sensors based on thick-film piezoresistors deposited on ceramic substrates are well established due to their low production cost. However, alumina is not optimal for piezoresistive sensing applications, as it is brittle, its elastic modulus is high and its strength rather low, compared to metallic substrates [1][2]. Moreover, ceramic pressure sensors usually require elastomer seals, which are not allowed in many demanding applications. Metallic alloy substrates therefore would me much better suited in many cases, especially relatively low-cost materials such as high-strength stainless steel, and have been investigated in in the past [3][4].

However, the standard thick-film firing temperature (850°C) is not compatible with most steels, owing to degradation of mechanical properties due to annealing or dimensional changes associated with martensitic transformation (which tends to destroy the thick-film layers). Special steels that do not suffer from these problems exist and give very good sensing properties with standard thick-film materials [5]. However, their cost remains very high. Therefore, a more realistic solution lies in lowering the firing temperature of the thick-film system below 700°C, a temperature that, in more common steels, avoids phase transformations and is still compatible with reasonably good strength retention.

In previous studies [5-8], we have developed and studied several such low-firing systems based on low-melting lead borosilicate glasses, where the glass composition, filler type and filler loading were changed to achieve compatibility with a wide range of substrates such as glass, aluminium metal and steels. An important aspect is the matching of the coefficients of thermal expansion (CTE), which ranges from 11 ppm/K for ferritic & martensitic steels to 17 ppm/K for austenitic ones and is much higher than that of alumina (7 ppm/K).

These systems exhibited promising properties, allowing the manufacture of prototype pressure sensors [5]. However they did exhibit unexpected problems at the silver-based resistor terminations: short resistors tended to have anomalously high values, indicating the presence of a highly resistive zone near the terminations. This so called "inverse size effect" has a deleterious effect on reliability and stability, and has already been observed by Prudenziati et al. [9], for RuO_2-based resistors with Ag-based terminations and with a glass matrix containing Al_2O_3, which exactly corresponds to our case.

This work also showed, that Bi_2O_3 additions, and lower interaction temperatures between the resistors and terminations could mitigate or even reverse the size effect back to normal, thereby improving reliability.

In this work, we therefore endeavour to systematically investigate the termination effect for our low-firing thick-film system, in order to eliminate the inverse size effect.

To this end, we studied the consequences of 1) changing the dielectric filler (alumina or quartz), 2) adding Bi_2O_3 to the dielectric, the resistor and/or the terminations, 3) reversing the firing order of the conductor and resistor and 4) post-firing a specially formulated conductor termination onto the resistor at a lower temperature than the resistor firing temperature.

Experimental

The following substrate materials were used: 96% pure alumina (Kyocera, Japan, A-476, CTE = 7 ppm/K) as standard thick film substrate and ferritic stainless steel 1.4016 (CTE = 11 ppm/K, comparable to the common precipitation hardening martensitic stainless steels). The steel substrates were pre-oxidised at 900°C during 1 hour in order to increase the adherence of the dielectrics.

The dielectric materials used in this work are based on the lead borosilicate glass ("V6") used in our previous dielectric and resistor studies [5][10]: 75% PbO + 10% B_2O_3 + 15% SiO_2 (mass%) with 2% Al_2O_3 added to inhibit crystallisation [11].

The examined filler materials were quartz (Nanostructured &Amorphous Materials Inc. 99.95%, D50:2.5-3 µm particle size) or alumina (Alfa Aesar, aluminium oxide alpha, 99.99%, 1 µm). In both cases, doping with 10% vol. Bi_2O_3 was also studied. The volume fractions and codes of the resulting dielectrics are given in table 1. All the dielectric layers were fired at a peak temperature of 625°C, with a 10 min dwell. Note: only the last dielectric layer (in contact with the resistors and the terminations) was changed; the first dielectric layers were always quartz-filled and without Bi_2O_3 (Alpha Aesar, bismuth(III) oxide, 99%). Together with the option of firing on bare alumina, this gives 5 dielectric + substrate variants.

A commercial Ag conductor material allowing a reduced firing temperature was applied: ESL 9912A, fired at 600°C / 10 min before (prefired) or after the resistors (postfired). Additionally, the possibility of using a very low-firing conductor, consisting of Ag powder (Nanostrutured & Amorphous Materials Inc. 99.9%, APS 30 nm) with a very low-melting glass binder ("V8", 85% PbO + 10% B_2O_3 + 5% SiO_2, mass%, + 2% Al_2O_3 [5]) was examined. As this conductor fires at 500°C, only post-firing on the fired resistors was examined in this case. As in the case of the dielectrics, doping with Bi_2O_3 was also examined. The resulting 6 conductors + firing order variants are given in table 1.

For the resistive composition, a commercial low-firing 10 kΩ paste, ESL 3114, was used, also with or without Bi_2O_3 doping (table 1). In this study, the resistors were always fired at 625°C / 10 min.

Samples for sheet resistance and TCR were 1.5 mm wide resistors of several lengths (figure 1), and were measured at 30°C, 65°C and 100°C.

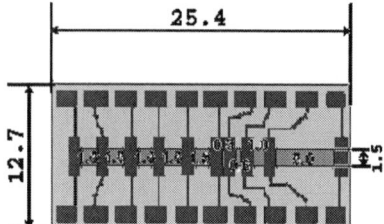

Figure 1: Layout of the test sample for measurement of electrical properties.

In order to have a quick assessment of the termination effects, we calculated a "length index" LI, defined as the ratio of the average sheet resistance of the three short resistors (0.3, 0.6 and 1.0 mm length) to the average of the 1.5 mm long ones. LI values greater than 1 imply the existence of highly resistive zones near the terminations, whereas the other case (LI < 1) corresponds to a locally decreased sheet resistance near the terminations.

Table 1: List of materials

Designation	Description
Substrate + dielectrics:	(alumina substrate)
*	40 V6 + 60 Quartz
Q	32 V6 + 58 Quartz + 10 Bi_2O_3
Q_{Bi}	50 V6 + 50 Al_2O_3
A	40 V6 + 50 Al_2O_3 + 10 Bi_2O_3
A_{Bi}	
Conductors & order:	
U	100 ESL 9912A prefired
U_{Bi}	85 ESL 9912A + 15 B_2O_3 prefired
V	100 ESL 9912A postfired
V_{Bi}	85 ESL 9912A + 15 Bi_2O_3 postfired
W	90 Ag + 10 V8 postfired
W_{Bi}	83 Ag + 7 V8 + 10 Bi_2O_3 postfired
Resistors:	
R	100 ESL 3114
R_{Bi}	90 ESL 3114 + 10 Bi_2O_3

Results and discussions

The results for length index LI, sheet resistance R_s and TCR are compiled in tables 2, 3 and 4 respectively.

On alumina substrate, the results show that conductors have to be fire before the resistor or after the resistor at low temperature. When the resistors are fired twice at 600°C (V), their sheet resistance increase, the TCR are damaged and the resistors become unreliable, confirmed by the LI index which increase with the V conductors.

On steel, the conductor and its firing temperature is clearly the dominant parameter for LI. Strong inverse size effects appear for conductors fired at 600°C (U & V), especially for post-fired ones (V). Conversely, this effect essentially

180

disappears in the case of conductors post-fired at 500°C (W). For this last group of "good" conductors, a smaller, secondary effect of the presence of Bi in the conductor and/or the resistor is evidenced. Best LI values (1 or slightly lower) are obtained in most cases except one: when the resistor contains Bi, but not the conductor, a slight increase of LI is observed, which could be due to diffusion of Bi from the resistor near the terminations into the glass frit of the conductor.

The sheet resistance decrease considerably when the resistor is doped

Table 2: Length index LI

Cond. & order Sub. + diel. + res.	U	U_{Bi}	V	V_{Bi}	W	W_{Bi}
* + R	1.19	1.10	1.41	1.47	0.87	0.90
* + R_{Bi}	0.97	0.93	1.49	1.59	1.07	1.09
Q + R	-	-	2.37	2.43	0.96	1.01
Q + R_{Bi}	1.86	-	2.61	2.68	1.02	1.02
Q_{Bi} + R	1.50	-	-	-	-	-
Q_{Bi} + R_{Bi}	-	1.67	-	-	-	-
A + R	1.59	1.55	2.59	3.09	0.98	0.97
A + R_{Bi}	2.48	1.51	2.56	3.05	1.12	0.95
A_{Bi} + R	1.71	1.44	2.70	2.80	1.02	0.96
A_{Bi} + R_{Bi}	1.76	1.26	2.74	2.87	1.19	1.08

Addition of Bi_2O_3 in the dielectric strongly decreases the sheet resistance, even if the resistor is not filled with Bi_2O_3 powder. The resistive terminations problem is completely solved with ESL 3114 on both dielectrics filled or not with Bi_2O_3: the sheet resistance becomes practically independent of length. However, filled the dielectric with Al_2O_3 and doped with Bi_2O_3 is favourable, because TCR is shifted towards 0 different from the dielectric filled with quartz. Doping ESL 3114 with Bi_2O_3 gives good results, but is not necessary to achieve good TCR and termination properties. The TCR results are even better with the ESL 3114 resistance.

Table 3: Sheet resistance R_s [kΩ]

Cond. & order Sub. + diel. + res.	U	U_{Bi}	V	V_{Bi}	W	W_{Bi}
* + R	9.23	8.79	16.68	22.3	12.4	12.39
* + R_{Bi}	3.34	3.09	5.68	5.29	3.82	3.80
Q + R	-	-	26.43	27.71	10.78	11.08
Q + R_{Bi}	8.44	-	22.23	26.42	7.26	7.08
Q_{Bi} + R	6.93	-	-	-	-	-
Q_{Bi} + R_{Bi}	-	9.47	-	-	-	-
A + R	11.10	11.97	50.99	41.03	8.24	7.46
A + R_{Bi}	7.35	7.21	22.86	21.19	3.80	4.45
A_{Bi} + R	7.55	6.33	16.99	22.07	3.58	3.75
A_{Bi} + R_{Bi}	7.09	5.82	12.80	10.25	2.33	2.71

Table 4: Temperature coefficient TCR [ppm/K]

Cond. & order Sub. + diel. + res.	U	U_{Bi}	V	V_{Bi}	W	W_{Bi}
* + R	-242	-231	-285	-286	-263	-292
* + R_{Bi}	-207	-203	-273	-269	-267	-264
Q + R	-	-	-147	-148	-161	-169
Q + R_{Bi}	-183	-	-271	-269	-232	-240
Q_{Bi} + R	-	-	-	-	-	-
Q_{Bi} + R_{Bi}	-	-201	-	-	-	-
A + R	-91	-80	-161	-147	-104	-137
A + R_{Bi}	-136	-116	-207	-199	-135	-142
A_{Bi} + R	-78	-60	-143	-157	-63	-82
A_{Bi} + R_{Bi}	-101	-70	-182	-158	-97	-112

The figure 2 illustrates the dependence of sheet resistance with the length of resistance. Size effects have been completely circumvented by firing the conductor after the resistance at lower temperature.

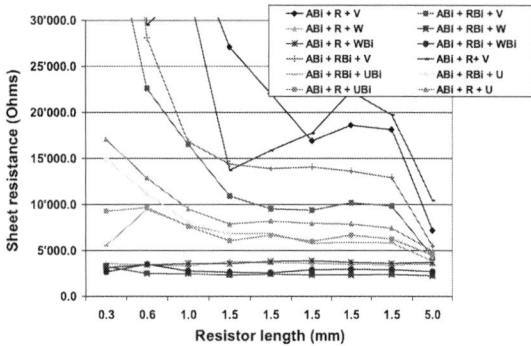

Figure 2: Evolution of sheet resistance with length of resistors.

Conclusion

The goal of this work was to study the resistive terminations between resistors and conductors on a compatible dielectric with a ferritic steel substrate. A serious problem with thick-film terminations is the increase of sheet resistance for the short resistors. In order to avoid this problem, several routes have been studies. The thick-film materials (dielectric, conductive and resistive materials) have been doped with Bi_2O_3 and the standard screen-printed sequence has been modified in order to modify the firing temperature of the conductor.

At the end of this study we have found a solution which eliminates the problem of high sheet resistance for the short resistors on dielectric for ferritic steel substrates. A dielectric filled with alumina doped with Bi_2O_3 combined with the commercial resistance ESL 3114 and a low firing temperature conductive composition proved to be up to the best combination to avoid the problem. By this way, we can guarantee the reliability of the resistors. The future work will be focus on the elimination of high sheet resistance on dielectrics compatible with austenitic steels.

References

[1] C. Jacq, T. Maeder, P. Ryser, "High-strain response of piezoresistive thick-film resistors on titanium alloy substrates", Journal of the European Ceramic Society, Vol. 24, No. 6, pp. 1897-1900, 2004.

[2] T. Maeder, H. Birol, C. Jacq, P. Ryser, "Strength of ceramic substrates for piezoresistive thick-film sensor applications", Proceedings, European Microelectronics and Packaging Symposium, Prague (CZ), pp. 272-276, 2004.

[3] N.M. White, "A study of the piezoresistive effect in thick-film resistors and its application to load transduction", thesis, University of Southampton, Faculty of Engineering & Applied Science, 1988.

[4] L. Fraigi, D. Lupi, L. Malatto, "A thick-film pressure transducer for cars propelled by natural gas", Sensors and Actuators A, Vol. 41, pp. 429-441, 1994.

[5] C. Jacq, T. Maeder, S. Martinerie, G. Corradini, E. Carreño-Morelli, P. Ryer, "High performance thick-film pressure sensors on steel", Proceedings, 4th European Microelectronics and Packaging Symposium, Terme Čatež (SI), IMAPS, pp. 105-109, 2006.

[6] C. Jacq, S. Vionnet, T. Maeder, P. Ryser, "Integrated thick-film hybrid microelectronics on aluminium substrates", Proceedings, European Microelectronics and Packaging Symposium, Prague (CZ), pp. 267-271, 2004.

[7] C. Jacq, T. Maeder, S. Vionnet, P. Ryser, "Low-temperature thick-film dielectrics and resistors for metal substrates", Journal of the European Ceramic Society, Vol. 25 (12), pp. 2121-2124, 2005.

[8] C. Jacq, T. Maeder, S. Menot-Vionnet, H. Birol, I. Saglini, P. Ryser, "Integrated thick-film hybrid microelectronics applied on different material substrates", Proceedings of the 15th European Microelectronics and Packaging Conference (EMPC), Brugge (BE), IMAPS, pp. S13.04, 319-324, 2005.

[9] M. Prudenziati, F. Sirotti, M. Sacchi, B. Morten, A. Tombesi, T. Akomolafe, "Size effects in ruthenium-based thick-film resistors: rutile vs. pyrochlore-based resistors", Active and Passive Electronics Components, Vol. 14, pp. 163-173, 1991.

[10] S. Vionnet-Menot, C. Grimaldi, T. Maeder, P. Ryser, S. Strässler, "Tunneling-percolation origin of nonuniversality: Theory and experiments", Physical Review B, Vol. 71, pp. 064201, 2005.

[11] M. Prudenziati, B. Morten, B. Forti, A.F. Gualtieri, G.M. Dilliway, "Devitrification kinetics of high lead glass for hybrid microelectronics", International Journal of Inorganic Materials, Vol. 3, pp. 667-674, 2001.

Nonferroelectric Lead-Free Ceramics and Thick Films with High Dielectric Permittivity - Synthesis, Sintering and Properties

J. Kulawik and D. Szwagierczak

Institute of Electron Technology, Cracow Division, Zabłocie 39, 30-701 Kraków

Phone +4812 6563144, Fax +4812 6563626, e-mail: jkulawik@ite.waw.pl, dszwagi@ite.waw.pl

Abstract

This work was focused on preparation and comparison of properties of the developed nonferroelectric lead-free high permittivity materials. Eight compounds of perovskite structure with compositions: $Bi_{1/2}Cu_{1/2}(Fe_{2/3}W_{1/3})O_3$, $Ca(Fe_{2/3}W_{1/3})O_3$, $Bi_{1/2}Cu_{1/2}(Fe_{1/2}Nb_{1/2})O_3$, $Ca(Fe_{1/2}Nb_{1/2})O_3$, $Bi_{1/2}Cu_{1/2}(Fe_{1/2}Ta_{1/2})O_3$, $Ca(Fe_{1/2}Ta_{1/2})O_3$, $Bi_{2/3}Cu_3Ti_4O_{12}$ and $Cu_2Ta_4O_{12}$ were synthesized. The obtained powders were used for fabrication of ceramic samples and pastes destined for dielectrics of plate and thick film capacitors. Capacitance and dissipation factor were measured at frequencies 10 Hz – 1 MHz in the temperature range from –55 to 700°C and –55° to 400°C for plate and thick film capacitors, respectively. Dc resistivity of the samples was investigated as a function of temperature in the range 20 - 500°C. Microstructure and chemical homogeneity of the ceramics and layers were studied using scanning electron microscopy and X-ray microanalysis. High permittivity values exceeding 10000, observed especially at low frequencies for all the developed ceramic samples and some of the thick films, are supposed to originate from the formation of internal barrier layer capacitors. Broad maxima of dielectric permittivity and dissipation factor versus temperature plots, shifting towards higher temperatures with increasing frequency, as well as a distinct dependence of dielectric properties on frequency were characteristic of the investigated materials. In view of their lead-free composition and relatively simple preparation procedure, yielding high capacitance and miniaturization, the developed ceramics seem to be promising dielectric materials for capacitive elements.

Key words: high permittivity, lead-free ceramics, thick films, capacitors

Introduction

Increasing demand for both environmentally benign materials and miniaturization is the reason of growing interest in the compounds exhibiting the so-called giant dielectric phenomena. Commercially used for many years high ε capacitor materials have been based on ferroelectrics or relaxors. Recently, some new nonferroelectric compounds, like $CaCu_3Ti_4O_{12}$, have gained great attention due to their very high dielectric permittivity, almost constant over a broad temperature and frequency range. Subramanian et al. [1] found that more than ten materials in the $ACu_3M_4O_{12}$ system exhibit large dielectric constant, exceeding 1000 at room temperature. $CaCu_3Ti_4O_{12}$ (CCTO) in the form of single crystals, ceramics and thin films has been the subject of several theoretical and experimental works [2 - 9]. The internal barrier layer capacitance model seems to be the most plausible explanation for the observed CCTO behavior. Cohen et al. [3] suggested that the large, temperature independent low-frequency dielectric constant exhibited by the CCTO single crystal stems from spacial inhomogeneities of local dielectric response. Probable sources of these inhomogeneities may be: twin, compositional ordering or antiphase

boundaries. Basing on impedance spectroscopy studies Sinclair et al. [5] stated that CCTO ceramic is a one step internal barrier layer capacitor with conductive grains and insulating grain boundaries. Zhang and Tang [7] showed that variable range hopping (VRH) mechanism is responsible for bulk dc conductivity in CCTO ceramics. Frequency response of CCTO thin films grown epitaxially on $LaAlO_3$ substrate by Tselev et al. [9] was found to be dominated by power-law, typical of localized hopping charge carriers.

Raevski et al. [10] proposed that very high values of dielectric permittivity of $AFe_{1/2}B_{1/2}O_3$ ceramics (A = Ba, Sr, Ca; B = Nb, Ta, Sb) observed in a wide temperature interval are due to Maxwell-Wagner relaxation.

Giant and nearly temperature independent dielectric constant values of 10^4 - 10^5 at low frequencies and/or at high temperatures were found also for single crystals of copper tantalum oxide $Cu_2Ta_4O_{12}$ by Renner et al. [11] and ascribed to surface barrier layer capacitors (SBLC) formation at the sample-electrode interface.

The large dielectric constant of $Bi_{2/3}Cu_3Ti_4O_{12}$ polycrystalline samples, independent of frequency and temperature below 150°C, was explained by Liu et al. [12] in terms of Maxwell-

Wagner relaxation. Two electrical responses in impedance and modulus formalisms were attributed to the grain and grain boundaries effects, respectively.

This paper focuses on the preparation and investigation of dielectric properties of a group of lead-free materials with perovskite-like structure characterized by high dielectric permittivity at low frequencies, prepared in the form of ceramics and as thick films.

Experimental

Syntheses of eight lead-free materials with perovskite-type structure: $Bi_{1/2}Cu_{1/2}(Fe_{2/3}W_{1/3})O_3$ (BCFW), $Bi_{1/2}Cu_{1/2}(Fe_{1/2}Nb_{1/2})O_3$ (BCFN), $Bi_{1/2}Cu_{1/2}(Fe_{1/2}Ta_{1/2})O_3$ (BCFT), $Ca(Fe_{2/3}W_{1/3})O_3$ (CFW), $Ca(Fe_{1/2}Nb_{1/2})O_3$ (CFN), $Ca(Fe_{1/2}Ta_{1/2})O_3$ (CFT), $Bi_{2/3}Cu_3Ti_4O_{12}$ (BCT) and $Cu_2Ta_4O_{12}$ (CT) were carried out by conventional solid-state reaction of oxides. Six of these compounds: BCFW, CFW, BCFN, CFN, BCFT and CFT have compositions analogous to relaxors with bismuth and copper or calcium substituted for lead ions, while the latter two are similar to $CaCu_3Ti_4O_{12}$.

BCFW, CFW, BCFN, CFN, BCFT and CFT were synthesized by the two-step coloumbite or wolframite methods. Small amounts of MnO_2 (0.5 – 1 mol %) were introduced to the batches in order to increase resistivity and decrease dissipation factor. During the first calcination step Fe_2O_3 reacted with WO_3, Nb_2O_5 or Ta_2O_5 at 1000°C for 4 h, while during the second step Fe_2WO_6, $FeNbO_4$ or $FeTaO_4$ reacted with Bi_2O_3 and Cu_2O or CaO at 800 - 850°C for 4 h. The syntheses products were ball-milled in alcohol, mixed with a water solution of polyvinyl alcohol, granulated and pressed into discs. The BCFW, BCFN, BCFT, CFW, CFN, CFT, BCT and CT pellets were sintered during 2 – 25 h at 870, 900, 1050, 1240, 1180, 1260, 1000 and 1230°C, respectively. Phase compositions of the sintered samples were detected by a Philips X'Pert diffractometer.

The compounds with lower sintering temperatures: BCFW, BCFN, BCFT and BCT were used for fabrication of thick film pastes. The pastes were prepared by mixing the synthesized powders, previously ball milled, with an organic vehicle - ethyl cellulose solution in terpineol. Thick film capacitors were screen printed on 96% Al_2O_3 substrates. The bottom and top electrodes made of Ag pastes were fired in a VI-zone BTU belt furnace according to a standard thick-film profile. The dielectric layers were deposited using a 260-mesh screen, dried and then fired during 45-minute cycle at a peak temperature of 820 - 870°C held for 10 minutes. The printing-drying-firing cycle was carried out three times. The thickness of the dielectric layers examined by means of a Hobson-Taylor profilograph was 20 - 30 μm.

Measurements of dc resistivity were performed by means of a Philips resistance meter in the temperature range 20 - 500°C and 20 - 400°C for ceramic samples and thick films, respectively. Capacitance and dissipation factor of capacitors were determined from –55 up to 700°C for ceramics and from –55 to 400°C for thick films, at frequencies from 10 Hz to 1 MHz. Preliminary impedance spectroscopy studies were also carried out by means of a LCR QuadTech meter.

A FEI scanning electron microscope and a Link Isis electron microprobe were used to study the microstructure and the chemical composition of thick films as well as the interaction between the ceramic layers and the electrodes.

Properties of ceramic samples

X-ray diffraction analysis confirmed that single phase, crystalline products were obtained as a result of the syntheses. Some of the developed materials are compounds with new compositions not avalaible in ICDD database.

The scanning electron microscope observations indicated that the microstructure of the ceramic samples was dense and fine-grained with grain size in the range 0.5 – 3 μm.

Figure 1 illustrates the temperature dependencies of dielectric permittivities at 1 kHz for all the studied ceramics. In the investigated temperature range the obtained materials exhibited broad maxima or humps on the plots. Their location shifted towards higher temperatures with increasing frequency. The maximum ε_r values were diminished significantly with increasing frequency. Flattening of $\varepsilon_r = f(T)$ curves took place both in the low temperature region and at high frequencies. At higher temperatures dielectric permittivity increased in monotonic way due to rising electrical conductivity. For BCT, BCFN, BCFT and CFW ceramics distinct maxima were observed which can be attributed to dielectric relaxation, while for BCFW, CFN, CFT and CT only the humps on $\varepsilon_r = f(T)$ curves occurred.

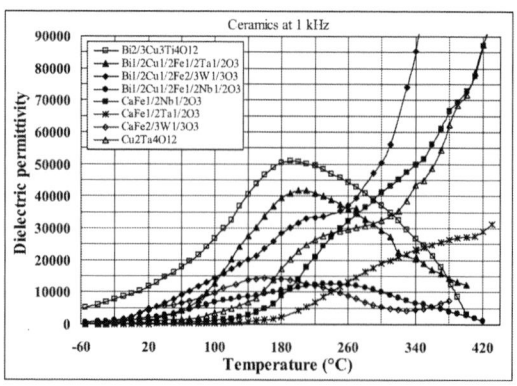

Figure 1: Temperature dependence of dielectric permittivity of lead-free perovskite ceramics at 1 kHz

The developed ceramics were characterized by very high maximum values of dielectric permittivity up to 100000. In the investigated frequency range the maxima and the humps were situated in the temperature range 70 – 440°C. The maximum ε_r values at 1 kHz corresponding to the peaks or humps as well as the resistivities at 20°C for BCFW, BCFN, BCFT, CFW, CFN, CFT, BCT and CT ceramics are listed in Table 1.

Shapes of the dielectric characteristics determined for CFW, BCFT and BCFW ceramics were close to those observed for other nonferroelectric materials with perovskite structure [1 - 12].

Table 1: Resistivity and dielectric permittivity ε_r of lead-free perovskite ceramics

Composition	Resistivity at 20°C, Ωcm	ε_r at the peak or the hump (1 kHz)
BCFW	10^7	34000
BCFN	10^6	13000
BCFT	10^6	42000
CFW	10^7	14000
CFN	10^{10}	37000
CFT	10^{11}	28000
BCT	10^7	51000
CT	10^{10}	30000

In Figure 2 the dielectric constant of CT ceramics is presented as a function of temperature at frequencies ranging from 10 Hz to 1 MHz.

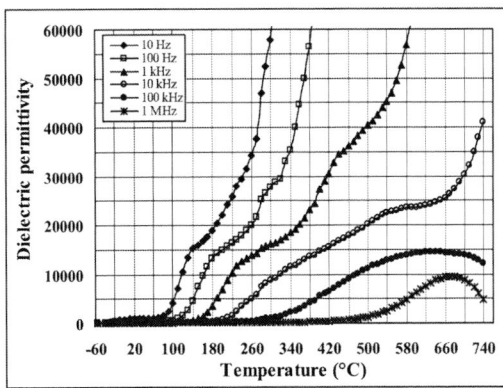

Figure 2: Dielectric permittivity of CT ceramic as a function of temperature in the frequency range 10 Hz – 1 MHz

As shown in Figure 3, there were found maxima in the dissipation factor versus temperature plots, shifting towards higher temperatures with rising frequency. Above the peak temperature the dielectric losses started to grow monotonically due to electrical conduction.

In Figure 4 the logarithm of relaxation times corresponding to the frequencies of dissipation factor peaks for the investigated ceramics is plotted as a function of 1000/T. The linear course of these plots indicate that the relaxation time τ obeys well the Arrhenius law:

$$\tau = \tau_0\, exp\,(E_r/k_BT)\qquad(1)$$

where E_r is the activation energy of dielectric relaxation, k_B – Boltzmann constant, T - temperature.

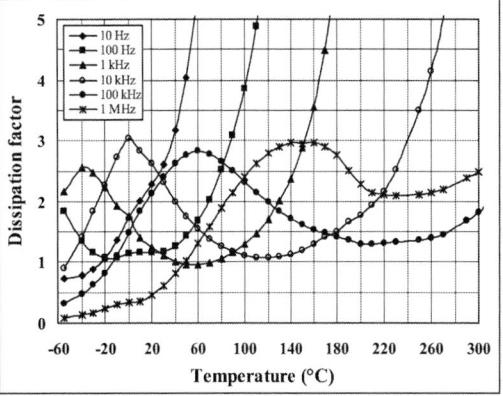

Figure 3: Temperature dependence of dissipation factor of BCFW ceramic in the frequency range 10 Hz – 1 MHz

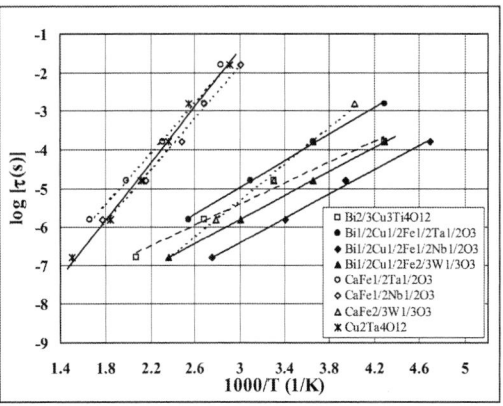

Figure 4: Logarithm of relaxation time of dielectric relaxation as a function of reciprocal of temperature for lead-free perovskite ceramics

In Figure 5 dc conductivities of the examined materials are presented as a function of 1000/T. The plots can be piecewise linearized according to the Arrhenius law:

$$\sigma = \sigma_0\, exp\,(-E_c/k_BT)\qquad(2)$$

where E_c is the activation energy of electrical conduction, k_B – Boltzmann constant, T - temperature.

The activation energies of dielectric relaxation E_r and electrical conduction E_c are listed in Table 2.

The E_r values range from 0.27 to 0.66 eV. The activation energies are lower (0.27 – 0.34 eV) and do not differ significantly for BCT, BCFW, BCFN and BCFT ceramics. The E_r values for CFN and CFT ceramics are the highest (0.66 and 0.64, respectively).

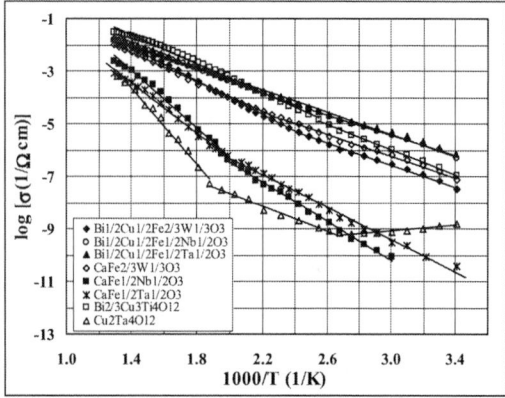

Figure 5: Logarithm of electrical conductivity as a function of reciprocal of temperature for lead-free perovskite ceramics

Table 2. Activation energies of dielectric relaxation and dc electrical conduction for lead-free perovskite ceramics

Composition	Activation energy of dielectric relaxation, eV	Activation energy of dc electrical conduction, eV
BCFW	0.31	0.41 (20-140°C)
		0.64 (150-500°C)
BCFN	0.31	0.38 (20-130°C)
		0.54 (140-500°C)
BCFT	0.34	0.42 (20-500°C)
CFW	0.47	0.40 (20-160°C)
		0.56 (170-500°C)
CFN	0.66	0.74 (60-230°C)
		1.09 (240-500°C)
CFT	0.64	0.61 (20-220°C)
		0.95 (230-500°C)
BCT	0.27	0.47 (20-110°C)
		0.54 (120-500°C)
CT	0.53	0.11 ((20-100°C)
		0.48 (110-260°C)
		1.43 (270-500°C)

The activation energies of electrical conduction E_c calculated in the lower temperature range were 0.11 - 0.74 eV, while those determined in the higher temperature range were 0.54 – 1.43 eV. The activation energies of electrical conduction

reported by Raevski et al. [10] for CFN ceramics, suggested by these authors to be determined by grain boundaries barrier height, were close to those obtained in this work at temperatures below 200°C (0.7 and 0.74 eV, respectively).

As seen from Table 2, the activation energies of relaxation do not differ significantly, although are not quite similar to those of electrical conduction. However, these discrepancies could result from the differences in the temperature ranges that are taken for fitting the straight lines in both type Arrhenius plots (Figure 4 and Figure 5).

Preliminary impedance spectroscopy studies confirmed the existence of semiconducting grains and insulating grain boundaries in the ceramics under investigation (Figure 6).

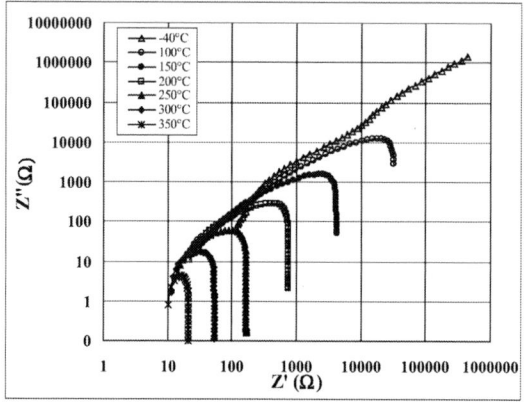

Figure 6: Log-log complex impedance plots Z" vs Z' for BCT ceramic at various temperatures ranging from 150 to 350°C

The resistances at 20°C of grains and grain boundaries determined for BCT ceramic on the basis of Z" = f(Z') plots were about 40 Ω and 10^7 Ω, respectively.

On the basis of the dielectric permittivity and impedance versus temperature and frequency studies we suppose, in agreement with the opinions of other authors reported for similar materials [1-12], that the dielectric characteristics of the compounds under investigation could be attributed to the formation of internal barrier layer capacitors.

Properties of thick films

The SEM observations revealed that the investigated thick films are characterized by dense, fine-grained microstructure with no cracks, blisters or delaminations at the electrode-dielectric layer boundary (Fig. 7).

In Figure 8 dielectric permittivities of the investigated thick films at the frequency of 1 kHz are plotted versus temperature. In Table 3 the maximum ε_r values and resistivities at 20°C of these layers are compared.

In Figure 9 temperature dependence of dielectric constant for BCFN thick film is depicted for the frequencies ranging from 10 Hz to 1 MHz. Broad maxima in the permittivity are observed decreasing and shifting to higher temperatures with increasing frequency.

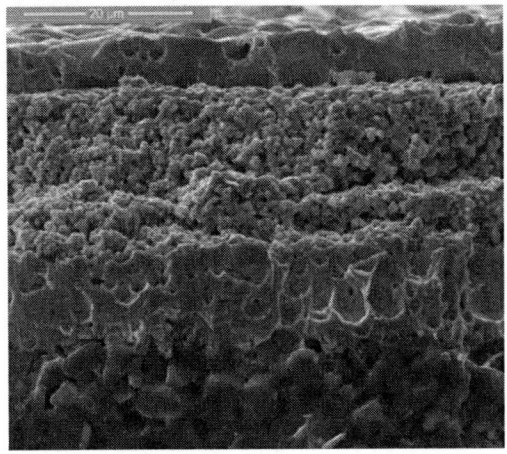

Figure 7: SEM microphotograph of a fracture of thick film capacitor with BCT dielectric layer,.

The highest permittivities, exceeding 11000, were reached for BCFW thick films. For BCFN and BCT layers the ε_r values were also high and characterized by advantageous, flat course of the $\varepsilon_r =$ f(T) curves entailing low temperature coefficient of capacitance over a wide temperature interval 100 – 300°C (Figure 10).

Table 3: Resistivity and dielectric permittivity of perovskite thick films

Composition	Resistivity at 20°C, Ωcm	ε_r at the peak or the hump (1 kHz)
BCFW	10^7	11000
BCFN	10^6	2300
BCFT	10^8	1100
BCT	10^7	5000

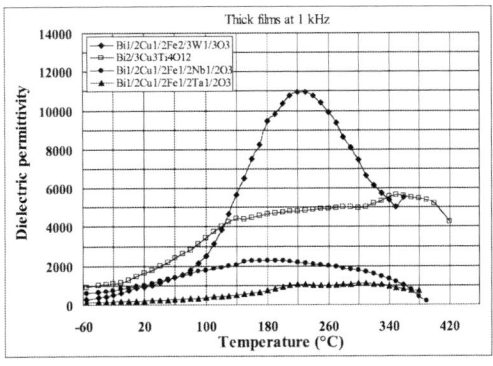

Figure 8: Dielectric permittivity of lead-free perovskite thick films at 1 kHz

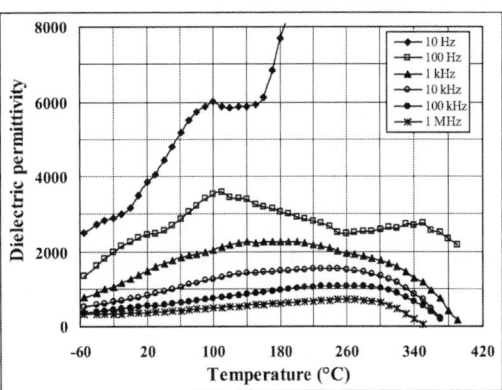

Figure 9: Temperature dependence of dielectric permittivity of BCFN thick films in the frequency range 10 Hz – 1 MHz

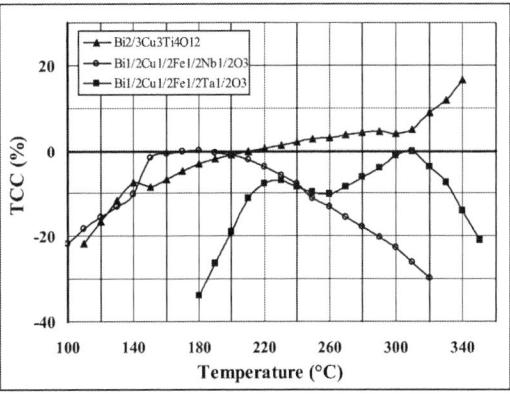

Figure 10: Temperature coefficient of capacitance of BCT, BCFN and BCFT thick films in the temperature range 100 – 340°C

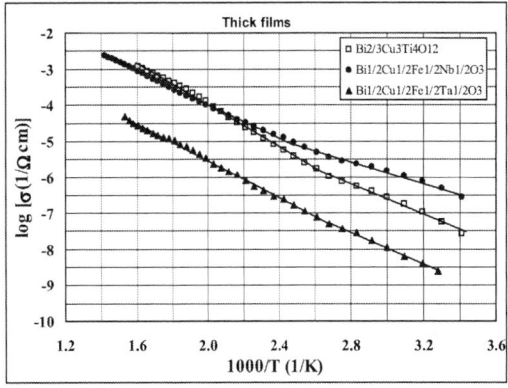

Figure 11: Logarithm of electrical conductivity versus reciprocal of temperature for BCT, BCFN and BCFT thick films

In Figure 11 the logarithm of dc conductivity of BCFN, BCFT and BCT thick films is plotted as a function of inverse temperature. The activation energies determined from the slopes of fitted straight lines are ranging from 0.3 to 0.45 eV in the lower temperature range (20 - 120°C) and from 0.46 to 0.57 eV in the higher temperature range (130 - 400°C). The E_c values for thick films are close to those determined for the ceramic samples with the same composition.

Conclusions

All the developed lead-free ceramics of perovskite structure with compositions: $Bi_{2/3}Cu_3Ti_4O_{12}$, $Bi_{1/2}Cu_{1/2}(Fe_{2/3}W_{1/3})O_3$, $Bi_{1/2}Cu_{1/2}(Fe_{1/2}Nb_{1/2})O_3$, $Bi_{1/2}Cu_{1/2}(Fe_{1/2}Ta_{1/2})O_3$, $Ca(Fe_{2/3}W_{1/3})O_3$, $Ca(Fe_{1/2}Nb_{1/2})O_3$, $Ca(Fe_{1/2}Ta_{1/2})O_3$ and $Cu_2Ta_4O_{12}$ exhibited high dielectric permittivity reaching 14000 – 50000 at 1 kHz. Three compositions chosen for thick film capacitor applications: $Bi_{2/3}Cu_3Ti_4O_{12}$, $Bi_{1/2}Cu_{1/2}(Fe_{1/2}Nb_{1/2})O_3$ and $Bi_{1/2}Cu_{1/2}(Fe_{1/2}Ta_{1/2})O_3$ were characterized by a dense microstructure, high dielectric permittivity and low temperature coefficient of capacitance at elevated temperatures (130 - 400°C).

Internal barrier layer capacitor formation is suggested to be responsible for the observed dielectric behavior of the investigated materials.

Acknowledgements

This work has been supported by Polish Ministry of Science and Higher Education under grant No. N507037 31/0906.

References

[1] M.A. Subramanian, Dong Li, N. Duan, B.A. Reisner, A.W. Sleight, J. Solid State Chemistry, Vol. 151, pp.323-325, 2000

[2] A.P. Ramirez, M.A. Subramanian, M. Gardel, G. Blumberg, D. Li, T. Vogt, S.M. Shapiro, Solid State Communications, Vol. 115, No. 5, pp. 217-220, 2000

[3] M.H. Cohen, J.B. Neaton, L. He, D. Vanderbit, J. Appl. Phys.,Vol. 94, pp.3299-3306, 2003

[4] Lixin Xe, J.B. Neaton M.H. Cohen, D. Vanderbilt,, C.C. Homes, Phys. Rev. B, Vol. 65, 214112, 2002

[5] D.C. Sinclair, T.B. Adams, F.D. Morrison, A.R. West, Appl. Phys. Lett., Vol. 80, No. 12, 2153-2155, 2002

[6] A.R. West, T.B. Adams, F.D. Morrison, D.C. Sinclair, J. Eur. Cer. Soc., Vol. 24, pp. 1439-1448, 2004

[7] L. Zhang, Z.-J. Tang, Phys. Rev. B, Vol. 70, 174306, 2004

[8] J. Li, A.W. Sleight, M.A. Subramanian, Solid State Communications Vol. 135, pp. 260-262, 2005

[9] A. Tselev, C.M. Brooks, S.M. Anlage, H. Zheng, L. Salamanca-Riba., R. Ramesh, M.A. Subramanian, cond. mat., 0308057, 2004

[10] I.P. Raevski, S.A. Prosandeev, A.S. Bogatin, M.A. Malitskaya, L. Jastrabik, J. Appl. Phys., Vol. 93, pp. 4130-4136, 2003

[11] B. Renner, P. Lunkenheimer, M. Schetter, A. Loidl, A. Reller, S.G.Ebbinghaus, J. Appl. Phys., Vol. 96, No 8, pp. 4400-4404, 2004

[12] J. Liu, C.-G. Duan, W.-G. Yin, W.N. Mei, R.W. Smith, J.R. Hardy, Phys. Rev. B Vol. 70, 144106, 2004

Development of Micro-Integrated Sensors and Actuators Based on PZT Thick Films

Sylvia E. Gebhardt, Thomas Rödig, Uwe Partsch and Andreas J. Schönecker

Fraunhofer Institut für Keramische Technologien und Systeme, Winterbergstr. 28, 01277 Dresden, Germany

Phone: +49 351 2553 694, Fax: +49 351 2554 160, E-mail Address: Sylvia.Gebhardt@ikts.fraunhofer.de

Abstract

Al_2O_3, Silicon (Si) and Low Temperature Cofired Ceramics (LTCC) are key functional materials, forming the substrate basis for microsystems technologies. They allow for three dimensional component integration, high robustness and excellent reliability. The combination of these substrate materials with piezoelectric films offer advanced sensor, actuator and ultrasonic transducer solutions which open new fields of application. Lead Zirconate Titanate (PZT, $Pb(Zr,Ti)O_3$) is the most common piezoelectric material because of its excellent ferroelectric and electromechanical properties. Integration needs engineered solution concerning design, PZT and electrode material selection as well as controlled processing. Most reliable interface solutions were achieved by screen printing and firing PZT thick films directly onto the substrate. Therefore, we developed a unique PZT thick film paste, recently. Sensors and actuators for microsystems applications like for example piezoelectric pressure sensors, ferroelectric printing forms and deformable mirrors have been developed successfully. The present paper deals with the feasibility of screen printable ultrasonic transducers and micro actuators on LTCC and Si-Wafer.

Key words: PZT, thick film, screen printing, sensors, actuators, ultrasonic transducers

Introduction

Microelectronic substrates like Al_2O_3, Silicon (Si) and Low Temperature Cofired Ceramics (LTCC) enable for three dimensional functional integration of electronic components and therefore for devices with high robustness and high reliability. The integration of piezoelectric transducers has already been considered, showing the interface formation, film densification and tailoring of functional properties to be the key issues.

Lead Zirconate Titanate (PZT, $Pb(Zr,Ti)O_3$) is the most common piezoelectric material because of its unique ferroelectric and electromechanical properties. Using thin film techniques like chemical solution deposition (CSD) or physical vapor deposition (PVD) PZT thin films with thicknesses between 20 nm and 2 μm can be applied attaining the desired characteristics for sensor applications [1]. The performance of bending actuators is limited to low power level.

Assembling methods, where piezoelectric plates are glued or joined onto a substrate material overcome this drawback. Problems arise from the weak interface, causing often failure by debonding.

PZT thick films with thicknesses between 20 μm and 150 μm fill the gap between PZT thin films and PZT plates. They are of great interest for microsystems applications where direct coating onto microelectronic substrates and high electromechanical performance are required.

The preparation of PZT thick films on various substrates has been an intensive research field for the last 15 years aimed at the application as information storage material in electrostatic printing machines and integrated sensors and actuators [2-6]. Preparation of high performance films turned out to be difficult. Problems arise from PbO evaporation, limited densification due to the constraint sintering conditions and reaction mechanisms between substrate material and PZT thick film.

We developed a unique screen printable PZT thick film paste having shown superior performance on substrates like Al_2O_3, Silicon and LTCC [2,3]. The present paper reports on our recent work dealing with the feasibility of screen printable ultrasonic transducers, and high performance micro actuators on LTCC and Si-Wafer.

Processing

Screen printing is a very flexible and cost effective technology for producing functional layers on green and fired substrates. Direct patterning of the functional layers is possible by the mask layout which is a great advantage of this method compared to spin coating processes. During screen printing the thick film paste is squeezed through a structured screen. Therefore a shear thinning or pseudo-plastic behavior of the thick film paste is needed. This allows for reduction of the viscosity under high shear rates whilst the paste passes through the mesh openings. After being deposited on the substrate the paste continues to level and thus smooth out surface irregularities before it stops flowing. Custom built pastes are required for the considered application. The pastes are basically composed of the functional powder and an organic print vehicle which is burned out during sintering.

We developed a PZT thick film paste based on a PZT-PMN formulation as described elsewhere [2,3,7].

As found experimentally, composition of the substrate material has an important influence on the dielectric and ferroelectric properties of the PZT thick film. The existence of Si or SiO_2 in the substrate material causes a reaction to lead based silicates, which deteriorate the ferroelectric behavior of the PZT thick film. This is relevant for PZT films on substrates like 96% Alumina, Silicon and LTCC.

In our experiments we used a commercial gold electrode (Heraeus 5789) for Al_2O_3 substrates. For Silicon and LTCC substrates a special gold electrode was developed, which also served as a diffusion barrier. Electrodes were deposited by screen printing and fired in a steel belt furnace at 850 °C.

PZT thick films with fired thicknesses of about 100 µm were built up by repeated screen printing. Sintering was performed at 900 °C or 950 °C for 5 h with shrinkage of approximately 35-45 % in the thickness direction. Lateral shrinkage was suppressed and no cracks were generated.

Top electrodes were screen printed onto the PZT thick film and fired in a steel belt furnace again.

Measurement Equipment

PZT films were poled at room temperature with E = 20 kV/cm for 5 min. Measurement of the dielectric parameters was done at least 24 h after poling.

The dielectric constants were measured at f = 1 kHz, using a Hewlett Packard 4194A Impedance Analyzer. The same device was also used for characterization of the impedance spectra. The resonance behavior and deflection of the thick film devices were characterized by a Polytec laser scanning vibrometer PSV 400.

The measurement of the piezoelectric coefficient d_{33} was performed at the Department of Physics of the Martin-Luther-University Halle at 130 Hz, using equipment based on a capacitive detector [8].

Ferroelectric hysteresis loops were determined by a modified Sawyer-Tower circuit. Internal resistance was measured by a Hewlett Packard 4339A High Resistance Meter.

Acoustic properties were characterized by pulse-echo measurements using a signal generator Agilent 33120A for driving samples and measuring the signals with a Hewlett Packard 54600B oscilloscope.

Properties of PZT Thick Films

Properties of PZT thick films strongly depend on the substrate material. Best results have been obtained for Al_2O_3 substrates with almost no silica content. As the Si content in the substrate material increases dielectric and electromechanical properties of the PZT thick film deteriorate as shown in Table 1 and Figure 1. Application of PZT thick films on LTCC and Silicon is only possible by the use of a diffusion barrier. We solved this problem by the development of an in-house gold electrode which served also as a diffusion barrier.

Table 1: Properties of PZT thick films on various substrates

Property	PZT on Al_2O_3 (99.7 %)	PZT on LTCC (DP 951)	PZT on Silicon
Dielectric constant $\varepsilon_{33}^T/\varepsilon_0$ at 1 kHz	1900	1500	1600
Dielectric loss tan δ at 1 kHz	0.038	0.033	0.055
Piezoelectric coefficient d_{33} [pC/N]	210	180	140
Remanent polarization P_r at 50 Hz [µC/cm²]	16	10	9
Coercive field E_C at 50 Hz [kV/cm]	15	13	12
Internal resistance R_{is} at 30 kV/cm after 30 s [Ω]	2×10^{11}	2×10^{11}	2×10^{11}

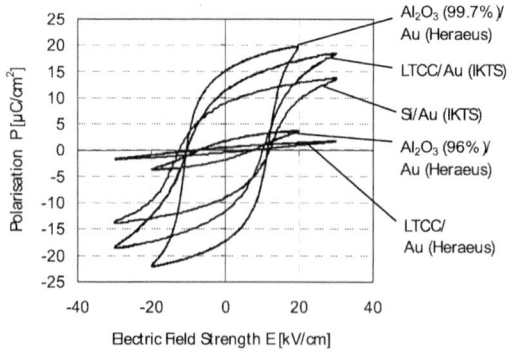

Figure 1: Ferroelectric hysteresis loops of PZT thick films on different substrates measured at f = 50 Hz

Actuators and Sensors Based on PZT Thick Films

Because of their high integration level and the strong inorganic bonding to the substrate material, PZT thick films are well suited for the development of actuators and sensors based on microelectronic substrates. We have already shown elsewhere, application of these films on Al_2O_3 for ferroelectric print rollers with a PZT thick film thickness of 100 µm, a width of 150 mm and an outer diameter of 200 mm [2], on 4 inch Silicon wafers for adaptive mirrors in the EUV (Extreme Ultra Violet) lithography [3] and on LTCC for pressure sensors in the range of 0-1.5 bars [9,10].

In the following, we describe our recent results on the feasibility of acoustic transducers and MEMS on Al_2O_3, LTCC and Si-Wafer.

Screen Printable Ultrasonic Transducer on Al_2O_3

A disc of a PZT thick film with a thickness of 120 µm was screen printed onto an electroded Al_2O_3 substrate as shown in Figure 2. The surface profile of the PZT disc after sintering is shown in Figure 3.

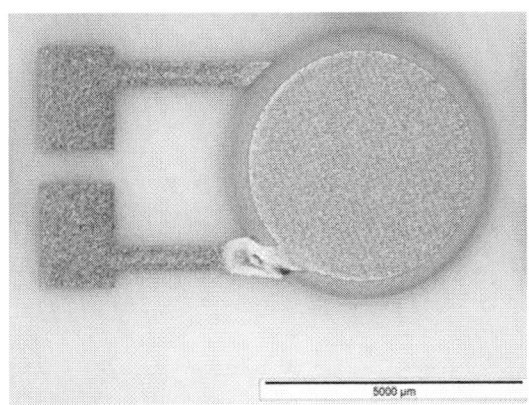

Figure 2: Ultrasonic transducer based on an electroded PZT thick film (thickness t = 120 µm) on Al_2O_3

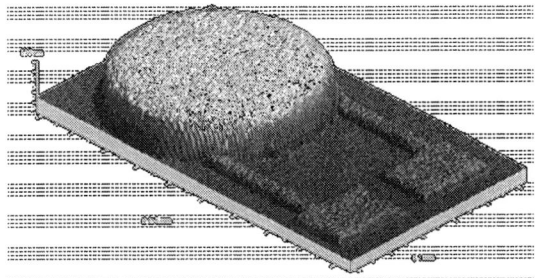

Figure 3: Surface profile of a sintered PZT thick film disc (Ø 5 mm, t = 120 µm) on electroded Al_2O_3 for application as an ultrasonic transducer

The resonance behavior of the transducer was characterized by measurement of the impedance spectra and the deflection. At a frequency of approximately f = 4.6 MHz a strong resonance occurred which was found to be a compound vibration of the PZT thick film and the substrate material. The Al_2O_3 substrate had a thickness of t = 630 µm. By driving the PZT thick film with U = 20 V the piezoceramic disc also excited the Al_2O_3 substrate to bend. Figure 4 shows a deflection picture of the compound vibration at f = 4.6 MHz.

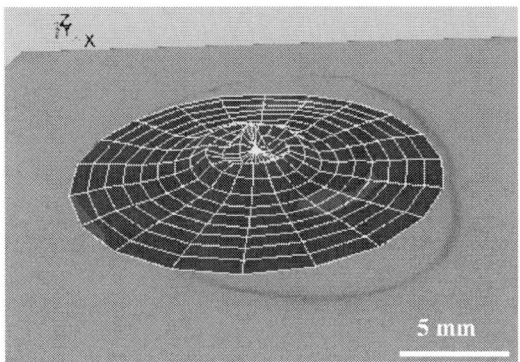

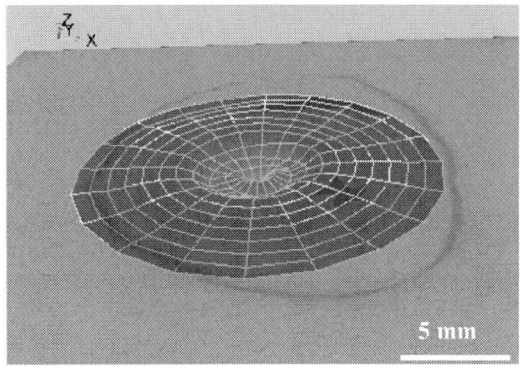

Figure 4: Bending mode of a PZT thick film disc on Al_2O_3 at f = 4.6 MHz

To study the feasibility of PZT thick films for ultrasonic transducer applications, pulse-echo measurements were carried out, too. Therefore, the transducer was glued on a plastic water tank and driven with a 5 periods burst signal of U = 3 V (peak-to-peak) at a frequency f = 4.6 MHz. The first back wall echo of the ultrasonic waves reflected at the water surface was received and measured by the transducer again. The fill height of the water tank was 38 mm.

As the ultrasonic wave sent by the transducer travels through the plastic into the water, it is reflected at the interfaces PZT/plastic and plastic/water. The same happens when the signal is reflected at the water surface and travels back to the transducer again. We used a test stand to provide evidence of transducer application. Figure 5 shows the pulse-echo signal of the ultrasonic transducer. The transmitted signal was cut off at U = 8 mV. Although no acoustic matching layers were used, the received signal had still a peak-to-peak amplitude of U = 3 mV.

The measurements clearly confirmed feasibility of our PZT thick films for application as ultrasonic transducers.

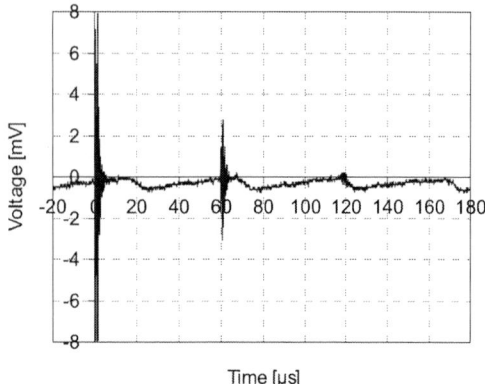

Figure 5: Pulse-echo curve of a PZT thick film disc on Al$_2$O$_3$ at f = 4.6 MHz

Structured Micro Actuator on LTCC

LTCC is a well known packaging material which can be easily shaped into three dimensional structures with membranes, channels and cavities. With these properties LTCC meets all requirements for demanding sensor and actuator solutions.

We combined LTCC technology with application of PZT thick films to develop integrated micro actuators.

DuPont LTCC 951 green tapes were used to build up substrates with a thickness of t = 660 μm with an enclosed membrane of t = 220 μm thickness. Then bottom electrodes and PZT thick films with a thickness of 80 μm were screen printed onto the membrane. After deposition of the top electrode, bending structures were cut out by laser machining. Figure 6 shows an example of a bending micro actuator based on structured LTCC.

Measurement of the resonant behavior of the bending micro actuator showed a strong resonance at f = 433 Hz. The deflection of the structure when driven with E = 20 kV/cm was characterized by a laser scanning vibrometer as seen in Figure 7. A deflection of up to approximately 14 μm could be achieved in one direction.

The results show one possible application of PZT thick films on LTCC and give an idea of a large field of MEMS applications which will be open combining both technologies.

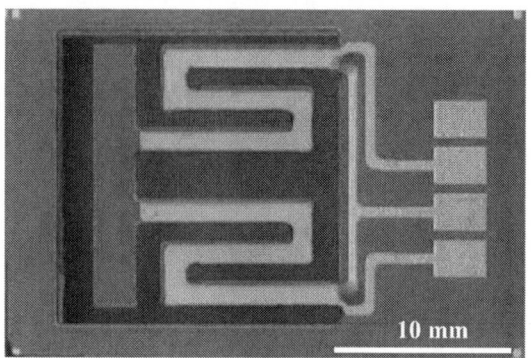

Figure 6: Bending micro actuator based on LTCC with PZT thick films

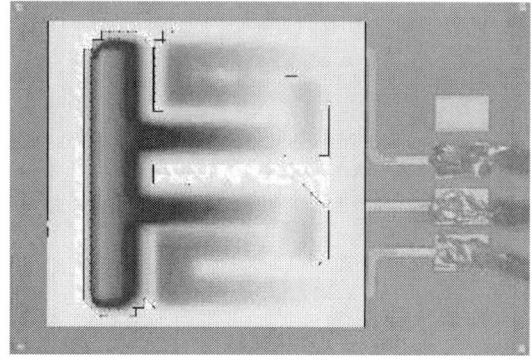

Figure 7: Deflection of the bending micro actuator shown in Figure 6 when driven with E = 20 kV/cm at f = 433 Hz measured by a laser scanning vibrometer

Micro Actuator Array on Si-Wafer

Sintering of PZT thick films on Silicon turned out to be the most challenging task to be solved. Because of the huge excess of Si, reaction to lead silicates is not easy to prevent. By using a gold electrode as a diffusion barrier reaction to lead silicates underneath is inhibited. Beside the bottom electrode, reaction of PbO which is evaporated during sintering with Si can still take place. This effect causes the development of glassy halos around the bottom electrode. The extension of the halos corresponds to the air flow in the oven. Therefore, application of PZT thick films at the back side of the Silicon wafer is reasonable.

PZT thick films on 4 inch Silicon wafer for adaptive optics have been introduced already as described in [2,3]. Recently, we enhanced the design to 6 inch Silicon wafers in co-operation with the Active Structure Laboratory of the University of Brussels (Prof. André Preumont). The aim of this work was to generate an active mirror for use in aerospace. Therefore 91 hexagonal PZT thick film structures with an edge to edge diameter of d = 8.58 mm and a thickness of t = 90 μm were screen printed onto the back side of an electroded 6 inch Silicon wafer as shown in Figure 8 and 9. The front side of the wafer is used as the mirror face. By driving each single PZT element, the mirror can be moved individually. This gives the basis for smart applications of adaptive optics.

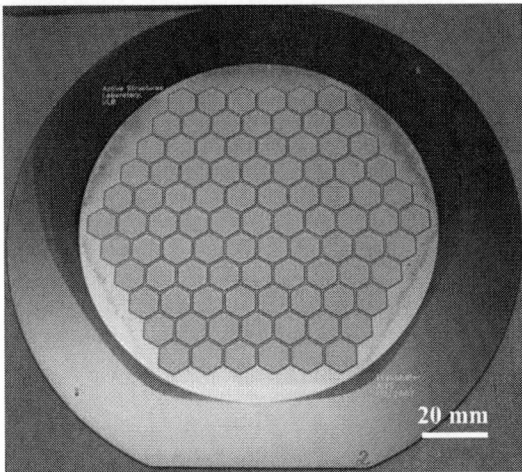

Figure 8: Active Mirror based on 6 inch Silicon wafer with honeycomb PZT thick film pattern (edge to edge diameter of hexagonal PZT structure d = 8.58 mm)

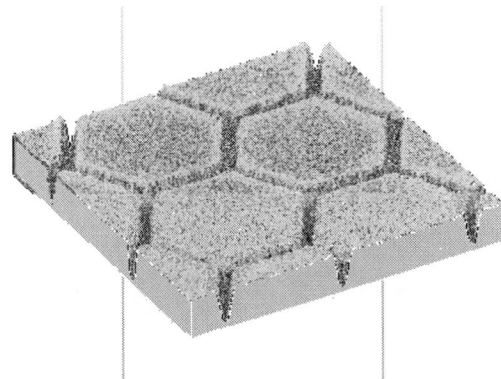

Figure 9: Surface profile of sintered honeycomb PZT thick film pattern on 6 inch Silicon wafer

Conclusions

PZT thick films have been applied on different substrates like Al_2O_3, Silicon and LTCC which are basis materials for microsystems technology. Therefore technological as well as material issues had to be solved. This allows now for the development of advanced microsystems with increased functionality. Three possible applications were presented in this article: an ultrasonic transducer based on Al_2O_3, a bending micro actuator based on LTCC and an active mirror based on a 6 inch Silicon wafer.

References

[1] J. Frey, A. Schönecker, F. Schlenkrich, "Self-Polarization and Texture of Wet Chemically Derived Lead Zirconate Titanate Thin Films", Integrated Ferroelectrics, Vol. 35, pp. 105-113, 2001.

[2] S. Gebhardt, L. Seffner, F. Schlenkrich, A. Schönecker, "PZT Thick Films for Microsystems Applications", Proc. of the 4th European Microelectronics and Packaging Symposium with Table-Top Exhibition (EMPS), Terme Catez, Slovenia, May 21-24, 2006, pp. 9-13.

[3] S. Gebhardt, L. Seffner, F. Schlenkrich, A. Schönecker, "PZT Thick Films for Sensor and Actuator Applications", to be published in Journ. Europ. Ceram. Soc., 2007

[4] M. Hrovat, J. Holc, S. Drnovsek, D. Belavic, J. Cilensek, S. Macek, M. Santo-Zarnik, M. Kosec, "Processing and Evaluation of Piezoelectric Thick Films on Ceramic Substrates", Proc. of the 4th European Microelectronics and Packaging Symposium with Table-Top Exhibition (EMPS), Terme Catez, Slovenia, May 21-24, 2006, pp. 3-8.

[5] L.J. Golonka, M. Buczek, M. Hrovat, D. Belavic, A. Dziedzic, H. Roguszczak, T. Zawada, "Properties of PZT Thick Films Made on LTCC", Microelectronics International, Vol. 22, No. 2, 2005, pp. 13-16

[6] E.S. Thiele, D. Damjanovic, N. Setter, "Processing and Properties of Screen-Printed Thick Films on Electroded Silicon", Journ. Am. Ceram. Soc., 2001, Vol. 84, No. 12, pp. 2863-2868.

[7] L. Seffner, H.J. Gesemann, K. Völker, "Process for the Production of PZT Coatings from a PZT Powder with a Low Sintering Temperature", German Patent DE 4,416,245, April 13, 1995.

[8] G. Sorge, T. Hauke, M. Klee, "Electromechanical Properties of Thin Ferroelectric-$PbZr_{0.53}Ti_{0.47}O_3$-Layers", Ferroelectrics, Vol. 163, 1995, pp. 77-88.

[9] U. Partsch, D. Arndt, U. Keitel, P. Otschik, "Piezoelectric Pressure Sensors in LTCC-Technology", Proc. of the 14th European Microelectronics and Packaging Conference & Exhibition (EMPC), Friedrichshafen, Germany, June 23-25, 2003, pp.

[10] U. Partsch, S. Gebhardt, D. Arndt, H. Georgi, H. Neubert, D. Fleischer, M. Gruchow, "LTCC-Based Sensors for Mechanical Quantities", Proc. of the 16th European Microelectronics and Packaging Conference & Exhibition (EMPC), Oulu, Finland, June 17-20, 2007

High Volume MEMS Packaging

Mark Shaw[1], Federico Ziglioli[1], Anne Marie Grech[2], Mario Cortese[3]

[1]ST Microelectronics, Corporate Packaging and Automation (CPA), Via Olivetti 2, Agrate, Italy

[2]CPA, Kirkop, Malta.

[3]MEMS Product Division, Castelletto, Italy.

Tel.: +390396035422, Fax: +3903964426930, E-mail: mark.shaw@st.com

Abstract

In this abstract we outline the critical aspects in the design of high volume packaging for MEMS applications. We will illustrate the evolution of MEMS packaging for devices now used in applications such as drop protection of lap tops and Cell phones and movement sensing in game controllers. In consumer applications the size and cost of the package are becoming increasingly important especially in the applications for Mobile telephones. The development and adoption of LGA style packages has opened up these new market segments to MEMS devices. The design of LGA based MEMS Packages are outlined and the problems encountered in packaging MEMS devices in this style of package structure discussed. In particular the aspects of the design that have influenced the performance of the MEMS devices are outlined. The requirements for size reduction discussed and the flexibility of this particular package platform is demonstrated. The Characteristics of the materials required in the assembly processes are discussed in order to optimise performance with the LGA platform. Extensive use of FEA simulations will show the influence of various aspects of the design and these will be correlated with actual devices results. Manufacturing processes will also be outlined showing how the LGA style package can be produced in a high volume environment.

Key words: MEMS, LGA, Accelerometer

Introduction

The original MEMS (micro electro-mechanical systems) applications such as collision sensors for automotive airbag, accelerometers and pressure sensors for engine manifolds, involved bulky packages using ceramic or pre moulded plastic packages. MEMS Packaging is currently evolving from these original formats into newer more low cost styles more suitable to consumer applications. New applications for example for accelerometers (such as free fall protection for hard disk drives on portable computers and cell phones, gaming interfaces, pedometers etc,) require lower cost and smaller packages in order to open up these markets [1,2]. The first MEMS packages were open cavity; however in standard electronics application lower cost packages are dominated by full moulded solutions either on lead frames or increasingly using substrates.

Figure 1: SO 24 and QFN lead frame packages

ST microelectronics first introduced a full moulded lead frame SO-24 (fig 1,2) and then a Quad Flat No Lead (QFN) package (fig1) in 2002. The SO style [3,4] and QFN style [5] packages are now widely used for MEMS devices.

Figure 2: MEMS sensor mounted on SO-24 lead frame

The MEMS devices produced in ST microelectronics are made on an 8 inch wafers using the Thelma process (Thick Epitaxial Layer for Microactuators and Accelerometers) standardized by exploiting state-of-the-art MEMS manufacturing

194

equipment. THELMA is a 0.8-micron process with a thick polysilicon layer for structures and a thin polysilicon for interconnections. This process platform is very mature and flexible enough to meet diverse market needs: from accelerometers, gyroscopes, megahertz oscillators and filters to dedicated confidential products.

A MEMS device, e.g. an accelerometer, being a mechanically moving part (fig3) must be protected from any particle or agent that could cause the mechanical surfaces to seize or jam. This requires the MEMS structure to be sealed within a protective cap in a controlled atmosphere. In plastic moulded packaged it must also be protected from any ingress of the moulding compound. In ST as in many companies this is done by wafer level hermetic sealing.

Figure 3: exposed MEMS sensor

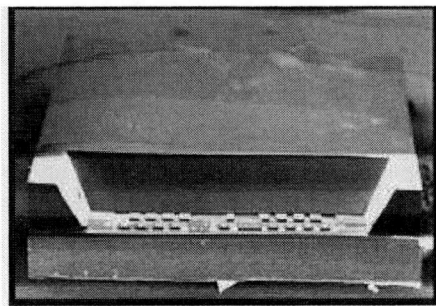

Figure 4: Section showing Hermetic Sealing and wirebond opening in cap

The most widely used method is that of sealing two wafers together. This can be done either with a glass frit or solder at wafer level [6]. If the cap design is such that it is the same size as the MEMS sensor with a 'cut out' for the wirebond pads, the wafer sandwich can be treated as a single wafer for any subsequent processing. This is important in the case of back lapping where in order to reduce the overall thickness of the package the wafer sandwich thickness needs to be reduced. The wafer sandwich is then diced using conventional saw dicing techniques, and no special precautions have to be taken as the MEMS device is sealed within its capped structure (fig 5). This is preferable to other

techniques where cap is bonded on the wafer and the cap layer is partially sawn to reveal the bond pads. In order to back lap the wafer a complicated process requiring filling tape in between the caps is required [7,8].

Recently a new family of packages for MEMS devices based on Land grid array packages, (LGA) has been introduced which give an increased new flexibility in manufacturing. In designing a MEMS package it is important that we remember that it is not only an electronic device with a package that protects the device and provides the standard electrical connections with associated electrical and thermal dissipation problems. A MEMS device is also a mechanical device where the package itself is the mechanical connection to the outside world. This means that the package has to take account of the physical interaction of the package on the MEMS device and its influence on the device performance in the same way as the electrical conductors resistance, inductance and capacitance does for the electronic part of the packaging[9].

LGA package and substrate design

The LGA package consists of a laminate substrate with a BT (Bismaleide. Triazine) resin Core, so far MEMS sensor devices have required only a top and bottom metal tracks with corresponding solder mask but as more complicated integrated designs develop these can be easily accommodated with extra layers. The exposed soldering pads (lands) and wirebond pads are suitably plated. One of the advantages of the LGA substrate design is it's flexibility with the ability to route the Cu tracks to permit different designs of sensor or ASIC to have the same pin out, or the same designs to have different pin outs with no change to the assembly process including even the wirebonding layout, only a different substrate design. This gives a large amount of flexibility which is not available in more traditional (e.g. QFN) designs.

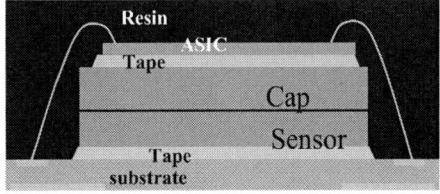

Figure 5: schematic of stacked MEMS LGA

In the case of an accelerometer the sensor device is usually co-packaged with a custom ASIC controller. This enables a more user friendly output for the customer either in analogue or in digital format. Depending on the size of die and package, different configurations of sensor and ASIC are

195

possible, they can either be bonded side by side or in a stacked mode (fig 5,6). The side by side mode (fig 6) obviously has a larger footprint than the stacked devices but can have advantages over lower profile devices, in that the separate die may not need to be back lapped to fit in the required package height. A stacked approach however with back lapped die will result in the lowest overall package size.

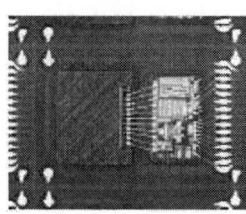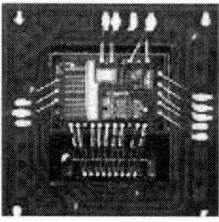

Figure 6: Photo of side by side bonded accelerometer and ASIC on LGA substrate (left) and stacked device (right)

The substrate design is made in standard strips in 4 blocks. This enables 240 devices per strip for a 7.5x4.4 design to 484 for a 4x4. The design and layout of the substrate is important in order to minimise the stress on the active area of the MEMS device and extensive FEA modelling is used in order to optimise the effects of the substrate.

Figure 7: FEA model of LGA NSMD land showing stress

One final aspect of the substrate design to consider is the design of the solder pads or lands for interfacing with the customers PCB. Two main configurations are used: SSMD (semi solder mask defined) and NSMD (non solder masked defined).

Table 1 Simulated Stress on various land designs

Stress	NSMD	Anchor trace		SSMD	
	MPa	MPa	%change	MPa	%change
Seqv	456	365	-20	276	-39
Sz	461	456	-1	261	-43
Syz	137	97	-30	59	-57
Sxz	204	202	-1	121	-41

In SSMD lands the final solder mask layer covers the edge of the land helping to fix it to the substrate.

An intermediate design uses only partial traces between the land and under the solder mask to anchor the lands. Without this extra clamping the land peel strength is limited to the adhesion of the pad to the substrate. Simulation and shear test results confirm the improvement with SSMD pads, with the fail mode during shear test for SSMD pads being always in the solder itself.

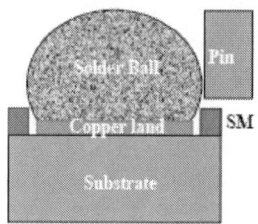

Figure 8: Schematic of shear test for NSMD LGA

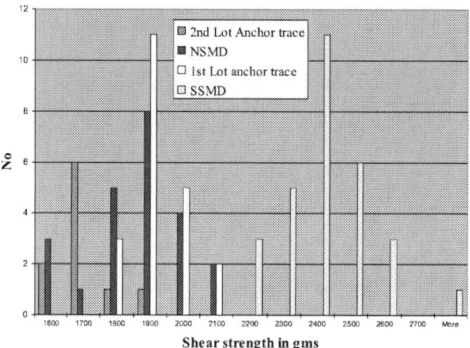

lands

Figure 9: Shear test of LGA lands 0.8mm Sn/Pb solder ball on LGA3x5 for NSMD, anchor trace and SSMD lands

Some concern has been expressed that SSMD designs could compromise the co-planarity of the lands or that the protrusion of the solder mask could cause problems during soldering but measurements have shown that the mask height above the land is no more than 30um.

It is important to understand the peel strength of the pads and also the effect of the stress induced by the soldering and final PCB design of the component (QFN [10] or LGA). If the PCB is not designed correctly stresses due to the placement position of other components on the PCB can cause warpage problems within the PCB which in turn gives rise to land peeling or stress on the sensor itself leading to drift in its performance or failure.

Assembly

The sensor die wafer is laminated with a combined die attach and dicing tape. The Die attach tape is combined with the dicing tape and applied to the wafer prior to dicing. The tape has characteristics such that for small die the adhesion is sufficient in the uncured state that the small diced chip remains in place during dicing and the die can then be picked with the die bonding tape adhered to it from the dicing tape. The dicing blade therefore cuts through the bonding tape but not the dicing tape. In selecting the die bond tape the bonding epoxy should flow sufficiently to take up any non planarity in the substrate and be sufficiently rigid after curing so that ultrasonic energy is not dispersed during wirebonding at elevated temperature.

Figure 10: SEM of stacked LGA accelerometer Die and wirebonded on substrate strip

The assembly operations, Die bonding, wire Bonding, moulding and device singulation are all carried out on standard automated machines.

Package stress

As mentioned previously the package is the mechanical linkage of the MEMS sensor to the outside world and its structure is therefore important in determining the stress on the MEMS devices and thus its final performance.

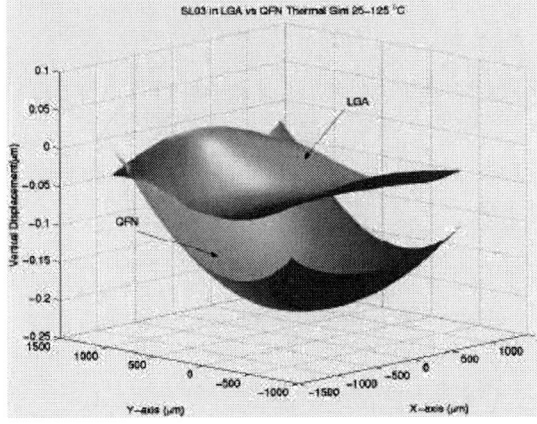

Figure 11: Simulation results of displacement in X and Y and Z axis for a QFN and LGA packaged accelerometer for the same accelerometer device over temperature.

Simulation of a QFN and LGA MEMS devices Fig (12,13) shows the difference in stress or mechanical movement for the MEMS structure due to stress. It can be clearly seen that with the LGA based device package the stress induced on a MEMS accelerometer is reduced. This is confirmed by actual measurement of the offset voltage at 0g of the different package styles SO-20, QFN and LGA.

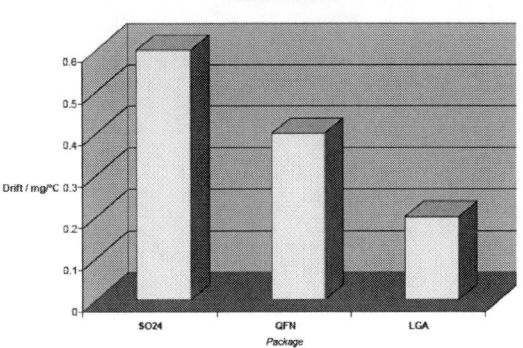

Figure 12: Offset drift for SO-24, QFN and LGA packages

However in an LGA package the stress can still be optimised. This stress is induced by the package design and the assembly tolerances. In fig (14) the variation of package warpage, (the bending of individual package due to stress), as a function of the package thickness is simulated. This shows that for a given package height the stress on the sensor can be optimised by varying the sensor die size for a given package height.

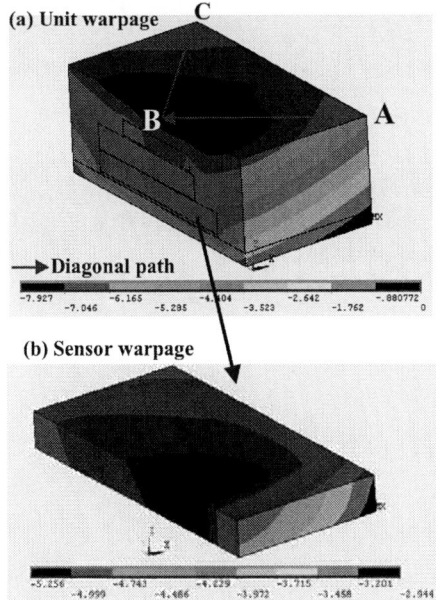

Figure 13: Unit warpage from simulation

The variation of the die attach thickness also effects the stress. Experimental verification of this can be seen in the results of the variation of die attach thickness with the offset drift.

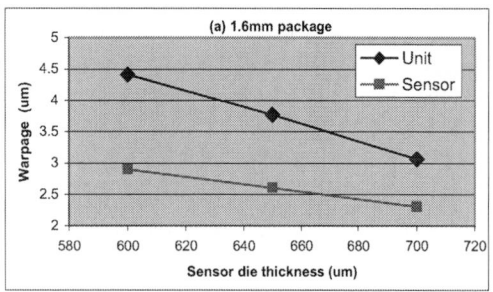

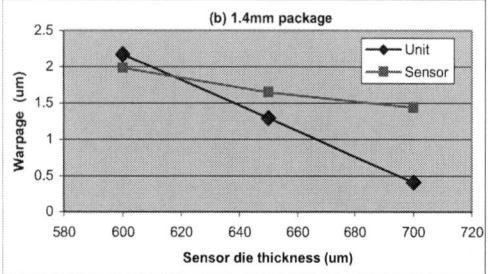

Figure 14: Simulation of the Diagonal path Warpage vs. Variation of sensor die thickness for a 1.4 and 1.6mm thick stacked package

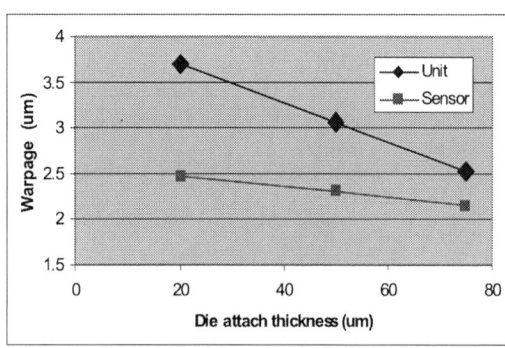

Figure 15: Simulation of the Diagonal Path warpage vs Variation of die attach thickness stacked package

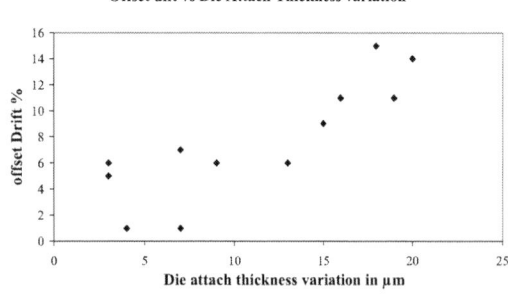

Figure 16: Experimental results of offset vs the variation of die attach thickness stacked package

The main factor in the stress induced is the thickness of the moulding compound above the die(s). This variation in package thickness and also the

correlation of warpage with the die thickness is due to the variation of the thickness of the moulding compound which has the most influence on package stress/warpage[11]. This is also shown in the simulation results for a 3x5x1mm side by side package.

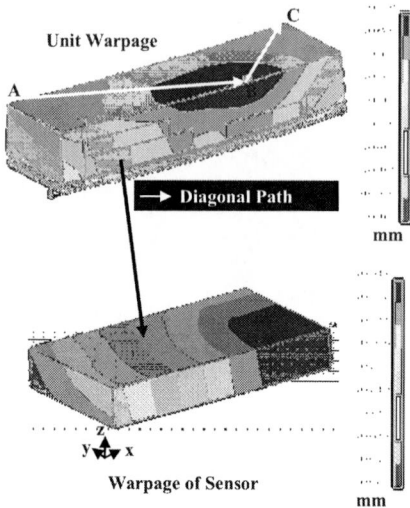

Figure 17: Simulation of side by side package

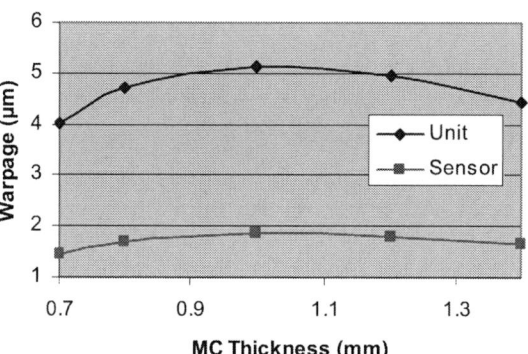

Figure 18: Moulding compound thickness vs Diagonal Warpage for side by side package

It therefore follows that the displacement of the die ASIC and sensor within the package structure can also effect the stress on the sensor device

The Individual devices together will also warp the whole block of devices on the strip which can cause problems in production with the tendency of die bond or wirebond machines to jam.

Device performance and Reliability

LGA packages with sizes of 7.5x4.4x1, 5x5x1.6mm, 3x5x1mm and 4x4x.1.5mm are now fully qualified to Jedec 3 passing stringent temperature cycling (-40+85°C) Humidity (100%, 120°C) storage and life tests, lead free and RoSH compatible. Typical performance over temperature (fig 19) shows the excellent low stress characteristics of this package.

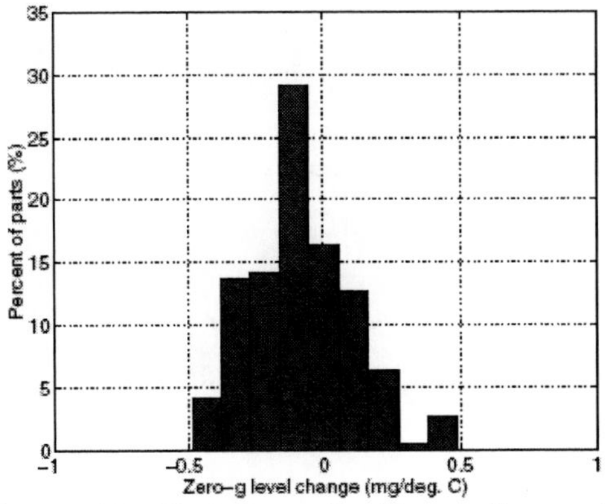

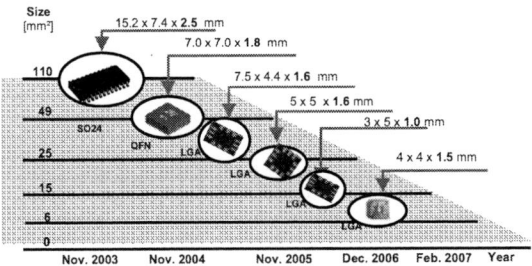

Figure 19: Variation of offset voltage with temperature (-40°C +85°C) for a LGA 5x5 accelerometer

Summary

We have outlined an LGA style packaging for MEMS devices which gives improvements in assembly and performance. This package style can be used for many types of MEMS devices including Accelerometers, Gyroscopes and pressure sensors.

Package Evolution for MEMS

Figure 20: Evolution of the full moulded package size in ST

The LGA package permits a flexible packaging line that can easily change between sizes and can result in an extremely small form factor (fig 20)

Acknowledgements

The authors would like to thank Xueren Zhang and Giovanni Frezza for the preparation of material used in this article.

References

[1] B.Vigna, "More than Moore: micro-machined products enable new applications and open new markets." Electron Devices Meeting, IEDM 2005 Technical Digest. Page(s):8

[2] B. Vigna, "Future of MEMS: An industry point of view", Proc. of 7th International Conference on Thermal, Mechanical and Multiphysics Simulation and Experiments in Micro-Electronics and Micro-Systems, 2006. EuroSime 2006. Page(s):1 – 8

[3] G. Li, A. Tseng, "Low Stress Package of a Micro-Machined Accelerometer," IEEE Transactions on Electronic Packaging Manufacturing, Vol. 24, pp. 18-24.

[4] G. Li, D. Mahadevan, M Chapman, "Packaging MEMS inertial sensors at Motorola", Lasers and Electro-Optics Society, 2003, Page(s):628 - 629 vol.2

[5] L.E Felton, N. Hablutzel, W.A. Webster, K.P Harney, "Chip scale packaging of a MEMS accelerometer", Proceedings, 54th Electronic Components and Technology Conference, 2004, Page(s):869 - 873 Vol.1

[6] S.A. Audet, K.M. Edenfel, "Integrated sensor wafer-level packaging", Proceedings International Conference on Solid State Sensors and Actuators 1997, Page(s):287 - 289

[7] E.S. Lacsamana, M.G. Mena, R.M. Navarro, N. Kuan, T.R Spooner, "Wafer Thinning Solution for Wafer-Level-Capped MEMS Devices ", Proceedings 56th Electronic Components and Technology Conference, 2006. Page(s):1118 - 112

[8] E.S. Lacsamana, R.M. Navarro, M.G. Mena, L.E. Felton, W. Webster, "A Very thin packaging of capped MEMS accelerometer device ", Proceedings of 7th Electronic Packaging Technology Conference, 2005. EPTC 2005. Volume 1, Page(s):5 pp

[9] X. Zhang, S.B. Park, R. Navarro, M.W. Judy, "Accurate assessment of packaging stress effects on MEMS devices", Proceedings of Thermal and Thermomechanical Phenomena in Electronics Systems 2006. ITHERM '06. Page(s):1336 - 1342

[10] T.Y. Tee, G. Frezza, M. Lim, H.S Ng, F. Ziglioli, Z.W. Zhong, "Design for Board Level Reliability of a Miniaturized MEMS Package: Stacked Die TQFN," International Journal of Computational Engineering Science (IJCES), 2003, Vol. 4(2), pp. 347-350.

[11] X. Zhang, T.Y Tee, J. Luan, "Comprehensive warpage analysis of stacked die MEMS package in accelerometer application", Proc 6th International Conference on Electronic Packaging Technology, 2005 Page(s):581 - 586

Screen-printed ruthenium dioxide pH-electrodes

Kathrin Reinhardt, Christel Kretzschmar, Claudia Feller

Fraunhofer-Institute for Ceramic Technologies and Systems

Winterbergstraße 28, D-01277 Dresden

Phone: +49(0)351/2553-837, Fax: +49(0)351/2554-338, E-mail: Kathrin.Reinhardt@ikts.fraunhofer.de

Abstract

The alternative use of metal oxide electrodes instead of glass electrodes for pH-measurements is reported. This study is focused on screen printed potentiometric pH-electrodes – which consists of a ruthenium dioxide - glass composite. The electrode will be used in a sensor for carbon dioxide measurements. This sensor will contain an electrolyte of alkaline $NaHCO_3$ solution, so the pH-properties of the electrodes in this buffer are from large interest. Former investigations have shown good results in pH-measurement, without detailed statements about the stability of electrode potentials. In this report the sensor properties, like pH-sensitivity, pH-response and especially the stability of potentials are characterized. The examined electrodes were made by thick-film technology. The ruthenium-glass inks were printed on alumina substrates. Two sorts of glass, which were selected for their behaviour in several solutions, were used. Different series of pH-electrodes were tested, one with lead oxide glass and one with lead free glass. The electrodes were treated in different solutions and measurements were taken in weekly time steps. The sensory properties are examined by potentiometric measurements. The dependence of electrode behaviour on surface structure was characterized by microscopic observation. The electrodes show a near-Nernstian dependence of potential upon pH in aqueous buffer between pH 4 and 9. Especially in alkaline buffer solutions the pH-data are very stable. Different behaviour of pH-response during the testing time could be explained with structural changing of the electrode surfaces. The best results have shown the ruthenium lead free electrodes in alkaline $NaHCO_3$-solution, which can be used in carbon dioxide sensors in the future.

Key words: ruthenium-dioxide, screen printed electrodes, pH sensor, thick film technology; potentiometry

Introduction

pH measurements are in many ranges necessary, for example in chemistry and food industry, in water and waste water treatment and in the medical technology. The glass pH electrode is the most commonly used pH sensor [1]. However, glass electrodes present several problems. For example these sensors are fragile, difficult to miniaturize, have high impedance and show an acid and alkaline error [2]. An alternative for glass electrodes are metal oxide electrodes like ruthenium dioxide sensors. Further results of several groups have shown that ruthenium oxide electrodes exhibit a nearly nernstian response to pH [3-7]. These electrodes are more mechanically robust, can tolerate operation under high temperature and have a simple structure. Another main advantage is the low electrical resistance of these electrodes, because no electrical shield is necessary to work with these sensors.

In this paper, screen printed ruthenium dioxide pH electrodes are presented. This technology permits the fabrication of miniaturized electrodes and a simple low-cost production with high reproducibility. In addition, thick film pH

electrodes are more rugged than the glass pH electrodes. The used electrodes consist of RuO_2-glass composites printed onto alumina substrates. Two sorts of electrodes were tested. One series consist of RuO_2 and lead oxide silicate glass (glass A), while the other electrodes were made up of a mixture of RuO_2 and lead free glass (glass B). The freshly screen printed pH sensors were aged over 70 days in a variety of solutions. In a weekly period the electrodes were tested by potentiometric pH measurements. The results were compared with the theoretical nernstian factor and the electrodes with the best results will used for carbon dioxide sensors in the future.

Experimental

Materials

The ruthenium dioxide powder was obtained from SAXONIA Edelmetalle GmbH (Halsbrücke, Germany). For the contacting a silver-based ink from Heraeus Holding GmbH (Hanau, Germany) was used. The polymer-based ink for covering the electrode conduction was bought by DuPont (Bristol, U.K.). Inks were printed on 96% alumina substrates (thickness 625 µm). Buffers (pH 4.01, 6.86, 9.18) for potentiometric measurements were purchased by Sensortechnik Meinsberg GmbH (Meinsberg, Germany). The used glasses A (lead silicate glass) and B (lead free glass) were selected out of five glasses after the testing of the chemical stability in different pH-solutions.

Electrode preparation

The layout of RuO_2 electrodes is shown in figure 1.

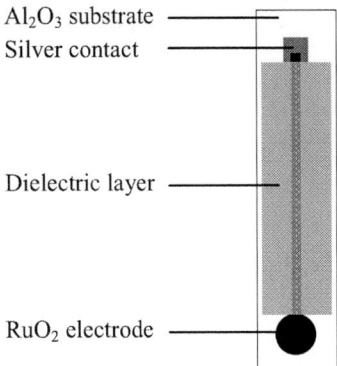

Figure 1: Layout of RuO₂ electrode

A semi-automatic screen printing machine, Microtronic II, from EKRA Automatisierungssysteme GmbH (Bönnigheim, Germany) was used to produce the screen printed pH-electrodes. First of all the silver based ink (conducting pad) was printed onto alumina substrates (10 x 45 mm²). Afterwards the silver paste was heated to 150 °C for 15 minutes to remove volatile organics and finally fired in an oven at 850 °C for ten minutes. After firing the silver contacts, the RuO_2-glass composite inks were printed. These inks were prepared by hand-mixing. Ink 1 (RuO_2-lead free glass composite = RLG) was a mixture of 76 Vol% RuO_2 and 24 Vol% glass A. The powders were mixed in an agate ball mill for 30 minutes. The mixture was ground together with a solution of ethyl cellulose in terpineol/ dibuthylphthalat to produce a printable paste. Ink 2 (RuO_2-lead free glass composite =RLFG) was prepared analogue to ink 1, except for using 69 Vol% RuO_2 powder and 31 Vol% glass B. Ink 1 and 2 were double printed on the end of the silver pad, dried at 150 °C and fired at 850 °C (ink 1) and 950 °C (ink 2). At last the electrode conductions were insulated with dielectric pastes. Ink 1 was covered with a lead silicate glass paste, which was coloured with $Co(AlO_2)_2$. This insulating layer was also dried at 150 °C for 15 minutes and fired at 850 °C for 10 minutes. For ink 2 the polymer based ink was printed over the RuO_2 conduction and heated at 200 °C for 20 minutes. The electrodes were contacted with a single cable.

Measurements

Fifteen electrodes per series were stored in five different media with three electrodes in each.

These were deionised water, dilute sodium hydrogen carbonate (0.01M, pH≈10), pH 4.01 buffer, pH 9.18 buffer and keeping dry at room temperature. The solutions were changed every week. The electrode potentials were measured versus a commercial silver/silver chloride reference electrode (Sensortechnik Meinsberg GmbH; E=0.197 V versus NHE) on a weekly basis over a 70 day period. The testing was carried out in three different pH buffers (pH 4.01, 6.86, 9.18) using a voltmeter. The buffers were thermostated at 25 °C and at each change of buffer, the electrodes were rinsed in deionised water. The potential-pH response of each electrode was measured after injection of the electrodes in the buffer solution for ten minutes. For each electrode following parameters were determined:
- pH-sensitivity (m)
- zero potential (U_0)
- responding behaviour

For surface characterisation of the electrode series measurements by FESEM (Gemini 982, co. Zeiss) and EDX analyze were done. Topographical and material differences and changes were examined.

Results and Discussion

The pH-sensitivity is calculated by the NERNST-equation:

$$U = U_0 - \frac{2,3 * RT}{nF} * pH = U_0 - m * pH \quad (1)$$

For a single electron mechanism (n=1) is the nernstian factor m = 59.2 mV at 25 °C. Table 1 shows the slope factors and zero potentials of each electrode series (15 electrodes per series) before aging in several media. The mean pH-sensitivity (-63.7 mV/pH) of the RLG electrodes was higher than the nernstian factor of a single electron mechanism (-59.2 mV/pH). This fact may stand for an exchange of less than one electron (n>1) at the electrode mechanism. It is possible that a ruthenium oxide, which has an oxygen-deficit ($RuO_{2-\delta}$) was formed by firing in the presence of lead oxide glass. After hydration of the oxygen-deficit ruthenium oxide,

$$Ru(OH)_{4-\delta} + (1-\delta)H^+ + (1-\delta)e^- \leftrightarrows$$
$$Ru(OH)_3 + (1-\delta)H_2O \quad (2)$$

the slope factor was around −55 mV/pH, which makes it possible that the pH-reaction follows a single electron mechanism.

The RLFG electrodes have shown a near nernstian manner of −57.5 mV/pH before starting long time storage. This relates to a single electron mechanism (n=1) like the following redox reaction:

$$Ru(OH)_4 + H^+ + e^- \leftrightarrows Ru(OH)_3 + H_2O \quad (3)$$

Table 1: Values of electrodes before beginning storage in five different media

	RuO₂-lead free glass electrodes	RuO₂-lead free glass electrodes
Slope factor [mV/pH]	-63.7 ± 5.1	-57.5 ± 1.1
U_0 [mV]	836 ± 61	674 ± 41

Long Time Storage

The averaged slope factors of the RLG electrodes are plotted in figure 2. The spread of results varied between the media which were used. The electrodes, which were stored in pH 4.01 buffer, deionised water and dry at room temperature, showed a very similar behaviour. The pH sensitivity changed from –64 mV/pH to approximately -55 mV/pH. The slight fluctuations could attribute to the aging of buffers. The alkaline stored electrodes showed a near nerstian manner over a period of nine weeks. After ten weeks the slope factor exhibited a clearly depression in the pH 9.18 buffer as well as in the diluted NaHCO₃ solution, so the electrodes failed after seventy days.

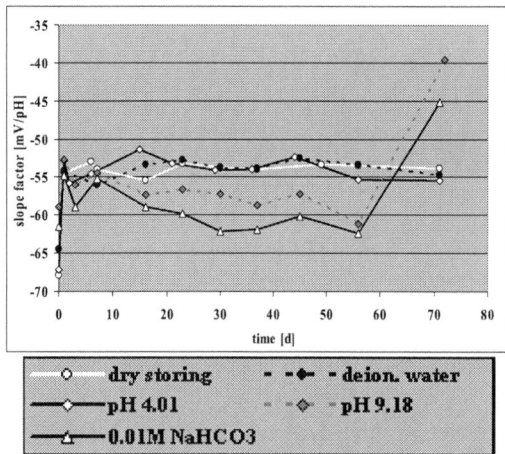

Figure 2: Averaged pH slope factors of RuO₂-lead free glass electrodes in several media

Figure 3 presents the pH sensitivity of the RLFG electrodes. The electrodes stored in acid media and deionised water showed at the beginning of the storage considerable fluctuations. After about four weeks the values stabilised around -52 mV/pH. The dry stored electrodes showed an average slope factor about -50 mV/pH over nine weeks. Afterwards the electrodes failed. The alkaline stored electrodes exhibited a constant pH sensitivity near the nernstian manner over the complete testing period.

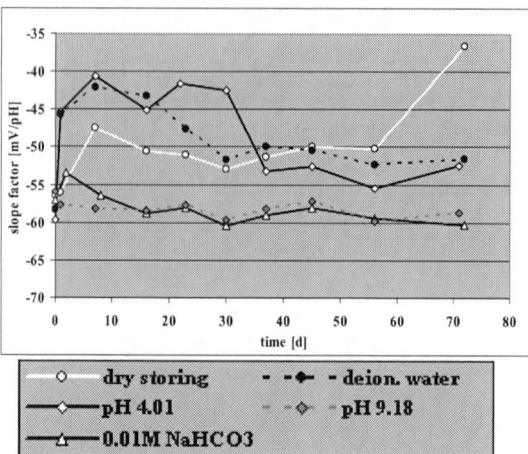

Figure 3: Averaged pH slope factors of RuO₂-lead free glass electrodes in several media

Surface characterisation

FESEM images of RLG electrodes are presented in figure 4. The freshly prepared electrodes (Fig. 4a) have a planar lead free glass surface with a few RuO₂ particles. Element identifications were done by EDX analysis. The light-gray region can be assigned to lead free glass, the dark-gray particles are RuO₂ grains and the black particles are SiO₂ particles, which are probably abraded of the agate ball mill. A corroded surface with many cracks at the electrode area was observed after 114 days in pH 4.01 buffer (Fig.4b). Electrodes in 0.01 M NaHCO₃ solution (Fig. 4c) exhibit no relevant changes by microscopic analysis.

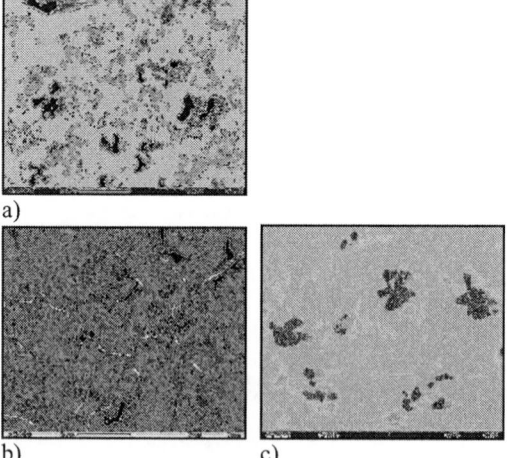

Figure 4: FESEM images of surfaces of RuO₂-lead free glass electrodes: a) before storage, b) 114 days in pH 4.01 buffer, c) 114 days in 0.01M NaHCO₃

The RLFG electrodes showed an uneven and porous surface with much more RuO₂ particles compared to RuO₂-lead free glass electrodes (Fig.

5a). By storing electrodes in pH 4.01 buffer (Fig. 5b) the number of pores increased. In alkaline medium a surface crystallisation could be noticed (Fig. 5c).

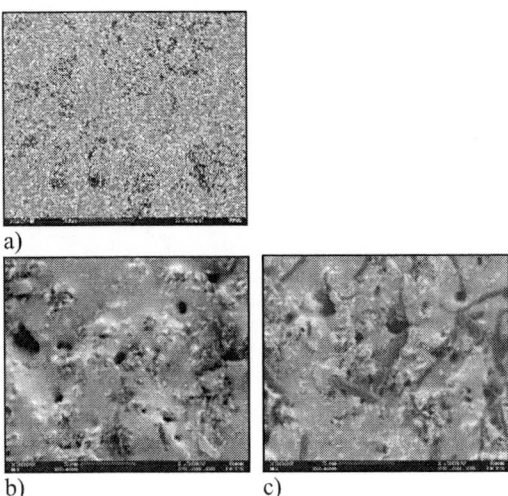

a)

b)   c)

Figure 5: FESEM images of surfaces of RuO₂-lead free glass electrodes; a) before storage, b) 114 days in pH 4.01 buffer, c) 114 days in 0.01M NaHCO₃

Response time

The response time of freshly prepared electrodes are compared with ten weeks testing electrodes, which were stored in five different media. Figure 6 show the response times of RLG electrodes before and after ageing in various solution. Before starting longtime use all RLG electrodes showed a small response time (Fig. 6a). After storing electrodes (Fig. 6b) in acid medium an increasing of response time was noticed. This could be attributed to the fact that cracks occurred at the electrode surface (Fig 4b). Potentiometric measurements are independent of the bulk of the active surface. It is important, how fast the buffer could be arrives the active area. That means the buffer needs much more time to get in contact with all the active surface of the electrode, when cracks and pores were occurred. The response time becomes longer. For electrodes, which were saturated in deionised water and dry stored, a small change in response time was observed. The potential of the alkaline stored electrodes (pH9.18, NaHCO₃) increased continuously, which could point to a pH-dependent secondary reaction.

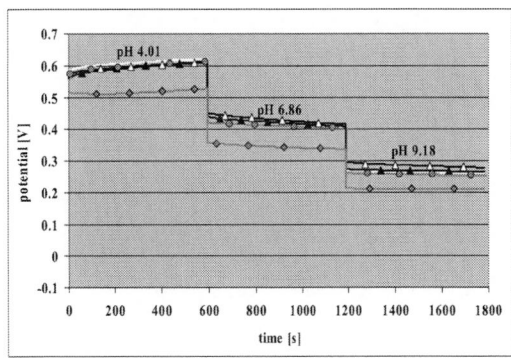

a)

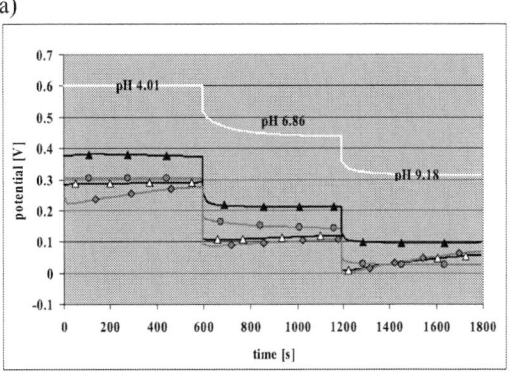

b)

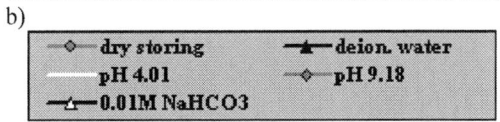

Figure 6: Response times of RuO₂-lead free glass electrodes a) before starting longtime storage b) after ten weeks testing in five different media

Figure 7 presents the responding behaviour of RLFG electrodes before and after the paging. Before the beginning of test the electrodes obtained a general delay of the potential (Fig. 7a). The response time decreased during the storing in alkaline solutions (Fig. 7b), which could be attributed to the crystallisation processes in the pores on the electrode layer. The buffer solution cannot intrude into the pores any longer. For this reason less time is necessary to get in contact with the whole active surface. The dry stored electrodes showed a non-equilibrium behaviour after 70 days, which may be caused by residuals of the buffer solutions in the pores. The acid aged electrodes showed an increase of the response time during the testing period. The potentials of the aqueous stored electrodes became not stable, so no potential equilibrium adjusted.

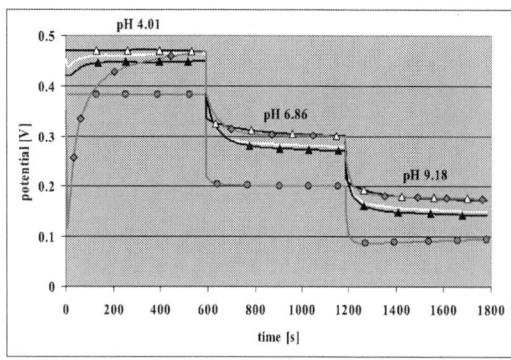

a)

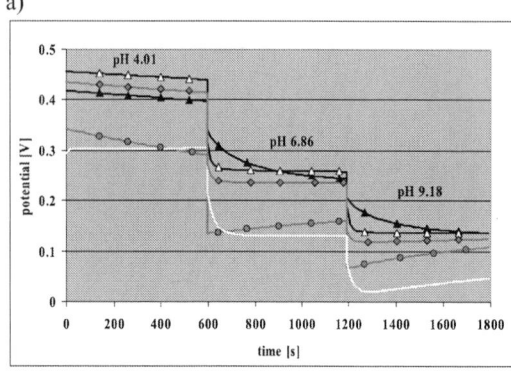

b)

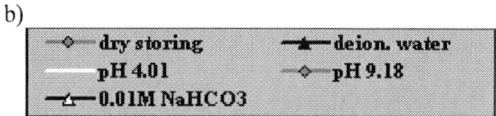

Figure 7: Response times of RuO₂-lead free glass electrodes a) before starting longtime storage b) after ten weeks testing in five different media

Conclusions

RuO₂-glass composite electrodes for pH measurements have been described in this paper. Two sorts of screen printed RuO₂-pH-electrodes were prepared (RuO₂-lead oxide glass; RuO₂-lead free glass), which differ heavily in their electrode surface. New RuO₂-lead free glass electrodes showed higher slope factors (-64 mV/pH), than the unaged RuO₂-lead free glass electrodes (-58 mV/pH), which could suggest a non single electron mechanism caused by an under stoichiometric ruthenium oxide. During the storage test the series showed different slope factors depending on the ageing solutions. The pH-sensitivities of both electrode series in alkaline solutions were around −59 mV/pH during the first nine weeks. After this the RLG electrodes failed. The response behaviour changed depending on the electrode composition and the storage conditions. The RuO₂-lead free glass electrodes showed the best results in acid solution, while the RuO₂-lead free glass electrodes were most stable in alkaline solutions over a period of more than ten weeks.

References

[1] R. Koncki, M. Mascini, "Screen printed ruthenium dioxide electrodes for pH measurements", Analytica Chimica Acta, Vol. 351, pp. 143-149, May, 1997

[2] M. J. Tarlov, S. Semancik, K. G. Kreider, "Mechanistic and Response Studies of Iridium Oxide pH Sensors", Sensors and Actuators, B1, pp. 293-297, 1990

[3] A. Fog, R. P. Buck, "Electronic semiconducting oxides as pH sensors", Sensors and Actuators, 5, pp.137-146, 1984

[4] K. G Kreider, M. J. Tarlov, J. P. Cline, "Sputtered thin-film pH electrodes of platinum, palladium, ruthenium, and iridium oxides", Sensors and Actuators, B 28, pp. 167-172, 1995

[5] H. N. McMurray, P. Douglas, D. Abbot, "Novel thick-film pH sensors based on ruthenium dioxide-glass composites", Sensors and Actuators, B 28, pp. 9-15, 1995

[6] J. A. Mihell, J. K. Atkinson, "Planar thick-film pH electrodes based on ruthenium dioxide hydrate", Sensors and Actuators, B 48, pp. 505-511, 1998

[7] L. A. Pocrifka, C. Concalves, P. Grossi, P.C. Colpa, E. C. Pereira, "Development of RuO₂-TiO₂ (70-30)mol% for pH measurements", Sensors and Actuators, B113, pp. 1012-1016, 2006

RF – Membrane Filter production and packaging challenges

M. Chatras*; N. Onda**; P. Nigg[1]; Wolfgang Tschanun

Altatec**; NTB IMT[1]; Xlim*; Reinhardt Microtech AG (RMT)

Aeulistrasse 10, CH – 7323 WANGS

Phone: + 41 81 7200456, Fax: +41 81 7200450 and E-mail: w.tschanun@reinhardt-microtech.ch

Abstract

The continuous development within the field of RF – technology is pushed to devices with higher performance and customer specific filter design. The basic design will be given with respect to understand the micromachined wafer stack of three Si – wafers. Individual solutions for some process steps are given. Main packaging issues were the ease of manufacturing and realisation of a cost optimised design. Packaging issues are addressed and several variants of packaging solutions are discussed and rated. The underfill solution is realised and discussed in detail. RMT will be able to realise on customers demand different MEMS – structures and packaging solutions. For better understanding see some of the practical work displayed at RMT's exhibition booth.

Key words: RF – MEMS Filter, Packaging, Customer Specific Design, BCB - Membrane

Introduction

To enlarge customer available technology base and fulfill customers needs a new project for RF – Filters was started. Main effort is the industrialization of a previously university lab shown for proof of principle fabricated device. This redevelopment includes a device modification and the redesign of process follow up for wafer level packaging.

Brief device description

The devices are built up by processing 3 different types of Si – wafers: Lower-, membrane (BCB)- and top wafer. The main functional principle is that of a resonating cavity containing a frequency selective electrode structure. The filter design itself was not altered.

Figure 1: Principal RF – Filter's layout

The realized filters are of Chebyshev type with a central frequency is 19.8 GHz [1]. The final device dimensions are 13.20 x 42.73 x 1.78 [mm] with a special opening for the RF - in – and output structure.

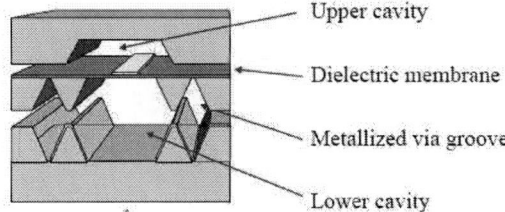

Figure 2: Principle cavities layout

The main functional issue is that the Au – electrodes are supported by a BCB –membrane with lowest RF – losses [2]. The basic substrates of the wafer stack are commercial Si – wafers.

Wafer processes

The wafers are processed by TMAH – etching and other standard thin films processes available at RMT[3]. Some of the complex middle wafer realizing processes can be seen at below

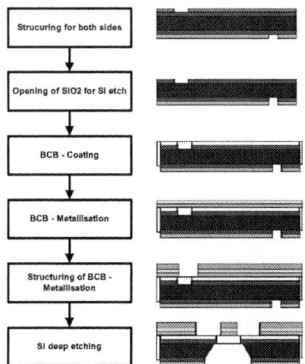

Figure 3: Principal membrane wafer processes

Changing of Device design

To enable a whole wafer processing the design of the individual devices had to be changed from a wafer part gluing process to a whole wafer process.

Main considerations were:

- no change of RF – properties
- change to a common ground
- protection of hot RF in and output
- use of underfill
- use of a conductive adhesive.

These design changes can be seen at Fig. 4 and 5.

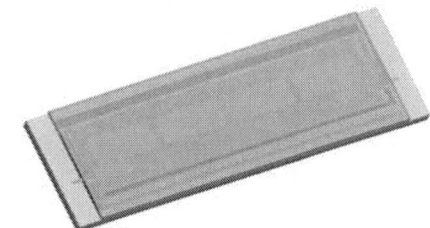

Figure 4: Initial device layout

Figure 5: Final device layout

Packaging process

The wafer design had also been adapted to the underfill requirement. After an investigation into several packaging technologies the conclusion was that the use of underfill technology will be the most economic and proposed reliable packaging of the investigated solutions.

A simple alignment wafer to wafer system was developed employing etched through windows for alignment under a microscope. The whole processing requires only hot plates and ovens to be performed.

The brief process follow up is:

- mount top- to membrane wafer
- mount 2 wafer stacks to bottom wafer
- full cure of underfill
- attach surface protection
- perform conducting adhesive processes
- prepare wafer stack for sawing
- saw process (3 depth cuts)
- remove sawing protection

- measure ground resistivity top – bottom
- final inspection

For quality assurance needs the process follow up generates also witness samples available after the device manufacturing.

As an example for the testability a picture of a incomplete underfill and a perfect underfill is shown at below:

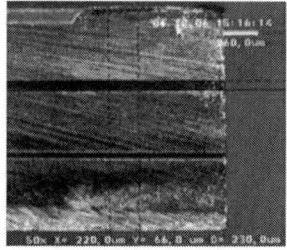

Figure 6: control incomplete underfill

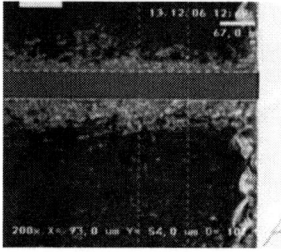

Figure 7: control of complete underfill

Conclusions

A new device packaging was introduced to build up RF – MEME Membrane Filters at GHz – range. The filters show a good compliance with the required RF – performance, environmental tests are ongoing; a space qualification is planned for a follow up project. The selected solution can easily transferred to other housing and sensor applications on customer demand.

Acknowledgement

The project partners thank the Swiss KTI and Swiss Space Office for project enabling funding.

References

[1] M. Chatras, P. Blondy, A. Vendier and J.L. Caseaux " A Surface Mountable Membrane Supported Filter"; 33rd European Microwave Conference, EUMW 2003, 6-10 Oct. 2003 Munich

[2] Please see Dow Chemical web page: http://www.dow.com/cyclotene/index.htm

[3] Please see RMT's web page: www.reinhardt-microtech.ch

New type of micro sensor with active nanoparticle surface

Radim Hrdý, Jaromír Hubálek, Kateřina Klosová

Brno University of Technology, Faculty of Electrical Engineering and Communication,

Department of Microelectronics, Údolní 53, 602 00 Brno, Czech Republic

E-mail: hrdyr@email.cz

Abstract

The using of template base methods is a very low cost technology in comparison with lithographic method and pore anodized alumina is one of templates base methods. Highly ordered nanopores in a thin-film alumina template have been prepared for electrochemical micro sensors fabrication with nanopatterned surface of electrodes. The template has been grown on the micro system composes by two gold comb-like electrodes. The template is used for nanopillars or nanotubes formation on the electrodes surface. The preparation highly ordered hexagonal pore [1,2,3] structure as template is done using anodization of Al film adhered on the micro system, which is the first phase of device fabrication. The second is the nanorods or nanotubes growing on the system. The anodized alumina (AA) as a template was used for its good dielectric properties and for ability of self ordering of aluminum under anodization [2,4]. The aluminum was directly evaporated on gold microelectrodes. The film thickness was obtained about 2 μm. The transformation thin aluminum film to nanopore alumina was preceeded by electrochemically controlled oxidation of aluminum. In this paper the two-step anodization [1] of thin Al film deposited on Si wafer was experimentally carried out with the aim to grow thin alumina on aluminum layer where alumina has double growing ratio causing high mechanical strain. Decreasing of the mechanical stress had to be done. The pore diameter and inter-pore distance vary in the range 15 – 30 nm and 20 – 50 nm respectively.

Key words: anodized alumina, nanopores, micro sensor, intermetallic layer,

Introduction

Since 1995, when the self-ordered porous alumina was discovered, [2] the effort to utilize of that extremely interesting material started. Up to now, self-ordered porous alumina with many different parameters was obtained, e.g, porous templates with inter-pore distance from 50 nm to 500 nm. [3,4,5] After this invention, many groups begun to find a use in many spheres of science like electronic, magnetic, optoelectronic devices. Also, the porous alumina has been found the use in orthopedic surgery like a component in bone implants with low coefficient of friction. [6,7] But the most interesting sphere is the applications of anodized alumina (AA) with various materials in sensors technology. Especially, the AA is used for increase areas of sensors active layers. The AA has been applied as template for deposition nanowires or nanotubes on substrate or as mask for etching tiny ordered motives. [8]

Various forms of AA has been studied for different application. [5,9] AA, which is fabricated from aluminum foils with thickness 250 μm, is one of them. This process of porous alumina fabrication is managed very well and it is described on Fig. 1. The disadvantage of this template lies on its difficulty manipulation. Better way is to deposit aluminum thin film directly on the substrate and after that to anodize porous alumina. [8,10,15]. However, it requires deposition equipment and clean rooms. This technique has made possible fabrication of AA and subsequently nanowires on various metallic layers without necessity to solve problems allied to manipulation and dissolving of remainder aluminum. Therefore, this technique is more suitable for industry.

Nowadays we can see a tendency to fabricate still smaller sensors elements but it is required the same or higher sensitivity.[8] One of many solutions could be enlargement of detection area by force of nanopillars or nanotubes deposited on surface. The utilization of different AA templates makes possible to produce nanoparticles with various sizes. Unfortunately, there are still many outstanding challenges in production multi metallic systems with perfectly ordered nanoparticles surface.

The aim of present work is to create a new high sensitive electrochemical sensor using AA template which is created from aluminum thin film.

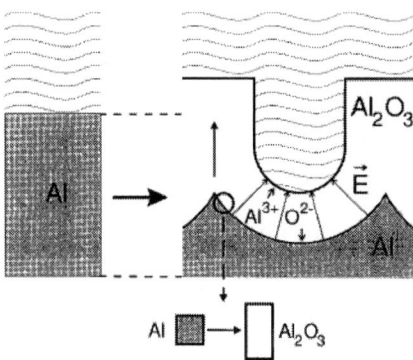

Figure 1. Scheme of nanopore growing. The process of anodization is based on drift mobility of Al^{3+} and O^{2-} through increasing barrier. Thickness of porous alumina is twice bigger than thickness of original aluminum film. [3]

Experimental

Here, the fabrication process of the porous alumina is described. First, we used to N-type of silicon substrate for sputtering of thin aluminum film. The thickness of film was 2 μm and the purity 99,999%. The first problem of the AA creation was appeared at once with deposition of aluminum film. The film was not homogenous and it was moulded by much bigger crystals than 100 nm. Crystal's boundaries were very apparent and deep, Fig.2. They avoided the creation of ordered structure. [11] Indeed, the porous AA structure was created, but it was not absolutely applicable as a template for arrayed deposition.

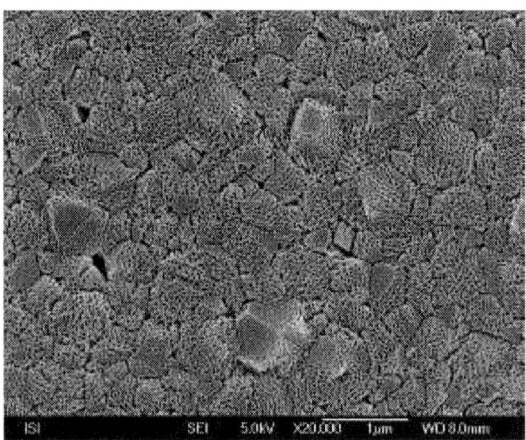

Figure 2 SEM image of the porous alumina surface anodized in sulfuric acid at 25V at 10°C. Anodization time was 13 minutes.

In the second experiment, the alumina film was deposited by vapour deposition (VD) method. The surface was glossier than sputtered film. In Fig. 3. the crystal's boundaries is evident but they did not affect the anodized process as much as in the first case. On the other hand, the problem with adhesion

was appeared. The evaporated aluminum films were cracked or involved by anodization in many cases.

In both experiments, samples with alumina film were cleaned and degreased in acetone at first. After that, they were put on a cell, where the sample was the anode and a stainless steel was the cathode. Sulfuric acid at 10% was used as electrolyte. The temperature of electrolyte was ranging from 0°C to 24°C. The voltage was set on value 25 V. The electrolyte was rigorously stirred. [3, 12, 13] Samples were anodized between 13 minutes and 15 minutes. During this time, current value decreased to value of 100 μA. After that, samples were etched in 5% phosphoric acid at 39°C in ordered to open the pores and dissolve the porous barriers. The etching time was between 2 and 3 minutes. After 3 minute of etching, the ceramic structure was started breaking down. The pore diameter and inter-pore distance vary in the range of 15 – 30 nm and 20 – 50 nm respectively, see Fig. 4.

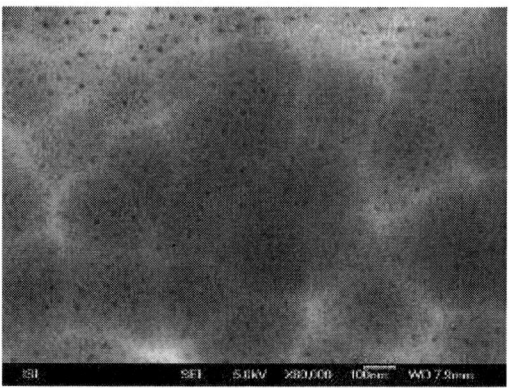

Figure 3 SEM image of the porous alumina surface anodized in sulfuric acid at 25V at 10°C. Anodization time was 15 minutes.

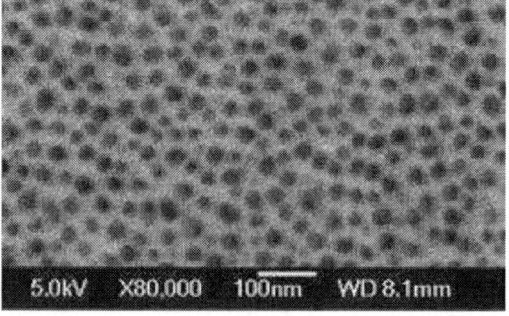

Figure 4 SEM image of the surface of the porous alumina anodized in sulfuric acid at 25V at 0°C. The sample was etched in 5% phosphoric acid. Anodization time was 15 minutes and etching time was 2 minutes.

Anodization of thin aluminum film in oxalic acid was done [2,5,14] However, the result was unsatisfactory. The layer did not apeared of porous structure. The result is shown in Fig. 5. The anodization of aluminum foil is easier to manage than the aluminum film.

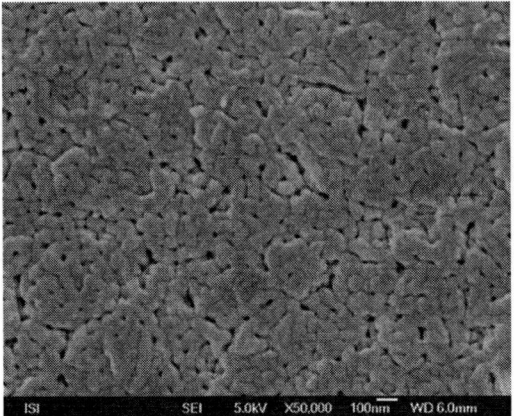

Figure 5 SEM image of the porous alumina surface anodized in oxalic acid at 40V at 1°C. Anoidzation time was 25 minutes.

Next step was preparation of AA on inter metallic layer. At first, 250 nm of nitride was fabricated on silicon wafer as non-conducting layer. On that, a photo-sensitive template was obtained with using a lithographic technique. The structure of designed template is shown in Fig. 6. The Lift-Off technique was used in order to create thin gold film with comb-like structure. The thickness of the gold layer was 300 nm. The template was dissolved in xylene ultrasonic bath. After that, the aluminum thin film was evaporated on the structure and the same conditions of anodization described above were applied. Unfortunately, the gold layer had very poor adhesion with non-conductive layer and the layer was often stripped. In all experiments, one-step anodization process was used.

Figure 6. New type of micro sensor It has length 9 mm.

Result and discussion

Fig. 2 and 3 show that the alumina structures are not enough self-organized. The pores have been created chaotically and they have different sizes. Nanocrystals of aluminum are the main cause of this issue. The size of nanocrystals grows is time dependent. Nanocrystals of aluminum are evident in Fig. 2, where many defects can be found between

nanocrystals. Using VD technique this problem was solved, see Fig. 3. The aluminum layer, which was deposited by sputtering, was turbid. On the other hand the layer, deposited by VD, had brilliant polished surface. Because the VD run in many steps and the alumina film has a thickness of 200 nm, nanocrystals do not grow up as much as nanocrystals in sputtered film. Therefore the alumina thin film deposited by VD is more quality.

Next issue is to improve the adhesion between silicon substrate, inter metallic layer and alumina film. One of the solutions is decreasing the stress, which is raised by the deposition process. The other solution is heating the substrate which results in better adhesion but it decreases the speed of deposition. Next solution is a using different intermetallic layers like titanium or titanium dioxide which have better adhesion and they can also create porous structure by special conditions. The using of anodic porous titanium oxide is going to bring benefit due to its conduction advantages. Titanium dioxide can be used as conductive matrix for deposition of nanopillars. [6]

Another task is finding conditions for creating ordered porous structure. In comparison, it is much difficult to apply anodization by Two-step method because it often terminates in dissolving or breaking up of aluminum films by anodization at second step. This problem is much dependent on adhesion of alumina layer. Anodization in one step was success managed and conditions were determined but then an initiation of second step is done, it is necessary to dissolve a first anodized layer of AA. This part of fabrications is in progress during higher temperature in range 40 – 60°C and the anodization is done at the same or lower a room temperature. This thermal difference causes very high mechanical stress between AA and bottom layer. The problem solving is in very slow increasing/decreasing of the temperature. The next very important condition is the electrolyte stirring. If the stirring is insufficient or it is stopped during anodization, free aluminum ions, which are usually stirred up in solution, begin to imbed on alumina surface. They cause the interruption of anodization process. [3] The stirring is very important for drawing-off rising hydrogen's molecules from AA surface. They usually stay on alumina surface, where they are reduced from water molecules, if a position is suitable or intensity of stirring is small. They cause non-uniform anodization of some alumina areas.

Conclusion

The conditions of the anodization process using aluminum thin films deposited by evaporation and sputtering were found. The optimal conditions of anodization processes have been determined for the first step. The self-organized process of thin films is dependent on high purity, homogeneity of aluminum film and adhesion between aluminum and silicon layers. Finally the VD method shows to be more suitable for deposition of aluminum thin film on silicon substrate than the sputtering method.

Acknowledgement

This research has been supported by Grant Agency of the Academy of Sciencies of the Czech Republic under the contract GAAV 1QS201710508 Impedimetric chemical microsensors with nanostructured electrode surface, and by the Czech Ministry of Education within the framework of Research Plan MSM 0021630503 MIKROSYN New Trends in Microelectronic Systems and Nanotechnologies.

References

[1] N.M. Yakovleva, A.N. Yakovlev, and E.A. Chupakhina, Structure of Al_2O_3 Films Prepared by Two-Step Anodization. *Inorganic Materials*, Vol. 34, No. 7, 1998, pp. 711-713.

[2] H.Madsuda, K. Yada, A Osaka, Selforderes of Cell cnofiguration of Anodic Porous Alumina with Large-Size Pores on Phosphoric Acid Solution, J.Apply.Phys,Vol 37,No11A,,1998,pp L1340

[3] O. Jessensky, F. Müller, and U. Gösele, Self-organized Formation of Hexagonal Pore Structure in Anodic Alumina. *J. Electrochem. Soc.* Vol. 145, No.11 (1998) 3735-3740.

[4] H.Madsuda, K.Fukuda, Ordered Metal Nanohole Arrays made by a two step replication of honey comb structures of anodic alumina. Science Vol.268,1995,pp,1566 8

[5] H. Masuda, F. Hasegwa, and S. Ono, "Self-Ordering of Cell Arrangement of Anodic Porous Alumina Formed in Sulfuric Acid Solution," J. Electrochem. Soc.,Vol. 144, No. 5, 1997. pp.L127-9

[6] E.P.Brings, M Karlsson et all, Formation of highly adhered nano-porous alumina on Ti-based substrates, a novel bone implant coating, Journal of materials science, Materials in medicine, 15, 2004, 1021-1029

[7] M.Karlsson a.R. Wilshawb, L.Di Silvio, Initial in vitro interaction of osteoblasts with nano-porous alumina, Biomaterials 24 (2003) 3039–3046

[8] A. Wang, R. M. White, "Thin-film anodized aluminum on an acoustic sensor," presented at the IEEE Ultrasonics Symposium, Seattle, WA, Nov. 7-10, 1995.

[9] Y. Kanamori, K. Hane, H. Sai and H. Yugami, 100nm period silicon antireflection structures fabricated using a porous alumina membrane mask. vol. 78, No. 2, 2001

[10] A.N. Govyadinov, S.A. Zakhvitcevich, Field Emitter Arrays Based on Natural Selforganized Porous Anodic Alumina. Technical Digest of IVMC'97, Kyongju, Korea, 1997, pp.735-738.

[11] R. Hrdý, J. Hubálek, Selfordered pore structure of anodized alumina thin film on Si substrate, Technical digest of EDS 05, Brno, Czech Rep. 2005 ISBN 80-214-2990-9

[12] K. Nielsch, J. Choi, K. Schwirn, R. B. Wehrspohn,* and U. Go1sele Self-ordering Regimes of Porous Alumina: The 10% Porosity Rule, the American Chemical Society, Vol. 2, Num 7, July 2002

[13] J. Choi, K. Nielsch, M. Reiche, R. B. Wehrspohn,a) and U. Gosele,J. Fabrication of monodomain alumina pore arrays with an interpore distance smaller than the lattice constant of the imprint stamp, Vac. Sci. Technol. B, Vol. 21, No. 2, Mar/Apr 2003

[14] W. Lee*, R.Ji, U. Gosele and K. Nielsch, Fast fabrication of long-range ordered porous alumina membranes by hard anodization, Nature materiále. Vol. 5 September 2006

[15] M.S. Sander, L.Than, Nanoparticle Arrays on Surfaces Fabricated Using Anodic Alumina Film as Templates. Adv. Funct. Material, 2003, 15, No.5 May

Creation of Nanostructured Metal Surface by Template-Based Electrodeposition Method and Its Employment in Sensor Technology

Kateřina Klosová[1], Jaromír Hubálek[2]

[1,2]Brno University of Technology, Faculty of Electrical Engineering and Communication, Department of Microelectronics, Czech Republic

Phone: +420 541 14 6163 and e-mail address: [1]k_kacka yahoo.com, [2]hubalek@feec.vutbr.cz

Abstract

Fabrication of an indented surface, i.e. nanostructures, can be used for an essential enlargement of the active area of microsensors. The microsensors are expected to gain high sensitivity (in spite of their miniaturization) due to the modification of their surface. The template-based electrodeposition method is an up-to-date low-cost technology for the nanostructure creation. It is based on using a non-conductive template which contains hexagonally arrayed nanopores. Metal nanostructures are created by galvanostatic electrodeposition of the required metal through the nanopores of the template. Metal ions are attracted to the cathode (a conductive substrate attached to one side of the template) leaving the non-conductive alumina. After the process of metal deposition, the alumina template is dissolved and the nanostructures are obtained. The nanostructures which can be created by the presented method are both nanotubes and nanowires. Nanotubes can also vary in the wall-thickness. The type of the structure depends on the diameters of the nanopores and on electroplating conditions under which the nanostructures are created.

Key words: nanostructures, nanowires, nanopillars, nanorods, nanotubes, nanopores, microsensors

Introduction

Modification of the surface by nanostructures is advantageous in all applications in which the area is considered to be a crucial parameter.

For example, the creation of an indented surface can solve the problem of reduction in sensitivity of the microsensors which is caused by miniaturization of the sensor. The nanostructures can be used in many other applications, such as: magnetic storage media [1], gene therapy [2], solar cells, [3], optoelectronics [4], cooling systems (micro-heatsinks) [5], etc.

The enlargement (i.e. surface modification) can be achieved by various methods like lithography [6], electrochemical step edge decoration method (ESED) [7], [8], controlled nanoparticle growth by diffusion [4], and deposition of metal through a nanoporous template. Nanostructures formation using the template can be accomplished by electrochemical deposition [9], [10], vapour deposition [4], or sputtering [1].

Lithographic techniques offer the best control over nanoparticle size, shape, and spacing; however, these techniques are rather expensive, limited to serial processing, and suitable to only a small number of material systems. Lithographic methods also fail to produce nanostructures with feature size smaller than 20 nm [4], [6].

The electrochemical step edge decoration method is based on metal electrodeposition at the edges of freshly cleaved graphite surfaces or electrodeposition of conductive metal oxides followed by hydrogen reduction [7], [8].

Creation of nanostructures or nanoparticles through the nanoporous template offers several advantages, including low price, the ability to fabricate large quantities of nanowires, and enables composition of two or more metals [4], [9]. The metal can be either sputtered or vapour deposited into the nanopores, although a problem arise due to non-uniform distribution of the metal and plugging of the nanopores [3], [11]. Therefore, the electrodeposition process appears to be the best option. Nano-sized particles can be prepared either from common electroplating baths or colloid solutions containing organic surfactants [3]. Even electroless metal plating can be used [5]. It is also possible to grow either single-crystal or polycrystalline nanowires by this method [11].

During the electrodeposition process, metal ions are attracted to the cathode (a conductive substrate attached to the template) leaving the non-conductive alumina. After the process of metal deposition is finished, the alumina template is dissolved (e.g. in NaOH or H_3PO_4) and the nanostructures are obtained (Figure 1).

The template can be produced by anodization of the thin aluminium film. Aluminium, as well as titanium, is well known for its self-assembling ability which happens during anodization under specific conditions. This ability enables to create the template which contains hexagonally arrayed nanopores [6], [12], [13]. However, a variety of templates (made out of various materials) with a range of pore sizes and pore densities is available, including polycarbonate membranes, mica crystals containing etched nuclear particle tracks, etc [9]. Compared to other types of the templates (or membranes) containing arrays of nanopores, anodic aluminum oxide has properties characterized by excellent uniformity in diameter and spacing of the pores. It has also good mechanical and thermal stability [1], [14].

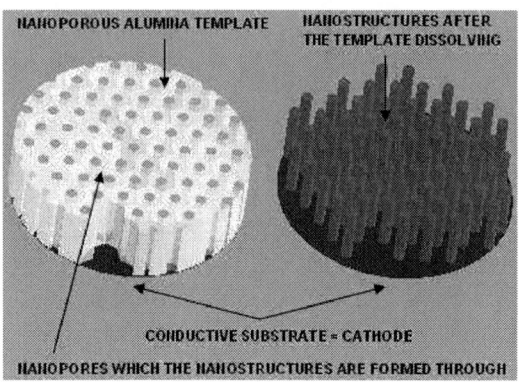

Figure 1: Principle of nanostructures creation

The dimensions, such as the diameter of a single wire and dispersion of the nanostructures, are given by the template and by the amount of metal deposited into the nanopores. The deposition speed is directly proportional to the time of the deposition and the current density, which is stated by Faraday's law, with a certain limitation of the mass transport in the nanopores. The amount of deposited metal also depends on the electron transfer, electrical potential, chemical potential, the processing temperature which can influence the mobility of ions, crystal growth, etc [15]. The nucleation and growth of metal can be classified into two categories: "interfacial (charge) controlled", in which the rate of the process is determined by the rate of incorporation of electroactive species into new phase, and "diffusion controlled" in which the rate of the process is limited by the mass transport of electroactive species either in the solution (solution diffusion) or on the substrate surface (substrate diffusion). The former type of nucleation is favoured at high concentrations of electroactive species and low deposition overpotentials, while latter is favoured at low concentrations of electroactive species and high deposition overpotentials [16].

The nucleation of nanowires on the electrode substrate during electrodeposition is influenced also by the crystal structure of the substrate, specific free surface energy, adhesion energy, lattice orientation

of the electrode surface, and crystallographic lattice mismatch at the nucleus-substrate interface boundary. The final size distribution of the electrodeposits, however, strongly depends on the kinetics of the nucleation and growth [17]. The current density and the temperature can affect the crystal growth of deposited metal and forming of fresh crystal nuclei [18]. The temperature also influences preferred orientation of electrodeposited nanowires (regardless of the pH of the electrolyte) [19]. Deposition of metal through the high aspect ratio nanopores can be considered as electrodeposition on arrayed nanoelectrodes and only diffusion as a way of mass transport is effective. It has several impacts, such as an increased ohmic drop of potential caused by raise in an ohmic resistance represented by the electrolyte volume in the pores, a current increase due to enhanced mass transport at the nanoelectrode boundary, changes in current density distribution etc. The dimensions of the nanopores change during the plating process (the aspect ratio decreases during deposition). This obviously results in changes in the mass transport (diffusion layers will differ) and a dynamic change of the current density distribution. In high aspect ratio nanopores where edge effect plays a predominant role, the diffusional flux toward the nanoelectrode is inhomogeneous over the nanoelectrode surface (quasi spherical diffusion layer of the pore is added on the top of the linear diffusion layer); it increases with decreasing distance from the nanoelectrode edge. The difference between electroplating on large planar electrodes and through the nanopores also depends on time. After a sufficiently long time, a steady state is established for electrodes of planar or spherical geometries. However, a steady state cannot be attained in the cases where the length of the nanopores tends to infinity [10] [20].

Regarding the diffusion, another very important parameter is the interpore distance. The diffusion quasi spherical layers may overlap and then affect one another. In such case the overall current measured is smaller than the sum of the currents passing through the nanoelectrodes when they operate independently. The extent of the diffusion layer overlap can be expressed in terms of the overlap factor derived from the ratio of the overall current to the sum of the currents at the independent individual nanoelectrodes. [10]. In order for the plating process to be the most effective the limiting current density (which corresponds to the maximum transfer rate that a particular species can sustain because of limitation of diffusion) has to be determined. If the current exceeds the value of the limiting current a sharp growth of the cathodic current can be observed due to the hydrogen evolution [21]. Therefore detailed knowledge of the diffusive behavior is crucial for the understanding and optimization of electrodeposition into the nanopores. Various attitudes to exploring of motions of molecules inside the nanopores have been used, such as fluorescence correlation spectroscopy (FCS) [22].

It has been found that metal nanostructures exhibit a variety of unique properties, such as quantization of the conductance, enhanced mechanical properties, etc [6]. The optical properties can be also modified by coating the nanostructures with fluorescent chromophores [3], [9].

Experiments

The metal which was used for the experiments was mainly nickel but also copper was used in some cases. However platinum and gold are considered for further experiments and practical applications. Watts Bath, containing 250g/l of $NiSO_4$, 50g/l of $NiCl_2$, and 34g/l of H_3BO_3, was used as an electrolyte. The nanostructures were created under various electroplating conditions, such as a wide range of current densities, various concentrations, various values of the pH, various diameters of the nanopores, usage of ultrasound waves. The temperature of the solution was usually 55°C. Dependence of the dimensions of the nanostructures on the time and the current density had been experimentally and numerically determined from Faraday's law and preliminary experiments carried out on macroscopics substrates. For the first experiments Whatman Anodiscs (ultrafine membrane filters developed by Whatman©) were used as the templates and then the Al_2O_3 templates (created only for the purpose of nanostructure fabrication) were employed. These templates differ from the Whatman Anodiscs, e.g. in the width and diameters of the nanopores. Using the Al_2O_3 templates is possible to create nanostructures of far smaller diameters. For the latest experiments standard stirring of the solution has been replaced by forced circulation of the electrolyte (by a pump). The pump supplies the fresh electrolyte continuously to the area where electrodeposition, or anodization in the case of the template fabrication, should proceed. After the electrodeposition, when the pores of the template are filled by the metal, the alumina template is dissolved in either NaOH or H_3PO_3 and desired nanostructures are obtained.

Results

The created nanostructures were examined by scanning electron microscopy (SEM). Both nanowires (Figure 2 & Figure 3) and nanotubes (Figure 4) were created. The type of the structure is given by the specific conditions under which the nanostructures are created. It has been found that the pH of the electrolyte, the diameters of the pores of the template, and probably the concentration and the ultrasound waves can influence the structure. The nanostructures which were created at low values of the pH or deposited through the nanopores of very small diameters (approx. 20 nm) turned out to be

nanowires while the nanostructures produced at high values of the pH or deposited through the nanopores of wide diameters (approx. 200 nm) have been nanotubes. The nanostructures fabricated at low concentration of the electrolyte tend to be nanotubes and nanowires were usually created in highly concentrated electrolytes. The nanostructures which were carried out in the ultrasound bath were often nanotubes, although dependence of the type of the structure on the usage of the ultrasound waves has not been fully proved. The nanotubes which can be formed through the templates can vary in the wall thicknesses. An example of thin-walled nanotubes is in Figure 5 and thick-walled nanotubes are in Figure 6. Unique phenomenon occurred on a few samples. Some of the nanowires were created in clusters (Figure 7) instead of uniform covering of the surface. The reason has not been made clear yet. The length of created nanowires is equivalent with the length determined by calculations and the preliminary experiments on macroscopic substrates, except of some nanotubes and only few nanowires. It is possible that the length differs due to the effect of diffusion and creation of the clusters – nonuniform distribution of the metal in the template.

Figure 2: Ni nanowires

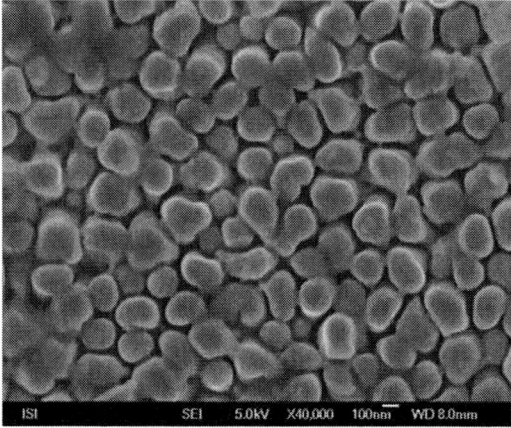

Figure 3: Ni nanowires

213

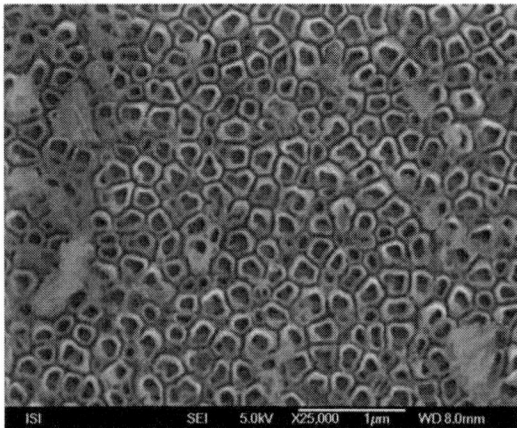

Figure 4: Ni nanotubes

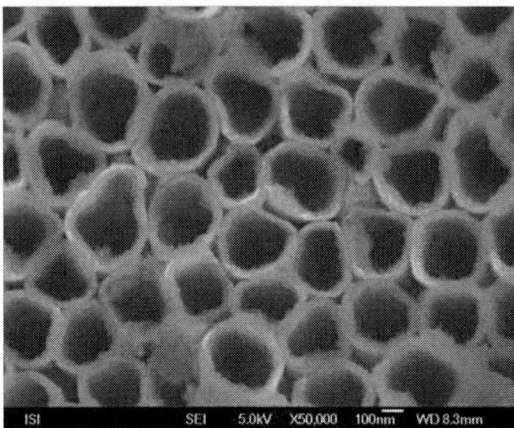

Figure 5: Thin-walled nanotubes

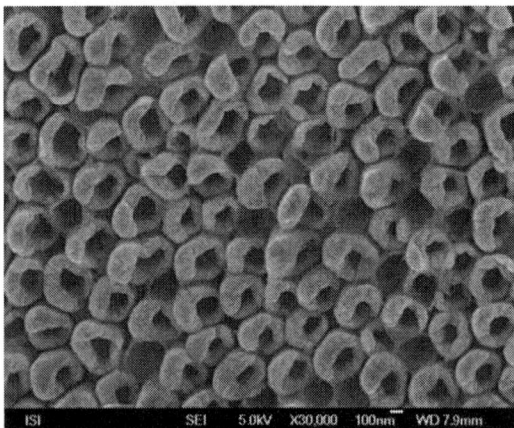

Figure 6: Thick-walled nanotubes

Figure 7: Clusters of nanowires

Conclusion

The procedure of creating nickel nanostructures has been put into practice. The nanowires of required proportions and other different nanostructures have been created.

The developing technique of the nanostructured surface creation can be applied when it is essential to enlarge the active area, like in solar panels, gas sensors, sensors of detection of heavy metals, sensors for solution conductivity measurement, biosensors etc. The nanostructures also exhibit changed physical properties due to their small dimensions and react with surrounding matter differently in comparison to macrostructures. This is why the nanostructures are considered as promising in the various fields of thin-film technology.

Acknowledgements

This research has been supported by Grant Agency of the Academy of Sciencies of the Czech Republic under the contract GAAV 1QS201710508 Impedimetric chemical microsensors with nanostructured electrode surface, and by the Czech Ministry of Education within the framework of Research Plan MSM 0021630503 MIKROSYN New Trends in Microelectronic Systems and Nanotechnologies.

References

[1] Z. L. Xiao, C. Y. Han, U. Welp, H. H. Wang, V. K. Vlasko-Vlasov, W. K. Kwok, D. J. Miller, J. M. Hiller, R. E. Cook, G. A. Willing, G. W. Crabtree, "Nickel Antidot Arrays on Anodic Alumina Substrates", Applied Physics Letters, Vol. 81, No. 15, pp. 2869-2871, October, 2002.

[2] Ch. Ji, P. C. Searson, "Fabrication of Nanoporous Gold Nanowires", Applied Physics Letters, Vol. 81, No. 23, pp. 4437-4439, October, 2002.

[3] D. J. Riley, "Electrochemistry in Nanoparticle Science", Current Option in Colloid & Interface Science 7, pp. 186-192, 2002.

[4] M. S. Sander, T. Le-Shon, "Nanoparticle Arrays on Surfaces Fabricated Using Anodic Alumina Films as Templates", Advanced Functional Materials, Vol. 13, No. 5, pp. 393-397, May, 2003.

[5] W. I. Son, J.-H. Hong, J.-M. Hong, "Fabrication of Micro-Heatsink by Nanotemplate Synthesis and Its Cooling Characteristic", Proceedings of the Spanish Conference on Electron Devices Conference (IEEE), Spain, Tarragona, February 2-4, pp. 439-442, 2005.

[6] M. Hernández-Vélez, "Nanowires and 1D Arrays Fabrication: An Owerview", Thin Solid Films 495, pp. 51-63, 2006.

[7] E. C. Walter, K. Ng, M. P. Zach, R. M. Penner, F. Favier, "Electronic Devices from Electrodeposited Metal Nanowires", Microelectronic Engineering 61-62, pp. 555-561, 2002.

[8] E. C. Walter, M. P. Zach, F. Favier, B. J. Murray, K. Inazu, J. C. Hemminger, R. M. Penner, "Metal Nanowire Arrays by Electrodeposition", Electrochemistry Special, CHEMPHYSCHEM 4, pp. 131-138, 2003.

[9] C. L. Chien, L. Sun, M. Taanase, L. A. Bauer, A. Hultgren, D. M. Silevitch, G. J. Meyer, P. C. Searson, D. H. Reich, "Electrodeposited magnetic nanowires: arrays, field-induced assembly, and surface functionalization", Journal of Magnetism and Magnetic Materials 249, pp. 146-155, 2002.

[10] K. Štulík, Ch. Amatore, K. Holub, V Mareček, W. Kutner, "Microelectrodes. Definitions, Characterization, and Applications", Pure Appl. Chem., © 2000 IUPAC, Vol. 72, No. 8, pp. 1483-1492, 2000.

[11] W. Schwarzacher, O. I. Kasyutich, P. R. Evans, M. G. Darbyshire, G. Yi, V. M. Fedosyuk, F. Rousseaux, E. Cambril, D. Decanini, "Metal nanostructures prepared by template electrodeposition", Journal of Magnetism and Magnetic Materials 198-199, pp. 185-190, 1999.

[12] N. M. Yakovleva, A. N. Yakovlev, E. A. Chupakhina, "Structure of Al2O3 Films Prepared by Two-Step Anodization", Inorganic Materials, Vol. 34, No. 7, pp. 711-713, 1998.

[13] R. Hrdý, J. Hubálek, "Selfordered pore structure of anodized alumina thin film on Si substráte", Technical digest of EDS 05, Czech Rep., Brno, September 15-16, pp. 300-304, 2005.

[14] Y.-H. Cheng, S.-Y. Cheng, "Nanostructures formed by Ag nanowires", Nanotechnology 15, pp. 171-175, 2004.

[15] L.-W. Pan, P. Yuen, L. Lin, "3D electroplated microstructures fabricated by a novel height control method", Microsystem Technologies 8, © Springer-Verlag, pp. 391-394, 2002.

[16] Ž. Petrović, M. Metikoš-Huković, Z. Grubač, S. Omanović, "The nucleation of Ni on carbon microelectrodes and its electrocatalytic activity in hydrogen evolution", Thin Solid Films, Vol. 513, Issues 1-2, pp. 193-200, August, 2006/

[17] D. Bera, S. C. Kuiry, S. Seal, "Synthesis of Nanostructured Materials Using Template-Assisted Electrodeposition", The Minerals, Metals & Materials Society – TMS, A Hypertext-Enhanced Article, 2004.

[18] C. Oropeza, "A New Approach to Evaluate Fracture Strength of UV-Liga Fabricated Nickel Specimen", Thesis written at Lousiana State University, pp. 16-32, May, 2002.

[19] I. Z. Rahman, K. M. Razeeb, M. A. Rahman, M. Kamruzzaman, "Fabrication and characterization of nickel nanowires deposited on metal substrate", Journal of Magnetism and Magnetic Materials, 262, pp. 166-169.

[20] N. Masuko, T. Osaka, Y. Ito, "Electrochemical Technology: Innovation and New Developments", Kodansha Ltd., pp. 173-174, 1996.

[21] Q. Juany, W. Wang, F. Jia, J. Zhang, "Electrochemical assembled p-type Bi/sub 2/Te/sub 3/ nanowire arrays", Thermoelectrics, 2003 Twenty-Second International Conference on – ICT, August 17-21, pp. 410-412, 2003.

[22] J. Hohlbein, M. Steinhart, C. Schiene-Fischer, A. Benda, M. Hof, Ch. G. Hübner, "Confined Diffusion in Ordered Nanoporous Alumina Membranes", Small 3, pp.380-385, 2007.

Capillary Flow Kinetics in the Presence of Uneven Gaps

Horatio Quinones

Asymtek Headquarters, 2762 Loker Avenue West, Carlsbad CA, USA

Tel: 1-760-431-1919; Fax: 1-760-431-2678; Email: info@asymtek.com

Abstract

Underfilling of various electronic components and packages for reliability improvement has been the plan of record for many years. The capillary of fluid between parallel plates namely, substrate and component/package including die, CSP, BGA by capillary kinetics is a mature process to a great extent. Present designs, however, often "violate" such parallelism, for instance, grooves on the substrate, vias, both blind and through-hole; present a challenge in the underfill process depending on their geometry relative to the component geometry and dimensions of the bump type used. This paper proposes a surface energy analysis mainly based in a perturbation approach using calculus of variations. Various geometries and configurations are solved numerically by this method including those instances where components are very closely packaged and their proximity my influence the actual capillary action during encapsulation. The paper also includes actual jet dispensing underfill data for capillary underfilling with various designs. The advantages of the capillary approach versus other methods such as molding or forced underfill are discussed.

Keywords: adhesion, capillary action, cohesion, differential geometry, dispense gap, dispensing, high density package,. jetting, London forces, perturbation theory, underfill, variational methods.

Introduction

The capillary action occurring in small ducts has been studied by several disciplines of science. Molecular forces of particles in a fluidic matrix are just an instance. In electronic packaging the fluid dispensing for various applications including, potting, filling, component underfilling is of common knowledge. The capillary kinetics plays an important role in several of these applications. Contrary to traditional injection molding, where the fluid is mobilized by an induced relatively high pressure differential at the surface of the fluid wave front, the capillary action is a result of adhesion forces overcoming the cohesive forces of the moving fluid. A Variational approach to determine the fluid-air-solid surface shape of the moving front will guide us in determining various geometric boundary conditions including gaps sizes, and steps occurring in the corresponding capillary ducts.

Theory and background

The analytical approach to solve the problem of surfaces is that of Mapertuis principle. These solutions, coupled with mechanical adhesive and cohesive forces that include van der Waals forces and London forces due to oscillation of electron clouds in molecules that are in close proximity, can give us a good description of the kinetics that takes place for slow fluid flow under capillary action. Given a definite integral with boundary conditions, its stationary value can be found by minimization of a functional using variational calculus tool [1].

$$\delta F(y, y', x) = F(y + \varepsilon\phi, y' + \varepsilon\phi', x) - F(y, y', x)$$

$$= \varepsilon\left(\frac{\partial F}{\partial y}\right)\phi + \left(\frac{\partial F}{\partial y'}\right)\phi' + (0)\varepsilon^2$$

The variational of the definite integral can be computed as follows:

$$\delta\int_a^b F(y, y', x) \cdot dx = \int_a^b \delta \cdot dx = \varepsilon \cdot \int_a^b \left(\frac{\partial F}{\partial y}\phi + \frac{\partial F}{\partial y'}\phi'\right) \cdot dx$$

Dividing by ε and integrating by parts the second term of the r.h.s. of above equations we obtain

$$\int_a^b \frac{\partial F}{\partial y'}\phi' dx = \left[\frac{\partial F}{\partial y'}\phi'\right]_a^b - \int_a^b \frac{d}{dx}\left(\frac{\partial F}{\partial y'}\right) \cdot \phi dx$$

We define I as the definite integral and since the $\phi(x)$ vanishes at the limits of integration (boundary conditions are satisfied exactly, $x=a$ and $x=b$)

$$\frac{\delta I}{\varepsilon} = \int_a^b \left(\frac{\partial F}{\partial y} - \frac{d}{dx}\left(\frac{\partial F}{\partial y'}\right)\right) \cdot \phi dx$$

We now define the function $\xi(x)$ as

$$\xi(x) \equiv \frac{\partial F}{\partial y} - \frac{d}{dx}\left(\frac{\partial F}{\partial y'}\right)$$

Combining above expression we can then write the stationary value of the corresponding definite integral as

$$\frac{\delta I}{\varepsilon} = \int_a^b \xi(x) \cdot \phi(x)dx = 0$$

It can be easily seen that the above expression would be satisfied for any arbitrary $\phi(x)$ if and only if $\xi(x)$ vanishes everywhere in the space [a,b]. We can, on the other hand make $\phi(x)$ vanish everywhere except in 'a small neighborhood around a point say, $x=\zeta$. Within this "small interval," $\xi(x)$ is "practically" constant and can therefore, bee taken out of the sum (integral operation) as a simple multiplier factor.

$$\frac{\delta I}{\varepsilon} = \xi(\zeta)\int_{\zeta-\mu}^{\zeta+\mu} \phi(x)dx$$

As the radius μ tends approaches zero, our "error' also tends to vanish. The first variation or linear term of the expression must vanish, hence we can write the expression known as the Euler-Lagrange Equation

$$\xi(x) = 0 = \frac{\partial F}{\partial y} - \frac{d}{dx}\left(\frac{\partial F}{\partial y'}\right)$$

About two centuries ago Poisson wrote the equation for the free energy of a solid elastic membrane

$$F = \left(\frac{k_c}{2}\right)\int_M (2H)\sqrt{2}dS$$

Where H and dS are the mean curvature and infinitesimal area element of the surface respectively, and k_c is the bending elastic modulus. The energy Euler-Lagrange equation corresponding to these functional can be written as

$$\nabla^2 H + 2H(H^2 - K) = 0$$

And the solution for such functional satisfying the minimization of surface energies is the critical curve known as the Willmore surface F, written as

$$W(f) = \int_M H^2 dS = \int_M (H^2 - K)dS + C$$

Numerical and Analytical Results

We carry out some solutions for the formulation presented above and using boundary conditions corresponding to the flow of a fluid in a capillary action, between two parallel plates, and in the presence of various surface topographies. The geometry to be treated consists of parallel plates as shown in Figure 1.

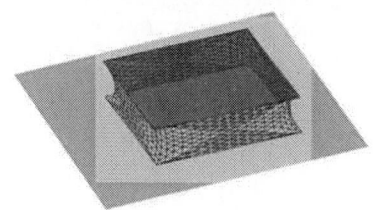

Figure 1. Drawing of the geometry of two parallel plates separated by a gap in the presence of a fluid flowing by capillary action between them.

The capillary motion, in the presence of a step, i.e., a sudden increase in the gap between the parallel surfaces (see Figure 2) resulting from the numerical analysis is depicted in the sequence of pictures shown in Figure 3.

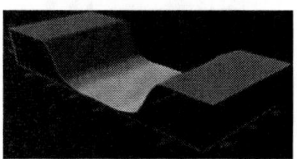

Figure 2. Sudden step on the organic substrate

There one can observe that the fluid front has a tendency to behave in a way that preserves state of symmetry about a virtual horizontal plane parallel to the surfaces, thereby avoiding the creation of voids in the fluid path. The assumption here of course is that all surfaces in contact with the fluid are wettable to it, and that the adhesion is about the same for all of them. This flow behavior is very different from that of the case where induced pressure differential (as it is in the case of injection molding) for instance, the propensity to create voids is rather high in similar geometries. For the case of holes, the flow around them is the primary cause of void formation and it is governed by the flow velocity field around the hole, tangent to the circumference of the hole, and the fluid velocity as the fluid goes to a larger gap.

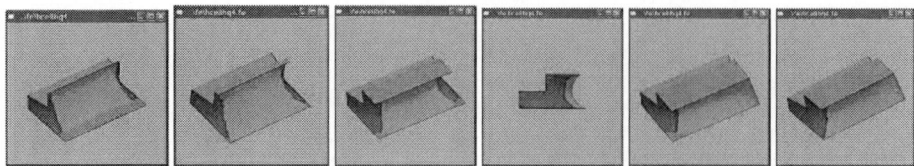

Figure 3. Capillary fluid flow in the presence of sudden step where the gap increases.

Other geometries analyzed are depicted in Figure 4, including grooves crossing and different groove directions.

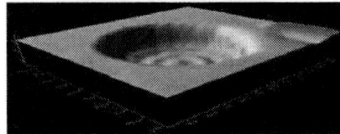

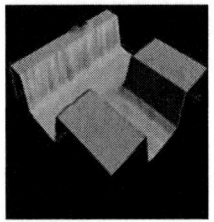

Figure 4. Various laser profilometer geometries treated as spatial boundary conditions: grooves, holes and groove intersections.

The case for underfilling components where the distance between them is comparable to the gap to be capillary underfill poses a whole new set of complications[3]. Aside from the fact that perhaps only jetting technologies can be of use (see Figure 5) given that the needle dispensing requirements do not physically permit its use, there are other limitations imposed by the physics of the underfill and capillary equilibrium principles.

Solutions to the above problem indicate that when the distance between the components is the same as the gap or less, then capillary underfilling does not take place; this is depicted in Figure 6.

Figure 5. Jetting fluid from a distance above the components allowing for small distance between components.

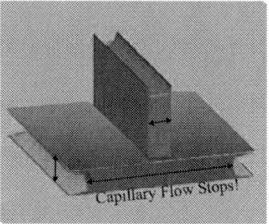

Figure 6. No capillary flow takes place for this geometry (the gap is the same as the distance of separation)

For cases where the distance separating the components is larger than the gap to be underfilled (see Figure 7), the solution indicates that capillary underfill occurs to completion, i.e., the fluid reaches the sides of the smaller of the two plates (in this case the glass plate).

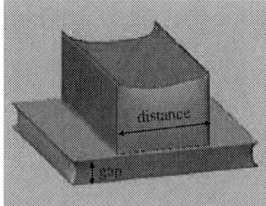

Figure 7. Two components side-by side separated by a distance larger than the gap to be underfilled.

As depicted in Figure 8, the fluid flows until it reaches equilibrium, and a fillet around the plate including that area in between the components is formed. A sequence of events through time is shown where the fluid is moving under the influence of capillary effects.

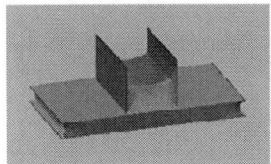

Figure 8. Sequence showing time slices of the fluid moving in a capillary manner in between the gap that is smaller than the distance separating the components.

Experimental Results

Glass plates were mounted on organic boards with gaps varying from 75 to 300 μm. The surfaces of the boards had different topographies including grooves of different depths and widths as well as holes drilled to different diameters and depths from 250 μm to 1.5 mm. Figure 9 depicts some of the samples used for the experiment.

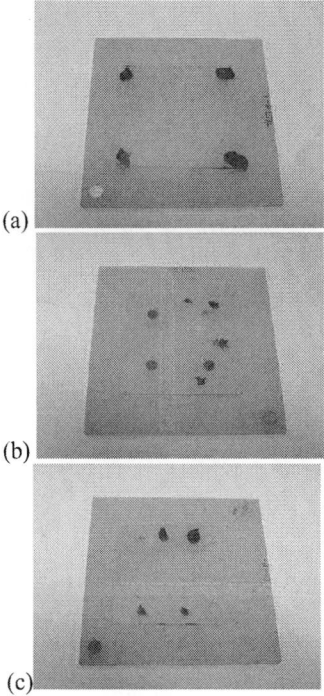

(a)

(b)

(c)

Figure 9. Samples used for capillary underfilling: (a) Drilled holes on organic substrate; (b) Grooves aligned parallel to the fluid flow direction; (c) Grooves normal to the capillary fluid flow direction.

Underfill material was jetted along a side of the glass plate using Asymtek's DispenseJet® DJ-9000 jet[4]. Figure 10 depicts the capillary flow of the underfill in the presence of a hole with diameter about three times larger than the gap and about 1 mm deep. It is noticed that a void is present; this resulted from the fact that the fluid flows along the peripheral of the hole following a near perfect tangential direction to its circumference.

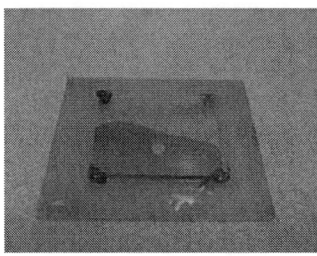

Figure 10. Capillary flow in the presence of a large hole on the organic substrate, void formation resulting from fluid flow tangential to the hole.

In the case of grooves present on the organic substrate, the fluid flows in a capillary fashion without allowing void formation. When the groove is parallel to the direction of the flow, voids could be formed if such groove is very narrow (compared to the gap) and in particular, if its depth varies. However, for the case of grooves normal to the fluid flow, voids are not formed independent of the groove geometry. For a very deep groove, the flow will simply cease to continue, this can be understood from the point of view of pure equilibrium mechanisms, i.e., it can be simply seen as a boundary condition similar to the starting edge of the glass plate. Figure 11 depicts the case where both grooves, normal and parallel to the fluid flow direction were present.

Figure 11. Grooves are present in the organic substrate in both directions, normal and parallel to the fluid flow direction; no voids are formed during the capillary underfill.

Conclusions

A comprehensive analysis of capillary fluid flow has been presented and validated by actual data. The coupling of molecular forces adhesion and cohesive in nature and the minimization of surfaces, as in the case of elastic membranes that yield Willmore critical surfaces in the differential geometry scheme gives an adequate way to model capillary fluid flow. These solutions were obtained essentially by using perturbation theory, where Variational calculus is applied to the functional derived from Poisson integral formulation. The results very closely matched the observed capillary fluid flow subjected to various boundary conditions including several different geometries. For dispensing processes being designed for spacing and high density packages, jetting technology lends itself in a very transparent and practical manner. Above results can be used as guidelines in situations where uneven surfaces and arrays are present and capillary flow is used. Design rules for packaging design need to include the capillary physics present during underfilling and fluid dispensing in general.

Acknowledgments

The author would like to thank Mrs. Roberta Foster-Smith for reviewing the present work; Mr. J. Klocke for his support and encouragement to complete this work, and Mr. A.Babiarz for his insights and guidance of this work.

References

[1] Martin Fuchs and Gregory Seregin, "Variational Methods for problems from Plasticity Theory and for Generalized Newtonian Fluids."

[2] Christoph Bohle, G. Paul Peters, "Bryant Surfaces with Smooth Ends", preprint, http://de.arxiv.org/abs/math.dg/0411480.

[3] Alec Babiarz, "Jetting Adhesives And Other Materials For Semiconductor And Electronic Component Packaging," Proceedings of the Pan Pacific Conference, SMTA, 2007.

[4] Horatio Quinones, Alec Babiarz, and Christian Deck, "Fluid Jetting for Next Generation Packages", Pac Tech, Berlin, April 2002.

Additional Stresses of ECA Joints due to Moisture Induced Swelling

Richard C. Löw[a,b], Ralf Miessner[b] and Jürgen Wilde[c]

[a] Robert Bosch GmbH, Automotive Electronics

[b] Robert Bosch GmbH, Corporate Sector Research and Advance Engineering
Alte Bundesstrasse 50, D-71301 Waiblingen, Germany
Phone: 0049–07151/503 2837, Fax: 0049–07151/503 2664, richard.loew@de.bosch.com

[c] University of Freiburg, Department of Microsystems Engineering, Laboratory for Assembly and Packaging
Georges-Koehler-Allee 103, 79110 Freiburg, Germany

Abstract

Surface Mounted Devices (SMD) are used as component parts of Electronic Control Units (ECUs) in the automotive industry. SMDs are either connected to the circuit with solder or with Electrically Conductive Adhesives (ECAs). Adhesives have some advantages over solder, like durability at elevated temperatures (~150°C) or the absence of creep during assembly of the circuit. Due to a mismatch in the Coefficient of Thermal Expansion (CTE) of the various materials used, mechanical stresses are induced in the adhesive. Additionally, the organic nature of the adhesives leads to an ingression of water into the polymer structure, which results in swelling. To correctly judge the influence of swelling on the reliability of the assembly, the swelling behaviour of the ECA was characterised under different test conditions. The results make clear that moisture induced strains are in the same magnitude as strains resulting from the CTE mismatch. Therefore using materials with different swelling properties causes mechanical stresses in the interconnections and decreases the effective lifetime. The investigation shows that humidity in the environment influences SMD behaviour and leads to strain and stress in the connection between circuit and device. This stress reduces the reliability and can not be neglected.

Key words: moisture, diffusion, swelling, simulation, FEA, visco-elasticity

Introduction

Electrically conductive adhesives (ECA) are used as conductive connection for Surface Mounted Devices (SMD) in automotive industry. Since the epoxy adhesive is an organic material and includes hydrophilic groups, it is susceptible to moisture attack. The mass gain due to water absorption can reach a magnitude of 10% [1], depending on the chemical structure, e.g. the percentage of hydroxyl-groups or flexibility of side-chains [2], the curing temperature and the glass transition temperature (T_g) [3]. It is the result of equalizing the concentration between dry adhesive and humid environment and has either a Fickian or a non-Fickian character [1]. Fickian character appears when the diffusion rate is much quicker than the relaxation rate. In these cases the mass gain kinetics agrees with Fick's law [4]. Non-Fickian relaxation occurs when the relaxation influences the diffusion properties, generally above the T_g [1].

As a result of the absorption of water into the polymeric network, the properties of the adhesive will alter. Water can increase chain mobility and plasticise the network by weakening the intermolecular interactions among the functional groups [2], [5]. It also can lead to chemical degradation of the matrix due to hydrolysis and therefore chain scission [6]. The constitutive behaviour of thermosetting plastics is known to be visco-elastic [7] and has therefore a general time and temperature dependence. The ingression of water reduces the cross-linked density of the network, which results in a lowering of T_g [8], reported to be 10°C to 15°C for every percent of mass fraction water for a particle filled epoxy based adhesive [9]. In this study, the visco-elastic properties of the adhesive used were determined for unaged and aged material, to perform correct computations in Finite-Elements-Analyses (FEA).

The absorbed water in the polymer is either bonded or unbonded. Both states were confirmed using different spectroscopic methods e.g. in [6] or [8]. In the former case, the water is hydrogen-bonded and causes swelling, while the water in the latter case is unbonded and free, filling the micro-voids [8]. To characterise the hygroscopic swelling one can use different test methods, as summarised in [10]. The result of the hygroscopic swelling measurement is a strain – mass concentration dependence. The change of the hygroscopic strain in relation to the amount of water is defined as Coefficient of Hygroscopic Expansion (CHE) in this study, according to the Coefficient of Thermal Expansion (CTE) and introduced in [5].

Hygroscopic swelling leads to stress, since there are materials with different swelling properties in a package. Automotive electronics, and therefore SMD joints, are widely exposed to humid environment and elevated temperatures during Highly Accelerated Testing or during the lifetime of the assembly. To predict the durability of the assemblies, one has to quantify the amount of stress. The authors in [11] presented a method for the computation of stresses in a package caused by moisture absorption using a thermal moisture analogy [4]. Since the material has visco-elastic behaviour and is therefore time- and temperature-dependent, no thermal – moisture analogy can be used. To correctly judge the stress of the polymer, a method was developed that is able to treat the moisture induced stress additionally to the stress resulting from the thermal mismatch.

To verify the method, a test assembly was built and the warpage, resulting from the swelling of the ECA, was determined and compared with the predicted ones.

Materials & Methods

In this study a one-component thermosetting ECA was used.

(a) Diffusion

The diffusion properties of the adhesive were taken up using a gravimetric analysis. Samples were made in forms of discs with a diameter of approximately 100 mm and a thickness of 1 mm. The samples were stored in a climate chamber and the mass gain was measured at constant time steps. As a result of the high aspect ratio one-dimensional diffusion occurs. The mass gain M(t) due to one-dimensional diffusion is given in equation (1). The solution consists of a trigonometrical-series type, of n series, that converges satisfactorily for a large number of n. The procedure of solving Fick's law is shown in [4].

$$M(t) = n_0 \left(1 - \sum_{n=0}^{\infty} \frac{8}{(2n+1)^2 \pi^2} \cdot \exp\left(\frac{-D(2n+1)^2 \pi^2 t}{l^2} \right) \right) \ (1)$$

Fitting the relative mass gain of the disc, with the thickness l, to equation (1) one gets the diffusion coefficient D and the maximum of the relative mass gain n_0. The diffusion coefficient is temperature-dependent and obeys an Arrhenius relation [12]. Using the maximum relative mass gain n_0 one can compute the concentration of water in the polymer. The vapour pressure of water, p, in the humid environment can be obtained using the law of Clausius-Clapeyron. If the environment is not fully

saturated with water vapour, the fraction of water is given as relative humidity. The pressure, which the gas would have if it occupies the volume alone, is known to be the partial pressure $p_{partial}$. Using the ideal gas law and the molar mass (M) of water molecules, one achieves the concentration of water in the humid environment, $c_{environment}$. The ratio of the concentration of water in the polymer to the concentration of water in the environment is known as solubility S and is temperature-dependent, following the Van't Hoffs law [12].

(b) Hygroscopic Swelling

Figure 1 shows the characterisation technique used. Introduced by [11], this technique offers some outstanding features, like usage of standard instruments coupled with a high accuracy. Combination of length change with mass loss leads to a strain – mass of water dependence. In this study the mass loss was computed using the three-dimensional solution of Fick's 2nd law, contrary to the measurement of the mass loss in [11], where a gravimetric analysis was used.

Strain

Hygroscopic swelling was measured using a Thermo-Mechanical-Analyser (TMA). Samples of the ECA were made in the form of cubes, with an edge length of approximately 5 mm. Each sample was placed in the TMA for 10 hours at 85°C and the reference curve of the dry material,

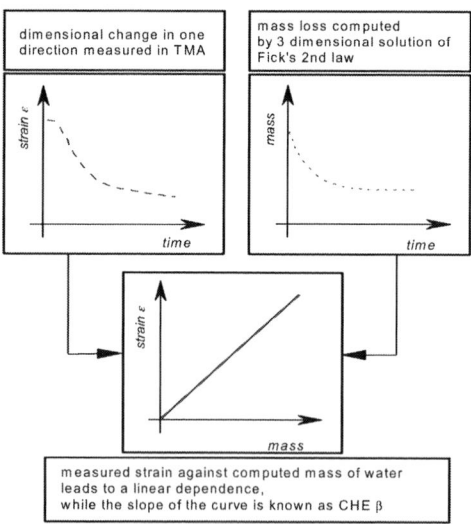

Figure 1: Characterisation technique for obtaining the Coefficient of Hygroscopic Expansion (CHE) introduced by [2].

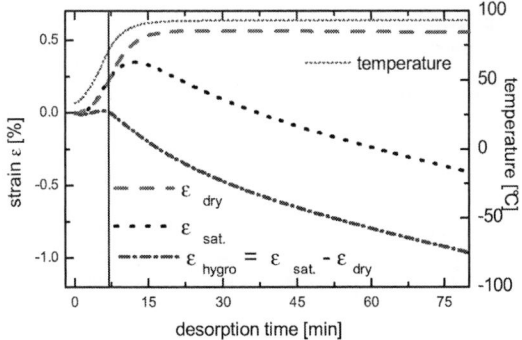

Figure 2: Obtaining hygroscopic strain ε_{hygro} by subtraction of the thermal strain ε_{dry} from the strain measured of the saturated sample ε_{sat}.

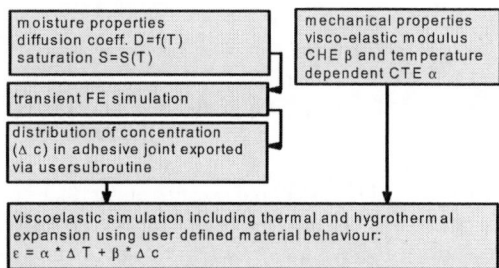

Figure 3: Procedure of coupling transient transport analysis and visco-elastic analysis.

containing length change due to thermal expansion, was measured. Thereafter the samples were stored in a climate chamber at different test conditions, namely 60°C / 95% relative Humidity (r.H.), 85°C / 85% r.H. and one sample was stored in a pressure cooker at 121°C / 100% r.H. Time to saturation of the samples had been calculated and the samples were transferred into the TMA, where drying at 85°C took place for at least 10 hours. Length change in the second run comprises thermal expansion as well as contraction during desorption of the sample. Therefore, the hygroscopic contraction could be obtained by subtracting the amount of the thermal strain, obtained in the first TMA run, from the strain of the saturated sample. In both TMA runs, the heating rate was 5 K / min and the accuracy was 1 micron.

An offset of the expansion at the beginning of the experiment was observed, because of the heating rate. Compensation was achieved by time-shifting the hygroscopic strain, shown in Figure 2.

Mass Change

The mass loss during drying M(t) was determined by using the three-dimensional solution of Fick's 2nd law [4].

$$M(t) = \frac{512 \cdot n_0}{\pi^6} \left(\prod_{i=1}^{3} \sum_{n=0}^{\infty} \frac{1}{(2n+1)^2} \cdot \exp\left(-\frac{D \cdot (2n+1)^2 \pi^2}{l_i^2}\right) \right) \quad (5)$$

For desorption of a cube with the side lengths l_1, l_2 and l_3 the solution is shown in equation (5). Therefore, the measured data of the mass gain were fitted with n=16 in each direction in space (i=1,2,3).

(c) Simulation

Simulations were performed using Abaqus 6.6.1. The simulation took place in two steps. In the first step the moisture concentration distribution was calculated in transient analyses using either the mass transport equations, provided in Abaqus, or the heat transfer equations. A subroutine had been programmed,

that saves the concentration at each node and time step. The second step comprises a visco-elastic analysis. The mechanical properties in forms of Prony coefficients were obtained by Dynamic Mechanic Analyses (DMA) for aged and unaged samples in tension mode. Aging of the samples was achieved by storing them in a climate chamber at 85°C / 85% r.H. for 24h, 48h and 96 hours before testing. The sample geometry was $50 \cdot 10 \cdot 1$ mm^3. More information about visco-elasticity and how to determine the Prony coefficients is given in [7], [13] and [14]. A second subroutine was programmed which reads the moisture concentration distribution from the file, saved in the first step, and adds hygroscopic and thermal strains. The procedure is shown in Figure 3.

(d) Verification

A glass die was assembled with ECA to a glass substrate, shown in Figure 4. Since the glass is inert to water, no absorption of water occurs in the glass. Therefore, diffusion of water into the ECA took place only via two dimensions, namely from the periphery. On this account, hygroscopic swelling started at the border area, while the centre of the glass die stayed unchanged. The die geometry was $18 \cdot 18 \cdot 0.15$ mm^3. The height of the ECA film was approximately 1 mm.

The Young's Modulus of glass was 68 GPa and Poisson ratio of glass was assumed to be 0.19. A sketch of the assembly is shown in Figure 4. The samples were stored in a climate chamber in 85°C / 85% r.H. condition up to 30 days and the warpage, due to hygroscopic swelling of the ECA was measured using a scanning laser-interferometer.

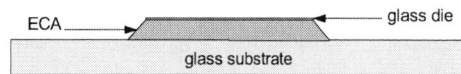

Figure 4: Assembly of the verification experiment. A glass die is placed on a glass substrate, joined by an ECA film. Warpage of the die was measured using a laser interferometer.

Results & Discussion
(a) Diffusion

Table 1 summarises the fitted values. Because of the good fit results, evidently Fickian diffusion took place for the absorption of water at temperatures up to 121°C. According to several studies e.g. [1],[9],[10] this means either that the relaxation time is much slower than the diffusion time or that there is a no relevant amount of micro-voids, since storage of water into micro-voids is related to non-Fickian diffusion. However, since the diffusion is Fickian, all absorbed water is bonded to the network and therefore leads to swelling [8].

(b) Swelling

Assuming that the contraction equals the swelling, the strain was plotted vs. the amount of water in the polymer, as shown in Figure 5 for 121°C / 100% r.H. condition. As reported by several other authors, there is a linear dependence of strain to the mass of water [10], [11], [15]. Data were fitted with the software package Origin using a Levenberg Maquardt optimization. Results are listed in table 2.

A fundamental assumption of this technique is that contraction and swelling are the same, reported in [10] and [11]. On the other hand, there are several studies that state different properties during sorption cycling, as an increase of diffusion coefficients with rising number of cycles [16], breaking of cross-links and leach-out [3], [6], [17]. The authors of [9] reported different maximum weight changes for sorption and desorption. Furthermore a decreasing maximum weight change was observed with increasing number of cycles, but there is no statement whether the absorbed water for each cycle is in bound or unbound state. Therefore, cycled sorption experiments have been carried out during this study and no change in the maximum mass gain was observed between sorption and desorption. Hence, the authors of this study decided to handle contraction equivalent to hygroscopic swelling.

One problem of using the TMA for hygroscopic swelling measurements is that the heating of the TMA requires considerable time and could not be loaded stepwise, leading to initial stage anomalies as reported in [10]. Compensation was achieved by a time shift of the data. Another possibility is to simulate the mass loss using FEA instead of using equation (5), but this procedure would be much more time consuming. As a result, the solution of Fick's 2nd law was used. The authors in [11] measured the weight loss using a Thermo Gravimetric Analyser (TGA). Using this technique, either one has to use the same samples for TGA and TMA or one has to produce identical samples. Since the diffusion properties of a polymer might alter during lifetime [16] the former case is not suitable. On the other hand, two samples won't be exactly the same. In contrast, the analytical equation could be fitted to the sample geometry used in TMA.

Table 1: Diffusion Properties of ECA

T [°C]	r.H. [%]	D $\left[10^{-12}\frac{m^2}{s}\right]$	n_0 [%]	solubility S
30	85	1.0952	0.15646	268.23
60	95	4.9961	0.2351	96.54
85	85	26.556	0.24959	38.99
100	100	43.479	0.4849	35.39
121	100	59.893	0.6263	24.07

Table 2: Hygroscopic Swelling Properties of ECA

Storage condition	CHE $\beta \left[\frac{cm^3}{g}\right]$	Coefficient of determination
60°C/95% r.H.	0.13841	0.82355
85°C/85% r.H.	0.27203	0.95609
121°C/100% r.H.	0.30172	0.98345

Therefore the method for determining CHE presented in this paper is much more precise.

Plotting hygroscopic strain vs. moisture content leads to a linear dependence, already reported in [10], [11] and [15]. The CHEs for different storage conditions differ between $0.1384 \cdot cm^3 / g$ and $0.30172 \cdot cm^3 / g$. Normally the CHE should be the same for all three test conditions. Since there are initial stage anomalies as reported in [10], the fit for the first storage condition at 60°C / 95% r.H. is poor and the CHE used for the simulations was $\beta = 0.272 \, cm^3 / g$.

Reducing the anomalies leads to a nearly constant signal of length change from the TMA that can be combined with a constant signal of mass loss, at all possible time steps. Using this technique, only one storage condition has to be tested that gives information for all moisture concentration states. Comparing to former studies [10], the CHE was reported to be approximately $\beta = 0.2 \, cm^3 / g$ for an epoxy moulding compound. Several packaging materials were characterised in [11] and the CHE was found to be temperature-dependent, varying between $0.1 \leq \beta \leq 2 \, cm^3 / g$.

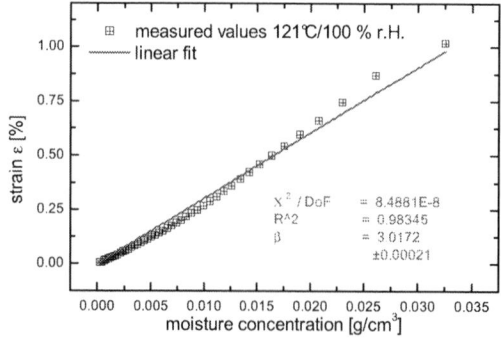

Figure 5: Linear dependence of hygroscopic strain from TMA and calculated moisture concentration. Data were fitted to a linear model.

(c)   Simulation of Stresses and Strains

The simulation was performed in two steps, whereas the first one computes the distribution of water concentration and the second one computes the resulting strains and stresses. Concentration distributions due to the transient diffusion process for different absorption times are shown in Figure 6. The second step comprises the visco-elastic analysis. The main advantage of this method is that moisture-induced stresses are handled additionally to the temperature induced ones in visco-elastic analyses.

The mechanical visco-elastic properties of the ECA will change during storage in humid environment as shown in Figure 7. The Prony coefficients of the aged material were determined and implemented in the FEA program. Storing the material in water leads to changes in the network, either reversible or irreversible as reported in [6]. Accordant to increasing moisture concentration in the polymer, the network density will decrease and chain mobility will increase. This results in a lowering of the T_g, reported to be 10°C to 15°C for every percent of mass fraction water for a particle filled epoxy based adhesive [9] and in a lowering of the storage modulus, especially in the rubbery state above T_g [3]. Both effects were noticed during this study.

The simulation method was verified using a simplified assembly (Figure 4) instead of a SMD. As a result of the diffusion process of water into the polymer, the ECA swells on the edges while the inner adhesive stays dry. This leads to a concave deformation profile furthermore called warpage of the die. The dependence of warpage on absorption time was predicted and measured. Unfortunately, placement of glass dies was not levelled, but it was compensated during evaluation of the data.

Measurement took place in one direction along a diagonal path of the die. Displacement of the die in out-of-plane direction was measured as difference between displacement in the middle of the path and the displacements of the edges, called $\Delta U3$, shown in Figure 8. As a result of the not-parallel assembling of the samples, different heights of the ECA film were received. Measurement using the interferometer resulted average heights with a maximum of 1581.29 µm and a minimum of 625.99 µm. Therefore, a wide standard deviation of the measured warpage was recorded. Figure 8 shows the assembly used in FEM and the simulated displacements in out-of-plane direction for different storage times in 85°C / 85% r.H. condition. Figure 9 shows the warpage of the glass die, including the measured and the predicted data for different storage times. One can see the high standard deviation due to the inclined planes of the different samples. Nevertheless, a general tendency of the measured values in accordance with the computed behaviour was observable in the verification experiment.

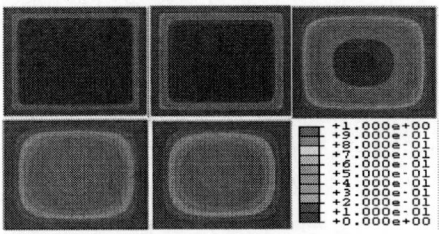

Figure 6: Moisture concentration distribution in the ECA during 85°C / 85% r.H. storage after 1d, 2d, 8d, 23d and 30 days of absorption.

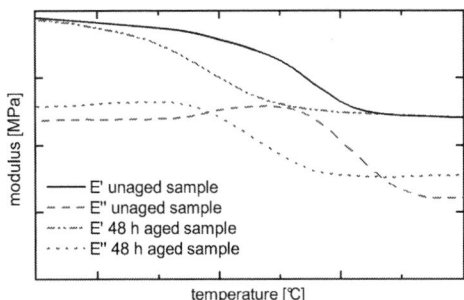

Figure 7: Visco-elastic properties measured in DMA: storage modulus in extension E' and loss modulus in extension E'' of unaged and 48 hours 85°C / 85% r.H. aged ECA.

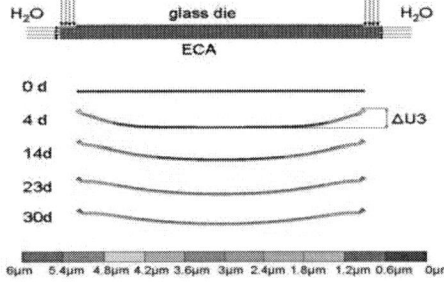

Figure 8: Simulated assembly and displacement in out-of-plane direction of the glass die due to water absorption in 85°C / 85% r.H. for different storage times.

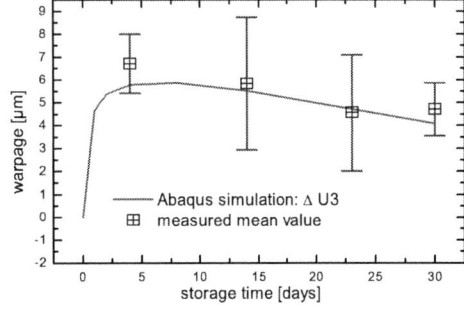

Figure 9: Transient diffusion process. The edges of the ECA swell while the inner adhesive stays dry and therefore leads to a concave warpage of the die, measured and predicted data.

Conclusion

The results show that humidity in the environment influences ECA behaviour and leads to strain and therefore to additional stress in the adhesive. Diffusion properties of the ECA were characterised and found to be Fickian in the case of water absorption up to 121°C for five different test conditions.

Since there is Fickian diffusion behaviour, all water molecules were hydrogen-bonded in the polymer matrix and lead to hygroscopic swelling. The hygroscopic swelling characteristics were determined by measuring strain using a TMA and computing mass change using the solution of Fick's 2nd law of diffusion. The hygroscopic strain was found to be $\varepsilon_{hygro} = 1\%$ in saturation. Since this strain induces additional stress to the thermal one, it is important for reliability analysis. For comparison only, the same strain is achieved due to a temperature increase from -50°C to 70°C.

Since there is additional expansion, moisture absorption will induce stress. This stress reduces the reliability and can not be neglected. To predict reliability of an assembly in FEA, the Coefficient of Hygroscopic Expansion (CHE) was determined and was found to be constant over moisture concentration.

A new method for simulation of visco-elastic materials in FEA, including time- and temperature-dependent data of the aged material as well as thermal and moisture induced stresses was presented. To verify, the warpage of a simplified test setup was predicted and measured, and the data were compared.

References

[1] W.K. Loh et.al, "Modelling anomalous moisture uptake, swelling and thermal characteristics of a rubber toughened epoxy adhesive", Int. Jour. of Adhesion and Adhesives 25 (2005), pp.1-12

[2] Liu, M. et al; „Contributions of side groups to the water diffusion in cured epoxy resins (2) study on physical aging", Jour. of Polymer Science, Part B., Polymer Phys., Vol. 41, No. 11, pp. 1135-1142

[3] K.I. Ivanova, R.A. Pethrick, S. Affross; „Investigation of hydrothermal ageing of a filled rubber toughened epoxy resin using DTMA and dielectric spectroscopy"; Polymer 41 (2000), pp. 6787-6798

[4] J. Crank, "The Mathematics of Diffusion", 2nd Edition, Oxford Science Publications, 1975 (ISBN 1560323256)

[5] Fei Su, Kerm Sin Chian, Sung Yi ; "An optical characterisation technique for hygroscopic expansion of polymers and plastic packages"; Microelectronics Reliability 46 (2006), pp. 600 – 609

[6] Shuangyan Xu, Dillard David A..; "Environmental aging effects on thermal and mechanical properties of electrically conductive adhesives", Jour. of Adhesion , Vol. 79 :699-723, 2003; ISSN 0021-8464

[7] John D. Ferry "Viscoelastic Properties of Polymers", 3rd Edition, John Wiley and Sons, (ISBN: 978-0-471-04894-7)

[8] Luo S., Leisen J., Wong C.P.; „Study on mobility of water and polymer chain in epoxy and its influence on adhesion"; Jour. of Applied Polymer Science, Vol. 85 (2002) No. 1, pg. 1-8, ISSN 0021-8995

[9] Michael G. McMaster, David S. Soane; „Water sorption in epoxy thin films", IEEE Transactions on Components, Hybrids and Manufacturing Technology, Vol. 12, No. 3 (1989)

[10] Haleh Ardebili, E. H. Wong, Michael Pecht; "Hygroscopic swelling and sorption characteristics of epoxy moulding compounds used in electronic packaging", IEEE Transactions on Components and Packaging Technologies, Vol. 26, No. 1, pp. 206-214 (2003)

[11] E. H. Wong, Y. C. Teo, and T. B. Lim, "Moisture diffusion and vapor pressure modeling of IC packaging," in Proc. 48th ECTC., 1998, pp. 1372-1378

[12] Wolf R. Vieth; "Diffusion In and Through Polymers: Principles and Applications", Hanser Gardner Publications (1991), (ISBN-13 978-2569901069)

[13] Abaqus Documentation 6.6.1

[14] Goehler, J., "Optimierungsstrategien zur Parameteridentifikation eines viskoelastischen Materialmodells für Epoxidklebstoffe", Proc. 2. Weimarer Optimierungs- und Stochastiktage, (2005)

[15] Eric Stellrecht, Bongtae Han, Michael G. Pecht; „Characterization of hygroscopic swelling behaviour of mold compounds and plastic packages", IEEE Transactions on Components and Packaging Technologies, Vol. 27, No. 3 (2004)

[16] Fernandez-Garcia, M. , Chian, M.Y.M. ; „Effect of hygrothermal aging history on sorption process, swelling and glass transition temperature in a particle filled epoxy based adhesive"; Jour. Appl. Polymer Science (USA), Vol. 84, (2002) no. 8, pp. 1581-1591,23.

[17] Shanahan, M.E.R., Xiao, G.Z. "Irreversible effects of hygrothermal aging on DGEBA/DDA epoxy resin", Jour. of Applied Polymer Science (UK). Vol. 69, (1998) ,no.2, pp. 363-369

Novel Packaging Technology for Combo Memory Package

Ville Pekkanen, Matti Mäntysalo, Jani Miettinen, Pauliina Mansikkamäki

Tampere University of Technology, Finland

P.O.Box 692, FI-33101 Tampere Finland

Phone: +358 3 3115 4506, Fax: +358 3 3115 3394, E-mail: ville.pekkanen@tut.fi

Abstract

Change from component based manufacturing to process based enables novel electronics integration methods. Printed electronics is an uprising technology for manufacturing electronic circuits. Different printing methods make it possible to manufacture a wide range of new products, for example low-cost displays, antennas, and smart packaging. Another issue is replacing today's PCB and wirebonding technology with inkjet printed interconnections. Key drivers for printed electronics are increased production flexibility and possibility for custom series and rapid prototyping. Digital printing methods can be used to form interconnections replacing traditional mask-etch technologies. Drop-on-Demand (DoD) inkjet printing combined with metallo-organic decomposition (MOD) can be used to form conductive traces accurately. MOD contains nano-sized metal particles which can be sintered in temperatures as low as 200°C. Additionally, fluids with dielectric properties can be printed with the same device. By combining conductive and dielectric materials, multilayer structures can be created. The manufacturing process is simplified and the digital process enables flexible production. The additive nature of the process also reduces the amount of waste produced. Inkjet printing was utilized as a method for interconnecting chip-scale package, replacing wire bonding and premanufactured interposer PCB. The novel method enables forming the wiring using inkjet deposited conductive and dielectric inks. Multiple layers of conductive ink and dielectric materials can be printed on epoxy-encapsulated components in order to achieve a multilayer interconnection structure.

Key words: printed electronics, inkjet printing, chip scale package, electronics manufacturing

Introduction

Lower cost, larger areas, simple manufacturing, and low amount of waste are the key drivers towards the rise of additive manufacturing technologies. Direct material deposition methods have reached a feature size of few microns, which enables the production of high-density circuit boards. The capability of inkjet printing for manufacturing high-speed electronics has already been demonstrated. Highly integrated SiP applications have been constructed as an evaluation of inkjet technology for electronics packaging and system integration.[1]

As the submicron level is reached, the number of logic functions per unit area is increased significantly. Enhanced performance combined with low cost per unit area is a factor that also attracts the use of technology in high speed products.[2]

A novel electronics packaging technique based on digital printing and nanosize metal particle behavior has been examined at Tampere University of Technology as an interconnecting method for chip-scale package (CSP). The package was manufactured in an additive manner using a materials printer. The package examined is based on

a combined memory device containing flash and pseudo static random access memory (PSRAM) integrated circuits in a TFBGA package designed to be used in low power wireless applications [3]. Integrating different memories in a single package occurs from the enlarged number of features in handheld devices, which obviously increases demands for the memory performance and capacity [4].

The novel method opens a way to create package and interconnections for modules in an additive manner. Design for manufacturing (DFM) is needed as electrical design must be done on production's terms. Design development becomes more flexible and prototyping can be done cost-effectively and relatively fast. Even so, testing the functionality and second level connections can be done using ordinary BGA package methods.

In this paper utilization of inkjet technology in electronics packaging is discussed. In addition, procedures preparing the way for printed electronics manufacturing are observed. That is, creating design rules and selecting materials for novel packaging approach. The goal of the study was investigate new manufacturing method for chip scale packaging and evaluate the possibilities of inkjet technology in electronics packaging and integration.

Design Considerations for Printed Electronics

Design for manufacturing is an important topic when discussing about the additive fluid deposition process. The nature of printed interconnections requires alternations in layout design. Connecting stacked structures is challenging with inkjet technology. Due to this, components are located side by side.

The firing routine of Drop-on Demand inkjet printheads places another demands for design phase. Electronics is usually designed using continuous vector graphics, but printing of fluids is done sequentially drop by drop, and therefore data conversion is needed. The digital nature of inkjet printing means, that a drop is either ejected or not on the specific location. The on-off signal information for the printhead nozzle is included in printfiles, and conversion to printed patterns is usually done with device-specific software. The process is binary and using small steps, i.e. high resolution, allows creating of good quality images. However, using high resolution images usually increases the printing time. On the other hand, printing low-resolution images a saw-tooth effect may occur, as seen in Figure 1.[5]

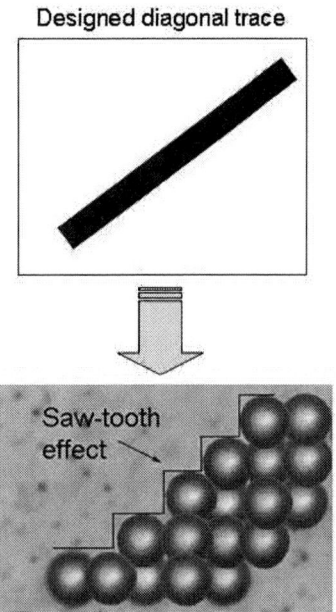

Figure 1. Saw-tooth effect in diagonal trace.

To avoid effects of the saw-tooth phenomenon, wiring is designed orthogonally, although it increases the length of traces and may cause challenges in reliability due to current accumulation in trace corners.

The performance of the circuits is determined by materials and processes. Therefore, the system design will be strongly related to the manufacturing process. The process parameters, such as temperature affect the behavior and reliability of materials. These issues need to be taken into account. To determine material suitability for certain printed electronics applications, modeling can be used. It has an important role when establishing the design guidelines for printable electronics and optimizing structures for the processes. In addition, printed electronics enables lots of opportunities which cannot be utilized with conventional manufacturing methods. For example, the number of layers deposited may vary depending on the location on the substrate. This provides much more freedom for designer whose role is to find out the design rules of the novel structures and architectures.

The sintered nano metal conductors are proven to have resistivity low enough for electronics applications, specific resistivity being 6 $\mu\Omega$cm [6]. In this application, the trace lengths must be observed, as a single trace total resistance should not exceed 3 Ω to ensure a decent signal level. Otherwise, the electrical design of the package follows common high-speed electronics design rules, such as minimal on-board noise generation and power noise, reduction of ground-bounce effect and simultaneous switching noise during operation, impedance matching and correct signal line termination [7].

Material Selection for Multichip Module

The conventional electronics materials supply will not totally satisfy the needs of printable electronics. The requirements for the printable substances and also for substances in the vicinity of printed materials are somewhat different. Inkjet printing is a liquid process and, therefore, requires drying and sintering phases between depositing different materials. In addition, materials must be compatible with each other and preferably cross-linked [8].

Manufacturing electronics with inkjet technology is based on the low temperature sintering of nanoscale metal particles, properties of particle protection materials, and solvents. Nanoscaled metal particles have a significantly lowered melting point resulting from the reduced size. The fundamental reason for the low melting point is the large number of surface atoms compared to inside atoms. Due to this, the interaction between the atoms diminishes gradually. [9]

During this experiment, a silver nanoparticle ink was used. The particle size in the carrier fluid was 2-7 nanometers. Agglomeration of neighboring nanoparticles was prevented by an organic dispersion agent. To enable the nanoparticles' sintering the agent needs to be evaporated. This can be done with heat. Forming of conductive traces can be achieved in temperature of 200-240°C. The temperature is almost equal to soldering process, but sintering requires elevated temperatures for longer periods of time. This makes the process more severe and material selection has to be reconsidered.

There are some fundamental requirements for materials used in printed electronics. The nanoparticles need to be electrically conductive and have certain dispersion stability in the liquid. Furthermore, fluid properties must be adequate to be used with specific printing apparatus. Silver nano particles used during this process have relatively low sintering temperature and low electrical resistivity.

Requirements for the printable dielectric material are wider than just adequate insulation and printability. To prevent shorts, the dielectric should not have pinholes. For subsequent processing and multilayer structures, material should be cross-linkable with other materials [8]. Additionally, printable dielectric must be compatible with surrounding materials. Requirements placed by printing equipment are also an issue that affects the material selection. Device manufacturers define the correct values for viscosity, surface energy and particle size for printable fluids.

Material Deposition Process

Forming interconnections with inkjet printing is an additive, self-evaporating process. The material layers are deposited in sequential fashion and no masking or etching procedures are done. Printed layers may have various different functionalities, but this implementation contains the two basic functions; conductivity and insulation. Conductive nanoparticle ink is used to form conductive traces and printable dielectric to insulate routed layers. The printing process (nozzle on-off cycle) is controlled with device-specific pattern files. The digital nature of the process offers an extremely flexible manufacturing environment. Smallest controllable unit is one drop of material that can be placed on surface with accuracy limits of printing system and theoretically every image after another can be different. In Figure 2 is presented the principle of inkjet printing utilization for interconnecting multilayer structures.

The surface of the module needs a proper treatment before material deposition. To ensure successful wetting and adhesion between the substrate and printed material, all impurities must be removed. Physical cleaning can be done either with oxygen plasma or UV-ozone.

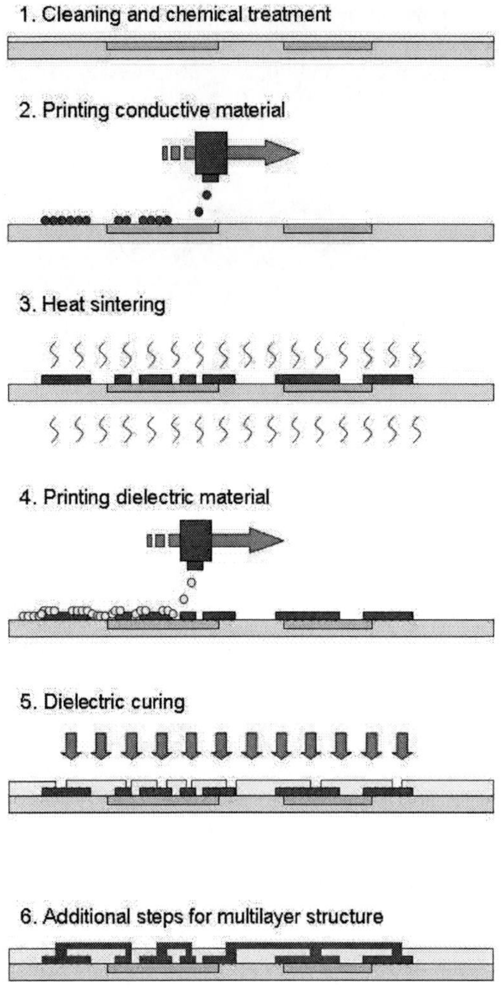

Figure 2. Principle of inkjet process utilized to interconnection forming

After cleaning, the surface energy of the substrate is very high, which has a significant effect on drop formation and printing quality. Surface energy is the measure of tendency of a surface to repel the molecules of another substance. That is, the disruption of chemical bonds that occur when a surface is created.[10]

Wetting must be kept within certain limits in order to achieve specific droplet sizes. As the energy after cleaning is very high, printing directly on the surface would result in total wetting, i.e. ink spreads and fine patterning becomes impossible. Correct wetting properties can be attained by treating the surface with sufficient chemical. 3M's surface treatment material EGC-1720 diluted with 3M HFE-7100 was used to decrease the surface energy. The material can be applied on the surface using various methods, depending on the application, such as wiping, spraying or dipping. Even slight amounts of surface treatment material prevent total wetting. However, by alternating the strength of the dilution droplet shape can be controlled. Wetting angle measurements show the basic difference in the shape of the drop with different dilutions. In the Figure 3 dilutions of 5 % and 20 % of 3M EGC-1720 have been applied on epoxy. Corresponding results of wetting are obvious.

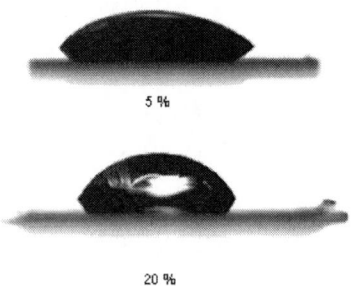

Figure 3. Differences in drop formation with different surface treatment material dilutions

Wetting angle measurements recorded on video provide useful information about drop behavior during landing on the substrate. However, imaging provides only side-view of the event. Further research of droplet profile was done in order to observe other properties, such as cross-sectional shape, possible "coffee-stain" effect and treatment material concentration effect on shape. Surface treatment concentrations of 0.5% to 20% were applied on polyimide foil and single droplets were printed on the treated surface. The effect on shape was observed optically. Analysis of the measurements proved the effect of the dilution strength on drop shape. Figure 4 shows the droplet shape when a 1% dilution of 3M EGC-1720 was used.

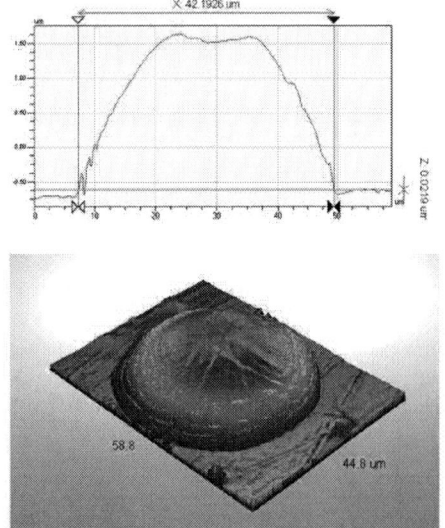

Figure 4. A single droplet profile with 1% surface treatment concentration

When the dilution strength was gradually increased up to 20 %, the drop shape changed respectively, becoming higher and smaller in diameter. The difference between different dilutions (1% and 10%) can be seen when comparing Figure 4 to Figure 5.

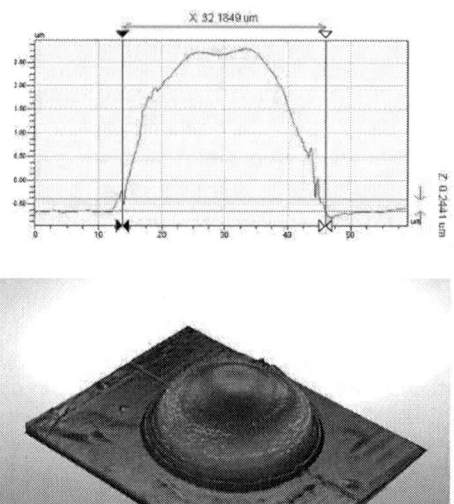

Figure 5. A single droplet profile with 10% surface treatment concentration

Although increasing the dilution strength makes the droplet diameter smaller, increasing the dilution over 10% has insignificant effect on drops of this size. The droplet shape and size indicates directly the minimum width and thickness of printed conductors. For multichip memory package, the trace width was defined by the requirements of connecting silicon die to the circuitry. During the design phase, this measure was specified to be 35 microns.

After surface treatment, the substrate was prepared for the first printed conductor layer. After printing, nanoparticle silver ink was sintered in an oven at 230°C for one hour to form a conductive path. After sintering, dielectric fluid was printed over the conductive layer and cured using a combination of UV and heat. For the second printed layer, this procedure was repeated. For the second level interconnections, the last print cycle includes printed I/O pads. The pads can be used for balling to enable conventional SMT connecting. However, the leaching of silver traces during the soldering process must be taken into account.

Experiment

The capability of the novel technology was proven by manufacturing a combined PSRAM/Flash multichip memory module. The memory type is widely used in handheld mobile applications and commonly available in chip scale stacked TFBGA package [3]. The goal was to manufacture the package using additive material deposition methods. The memory dies were separated and placed next to each other. The layout was designed in order to achieve a routing suitable for digital printing. The structure was made as thin as possible and routing design accomplished with two layers. Components were assembled on a heat and chemical resistant adhesive carrier and molded in an epoxy encapsulant. Printing process of multilayer structures includes repeated sintering processes and exposure for chemicals, such as surface treatment material and solvents, which the substrate material must last. In addition, thermal, mechanical, and electrical properties should be adequate and remain the same during the process phases.

After molding, the adhesive carrier was removed to uncover the active side of the silicon chip for printing. The quality of the interface between the chip and molding material has to be controlled carefully to avoid shorts in the circuitry. In addition, the molding process parameters must be optimized to achieve a smooth printing surface and to avoid warping, which is considered to be an important issue in device's reliability. Warping has been modeled in order to find suitable material combinations and process parameters [11]. After molding the printing process could be started. In order to achieve good print quality, a proper printing algorithm must be used. An optimized algorithm is essential when printing liquid in sequential fashion to avoid fluid bulging, which results in poor print quality.

The printing substrate contains two types of materials, components and the encapsulating material. Dielectric and conductive materials were printed on the module surface to form the first wiring layer to connect the die I/O's. Materials were treated after printing at 230°C, and the result can be seen in Figure 6 and Figure 7. It can be seen from the picture that inkjet printing proved to be a suitable method for connecting silicon chips with pad size of 45 microns.

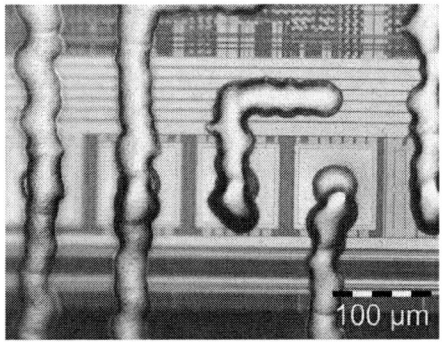

Figure 6. A silicon chip connected with inkjet printed silver conductors

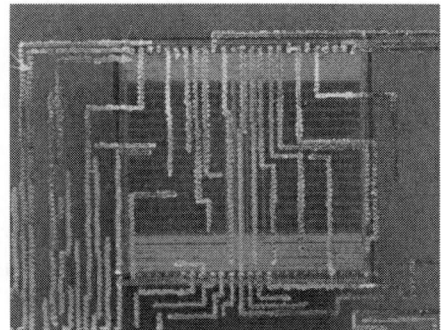

Figure 7. Inkjet printed interconnections

Vias between two conductive layers were created with printed layers of conductive ink. Vias act as contact point for second layer wiring and together with dielectric layer form a smooth surface for second layer conductive wiring. Second layer was formed by printing dielectric and conductive materials alternately.

Over the second wiring layer the higher level interconnection pads are printed. These pads are for connecting the module in larger device. The pads are sized to match with chosen interconnection method, for example with conventional surface mount technology. Using conventional connecting methods eases the functional testing as new apparatus is not needed.

Conclusions

In this paper we evaluated the capability of additive deposition of functional fluids as a packaging method for microelectronics by means of inkjet printing. Factors affecting the successful process for fine printing quality were discussed as well as material requirements for printed interconnections.

Printing fine accuracy interconnections using the inkjet technology with the metallo-organic decomposition was proven to be possible. The conductive and dielectric layers were successfully printed on the surface despite various different materials on the surface. It was also discussed, that it is possible to decrease the resistivity of printed traces with optimized printing process, drop placement and by means of optimized sintering process.

The demonstrator of the technology, a multichip memory device was manufactured using additive methods. Furthermore, the equipment needed for the module manufacturing was minimal and it was possible to complete the entire manufacturing phase in single location.

Acknowledgements

The authors would like to thank the Finnish Funding Agency for Technology and Innovation (TEKES) and Nokia Research Center for support.

References

[1] Mäntysalo, M., Mansikkamäki, P., Miettinen, J., Kaija, K., Pienimaa, S., Rönkkä, R., Hashizume, K., Kamigori, A., Matsuba, Y., Oyama, K., Terada, N., Saito, H., Kuchiki, M., Tsubouchi, M. "Evaluation of Inkjet Technology for Electronic Packaging and System Integration", accepted to be published at 57th Electronic Components and Technology Conference (ECTC) in Reno, Nevada USA, May 29-June 1, 2007.

[2] Molesa, S. 2006. "Ultra-Low-Cost Printed Electronics" Technical Report No. UCB/EECS-2006-55. University of California at Berkeley, Electrical Engineering and Computer Sciences.

[3] ST Microelectronics. 2005. "NOR Flash memories – Advanced solutions for wireless applications" 4 p. Available at http://www.st.co m/stonline/products/promlit/pdf/flnormob1005. pdf. Referred Feb.23, 2007

[4] Yokotsuka, M. 2004 "Memory Motivates Cell-Phone Growth" Electronics Design: Wireless Systems Design, April 2004. Penton Media, Inc.

[5] IdTechEx Printed Electronics Review. "Inkjet Printing Electronics" Nov. 2nd, 2004. Available at http://www.idtechex.com/printelecreview/en/ articles/00000100.asp. Referred March 26, 2007.

[6] Lee, K. J., Jun, B. H., Kim, T. H., Joung, J. 2006 "Direct Synthesis and Inkjetting of Silver Nanocrystals Toward Printed Electronics" Nanotechnology, number 17, April 2006

[7] Freescale semiconductor 2006. Application Note 2536. "High Speed Layout Design Guidelines" Rev.2, 04/2006.

[8] Wu, Y., Liu, P., Li, Y., Ong, B. S. 2006 "Enabling materials for printed electronics" IEEE Lasers & Electro-Optics Society. Oct. 2006. pp. 434-435.

[9] Buffat, P., Borel, J. 1976. "Size Effect on the Melting Temperature of Gold Particles" Physical Review A, vol. 13, number 6. June 1976.

[10] DuPont Antron web page. Definition of surface energy. Available at http://antron.dupont.com/c ontent/resources/carpet_glossary/ant06_03_19.s html. Referred Feb. 22, 2007.

[11] Kaija, K., Miettinen, J., Mäntysalo, M., Mansikkamäki, P., Kuchiki, M., Tsubouchi, M., Rönkkä, R. "Modeling the Effect of Assembly Parameters on Warpage and Stresses of Molded Package for Ink Jet Printing", accepted to be published at 16th European Microelectronics and Packaging Conference & Exhibition in Oulu, Finland, June 17-20, 2007.

Cost Efficient Quality and Production Strategies for Electronics Production

Martin Oppermann, Wilfried Sauer, Klaus-Jürgen Wolter, Thomas Zerna

Technische Universität Dresden, Electronics Packaging Lab.
D-01062 Dresden, Germany

Phone: +49 351 463 35051, Fax: +49 351 463 37069, E-mail: Oppermann@zmp.et.tu-dresden.de

Abstract

The main goals of quality management in all industries (and so in electronics productions) are customer satisfaction by delivery of defect-free products, the radical reduction of defect rates and also of quality costs and production costs. Controlled technological processes are the most important way to reach these goals. In the last years the authors developed new and powerful quality cost models to optimize quality strategies in electronics production. These models are successful in use and were published in some books, tutorials, articles and papers. The basic quality cost model and its extensions (i.e. for sampling inspection) looks for the quality processes only to find out the optimized quality strategy by the minimum of quality costs. The question is testing or no testing after a technological process dependent on the quality costs of the necessary inspection and repair processes. The next step was the development of an extension of this model to include the production process with its costs and its quality dependent properties into the cost model. The question is: How influences the improvement of a production process by investment the quality and the costs of the products? How much time does it need to reach a better quality level by lower costs? The new extension of the quality cost models gives help to the production engineers to make a decision about the right quality strategy.

Key words: quality cost, production cost, cost optimization, process optimization

Introduction

Controlled technological processes are the most important way to reach a high quality level in mechanical engineering industries. Statistical Process Control (SPC) and Failure Mode Effect Analysis (FMEA) are more methods to reach this goal. It is possible to use these methods also in electronics production in the case of producing batches with a high number of same PCBs. But in a production process with many different products and small batch sizes (low volume, high mix) we expect relatively high defect rates, and some technological processes may be uncontrolled. The objective of this publication is to show mathematical models to decrease the quality costs of such technological processes. The last part of this publication describes the influence of investment in production processes to decrease the defect rate and so to influence the summary costs of an electronic product.

Optimizing the Quality Strategy – The Basic Quality Cost Model

Starting point of our consideration is a chain of technological processes - for instance a SMT production line as in Figure 1. The SMT line consists of PCB storage and transport system (1), solder paste printer (2), placement machines (3), reflow oven (4) and test facility at the end (5).

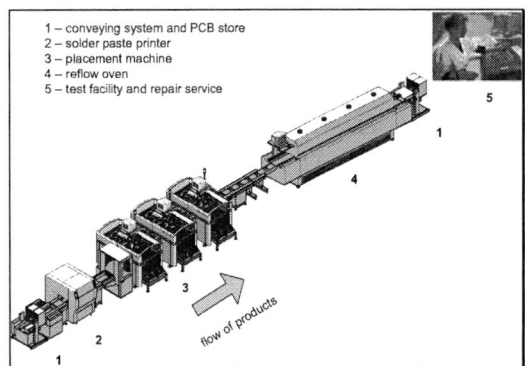

Figure 1: Typical SMT production line

In this example the process of solder paste printing is not well controlled and so there are some poorly printed PCBs. These PCBs are represented by the defect rate p. What is the right way to fix this problem? It is the insertion of a test and a repair process in the line after this uncontrolled technological process. The new strategy is shown in Figure 2. This picture shows the abstraction of the processes too: T indicates a technological process and Q a quality process. A quality process consists of an inspection and a repair process.

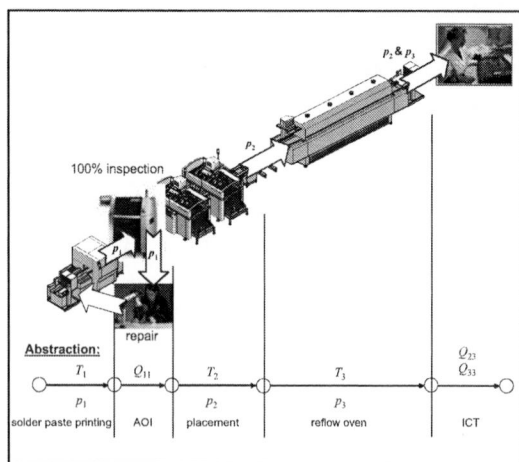

Figure 2: SMT production line with quality process and abstraction

But is it sensible to do this quality process Q in the context of quality costs? The use of quality cost models helps to find the answer. The basic model is described in this chapter. Figure 3 shows the abstraction of one technological process with and without a quality process.

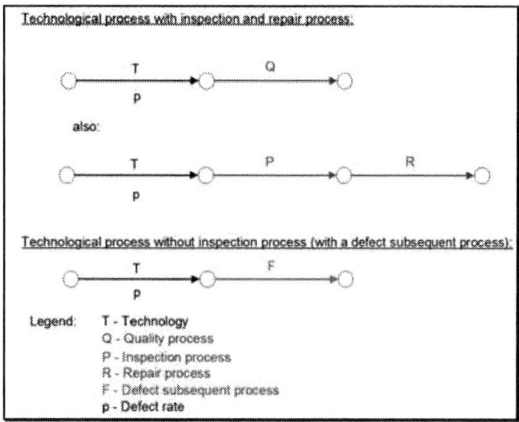

Figure 3: Model for one technological process

The technological process T (see Figure 3) has a defect rate p ($p \neq 0$). In the first case a quality process Q (consisting of the inspection process P with 100% inspection of the PCBs and the repair process R) is added to detect and to eliminate these defects. The inspection and repair processes cause quality costs. In the second case there is no quality process to detect and to remove the defective parts directly after the technological process. But there is also a defect rate p ($p > 0$) and the defects will be "processed" in the next technological processes. It means the defects also cause quality costs, but later in the technological line (latest in the final inspection process). With the view to these quality costs it is possible to define a virtual process called defect subsequent process F. This process describes

the necessary quality process anywhere later in the line. Now it is possible to answer the question: Which case is cheaper – with or without a quality process right after the technological process T? To answer this question the following assumptions are necessary:

- all of the produced PCBs are inspected (100% test)
- all PCBs with defects are identified and repaired
- after the last technological process in the line is a 100% test of the PCBs installed (for example an in-circuit test)
- quality costs are calculated as costs per unit (for example per PCB).

To describe the quality costs of the above discussed two cases the following equations are used:

- costs without an inspection after T:
$$k_0 = k_T + p \cdot k_F \qquad (1)$$
- costs with an inspection after T:
$$k_1 = k_T + k_P + p \cdot k_R \qquad (2)$$

with:

k_T – costs of the technological process T per unit
k_F – defect subsequent costs per unit
k_P – inspection costs per unit
k_R – repair costs per unit
p – defect rate of T.

Of course in this model the technological costs k_T are the same with and without the quality process Q and so they are not a necessary part of the calculation. If the defect subsequent costs k_F are higher than the sum of inspection and repair costs ($k_P + k_R$), a point of intersection of the two lines (the graphs of the equations (1) and (2)) exists. This point (so called break even point) is described by a defect rate p^* and the costs k^*. The equation of p^* is:

$$p^* = \frac{k_P}{k_F - k_R} \qquad (3)$$

with: $\begin{aligned} k_0 = k_1 = k^* \\ p = p^* \end{aligned}$ if $k_F > k_P + k_R$.

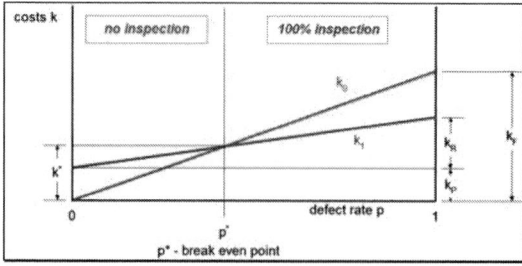

Figure 4: Quality costs with and without quality process

For low values of p (see Figure 4 - left from p^*) the use of a quality process Q at this time is

234

more expensive than to inspect the PCBs later on during the final inspection. With this simple model it is possible to design a technological chain (with technological and quality processes) only dependent on the defect rate p, if the quality costs are known.

The following extensions of the basic quality cost model are published in (e. g.) [1], [2] and (especially) [3]:

- extension for more than one technological process
- extension for the influences of the inspection process (the influence of pseudo defects and defect flow) to the cost
- extension for the use of sampling inspection.

The first practical use of the quality cost models was at the electronics production business unit of Heidelberger Druckmaschinen AG (HDM) in Wiesloch (Germany). Together with our partners we adapted the models and we analyzed the quality data of complex assembled PCBs with relatively high defect rates. After this analysis two products were selected and the quality processes of these boards were optimized. Figure 5 shows the result for one visual inspection step.

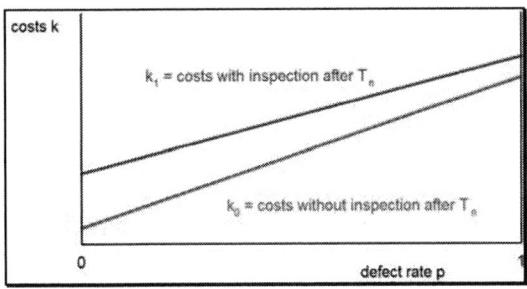

Figure 5: A practical result of using quality cost models

What does Figure 5 mean? For any value of the defect rate of the considered board and technology it is cheaper to have no inspection after the technology T_n than to have it. The reason for this result was a high defect flow of the investigated visual inspection process. After that statement the management renounced the considered inspection process and saved about 2.50 EUR on every board of the considered products. More about is written in [1]. Calculated examples are published in [4], [5] and [6].

But what is the disadvantage of the basic quality cost model and its extensions? These models look for the quality processes only and give recommendations to change these processes. There is no changing of the quality of the technological processes and so no influence to the defect rate p of the processes.

Changing the Defect Rate – The New Extension

Process Improvement by Investment

In the previous discussions, the fluctuations of the defect rate p did not have any effect on the costs k_T of the technological process T because the fluctuation is caused by random processes. If, however, technological, control-technical, organizational and other measures are intentionally used in the technological process with the goal of systematically influencing the defect rate, this causes costs when the defect rate will be reduced.

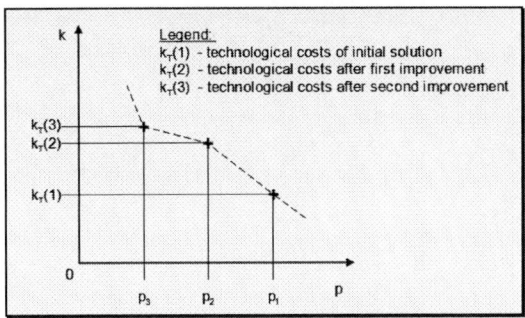

Figure 6: General dependency of the technological costs on the goal defect rate

The specific process costs increase with a reduced goal defect rate (envisaged defect rate). This results from the investment that must be made in order to achieve this goal. Figure 6 shows an example of this dependency. The lower the envisaged defect rate, the higher are the technological costs.

The costs for the quality strategy remain unaffected by the change of the technological costs. The effect of the increase of the technological costs on the complete costs of the process (sum of the technological costs and quality costs for the envisaged defect rate) must now be evaluated. The cost development is shown in Figure 7.

The starting point for the considerations in the example shown in Figure 7 are the technological costs $k_T(1)$ of the initial solution for the technology T for the average defect rate p_1. The quality strategy resulting by this given defect rate has a quality process directly after the technology T ($p^* < p_1$). To achieve a significantly lower defect rate p_2 for the considered process, new equipment must be obtained for the technological process T. This investment leads to an increase of the technological costs per unit by $\Delta k_T(2)$. The associated quality strategy for the defect rate p_2 results from the basic model and means in this case the non-inspection after the technology T (see Figure 7). If the change of the costs for the improved technological process $\Delta k_T(2)$ is now added to the quality costs for defect rate p_2, this produces the cost value $k(2)$. In the example shown in Figure 7, this value is higher than the cost $k(1)$ of the initial solution. This means that in despite of a significant reduction of the defect rate, the summary costs per unit for this technology

235

with the new solution are higher than those for the old solution.

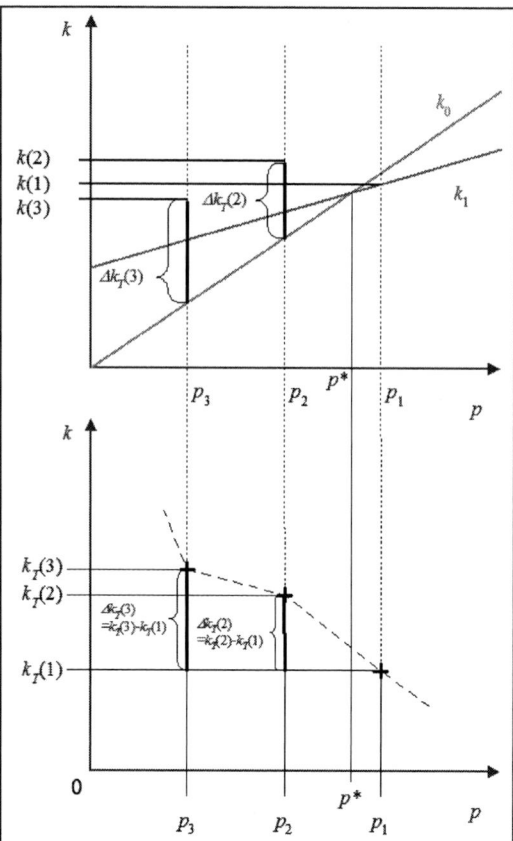

Figure 7: Development costs for the achieved reduction of the defect rate

In practice, it is often the case that another, smaller investment can further positively affect the properties of the technological process in direction of an additional reduction of the defect rate. This case is represented with the defect rate p_3 in the example. In this case, the cost component $\Delta k_T(3)$ for the improvement of the technological process compared with the initial solution must be added to the quality costs. The graphs (Figure 7) show that the resulting costs $k(3)$ are lower than the costs $k(1)$ of the initial solution. The optimization goal of the reduction of the defect rate and the unit costs for this technological process has been achieved.

The example shown in Figure 7 certainly occurs in practice, e.g., for the automatic component placement. An automatic placement machine operates at (or even outside) the limits of its specification, and so it has a relatively high defect rate for the placement of fine-pitch components. An investment is made in a new automatic placement machine that is specified with a higher placement accuracy (high investment → higher technological costs $kT(2)$). The operation shows that the placement of chip components of the size 0201 still has a high

defect rate. The manufacturer of the new automatic placement machine recommends the installation of a vision system with a higher resolution for the inspection and positioning of the components (additional, but lower investment → higher technological costs $kT(3)$). This solution can be used for the placement of the complete component spectrum for a smaller process defect rate. Whether the optimization goal of the reduction of the defect rate and summary costs has been achieved, depends on the actual costs.

The Influence of Time

The model in the last chapter was discussed on the assumption that the effect of the investment in a new machine (the reduction of the defect rate and so the cost) acts immediately. This perfect action shows Figure 8. After the first investment the total costs per unit are a little higher than before. After the second investment the total costs are lower.

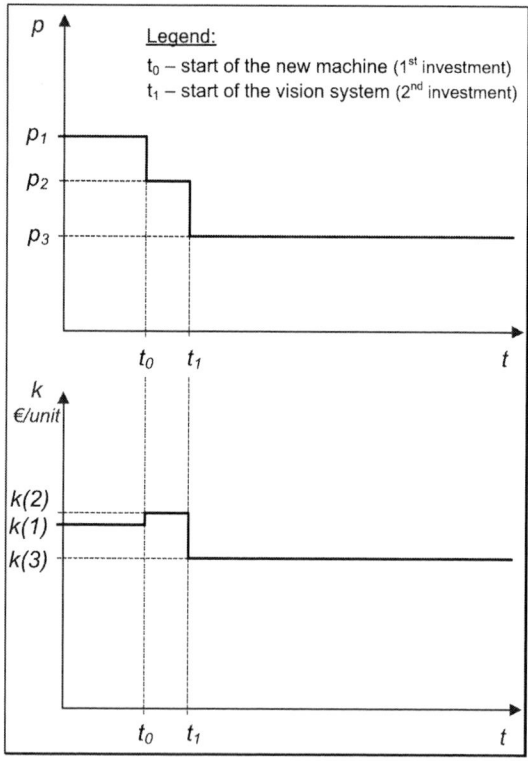

Figure 8: Defect rate and costs over time (ideal)

But in reality there is no step function. The modifications of the defect rate p and the costs k proceed slowly. A practical example for that is the shrinking process in semiconductor production. After the shrinking the yield is lower than before (the defect rate is higher), but after a short time the technological processes become stable and the yield increases. Figure 9 shows a possible behaviour (based on the example from the previous chapter).

The transition range of the costs can be described as

$$\int_{t_0}^{t_0+\Delta t}(k(t)-k(3))dt \triangleq A_k \qquad (4).$$

What is the reason for this behavior? At the time of starting the use of the new and better technology the quality strategy is changed with the view to the expected (lower) defect rate. But the defect rate has not reached this value yet and so the wrong quality strategy is in use and the quality costs are higher than expected.

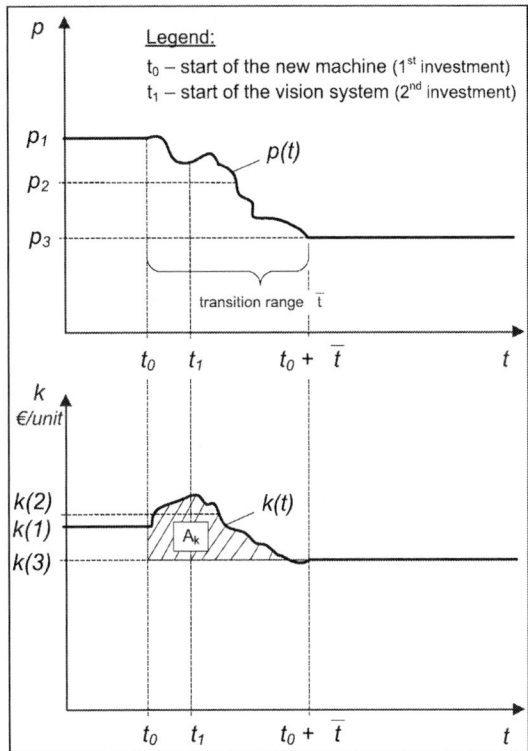

Figure 9: Defect rate and costs over time (real)

Figure 10 shows this connection at time t_1, the time of starting the new vision system as a part of the new machine. This time describes the end of the investment to improve the technological process. Now it needs time to reach the planned defect rate and so the expected lower summary costs per unit. The goal is the minimization of this time to reach the expected costs. This goal is attainable by the persistent use of quality management tools and by training of the employees.

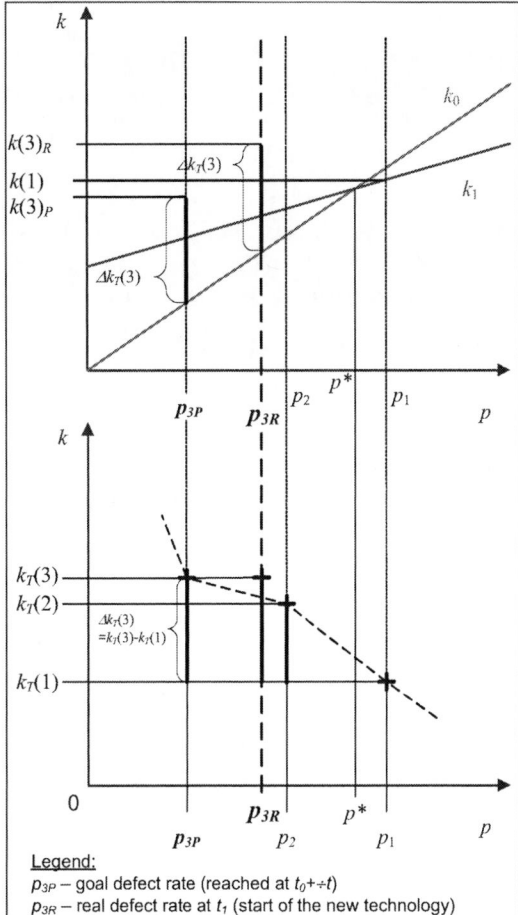

Figure 10: The reason for the higher costs at t_1

Conclusion

This paper presents an overview to realize optimal test strategies for electronics and Microsystems production lines with a view to the costs. It shows the mathematical background of the basic model and the influence of investment in better technologies to the summary costs. The quality cost models are in use in an electronics production. The necessary steps in practice depend on the specific quality of the investigated processes and on the complexity of the investigated products. The improvement of technological processes by investment is a good way to get better products. It is possible to evaluate the summary costs of such a technological process by using the new extension of our models. But it needs time the reach the expected quality and costs. A necessary way to shorten this time is training of the employees and a good quality management system.

References

[1] M. Oppermann, W. Sauer, H. Wohlrabe, T. Zerna, "New Quality Cost Models to Optimize Inspection Strategies", IEEE Transactions on Electronics Packaging Manufacturing, Vol. 26, No. 4, pp. 328-337, a publication of the IEEE CPMT Society, 2003.

[2] M. Oppermann, W. Sauer, H. Wohlrabe, "Optimization of Quality Costs", Robotics and Computer Integrated Manufacturing – An international journal of manufacturing and product and process development, Vol. 19, No. 1-2, pp. 135-140, Elsevier Science Ltd., 2003.

[3] W. Sauer, M. Oppermann, G. Weigert, S. Werner, H. Wohlrabe, K.-J. Wolter, T. Zerna, "Electronics Process Technology – Production Modelling, Simulation and Optimisation", Springer-Verlag, London, Chapter 6, pp.303-347, 2006.

[4] M. Oppermann, "Modellierung und Optimierung des Qualitätsverhaltens von Fertigungsprozessen in der Elektronik", Verlag Dr. Markus A. Detert, Templin, 2002.

[5] G. Linß, *et al*, "Training Qualitätsmanagement", Fachbuchverlag Leipzig im Carl Hanser Verlag, Munich Vienna, Chapter 3.5, pp.125-150, 2003.

[6] G. Linß, *et al*, "Statistiktraining im Qualitätsmanagement", Fachbuchverlag Leipzig im Carl Hanser Verlag, Munich Vienna, Chapter 5.5, pp.367-378, 2006.

Towards Low Cost Coupling Structures for Short-Distance Optical Interconnections

N. Hendrickx[1], J. Van Erps[2], T. Alajoki[3], N. Destouches[4], D. Blanc[4], J. Franc[4], P. Karioja[3], H. Thienpont[2], P. Van Daele[1]

[1]Ghent University, TFCG Microsystems, Dept. of Information Technology (INTEC),
Technologiepark 914A, B-9052 Ghent, Belgium
[2]Vrije Universiteit Brussel, Dept. of Applied Physics and Photonics (TONA-FirW),
Pleinlaan 2, B-1050 Brussels, Belgium
[3]VTT,Kaitoväylä 1, 90570 Oulu, Finland
[3]Laboratoire Hubert Curien, UMR 5516, Université Jean Monnet
18 Rue Benoit Lauras 42000 Saint-Etienne, France
Contact: nina.hendrickx@intec.ugent.be, Phone: +3292645370, Fax: +3292645374

Abstract

The performance of short distance optical interconnections in general relies very strongly on coupling structures, since they will determine the overall efficiency of the system to a large extent. Different configurations can be considered and a variety of manufacturing technologies can be used. We present two different discrete and two different integrated coupling components which can be used to deflect the light beam over 90° and can play a crucial role when integrating optical interconnections in printed circuit boards. The fabrication process of the different coupling structures is discussed and experimental results are shown. The main characteristics of the coupling structures are given. The main advantages and disadvantages of the different components are discussed.

Key words: Coupling Structures, Deep Proton Writing, Laser Ablation, Optical Interconnections, Printed Circuit Board, Diffraction Grating

Introduction

The performance of short distance optical interconnections in general relies very strongly on coupling structures, since they will determine the overall efficiency of the system to a large extent. The integration of the optical interconnection to the board-level is done with the use of a polymer optical layer. The optical layer contains multimode optical waveguides, which guide the light in the plane of the optical layer. The availability of VCSELs and photo-detectors at the targeted wavelength of 850nm, which are placed on top of the optical layer, requires the use of coupling structures. These coupling structures deflect the light beam over 90°, enabling light to be coupled from the VCSEL towards the optical waveguide or from the waveguide towards the photo-detector. Different configurations can be considered to obtain the 90° beam deflection. We will discuss two integrated and two discrete coupling structures. The main advantage of the integrated structures is the fact that the alignment between the waveguides and the coupling component is arranged during the fabrication itself whereas the discrete components have to be inserted into the optical layer and require passive or active alignment for placing the component at the right position. The main disadvantage of the integrated structures is the fact that they have to be compatible with the entire fabrication process, which may include elevated pressures and temperatures, whereas the discrete ones can be fabricated in a separate step and inserted into the optical layer in a later phase. In our case, active alignment is used to arrange the alignment between the couplers and the waveguides. In the next sections, the different coupling structures are discussed. The fabrication process is described and experimental realizations are shown. The main characteristics are also given. The different components are finally compared in an objective way, giving the main advantages and disadvantages.

Laser ablated micro-mirrors

Laser ablation is a versatile micro-structuring technology that can be used to pattern the main building blocks of the optical interconnection into the polymer optical layer [1]. The ablation set-up available at Ghent University contains three different laser sources: a KrF excimer (248nm), frequency tripled Nd-YAG (355nm) and a CO_2 (9.6μm) laser. The excimer laser beam can be tilted with respect to the sample, which eases the fabrication of angled features. During the processing, the sample is placed on a computer-controlled translation stage which has an accuracy of 1μm.

The laser beam is send through an optical projection system and projected onto the sample. The photon energy is absorbed by the polymer material. As soon as the photon density inside the material exceeds a certain threshold, the photon energy can be used for material decomposition. This includes both photo-thermal and photo-chemical processes and results in the ejection of the decomposed material in the form of an ablation plume. This plume contains both evaporated and non-evaporated particles. The solid particles will fall back to the surface and cause debris, which is off course highly undesirable since it increases the surface roughness of the ablated area. Polymer materials typically show a high absorption in the UV-range allowing for an ablation with a low deposition of debris. The optical layer is thus patterned by physically removing material.

The material used for the optical layer, Truemode Backplane™ Polymer, is highly cross-linked acrylate-based polymer which has excellent optical and thermal properties. The optical layer consists of a cladding-core-cladding stack, where the cladding material has a slightly lower refractive index than the core material. The light can in this way be trapped inside the core layer by means of total internal reflection (TIR). Multimode waveguides are used to guide the light in the plane of the optical layer. The waveguide cores are ablated into the core layer with the KrF excimer laser. Material is removed on both sides of the resulting waveguide core. The waveguides have a cross-section of 50μm×50μm and are on a pitch of 125μm. The propagation loss of the ablated waveguides is 0.12dB/cm at 850nm [2]. A cross-section of an array of ablated waveguides is given in Fig. 1.

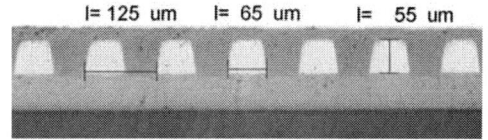

Figure 1: cross-section of an array of laser ablated multimode waveguides.

45° micro-mirrors are good candidates for the 90° beam deflection. They are wavelength independent, highly reproducible and can be fabricated with a number of technologies. These 45° micro-mirrors can be ablated into the optical layer with use of the tilted KrF excimer laser beam. There is always a certain degree of tapering during the ablation, which can be measured experimentally and corrected for.

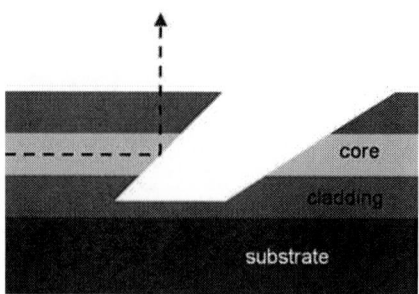

Figure 2: schematic showing the principle of the TIR mirror.

The ablated trench contains two interfaces which can both be used to deflect the light beam over 90°. Because of the tapering, only one of the two interfaces will have a 45° angle. In case the core-air interface has a 45° angle, the beam deflection is based on TIR at the polymer-air interface. In the case of the air-core interface two additional processing steps are required. In order to function as a mirror, the facet has to be metal coated and the trench has to be filled with cladding material. We present the results on the TIR mirrors. The schematic is shown in fig. 2.

The surface roughness of this facet can not be measured because of the fact that it is enclosed. The average RMS surface roughness of the other interface is 53nm on a scan area of 52μm×174μm measured with an optical profiler (Wyko NT3000).

Measurements have been carried out on a demo board which contains an array of multimode optical waveguides, integrated on the PCB, and an ablated TIR mirror at a wavelength of 850nm. The light is coupled into the optical waveguides with a multimode optical fiber with a core diameter of 50μm and a numerical aperture (NA) of 0.2. The light that is coupled out vertically at the output facet is detected with a multimode optical fiber with core diameter 100μm and NA 0.29. The Truemode waveguides have NA 0.3. The TIR is not coated with a thin Au layer; this could however increase the coupling efficiency. The first measurement results give an average coupling loss of 3.6dB for the entire link (coupling into the Truemode waveguides, propagation through the waveguide, 90° deflection at the 45° facet and outcoupling at the output facet towards the fiber). The vertical distance between the output facet and the fiber has however not been optimized. Optimization of the ablation parameters (pulse energy and repetition rate) and the use of a Au coating could however improve the efficiency and will be investigated in the near future.

Resonant grating coupler

The solution proposed by Laboratoire Hubert Curien is the use of a resonant grating structure at the bottom of the guiding layer as coupling element, in the place of a 45° micro-mirror. Such a structure can theoretically diffract more than 90% of the incident light in the 1st diffraction order [3], which angle can be adjusted to couple light into the optical waveguide. The resonance effect is obtained by combining a metallic diffraction grating and a thin dielectric layer [4] which refractive index is higher than the one of the covering multimode waveguide. The whole component is made thanks to standard planar technologies.

The component includes two resonant gratings (Fig. 3), 3 cm apart, to couple light respectively towards and from the optical waveguide. The grating corrugation in the pyrex substrate is obtained by first exposing a 300 nm thick positive resist to an interferogram at the HeCd laser wavelength of 442 nm, then by reactive ion beam etching into the pyrex surface. The two gratings are made simultaneously and are rigorously parallel. Then a 150 nm thick gold layer is evaporated on the corrugated parts of the substrate in order to create the metallic reflecting grating and a 176 nm thick HfO$_2$ layer is deposited by sputtering on the gold layer to complete the resonant structure. Finally the whole pyrex plate is dip-coated with a 50 µm PMMA layer which forms the multimode waveguide.

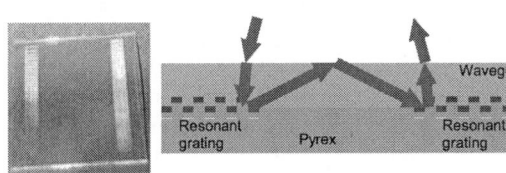

Figure 3: Left: top view of the 5x5mm² pyrex plate on which two resonant gratings are etched (colored bands) to couple the light to and out the waveguide deposited on the whole plate. Right: sketch of the cross-section of the component with two coupling gratings.

The angular width of the designed resonance is few degrees large and allows maintenance of high diffraction efficiency with a focused beam. The measurement set-up used for the optical characterization of the component is sketched on Fig. 4. The laser wavelength is 850 nm, the coupling grating is placed in the focal plane of a lens of 60 mm focal length and the main detector measures the intensity of the light diffracted by the out coupling grating. A reference detector records simultaneously the fluctuations of the incident light. A rotation stage not drawn on the sketch allows us to vary the incidence angle of the focused beam on the coupling grating. In such a configuration the overall efficiency of the component, ie the ratio of the out coupled power and the incident power, reaches more than 60% for incidence angles in the range [-3°; -3.5°]. This means that the overall losses, which include the light reflected at the air-waveguide interfaces, the light absorbed by the metal layer, the light scattered during the propagation in the

multimode waveguide and on the diffracting structures, do not exceed 2.2 dB.

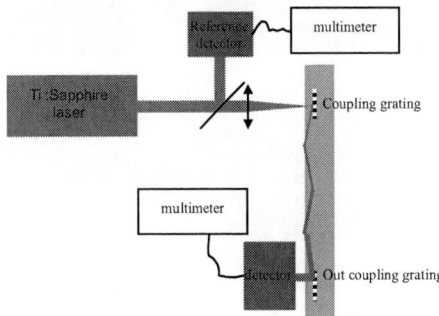

Figure 4: measurement set-up used to characterize the component.

Some remarks have to be add to show the versatility of such a grating coupler. It could also be used to couple free space light towards channel waveguides perpendicular to the grating lines. In order to simplify the making process of the grating, the latter can be made without etching by using a photosensitive hybrid sol-gel thin film deposited on the pyrex substrate [5].

Pluggable out-of-plane coupler

At the Vrije Universiteit Brussel, we investigate the use of a pluggable micro-optical component to couple the light to or from PCB-integrated waveguides by inserting it into a micro-cavity fabricated in the board. The insert contains a 45° micro-mirror which is used to deflect the light beam over 90° and which can be Au-coated to increase the coupling efficiency. For the fabrication of the pluggable out-of-plane coupler, we make use of our in-house rapid prototyping technology of Deep Proton Writing (DPW) [6]. It consists of the following processing steps, as illustrated in Fig. 5. First a collimated 8.3MeV proton beam is used to irradiate an optical grade PMMA sample according to a predefined pattern by translating the PMMA sample, changing the physical and chemical properties of the material in the irradiated zones. As a next step, a selective etching solvent is applied for the development of the irradiated regions. This allows for the fabrication of (2D) arrays of micro-holes, optically flat micro-mirrors and micro-prisms, as well as alignment features and mechanical support structures. On the other hand, an organic monomer vapour can be used to expand the volume of the bombarded zones through an in-diffusion process. This enables the fabrication of spherical (or cylindrical) micro-lenses with well defined heights. If necessary, both processes can be applied to different regions of the same sample, yielding micro-optical structures combined with monolithically integrated micro-lenses.

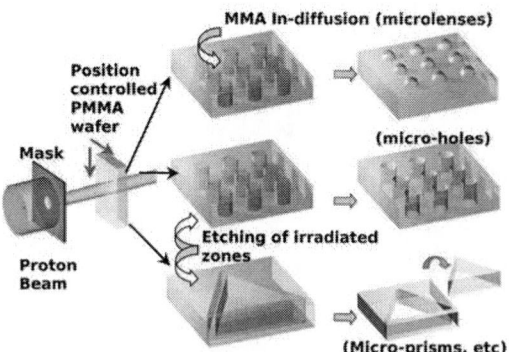

Figure 5: Deep Proton Writing: basic processing steps. Proton beam exposure of PMMA is followed by selective etching and/or swelling

We use high molecular weight PMMA with a thickness of 500μm, which allows the 8.3MeV protons to completely traverse the sample. During the irradiation step, the PMMA sample is semi-continuously translated perpendicularly to the beam in steps of 500nm using high-precision translation stages with an accuracy of 50nm, according to the pre-defined pattern shown in the bottom part of Fig. 6. Optimal surface roughness results are obtained by using a proton dose of 50pC per step of 500nm, with a proton current of 160pA. This current is monitored by measuring the charge that the protons induce on a target located directly behind the sample. The deposited dose can then be determined by integrating this proton current during the irradiation using a precision-switched integrator trans-impedance amplifier that aims at compensating any current fluctuations of the proton source.

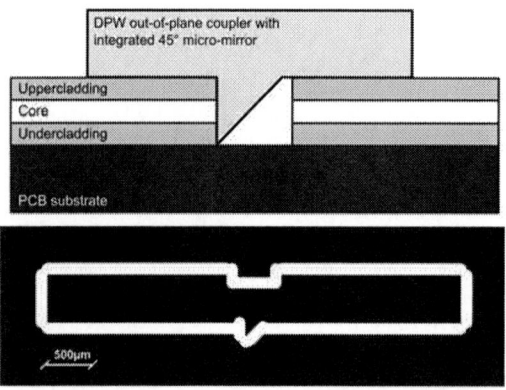

Figure 6: Schematic operation principle (top) and design lay-out (bottom) of a pluggable out-of-plane coupling component

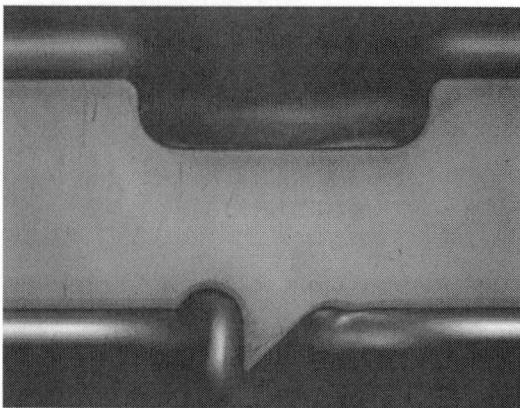

Figure 7: Fabricated out-of-plane coupling component with integrated 45° micro-mirror

After the exposure step, the irradiated zones can be selectively etched, resulting in micro-components with high-quality optical surfaces. An optical microscope picture of a fabricated pluggable out-of-plane coupler is given in Fig. 7.

It is obvious however that DPW is not a mass fabrication technique as such. However, one of its assets is that, once the master component has been prototyped with DPW, a metal mould can be generated from the master by applying electroplating. After removal of the plastic master, this metal mould can be used as a shim in a final micro-injection moulding or hot embossing step [7]. This way, the component can be mass-produced at low cost in a wide variety of high-tech plastics.

For the characterization of the critical optical surfaces of the component, namely the flat top exit facet and the 45° mirror facet, we use a WYKO NT-2000 non-contact optical surface profiler (Veeco). Since the entrance facet is not accessible with the microscope objective, this surface was not measured, but its surface roughness will be analogous to the two others. The surface roughness analysis reveals that the flat top part has an average local RMS surface roughness Rq of 14.1nm ± 2.7nm measured over an area of 60μm by 46μm. We averaged at least 5 measurements of randomly chosen positions. Applying the same measurement method to the 45° angled facet reveals an RMS roughness of 17.1nm ± 4.2nm. The flatness R_t or peak-to-valley difference along the depth of 500μm of the component is measured to be smaller than 2.5μm. This is due to the scattering of the protons during the interaction with the PMMA. As a conclusion, we can say that our developed DPW surfaces have a very good and reproducible optical quality: almost flat and a very low RMS roughness.

Measurements have been carried out on an optical board, containing Truemode™ optical waveguides and a laser ablated micro-cavity. The out-of-plane coupler is inserted passively into the micro-cavity. Again, 850nm light is coupled into the waveguides using a MMF with core diameter 50μm and NA 0.2 and the power coupled out by the DPW pluggable coupler is detected by means of a MMF with core diameter 100μm and NA 0.29. The coupling efficiency that is measured for the entire link is -5.68dB. It should be noted that when the DPW out-of-plane coupler is measured in a direct fiber-to-fiber coupling scheme, using the same input and output MMF as described above, efficiencies up to -0.72dB are measured.

Another important advantage of the pluggable out-of-plane coupler fabricated with DPW is the fact that micro-lenses can be monolithically integrated in the component, and that the concept can easily be extended towards multilayer optical waveguide structures integrated on printed circuit boards, as reported in [8] and [9].

Glass micro-mirror

At VTT, micro-mirrors for 90° beam turning were fabricated by grinding and polishing one edge of 100μm thick glass substrate in such as way that a 45° bevel was formed. Several substrates were grinded by stacking them into a jig which made it possible to mount them in a 45° angle precision lapping and polishing machine. After polishing, the substrates were diced with a dicing saw. An aluminum coating layer was evaporated on the glass surfaces for high reflectivity. Also, a thin layer of chrome was evaporated in order to achieve better adhesion of the aluminum coating.

Optical loss measurements have been carried out on an optical board with an embedded glass mirror. The top view of the demoboard is shown in Fig. 8. The measurement set-up is schematically shown in Fig. 9. MMF with core diameter 50μm and NA 0.22 is used to couple light into the multimode optical waveguides. The light beam that is deflected by the glass mirror is detected with a MMF with core diameter 200μm and NA 0.22, which is equipped with a coupling lens. Excess loss caused by the micro-mirror coupling was estimated from the measured losses by subtracting the waveguide loss and the fiber-coupling losses, which were obtained from the measurements done using butt-coupling to waveguides of the same length at both ends. The excess loss calculated thus includes losses due to the quality (roughness, angle) of the mirror, due to refraction and scattering from the under cladding surface, and due to the possibly poorer quality of the lithographically patterned waveguide facet compared to a sawed facet.

Four channels (i.e. waveguides) were measured on two different demo boards, resulting in eight measurement results. The excess losses were between 4.0dB and 6.5dB with an average of 5.1dB [10]. The variation is partly due to the varying quality of the sawed, not polished, waveguide facet at the in-coupling. It should be also pointed out that the excess loss is probably slightly different in the fully assembled demonstrator, since the micro-lens based collecting optics performs differently than the

fiber-coupling optics used in this experiment. It is clear that the edge of the under-cladding layer blocks part of the optical beam between the waveguide end facet and the micro-mirror. This part of the optical power is lost due to scattering and reflections to wrong directions, thus significantly increasing the total path loss.

The average RMS surface roughness of the mirror facet, measured on a scan area of 59μm×45μm, is 42.3nm. The coupling efficiency may be improved by placing the mirror into a micro-cavity in the optical layer. The distance between the output facet of the waveguides and the mirror facet can in this way be decreased. Tests where the mirror is placed in the same demo board as the pluggable out-of-plane coupler will be carried out in the near future.

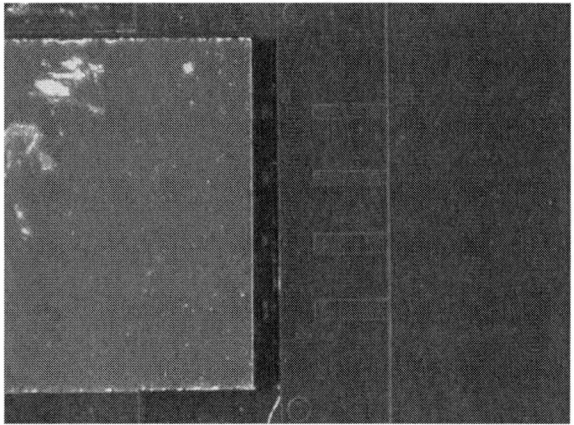

Figure 8: top view of the demo board with the inserted glass mirror.

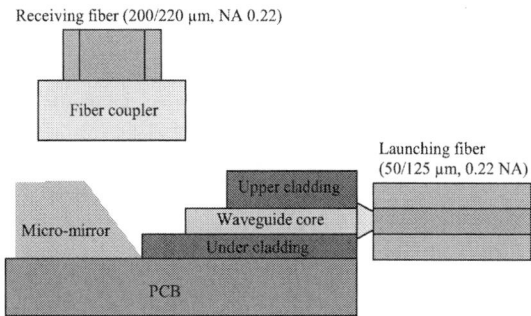

Figure 9: measurement set-up used to characterize the excess loss.

Conclusion

Main advantage of the ablated micro-mirrors is the fact that they are integrated with the optical waveguides. We can in this way achieve high alignment accuracies with use of passive alignment. The performance of these mirrors can be improved by optimizing the ablation parameters and with the application of a Au coating on the mirror facet. This way, the trench could be filled with cladding material to protect the mirror from dust and moist intrusion.

The advantage of the grating coupler is the very high coupling efficiency that can theoretically be achieved. Drawback is the fact that the device is wavelength dependent, and the strong requirements on the input side (input angle, low beam divergence).

The pluggable out-of-plane coupler is very versatile because it can easily be extended towards multilayer structures, where multiple optical layers are stacked. It is also possible to monolithically integrate micro-lenses which can ease the alignment tolerances. Although PMMA is not suitable for standard PCB processing, the DPW prototypes are compatible with low-cost mass-fabrication techniques in high-tech plastics capable of withstanding high temperatures.

The glass mirrors have excellent optical and thermal properties and are fully compatible with standard PCB manufacturing and soldering processes. We believe that the efficiency can be improved by putting the output facet of the waveguides closer to the mirror facet.

Acknowledgements

Nina Hendrickx would like to acknowledge the Institute for the Promotion of Innovation by Science and Technology in Flanders (IWT Flanders) for financial support. Jürgen Van Erps acknowledges the FWO for financial support. This work was carried out within the framework of the network of excellence on micro-optics (NEMO), supported by the European Commission through FP6 program.

References

[1] G. Van Steenberge, P. Geerinck, S. Van Put, J. Van Koetsem, H. Ottevaere, D. Morlion, H. Thienpont, P. Van Daele, "MT-Compatible Laser-Ablated Interconnections for Optical Printed Circuit Boards", Journal of Lightwave Technology, Vol. 22, No. 9, pp. 2083-2090, September, 2004.

[2] G. Van Steenberge, N. Hendrickx, E. Bosman, J. Van Erps, H. Thienpont, and P. Van Daele, "Laser Ablation of Parallel Optical Interconnect Waveguides", Photonics Technology Letters, Vol. 18, no. 9, May, 2006.

[3] J. Franc, N. Destouches, D. Blanc, J.-C. Pommier, S. Tonchev, G. Van Steenberge, N. Hendrickx, A. Last, O. Parriaux "High-efficiency diffraction grating coupler for multimode optical interconnect", Proc. SPIE Vol. 6185, 61851F (Apr. 21, 2006) Photonics Europe

[4] A.V. Tishchenko and V.A. Sychugov, "High grating efficiency by energy accumulation in a leaky mode", Optical and Quantunm Electronics 32, 1027-1031, 2000.

[5] D. Blanc, S. Pelissier, K. Saravanamuttu, S.I. Najafi, M.P. Andrews, Adv. Mater. 11, 1508–1511, 1999.

[6] C. Debaes et al., "Deep proton writing: a rapid prototyping polymer micro-fabrication tool for micro-optical modules", New Journal of Physics, vol. 8, 270, 2006.

[7] M. Heckele and W.K. Schomburg, "Review on micro molding of thermoplastic polymers", Journal of Micromechics and Microengineering, vol.14, pp. R1-R14, 2004.

[8] J. Van Erps, N. Hendrickx, C. Debaes, P. Van Daele and H. Thienpont, "Prototyping of pluggable out-of-plane coupling components for multilayer board-level optical interconnections", Poster session 1, EMPC 2007.

[9] N. Hendrickx, J. Van Erps, H. Thienpont and P. Van Daele, "Intra-plane coupler for PCB-integrated multilayer optical interconnects", Poster session 1, EMPC 2007.

[10] M. Karppinen et al., "Parallel Optical Interconnect between Ceramic BGA Packages on FR4 board using Embedded Waveguides and Passive Optical Alignments" Proc. of 56st Electronic Components & Technology Conf. (ECTC), pp. 799-805, 2006

Surface mounted coupling elements for PCB embedded optical interconnects

Thomas Kühner, Marc Schneider

Forschungszentrum Karlsruhe GmbH, Institute of Data Processing and Electronics
Hermann-von-Helmholtz-Platz 1, D-76344 Eggenstein-Leopoldshafen, Germany

Phone: +49 7247 828757, Fax: +49 7247 825594, marc.schneider@ipe.fzk.de

Abstract

Optical interconnects at printed circuit board (PCB) level facilitate improvements regarding data rates, packing density and electromagnetic compatibility (EMC) not only for telecom applications, but also for data acquisition systems and sensor applications. Preconditions for the successful implementation are a low attenuation of the optical waveguides, the compatibility of the optical layer to the established processes of the printed circuit board manufacturers and the feasibility of automated placing of the electro-optical coupling components onto the printed circuit board. In this paper we present a board level optical interconnect based on embedded standard multimode glass fibers and novel surface mounted coupling elements for passive alignment. The coupling elements combine both the optoelectronic component (VCSEL or pin-diode) and a tailored microstructure which supports the solely passive alignment of the optoelectronic chip to the glass fiber.

Key words: optical interconnect, printed circuit board, coupling element, passive alignment, multimode fiber

Introduction

Integration of optical interconnects into printed circuit boards becomes a major issue when very high data rates or a large number of data channels are required. Furthermore optical signal transmission provides good means to reduce electromagnetic interference (EMI), not only important for mixed-signal systems, where high accuracy analog electronics for sensor applications meets high speed digital electronics for data acquisition and signal processing.

The success of integration of optical interconnects into printed circuit boards (PCB) will be largely determined by the compatibility of the fabrication steps and material properties of the waveguides with the established manufacturing and assembly techniques of printed circuit board industry. The availability of coupling elements for coupling the optical signals into and out of the waveguides is important as well. Most suitable will be elements, which can be mounted by common pick-and-place machines without active alignment.

Numerous approaches focus on the integration of waveguides using structured polymer layers [1]. Main topics are the thermal and mechanical stability of the used polymers against the process conditions at the PCB fabrication as well as the optical attenuation of the polymers to realize suitable waveguides for distances up to two meters. The methods of waveguide coupling sometimes show a high complexity of the coupling components using mirrors and focusing optics. Due to the close tolerances and inexact position of the waveguide within the PCB, the mounting of the coupling elements may require an active alignment.

To reduce such problems we developed a system that uses standard multimode glass fibers to be integrated into the layer stack of a PCB. Because of the high thermal and mechanical stability of glass fibers the proven processing steps of the PCB manufacturer can be applied to embed the fibers into the multilayer PCB. In addition to the very low optical attenuation the geometrical accuracy of the glass fiber is of great importance for the used coupling method. The active optoelectronic components (VCSEL, pin-photodiode) are directly adjusted at the fiber stubs accessible by cut-outs in the PCB. This passive alignment is achieved by a specific microstructure combined with the optoelectronic chips. No mirrors or lenses are needed for waveguide coupling.

Optical Printed Circuit Board

The standard lamination process of printed circuit boards will be done under pressure of up to 15 bar at a temperature of up to 180 °C. Glass fibers have the favorable characteristics to withstand such conditions without damage or decline of their optical transmission properties. In addition glass fibers are very accurate to contour and also inexpensive.

The fibers are embedded between the top and bottom layers of the printed circuit board. These layers may consist of several separate layers. Figure 1 shows the cross section of a PCB embedded fiber.

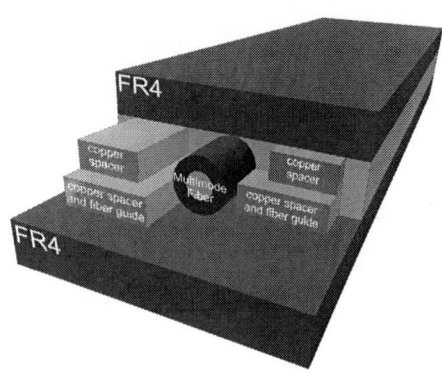

Figure 1: Cross section of printed circuit board with embedded glass fiber.

To ensure reliable results when integrating the optical fibers into the layer stack the following processing steps have to be carried out.

Each single layer of the layer stack has to be processed before the insertion of the fibers. As the end facets of the fibers are accessed through openings within the PCB, all layers have to be milled accordingly. Along the path of the fiber copper pads are placed onto the top and bottom layers to guide the fiber and to ensure the proper distance between the layers during the lamination process. Areas around and between the fibers are leveled with a prepreg layer to obtain a plane surface of the multilayer.

To fabricate optical quality end facets the fiber is cleaved using a conventional fiber cleaver. Then the fiber is laid between the copper pads of the bottom PCB layer and fixed by epoxy resin. After that the layers and the prepreg are stacked together and laminated using a standard lamination process.

Figure 2: Prototype board of optical PCB with embedded multimode glass fibers.

Electrical tracks, bores, throughplatings, and solder stop are fabricated subsequently. At these processes care must be taken not to damage the fiber

stubs in the openings. Figure 2 shows a prototype board with embedded glass fibers, throughplatings, and electrical tracks for contacting the transmitter and receiver electronics.

The openings are quite large with 4×5 mm² for easy inspection of the placed coupling element. Figure 3 shows a close-up view of the openings with single glass fiber stubs. In a final board the openings will be much smaller, even if they house multiple fibers in parallel.

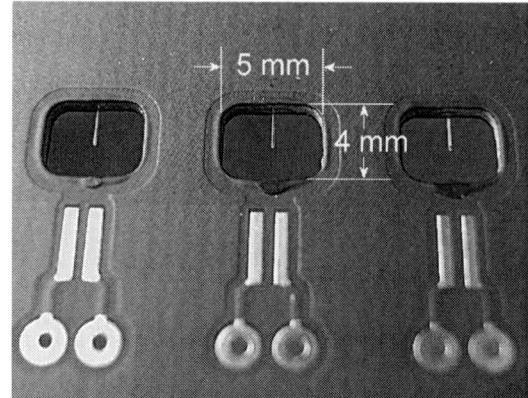

Figure 3: Close-up view of openings with single glass fiber stubs.

An alternative process to achieve optical quality facets at the embedded fibers is to cleave the fiber in the PCB after the PCB fabrication. This will reduce the risk of damaging the fiber stub or end facets during the PCB fabrication processes. For this the glass fibers cross the opening and are scribed either by a diamond knife or a laser and subsequently cleaved [2]. This process has to be further developed to get reliable results.

Surface Mounted Coupling Elements

Coupling elements are needed to couple light into and out of the fiber stub. Our approach originates from the demand that the mounting of the optoelectronic transmitters and receivers has to fit into standard pick-and-place processes as far as possible. Compared to waveguides in polymer foils the embedded glass fibers facilitate the adjustment of the optoelectronic transmitter and receiver components by reason of their contour accuracy. Therefore we developed novel optoelectronic coupling elements, which enable a solely passive alignment through an advanced microstructure to achieve low coupling losses without active position optimization.

Figure 4 shows a schematic diagram of the coupling element. The optoelectronic chip (VCSEL or pin-photodiode) is bonded onto a substrate made of standard PCB material with electrical tracks and leads. The dimensions of the chip depend on its type and are between $250 \times 250 \times 150$ µm³ and

$250 \times 480 \times 150~\mu m^3$, the substrate dimensions are $2.54 \times 3.00 \times 0.81$ mm^3. To achieve a correct adjustment of the coupling element, it has an optically clear alignment structure with a large V-groove by which it is placed on the free standing fiber stub. The alignment structure covers the optoelectronic chip and the gap between both is filled with optical UV-curable glue. Thus the chip is encapsulated and protected from environmental influences. To achieve low coupling losses, the emitting area of the VCSEL or the sensitive area of the photodiode has to be placed with very tight tolerances in the range of a few micrometers relative to the V-groove. Thereby the large V-groove enables a coarse adjustment of the component on the fiber with a tolerance of about 100 µm. The final passive self-alignment of the active chip surface in front of the fiber's end facet is better than 10 µm.

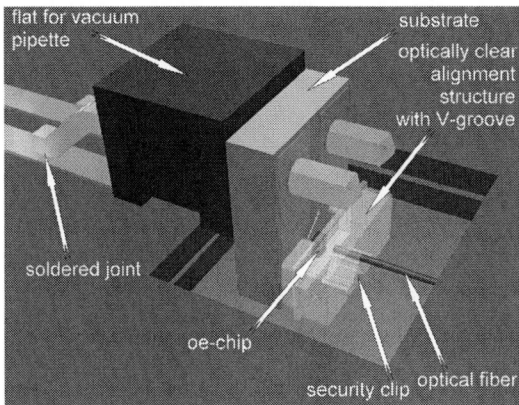

Figure 4: Schematic of mounted coupling element

The mechanical support of the coupling element is improved by the leads for the electrical contacts. The leads are partly in a plastic block with a flat top so that the coupling elements can be picked up by a vacuum pipette of a common pick-and-place machine. The correct position of the fiber in the V-groove is ensured by a security clip, which is clipped into appropriate mounting structures. Further application of optical glue helps to improve the long term durability of the connection and to reduce reflections on the interfaces.

The alignment structure of the prototype coupling element consists of three PMMA parts. Figure 5 a) shows the parts: a frame, which acts as spacer and first part of the chip encapsulation, a foil, which acts as second part of the encapsulation and builds an optically clear window for the fiber end facet and the V-groove part for the fiber alignment. Additionally the security clip is shown, which holds the fiber in place. All parts, except the foil, are fabricated with deep X-ray lithography (LIGA) [3]. After joining the parts the complete alignment structure results as shown in Figure 5 b).

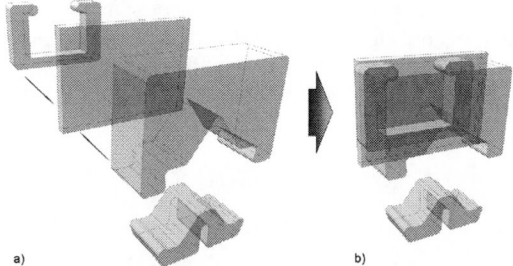

Figure 5: Design of LIGA alignment structure

This alignment structure is on the one hand quite complicated to fabricate and on the other hand not temperature stable due to the used PMMA. Therefore we use a centrifugal casting process with silicone molds to replicate a master made from PMMA. The casting material must meet several conditions. It has to be optically clear in the 850 nm region, temperature stable for soldering, dimensionally stable while curing and soldering, and fast curable. It must not adhere to or diffuse into the mold or corrode or swell its surface. After a number of experiments we chose an UV curable optical encapsulant for semiconductor devices. Using the casting technique, a larger number of alignment structures can be fabricated in limited time, therefore suitable to fabricate a larger number of prototypes to test the coupling element in real-life.

A fully functional device with LIGA alignment structure mounted on an embedded fiber of the optical PCB is shown in Figure 6.

Figure 6: Coupling element mounted on PCB embedded fiber

For future requirements of multiple parallel optical interconnects, the concept for the optoelectronic coupling element can be extended to multiple emitters and receivers using VCSEL- and photodiode-arrays.

Figure 7 shows a sketch of such an element for four parallel optical links. The apex of the V-groove is replaced by a saw-like structure with dents for every fiber. The sides of the groove provide an initial coarse positioning of the element on the outer fibers and the saw-like structure the final fine positioning. Bidirectional coupling elements with VCSELs and photodiodes combined are also possible in this concept.

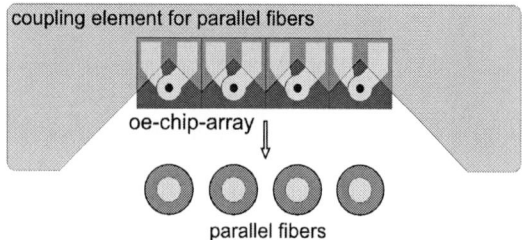

Figure 7: Coupling element for multiple parallel fibers

Test Environment

Within the coupling elements single VCSEL and pin-photodiode chips are applied as transmitter and receiver respectively. Both types use the same substrate and alignment structure design.

For a start we aim at data rates of up to 2.5 Gbit/s. The data rate is not limited by the concept in principle but by currently available low cost components for the back end electronics. The short multimode fibers used here are able to transmit more than ten gigabits per second at one wavelength channel. In fact the speed limiting factors are electronics and VCSELs.

To proof the quality of the embedded fibers in our test boards and to verify the coupling concept we use a modular system to drive and evaluate the optoelectronic coupling elements. This allows testing of all optical paths within the test boards without the need to fully equip each channel with the required electronics for data transmission.

The restriction of the aimed data rate to 1 ... 2.5 Gbit/s offers several advantages. Various semiconductor manufacturers offer integrated VCSEL-drivers as well as transimpedance and limiting amplifiers for the pin-diode receiver in this speed range at a reasonable price. The choice of packaged devices not only of bare chips enables the production of low-priced SMT-modules. Further-more the demands on PCB-Layout concerning high speed board design can be met with reasonable efforts. Data to and from the modules are transferred via differential links with standard signaling (PECL, CML).

The optical and electrical paths of the test boards are designed in such a way that both the modules for driving the VCSELs and for readout of the pin-diode can be mounted on each side of the waveguide. Whereas the coupling elements have to be fixed for all waveguides, the electronic modules can be plugged from one channel to be tested to the next. Figure 8 shows the optical PCB equipped with components for one optical link.

Figure 8: Optical PCB with one equipped optical link.

A simple test setup consists of a VCSEL-driver module controlled by a clock or pattern generator, the optical waveguide with transmitting and receiving coupling elements, and a pin-diode-amplifier module connected to an oscilloscope or to a digital signal analyzer (DSA).

To expand our test facilities and to aim for further applications we designed a universal test bed based on a current Gigabit Transceiver. This chip combines two independent 1.25 Gbit/s Serializer/ Deserializer (SerDes) channels, internal 8B/10B encoder/decoder, clock recovery, and parallel data interfaces.

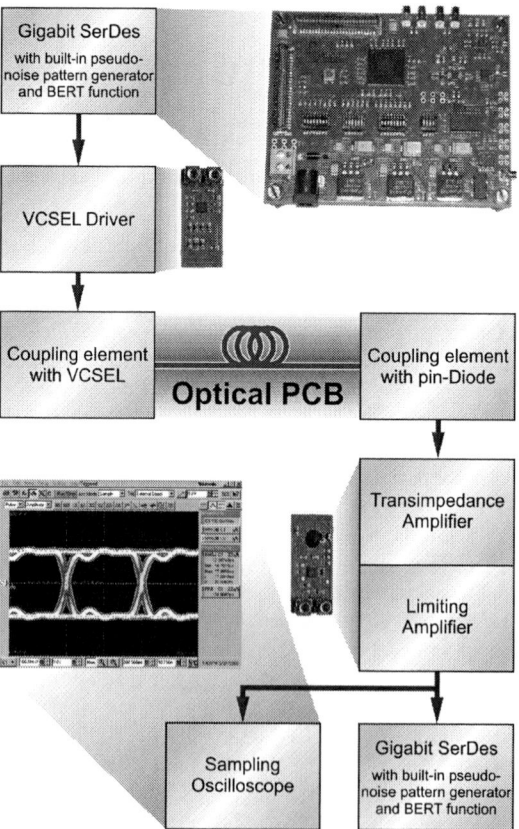

Figure 9: Test setup with Gigabit SerDes Board

The differential data links are routed to MMCX sockets to connect the VCSEL-driver and the pin-diode amplifier modules by coaxial cables. In addition the transceiver chip has an integrated pseudo-noise pattern generator and a built-in bit error rate test (BERT) function. Thus the test board can be used as a stand-alone test system for data-eye signal generation and observation with a sampling oscilloscope and for bit error rate testing of the electro-optical link. The test setup is shown in Figure 9.

Characterization

To specify the quality of the optical links we used the setup as described above, testing the fibers one after the other applying the same coupling elements and driver/receiver modules.

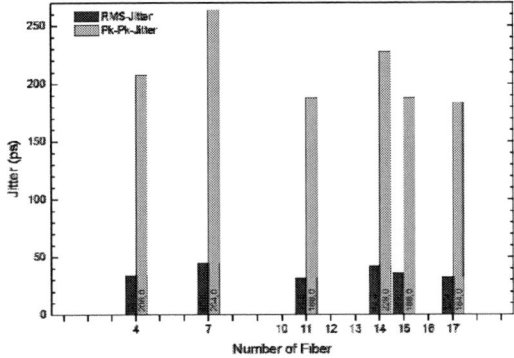

Figure 10: Peak- and RMS-jitter of tested links

The coupling elements are solely passive aligned, the data rate is 1.25 Gbit/s. Peak-to-peak- and RMS-jitter of six tested links are resumed in Figure 10.

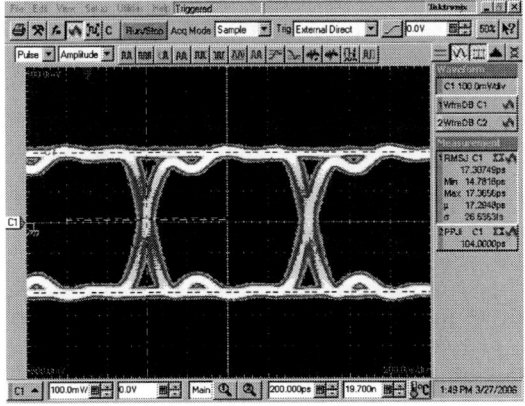

Figure 11: Measurement result of optical link with soldered driver/receiver modules

The chart shows resembling results for all tested channels. The corresponding clear open eye-diagrams indicate reliable data transmission via the embedded optical links.

By soldering the driver/amplifier modules onto the optical PCB instead of using board-to-board interconnects, jitter significantly decreases (17 ps RMS), showing the potential of the embedded fiber concept for still higher data rates (Figure 11).

Conclusions

In this paper a concept and evaluation setup of an optical interconnection system, consisting of several optical, optoelectronic, and electronic components has been presented. The optical PCB comprises embedded multimode glass fibers as well as electrical tracks and throughplatings. Light is coupled into and out of the fibers by special coupling elements. These elements include optoelectronic chips and alignment structures for a solely passive alignment. This concept is also expandable to build up multiple parallel optical interconnects. To characterize the embedded optical links, transceiver electronics for a multifunctional test bed has been developed.

Acknowledgements

We would like to thank the Institute of Microstructure Technology (IMT) of Forschungszentrum Karlsruhe for fabricating the LIGA parts.

Parts of this work are done within the European Network of Excellence on Micro-Optics (NEMO).

References

[1] Hyun-Shik Lee *et al.*, "Fabrication of a 2.5 Gbps × 4 channel micro-module for O-PCB application", Microelectronic Engineering 83, 2006, pp 1347-1351.

[2] G. Van Steenberge, P. Geerinck, S. Van Put, J. Watté, H. Ottevaere, H. Thienpont, P. Van Daele, "Laser cleaving of glass fibers and glass fiber arrays", J. Lightwave Technol. 23 (2) 2005, pp 609-614.

[3] J. Mohr, "LIGA - A technology for fabricating microstructures and microsystems", Sensors Mater. 10, 1998, pp 363-373.

Optical Coupling and Optoelectronics Integration Studied on Demonstrator for Optical Interconnects on Board

Teemu Alajoki[1], Nina Hendrickx[2], Jurgen Van Erps[3], Samuel Obi[4], Mikko Karppinen[1], Hugo Thienpont[3], Peter Van Daele[2], Pentti Karioja[1]

[1] VTT Technical Research Centre of Finland, Kaitovayla 1, P.O. Box 1100, FI-90571 Oulu, Finland
teemu.alajoki@vtt.fi, tel. +358 20 722 2149

[2] Ghent University (UGent), TFCG Microsystems, Dept. Information Technology (INTEC), Ghent, Belgium

[3] Vrije Universiteit Brussel (VUB), Dept. of Applied Physics and Photonics (TONA), Brussels, Belgium

[4] Centre Suisse d'Electronique et de Microtechnique (CSEM) SA, Zürich, Switzerland

Abstract

We have studied methods to design and fabricate coupling structures for short distance optical interconnects on printed wiring boards (PWBs). In order to demonstrate the developed technologies, a parallel optical interconnect was integrated on a standard FR4-based PWB by using surface-mounted component packages and board-embedded optical waveguides. The aspect of particular interest in this study was the in-coupling efficiency and alignment tolerances of the transmitter module, since they will determine the overall performance of the system to a large extent. The transmitter was built on multilayer ceramic (LTCC – low-temperature co-fired ceramics) substrate, which was then surface mounted onto the O/E-PWB using ball grid arrays (BGA). A 4-channel VCSEL array was used as source. In order to reduce the farfield divergence, the VCSEL had integrated microlenses UV-moulded directly on top of the VCSEL emitting area made in a wafer-scale process. Optical waveguides were made of optical polymer material and patterned by laser ablation. A discrete optical out-of-plane coupler component was realized by deep proton writing (DPW) technology and inserted into a laser ablated cavity to bend the light beam by 90° and to couple it to the waveguide. The results of the alignment tolerance and coupling efficiency measurements of the coupling structure are presented.

Key words: optical interconnect, micro-optics, LTCC, polymer optical waveguide

Introduction

With ever-increasing processor clock frequencies and telecom data rates the electrical interconnects inside equipment are becoming a bottleneck. The performance of electrical interconnects has many physical limitations, for instance, due to dispersion, electromagnetic emission, and susceptibility against electromagnetic radiation. Optical transmission with high bandwidth, low loss and cross-talk, insensitivity to EMC/EMI and low heat dissipation could offer several advantages in off-chip I/O connections of fine-pitch packages and chip-to-chip interconnects.

In recent years, a number of optical chip-to-chip interconnections has been demonstrated, e.g. [1,2]. At the board-level, embedded multimode waveguides with similar physical topography and dimensions to electrical strip-lines can be regarded as the most promising approach. Polymeric waveguides as additional optical layer in PCBs have the potential of fulfilling integration and compatibility requirements.

Background – Evolution of Optical Interconnection Demonstrators at VTT

Since 2002, VTT has been running research projects in which we have studied technologies to design and fabricate high-bit-rate optical interconnects on PWBs using board-embedded polymer waveguides and surface-mounted component packages. The work has been carried out with several collaborators, including Helsinki University of Technology to name one. The aim has been that the developed technologies should be compatible with existing PWB technology. One key element in this research has been VTT's in-house LTCC pilot production line. LTCC substrate technology was selected for high-speed optoelectronic module platform thanks to the possibility to implement precision structures for optical alignment directly onto the electronics packaging substrate. LTTC also enables high packaging density due to its multilayer structure and possibility for passive-component integration and bare-chip encapsulation. In addition, good high frequency and thermal properties as well as stability, reliability, and compatibility with hermetic sealing are advantageous for high-speed optoelectronics modules. A review of the LTCC photonics integration at VTT can be found in Ref. [3].

Our first optical interconnection demonstrator, shown in Figure 1, utilized modular structure [4]. The system consists of three separate units: transmitter, receiver, and optical waveguide board. The optical front-ends of the transmitters and the receivers are based on LTCC modules, which were assembled using BGA connection on transmitter and receiver PWBs. The modular approach allows testing and characterization of alignment tolerances with different waveguides and optical coupling structures. Three different coupling concepts were compared.

Figure 1: Modular demonstration platform.

Our second demonstrator was 4x10 Gb/s optical interconnect completely integrated on a standard FR4 PWB [5]. The optical link demonstrator consists of 4-channel BGA-mounted transmitter and receiver modules built on LTCC substrates as well as of four parallel multimode optical waveguides fabricated on the PWB. A photo of the demonstrator is shown in Figure 2.

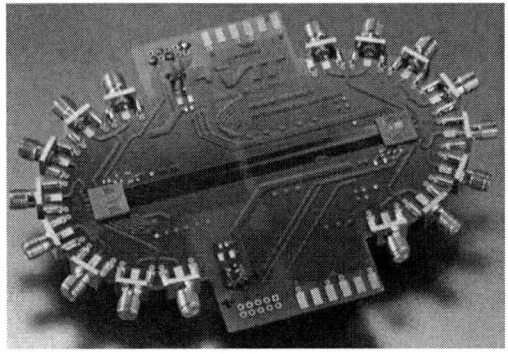

Figure 2: Photo of the 2nd optical interconnect demonstrator on FR4 PWB. Optical waveguide layer is on the PCB between the (blue) transmitter and receiver modules.

The schematic structure of the transmitter and receiver modules and the optical coupling are illustrated in Figure 3. Coupling between the VCSEL/PD array and the waveguide array is based on two microlens arrays and a micromirror. One lens array is mounted into a cavity on the BGA side of the LTCC substrate, whereas the other one and the mirror are mounted on the PCB. With this design, an expanded and collimated beam is obtained between the two microlens arrays, i.e. between the module

and the board; thus, relieving the sensitivity to the potential misalignments due to the BGA board assembly. The characterized transversal tolerances with 1 dB loss margin were ±40...60 μm and total loss of the link estimated to be around 19 dB.

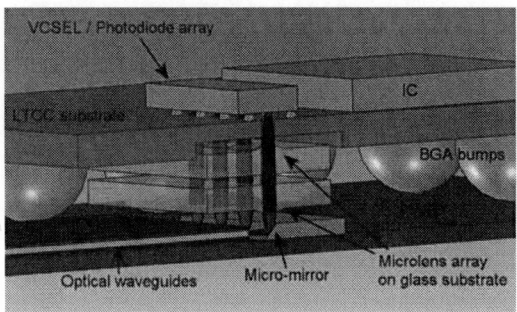

Figure 3: Schematics of the optical coupling and module structures of the 2nd demonstrator. Expanded beam concept between O/E-module and PWB loosens alignment requirements.

Design and fabrication of the modified optical in-coupling structures

Within the framework of the Network of Excellence in Micro-Optics (NEMO) supported by the European Commission through the FP6 program, project partners can access various micro-optical technologies not yet available on commercial markets. To enhance the performance and to reduce the amount of optical components in the 2nd VTT demonstrator described in the previous chapter, altogether three novel technologies were introduced to this demonstrator platform.

Assembly of the discrete microlens array that collimates the VCSEL beam requires high accuracy and is thus costly. The Centre Suisse d'Electronique et de Microtechnique (CSEM) has developed a wafer-scale replication process, which combines UV-casting and lithography, to realise microlenses directly on top of the VCSEL wafer [6]. These i-VCSELs can reduce the farfield FWHM–divergence to around 6° with 7 mA driving current and 1 mW of emitted optical power, according to our measurements. Figure 4 shows 4-channel i-VCSEL array chip with 250-μm pitch wire-bonded onto the bottom of a cavity on an LTCC substrate.

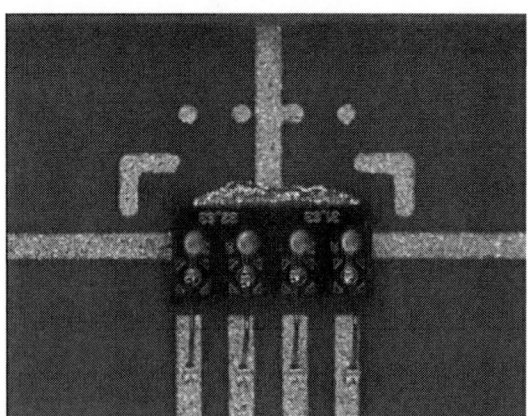

Figure 4: i-VCSEL array chip in the bottom of the cavity on LTCC substrate.

Multimode optical waveguides were fabricated on top of 2nd VTT demonstrator PWB at the Ghent University, Belgium. The optical material, Truemode Backplane™ Polymer, was spin-coated on the PWB. It is a commercially available acrylate based polymer (Exxelis Ltd.) which shows excellent optical and thermal properties and is fully compatible with standard FR4 processing, according to the supplier. Waveguides were patterned by excimer laser ablation and had a cross-section of 50 µm x 50 µm and NA of 0.3. After spin-coating of the top cladding, a micro-cavity was ablated (shown in Figure 5). A detailed description of the laser ablation process in optical waveguide fabrication is given in [7].

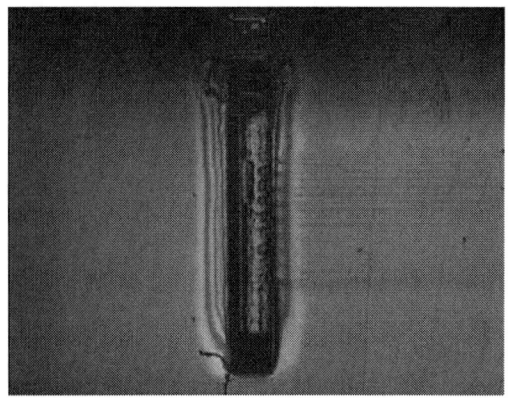

Figure 5: Laser ablated cavity on PWB.

In order to couple the light to the optical waveguides, the light beam has to be bent by 90°. In this study, we used a pluggable micro-optical component incorporating a 45° micromirror fabricated with DPW technology at the Vrije Universiteit Brussel, Belgium. Photo of the realised component is shown in Figure 6. The fabricated coupling components are potentially suitable for low-cost mass production since the DPW technology enables fabrication of moulds for standard replication techniques, such as hot embossing and injection molding [8]. In addition, the component

should be made of another material than the currently used PMMA in order to enable standard surface-mount assembly of the module with reflow soldering process.

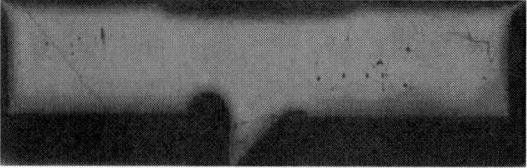

Figure 6: Micro-optical out-of-plane coupler.

The principle of the optical coupling and the structure of the transmitter module are schematically illustrated in Figure 7. The module is built on an LTCC substrate with a cavity for the i-VCSEL array. The cavity enables to adjust the separation between the VCSEL and the PWB so that the optical components can fit in between them. The array chip is mounted with conductive adhesive and wire-bonded to contact pads. The DPW coupling component has a mechanical alignment structure with laser ablated cavity edge and can be assembled either by a die bonder or by a pneumatic gripper on a precision active-alignment station. The component is fixed with UV-cured adhesive. The transmitter module can be mounted to PWB with BGAs by using flip-chip bonder, or alternatively, it might be assembled with a high-precision SMT pick-and-place-machine.

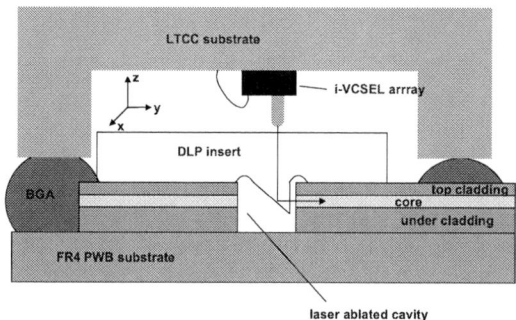

Figure 7: Schematic of the optical coupling and module structures.

Characterization of the optical performance

Optical loss measurements were made according to Figure 7, where the ends of the 3.5 cm long waveguides were polished and the output light was butt-coupled to a 200/220-um MM fiber with 0.22 NA. Preliminary results of this measurement gave average insertion loss of 10.4 dB. This includes in-coupling, waveguide propagation and out-coupling losses. The propagation loss of the ablated waveguides is 0.12dB/cm [7].

Alignment tolerance measurements were done (before the BGA assembly) by attaching the transmitter module into a moving stage of Newport AutoAlignment Station 8100 and by measuring the

252

out-coupled optical power as a function of misalignments in the lateral directions (x- and y-axes). Axis nomination is shown in Figure 7. Figure 8 shows the obtained alignment tolerance curves.

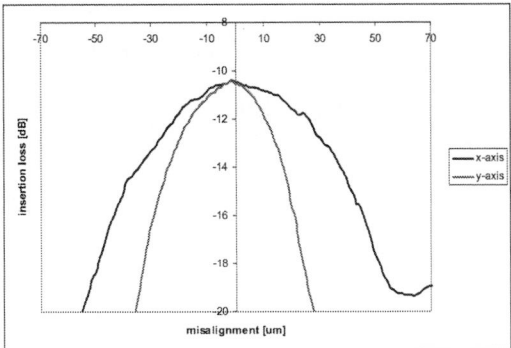

Figure 8: Alignment tolerances of the transmitter module.

-3dB alignment tolerance is ±33 μm along x-axis and ±20 μm along y-axis. -1 dB alignment tolerances are ±16 μm and ±10 μm along x- and y-axes, respectively.

When compared to VTT's 2nd demonstrator, this coupling structure demands higher alignment accuracy, because it does not provide expanded beam concept. However, the structure includes less components. Thus, the assembly process is simpler and potentially more cost-effective. Optical coupling efficiency performance between VTT's 2nd demonstrator and this study can be considered quite similar, although measurement results are not entirely comparable.

Conclusions and future perspectives

A novel optical coupling scheme was presented, consisting of transmitter module built on an LTCC substrate and equipped with a VCSEL array with integrated microlenses; a pluggable micro-optical out-of-plane coupling component made with the DPW technology; and laser ablated polymer waveguides and cavities. Promising results were obtained from the characterization of the optical performance, indicating that the technology could be feasible for optical interconnects on PWBs.

Further study is planned to include the making of waveguide facets with 45° micromirrors by laser ablation. In the future, we will also investigate in more detail the mounting process of the DPW coupling component and the transmitter module in order to optimize the optical coupling efficiency.

Acknowledgements

Part of this work was carried out within the framework of the NEMO supported by the European Commission through the FP6 program. The authors wish to thank several colleagues, especially Airi Weissenfelt and Risto Karjalainen for their contribution to the assembly process. In addition,

Aspocomp Group Plc and Helsinki University of Technology are acknowledged for their contribution to the implementation of the PWBs.

References

[1] R. T. Chen et al., "Fully embedded board-level guided-wave optoelectronic interconnects", Proc. of the IEEE, Vol. 88, No. 6, pp. 780-793, 2000

[2] D. Krabe et al., "New technology for electrical/optical systems on module and board level: The EOCB approach," Proc. of Electronic Components & Technology Conf. (ECTC), pp. 970-975, 2000

[3] P. Karioja et al., "LTCC Toolbox for Photonics Integration" IMAPS / ACerS 2nd International Conference and Exhibition on Ceramic Interconnect and Ceramic Microsystems Technologies (CICMT) 2006

[4] M. Karppinen et al., "Embedded Optical Interconnects on Printed Wiring Board", SPIE Photonics Europe, Proc. of SPIE, Vol. 5453, pp. 150–164, April 2004

[5] M. Karppinen et al., "Parallel Optical Interconnect between Ceramic BGA Packages on FR4 board using Embedded Waveguides and Passive Optical Alignments" Proc. of 56st Electronic Components & Technology Conf. (ECTC), pp. 799-805, 2006

[6] C. Gimkiewicz et al., "Wafer-scale replication and testing of micro-optical components for VCSELs", Proc. of SPIE Vol. 5453, pp. 13-26, 2004

[7] G. Van Steenberge et al., "Laser Ablation of Parallel Optical Interconnect Waveguides", Photonics Technology Letters, Vol. 18, no. 9, May 2006

[8] J. Van Erps et al. , "Deep lithography with protons to prototype pluggable micro-optical out-of-plane coupling structures for multimode waveguides", Proc. of SPIE, Integrated Optics: Theory and Applications, Vol. 5956, pp. 52-63, October 2005

Study of Thermal Behavior in a Multi-Chip-Composed Optoelectronic Package

J. Tian, S. Sinaga and M. Bartek

Delft University of Technology, Laboratory of High-Frequency Technology and

Components/DIMES, Mekelweg 4, 2628 CD Delft, the Netherlands

Phone: +31-15-2789421, e-mail: j.tian@tudelft.nl

Abstract

In this paper, thermal behavior of a multi-chip-composed optoelectronic package is analyzed. The optoelectronic module consists of a common silicon platform on which an InP optoelectronic chip and one or more silicon control circuitry chips are mounted. The InP chip contains a semiconductor optical amplifier (SOA) which optical gain is temperature sensitive and therefore requires thermal stabilization. The SOA itself generates significant heat flux that need to be properly dissipated. Two packaging schemes (wire bonding and flip chip) are compared. The factors that may affect the SOA operating temperature and thermal stability are analyzed with extensive thermal simulations using Ansoft ePhysics^{TM}.

Key words: semiconductor optical amplifier (SOA), thermal behavior, multi-chip module (MCM), optoelectronic packaging

Introduction

Optoelectronic packaging based on a multi-chip module (MCM) with active or passive silicon platform is a promising approach for realizing complex optoelectronic systems [1]. MCM or System in Package (SiP) approach allows achieving of high integration levels without fabrication compatibility problems when single chip solutions would be considered. In MCM optoelectronic system, however, thermal management becomes a critical issue due to reduction of system dimensions and strict requirements of photonic ICs on temperature stabilisation. Application of silicon platform can provide additional infrastructural functions in the form of adaptive temperature control subsystem.

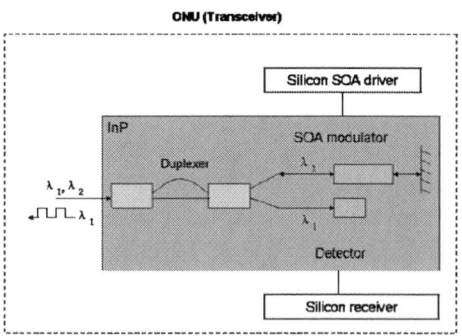

Figure 1: Optical network unit consisting of an InP integrated transceiver and Si control circuitry.

In our application, which is an optical network unit (ONU) consisting of a semiconductor optical amplifier (SOA) on InP substrate, a silicon SOA driver and a silicon receiver (see Figure 1), the optical gain provided by the SOA is temperature sensitive. Due to the increased carrier mobility, the gain of the SOA decreases with the increasing temperature. In general, the operating temperature of the SOA is required to be stabilized around room temperature. The SOA, however, generates significant heat flux, which causes an excessive temperature rise in the active region due to the poor thermal conductivity of InP [1]. This would potentially affect the SOA temperature, and consequently, its performance.

The goal of this work is to design a hybrid optoelectronic packaging solution, which would produce stable operating temperature (around room temperature) for the SOA. In this contribution, the thermal behaviour of two solutions, InP chip face-up (wire bonding) and -down (flip chip), are characterized by extensive simulations using Ansoft ePhysice^{TM}. The main effort is put on the SOA, which generate most of the heat in the system, on the InP substrate.

In this paper, the characterization results obtained for two different packaging schemes are presented and discussed.

Packaging schemes

Figure 2 shows the two packaging schemes evaluated in this study. In the first scheme, all the modules in the ONU are connected with bonding wires. The heat generated is transferred to the surrounding air by convection, and into the InP substrate and the bonding wire by conduction. This is the most straightforward and cost effective solution. It, however, gives higher electrical parasitics, which may limit the high-frequency performance. In the second scheme, all the modules are flip-chip bonded onto a common silicon substrate with Au/Sn bumps. Cu transmission lines for the electrical connection between modules are fabricated on the silicon substrate with PECVD silicon dioxide underneath. For better electrical isolation between the transmission lines, high resistivity silicon (HRS, resistivity 2 kΩ·cm) is used. In this solution, the heat is mainly conducted directly through the bumps to the silicon and the InP substrate. This solution is more complicated than the first one, but the electrical path in this solution is much shorter with all benefits for high-frequency performance.

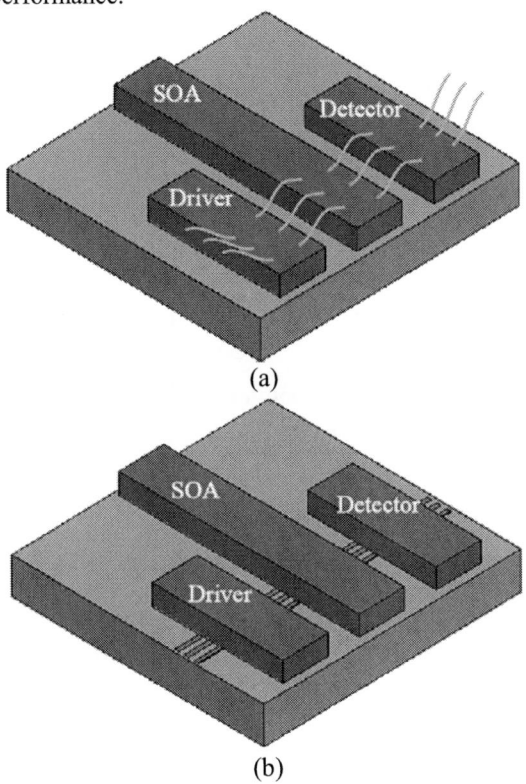

(a)

(b)

Figure 2: Packaging schemes studied in this work, (a) wire-bonding solution and (b) flip-chip solution.

In both of the two schemes, in order to maintain the SOA working at room temperature, Peltier element (set point 20°C) is considered to be attached to the backside of the InP chip (for the first scheme) or the silicon substrate (for the second scheme).

Electrical current injected into the SOA generates photon emission power, as well as joule heat that produces large heat flux and consequently temperature rise in the active region of the SOA. In order to drain the heat generated away from SOA

active region, heat sink should be included in the packaging solution.

The designed packaging solution should be able to conduct as much heat flux as possible to the heat sink. And this heat flux conducted, q_k, is governed by Fourier's law:

$$q_k = -kA\frac{\partial T}{\partial n}, \qquad (1)$$

where k is the thermal conductivity; A is the cross-sectional area of the conduction path; T is the temperature and n is conduction distance [2]. It can be observed from Eq. (1) that the heat flux is proportional to the cross-sectional area and conductivity of the conduction path; and it is inversely proportional to the heat travel distance. These factors have to be considered in the packaging solution design.

Finite element modeling

The proposed packaging designs comprise a lot of different materials: silicon, InP, Cu, etc. They represent composite structures with a complex geometry. When activated thermally, they undergo complex thermal behaviors, which cannot be described using a simple analytical method [3]. In this work, finite element (FE) modeling is used to evaluate performance of the two packaging schemes. The finite element software package Ansoft ePhysics™ is used for the modeling. The packaging models built for the FE analysis are shown in Figure 3.

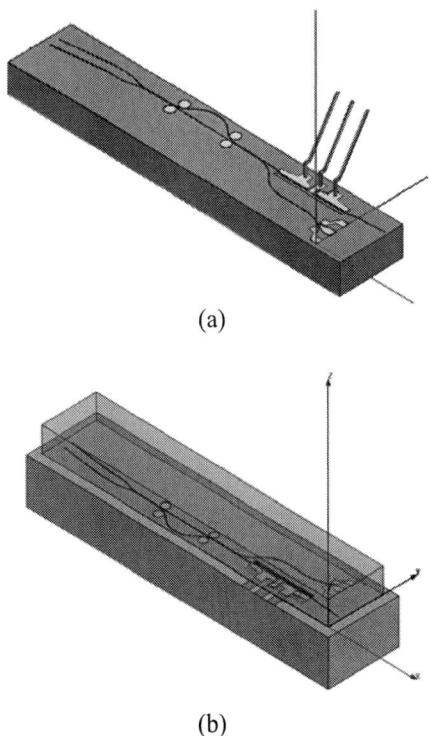

(a)

(b)

Figure 3: Simplified simulation models, (a) wire bonding solution, (b) flip-chip solution.

A. Boundary conditions and geometry

Figure 4 shows the packaging configuration modeled. The detailed component dimensions are listed in Table I. The number of bumps is one of the evaluation parameters in this study. One to three pairs of bumps on the ground pads are evaluated.

Top view

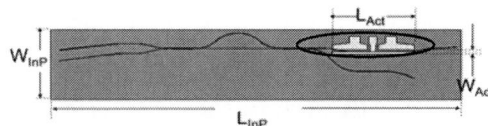

Side view of the flip-chip scheme

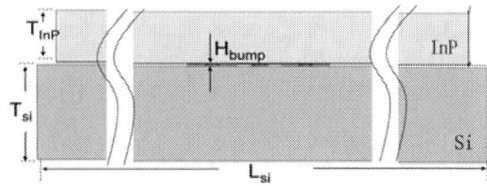

Figure 4: Dimensions of the components in the packaging schemes (the SOA active region is encircled in the top view).

In the modeling, the device is placed in the air. For preliminary study, free convection is assigned to all the object faces that are in contact with the surrounding air. The calculation of the convection coefficient is described in the following section.

The heat source in the model is the active region of the SOA. This becomes a thermal source in the modeling. The detailed calculation on the heat source is given in the following section as well.

The ambient temperature in this study is an evaluation item, in order to find out the temperature stability of the SOA active region with respect to the ambient temperature change. The thermal performance of the packaging schemes under both 27 and 37°C are studied with the Peltier element temperature set to 20°C.

B. Material properties

As a preliminary study, the present modeling assumes that material properties are constant and temperature independent [4, 5]. The material properties used in this work are listed in Table II.

C. Heat transfer in the air

In conduction transfer, thermal energy is transferred through a stationary medium through the vibratory motion of atoms and molecules. Heat conduction in air is relatively important when the air gap is small, say several microns. It happens significantly over a conduction layer in a close proximity to the hot device [3].

Table I: Dimensions of the packaging schemes.

Dimenstions	Value (µm)
Length of the InP chip (L_{InP})	3700
Width of the InP chip (W_{InP})	700
Thickness of the InP chip (T_{InP})	300
Length of the active area in the SOA (L_{Act})	750
Width of the active area in the SOA (W_{Act})	26
Depth of the active area in the SOA	3
Length of the silicon substrate (L_{Si})	3900
Width of the silicon substrate (W_{Si})	900
Thickness of the silicon substrate (T_{Si})	550
Thickness of the Au pad on the InP chip	1.5
Thickness of the Cu pad on the silicon substrate	1.5
Diameter of the Au/Sn bump	60
Height of the Au/Sn bump	10

In convection transfer, thermal energy is transferred through mass movement (such as a gas or liquid), which is flowing around the heat generating object. In radiation transfer, the thermal energy is converted into electro-magnetic radiation, which is then absorbed by the surrounding environment [8]. At a low operating temperature (e.g., <300°C), micro-scale heat transfer by radiation in air is small as compared to that by conduction or convection over a wide range of dimensions [3]. In electronic packaging, radiation effects are seldom sufficient to cause a noticeable change [9]. Therefore, only heat conduction and convection in air are discussed in the present modeling.

Table II: Material properties [6].

Material	Thermal conductivity, W/mK	Mass density, kg/m^3	Specific heat, J/kgK
InP	97	4790	310
Cu	386	8960	380
Au	318	18900	130
80%Au/20%Sn	57	15000	15
Polyimide	0.11	1400	1300
AlN	140	3260	740
SiO$_2$	0.5	2200	745
High resistivity Si	148 [7]	2330	712

Based on the packaging schemes shown in Fig. 3, the most significant heat conduction can be found in the flip-chip solution, where the heat is conducted from the active area in the SOA in the InP chip through the air gap (distance determined by the bump height) to the silicon substrate. It is approximated [10] to have a heat conduction coefficient h_{cond},

$$h_{cond} = \frac{k_{air}}{d}, \qquad (2)$$

where k_{air} is the air thermal conductivity of $0.027\ \text{Wm}^{-1}\text{K}^{-1}$ [11]. In this case, the effective heat transfer coefficient by air conduction is around $2077\ \text{W m}^{-2}\text{K}^{-1}$.

The heat transfer by convection is given by:

$$Q = h_{conv}A(T-T_{amb}), \qquad (3)$$

where the h_{conv} is the convective heat transfer coefficient, A is the area.

The average convective heat transfer coefficient is given by:

$$\overline{h}_{conv} = \frac{k_{air}\overline{N}_u}{L}, \qquad (4)$$

where L is the characteristic length and Nu is Nusselt number. In Eq. (4), the bar above quantities signifies an average value. The Nusselt number is related to Rayleigh number, which is the product of the Grashof and Prandtl number:

$$Ra = Gr \cdot Pr \qquad (5)$$

Grashof number, Gr, is given by:

$$Gr = \frac{L^3 \cdot g \cdot \beta \cdot (T - T_{amb})}{\nu^2}, \qquad (6)$$

where g is gravity, ρ is the density of the fluid, β is the thermal expansion coefficient , μ is the dynamic viscosity, ν is the kinematics viscosity.

Prandtl number is given by:

$$Pr = \frac{c_p \cdot \mu}{k_{air}}, \qquad (7)$$

where c_p is the fluid specific heat.

According to the dimensions listed in Table I, the average conductive heat transfer coefficient is estimated to be $41\ \text{Wm}^{-2}\text{K}^{-1}$.

D. Heat source

Electrical current (130 mA) is delivered by the burst mode silicon driver to the SOA. According to Hudgings, Pipe and Ram [12], the bias power that is electrically injected into the element is almost completely dissipated near the active region. The power loss via photon emission in the SOA can be described by:

$$Q(x) = \left(gP_0 e^{gx}\right) \cdot \Delta x, \qquad (8)$$

where P_0 is the injected optical power and g is the model gain coefficient [12]. So the maximum heat dissipation in the device at a given current is when

the power loss via photon emission is assumed to be zero, which means there is no light coming out of the SOA. From electrical measurement on test samples, it has been determined that the electrical resistivity of the active region of the SOA is around $10\ \Omega$. So the maximum heat dissipation in the system is estimated to be 170 mW.

Results and discussions

In this work, both static and transient thermal behaviors are studied. In the static thermal characterization, we focus on the temperature distribution along line AB (shown in Figure 5), which is on the device surface of the InP chip along the length direction across the active region of the SOA.

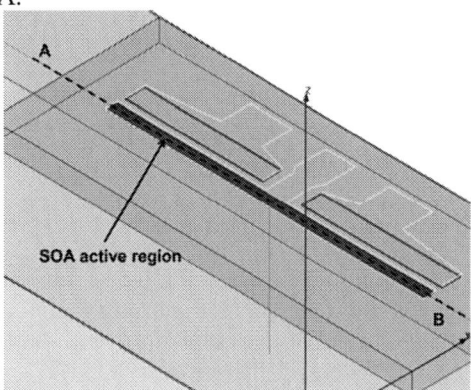

Figure 5: The temperature distribution along line AB is studied.

A. Influence of the flip-chip bumps

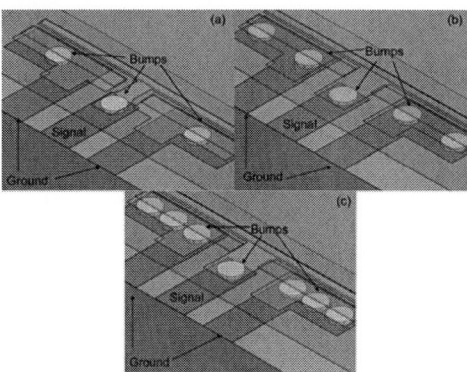

Figure 6: (a) one pair of bumps, (b) two pairs of bumps and (c) three pairs of bumps on the ground of the transmission lines are evaluated.

In the flip-chip packaging solution, major part of the heat generated passes through the bumps into the silicon substrate. The number and the dimensions of bumps determine the cross-sectional area of the heat path and the heat travel distance. According to Eq. (1), the heat flux conducted increases with the increasing area and the decreasing heat travel distance. So the number of the bumps, bump height and diameter are some of the key factors influencing the thermal performance of the SOA. These factors are studied in the modeling.

Here, only the influence of the number of bumps is discussed.

In order to avoid electrical parasitic effect introduced by the additional bumps, the extra bumps are all placed on the ground of the transmission lines. This is shown in Figure 6. In this case, the ambient temperature is set to 27°C, and one to three pairs of bumps are evaluated and compared with the bond wire solution. Here, the dielectric layer on the silicon substrate is 1 µm silicon dioxide. The simulation results are plotted in Figure 7.

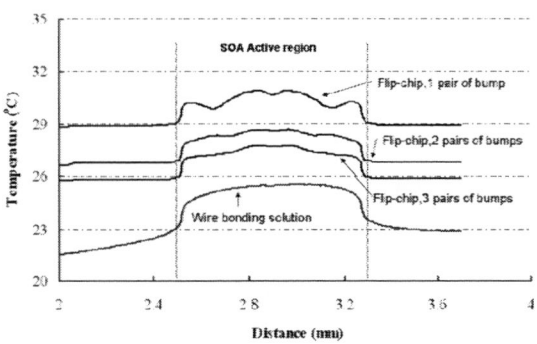

Figure 7: Temperature distribution vs. number of bumps on the ground of the transmission lines (the temperature distribution of the wire-bonding solution has been verified by experiment).

From Figure 7, it can be seen that the lowest active region temperature is achieved by wire-bonding solution. This result is expected. Because, in the wire-bonding solution, right after the SOA is switched on, the heat transferred away from the SOA active region is composed of the heat conduction through the InP substrate to the Peltier element at the back, the heat convection from the Au pads above the active region on the InP chip to the air and the heat conduction through the Au pads to the bond wires. In the flip-chip solution, the SOA active area faces the narrow gap (determined by the bump height) in between the InP and the silicon chips, where the heat conduction plays a more important role than convection [3]. There are two major heat conduction paths in parallel in the flip-chip solution. One is from the active region through the Au pad on the InP to the bumps, which is similar to the heat conduction situation in the wire-bonding solution. The other is from the active region through the air gap to the bumps. Calculated from Eqs. (1) and (3), the heat flux passing through these conduction paths is smaller than that in the wire-bonding solution, and consequently results in a higher temperature.

Because the heat conduction area, which refers to active region-to-air-to-bumps situation (the increased bumps are not connected to the active region directly), is increased, from Figure 7, it can also be observed that increasing the number of bumps decreases the active region temperature. From the bump dimensions study, it is also known

that the SOA temperature decreases with decreasing the bump height, or increasing the bump diameter.

B. Influence of the insulating layer

In the flip-chip solution, after the heat being transferred from the SOA active region to the bump, the majority of the heat is conducted through the bump, insulating dielectric layer under the bumps and silicon substrate to the Peltier element. In this conduction path, the insulating dielectric layer may also have significant influence on the heat conduction. Determined from Eq. (1) and Table II, the dielectric layer used in the above discussed situation, SiO_2 might limit the heat conduction. Reducing the thickness of the SiO_2 or using some other dielectric materials with higher thermal conductivity may help to increase the heat flux conducted to the silicon substrate. These two factors are also studied in this work. In this section, we discuss the influence to the SOA temperature by using other dielectric material, AlN. AlN is a ceramic material with thermal conductivity of $140\ Wm^{-1}K^{-1}$ (refer to Table II), which is also possible to be used as dielectric layer on the silicon substrate.

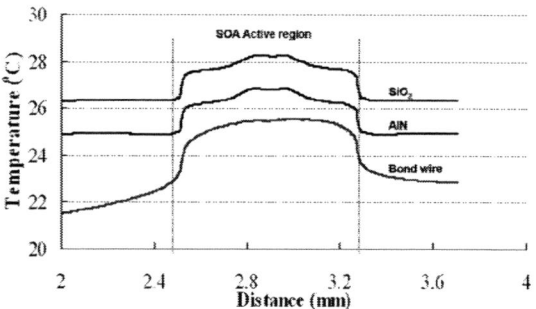

Figure 8: Temperature distribution vs. different dielectric layers used under the bumps on the silicon substrate (three pairs of bumps at ambient temperature 27°C).

Figure 8 shows the temperature distribution comparison between AlN and SiO_2 with the same thickness (1 µm) as the dielectric layers on the silicon substrate. As a reference, the temperature from the wire-bonding solution is also plotted.

As it is analyzed above, the implementing of AlN as the dielectric layer decreases the temperature in the active region. A similar result can also be achieved by decreasing the SiO_2 layer thickness.

C. Influence of the ambient temperature variation

According to Eqs. (4-6), the convective heat transfer coefficient is ambient temperature dependent. And it has been pointed out that the thermal conductivity of most materials decreases with increasing temperature [13]. Furthermore, the ambient temperature variation affects the SOA working temperature directly. So the ambient temperature variation is one of the factors that could possibly affect the SOA working temperature.

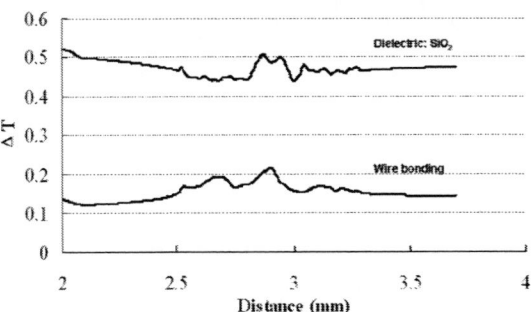

Figure 9: Temperature change along the AB line, after the ambient temperature rises from 27 to 37°C.

In this case, the ambient temperature rising from 27 to 37°C is both simulated with wire-bonding solution and flip-chip solution with SiO_2 as dielectrics (three bump pairs) respectively. All other boundary conditions and the Peltier element temperature setting are not changed for both solutions. Fig. 9 plots the temperature rise on the InP chip along the AB line, after the ambient temperature rises from 27 to 37°C, which is:

$$\Delta T = T_{SOA@Tamb=37C} - T_{SOA@Tamb=27C} \quad (9)$$

It is found in Fig. 9 that after the ambient temperature changes, the temperature rise on the InP chip surface from flip-chip (SiO_2) is more significant than the wire-bonding solutions. In general, the temperature rises in both solutions are smaller than 0.5°C. This suggests that both packaging solutions be thermally stable with respect to the given ambient temperature variation.

D. Transient thermal behavior

The results of transient thermal simulations show that all the packaging schemes can reach their equilibrium state within a short period of time (<0.5 s). The flip-chip scheme with SiO_2 insulation layer yields slightly longer stabilization time than the rest.

Conclusion

In this work, the thermal behavior of an MCM optoelectronic package solution is studied by extensive thermal simulations with Ansoft ePhysics™. Different packaging schemes, packaging parameters are analyzed and compared. The results suggest that the thermal behavior of such a system is relatively sensitive to and therefore can be easily tuned by the dimensions of the flip-chip bumps, as well as bump count and dielectric layer used. The proposed solutions are, in general, stable with respect to environment temperature change. The results achieved in this work serve as a starting point for the MCM package design.

Acknowledgements

This work is done in the frame of Freeband Broadband Photonics project [14], where consortium of industrial and academic partners works on development of photonic integrated circuits for low-cost dynamically reconfigurable optical access networks.

References

[1] A.R. Mickelson, N.R. Basavanhally and Y.C. Lee, "Optoelectronic packaging", Wiley Interscience, ISBN 0-471-11188-0, pp. 59-60.

[2] F. Kreith et. al, "The CRC handbook of mechanical engineering", 2nd edition, chapter 4, CRC press LLC, 2005.

[3] G.K. Lau, T. Chu Duc, J.F.L. Goosen, P.M. Sarro and F. v Keulen, "In-plane thermal unimorph using confined polymers", J. Micromech. Microeng, 17 (2007) pp. 1-10.

[4] C.S. Pan and W. Hsu, "An electro-thermally and laterally driven polysilicon microactuator", J. Micromech. Microeng. Vol. 7, 7–13, 1997.

[5] Q.A. Huang and N. K. S. Lee, "Analysis and design of polysilicon thermal flexure actuator", J. Micromech. Microeng. Vol. 9 64–70, 1999.

[6] J.E. Sergent and Al Krum, "Thermal management handbook for electronic assemblies", McGraw-Hill, pp3.8-3.16, 1998.

[7] M.G. Buozo, P.L. Komarov, P.E. Raad, "Non-contact thermal conductivity measurement of P-doped and N-doped gold covered natural and isotopically-pure silicon and their oxides", EuroSimE, pp269-276, 2004.

[8] X.G. Liang and Z.Y. Guo, "The scaling effect on the thermal processes at mini/micro scale", Heat Transf. Eng. 27 30–40, 2006.

[9] A.B. Lostetter, F. Barlow, A. Elshabini, "An overview to integrated power module design for high power electronics packaging", Microelectronics Reliability, vol. 40, pp. 365-379, 2000.

[10] R. Hickey, D. Sameoto, T. Hubbard and M. Kujath, "Time and frequency response of two-arm micromachined thermal actuators", J. Micromech. Microeng. 13 40–6, 2003.

[11] F.P. Incropera and D.P. DeWitt, "Free convection Fundamentals of Heat and Mass Transfer", 4th ed (NewYork: Wiley) chapter 9, pp 248–333, 1996.

[12] J.A. Hudgings, K.P. Pipe and J. Ram, "Thermal profiling for optical characterization of waveguide devices", Vol. 83, no. 19, pp 3882-3884, Applied Physics letter, 2003.

[13] G.R. Blackwell, "The electronic packaging handbook", chapter 11, CRC press LLC, 2000.

[14] Freeband Broadband Photonics project (http://www.freeband.nl/project.cfm?id=526).

The Market Situation of Ceramic Micro Circuits

Erwin Effenberger

Chairman of the ZVEI Ceramic Micro Circuit Committee

Introduction

There is a correlation between the GPD and the usage of consumer goods like Personal Computers, Mobil phones, TV-sets, Vehicles etc.

This correlation is an indicator of the prosperity of a specific country. Nevertheless in Western countries we have seen a change during the past years. Due to the transfer of production to low cost countries the standard of living has decreased in the exit countries and high unemployment rate has caused problems in the society.

With this transfer, know-how and possibly patent rights may become lost.
Microelectronic parts, including their use are the key elements in this process and therefore an important criterion for such a development. This situation represents a big challenge to politics.

The knowledge and the evaluation of market trends with view to the future are necessary. This presentation should contribute to give a support to your planning and your actions.

First of all I'll give an overview of the Microelectronic market form 2005 through 2010.
Afterwards I'll present the market for Ceramic Micro Circuits in Europe, Germany, and worldwide. Automotive Electronics will be the main subject of the presentation, because this segment is well established in Europe.

The presented data, trends and prognosis about the past and the future development are based on own research and sources of national and international statistics from organizations and committees.

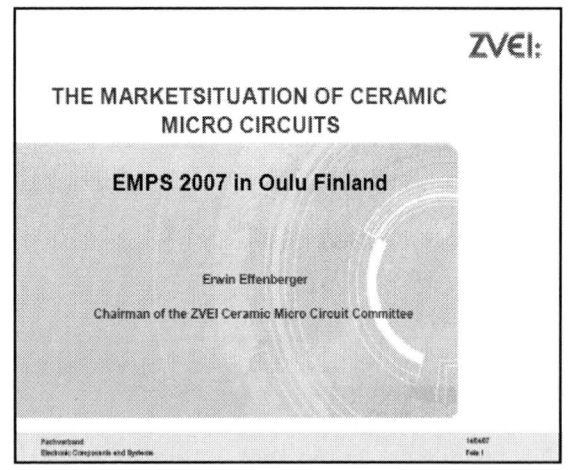

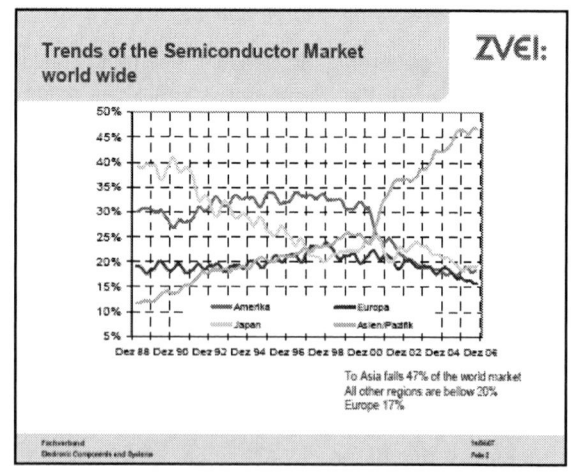

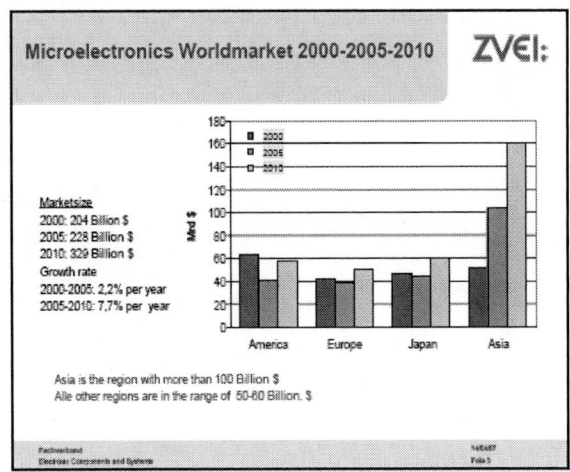

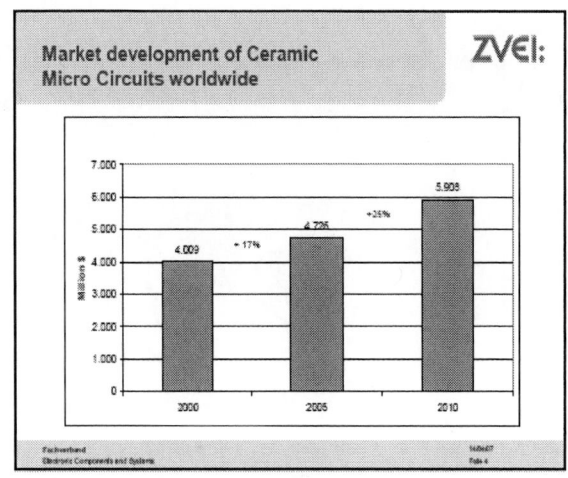

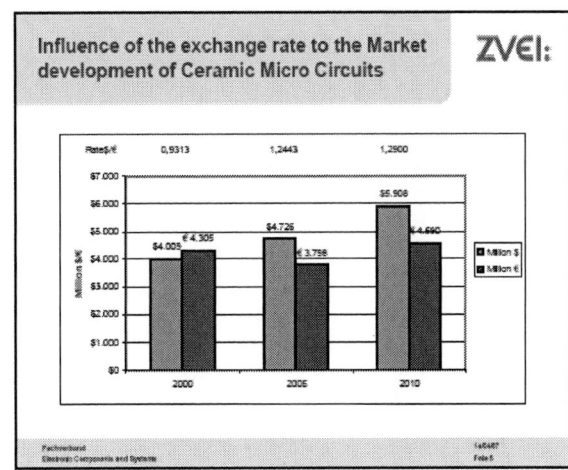

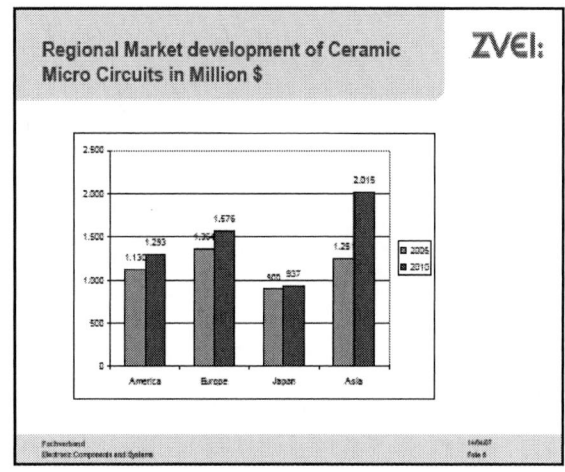

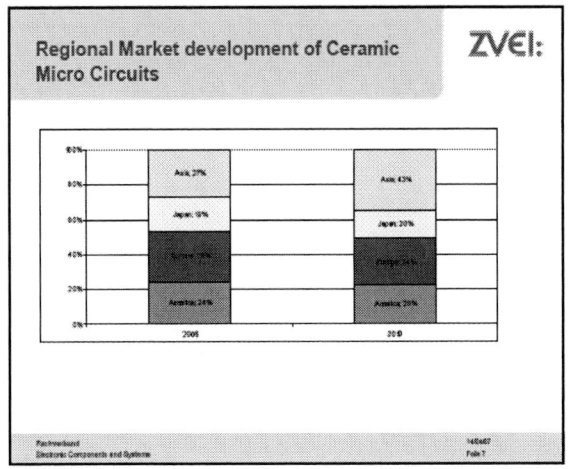

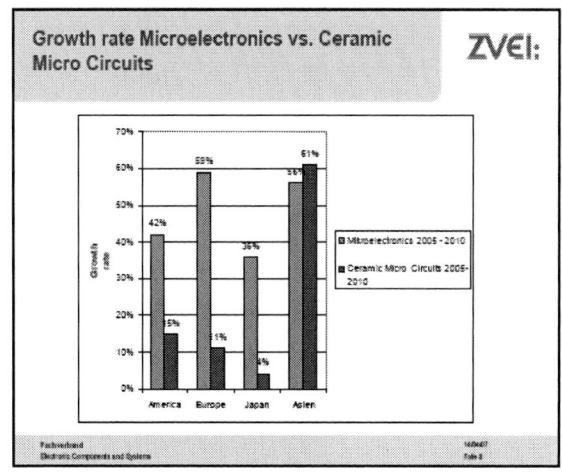

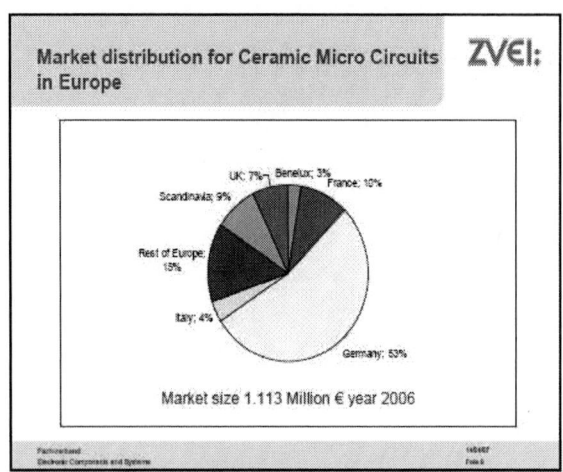

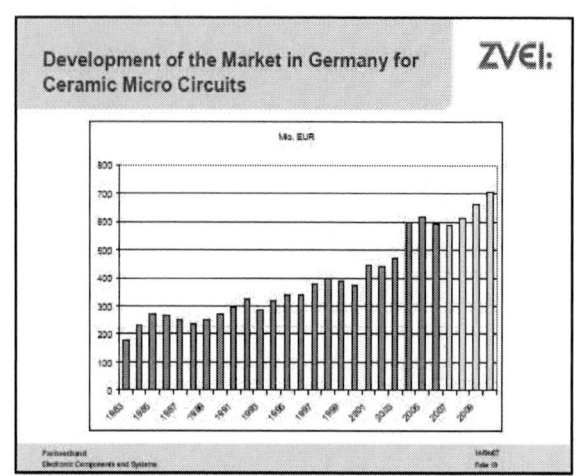

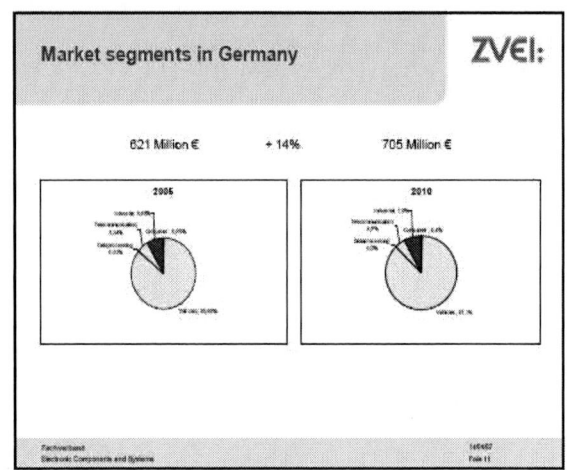

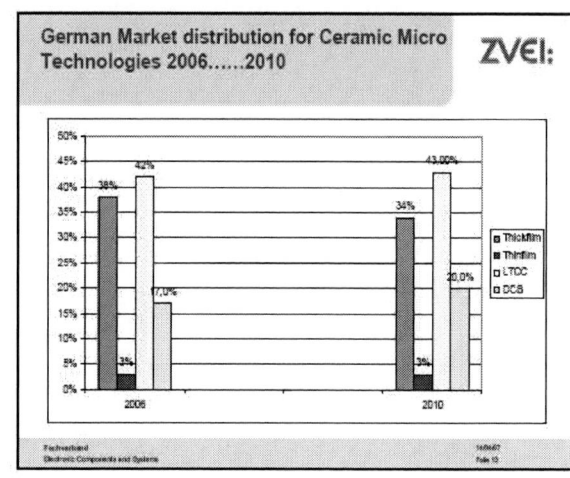

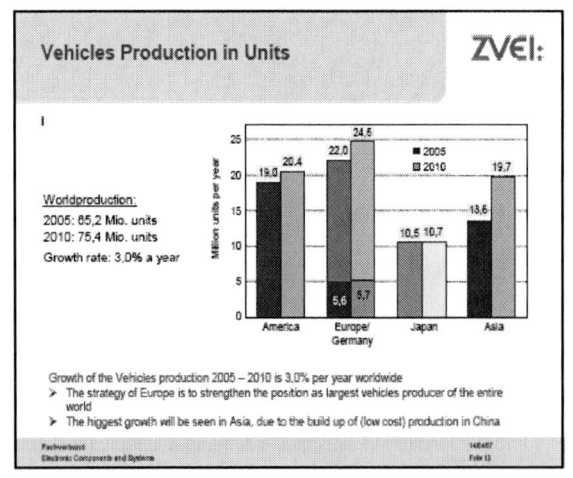

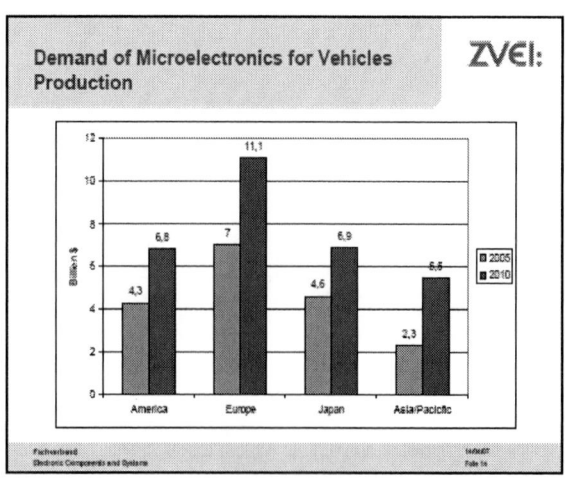

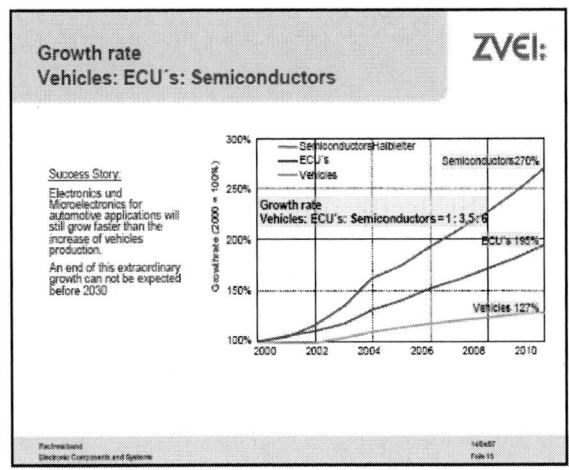

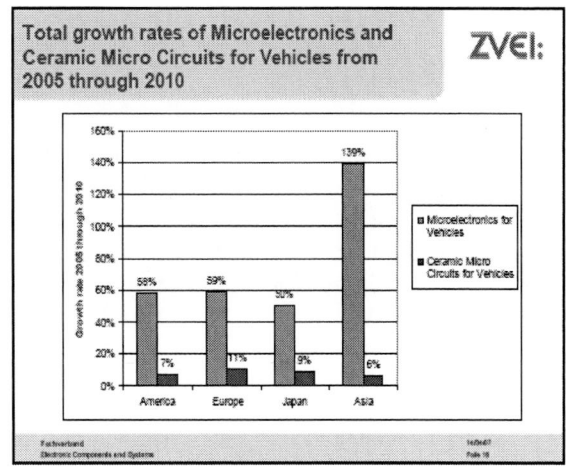

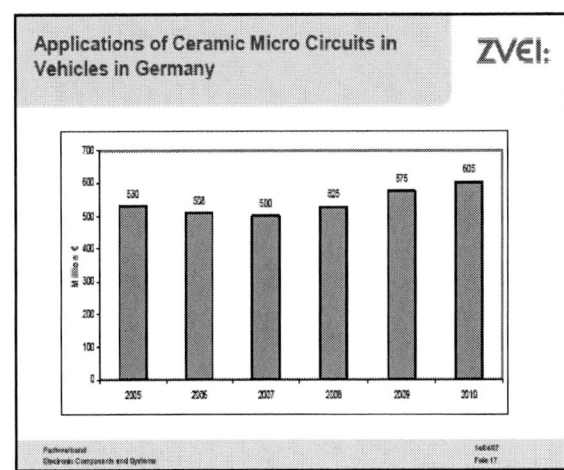

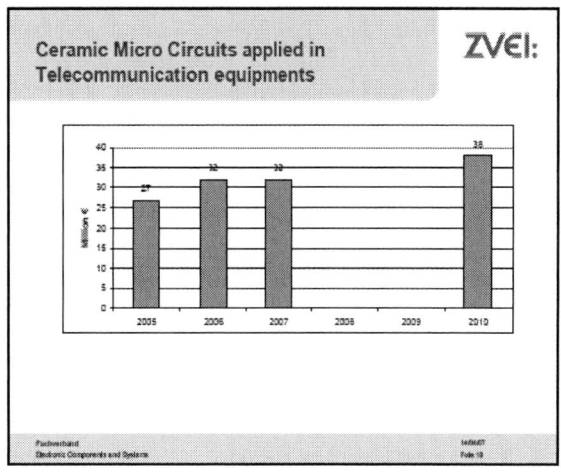

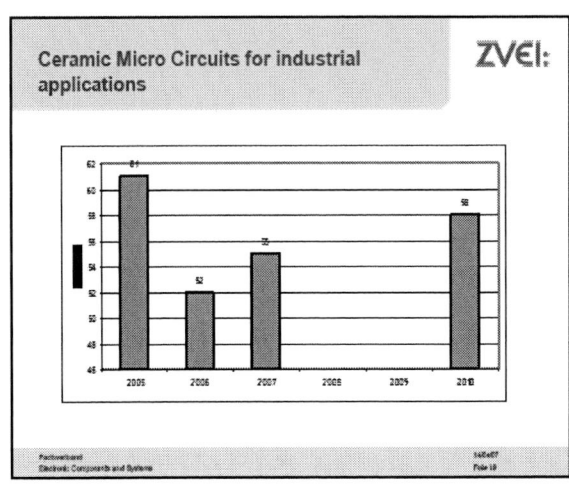

Electronics Manufacturing in Europe – Competing in a Global Market

Dr Indro Mukerjee - C-MAC MicroTechnology

In today's global and highly competitive business environment, electronics companies are facing continual pressure to improve the quality of their products and services, deliver shorter cycle times and at reduced cost. The electronics industry is driven by consumers that demand products that are more reliable, smaller and cheaper than the ones they replace. These demands have led to the electronics industry moving to lower cost regions with China and other East Asian countries, particular beneficiaries.

As the CEO of C-MAC MicroTechnology, Indro Mukerjee has a first hand understanding of the challenges facing electronics design and manufacturing companies in Europe competing in a Global market. Indro will address the conference and outline what steps C-MAC is taking to rise to these challenges:

What are companies like C-MAC doing to compete in an ever more competitive global market?

What future does C-MAC see for its core business, high reliability electronics and how does it plan to build on its core expertise in Thick Film Hybrids?

How is C-MAC modifying the approach it takes to providing standard vs custom products and how does it differentiate its service offerings for its key markets Automotive and Aerospace/Defence electronics?

What steps is C-MAC taking to ensure that it has the human resource skills and talent it needs for the 21st Century?

Indro Mukerjee was appointed CEO of C-MAC MicroTechnology at the end of 2005 from Philips Semiconductors, where he was Executive Vice-President & General Manager of the Automotive & Identification Business Unit.

SiP, PoP, MCP, and Such – Trends in Advancing Cell Phone Capabilities Through Package Integration

Jeff Brown and Niels Kellerhoff - Portelligent, Inc.

The cell phone continues to drive the electronics industry, delivering major functional improvements while reducing size, power, and cost. Multiple cellular protocols, Bluetooth, Wi-Fi, GPS, DVB, mega-pixel cameras, multi-media playback, and more are appearing in these mobile platforms at an ever increasing rate, while their size and cost continue to shrink. These dramatic advances in cell phone features have been made possible by new packaging technologies including stacked die, package-on-package, flip chip, and system-in-package. Using Portelligent's Product Profile Database of over 100 UMTS, W-CDMA, and GSM/GPRS cell phones over the last three and one half years, we will demonstrate how packaging technologies have enabled new capabilities in cell phones without increasing the size of the device. Trends in packaging technologies as seen in recent teardowns will be reviewed along with the implications of these trends for future products.

How Packaging Technologies Can Add Value to Portable Devices

Kari Kulojärvi - NOKIA

The presentation will discuss about ways how new advanced packaging technologies add value to original device manufactures (ODM) products.

NXP System-in-Package vision and latest 3D technology developments

J.-M. Yannou, F. Neuilly, J.-O. Moreno, M. Pommier, S. Bellenger, P. Biermans

NXP Semiconductors, 2 esplanade Anton Philips, Campus EffiScinece,
Colombelles, BP2000,
14906 Caen Cedex 9 - France

+33 (0)231452249
jean-marc.yannou@nxp.com

Abstract

Today, the System-in-Package approach offers a new dimension to system integration, far beyond mere dense micro-packaging of existing SoC solutions. Not only does SiP offer the capability to integrate almost any kind of companion passive component with a given active circuit, but it also enables flexible combinations of analogue circuits, RF functions or even micro-electro-mechanical (MEMS) components with digital integrated circuits. SiP widens the degree of integration and allows ultimate miniaturization as it introduces the third dimension to microelectronics. To any given SoC technology node, SiP brings yet another step in system integration for a wide range of applications.

Due to its modular and flexible approach and its multi-technology nature, SiP poses a serious question of choice when it comes to defining future technologies. Too many different technology developments will lead to an overall weakening in terms of "breakthrough" innovation.

The first part of the presentation will emphasize on NXP's SiP vision, both in terms of technology and expected application benefits. Then the SiP-specific technology challenges will be listed. Several examples of NXP's on-going 3D vertical integration innovation projects will be given through the presentation to showcase the many aspects to consider in a new SiP technology platform development, from packaging to passive integration, from test to Computer-Aided-Design and virtual prototyping, how they relate to one another, and how the apparent wide diversity of constituent technology building blocks can lead to a coherent integral roadmap with limited complementary "SiP platforms".

Key words: 3D vertical integration, SiP, Through silicon vias, passive integration, Wafer-level Packaging

Introduction

Over the past 30 years, system integration of electronic functions has consistently focused on single die integration, following the System-on-Chip route driven by Moore's law (the density per surface area of transistors doubles every one and a half years).

Today, the System-in-Package approach offers a new dimension to system integration, far beyond mere dense micro-packaging of existing SoC solutions. Not only does SiP offer the capability to integrate almost any kind of companion passive component with a given active circuit, but it also enables flexible combinations of analogue circuits, RF functions or even micro-electro-mechanical (MEMS) components with digital integrated circuits. SiP widens the degree of integration, and also allows ultimate miniaturization as it introduces the third dimension to microelectronics. To any given SoC technology node, SiP brings yet another step in system integration for a wide range of applications.

For the monolithic SoC approach, technology innovation is led by the 3 most advanced semiconductor manufacturers, who specify the next generation equipment for the whole industry. Due to its modular and flexible approach and its multi-technology nature, SiP poses a serious question of choice when it comes to defining future technologies. Too many different technology developments will lead to an overall weakening in terms of "breakthrough" innovation. The diversification of technologies between companies, increases the risks of poor customer acceptance, and poses serious difficulties for standardization, industrialization and cost-reduction roadmaps.

Based on a wide application spectrum analysis of customer needs, NXP's focused market-driven approach of new SiP technology innovation allowed for the making of a differentiated yet understandable and coherent as a whole SiP technology portfolio. Its passive integration technologies have proven successful in high volume production across many different applications and combined with many different packaging platforms.

New test techniques and system design approaches allow for the planning of further hardware technology improvements, so as to achieve ultimate miniature integration from sensors to the actuators of a smart system.

The first part of the presentation will emphasize on NXP's SiP vision, both in terms of technology and expected application benefits. Then the SiP-specific technology challenges will be listed. Several examples based on some of NXP's on-going innovation projects, such as 3D vertical integration, will be given through the presentation to showcase the many aspects to consider in a new SiP technology platform development, from packaging to passive integration, from test to Computer-Aided-Design and virtual prototyping, how they relate to one another, and how the apparent wide diversity of constituent technology building blocks can lead to a coherent integral roadmap.

SiP is more than just advanced packaging

Each new System-in-Package design is a unique configuration of mixed technologies and materials which requires a wide technology competence framework to develop and qualify. Of course, new SiP products embed existing System-on-Chip integrated circuits, supposed robust by the SiP architect or project leader, when qualified as stand-alone, and, of course, SiPs are generally quite easier to prototype than Integrated circuits. However, qualifying and productizing a new innovative SiP product is yet another story and requires optimal project lead and team work to anticipate issues and keep the time-to-market advantage brought forward by SiP easy prototyping. Indeed, making a new SiP product does not consist in placing several existing circuits side-by-side in a known package platform, but it supposes that the following SiP-inherent challenges will be addressed with anticipation:

- New package design constraints emerge with SiP which must be listed as as many design rules, such as:
  - Minimum stand-off distance between 2 side-by-side integrated circuits, maximum wire bond loop height in stacked dies configurations…
- An SiP product is a new system, with functionalities exceeding that of the embedded SOCs. System design must be planned, including passive integration and system-level ESD protection, to ensure high system performance.
- SiP is bringing the third dimension to micro- and nano- electronics, in a miniature environment, with associated subsequent challenges of electromagnetic compatibility between the various embedded components, thermo-mechanical robustness and management of the thermal budget.
- Final test of the SiP module is tricky since Design-for-test is provided for each embedded integrated circuit. Since SiP allows for many interconnections in between the embedded circuits and components, few Integrated circuit pins are accessible from outside the SiP module for test, and test of the SiP module must be thought in a complete different way, i.e. performing "system test", as opposed to "structural test".
- SiP uses new emerging technologies such as fine-pitch bumping, passive integration, redistribution, silicon substrates. Complete new library cells for design need to be written for these new technologies
- Because SiP products include many components of various origins, supply chain is particularly tricky to manage.

All these challenges account for the particular care needed in the SiP design phase, with the objective to make a reliable system with robust system performances, and to make it first time right. Only then, the promised benefits of SiP to make a fast time-to-market integration of a complete system can be delivered.

To achieve such anticipation, cross-fertilization of multiple competences is highly recommended. In practice, a SiP design and industrialization project team needs to be constituted including at least a project leader in charge of coordination, timeframes, prototyping and sampling, and customer relationships, an SiP architect with system competences, one IC designer or field application engineer for each embedded integrated circuit, a package designer with competences in thermo-mechanical simulations for virtual prototyping for reliability, designers for the passive integration substrate if any, and a test engineer and a product engineer for production ramp-up.

The NXP technology portfolio and vision

SiP technology innovation is yet a bigger challenge. Here again, the classical organization charts of the semiconductor industry are contradicted as a multi-skilled team needs to operate in the same direction to bring from innovation to manufacturing technologies which aim at bridging wafer front-end and assembly back-end techniques, the system and its constituent components, the nanometer scale with the micron and the millimeter scales, all by constantly caring for system electrical performance and robustness, reliability and manufacturability. Placing ourselves at the curfew of these existing technologies, our vision is that NXP will maximize SiP value by joining these competences together and by orienting them as much as possible towards our core competence as a semiconductor manufacturer: front-end silicon technologies. This means that we think that SiP not only follows but also leads this other major trend of IC packaging: wafer level packaging (WLP). WLP consists in carrying out packaging operations at the wafer level, before dicing the wafer. WLP techniques include, but are not limited to bumping, under bump metallurgy, redistribution, wafer molding, wafer to wafer bonding, the making of through silicon vias, wafer-level test and burn-in, wafer thinning. Thanks to the combination of SiP and WLP altogether, the following benefits for the application are expected:

- Higher electrical performance thanks to shorter interconnections
- Cost management thanks to more collective operations at the wafer level at the expense of fewer operations at the die level
- Supply chain simplification as we refocus on our core silicon competences and decrease the component count
- Optimal miniaturization thanks to denser routing substrates, and in-silicon vertical integration

Figure 1, although only a "conceptual view" still far from technology specification, illustrates our SiP technology vision.

Two wafer level platforms for 3D vertical SiP integration

The analysis of our customer needs led to the definition and specification of 2 new technology platforms so as to propose a complete technology portfolio for 3D vertical SiP integration. In

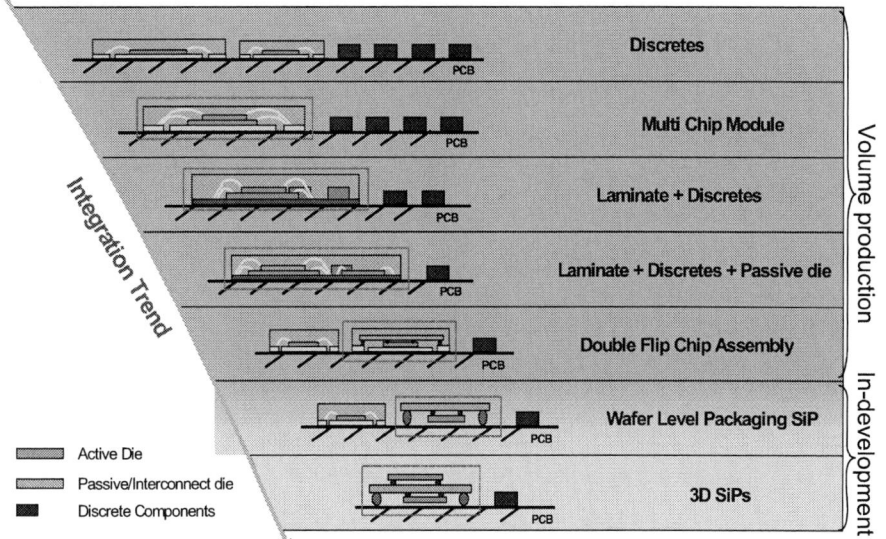

Figure 1: NXP SiP Technology Vision

agreement with our SiP technology vision, both these platforms mostly use wafer-level technology bricks, and both allow for 3D integration without wire bonds, which are space consuming and introduce parasitic phenomena [2]. Both these platforms are currently under development with promising results. The first one, named WL-SiP, mainly targets high-end RF applications with low pin counts and stacks of 3 circuits maximum, whereas the second one, 3DWLP, is meant for high pin count digital and memory devices, with no technology-constrained limit of the amount of

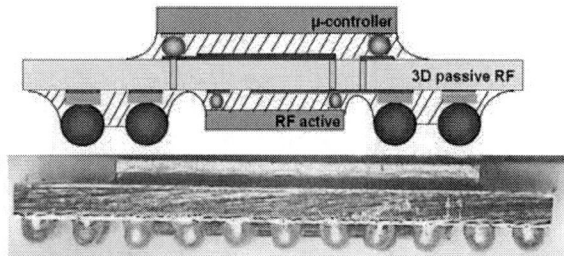

Figure 2: conceptual cross-section drawing and picture of WL-SiP demonstrator

stacks. Thanks to their complementarities, we expect these two new technologies to cover most of our customers' 3D vertical integration needs in the coming years.

Thanks to through silicon vias in the NXP PICS (standing for Passive Integration Connecting Substrate) Integrated Passive Device (IPD) [1], WL-SiP (Figure 2) [4] offers record short connections between active and passive components and to, possibly, a third circuit, like an associated transceiver (modulator/demodulator) mixed-signal device. We recently and successfully demonstrated such a construction. To do so, many challenges had to be faced, relating to various technology sectors, like:

- for silicon front-end technologies: substrate isolated metal-filled through silicon vias or bumping.
- For assembly technologies: dual-side pick, flip and placement of bumped ICs , i.e. on each side of the 3D silicon and passive integration connecting substrate
- Modeling of new library cells, definition of a design flow, and characterization.

Based on the emerging technologies of wafer reconfiguration and wafer molding (Figure 4), redistribution on reconfigured molded wafers [2,3], 3DWLP extends the so-called fan-out Wafer-level Chip Scale Packaging (fan-out WLCSP) concept to the third dimension (Figure 3).

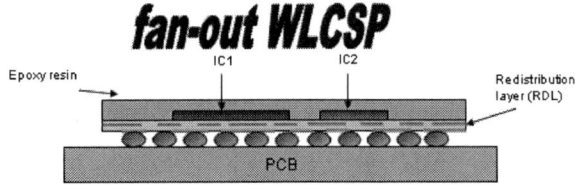

Figure 3: conceptual cross-section drawing of fan-out WLCSP and 3DWLP

Because of its expected higher reliability at 2nd level (solder joint to Printed Circuit Board) than regular

CSP [4], fan-out WL-CSP is expected to compete with Ball Grid Array (BGA) packages, and to overcome them in terms of cost, routing density and low thickness. Fan-out WLCSP is a typical example of mixing IC technologies: usual packaging materials like epoxy resins are used to rebuild wafers and process them back on silicon front-end equipments for packaging

purposes (redistribution and bumping).

Figure 4: picture of a molded reconfigured wafer made at NXP

3DWLP adds one more wafer level operation: wafer stacking: indeed, reconfigured molded wafers are made in such a way as to make equal size fanned-out circuits on each layer. Therefore wafers can be stacked and glued on top of one another before sawing the wafer stack. Each layer is expected to be as thin as 100μm, including silicon, epoxy resin, redistribution and glue. Then the layers are interconnected to one another thanks to vertical lines on the external vertical sides of the multi-stack devices. This can be achieved thanks to metal plating on the external sides of the device. Then, laser patterning is applied to remove electrical shorts.

Expected applications are:
- Memory stacks: up to GigaBytes flash memory can be obtained in less than 500μm with 4 layers containing each half a GByte flash memory. On a fifth layer, the memory microcontroller can be added, possibly together with an integrated passive die.
- Replacement of "Package-on-Package": memory and digital fan-out layers can be prepapred and tested separately at the wafer level, and then assembled on top of one another. Huge cost benefits and record low

thicknesses can then be obtained, with respect to current "package-on-package" solutions.

Conclusion

Forthcoming SiP technology innovations need to be diverse enough to address as many customer needs as possible. However, the technology roadmap must be kept as simple as possible: there are so many possible combinations of building blocks that the roadmap must be driven by a precise technology vision otherwise efforts will be diluted in too many concurrent developments, and the application mapping of these technologies may become illegible. Each new SiP platform development requires cross-fertilization of teams with multiple competences ranging from silicon front-end technologies to assembly and packaging techniques, from test to design environment and modeling. Adequate project management is another key success factor. Thanks to this focused approach, the technology innovation examples here presented are expected to far exceed existing technologies in miniaturization benefits by truly bringing the third dimension to Integrated Circuits.

References

Articles from conference proceedings
[1] D. Chevrie, "A silicon-based System-in-Package technology", 15th European Microelectronics and Packaging Conference (EMPC), Brugge, Belgium, June 12-15 2005.

Journal article
[2] A. Mangrum, "Packaging Technologies for Mobile Platforms", 2006 Electronics Packaging Technology Conference (EPTC), Singapore, December 6-8 2006.

Book
[3] M. Brunnbauer, "Embedded Wafer Level Ball Grid Array", 2006 Electronics Packaging Technology Conference (EPTC), Singapore, December 6-8 2006.
[4] N. Strusevich, "Modeling the behaviour of solder joints for Wafer-Level SiP", 2006 Electronics Packaging Technology Conference (EPTC), Singapore, December 6-8 2006.

Conductive Adhesives for Blue LEDs

Noritsuka Mizumura, Sen-ichi Ikarashi, Michinori Komagata and Yukio Shirai*

Technical R&D Division, NAMICS Corporation

*Sales Division, NAMICS Corporation

3993 Nigorikawa, Kita-ku, Niigata-city. 950-3131, Japan

Phone: +81-25-258-5577, Fax: +81-25-258-5511, E-mail: nmizu@namics.co.jp

Abstract

LEDs have superior features such as high-speed response, low electric consumption, long life, small size, light weight and so on. It has been used for light bulbs, head lumps for automotives, the interior, back light for LCD and some other applications. In spite of all these superiority, epoxy conductive adhesive which is used to adhere LED element to the flames induce the change of color with UV light and heat emitted from the element of blue/white LEDs in particular. It causes the degradation of LEDs luminance which is the serious problem. In order to solve this problem, Benzene ring in epoxy resin/cure agent/cure accelerator which is the prime cause of the degradation of UV/heat resistance is paid attention and cure system without Benzene ring was investigated. As a result, conductive adhesive of less degradation of reflectance was found in the cure system of hydrogenation epoxy resin and polyurethane system resin at which the wave length is 450 nm. In addition, this conductive adhesive satisfied the reliability properties required for LEDs such as MRT reflow, Thermal Cycle and so on.

Key words: Die Attach, Hydrogenation Epoxy Resin, Urethane, Reflectance, Light Stability, Resistance

Introduction

LEDs have superior features such as high-speed response, low electric consumption, long life, small size, light weight and so on. It has been used for light bulbs, head lumps for automotives, the interior, back light for LCD and some other applications.

Improved luminance efficiency is indispensable for the development of LEDs in the future. Long life of LEDs is also necessary for the development by controlling the luminance degradation in long time lighting of blue/white LEDs. The primary cause of the luminance degradation is yellowing of encapsulant resin or the Die Attach (DA) for LEDs.

Generally, adhesives are composed of epoxy resin which has the properties of high elasticity and adhesion. It is acknowledged that benzene ring in epoxy resin change into yellow, which is affected by the wavelength of 400 to 500 nm. 1) 2) Therefore, in recent years, investigation of a long-life is made by using hydrogenation epoxy resin or silicone resin for die attach. Now silicone resin is used more often for encapsulant resin. Furthermore, silicone resin is also beginning to be used for die bonding attach to connect two wire type LEDs, though, it has a weakness of low adhesion. Therefore, integrated type of hydrogenation epoxy resin (H-BPA) and silicone resin is used.

Actually, Conductive Die Attach (CDA) for one wire type, as the name tells, give a direct influence on LEDs luminance of which electrical property is important. There exists no conductive die attach material which has the properties of both electrical reliability and light stability (i.e. stable reflectance ratio) as the ingredient materials are limited.

Conductive Die Attach can control the degradation of reflectance ratio, however, it can not attain enough reliability in electrical property.

Even if alicyclic epoxy resin or hydrogenation epoxy resin is used, yellowing and degradation of reflectance ratio occur in case materials with benzene ring are used as curing agent or curing accelerator. Use of acid anhydrate such as MeHHPA enables to make conductive die attach with less reflectance degradation, though, pot life is remarkably short. 3) Moreover, degradation is easy to happen due to moisture absorption of acid anhydrate.

As a result, after a number of investigations, we found that it is possible to develop conductive die attach which possess both adhesion reliability and light stability (stable reflectance ratio) by using hydrogenation epoxy resin and curing agent with urethane structure. We would like to describe the properties of the conductive die attach.

Materials

Table 1 shows the compositions for four conductive die attach in comparison with XH9680-43 which was developed for blue LEDs.

Table 1 Compositions of CDAs

	XH9680-43	Sample A	Sample B	Sample C	Sample D
Conductive Filler	Silver Flake	←	←	←	←
Base Resin	H-BPA	H-BPA /ACE	silicone Epoxy	BPA	BPA
Curing agent	Urethane	MeHHPA	MeHHPA	Phenoric resin	aromatic Amine

Test Method

In this study, we confirm the property change with irradiation of light, and resistance change and preservation stability with heat.

Reflectance Test
Irradiation of Blue HID Lump
* Irradiated plane:
  Glass plate (measured reflectance)
* Distance of test piece to the lamp:
  Approximately 50mm
* Beam/ iluminance:
  8000 lm/ Approximately 20TH Lx
* Temperature of the test piece while irradiating :
  Approximately 150 °C
* Wavelength: Shown in Figure 1 .

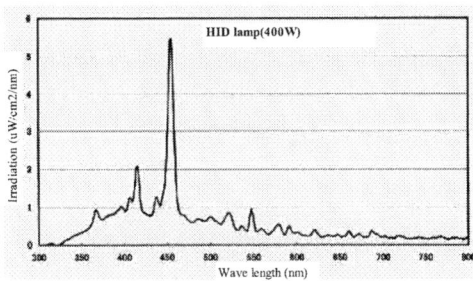

Figure 1 Wavelength of Blue HID Lamp

Procedure
1. Conductive die attach is applied on the glass plate and cured.
2. 45 degree angle reflectance is measured by color meter from the backside.
3. After the measuring, the light at 400/450 nm is irradiated by the blue HID (400W).
4. After irradiating for certain amount of time, reflectance is measured in the same way as No.2 above.

Delta vf Simulation Test
1) The change of the contact resistance are measured on thermal cycle test which is ramped up to 300 °C and kept for 15 seconds and ramped down to room temperature. It is conducted three times.

2) After moisture test of 30 °C /70% for 168 hours and reflowed at 260°C by three times, resistance change is measured.

Test Apparatus
FR4 Substrate with Cu/Ni/Au electrode (shown in Fig. 2)
Alloy 42/Ag plated Pad (3.0mm x 2.0mm)

Procedure
1. The paste of 0.5mm square / 70um thickness is applied on the inside electrodes. (shown in Fig. 3)
2. The pad is set on the paste and pressed into 50um thick. (shown in Fig. 4)
3. The test piece is warmed to room temperature after curing at the recommended condition.
4. Transparency encapsulant is pored after Dam paste was applied, and it was cured for 2 hours at 180 °C, then it was warmed to room temperature. (shown in Fig. 5)
5. Probe is put on the outside electrode and the voltage is checked while 10mA electrode flowing, then resistance voltage is calculated. (n=4)
6. The test piece above No. 4 is left on the hot plate at 300 °C for 15 seconds and it is warmed to room temperature, then resistance voltage is calculated in the same method as No. 5.
7. The process of above No. 6 is repeated three times.

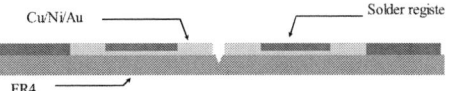

Figure 2 Substrate

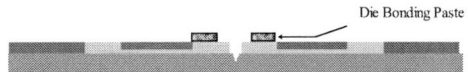

Figure 3 Application

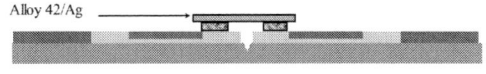

Figure 4 Pad Mount

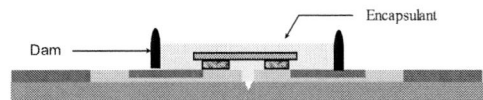

Figure 5 Encapsulant

Shear Strength Test
Test Apparatus
  Si Chip of 2 square mm
  Substrate: Ag plated FR4

Test Method
  Shear Strength was measured at room temperature and 200 °C .

Refer to Figure 6

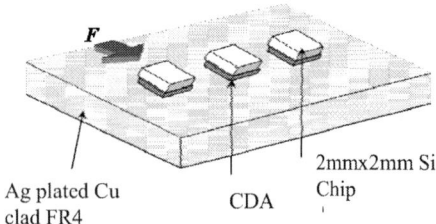

Figure 6 Shear Strength Test Method

Irradiation Test

Confirm the strength change by irradiating Blue HID Lamp. Glass substrate is used when lighting on Conductive Die Attach. Refer to Figure 6 on strength measurement.

Pot Life Test

Confirm the viscosity change of test sample left under the conditions of 25 °C /50%.

Results and Discussion

Reflectance

Reflectance Test

Figure 7 is the result of reflectance change. Sample C and D are composed of BPA epoxy resin with benzene ring and, phenoric resin and aromatic amine as curing agent also contain benzene ring. This causes the degradation of reflectance from early stage and generating delamination between chip and glass substrate. (i.e. chalking phenomenun)

On the other hand, the degradation of reflectance gradually happens on sample A which contains no benzene ring. It is presumed that resin yellowing due to oxidization from heat lowers the reflectance ratio. As the figure shows, XH9680-43 and sample B with silicone modified epoxy resin keep stable for a long- time irradiation.

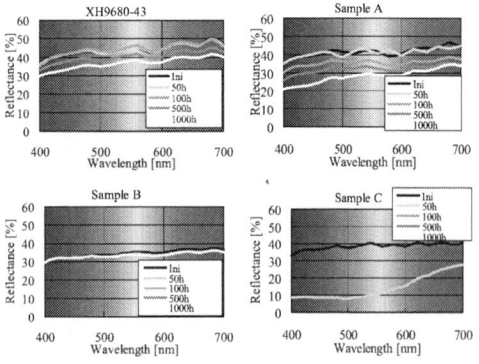

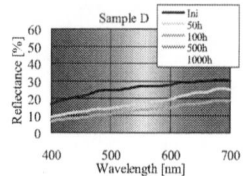

Figure 7 Change of Reflectance

Delta vf Simulation Test

Change of the contact resistance was measured under the thermal cycle from 300 °C for 15 seconds down to room temperature by three times.

Figure 8 shows the result of thermal cycle test at reflow simulated temperature between 300 °C and room temperature.

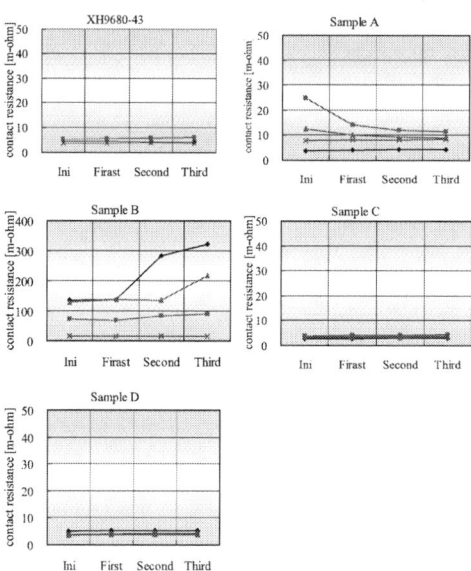

Figure 8 Change of the Contact Resistance

After moisture test of 30 °C /70% for 168 hours and reflowed at 260°C by three times, resistance change is measured. Figure 9 is the result for it.

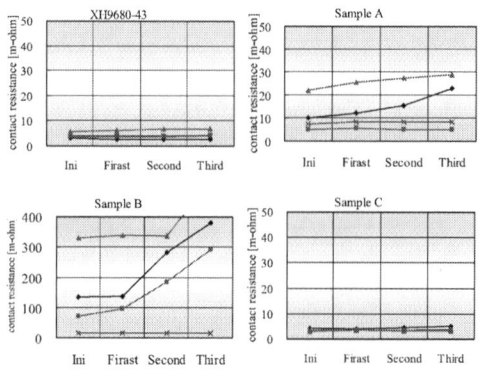

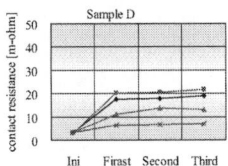

Figure 9 Change of the Contact Resistance (JEDEC Level 3)

Considering Figure 8, use of acid anhydrate might cause the degradation of contact resistance as sample A. While variable contact resistance is observed on sample B in early stage, and also contact resistance increase is comfirmed. This is because of big influence by heat and degraded strength due to siloxane structure in silicone.

As Figure 9 shows, under severe conditions of moisture reflow different from simple temperature cycle as Figure 8, a slight change occurs even on sample D. XH9680-43 keeps stable contact resistance under such a harsh environment.

Shear Strength Test

Measurement was done at room temperature and 200 °C.
The result is shown in Figure 10

Sample A and D have high strength at room temperature. Sample C and XH9680-43 show high strength at higher temperature compared with room temperature.

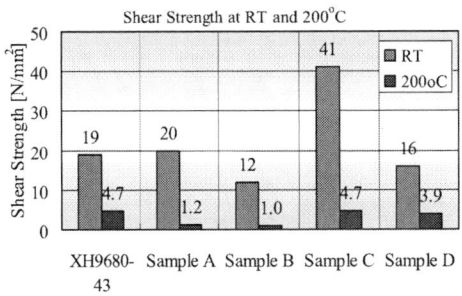

Figure 10 Shear Strength at Irradiation test

Irradiation test

Figure 11 shows the result of shear strength change when Blue HID Lamp is irradiated.

On sample C and D, the shear strength decreases in proportion to the degradation of reflectance ratio.

On the other hand, sample B and XH9680-43 show stable shear strength. This might give us the assumption that resin system of XH9680-43 is hardly influenced by the light of 450nm along with silicone modified epoxy.

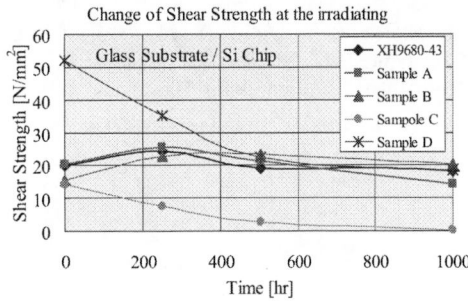

Figure 11 Shear Strength at Irradiation Test

Pot Life Test

Figure 12 is the result of pot life measurement.
The viscosity of sample A and B doubles after 24 hours. Therefore, keeping stable pot life for acid anhydrate might be difficult. The viscosity of other samples keeps stable and XH980-43 has pot life in two days.

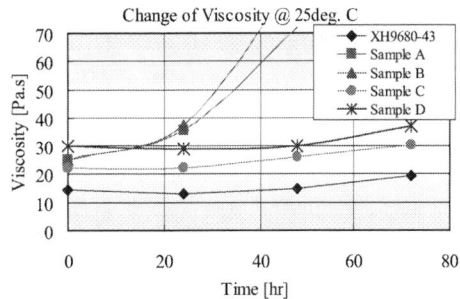

Figure 12 Change of Viscosity at 25 °C

Conclusion

We have succeeded in developing Conductive Die Attach for blue LEDs which is stable with irradiation of the light of 450nm and also stable with moisture reflow with the property of high temperature resistance by using both hydrogenation epoxy resin and curing agent with urethane structure.

Since urethane bonding is generated, it is thought that there would be low resistance change due to high stability with light and high temperature resistance.

References

[1] Atsushi Okuno, "Resin for White LEDs and Latest Technology for LEDs Packaging" The Japan Society of Epoxy Resin Technology, pp. 2-3, March 2006

[2] Yoshinobu Oonuma, " The Latest Trend of Epoxy Resin for LEDs Encapsulant and Curing Agent" , pp. 9, The Japan Society of Epoxy Resin Technology, January 2006

[3] Masafumi Tanaka, "Epoxy Resin Composition and Adhesive for Optical Semiconductor Using The Same, Patent Publication 2004-256603

[4] Keiji Iwata, "Polyurethane Hankbook", pp. 38-40, Nikkan Kogyo Shinbunsha, September 1987

[5] C.W. Bunn: J. Polym. Sci. 16, 323(1955)

Short and long term reliability of in-mould sealed bare and glob-top shielded LED devices

K. Keränen[1], M. Silvennoinen[2], A. Lehto[3], J. Ollila[1], T. Salmi[4], J.-T. Mäkinen[1], A. Ojapalo[5], M. Schorpp[6], P. Hoskio[7] and P. Karioja[1]

[1]VTT, Kaitoväylä 1, 90570 Oulu,
+358 20 722 2272, +358 20 722 2320, Kimmo.Keranen@vtt.fi
[2]Perlos Corp., Menotie 1, 33470 Ylöjärvi,
[3]Helsinki University of Technology, Laboratory of Physical Metallurgy and Materials Science,
P.O.Box 6200, 02015 HUT,
[4]VTT, Metallimiehenkuja 6, 02044 VTT,
[5]Perlos Corp., Äyritie 8A, 01510 Vantaa,
[6]Nokia Corp., P.O. Box 1000, 33721 Tampere,
[7]Electroplast Corp., Kempeleentie 46, 90440 Kempele, FINLAND

Abstract

In-mould integration of bare and glob-top shielded LED chips was demonstrated in an injection moulding process. The LED chips were first attached on conventional FR4 substrates and the substrates were used as inserts in the injection moulding process. Test series of 39 modules containing four green LED chips and 80 modules containing one blue chip were in-mould sealed using a polymethyl methacrylate (PMMA) polymer. When the holding pressure used in the injection moulding processing was 1000 bar or lower, 100% yield was obtained for the injection moulded LED modules. LED modules, which were operational just after in-mould sealing, remained operational after six months storage in normal office room conditions. Sealed LED long term reliability was estimated by accelerated environmental stress tests with elevated temperature and humidity. The test results showed that the in-mould sealing is a very promising manufacturing method for cost-effective module production. In order to get more reliable estimate for volume-production yield, however, larger series of test modules should be fabricated and tested. The required production tests are performed in the next phase of the manufacturing concept study.

Key words: in-mould sealing, module integration, bare chip, cost-effective production

Introduction

Photonic module manufacturers pursue miniature, long term stable and precise module realisation and assembly technologies in order to achieve cost-effective and reliable solutions for market demands. The integration of optical, electrical and mechanical functionalities into one system, can greatly improve the cost efficiency of systems, due to the fact that the packaging cost is reduced [1],[2]. In addition, the very tight alignment tolerance requirements of discrete photonic devices are mainly avoided through high precision lithographic or comparable processing technologies, which further enhance the long-term performance and reliability of the systems [3],[4].

Monolithic integration of signal processing and photonic devices in a single chip, however, is not yet cost-efficient or technologically possible and hybrid integration is still an applicable approach [5].

The best long term reliability of the photonic modules is achieved by sealing the sensitive devices hermetically [6],[7],[8]. The hermetic sealing, however, is not necessarily needed in all applications. The mainstream packaging method for high-volume production of light emitting diodes (LEDs) is device sealing with epoxy moulding. The moulding offers the required long-term reliability for devices. In addition to the reliability, the sealing material provides refractive index matching between the semiconductor material of the LED chip and air.

We studied possibilities to utilize in-mould hybrid integration technology in photonic module manufacturing. The concept idea is to assemble bare chips on a substrate or circuit board, such as, a conventional FR4 or low temperature co-fired ceramics, LTCC, substrate, and use thereby formed subassembly as an insert in injection moulding. Mechanical and optical structures are formed and bare chips are sealed in a single or multi-step injection moulding process in such a way that required mechanical and optical interconnects and required device encapsulation are provided for the module. The in-mould injection moulding of photonic devices using optical thermoplastics seems to be a similar and cost-efficient method as epoxy moulding to obtain the required long-term reliability of photonic devices.

In this paper, we demonstrate the in-mould integration concept of photonic modules based on injection moulding of optics and mechanics as well as optoelectronic subassemblies as in-mould inserts in a conventional injection moulding process. The manufacturing concept is demonstrated with bare and glob-top shielded LED devices. The manufacturing yield and long-term reliability of packages are evaluated.

In-mould Integration of LED chips

The in-mould integration process of LED integration is the following. Electronic substrates, such as, FR4 or LTCC substrates are processed and LEDs and other possible active devices as well as semiconductor chips, are assembled onto the substrates. The produced subassemblies are used as inserts in the injection moulding process, in which the module optical and mechanical interconnections are implemented [9],[10]. The encapsulation of the active devices and refractive index matching between chip semiconductor material and air is achieved in the same injection moulding process.

The design of electronic subassemblies suitable for in-mould insertion process is a very critical issue. The electronic insert should provide accurate alignment between the devices on the substrate and optical structures processed in the injection moulding process. In practice, the achievable alignment accuracy between chips and optics is defined by the chip alignment accuracy on the substrate, substrate alignment accuracy within the mould, mould tooling accuracy and injection mould tooling accuracy. These factors together generate the tolerance chain for the module implementation. By using modern flip-chip assembly machines, chips can be assembled on the substrate below +/- 1 µm horizontal alignment inaccuracy. The achievable alignment accuracy in chip assembly typically depends of the available cycle time. In mass production, a short cycle time is pursued, which results to a lower alignment accuracy than with a long cycle time. Dimensional accuracy of the substrate and the mould define the best possible accuracy between the devices and optical structures, when mechanical alignment structures are used in the mould for the substrate alignment. In practice, the manufacturing errors processed in the injection moulding make the final contribution to the tolerance chain.

It is very important to adequately notice in the design of the electronic inserts and assembly of the devices that the insert and the devices should tolerate the injection moulding process conditions. These conditions are very challenging: the temperature of the melt thermo plastic material can be 300°C (573 K) and holding pressure over 1000 bar (equals 100000 kPa) for several seconds. Thermal shock induced to the sub-substrate and devices combined with high pressure seemed to offer a formidable challenge for the devices and assembly technologies.

We studied the process practices in device assembly and parameters in injection moulding in order to obtain high yield in packaging. Operation of the components after packaging can be considered to be based to the overall capability to produce light output. In a case of LED device, the light output of an active device can be higher after injection moulding encapsulation compared to non-encapsulated devices due to better refractive index matching, which allows larger optical power to be extracted from the device.

The achieved yield forms a very critical parameter value, when evaluating the packaging processes overall feasibility and cost. The lifetime of a photonic device is normally defined by the excessive light output degradation. Typically, defined lifetime criterion is met, when the device optical power is decreased to 80% or 50% level compared to the initial value. The lifetime of an encapsulated device can be considered to represent its long-term reliability. Naturally, device total operation failure is possible, also. The total operation failure causes are typically introduced by contact wire break or shortage or by device catastrophic optical failure (COD).

Injection Moulding Experiments

The LED devices used in the demonstrator were based on the G•SiC substrate technology combined with efficient InGaN materials and manufactured by CREE. The type of the devices was CxxxXT290-Sxx00-A. The measures of the chip were 300 µm x 300 µm x 105 µm (LxWxH).

Electronic substrates using flame resistant 4 (FR4) fibre glass reinforced epoxy and low thermal co-fired ceramics (LTCC) materials were designed and manufactured for a demonstrator implementation. The demonstrator substrate metallization design is shown in Figure 1.

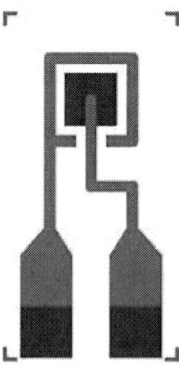

Figure 1: Substrate Design for LED Demonstrator

The LED device was dye bonded to the FR4 substrate using Epotek H20E conductive epoxy. The upper contact was wedge bonded using 25 µm gold wire.

The LED demonstrator system was planned to be processed with a Fanuc α100iA injection moulding machine. The mould was designed so that by changing the other half of the tool, another demonstrator system was possible to implement. Designed tool halves for integrated injection moulding of the LED demonstrator are shown in Figure 2 and 3.

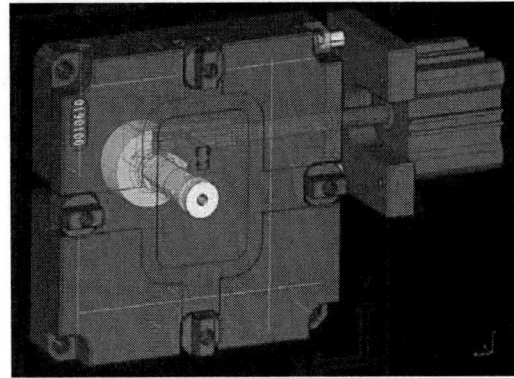

Figure 2: Fixed Tool Half

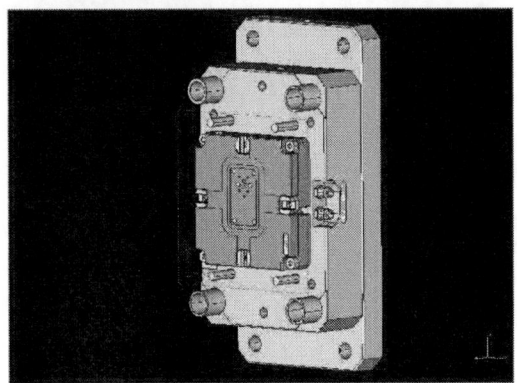

Figure 3: Moving Tool Half

Adjustable shut-off ensures that the substrate contact area is not overmoulded. In addition, the adjustable shut-off ensures that the thickness of the polymer layer keeps constant despite substrate thickness variation. A diamond turned insert could be assembled into the moving half tool to produce high quality lens surface. In processing of the LED demonstrator, however, a hand polished insert was used. The basic idea was to implement a clear window above the LED device. The clear window enables evaluation of bonding wire attachment both to the device and to the substrate bonding pad by a microscope.

The prototypes were implemented in the injection moulding process in the following way. Electronic inserts were put on the fixed half by hand. Two pins in the tool and matching semicircles in the insert were used for the substrate alignment with the mould. In the test runs the operation of the LEDs were tested just before inserting the substrate in the mould. The FR4 LED substrate containing four

green devices placed in the mould is shown in Figure 4.

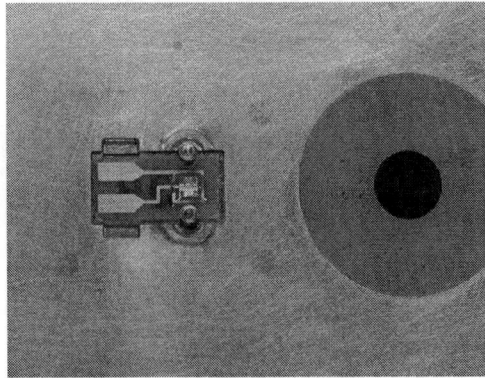

Figure 4: FR4 Substrate Placed in the Mould

Prototypes were injection moulded using thermoplastic PMMA, type Plexiglas 6N. Peak injection pressure during moulding was around 700 bar and holding pressure varied between 600 and 1200 bar. The temperature of the melt material was 245°C and the temperature of the mould 75°C. The implemented FR4 LED demonstrator is shown in Figure 5.

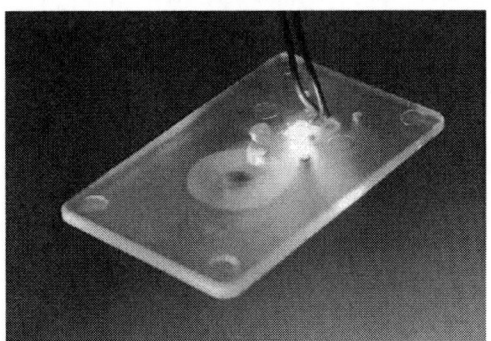

Figure 5: Implemented FR4 LED Demonstrator

The overall moulding experiences achieved with LTCC substrates were quite similar to the results with FR4 substrates with two main differences. The LTCC multi layer ceramic substrate is very rigid and it is dimensionally more accurate than FR4. The dimensional accuracy allows more accurate alignment of the substrate with the mould, when using alignment structures in the mould. On the other hand, the substrate hardness can be a negative characteristic, if the substrate alignment feature dimensions are not implemented with adequate tolerance. Breaking of the LTCC substrate happened in cases, when substrates with too small semicircle radius alignment features were inserted to the alignment pins of the mould and used in-mould injection tests. The implemented LTCC LED demonstrator is shown in Figure 6.

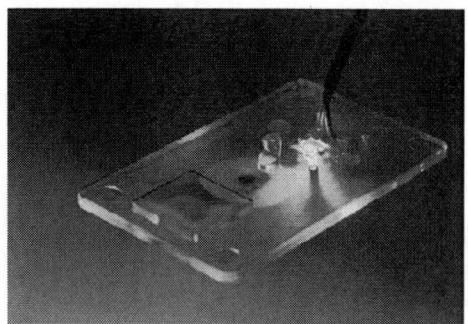

Figure 6: Implemented LTCC LED demonstrator

Testing of the First Run Demonstrators

The operation of the encapsulated devices was tested just after the product surface temperature was decreased to near room temperature. Immediate observation target in the test runs was to follow the device failure dependency of the used holding pressure in the injection moulding. Both bare and glob-top shielded devices were used in the test processing.

In the first test run a series of 39 LED modules each containing 4 green LEDs were implemented. LEDs were glob-top shielded in 24 modules. In 15 modules, bare LEDs were encapsulated in injection moulding. All glob-top shielded LEDs (44/44) were operational just after processing, when holding pressure equal or below 1000 bar was used in the processing. Three LEDs of 32 were non-operational, when bare LEDs were encapsulated using equal or below 1000 bar holding pressure. As a conclusion the yield with glob-top shielded devices was 100% and with bare devices 91%, while equal or below 1000 bar holding pressure was used in the processing. It seemed that the module yield was possible to increase by decreasing the holding pressure used in the processing.

The fault analysis of the non-operational devices showed that the main reason for device failure was loosening of bonding wires during the injection moulding process. In Figure 7, a close-up of a processed module with bare LEDs is shown.

Figure 7: Bare Bonding Wires after the Injection Moulding Process

As one can see from Figure 7, the thermoplastic material flow has formed the bonding wires distinctively. It seemed that the bonding wire loosening from substrate metallization or device pad was especially possible, when the wire direction was transverse to the material flow. On the other hand, a short circuit was possible, when the wire direction was parallel to the material flow. The short circuit seemed to be possible to avoid by layout modification by locating device to the margin of the metallization and using as short wire as possible.

The forming of the bonding wires by the material flow can be reduced by shielding the bonding wires by a glob-top material. In Figure 8, a close-up of a processed module with glob-top shielded LEDs is shown. As one can see from Figure 8, the glob-top shield has prevented wire forming and possible loosening during the injection moulding process.

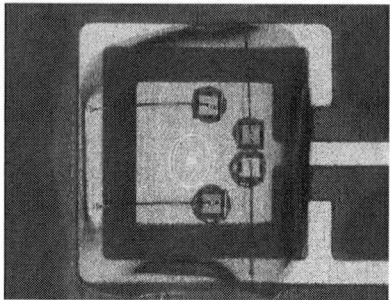

Figure 8: Glob-top Shielded Bonding Wires after the Injection Moulding Process

Processing and Testing of the Second Run Demonstrators

In the second test run 80 demonstrator modules, each containing one LED device was processed. LEDs were glob-top shielded in 40 modules and bare LEDs were used in 40 modules, correspondingly. The used injection pressure and holding pressure was 1000 bar in the processing. The temperature of the melt material was 245°C and the temperature of the mould 75°C. All the encapsulated LEDs were operational after the processing, so the resulted module yield was 100%. In Figure 9 an example of processed LED module in the second test run is shown.

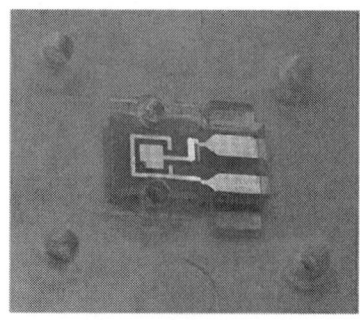

Figure 9: Example of a Processed LED Module in the Second Test Run.

The long-term reliability of the demonstrators was preliminary tested by accelerated stress tests in elevated humidity and temperature. Ten pieces of both shielded and non-shielded modules were placed in test conditions of 80%RH and 60°C. The total test time was 2790 hours for the modules. All modules were operational after the test period and insignificant optical power drop was monitored from the modules during the test.

The effect of thermal cycling was preliminary tested with the modules. When thermal cycle of -40 ... + 70°C were used for the LED demonstrator, the monitored optical power of the LED varied as a function of temperature. The variation was at the same level than the LED device inherent performance in equivalent temperature variation. The encapsulated LED device remained operational after several hundreds of cycles. In addition, all encapsulated LEDs were operational after six months storage in normal office conditions.

Discussion

We demonstrated LED device encapsulation using electronic substrates containing bare and glob-top shielded LED devices as inserts and encapsulated the substrates with PMMA thermoplastic in injection moulding process. According to the results achieved with demonstrator series, it is possible to achieve high yield in the processing. The long-term reliability of the encapsulated devices seemed to be good according to the performed accelerated aging tests using elevated humidity and temperature. The number of implemented prototypes in this case, however, was very limited. Test series containing at least hundreds, preferably thousands of prototypes, should be implemented in the future in order to evaluate large-scale production feasibility and attainable yield more definitely.

Conclusions

Cost-efficient integration of mechanical and optical structures and encapsulation of bare LED chips in the standard injection process proved out to be realisable. Using the in-mould integration method, we encapsulated optoelectronic subassemblies containing bare and glob-top shielded LED chips as inserts and encapsulated the subassemblies with the PMMA thermoplastic material in the injection moulding process. The reliability tests carried out for the prototype series show promising results. In order to confirm the high-volume production capability of this technology, however, large scale prototype runs and reliability tests should be carried out.

Acknowledgements

Financial support received from National Technology Agency of Finland (TEKES), VTT and Finnish industry is acknowledged. Furthermore, the authors want to acknowledge Sari Kivelä and Airi Weissenfelt who carried out assembly of the devices to the substrates.

References

[1] R. Lestra and J.-Y. Emery, "Monolithic integration of Spot-Size Converters with 1.3 μm lasers and 1.55 μm polarization insensitive semiconductor optical amplifiers," IEEE J. of Selected. Topics in Quantum Electron., vol 3, pp 1429-1440, December, 1997.

[2] H. Cho, K.-M. Chu, S. Kang, S. Hwang, B. Rho, W. Kim, J.-S. Kim, J.-J. Kim and H.-H. Park, "Compact packaging of optical and electronic components for on-board optical interconnections," IEEE Trans. Adv. Packag, pp. 114-120, February, 2005.

[3] D. Bartelink, "Integrated systems," IEEE Trans. on Elect. Dev., vol 43, pp 1678-1687, October, 1996.

[4] M. Aoki, M. Komori, H. Sato, T. Tsuchiya, A. Taike, M. Takahashi, K. Uomi and S. Tsuji, "Reliable wide-temperature-range operation of 1.3 μm beam-expander integrated laser diode for passively aligned optical modules," IEEE J. of Selected. Topics in Quantum Electron, vol 3, pp. 1405-1412, December, 1997.

[5] E. Towe, "Heterogeneous optoelectronic integration,", SPIE, Bellingham, Washington, pp. 1-6, 2000.

[6] T. Sakai, S. Okamoto, T. Iikawa, T. Sato and Z. Henmi, "A new laser hermetic sealing technique for aluminium package" IEEE Trans.on Comp., Packag., and Manufact. Tech., vol 10, pp 433-436, September, 1987.

[7] J. Ollila, K. Kautio, J. Vähäkangas, T. Hannula, H. Kopola, J. Oikarinen and M. Sivonen, "Hermetic diode laser transmitter module", In Proceedings of SPIE, vol 3626, pp 123-126, 1999.

[8] D. Schleuning, R. Dato, G. Frangineas, M. Jansen, C. King, R. Nabiev and C. Nabors, "Packaging multiple active and passive elements in a hybrid optical platform", J. of Lightwave Tech., vol 22, pp. 1320- 1326, May 2004.

[9] J.-T. Mäkinen, K. Keränen, J. Hakkarainen, M. Silvennoinen, T. Salmi, S. Syrjänen, A. Ojapalo, M. Schorpp, P. Hoskio and P. Karioja, "Inmould integration of a microscope add-on system to a 1.3 Mpix camera phone", In Proc. of SPIE, vol 6585, 2007.

[10] K.Keränen, T. Saastamoinnen, M. Silvennoinen, J.-T. Mäkinen, I. Mustonen, P. Vahimaa, T. Jääskeläinen, A. Lehto., A. Ojapalo, M. Schorpp, P. Hoskio and P. Karioja, "Injection moulding integration of a red VCSEL illuminator module for a hologram reader sensor", In Proc. of SPIE, vol 6585, 2007.

EMBEDDING A THIN POLYMER VOLTAGE ESD SUPPRESSING CORE IN A CHIP PACKAGE - ALTERNATIVE TO ON CHIP ESD PROTECTION

Karen Shrier, Electronic Polymers and Paul Collander, Poltronic

Electronic Polymers, Inc.

525 Round Rock West Drive Suite 200, Round Rock, TX 78681

Phone +1 (512) 583-8300, kshrier@electronicpolymers.com

Abstract

Today's portable cell phones and commercial electronics have the potential to impact the quality of our daily lives by the reliability of their performance. One of nature's hidden threats to electronics is electrostatic discharge (ESD). As the electronics content of our daily lives continually expands, we have become more and more dependent on electronics that contain digital chips whose ESD reliability is getting worse. Engineers are being forced to sacrifice reliability for performance in the competitive race to provide more features in less space.

This paper describes the ESD problem and how it has been solved with new nanomaterial filled polymers. It describes how the ESD suppressors are tested and how the new material enables designers to design in protection in the packaging.

A cell phone is used as a well known example of a device which has ESD sensitive I//O channels. EPI's surface mount technology and EPI-Core embedded in an IC package can be applied as an easy and cost effective way to attain ESD protection.

Key words: ESD-protection, embedded passives, reliability testing

Introduction

The drive for new features has been supported by advances to 65nm CMOS chips and GaAs chips which are far more sensitive to ESD than their predecessors of 5 years ago. In 2005 the National ESD Association published a roadmap showing ESD sensitivity of CMOS chips has not only taken a huge nose dive over the last five years, but also the map is predicting ESD sensitivity is going to get even worse between now and 2010. [1] ESD survivability in 1995 for Human Body Model (HBM) ESD was at 2000 V, today, in 2006 HMB survivability has dropped to 200V, and is heading to less than 200V.

Further exasperating the ESD problem is the trend towards pervasive use of portable electronics, which led to the introduction in early 2000 of a system level ESD standard that anticipates the amount of current a digital chip can be exposed to at the system level is going to increase from HBM, worst cast, of ~ 6 Amps to System Level Current of ~ 30 Amps. The duration of the ESD event is extremely short, typically less than 100 ns, but the speed of System Level ESD is so fast that it can reach 8KV in 1 ns. In 2002 Scientific American published an article written by an IBM expert of on-chip ESD design stating "Electrostatic discharges threaten to halt further shrinking and acceleration of electronic devices in the future" indicating ESD protection on semiconductor components is a barrier to Moores Law. [2] The key problem is digital chip designers need low capacitance ESD protection structures on chips in order to run at high frequencies without interfering with the signal transmission. Since no cost effective solution has been readily available to the ESD Engineer, the only alternative has been to trade reliability for performance. It is estimated that 27-33% of customer returns are due to ESD returns. [3]

Polymer voltage suppressors

At Electronic Polymers Inc., EPI, we have developed EPI-FLO™ ESD polymer voltage suppressors specifically for ESD protection of sensitive digital chips. Initial product developed was standard foot print 0402 Surface Mount components for protection of IC's from HBM and System Level ESD. The Surface Mount product led to EPI-FLO™ Connector press fit arrays capable of multiple line ESD protection.

The EPI-FLO™ polymer is manufactured as laminate 12x18" sheets, with the voltage sensing polymer sandwiched between copper sheets. The typical sheet thickness is less than 2 mils, composed of .75 mil copper. Using standard printed circuit board circuitzing processes the laminate is transformed into EPI-FLO™ Surface Mount and Connector Arrays devices. **Figure 1** is a cross sectional diagram of an EPI-FLO™ SMT. **Table 1** lists typical SMT data sheet parameters.

Table 1: SMT Data Parameters

SMT Data Parameters

- Typical Capacitance <200fF
- Pico second response time
- Wide Bandwidth > 10Ghz
- Low leakage current < 1 micro amp
- Protects single or multiple I/Os
- Bidirectional
- Low Profile < 10mils
- Up to 40 line ESD Protection

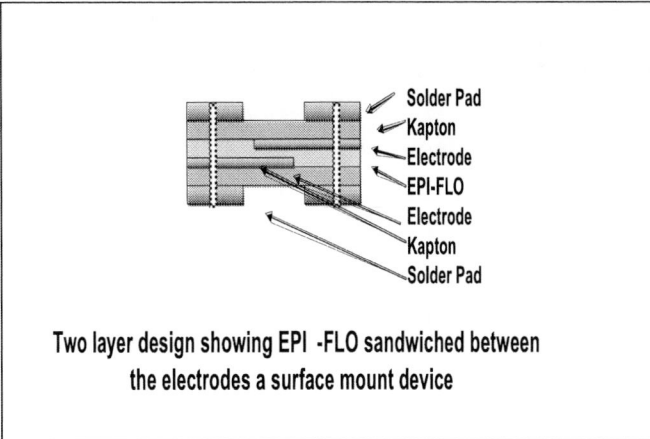

Two layer design showing EPI -FLO sandwiched between the electrodes a surface mount device

Figure 1: EPI-FLO^{TM} is between the two electrodes of the surface mount device.

EPI has also developed product design to accommodate the build of a 4 layer package with embedded EPI-FLO^{TM}. Specifically, EPI has designed a method to embed EPI-FLO™ in the package of a cell phone GaAs power amplifier. This design involves extending the through hole via process and etch process that EPI used previously for two layer Surface Mount and Connector Array devices. **Figure 2** is a diagram of the etched EPI-FLO™ laminate embedded in a 4 layer Power Amplifier board. The unique facture of the SMT that leads to the concept of embedding EPI-Core™

in IC Packages is the fact that there EPI-Core material is the substrate combining the ESD voltage suppression with the Femto Farad capacitance of EPI-Core™, opens the door for off chip ESD suppression in the IC package. EPI-Core™ embedded in the package provides I/0 as well as pin to pin protection in cooperation of EPI-Core™ laminate in an IC package adds another layer in the package board, transforming a 2-layer into a 4 layer board. The EPI-Core™ increased the package thickness by 2 mils (50_m)

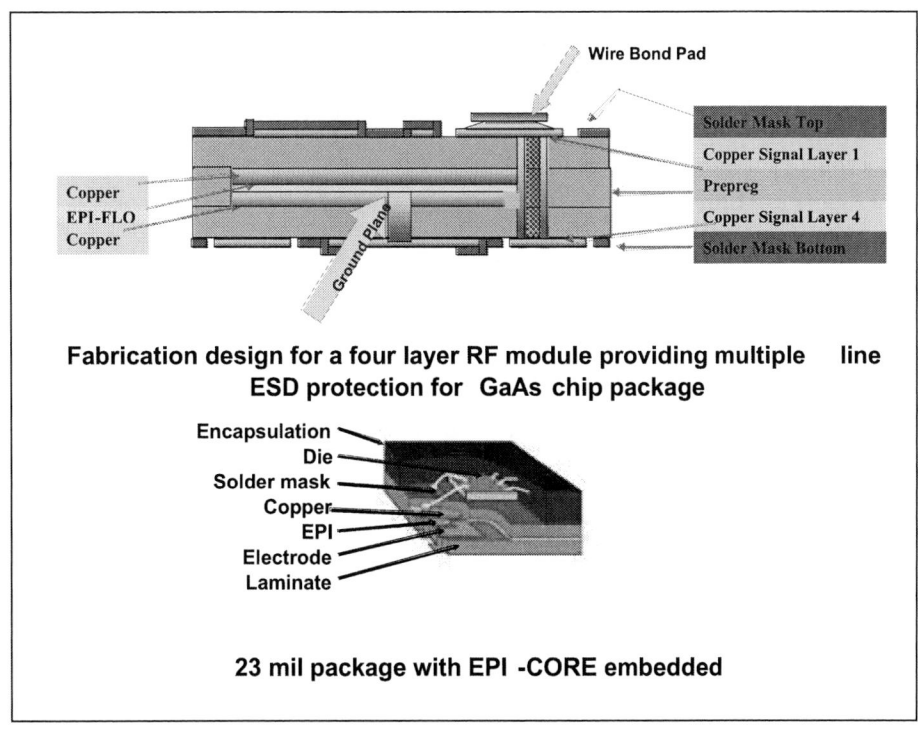

Fabrication design for a four layer RF module providing multiple   line ESD protection for GaAs chip package

23 mil package with EPI -CORE embedded

Figure 2: EPI-CORE can be embedded in semiconductor packaging.

Figure 3 shows an EPI-FLO™ connector array that has been press fitted onto a microprocessor chip for prototype evaluation prior to building EPI- Core™ into the package from the EPI- Core™ laminate. The artwork for connector array allows for direct attachment of ground pins to the array ground with a press pin fit design. Similarly sensitive signal pins use a press fit washer to provide an electrode for shunting an ESD pulse to ground.

Figure 3: EPI-FLO™ connector array is made to be press fitted onto microprocessor chip.

Cell phones ESD sensitivity and protection

All electronic devices are ESD sensitive on their I/O channels. In a mobile phone these are antennas, keyboard and microprocessor. In addition to protecting the I/Os, components next to the I/O must also be protected. Such components include the PA (Power Amplifier) and the LNA (low Noise Amplifier). These are the components may be degraded or killed by the ESD pulses. Cell phone services providers identifying a large number of dropped calls caused by ESD degraded components, especially PA's. Thus ESD protection is a common interest for all in the cell phone supply chain, the operator and the user.

Protection of the antenna and human interfaces in a cell phone is fairly easy, but it needs a holistic approach, independent of component suppliers. **Figure 4** details the most ESD sensitive components of a high end mobile phone. To reap the benefits of reduced weight, space, height and reliability requires designing in single or multiple IO protection on the PCB or in the IC package during the cell phone design phase. **[4]**

Areas detailing the use of SMT, connector, and EPI- Core™ embedded in the package are shown in Red.

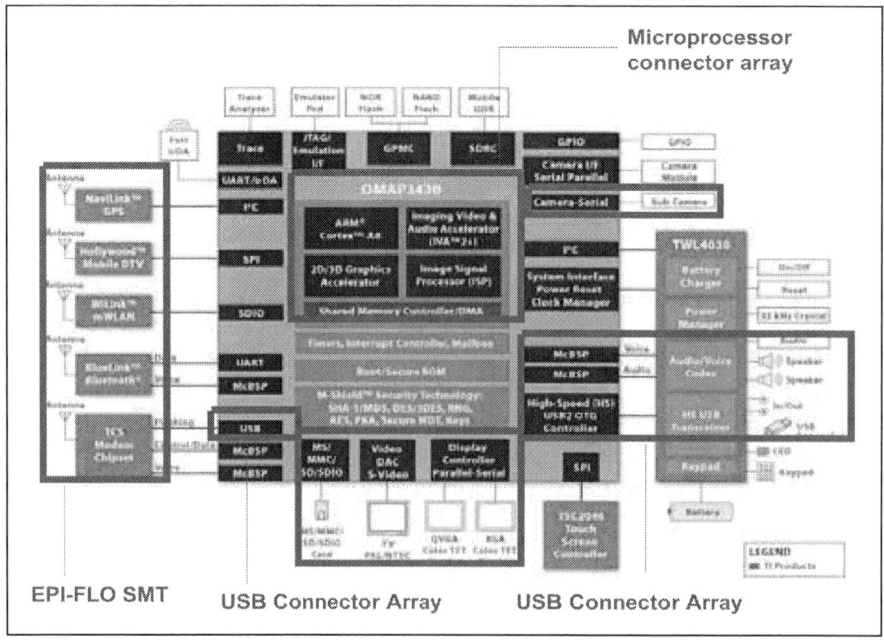

Figure 4: ESD sensitive areas of a mobile phone are outlined in red as well as EPI-FLO™ protective solutions.

Test protocols for ESD protection of chips

At Electronic Polymers test protocols and test equipment have been developed to enable ESD protection of sensitive chips. We have used our test protocol successfully to protect chips from Human Body Model, Machine Model, Charged Device Model, System Level IEC 61000-4-2, and the new Cable Discharge Event ESD. To provide ESD solutions we first determine the failure voltage of the chip using the ESD standard specified, and then using our System Level/IC, Very Fast Transmission Line Pulser with a 126Hz, 40Giga samples/sec sampling, 1.5, 12GHz rate, we establish the voltage and current that damages the chip. The EPI-FLO™ is then specified to turn on at a lower voltage. Then the IC fails to dissipate current which would cause IC failure. The protection is verified initially by building test fixtures and mounting EPI-FLO™ surface mount 0402 units in front of the pins to be protected. Protection is demonstrated by showing the chip can survive multiple ESD hits as required by the specification with both positive and negative pulses. Typically devices are most sensitive to negative pulses. Failure of the chip is measured by a 10- 20% change in leakage current
under power. Most PAs will continue to increase the leakage current form µA to milliamps long after the 20% change in leakage has occurred. It is the damaged unit that can cause dropped calls, and increased battery use.

Figure 5 Shows a typical Test setup for a USB VI/IVL curve trace system.

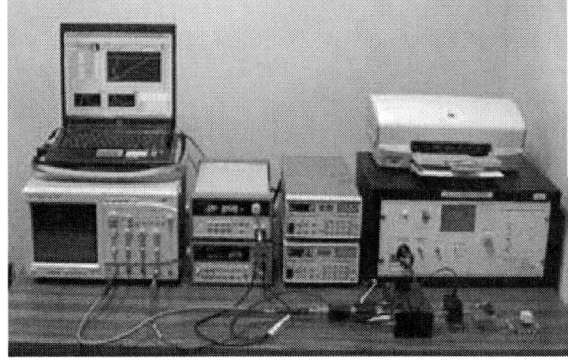

Figure 5 shows a typical test set up

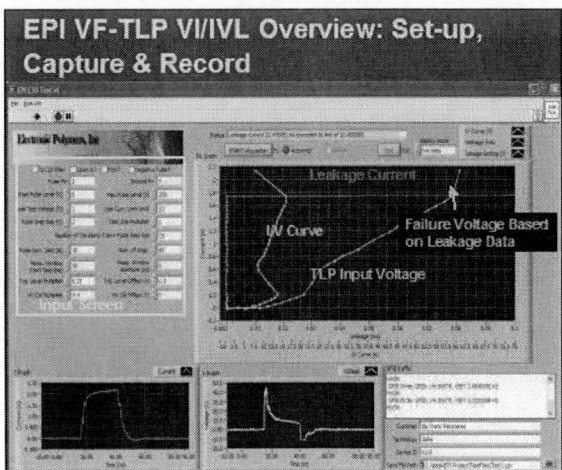

Figure 6: Test set-up panel

Figure 6 shows a VI/IVL curve trace test set-up panel showing input window and cumulative test results from subsequent test pulses.. This system is instrumental in EPI's failure analysis studies.

Once failure specification is determined, artwork is created for the multi-layer package with ESD protection. The unique benefit of embedding the EPI-FLO directly in the package is multiple protection of signal pins as well as pin-to-pin combinations. As an example, in a 6 pin GaAs power amplifier, 6 pins and 21 pin combinations are ESD protected.

Following is data showing the EPI-FLO enhanced ESD survivability of a GaAs cell phone Power Amplifier and a USB IC to System Level ESD.

Protection of IC

Figure 7 is I/V trace of a cell phone USB IC protected with an 0402 surface mount device, in parallel to ground. The scope trace shows no change in leakage current with 680V and 25.5Amp in the IC. Without EPI-FLO™ protection the device failed at 11 Amps, 440 Volts. To view the current through the chip a current probe was attached to the IC input. To view the current through the chip a current probe was attached to the IC input.

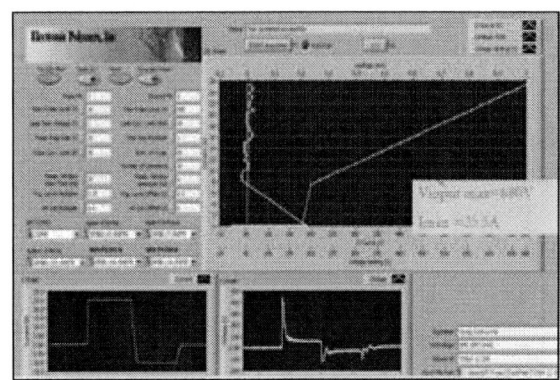

Figure 7: 0402 protection of USB IC to 23Amp.

Figure 8: Below shows the majority of the current goes through the EPI-FLO™. The scope trace shows that with 610 Volts input, 21Amps went through the 0402 and only 1.9Amps went through the USB IC.

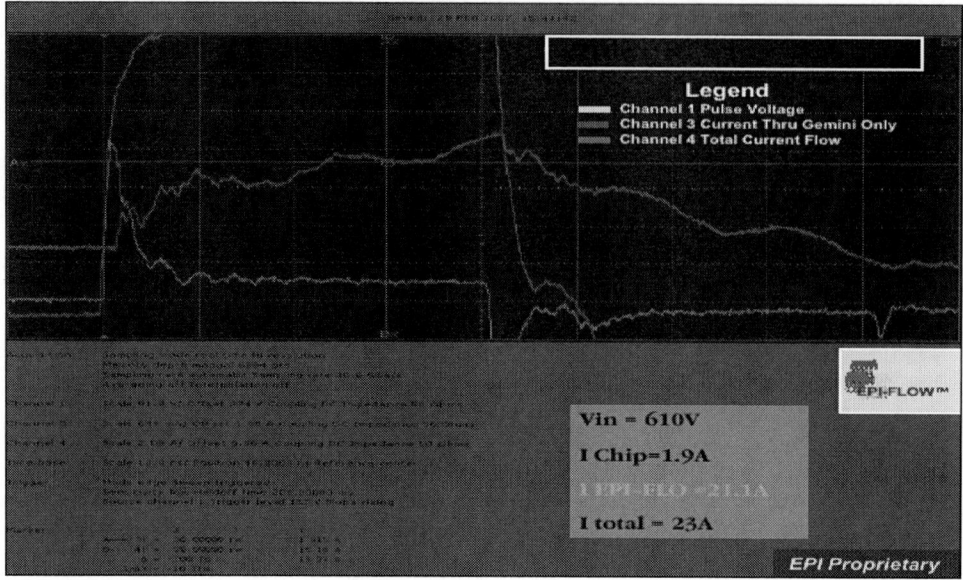

Figure 8: Scope trace shows total current flow on Channel 4, and flow thru the IC only on Channel 3, channel 1 is the pulse voltage

Protection of Ethernet Chip

Like the EPI-FLO™ surface mount, the EPI-Core embedded solution has also show protection of PA's. Figure 9 is a typical I/V trace of the protection of the PA with an EPI-Core embedded package. The PA is able to withstand >9A, 250V with EPI-Core. 9Amps correlates to an 8KV HBM ESD, without EPI the PA fails at 500V HBM, and 1.5A, 85V on the VI/IVL TLP.

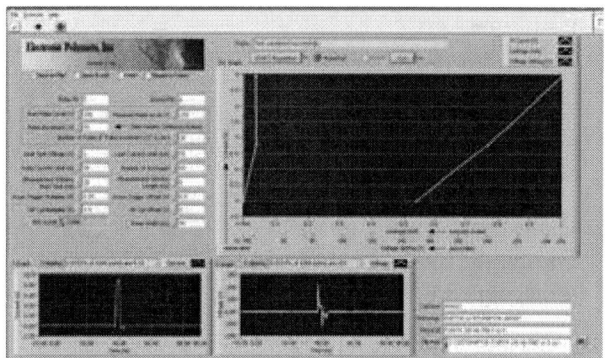

Figure 9: Failure of PA protected with EPI-FLO™ in package increased from 1.5A to > 9A

As shown in the picture, the EPI-FLO™ SMT shunts 21A of current. Therefore, the Ethernet chip is exposed to only 1.9A, and protected from this ESD event.

Conclusions

The reliability of electronic components has been in decline for the last five years. Due to the higher sensitivity and the increasingly small dimensions of electronic components, the risk of ESD damage will only become worse. There are very fast switching protective devices which can protect almost any electronic component. However, most ESD protective components have a detrimental impact on high speed performance and interfere with operating signal.

EPI-FLO^{TM} ESD polymer voltage suppressors have low capacitance, and provide ESD protection without interfering with the IC's operating signal. Further, EPI's various test protocols have provided a means to test certain types of components with different circuitry and environments. It is important to match actual use to specific situations. As a result of testing, EPI has been able to develop EPI-FLO^{TM} polymer voltage suppressors shown to provide adequate as well as cost effective solutions to ESD sensitive devices.

References

[1] ESDA Association Technology Roadmap, http://www.esda.org/documents/ElectrostaticDischargeRoadmap-March42004--ESDA--Final.pdf

[2] Steven H. Voldman IBM, Lightning Rods for Nanoelectronics, October 2002 Scientific American

[3] Karen Shrier, Tuyen Troung, and Jimmie Felps, "Transmission Line Pulse Test Methods, Test Techniques and Characterization of Low Capacitance Voltage Suppression Device for System Level Electrostatic Discharge Compliance," Proceedings of the EOS/ESD Symposium, 2004.

[4] Karen Shrier, Chi Jia, "Cell Phone GaAs Power Amplifiers: ESD, TLP, and PVS Devices", Proceedings of the EOS/ESD Symposium, 2005.

[5] Stephen Halperin, "Guidelines for Static Control Management," Eurostat, 1990.

[6] Stephen Halperin, Guidelines for Static Control Management, ESD Association, 2001

Tunable Dielectric Material Embedded in LTCC for GHz-Frequency-Range Applications

Sven Rentsch[1], Tao Hu[2], J. Müller[1], R. Stephan[3], M. Hein[3], H. Jantunen[2]

[1]TU Ilmenau, Functionalised Peripherics Group, ZIK MacroNano®
P.O. Box 10 05 65, D-98684 Ilmenau/ Germany
Phone: +49 3677 69 3372, Fax: +49 3677 69 3379, E-mail : sven.rentsch@tu-ilmenau.de, www.macronano.de

[2]University of Oulu, Microelectronics Laboratory and EMPART Research Group of Infotech Oulu
P.O. Box 4500, FIN-90014 Oulu/ Finland
Phone: +358 8 553 2710, Fax. +358 8 553 2728, hutao@ee.oulu.fi

[3]TU Ilmenau, Department of RF and Microwave Techniques
P.O. Box 10 05 65, D-98684 Ilmenau/ Germany
Phone: +49 3677 69 2831, Fax: +49 3677 69 1586, matthias.hein@tu-ilmenau.de

Abstract

Tunable dielectric materials offer an opportunity to realize novel wireless broadband applications such as antenna arrays with electrically controlled propagation delays. This kind of devices with good performance are already realized using discrete components such as diodes or barium strontium titanate (BST) thin films [1-2]. In the first case, however, extra assembling is required whereas with thin films, the production of the component can be expensive in large volumes. One promising method to exploit tunable dielectric materials is to make them compatible to LTCC (Low Temperature Co-fired Ceramics) process, thus enabling inexpensive and flexible manufacturing of functional multilayer ceramic modules (MCMs) for telecommunication industry. Barium strontium titanate with low sintering temperature (~900°C) has already been investigated for the LTCC process [3-4]. Additionally, first attempts to integrate it with other LTCC materials have been reported [5]. However, further research to enable a reliable co-firing process is needed since different materials have different shrinkages and thermal expansions easily causing delamination of the layers or warping of the whole module during sintering. In this work, a detailed research of the integration and interaction of the LTCC compatible BST layers with dielectric LTCC layers is presented. For process development, Pressure Assisted Sintering (PAS) is considered to enable flat multilayer multimaterial modules. Additionally, the effects of co-firing on relative permittivity and loss of the BST are reported as well as the temperature dependence of the relative permittivity (20°C-400°C).

Key words: —   BST, LTCC, tunable dielectric

Introduction

Low Temperature Co-fired Ceramics (LTCC) provides good possibilities for multilayer circuits incorporating buried passives for microwave applications along with relatively low production costs. One interesting option in the future would be the integration of a tunable layer into multilayer ceramic modules enabling more functionality. LTCC BST (barium strontium titanate) material had been already developed and first attempts to co-fire it with commercial LTCC dielectric tapes have been performed [4]. Co-firing of different ceramic materials is a challenging work because of their various sintering behaviors and different coefficients of thermal expansion (CTE). A common free sintering process always requires matched sintering behaviors and an adjustment of CTE for mixed material systems. Cracking, delamination or warping often occurs. Fig. 1 shows a sample warping when

Figure 1 Warping sample after sintering due to mismatched materials

prepared of tapes with two different materials and co-fired without special arrangements.

In this work, a co-firing of LTCC BST tape with DuPont® tapes (DP951 and DP943) was investigated. A pressure assisted sintering (PAS) as well as pressureless assisted sintering (PLAS) was used for the integration of BST layer into the DuPont LTCC layers [1]. The feasibility of co-firing and the microstructure of the multilayer was studied.

Test setup

The LTCC ferroelectric BST tape was made by tape casting using earlier developed material $Ba_{0.55}Sr_{0.45}TiO_3 - 5$ wt% $(B_2O_3... Li_2CO_3)$ [4]. The basic BST powder was provided by Filtronic Comtek UK. The LTCC BST powder was mixed with solvents (ethanol and xylene) and dispersant (blown Z-3 menhaden fish oil) in a ball mill for 24 hours, after which a binder (polyvinyl butyral) and plasticizers (butyl benzyl phthalate and polyalkylene glycol) were added and mixing for another 24 hours was done to obtain the slurry [6]. The organic additives, all commonly used in the tape casting, were supplied by Richard E. Mistler, Inc., USA. The tape casting was done with a laboratory caster (Unicaster 2000, University of Leeds, Leeds, UK). The green tapes were cut into 5x5 cm² size with thickness of 73 ± 3 μm.

The LTCC BST green tapes were laminated with DuPont tapes in an isostatic press at 70 °C / 180-250 bar / 10-30 min. Then the green body was fired in a HP 603-LTCC sintering press for PAS and a PEO601 furnace for PLAS, AVT Technologie GmbH, Germany. The whole layer stack (Fig. 2) was kept under pressure during the pressure assisted sintering process by porous plates allowing only Z

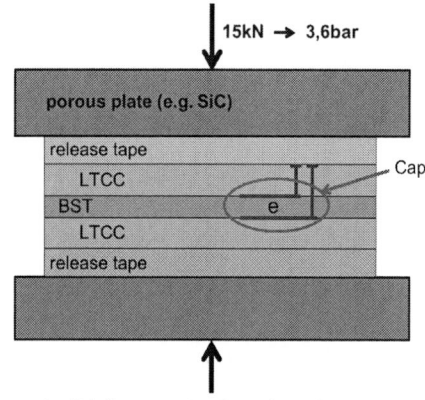

Figure 2 PAS setup during sintering

direction shrinkage. Furthermore release layers, which have to be become porous during firing (required for the burnout of the LTCC body), are needed to avoid bonding with the pressure plates [1]. The release layers consist mainly of Al_2O_3 powder, binder and solvents, briefly high temperature co-fired ceramic (HTCC) tapes. The whole zero shrinking process can be obtained without pressure as well (PLAS) by previously laminated specific tapes which keep the green ceramic body in YX-shape.

In this research, multilayers composed of DP951-BST-DP951 and DP943-BST-DP943 were examined. The considered dimensions of the substrate were 50 mm in width and length and the thickness of the embedded BST was 63 μm after firing (PAS). For sufficient mechanical stability of the substrate, thick dielectric layers on top and bottom were used. Two layers of 951 PX and four layers of 943 PT with an overall fired thickness of ~420 μm and DP951 RT (release tape) on both sides of the BST tape were used. The whole assembly was isostatically laminated under standard conditions (70 °C, 200 bar, 10 min). The pressure during the PAS was 3,6 bar. After removing the release tape, the thickness of the substrate amounted to about 900 μm.

For an electrical measurement embedded plate-capacitors were manufactured using silver paste (DP6148) and gold paste (HF522) with the DP951 and DP943, respectively. The relative permittivity and loss of BST in each multilayer module up to 1 MHz were measured by a LCR meter (HP Agilent 4284a). Microstructures were examined by SEM (Jeol JEM-6400, Tokyo, Japan).

Material analyses

The cross section of one co-fired sample by PAS is presented in Fig. 3. It clearly shows that the

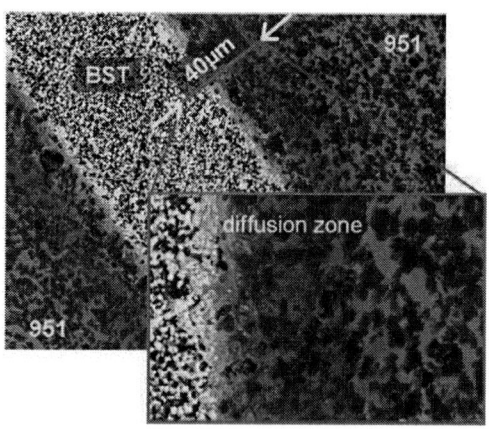

Figure 3 SEM-picture of the combination DP951-BST-DP951

DP951 and BST layer were firmly bonded without delamination. The diffusion zone between BST and DP951 was ~20 μm. Co-firing of the LTCC BST with DP943, however, was not successful, starting with cracks through to complete delaminations (Fig. 4). DP943 and BST showed an inert behavior in all test configurations. A firmly bonded compound couldn't be achieved.

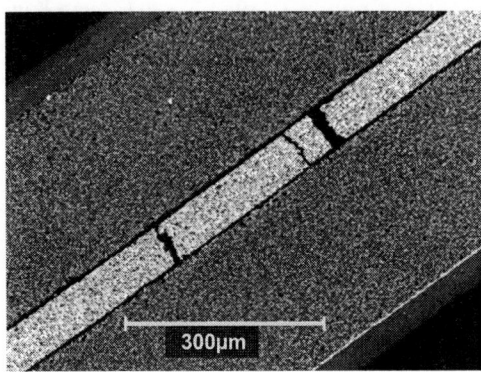

Fig. 4   SEM-picture of DP943-BST-DP943

Measurements

For the capacity measurements up to 1 MHz the prepared LTCC assemblage was put onto a computer controlled heating plate. The measurement starts at room temperature. Following the temperature was increased 5K at a time until 400°C. One must consider that the whole system has to achieve stable conditions after temperature changes, consequently the capacity remains steady. Gathered all capacity values against the temperatures, the permittivity was calculated in accord with the formula

$$\varepsilon_r = \frac{C \times d}{A \times \varepsilon_0},$$

where C is the Capacity measured for different temperatures and frequencies (50kHz, 100kHz, 1MHz), d represents the distance between the electrodes (57μm, process dependent), A the area of the electrodes (10x10 mm²) and ε_0 the electric constant. Due to the fact that the electrodes are relatively large-scaled potential field effects at the edges of the electrodes were not considered.

Fig. 5 presents the measured temperature

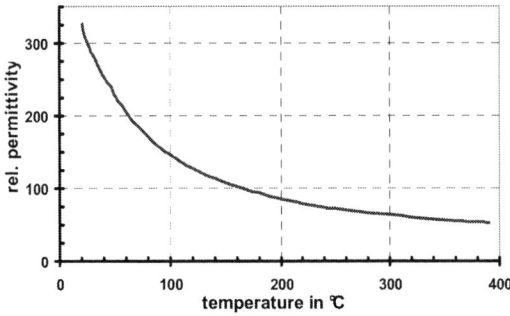

Figure 5 Permittivity vs. temperature of LTCC BST

dependence of the relative permittivity of the embedded BST. The relative permittivity decreases gradually from 320 at 20°C down to 50 at 400°C. Especially in the range of room temperatures a significant temperature dependence is recognizable while the value of the permittivity is relative high. However, at high temperatures the relative permittivity is barely affected by temperature changes. With regard to the considered frequency range from 50 kHz up to 1MHz the permittivity remains stable.

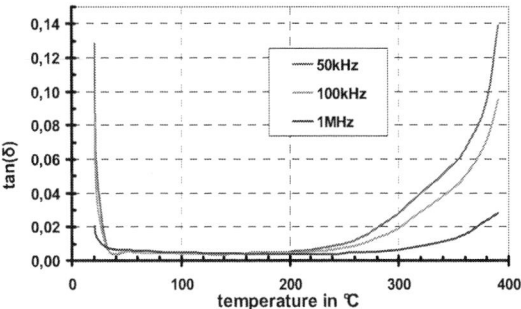

Figure 6 Dissipation factor vs. temperature of LTCC BST

The dissipation factors (tan(δ)) at different frequencies versus temperature are shown in Fig. 6. The dissipation factor was measured simultaneously with the capacity. It follows from the quality-factor Q,

$$\tan(\delta) = \frac{1}{Q}.$$

The losses are applying not only for the BST itself but also for the whole measured structure. However, using short lines to the electrodes and working at relative low frequencies with respect to the relative high permittivity of the BST the additional parts are negligible.

From 20°C to ~35°C the dielectric losses of the BST are decreasing. Due to the specific composition of the ferroelectric ceramic the Curie point is shifted below room temperature, hence all measurements are carried out in the paraelectric phase of the BST. Actually the losses at room temperatures should be much lower. But they can be suppressed by fine tuning the composition of the BST (Curie temperature).

The losses in the temperature range from 40°C to 160°C are constant at tan(δ)=0,006. At 1MHz they are slightly higher compared to the losses at 100kHz and 50kHz. Above 160°C the losses increase, especially at lower frequencies.

Conclusion

The results showed that the co-firing of LTCC BST tape with DP951 tapes by pressure assistant sintering succeeded. The SEM image of the sintered multilayer showed a dense microstructure without cracks or delaminations, although some diffusion layer formed. The measured dielectric values were promising and additional work is required for a further optimization of the co-firing process.

In the range of room temperatures the permittivity of the embedded BST is highly affected by temperature changes. If a stable permittivity is required a capable temperature regulation is necessary. On the contrary the very good temperature sensitivity of the BST can be used for sensor applications or for tuning the permittivity instead of using high electric field strengths. The tunability by an applied electric field is presently being investigated for different frequencies.

The losses are constant $\tan(\delta)=0,006$ within a temperature range of more than 100°K. There are no significant differences between 50kHz and 1MHz recognizable. One can expect, that this applies for higher frequencies as well and the losses are remaining at a relative stable low level. Further investigations of the high frequency properties of the LTCC BST have to be done.

Acknowledgement

The authors acknowledge Mr. Timo Vahera for preparing the LTCC BST tapes. Many thanks go to Mrs. Ina Koch for the preparation of the samples.

This work was supported by D.A.A.D. (D/05/51668) and the BMBF Project "MacroNano®"

References:

[1] K.R. Mikeska, D.T. Schaefer, R.H. Jensen "Method for Reducing Shrinkage During Firing of Green Ceramic Bodies", United States Patent US005085720A, USA, 1991, EP 0 511 301 B1, DE 691 06 345 T2

[2] V. Sherman, K. Astafiev, N. Setter, A. Tagantsev, O. Vendik, I. Vendik, S. Hoffmann-Eifert, U. Böttger, and R. Waser: Digital Reflection-Type Phase Shifter Based on a Ferroelectric Planar Capacitor, IEEE Microwave and Wireless Components Letters, Vol. 11, No. 10, (Oct. 2001)

[3] Sang-Min Han, Chul-Soo Kim, Seong-Soo Lee, Dal Ahn, and Tatsuo Itoh: Higher Phase-Tunable Phase Shifters Using DGS Termination Loads, 2005 European Microwave Conference, 4-6, Volume: 2, 4 (Oct. 2005)

[4] Tao Hu: "BST-Based Low Temperature Co-Fired Ceramic (LTCC) Modules For Microwave Tunable Components", Dissertation, Department of Electrical and Information Engineering and Infotech Oulu, University of Oulu, University of Oulu, ISBN 951-42-7292-7, (2004)

[5] Matjaz Valant and Danilo Sovorov: "Low-Temperature Sintering of $(Ba_{0.6}Sr_{0.4})TiO_3$", J. Am. Ceram. Soc., 87 [7] 1222-1226 (2004)

[6] Heli Jantunen, Tao Hu, Antti Uusimäki, Seppo Leppävuori: "Tape Casting of Ferroelectric, Dielectric, Piezoelectric and Perromagnetic Materials", Journal of the European Ceramic Society 24 1077–1081 (2004)

Vacuum Package design for a MEMS based IR Detector Array

J. Ollila[1], M.F. Toy[2], O. Ferhanoglu[2], P. Karioja[1] & H. Urey[2]

[1]VTT Technical Research Centre of Finland, Kaitoväylä 1, PO BOX 1100, FI-90571 Oulu, Finland

Tel. +358 20 722 2250, Fax +358 20 722 2320, jyrki.ollila@vtt.fi

[2]Koç University, Optical Microsystems Laboratory, Rumelifeneri Yolu, 34450 Sariyer, Istanbul, Turkey

Abstract

A vacuum package was designed and manufactured for the testing of a MEMS-based thermal imaging detector array. Optimum operation of the pixel array requires low pressure and stabilized temperature. The pixel array devices are under development; therefore, the package needs to be reusable as a test fixture for chip revisions. Two optical windows are required: the top window to be transmissive at the long-wave infra-red (LWIR) wavelengths and the bottom window at visible (VIS) readout wavelengths. Thermal noise is a fundamental source deteriorating the performance of the pixel array; hence, a thermocoupler, thermo electric cooler (TEC) and heatsink were included in the package to allow thermal stabilization. To minimize the thermal noise, the TEC and heatsink were located between the VIS window and the pixel array. Such a configuration for thermal stabilization resulted in the use of special TEC and heatsink designs that include holes matched with the VIS window size. Simulations on pixel array performance indicate a 1 mTorr vacuum pressure to provide the optimum performance of the pixel array. In order to meet the vacuum requirement, a Kovar package with the minimum possible size, considering the size of elements assembled in the package, was chosen. The removable ZnSe IR window was hermetically sealed with an O-ring and vacuum grease. The VIS window was glass soldered to the bottom of the package. Active pressure control was obtained by a vacuum pump.

Key words: MEMS, thermal testing, vacuum packaging, IR detector packaging, IR testing

Introduction

In an IR detector array package, it is necessary to minimize convective heat transfer to the detector chip in order to prevent crosstalk between pixels and reduce detector noise. Therefore, operation in vacuum is a critical requirement for the package design. Furthermore, temperature tuning and stabilization provide additional means for optimizing the performance of the detector array.

In this paper, we describe the structure of a bimaterial MEMS detector optimized for IR detection. We show the design and implementation of a vacuum package customized for the MEMS device testing. In addition, we show vacuum package test setups and test results.

MEMS-based IR Detector

The IR detector mentioned here is a thermo-mechanical detector with optical readout [1] [2]. The design of the detector is shown in Fig. 1. The operation principle of the detector is the following: Firstly, IR radiation is absorbed in the membrane of the detector causing temperature increase in the membrane. Secondly, the heat induced in the detector as the result of the absorbed radiation is converted into mechanical deflection by the help of bimaterial legs that support the radiation absorbing membrane. In Figure 1, the temperature increase in the membrane is shown with red color, and the

bending of the bimaterial legs is clearly depicted, also. The displacement of the membrane can be precisely measured from underneath the device using a diffraction grating interferometer setup [3]. Thus, two windows are needed on the opposite facets of the package: a long-wave infra-red (LWIR) window on top of the package and a visible (VIS) readout window in bottom of the package.

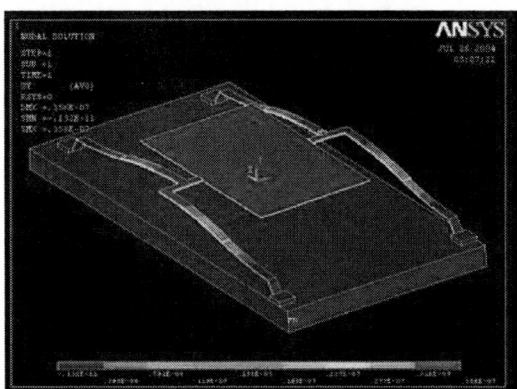

Figure 1: Thermo-mechanical simulation of the operation of the IR detector illustrating bimaterial bending in the supporting legs due to the heat load caused by IR radiation absorption in the membrane.

According to simulations, thermal noise is a fundamental source deteriorating the performance of the pixel array; therefore, the optimum operation of the pixel array requires good vacuum and temperature stabilization. Simulations indicate that a 1 mTorr vacuum pressure is needed to provide the optimum performance. The pixel array devices are under development; therefore, the package needs to be reusable as a test fixture for chip revisions.

Package Design

The package design was carried out with the mentioned considerations. A schematic illustration of the package is shown in Figure 2. The removable ZeSn window transmitting at IR wavelengths is attached on top of the Kovar package using an O-ring. The top window allows the substitution of the detector chip for multi-use purposes. The readout is performed through the bottom window.

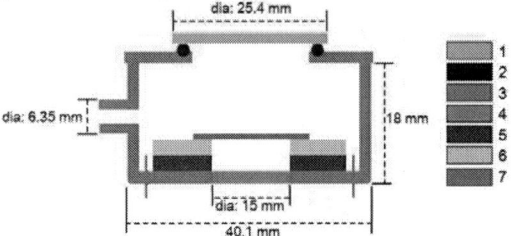

Figure 2: Schematic drawing of the package. The captions and explanations are the following: 1. Removable ZeSn window for LWIR transmission; 2. O-ring; 3. Kovar package; 4. B270 glass window for optical readout at VIS wavelengths; 5. Heat sink; 6. TEC for thermal stabilization; 7. Detector array die.

The IR detector is an uncooled detector; therefore, the operation does not require expensive cryogenic cooling. However, thermal stabilization for the package is necessary to reduce temperature fluctuations and thermal noise. Moreover, the sensitivity of the detector can be tuned by adjusting the operation temperature. In the package, temperature stabilization and control is accomplished with a thermoelectric (TEC) cooler and heat sink.

Assembly

The vacuum package consists of a thermo electric cooler (TEC), thermistor and heat sink that were included in the package to allow the thermal stabilization of the IR detector. To minimize the thermal noise of the detector, the TEC and heat sink were located inside the package in the bottom side of the IR detector, i.e., between the VIS window and the pixel array element.

Figure 3: Vacuum package and Kovar lid from Technotron.

Figure 3 shows the package and corresponding lid that were ordered from a commercial supplier, Technotron. The inside dimensions of the package are 17.0 mm (H) x 38.1 mm (W) x 38.1 mm (L). The open hole diameter at the bottom of the package is 15 mm.

The heat sink material is AISI 430F. The heat sink was tooled by the use of a 5-axis machine tooling. The VIS window (B270) was designed to form hermetically sealed interface in the bottom of the package. The hermetic seal for the window was processed into the heat sink using a glass perform (Electro Glass Products; Material 7572). The 7572 solder glass is a low temperature crystalline glass material specifically designed for sealing windows hermetically in metal packages. The glass has the following key properties: CTE 9.2 ppm / °C, T_g 315°C, density 6.33 g/cm^3 and sealing temperature 480…500°C.

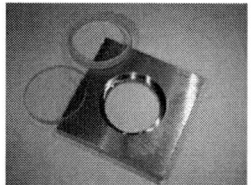

Figure 4: Visible window, the solder glass perform and the heat sink before and after glass sealing.

The VIS window was sealed to the heat sink using the solder glass preform at about 500 °C, see Figure 4. The sealing temperature is below the transformation temperature of the window glass; therefore, the window is not deforming during the sealing process. [4]

Figure 5: Heat sink solder jointed on the bottom of the package.

290

After hermetic window sealing, the heat sink was electro-galvanized with nickel. The heat sink was solder jointed into the package using AIM NC254 solder paste (SAC; Tin-Silver-Copper). After soldering, the flux was cleaned with isopropyl alcohol. The package with the heat sink is shown in Figure 5.

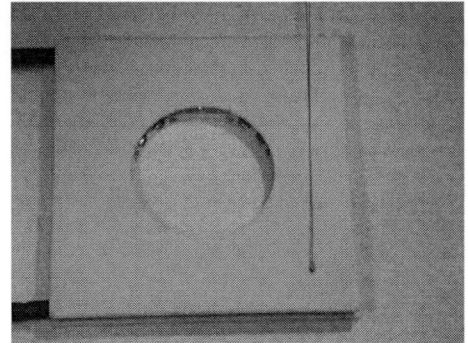

Figure 6: TEC (Melcor SH1.0-95-05L and thermistor (Shibaura Electronics PB7-43-SP2).

Figure 6 shows the thermoelectric cooler (Melcor SH1.0-95-05L) and thermistor (Shibaura Electronics PB7-43-SP2) before assembliy. The hot side of the TEC was bonded onto the heat sink with a heat-conductive silicone (Dow Corning Q-9226). After that, the TEC was contacted to the electrical pins of package using soldering. The thermistor was attached to the top plate with a heat-conductive silicone and connected to the electrical pins using a micro welding equipment. The final IR detector package is shown in Fig. 7.

Figure 7: On the left, the open package with the heat sink, TEC and thermistor. On the right, the bottom side from package showing the hermetically sealed VIS window.

Testing Results

The vacuum level of the package was tested using the vacuum pump setup shown Figures 8 and 9. The package lid was attached by gluing onto the package and the IR window was sealed with an O-ring. A vacuum pipe was connected to the pump system with a Swagelok fitting.

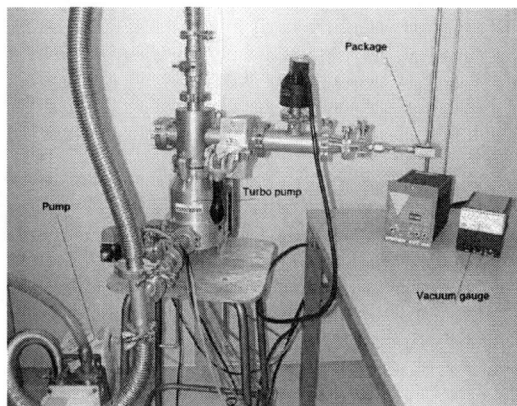

Figure 8: Vacuum pump setup used for the testing of the vacuum level.

The measured vacuum level of the package was 2×10^{-6} mbar (1.5×10^{-3} mTorr) that perfectly fulfils the 1 mTorr vacuum level specified for the package. The system can be used to test different version of IR detector arrays.

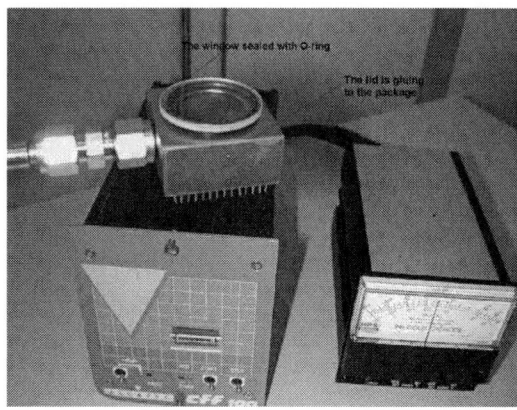

Figure 9: Package in the vacuum test.

Conclusions

A vacuum package was designed, manufactured and tested to be used for a MEMS based IR detector testing. The measured vacuum level show that a detector array will safely operate in the required vacuum without any performance degradation due to convective heat transfer. Future work involves detector array testing using the developed vacuum package as a reusable fixture to encapsulate the devices under test.

Acknowledgements

The authors acknowledge the European Network of Excellence in MicroOptics (NEMO, http://www.micro-optics.org/) for its support in the development of the read head

References

[1] P.G. Datskos, N.V. Lavrik, and S. Rajic, "Performance of uncooled microcantilever

thermal detectors", Review of Scientific Instruments 75, 1134–1148, 2004.

[2] Y. Zhao, M. Mao, R. Horowitz, A. Majumdar, J. Varesi, P. Norton, and J. Kitching, "Optomechanical uncooled infrared imaging system: design, microfabrication, and performance", J. Microelectromechanical Systems 11, pp. 136–146, 2002.

[3] H. Torun, H. Urey, "Uncooled Thermo-mechanical Detector Array with Optical Readout," Opto-Electronics Review, Vol. 14 (1), pp. 55–60, 2006.

[4] J. Ollila *et al.,* "Hermetic diode laser transmitter module", Proceedings of SPIE, Vol. 3626, pp. 123-126, 1999.

Capacitive Micro Force Sensors Manufactured with Mineral Sacrificial Layers

Y. Fournier, S. Wiedmer, T. Maeder, P. Ryser

Ecole Polytechnique Fédérale de Lausanne, EPFL-LPM, Station 17, 1015 Lausanne, Switzerland

Phone: +41 21 693 78 46; Fax: +41 21 693 38 91; http://people.epfl.ch/yannick.fournier; http://lpm.epfl.ch

Abstract

In this work a prototype of micro force sensor of range µN...mN is presented. Instead of the traditional piezoresistive strain sensing through thick-film resistors used for higher forces, a more effective principle is used: measurement of beam displacement rather than strain. A design of a cantilever sensor with capacitive electrodes, optionally coupled with an electrostatic force cancelling to achieve higher sensitivity, is proposed and discussed. The structuration of the device is achieved in alumina with thick-films and mineral sacrificial layers (MSL), and the resulting sensor properties are promising. The composition and dissolution of the sacrificial paste is of high importance, and is the cornerstone of the presented research.

Key words: Force sensor, Capacitive, Thick-film, Sacrificial layers, Mineral

Introduction

Cantilever beam force sensors based on LTCC and thick-film technology commonly apply piezoresistive strain sensing through thick-film resistors, and allow measurement of forces down to ca. 100 mN [1]. Extension of this range down to µN...mN forces is attractive, because these sensors are robust and low-cost, but is no longer compatible with piezoresistive sensing (Figure 1). For such forces, it is more effective to measure the beam displacement rather than the strain.

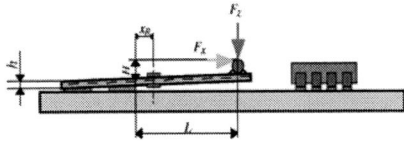

Figure 1: Schematic side view of the MilliNewton piezoresistive cantilever force sensor

Beam displacement is best measured with capacitive electrodes. This allows for purely passive measurements, but also active with electrostatic force cancelling to achieve higher sensitivity (with the help of extra electrodes, which can be combined). The capacitance involved is relatively low (~pF), but today's dedicated chips such as the *Analog Devices AD7745/46* render the task easy, with multiple available configurations (differential, reference, temperature compensation, etc.).

The challenge in creating this "variable capacitor" resides in the structuration of the cantilever – in our case, by the mean of sacrificial mineral layers (MSL) fired like standard thick-films, later removed by selective acid dissolution that leaves the cantilever free-standing.

This work first presents the different structuration techniques, then the developed solution, followed by the prototypes we have realized in alumina (passive measure), and discusses in depth the sacrificial paste composition and dissolution. The discussion will focus on design geometries, foreseen improvements to facilitate dissolution, as well as migration of paste to be LTCC-compatible.

Sacrificial Layers

Structuration through sacrificial layers removed at the end of processing is standard in micromachined silicon MEMS, but is more seldom used in thick-film and LTCC technology, with the exception of carbon-based fugitive phases for LTCC.

Pioneering, but little-known work on carbon and other sacrificial layers was also carried out on classical thick-film technology, already in the early 1980's at Bosch by G. Stecher et al. [2][7][8]. The processing steps, however, were quite complex, involving a succession of depositions and firings in nitrogen and air, and therefore did not achieve large-scale manufacturing.

Carbon sacrificial layers that burn out during firing are useful for structuration of LTCC, but using them for delicate structures such as thin membranes can be difficult, due to a very high process sensitivity [3][5]. Also, carbon must be used in rather "closed" environments. If not it can oxidize before the structure is sintered, leading to sagging and/or lateral deformation.

Mineral-based Sacrificial Layers

For more "open" structures, structuration with a sacrificial material that survives the firing step and supports additional layers, then is etched away by an acid as is MEMS technology, would be much more straightforward. Unlike setter tape [6], such a structuration method also leaves no particles, which are highly undesirable in microstructures.

Previously, we have explored chemically dissolvable mineral sacrificial pastes based on mixtures of CaO (refractory) and B_2O_3 (melts at ca. 450°C) system [4]. Unexpectedly, shrinkage was found to be small, ca. 7-8%, regardless of the B_2O_3 content, and the films remained porous. This was ascribed to the high volatility of free B_2O_3 and to the high melting points of its compounds with CaO, hindering sintering. Although the open porosity made for very easy etching with acids, the low shrinkage did not allow successful co-firing with LTCC.

Attempting to improve this situation is the subject of this work: replacing B_2O_3 with borax ($Na_2B_4O_7$, sodium tetraborate), because it is known to reduce volatility and enhance sintering. This will be presented in the following chapters.

Sensor Principle

In order to setup and validate our new mineral sacrificial paste (MSP), we decided to initially work with thick films on alumina substrate; it will be ported on LTCC later, once the process is under control. The goal is to make a micro force sensor of 1-2 mN range full scale behaving like a variable capacitor, and to determine the smallest force measurable. The principle is simple (Figure 2): the following succession of pastes is screenprinted and fired on alumina: conductor (lower electrodes + GND), MSP, conductor (upper electrodes), and dielectric (i.e. the cantilever). The MSP must then be dissolved by acids (HCl or H_3PO_4). The challenge is twofold: the acid must act selectively (only the MSP must be affected and removed totally), and the MSP must not react with adjacent pastes.

Once the MSP removed, contacts are established on the electrodes by gluing standard DIL pins. The displacement of the cantilever (due to an external force) reduces the distance between the capacitive electrodes, and the variation of capacity allows the measurement of the applied force.

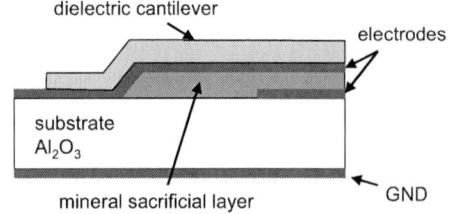

Figure 2: Schematic cross section of capacitive cantilever force sensor with sacrificial paste

By adding a pair of reference electrodes and a non-moving cantilever of same characteristics, the capacitance measurement can be done in differential

and get rid of perturbations induced by temperature, humidity etc.

Theory of the Capacitive Force Sensor

The force to measure (F_a) is applied at the tip of the cantilever, which bends and acts like a spring with a force of recall $F_r = -k \cdot d$, where k is the spring constant and d the displacement.

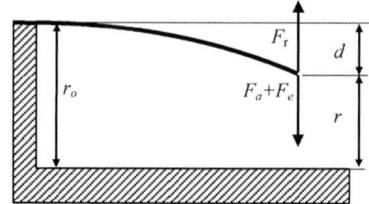

Figure 3: Schematic of forces for capacitive measure (passive and active)

Figure 3 shows the schematic of the forces. The capacity between electrodes can be written as $C = \dfrac{\varepsilon_0 \varepsilon_r \cdot A}{r}$, with $\varepsilon_0 = 8.854 \cdot 10^{-12}$ [F·m^{-1}] the dielectric constant, $\varepsilon_r = 1$ (in air) the relative permittivity, and A the electrodes area. By adding (r_0) r the (initial) distance between electrodes, the non-linear relation between applied force and measured capacity is:

$$C(F_a) = \frac{\varepsilon_0 \varepsilon_r \cdot A}{r_0 - \dfrac{F_a}{k}}$$

Active Measurement with Electrostatic Forces

The last force of Figure 3, F_e, is an optional electrostatic force used in the case of active measurement. An extra pair of electrodes can be used (or the existing pair if the measuring and acting signals are mixed) to attract the cantilever in a middle idle position; any external applied force will modify the equilibrium, and the electrostatic voltage has to be decreased by the close-loop feedback in order to maintain the idle position constant.

The electrostatic force is stated as: $F_e = \dfrac{1}{2}\varepsilon_0 \varepsilon_r A \cdot \left(\dfrac{V}{r}\right)^2$. By equalizing the distance r from the capacitance and doing a few transformations, the applied force F_a can be directly extrapolated from the value of the electrostatic voltage V, or the distance r: we have $V(F_a) = \sqrt{2\dfrac{F_r - F_a}{\varepsilon_0 \varepsilon_r A k}} \cdot r$, and

$V(r) = \sqrt{2k\dfrac{r_0 - r}{\varepsilon_0 \varepsilon_r A}} \cdot r$. The behavior is more linear than in the passive case, but one has to be careful when the electrodes are close to each other as the system is instable.

Unfortunately the active measurement has not yet been tested, as it requires developing specific electronics. However dedicated electrodes are available on our designs for future tests.

Sensor Designs and Geometries

Two designs have been tested: the simple cantilever and the bridge, as depicted on Figure 4. Both geometries present pros and cons, and the best design will be chosen after extensive testing.

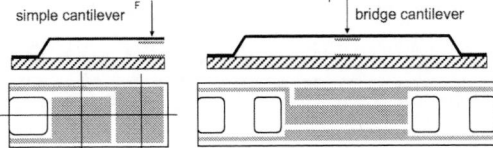

Figure 4: Top: schematics of the simple cantilever and bridge designs. Bottom: location of electrodes (gray) and of inner cuts (rounded rectangles).

The electrical pins and electrodes are placed as depicted on Figure 5. This allows using the same electronic setup for all variants, and in the future to add the active electrostatic feedback without hassle.

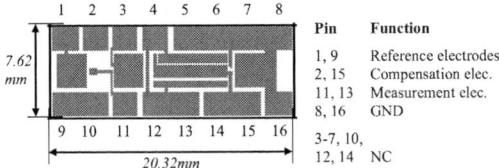

Figure 5: Disposition of pins and electrodes on the top surface of the alumina substrate.

Six variants have been selected, with variable overall lengths (7-15mm), presence or not of inner cut(s) and their lengths (1-6mm). The width is 3mm for all variants (cf. Figure 6). The estimated capacitance is around 1 pF without force, and 2 pF at 50% displacement.

Figure 6: 4"x4" alumina substrate with ten samples for each six retained geometry variants. Dielectric is blue and silver-palladium grey.

Thick-films Pastes Selection

Apart from testing different cantilever geometries as seen above, it is also possible to change the order, the type and number of screen-printed pastes. The following considerations have lead to eight screenprinting variants (Table 1):

- It is mechanically favorable to have a sandwich cantilever (diel-cond-diel) with the "soft" layer in the middle, to avoid hysteretic behavior.
- Passive measurement accommodates this sandwich easily. However for active electro-static compensation, an additional dielectric layer in addition to air is unfavorable.
- The cantilever being mainly governed by the dielectric layer, it is wise to print two or more dielectric layers on top of each other to ensure good mechanical stability and homogeneity.
- The goal is to measure the smallest possible forces; one variant with only one dielectric layer to lower stiffness is hence interesting.

Explanation of layers shown in Table 1:
- Sub: 96% alumina substrate, 0.635mm
- Ag: silver-palladium AgPd, *ESL 9635B*
- MSP: mineral sacrificial paste (see later)
- Diel: dielectric, *ESL 4913*
- Au: gold, *DuPont 5744*
- R: resistor paste, *DuPont 2031*

The dielectric has been chosen among three (DuPont 4903, 4913, 4924) after early dissolution tests (cf. next chapter). While the 4903 resisted to acids, it formed bubbles with the MSP; the 4924 was totally dissolved by HCl.

Table 1: Screenprinting variants

V1	V2	V3	V4	V5	V6	V7	V8
			Diel				
			Diel				
Diel	Diel	Diel	Au	Diel	Diel	Diel	
Ag	Au	R	Diel	Diel	Diel	Diel	Diel
Diel	Diel	Diel	Diel	Ag	Au	R	R
MSP	MSP	MSP	MSP	MSP	MSP	MSP	MSP
MSP	MSP	MSP	MSP	MSP	MSP	MSP	MSP
Ag	Ag	Ag	Ag	Ag	Ag	Ag	Ag
Sub	Sub	Sub	Sub	Sub	Sub	Sub	Sub
Ag	Ag	Ag	Ag	Ag	Ag	Ag	Ag

The screens used were 325/30 [mesh/µm] for conductors, 200/40 for dielectric and 200/50 for the MSP.

Sacrificial Paste Preparation

Continuing the work of Dr. Hansu Birol [4], we decided to modify his initial MSP recipe of $CaO+B_2O_3$, because it gives a very porous paste. It would have been a bad basis for future fine screen prints; if the paste solvent get absorbed by the under-lying MSP during the pass of the squeegee, adverse effects can occur.

The role of the CaO (*Alfa Aesar 99.95%, 010923*) is to be the skeleton of the paste, as its melting point is high (2572°C). The B_2O_3 acts as a binder during firing; it melts around 450°C. Thus we tried to substitute the bore oxide with borax (anhydrous sodium tetraborate $Na_2B_4O_7$, *Riedel-*

deHaën granulated dry 98%, 11648), known for its good vitreous and melting properties, in order to better join CaO grains. Borax alone dissolves well in HCl or H_3PO_4; and as we are working on alumina, we are less subject to shrinkage issues than on LTCC – for the time being!

Table 2: Homemade organic binder composition

Component	Mass [g]	Mass [%]
Terpineol $C_{10}H_{18}O$ *Fluka 86480, anhydrous*	32.00	64.5
Dibutylcarbitol *Aldrich 99+%, 20,562-1*	16.00	32.2
Ethylcellulose *Aldrich 300 [cps], 20,065-4*	1.64	3.3

To determine the best CaO-borax mix, initial tests have been carried out with varying proportions. Both powders have been mortar grinded (*Retsch RM100*) for at least 1 hour before careful weighting (*Denver Instruments*) and mixing. Then a homemade organic binder was added to the mix before passing through a tricylinder grinder (*EXAKT*) to further reduce grain size and to get a homogeneous paste suitable for screenprinting. The organic binder had the composition listed in Table 2 (magnetically stirred for 1+ hour at 60-80°C).

Sacrificial Pastes Tested

Seven pastes have been prepared, with these proportions of borax compared to CaO (in % vol. / vol.): 10, 20, 40, 50, 60, 70, and 80%. Rapid tests of firings (*SierraTherm* belt air furnace, 850°C-45min) on alumina and all envisioned pastes showed the following behaviors:

- **<40%**: very white, very powdery, porous, unbundled; strongly hydroscopic; not suitable.
- **40%**: white, well bundled, slightly powdery, hydroscopic (decomposed after 1 week).
- **50%**: white, slightly porous, not powdery, not hygroscopic, not easily strippable with nails; bubbles with resistor paste *DuPont 2031*.
- **60%**: less white, vitreous surface, strippable with hard tool, bubbles with dielectric pastes.
- **>70%**: transparent, nearly impossible to strip away, strong chemical reaction with dielectrics.

Final Sacrificial Paste

From the above experiments it turns out that the most suitable compositions are the 40% and the 50% of borax. The 40% decomposed within a few days, so we decided to go with the 50%. This was at posteriori a mistake, as it proved difficult to dissolve. Using the 40% borax would be smarter, but the subsequent screenprinting operations need to take place immediately afterward in order to avoid short-term destruction by air humidity.

The 50% powder mix was stirred with organic binder in proportions 60-40% (mass-mass)

respectively, as listed in Table. The same procedure as described before was used (mortar and tricylinder grinding).

Table 3: Composition of final mineral sacrificial paste with 50% borax-50% CaO (% volume)

Component	Mass proportions	Used quantity [g]
Borax	24.8%	7.81
CaO	35.2%	11.06
Organic binder	40.0%	12.58
Total	100%	34.82

However, no dispersant (acetylacetone, *Aldrich 99+%*) was used in the final paste. It is usually added as 5% of the powder mass to enhance screenprinting pastes, but a test turned to disaster: upon stirring the mix suddenly heated, and after firing the resulting paste was yellowish, completely cracked and easily removable.

Results – Pastes Compatibilities upon Firing

The screenprinting went relatively well, despite some tridimensional problems (vias filled by adjacent layers, tight margins). The MSL fired thicknesses were 30um after first pass and 50um after second pass, enough to ensure a good displacement range, but just enough to allow for good circulation of acid during dissolution.

Figure 7: Photographs of bubbles on cantilevers due to paste incompatibilities.

At this stage some paste combinations can already be disregarded (Figure 7):

- Resist-diel (V3, V7, V8) and MSP-Au-diel (V6) lead to bubbles formation, strongly affecting mechanical properties for the resistance variants. Despite this, Au variants (V2, V4) gave overall the best results.
- Strange black stains spread around the MSP for resistance variants; however no other complications were encountered.

Results – Sacrificial Layer Dissolution

This step is indeed not yet under control. The sodium of borax tends to react with adjacent layers and renders dissolution of buried MSP difficult. In particular, the AgPd lower electrodes stick to the MSL and get ripped off with it. First tests were done with diluted HCl (*Fluka 32%*), but it was too strong and we switched to H_3PO_4 (*Aldrich, 98+%*) to (hopefully) increase selectivity.

Systematic tests have been carried out with alternating dissolution (1 hour) and rinsing (1 hour in DI water), in order to characterize the variants. This loop was repeated until all cantilevers of a variant were enough detached. Figure 8 shows the result of variant 2 after 5:30 passed in H_3PO_4.

Figure 8: Cantilevers of Variant 2 (Au-diel-Au) after 5:30 of dissolution in H_3PO_4. Note the detached lower AgPd electrodes.

The best results were obtained on sensors left 30 hours in 80-90°C phosphoric acid, then thoroughly rinsed (DI H_2O for 1-8 hrs, IPA) and air dried at 100°C (150°C bends and cracks the cantilever). More should be learned with the next generation of MSP tested in near future.

Results – Force Measurements

A few sensors were good enough to be fitted with DIL pins and tested with the demo-board of the *Analog Devices AD7746EB* (Figure 9). The achieved performance shows a force range of ~2 mN with a resolution of ~1% of FS, which is quite remarkable for passive measurement. Active measurement with electrostatic counteraction has not yet been implemented. Absolute or differential measurement show little differences; in differential, humidity affects the reference electrodes too…

The *AD7746EB* chip demo-board presents astonishing performances. Its resolution is 4 fF and linearity ±0.01%; it is easily programmed by USB.

Figure 9: Photograph of prototype of thick-film capacitive cantilever force sensor on alumina.

As can be seen on Figure 10, the first estimation of capacity in function of applied force was too pessimistic regarding sensitivity (dashed line). Although the sensitivity is indeed lower, the capacitive *AD7746EB* chip has a range of ±4 pF (or 0-8 pF, displaceable up to +17 pF); thus, it would not be favorable to increase the sensitivity too much.

Based on actual dimensions and measures (squares), the estimation was adapted and now fits reality rather well (except for the saturation which

was not modeled, when the electrodes touch each other). The waves at the beginning of the curve are due to bad experiment conditions (door and window opened and closed, measures made over 2 hours).

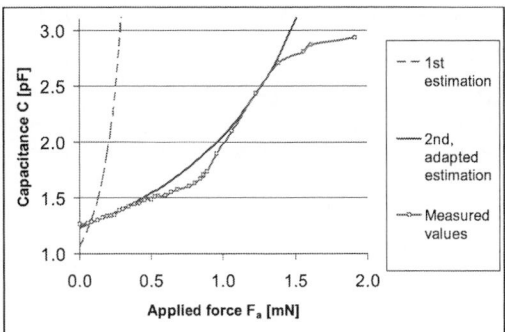

Figure 10: Measurements of prototype thick-film capacitive force sensor along with estimations.

Analysis – Design Variants

The different layouts proved to be relatively good. Nevertheless, variants with very long cantilevers and long inner cuts tend to be unusable because they twist and tear easily.

Surprisingly, both sandwich and asymmetric cantilever designs were successful.

Improvement of MSL and Migration to LTCC

As we recently tested, a promising solution is to reduce the sodium content by making a paste with 60% vol. CaO, 15% borax and 25% H_3BO_3 or B_2O_3. To suit LTCC, the shrinkage has to be increased (from 7-8 to 15%), per example by adding bismuth borate, known to melt at 726°C (congruent).

Conclusions

The manufacturing of a capacitive microforce sensor has proven to be feasible in thick-films on alumina with mineral sacrificial layers, but not yet on LTCC, due to shrinkage and dissolution issues. The achieved performance shows a force range of ~2 mN with a resolution of ~1% of FS, which is quite good for passive measurement. Active measurement with electrostatic counteraction has not yet been implemented.

Resistor paste has seen strong interactions with dielectric; gold conductors were the best choice despite some bubble deformations. Both sandwich and asymmetric cantilever designs proved to be successful. Simple cantilevers tend to be easier to dissolve than bridges; however no extensive tests could be done on mechanical aspects.

Regarding the mineral sacrificial paste (MSP 50% CaO - 50% $Na_2B_4O_7$), the sodium contained in borax tends to react with adjacent layers and renders dissolution difficult. A promising way is to make a paste with 60% vol. CaO, 15% borax and 25% boric acid or oxide.

Acknowledgements

The following people are warmly thanked: Mr. G. Corradini, T. Haller and M. Garcin for screenprinting and gluing operations; Mrs. C. Jacq for technical advices on sacrificial paste preparation.

References

Articles from conference proceedings

[1] H. Birol, M. Boers, T. Maeder, G. Corradini, P. Ryser, "Design and processing of low-range piezoresistive LTCC force sensors", Proceedings, XXIX International Conference of IMAPS Poland, Koszalin, pp. 385-388, 2005.

[2] G. Stecher, "Free supporting structures in thickfilm technology: a substrate integrated pressure sensor", Proceedings, 8th European Microelectronics Conference, pp. 421-427, 1987.

[3] H. Birol, T. Maeder, C. Jacq, P. Ryser, "3-D structuration of LTCC for sensor micro-fluidic applications", Proceedings, European Microelectronics and Packaging Symposium, Prague (CZ), 366-371, 2004.

[4] H. Birol, T. Maeder, P. Ryser, "Preparation and application of minerals-based sacrificial pastes for fabrication of LTCC structures", Proceedings, 4th European Microelectronics and Packaging Symposium, Terme Catez (SI), IMAPS, 57-60, 2006.

Journal article

[5] H. Birol, T. Maeder, P. Ryser, "Application of graphite-based sacrificial layers for fabrication of LTCC (low temperature co-fired ceramic) membranes and micro-channels", Journal of Micromechanics and Microengineering 17, 50-60, 2007.

[6] K.A. Peterson, K.D. Patel, C.K. Ho, S.B. Rohde, C.D. Nordquist, C.A. Walker, B.D. Wroblewski, M. Okandan, "Novel microsystem applications with new techniques in low-temperature co-fired ceramics", International Journal of Applied Ceramic Technology 2 (5), 345-363, 2005.

Patent

[7] G. Stecher, K. Spitzenberger, K. Müller, "Pressure sensor", USA patent 4'382'247, 1983.

[8] G. Stecher, H. Zimmermann-H, "Electrical thick-film free-standing, self-supporting structure, and method of its manufacture, particularly for sensors used with internal combustion engines", USA patent 4'410'872, 1983.

Cavity Formation in a Silicon Cap Wafer
Using Aluminum Etch-Stop Layer

S. Sosin, J. Tian, and M. Bartek

Laboratory of High-Frequency Technology and Components/DIMES, Delft University of Technology,
Mekelweg 4, 2628 CD Delft, the Netherlands

Phone: +31(0)152789421, fax: +31(0)152623271, e-mail: s.sosin@ewi.tudelft.nl

Abstract

In this paper, our progress in development of a hybrid wafer level packaging (HWLP) scheme suitable for RF-MEMS applications is presented. The scheme is based on a prefabricated high resistivity silicon (HRS) substrate that provides vertical interconnects, protection of MEMS structures, electrical interconnect redistribution and through-substrate cavities for flip-chip bonded additional ICs. Simultaneous formation of through-substrate circular vias for electrical interconnect and large rectangular cavities for flip-chip die placement (i.e. structures with highly different aspect ratios) is rather demanding. Previously we have shown that this can be achieved by first etching only non-continuous narrow trenches at the periphery of the future cavities that after wafer thinning leave temporary beams supporting a mass of silicon inside the cavity. During ultrasonic cleaning of the wafer these beams are broken leaving a clean cavity. In this contribution we present a novel, further simplified method for via, cavity and recess formation with sizes ranging from tens of micrometers up to millimeters that is based on conductive aluminum etch-stop and mechanical-support layer.

Key words: hybrid packaging, wafer-level MEMS packaging, cavity fabrication.

Introduction

In the last decade, starting from a research subject, wafer level packaging (WLP) has become a widely used technology, mainly due to the continuous demand of further miniaturization from the high-volume hand-held product market. Another reason for its wide use is the added value in radio-frequency (RF) and MEMS applications given by its capability to realize additional functionality like integration of RF passives or MEMS device protection at a very limited additional cost.

We are presenting an update on our work on developing a hybrid wafer-level packaging (HWLP) scheme. The scheme consists of a HRS capping and electrical interconnect redistribution wafer and a to-be-packaged MEMS wafer. The packaging scheme shown in Figure 1 provides through wafer vertical interconnects to the MEMS wafer, protection of MEMS device using a shallow recess etched on the capping wafer and through substrate cavities for additional flip-chip ICs.

Capping Substrate Fabrication

The fabrication sequence consists of several steps. First, the capping wafer is completely prefabricated and then aligned and bonded to the device wafer. In the last step (hybridization step) additional ICs are placed using flip-chip technology in the prefabricated cavities that serve also as pre-alignment for the solder reflow.

Fabricating structures having large variations of lateral dimensions is rather challenging although there are many available technologies for bulk micromachining. The capping wafer makes no exception as it requires simultaneous formation of opening with dimensions ranging from 20-100 μm for the through-substrate circular vias for electrical interconnect; few by few millimeters for large rectangular cavities for flip-chip die placement; and finally hundreds of microns for shallow recesses for MEMS device protection. As illustrated in Figure 2b, the etch rate might significantly depend on the dimensions of the etched structures.

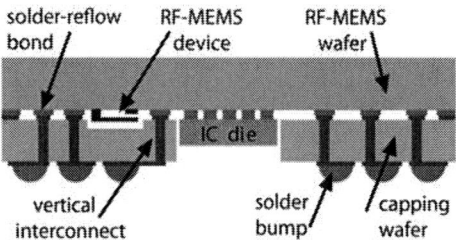

Figure 1: Hybrid wafer-level packaging scheme proposed in this work.

To determine the optimum trench width we have designed structures for measuring the dependency of the DRIE plasma etch rate on the etched trench width as shown in Figures 2 and 3. For this, a set of trenches with constant length and increasing width from 2 μm up to 50 μm was used (Figure 2).

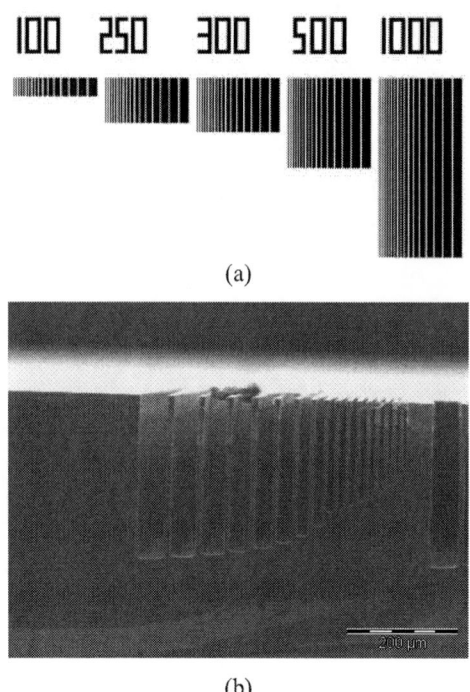

(a)

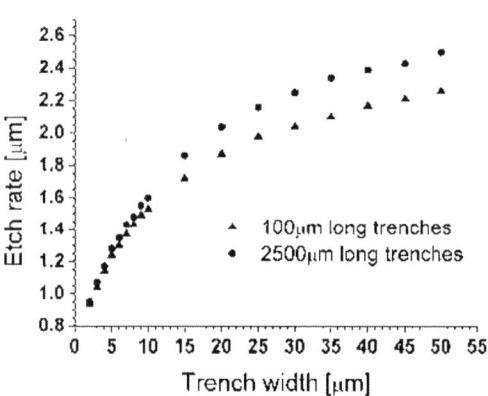

(b)

Figure 2: Test structures with increasing trench width for etch rate measurements: a) layout; b) SEM cross-sectional photograph showing significant dependency of the DRIE plasma etch rate on the etched trench width.

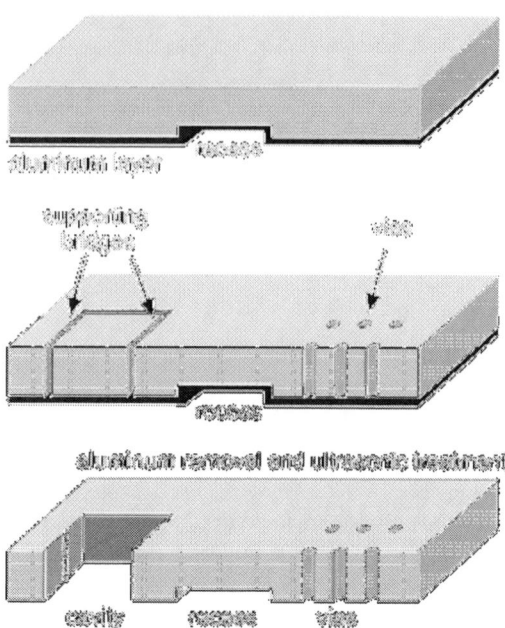

Figure 4: Capping substrate structuring sequence: shallow recesses are formed and Al etch-stop and mechanical-support layer is deposited; vias and trenches defining cavity periphery are etched; Al layer is removed and optional ultrasonic step breaks the temporary supporting bridges.

Fabrication of the capping wafer starts with silicon wafers already thinned to the required thickness of 250 μm. On the wafer backside, recesses with depth as needed for protection of (RF-) MEMS structures are first formed by deep reactive ion etching (DRIE). This wafer side is then covered with an aluminum layer (thickness 2-4 μm) that will act as a conductive etch-stop layer in the second DRIE step and also as a mechanical-support layer. Subsequently, from the wafer front side, vias and continuous narrow trenches at the periphery of future cavities are etched down to the Al etch-stop layer in the second DRIE step.

The thickness and mechanical strength of the Al layer is sufficient to support the silicon mass inside the cavity, thus avoiding the need for temporary supporting silicon bridges. After aluminum layer removal, the 3D structuring of the capping wafer is completed and the capping wafer is ready for further processing as described previously [4]. If temporary supporting bridges have been used, these can be fractured using ultrasonic step or directly by applying a mechanical force.

For testing purposes, cavities with lateral dimensions from 250x250 μm² up to few square millimeters, supporting beams of different width (2 μm up to 25 μm) and two topologies (see Figure 5) have been designed and fabricated. This will allow extracting the 'safe' processing window

Figure 3: Dependency of the DRIE plasma etch rate on the etched trench width for 100 μm and 2500 μm long trenches.

The complicated processing (e.g. application of photoresist, vacuum clamping, etc.) after the through-substrate structures have been formed is even more critical. We have developed a novel technique for the capping substrate structuring. The schematic fabrication sequence is shown in Figure 4.

300

for using this process for reliable through-substrate cavity fabrication.

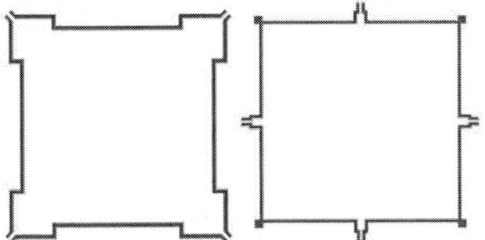

Figure 5: Test structures showing the two temporary supporting beam topologies.

This design contains also cavities with continuous trenches at the periphery of the future cavities for the improved version of our previously reported process [4], [5].

To allow detailed analysis of the beam fracture dynamics, structures for tensile test measurements have also been included. These contain simple and notched beams as shown in Figure 6. Previously built tensile test measurement setup [3] will be used for beam fracture studies (see Figure 10).

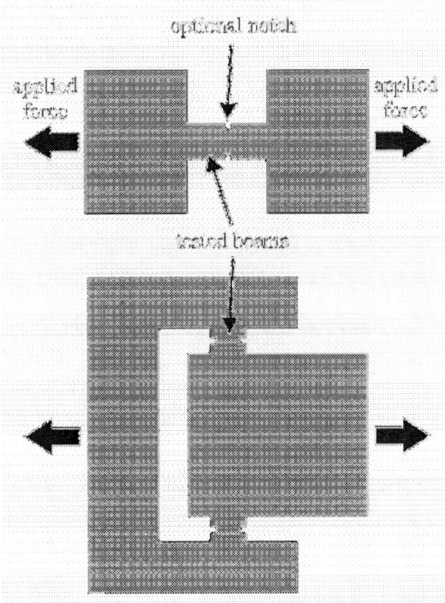

Figure 6: Test samples used for beam fracture studies (quantitative measurements).

Results

We have investigated three aspects: formation of cavities using non-continuous trenches [5] using beams from 2 µm to 25 µm, cavity formation using continuous trenches and fracture of narrow beams.

By fabricating cavities of 1500µm x 1000µm having 1, 2 and 4 supporting beams with 2, 5, 10,

15, 20 and 25 µm in width, we have defined an interval in which both versions of the process can be reliably used. In Table 1 shows this process reliability interval. The narrow beams (2 µm and 5 µm) were not strong enough to support the central part and the wide beams (20 and 25 µm) were too strong and did not break during ultrasonic treatment. From Table 1 we can conclude that beams having the width between 10 and 15 microns can be reliably removed during ultrasonic treatment, leaving a cleared cavity.

Table 1: Process reliability interval for cavity formation using narrow beams.

Number of beams	1	2	4
Beam width [µm]			
2	—	—	—
5	—	—	OK
10	OK	OK	OK
15	OK	OK	OK
20	—	—	—
25	—	—	—

Fabrication results can be seen in Figure 7, showing a four 10 µm beams cavity before and after ultrasonic treatment.

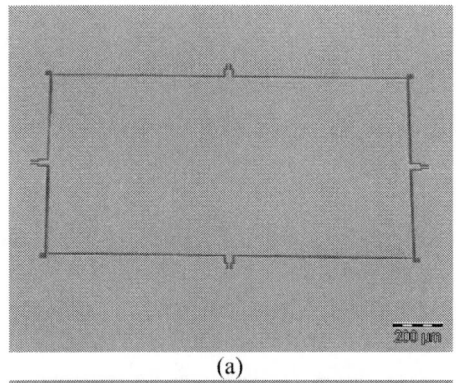

(a)

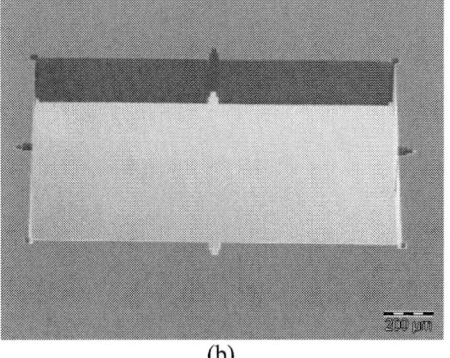

(b)

Figure 7: Cavity (a) before and (b) after ultrasonic treatment.

By using continuous trenches and an etch-stop and mechanical-support aluminum layer we can avoid using the temporary beams and the ultrasonic treatment step. As shown in Figure 8 large area cavities can be formed in the same process with small diameter vias.

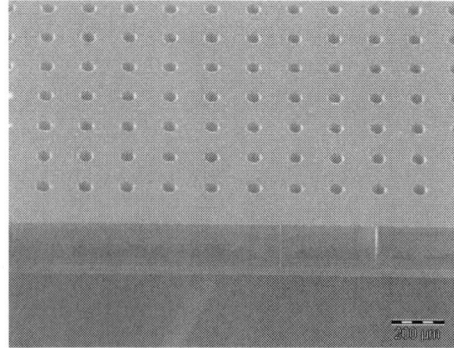

Figure 8: Close-up of 4x4 mm² cativy surrounded by 50 μm diameter vias.

Insertion tests were performed using dummy dies and the cavities for pre alignment. The result can be seen in Figure 9.

Figure 9: Cavity after dummy die insertion.

Tensile tests to determine fracture strength of narrow beam were performed using the tensile setup shown in Figure 10. The measurement results are summarized in Table 2. Figure 11 shows an example of a test structure before and after measurement.

Table 2: Fracture force measurement results.

Beam width [μm]	Fracture force [N]
10	2.04
20	2.37
50	7.1

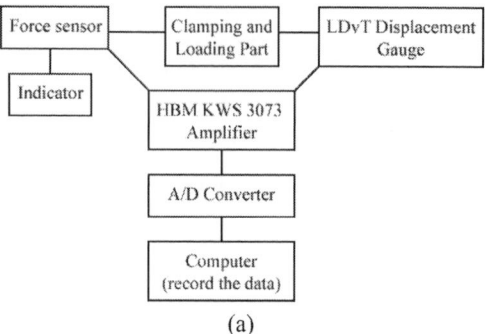

(a)

(b)

Figure 10: (a) Schematic block diagram and (b) photograph of the tensile test setup.

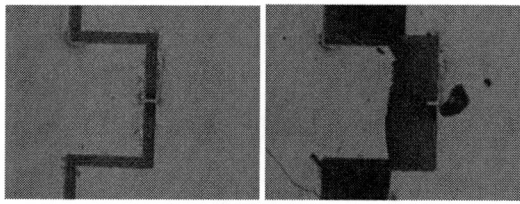

Figure 7: Photograph showing test structure before and after tensile measurement.

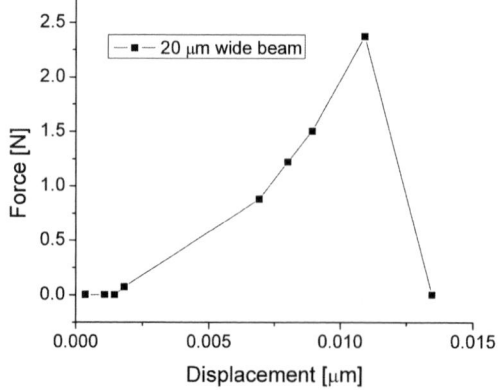

Figure 8: Graph showing fracture force measurement for 20 μm wide beam.

302

Conclusions

We have demonstrated a new fabrication technique for 3-D structuring (simultaneous through-substrate cavity and via formation) of a capping silicon wafer for hybrid wafer-level MEMS packaging. Using continuous trenches it is easy to fabricate structures having large area variations. By using the proposed method cavities with sizes from $250x250\ \mu m^2$ up to $4x6\ mm^2$ were successfully fabricated on the same wafer, all having the same sidewall profile and accurate edges. We have also determined a processing interval for reliable formation of through-substrate cavities using non-continuous narrow trenches on the periphery of the future cavity and temporary supporting bridges.

For further understanding of the fracture mechanism of narrow beams new optimized test structures have to be designed, which will be easier to handle and test. Also improvement of the measurement setup has to be implemented.

References

[1] A. Polyakov, S.M. Sinaga, M. Bartek, et al., "WLP for RF Applications Using High-Resistivity Polycrystalline Silicon Substrate Technology", Proceedings of IMAPS 2004, Long Beach, Nov. 2004, pp. WP1 7/1-8.

[2] J. Iannacci, J. Tian, S.M. Sinaga, R. Gaddi, A. Gnudi, M. Bartek, "Electromagnetic Optimization of a Wafer-Level Package for RF-MEMS Applications", Proceedings of EMPS 2006, Terme Catez, Slovenia, 2006.

[3] L. Wang, K.M.B. Jansen, M. Bartek, A. Polyakov, L.J. Ernst, "Bending and stretching studies on ultra-thin silicon substrates", Proceedings of the 6th International Conference on Electronics Packaging Technology, Shenzhen, China, Aug.-Sept. 2005, pp. 682-686.

[4] J. Iannacci, M. Bartek, J. Tian, S. Sosin, A. Akhnoukh, R. Gaddi, A. Gnudi, "Hybrid Wafer Level Packaging for RF MEMS Applications", Proceedings of the 39th International Symposium on Microelectronics (IMAPS 2006), San Diego, USA, October 9-13, pp. 246-253, 2006.

[5] S. Sosin, J. Tian, L. Wang, M. Bartek, "Through-Substrate Cavity Fabrication for Hybrid Wafer-Level Packaging, Technical", Proceedings of the 17th Micromechanics Europe Workshop (MME 2006), Southampton, UK, September 3-5, pp. 181-184, 2006.

Structural and Electrical Investigation of PZT Films on Different Substrates

Darko Belavič[1], Marko Hrovat[2], Hana Uršič[2], Silvo Drnovšek[2], Mitja Jerlah[1], Jena Cilenšek[2], Janez Holc[2], Marina Santo Zarnik[1], Marija Kosec[2]

[1] HIPOT-R&D d.o.o., Trubarjeva 7, 8310 Sentjernej, Slovenia

[2] Jozef Stefan Institute, Jamova 39, 1000 Ljubljana, Slovenia

Phone: ++386 1 4773479, Fax: ++386 1 4773887, E-mail: darko.belavic@ijs.si

Abstract

Lead zirconate titanate (PZT) is a piezoelectric material that can sense or respond to mechanical deformations and can be used in ceramic micro-electro-mechanical systems (C-MEMS). A thick-film paste was prepared from a pre-reacted PZT powder ($PbZr_{0.53}Ti_{0.47}O_3$) and thick-film technology (screen-printing and firing) was used to deposit the PZT layers on LTCC tapes and on alumina substrates. The microstructural, electrical and piezoelectric characteristics of the thick PZT films on relatively inert alumina substrates and on LTCC tapes were studied. Preliminary experiments indicated that due to the interaction between the printed PZT layers and the LTCC substrates during firing the electrical characteristics deteriorate significantly. To minimise the influence of substrate-film interactions different electrode materials and the use of additional intermediate layers as a barrier were evaluated. The dielectric permittivities, dielectric losses, and piezoelectric coefficients (d_{33}) were measured. The dielectric permittivities of the thick films fired on LTCC substrates were lower (210 with gold electrodes and 430 with silver electrodes) than those measured on alumina substrates (500). The piezoelectric coefficients d_{33} were measured with a Berlincourt piezometer. The d_{33} values measured on the LTCC substrates were relatively low (60-80 pC/N) compared with the values obtained for the alumina substrates (around 140 pC/N). The lower dielectric constants and piezoelectric coefficients d_{33} of the films on LTCC substrates are attributed to the formation of phases with a lower permittivity. This was a result of the diffusion of SiO_2 from the LTCC into the active PZT layer. The diffusion of silica was confirmed by the SEM and EDS analyses.

Key words: thick films, thick-film PZT, piezoelectric characteristics, alumina substrate, LTCC substrate

Introduction

Piezoelectric ceramics are used in a wide range of sensors, actuators and transducers that are important in diverse fields such as industrial process control, environmental monitoring, communications, information systems, and medical instrumentation. Thick-film technology, i.e., the deposition of thick-film pastes by screen printing, primarily on alumina substrates, is a relatively simple and convenient method for producing piezoelectric layers with a thickness of up to 100 µm. The characteristics of thick-film ferroelectrics are similar to those of bulk materials [1]. The compositions of piezoelectric thick-films are almost exclusively based on $Pb(Zr_{1-x}Ti_x)O_3$ solid solutions, often referred to as PZT. The PZT material for sensors and actuators was a ferroelectric thick-film paste based on PZT 53/47 powder ($PbZr_{0.53}Ti_{0.47}O_3$). The material was made at the Jožef Stefan Institute.

The substrates for thick PZT films are mainly alumina or silicon [2-5]. However, LTCCs (low-temperature co-fired ceramics) have some advantages over alumina substrates: mainly a lower Young's modulus (alumina 215–414 GPa, LTCC 90–110 GPa), which is important for sensor and actuator applications. Some of the characteristics of fired LTCC laminates in comparison with alumina are presented in Table 1.

LTCC materials are sintered at the low temperatures typically used for thick-film processing, i.e., around 850°C. LTCC technology is a three-dimensional ceramic technology utilizing the third dimension (z) for the interconnects layers, the electronic components, and the different 3D structures, such as cavities, channels, diaphragms, cantilevers, bridges, or other structures. These possibilities are widely used for the production of high-density ceramic interconnections and increasingly also for MEMS (micro-electro-mechanical systems) [6-7].

The LTCC tape consists of alumina and glass particles suspended in an organic binder. The materials are either based on crystallisable glass or a mixture of glass and ceramics, for example, alumina, silica or cordierite ($Mg_2Al_4Si_5O_{18}$) [8]. The composition of the inorganic phase in most LTCC tapes is similar to, or the same as, materials in thick-film multilayer dielectric pastes. To sinter to a dense and non-porous structure at these, rather low, temperatures, it has to contain some low-melting-point glass phase. This glass could presumably interact with other thick-film materials (in our case thick-film PZT), leading to changes in the electrical characteristics [8-11].

Table 1: Some characteristics of fired LTCC and 94.0–99.5% Al_2O_3 ceramics.

Characteristic	LTCC	Al_2O_3
TEC ($\times 10^{-6}$/K)	5-7	7.6-8.3
Density (g/cm^2)	2.5-3.2	3.7-3.9
Flexural strength (MPa)	170-320	300
Young's modulus (GPa)	90-110	215-415
Thermal cond. (W/mK)	2.0-4.5	20-26
Dielectric constant	7.5-8.0	9.2-9.8
Loss tg. (x10^{-3})	1.5-2.0	0.5
Resistivity (ohm.cm)	10^{12}-10^{14}	10^{12}-10^{14}
Breakdown (V/100 μm)	>4000	3000-4000

Due to the chemical composition of an LTCC material and the shrinkage of an LTCC tape during sintering, special thick-film materials were developed. On the other hand, some special thick-film materials, such as PZT thick-film material, have not yet been developed for applications on LTCC tape. As a result, these materials can only be used on a pre-fired LTCC laminate. But before their application, the compatibility and the characteristics of these materials must be carefully investigated and evaluated.

Our preliminary investigations [12-14] and data from the literature [15-16] indicated that due to the interaction between the printed thick-film PZT layers and the LTCC substrates during firing the electrical and piezoelectic characteristics deteriorate significantly.

Experimental

PZT 53/47 powder (PbZr$_{0.53}$Ti$_{0.47}$O$_3$) with an excess of 6 mol.% PbO was prepared by mixed-oxide synthesis at 900°C for 1h from high-purity oxides: PbO (litharge) 99.9% (Fluka), ZrO$_2$ 99% (Tosoh), and TiO$_2$ 99% (Fluka). To this was added 2 wt.% of lead germanate, with the composition Pb$_5$Ge$_3$O$_{11}$ (melting point 738°C), as a sintering aid. Lead germanate (PGO) was also prepared by mixed-oxide synthesis from PbO and GeO$_2$ 99% (Ventron) at 700°C. After the synthesis, both compositions were ball milled in acetone for 1 h and dried. A thick-film paste was prepared from the PZT (2% PGO) and an organic vehicle (ethyl cellulose, alpha-terpineol and butil carbitol acetate) by mixing on a three-roller mill.

For the electrical and piezoelectrical characterisation a special test pattern was designed. This test pattern consists of 16 elements (capacitors) with lateral dimensions 4.7 × 4.7 mm and a thickness of about 50 μm. The test sample is shown in Figure 2. The thick-film structures shown in Figure 3 were made on alumina (which was used as a reference) and on pre-fired LTCC substrates. The dimensions of the alumina substrate are 30.0 × 25.4 × 0.25 mm and the dimensions of the pre-fired LTCC substrate are 30.0 × 30.0 × 0.4 mm. The pre-fired LTCC substrates were prepared by laminating two layers of the LTCC tape (Du Pont, 951) at 70°C and at a pressure of 200 bar. This laminate was then sintered in a one-step process with a special burnout-and-firing temperature profile with a peak temperature of 875°C.

Figure 1: The test sample of 16 elements – capacitors on the alumina substrate (left) and the LTCC substrate (right)

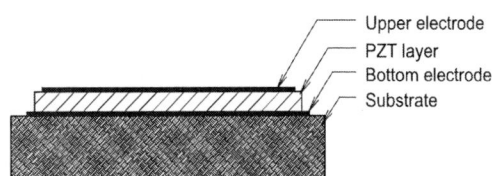

Figure 2: The basic PZT-on-Substrate structure (schematic – not to scale)

The thick-film structures were prepared by first printing and firing the gold or the silver conductor layer as bottom electrodes of the capacitors for 10 minutes at 850°C. Over these electrodes the active PZT film was printed twice and fired, and then again printed twice and fired for 18 minutes at 850°C. On the top of this structure the gold or the silver conductor layer, as the upper electrodes of the capacitors, was printed and fired for 10 minutes at 850°C. In some test samples a barrier layer, based on PZT, as a first layer on the LTCC substrate was used. This layer acts as an intermediate layer between the LTCC substrate and the active PZT structure. The function of this layer is to prevent or to minimize the chemical interactions between the printed thick-film PZT layers and the LTCC substrates during the firings.

The approximate thicknesses of the test structures after the thermal treatments were as follows: gold electrode, 3 μm; silver electrode, 20 μm; active PZT layer, 50 μm; and intermediate PZT layer, 15 μm.

Six different types (combinations of substrate, electrode and barrier) of test samples were fabricated for the structural and electrical investigations. All the types were made with the same thick-film PZT material and with the same technological process. The list of test samples is shown in Table 2.

Table 2: List of test samples

Test samples	Substrate	Electrode	Barrier
Type 1	Al₂O₃	Au	NO
Type 2	Al₂O₃	Ag	NO
Type 3	LTCC	Au	NO
Type 4	LTCC	Au	YES
Type 5	LTCC	Ag	NO
Type 6	LTCC	Ag	YES

For the microstructural investigation the samples were mounted in epoxy in a cross-sectional orientation and then cut and polished using standard metallographic techniques. A JEOL JSM 5800 scanning electron microscope (SEM) equipped with an ISIS 300 energy-dispersive X-ray (EDS) analyzer was used for the overall microstructural and compositional analyses. The EDS spectra were quantified using the ZAF (Z – atomic number correction, A – absorption correction, and F – fluorescence correction) method and a library package of virtual standards. The library contains pre-recorded standard-element profiles under the same experimental conditions.

The electrical characteristics of all the capacitor test samples were measured. The dielectric permittivity and the dielectric losses were measured with an HP-4284 Precision LCR Meter at a frequency of 1 kHz. After this measurement the samples were heated to 160°C and polarised with an electrical field of 100 kV/cm for 20 minutes and then cooled to room temperature.

The values of the piezoelectric coefficient d_{33} were estimated by using the conventional Berlincourt method at 100 Hz with a "Piezometer system PM 10". The standard samples for this method are bulk pellets with defined dimensions. As our thick-film samples fired on rigid substrates are non-standard, the obtained values are only used for the benchmarking of different technologies.

Results and discussion

The microstructures of the cross-sections of the samples with and without the PZT barrier on LTCC substrates are shown in Figure 3 (gold electrodes, without barrier), Figure 4 (gold electrodes, barrier), Figure 5 (silver electrodes, without barrier) and Figure 6 (silver electrodes, barrier). The LTCC substrate is on the bottom. The

LTCC material is a mixture of a darker alumina-rich phase and a lighter silica-rich phase. The thickness of the PZT barrier layer is around 15 μm. During firing the PbO diffused from the PZT films, either from the active layers or from the barrier layers, into the LTCC. The lighter layers on the top of the LTCC substrates are rich in PbO. The estimated depth of the PbO diffusion layer is between 20 and 30 μm.

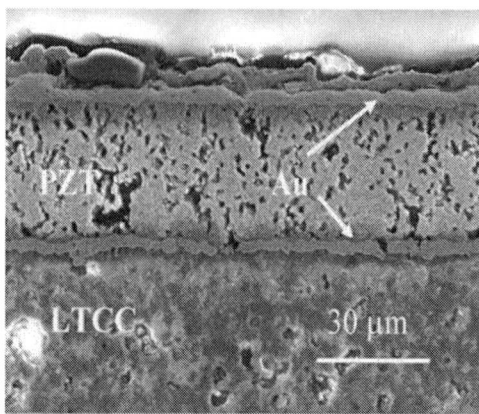

Figure 3: Microstructure of the cross-section of the LTCC/Au-electrode/PZT/Au-electrode structure

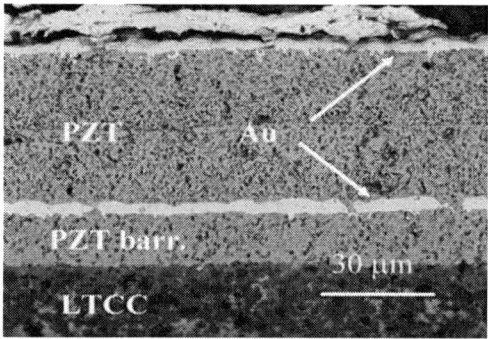

Figure 4: Microstructure of the cross-section of the LTCC/PZT-barrier/Au-electrode/PZT/Au-electrode structure

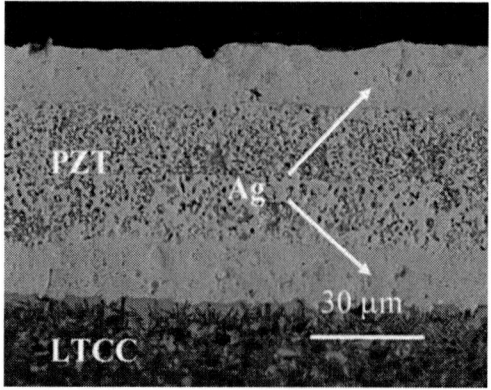

Figure 5: Microstructure of the cross-section of the LTCC/Ag-electrode/PZT/Ag-electrode structure

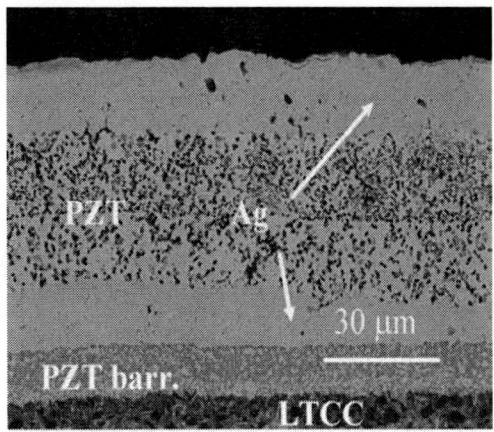

Figure 6: Microstructure of the cross-section of the LTCC/<u>PZT-barrier</u> /Ag-electrode/PZT/Ag-electrode structure

The EDS microanalysis of the PZT layers showed, besides Pb, Zr and Ti, a relatively high concentration of Si, indicating the diffusion of the silica-rich glassy phase from the LTCC into the PZT. The concentrations of the oxides in the PZT layers with silver and gold electrodes, and with or without the barrier, are shown in Figure 7. The concentrations of SiO_2 are higher for structures with gold electrodes (around 15%) than for silver electrodes (around 12%). The difference is attributed to the thicker silver electrodes, which act as a barrier to the diffusion. The intermediate PZT layer, as an added barrier, also slightly decreased the SiO_2 concentration.

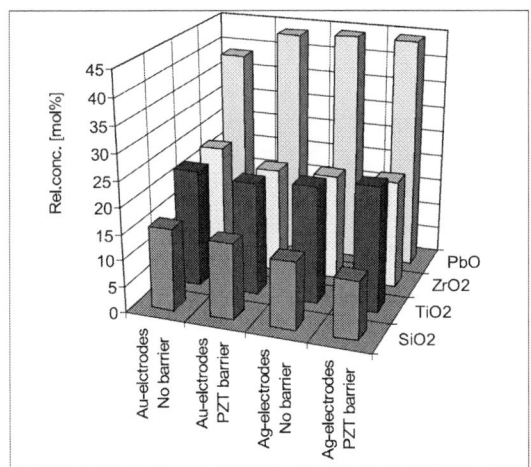

Figure 7: Relative concentration of the oxides in the PZT layer on the LTCC substrate with different electrodes

The dielectric permittivities (ε), dielectric losses (tan δ), and piezoelectric coefficients (d_{33}) were measured for the capacitors on all types of substrates and are presented in Table 3 and Figures 8 and 9. Two main influences on the measured properties can be observed, i.e., the substrate and the electrodes, while the additional intermediate PZT layer does not have a significant influence on the properties.

Table 3: Electrical characteristics (dielectric constant ε', dielectric loss tan δ, and piezoelectric constant d_{33}) of the PZT layers on the alumina and LTCC substrates. The samples with the PZT barrier are denoted Au+B or Ag+B.

Substrate	Electrode	(pC/N) $d33$	(@1kHz) ε	tan δ
Al_2O_3	Au	138	490	0,020
	Ag	96	515	0,028
LTCC	Au	78	200	0,016
	Au+B	81	225	0,011
	Ag	61	410	0,017
	Ag+B	58	445	0,011

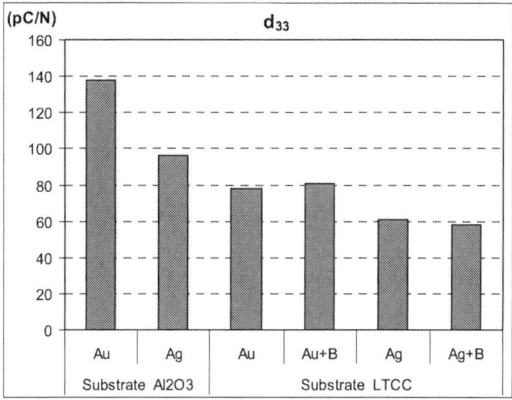

Figure 8: Piezoelectric constants d_{33} of the PZT layers fired on alumina and LTCC substrates

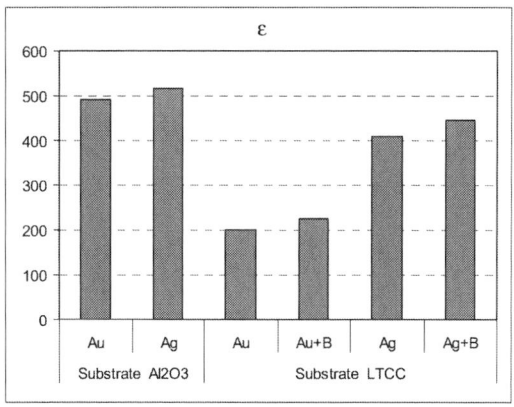

Figure 9: Dielectric permittivity (ε) of the PZT layers fired on alumina and LTCC substrates

The influence of the additional intermediate PZT layer on the piezoelectric coefficients (d_{33}) is within the measurement uncertainty. In contrast, the structures on the LTCC substrate with the additional intermediate PZT layer have between 10 and 15% higher values of the dielectric permittivity.

Conclusions

The compatibility of thick-film piezoelectric (lead zirconate titanate – PZT) material with LTCC substrates (Du Pont 971) was studied. Thick-film PZT material was evaluated as a possible actuator or sensor for C-MEMS. For the investigation of the properties the thick-film PZT $(Pb(Zr_{0.53}Ti_{0.47})O_3)$ paste was screen-printed and fired on LTCC substrates and alumina as a reference. The values of the dielectric permittivity (ε) and the piezoelectric constant (d_{33}) of the PZT layers on the LTCC substrates were about half those of the values on the alumina. The additional PZT intermediate layer between the LTCC substrate and the active PZT layer increases the values of the dielectric constant by about 10-15%. For thicker bottom and upper silver electrodes the values of the dielectric permittivity (ε) are higher and the values of the piezoelectric constant (d_{33}) are lower.

Acknowledgements

The financial support of the Slovenian Research Agency in the frame of the project L2-6462 and the European Commission's support of the project MINUET (FP6-505657) and NoE MIND (FP6-515757) are gratefully acknowledged.

References

[1] P. Tran-Huu-Hue, F. Levassort, F. V. Meulen, J. Holc, M. Kosec, M. Lethiecq, "Preparation and electromechanical properties of PZT/PGO thick films on alumina substrates", Journal of European Ceramics Society, Vol. 21, pp. 1445-1449, 2001.

[2] M. Kosec, J. Holc, B. Malič, V. Bobnar, "Processing of high performancelead lanthanum zirconate titanate thick films", J. European Ceram. Soc., Vol. 19, pp. 949-954, 1999.

[3] M. Kosec, D. Murko, J. Holc, B. Malič, M. Čeh, T. Hauke, "Low-temperature processingof (Pb,La)(Zr,Ti)O3 thick films on alumina substrates", Z. Metallkd., Vol. 92, pp. 97-104, 2001.

[4] M. Kosec, J. Holc, M. Hrovat, D.Belavič, B. Malič, "Preparation of piezoelectric thick films on different substrates", Proceedings of the 37th International Symposium on Microelectronics (IMAPS 2004), Long Beach, California, USA, November, 2004.

[5] M. Kosec, J. Holc, M. Hrovat, D.Belavič, "Piezoelectric Pb(Zr,Ti)O$_3$ (PZT) – based thick films: from materials to devices", Proceedings of the 28th International Conference of International Microelectronics and Packaging Society – Poland Chapter, Wroclaw, Poland, September 26-29, pp. 49-56, 2004.

[6] T. Thelemann, H. Thust, M. Hintz, "Using LTCC for microsystems", Microelectronics International, Vol. 19, No. 3, pp. 19-23, 2002.

[7] L.J. Golonka, A. Dziedzic, J. Kita, T. Zawada, "LTCC in microsystem application", Informacije MIDEM, Vol. 32, No.4, pp. 272-279, December, 2002.

[8] C.J.Ting, C.S. Hsi, H.J. Lu, "Interactions between ruthenium-based resistors and cordierite-glass substrates in low temperature co-fired ceramics", Journal of the American Ceramics Society, Vol. 83, pp. 23945-2953, 2000.

[9] A.A. Shapiro, D.F. Elwell, P. Imamura, M.L. McCartney, "Structure-property relationships in low-temperature cofired ceramic", Proceedings of the 1994 International Symposium on Microelectronics ISHM-94, Boston, Massachusetts, pp. 306-311, 1994.

[10] J.H. Jean, C.R. Chang, "Camber development during cofiring Ag-based low-dielectric-constant ceramic package", Journal of Materials Research, Vol. 12, pp. 2743-2750, 1997.

[11] R.E. Doty, J.J. Vajo," A study of field-assisted silver migration in low temperature cofirable ceramic", Proceedings of the 1995 International Symposium on Microelectronics ISHM-95, Los Angeles, California, pp. 468-474, 1995.

[12] M. Hrovat, J. Holc, S. Drnovšek, D. Belavič, J. Bernard, M. Kosec, L. Golonka; A. Dziedzic; J. Kita, "Characterization of PZT thick films fired on LTCC substrates", J. Mater. Sci. Lett., Vol. 22, pp. 1193-1195, 2003.

[13] M. Kosec, J. Holc, F. Levassort, L.P. Tran-Huu-Hue, M. Lethiecq, "Screen-printed Pb(Zr,Ti)O$_3$ thick films for ultrasonic medical imaging applications", Proc. 34th Int. Symp. on Microelectronics IMAPS-2001, Baltimore, pp. 195-200, 2001.

[14] M. Buczek, L.J. Golonka, M. Hrovat, D. Belavič, A. Dziedzic, H. Roguszczak, T. Zawada, "Electrical and microstructural properties of PZT thick films made on LTCC", Proceedings of the 28th International Conference of International Microelectronics and Packaging Society – Poland Chapter, Wroclaw, Poland, September 26-29, pp. 187-190, 2004.

[15] R.N. Torah, S.P. Beeby, N.M. White, "Experimental investigation into the effect of substrate clamping on the piezoelectric behaviour of thick-film PZT elements", J. Phys. D: Appl. Phys., Vol. 37, pp. 1-5, 2004.

[16] B. Y. Lee, C. I. Cheon, J. S. Kim, K. S. Bang, J. C. Kim, H. G. Lee, "Low temperature firing of PZT thick films prepared by screen printing method", Materials Letters, Vol. 56, pp. 518-521, 2002.

Reliability Aspects of Embedded Chips

A. Ostmann[1], D. Manessis[2], M. Cauwe[3], and Johann-Peter Sommer[4]

[1]Fraunhofer Institute for Reliability and Microintegration (IZM)

[2] Microperipheric Research Center, Technical University Berlin (TUB)

[3] University of Gent – Interuniversitair Microelektronica Centrum (UG-IMEC), Gent

[4]Chemnitzer Werkstoffmechanik GmbH (CWM)

Tel: +49-30-46403187, E-mail: ostmann@izm.fhg.de

Abstract

In the framework of the European project "HIDING DIES" an innovative chip embedding technology for integration of active component into printed circuit boards (PCBs) has been developed. Besides the successful development of a technology platform for chip embedding in the PCB manufacturing flow, the reliability of test vehicles was assessed. In this paper experimental results of reliability tests will be discussed. The tests were performed using thin chips, embedded into build-up layers of PCBs. The chip thickness was 50 µm and the sizes varied from 2.5x2.5 mm² to 10x10 mm². The performed tests were high temperature storage, humidity storage, thermal shock and reflow tests. Additionally experiments with embedded Si chips without interconnects were performed in order to investigate the polymer interface behaviour after stress conditions. Finite element simulations were used for thermo-mechanical pre-optimization with respect to the desired functionality and operating conditions. Reliability aspects were also investigated numerically before first tests with real devices were available, revealing points of potential defects.

Key words: Embedded Chips, Reliability, finite element modelling

Introduction

In the European Hiding Dies project /0/ seven partners from industry and research centers work together on the development of a technology for embedding of active chips into printed circuit boards (PCB). The goal to achieve a generic technology platform for chip embedding in a PCB manufacturing flow has been successfully achieved /0/. Technology development was performed by the use of different test vehicles. They contained chips of 50 µm thickness and sizes in between 2.5x2.5 mm² and 10x10 mm². The process flow of the Hiding Dies technology is:

1. die bonding of thin chips onto a core substrate (e.g. multi-layer FR4) by printed adhesive or die attach film
2. embedding of chips by vacuum lamination of a RCC (resin coated copper) layer
3. drilling of micro-vias to the embedded chips and to the core layer
4. metallization of vias by electroless Cu and electroplating of Cu
5. structuring of the Cu layer

A detailed discussion of the process steps has been published in /0/. For the performance of reliability the main test vehicle of the Hiding Dies project was used (see Figure 1). In case of a complete population it contains 72 embedded chips, varying from 2.5x2.5 mm² and <20 I/Os up to 10x10 mm² and 184 I/Os. The chips had pitches of 300 and 200 µm. Test vehicle boards was evaluated by high temperature storage, humidity storage, thermal cycling and reflow tests. Conditions and results of the tests are detailed in the following section

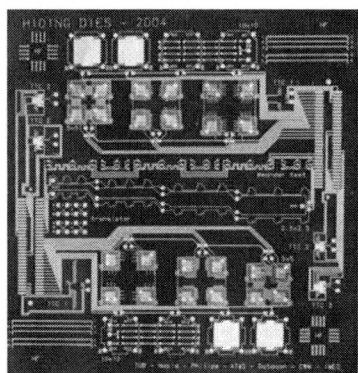

Figure 1: Picture of Hiding Dies test vehicle with embedded chips (size 10x10 cm²).

In a further set of experiments the interface behaviour of substrates with embedded chips under stress conditions (reflow tests, thermal cycles) was investigated. Finally a short example of the accompanying thermomechanical modelling and simulation work is presented.

Reliability Tests

Conditions

For the reliability tests samples of the above described Hiding Dies test vehicle board were used. The samples originate from different process runs over a longer period of time, giving an overall status of the technology under development. Even samples from an early stage of the process were included. Test conditions are listed in Table 1, followed by a more detailed description below. In regular periods electrical measurements of the daisy chains were done. After completion of the test cross-sections were prepared. Failure criteria were resistance changes > 20 % or delaminations.

Table 1: Overview of reliability tests

test	conditions
humidity storage	1000 h @ 85 °C / 85 % RH
thermal storage	1000 h @ 150 °C
humidity storage under bias	1000 h @ 85 °C / 85 % RH, 10 mA bias current
thermal cycling	1000 cycles @ –40/+125 °C 4000 cycles @ -55/+125°C
reflow	3 x reflow @ 260 °C

Humidity storage (under bias)

Samples were stored at 85 °C and 85 % humidity. The duration of the test is specified at 1000 hours, but some samples were tested more than 2000 hours. Additionally, a total of three test vehicle boards were tested under bias. Normally this is done by putting a bias voltage onto the circuit under test. This was not sensible for the test designs used, so instead it was opted to stress the daisy chains with a bias current of about 10 mA.

Thermal storage

Samples were stored at an elevated temperature of 150 °C for a minimum of 1000 hours. This test was mainly used for initial reliability testing, since it is a fairly light-weight test.

Thermal cycling

During a 1 hour cycle, the samples were submitted to a temperature range from -40 °C to +125 °C. This cycle was repeated a 1000 times. Some samples were tested under thermal shock conditions -55 °C/+125 °C and a cycle time of 30 min, with a dwell time of 10 min. at upper and lower temperature. For this test condition electrical measurements were done up to 4000 cycles and the test is still ongoing.

Reflow

Reflow tests were performed according to the JEDEC standard. This standard describes different levels of testing. All levels include three reflow runs with a Pb-free profile (Tmax = 260 °C), but are differentiated by the humidity storage prior to the first reflow. For level 1 the boards are stored for 168 h at 85 °C and 85 % R.H.; level 3 requires 192 h at 30 °C/60 % R.H. before the first reflow. JEDEC level 1 is the most severe test, which fails for most types of plastic packages.

Results

Table 2 summarizes results of all reliability test runs performed. Overall a significant number of contacts to the chip were investigated, although only a limited amount of samples were available for testing. Only a few embedded chips 2.5x2.5 mm² with 64 measurable I/Os together were tested for 400 cycles at thermal shock conditions - 55 °C/+125 °C. No electrical defect was measured.

Table 2: Summary of reliability results

test	contacts	electrical defects	delamination
humidity storage	3002	6	1 sample
thermal storage	1232	0	No
humidity storage under bias	1704	0	No
thermal cycling	4712	8	No
reflow	1018	0	1 sample

A more detailed look to the results shows, that the electrical defects during thermal cycling humidity storage occurred on boards from an early development phase. They initially already showed a large number of failed daisy chains. Cross sectioning revealed that this was due to a faulty metallization process, resulting in large voids inside the plated vias (see Figure 2).

Figure 2: Cross-section of via to an embedded chip: with insufficient via metallization.

Figure 3: Cross-section of a via to an embedded chip: with proper Cu via metallization

Embedded chips with proper via metallization (as shown in Figure 3) have never shown electrical failures at any test.

Delamination occurred for two samples after humidity storage and thermal cycling. In both cases a separation between chips and the silver-filled epoxy adhesive layer occurred (cross section see Figure 4). It should be noted that the interconnection to the chip did not fail electrically.

Figure 4: Delamination between adhesive and chip after reflow test.

Summary

In summary results of the reliability tests are:

- No failure occurred after thermal storage.
- Improper (weak) Cu metallization can lead to electrical failures during humidity storage and thermal cycles due to Cu break
- No electrical defects occurred at samples with properly metallized vias. This was always the case in the final process runs after optimization of plating parameters.
- Delamination between chip and die attach material was observed for two samples after humidity storage and reflow test at JEDEC level 1 conditions.
- A low number of chips were tested for 4000 cycles without showing electrical defects.

Break of via interconnects is considered as a solved problem, since it was clearly related to an immature process at an earlier development phase. Delaminations however are a basic issue of polymeric materials, not induced especially by chip embedding into the PCB substrate structure.

Interface Stress Test

After the reliability test described above a further set of experiments was carried out, which aimed especially at the investigation of polymeric interfaces after different stress conditions. In order to concentrate on the relevant interfaces and to keep sample preparation simple, embedded Si chips without electrical interconnects were used only. A comprehensive experimental matrix was built in order to study the effects of lamination preessure, heating rate, RCC resin type, and adhesive type on the interfacial perfromance of samples with embedded chips. In specific, 10x10 cm² FR4 panels of 676 µm thickness coated on both sides with 16.8 µm Cu were used. They. Si wafers were thinned down to 50 µm and cut in 2.5 x 2.5 mm² dies for attachment on the boards. A silver-based adhesive was used for die attachment as well as a die adhesive film. The thickness of the printed silver-adhesive was about 20 µm. Two b-stage RCC materias were used for the experiments from different suppliers; one with a non-filled epoxy (A) 100 µm/9 µm Cu and one with filled epoxy (B) 90 µm thick/5 µm Cu. The silver-based attach material was stencil-printed and the dies were attached. Subsequently, the silver paste was cured at 150 °C for 3 h. The lamination experiments were performed using a PCB vacuum press system. Table 3 summarizes the RCC materials and test conditions of the experimental matrix.

Table 3. Lamination experimental matrix

#	RCC type	adhesive type	heating rate (°C/min)	pressure (bar)
1	A	paste	3.6	5
2	A	paste	3.6	10
3	A	paste	3.6	20
4	A	paste	7.3	5
5	A	paste	7.3	10
6	A	paste	7.3	20
7	B	paste	3.6	5
8	B	paste	3.6	20
9	B	paste	7.3	20
10	B	film	3.6	20

Reflow Test

Reflow tests at JEDEC level-1 (IPC/JEDEC J-STD 020C) were performed by holding at 125 °C for 24 hrs before test and then moisture soaking at 85 °C/80 % RH condition for 168 hrs; hold for 1 h and then reflow three times with JEDEC level-1 reflow profile.

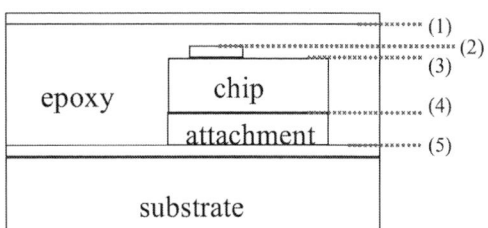

Figure 5. Delineation of delamination interfaces: (1) Cu/epoxy, (2) epoxy/pad, (3) epoxy/chip, (4) chip/adhesive, (5) adhesive attachment/Cu from FR4 substrate.

After exposure the samples in JEDEC level 1 condition voids can be found which originate from the lamination process as mentioned before and not from the actual JEDEC level 1 test. Figure 5 shows all potential delamination interfaces resulting from reliability testing.

There are three different delaminations modes observed by JEDEC level 1 testing at interfaces (1), (3) and (4) according to notation in Figure 5. For defect statistics after JL1, nine samples for each lamination condition were tested. Delaminations at (2) and (5) interfaces were not observed. Delamination at interface (1) occurs rarely for "A" RCC and sometimes was observed for "B" RCC epoxy under all conditions. For "A" RCC appears only under high heating rate and small pressure of 5 bar experiment and it implies the necessity of high pressure (at least 10 bar) for lamination because it promotes the Cu adhesion on the epoxy. Figure 6 shows such Cu delamination from the epoxy because of low pressure for "A" RCC. Such defect at interface (1) may further imply that the adhesion between Cu and the epoxy resin of the RCC should be improved from the RCC provider. Delamination between chip and die attachment material (interface 4) is shown in Figure 6. It has been found frequently and is considered normal for the harsh environment of JL1 and may be also attributed to partial voiding in the printed and cured adhesive paste. These voids largely account for the delamination phenomenon. Nevertheless, defect at interface (4) shows that the lamination quality is good and reliable and only the attachment material should be improved or the printing should be void-free in order to avoid failures at interface (4). Delaminations in the chip/epoxy interface (3) appear only in "A" epoxy for 5 bar lamination with low heating rate and for all laminations with high heating rate and especially with 5 bar pressure, as shown in Figure 8. Obviously for "A" RCC, the low pressure of 5 bar and the increased fluidity of the epoxy due to high heating rate have an adverse effect on the epoxy adhesion on the chip. "B" RCC does not show any problem at interface (3) compared to "A". This delamination can be caused also by the epoxy expansion and

bending during the repeated reflow in JL1. Such bending can create high stresses especially in the sample of this study where no lamination was planned from the FR4 bottom side in order to counterbalance the stresses. Furthermore, moisture absorption and mechanical strength of the epoxy play a significant role. The superior performance of "B" RCC than "A" especially for delamination at interface (3) may be attributed to the lower moisture absorption of "B" than "A" and to its higher mechanical strength owing to its filler content.

Figure 6: Delamination at interface (1) (Cu/epoxy) after JL 1 testing for lamination with "A" RCC at 5 bar

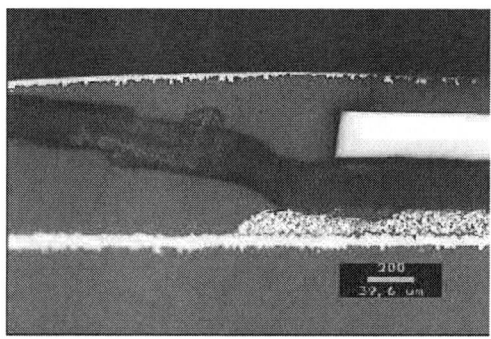

Figure 7: Delamination at interface (4) (Chip/ adhesive) after JL1 testing for lamination of "B" RCC at 20 bar

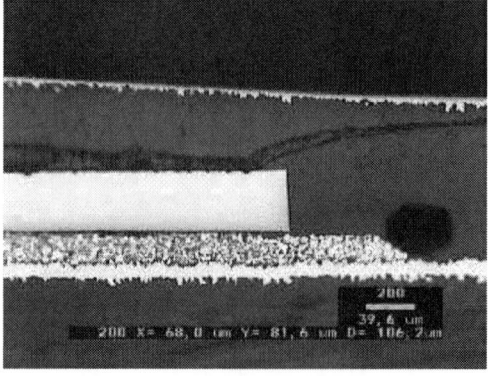

Figure 8: Delamination at interface (3) (epoxy/ chip) after JL1 testing for lamination of "A" RCC at 5 bar

Thermal Cycling

Thermal cycling test exposed the samples up to 1000 cycles at -55 °C / +125 °C (10 min at -55 °C, 5 min at 25 °C, 10 min at +125 °C, 5 min at 25 °C and then back to -55 °C for 10 min

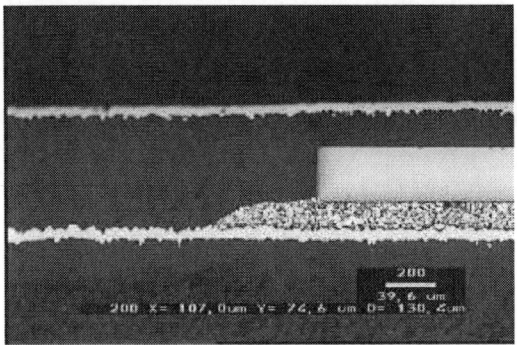

Figure 9: Delamination-free embedded die after 1000 cycles at -55 °C/+125 °C temperature range ("A" RCC, 20 bar)

All samples have passed successfully 1000 thermal without indication of delaminations at all interfaces or other defects. Some voids exist but stem from the lamination process. Experiment 10 has also revealed good reliability for the adhesive film under thermal cycling conditions. Figure 9 indicatively shows an embedded silicon die after 1000 cycles.

Summary

- Three different delaminations take place after JEDEC Level 1 testing; at interfaces (1), (3) and (4). Delamination between chip and die attach material (interface 4) is mainly due to voids pre-existed during screen printing of the adhesive paste. Delamination at interface (1) (Cu/epoxy) can be improved by better adhesion of the Cu on the epoxy. Delamination at interface (3) (epoxy/chip) can be attributed to thermal mismatch coefficients of the epoxy and the chip and is also due to excessive unsymmetrical bending during reflow. Lamination also from the bottom side is recommended.
- All samples have successfully passed 1000 cycles in thermal cycling test from -55°C to +125°C without showing any delamination.

Simulation and Modeling

An example of the extensive modelling and simulation in the Hiding Dies project is a study of the thermal and the thermo-mechanical behaviour in the vicinity of test vehicle. Three variants with different actions implemented for leading the dissipative heat off were realized in a single 3D FE (finite element) model. Figure 10 represents the version with continuous copper at the top and thermal vias for a best thermal performance.

As a disadvantage, the shape of the thermal vias had to be estimated, and it could be shown, that the real shape is of essential influence on the maximum stresses around the vias. But the interaction between neighbouring vias was found to be negligible. After having first real parts, the vicinity of a single via was modelled in detail.

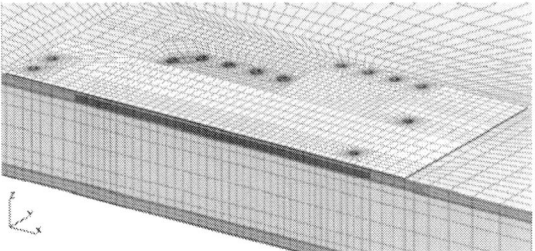

Figure 10: 3D FE-model of an embedded chip

Figure 11 gives an impression of the stress distribution around the via. The highest stresses are expected at the interface chip to die pad and the transition from the die pad to the via, not at the upper interconnection to the outer lead. Smoother transitions between chip, die pad, via, and outer lead are found real parts compared with the idealized ones in the model. Therefore, the computed stress maxima will act as upper bound values.

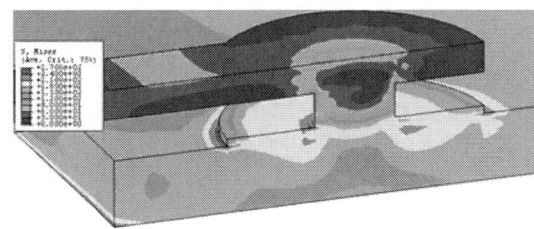

Figure 11: Mises stress distribution at the symmetry plane through a Cu micro via

As the most important result, information about the differences in strength are expected for the dies and the interconnects in 1- resp. 2-directions.

In summary it can be stated that after embedding a compression stress state around the embedded dies exists. Tensile stresses occur only at the edges but can be neglected due to the relative low stiffness of the RCC resin around. Long copper lines inside the package and outside, which are bridging the chip edges, are relatively intensively stressed during thermal cycling. A plastification was detected only at the first simulated thermal cycle. After that, a deformation reserve helps to suppress plastic cycling which is necessary for a good reliability. Further details of the thermo-mechanical modelling are published in /0, 0, 0/.

Conclusions

In the Hiding Dies project a process technology for embedding of chips into PCB build-up layers was developed and the reliability performance evaluated.

No defects were observed after thermal storage at 150 °C.

Samples from an early phase of the project failed during thermal cycles and humidity storage. The defects were caused by break of improperly metallized vias.

No electrical defects or delaminations were observed for chips with fully metallized vias. A low number of samples were even tested for 4000 thermal shocks without failure.

After humidity storage occasionally delaminations between chips and die bond adhesive were observed. In order to improve this, a screening for better performing die attach materials is planned.

After reflow tests according to JEDEC level 1 delaminations frequently occurred. This however can be generally observed for many types of plastic packages. Reflow tests according to JEDEC level 3 did never lead to delaminations.

In a further experiment only Si chips without electrical connections were embedded, using different lamination parameters, in order to investigate the polymer interfaces after different stress conditions. Also here no delaminations could be forced by thermal shocks. Reflow tests at JEDEC level 1 conditions again caused delaminations at different interfaces.

The technology developments of the project were always accompanied by modelling and simulation activities. This activity supported the design of device structures and reduced the overall effort for experiments.

Acknowledgments

The authors would like to thank the European Commision for the financial support. The Hiding Dies project is part of the 6. framework programme. Special thanks to Arno Kriechbaum, Rainer Patzelt and Alexander Neumann for preparation of samples.

References

[1] Project website www.hidingdies.net

[2] A. Ostmann, .J.De Baets., A. Kriechbaum, H. Kostner, A. Neumann, "Technology for Embedding Active Dies", Proc. of 15th European Microelectronics and Packaging Conference, June 12-15, Brugge, Belgium

[3] Kriechbaum, A., "Manufacturing of demonstrator printed circuit boards with embedded active components", Proc. of 16th European Microelectronics and Packaging Conference, Oulu, Finland, June 17-20, 2007.

[4] Sommer, J.-P., Döring, R., Dost, M., Michel, B.:"Advanced Packages with Buried Dies: Design Support by Means of FE Analysis and Deformation Measurement in Micro Scale", Proc. of 4th European Microelectronics and Packaging Symposium (EMPS 2006), Terme Catez, Slovenia, May 22-24, 2006, 139-145.

[5] Sommer, J.-P., Michel, B., Ostmann, A.: "Electronic Assemblies with Hidden Dies – Design Support by Means of FE Analysis", Proc. 1st Electronic Systemintegration Technology Conference, Dresden, September 5 - 7, 2006, 1088-95.

[6] Sommer, J.-P., Michel, B., Ostmann, A.: "Numerical Characterisation of Electronic Packagiyng Solutions based on Hidden Dies", Proc. ICEPT 2005 (6th International Conference on Electronics Packaging Technology), August 30 - September 2, 2005, Shenzhen, Chinay, 300-306.

Manufacturing of demonstrator printed circuit boards with embedded active components

Arno Kriechbaum

AT&S AG Austria Technologie & Systemtechnik Aktiengesellschaft, Fabriksgasse 13, 8700 Leoben, Austria

Phone:+4338422005719, Fax:+433842200216, a.kriechbaum@ats.net

Abstract

The manufacturing and challenge in processing of three different demonstrator printed circuit boards (PCBs) with embedded dies is presented. In frame of the European research project "HIDING DIES" a power RF demonstrator, a smart card module and a sensor demonstrator was specified for the validation of the developed chip embedding technology. It is the goal of this paper to present the different manufacturing conceptions for each demonstrator and potential process optimisations. The basic concept is to embed thinned dies (50µm) with RCC foils into build up layer by vacuum lamination. Electrical contacts to the chip are made by laser drilled microvias followed by PCB compatible Cu plating. Due to the different applications and requirements of the demonstrators varying and modified process steps are necessary. The smart card modules are used for bankcards and identification cards. This application demanded an ultra thin and cost effective build up. For the RF power amplifier demonstrator is excellent electrical properties and good thermal conductivity necessary. The sensor demonstrator is a 2D compass and the miniaturization of the system and the reliability aspects of different die types are important.

Key words: chip embedding, smart card module, power RF amplifier, sensor demonstrator

Introduction

Due to the different requirements of miniaturization in applications of the electronic industry, it's necessary to embed passive components (Resistors, Capacitors) and active components into printed circuit boards. In addition to the miniaturization the integration of active components has different functional reasons like improved radiofrequency characteristics by shorter signal tracks, better thermo mechanical reliability and the possibility of chip stacking.

The presented results are outcomes from the project "HIDING DIES" which was started January 2004 and is funded by the European Commission. The consortium of this project consists of seven partners from five different countries. The project partners are IMEC (Belgium), Datacon (Austria), AT&S (Austria), CWM (Germany), Philips (Netherlands), Nokia (Finland) and TU-Berlin (Germany) [1].

The schematic process flow of embedded thinned dies is displayed in Figure 1. At first the thinned die (thickness≤50µm) is bonded on the structured core with an adhesive. Afterwards the chip is embedded with an RCC foil (Resin Coated Copper) and a vacuum lamination process. The electrical contact between active component and outer layer is manufactured by laser drilling and following copper plating process. The last step is the copper structuring of the outer layer. Figure 2 shows the build up of an embedded active component.

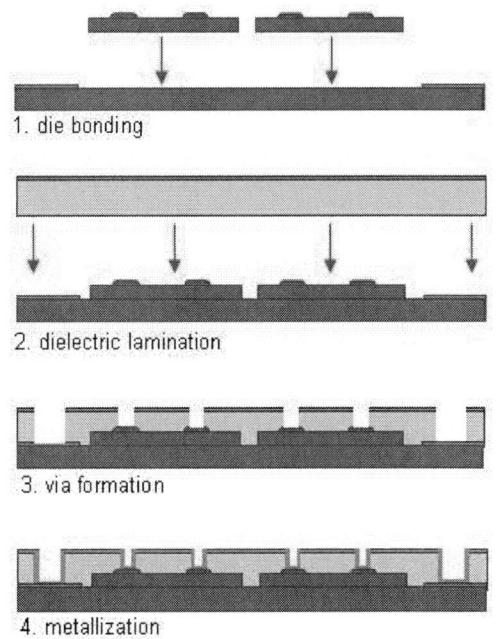

Figure 1: Process flow for chip embedding technology

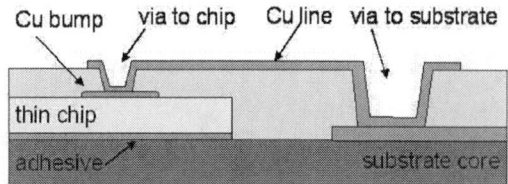

Figure 2: Build up of PCBs with embedded active components

Die bonding

Before die bonding process, the wafer must be prepared especially for this chip embedding technology. The contact pads are usually consists of aluminium. This kind of pad is incompatible with conventional PCB process steps like laser drilling and metallization. Therefore it's necessary to cover the aluminium pads with a copper metal layer. In case of very small distances of chip pads it's essential to apply a redistribution layer (RDL), because the accuracy of current processes is not high enough. With RDL it's possible to convert the small pads in border area to a processable pad size. After the copper layer application the wafer is thinned to ≤50μm and sawed into single chips.

For bonding the thinned dies on the core substrate a liquid adhesive or a special adhesive tape is used. The application of the liquid adhesive can be done with screen printing, stencil printing or dispensing and the challenge is to realize an adhesive layer with ≤20μm thickness, uniform thickness distribution and without bubbles. With adhesive tapes a uniform thickness distribution for layer thicknesses down to 10μm is viable.

After the adhesive application the chip assembly is done on fully automatic die bonder with tools especially developed for thin die processing. Currently the chip assembly on panel sizes (e.g. 18x24 inches) which are commonly used in PCB manufacturing is not possible with the required accuracy of less than ±15μm. Up to now the chip assembly is done on substrate sizes of 12x18 inch and were put into a frame with size of 18x24 inch for further processing. For the required high accuracy at chip assembly local fiducials close by the chips are used.

Chip embedding

The embedding of dies occurs with a vacuum lamination process of the structured die bonded cores with an RCC foil. Vacuum ensures a void free distribution of the dielectric during the lamination press. For further process steps the planarity of the surface after vacuum lamination process is very important. Therefore a special lay up of the press book for chip embedding has been developed. A large quantity of craft-paper effected a uniform pressure distribution across the whole panel and with thick polished steel panel a very good planar embedding of dies has been achieved. The lamination process is done at a maximum

Temperature of 235°C at maximum pressure of 20bar and takes 120min. For minimizing the warpage the panels are cooled down under pressure. For the chip embedding was selected a RCC foil with high filler content, high glass transition temperature and with good flow behaviour. Figure 3 shows an embedded die.

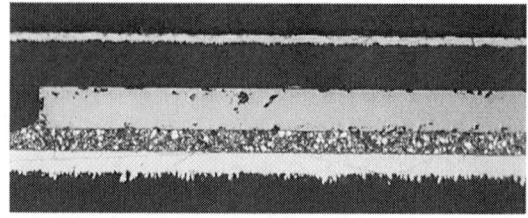

Figure 3: Embedded thinned die with vacuum lamination process

For the laser drilling step afterwards the thickness and the uniformity of the dielectric are very important. The dielectric thickness affected the diameter of the Microvias to the chip and the uniform thickness distribution is important for the process stability of laser drilling. These two dimensions are depending on adhesive thickness respectively distribution of adhesive thickness, chip thickness respectively distribution of chip thickness, numbers of embedded chips and the lamination process. At Micro via diameters of 50μm a dielectric thickness in the range from 20μm to 30μm is aimed. Figure 4 shows a uniform thickness distribution of dielectric above the chip.

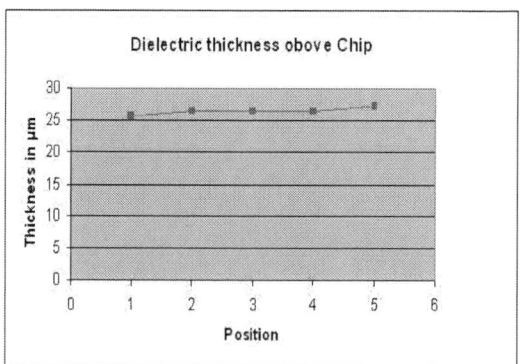

Figure 4: Uniform distribution of dielectric thickness above embedded die

Laser drilling

For drilling of the Microvias to contacting the chip pads, a machine with combined laser system is used. At first the copper is opened by an UV laser (Figure 5) and afterwards the resin is removed by a CO_2 up to the chip pads.

The advantage of combined systems is that the copper pads are not damaged by the CO_2 laser. At machines with only UV laser is the risk to damage the chip pads at varying thickness distributions or varying process parameters very high.

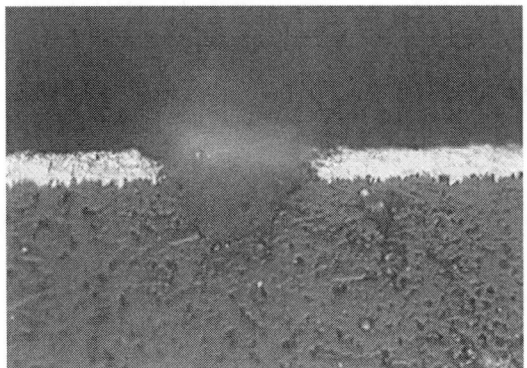

Figure 5: Opened copper layer by UV laser

Due to the very small diameter of the chip pads of 150μm, the common alignment methods in PCB manufacturing are too inaccurate. Therefore the local alignment occurs on four local fiducials on the inner layer where the chip is bonded, that means the copper and the resin has to be removed above the fiducials before alignment (Figure 6).

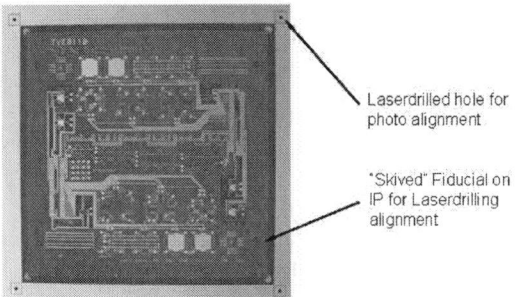

Figure 6: Local alignment for laser drilling and photolithography process

Metallization

After the laser drilling of the Microvias a hole cleaning step, followed by metallization is done. At first the hole is cleaned in a desmearing step with potassium permanganate ($KMnO_4$). Subsequently the surface is activated by Palladium and a layer 2μm electro less copper is superimposed. Last step is the filling of the 50 μm vias with pulse plating technology. Figure 7 shows a filled Microvia for contacting the chip pads.

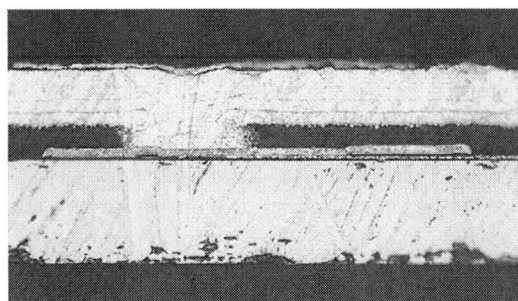

Figure 7: Filled Microvia for contacting the chip pads

Structuring of outer layer

The structuring of the outer layer is realized by laser direct imaging and spray etching afterwards. In order to achieve the required high accuracy of ±15μm, it is necessary to use local fiducials for the alignment (Figure 6). Figure 8 shows the top view of structured outer layer and connected chip.

A standard resist with 30μm thickness is not capable for the realisation at track to track spacing of ≤50μm which are common at this technology. Therefore the use of resist types with 15μm to 20μm thickness is required additional to copper thickness in the range of 25μm at the outer layer.

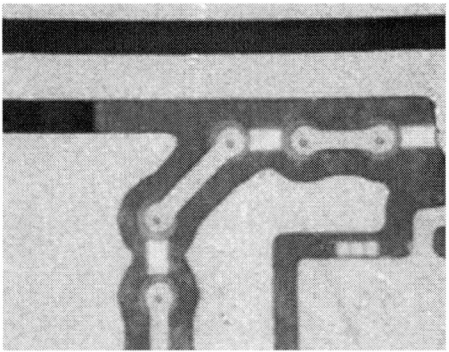

Figure 8: Top view of structured outer layer and connected embedded chip

The manufacturing of the demonstrators is done with previous described process steps. The following chapter describes the demonstrators, special process steps for manufacturing and process adjustments by reason of the requirements for each demonstrator.

Smart Card Module

This is a straightforward demonstrator as used for bankcards and identification cards. The Module size is 12,6x11,4 mm and it's one controller chip with 3,2x2,9 mm embedded. Goal is to realize a very thin module with thickness of 0,35 mm. The die is bonded on the core with a special tape and the standard 80μmx80μm Al pads of the chip are covered with copper and enlarged to 150μmx150μm.Figure 9 shows the top view of the manufactured Smart Card Module.

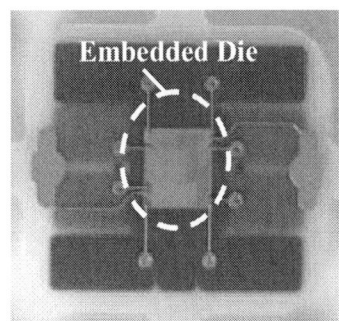

Figure 9: Top view of Smart Card Module with embedded die

Lamination cycle for ultra sensitive dies

Lamination of the Demonstrator with the above mentioned parameters showed extensive cracking during processing. Reasons were to high copper bumps (23µm) which make the chip very sensitive upon pressure application and of course the lamination pressure of 20 bar. Therefore it was essential to develop a new press cycle for ultra sensitive chips.

At first the actual temperature in the product has been measured during the lamination cycle because it's important for the curing of the resin and the pressure applying.

An outcome of the temperature measurement of the product and the viscosity curves of the RCC was to start with 4 bar (minimum possible pressure) and to increase the pressure after 40 min from 4 bar to 10bar.The increase up to 10 bar leads to better and uniform distribution of the resin. For better pressure distribution more craft paper were used, that means in the middle of the press book 4 sheets craft paper were added. The build up we used before has only 8 craft papers top and bottom of the press book. The temperature cycle is the same as we used in test vehicle manufacturing.

Figure 10 shows a cross section of the Smart Card Module laminated with new press cycle, RCC foil with 90µm resin thickness and reduced thickness of chip pads of 5µm

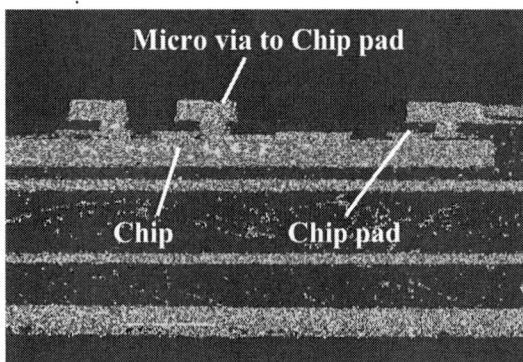

Figure 10: Cross section of Smart Card Module with embedded die

The combination of adjusted press cycle and thinner chip pad thickness led to an essential improvement in embedding performance. Optical inspection showed only 2 cracked dies during lamination by total number of 770 tested chips.

RF Power Amplifier

This demonstrator is part of an RF Power Amplifier and objective is to make a comparison in performance with current techniques. It's a test amplifier chip with 1mmx1mm die size embedded. The chip has a small pitch of 90µm, but with additional copper plating on top of the bond pads, a pitch larger than 150µm can be realized.

The performance of the die is mainly limited by the temperature and therefore low thermal resistance of the whole build up is required. That means base material with high thermal conductivity is used, the chip is bonded on the core with a sliver filled adhesive and in the core directly under the die are laser drilled and with copper filled thermal vias.

Processing of thermal vias

Due to the requirement of good thermal conductivity of the whole build up, it is a need for thermal vias directly underneath the chip. The use of mechanically drilled PTH would require these holes to be filled and the copper above would need to be smoothed to achieve good die bonding results. An easier solution is to use a thin core and drill the vias from the backside using laser drilling. These vias can then be filled using extra copper plating. AT&S decided to do the laser drilling with a combined UV/CO2 system. Different laser drilling parameters have been tested and Figure 11 shows the result of optimized laser drilling parameter for 200µm hole diameter and 150µm thick core.

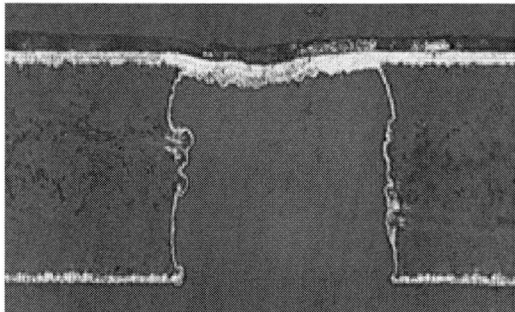

Figure 11: Laser drilled thermal via

Thermal measurements and simulations at test vehicles with thermal vias showed that's a copper thickness in the hole and on the surface of 20-25µm is required for sufficient heat dissipation.

The first idea were not completely filled holes with minimum copper thickness of 25µm in the hole and on the surface, but due to the high density of large laser holes with 200µm diameter, there are unfavourable flowing conditions during copper plating and we got areas with no copper between the thermal vias on the panel back side.

Completely filled vias provides no surface defects but the filling of thermal vias results in non unequal copper thickness on the panel surface and in the area of the thermal vias and affects the conditions at copper structuring afterwards. There are two additional steps necessary to achieve the specified copper thicknesses: First, a differential etching step of the whole panel to achieve 20-25µm in the area of thermal vias back. Second step is to make an etching compensation. That means to adjust the original dimensions of the design respective to the copper thickness in order receive the specified dimensions after etching.

Figure 12 shows the completely filled thermal vias with 200μm diameter

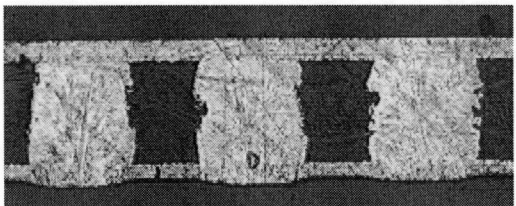

Figure 12: Filled laser via for thermal management of RF Power Amplifier

Sensor Demonstrator

The sensor application is a 2D compass based on a control ASIC and a magnetic field sensor die. The control ASIC die and two daisy chain dies are embedded, one with copper pads on bare silicon and the other with Redistribution layer. The control ASIC is originally designed for wire bonding and has a minimum pitch of 78μm and therefore it requires also a Redistribution layer to apply the "Hiding Dies" technology. The sizes of the embedded dies are 2,2mmx2,3mm, 10mmx10mm and 7,2mmx6,7mm.The minimum pad size is 150μm. The dies are bonded with a special adhesive tape on the core.

The demonstrator can be used both to verify the applicability of embedding technology for manufacturing functional modules and to compare the reliability aspects of embedding different die types.

Conclusion

In frame of the European research project "Hiding Dies" a production capable technology for chip embedding into PCB's is developed. At first the thinned die (thickness≤50μm) is bonded on the structured core with an adhesive. Afterwards the chip is embedded with an RCC foil (Resin Coated Copper) and a vacuum lamination process. The electrical contact between active component and outer layer is manufactured by laser drilling and following copper plating process. The last step is the copper structuring of the outer layer.

Three different demonstrators, their processing and their special requirements have been described.

The smart card module is a straightforward demonstrator. Goal is to realize a very thin and low cost build ups. Because of chip cracking a lamination cycle for ultra sensitive dies has been developed which results in an essential improvement of lamination yield.

The performance of the RF power amplifier demonstrator is mainly limited by the temperature and low thermal resistance of the whole build up is required. Therefore a manufacturing technology of filled laser holes for the thermal management has been introduced.

The sensor demonstrator is a 2D compass with three embedded dies and can be used both to verify the applicability of embedding technology for manufacturing functional modules and to compare the reliability aspects of embedding different die types.

Acknowledgements

The author would like to thank the European Commision for the financial support. The HIDING DIES project is part of the 6. framework programme.

References

[1] HIDING DIES website www.hidingdies.net

Embedded Actives Technology, from Functional Densification to Fan-out Redistribution

A. van der Lugt[1], W. Peels[2]

[1] Philips Applied Technologies, Eindhoven, Netherlands

[2] NXP Semiconductors, IS&O CSC-Inno, Nijmegen, Netherlands

Abstract

Cost reduction and miniaturization were for years the main drivers of new package development. Lead frame based packages are still the major number of SMD packages on the market today but with the introduction of leadless packages and system-in-package (SiP), new IC packaging solutions are explored and introduced. The rise of SiP is driven by functional densification, while leadless packages require fan-out redistribution to bridge the interconnect gap between die and PCB; both introduce a new set of requirements. Traditionally 'IC manufacturing and packaging' and 'board and system assembly' are two separated worlds. Today various new packaging technologies exploring the IC manufacturing and packaging techniques to realize the SiP requirements are seen. Also the embedding of passives and actives, moving down in the packaging and assembly food chain from main board to module board and further are studied. One of these embedding techniques is explored in the European project 'Hiding Dies' where printed board build-up technology is re-used for the embedding of naked dies in the inner layers of a printed circuit board. This paper presents the attractiveness and challenges of this technique for use in IC packaging and more particularly discretes packaging. Results of first fabricated samples are very promising.

Key words: Embedded actives, discretes, build-up, and packaging

Introduction

For years the main function of IC packages was interconnect fan-out redistribution and IC protection. The IC package had to close the interconnect gap between IC and PCB. Wafer level processes, packaging and board level technologies were separated worlds, each with their own scale of interconnect dimensions. IC's have dimensions a factor 10 to 100 times smaller than printed circuit boards and packages have dimensions in-between. Printed board technology realized a factor 10 reduction in line pitch dimensions in 25 years, but IC technology realized a reduction factor of 100 in the same time period. The interconnect gap enlarged.

Integration at chip level or board level

Miniaturization and functional diversification are strong drivers for innovation in electronic industry. Function integration and densification was traditionally realized on chip level or on printed board level. The System on Chip (SoC) approach is the way to realize ultimate densification and often also results in lowest overall cost level. (Yet) non-integrate-able functions or long development times however limit this SoC approach. The System on Board (SoB) approach was the fastest way to realize complex systems, but, although the assembly of miniaturized components, new HDI board technologies and embedding of passives are introduced, miniaturization is limited.

The way between these two system approaches has always been the use of sub-systems or modules. By dividing the system in sub-systems and optimizing these sub-systems in functionality and miniaturization, the real optimum between conflicting system requirements can be realized.

Integration at package level

The need for highly integrated and miniaturized sub-systems introduced new system approaches like System in Package (SiP) and System on Package (SoP) [1]. Electrical, mechanical, optical and thermal functions are to be integrated in one module. System in Package became the buzzword for miniaturized sub-systems.

One could say that with SiP the chip package evolved from fan-out redistribution of chip interconnect to functional integration of complete sub-systems. Today one find packages with 3D chip stacking, including IC and SMD components, packages with mechanical and optical parts and packages with components embedded in the substrate. Package level integration is a new integration field being widely explored (Figure 1).

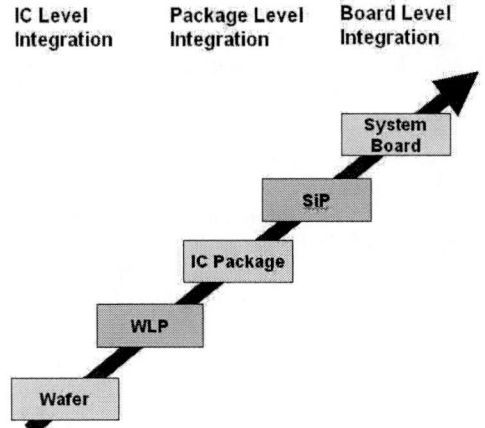

Figure 1: Integration levels in the package and assembly food chain

At the same time also IC level integration and board level integration evolves. Wafer level packaging and embedded component printed circuit boards are just two areas under exploration.

Embedded actives technology

Embedded actives technology started at SiP-level, where the need for functional densification was felt most heavily. Companies like Sanyo, Matsushita, and Imbera Electronics Oy embed dies in the substrate.

Today these technologies are re-used for IC-packaging. On packaging level the short-term objective is improved fan-out redistribution. Infineon with molded reconfigured wafer technology, Freescale with Redistributed Chip Package (RCP) technology, and Casio with Embedded Wafer-Level Packages (EWLP) are some examples of activities in this area.

Even in wafer level packaging initiatives develop. IME developed their Polymer Embedded Technology, IMEC the Ultra-Thin-Chip Stacking technology.

Chip embedding with HD technology

A technique for embedded chips in printed circuit board is developed within the European funded project 'Hiding Dies' [2]. For contacting the chip I/O's in chip embedded products, build-up technology as used in printed circuit board manufacturing is applied. A short overview of the process is given in figure 2.

At wafer level 5μm thick Cu is added to the chip I/O pads and the wafer is thinned to 50 μm before dicing. The additional Cu on chip I/O pads is necessary to achieve laser-drilling selectivity and to realize good chip contact after via metallization. The thin chips are then die bonded on a substrate core after which the chips are embedded in a polymer layer by vacuum lamination of resin coated copper foil (RCC). Next the vias are laser drilled from the outer layer to the chip and to the core substrate,

followed by via metallization and structuring of the Cu at the outer layers.

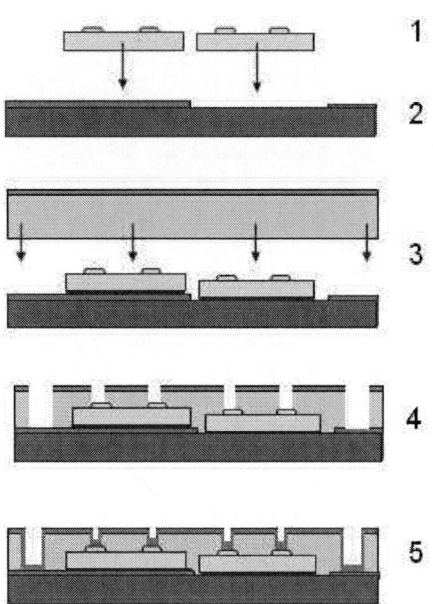

Figure 2: Process flow of chip embedding

The Hiding Dies (HD) way of embedding chips is attractive because it is very compatible with mature PCB manufacturing, which is low cost. On the other hand minimal I/O pitch of 150μm is needed for alignment of micro via to chip. This I/O pitch is needed because manufacturing takes place on large panels. Tolerances of chip placement, via drilling and Cu structuring and minimal Cu line space contribute to this minimal I/O pitch. Most chips on the market are designed for wire bond technique and the size of chips is often I/O limited, limiting the use of the HD technology. Therefore redesign of the chip or an additional redistribution layer could be necessary for the adoption of the HD technology. The embedding of flip chip can be a way out to integrate chips with pitch lower than 150μm [3].

A challenge for all embedding technologies is the process yield in combination with known good die percentage. Because of high added value, even a small yield loss or non-good die percentage ends with non acceptable board yield when multiple chips are to be embedded in one board (figure 3). Therefore use of chip embedding is expected to start in module- / SiP-sized substrates where the chip manufacturer takes ownership for the supply chain from chip to SiP component.

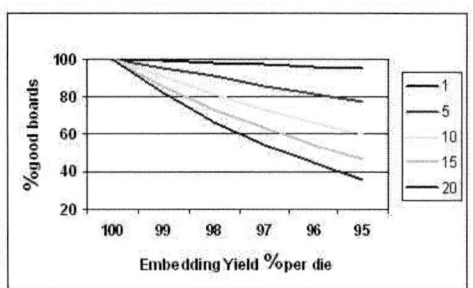

Figure 3: Board yield versus yield per chip for different number of chips per board

Chip embedding on coreless substrate

The HD embedding technique gives options to work with core-less substrates. In that case the process starts with chip placement on a single Cu sheet and lamination of RCC foil will be single sided. Very thin laminates with embedded chips can be realized with two sided Cu-structures. This way the HD technology can be used for fan-out redistribution to create IC-packages.

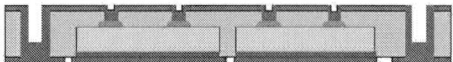

Figure 4: Embedded chip on coreless substrate

Discrete, Vertical Devices

The business portfolio of NXP MultiMarket Semiconductors mainly consists of:
- Standard products, such as small signal transistors and diodes, standard MOS, logic, and RF products and
- Application specific standard products (ASSP), such as micro controllers, sensors, integrated discretes, integrated power, interface products, and data converters

The major part of the standard products is so called "discretes". Most have a vertical device inside. Vertical devices have to be contacted both on the top as well as on the bottom. This makes discretes different from IC's that can be contacted from a single side only. Another difference with IC-packages is that the dies can be very small, down to 250x250µm. For a transistor with two contacts at the top this will result in a gap smaller than 200µm between gate and source.

Since most of these standard products are commodity products there is a heavy price pressure in this market. The combination of a vertical device with small size and low price makes packaging very challenging, particularly when taking into account that the die-free package costs as part of the total product costs can be as high as 80% for transistors and diodes.

Two demonstrators have been chosen to evaluate the opportunities of the "Hiding Dies" technology for the packaging of discretes:
- A power MOS transistor
- A complex discretes circuit, being a combination of multiple transistors and diodes in one package.

The major specification issue for a power MOS product is the $R_{DS(on)}$. This is the resistance between the drain and the source in the on state. $R_{DS(on)}$ is the sum of the drain-source in transistor on state and the package resistance. Today the drain-source resistance is still dominant, but this will change with the introduction of next generations' silicon in the near future. The package resistance will become more important. Multiple wires, thick wires and clip bonding are ways to reduce the package resistance.

Embedded active technology is offering another possibility to reduce the package resistance. Moreover faster switching speeds could be realized as well as better thermal performance (with 2-sided cooling), and improved reliability.

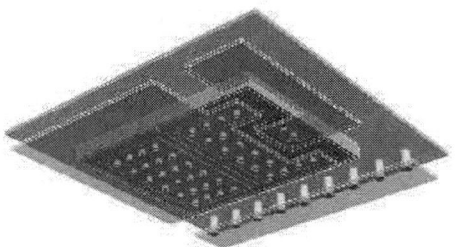

Figure 5: Power MOS Demo (bottom view)

For complex discretes there is another consideration, the route-ability. With up to 4 dies in a SOT457 package, making connections becomes difficult and wire bonding is a real challenge.

Embedded active technology is offering much more freedom in contacting the individual dies. Smaller package dimensions would even become possible.

Figure 6: Complex Discretes Demo (top view)

From the "Hiding Dies" technology a coreless solution was adopted for these demonstrators.

The dies are 150 respectively 200µm thick for the power MOS transistor and the complex discretes circuit demonstrator. All dies had an aluminum top metallization. 4-5µm copper (laser stop) is plated on the bond pads. The size of the bond pads for the transistors is enlarged to be less critical in position accuracy of the laser vias.

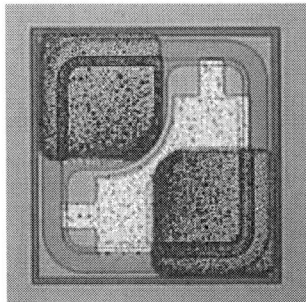

Figure 7: Die with copper plating

Starting material for die attach is a 1 oz. electrodeposited copper foil that is commonly used in the printed circuit board industry. The dies are attached to the treated side of the foil. With the treated side faced towards the laminate a good adhesion is guaranteed later on. The power MOS dies are soldered and for the complex discretes dies a conductive adhesive is used for bonding to the base foil.

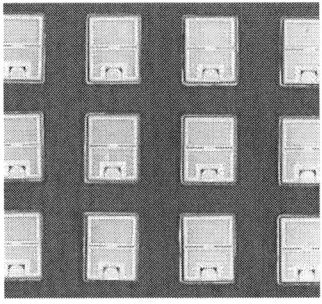

Figure 8: Power MOS after die attach (soldering)

A prepreg with holes corresponding to the dies and with the same thickness is aligned onto the foil assembly followed by a resin coated (½ oz.) copper foil. This stack is laminated under vacuum. Optimization of the lamination conditions (pressure and temperature) is necessary to prevent die crack.

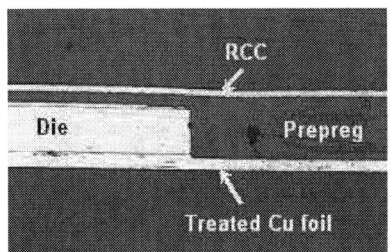

Fig. 9: After lamination

Vias are laser drilled and make contact to both the die and the bottom copper foil. The via diameter is 75 respectively 100µm for the die and Cu foil via.

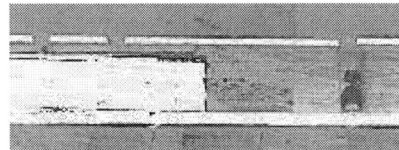

Figure 10: After laser drilling

Next the vias are plated. The aspect ratio of 2 for the blind vias to the bottom copper required optimization of the plating conditions.

After via filling the top and bottom copper layers are structured by photo etching. If necessary a solder resist could be applied to define the footprint of the package or to isolate the top surface. Finally sawing separates the packages. Since there is no metal in the sawing lane, sawing does not pose any issues and sawing speeds can be high.

Conclusions

Functional densification and fan-out redistribution are the main drivers for embedded actives technology on all levels of the package and assembly food chain. First SiP products with embedded actives are being introduced in the market and many new initiatives will follow in due time.

In discretes packaging, sampling and characterization is still ongoing. Nevertheless from the results so far one can conclude that embedded actives technology seems to be a feasible alternative for conventional packaging.

Implementation will put the discretes packaging industry to its biggest challenge ever: how to control the supply chain if subcontracting of a major part of the fabrication to PCB-fabs seems inevitable. This is a major drawback for embedded actives technology and could endanger its future in discretes packaging.

In IC-packaging wafer level embedded active technologies are under development to overcome this disadvantage.

Acknowledgements

The authors would like to thank TU Berlin for sampling support. The project Hiding Dies is financially supported by the European Union under contract IST507759.

References

[1] Rao R. Tummala, "SOP: What is it and Why?", IEEE Transactions on advanced packaging Vol. 27, no 2 May 2004.
[2] Hiding Dies: http:\\www.hidingdies.net
[3] T. Löher, "Laminate Concepts for Chip Embedding", Proceedings of the 4th European Microelectronics and Packaging Symposium, Terme Čatež, Slovenia pp. 135-138, 2006.

Lamination Process Studies for Realisation of chip Embedding Technologies-Current Applications and Technical Challenges

D. Manessis[1], A. Ostmann[2], S-F.Yen[2], R. Aschenbrenner[2], and H. Reichl[1]

[1] Microperipheric Research Center, Technical University Berlin (TUB)

[2] Fraunhofer Institute for Reliability and Microintegration (IZM)

Gustav-Meyer-Allee 25, 13355 Berlin, Germany

E-Mail: manessis@izm.fhg.de, Tel: +49-30-46403229

Abstract

This paper focuses on the lamination technology employed for the development of a production-capable technology for embedding active chips into printed circuit boards (PCBs). The work is jointly performed by a consortium of partners from industry and research within the frame of the European research project "HIDING DIES". In specific, Resin-Coated-Copper (RCC) films can be laminated on assembled chips and components providing the polymer dielectric matrix for further 3D-SiP package processing. Lamination of RCC films can superbly replace the spin-coating processes. Filled and no-filled epoxy RCC's have been successfully used to laminate very thin (~50μm) as well as relatively thick chips up to 200μm. Pressures from 5 up to 20 bars and heating rates of 3°C/min and 8 °C/min have been used to study the integrity of the resultant interfaces. Lamination at the highest heating rate and pressures of 5 and 10 bar yields interfaces with many voids and a thick epoxy thickness above chip compared to other lamination conditions. Based on shear test results with a shear speed of 100μm/sec and shear height of 40μm, the low pressures of 5 bar and 10 bar and the highest heating rate result in lower shear strength values than the slowest heating rate of 3 °C/min. Lamination at a pressure of 20 bar yields embedded structures with the highest strength regardless the lamination heating rate chosen. Lamination of a combination of 2-prepreg layers and RCC's with 25μm epoxy thickness can achieve embedding of chips with even 200μm thickness. Reliability testing of the laminated embedded chips has shown very promising results. Lamination related issues are discussed and lamination process tips are provided for successful chip embedding and further 3D package processing.

Keywords: chip embedding, lamination, Resin-Coated-Copper (RCC), chip in polymer (CiP)

Introduction

The coming generations of portable products will require significant improvements of integration and packaging technologies, mainly due to increasing signal frequencies and the demand for higher density of functions at acceptable cost. The current technology provides organic substrates with high-density build-up layers and microvias, assembled on both sides with surface mount passive and active components. The lateral space requirement of active components can be reduced to a minimum by using CSPs (chip size packages) or flip chips. A further miniaturisation however requires 3-dimensional integration of components. The obvious benefits to name only a few are (a) simplified and shortened wiring, (b) increased reliability by reduction of solder joints, and (c) better electrical performance by reduction of parasitics especially for high frequency applications in the order of several GHz [1-3].

A new concept for a high level of 3D integration of active components is the "Chip in Polymer" technology, introduced first by Fraunhofer IZM and TU Berlin [[2]]. It is based on the embedding of ultra-thin chips into build-up layers of printed circuit boards (PCBs). The basic interconnect structure, which is neither a flip chip nor a wirebond, is shown in Figure 1.

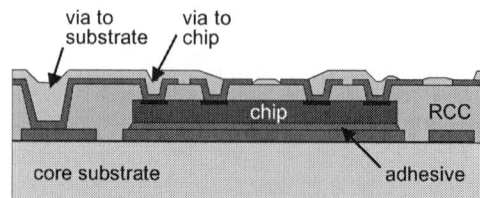

Figure 1: Interconnect principle of an embedded chip in a PCB build-up layer.

The present work will explain briefly the process steps of the "Chip in Polymer" embedding technology and will elaborate mainly on the lamination technology which has been greatly further developed in this work to achieve not only embedding of very thin chips (<50µm) but also "thick" chips in the order of 200µm thickness for power RF applications. The role of the lamination parameters on the integrity of embedded structures will be discussed.

Technology Description

The "Chip in Polymer" technology [1-3] is further developed and optimised through a European project "Hiding Dies" which has started since 2004. The project intends to develop the chip embedding towards a production technology. The main manufacturing process steps developed in the project are shown in Figure 2. For chip embedding, a die is bonded on the PCB (step A) using printed adhesive paste or adhesive film then the assembly is vacuum laminated with Resin-Coated-Copper (RCC) polymer films (step B). Subsequently, a laser drills via openings to chip and substrate pads (step C). Copper metallisation of the vias and Cu structuring follow (step D) to create the package interconnect structure. Detailed discussion of the technical challenges and solutions for the steps A-D can be found in literature [1-4]. This study describes in detail the vacuum lamination of RCC (Step B) for chip embedding and 3D package development.

Lamination Technology

The actual embedding of the chips into the PCB structure occurs in the vacuum lamination step. The core substrate with the die bonded chips is covered from both sides with a RCC layer. A press stack consisting of separation films, pressure distribution material and steel plates is used. The vacuum in the press chamber ensures a void-free distribution of the RCC dielectric. The lamination pressure profile should be adjusted in such a way to protect the thin chips and on the other hand to provide sufficient flow, curing and good Cu and epoxy flatness. A high flow, high Tg RCC is used to obtain optimal results. The pressure needs to remain low to prevent introducing cracks in the thin silicon, however a minimum level is required to ensure sufficient flow in order to avoid voids around the chips. Typical pressures above 12 bar are used although a most common pressure of 20 bar has been established in PCB manufacturing. Vacuum is maintained throughout the lamination experiment.

The temperature profile and heating rate are also very important lamination parameters. The higher is the heating rate the lower is the minimum epoxy fluid viscosity (MV) and the higher is the temperature where the (MV) is reached. Heating rates of 3°C/min up to 8 °C/min can be used.

The temperature and the respective time range where the pressure has to be raised from 5 bar to the final lamination pressure are also important factors for good lamination results. The pressure rise

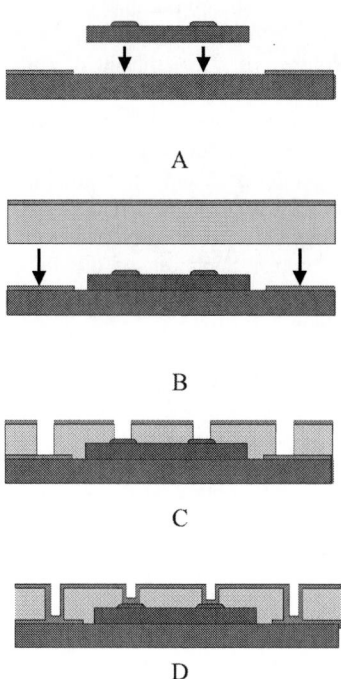

Figure 2. Process steps of chip embedding: (a) die bonding, (b) embedding in a polymer layer by vacuum lamination, (c) laser drilling of vias to chip and substrate, (d) metallisation of vias and Cu structuring.

should take place well before the minimum viscosity point (MV) temperature to avoid excessive flow of the epoxy which can potentially result in turbulent flow and leave uncovered areas around the chips. The usual temperature range for most RCC`s ro rise the pressure is from 80 °C to 105 °C. It is advisable to obtain or request from the RCC provider viscosity vs temperature (fluidity) curves and Differential Scanning Calorimetry (DSC) results for every epoxy resin of RCC's used before the selection of the temperature and pressure lamination profile. Typical temperature and pressure lamination profiles are shown in Figure 3. As experienced in the "Hiding Dies" project, due to the very different properties (especially CTE) of the materials, warpage of the board is a concern. A symmetric build-up is a first requirement to avoid excess warpage especially if core thickness is less than 0.5mm. In order to minimize warpage, the boards are cooled under pressure at the end of the cycle. An important dimension is the resultant epoxy thickness on top of the embedded die, after lamination. This thickness will determine the possibility for making small microvias for contacting the I/O pads of the chip. The uniformity of the covering epoxy layer is important for the process window of the laser drilling. Different factors as die bond layer thickness, thinned silicon die thickness and planarity of the lamination press will influence the uniformity. Because the bare dies are embedded in the build-up layer, the lamination thickness will also depend on the number of embedded dies, or the total surface taken by silicon. A 15-20 µm epoxy layer on top of the embedded die is considered a good compromise between laser drilling margin and ability to create small microvias. Based on these considerations, the appropriate epoxy thickness of the RCC material should be chosen from the available inventories of RCC suppliers.

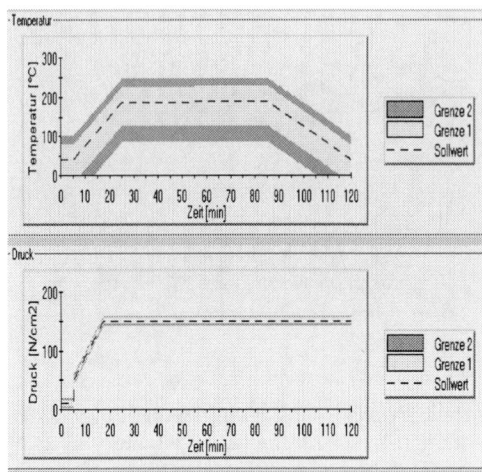

Figure 3. Typical temperature and pressure lamination profile.

Experimental Methodology

This study used Cu/FR4/Cu panels as core substrate for lamination of thin silicon chips. Cu/FR4/Cu boards were cut in 10cmx10cm dimensions and their thickness was 676µm. The thickness of the Cu layers in the stack was 16.8µm. Silicon wafers were thinned down to 50µm and cut in 2.5mmx2.5mm dies for attachment on the Cu/FR4/Cu boards. A silver-based adhesive was used for die attachment as well as an adhesive film. The thickness of the printed silver-adhesive was about 20µm. Two B-stage RCC materias were used for the experiments from different suppliers; one with no-filled epoxy (A) 100µm/9µm Cu and one with filled epoxy (B) 90µm thick/5µm Cu. The silver-based attach materials were screen-printed and the dies were attached using a DATACON machine.

Subsequently, the silver paste was cured at 150°C for 3 hrs. Lamination parameters included two heating rates and three lamination pressures in order to investigate the role of heating rate and pressure to lamination results. The lamination experiments were performed using a Lauffer LC40/2E HAT Vacuum PCB-Laminating Press System. Table 1 summarizes the RCC materials and test conditions of the experimental matrix.

Shear tests were performed on the embedded chips by using a Dage 4000 shear tester at room temperature. Before shear test, the chip-laminated area was cut through in 6mmx10mm dimensions and subsequently for the shear test a laser cut about 300µm from the chip edge revealed the laminated chip. Additionally, the laser-cut chip edge was polished with 1um diamond paste. The samples were attached to glass slides with a double-side adhesive tape. The shear tests were performed at 40µm and 70µm from the top Cu foil in the chip and in the epoxy cutting areas.

Table 1. Lamination experimental matrix

#	RCC	Heating rate (°C/min)	Pressure (bar)
1	A	Paste/3.6	5
2	A	Paste/3.6	10
3	A	Paste/3.6	20
4	A	Paste/7.3	5
5	A	Paste/7.3	10
6	A	Paste/7.3	20
7	B	Paste/3.6	5
8	B	Paste/3.6	20
9	B	Paste/7.3	20
10	B	Film/3.6	20

Lamination results

The "A" RCC has shown small voids under lamination at the low heating rate and pressure of 5 bar as shown in Figure 4. No voids were found in cross sectioning analysis after lamination at 10 and 20 bar as shown in Figure 5. Obviously a minimum pressure of 10 bar is necessary for optimum lamination of a non-filled B-stage epoxy resin without voids. The number of voids was drastically increased for the "A" RCC with increasing heating rate and for pressure of 5 bar, as shown in Figure 6. Increasing the pressure to 10 bar, the voids were fewer and smaller than for 5 bar pressure. Finally for lamination at 20 bar, no voids were found. Figure 7 shows the laminated chips. The filled "B" epoxy has shown better laminating characteristics than the non-filled "A" epoxy. No voids have appeared even at low pressures of 5 bar with low and high heating rates as shown in Figure 8. The laminated chips which haven been attached with a polymer adhesive film show great flatness and no appearance of voids. Table 2 cites the void appearance for all experiments listed in the Table 1 matrix. The role of pressure is very significant for good lamination along with the heating rate. However, higher heating rates create more and larger voids and then pressure should be at least 20 bar for good lamination. A much better lamination quality is demonstrated by the filled "B" which shows a more independent lamination behavior than

"A" RCC. That implies the larger process window provided from filled epoxies than for non-filled epoxies. The epoxy thickness over the chip was measured by cross sectioning the laminated chips. All chips laminated with the no-filled "A" epoxy of

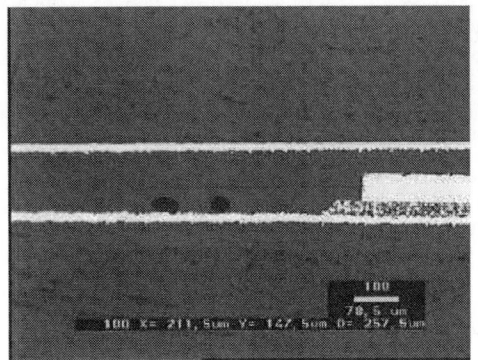

Figure 4. Lamination of silicon chips with no-filled "A" epoxy with low heating rate and 5 bar pressure. Small voids are created. (Stack: Cu/epoxy/chip/adhesive/Cu substrate)

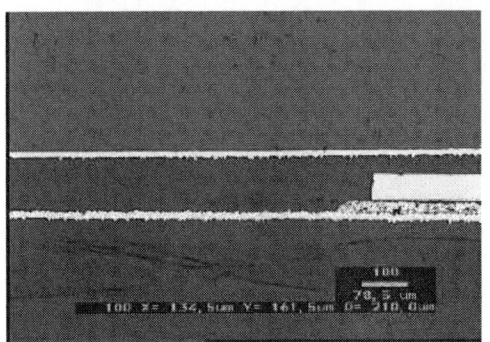

Figure 5. Lamination of silicon chips with no-filled "A" epoxy with low heating rate and 10 bar pressure. Complete elimination of voids as shown in Figure 4.

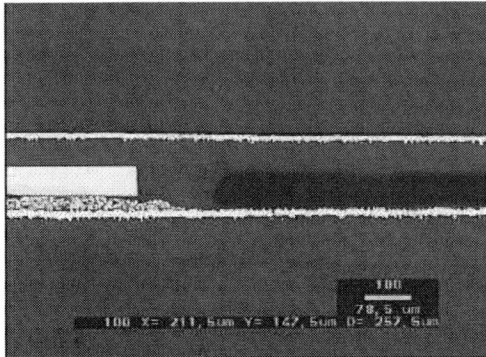

Figure 6. Lamination of silicon chips with no-filled "A" epoxy with high heating rate and 5 bar pressure. Appearance of many and large voids. (Stack: Cu/epoxy/chip/adhesive/Cu substrate)

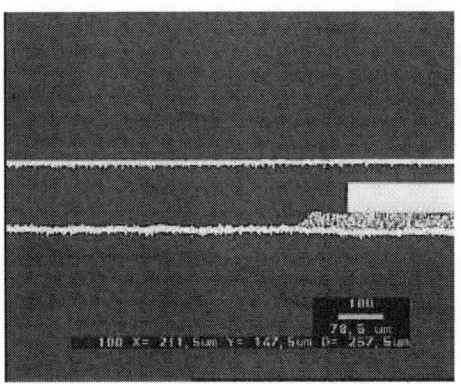

Figure 7. Lamination of silicon chips with no-filled "A" epoxy with high heating rate and 20 bar pressure. Increase in pressure eliminates the voids in Figure 6.

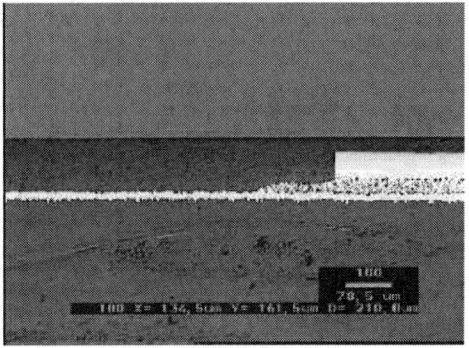

Figure 8. Lamination of silicon chips with filled "B" epoxy with low heating rate and 5 bar pressure. No voids even at low pressure.

Table 2. Lamination defects

#	RCC	Heating rate (°C/min)	bar	voids
1	A (no-filled)	3.6	5	yes
2	A	3.6	10	no
3	A	3.6	20	no
4	A	7.3	5	yes
5	A	7.3	10	yes
6	A	7.3	20	no
7	B (filled)	3.6	5	no
8	B	3.6	20	no
9	B	7.3	20	no
10	B	3.6	20	no

100μm had a thicker epoxy over the chip active surface than for the filled "B" epoxy of 90μm for all lamination conditions. The higher heating rate for the no-filled "A" epoxy resulted in slightly thicker epoxy over the chip than the lower heating rate for 5 and 10 bars pressure as shown in Figure 9. The increase in pressure reduces more the thickness over the chip as expected. As opposed to "A", the filled epoxy "B" shows a reduced thickness over the chip surface with increasing heating rate with 20 bar pressure. The opposing effect of the heating rate on the "A" and "B" may be attributed to the higher original thickness of the "A" epoxy than the "B" epoxy and also the different flow behaviors of "A"and "B" resins.

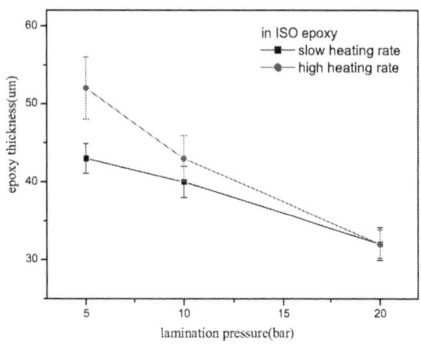

Figure 9. Effect of heating rate and pressure on epoxy thickness over the chip for no-filled "A" RCC.

Three different delaminations take place after Jedec Level 1 testing; delamination between chip and die attach material is mainly due to voids pre-existed during screen printing of the adhesive paste. Delamination at interface (Cu/epoxy) can be improved by better adhesion of the Cu on the epoxy. Delamination at epoxy/chip interface can be attributed to thermal mismatch coefficients of the epoxy and the chip and is also due to excessive unsymmetrical bending during reflow. Lamination also from the bottom side is recommended.

All samples have successfully passed 1000 cycles in thermal cycling test from -55°C to +125°C as well as JEDEC level 3 without showing any delamination.

Applications & technical challenges

CASE I: Chip Card Modules

The main application of the Chip in Polymer technology is expected in the manufacturing of small packages like stackable chip packages, SiPs or small modules with only a few chips [3]. The chips should be tested before embedding because repair is not possible. Thin chips of 50 have been embedded successfully using "C" RCC with a thickness of 80μm/5μm Cu. For the lamination experiment a typical final pressure of 20 bar was applied without any problem of chip cracking. Nevertheless, by the chip card module lamination in Hiding Dies project, the technical challenge was that chips were cracked because the chips had relatively high Cu bump metallisation and the whole stack thickness including the adhesive was 90μm. By applying a pressure of 20 bar and with a RCC of 80μm thickness, the thin chips have cracked as shown in Figure 10. The solution to this problem can be the usage of thicker RCC if this is available in the market or the reduction of the lamination pressure. A "C" RCC was used with 90μm/5μm Cu thickness and experiments were conducted with 5 bar and 10 bar lamination pressure.

Figure 10. Chip card breakage due to high pressure and inappropriate RCC thickness.

The problem of thin chip cracking was solved with the usage of thicker 90μm/5μm RCC and lower pressure. At 5 bar, all chips were embedded successfully without cracking. Some voids appear in the epoxy resin, therefore it was decided to use 10 bar. Even at 10 bar, neither chip cracking nor voids were observed. The pressure of 10 bar was finally selected for the lamination of the thin chips with the thicker 90μm/5μm RCC. A compromise should be reached between high pressure for good coverage and epoxy adhesion on the FR4 and low pressure for avoidance of chip cracking. The results of Case I study have shown that for embedding of thin chips lower pressure than 20 bar can be applied without significant compromise on chip coverage, epoxy flatness and void formation.

The functional chip card module is shown in Figure 11; the further steps of microvia opening and Cu deposition with subsequent structuring are also delineated. It contains a Philips chip with 3.2 x 2.9 mm² size and 10 connected contacts. The 50 μm chips were bonded on a 100 μm FR4 core, then RCC was laminated on both sides, giving the module with 4 Cu layers. Total module thickness is about 300 μm. The functionality of the controller chips was successfully tested after module manufacturing [3, 4].

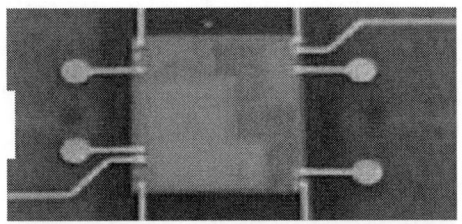

Figure 11. Module with embedded Chip Card controller: (a) top view, (b) cross-section (3D view).

CASE II: Package for Power MOSFET chips

An internal Fraunhofer IZM project aims to develop a low-cost package technology for power MOSFET chips. The chips are designed to switch currents of 80 A and more. They have a size of 6 x 6 mm² and a thickness of 200 μm. Their I/Os are Al contacts for gate and source at the top and a solderable contact with Ag finish for drain at the bottom. The basic strategy for the project was to generate a SMD package for the power chips and to keep this as simple and thin as possible. Therefore no core substrate was used. Instead the chips were soldered by their drain contact (backside) to a 36 μm Cu foil [3]. In Case II study, relatively "thick" chips of 200μm thickness are soldered on the rough side of Cu foils and then have to be laminated. Thick RCC's are not standard products in the market and they can be only found after dedicated development of the RCC suppliers. The challenge of thick chip embedding was approached by using 2 layers of prepreg material (180μm thickness each) which have openings little larger than the chip size (5mmx5mm) and surrounded the chips. On top of the prepregs, a Cu foil of 35μm thickness with its rough side towards down is laminated together with the prepregs. A relatively slow heating rate of 3.2°C/min was used along with a normal pressure profile of 20 bar. The number of the prepreg layers used is dependent on the chip thickness and the reduction ratio under the given lamination conditions. The results have been very promising and the 200μm chips could be successfully embedded without using RCC material. This approach offers obvious economical benefits since expensive RCC material has not been used in this case for embedding. The results are shown in Figure 12 where a total epoxy thckness of 30μm covers the chip top surface. The flatness of the embedded chips was very good above the chip surface and away from the chip. For high

heating rates, it has been observed ununiformity of the embedded structures due to excessive flow of the epoxy resin of the prepreg during the lamination heating stage over the chip top surface.

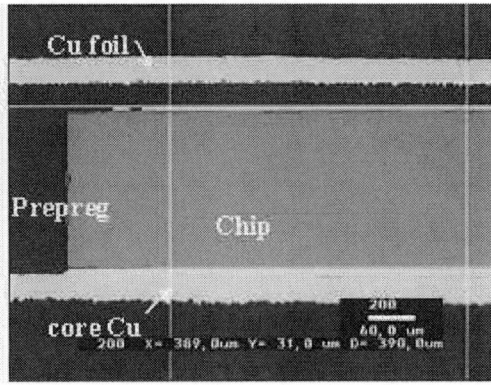

Figure 12. Embedding of 200μm "thick" chips by lamination of 2 layers of prepreg and Cu foil. The epoxy thickness over the chip is ~30μm.

Summary

Experimental studies on lamination process have shown that a minimum pressure of 20 bar and low heating rate can yield good lamination surfaces and embedded structures without voids for no-filled RCCs. The process window becomes wider for no-filled RCC where 10 bar can be sufficient pressure. For thick chips due to lack of thick RCC a combination of B-stage prepregs and thin RCC can be employed successfully for chip embedding.

Acknowledgments

Financial support under the EU "Hiding Dies" project is greatly appreciated. Dr. Shiu-Fang Yen would like to thank the DAAD educational exchange program.

References

[1] Aschenbrenner R., Ostmann A., Neumann A., and Reichl H., "Process Flow and Concept for Embedding Active Devices", Proc.EPTC 2004, Singapore, 2004.

[2] Ostmann A. Neumann A , J. Auersperg, C. Ghahremani, G. Sommer, Aschenbrenner R., and Reichl H., "Integration of Passive and Active Components into Build-Up Layers", EPTC 2002, Singapore, 2002.

[3] Ostmann A., Manessis D., Neumann A., Reichl H."Lamination Technology for System-in-Package Manufacturing", IMAPS UK, Microtech 2007, Daventry UK, March 2007.

[4] Manessis D., Yen S-F., Ostmann A., Neumann A., Aschenbrenner R., Reichl H., "Lamination technology of Resin-Coated-Copper (RCC) films for chip and component embedding in printed circuit boards", Proc. IMAPS Nordic 2006, Gotenborg, September 2006.

High-frequency modeling and measurements of tracks running on top of active components embedded in printed circuit boards

Maarten Cauwe and Johan De Baets

TFCG Microsystems, Electronics and Information Systems Department, Faculty of Engineering, Ghent University, Technologiepark 914A, B-9052 Gent - Zwijnaarde, Belgium

Tel. +32-9-2645353, Fax. +32-9-2643594, E-mail Maarten.Cauwe@elis.UGent.be

Abstract

Embedding active components in the build-up layers of a printed circuit board is a cost-effective technology, providing a high degree of integration for future packaging requirements. Since this technology offers clear advantages for high-speed applications, a high-frequency characterization is essential. This paper studies the influence of embedded chips on high-speed interconnects running on top. A similar geometry found in integrated circuits is used as a start for the research. Investigation of this Metal-Insulator-Semiconductor transmission line reveals three propagating modes, depending on frequency and silicon conductivity: a dielectric mode, the skin-effect mode and the slow-wave mode. A mode analysis of the parallel-plate approximation is used to formulate detailed expressions describing the frequency dependent behavior of the real and imaginary part of the effective dielectric constant. 3D electromagnetic simulations confirm that this model gives a good description of the behavior of embedded dies. The specific geometry of substrates with embedded components requires adaptations to the formulae predicted by the parallel-plate approach. Especially the presence of a die bonding adhesive layer and the high track thickness needs to be taken into account. The parameters extracted from the measurement show a good correspondence to the model as far as the real part of the dielectric constant is concerned. Due to the limitations of the available test structures, an accurate quantitative analysis of the losses is not possible; however the overall behavior matches with the model. This paper shows that the MIS model can act as a good starting point for an in depth characterization of microstrips running on top of embedded components.

Key words: High-frequency characterization, Chip embedding, Maxwell-Wagner permittivity

Introduction

Miniaturization has been one of the driving forces in modern day electronics. This miniaturization is starting to shift from the chip level to the printed circuit board connecting the different chips. Embedding active components into printed circuit boards or even flexible substrates is a three dimensional solution to this interconnection problem. The benefits of 3D packaging in general and embedding active components in particular are improved electrical and thermal performance, a higher degree of miniaturization, and more design flexibility.

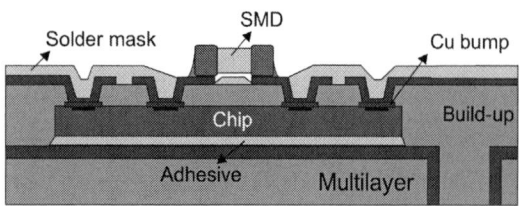

Figure 1: Schematic drawing of an embedded chip

The three dimensional nature of this technology makes it very promising for high-frequency applications. Short signal paths and very low restrictions on track routing are clear benefits over standard wire bonding techniques. However, embedded silicon dies have a distinct influence on high-frequency tracks running over the chips, altering the effective dielectric constant of the microstrip and increasing losses. High-frequency measurements of tracks running on top of embedded dies were performed in order to qualify the difference in effective dielectric constant, the change in impedance and the increase of losses.

The high-frequency problem of tracks running on top of embedded dies shows quite some resemblance to the problem of microstrip interconnects on Si-SiO$_2$. This problem has been described at length in literature with Guckel et al. [1], and Hasegawa et al. [2] as base references. The former found that depending on the silicon conductivity there are different types of propagating modes: a quasi-TEM mode for high resistivity silicon, a skin-effect mode for highly conductive silicon and a slow-wave mode characterized by a very low propagation velocity and a considerably smaller attenuation. Hasegawa et al. give a theoretical and quantitative analysis of these different modes and the transitional regions in between. Based on the resulting formulae, they create the resistivity-frequency domain chart defining the regions that apply to these modes. These two papers, and most of the subsequent papers [3] – [5] use the parallel-plate waveguide approach as an approximation of the problem. The fringing effect of the microstrips is regarded as a small deviation from this model, as long as the width of the strips is much larger than the total height of the substrate.

This paper compares the metal-insulator-semiconductor waveguide model, based on interconnects on integrated circuits, to measurements of microstrips running on top of embedded components. Section 2 starts with a description of the parallel-plate approach and the corresponding mode analysis. These formulae are compared to finite element method simulations of zero thickness microstrips with a silicon-dielectric substrate. The high-frequency test structures are described at the beginning of section 3. This section continues to reveal the effective dielectric constant and the attenuation constant extracted from the measurements and discusses the similarities and differences compared to the model.

Mode analysis and FEM simulations

A theoretical approach to determine the high-frequency behavior of a waveguide is to calculate the propagation constant by performing a mode analysis on the cross-section of the waveguide. This can be done analytically for simple waveguides as the parallel-plate waveguide or the rectangular waveguide. More complex waveguides require advanced numerical techniques to determine the propagating modes and often 3D electromagnetic simulators are used to determine a solution.

An exact analytical solution for a microstrip waveguide requires solving an extremely complicated eigenvalue problem. For this reason the microstrip is often treated as a perturbation of the parallel-plate waveguide, which is a valid approximation if the width of the strip is much larger than the height of the substrate. This approximation is valid for interconnects on integrated circuits, where wide strips on very thin oxide layers are used. A suitable model for a microstrip on a Si-SiO$_2$ substrate is a two-layer parallel plate structure filled with a lossless dielectric representing the oxide and a conductive dielectric medium replacing the silicon. As described in [2], the behavior of this structure can be divided into three regions, depending on frequency and silicon conductivity. For high frequencies and low silicon conductivity the waveguide can be regarded as a sandwich of two dielectric layers. When the silicon conductivity is high, the depth of penetration at high frequencies will become equal or smaller than the substrate

thickness. The wave propagates almost entirely through the lossless dielectric, with the silicon acting as an imperfect ground plane. The third region is referred to as the slow-wave mode. Due to strong interfacial polarization between the conducting and non-conducting media, the effective dielectric constant increases to values far above those of the two materials. This effect can be described by the Maxwell-Wagner mechanism. Typical for this slow-wave mode is the very slow propagation velocity and a minimal attenuation constant for a given conductivity.

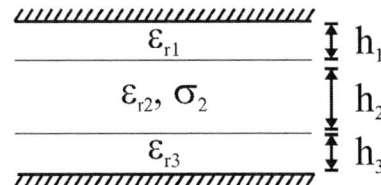

Figure 2: Geometry for the parallel-plate approximation

From the mode analysis, the frequencies defining the transitions between the different regions for a given silicon conductivity can be determined. The transition from the slow-wave mode to the dielectric region is characterized by the dielectric relaxation frequency of the silicon layer (Material parameters as defined in figure 2).

$$f_e = \frac{1}{2\pi} \frac{\sigma_2}{\varepsilon_0 \varepsilon_2} \qquad (1)$$

The transition frequency for the skin effect region can be calculated by comparing the depth of penetration to the thickness of the silicon layer.

$$f_\delta = \frac{1}{2\pi} \frac{2}{\mu_0 \sigma_2 h_2{}^2} \qquad (2)$$

The thickness of the silicon used for embedding components in printed circuit boards is around 50 μm, with smaller thicknesses expected in the future. Even for high conductivity silicon ($\sigma_2 = 10^3$), the transition frequency f_δ for thin chips is above 100 GHz, so beyond the frequency range targeted in this paper. The main propagating modes will be the slow-wave mode and the dielectric mode. The formulae for the transition between these two regions can be found in [2], and in [6] simulations indicated that these were also valid for a wide microstrip on a thin silicon and dielectric layer.

As stated above, the MIS model was developed to describe interconnects on integrated circuits. The geometry for embedded active components in printed circuit boards is however quite different from the situation on integrated circuits. The insulating layer of the metal-insulator-semiconductor transmission line, usually consisting

331

of SiO_2 is very thin (about 1 µm) compared to the silicon thickness (up to 600 µm). This is however not the case for embedded active components, due to the use of thinned chips (about 50 µm thickness) and laminated RCC as dielectric (in the order of 20 µm above the chip). Between the chip and the underlying copper layer is a die bonding adhesive, which can be a polymer die attach tape (DAT) but also a conductive paste. Typical printed circuit board characteristics as high track thickness, Ni/Au finish and solder mask have to be regarded in the final analysis.

The first major difference that needs to be studied is the influence of the third material layer, since there are now two interfaces at which the Maxwell-Wagner polarization occurs. The formula for the optical value of the Maxwell-Wagner permittivity $\varepsilon_{r\infty}$ is based on the calculation of the effective dielectric constant of a two-layer dielectric mixture and can easily be expanded to three layers.

$$\varepsilon_{r\infty} = \frac{h_1 + h_2 + h_3}{\dfrac{h_1}{\varepsilon_{r1}} + \dfrac{h_2}{\varepsilon_{r2}} + \dfrac{h_3}{\varepsilon_{r3}}} \quad (3)$$

The static value of the Maxwell-Wagner permittivity is calculated from the interfacial polarization between the dielectric and the silicon. If the dielectric is assumed to be non-conductive ($\sigma_1 = 0$), only the dielectric constant of the RCC is included in the formula for ε_{rs}. Adding the die bonding adhesive layer, this is also assumed to be non-conductive, the new value of ε_{rs} becomes an averaging of the dielectric constant of the RCC (ε_{r1}) and of the die attach tape (ε_{r3}) [8].

$$\varepsilon_{rs} = \frac{\dfrac{h_1}{h}\varepsilon_{r1} + \dfrac{h_3}{h}\varepsilon_{r3}}{\left(\dfrac{h_1}{h} + \dfrac{h_3}{h}\right)^2} \quad (4)$$

Where h is the total height of the substrate.

The transition from the slow-wave mode to the dielectric region needs to be adapted to incorporate the effect of the second interface and the related relaxation frequency.

$$\varepsilon'_{reff} = \varepsilon_{r\infty} + (\varepsilon_{rs} - \varepsilon_{r\infty})\sum_i \frac{R_i}{1 + k_i^2 \dfrac{\omega^2}{\sigma_2^2}} \quad (5)$$

$$\varepsilon''_{reff} = (\varepsilon_{rs} - \varepsilon_{r\infty})\sum_i R_i \frac{k_i \dfrac{\omega}{\sigma_2}}{1 + k_i^2 \dfrac{\omega^2}{\sigma_2^2}} \quad (6)$$

where

$$k_i = \varepsilon_0 h_2 \left(\frac{\varepsilon_{ri}}{h_i} + \frac{\varepsilon_{r2}}{h_2} \right) \quad i = 1,3 \quad (7)$$

$$R_i = \frac{\dfrac{\varepsilon_{ri}}{h_i}}{\left(\dfrac{\varepsilon_{r1}}{h_1} + \dfrac{\varepsilon_{r3}}{h_3} \right)} \quad i = 1,3 \quad (8)$$

To verify the effect of the third layer, a 3D finite element method (FEM) electromagnetic simulation was used to calculate the propagating mode. These simulations were performed using the RF module of Comsol Multiphysics. A mode analysis revealed the propagation constant for a given frequency and this was used to calculate the real and imaginary part of the effective dielectric constant.

$$\gamma = \alpha + j\beta \quad (9)$$

$$\alpha = \frac{1}{2}\left(\frac{2\pi f}{c} \sqrt{\varepsilon'_{reff}} \frac{\varepsilon''_{reff}}{\varepsilon'_{reff}} \right) \quad (10)$$

$$\beta = \frac{2\pi f}{c} \sqrt{\varepsilon'_{reff}} \quad (11)$$

The first geometry consists of a simple parallel plate with the following materials: a 20 µm dielectric layer (RCC: $\varepsilon_{r1} = 3.8$), a 50 µm thick silicon layer ($\varepsilon_{r2} = 12.1$) and a 20 µm dielectric layer (DAT: $\varepsilon_{r3} = 3$). Figure 3 shows the comparison of the simulation results with formula (5) for the real part of the effective dielectric constant. The behavior of the simulated waveguide corresponds well to the formula. Varying the conductivity of the silicon shifts the curves according to the change in the relaxation frequency f_e, but the step-like shape remains the same. Figure 4 shows the results for the attenuation constant which is independent of frequency above f_e and increases quadratic with frequency for frequencies below f_e.

Using the same material sandwich, but this time for a microstrip waveguide, a new series of simulations was performed. The geometry used for the simulation consists of a perfect electric conductor as ground plane and a 200 µm wide perfect electric conductor as microstrip. Figure 5 shows the results for the real part of the dielectric constant extracted from the simulations. The frequency dependency is less steep compared to the parallel-plate. At high frequencies the behavior of the dielectric mode corresponds well to that of a microstrip on multiple dielectric layers. The value for the real part of the dielectric constant at high frequencies can be predicted by calculating the effective value of the optical value of the Maxwell-

Wagner permittivity εr∞, using the classic formulae [7]. At the low end of the frequency spectrum, the value of the dielectric constant is much higher than predicted, even higher than the static value of the Maxwell-Wagner permittivity for the parallel-plate waveguide. This, along with the smoother dependency of frequency, indicates that at lower frequencies an additional effect is present. Possible explanations are the limited width of the microstrip, where the current distribution on the strip in combination with the conductivity of the silicon results in an inductive effect, raising the value at lower frequencies. This effect needs to be investigated further to incorporate it in the formulae above.

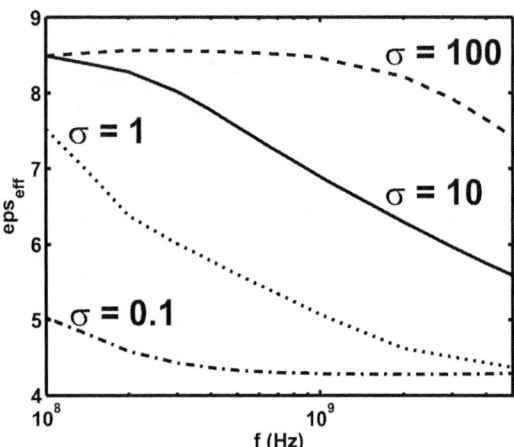

Figure 5. Simulation results for the real part of the effective dielectric constant for the microstrip geometry

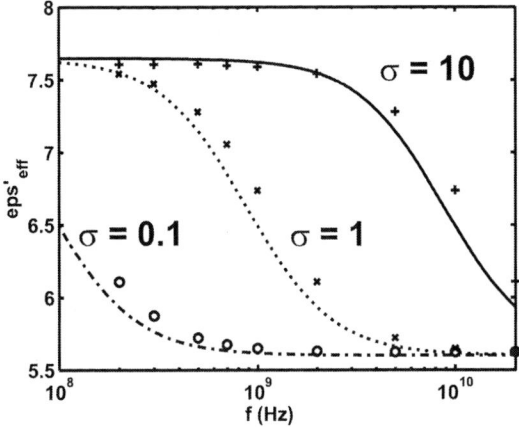

Figure 3: Comparison of the simulation results (markers) for the parallel-plate with formula (5) for the real part of the effective dielectric constant

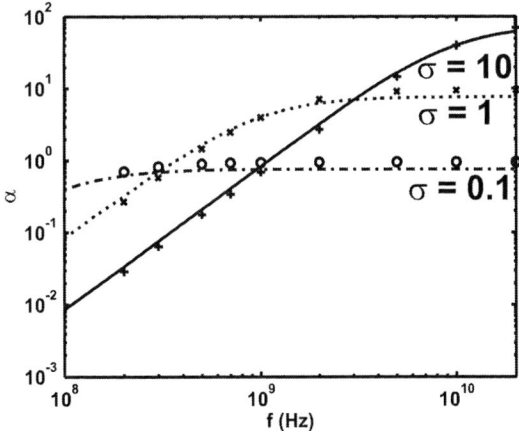

Figure 4: Comparison of the simulation results (markers) for the parallel-plate with formula (10) for the attenuation constant

Measurement results and discussion

In this section the waveguide model above is compared to measurements of the high-frequency test structures on the Hiding Dies test vehicle. The measurements were performed using an Agilent PNA 8364B vector network analyzer in a frequency range from 100 MHz up to 20 GHz. The test structures were contacted using coplanar microwave probes (Picoprobe 50A-GS-500-P) and a dedicated printed circuit board probe table. The Hiding Dies test vehicle includes several test structures to characterize the high-frequency behavior of interconnections running on top of an embedded die, as well as the connections to the embedded chips using microvias. The modeling of the microvias will not be discussed in this paper as the focus is on the influence of embedded components on the tracks on top. The test structures consist of 2 cm microstrips with varying width on top of a 10 mm x 10 mm embedded die. The embedded dies used for the test vehicles were simple test chips containing only an aluminum metallization test pattern. The widths used for the test structures were calculated to obtain an impedance of 50 Ω by simulating the effective dielectric constant for three different cases. The first width was calculated for a substrate containing only RCC (w = 275 μm), the second substrate consisted of both RCC and silicon (w = 233 μm) and finally the effective dielectric constant was calculated for the case with RCC, silicon and die bonding adhesive (w = 221 μm). Some test structures include a step in width above the chip to compensate for the change in impedance. Figure 6 gives a detailed overview of the different configurations. The reference measurements use the same structures without embedded dies.

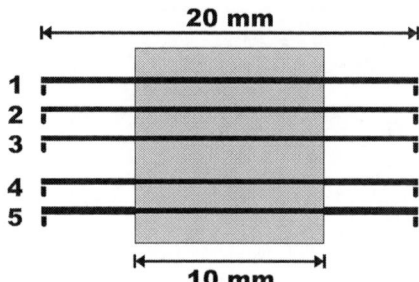

Figure 6: High-frequency test structures on top of an embedded die. Line 1: w = 275 μm, Line 2: w = 233 μm, Line 3: w = 221 μm, Line 4: w1 = 275 μm and w2 = 233 μm, Line 5: w1 = 275 μm and w2 = 221 μm

The high-frequency parameters were extracted from the S-parameter measurements and compared to the results from the formulae based on the parallel-plate aproximation. The effective dielectric constant can be determined from the phase shift of the S21 parameter. The high-frequency test structures are composed of a part above the chip and two short parts next to the chip. To compensate for this, the phase shift caused by the two short parts was calculated based on the reference structures and subtracted from the measurements of the microstrips running over the embedded dies. The apparent increase of the dielectric constant at low frequencies caused by the skin-effect was also compensated and the resulting effective dielectric constant was transformed using the classical formulae for microstrips [7]. Figure 7 shows the results for three different widths on two samples. The parameters in formula (5) were calculated based on the following geometry: 20 μm die bonding adhesive, a 60 μm thick chip and a dielectric layer of 20 μm above the chip.

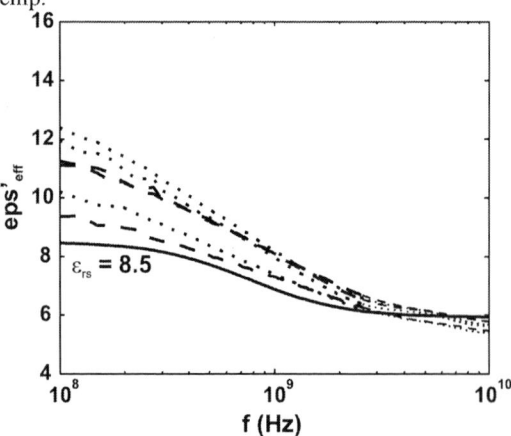

Figure 7: Comparison of the dielectric constant extracted from the measurements (dotted and dashed line) to the model (thick line) with following parameters: h1 = 20 μm, h2 = 60 μm, h3 = 20 μm, εr1 = 3.8, εr2 = 12.1, εr3 = 3, σ2 = 0.6

The measurement values are displaying a strong dependence on the frequency. This will make it hard to calculate a broadband value for the characteristic impedance. The high-frequency values correspond well to the calculated value of $\varepsilon_{r\infty}$. Part of the difference can be explained by the dispersion of the dielectric constant of the RCC and die attach tape, which was not included in the model. At low frequencies, the spreading of the measurement results is higher. This is because the parameter extraction method based on the phase shift is less accurate at low frequencies. More advanced test structures and measurements at lower frequencies would allow for a better characterization. In accordance to the simulations, the value at low frequencies is higher than predicted by the model. Overall we can state that the MIS model for the real part of the dielectric constant can be used over a broad frequency range for the characterization of embedded components; however the additional effects at low frequencies need to be studied further in detail.

As usual, modeling the losses is more difficult. A similar method as for the dielectric constant is used to extract the attenuation constant. This method disregards the reflections caused by the change in geometry at the edge of the chip, and is therefore less accurate. The losses of the parts next to the chip are again extracted from the reference test structures. However, the losses of the reference structure are higher then expected, caused by the presence of a conductive die bonding adhesive (Hereaus PC 3001 silver filled paste), which was applied to all the die bonding positions by screen printing. This conductive layer could be modeled as a lossy ground plane, but large changes in thickness prevent from extracting an effective thickness for this layer. Figure 8 shows the extracted attenuation constant for different widths and samples. The frequency dependent part of the conductive loss was calculated using the conductivity of nickel in stead of copper, because of the high thickness of the Ni/Au finish along the entire track. The value for the tanδ of the RCC in the dielectric loss was estimated based on the reference measurements and corresponds to the value of 0.03 found in the datasheet. The frequency dependence of the attenuation constant in figure 8 does not correspond to the expected value predicted by the model, as the increase with frequency does not diminish at high frequencies. Since the imaginary part of the dielectric constant, responsible for the losses, is connected to the real part by the Kramers-Kronig relations and the latter does show good correspondence with the measurements, one can conclude that the high-frequency test structures used here do not allow the correct characterization of the losses of tracks running on top of embedded components.

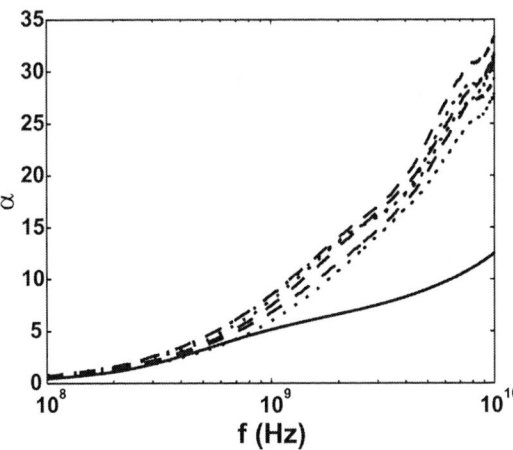

Figure 8. Comparison of the attenuation constant extracted from the measurements (dotted and dashed line) to the model (thick line) with following parameters: h1 = 20 μm, h2 = 60 μm, h3 = 20 μm, εr1 = 3.8, εr2 = 12.1, εr3 = 3, σ2 = 0.6

In order to quantify all of these phenomena, new measurement structures are designed that allow for the exact determination of the propagation constant [9]. The reference structures should be manufactured without die bonding adhesive and with the same track thickness as the microstrips on top of the embedded dies. This will allow a more accurate characterization of the different materials, which can then be applied to the tracks running over the embedded components.

Conclusions

This paper presented the Metal-Insulator-Semiconductor transmission line as a model for microstrips running on top of embedded components. Both simulations and measurements show that the general high-frequency behavior of these tracks can be described using the parallel-plate approximation. The real part of the dielectric constant becomes independent of frequency at both low and high frequencies. This Debye-type relaxation is caused by the Maxwell-Wagener mechanism for interfacial polarization. Both the static and optical value of the simulated Maxwell-Wagner permittivity are lower than predicted by the MIS model because of fringing effects. It is important to keep in mind the large geometrical differences between a microstrip waveguide on integrated circuits and a microstrip running on top of embedded dies. The presence of the die bonding adhesive, the low thickness of the silicon, the higher thickness of the dielectric layer above the silicon and the increased track thickness are some of the key differences, requiring adaptations to the model and the corresponding formulae. Some of these adaptations can be calculated quite easily, as for example the three layer equivalent of the optical value of the Maxwell-Wagner permittivity. Others,

as the influence of the extra interface on the static value and the frequency dependent behavior in general, are less straight-forward. With these adaptations and more advanced high-frequency test structures an accurate characterization of the high-frequency behavior of embedded active components is possible, leading to design rules and formulae similar to those available for standard microstrip transmission lines.

Acknowledgments

The author would like to thank all the partners of the HIDING DIES consortium for sample preparation, and the European Commission for funding this FP6 project. For more information see the HIDING DIES project website at www.hidingdies.net.

References

[1] Guckel, H., Brennan, P. A., Palocz, I., "A Parallel-Plate Waveguide Approach to Micro-miniaturized, Planar Transmission Lines for Integrated Circuits," IEEE Trans. Microw. Theory Tech., Vol. 15, No. 8 (1967), pp. 468-476.

[2] Hasegawa, H., Furukawa, M., Hisayoshi, Y., "Properties of Microstrip Line on Si-SiO$_2$ System," IEEE Trans. Microw. Theory Tech., Vol. 19, No. 11 (1971), pp. 869-881.

[3] Jäger, D., "Slow-Wave Propagation along Variable Schottky-Contact Microstrip Line," IEEE Trans. Microw. Theory Tech., Vol. 24, No. 9 (1976), pp. 566-573.

[4] Gilb, J., Balanis, C. A., "MIS Slow-Wave Structures Over a Wide Range of Parameters," IEEE Trans. Microw. Theory Tech., Vol. 40, No. 12 (1992), pp. 2148-2154.

[5] Williams, D. F., "Metal-Insulator-Semiconductor Transmission Lines," IEEE Trans. Microw. Theory Tech., Vol. 47, No. 2 (1999), pp. 176-181.

[6] Cauwe, M., De Baets, J., Van Calster, A., "High-frequency characterization of embedded active components in printed circuit boards," Proc 8th Electronics Packaging Technology Conference, Singapore, Dec. 2006, pp.643-650.

[7] Bahl, I. J., Ramesh, G., "Simple and accurate formulas for a microstrip with finite strip thickness," Proc. IEEE, Vol. 65, (1977), pp. 1611-1612.

[8] Kita, Y., "Dielectric relaxation in distributed dielectric layers," J. Appl. Phys., Vol. 55, No. 10 (1984), pp. 3747-3755.

[9] Janezic, M. D., Jargon, J. A., "Complex Permittivity Determination from Propagation Constant Measurements," IEEE Microw. Guided Wave Lett., Vol. 9, No. 2 (1999), pp. 76-78.

Cost modelling for embedded component technology

J. De Baets[1], G. Willems[1], A. Ostmann[2], A. Kriechbaum[3], H. Kostner[4]

[1] University of Gent - Interuniversitair Microelektronica Centrum (UG-IMEC), Gent, B

[2] Fraunhofer Institute for Reliability and Microintegration (IZM), Berlin, D

[3] AT&S, Leoben, A

[4] Datacon, Radfeld, A

E-Mail: Johan.DeBaets@imec.be, Tel: +32-9-2645361

Abstract

A generic yield and cost model has been developed for a technology to embed silicon dies into printed circuit boards. The process is split into different process blocks, each with its associated cost, yield and test coverage figures. The developed model is a useful tool to analyse the cost factors, and the influence of process yield and testing on the final module yield and cost. Especially for complex boards with expensive dies, process yield is extremely important.

Keywords : embedding technology, printed circuit board, yield, cost model

Introduction

New developments in electronic interconnection and packaging technology have lead to embedding of components into printed circuit board layers. The embedding technology developed inside the EC-FP6 project "Hiding Dies" uses sequential build-up layers for embedding thinned Si dies. The technology offers new possibilities for compact 3-D packaging. Through the use of inexpensive printed circuit board processes, it is also a potentially low cost technology. But as the manufacturing flow with embedded components is different from a normal printed circuit board manufacturing, care has to be taken.

By adding high value components (Si dies) during the board manufacturing process, the yield control is increasingly important in order to limit scrap cost.

In the project a generic cost model is being made to investigate and quantify the impact of process yield on the final cost of an electronic package or module. The cost model takes into account testing and test coverage, and yield at different stages of the processing. The yield of the final module is calculated from DPMO (number of defects per million opportunities) figures for components and manufacturing processes of the module. Test coverage and test cost at intermediate fabrication stages and of the final product are included. The final outcome is a cost distribution between the different parts/processes of a embedded components module. By varying the different input parameters, the user of the model can see the influence of yields, die cost, testing, manufacturing choices… on the final cost.

The die embedding process

The chip embedding manufacturing process differs significantly from the standard printed circuit board and component assembly manufacturing processes we know today.

In standard multilayer printed circuit board technology double sided laminate layers are fabricated, inspected and tested separately, so that the lamination into a multilayer stack is done with only known-good layers. Sequential build-up

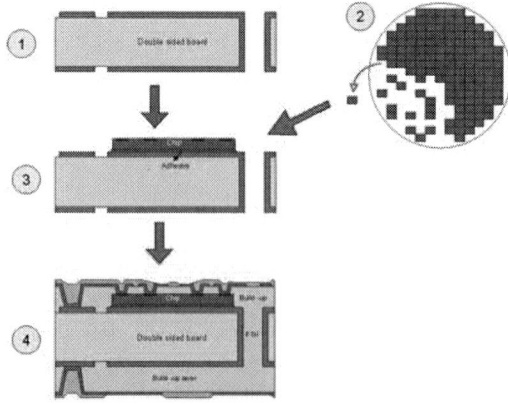

Figure 1 : Process steps for die embedding (1 – double sided board manufacturing; 2 – wafer back end processing and singulation; 3 – Die placement; 4 – embedding in SBU layer with microvia contacts)

technology deviates from that concept by building up onto an existing substrate. Any defect introduced in that sequential step is also making the core substrate worthless. In embedded-die technology, that core substrate is particularly valuable through the assembly of bare dies, increasing the scrap cost tremendously, in case of defective sequential build-up layers.

Basically the die embedding process has a mixed process flow, which involves printed circuit board manufacturing, die bonding and sequential build processes in a non-standard order[1]. First a core substrate is manufactured (Figure 1 – step 1), then bare die bonding on the substrate is done (Figure 1 – step 3), followed by another printed circuit board process, the sequential build-up of the embedding layers (Figure 1 – step 4). The cost calculation has also to take into account the wafer back-end processing, including redistribution and Cu pad finish, adhesive layer application and die singulation (Figure 1 – step2).

Cost and yield model

It is essential that in a technology cost model, yield estimates and calculations are included. Especially in the case of embedded active components the yield of process steps after the dies have been bonded onto the substrate is important for

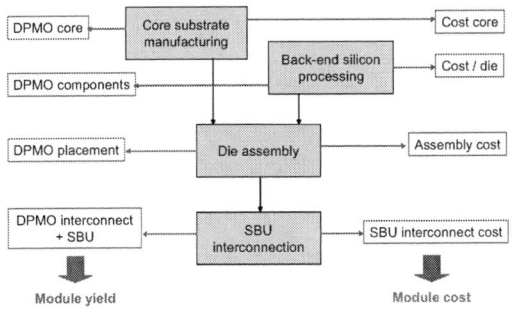

Figure 2 : Schematic representation of DPMO figures and cost

keeping the cost at an acceptable level.

We have chosen to use a yield calculation based on the principles developed in IPC standards IPC-7912 and IPC-9261[2], by introducing DPMO (= Defects Per Million Opportunities) levels. The die embedding process is basically seen as an assembly, with an additional process step (the SBU layer).

A defect opportunity is any occasion for a defined type of defect to occur. For the embedded die process we have determined the following defect types:

- Silicon die DO : a defect component, delivered as bare die, after back-end processing and singulation
- Core substrate DO : a defect core substrate, as it is supplied by the PCB manufacturer

- Placement DO : wrong placement of the component
- Interconnection DO : failing connection to a chip
- SBU DO : any other defect in the build-up layer

Other categories could be defined or added, if more differentiation in defects is needed. The actual occurrence of defects, is given by the DPMO figure per defect opportunity class, known on a statistical basis, e.g. from suppliers' data, or from process analysis…

An electronic circuit or module with embedded dies has then a final yield (in %) calculated from the DPMO figures, as follows :

$$Yield_{module} = 100 \cdot \prod_{i=1}^{n} (1 - DPMO_i / 10^6)$$

with

$DPMO_i$ = the defect level for the ith defect opportunity;
n = total number of defect opportunities

When we take also testing into account, we can describe each stage of the manufacturing of a module with embedded dies as a generic block with its own cost, test coverage and yield figures.

In Figure 3 an elementary block is shown, with indication of the important parameters :
- N = number of modules entering that stage
- Ng = number of modules considered "good" to go the next stage, but with a certain defect level (DPMO(i))
- Ns = number of modules that is scrapped in that stage
- DPMO(i) = number of defects per million opportunities for modules leaving stage i
- Y_i = yield of the process step i
- TC_i = test coverage of the test procedure after process step i (TC_i = 0% when no testing is done at that stage)
- Process cost, test cost, scrap cost = costs associated with stage i

Testing of the module at an intermediate stage can be interesting in order to minimize additional costs. For example core substrate testing will be mandatory to obtain "known-good" modules before mounting expensive bare dies.

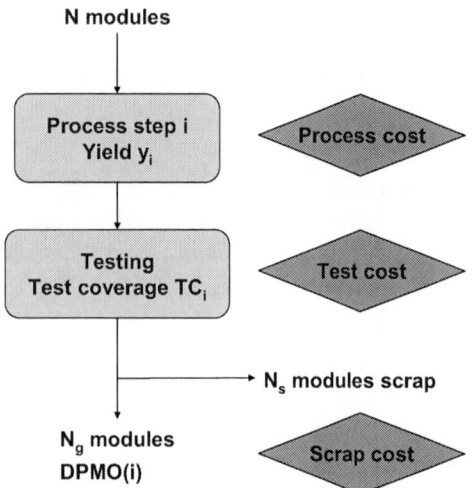

Figure 3 : Elementary block of the cost/yield model

Testing involves a certain additional cost, and provides a selection of good modules that continue to the next stage, and defect modules that are scrapped. Usually a test – or a combination of different tests – is not capable of testing the complete module. Some types of errors are not detected because they are not accessible with the test methods used. This is expressed with test coverage (TC$_i$ - in %).

The model as described above has been implemented into a MS Excel-workbook. Cost, yield and testing figures have been combined in different sheets, according to their processing stages (die manufacturing, core substrate manufacturing, assembly, sequential build-up).

The actual cost model takes into account cost parameters as equipment capital investment, operating cost, maintenance cost, materials, NRE cost... Also throughput times, production batch sizes, machine parameters are covered. Although the different processing steps are not implemented to the same detail, as the input comes from different partners and backgrounds.

Results and examples

The model has been applied to a practical case, for which the effect of variations in the input parameters have been investigated. The case chosen uses a double sided core substrate and two SBU layers, of which one layer has active dies inside. The production panel has a large amount of small modules (12 x 12 mm) with 1 chip having 12 I/Os.

A typical cost breakdown for this single chip package is shown in Figure 4. As expected, the core

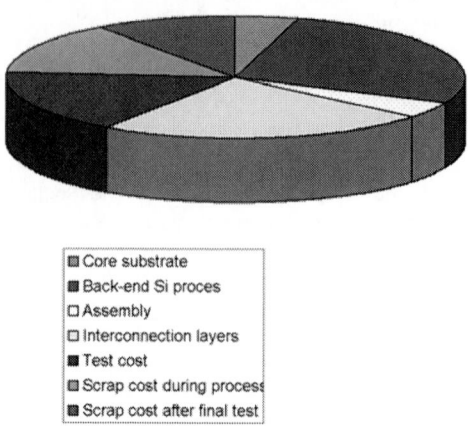

Figure 4 : Typical cost breakdown for an embedded die package

substrate is only a small part of the cost. Back-end silicon processing is covering the redistribution pattern on the chip, with Cu pad finish. For very large volumes, this cost could be reduced, by adapting chip design and layout, and the silicon processing to this type of package. Also the cost associated with attaching the die-bond adhesive

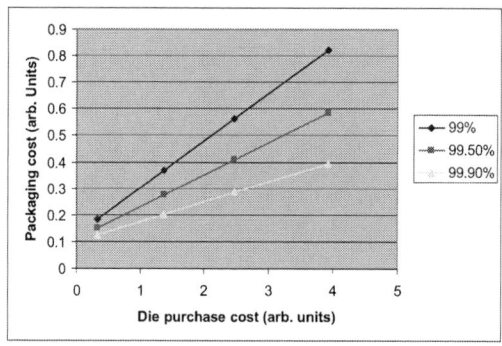

Figure 5 : Packaging cost vs. die purchase cost, as a function of interconnection yield

layer to the back of the silicon wafer is included in this back-end processing cost. A large part of the cost is associated with the interconnection through the SBU layer, and the laser drilled microvias. Also the external pad finish (NiAu in this case), solder mask and machining, are included in this cost. Testing includes intermediate testing and final functional testing. The amount of scrapped modules is depending on the yield of the process. Modules can either be scrapped after intermediate testing or after final functional testing.

In Figure 5 one sees the influence of the interconnection yield (= errors caused by laser

338

drilling and Cu plating process) on the packaging cost. The influence becomes higher with increasing die cost. The latter is easily explained by the scrap cost, taking an increasing part of the total cost. The

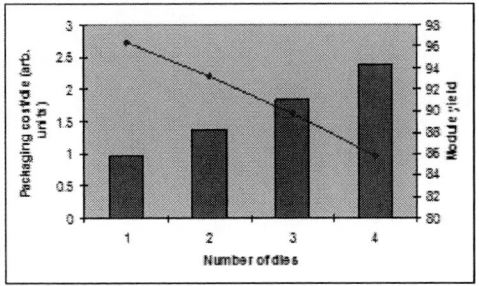

Figure 6 : Decreasing module yield (line graph), and increasing module cost (bar graph), with increasing number of embedded dies

model allows analysing the influence of different yield factors on the final module yield and the module cost.

Embedded die technology will mainly be applied when the processing yield is high enough. A calculation example is showing the decreasing yield when embedding more dies in one module. In **Error! Reference source not found.** the line graph is showing the decreasing yield for the module (from 96 % for 1 die, down to 86 % for 4 dies). Although the yield decrease is not dramatic, the (embedding) packaging cost is increasing dramatically with a factor of 2.5. The packaging cost is calculated per chip, so that the embedding area and complexity are comparable for all 4 cases.

Conclusion

A cost and yield model has been developed for embedded die in printed circuit board technology. The model allows to analyse the impact of the yield and cost factors in the different technology steps, on the final yield and cost. From real cases that have been studied inside the Hiding Dies project, it is clear that embedding die technology will be a challenge for the printed circuit board manufacturing. The processes have intrinsic yields that are sufficient for cost-effective production, but tight manufacturing process control is needed to keep the yields high enough.

Acknowledgement

This work has been funded by the European Commission through the FP6-IST project "Hiding Dies".

References

[1] A. Ostmann, J. De Baets, A. Kriechbaum, H. Kostner and A. Neumann, "Technology for Embedding of Active Dies", Proceedings of the EMPC 2005 (Bruges), June 2005, pp. 101 – 106.

[2] IPC-7912 "Calculation of DPMO and Manufacturing Indices for Printed WiringAssemblies", and IPC-9261 "In-Process DPMO and Estimated Yield for PWAs"

Double and triple stacked solder joint technology for further miniaturization of 3D-SIP

Serguei Stoukatch, Christophe Winters, Tom Torfs, Walter De Raedt, Eric Beyne,

Chris Van Hoof

IMEC, Kapeldreef 75, 3001 Leuven, Belgium

Tel.: +32-(0)16.28.83.15. Fax.: +32-(0)16.28.15.01, stoukat@imec.be

Internet: http://www.imec.be

Abstract

In this paper we demonstrate the use of a lead-free solder technology to achieve 3D-SIP integration. In addition, we acquire a larger separation or standoff between the individual SIP layers utilizing a new method of triple ball stacking. Increased stand-off enables the use of larger SMD components and broadens the selection of passive and active components. Such increased stand-off permits sequential stacking of functional layers populated on both sides with components in formats up to 0603. As a result of the successfully developed lead-free triple ball stacking technology we realized a fully functional a wireless bio-electronic low power sensor node. The product is a portable electroencephalogram (EEG) system for brain activity monitoring.

Key words: 3D SIP integration, Package-on-package stacking, double and triple solder ball joint, autonomous sensor nodes.

Introduction

Previously [1, 2, 3] we reported on a wireless bio-electronic low power sensor node which was realized using a 3D SIP integration approach. This system includes a commercial Nordic nRF2401 2.4 GHz transceiver, a TI MSP430F149 low-power microcontroller, a custom designed matched antenna, a custom designed EEG ASIC, crystals and all necessary passives. The result was a portable, electroencephalogram system for brain activity monitoring. The selection of the passives (choice between 0201, 0402 and 0603 type) was driven by miniaturization and capability of providing adequate values in the smallest available format.

The 3D SIP was constructed using a conventional soldering technology [4, 5] where the solder paste and solder balls comprise lead as a part of their composition. High-lead solder (90% lead, 10% tin composition) balls were attached to each of the individual layers. An eutectic solder paste (183° melting point) was deposited on top of the solder balls. The corresponding solder balls were aligned to each other and the assembly was finally reflowed at peak temperature of 223°C, which is common to the eutectic soldering process. In this case only the solder paste was molten, while the high-lead solder balls remained non-collapsible because of their higher melting point (280°C). As a result, the required stand-off of 1.5mm was reached. This stand-off requirement was defined by the demand to accommodate a set of large SMD components.

In response to the European RoHS (2002/95/EC) and WEEE (2002/96/EC) Directives the task was set to convert the product in compliance to these guidance. The purpose is to exclude banned lead from the assembly, both from the solder paste and solder spheres. When utilizing lead-free solder paste and lead-free solder balls, the separation between the complementary layers must remain adequate to accommodate large SMD components. Practically, the stand-off must be similar to what was achieved using high-lead solder balls, 1.5mm or larger.

We carefully explored an alternative route, by replacing larger components by smaller (especially thinner) alternatives. The component selection for the system is driven by its functionality to meet the low power consumption requirements and miniaturization. Among IC's the system comprises a significant amount of surface mount type components (SMD) such as crystals and necessary passives. In total 120 pieces of surface mount type components have been integrated, which occupy up to 95% of the total volume of the 3D-SIP stack. Besides the PCB (0.8mm thick) which serves as a carrier for the assembly, they are the main contributors to the total thickness of the 3D-SIP stack. Based on this we identified that the main scope of efforts towards further assembly miniaturization would be related to SMD integration. Therefore the most logical way to achieve further miniaturization would be just to replace current SMD's by their smaller analogs. However, for the demonstrated assembly module the smallest available SMD type of components have been already utilized. Further SMD's dimension reduction is in contradiction to their electrical properties. For example, the required capacitance value of 10μF can only be met by utilizing a chip capacitor in format 0603 at smallest. A 0402 capacitor will provide a value of 2μF maximum that is 5 times less as required. The 0201 capacitor provides only 0.1μF capacitance.

Therefore, the task remains to achieve a stand-off of at least 1.5mm between the layers using the lead-free assembly option.

Method

The interconnection, both electrical and mechanical, between adjoining modules, or packages in case of PoP (package-on-package), can be achieved by solder balls. In principle the interconnection can be implemented by a conventional technology using a single row of solder balls (the single row of solder balls leads to a single-ball solder joint assembly scheme, further named as single-ball solder joint). In order to maintain an appropriate interconnect density an interconnect pitch of 1.27mm was selected, which is in agreement to the JEDEC standard. The standard recommends a solder ball diameter of 0.75mm for the selected interconnect pitch. As a result the stand-off is in the range of 0.50-0.75mm depending on the bond pad geometry and assembly scheme (more details will be given below). Using a double-row solder ball the interconnect scheme (further referred as double-ball joint) we end-up with 1.00mm in the best case. This is a significant lower stand-off than the targeted value of 1.5mm that was reached using the high-lead non-collapsible solder balls. In order to get the interconnection, 2 lead-free balls are co-molten together. It differs from the high temperature lead-containing solder balls process where the balls stay non-collapsible and only the eutectic solder paste melts and forms the joint between 2 solder balls. A possible solution is to increase the solder ball diameter, but this would require a larger interconnect pitch which conflicts to the interconnect density. An other solution we propose here is the use of a triple row solder balls (triple-ball solder joint).

Based on the initial observation in the first instance a model was built that describes the relation between an initial solder ball diameter and the resulting solder joint height. This solder joint height defines the module stand-off. We find out that another parameter, the dimensions of the landing pad for the solder balls, also has an impact on the stand-off. The solder ball joint's forming method is an essential factor to create a required stand-off. More specifically, the interconnection between adjoining modules, can be achieved by utilizing single row solder balls (further single-ball solder joint), double-row (double-ball) or triple row solder balls (triple-ball solder joint).

Within each of the methods there is a certain degree of freedom to tune the stand-off towards the required value. In respect to that, we studied here two main cases. First is the impact of the landing pad dimension on the solder bump height. The solder bump is the main contributor to the PoP stand-off. The mechanism of the impact of the landing pad dimensions on the bump height is simple and well understood. We vary the landing pad's diameter; meanwhile the total amount of solder dispensed on the landing pad stays the same and is defined by the solder ball diameter. The landing pad finished by a fully solder wet-able layer determines the spreading area for the molten solder during reflow and the solder mask restricts further the solder flow. Based on that, we performed the stand-off calculation. We also presume here that there is no wetability issue and solder equally covers the total area of landing pads. This presumption is valid for properly prepared landing pads, which are free of native oxide and any other foreign contamination and particles. Such conditions are normally achieved on an SMD assembly line by controlling the environment on one side (particles control) and applying flux which reduces eventual oxide.

The second factor involved in the stand-off management is a partial encapsulation of the solder bumps. In respect to that we considered the following cases: no encapsulation, partial encapsulation and full encapsulation. Also, the encapsulant material might have some effect on the final solder joint. Without encapsulation the product of 2 co-molten solder ball is a solder joint whose dimension is defined except by fixed parameters (solder ball and landing pad diameter), by solder surface tension, which is constant for given solder composition. As a consequence there is no room for stand-off management, except maybe the top module weight in certain cases. The module weight if it is light has no effect on the stand-off, however in case of a heavy module there is a load applied due to the module's weight which is in the same magnitude as the solder surface tension force. The resulting stand-off may be decreased. The stand-off management by varying top module weight was not investigated in this study.

The mechanism of stand-off management by different degree of encapsulation is as following. The encapsulant applied around the solder bump prevents solder from spreading in lateral direction during sequential reflow. The molten solder on each of the adjoining modules stays in a cavity formed by the encapsulant and after solidification remains confined. Only solder recess is involved in solder joint formation. The result of co-melting two confined solder balls is a solder joint in an 8-like shape.

341

Based on that we calculated analytically the solder joint height in each described case. A universal correlation between the AR (aspect ratio) defined as a ration of an interconnection height to a ball diameter and the ratio of the pad diameter to the ball diameter was found. The correlation is universal in sense that it does not depend on the actual solder ball diameter and landing pad dimension. Based on this we can conclude that the ball stacking scheme is the major factor which drive the height of solder interconnection.

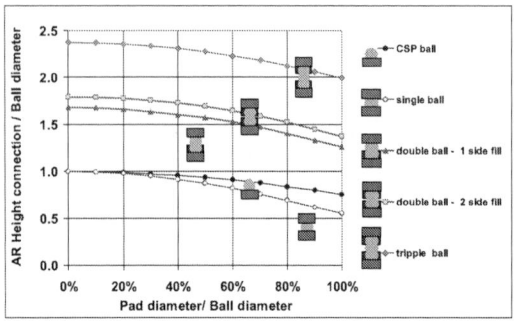

Figure 1: Aspect ratio versus pad/ball diameter and depends of the number balls in stack.

The universal correlation helps us to define the solder ball diameter required to achieve the targeted stand-off. The value of the stand-off is used as an input parameter, and as output is read-out the ball diameter and the ball assembly scheme (choice between single-, double- and triple-ball) correspondingly.

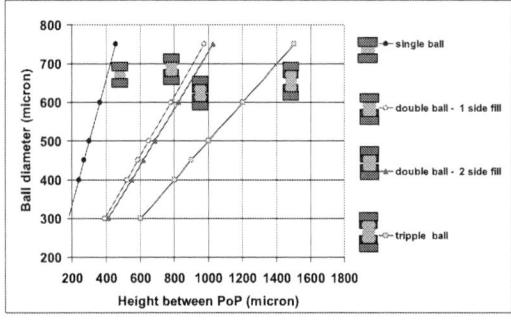

Figure 2: Relationship between separation substrates and diameter of the solder balls used versus the solder ball stacking scheme.

In our specific case the figure 2 suggests that the targeted stand-off of 1.5mm, requires utilizing a solder ball of 0.76mm diameter and can be realized by triple ball stacking. An important remark is that the 0.75mm diameter solder is the maximum solder ball format according to JEDEC standard.

Experiment

The test boards have been designed for single-, double- and triple- solder ball joint process formation optimization. The PCB was commonly used commercially available FR4 of 0.8mm thick.

The landing pads (LP) have a circular shape and were designed in different diameters, thus 20, 40, 60, 80 and 100% of the ball size. The ball size (BS) investigated is 760μm (30mil) and interconnect pitch (the distance between the neighboring landing pad centers) is 1270μm (50mil). The landing pad geometry and solder ball diameter is in compliance with the JEDEC recommendations.

Solder ball attachment is a conceptually simple process and consists of the following sequential steps: flux dispensing, solder ball placement and finally a solder reflow step. The flux provides the required hold on force to retain the ball at designated position (on top of landing pad) prior to reflow. The reflow profile is a recommended profile for commonly used lead-free solder composition (96.5%Sn3.5%Cu0.5%Ag). The profile features a dwell time of 60 second at peak temperature of 250°C. During reflow molten solder spreads on the landing pad surface and forms a solder bump, often called a CSP (chip scale package) ball. The solder bump height and diameter has been recorded.

Table 1: Solder bump diameter versus landing pad dimension.

	LP size, μm	LP/SB, %	Solder bump diameter, μm
1	152	20	740
2	304	40	730
3	456	60	680
4	608	80	660
5	760	100	630

As presented in the table the maximum solder bump height achieved on the test PCB is 740 μm. The height was observed on a minimal landing pad. As the landing pad becomes larger the bump height decreases. The smallest bump height of 630μm was registered on 760μm landing pad. We also observed here that a number of processing mistakes, such as a ball misplacement and/or double-ball formation increased dramatically once the threshold of 456μm is crossed. Thus the ball attach process on 304 μm and 152μm was not reproducible in our case. The process on 456μm has reduced yield. The ball attach on 608 μm and 760μm landing pads was robust and highly reproducible, however the bump height gain of 30 μm is less than 3%, which is not significant in our case. All processing described below has been performed on 608μm diameter landing pads.

The package with CSP ball, or simply CSP, is usually mounted on the board and the resulting stand-off is always less than the initial solder ball diameter In our experiments it was in the range 0.5-0.7mm depending on the conditions.

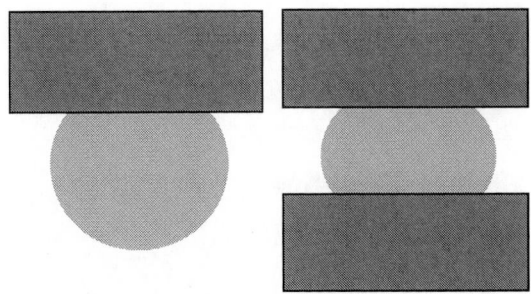

Figure 3: Schematic cross-sectioning view CSP ball (left) and single-ball solder joint (right).

We attached sequentially 2 solder balls on top of each other. In this case two contiguous solder balls were co-melting together and form the so-called 8-type shape solder bump. To prevent the formation of a larger solder bump,to maintain the targeted stand-off and to secure a higher interconnect density the first row solder balls have been encapsulated by epoxy based encapsulant. The encapsulation is an essential process feature.

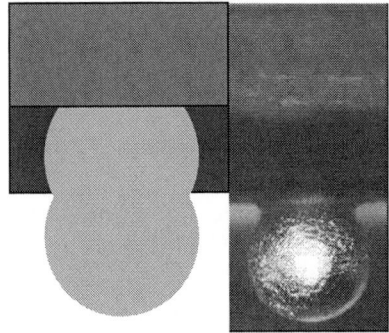

Figure 4: Double row solder ball: Schematic cross-sectioning view (left) and side view (right).

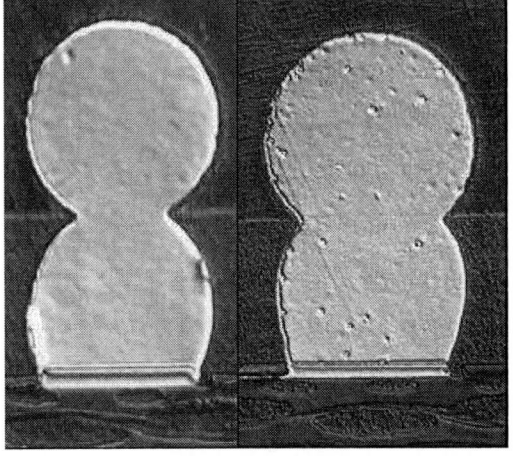

Figure 5: Example of double-stacked solder balls (cross-sectioning view-8-type shape).

To form the double-ball solder joint a module with pre-formed 8-shape solder bumps is attached to the adjoining stackable package. The resulting double-stacked solder ball joint has a height which varies in a certain range, in our specific case it is 0.9-1.0mm at most.

For a triple-ball solder joint formation we developed a process scheme where on one side there is a module with the CSP ball already attached and on the adjoining module a double-row solder ball is pre-formed. The process of the attachment of two adjoining packages is similar to the single- and double-stacked solder ball joint. It comprises a flux deposition, an alignment and is finalized by reflowing.

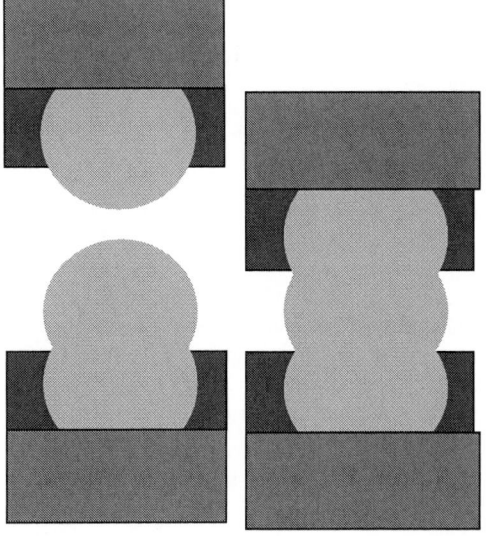

Figure 6: Schematic view on triple-ball solder joint formation.

To preserve the required interconnect density (interconnect pitch of 1.27mm) and to meet the required stand-off specification (1.5mm) we used encapsulation for both CSP-ball and double-row solder ball. As a result the triple-stacked solder ball joint has a height of 1.55mm for the described assembly scheme, which was enough to accommodate the large SMD components. The quality assessment procedure for the triple-solder joint is similar for a conventional SMT and BGA assembly and may contain X-ray inspection and cross-sectioning techniques.

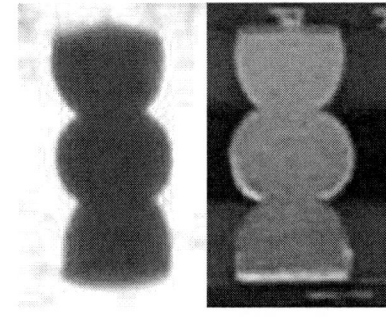

Figure 7: X-ray image (left) and cross-section view of the triple-ball solder joint

343

To conclude the experimental charter the on solder joint formation study: for an interconnect density of 1.27mm pitch using solder balls of 0.76mm, it is possible to reach a stand-off between stackable packages of 0.7mm for single-ball solder joint, 1.0mm for double-ball and 1.55mm for triple-ball solder joint depending on the solder joint formation scheme. It should be remarked that the stacking process is a similar process than the conventional BGA and CSP placement process, comprising of flux dispensing, alignment of adjoining packages and a reflow step.

3D-SIP module using triple ball solder joint technology

The developed double- and triple solder ball joint technology was utilized for the assembly of a 3D-SIP module. The 3D-SIP integration approach features novelty of stacking functional layers or a package-on-package (PoP) where each of the packages in the stack are sub-system SIP circuit blocks. As a result of using this approach, the required number of connections between the SIP packages is greatly reduced and a PoP can offer sufficient 3D-interconnect density.

A particular interesting application area for 3D-SIP is the realization of distributed, fully autonomous systems for realizing so-called "ambient intelligence" systems. Such systems can be divided into clear subsystems: the radio (antenna, rf-front-end base band), the main application (processor, sensors, actuators) and the power management (regulation, storage, generation). Each of these functions can be realized as a SIP-subsystem.

Using double- and triple- solder joint technology we built the 3D-SIP fully functional demonstrator. The demonstrator [1, 2, 3] is a wireless bio-electronic low power sensor node. The system includes a commercial Nordic nRF2401 2.4 GHz transceiver, a TI MSP430F149 low-power microcontroller, a custom designed matched antenna, a custom designed EEG ASIC, crystals, and all necessary passives. The result was a portable, electroencephalogram system for brain activity monitoring.

Because of triple-solder joint interconnect the required stand-off of 1.5mm was achieved and actually exceeded (1.55mm) to keep a little margin that allows to accommodate a larger SMD (format 0603).

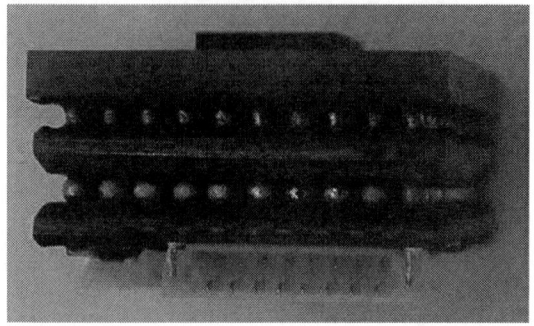

Figure 8: Perspective view: Triple-ball double stack 3D-SIP

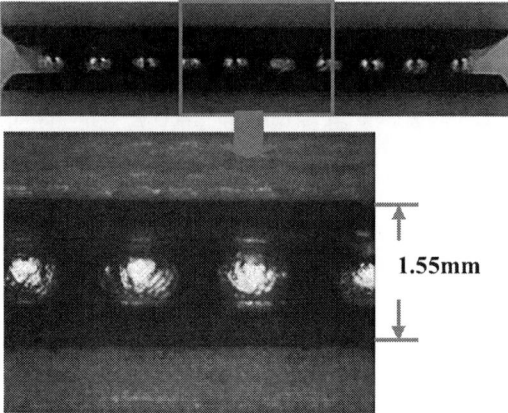

Figure 9: Side view: Triple ball double stack embodiment for increased stand-off.

For assembly quality control a conventional SMT and/or BGA metrology tools such as X-ray inspection and cross-sectioning technique can be used.

As a non-destructive method for analyses of the module an X-ray inspection is a quick and a trustful method for an eventual defect detection and can be recommended for all samples. A rotation of the sample and changing the angle between sample's axis and X-ray beam allows to obtain a comprehensive view for each individual solder joint.

Figure 10: X-ray images on 3D-SIP: top view (left) and angle view (right).

Cross-sectioning the sample might be useful on a very limited amount of samples. We performed the cross-sectioning on four sides of the assembly. The solder joints have no visible defect and have repeatable shape.

Figure 11: 3D-SIP cross-sectional image.

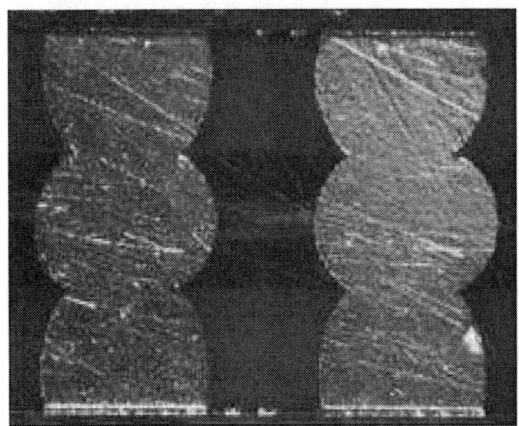

Figure 12: Cross-sectioning view of two neighboring triple-ball solder joint.

Conclusion

We demonstrated the use of a lead-free solder technology to achieve 3D-SIP integration. The required separation between SIP layers of 1.5mm, defined by the demand to integrate large passives of 0603 format is achieved without using any spacers. As a technology 'know-how" a novel solder ball stacking concept is developed (double- and triple-solder joint). The analytical model was built which describes a correlation between the solder ball diameter and resulting separation between modules. The series of experiments conducted to validate the model and convergence between the analytic calculation and the experimentally achieved separation between SIP layers is demonstrated.

The high-aspect ratio triple-ball lead-free integration method permits component assembly on SIP layers for increased integration density. As a result of the successfully developed lead-free triple ball stacking technology we realized a wireless bio-electronic low power sensor node. The product is a portable, electroencephalogram system for brain activity monitoring.

It should be remarked that the modules' stacking utilizing double and triple solder ball joint concept and final mounting of 3D SIP are compatible with existing SMT and BGA assembly processes, improving the 3D-SIP chances for commercialization and large-scale production.

Acknowledgements

The authors would like to thank Danny Frederickx for his contribution in the 3D assembly process, Tomas Webers for his help with the substrate design, Steven Sanders for his input on the system built-up, Steven Brebels, Kristof Vaesen for antenna design and measurements, Human++ program director Bert Gyselinckx and last but not least Frederic Duflos for assembly analysis.

References

[1] S. Brevels, S. Sanders, C.Winters, T.Webers, K.Vaesen, G.CArchon, B.Gyselinckx, W.De Raedt. "3D SoP Intergration of A BAN Sensor Node" *Proc. 55th Electronic Components and Technology Conf.*, Lace Buena Vista, Florida, May 31-June 4 2005, pp. 1602-1606.

[2] R.F. Yazicioglu, P. Merken and C. Van Hoof, "Integrated low-power 24-channel EEG front-end", Electron. Lett. Vol. 41(8) pp. 457-458 (2005).

[3] C. Van Hoof, E.Beyne, W. De Raedt, S.Stoukatch, T.Torfs, C. Winters. "3D System-in-Package Integration of Wireless Sensors Nodes". ISSCC2007, San Francisco, Ca., Feb 11-15 2007.

[4] S. Stoukatch, T. Webers, C. Winters, P. Ratchev, I. De Wolf, K. Baert, E. Beyne, Y. Oya, A. Okubora. FCOB: Packaging issues for RF-Mems applications and reliability study. Proceedings of the 15 th EMPC (European Microelectronics and Packaging Conference & Exhibition Brugge, Belgium, June 12 - 15, 2005,), pp. 195-200.

[5] S. Stoukatch, C. Winters, E. Beyne, W. De Raedt, C. Van Hoof. "3D-SIP Integration for Autonomous Sensor Nodes". Proc. of the Electronic Components and Technology Conference (56nd ECTC), (San Diego, California, USA), pp. 404-408. May-June 2006.

Challenges Of 3D/ Stack Die Integration for Thin Large Die

Gaurav Mehta, Tan H Hong, Wilson Ong, YC Koh, John D Beleran, Ravi Kolan

United Test and Assembly Center Ltd, 5 Serangoon North Avenue 5, Singapore 554916

Tel: (65)-65511585, Fax: (65)-65518711, Email: gaurav.mehta@sg.utacgroup.com

Abstract

The focal driving force in today's semiconductor industry is the consumer market. And the key-driving factor for this tremendous growth in this market is the continuous change in the way we do things. Mobility and Portability is key to achieving these goals. The most fundamental technique to achieve this is to incorporate multifunctional devices within a small form factor. And the simplest way to achieve this is via 3D integration. This easiest form of 3D integration readily available to all is to literally stack-up dies within a package structure, thereby increasing the functionality and/or the capacity of a product while maintaining the same footprint. This paper review the most popular stacking options currently available to achieve the desired package structure and performances. However, since the growth in the consumer market dictates the development for most of the packaging houses, the aim will always be advances to reduce the form factor while at the same time integrating/ customizing more multifunctional devices! This paper will give an insight into some of the Stack Die 3D Packaging work that is being developed at UTAC. The articles main focus is also to review the key technologies (Wafer Thinning, Die Attach and Molding) and concepts from a backend assembly point of view that are crucial to meet these new challenges in device packaging. Finally this paper will also discuss the current limitations in 3D integration, like high cost, non-optimal performance, manufacturability and some possible solutions to counter these issues.

Key Words: Stack Die Strength Packaging Thinning Slider

Motivations for Stack Die/ 3D Integration

As the semiconductor industry grows and gadgets become more feature-rich, the clear goal for everyone will be to improve on their Silicon Packaging Efficiency (SPE) of an IC package. "Silicon efficiency" is defined as the ratio of the total silicon (Si) or die area(s) to the associated package area. A very obvious, but gradual shift has been seen from the traditional PCB assembly towards a more hybrid like MCM type package or 2D system on chip solutions. These solutions will usually have a SPE of anywhere from 25-70%. Fusing both 2D together with elements of 3D MCM can further improve the SPE to around 70-90%. However a SPE of >100% can be achieved using a wafer/chip level package integration solution. This type of packaging integrates multiple chips of either similar or different functionality and stacks them in the vertical configuration. For the foreseeable future, the technology trend will move from conventional 2D to 3D die stacking (using conventional WB/FC bumps) and eventually to pure 3D ICs using Thru Silicon Via (TSV) technology to achieve a paradigm shift in the SPE. Some of the common stack-die pkgs under development in UTAC are shown below.

Figure 1a/b: Common Stack Die Pkg

The structures shown Fig 1 (a), (b) are very common in the memory applications where same die stacks (doubles the memory capacity) are achieved with either a spacer technology or via back-to-back pkg. As 3D packaging gain in popularity, more innovative ways are used to incorporate the different functional dies within a package. The structures below show some examples of structures used in the like automotive, communications or the medical industry.

Figure 1c/d: Common Stack Die Pkg

Also popular and growing in demand is the packaging of Flip Chip together with WB stack dies. Additionally some applications require a strict control of the wire length for specifically for RF devices or where wire encapsulation is required. The structure in fig 1 (e)/(f) below show some concepts that could satisfy these requirements perfectly.

Figure 1e/f: Common Stack Die Pkg

The limits of these types of stacking are endless and only restricted by the capability of shrinking various components of the IC Pkg. The key requirements in almost all of the above structures shown, require a reliable wafer thinning capability that can meet the necessary requirements of the mold cap while meeting the moisture/ stress conditions that a typical IC device undergoes. Also required is suitable assembly process like Die Attach stacking with compliant adhesives materials applicable for very thin substrates/ wafers handling. The challenge is then for the wire bonding process to meet the necessity for ultra long low wire loops, bonding on overhang Si and the fine pitch capability. Additionally, to meet the growing demands of shrinking form factor, a robust and reliable encapsulation process is necessary. This is where Vacuum Molding provides the necessary process margins while satisfying the package performance requirements. The following sections describe these processes in more details.

Wafer Thinning (BG) and Die Reliability

Wafer thinning is a crucial requirement to achieve the necessary form factor required in today's market. However wafer thinning can lead to a significant increase in wafer warpage and "bowing" effect, which intern can lead to mechanical tension on the active layer of the wafer. This tension within the layers can then very easily propagate through any one of the many mechanical defect left behind by the grinding process, leading to eventual chip failure. Conventional backgrinding processes with coarse (Z1) and fine (Z2) grind are just not sufficient for semiconductor device packaging that requires ultra thin die thickness. An additional "stress relief" process is crucial to eliminate or minimizing these defect on the wafer back that can lead to flexural damage. Common "stress relief" methods used and available in the industry are CMP (Chemical Mechanical Polishing), Dry Polishing or Plasma Etch or Wet Etch (shown below).

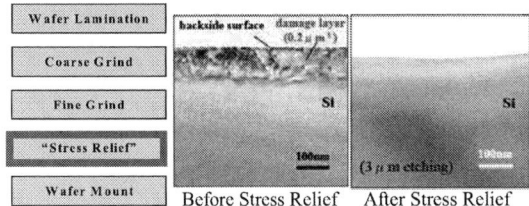

Figure 2: Impact of Stress Relief [3]

In this paper we will evaluate and optimize the Dry Polish process for 50/75um die thickness. The BG system utilized is the fully integrated DICSO DGP8760 Backgrinding/ DFM2700 Wafer Mount system. Due to the high warpage induced by 50/75um wafers, a fully automated system is *mandatory*. All Z1/Z2 Grind parameters are kept constant and only the dry polish parameters are evaluated. Output parameters to be scrutinized are:

- ✓ *Wafer Roughness,* using AFM/Laser system
- ✓ *Die Strength,* uses a Instrom strength tester
- ✓ *Pkg Die Reliability,* using MSL/Etest Data

The sampling plan for the wafers roughness measurements was 5pt/wafer/leg and the sampling plan for die strength characterization is 60pt/wafer. The evaluation matrix for the study is shown in Table 1 below. The parameters ranges were obtained

Table 1: Dry Polish Parameter Study

Thin Die Charactrisation Matrix				
Dry Polish Parameters	Leg 1	Leg 2	Leg 3	Leg 4
Load (N)	Low	Low	High	High
Time	Low	High	Low	High

by a preliminary screening DOE.

Wafer Roughness measurements using a laser system was found to be severely inadequate as its resolution was found to be too low. Wafer roughness, R_a, was found to be in the range of 100 angstroms and this is more suitable for the pre-polish roughness measurement. Normal grinding process can have a R_a in the range of around 100-300 angstrom. Subsequent measurement using the AFM showed most accurate results and Leg 1 was found to have the lowest roughness, R_a (0.308nm) compared to all the other legs. The table below

Table 2: Wafer Roughness Data for 75um Die

Thin Die Characterization Matrix (nm)				
Dry Polish Parameters	Leg 1	Leg 2	Leg 3	Leg 4
Roughness	0.308	0.381	0.358	0.344

shows the wafer roughness data for all the eval legs.

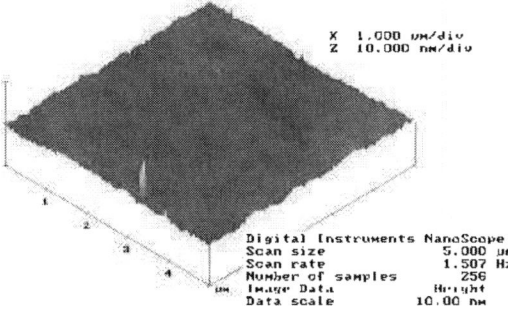

Figure 3: AFM image of Si wafer after Dry Polish

Die strength analysis was then performed using the following bend tests. Sawing parameters were kept constant for all legs. The die sizes and the test jig setup details are shown in the figure below,

	Type of test	Wafer Type	Die size (mm)	Load span / Dia (mm)	Support span / Dia	Test Area (mm²)
A	3-Point Bend	RDL	9.712x5.012	NA	4.0	20.00
A	3-Point Bend	Non-RDL	9.712x5.012	NA	4.0	20.00
B	4-Point Bend	Mirror	20.0x5.5	7.0	15.0	38.50
C	Ring-on-Ring	Mirror	20.0x20.0	7.0	15.0	38.50

Figure 4: Die Strength Tests Config/Methods

347

The highest average die strength for the all the 3 tests above was consistently achieved for leg 1, as shown by the data summary below. Also shown is the typical force/extension graph obtained

Thin Die Charactrisation Matrix for 75um Std (N)				
Dry Polish Parameters	Leg 1	Leg 2	Leg 3	Leg 4
3Pt Test	1.83	1.03	1.56	1.74
4Pt Test	1.15	1.10	1.12	1.12
Ring/Ring	11.26	7.65	9.63	8.03

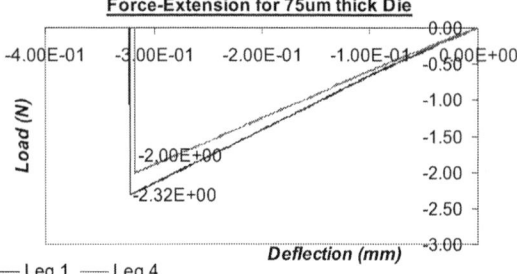

Figure 5: Force-Extension for 75um Thick Die

Although the test data for 4pt vs. ring on ring cannot be compared identically, they do however give a good indication of the impact of the different types of defects within the chip. Additionally for this study, the test area was intentionally kept identical by design so that no correlation study needs to be performed to compare the 2 test results. The 4pt bend test incorporates the effects of both surface defects (wafer thinning process) and also edge defects (wafer saw related), while ring on ring only correlates the surface defects. Comparing the data of the 4pt vs. ring on ring, it is obvious that the edge defect of the chip have a greater influence on die strength when using bend tests (for ultra thin die).

Additionally, data was also collected for 50um thick dies (using 3pt bend test), shown below in table 3 and a significant drop in the average die strength were seen. However a consistent trend for the best sets of parameter was not achieved for this and the other two tests. It was deduced that the load cell of 10N used for the 75um dies, was not suitable when testing 50um dies. A very high systematic error was induced due to the fact that the measured die strength values were only about 5% of the total load cell. We would recommend that a much lower load cell be used, preferably a 2N or 5N load cell if error free data is to be achieved.

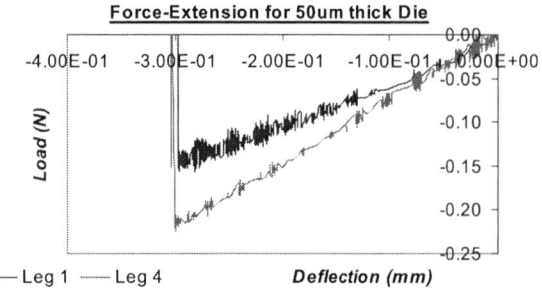

Figure 6: Force-Extension for 50um Thick Die

Table 3: 50um Die Strength Data

Thin Die Charactrisation for 75-50um Std Die (N)				
Dry Polish Parameters	Leg 1	Leg 2	Leg 3	Leg 4
75um Die	1.83	1.03	1.56	1.74
50um Die	0.70	0.61	0.77	0.59

Next, the impact of RDL (Redistributed Layers) process on the Die strength was investigated. RDL are becoming more and more common in the 3D packaging as designers incorporate an RDL process into a wafer fabrication process to ease the impact on wire bonding of an already complicated wire layout and to further increase the performance of an IC. The RDL's evaluated in this study were ~10-15um in thickness. The net impact of this is that the Si thickness for a typical 75um die is reduced to ~55-60um only! The effect of this is a direct reduction of the strength of the die. The table below shows the comparison of die strength of a 75um Std and RDL die. However more importantly, as the results show, Leg 1 still had the best strength results.

Table 4: Die strength Data for Eval Legs

Thin Die Charactrisation Matrix (3pt Test,N)				
Dry Polish Parameters	Leg 1	Leg 2	Leg 3	Leg 4
Std Die	1.83	1.03	1.56	1.74
RDL Die	1.03	0.99	0.89	0.85

The key process development challenge for the backgrinding process was to reduce the warpage induced on these thin wafers, especially for the RDL wafers where actual thickness are 35-40um for a 50um die and 60-65um for a 75um die, and thereby to improve the robustness of the wafer handling/ transfer processes itself. It was found that even though the dry polishing had an improvement on the wafer warpage/TTV (reduction), the key factor was the right selection of the lamination tape used. This becomes more critical as the trend for wafer diameter is moving more towards 12". Additionally this issue is significantly aggravated when RDL wafer have to be processed. The challenge of processing RDL wafers was successfully overcome by using a special *"anti-warp"* tape. Results below show the difference in warpage for a 75um RDL wafer with STD vs. Anti-warp tape. Even with a fully automated BG/Mount system, the RDL wafers with the Std lamination tape could not be processed due to vacuum leak on the process table. Shown below is the effect of Std tape on RDL wafers that were *"spring-rolled"* vs. those using anti-warp tape.

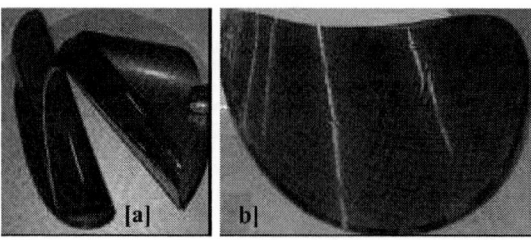

Figure 7: Std [a] vs. Anti-warp [b] tape on 75um

Critical Assembly Processes

To meet the requirement of Ultra Thin packaging, all components within an IC package have to shrink. BGA substrates have been reducing in thickness from 0.3mm to 0.1mm and may even go lower. The use of a thin core of 60um is ever more increasing. To incorporate more functionality within an IC, the ultra thin wafer need to be packaged and this creates the next big challenge for IC assembly. The challenges and how to handle these thin substrates/ wafers and process them is discussed in the next few sections. Also discussed are some novel solutions to meet these requirements.

Die Attach Process: In most cases, stack die configuration dictates the type of direct materials applicable. Key factors in this decision-making are:
- ✓ Die Thickness/ Size
- ✓ Stack-up Structure (Pyramid or Same die)
- ✓ Critical Process Design rules

Conventional Die Attach adhesives like soft solders, epoxy or pastes materials have a distinct disadvantage in terms of BLT control, Die tilting and coverage. Epoxy bleed out is a common occurrence and a major cause for concern when 2-stacked dies have small clearances at the WB level. This is where DAF (Die Attach Film) provides an ideal alternative to the conventional die attaching process. As shown by Song [1], DAF material exhibit a lower reaction rate and lower dynamics modulus, which is better in ensuring its fluidity and gap filling characteristic for better adhesion and reliability performance in the package. DAF also provides a consistent and uniform BLT/ Die Tilt with 100% coverage. Additionally as is its property, it does not in general have a resin separation leading to bleed out, thus providing the ideal materials for small clearances application in a pyramid stack-up.

However the real advantage of DAF materials is seen when processing very large thin (50/75um) dies (with possibly high aspect ratio), where it would not be possible to establish a stable process control using conventional epoxy materials. Additionally, another crucial benefit of DAF materials is the elimination of epoxy cure processes, thereby reducing the overall substrate warpage induce by any heat treatment. This provides an additional processing benefit for stack die packages that require a multi-pass DA/WB process flow. The higher cost of DAF materials is easily countered by the many potentials gains in terms of ease of mfg, wider process window and an improved IC design capability. The biggest challenge for the Die Attach process is stable and reliable die pickup. As dies get thinner conventional needle pickup is not recommended as it can induce an impact force on the die back, which is the weakest point in the die. There are many new and upcoming pickup methods available in the market, some that utilizes varying configuration of needle /plunger to achieve impact free pickup or using a combination of needles and

mechanical offset to separate tape from die. Some concepts involve using vibration (or ultrasonic) to separate the die from film to achieve separation.

Very few of the methods stated above can be classified as truly robust and "impact free". In this paper we will review the patented slider method of die pickup used by the Canon Die Attach System and qualified in UTAC for ultra thin die pickup.

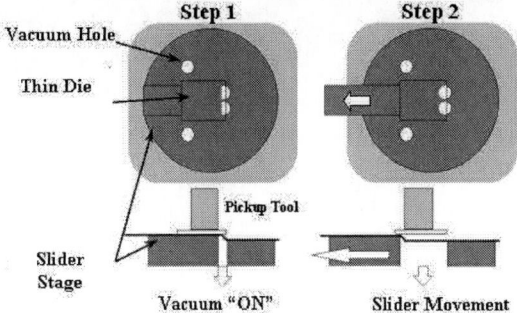

Figure 8: Slider Pickup Concept

The pickup process utilizes a "slider tool" that moves under the die to be picked. Separation of the die from the dicing tape is achieved by the release of vacuum under the slider as it moves across the die. This movement separates the dicing tape and allows the pickup tool to lift the die. As all the movement is in the xy-axis, no impact is applied on the die. The figure 8 shows the schematic of this pickup concept in more detail.

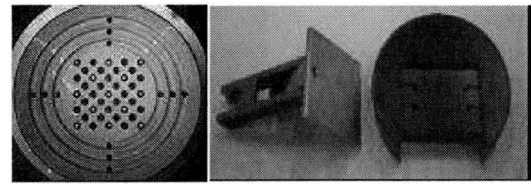

Figure 9: Needle Configuration vs Slider Tool

The key to this pickup process is the fact that the chip is in an almost stationary mode during the pickup process. Once the pickup tool comes into contact with the die surface, the chip vacuum is triggered, holding the chip stationary. This is followed by the initiation of the tool vacuum synchronized together with the slider movement, leading to dicing tape peeling away from the DAF/VU tape material. Once full separation is achieved, die bond can proceed. A study was performed to evaluate and quantify the main the difference of using the slider pickup concept versus the traditional multi-needles ejection system. Setup using the multi-needles type was found to be un-reliable and tedious. Needle configurations using up to 16 needles were necessary for clean pickup. Tight needle planarity is required to ensure that a uniform ejection force applied at the die back. Nonlinear needles can lead to die mis-pickup and possibly die cracks. Needles layout on the pepper pot was very critical and dedicated tooling were required. The setup time for the slider tool was <5min versus the

349

traditional multi-needles ejection type of ~20min. Pickup time was also significantly faster for the slider tooling. This pickup methodology has been successful tested on 25/50um dies, up to 18mm in size using various DAF and UV films. The structure below shows a successful 50um same-die stacking using DAF and Si spacer technology.

Figure 10: 7 Die FBGA Pkg using DAF

Wire Bond: Having a Wire Bond process that can bond on thin overhang dies, or bond long low loop wires is critical for 3D packaging. Ultra Low loops of 50um are a necessary requirement in these days of same die stacking structures.

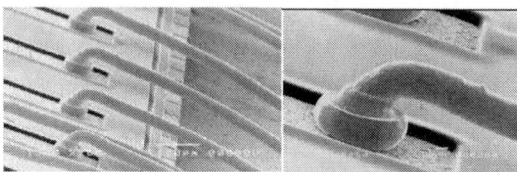

Figure 11: Ultra Low Loop mode (ULL)

For this study, we use the the structure shown below with a die thickness on 75um and an overhang length of 0.8-1.0mm. The most improtant control during this WB process is the amount of die deflection and is a very critical aspect of package design and process optimization. Since there is a lower die with wire bonds present, the deflection of the top die cannot be excessive or the wires beneath can be damaged. Therefore either the die deflection is kept to the min or the loop height for the loop below is minimized. The results achieved are shown in the figure/table below: -

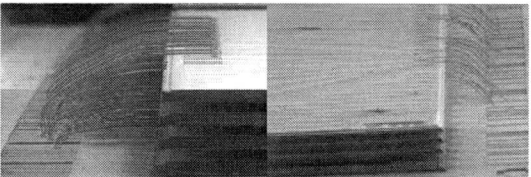

Figure 12: ULL and Reverse Loop modes

The Ultra low loop together with the folded loop modes were successfully able to meet the requirements for the long low loops and overhang WB required for the package structure shown above.

Table 5: WB output data on Overhang dies

Output Responses		Die 2 ULL	Die 3 ULL	Die 4 ULL
Ball Shear [gf]	Average	26.0	26.2	26.7
	Std	1.9	2.6	3.3
	Min	20.7	20.1	21.5
	Max	30.9	34.3	36.0
Bond Pull [gf]	Average	5.7	5.8	5.8
	Std	0.5	0.9	1.1
	Min	4.6	4.2	4.2
	Max	6.7	7.3	8.0
Loop Height [um]	Average	53.7	55.7	50.5
	Std	4.5	6.8	6.5
	Min	45.0	42.0	40.0
	Max	63.0	65.0	64.0

Molding: This process becomes one of the most critical and one of the most challenging when applied for stack die packages. Key challenges are to incorporate large thin dies within the required thin form factor with low die-mold clearances while incorporating overhang dies and achieving a void free encapsulation. The standard mechanism of voids formation, due to the fast/slow flow of materials within the substrate panel as shown below.

Figure 13: Voids Formation in a Stack Pkg

Although Vacuum molding has been in the industry for quite a number of years, its benefits has not really been fully utilized until now, when process capability for DA/WB have reached a point that stack packages have die-mold clearances as low as 70 to 120um. Options such as high spiral flow materials or small filler size materials help to solve these constraints are used, but these come with a significant increase in the package cost due to high materials cost. Investing in a vacuum molding system therefore seem a more ideal options in the long run for such conditions. Multiple Die stacking also leads to a significant increase in stack-up tolerances induced due to the wafer thinning process. A normal thinning process can have a tolerance of ±5-10um. In a typical 4 die stack package, even with a tolerance of ±5um in wafer thinning, there is a good possibility of having a variation in mold clearance of up to 20-30um, which can lead to voids in conventional molding processes.

Many mold flow simulation software tools available in the market that can help predict potential issues that may be encountered for a specific stack-up structure. However no software is yet able to incorporate the effect of vacuum molding within its sets of parameters. An 18leg DOE was performed to analysis the key parameters/ factors that would affect the output response for the mold process. The main factors considered are: -

✓ Stack-up Height and Mold Clearance
✓ Die Layout (unit matrix, shown below)
✓ Die Size (X/Y), Die/Pkg ratio
✓ Conventional vs. Vacuum Molding

Table 6: Mold DOE Summary Table

Leg #	Die matrix	Die : Pkg ratio	Clearance	Vacuum	External voids	Internal voids	Total voids
1	2 x 2	0.33	100	on		0	
2			200	off	3	0	3
3			300	off	0	0	0
4		0.67	100	on	1	1	1
5			200	off	3	0	3
6			300	off	0	0	0
7	3 x 3	0.33	100	on	1	0	1
8			200	off	3	0	3
9			300	off	1	0	1
10		0.67	100	on	1	0	
11			200	off	3	0	3
12			300	off	0	0	0
13	4 x 4	0.33	100	on	0	0	0
14			200	off	0	0	0
15			300	off	0	0	0
16		0.67	100	on	0	0	0
17			200	off	2	0	2
18			300	off	0	0	0

Clearances of 100, 200 & 300um were used. 3 Die layouts as shown below were used and die/package ratios of 0.33 and 0.67 were run. The main output responses were internal/external voids, delamination and wire sweep. Mold compound was kept constant. The table above shows the DOE matrix and the voids results. The main conclusions derived from this study are as follows:

- ✓ Vacuum molding and die clearance are the 2 most significant factor
- ✓ Die Matrix has some influence on the overall voids performance. Indirectly this refers to the top die size over which mold flow takes place.
- ✓ No significant contribution on wire sweep was seen for the various legs run.

Although the results shown above were found to be consistent with other stack die development work, we would like to highlight the additional impact of stack die structure, overhang length, substrate size, vacuum mold technology/ capability, mold compound, could lead to a different result.

Incorporating all the learning described in the above sections, a MSL assessment study was run to validate the all the different process conditions/ parameters derived from the various studies. The test vehicle chosen was a D2-wCSP package. Output response was process manufacturability and package reliability. Shown below is the evaluation matrix.

Leg #	Die2 Thickness	Die Type
Leg 1		STD
Leg 2	75um	RDL
Leg 3		STD
Leg 4		RDL
Leg 5	60um	STD

Figure 14: Reliability Matrix for Thin Die

The legs shown in the table above were subjected to MSL3@260 condition and environmental conditions up to TC1000x and HTS1000hrs. No die related failure was observed. Shown below are the SAT images Post TC1000x and HTS1000hrs for leg 5.

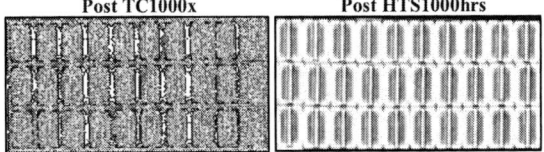

| Post TC1000x | Post HTS1000hrs |

Figure 15: Reliability Results for Thin Die

These results successfully established the positive reliability results achieved through the implementation of the various materials and process controls put in place during the course of this study.

Conclusions

The various Process, Materials & Equipment evaluations and studies presented in this paper have led to a successfully development of 3D Stack Die packages with the desired reliability and electrical test performance achieved. Additionally significant groundwork has been set for future development of

Multi Chip Packages. The Wafer Thinning process was successfully characterized for 75/60um wafer thickness using Dry Polishing as the "stress relief" technology and the reliability of these thinned wafers was successfully demonstrated. Wafer RDL process was found to be a critical factor for the thinning process and detail study of impact of different RDL design need to be considered when setting up an ultra thin wafer processing line.

The challenges that are normally faced by using conventional epoxy materials were successfully countered by using DAF materials. Significant process margins and yield improvement were obtained using this type of material. Reliability performances were met up to MSL2 @ 260degC. Additionally, tighter package design could be processed due to the stability/repeatability of the DAF material. Vacuum molding process was seen to be a necessary requirement to provide the extra tolerances when processing multi stack-up structures with ultra low clearances or with deep overhang situations. The desire for portability, mobility and increased functionality will continue to drive the semiconductor packaging and UTAC will continue to pursue further enhancements to improve our process capability and product portfolio.

Acknowledgment

The authors would like to thank the management of UTAC for giving the opportunity to work on these stack die projects. We would also like to thank everyone in the UTAC Packaging & Assembly Technology Team for all their support during the development during of these Stack Die Packages. We would also like to thank the UTAC Design (especially BK Lim), Reliability and FA Gp for all their support in the endless job requests! Additionally we would like to thank Qimonda and its Development Gp based in Dresden, Germany, for all their support, suggestions and help during the co-development of the Joint Stack Die ToolBox project, especially to Kimyung Yoon and Christine Hinz.

Finally we would like to acknowledge all the equipment vendors with whom we collaborated on the process development for these stack die packages, especially to Canon Machinery Inc for their endless support for the Thin Die Die-Attach Process.

References

[1] S.N. Song, H.H Tan, P.L. Ong, "Die Attach Film Application in Multi Die Stack Package" Proceeding of Semicon Singapore, May 2005

[2] Paydenkar. C Poddar. A, Chandra. H, "Wafer Sawing Process Characterization for Thin Die (75 micron) Application," IEEE/CPMT/SEMI 29th International, pp. 74-77, Jul. 2004

[3] Dr Kiyoshi Arita, "Plasma Stress Relief Technology for Wafer Thinning Process" 11th Annual International KGD Packaging and Test Workshop, September 12-15, 2004

Through Wafer Interconnect Technologies for 3D System-in-Package

E. van Grunsven[1], F. Roozeboom[2], F. Sanders[1], F. van den Heuvel[3],

M. Burghoorn[1], T. Grob[1]

[1] Philips Applied Technologies , [2] NXP Semiconductors Research, [3] Philips Research

High Tech Campus, 5656 AE Eindhoven, The Netherlands

Abstract

The next generations of cellular RF transceivers require a higher degree of integrations and three dimensional interconnect. This paper describes technologies that we recently studied and which have found or may find their implementation in RF and other System-in-Package (SiP) applications. First, a brief introduction is given on the applications that use 3D packaging. The focus is on NXP technology road map. Next, the options for through wafer interconnects will be discussed, with special focus on via forming by means of Deep Reactive Ion Etching (DRIE). Third different concepts for through wafer interconnect, that are in development at Philips and NXP, will be presented. A concept based on copper paste printing for via filling will be elucidated in more detail.

Key words: system-in-package, SiP, 3D packaging, through wafer interconnect, TWI

Background

Recently a new type of highly integrated cellular RF transceiver modules was launched using NXP Semiconductors' silicon-based *System-in-Package* (sbSiP) technology [1-3]. This new technology utilizes back-end silicon processing to integrate passives onto a silicon substrate. This technology is called PICS (*Passive Integration Connecting Substrate*). It serves as a platform for the heterogeneous integration. As an example, a transceiver IC can be flip-chip mounted onto this passive component silicon substrate, thus minimizing inter-connect parasitics and footprint area. This sub-assembly stack is next flipped into standard IC-sized lead frame package (Fig. #1).

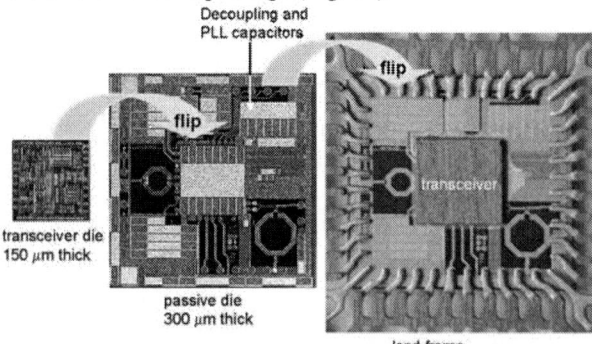

Fig 1. Bluetooth SiP radio module of active RF-transceiver on passive die in an open lead frame.

For the next generations of such sbSiP devices an interconnect through the PICS substrate is envisaged (Fig.#2). This would allow a low inductance interconnection from one wafer side to the other, e.g., for electrical grounding or signal redistribution.

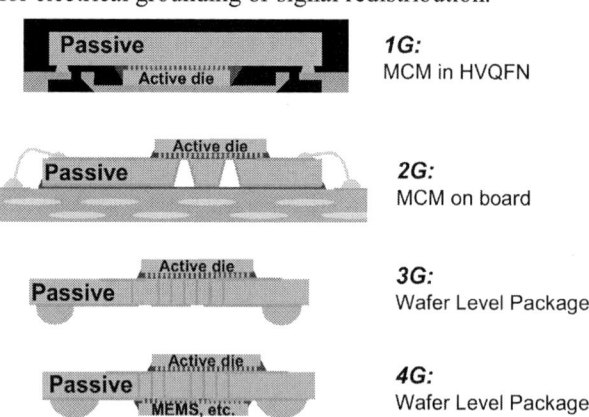

Fig 2. Different generations sb-SiP modules.

Through Wafer Interconnect Technologies

Through wafer interconnection (TWI) requires the following basic process steps:
1. via formation
2. via isolation
3. seeding and barrier coatings
4. via filling with conductive material
5. redistribution and finishing

Depending on the specific application these process steps need to be compatible with the pre- and post processes. Considerations on "vias formed in front-end process" or "vias formed in back-end process" do determine the applicability of materials and processes.

In this presentation a short overview of the various TWI technologies and conceptual considerations will be given. In particular the characterization of the via etching process using DRIE, will be elucidated. In this characterization the effect of via geometry and loading on the etch rate is presented. Circular and elongated vias of various diameters were etched using the so-called *"Bosch"* process. The effect of the loading (% area of exposed silicon to be etched) on these different structures is determined. (Fig. #3)

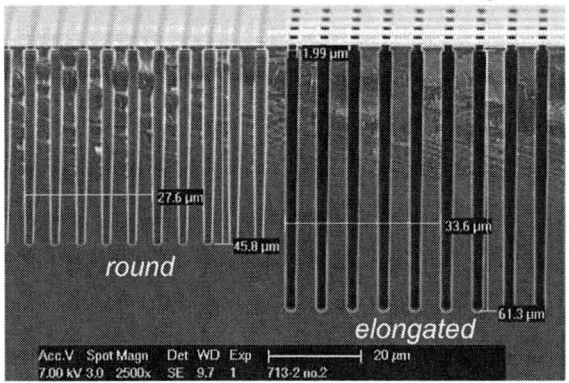

Fig 3. Cross sectional view; etch depth for a round shape 2µm (left) and a elongated via 2*16µm (right)

Concepts for Through Wafer Interconnect

Philips and NXP are developing different concepts for through wafer interconnect. Some of these concepts are realized in front-end technologies, using narrow vias with a size of approximately 3µm and filled with tungsten or polycrystalline silicon. Other concepts are made in back-end technologies and consist of wider vias, with a diameter of 10µm to 100µm, and use copper as a via filling material.

In the presentation a short overview of the different concepts shall be given. Also the main properties of these concepts will be listed.

Finally, one concept based on copper paste filled vias, will be extensively elaborated. This concept is good example of back-end processing and aims at vias with a diameter of 30µm to 100µm. A first technology demonstrator is made and evaluated. The filling paste consists of silver coated copper grains, with a size of 2µm to 3µm, embedded in an organic binder. This paste shows excellent properties for filling vias with a high aspect ratio. Evaluation with SEM revealed that even in a 10µm wide via gets filled with this copper paste. (Fig.#4).

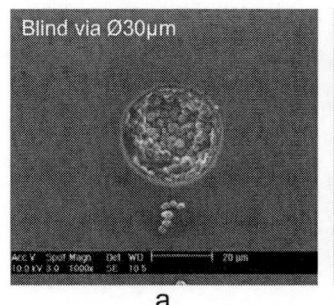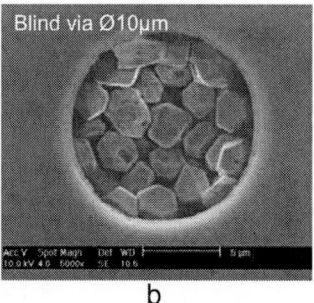

Fig 4. Top view of "blind' vias after paste printing, a) φ via =30µm, b) φ via =10µm

The filling properties were confirmed by angular X-ray inspection and cross section analyses. Figure 5 shows a result of the latter.

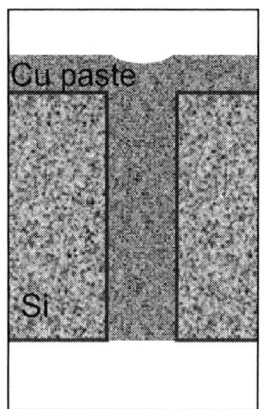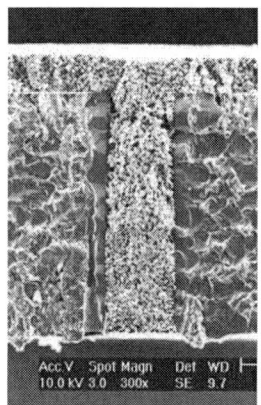

Fig 5. Cross sectional view of a via filled with copper paste; φ via =30µm.

Electrical evaluation of the cured copper paste was done by measuring with a "4 point in-line probe". These measurements showed that the specific resistivity of this material is approximately $3*10^{-5}$ Ωcm, which is still significantly higher than that of bulk copper. Further analyses on chain structures of these vias are ongoing.

References

[1] F. Roozeboom et al., 'Passive and heterogeneous integration techniques for 3D System-in-Package applications',12th Annual Pan Pacific Microelectronics Symposium, Jan. 30- Feb. 1, 2007

[2] F. Roozeboom et al., 'Passive and heterogeneous integration towards a silicon-based System-in-Package concept', Thin Solid Films, 504 (2006) 391-396, and refs. therein.

[3] D. Chevrie, et al., 'A silicon-based system in package technology', 15th European Microelectronics and Packaging Conf., June 12-15, 2005, Bruges, Belgium

Thermal Performance of Embedded IC Structure

Tanja Karila

Imbera Electronics Oy, PO Box 74, FIN-02151 Espoo

Phone: +358-40-540 7278, Fax: +358 207 400 258, E-mail: Tanja.Karila@imbera.biz

Abstract

Thermal performance of electronic packages is critical in terms of reliability of the entire electronic application. The miniaturization trend along with constantly increasing power density poses an issue of overheating. In addition, as embedded ICs are in question, the thermal management of the packages is typically even more challenging: gained benefit in size and weight produces fewer options for thermal management methods. However, despite of lost volume, embedded IC technology developed by Imbera Electronics offers also some degrees of freedom to manage the thermal load produced by ICs. This paper presents results about the studies of the thermal performance of Integrated Module Board (IMB) technology. System-in-Board (SIB) and System-in-Package (SIP) types of packages have been manufactured and measured in standard natural convection environment and modelled with FloTherm® thermal simulation software. The results have provided fundamental information about the thermal behaviour of the IMB structure - e.g. main heat flow paths have been determined and the efficiency of various thermal enhancement methods has been evaluated.

Key words: IMB, embedded IC, thermal properties, thermal via, thermal design

Introduction

One of the main challenges in current electronics manufacturing and packaging development is how to integrate more functions into the same or even smaller size. With traditional technologies it has become more difficult to increase the packaging density. To meet these requirements Integrated Module Board (IMB) technology has been developed. IMB technology utilises standard PCB manufacturing processes to integrate active and passive components inside organic materials (e.g. PCB core layer). The process combines numerous separate production phases into a single process, enhancing overall efficiency and offering a range of new capabilities. [1]

Increased packaging density has led to increased power densities raising the potential for reliability risk. Thus, thermal design for effective heat transfer is essential to ensure proper functionality of the final product. IMB technology enables various methods to keep the product operating temperature within specifications. Thermal vias and other copper structures can be manufactured to enhance thermal conduction from IC junctions, where the heat is generated, to the edges of the package. The critical areas can be protected and hot spots can be managed by proper thermal design.

To get an understanding of the thermal performance of IMB technology, thermal modelling and measurement of samples have been performed adapting JEDEC 51 standards. The power level of

the study is in line with the power levels of typical hand-held application. The summary of the results is presented in this paper. [2]

Sample Description

To define the thermal performance of IMB technology two kinds of structures were manufactured, measured and modelled – system in board (SIB) type of structure and ball grid array (BGA) type of SIP modules. Figure 1 illustrates the SIB structure and the basic structure of BGA mounted on test PCB is depicted in figure 2.

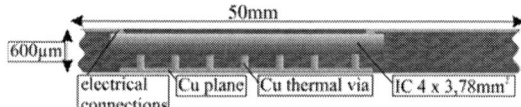

Figure 1. Principle of SIB sample. IC embedded in FR4. Thermal vias reach from the IC back to the heat spreading Cu-plane.

SIB test series consisted of 10 samples each having a different structure for thermal management. Outer dimensions of the samples were 50 mm x 50 mm x 600 μm. The front sides were similar, but back sides varied as regards to thermal vias (number, size) and copper plane (size, geometry). Copper volume was chosen to correspond to different applications - varying from zero to maximum volume. E.g. electrical design, reliability and cost are limiting the thermal design. Thus, the copper volume and geometry must be optimised application-specific. Test series was

354

designed to evaluate several thermal management methods and to be able to apply the methods effectively in actual modules.

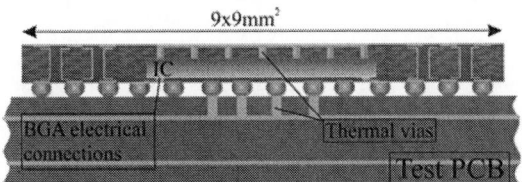

Figure 2. The schematic cross-section of 9 mm x 9mm BGA mounted on test PCB.

BGA test series composed of 4 dissimilar samples – two without thermal structures and 2 with thermal vias reaching from the IC back side to the edge of the package. The orientation of IC was a variable as well – active side facing the top and bottom (as seen in figures 2 and 3) of the package.

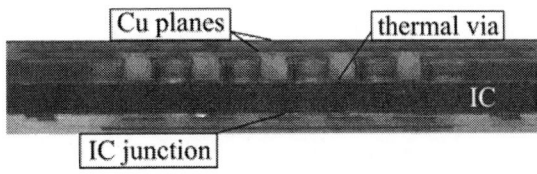

Figure 3. The cross-section of BGA in IC area. Thermal vias reach from the IC back side to the package top.

By contrast to SIB samples, room for varying copper structures in BGAs was limited. The main interest was to understand the effect of IC orientation, the presence of back side thermal vias and the effect of mother board thermal design. IC facing the bottom of the package (in close proximity to solder balls) is usually recommended due to better electrical performance, e.g. shorter signal path. Also, from design point of view the number of through-holes may limit the freedom to choose IC orientation. From thermal standpoint the shorter distance from back side thermal vias to ground solder balls is the optimal solution.

BGAs were mounted on three types of thermal test boards: 1) JEDEC standard test board conforming to JESD51-9 specification, 2) modified thermal test board (to measure the packages mounted on today's multilayer PCB) and 3) PCB without thermal management.

Thermal test IC (4 mm x 4 mm) was embedded in both SIB and SIP structures. The electrical functionality of the IC is shown in figure 4. The IC enables the controlled heating over its surface. Diode in the center is for measuring the surface temperature.

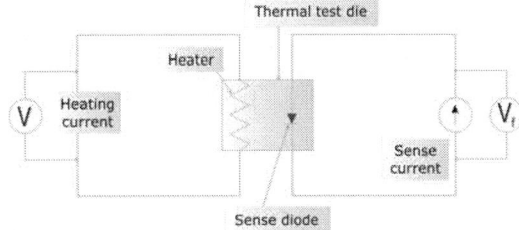

Figure 4. The schematics of test IC functionality.

Measurement Setup and Theory

Thermal resistance from IC junction to ambient air was measured utilising MicReD's thermal transient tester, T3Ster. Measurement system consists of hardware and software elements for measurement, data acquisition, subsequent evaluation and display. The surroundings for the measurements equalled still air environment specified in JEDEC JESD51-2 standard. Ambient air temperature was measured with thermocouple and the IC temperature obtained from the sense diode.

The measurement results were in form of thermal impedance curves – showing the normalised IC temperature rise as a function of time. Thermal resistance is equal to thermal impedance once the thermal equilibrium is reached. Typically, thermal resistance from junction to ambient (R_{JA}) is presented as shown in equation 1.

$$R = \frac{T_j - T_a}{P} \quad , \qquad (1)$$

where T_j is the junction temperature, T_a equals the ambient temperature and P is the dissipating power.

Also, the cumulative structure functions gained from the measurement data were analysed to better understand the thermal behaviour. Cumulative structure functions are the graphic representations of one-dimensional equivalent thermal RC-network of the system, giving the sum of the thermal capacitance as the function of the thermal resistances – measured from IC junction towards the ambient. The micro scale thermal behaviour can be interpreted from the structure functions. [3, 4]

Thermal Modelling

The detailed thermal models of both SIB and SIP structures were constructed with FloTherm simulation software. They corresponded to the measurement setup as regards to samples and surroundings. Table 1 shows the thermal conductivities of the materials used in models.

Table 1. Thermal conductivities of materials utilised in modelling.

Material	Thermal conductivity (W/mK)
FR4 core	0,30
Build up layers	0,36
Solder resist	0,27
Cu traces/vias/planes	380
IC	120
Solder balls	50
Ambient air	0,027

The BGA structure in FloTherm drawing board is illustrated in figure 5. Packages are modelled accurately with accurate dimensions. Simplifications had to be utilised in thermal vias. Combination of the thermal conductivities of epoxy and copper was calculated for the conductivity of thermal vias. Solder balls and pads were modelled as combined cuboids with thermal conductivity of 70 W/mK. Generally, the model was detailed and the simplifications concerned mainly round shapes – solder balls and thermal vias.

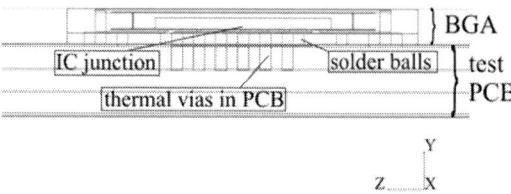

Figure 5. The details of the BGA model.

The grid for calculation was defined finer in the IC area and coarser in the edges of the system. Simulation rounds were started from the JEDEC standard enclosure. The resulting air flow – PCB boundary conditions were imported into more detailed model including PCB geometries and module geometries. Radiation was included in the simulations. The contribution of the radiation to the heat transfer depends on the thermal environment of the sample under test or in application. For a package on a test board with high temperatures, the radiation exchange with the ambient can not be neglected.

Results

Both measurements and modelling gave fundamental information about the package level thermal behaviour of the IMB structure in standard still air environment. Summary of the results are presented in this section.

Thermal resistances from junction to ambient of SIB structures are shown in table 2. The equivalent thermal impedance curves are presented in figure 6.

Table 2. Thermal resistances of SIB samples.

Sample number	R_{JA} (K/W)
1	265
2	265
3	180
4	135
5	117
6	103
7	100
8	99
9	45
10	44

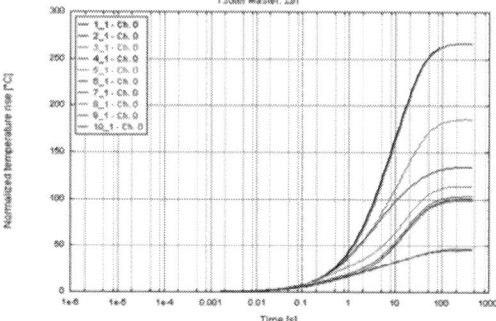

Figure 6. Thermal impedance curves of the 10 SIB samples.

In addition to thermal resistance values, the cumulative structure functions were studied to fully understand the thermal behaviour. The curves are presented in figure 7. The curve profiles were analysed to specify the fine details of the heat flow paths in each structure.

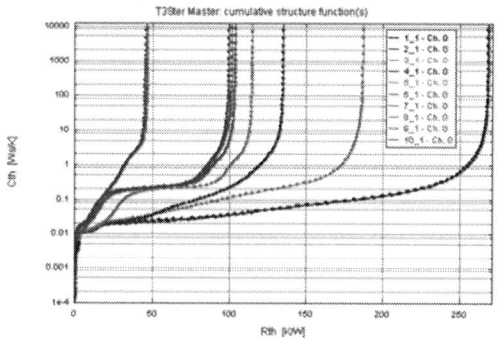

Figure 7. Cumulative structure functions of SIBs.

Several measurements and simulations were performed for BGA setup. The summary of the results is presented in table 3. The effect of mother board thermal design is shown with 8 mm x 8 mm BGA. "Standard" corresponds to test board in accordance with JEDEC JESD51-9 specifications, "worst case" stands for the mother board without thermal management and "enhanced" means typical thermal via layout of today's multilayer PCB.

Package level thermal management is shown with 9 mm x 9 mm BGA.

Table 3. Thermal resistances of BGAs.

Sample	R_{JA} (K/W)
8x8BGA worst case	64
8x8BGA standard	44
8x8BGA enhanced	28
9x9BGA no vias, face up	57
9x9BGA no vias, face down	45
9x9BGA vias, face down	35
9x9BGA vias, face up	29

The temperature distribution of the BGA in standard test board is presented in figure 8 and cumulative structure functions of 9 mm x 9mm BGA with 4 different power levels are shown in figure 9.

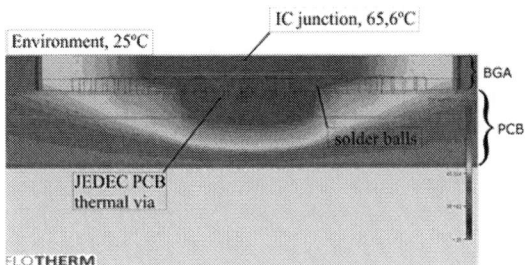

Figure 8. Temperature distribution of the BGA cross-section as powered with 1 W.

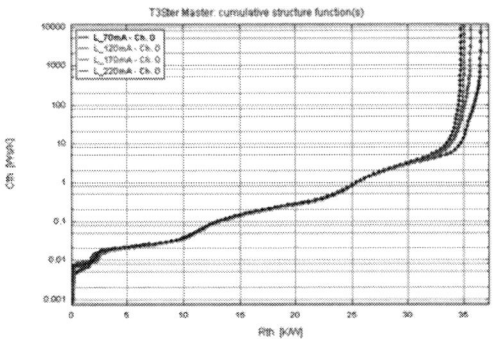

Figure 9. Cumulative structure functions of BGA with thermal vias. IC facing the bottom of the package.

Thermal modelling accuracy was around 95%. As detailed models are concerned, inaccuracy is a result of variation in copper volume, silicon-copper interface and the thicknesses of dielectric materials between the model and the real world. The contact thermal resistances within IMB structures are not totally defined. Also, the changes in copper volume (due to plating process tolerances in test series manufacturing) induce error, since even a minor change can contribute to the thermal behaviour of the structure.

Discussion

The thermal resistances of SIB structures are higher due to the measurement setup: samples are not mounted on test PCB, which would take some of the thermal load. Regardless of non-standard configuration of SIB in package level, the results are useful for comparing the different thermal structures. As can be seen, the highest value is about 6-fold to the lowest.

Sample 1 represents the structure without thermal management. Cumulative structure function shows good thermal performance only within an IC, where heat is conducted rapidly. Somewhat better conduction is noticed via electrical interconnections. Small copper plane in sample 2 does not improve the thermal performance, since there is no path to conduct heat to the periphery of the module.

Additional fins from small copper plane decreases the thermal resistance of sample 3 considerably. The volume only increases 10%, but increased copper area is enough to promote the heat transfer from module to ambient via copper fins. Copper grid over the back side of the module (representing a typical amount of copper in interconnections) having less volume of copper as in sample 2 decreases the thermal resistance of sample 4 substantially due to continuous thermal paths from the hottest areas to the edges of the module.

Larger copper plane improves thermal performance of sample 5, and the direct contact of copper plane and IC (in sample 6) further decreases the thermal resistance. The cumulative structure function shows that in close proximity of IC thermal conductivity is improved compared to samples 1-5 (without IC – copper plane contact). The copper volume is only increased by few percents, but the steep sections in the curve show more effective thermal conductivity close to the IC. The structures without a contact have flat regions before the steep phase presenting the worsened heat conduction through pure FR4.

Two different thermal via matrixes of samples 7 and 8 are both slightly more effective than large contact of sample 6. The copper volume increase of around 0,18 mm^3 in IC area (sample 7) is beneficial, but the further increase of copper in sample 8 (around 60% more volume) is no longer that beneficial. It is noticed that additional vias do not affect thermal performance once the adequate contact is reached. This is confirmed by samples 9 and 10. After sufficient thermal conduct the dominant factor in heat transfer is the copper plane geometry. Maximum sized copper plane results low thermal resistances.

Wide range of simulation and measurements with BGA structure showed the importance of proper thermal management. The main issues are copper – IC contact, sufficient copper volume and mother board thermal design. The thermal resistance of 8mm x 8 mm BGA mounted on test board without any thermal contact was nearly one and a half fold to standard test board, in which the thermal vias connect the middle solder balls to the upper ground plane of PCB. The enhanced version is again considerably better due to extended thermal vias in PCB reaching from solder balls to the bottom copper plane.

The thermal resistance values of 9 mm x 9mm BGAs were expected: the highest value is for package without thermal vias, IC facing up. Poor thermal conduction from IC through epoxy and signal traces results the highest thermal resistance. Shortened distance from IC to the solder balls decreases the thermal resistance even without proper thermal contact. Thermal vias improve the performance: heat is most efficiently conducted directly from the IC back side to the solder balls. Thermal resistance of the most probable actual design (IC active side facing the bottom of the package) is also substantially lower. The cumulative structure function of it indicates that the main heat flow path is through the top of the package via through holes to the solder balls. The same phenomenon is seen in simulation data.

The presence of thermal solder balls in the middle of the package was studied with simulations. It was shown that even few thermal contacts are beneficial compared to structures without thermal solder balls conducting the heat to the PCB ground planes.

Both SIB and SIP samples had the same feature as regards to comparison of different power levels. The phenomenon is seen in figure 8 and also noticed in thermal impedance curves. The highest power level of the sample gives the lowest thermal resistances due to the best cooling capacity. Larger temperature gradient results higher air speed transferring heat more effectively.

All the results of the SIB and SIP are not directly comparable as such, but as combined they increased the knowledge of the thermal behaviour of an embedded IC. The effect of environment, significance of copper contact and the copper heat spreading capacity is defined with IMB thermal samples. According to these results it is possible to give general thermal design guidelines for the IMB module, if the environment, power level per IC area and dimensions are known.

Conclusions

Thermal properties of IMB technology have been modelled and verified. According to the results, the thermal functionality of the various IMB structures is well controllable and gained thermal resistances showed good performance in power levels of hand-held electronics. Optimisation of thermal design is a necessity in case of increased power density. Significant differences in thermal resistances with minimal and maximal copper volume were shown – depending on a case even sixfold resistances were achieved with poor thermal

management. Also, the importance of the thermal design of package surroundings was shown. E.g. thermal vias in the package are not effective, if the mother board is not designed to take advantage of them.

It must be emphasized that gained thermal characteristics will differ from the characteristics of 3D packages with multiple ICs embedded in the structure. Along with characterisation of 3D structures the front side thermal via process is studied. With existing thermal via processes the control of local hot spots is satisfying, but to gain the best performance, e.g. with GaAs applications, the thermal via process for IC front side must be optimised.

Acknowledgements

Author would like to thank MicReD (part of Flomerics Group) for thermal transient measurements.

References

[1] P. Palm, R. Tuominen, A. Kivikero, "Integrated Module Board (IMB); an Advanced Manufacturing Technology For Embedding Active Components Inside Organic Substrate", Proceedings of the 2004 Electronic Components and Technology Conference (ECTC), Vol. 2, June 1-4, pp. 1227-1231, 2004.

[2] JEDEC, the JEDEC standards for thermal characterisation are available from JEDEC at http://www.jedec.org/

[3] Rencz, M.; Szekely, "Structure function evaluation of stacked dies", Semiconductor Thermal Measurement and Management Symposium, March 9-11, pp. 50-54, 2004

[4] MicReD Ltd., "Properties of the structure function and its use for structure identification and for compact model generation", Available: http://www.micred.com/applications/documents/t3ster/t3ster-STRFUNCT_detailed.pdf, 2000

Application of Metallo-organic Pastes on LTCC Substrates

Jaroslaw Kita, Ralf Moos

University of Bayreuth, Functional Materials Group, D-95440 Bayreuth, Germany

Phone +49 921 55 7401, Fax +49 921 55 7405, Email: Jaroslaw.Kita@Uni-Bayreuth.de

Abstract

Metallo-organic pastes (resinate pastes) have been well-known for many years. They may consist of precious metal compounds (gold, platinum) that are dissolved in organic oils. Although the described materials are well known, no information about possible applications of metallo-organic pastes on LTCC is available. In our opinion, the pastes can be advantageously used for LTCC devices. Their low film thickness (one layer less than 1 µm) allows improving the flatness of the structure. Due to their lower thickness, the sheet resistance of resinate films is higher than those of typical thick-film conductive materials. This may be advantageous, for example for designing ohmic resistors for heater purposes. This contribution presents results of a study about resinate pastes on LTCC ceramics. In our study gold resinate paste that were printed on different LTCC tapes were characterized. The test structures were arranged both as surface and as buried elements. The prepared structures were examined by SEM and EDX-analysis. Electrical properties of the fabricated structures are discussed and compared with properties of similar elements manufactured on standard alumina substrates.

Key words: LTCC, low temperature co-fired ceramics, resinate pastes, metal-organic pastes

Introduction

Miniaturization of active and passive devices as well as of circuit boards is a strong trend in electronics industry. At the moment, the typical line width of thick-film structures is about 100 µm. In order to realize smaller structures, other processes are necessary. Miniaturization can be achieved by photolithographic methods. Appropriate materials are Fodel from DuPont or the KQ-System from Heraeus. Another method is laser direct writing fired or unfired films. Application of frequency-tripled (355 nm) Nd:YAG laser allows to achieve line widths as small as 20-30 µm.

Our previous investigations showed that structuring of unfired conductive films gives somewhat better results. In this case however, a shrinking of structured lines on alumina substrates occurs during the firing process. This phenomena is understood on typical LTCC ceramics, but the same was observed on alumina substrates as well [1]. Laser direct-writing results depend strongly on the film composition. Films with large grains are difficult to pattern and the quality of the lines is worse. A glass from the film melts and can burn-out in an uncontrolled way. Fritless pastes, mostly platinum ones, are more appropriate for laser patterning but in some cases large grains and high porosity do not allow to manufacture as small lines as desired.

Typical for thick-film technology are pastes with glass and/or metal-oxide additives. These pastes are sometimes referred to as "cermet" pastes due to their high temperature reliability. In terms of high temperature reliability, some particular cermet pastes can be metallo-organic materials.

Metallo-organic pastes for technical applications

Metallo-organic pastes consist of precious metal compounds (mostly gold or platinum) that are dissolved in organic oils [2]. Metallo-organic pastes are used in some technical products, like thermal printheads, pressure sensors, temperature sensors and touchpanels [3].

Resinate pastes do not contain glass compounds but some additives can be built-in as bonding agents. Resinate pastes can be deposited in different ways but usually they are screen-printed. Resinate pastes can be used on all substrates known from thick-film technology, i.e. alumina, steel, or glass. The screen printing process does not differ compared to typical thick-film materials. Screens with 300-350 Mesh can be applied. After printing, the film thickness is about 10…25 µm. The firing temperatures of resinates are about 800 - 900 °C. The similarity to typical thick-film pastes is obvious. The process differences between these two materials occur after screen printing. The drying time is much longer as usual due to high content of organic materials within resinate film. According to manufacturers recommendation, about 50% of organics should be removed during the drying process. Loss of organic solvents during drying and firing causes the layer thickness to decrease to 0.1…1 µm. To achieve a thicker film, the structure should be subsequently printed and fired.

As mentioned above, the resinate pastes can be used on different substrates but at the moment no public information is available on the application of resinate pastes on LTCC substrates, especially about co-firing of resinates and LTCC.

The application of resinate pastes on LTCC is very interesting because of some properties of this material. First of all, application of resinate pastes is possible to produce conductive films with thicknesses typical for thin-film technology with the same equipment as for thick-film technology, i.e., the cost level maintains as low as for thick-film technique.

Compared to a typical thick-film layer on LTCC (thickness between 7...10 µm), thickness of single resinate layer (0.1...1 µm) has only a marginal effect on the flatness of the LTCC multilayer. This is particularly important, if the thickness of an LTCC tape is low and the conductive lines are placed inside the substrate.

Here, two important properties of resinates should be noticed. The necessary firing of each layer after deposition means that, in the case of co-firing with LTCC, only one single layer of resinate can be applied. The low layer thickness affects the sheet resistance. The R_{sq}- value of single layer is, as for conductive materials, high, between 200 mΩ (Au-paste) and 5 Ω (Pt-paste - single layer). It limits the application of resinate films as typical conductive lines, but in some applications like heating systems or low-ohmic resistors, it allows to reduce the number of squares, i.e. the minimization of the structure. Moreover, metallo-organic pastes can be used and joined with standard thick-film pastes.

In Germany, Ceramic Colour Division of W.C. Heraeus GmbH produces resinate gold and platinum pastes for technical applications. They are available in a wide range of metal contents from 10 % to 22% Au and 12.5 % to 15% content of platinum, respectively.

Experimental

In our experiments, the resinate layers were deposited on 96 % alumina substrates and on two unfired LTCC tapes: HL 2000 zero-shrinkage tape (Heralock® System, Heraeus, Germany) and a conventional non-zero-shrinkage tape. Due to our precondition – co-firing with LTCC ceramics - only single layer was printed using 325 Mesh steel screen. Printed films were dried for 30 min at 100 °C. In case of LTCC the tapes were laminated with additional tapes for mechanical stability. Samples were prepared as surface elements as well as buried ones, i.e. between LTCC tapes. The structures were fired with a peak temperature 865 °C. The LTCC firing profile had an extended burn-out phase to remove all organic components from resinate films and LTCC tapes.

For the initial tests, two resinate pastes have been chosen: RP20003 with 20% gold content and RP10003 with 12,5% platinum content (both pastes Ceramic Colour Division, Heraeus, Germany). Since not all investigations have been finished yet, the results presented in this paper have to be considered as preliminary.

Resistance vs. temperature characteristics $R(T)$ was measured according to German DIN standard 41850 Part 2 [4]. The test structure contains 100 squares. SEM and EDX investigations were made with LEO 1450VP (Zeiss, Germany) scanning microscope.

Single Resinate-Layer on Alumina

Metallo-organic pastes for technical applications were designed to use, most of all, on alumina ceramics. Therefore, the structures printed on 96 % alumina were used as reference material.

Figure 1. Test structures according to German DIN standard 41850. Line width 500, 300, 200 and 100 µm

Fig.1. shows test structures printed on alumina. All printed lines were of good quality, without shortcuts between lines. Minimal obtained line width was 100 µm. Simple measurement of layer thickness with a profilometer gave no information about the layer thickness.

The structures were characterized in a tube furnace in the temperature range up to 800 °C. The measurements were done in four-wire configuration. Changes of resistance over time are shown in Fig. 2.

The temperature characteristics were almost linear above 150 °C (Fig. 3). The sheet resistance of the film was about 230 mΩ/□ (at room temperature). HTCR was about 2580 ppm/K (25...125 °C). These values can be used as reference values for one-layer-printed gold resinate pastes.

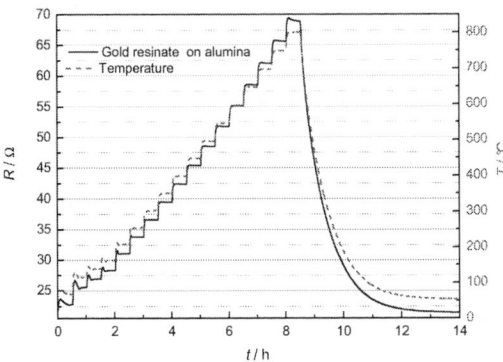

Figure 2. Resistance changes over the time for gold resinate on 96 % alumina when exposed to stepwise increased temperature.

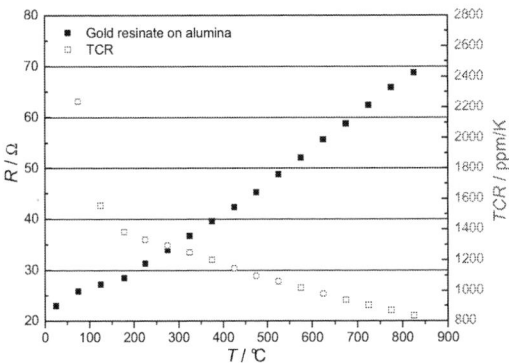

Figure 3. Resistance vs. temperature characteristics of gold resinate on 96 % alumina

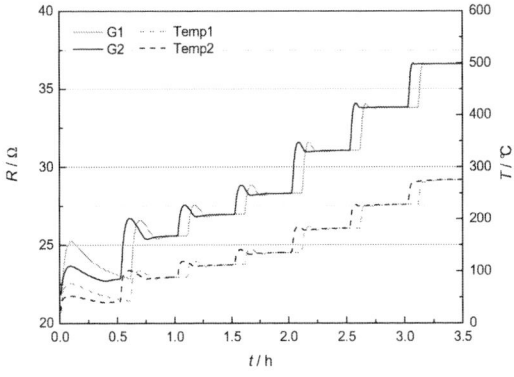

Figure 4. Comparison of $R(t)$ characteristics for different samples printed on alumina substrates

The gold resinate film printed on alumina can be characterized by very good reproducibility as shown in Fig. 4. The differences in sheet resistance between samples as lower as 2 %.

Application of resinate pastes on LTCC

Resinate layers on LTCC were prepared in the same way as on alumina. As mentioned above,

the firing profile was a little bit extended to be sure that all organics from resinate layer and from LTCC were removed. Figures 5 and 6 show test structures after firing. The structures printed on zero-shrinkage tapes were of very good quality, despite of only one single layer was printed. No cracks or holes in the conductive layer were found. Similar to samples on alumina, it was impossible to obtain information about the layer thickness by a simple profilometer measurement.

Figure 5. Test structure with gold resinate paste (single layer) co-fired with Heralock tape. Line width 100, 200, 300 and 500 μm

Results obtained from structures printed on typical non-zero-shrinkage LTCC tapes were much more worse (Fig. 6). The layer consisted of small clumped gold particles. There was no contact between adjacent particles. The reasons of this behavior of resinate paste on typical non-zero-shrinkage tape is presently under investigation. For this reason, only gold resinate layers on Heralock tape were electrically measured.

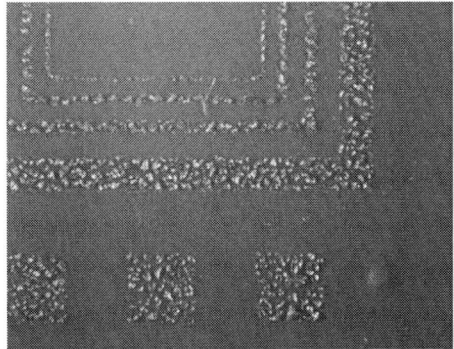

Figure 6. Gold resinate layer after co-firing on a typical non-zero-shrinkage tape.

Structures with buried elements were examined by scanning electron microscopy (SEM). Figure 7 presents an example of such a buried element. The lines were properly laminated and fired. Neither delamination nor crack formation can be found.

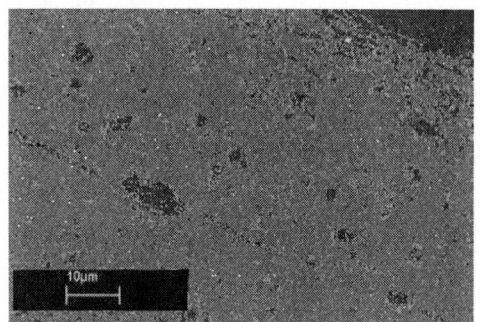

Figure 7. Buried structures made of gold resinate paste.

The visible voids in the ceramics, characteristic for Heralock tape, are small, mostly with a diameter of only 2-3 μm. The thickness of the buried resinate-based gold layer itself was about 600...700 nm. The EDX analysis showed no remarkable interactions between resinate film and LTCC ceramics.

The resistance measurement conditions were the same as for structures on alumina. Figure 8 shows the resistance characteristics over the time of two selected samples. In both cases, a very good stability over the time was observed, only above a temperature of 750 °C, some samples exhibited a resistance drift.

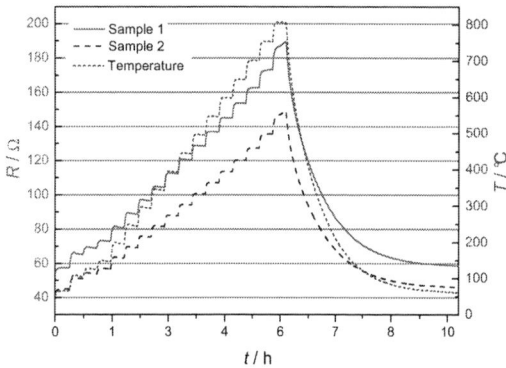

Figure 8. R(t) characteristics of surface resinate gold layer on Heralock tape.

The resistance variations were much worse compared to samples on alumina - up to 50%. This point to differences in the film thickness between structures that can be caused by improper printing parameters. Another reason can be the roughness of Heralock tape, which is higher than alumina. Due to the low thickness of the resinate film it is possible that the high roughness of substrates extent the surface of the metal layer. To prove this assumption, additional investigations are necessary.

The measured surface resistance of gold single layers was between 420 and 560 mΩ/□ (at room temperature). This value is much higher compared to test structures on alumina (230 mΩ/□). This phenomenon is well known and almost all "non-LTCC" conductive or resistive pastes exhibit on LTCC higher sheet resistances than the same elements on alumina [5].

Both samples exhibit almost the same TCR characteristics (Fig. 9). While the TCR is a material property, this proves that the difference in resistance is caused by different film thicknesses or microstructures of the layer. The HTCR value (2030 ppm/K) is about 20% lower compared to the structure on alumina (2580 ppm/K). This could indicate some kind of interactions on the interface resinate film - LTCC, although EDX showed no interactions. This should also be subject of further investigations.

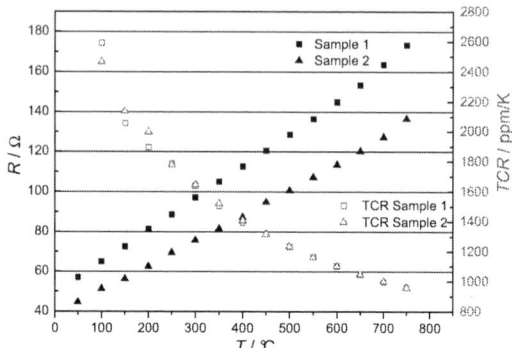

Figure 9. Temperature characteristics and TCR values of surface resinate gold structures on Heralock tape.

Patterning the gold resinate layers

As mentioned above, the minimal line width in the standard thick-film technology is 100 μm. Different photolithographic methods allow to obtain lines as narrow as 20...30 μm. Similar dimensions can be achieved by laser patterning. Laser direct-writing compared to photolithography provides advantages like flexibility and rapid prototyping, maintaining high accuracy and high resolution.

Our preliminary investigations showed that resinate pastes can be easily laser-structured in the unfired as well as in the fired state. The cut is clean and of very good quality, better compared to typical thick-film materials (Fig. 10).

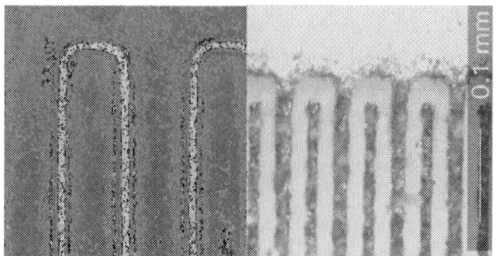

Figure 10. Quality of cut of unfired films; left: resinate gold paste, right: typical gold thick-film paste, cut width 20 μm

For patterning resinate films, a frequency-tripled (355 nm) Nd:YAG laser (Microline Cut 350ML, LPKF) was used. The decrease of the wavelength let to an increased absorption coefficient for many materials that are used in thick-film technology. Figure 11 presents conductive lines patterned in gold resinate layer. The minimal obtained line width was 20 µm. Here, a special programmable routine was applied to remove the metal film from the substrate. In contrast to other methods, it is possible to remove all unnecessary material without remarkable injury of the substrate.

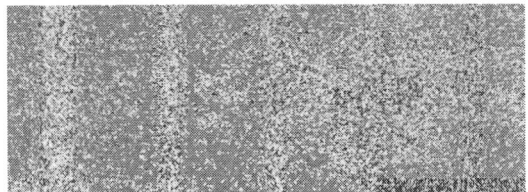

Fig. 11. Conductive lines patterned in gold resinate layer. Line thickness: 75, 50, 40 30 and 20 µm

Conclusions

Our tests showed that resinate pastes can be applied on zero-shrinkage LTCC tapes. The quality of the co-fired conductive layers is very good, without holes or cracks. The variations in sheet resistance were probably caused by film thickness difference. However, this assumption needs additional investigations.

The sheet resistance of structures on LTCC is higher than on alumina. The resistance characteristics $R(T)$ is measured up to 800°C. The resistance increases almost linear with temperature. The high TCR value and stability over the time are very promising. In our opinion, resinate pastes could be applied as a material for micro-heaters.

Laser direct-writing of resinate layers gave very promising results. A minimal line width of 20 µm was achieved. The quality of the patterned lines was very good.

Our pre-conditions and expectations regarding the application of resinate pastes on LTCC ceramics were met, although the experiment with typical LTCC tapes were as successful as the tests performed with zero-shrinkage tapes.

Some technological processes, particularly screen printing, should be optimized. The origin of resistance variations of structures should be further investigated.

In upcoming investigations, we will try to use resinate paste in practical applications, first of all, for designing a micro-heaters for new types of gas sensors with very low power consumption..

Acknowledgements

The authors wish to thank Mrs. Christina Modes from Thick-Film Circuit Division and Dr. Herbert Fuchs from Ceramic Colour Division, W.C. Heraeus, Germany for providing LTCC tapes and resinate pastes used in investigations. We will also wish to thank Mrs. Monika Wickless for sample preparations and SEM and EDX investigations.

References

[1] Gollner E., Kita J., Moos R.: Frequency-tripled Nd:YAG-laser in thick-film and LTTC applications, Proceedings of XXX IMAPS Poland Conference, Krakow, Poland, 24.-27.9.2006, p.147-15

[2] Reichl H., Hybridintegration, Springer Verlag, 1980

[3] Fuchs H., Metallo-Organic Preparations for Electronic Components, cfi/Ber. DKG 82 (2005) No. 4

[4] German Standard DIN 41850-2, Film and hybrid integrated circuits; materials, methods for the assessment of conductive pastes, September 1985

[5] Dziedzic A., Golonka L., Kita J., Kozłowski J. Macro- and microstructure of LTCC tapes and components, XXIV International IMAPS - Poland Conference, Rytro 2000

Fabrication of High Performance RF-MEMS Structures on Surface Planarised LTCC Substrates

Massimiliano Dispenza[1], Roberta Buttiglione[1], Anna Maria Fiorello[1], Jarkko Tuominen[2], Kari Kautio[2], Jyrki Ollila[2], Pentti Korhonen[2], Manu Lahdes[2], Kari Rönkä[2], Simone Catoni[3], Daniele Pochesci[3], Romolo Marcelli[3], Vittorio Foglietti[4], Elena Cianci[4], Andrea Coppa[4]

[1]Selex Sistemi Integrati, Rome, Italy

[2]VTT Technical Research Centre of Finland, Oulu, Finland

[3]CNR-IMM (Institute for Microelectronics and Microsystems), Italy

[4]CNR-IFN (Institute for Photonics and Nanotechnologies), Italy

Abstract

Fabrication of high performance thin-film RF-MEMS structures directly on surface treated LTCC substrates has been investigated and demonstrated. After extensive testing of different LTCC materials as well as the characterization of lapping and polishing conditions, Heraeus CT707 was chosen as substrate material. The fabrication process for high performance thin film patterning and RF-MEMS structures has been developed, optimised and demonstrated. The release process for RF-MEMS structures has been optimised taking also into account the compatibility with LTCC materials. Also, hermetic sealing for the MEMS structures using brazing both with Kovar and ceramic lids has been designed, optimised and demonstrated.

Keywords: Thin film, MEMS, LTCC, polishing, True Time Delay

Introduction

In electronically scanned antennas the steering of the beam can be defined by changing the amount of phase shift imposed between adjacent T/R modules. If large instantaneous bandwidth signals are used the phase shift would not be the same for all the frequency components of the spectrum, thus giving rise to the beam squint effect. This problem may be counter measured by imposing a net time delay between adjacent T/R modules which would be intrinsically independent of frequency. Long delays are required to be implemented for this scope.

In such context, it would be quite fruitful to develop a fabrication process able to integrate RF-MEMS switches directly onto LTCC multilayer substrate. This would, in fact, combine the well known advantages of MEMS switches (wide bandwidth, low losses, linearity ...) with the ease of building-up long lines in a compact form, thanks to multilayer LTCC (Low Temperature Cofired Ceramic) technology.

LTCC is a well established packaging platform competitive especially in harsh environments and very high frequency applications [1]. For RF-MEMS applications LTCC can offer a platform with low electrical loss and capability of hermetic sealing.

One of the most critical problems encountered in thin film fabrication process consisted on the high co-fired LTCC substrate roughness related to the design, the materials and the manufacturing process of LTCC substrate, such as, sintered dielectric homogenity, metallization distribution and firing setter plate waviness.

A substrate roughness comparable to the suspended membrane, sacrificial layer thickness and dimples depth could cause a negative impact on the correct mechanical behaviour of the device. Excess of roughness could also reduce the intimate contact in the dimples region with a lack of reproducibility in the resistive contact. Thus, it emerged the necessity to surface treat the LTCC substrate by lapping and polishing and/or thin-film planarization [2].

Design & Material Characterization

As a Single-Pole-Single-Thru (SPST) building block, a series-ohmic-bridge RF-MEMS switch configuration was selected. It is based on a metal membrane (the bridge) suspended, as a series switch, above an interrupted RF microstrip line.

In the "on" state (down position) the bridge is lowered by electrostatic forces applied through actuation pads connected to a pull-down electrode, placed beneath the bridge itself. The choice of series configuration is aimed to achieve high enough isolation in the "off" (up position) state.

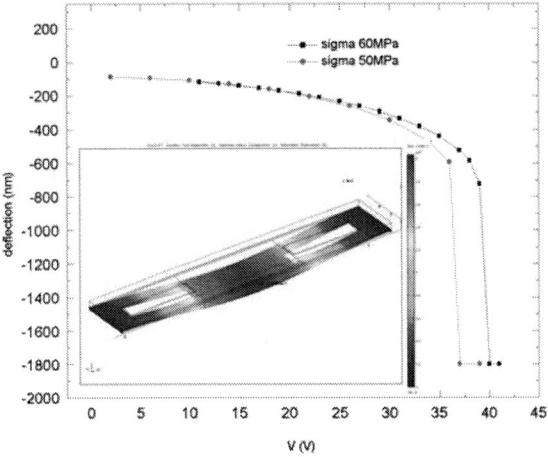

Fig. 1 - Deflection vs. applied voltage from the mechanical simulations with two different values of residual stress σ.

As a first issue, the mechanical behaviour was considered. A set of mechanical simulations (Fig. 1) by COMSOL multiphysics were performed in order to evaluate the impact of critical parameters, such as membrane thickness and gap height, onto the actuation voltage.

Performing such analysis with a realistic approach requires a clear characterisation of the material used to realise the moving membrane, at least in terms of residual stress σ and Young's modulus E. In order to identify the value of σ, test structures were realised which, once released, undergo deformations as a result of their internal residual stress. These passive strain sensors consist of double-clamped narrow beams, V-shaped with the V-tips towards each other, and supporting a Vernier scale [3].

In order to determine also the Young's modulus of the deposited films, membranes were also fabricated, whose deflection, as a function of the force applied by a stylus probe, was finally measured.

Several material options were characterized, such as electroplated gold, sputtered gold, Ti/Au 2-fold layer membrane. Simulations were carried out with such materials. Following the results of such analysis, sputtered gold was selected as a reasonable choice. Its tensile stress and Young's modulus were found to be 7 MPa and 60 GPa, respectively.

Following the inputs collected from the aforementioned simulations, a set of test bridge structures were designed (Fig. 2) and realised. The main features of such bridges are listed in the following.

The electrical contact between the bridge and the two ends of the signal line is established in few points, named dimples. The small size of such contact points addresses increase of metal-to-metal pressure and thus reduction of contact resistance.

The actuation lines provide the driver signal to the moving membrane. They should act as an RF block to the microwave signal travelling along the transmission line. These requirements are satisfied with the resistance value of about 100 kΩ for each line. To avoid any contact between the membrane and the actuation lines a dielectric passivation layer is used.

As mentioned above the aim of such a work is to develop RF-MEMS components to be incorporated in a TTD (True Time Delay) module for wideband antennas. Thus the RF design addressed optimisation of EM behaviour up to 35GHz. RF simulations show acceptable down-state losses within the band of interest with isolation in up-state better than 20dB (see Fig, 3).

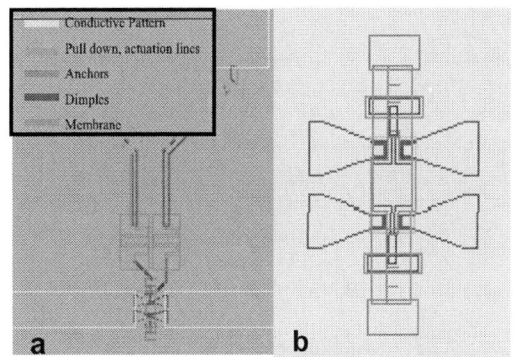

Fig. 2 - CAD lay-out of the switch structure: a) complete view including RF and actuation lines, and b) blow-up view

Accurate electromagnetic simulation require precise knowledge of the electrical properties of the LTCC substrate and metallisation. Since such data at microwave and millimeter wave frequencies is not readily available, surface (microstrip) ring resonator structures on Heraeus CT707 were designed, manufactured and measured. The resonators were measured with on-wafer S-parameters system. From the resonance frequencies and widths of the resonance peaks, the effective relative dielectric constant ($\varepsilon_{r,eff}$) and the atteanuation constant (α) can be extracted at the resonance frequencies.

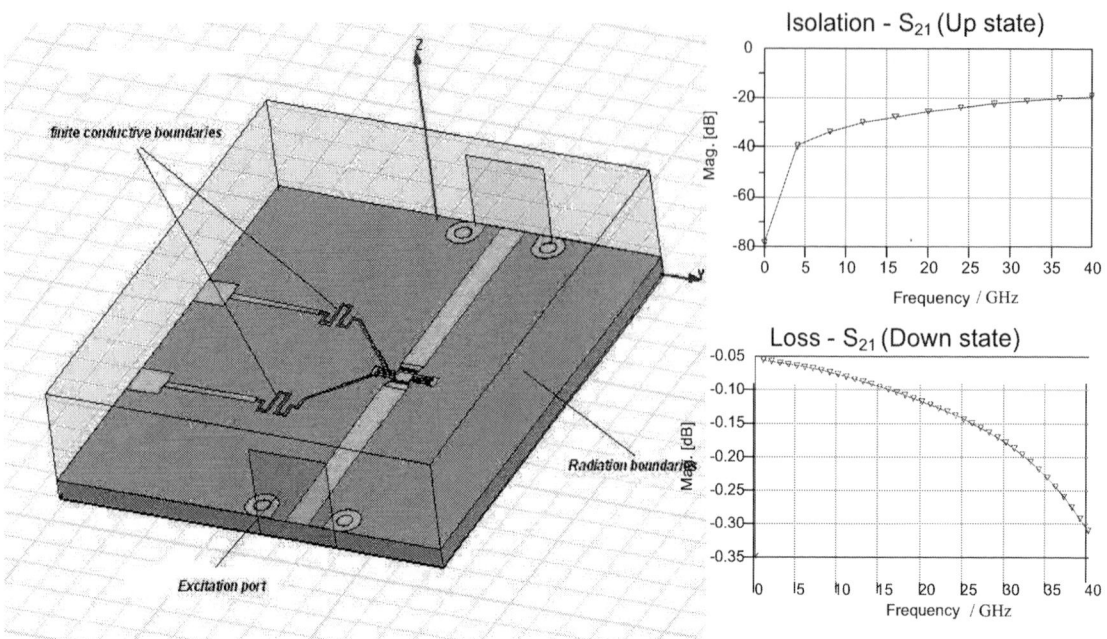

Fig. 3 - RF response of the switches as expected from the mechanical simulation

The extraction of the $\varepsilon_{r,eff}$ is rather straightforward requiring the exact knowledge of the ring radius and the resonance frequency. The surface microstrip resonators give the $\varepsilon_{r,eff}$ and the ε_r must be extracted from this. This was done with a commercial transmission line calculator by using the transmission line geometry and the measured $\varepsilon_{r,eff}$. The attenuation constant was calculated by observing the 3 dB width of the resonance peaks resulting the loaded Q_l value, which was then corrected with the coupling coefficient to get the unloaded Q_0.

The attenuation constant can be calculated from Q_0 and the resonance frequency. This is, however, not straightforward since the measurement method cannot distinguish different loss mechanisms. For that purpose the commercial transmission line calculator was again used. It is also important to know the effect of the relatively high roughness of LTCC printed conductors, and therefore, a 0.8 μm RMS roughness value was used in the calculations. After subtracting the conductor losses, the dielectric loss tangent, tanδ can be calculated from the dielectric losses. The test was conducted with un-lapped and un-polished LTCC substrates.

The test structures were measured with a vector network analyzer and probe station. The measurements were conducted over a frequency range of 4 - 100 GHz. The calibration was done with TRL-set manufactured on the LTCC, to minimize the effects of the feed lines and to move the reference plane as close to the rings as possible. Fig. 4 shows the S21 of the ring (radius 1220 μm), the

resonance peaks occur at 17.36 GHz, 34.80 GHz, 51.83 GHz, 68.55 GHz, 85.05 GHz. Table I shows the calculated $\varepsilon_{r,eff}$ and ε_r values and attenuation constant (microstrip line w=126 μm, h=102 μm, t=5 μm).

Table I. Overview of the calculated material parameters

Freq (GHz)	N	$\varepsilon_{r,eff}$	ε_r	α_{tot} (dB/m)
17.36	1	5.07	7.21	66.0
34.8	2	5.03	7.19	114.4
51.83	3	5.13	7.23	175.4
68.55	4	5.18	7.13	228.3
85.05	5	5.29	7.16	272.7

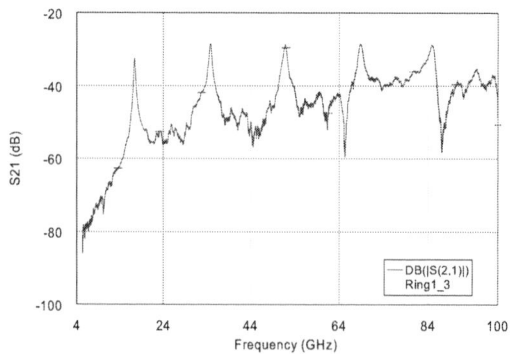

Fig. 4 - S21 response of a microstrip ring resonator with radius of 1220 μm.

Fabrication & Experimental Results

The extensive work for finding out the most suitable substrate material for TTD RF-MEMS structures has been carried out earlier, and the experimental methods are explained in more detail in [2]. For the chosen material system, Heraeus CT707, the following sintering parameters were used: SiC setter plate, sintering temperature of 850 °C and lamination pressure of 3000 psi.

The fabrication of delay line test circuit using Heraeus CT707 tape system was straight forward. Via holes were punched to the tape sheets (6 layers) and filled using stencil printing. Only the buried conductor layers were printed. The surface layer had only via connection to the inner layers. Substrate size was 55mmx55mm and thickness 630 µm.

The substrates flatness was good and no particular problems were seen with the metallisation system (both Au and Ag were used), either. The co-firing shrinkage tolerance of the fabricated test substrates was typically +/- 0.03% in x and y. This is very low value using a free shrinkage LTCC process. The choice of metallisation system, either silver or gold, has some effect on the co-firing shrinkage but not on the dimensional tolerances.

As to lapping conditions, the following parameters were used: 3 µm particle size Al_2O_3 abrasive slurry, 70 rpm plate speed, 2.8 kg load, 45 min. processing time; and as to polishing conditions, the following parameters were used: colloidal silica polishing fluid with Chemcloth pad, 70 rpm plate speed, 2.8 kg load, 1h 20 min. processing time.

The surface quality of the processed substrates was characterized by using an optical profiler (Veeco NT3300) and by taking FESEM micrographs (JEOL JSM-6300F). Demanding acceptance criteria for the surface defects were set; defects should not be deeper than 300 nm and not bigger than 1 µm² in size. If both of the criteria were not fulfilled, the number of defects per mm² should not be higher than 10.

The waviness of the as-fired Heraeus CT707 substrates sintered on alumina setter plate was typically ± 5…10 µm. This was reduced to ± 2...4 µm on the LTCC surface, facing the flat SiC setter plate, whereas the waviness of the LTCC surface facing the sintering oven atmosphere was not reduced. In addition, a homeycomb structured cordierite setter was tested to improve the organics burnout and to reduce the porosity of the dielectric. However, the support was not adequate causing pattern build-up and substantial waviness to the substrate. Thus, the SiC setter plate was seen more favourable [2].

After lapping and polishing, the surface roughness values of Rq and Rt of CT707 were reduced down to 14 nm and 200 nm, respectively, thus, fulfilling the set requirements for the thin film process The value Rq is the RMS roughness calculated over the entire measured array, and the value Rt is the peak-to-valley difference calculated over the entire measured array. The surface texture of both the as-fired and the polished CT707 substrate is displayed in Fig. 5.

The variations of the LTCC manufacturing process, such as lamination pressure and sintering peak temperature and profile, were not found to have any major impact on the quality of the lapped and polished surfaces.

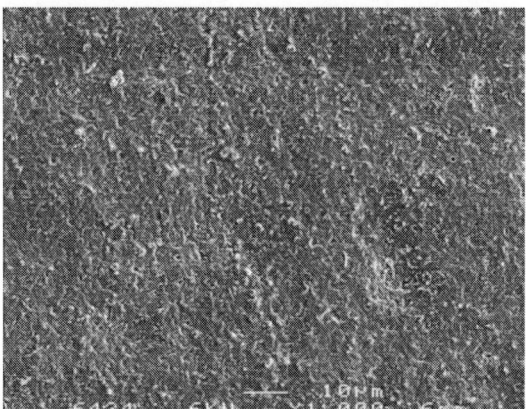

Fig. 5 - FESEM micrographs of (a) as-fired and (b) lapped and polished CT707 substrates.

Regarding the metallised vias, the results after the polishing step revealed that the vias were posted by 1 to 5 µm. One possible explanation to this phenomenon could be that, during the polishing step, the surface layer of LTCC material is removed with different rate than via fill material (gold or silver).

This result might present problems during high-precision photolithographic post-processing, especially if contact exposure is used. Etching of the posted vias could serve as one solution to this problem. The preliminary tests showed that etching with potassium iodide (KI) for 3 minutes was enough to planarise the vias. There was also a short polishing step needed (10 min.) to finalize the surface quality.

A seven masks thin film fabrication process flow has been developed (see Fig. 6).

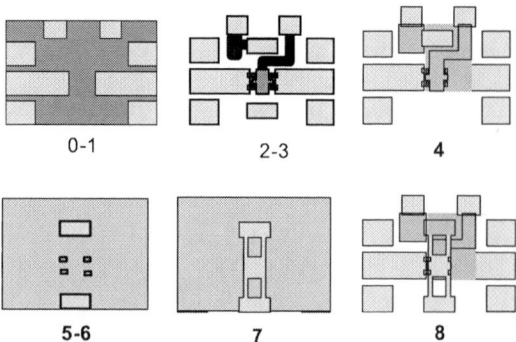

Fig. 6 - Process flow: main steps.

As a preliminary step a purposely defined treatment is applied to the substrate, in order to make it inert with respect to some of the next process steps, which would be chemically aggressive for LTCC compounds.

Then TaN/NiCr/Pt/Au thin film deposition starts the fabrication process. The first photolithography step is performed in order to define the conductive pattern, this is realised by a sequence of gold electroplating and partial thin film ion etching (1). After that a second photolithography step followed by an ion etching process define the pull down electrode and the DC pattern (2), while a further ion etching defines the high resistive TaN actuation lines (3). Two steps (deposition, photolithography) are then performed to realise the Ta_2O_5 passivation film placed upon the pull-down electrode and the actuation lines (4). A 1.8 μm thick sacrificial layer is deposited and anchors and dimples are defined by reactive ion etching (5-6). The sacrificial layer acts as a spacer to fix the right distance between membrane and pull-down electrode.

Afterwards a sequence of 2 μm thick gold deposition, photolithography (7) and ion etching defines the membrane structure. Finally, the sacrificial layer release allows completing the suspended and moving air bridge (8). The whole process flow summarised above includes a total of about 25 single processing steps.

The series switches (Fig. 7) fabricated on polished CT707 substrates have been electrically characterised in vacuum environment, and the best measured actuation (deactuation) voltage turned out to be 30V (25V). Few hundreds of switching voltage cycles were applied and a repeated up/down response was observed. Further life-cycle tests are expected to be performed, in order to evaluate reliability and lifetime of fabricated components.

Fig. 8 - Kovar lids on LTCC substrate.

Brazing with Au80Sn20 solder was chosen as the preferable sealing method. The sealing process on the polished LTCC surface was demonstrated both with metallic Kovar lids as well as with LTCC lids (outside dimension: 20.3mm). The peak temperature of the reflow profile and the melting time were 340 ºC and 60 seconds, respectively. Reflow was done under nitrogen atmosphere. In order to achieve successful sealing the maximum curvature of the LTCC substrate needs to be less than 30 – 40 μm in the lid length, since the solder preform thickness is 50 μm before sealing. In some of the samples leaks in the lid solder joints were noticed during sealing tests. The substrate bows were measured, which were around 60 – 200 μm for Kovar lid samples and around 10 – 60 μm for LTCC lid samples (sample dimensions 50 mm x 50 mm). The possible solutions to this problem include selecting a lid material with the same CTE as

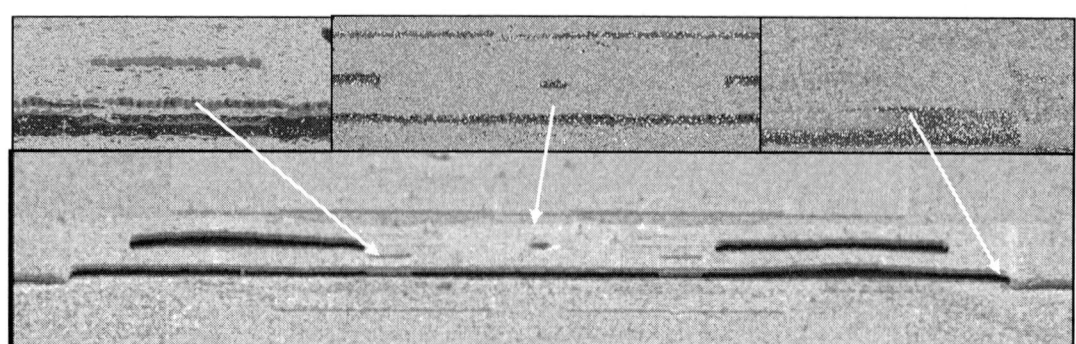

Fig. 7– SEM photograph of fabricated RF MEMS switches

CT707, and optimizing lapping and polishing parameters and times.

Conclusions

LTCC material system selection is the most important factor for the most demanding photolithographic thin-film post-processing applications and requirements. With Heraeus CT707 material system the eligible demanding surface roughness requirements were fullfilled.

It was shown that the surface treatment methods of LTCC substrates can greatly improve the quality of the substrate surface. The achievable surface roughness is most likely determined by the material properties, such as grain size and material system chemistry.

The lapping and polishing process needs to be optimised in more detail, and compromises may be necessary between the high frequency requirements, the achievable surface roughness and processing time. Attention has to be paid on the substrate bow, which needs to be reduced before reliable sealing can be done.

RF MEMS switches based on a series-ohmic-bridge configuration were successfully fabricated upon a purposely prepared LTCC substrate. Repeated switching behaviour of the switches was demonstrated with an actuation voltage of 30 V.

References

[1] Jaakko Lenkkeri et al., "Prospects and limits of LTCC technology", Advancing Microelectronics, Vol. 32, No. 3, pp. 9 – 12, 2005.

[2] Jarkko Tuominen et al., "Surface planarisation of LTCC for high performance thin-film applications", The IMAPS Nordic Annual Conference, Göteborg, Sweden, September 17 - 19, IMAPS Nordic, pp. 166 – 170, 2006.

[3] Y. Gianchandani and K. Najafi, "Bent-beam strain sensors", Journal of Microelectromechanical Systems, Vol. 5, No. 1, pp. 52 – 58, 1996

LTCC ultra high isostatic pressure sensors

Th. Maeder, B. Afra, Y. Fournier, N. Johner and P. Ryser

Laboratoire de Production Microtechnique, Ecole Polytechnique Fédérale de Lausanne (EPFL)
Station 17, CH-1015 Lausanne, Switzerland

Phone: +41 21 693 38 17; fax: +41 21 693 38 91; e-mail: thomas.maeder@epfl.ch

Abstract

Piezoresistive pressure sensors using classical deformable structures such as membranes progressively run into difficulties at very high pressures (above 1 to 2 kbar), due to the decreasing ratio between the available sensing strain on the outer membrane surface and the maximal materials strain experienced on the pressurised side. Therefore, a novel "isostatic" pressure sensor concept, whereby the sensing element is immersed in the pressure fluid, has been developed. This method in principle allows the measurement of very high pressures, because materials stresses within the sensing beam are all compressive. LTCC (Low-Temperature Cofired Ceramic) beams with hermetically embedded thick-film piezoresistor bridges have been fabricated, packaged and characterised as sensor elements. The observed sensitivity is comparable to the response expected from the LTCC and piezoresistor materials properties.

Key words: piezoresistivity, thick-film, LTCC, high pressure sensor.

Introduction

Initially designed for high-frequency, high reliability electronics and multi-chip modules [1-4], low-temperature cofired ceramic (LTCC) has recently drawn considerable attention in the field of sensors, actuators and fluidics [5-9], due mainly to its ease of 3D structuration, combined with excellent chemical and thermal stability, hermeticity, and moderate cost.

Piezoresistive sensor technology using thick-film resistors can be applied to LTCC mechanical sensors, as reasonable gauge factors [5] can be obtained, and the long-term design strain (e.g. allowable mechanical "signal") of LTCC is roughly equivalent to that of commonly-used alumina substrates [10]. These results were applied successfully to a low-range force sensor [11].

For classical piezoresistive sensors operating in substrate bending mode (cantilevers, membranes, etc.), the properties of LTCC (low strength, low elastic modulus, very low available sheet thicknesses and ease of making fine structures) make it better suited to low-range sensors. However, for measuring very high pressures, one can also make use of the direct response of a resistor lying on a substrate (figure 1) to ambient hydrostatic pressure. This work demonstrates the application of LTCC to a very high-pressure "isostatic" sensor.

Piezoresistive "isostatic" sensing model

In general, the relative variation under strain of a resistor of an isotropic material, with current flowing along the x axis (figure 1), can be expressed by relation (1), incorporating the variation of the resistivity (2) and of the geometric dimensions (the mechanical strains) [12-14]:

$$(1) \quad \frac{dR}{R} = \frac{d\rho}{\rho} + \varepsilon_x - \varepsilon_y - \varepsilon_z$$

$$(2) \quad \frac{d\rho}{\rho} = \Gamma_L \cdot \varepsilon_x + \Gamma_T \cdot \left(\varepsilon_y + \varepsilon_z \right)$$

Here, R is the resistor value, ρ the resistivity and ε_x, ε_y, and ε_z the mechanical strains along the longitudinal axis x and the in-plane transverse axis y and the out-of-plane transverse axis z. The piezoresistive response of thick-film resistors is dominated by the rather large piezoresistive coefficients arising through interparticle tunnelling effects [12][13], denoted Γ_L (longitudinal) and Γ_T (transverse). Upon application of an isostatic pressure P on the substrate-resistor combination, we get, combining (1) & (2):

$$(3) \quad \frac{1}{R} \cdot \frac{dR}{dP} = (\Gamma_L + 1) \cdot \frac{d\varepsilon_x}{dP}$$

$$+ (\Gamma_T - 1) \cdot \left(\frac{d\varepsilon_y}{dP} + \frac{d\varepsilon_z}{dP} \right)$$

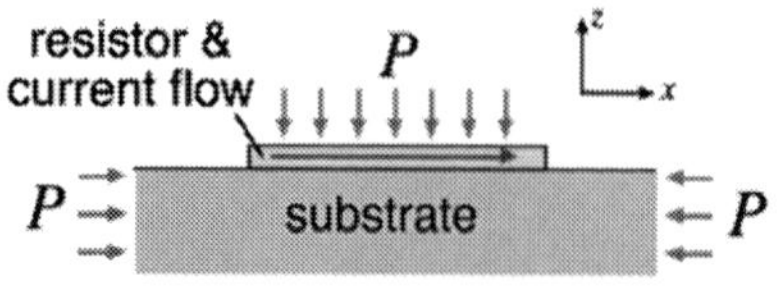

Figure 1: Resistor on a substrate, subjected to ambient isostatic pressure P

The strain response will not be isotropic in the general case, because the substrate (elastic modulus E_S and Poisson's coefficient v_S) dominates the mechanical properties in the plane (lateral clamping of the resistor), and its mechanical properties are in general different from that of the resistor (E_R and v_R). Neglecting the influence in the substrate plane of the resistor, we calculate the in-plane strains:

$$(4) \quad \varepsilon_x = \varepsilon_y = -\frac{1-2v_S}{E_S} \cdot P$$

The out-of-plane resistor strain ε_z can be obtained by superposing the direct effect of vertical pressure and that of the in-plane strains. It is best expressed in the same form as (4), with a corrective term w due to the difference in mechanical properties between resistor and substrate:

$$(5) \quad \varepsilon_z = -\frac{1-2v_S}{E_S} \cdot P \cdot (1+w)$$

$$(6) \quad w = \frac{1+v_R}{1-v_R} \cdot \left(\frac{E_S}{E_R} \cdot \frac{1-2v_R}{1-2v_S} - 1 \right)$$

These relations are valid when the substrate is much stiffer than the resistor, which is true for most cases, and are somewhat at difference with more simplified ones in previous work [15]. Combining (3 - 5), we get the piezoresistive response under isostatic pressure of a resistor with in-plane current:

$$(7) \quad \frac{1}{R} \cdot \frac{dR}{dP} = -\frac{1-2v_S}{E_S}$$
$$\cdot \left[\Gamma_L + \Gamma_T + (\Gamma_T - 1) \cdot (1+w) \right]$$

The piezoresistive coefficients Γ_L & Γ_T are related to the more often used gauge factors of bending cantilevers K_L (longitudinal) and K_T (transverse) as follows, using (1), (2) and [13]:

$$(8) \quad K_L = 1 + \Gamma_L - (\Gamma_T - 1) \cdot \frac{v_R}{1-v_R}$$

$$(9) \quad K_T = (\Gamma_T - 1) \cdot \frac{1-2v_R}{1-v_R}$$

Note that the plane strain approximation is used here (lateral strain = 0), which matches most cases involving thick-film substrates (where width >> thickness), in contrast with other work positing plane stress (lateral stress = 0) [14][15].

Previous work on isostatic pressure loading

Compared to substrate bending experiments involving cantilevers, membranes, etc., there is relatively little experimental data on isostatic loading of thick-film resistors on ceramic or LTCC substrates. Combined substrate bending, hydrostatic pressure and temperature studies were carried out by Fawcett & Hill [15] on alumina (with some issues with the analytical model). A large volume of data, on alumina and LTCC, was published by Dziedzic et al. for both cermet thick-film resistors [16] and polymeric ones [17], without detailed theoretical analysis.

These experiments nevertheless show very interesting results: a resistor response of 0.3%, which is typical for cermet thick-film resistors, is already obtained at pressures in the 100 bar (10 MPa) range, whereas the corresponding design stresses for alumina-based thick-film sensors lie in the 100 MPa range (in the alumina). Therefore, direct immersing resistors is a very promising principle for high-pressure sensors.

Immersed-resistor pressure sensor concept

The pressure sensor concept is depicted in figure 2 (cross section). A metal body compatible with any type of mounting (screwing, welding, etc.) is fabricated, and two cavities (pressure and reference) are drilled at each end, linked by a hole just large enough to accommodate the sensing element. This element, best fabricated as a stick-shaped LTCC module with buried resistors in a Wheatstone bridge configuration, is then inserted into the hole, and bonded through a strong adhesive such as epoxy. To carry out this operation, the glue is best injected through a narrow lateral hole (not shown) in the centre of the body.

This design has several advantages. First, the sensing materials are nominally in compression – pressure capacity is mainly limited by the adhesive feedthrough. Moreover, direct vertical loading of the resistor gives a high sensitivity, due to it having a much lower elastic modulus (60...85 GPa [18][19]) than the substrate (110 GPa for LTCC [18] and 300...340 GPa for Al$_2$O$_3$ [18][19]). This is also seen in force sensors where piezoresistors are directly loaded in compression [20].

LTCC has two advantages over alumina. Using a buried-resistor configuration reliably protects the bridge against aggressive media, and its much smaller elastic modulus gives a larger output signal.

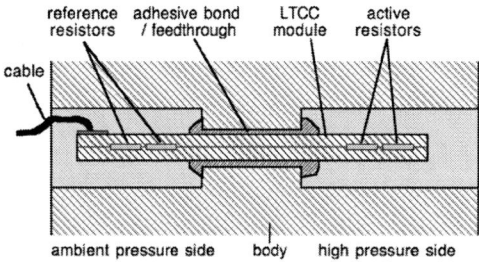

Figure 2: Immersed-resistor pressure sensor

LTCC modules immersed in high pressures have been used successfully for applications such as a flow sensor for high-pressure automotive engine fuel injection [21].

Pressure sensing LTCC module

A LTCC pressure sensing module was designed and fabricated according to the abovementioned principle, with length, width and height ca. 58, 2.2 and 0.85 mm respectively. A four-layer configuration was used : 1) bottom lid ; 2) bottom sensing half-bridge ; 3) top sensing half-bridge ; 4) top lid. Thus, the sensing resistors are fully buried in the LTCC, and the resulting "stick" is very narrow. This is important, as a small feedthrough hole is more reliable and allows operation at higher pressures.

The layout of the resistive half-bridge is shown in full in figure 3, and the full sensing circuit is schematised in figure 4. Here, "+" & "-" denote the positive and negative supplies, and "s+" & "s-" denote the positive and negative bridge outputs.

The materials used for the fabrication of the LTCC module were DuPont 951 LTCC, 6146 AgPd conductor and 2041 10 kΩ resistor. After screen-printing and drying, the LTCC sheets were stacked and laminated at 21 MPa and 72°C for 10 min, the recommended conditions. Finally, the modules were fired at a peak temperature of 850°C, yielding the LTCC sensing "sticks" depicted in figure 5.

Figure 3: Layout of one resistive half-bridge

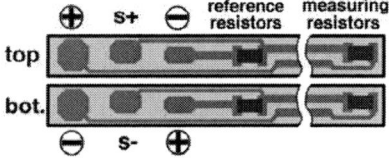

Figure 4: Complete sensing circuit.

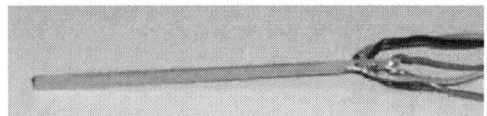

Figure 5: Fired LTCC sensing module

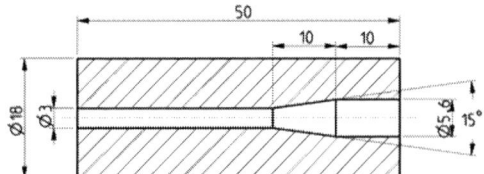

Figure 6: Supporting structure for tests

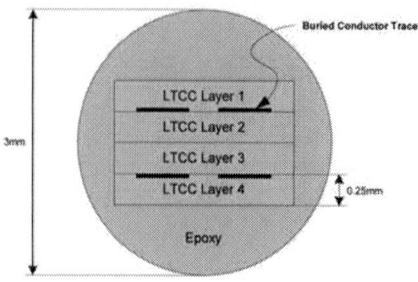

Figure 7: Cross-section of feedthrough

Assembly of complete sensor

For our tests, we used a simplified design compared to that of figure 2. Standardised simple bodies (cross-section in figure 6), which can be fitted with a jig in our pressure testing apparatus, were machined out of austenitic stainless steel (1.4305 - X10CrNiS18-9) and cleaned. The LTCC modules (figure 5) were then inserted in the bodies, and epoxy (Stycast) was used to fill the feed-through, and hardened ca. 24 h at room temperature. A transverse cross-section of the LTCC module in the feedthrough is depicted in figure 7.

Results – resistive bridge characterisation

Statistics on the resistor values and the bridge offset are given in Table 1. In order to assess the effect of burying ("covered" samples) DuPont 2041, which is a standard thick-film resistor not developed for LTCC, some sensors ("open") were manufactured with openings cut in the top and bottom lids, giving co-fired (but not buried) resistors.

Compared to the nominal sheet resistance of 10 kΩ, simply co-firing strongly reduces the value, whereas burying increases it. Nevertheless, both the values and the bridge offsets (given by value matching of the resistors) remain acceptable.

Results – pressure response of the sensors

The pressure response was measured by monitoring one resistor or the bridge output voltage (excitation voltage = 5.00 V) as a function of the pressure, which was generated by a DH-Budenberg 540 VHX manual hydraulic press, with the samples placed in oil.

Table 1. Statistics of resistor values

Type	Covered	Open
Resistor value		
Average [kΩ]	24.1	3.55
Standard deviation [%]	5.9	6.0
Bridge offset		
Average [mV/V]	7	78
Standard deviation [mV/V]	21	33

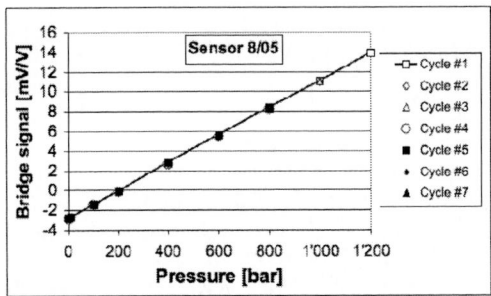

Figure 7: Sensor response under pressure

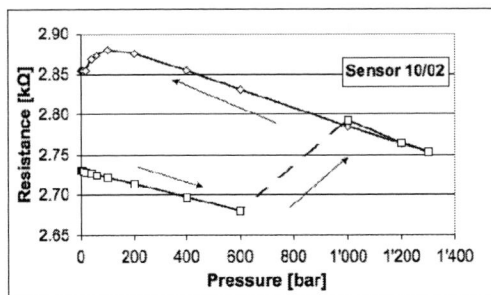

Figure 8: Defective resistor under pressure

Figure 7 shows the response of a "good" sensor with buried resistors, over 7 pressure cycles. Drift of the unloaded bridge offset after 7 cycles is below 0.1% of the sensor response.

The response over 1'000 bar is ca. 14 mV/V, which corresponds to a shift of the measuring resistors of -2.8%, because only half of the bridge is active. This variation agrees with direct measurements of the resistor values, and with the response of a second nominally identical sensor.

Sensors with "open", uncovered resistors gave slightly lower responses at 1'000 bar, corresponding to a resistor shift of -2.6%. This difference is likely due to a slight change of the piezoresistive properties through chemical resistor-substrate interactions during firing, especially with resistors buried in LTCC.

Our results are quite comparable to that of Dziedzic et al. [16] on alumina for the same resistive composition (ca. 1.8%): a higher response is expected on LTCC due to the lower substrate elastic modulus.

While these sensors seem very promising, some failures were also encountered, such as shown in figure 8. While no detailed failure analysis was carried out, the collapse of a void inside the ceramic sensing element or inside the glued feedthrough is the likely cause, as such voids were sometimes observed during manufacturing of the samples.

Discussions

It is interesting to compare our results and those in the literature with the model given at the start of this work. This is attempted in table 2.

Table 2. Calculated pressure responses

Substrate	LTCC	Al$_2$O$_3$
E_S [GPa]	110	315
E_R [GPa]	70	70
ν_S	0.17	0.23
ν_R	0.25	0.25
w	0.32	5.28
K_L	11.6	11.6
K_T	8.1	8.1
Γ_L	15	15
Γ_T	13	13
$\Delta R/R$ @ 1 kbar	-2.6%	-1.8%

The values for the data were taken from various sources. Mechanical properties of LTCC were taken from Santo-Zarnik et al. [18], and that of alumina from Kyocera [22]. For the resistor, a compromise between two sources [18][19] was taken. Finally, the gauge factors of the resistors were taken from our previous studies on alumina [23].

The resulting calculated responses for DuPont 2041 at 1'000 bar agree very well with our results (-2.8% or -2.6%) on LTCC and that of Dziedzic et al. (-1.8%) on alumina. This indicates that the alteration of the gauge factor on LTCC is not very large.

In principle, this sensing method is very powerful, as sensor breakage is much less likely to occur in compression than in bending, where tensile stresses are experienced. In practice, however, the presence of voids that can implode at very high pressures and the feedthrough will limit the maximum pressure. In the latter case, several techniques can be used to improve the pressure capacity of the feedthrough [24][25]. This may often necessitate the break-up of the sensing element in two parts linked by small wires, which is much less convenient to assemble. Nevertheless, this can be interesting for measurements that require good sensitivity at very high pressures, as standard high-pressure piezoresistive foil gauge-type sensors that measure strain on the outer wall of a thick-walled tube have a rather small response of only 1 mV/V [26].

Conclusions

A novel high-pressure immersed-resistor pressure sensor concept was introduced, modelled and tested in this work. Compared to sensors using bending or stretching of walls, much higher output signals are achieved at high pressures. Moreover, the compressive loading of the sensing element theoretically allows measurement up to very high pressures. These sensors are also simple to fabricate and assemble.

However, the pressure the sensors actually withstand can be limited in practice by defects such as bubbles and voids inside the materials, which can collapse under pressure and lead to destruction of

373

the device. Particular care must therefore be given to the manufacturing process.

Acknowledgements

The help of G. Corradini, I. Saglini, M. Garcin and T. Haller in fabricating the samples is gratefully acknowledged.

References

[1] S. Nishigaki, J. Fukuta, "Low-temperature, cofireable, multilayered ceramics bearing pure-Ag conductors and their sintering behavior", Ceramic Substrates and Packages for Electronic Applications (Advances in Ceramics vol 26) Vol. 26, pp. 199-215, 1989.

[2] C.J. Sabo, W.A. Vitriol, C.L. Slaton, D.L. Rychlick, "The use of low-temperature, cofired ceramic technology for the fabrication of high-density, hermetic, multicavity modules", Ceramic Substrates and Packages for Electronic Applications (Advances in Ceramics vol 26) Vol. 26, pp. 217-228, 1989.

[3] R. Kulke, W. Simon, C. Gunner, G. Mollenbeck, D. Köther, Rittweger, "Vergleichstest verlustarmer LTCC-Systeme bis 40 GHz", PLUS - Produktion von Leiterplatten und Systemen, No 8, pp. 1-7, 2003.

[4] S. Nishigaki, U. Goebel, W. Roethlingshoefer, "LTCC (LFC) material systems and its application in automotive ECU's", Proceedings, 2004 IMAPS Conference, Ceramic Interconnect Technology, Denver, USA, pp. WP32, 2004.

[5] D. Belavič, M. Hrovat, A. Benčan, J. Golonka, A. Dzedzic, J. Kita, "Thick-film resistors on LTCC for possible sensor applications", MIXDES 2002 : proceedings of the 9th International Conference Mixed Design of Intergrated Circuits and Systems, Wroclaw, PL, pp. 521-526, 2002.

[6] J.J. Santiago-Avilès, M. Gongora-Rubio, P. Espinoza-Vallejos, L. Sola-Laguna, "Sensors, actuators and other non-packaging applications of LTCC tapes", Proceedings, 2004 IMAPS Conference, Ceramic Interconnect Technology, Denver, USA, INV12, 2004.

[7] T. Thelemann, H. Thust, M. Hintz, "Using LTCC for microsystems", Proceedings, European Microelectronics Packaging and Interconnection Symposium, IMAPS, Cracow, Poland, pp. 187-191, 2002.

[8] J. Youngsman, A. Moll, D.G. Plumlee, M. Schimpf, "Mini- and micro-channel devices in LTCC", Proceedings, 1st International Conference on Ceramic Interconnect and Ceramic Microsystems Technologies (CICMT), Baltimore, USA, WA33, 2005.

[9] K.A. Peterson, K.D. Patel, C.K. Ho, S.B. Rohde, C.D. Nordquist, C.A. Walker, B.D. Wroblewski, M. Okandan, "Novel microsystem applications with new techniques in low-temperature co-fired ceramics",

International Journal of Applied Ceramic Technology, Vol. 2, No. 5, pp. 345-363, 2005.

[10] H. Birol, M. Boers, T. Maeder, G. Corradini, P. Ryser, "Design and processing of low-range piezoresistive LTCC force sensors", Proceedings, XXIX International Conference of IMAPS Poland, Koszalin, pp. 385-388, 2005.

[11] H. Birol, T. Maeder, I. Nadzeyka, M. Boers, P. Ryser, "Fabrication of a millinewton force sensor using low temperature co-fired ceramic (LTCC) technology", Sensors and Actuators A, Vol. 134, pp. 334-338, 2007.

[12] C. Grimaldi, P. Ryser, S. Straessler, "Longitudinal and transverse piezoresistive response of granular metals", Physical Review B , Vol. 64, pp. 064201, 2001.

[13] S. Vionnet-Menot, C. Grimaldi, T. Maeder, P. Ryser, S. Strässler, "Tunneling-percolation origin of nonuniversality: Theory and experiments", Physical Review B , Vol. 71, pp. 064201, 2005.

[14] Santo-Zarnik-M Belavič-D Wymyslowski-A Friedel-KP, "A numerical analysis of the piezoresistive properties of thick-film resistors in pressure-sensor applications", Proceedings, European Microelectronics and Packaging Symposium, Prague, pp. 601-606, 2004.

[15] N. Fawcett, M. Hill, "The electrical response of thick-film resistors to hydrostatic pressure and uniaxial stress between 77 and 535 K", Sensors and Actuators A , Vol. 78, pp. 114–119, 1999.

[16] A. Dziedzic, R. Poprawski, A. Kolarz, "Thick-film and LTCC resistors under high hydrostatic pressure", Proceedings, XXVII International Conference of IMAPS Poland Chapter, Podlesice, Poland , Vol. 27, pp. 140-142, 2003.

[17] A. Dziedzic, A. Magiera, R. Winsiewski, "Hydrostatic high pressure studies of polymer thick-film resistors", Microelectronics Reliability , Vol. 38, pp. 1893-1898, 1998.

[18] M. Santo-Zarnik, D. Belavič, S. Macek, J. Kita, "Modelling of a piezoresistive ceramic pressure sensor", Proceedings, MIDEM 2003, Ptuj, SI, 2003.

[19] N.M. White, "A study of the piezoresistive effect in thick-film resistors and its application to load transduction", thesis, University of Southampton, Faculty of Engineering & Applied Science, 1988.

[20] B. Puers, W. Sansen, S. Paszczynski, "Assessment of thick-film fabrication methods for force (pressure) sensors", Sensors and Actuators, Vol. 12, pp. 57-76, 1987.

[21] U. Schmid, "A robust flow sensor for high pressure automotive applications", Sensors and Actuators A93-98, pp. 253-263, 2002.

[22] Kyocera, A-476 96% alumina datasheet.

[23] S. Vionnet, T. Maeder, P. Ryser, "Firing, quenching and annealing studies on thick-film

resistors", Journal of the European Ceramic Society, Vol. 24, No. 6, pp. 1889-1892, 2004.

[24] J. Brielles, D. Vidal, P. Malbrunot, "Passage de courant pour mesures précises de température sous hautes pressions avec des couples thermoélectriques", Journal of Physics E: Scientific Instruments, Vol. 6, No. 7, pp. 609-610, 1973.

[25] M.J. McNamee, W.R. Wawersik, M.E. Shields, D.J. Holcomb, "A compact coaxial electrical feedthrough for use up to 400 MPa", Review of Scientific Instruments, Vol. 62, No. 6, pp. 1662-1663, 1991.

[26] A. Schäfer, W. Viel, C. Rapp-Hickler, K. Mkulecki, "A new type of transducer for accurate and dynamic pressure measurement up to 15000 bar using foil type strain gauges", Proceedings, XVII IMEKO World Congress, Dubrovnik, HR, 2003.

LTCC-Based Sensors for Mechanical Quantities

U. Partsch*, S. Gebhardt*, D. Arndt **, H. Georgi**,
H. Neubert***, D. Fleischer***, M. Gruchow***

* Fraunhofer Institute Ceramic Technologies and Systems, Winterbergstraße 28, 01277 Dresden, Germany
** ADZ Nagano Gesellschaft für Sensortechnik mbH, Bergener Ring 43, 01458 Ottendorf-Okrilla, Germany
*** Dresden University of Technology, Institute of Electromechanical and Electronic Design, 01062 Dresden, Germany

Phone: +49-351-2553 696, Fax: +49-351-2554 161, mail: uwe.partsch@ikts.fraunhofer.de

Abstract

LTCC (Low Temperature Co fired Ceramic) is a ceramic multilayer technology for the manufacturing of highly integrated, reliable, high frequency suited, high temperature stable 3-D electronic packages. The LTCC-technology is commonly used for packages for mobile communication, automotive, airborne/ space or medical applications. Because of its multilayer structure LTCC is also an suitable technology for the fabrication of 3-D components of ceramic based microsystems such as sensors for mechanical quantities. The paper describes the development of sensors for the detection of pressure and acceleration. The required materials for the screen printing of the sensor layers were selected (piezoresistive pastes) or developed (piezoelectric paste). Before manufacturing different simulations (FEA) were done to calculate and optimize sensor characteristics. A manufacturing technology for the sensors was developed, the sensor behaviour was measured.

Key words: LTCC, Low Temperature Co fired Ceramic, acceleration sensor, pressure sensor, Thick-Film technology, piezoelectric thick films

Introduction

Besides it's usage for 3-D electronic packages LTCC is a well suited technology for the integration of Microsystems applications. The stacking and lamination of different structured layers allows the manufacturing of a wide range of 3-D geometry elements such as beams, diaphragms, cavities or channels.

LTCC-based components are high temperature stable and show a ceramic-like linear stress/ strain behaviour. Using LTCC for mechanical sensors leads to components with a very linear and nearly hysteresis free characteristics. LTCC-based components are very reliable and stable against thermal and mechanical shock.

The geometric configuration of LTCC-based 3-D multilayer is variable. In X-Y-direction many types of shapes can be produced e.g. by punching, micro milling or Laser cutting/ ablation. In Z-direction there is the possibility to combine different layers to create complex structures.

Not at least LTCC is a high density electronic packaging technology which allows the integration of electronic components which are needed for signal condition functions.

All these facts leads to the conclusion that LTCC-technology is very suited for the cost effective fabrication of high performance sensors e.g. for mechanical quantities.

Materials

In the materials section some tests were done to verify mechanical characteristics of the used LTCC-systems. This allows comparing the datasheet values (Young's Modulus, bending strength) with the characteristics of components fabricated with real technology steps (not optimal surfaces or edges). These values are important for the sensors modelling and design. Mechanical sensors have to transmit mechanical strain due to the measured quantity into an electrical signal. For this purpose different thick film pastes can be used. For the fabrication of sensor layers on LTCC piezoresistive and piezoelectric pastes were developed and/or characterized.

LTCC

It's the linear stress/ strain behaviour of LTCC which makes it suited for the manufacturing of bending elements. For the development of pressure and acceleration sensors Du Pont's Green Tape™ 951 system was used. It offers different tape thicknesses (50 – 254 μm unfired) and some of the required pastes (inner/outer conductors, via, outer/inner resistors).

The used LTCC system is a good balance between Young's Modulus (Sensibility) and Fracture Strength compared to other ceramic materials (table 1).

Table 1: Mechanical Properties of Different Ceramics.

	Unit	Alu-mina *	LTCC*	ZrO$_2$-ZTA*
Firing Temperature	[°C]	> 1600	900	1450
Young's Modulus	[GPa]	320	120	256
Fracture Strength	[MPa]	350	320	1399

* Kyocera datasheet A-476 (alumina 96%), LTCC: Du Pont datasheet DP 951; ZrO$_2$-ZTA [2].

For the verification of the datasheet values of Young's Modulus and Fracture Strength 3-point-bending tests for Du Pont's LTCC have been carried out. It could be shown that the datasheet values are approx. 15% higher than the measured values (Fig. 3). The reason for this fact could be the manufacturing technology for the used bending cantilevers.

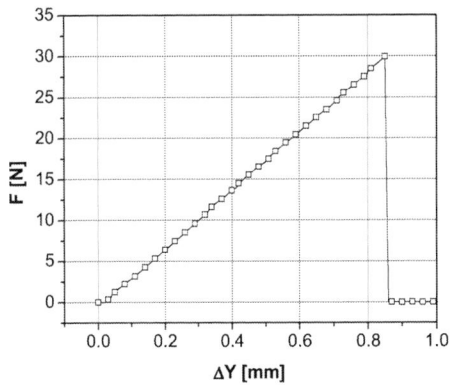

Fig. 1: 3-knife bending test of a DP 951 cantilever.

The cantilevers (50x9x0.88 mm^3) were structured using Laser cutting. In this case the edges are not optimal which can cause cracks earlier.

Piezoresistive Pastes

Piezoresistive layers can be used for the transmission of a mechanical strain into an electrical signal. Piezoresistors on LTCC made by screen printing of special pastes have to be as strain sensitive as possible. Furthermore they have to have a low TCR, low current noise and a high stability against aging. Most of the pastes which are known for high strain sensitivity are designed for alumina substrates. In the case of using these pastes as post fired strain gauges at LTCC a change of the resistor characteristics is possible. Causes for this fact are the different CTE of LTCC in comparison with alumina and chemical interactions between resistor and LTCC while firing of the pastes.

Different measurements with 5 different 10 kOhm Pastes on DP 951 have been carried out to obtain the mentioned characteristics.

Table 2 shows the results of the measurements. It could be shown that the paste characteristics on DuPont's LTCC 951 are in the range of the data sheet values (characteristics on alumina).

Table 2: Characteristics of different 10 kOhms-Pastes on LTCC. Rsqr – sheet resistance [kOhm/sqr], HTCR – Hot TCR [ppm/K], CTCR – Cold TCR [ppm/K], Kl – Gauge Factor longitudinal, Kt – Gauge Factor transversal, N - current noise [dB], π R/R0 [%] –aging stability.

	DP 1541	DP 1641	DP 2041	DP QT84	ESL 3414
R$_{sqr}$	8.3	11.4	7.3	11.1	6.9
CTCR*	-112	-118	5	-80	-152
HTCR**	-51	-23	60	-12	-85
K$_l$**	18.1	14.4	12.4	12.1	18.3
K$_t$**	13.6	11.1	10.2	10.8	11.9
N**	4.2	-10.1	-18.1	-6.9	-2.8
π R/R$_0$ @150°C	0.17	0.14	0.11	0.07	0.32
π R/R$_0$ @85% H	0.22	0.02	0.01	0.01	0.14

*   T = -55 .. 25 °C, 16 resistors 1x1 mm^2
**   T = 25 .. 150 °C, 16 resistors 1x1 mm^2
***   3-point bending method, 25 °C, 16 resistors 2x2mm^2
**** Used equipment: QuanTech 315 C, 16 resistors 1x1 mm^2

Different analytical analysises (SEM, EDX and XRD) have been carried to obtain the micro structure of the pastes. As in the literature described [4], [5], [6], [7] pastes with large conductive grains show a higher strain sensitivity on one hand but also a higher current noise and a lower stability on the other hand. Different interfaces resistor/ LTCC were observed caused by glass exchange between the resistors and LTCC. EDX measurements detected a high amount of lead oxide of the resistor pastes as one possible reason for this fact (Fig. 2).

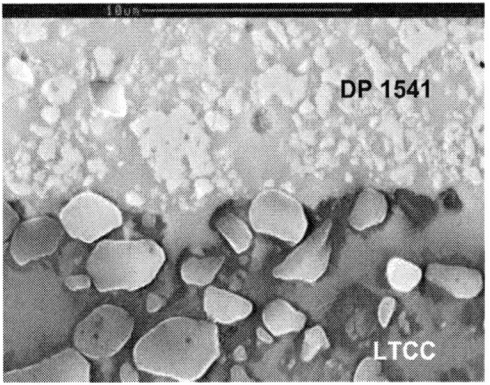

Fig. 2: SEM-picture of the interface LTCC/DP 1541.

The measurements make clear that the choice of a paste as strain gauge for a sensor is a compromise between strain sensitivity on one hand and current noise and stability on the other hand (Fig. 3).

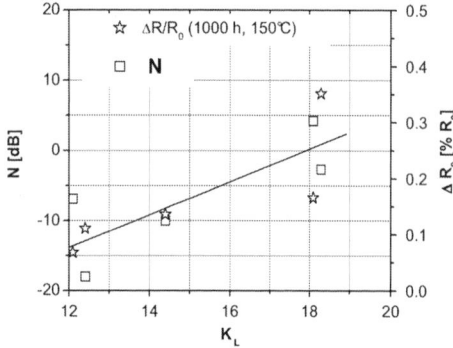

Fig. 3: Longitutinal Gauge Factor (K_L) vs. Current Noise (N) and Stability at 150°C

Piezoelectric Paste

Piezoelectric layers which are screen printed on LTCC can be used as sensors or actuators. Actually PZT (Lead-Zirconate-Titanate) is the most common material for the fabrication of piezoelectric components.

The use of piezoelectric thick-films on LTCC was prevented so far by different facts. The achievement of nonporous PZT layers is difficult because of the low firing temperatures (LTCC-compatibility: < 900°C) as well as the lateral fixing on the substrate. Additionally the contact of silicon phases contained into the LTCC with PZT leads to a massive degradation of the ferro- and piezoelectric characteristics. To prevent this case a dense diffusion barrier layer has to be placed between PZT and LTCC. To decrease the sintering temperature of PZT to a LTCC compatible level a sintering aid was used. A mixture of Bi_2O_3/ ZnO screen printed on the dried PZT-layer leads to a liquid phase while firing densifying the PZT.

Within the work a low sintering PZT paste/ electrode system developed at IKTS [8] was tested on Du Pont's 951 LTCC to check the functionality as sensor layer.

Fig. 4: Microstructure of a PZT layer screen printed on LTCC.

The PZT-paste was screen printed on a pre-fired Au-electrode (situated on LTCC). The Au-layer acts as electrode as well as the required diffusion barrier. The PZT-layer was fired at 875°C for 2 hours. Fig. 5 shows the microstructure of the fabricated layer.

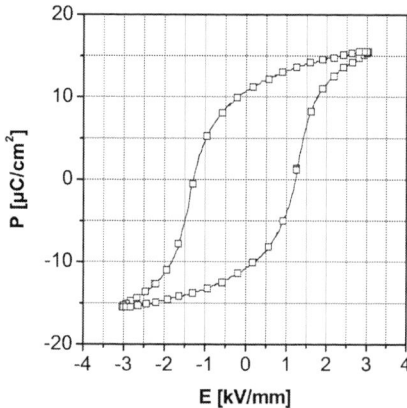

Fig. 5: Ferroelectric hysteresis loop of PZT thick film on LTCC (DP 951).

Relevant characteristics of the PZT-layer were measured. The layer shows proper characteristics and is very suited for the application in LTCC-based sensors [Tab. 3].

Table 3: PZT-Thick-Film properties on LTCC.

Property	PZT on DP 951
Dielectric constant $\varepsilon_{33}^{T}/\varepsilon_0$ @ 1 kHz	1500
Dielectric loss tan δ @ 1 kHz	0.033
Piezoelectric coefficient d_{33} [pC/N]	140
Remanent polarization P_r @ 5 Hz [μC/cm^2]	10
Coercive field E_C @ 5 Hz [kV/cm]	13
Internal resistance R_{is} @ 30 kV/cm[Ω]	> 10^6

Analytic Measurements using EDX verified that no Si-based phases penetrated the PZT layer (Fig. 6).

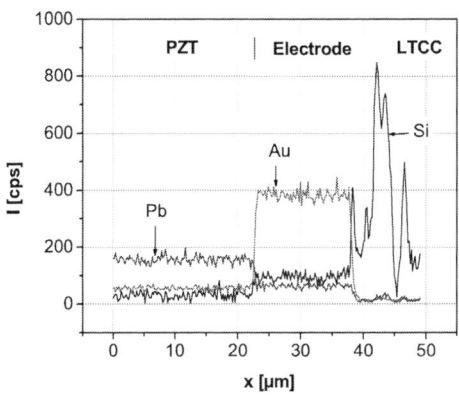

Fig. 6: EDX-line scan PZT /Au electrode/ LTCC (DP 951).

LTCC-based Pressure Sensors

Sensors for pressure detection mostly base on the usage of a diaphragm. The pressure-caused deflection is transformed by different measuring principles (e.g. piezoresistive, capacitive or piezoelectric principle) into an electrical signal.

LTCC-based pressure sensors have many advantages in comparison with classic steel or ceramic-based pressure sensors [3]. Different diaphragm thicknesses can be achieved using different tape thicknesses. The arrangement of the diaphragm is free. All types of pressure sensors (relative, absolute, differential) can be built up in one step. Furthermore all components of the sensor system (sensor body, electronics) can be integrated in one LTCC-based multilayer substrate.

For the mechanically decoupling of the sensor cells a new LTCC-based concept was developed (Fig. 8). The cell is fixed by 4 thin LTCC-beams containing micro channels for the pressure connection of the sensor cell.

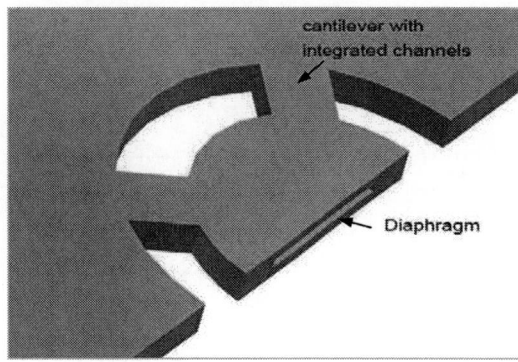

Fig. 7: Sectional drawing of the LTCC-pressure sensor body.

The described new sensor design prevents mechanical strain from the diaphragm which guarantees the proper function and stability of the sensor.

Piezoelectric Pressure Sensors

The piezoelectric pressure sensor consists of a LTCC-diaphragm, the ground electrode, the PZT layer and a top electrode. In dependence of the geometrical design and the Young's Modulus of the diaphragm there are different frequencies of resonance (Fig. 9).

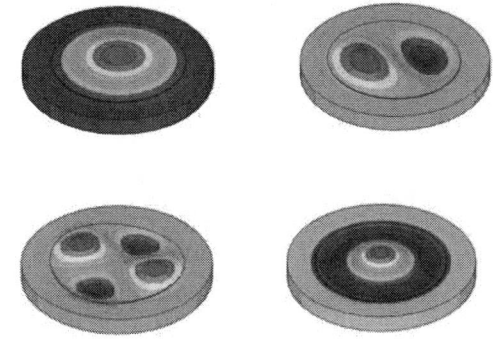

Fig. 8: FEA simulations results: Different oscillation modes of a LTCC diaphragm.

The piezoelectric layer works both sensor and actuator. It drives the diaphragm on its resonance frequency (first axial-symmetrical mode). The measuring principle bases on the pressure pro-portional shifting of the frequency of resonance (Eq. 1) and can be assumed to be linear in small intervals (with C1 .. geometric constant, f_0 .. frequency of resonance without pressure, C_1.. geometric dependent constant, ω_0 .. diaphragm deflection caused by pressure, d .. diaphragm thickness) [9]

$$f = f_0 \sqrt{1 + C_1 (\omega_0 / d)^2} \qquad \text{(Eq.1)}.$$

The first axial-symmetrical oscillation mode has the strongest dependence on pressure. Fig. 10 shows the dependence of the frequency of resonance from diaphragm thickness (FEA simulations results).

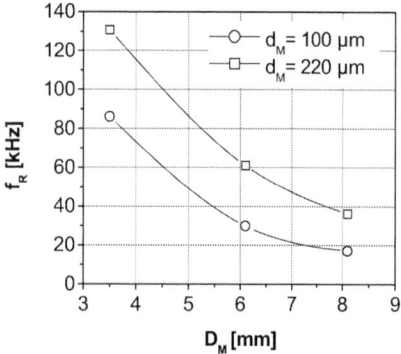

Fig. 9: Results of FEM simulations: Frequency of the first axial symmetrical mode (f_R) vs. diaphragm diameter (D_M) and thickness (d_M). LTCC-material: DP 951.

After fabrication of the sensors different measurements were performed to obtain sensor characteristics. The resonance frequency was measured by an impedance analyser (HP 4194-A) and about 34 kHz.

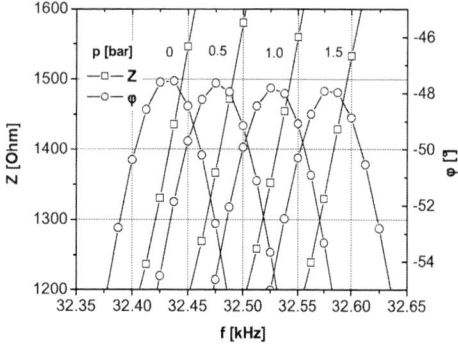

Fig. 10: Shifting of frequency of resonance in dependence of pressure.

Fig.10 shows |Z| and φ vs. frequency in dependence of pressure. It could be shown that there is a linear behaviour of the resonance frequency in dependence of pressure (Fig. 11). Table 4 illustrates sensor characteristics.

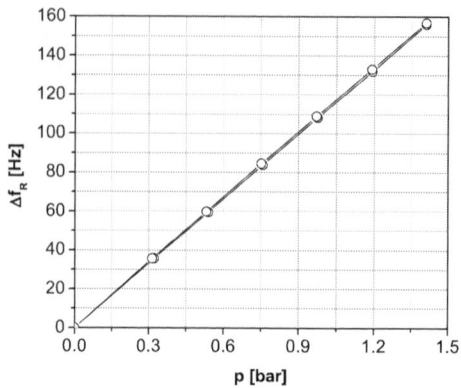

Fig. 11: Characteristic curve f_R vs. p piezoelectric pressure sensor.

Table 4: Sensor characteristics of the piezoelectric pressure sensors (Diaphragm diameter 8 mm, thickness 200 µm), T=25°C, tolerance: ± 3σ$_{n-1}$

Sensitivity	S	105±13.7	Hz/bar
Resonance frequency	f_R	32.4±0.4	kHz
Linearity	L	0.25±0.3	%FS
Hysteresis	H	0.25±0.3	%FS

Piezoresistive Pressure Sensors

In the case of using the Piezoresistive measuring principle for pressure detection the diaphragm deflection caused strain is measured by thick-film resistors which are screen printed at the surface of the LTCC diaphragm. The thick film resistors are connected to a Wheatstone bridge. The bridge voltage can be calculated as follows (Eq. 2, with U_b .. bridge voltage, U_v .. supply voltage):

$$U_b = \frac{U_V}{R_2 + R_4} \cdot R_2 - \frac{U_V}{R_1 + R_3} \cdot R_1 \qquad \text{(Eq. 2)}.$$

While pressure impact two resistors are under strain and two resistors under compression. In this case there is a pressure proportional shifting of the bridge voltage.

Several FEA analyses were carried out to calculate the sensor sensitivity caused by the pressure impact in dependence of diaphragm thickness.

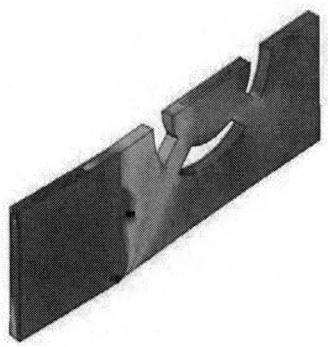

Fig. 12: FEA-modeling result (mechanical stress caused by the bonding of a metal block on LTCC at 200 °C after cooling down to room temperature).

Furthermore there is the need to verify the mechanical decoupling of the sensor cell. Fig. 13 shows one result of these calculations.

Sensors for different pressure ranges were characterized with respect to their sensitivity, offset voltage, linearity and hysteresis.

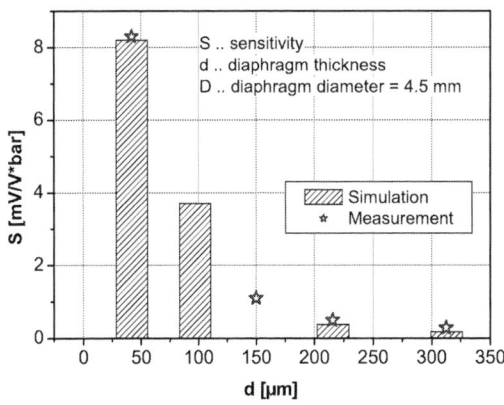

Fig. 13: Sensor sensitivity (calculated vs. measured values) in dependence of diaphragm thickness.

The measurements follow the DIN/ ISO 16086 and were accomplished between -20 and 120°C. A measuring system consists of a pressure controller (DPI 520, 0 ... 30 bar), a digital multimeter with an integrated 12 channel multiplexer (Keithley 2700) and a programmable temperature chamber (Weiss temperature and humidity test chamber) was developed and installed. The measurements of the sensitivity show a good matching to the simulation results (Fig. 12).

All sensors show a very linear behavior and are nearly free of hysteresis. Fig. 14 shows 3 characteristic curves.

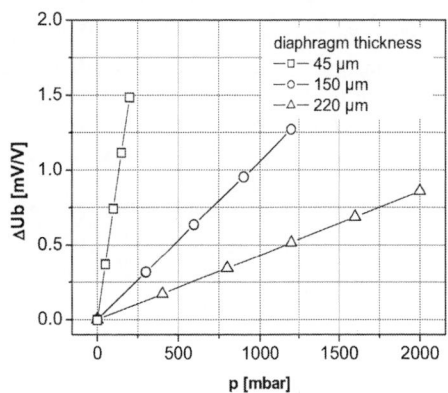

Fig. 14: Characteristic curves π U_b vs. p of different sensors (pressure ranges)

Table 4 shows the overall characteristics of the sensors for different pressure ranges (0.2, 1.5, 5 bar). The operation pressure of the different sensors was defined by additional burst tests. All sensors have a 4-times overload safety.

Table 4: Characteristics of different sensor types after DIN/ISO 16086 (Offset Voltage < 25 mV/V, Bridge resistance ~ 25 kOhm,T=25°C). D – diaphragm diameter, d – diaphragm thickness, p_{op} – operation pressure, S – sensitivity, L – linearity, H – hysteresis; FS – full scale.

Diaphragm		p_{op} [bar]	S [mV/V *bar]	L [%FS]	H [%FS]
D [mm]	d [µm]				
4.5	0.22	5.0	0.4	0.06	0.05
4.5	0.15	1.5	1.1	0.04	0.01
4.5	0.04	0.2	4.6	0.07	0.08

As an additional feature sensors were manufactured with an ASIC for signal conditioning (amplifier, temperature compensation, Fig. 15).

Fig. 15: LTCC-pressure sensor with electronics for signal condition.

LTCC-based Piezoresistive Acceleration Sensors

The principle of acceleration measurement is a mass element on a string. An acceleration effects an elongation of the mass and a deformation of the string. This is measured by a transducer, e.g. a capacitive or a piezoresistive. Today's acceleration sensors made of silicon offers a sufficient functionality in a cost effective way. In contrast acceleration sensors made of LTCC work under elevated temperatures, or as an additional feature integrated in LTCC substrates or electronic assemblies.

The LTCC multilayer technology suggests a sensor layout with leaf springs in a layer stressed for bending or for torsion. Thus an acceleration perpendicular to the layer effects a deformation of the springs which can be measured by piezoresistive thick film resistors printed on the springs. The layout has to consider some conflictive requirements, e.g. high resonance frequency, i.e. high stiffness, versus high sensitivity, i.e. lower stiffness, or uniform strain in the benders for a high reliability versus high strained areas of the springs for a high sensitivity.

We used analytical and finite element models to optimize the sensor design. Two parallel trapezoidal benders have advantages compaired to rectangular benders or torsion springs or the bridge and cantilever structures proposed in [10]. A FEA model of the optimized layout fabricated in LTCC is shown in fig. 16.

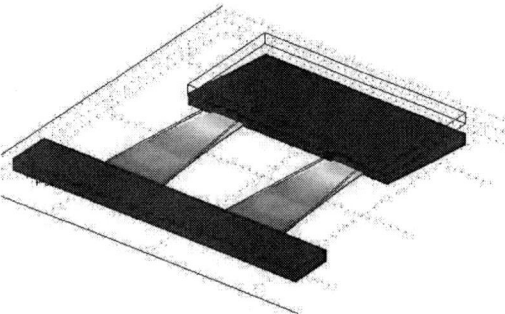

Fig. 16: FEA model of an acceleration sensor with two trapezoidal benders (equivalent strain).

The first resonance frequency has a strong influence to the working range and is obtained from FEA models or from the approach in Eq. 3 (c .. spring rate of the mass elongation perpendicular to the layer, m .. mass of the sensor element):

$$\omega_e = \sqrt{c/m} \ \text{(Eq. 3)}.$$

A linear relation between acceleration and deformation can be expected about up to the half of ω_e. Resonant excitation would destroy the sensor. Four thick film resistors, two on the benders and two on the frame are connected to a Wheatstone bridge like it's been done to the pressure sensors. Fig. 17

shows a LTCC acceleration sensor with a resonance frequency of 250 Hz. Excited by 100 Hz the sensor gives a slightly higher signal due to the closer distance to the resonance frequency compared to 50 or 75 Hz (fig. 18).

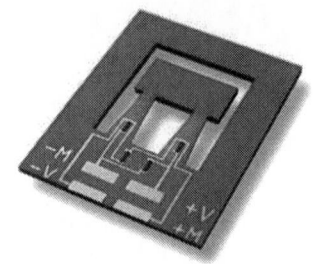

Fig. 17: LTCC-acceleration sensor with piezoresistors.

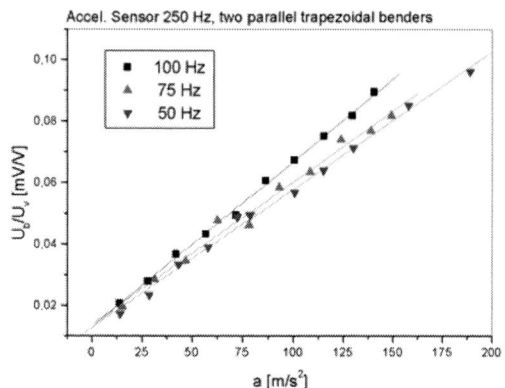

Fig. 18: LTCC-acceleration sensor characteristic curve, bridge voltage U_b over supply voltage U_v vs. acceleration a under several excitation frequencies.

However the tested sensors are sensitive for large accelerations smaller effective ranges can be designed.

Conclusions

Within the work it could be shown that LTCC is very suited as an integration platform for sensors for mechanical quantities. Different types of pressure sensors could be demonstrated as well as a new type of a LTCC-based acceleration sensor. Piezoresistive and piezoelectric pastes were characterized for the usage as sensor layers.

The main advantage of the demonstrated sensors is the quasi-monolithic design which prevents mechanical strain caused by the using of adhesive layers or the combination with other materials. Using the monolithic design LTCC-based sensors for mechanical quantities shows a linear and hysteresis free behaviour and offers an additional functionality for LTCC-based microsystems.

References

[1] Pfeifer, G.; Werthschützky, R.: Drucksensoren, Verlag Technik, Berlin, 1989

[2] Dörfel, I. ; Effner, U.; Glaubitz, S.; Österle, W. and Rudolph. E.: Characterization of high-performance ZrO_2 materials; Conference Proceedings "Werkstoffwoche '96", Stuttgart, 28–31 May 1996

[3] Slosarcik, S; Bansky, J.; Dovica, M.; Tkac, M.: Unconventional Application of Low Temperature Cofired Ceramics for (Under-) Pressure Sensors, J. Electrical Engineering, (48) 1997, Nr. 9-10

[4] Prudenziati, M.; Morten, B.: Piezoresistive Properties of Thick Film Resistors - An Overview, Hybrid Circuits, 10/1986

[5] Hrovat, M.; Drazic, G.; Holc, J.: Correlation between Microstructure and Gauge Factors of Thick Film Resistors, Journal of Materials Science Letters, 14/1995

[6] Forlani, F.; Prudenziati, M.: Electrical Conduction by Percolation in Thick Film Resistors, Electrocomponent Science and Technology, 1976, Vol. 3

[7] Abe, O.; Taketa, Y.: Strain Characteristics of Thick Film Resistors and its Application to a Strain Sensor, IMC Proc. 1986

[8] Seffner, L.; Partsch, U.: Niedrigsinternde PZT-Schichten auf LTCC", annual report 1998, Fh-IKTS Dresden

[9] Smiths, J.; Tilmans, H.; Lammerick, T.: Pressure Dependence of Resonant Pressure Sensor, Transducers 85, Proceedings, Philadelphia 1985, S. 93.

[10] Thelemann, T.: Die LTCC-Technologie als Basis von sensorischen, aktorischen und fluidischen Komponenten für Mikrosysteme, Diss. thesis of Ilmenau University of Technology 2005

Epoxy-based polymer for capacitive chemical sensors

Marijan Maček[1], Marta Klanjšek Gunde[2], Nina Hauptman[2]

[1]University of Ljubljana, Faculty of Electrical Engineering, Tržaška 25, SI–1000 Ljubljana, Slovenia

[2]National Institute of Chemistry, Hajdrihova 19, SI–1001 Ljubljana, Slovenia

E-mail: marijan.macek@fe.uni-lj.si

Abstract

A capacitive micro–machined sensor module, fully capable of integration with a generic double metal CMOS technology, covered with photosensitive layer of SU–8 high functionality epoxy, was developed. Sensitivities to different volatile organic compounds were studied after curing the polymer at different temperatures between 200 °C and 340 °C. By FTIR spectroscopy also the changes in the chemical structure of polymer were studied. Heat treatment at temperatures above 300 °C results with prominent changes inside epoxy groups and aromatic rings and can be explained as destruction of the crosslinks and formation of uncrosslinked monomers. Therefore, initial high functionality of epoxy groups is decreased. As a result, a prominent increase in the sensitivity of sensors is observed. This increase may be related to the changes in the solute–solvent interaction, and therefore in the partition coefficient K, or to the changes in the polymer susceptibility to swelling.

Key words: micromechanical capacitive sensor, *SU–8*, FTIR

Introduction

Chemical sensors for gases and vapours are at the forefront of the information acquisition chain in environmental applications. A special branch of them are the capacitive chemical sensors, whose response is based on the variation in their capacitance. The change in capacitance due to the sorption of the analyte is related to three distinctive mechanisms [1]: adsorption of the analyte on the polymer surface, absorption into the polymer and swelling (the change of thickness) of the polymer layer.

$$\Delta C = \Delta C_{ad} + \Delta C_{\varepsilon} + \Delta C_h \qquad (1)$$

The adsorption term is important only for sensors with polymer thickness much smaller then inter–electrode spacing and is usually neglected.

The second term is the most important term for analytes with $\varepsilon_a >> \varepsilon_p$. It will be shown latter, that this contribution can be either positive or even negative,

The third, swelling, term is always positive and depends on the electrode geometry, polymer thickness as well as on the polymer analyte interaction. The last two terms in Eq. (1) are inter–related, since both depend on the absorption of the analyte into the polymer.

Description of the absorption process of the analytes into the polymer and their interaction is similar to that used in gas chromatography [2]. In thermal equilibrium the absorption of the analyte in the polymer depends, in first approximation, on the

saturated vapour pressure p_s, and absolute temperature T. The ratio between the concentration of the analyte in the gas phase c_g and in the polymer phase c_p is known as the partition coefficient K given by the following equation

$$K = \frac{c_p}{c_g} = \frac{RT}{p_s V_p^{mol}} \qquad (2)$$

V_p^{mol} denotes molar volumes of polymer, and R the gas constant. Partition coefficients for the individual polymer are inversely proportional to the saturated vapour pressure and are in the range from $100 – 100.000$. Eq (2) is only an approximation, valid for low concentrations of solute, where is expected that Henry's law holds. It doesn't include the nature of the analyte (solute)–polymer (solvent) interaction. Generally, the partition coefficient K for the particular polymer and for a certain chemical group like alcohols is inversely proportional to p_s [3].

It was shown in [4] that the second term ΔC_{ε} in Eq. (1) is proportional to the concentration of analyte in air and to the ratio of dielectric constant.

$$\frac{\Delta C_{\varepsilon}}{C_0} \propto \cdot K \left(\frac{\varepsilon_A}{\varepsilon_p} - 1 \right) \cdot p_A \qquad (3)$$

The partition coefficient K is inversely proportional to the saturated vapour pressure, $K=K(1/p_s)$, Eq. (2). Therefore, the relative change in capacitance is proportional to relative concentration of analyte in air, $p_a/p_s(T)$, or in case of humidity detectors to the relative humidity. A very similar dependence gives also Maxwell and Clausius–Mosotti models [1]. According to Eq. (3) the second term in Eq. (1) is negative for analytes with dielectric constant lower then that of sensing polymer layer.

Modeling of the swelling contribution is rather complicated. The change in the capacitance depends on the geometry, applied sensing polymer and analyte. It follows from analytical model published in [1] that this term is important for combinations when $\varepsilon_p < \varepsilon_a$., ($\varepsilon$ toluene = 2,38, ε n-hexane = 1.9, or $\varepsilon_p \sim \varepsilon_a$. With a proper combination of absorbing layer thickness and inter–electrode spacing, even the null response can be achieved for a certain polymer–analyte combination.

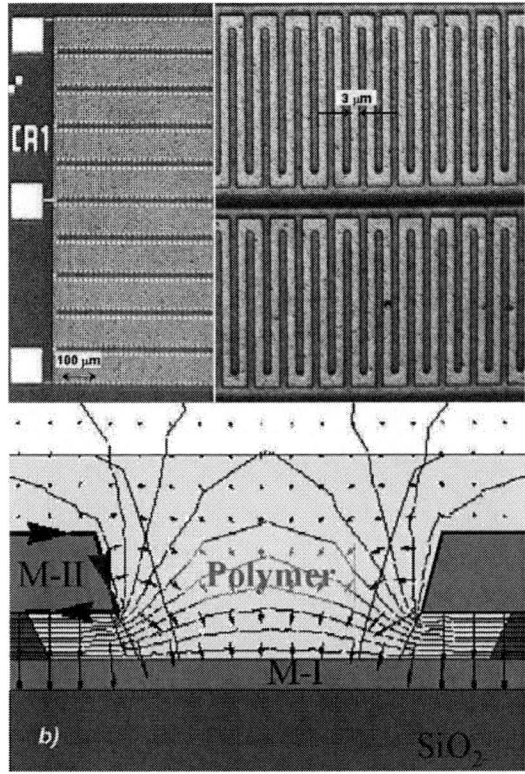

Figure 1. a) Photomicrograph of the capacitor with the interdigitated top electrode. b) cross–section of the structure with equipotential lines as used for 2D calculations.

In previous papers [4,5] we compare the micro–machined capacitive sensors realized with *SU-8* photosensitive epoxy with the common polyethersulphone (*PES*) polymer as the humidity detectors. It was found out that *SU-8* is very promising material, partly due to the simplified application (photo definable) and partially due to up to two times higher response after proper curing. Preliminary FTIR measurements [6] indicated changes in the epoxy polymer structure, which can be related to the response of the detectors.

In this contribution we will show the possible application of *SU-8* negative resist as a chemical sensor for volatile organic compounds with emphasis to the change in response characteristics after different curing in the temperature range from 200 – 340 °C.

Experimental

During the design of the micro–machined sensors which are the subject of this contribution, all the limitations given by the *CMOS* processes, which are used during further fabrication, of the integrated sensors were taken into account. The post–processing which enables us to etch the passivation layer and inter–metal sacrificial dielectric, as well as subsequent polymer deposition and curing, was applied. More details are given in [5]. Photomicrograph of the sensor and its shematic cross section whit equipotential lines are shown in Fig. 1.

SU–8 is epoxy–based negative photo resist with the highest commercially epoxide available functionality (f = 8). Its structural formula is shown in Fig. 2. It was initially developed by IBM for optical lithography of ultra thick applications, giving structures with high aspect ratio (>50) [7]. Once it is fully crosslinked, the material becomes thermally stable and practically insoluble in most chemicals. It is extensively used in photolithographic fabrication of micro–electronics machine systems (*MEMS*) or in packaging applications.

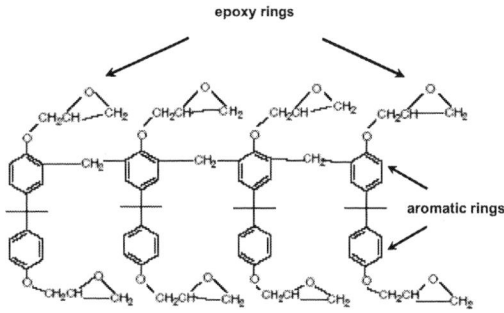

Figure 2. Chemical structure of the epoxy *SU–8* monomer.

This resist was deposited by a standard spin-on technique, with subsequent drying on the hot plate at 95 *C* for 1 min. The target thickness was 2.2 μm, as measured by a reflectometer on flat test wafers. At this thickness, the best linearity of humidity sensors was obtained [5]. More details regarding *SU–8* curing were given elsewhere [4,5].

Capacitance measurements were performed by a computer controlled CV-analyser at 1 MHz at room temperature. The accuracy of the measuring set–up was below 3 fF. Measurements were performed in a gas mixture of pure nitrogen and controlled amount of the analyte vapour from a bubbler kept at the constant (± 0.05 °C) temperature, lower than the room temperature to prevent unwanted condensation. The gas flow was controlled by MFC. The carrier flow was kept constant at 150 ml/min to minimise differences in evaporation.

For the purpose of *FTIR* analysis, thin layers of *SU8* were deposited on device-grade silicon wafers by spin-coating in thickness of 0,86 μm. Samples were cured on the same way as sensors.

The IR transmittance spectra of all samples were measured using a Perkin-Elmer System 2000 *FTIR* spectrophotometer. IR transmittance of the silicon wafer used for the deposition of layers was measured as a reference. During all IR spectral measurements the sample compartment was purged with dry nitrogen to diminish the amount of water vapour and CO_2.

Results and discussion

Fabricated sensors react very fast on to the change in the analyte concentration as shown in Fig. 3. In this figure the response of the sensor realised with *SU–8* cured at 340 °C on toluene is shown. Typical time constants obtained from the best fit by the following equation (we assume two different mechanisms in sorption of the analyte)

$$\Delta C/C_0 = \cdot C_1\left(1 - \exp(-t/\tau_1)\right) + C_2\left(1 - \exp(-t/\tau_2)\right) (4)$$

are between 20 s and 50 s for all studied analytes with the only exception to be methanol. In this case the time constants are of the order of 300-500 s.

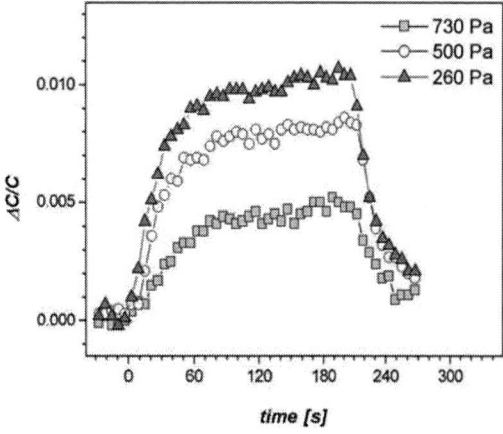

Figure 3. Relative response $\Delta C/C_0$ **of** *SU–8* **capacitive sensors cured at 340 °C vs. time for three different partial pressures of toluene.**

The response is linear with the partial pressure of analyte up to 500 Pa. The reason for the

non–linearity at higher pressures should be either the experimental one (error in gas flow, the differences in evaporation), or the concentration of analyte excided the limits of the region where the theory, Eq. (1–3), based on Henry low holds. It is worth to note, that the response of the same sensors to the water was linear from 10% – 90% relative humidity, or almost 3 kPa [4].

In Fig. 4, the relative sensitivities, $C^{-1}(dC/dp_A)$, of sensors cured at different temperature are shown. The sensitivity steadily increases for the temperatures between 200 °C and 300 °C. Above this temperature a steep increase was observed, the same as in the case of humidity detectors [4]. N-hexane characteristic is different, without prominent changes at temperature above 300 °C. But the sensitivity is only slightly higher than the sensitivity of measuring set up.

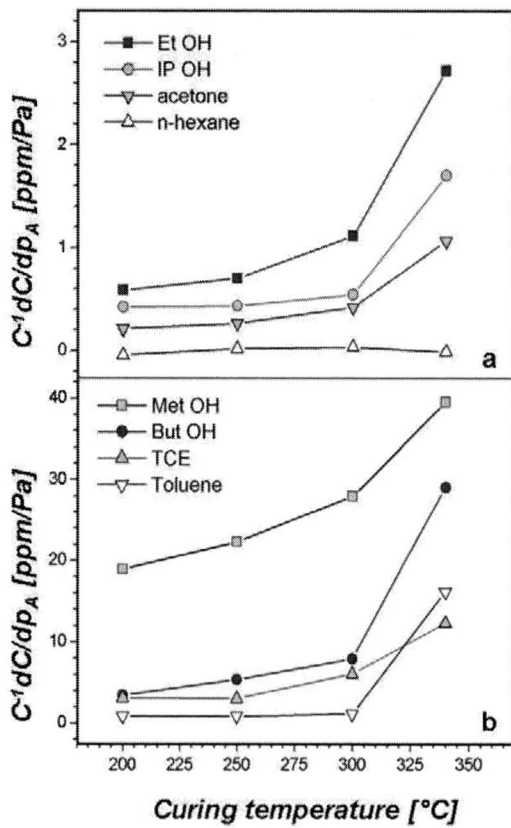

Figure 4. Relative response $C^{-1}(dC/dp_A)$ **of** *SU–8* **capacitive sensors on to different analytes as function curing.**

From Eq. (2, 3) we see that the sensitivity of capacitive sensors is proportional to

$$C^{-1}\left(dC/dP_a\right) \propto \left(\frac{\varepsilon_A}{\varepsilon_p} - 1\right) \cdot p_s^{-1}(T) \qquad (5)$$

Therefore, for n-hexane and toluene the sensitivity should be negative since dielectric constants (1.9 and 2.38 respectively) are lower than $\varepsilon_{SU8} = 3$. Only for n-hexane we measured negative sensitivity. For toluene, a low positive sensitivity was measured for curing temperatures up to 300 °C. This indicates moderate effect of swelling. For specimens cured at 340 °C the sensitive increase for an order of magnitude. This can be related to the effect of swelling due to the changes in polymer structure and/or to different mechanism of interaction between solvent and solute. It was already stated that the definition of partition coefficient neglect the different nature of solute–solvent interaction.

IR spectra of exposed *SU–8* layers consist of two wavenumber regions, above and below 2000 cm^{-1}, showing vibrations of molecular groups (such as OH and Si–H$_n$) and the so-called fingerprint of the whole molecule, respectively. The fingerprint region of IR spectra of layers heated at 200, 250, 275, 300, 320 and 340 °C are shown in Fig. 5. Due to relatively large *SU–8* molecule, the fingerprint region contains large number of mostly overlapped peaks. For the purpose of our analysis we followed the three characteristic peaks of the epoxy ring (appearing at 1250, 831 and 915 cm^{-1}) and the formation of carbonyl (C=O at 1730 cm^{-1}) [8]. The relative intensities of the considered peaks in dependence on curing temperature of exposed layers are shown in Figure 6. The vibration of aromatic ring at 1608 cm^{-1} was taken as a measure of the invariant aromatic content of the resin.

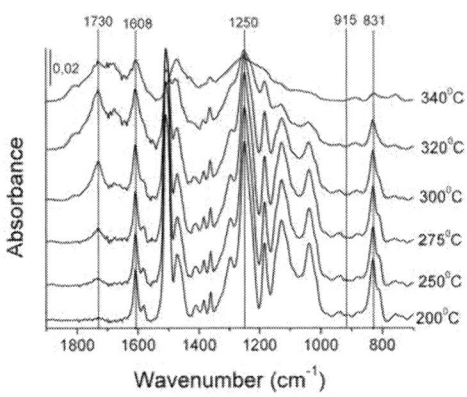

Figure 5. IR spectra of exposed *SU–8* layers after heating at 200, 250, 275, 300, 320 and 340 °C.

The chain-terminating epoxy group is involved in crosslinking of the polymer. The peak at 915 cm^{-1} corresponds to stretching of CH$_2$ group on epoxy ring. Its intensity is usually applied as a measure of degree of polymerization of the resin [9]. In our samples, this peak is very small at 200 °C and completely disappears at higher temperatures. It shows that the exposure alone produced almost fully crosslinked layer and it is completed with post–exposure backing at temperatures above 200 °C. The peaks at 831 and 1250 cm^{-1} correspond to symmetric and antisymmetric stretching vibrations of C–O–C group in monosubstituted epoxy rings [9]. Their normalized intensities quickly diminish at temperatures above 275 °C whereas the carbonyl peak intensity (at 1730 cm^{-1}) increases simultaneously. These effects may be explained as a decay of epoxy rings, giving rise to destroying of the crosslinks between monomers, formation of uncrosslinked monomers and diminishing their functionality.

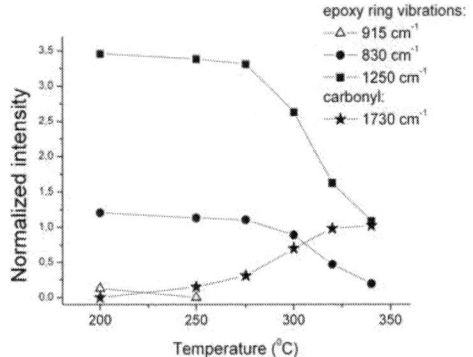

Figure 6. Normalised intensities of the considered vibrations.

Weather the decay of epoxy rings can change the nature of the interaction solute–solvent and thus increases the partition coefficient or the destruction of crosslinked structure at temperatures above 340 °C change the polymer to be more prone to swelling is a question. The only way to distinguish unambiguously between these two mechanisms is to fabricate the sensors with different polymer thicknesses. From the analytical solution of the similar problem [1] follows, that the response of thick polymer layers is almost solely composed of the pure ΔC_ε term.

Conclusions

A capacitive sensor module, fully capable of integration with a generic double metal *CMOS* technology, was developed. The sensing layer of photo definable epoxy based *SU–8* resin was successfully implemented.

Sensors react fast ($\tau = 20 – 50$ s) to the variations in partial pressure of most of the studied analytes and all heat treatments. The response to the methanol vapours is rather slow ($\tau = 300 – 500$ s). On the other hand, the sensitivity to methanol is much higher then expected from the relationship given by Eq. (5). The reason is not known, but most probably is related to the mechanism of the polymer-methyl group interaction. All other tested alcohols follow this relationship.

Swelling mechanism plays an important role for the fabricated sensors covered with 2.2 μm thick *SU–8* layer. This is reflected on the sensitivity to toluene ($\varepsilon = 2.38$) and n-hexane ($\varepsilon = 1.9$). The part corresponding to the change in dielectric constant, Eq. (5) should be negative. For toluene for all samples we measured positive response, while for the n-hexane, the response was slightly below zero.

Curing of *SU–8* at temperatures between 200 °C and 340 °C increases the sensitivity of the detectors for all studied volatile organic compounds. The most prominent changes are seen at the temperatures above 300 °C. After polymer curing at 340 °C the sensitivities increase for factor 2.1 (methanol) to 8.5 (butanol) with exception of toluene. Its expected low sensitivity (0.8 – 1.0 ppm/Pa for curing between 200 °C and 300 °C), increases for factor 18.5 after curing at 340 °C.

Analysis of *FTIR* spectra shows the two–step changes that may be interpreted as follows:

i. 250 – 275 °C; the beginning of formation of carbonyl groups, no appreciable change of aromatic rings and epoxy groups,

ii. 300 – 340 °C; further formation of carbonyl groups, with prominent changes inside epoxy groups and aromatic rings. These effects may be explained as a decay of epoxy rings, giving rise to destroying of the crosslinks between monomers, formation of uncrosslinked monomers and diminishing their functionality from 8 to lower values.

An increase of the sensitivity may be related to the destruction of crosslinks between epoxy monomers for curing at high temperatures which changes the interaction solute–solvent and consequently partition coefficient K. On the other hand, destruction of crosslinks may affect the susceptibility of the polymer to swelling. Additional studies are necessary to find out the correct explanation and understanding the chemical nature of these effects caused by curing.

Acknowledgment

This work was supported by the Slovenian Research Agency under the Contract No. J2-9455-0104-06. One of the authors (N.H.) thanks the Slovenian Research Agency for financing her training research as well as her postgraduate study program.

References

[1] R. Igreja, C.J.Dias, Sensors and Actuators, B 115, 69–78 (2006).

[2] Koll, E. Leibnitz, H.G. Struppe, „*Handbuch der Gaschromatographie*", W&P Buchversand f. Wissenscaft u, Praxis, Weinheim, 3rd ed (1984) (1999).

[3] Koll, Diss ETH, No 13460 (1999).

[4] M. Maček, N. Hauptman, EMPS 2006 - 4th European Microelectronics and Packaging Symposium with Table-Top Exhibition, May 21-24, 2006, Terme Čatež, Slovenia. Proceedings. Ljubljana: Midem, cop. 2006, pp. 27-32.

[5] M. Maček, R. Osredkar, Proc. of 41st International Conference on Microelectronics, Devices and Materials, (2005), pp. 321 – 326.

[6] N. Hauptman, M. Klanjšek Gunde, M. Maček, M. Kunaver, Slovenski kemijski dnevi 2006, Maribor, 21. in 22. September 2006. Maribor: FKKT, 2006.

[7] J.M. Shaw, J.D. Gelorme, N.C. LaBianca, W.E. Conley, S.J. Holmes, "Negative photoresists for optical lithography", IBM J. Res. Develop. 41, 81–94 (1997).

[8] D.I. Bower, W.F. Maddams: *"The Vibrational Spectroscopy of Polymers"*, Cambridge University Press 1989.

[9] Norman B. Colthup, Lawrence H. Daily, Stephen E. Wiberley: *"Introduction to Infrared and Raman spectroscopy"*, Academic Press, Boston, San Diego, New York, London, Sydney, Tokyo, Toronto 1990.

Leakage Current, Noise and Reliability of NbO and Ta Capacitors

V. Sedlakova[1], J. Sikula[1], H. Navarova[1], J. Pavelka[1] and J. Hlavka[1], Z. Sita[2]

[1]Brno University of Technology, Department of Physics, Technicka 8, 616 00 Brno, Czech Republic

[2]AVX Czech Republic s.r.o., Dvorakova 328, 563 01 Lanskroun, Czech Republic

Abstract

An analysis of charge carrier transport and noise in NbO and Ta capacitors was performed to prove the stability, reliability and non-burning performance of NbO and Ta capacitors. VA characteristics in normal and reverse mode at room and elevated temperature have been measured to determine the mechanism of current flow and current noise sources. The charge is accumulated not only on NbO/Ta and MnO_2 or conducting polymer (CP) electrodes, but also in the Nb_2O_5 or Ta_2O_5 insulating layer. Capacitance is given by a charge stored on both NbO/Ta and MnO_2 electrodes and also by a charge stored on localized states in Nb_2O_5 or Ta_2O_5 insulating layers. The charge carrier transport is determined by the Ohmic, Poole-Frenkel and tunnelling mechanisms in the normal mode (for NbO/Ta electrode positive). The dispersion of leakage current values at given applied voltage is due to interface dipole layers and charge carrier transport in the insulating layer flaws and defects. The dominant mechanism in reverse mode is: (i) for low voltage (U<0.4 V) the Ohmic charge carrier transport and (ii) for higher voltage - Schottky over barrier charge carrier transport. The electron transport has stochastic component, which is analyzed by noise measurement. For low electric field intensity (E< 100MV/m) shot and g-r noise are dominant. These sources have white noise spectral density and physical processes are reversible. They correspond to high quality and reliable technology. If 1/f noise spectral density is dominant then charge carrier transport contains irreversible processes which are related to low reliability. Reliability testing method based on noise and potential barrier evaluation is proposed.

Key words: NbO capacitor, Ta capacitor, MIS structure characteristics, Nb_2O_5, Ta_2O_5, reliability, noise

Introduction

A charge carrier transport and noise mechanism analysis has been performed on NbO and Ta capacitors to determine the sources of irreversible processes responsible for device quality and reliability. The method for quality and reliability of NbO and Ta capacitors is based on assessment of defects in active region from evaluation of VA and noise characteristics and theirs temperature dependences. For the capacitor polarized in the "normal mode", (with the NbO and Ta electrode positive), ohmic, Poole-Frenkel and tunnelling are the dominant conduction mechanisms. Insulating layer in these components has 30 to 100 nm, so they belong to nanoscale electronic devices in which quantum effects play important role. The electron conduction can occur by thermally activated hopping in impurity band and tunnelling between deep impurity states.

The model of the MIS structure can be used to give a physical interpretation of the NbO and Ta capacitors VA and CV characteristics and theirs temperature dependences. The MIS structure consists from metallic NbO or Ta, insulating layer made from Nb2O5 or Ta2O5 and semiconductor: MnO2 or conducting polymer (CP).

VA characteristics

To find more information about the current flow processes, the VA characteristics were measured both in normal and reverse mode at temperature T = 300 K and 100 K. For temperature lower than 200K, Ohmic and Poole-Frenkel components are lower, than the tunnelling one, and VA characteristics is described only by charge carrier tunnelling. At room temperature VA characteristics in normal mode and reverse mode differ significantly while at low temperature match each other. These capacitors can be used as bipolar ones at temperatures lover than 200 K [1,2,4 and 6].

The VA characteristics of a NbO sample for normal and reverse mode is given in Fig.1. For voltage lower than 0.3 V the VA characteristics is symmetric. The current carrier transport is Schottky and Poole-Frenkel with very low Ohmic component. This NbO capacitor has in reverse mode saturation current I_0=9x10^{-11}A and β= 14.2 V^{-1}. There is "threshold" reverse voltage U_N in the range of 1.0 to 1.4 V for NbO technology and U_N in the range of 1.8 to 2.4 V for Ta technology (Fig.2).

When voltage higher than "threshold" reverse voltage is applied Nb_2O_5 and Ta_2O_5 layers become conductive. In this case the reverse voltage is equal to potential barrier between NbO/Ta and MnO_2 electrodes.

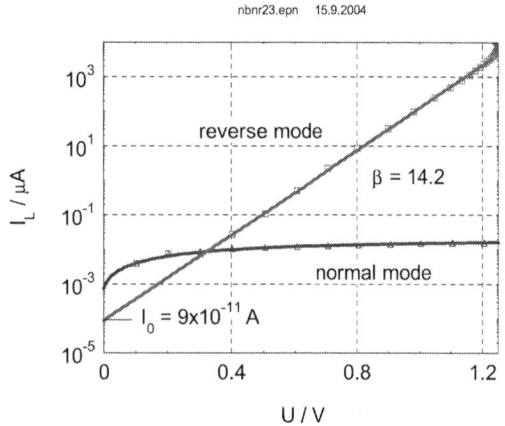

Figure 1. VA characteristics for NbO capacitor at temperature T = 300 K

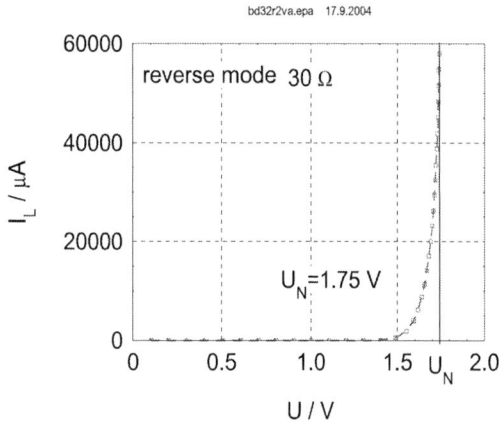

Figure 2. Reverse mode VA characteristics in linear scale for Ta capacitor

More information follows from VA characteristics in lin-log. scale, where ohmic and potential barrier current components can be separated as is shown in Fig.1. VA characteristics can be approximated by exponential dependence of current on voltage:

$$I = I_0 \ (exp(\beta.U)-1) \qquad (2)$$

where $\beta = 10$ to 25 V^{-1}. This value of parameter β corresponds to high ideality factor $n \geq 2$.

Schottky barrier can be find from saturation current I_0, given by [3, 4]

$$I_0 = ART^2 exp(-e\Phi/kT), \qquad (3)$$

where $e\Phi$ is Schottky potential barrier.

For capacitor area $A = 100$ cm^2 and $T = 300$K, assuming Richardson constant $R = 100$A/cm^2 we have the potential barrier

$$e\Phi_{Nb} = 1.0 \text{ to } 1.4 \text{ eV for NbO capacitor,}$$

$$e\Phi_{Ta} = 1.7 \text{ to } 2.2 \text{ eV for Ta capacitor.}$$

At temperature T = 100K Poole-Frenkel and Schottky mechanisms current transport are negligible and tunnelling is a dominant mechanism (see Fig. 3 and 4).

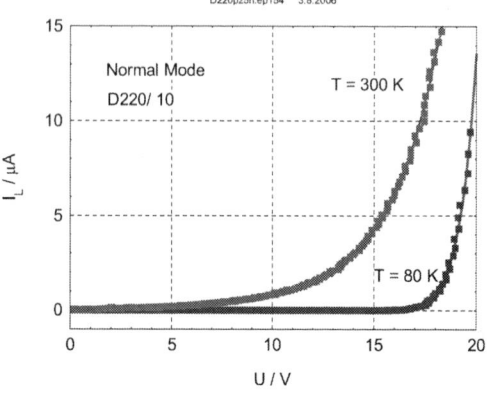

Figure 3. VA characteristics of Ta capacitor in normal mode at T = 300 K and 80 K

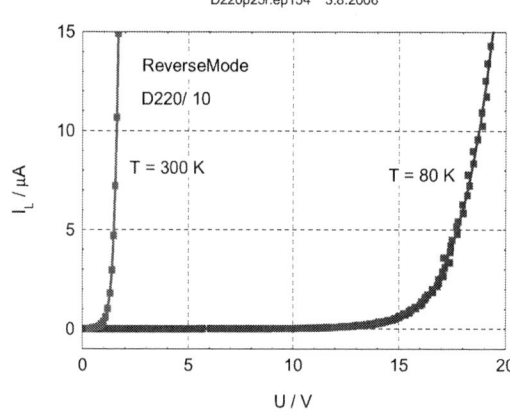

Figure 4. VA characteristics of Ta capacitor in reverse mode at T = 300 K and 80 K

VA characteristics at T = 80 K (Fig.3. and 4.) can be fitted by tunnelling component only:

Tunnelling current is given by

$$I_T = I_{0T} \ exp(-U_T / U), \qquad (4)$$

which is in the first approximation temperature independent. Here I_{T0} and U_T are tunnelling current constants.

We found that VA characteristics of Ta capacitor D220 μF/10V (Fig.3 and 4) can be fitted in normal mode by Poole-Frenkel components at T = 300 K:

In normal mode by:

$$I = G.U.\exp(b_{PF}\,U^{1/2}), \qquad (5)$$

where conductivity

$$G = 1.3\times10^{-6}\,S \text{ and } b_{PF}=3.1\,V^{-1/2}$$

In reverse mode:

$$I = I_0.\exp(\beta U - 1), \qquad (6)$$

where saturation current

$$I_0 = 1.4 \text{ nA and } \beta = 5.6\,V^{-1}.$$

At temperature T= 80 K we have:

In normal mode:

$$I = I_{T0}.\exp(-U_T/U) \qquad (7)$$

where tunnelling constants are

$$I_{T0} = 5\times10^5 \text{ A and } U_T = 468 \text{ V}$$

In reverse mode:

$$I = I_{T0}.\exp(-U_T/U), \qquad (8)$$

where

$$I_{T0} = 194 \text{ A and } U_0 = 228 \text{ V}.$$

Tunnelling parameter U_T (see Fig.5) is given for barrier Φ_0 with thickness t_0 by

$$U_T = (8\pi\sqrt{2m^*}\,/3eh)(e\Phi_0)^{1.5}t_0 \qquad (9)$$

m^* is effective electron mass, h= 6.6×10^{-34} Js is Planck constant and $e\Phi_0$ is the barrier energy.

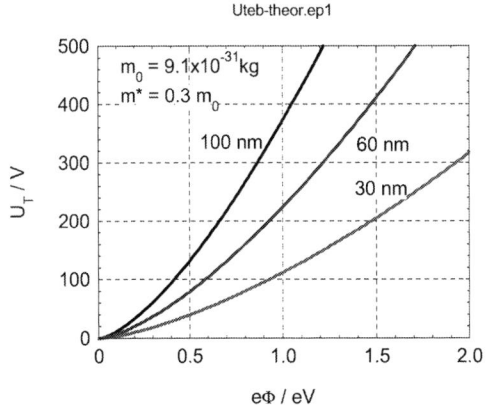

Figure 5. U_0 vs. barrier $e\Phi_0$ for layer thickness t_0 =30 to 100 nm

Tunnelling process has in the first approximation very low temperature dependence and then leakage current is constant in temperature range 80 to 200 K. Band diagrams for electron tunnelling in NbO capacitor in normal mode is in Fig.6 below. Similar diagrams are valid for Ta capacitors.

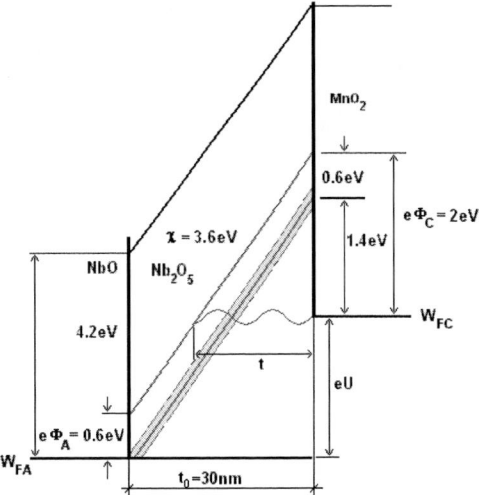

Figure 6. Band diagram of electron tunnelling in NbO capacitor in normal mode

On the basis of VA characteristics experimental data evaluation we came to the conclusion that tunnelling component is dominant at high electric field and temperature T = 100 K in both normal and reverse mode.

At temperature T = 300 K the VA characteristics in normal mode are given by Poole-Frenkel and Schottky mechanisms current transport. For low leakage current a high potential barrier is needed at the interface of Nb_2O_5 or Ta_2O_5 insulating layers and MnO_2 electrode. Barrier height is shown for 7 samples of NbO capacitors in Fig. 7.

Devices reliability is related to the stability of the technology and also to low parameters dispersion. Low potential barrier variation is related to low leakage current dispersion.

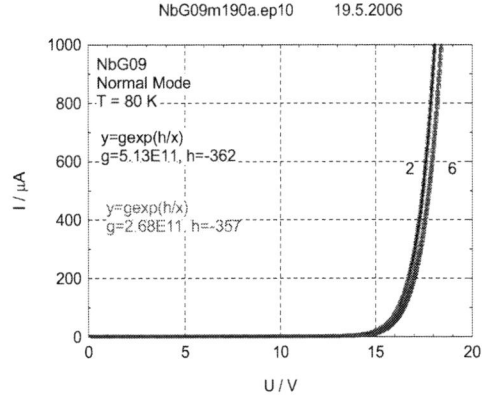

Figure 7. Normal mode characteristics for NbO capacitors, where barrier $e\Phi_C$ = 1.4 eV

Potential barriers and energy band diagram

Physical parameters of the NbO and Ta MIS structure before thermodynamic equilibrium are in Fig. 8. The real band model of the NbO and Ta MIS structure in thermodynamic equilibrium (no applied voltage) is illustrated in Fig.9. Potential barrier model parameters are summarized in Tab. 1 and 2.

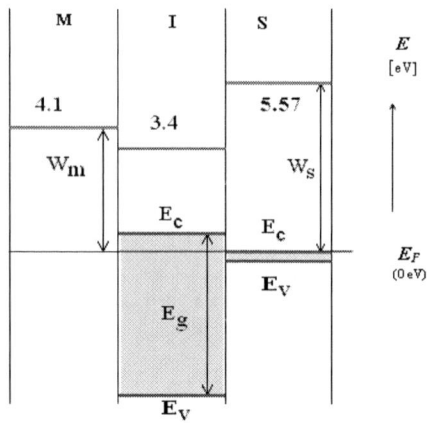

Figure 8. Band diagram of NbO/Ta capacitor

Table 1.

Material	W / eV	E_g / eV
NbO	4.1 to 4.3	
Ta	4.1	
Nb_2O_5		3.5 to 4.5
Ta_2O_5		4.5
MnO_2	5.6	0.2 to 0.4

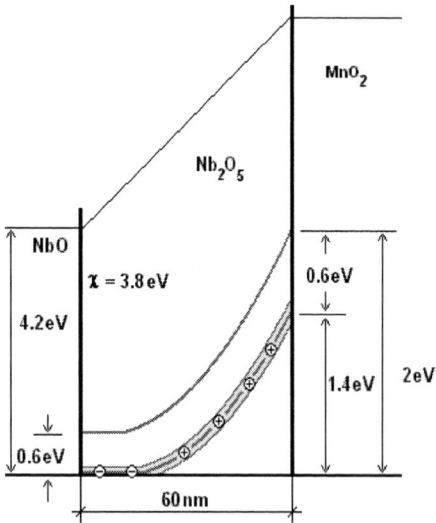

Figure 9. Potential barriers of NbO capacitor for applied voltage U = 0

Table 2.

Capacitor	$e\Phi_M$ / eV	$e\Phi_S$ / eV
NbO	0.4 to 0.6	1.0 to 1.4
Ta	0.7 to 0.8	1.7 to 2.2

The band structure for thermodynamic equilibrium is in Fig. 9 and for flat band insulating layer is in Fig. 10. Values of these parameters must be for given technology stable and they must be in the ranges given by Tab. 1 and 2.

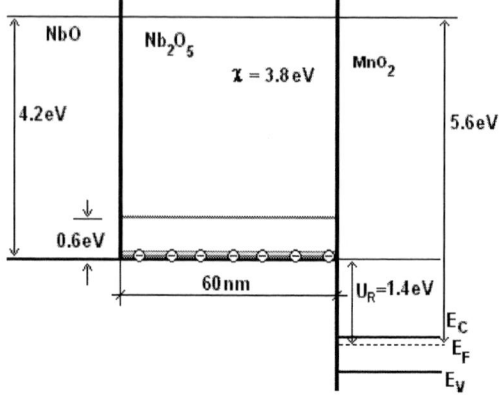

Figure 10. Potential barriers of NbO capacitor for $U_R = 1.4$ V

Noise

Noise characteristics can be explained on the basis of following noise sources: 1/f noise, burst noise and hot carrier noise. The noise spectral density is 1/f type in the whole measured frequency range for the most of samples both in normal and reverse mode (see Fig.11.). For the higher frequencies thermal noise of the load resistance is superposed yielding constant value of noise spectral density.

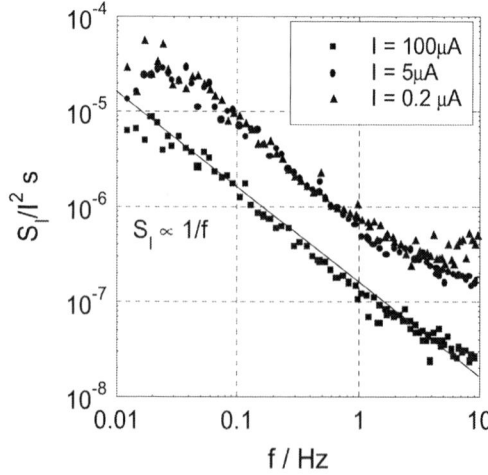

Figure 11. Normalized current noise spectral density in reverse mode

Due to defects in Ta_2O_5 layer the metal semiconductor (MS) diodes of $Ta - Mn_2O_3$ contacts are created during self-healing. Burst noise is experimentally observed if one elementary MS diode burst noise source is dominant. In Fig.12 the typical burst noise voltage signal in time domain is given, showing bistable fluctuation of current driven by carrier capture and emission processes at the defect spot.

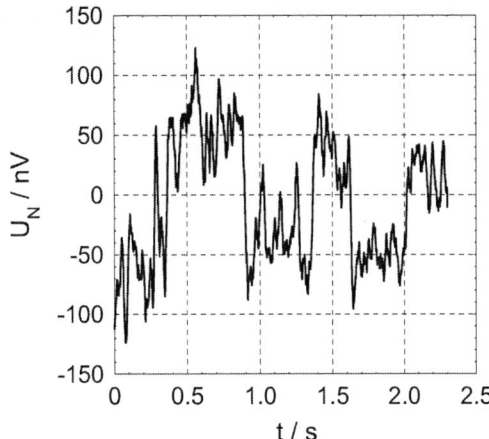

Figure 12. Burst noise voltage time dependence

The crystal structure of insulating film is very amorphous and we can expect a large number of randomly distributed localized states in forbidden band and at interfaces. Then 1/f noise spectral density can be generated as a superposition of Lorentzians with exponentially distributed relaxation time constant. We are surprised, that 1/f noise component has frequency exponent strictly equal to 1.

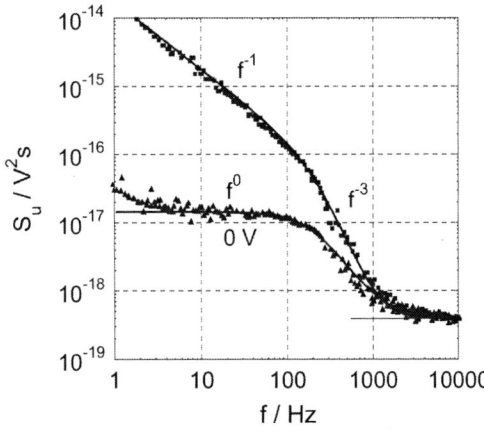

Figure 13. Noise spectral density vs. frequency

The frequency dependence of noise spectral density is affected by the changing the value of load impedance. The preamplifier input is shunted by the parallel combination of sample capacitance C_X and load resistance R_L and then the measured signal is attenuated with frequency. In the frequency region, where $\omega \ll 1/R_LC_X$ the noise spectral density is proportional to 1/f, while for frequency $\omega > 1/R_LC_X$ it decreases as f^{-3}, as is shown in Fig.13. The background thermal noise of load resistance $R_L = 1$ kΩ for zero applied voltage is also decreasing for higher frequencies until it reaches the value of amplifier background noise. To obtain undistorted results, numerical correction according to following equation should be done

$$S_I = \frac{S_U}{R_L^2}(1 + \omega^2 R_L^2 C^2)$$

where S_U is measurable quantity – voltage noise spectral density on the amplifier input. In Fig.14 the theoretical circuit characteristics is fitted in the experimental data.

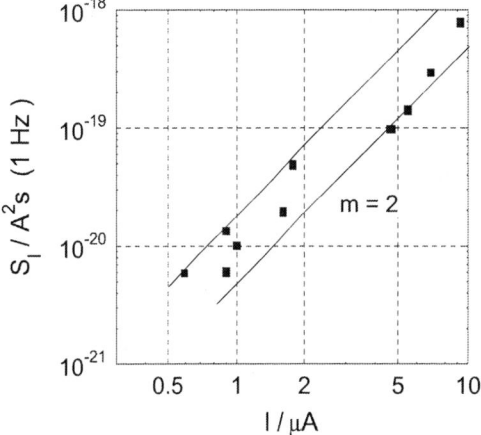

Figure 14. Noise spectral density value at 1 Hz versus leakage current for one sample at various applied voltage

Noise in normal mode results from current fluctuations and then current noise spectral density can be obtained from

$$S_I = S_U (1+\omega^2\tau^2) / R_L^2$$

where S_U is a measurable quantity – voltage noise spectral density on load resistance R_L and τ is the time constant characterizing signal attenuation due to capacitor shunting amplifier input:

$$\tau = R_LC = 0.1 \text{ s } (C = 100 \text{ } \mu F, R_L = 1 \text{ k}\Omega).$$

For $\omega\tau = 1$ the cut-off frequency is 1.5 Hz, and for $\omega\tau \gg 1$ it holds:

$$S_I = S_U \cdot \omega^2 C^2$$

The current noise spectral density S_I does not depend on load resistance.

Noise spectral density is proportional to square of current for low and medium value of electric field intensity both in normal (see Fig.14.) and reverse operation mode. For high electric field intensity normalized noise spectral density increases with increasing field intensity in normal mode probably due to avalanche process. In contrast normalized noise spectral density decreases in reverse mode probably due to formation of hot conducting channels, which change transport mechanism due to temperature dependence of electrical conductivity. There exists a feedback between Joule heating and electron transport. In this reverse mode VA characteristics, a negative resistance region was observed.

Leakage current has one component given by defects and then its value is used to estimate the dielectric layer quality. We observed that the current noise spectral density is also related to the technology and it constitutes the reliability indicator of capacitors. In Fig.15 the correlation between leakage current at rated voltage and noise spectral density value at frequency 1 Hz (in the 1/f region) is shown for ensemble of 80 samples. The majority of samples follow the quadratic law, although there are also some samples characterized by low noise and high leakage current. We suppose that DC current is a sum of at least two independent current flow mechanisms, which have not the same noise intensity.

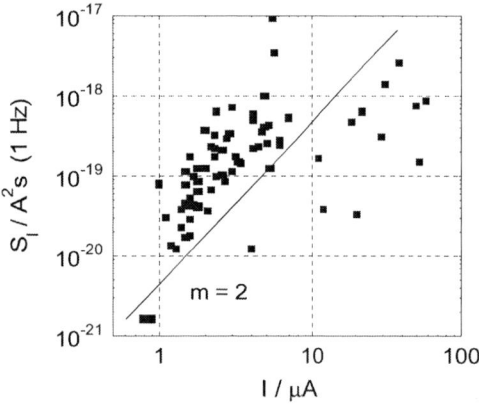

Figure 15. Noise spectral density value at 1 Hz versus leakage current for ensemble of 80 samples at rated voltage

Conclusion

1. Insulating layer thickness in these devices is from 20 to 100 nm and then these devices belong to nanoscale electronic devices in which quantum effects play important role.

2. Insulating layer is not perfect structure and it contains oxygen vacancies and other defects. They create deep localized states acting as donors with concentration about 10^{18} - 10^{19}cm^{-3}. At such level of donors a concentration impurity band can be created.

3. Theoretical barrier height for electron transport in normal mode is given by the difference between Ta/NbO and MnO$_2$ work functions. In these cases the potential barriers are about 1.5

eV. This value is changed if the traps are present at the Nb$_2$O$_5$ - MnO$_2$ or Ta$_2$O$_5$ - MnO$_2$ interfaces.

4. For electric field intensity higher than 100MV/m the second tunneling component appears. In this case higher barrier about 1 eV was found.

5. We have three methods to determine barrier height: VA characteristics in reverse mode at T = 300 K, the tunnelling current transport in normal and reverse mode at T = 80 K and CV characteristics analyses.

6. Tantalum and niobium oxide capacitors have identical conductivity mechanisms in dielectric that can be described by the same physical characteristics.

7. Knowledge of the theoretical model and the basic physical parameters is crucial for understanding the conductivity and degradation mechanisms in dielectric and subsequently for controlling and tuning the technology, so that optimum parameters of final capacitors are achieved.

Acknowledgment

This research has been partially supported by GACR No.102/05/2095 and under the project MSM 0021630503.

References

[1] J.Sikula et al., Charge Carrier Transport in NbO and Ta Capacitors in Temperature 100 to 300 K CARTS EU, 2006, 189-196

[2] J.Sikula et al., Tantalum Capacitor as a MIS Structure; CARTS USA 2000, 102-106

[3] K.W. Boer, Survey of Semiconductor Physics, Van Nostrand Reinhold (1990)

[4] S.M. Sze, Physics of Semiconductor Devices, Wiley-Interscience, New York, 1981

[5] C. A. Mead, Phys. Rev. 128, 2088, 1962

[6] A. Teverovsky, Reverse Bias Behavior of Surface Mount Solid Tantalum Capacitors; CARTS USA 2002, 105-123

Analysis methods for characterization of unleaded solderable surfaces

Thomas Hetschel[1,2] and Klaus-Jürgen Wolter[2]

[1] Robert Bosch GmbH, Stuttgart, Germany

[2] Electronic Packaging Laboratory, Dresden University of Technology, Germany

Phone: +49 711 811-42187, Fax +49 711 811-5112722, thomas.hetschel@de.bosch.com

Abstract

Due to the conversion to lead-free assemblies in electronics packaging industry, the solderability of the substrate surfaces play a decisive role for durable electronics products. The ageing behaviour of these surfaces affects the wetting properties of the solder material during the assembly process. A good wetting is very important for ensuring reliable solder interconnections. Different effects on the solder joint reliability are assumed to depend on the ageing conditions of these surfaces. For analysing the ageing variances complex analysis methods are required. The aim of this paper is to evaluate suitable methods for characterising lead-free solderable surfaces and their ageing behaviour. Different methods are examined and rated by their selectivity between unconditioned and aged surfaces. Every variance in the surface properties in a certain period of time can be seen as surface ageing. For Printed Circuit Boards, surface ageing can be classified in process ageing (e.g. reflow soldering) and storage ageing prior the assembly. To understand both ageing systems, surfaces have to be characterised accurately. For the evaluation of the analysis methods several surfaces (immersion tin, nickel-gold, organic solderability preservative layer) are investigated under different ageing conditions. Three characteristic properties are determined: the layer thickness, the chemical composition of the surfaces and the chemical layer composition. For measurement of these characteristic quantities, the following analysis methods are selected, X-ray photoelectron spectroscopy, time-of-flight mass spectrometer, auger spectroscopy, sequential electrochemical reduction analyses and focus ion beam.

Key words: analysing methods, lead free surface finish, immersion tin, immersion tin oxides

The printed circuit board (PCB) market shows a slow change to new lead free surface finishes, which will be used increasingly for new innovative products. Remarkable advantages like flatness, healthy and environmentally friendly manufacturing process are leading to this changeover which is also required by the legislation.

The introduction of new solderable surface finishes gives new challenges to the electronic industry, since there is no uniform standard of these solderable surface finishes defined until now. Different surface finishes are used in different countries, e.g.:

- North America - immersion silver
- Europe - immersion tin
- Asia - organic solderability preservative

The major function of the surface finishes is the protection of the copper on the PCB against oxidation and contamination with other substances. In order to determine the solderability and the ageing behaviour of these new surface finishes will be investigated and analysed in detail.

Previous standard analysing methods of the PCB industry are based on x-ray fluorescence spectroscopy (XRF), light microscopy and wetting tests with solder material. In special cases scanning electron microscopy (SEM) will be applied.

Due to the fact, that the thickness of the layer will be constantly decreased (e.g. 20 µm hot air levelling down to 1 µm immersion tin), alternative analysing methods are required to characterise lead free surface finishes.

Within the scope of this paper, different alternative analysing methods will be compared. A special focus is laid on the immersion tin surface since this is the mainstream in Europe. Immersion nickel-gold and organic-solderability-preservative surface finishes will be considered marginally.

Methodology

There are many methods to characterise thin layers or chemical compositions. The analysis methods are selected, depending upon their particular surface and the objectives of the investigation.

For the characterisation of thin layers and their chemical composition, high resolution measuring procedures are needed. Therefore, the necessary analysing methods are very complex and expensive. Possible analysis methods are following listed:

- X-ray photoelectron spectroscopy
- time-of-flight mass spectrometer
- Auger spectroscopy
- sequential electrochemical reduction analysis
- X-ray fluorescence spectroscopy
- energy dispersive X-ray analysis
- glow discharge optical emission spectroscopy
- focus ion beam

Principally, most of the methods are working similar. The surface will be bombarded by using X-ray, electron- or plasma beams. The response will be detected and evaluated by using spectroscopy detectors. Another procedure is based on the electrochemical reduction effect (SERA). In this paper is mainly examined:

- Auger spectroscopy (AES)
- X-ray photoelectron spectroscopy (XPS)
- time-of-flight mass spectrometer (ToF-SIMS)
- sequential electrochemical reduction analyses (SERA)
- focus ion beam (FIB)

The methods were selected under the consideration of their resolution, chemical composition of the surface finishes, detectability of wanted chemical elements and the availability of the needed equipment. The next chapter describes the function of the examined analysing methods.

Function of examined analysing methods

In Auger spectroscopy, the excitation source is a finely focused electron beam. Upon sample bombardment, a transfer of energy occurs which excites a core electron into an orbital of higher energy. Once in this excited state, the atom has two possible modes of relaxation: emission of an X-ray, or emission of an Auger electron. In both processes, the emitted particle will have an energy characteristic of the parent element. An energy spectrum of the detected electrons shows peaks assignable to the elements present. The ratios of the intensities of Auger electron peaks can provide a quantitative determination of surface composition. AES is sensitive to low atomic number elements and all elements save hydrogen and helium. Information in deeper layers is detected by combination of sputtering with an Argon gas and surface measurement.

In X-ray photoelectron spectroscopy (XPS) also called electron spectroscopy for chemical analysis (ESCA) - X-rays excite photoelectrons, and the emitted electron signal is plotted a spectrum of binding energies. The functionality is like as AES, only the stimulation source is an X-ray beam. Differing chemical states resulting from compound formation are reflected in the photoelectron peak positions and shapes. Spectral information is collected from a depth of ~5 nm layers, depending on the material studied. Information about sample regions lying deeper below the surface can be obtained by depth profiling with Argon ions. In these experiments, energetic Argon ions are used to etch the surface of the sample. The new surface created by the etching process is analysed by XPS.

Time-of-Flight Mass Spectrometer (TOF-SIMS) is a technique to determine elemental composition of samples, based on measurement of mass of atoms and molecules. The sample is bombarded by high energy gallium ions, they sputter secondary ions from the surface, and the masses of the secondary ions are measured in the mass spectrometer. In this work a time-of-flight detector of secondary ions was used. It has high mass resolution above 7000, high sensitivity and unlimited mass range.

Sequential Electrochemical Reduction Analyses (SERA) is an electrochemical process used to determine a variety of coating parameters that can predict solderability of PCB surface finishes. A small defined area is isolated on a test piece and a current is applied to oxidise or deoxidise a surface species. Potential is recorded over time yielding a series of plateaus corresponding to the appearance of oxidation. The voltage levels identify the species present and the time at each level measures the amount present.

Focus ion beam (FIB) is a combination of etching and scanning electron microscopy. A gallium ion beam cut a small partial ditch in the surface. The gallium ion beam is focused by a system of electrostatical lenses on a defined area of the sample surface. The interaction of the ion beam with the surface leads to the emission of secondary electrons and ions which can be used independently of each other for the generation of an image. The secondary electrons provide a clear material contrast (e.g. the difference between materials such as Sn, Cu_3Sn etc.). By using the so-called "channeling effect", the grain structure of poly-crystalline metals can be analysed.

Test specimen

A key parameter for good results is a suitable test sample. The design and assembly of the test specimens are decisive for the success of the investigation. An important parameter concerning the manufacturing process is the precipitation of the surface finish. Further fundamental parameters are "handling" and "storage" of test specimens during the investigation. In this investigation one-inch printed circuit boards (PCB) are used which are completely coated around the PCB geometry. There are three variants of PCB's with different surface finishes; Figure 1 shows the basic configuration of all variants.

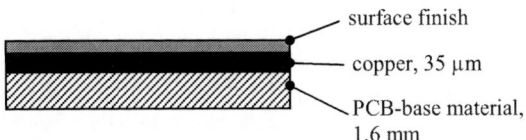

Figure 1: Cross section of a typical PCB configuration.

All specimens consist of the same base epoxy PCB material. The copper layer thickness is around 35 µm and the surface finish is directly deposited on the copper. Table 1 displays the important manufacturing parameters of the above mentioned three surface finishes, measured by the supplier with X-ray fluorescence spectroscopy (XRF).

Table 1: Manufacturing parameters of surface finishes.

surface finish	parameters	
	thickness	element concentr.
Immersion tin	1,1 µm	100 % Sn
Immersion nickel-gold	Ni : 5 µm / Au : 50-60 nm	100 % Au / 90 % Ni and 10 % P
Organic solderability preservative	300-500 nm	-

This investigation will focus upon the immersion tin surface. Immersion nickel-gold and organic-solderability-preservative will be also investigated but not discussed within this paper.

Ageing parameters of surface finishes

It is not sufficient to examine only the new condition of surface finishes. Finally, the point of interest is the alteration of the surface finishes after ageing under e.g. defined storage conditions. Hence, above mentioned analysing methods will be evaluated for different ageing conditions. Until now, there is no defined standard for accelerated ageing which is simulating the alteration of surface finishes during life time. Many investigations are using different ageing methods to simulate warehouse storage or manufacturing ageing (e.g. by multiple reflow procedures). Accelerated ageing will be performed under high temperatures and a high humidity [3], [4].

Refer to Table 2; three different variants of ageing conditions will be simulated. The ageing conditions correspond to the standard ageing conditions for tin lead systems according to IPC/EIA J-STD-003 [3].

Table 2: Conditions of ageing parameters and their identifications.

ageing states		label
new	-	x1
temperature storage 4h@155°C	TS	x2
temperature storage 4h@155°C+2 times reflow	TS	x10
temperature and humidity storage 24h@85°C/85% r.H.	TS/HS	x17

Results and Discussion

According to the current knowledge, Inter-metallic compound (IMC) will be accumulated by immersion tin at room temperature. A fast growth will be achieved by higher temperatures and a high humidity. Two types of inter-metallic layers are formed. At first, the growth of the Cu_6Sn_5 layer can be identified followed by the Cu_3Sn layer. At the tin surface, tin Oxide SnO and SnO_2 can be observed. The thickness of the monitored oxide layer is in a range between 2-4 nm (new condition). After ageing the thickness of the oxide layer grows slowly [6].

With the focus ion beam (FIB) analysing method the growth of the immersion tin inter-metallic layers can be seen; Figure 2 shows a picture of the new condition and Figure 3 a picture of an ageing condition at temperature storage (4 h@155 °C).

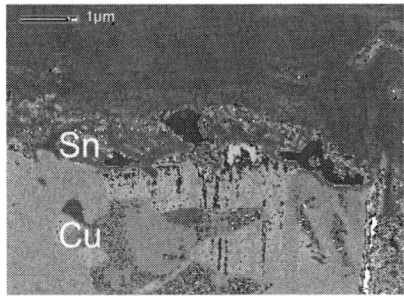

Figure 2: New condition of immersion tin surface finish.

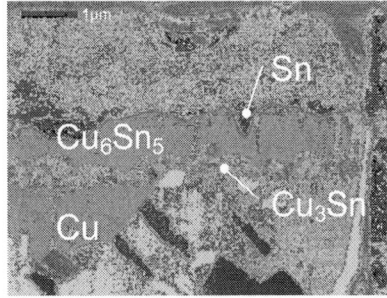

Figure 3: Ageing condition temperature storage (4h@155°C) of immersion tin surface finish.

Figure 3 shows that parts of the inter-metallic layer Cu_6Sn_5 touch the surface. The measured growth speed of the inter-metallic layer is in agreement with the results of other papers [5].

Figure 2 and 3 are made with a single beam system. New generation of focus ion beam analysing systems can determine chemical elements with an additional EDX unit. Since the resolution of scanning electronic microscopy is not sufficient, the growth of oxides can not be observed.

For analysing the elementary composition AES, XPS and TOF-SIMS are used. XPS and AES are similar analysing methods. Both methods can detect concentrations of chemical elements. All results of XPS and AES measurements are related to 100 atom percent. In comparison to the XPS results (Figure 5) the AES results shows a less concentration of carbon. During the AES investigation a colour change of the measuring point can be observed (refer to Figure 4).

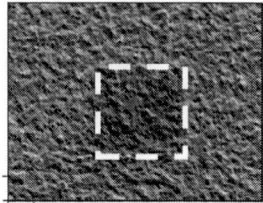

Figure 4: Measuring field after the AES surface investigation of immersion tin surface finish.

This effect is related to the electron beam, which leads to an undefined alteration of the surface. Due to this effect it is not possible to get reproducible results. In comparison to above mentioned result (AES), the XPS measurement does not show any modification of the surface.

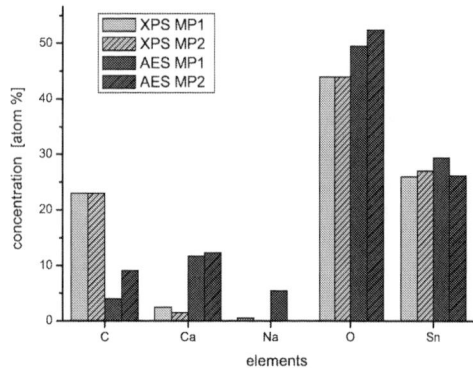

Figure 5: Comparison of XPS and AES surface analysing, specimen new condition and two measuring points MP1 and MP2.

Calcium and sodium can be seen as a contamination on the specimen. A remarkable higher concentration of calcium and sodium is observed by using the AES measurement. The exact thickness of the tin oxide can not be determined out of figure 5, since the maximum resolution concerning the depth is around 5 nm for the AES- and XPS- measurement system. Related to the latest technical expertises, the thickness of oxide layer is in the range between 2-4 nm [4]. Hence, the results of the AES and the XPS

measurements show only the spectra of the total Sn, SnO and SnO2 composition. The XPS analysis of aged specimens displays an interesting characteristic regarding the carbon concentrations which is obviously increased by high temperature ageing (refer to Figure 6). It is remarkable, that carbon can be still detected on the surface at 270°C. Under the condition x17 (24h@85°C/85% r.H.), the carbon concentration is less detectable due to a thin film of humidity which avoids the deposition of carbon on the tin surface. In addition, a higher oxygen concentration can be observed which probably leads to a growth of the tin oxide layers.

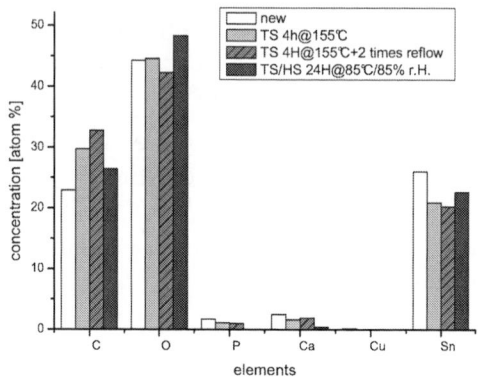

Figure 6: Comparison different ageing condition immersion tin by XPS surface analysing.

The compositions of layers are investigated by using AES and XPS depth profiles. The layers will be removed by using a sputter procedure with Argon gas. There is a correlation between the sputter time and the removed layer thickness. The result depends on the settings of the used equipment which is different for each system. Thanks to calibration options (e.g. sputter time versus deepness) quantifiable values can be achieved. Within the scope of these investigations the calibration has not been considered. Figure 7 shows an AES depth profile with x1 (new condition) and temperature storage x2 (4h@155°C).

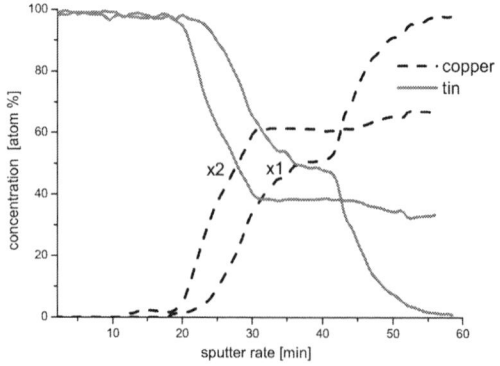

Figure 7: AES depth profile of immersion tin surface finish x1 new condition and x2 ageing condition (4h@155°C).

A shift of the intercepting point between copper and tin to the right side can be clearly observed in Figure 7. This is related to the growth of an inter-metallic layer (refer to Figure 7).

After approximately 30 min sputter time, the stoichiometric inter-metallic layer Cu_6Sn_5 can be detected in curve x2 (refer to Figure 7). The tin line achieves approximately 40% and the copper line achieves approximately 60%. This can be seen as the first inter-metallic phase. Figure 8 shows XPS depth profiles of immersion tin surface finishes comparing curve x1 (new condition) and curve x2 (4h@125°C+2 times reflow). The change from the the Cu_6Sn_5 layer to the Cu_3Sn layer can be clearly observed.

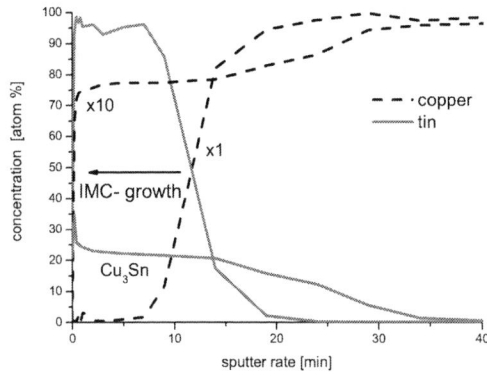

Figure 8: XPS deep profile of immersion tin surface finish x1 new condition and x10 ageing condition (TS 4h@155°C+2 times reflow process).

Based on these facts, the following conclusion can be made. It is possible to obtain the chemical composition and the corresponding depth information. The depth profiles of XPS and AES are comparable and show the same results.

Another examined measurement for surface composition is ToF-SIMS. Upon the surface a lot of hydrocarbons and tensides can be found. In addition, traces of inorganic compounds e.g. iodine, chlorine and bromine are detected. ToF-SIMS analysis shows only qualitative values but it is very sensitive concerning organic substances. Many measured substances point out traces of the PCB base material. Most substances are remaining on the surface after the manufacturing of printed circuit boards.

It is not possible to find significant differences of the specimens with different ageing states. Based on these investigations the measurement results present no statements concerning the surface oxidation or the surface contamination with other substances. Therefore, it is recommendable to use ToF-SIMS as an additional analysing method but not as the main investigation method. It is possible to detect some chemical elements but without any quantified information.

In order to measure the thickness of oxide layers, the sequential electrochemical reduction analysis (SERA) should be used since above mentioned analyzing methods are marginally practicable for this kind of thickness measurements.

The thickness of the immersion tin oxide layers will be measured in all different ageing states by using the SERA analysing method. Specimens under new condition and temperature storage (x2 and x10) do not show a significant oxide growth. Only the specimen with the x17 temperature and humidity ageing shows the growth of a SnO layer (refer to Table 3). A SnO_2 oxide layer can not be detected within the detection range of 1,5 nm.

Table 3: Results of SnO thickness measurements , measured by using SERA at different ageing states.

Ageing states		Thickness of SnO [nm]
New	X1	2,3
TS 4h@155°C	X2	2,5
TS 4h@155°C+2xreflow	X10	2,8
TS/HS 24h@85°C/85%r.H	X17	4,4

The accuracy of the analysing system SERA has to be checked, but the measured values are in agreement with the values of the literature [5]. Only a few methods are known to measure such thin layers. Possible alternatives are X-ray photoelectron spectroscopy (XPS) with different incident angles or transmission electron microscopy (TEM). Further examinations are planned with TEM.

Interestingly, no copper can be found on the surface due to the growth of inter-metallic compounds. The penetration depth of XPS is approximately 5 nm and the diameter of the measuring point is approximately 100 μm. The measured chemical elements are displayed in figure 6. The spectra of the tin compounds indicates that tin oxide and purely tin are existing. Related to these facts, following conclusion can be made: The inter-metallic layers grow up to the oxide layers, but do not penetrate the oxide layers. An academic comparison of the bond enthalpy between SnO, Cu_6Sn_5, and Cu_3Sn shows that the SnO compound is very stable in comparison to the inter-metallic layers [7], [8].

$\Delta H_B(SnO)$ = -285,8 KJ/mol

$\Delta H_B(SnO2)$ = -580,7 KJ/mol

$\Delta H_B(Cu_6Sn_5)$ = -67,1 KJ/mol

$\Delta H_B(Cu_3Sn)$ = -32,8 KJ/mol

(Values at 25°C)

Hence, the wetting takes place on the oxide layers and not on the inter-metallic layers. Further investigation will examine the wetting behaviour of different oxide states on immersion tin surface finishes.

Conclusion

There are a lot of analysing methods for the characterisation of the lead free surface finishes. Within the scope of this work different high-end analysing methods will be compared by using an immersion tin surface. Under consideration of boundary conditions like different measurement points of specimens and settings of the used analysing equipment, XPS shows the best results for immersion tin surface investigations. In comparison to AES no surface changes can be observed by using XPS. The quantifiable results show a good selectivity between test samples under new condition and ageing condition. Tin oxide layers can be detected with XPS, but it is not possible to determine the thickness of these layers. Based on the investigated surface finishes, AES and ToF-SIMS do not have any advantages compared to XPS.

In order to determine the very thin tin oxide layers, SERA is a possible investigation method. The thickness of tin oxide layers depends on the decreasing potential during the measurement.

To get a pictorial impression of the layer construction, FIB is a suitable method. Based on the pictures, the layers can be simply identified and the thickness of the layers can be easily measured. Needed information of the layer composition will be gained by using a depth profile measurement with XPS or AES. Referring to the results of the surface measurements, XPS is preferred for analysing depth profiles.

Related to above mentioned facts, XPS and SERA in combination present the best results within this investigation.

References

[1] M. Henzler and W. Göpel, "Oberflächenphysik des Festkörpers", Teubner Studienbücher, second edition, Stuttgart 1994.

[2] R. Biedorf, "Analytische Praxis in der Elektronikfertigung", Eugen G. Leuze Verlag, Bad Salgau 2005.

[3] IPC-4554, "Specification for Immersion Tin Plating for Printed Circuit Boards", Final Draft 3, May 2006.

[4] Sungil Cho, Jin Yu, Sung K. Kang, Da-Yuan Shih; "Oxidation Study of Pure Tin and Its Alloys via Elektrochemical Reduction Analysis", Journal of Electronic Materials, Vol. 34, No.5, 2005.

[5] Erika Gehberger, "Lötversuche auf bleifreien Oberflächen", VTE 15, Heft 2, 2003.

[6] Manfred Jordan, "Die galvanische Abscheidung von Zinn und Zinnlegierungen", Eugen G. Leuze Verlag, Bad Salgau 1993.

[7] H. Flandorfer, U. Saeed, C. Luef, A. Sabbar, H. Ipser, accepted for publication in "Thermochimica Acta 2007".

[8] K. Loosen, Thomas Claßen, Michael Müller; Thermodynamische Daten; http://www.chempage.de/ (20.01.2007).

Influence of Environmental Conditions on Pb-free Solder Joints Quality

A. Skwarek, K. Witek

Institute of Electron Technology, Cracow Division, 30-701 Cracow, Zablocie 39, Poland

+48126563144, askwarek@ite.waw.pl

Abstract

Pb-free solder joints and finishes, usually based of tin, applied in many electronic devices have been studied recently at the angle of their reliability in different working conditions. There are two types of phenomena, which can occur in different tin alloy influencing on electronic circuits reliability:

- *Tin pest – an allotropic transformation of white β - tin (body centered tetragonal structure) into gray α - tin (diamond cubic structure) in the temperature under + 13.2 °C*
- *Tin whiskers – single crystals of a few micrometers in diameter and a few millimeters in length growing spontaneously from the pure tin surface.*

Both, tin pest and tin whiskers, can lead to the deterioration of circuits reliability. Literature data indicate that these phenomena can occur in the alloys consisted of 95% of tin. In many commercial available Pb-free solder joints (SAC) the tin content is just at this level. It is also known that Pb addition at the level of 3-5% can prevent the changes in Pb-free solder joints, however European RoHS Directive restricts this value to 0.1%. Also pure tin is often using as an economical Pb-free plating option. Taking it under consideration, it is important to estimate the influence of environmental conditions and some factors on of these two phenomena occurrence.

Key words: tin pest, tin whiskers, SAC, environmental conditions

Introduction

The RoHS Directive introduced in July 2006 has had a great influence on electronic market. Many electronic manufacturers replaced traditional the Pb alloy by unleaded alloys, usually based on tin (SnAgCu – *abrev.* SAC). Plating composed of pure tin has been also widely used in many electronic applications as an agent enhancing solderability and preventing metal oxidation (in consequence element corrosion). The greatest advantage of this finish is a possibility of excellent control and uniformity of plating thickness. Tin finish possesses also additional feature enhancing its popularity; gives shiny surface which can be demanded in certain applications [1].

Wide use of tin finishes renewed concern over two phenomena occurring in tin rich materials - tin whiskers and tin pest. Both phenomena can occur in the alloys containing 95% of tin. Some additions can inhibit, the other ones on the contrary can promote the processes.

1. Tin whiskers

Tin whiskers are single crystals of a few micrometers in diameter and a few millimeters in length spontaneously growing from the alloy surface. Typical growth rates are about $0.01 - 0.1$ Å s^{-1} [2]. The phenomenon is believed to take place for compressive stress relief [3]. The whiskering process is strongly dependent on environmental conditions (Tab. 1). Temperature of +50°C, thermal cycling –40°C/+90°C and mechanical stress (scratching) are the strongest factors promoting whisker formation. Neither barometric pressure nor electric field are necessary conditions for whisker formation. The moisture influence on whisker formation is not still unequivocally determined [1].

Literature data indicate also the influence of factors connected with plating chemistry and process on phenomenon occurrence; e. g. plating thickness, grain size and shape. Another theory indicates the role of screw dislocation or Frank-Read source in whiskering process but for now there are no direct data confirming this hypothesis [4].

Many researchers have reported the promotion of tin whiskers by some impurities, especially copper, but also lutetium or zinc [2, 3].

Generally, the process intensifies with both increasing compressive stress and diffusion process rate.

Whiskers can be dangerous for circuit reliability because of short circuits or device littering. Due to circuit geometry reduction the problem became very important. Whiskers are responsible for many system failures in military, medical and telecommunication industries. Three earth satellites were completely destroyed as a result of tin whisker short originating from pure tin plated relays [1].

Table 1. Chosen factors promoting whisker formation [1, 3]

Factor	Description
Environmental stresses	▪ Temperature of +50°C ▪ Thermal cycling ▪ Moisture (?)
Mechanical stress	▪ Stretching or blending
Intermetallic formation	▪ Cu_6Sn_5
Impurities	▪ Copper, zinc
Plating parameters	▪ Plating thickness 2μm – 10 μm ▪ Grain size <1 μm
Dislocation mechanisms (?)	▪ Screw dislocation ▪ Frank-Read source
Recrystallization mechanism (?)	▪ Formation of new grain sets
Diffusion mechanisms	▪ Movement of material from the substrate to the plating layer
Tin oxides	▪ Oxidation of tin layer

(?) - ambiguous data

Many scientists try to elaborate whisker mitigation methods [1, 5] . Generally all methods lead to whiskers reduction but not elimination. The simplest method is pure tin plating avoidance. This includes the application of alloys with a second metal as well as usage of "matte" tin plating of thickness less than 1μm or greater than 20 μm (grains in the range of 1 – 3 μm are less prone to whisker formation). Since recent experiments have reported that 3% of Pb addition can efficiently limit the danger of whiskering, application of the alloy with such a lead content in military was accepted in framework of RoHS Directive exemptions. Another way of action is striping and replacement of already existing plating by SAC finish, but only in the case when SAC paste and process will be used for assembly. An application of physical barriers (washers, ceramic spacers, etc.) is also widely used. Where it is possible, hot solder dipping is recommended but the limitation of this method is connected with Pb application in some electronic equipment (RoHS Directive) [1].

The second mitigation method recommended for reduction of whisker formation is conformal coating usage. For this purpose urethanes can be applied but the data concerning other conformal coating (silicones, acrylic, parylene and epoxy) are very poor. As mentioned, conformal coatings protective layers only reduce the risk of whisker formation (not prevent). NASA data indicate that a layer of thickness of 50 μm protects the circuits from whisker formation after 3 years ambient and 50°C storages.

For reduction risk of tin whisker formation as a result of tin oxidation two processes are possible: (i) performing burn-in in an inert environment or (ii) tin surface reflow. Both processes lead to tin oxide reduction and in consequence a compressive stress decrease.

Table 2. Mitigation methods of tin whiskers formation [1, 5]

Mitigation method	Description
Pure tin avoidance	▪ Where it is possible, usage of tin alloy with second metal addition ▪ Hot solder dipping
Plating process modification	▪ Grain size in the range 1 – 3 μm ▪ Plating thickness < 1μm or > 20 μm
Conformal coating usage	▪ Whisker formation slowering not elimination
Prevention of tin surface oxidation (compressive stress reduction)	▪ Burning-in in an inert environment ▪ Tin surface reflow
Prevention of intermetallic formation	▪ Physical barriers (e. g. washers) ▪ Underplating (Ni layer)

2. Tin pest

The basis of the process is an allotropic transformation of white β- tin (body centered tetragonal structure) into gray α - tin (diamond cubic structure), which additionally causes the volume increase of about 26 percent. Tin pest (tin plague or tin disease) occurs in the temperature under 13.2 °C. This phenomenon can be harmful for SAC, often used instead of Pb solder joints. Tin pest intensifies with temperature decreasing and reaches the maximum at about –50 °C [6]. Characteristic feature of this process is its contagiousness. The transformation into α - tin initiates at the surface and expands on it [7]. The allotropic forms differ in density and conducting properties. β- tin exhibits density of 7.31 g/cm^3 and metal conducting properties, whereas density of α-tin is only 5.77 g/cm^3 and the conductivity shows semiconducting character [8]. β- tin is also characterized by malleability (capacity of being hammered into sheets) and ductility (capacity of being drawn into a thin wire). Becoming α-tin, this element transforms into gray powder [9].

$$Sn_{cub} \xrightarrow{13.2°C} Sn_{tetr} \xleftarrow{161°C} Sn_{r\,hom} \xleftarrow{231.8°C} Sn_{lq}$$
$$\quad\; \alpha \text{ - tin} \qquad\quad \beta \text{ - tin} \qquad\quad \gamma \text{ - tin}$$

where: cub – cubic structure, tetr – tetrahedral structure, rhom – rhombohedral structure, lq – liquid phase (according to Bielański.A. [6])

The process can be accelerated by certain solvents or addition of some elements (such as aluminum, zinc, germanium), trace impurities as well as by deformation or strain in the solid and retarded by another elements, such as antimony, bismuth, cadmium or lead.

The biggest problem with tin pest is the lack of unequivocal detection methods. For this purpose XRD analysis is recommended, but there are some limitation connected with this method. In literature there exist the cases where tin pest was identified although the results of x-ray analysis were negative. Tin pest appearance in the French organ in Bordeaux was deduced by Bouchy and Roland on the basis of the morphology and microstructure of the deteriorated materials, although the XRD analysis did not detect gray tin [10].

Experimental

For evaluation of the influence of environmental conditions on Pb-free circuit reliability, two groups of samples were exposed at temperature of –55°C or +55°C. The circuits were prepared with PCB technology using ImmSn and ImmAu finishes, PLCC and QFP fine pitch integrated circuits and commercially available SAC pastes (Tab. 3). Then the samples were examined at the angle of tin pest (those exposed at -55°C) or tin whiskers (those exposed at +55°C) presence. Additionally, one part of the samples was protected with conformal coating for verification of the hypothesis that this type of protection can reduce the risk of tin whiskers.

The results of investigation at the angle of the tin pest occurrence showed significant changes in the shear force values after exposure at -55°C. The differences between various SAC solder joints are shown in Figure 1 and Figure 2. One can notice the differences in the shear force values dependent on the finish type (ImmSn and ImmAu).

The biggest decrease of the shear force values after 1000 h of temperature stress was observed for the samples with Indium solder joints and ImmSn finish (Fig. 1). In case of 3500 h of

temperature stress the worst results were obtained for Cobar solder joint with ImmSn finish. Preliminary studies exhibited that the decrease of the shear force values can be caused by α tin presence. Many blisters which were observed on the finish surface may confirm the phenomenon occurrence (Fig. 3).

Table 3. Chemical composition of SAC pastes

Commercial name	Sn (wt.%)	Ag (wt.%)	Cu (wt.%)
Indium 241	95.5	3.8	0.7
Cobar CUAG-XM 3S	95.5	4.0	0.5
Multicore LF300	95.5	3.8	0.7
Omnix 5100	96.5	3.0	0.5

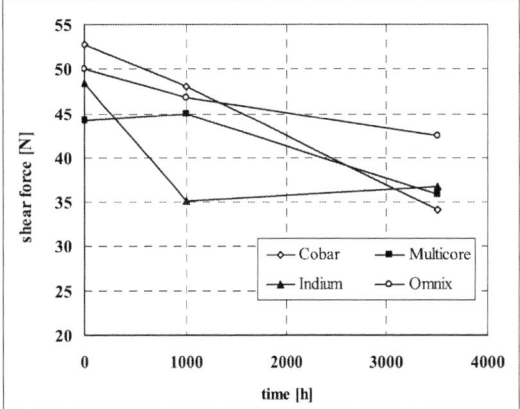

Figure 1. Shear force versus time of exposure at –55°C for the samples with ImmSn finish

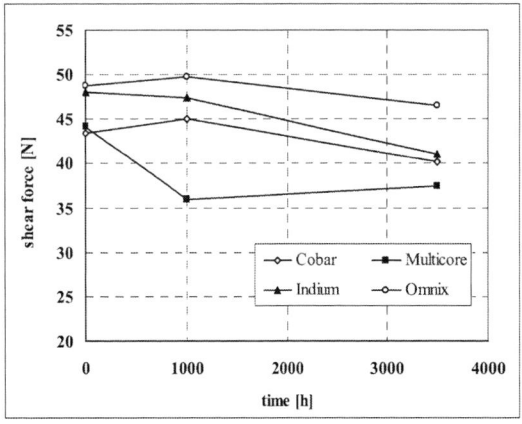

Figure 2. Shear force versus time of exposure at –55°C for the samples with ImmAu finish

Appearance of α tin gray powder in solder joints can lead to their mechanical properties degradation and as a result joints weakening. However, this hypothesis requires confirmation.

For this purpose X-ray analysis can be used but the results could not give a clear response by the reasons given above (see Introduction).

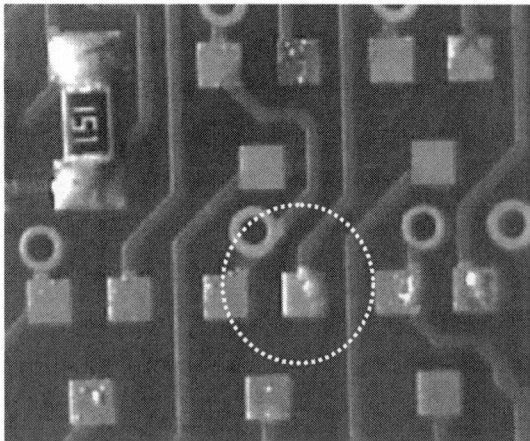

Figure 3. Appearance of a sample with ImmSn finish after 3500 h of exposure at –55°C

The tests with the samples stored at +55 °C were performed for estimation if tin whiskers are a real danger in the circuits made with unleaded technology. SEM images realized using scanning electron microscope Nova NanoSEM 200 (FEI Company) are presented in Figures 4, 5 and 6. The tests showed that after 3000 h of temperature stress of +55 °C the phenomenon of tin whiskers formation took place on the tin finish surface. As illustrated in Figs. 4 and 5, the tin whiskers found on finish surface were needle-shaped.

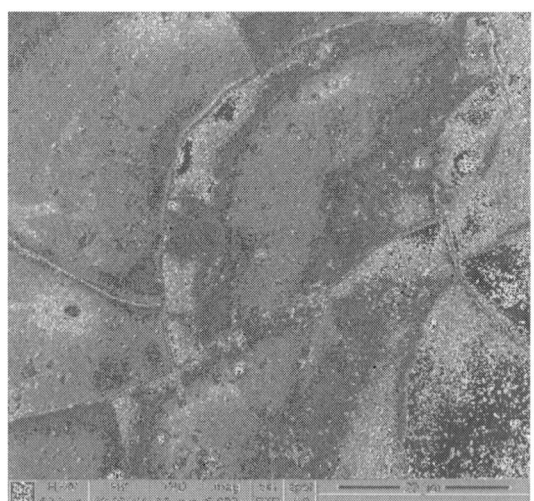

Figure 4. SEM image of the tin finish covered with conformal coating

The samples with conformal coating did not exhibit any noticeable form of whiskers (Fig. 4). It can be because the whiskers observed were shorter (2 µm) than the coating thickness (20 µm), so they were invisible in scanning microscope. Their

disclosure requires a longer time of exposure to the temperature (for whisker elongation).

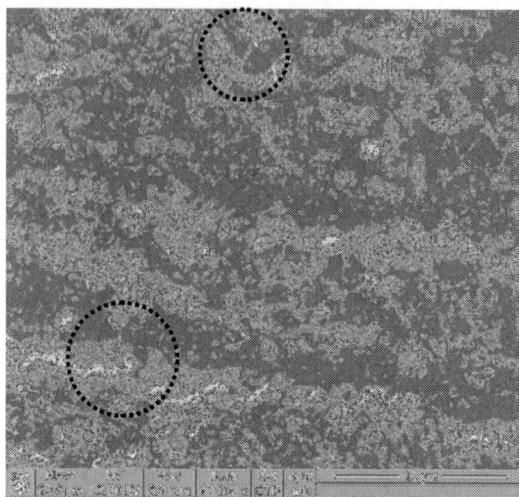

Figure 5. SEM image of the tin finish surface with visible tin whiskers (magnification 10 000x)

Figure 6. SEM image of the tin finish surface with a single tin whisker (magnification 50 000x)

Conclusion

The knowledge acquired during the research task, both of the literature and the experimental nature leads to the following conclusions:

- Both phenomena, tin pest and tin whiskers, are a real danger occurring in circuits realized with unleaded technology
- Evaluation of the factors inhibiting these two phenomena requires more time and further tests
- The recommended test methods could not give a clear response on the asked questions concerning the tin pest and tin whiskers detection

References

1. J. A. Brusse, J. G. Gary, J. P. Siplon " Tin Whiskers: Attributes and Mitigation" 16 [th] Passive Components Symposium, pp. 221-233, 2002.

2. T. H. Chuang, H. J Lin, C. C. Chi " Rapid growth of tin whiskers on the surface of Sn-6.6Lu alloy" Scripta Materialia, vol. 56, pp. 45-48, 2007

3. W. J Choi, T. Y. Lee, K. N. Tu, N. Tamura, R. S. Celestre, A. A. MacDowell, Y. Y. Bong, L. Nguyen, G. T. Sheng " Structure and Kinetics of Sn Whisker Growth on Pb-free Solder Finish", Notional Semiconductor website http://www.national.com/packaging/files/sandk ofsn.pdf

4. R. Gedney, J. Smetana, N. Vo "NEMI Tin Whisker Projects" http://thor.inemi.org/webdownload/newsroom/P resentations/Amsterdam04_tin_whiskers.pdf

5. M. Osterman "Mitigation Strategies for tin whiskers", CALCE website http://www.calce.umd.edu/lead-free/tin-whiskers/TINWHISKERMITIGATION.pdf

6. A. Bielański "Chemia ogólna i nieorganiczna", PWN pp. 487, 1973 (in Polish)

7. Y. Katiya, C. Gagg , W. J. Plumbridge "Tin pest in lead-free solders", Soldering & Surface Mount Technology, vol. 13/4, pp. 39-40, 2001

8. R. Larsky "Tin Pest: Still a Forgotten Concern in Lead Free Assembly" www.indium.com/drlarsky/entry.php?id=310

9. www.chemistryexplaind.com/elements/T-Z/Tin.html

10. C. Chiavari , C. Martini, G. Poli , D. Prandstraller „Deterioration of tin-rich organ pipes" J. Mater. Sci, vol. 41, pp. 1819-1826, 2006

Forward Compatibility Assessment for Aeronautical and Military Communication Systems (GEAMCOS project)

O. Maire[1], A. Chaillot[1], C. Munier[1], I. Lombaërt-Valot[1], S. Bousquet[2], C. Chastanet[2],
D. Plouseau[3], E. Munier[3], D. Maron[4], P. Raynal[4], S. Villard[5], R. Dumonteil[5]

[1] EADS France, 12 rue Pasteur BP 76, 92152 SURESNES Cedex, France

[2] AIRBUS France, [3] EADS Secure Networks, [4] ACTIA, [5] TECHCI

E-mail: olivier.maire@eads.net, Phone: +33 (0)1 46 97 37 61

Abstract

According to the RoHS European directive, electronics industry moves to lead-free soldering. Aerospace and military applications are not directly concerned by this evolution, nevertheless they have to face supply chain issues, obsolescence of components and they have to guarantee reliability of their equipment. In order to assess lead-free assemblies in harsh environments, the European LIFE project GEAMCOS, for Green Electronics in Aeronautical and Military Communication Systems, was launched at the end of 2005. The first stage of the project concerns the reliability and compatibility of leaded components soldered with a lead-free process. This step is called the forward compatibility. This paper presents the results concerning: the solder paste selection, the influence of process on delamination, and the results of ageing tests on reliability of the solder joints.

Key words: lead-free, SIR test, reliability, microstructure, ageing test

Acronyms and symbols

CTE   Coefficient of linear Thermal Expansion
ENIG   Electroless Nickel Immersion Gold
IMC   InterMetallic Compounds
SAM   Scanning Acoustic Microscope
SEM   Scanning Electronic Microscope
T_g Glass transition temperature [°C]

Introduction

The commercial electronics industry has implemented lead-free assembly since the 1st of July 2006, but concerning lead-free electronics, specifications requested for aeronautical and military communication applications (reliability, life-time, security, maintainability) do not permit to use simply alternative solutions already proposed for consumer electronics without some evaluations [1]. The main objectives of this three-and-a-half-year project, supported by the European Commission, are thus to evaluate lead-free technologies in harsh environments and to transfer these new technologies to other sectors with the same reliability needs. The tasks of this project are:

- to define lead-free processes (materials, soldering profiles, design rules),
- to assess reliability of lead-free assemblies in harsh environments,
- to assess compatibility and reliability of leaded components soldered in lead-free process, the usually called step: forward transition.

Due to the life cycles of products, leaded components are stored for long-term period to maintain equipment, to assure reparability and to face to component obsolescence. Mixed assemblies will be thus encountered in a near future, with lead-free process and leaded components. A component with lead (Pb), with SnPb pin coating, is said to be forward compatible if it can be reliably attached to a printed circuit board (PCB) using lead-free solder paste with a lead-free profile (typical peak reflow temperatures ranging from 240°C to 260°C).

This paper presents the first results of the experiments carried out on a dedicated test board focusing on risks of forward assembly, metallurgical compatibility and failures.

Presentation of the GEAMCOS project and the test boards

The present work is part of the European LIFE project GEAMCOS supported by European commission. This acronym stands for Green Electronics in Aeronautical and Military Communication Systems. The project leader is EADS. The partners of this project are: AIRBUS and EADS Secure Network (end users), TECH CI (PCB manufacturer) and, ACTIA (equipment manufacturer).

Its first objective is to qualify mixed assemblies (lead-free alloy and leaded components) encountered during board repair or maintenance. For that purpose, a test board (Figure 1) was designed with the characteristics specified in the Table 1.

Two sets of boards were manufactured:

- the test boards which were soldered with convection oven, called convection boards,
- the test boards which were soldered with repair process, called repair boards.
-

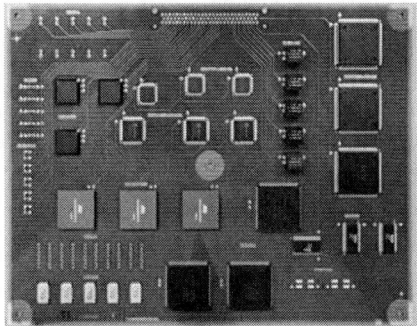

Figure 1: GEAMCOS forward test board without PTH connector

Table 1: Main characteristics of the forward test board

Components	FBGA, CICGA, SQFP, PQFP, TQFP, PLCC, LCC, SSOP, C0709, DO27A, DO213, DIL	
PCB	Laminate material	- high T_g (150℃) and low CTE FR4 - current FR4 (T_g = 150℃)
	Finish	ENIG, Immersion Sn, HAL
Solder material for reflow and wave	Sn-3Ag-0.5Cu (SAC305)	
No clean flux used with cleaning		

A full lead-free test board will then be used to assess reliability in harsh environments in another part of this project, and to adjust process parameters. The final objective is to validate the results with a full lead-free functional demonstrator used in military communications. In parallel, finite element modelling (FEM) simulations complete this work.

Experimental procedure

First of all, Surface Insulation Resistance (SIR) tests were carried out on samples to choose a solder paste and define a cleaning procedure.

The components used for this project are functional or daisy-chained. The content of Pb in pin coating is in the range of 5 wt% to 25 wt%. Ni or Cu was identified as under layer barrier according to the different packages. The lead-frame material is Cu-based or Fe-42Ni alloy. The composition of the FBGA balls is Sn-37Pb (eutectic composition) and for the CICGA, the balls are made with 90Pb-10Sn hard solder material.

After delivery, the components were examined by SAM from KSI with mainly 25MHz transducer. The nature of pin coating material was analysed with JEOL 6490LV SEM coupled with Energy Dispersive X-ray (EDX). Visual inspection and electrical continuity measurements were performed after soldering and ageing tests. The test boards were inspected with X-Rays to detect the percent of voids and potential bridges between pins or balls.

Then, they were submitted to thermal cycles or temperature/humidity tests. In order to study the reliability of the solder joints, daisy-chained and passive components were monitored during thermal cycles to detect electrical failures and electrical discontinuities. This monitoring allows to identify failure location for cross-sections. The cross-sections were then observed with SEM/EDX to analyse the microstructure, the IMC within the joints, the crack path, etc. Previously, X-Ray, SAM and optical inspections were also performed before DPA.

Furthermore, some shear tests were performed on the smallest components to evaluate the decrease of shear strength with the ageing time. The test boards were also submitted to vibration tests but the results of this part are not presented in this paper.

Solder paste evaluation

First of all, two solder pastes were tested for the convection soldering process: noted paste A and paste B. Different cleaning methods were also tested: cleaning or no cleaning.

SIR tests were carried out on samples with modified patterns compared to the IPC-TM-650 2.6.3.3 standard. It is composed of 4 interdigital comb patterns:
- one classic SIR comb (pattern A): a 400µm width of the lines with a 500µm spaces between interleaved combs (like the pattern IPC-B-24 of the norm IPC-TM-650),
- two modified SIR combs (patterns C and D): a smaller space between the interleaved combs than for the pattern IPC-B-24 (that is a 400µm width of the lines with a 200µm space between the interleaved combs),
- one SIR comb (pattern B) with a solder mask in order to protect the copper conductor track from oxidation and to prevent solder bridging during the soldering process. This pattern is also characterized by a 400µm width of the lines and 200µm space between the interdigital combs.

These SIR board modifications are made in order to consider the technical evolution of printed circuits [2]. The A pattern is the one defined by the IPC standard. The others are closer to the circuits used nowadays.

The SIR test conditions are described in Table 2. Test samples were observed at initial time and then every 168h until 504h.

Table 2: SIR test conditions

Ageing conditions			Measurements	
T	Bias	Humidity	Frequency	Bias
40°C	5V (DC)	93 % RH	5min	5V (DC)

A few comments on SIR boards during or after the soldering phase may be done:

- The SIR boards processed with the solder paste A gave the best results. Only one short-circuit was touched up after reflow on the thinnest SIR pattern (B pattern). The behaviour of the paste after the screen process was good.
- With the solder paste B, many short-circuits were touched up after reflow (B pattern).

The optical observations of the SIR test boards did not reveal any major defect (dendrites or corrosion of metallisation). The main remarks are:

- Residues of flux are observed on most of the SIR test boards. They become more crackled after 500 h of ageing tests (Figure 2). Nevertheless, they do not represent any risk for the insulation resistance.
- Some metallic tracks of boards soldered with paste B have granular aspect but they have the same before and after 500 h of ageing tests.
- Some "wires" were included with the flux but they are not metallic and they have the same aspect before and after 500 h of ageing tests.

The SIR electrical measurements confirm that there was no major damage during the 500h of ageing (Figure 3). Measurements were always up to $10^8\Omega$ i.e. no leakage current or isolation resistance degradation was observed independently of the solder pastes tested. Some typical curves of insulation resistance versus time are given in Figure 3. A few remarks on SIR results can nevertheless be done:

- Insulation resistance drops were observed about 1 and 2 weeks due on the one hand to the short stop of measurements for optical observations and on the other hand to a stability problem of humidity of the climatic chamber.
- There is no important impact of the geometry of SIR patterns (different widths between the lines).
- The use of a solder mask, used to protect the copper conductor track from oxidation and to prevent solder bridging during the soldering process, has also a very small impact on resistance data.
- The cleaning method did not influence the results.

According to this solder paste selection, the solder paste A was used for the test board soldered in the convection oven.

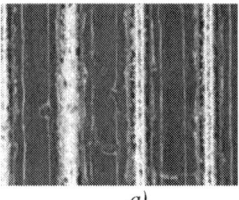 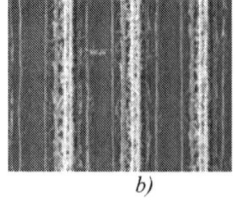

a)   b)

Figure 2: Flux residues between tracks after soldering (a) and, after 3 weeks of ageing (b) – (solder paste A, immersion Sn PCB finish, cleaned)

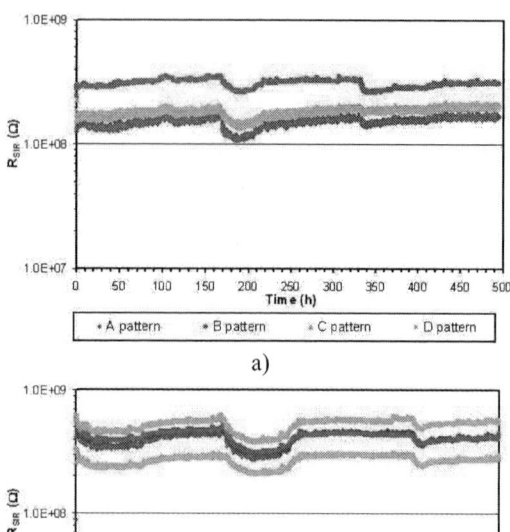

a)

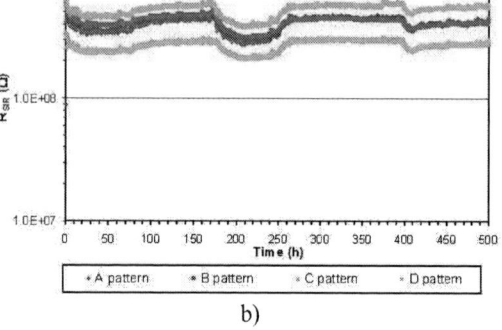

b)

Figure 3: SIR measurement curves with ENIG PCB finish and cleaning, for a) A (a) and B (b) solder pastes

Cleaning method assessment for wave soldering

Different cleaning methods were also assessed for wave soldering process:

- cleaning with saponified water,
- cleaning with dedicated product.

The soldering material used was: the SAC305 with a no-clean flux. The SIR tests were only performed on samples with ENIG PCB surface finish.

After soldering, on boards cleaned with dedicated product as well as with those cleaned with saponified water, only the presence of solder residues and microballs can be detected. Boards cleaned with dedicated product did not change during and after the ageing period (500 h) and a stable SIR was measured. On the contrary, boards cleaned with saponified water have presented dendritic growths after 1 week at 40°C/93% RH (Figure 4) and, a decrease of the SIR values was noticed on the patterns C and D (Figure 5). Furthermore, SIR curves presented more noise representing small leakage currents. The metallic migration was favoured by the presence of surface impurities at the surface of the board. The cleaning with dedicated product is then recommended for our test board soldered with wave soldering process.

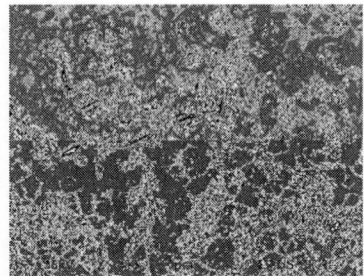

Figure 4: Dendritic growth after 1 week of ageing test at 40°C/93% RH on samples cleaned with saponified water

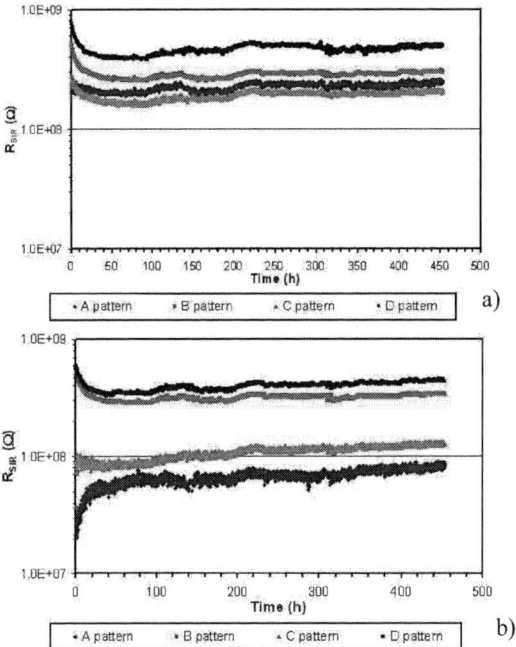

a)

b)

Figure 5: SIR measurement curves with dedicated product for cleaning (a) and saponified water(b)

SAM results

One of the main drawbacks with forward assembly is the risk of package failures when the components are soldered with higher temperature [3]. Though delaminations may not be a direct cause of failure, their presence may induce several modes of failures and, must be considered as not acceptable failures as: penetration of moisture, rupture of interconnects inside the package or, increase of package thermal resistance followed by overheating. The components were thus inspected with SAM as received and after soldering. Basically, five samples of each plastic encapsulated component were investigated.

Except two minor particulars for one SQFP44, with a very little suspicious area, and for four SSOP56, with a slight delamination at the outer tip of the lead frame fork, all other components have been checked with no default found at delivery.

After soldering, all the components were not observed due to a limited scan area and large size of the test board. Nevertheless, the investigated components did not show any defaults, except two PLCC84, which showed abnormal phase reversal of the acoustic echo. It indicates a probable delamination default. The delamination is located on the top of the die and, the echo returning from the glue under the die is also abnormal.

X-Ray inspection results

No soldering default, like bridge between interconnection, was visualized. Nevertheless, X-Ray inspections allowed quantifying voids within solder joints. For the FBGA components, the number of balls without any void as well as the number of balls with a surface of void superior or inferior to 25% of the global ball surface, were calculated. The Figure 6 and Figure 7 report these percentages. Solder leaded BGAs with lead-free solder paste results in increased voiding [4].

The repair process decreases the percentage of voids. The balls without any void are more important and the size of voids is reduced compared to convection soldering. The temperature was more important in this process allowing more gas vaporization.

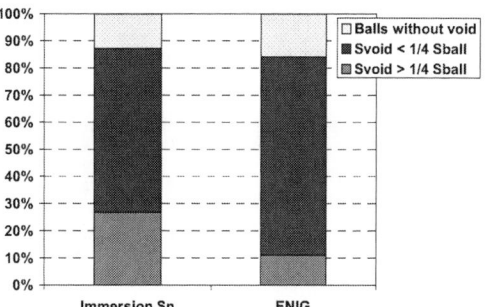

Figure 6: Percentages of FBGA balls without void, with a surface of void superior or inferior to 25% of the total ball surface (convection process)

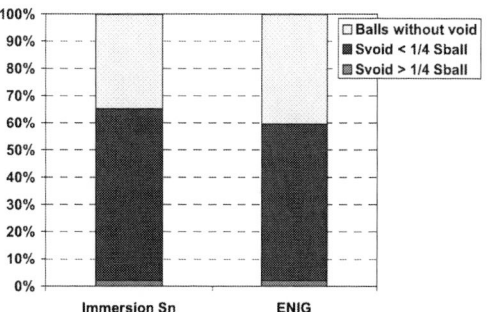

Figure 7: Percentages of FBGA balls without void, with a surface of void superior or inferior to 25% of the total ball surface (repair process)

Metallurgy of mixed solder joints

Cross-sections of the components were carried out after soldering in order to assess metallurgical compatibility between pin coatings and SAC (Sn-3Ag-0.5Cu) solder material. The lead (Pb) from pin coating spreads in all the solder joints excepted for through-hole ones. In that case, the lead (Pb) is localized in some areas (Figure 8). For BGA components, complete mixing is achieved (Figure 9). The interfacial reactions are similar to the ones observed with full leaded solder joints. Cu_6Sn_5 IMC forms when Sn reacts with Cu from the component pin or the PCB pad. Ni_3Sn_4 IMC is observed at the pad level when ENIG PCB surface finish or diffusion barrier from the component is used. No particular issue was revealed after soldering according to metallurgical point of view.

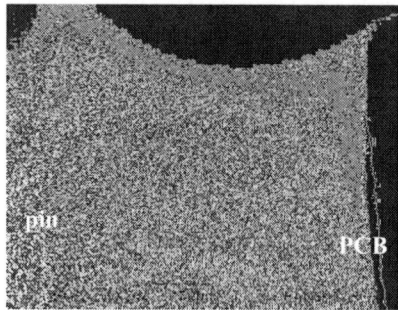

Figure 8: Localization of Pb observed in through-hole solder joints (diode component)

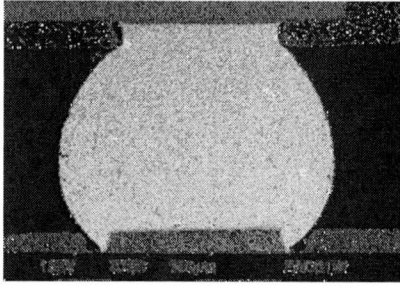

Figure 9: Microstructure of BGA balls after soldering, showing a uniform distribution of Pb within the solder joint

Otherwise, important differences were observed for the solder joints of through-hole components between the convection soldering profile and the repair process. In the first case, the microstructure was composed of large Sn dendrites separated by the precipitation of small Ag_3Sn and Cu_6Sn_5 IMC. Furthermore, large platelets or needles of Ag_3Sn were observed. In the second case, Cu_6Sn_5 IMC as star shape was identified (Figure 10). They were localized preferentially in the bottom face of the board where the temperature was higher. This morphology could be explained by an excess of temperature and a rapid diffusion or dissolution of Cu to form this kind of IMC.

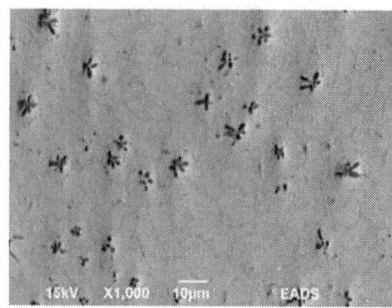

Figure 10: Solder joint microstructure of through-hole components soldered with repair process showing Cu_6Sn_5 IMC with star shape

Forward soldering defects

The main problems encountered on through-hole components were: pad lifting, fillet lifting and fillet tearing (Figure 11 and Figure 13). This defects were also reported in literature with lead-free soldering process [4][5].

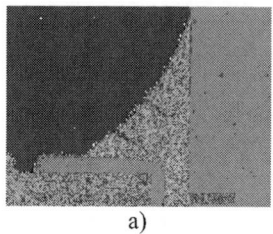 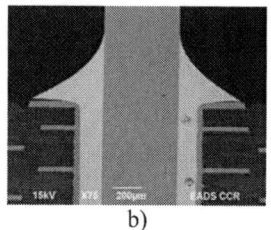

a)   b)

Figure 11: Defect on through-hole components, a) fillet tearing, b) fillet lifting, observed on the top of the boards

Pad lifting, fillet lifting and fillet tearing are effects resulting from the differences in thermal expansion coefficients of the PCB base material, the epoxy/glass FR-4 laminate, as well as the copper barrels and copper tracks on the PCB. During contact with the solder, the board material expands more than copper hole-wall metallisation in the Z-direction. After the board has left the solder bath, the connection begins to cool down to ambient. The board material cools down and returns to its original planar shape when the joint begins only solidifying. This movement introduces considerable stresses to the solder joint, which is still very weak at this stage. These stresses may cause pad lifting; or, if the adhesion between pad and board is at that point stronger than the solder, it can cause cracks in the solder surface, known as fillet tearing, or fillet lifting (Figure 12). Pad lifting was always observed on bottom side of the boards, whereas fillet lifting was noticed on their top side.

Fillet tearing, or small cracks at solder joint surface, and fillet lifting on plated through-hole solder joints are commonly considered cosmetic issues, having no detrimental effect on solder joint reliability [4]. However, pad lifting may compromise joint reliability. Some solder joint components submitted to the repair process

410

presented un-acceptable cracks and delaminations after soldering without any ageing test stress. Pad lifting usually conducted to some internal delaminations and copper track cracks (Figure 13).

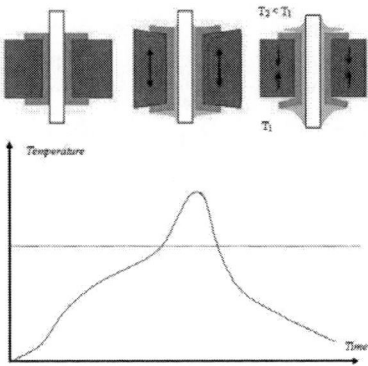

Figure 12: Schematic representation of fillet and pad lifting during cooling, due to CTE mismatch

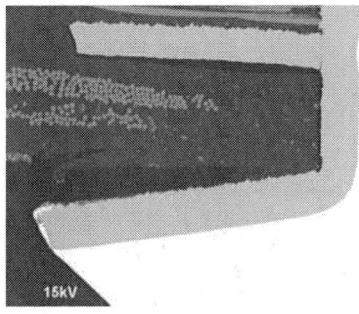

Figure 13: Pad lifting, due to CTE mismatch, leading to PCB delamination and copper crack

Thermal cycling tests

The test boards were submitted to thermal cycling tests to accelerate thermo-mechanical fatigue of solder joints due to CTE differences. The range of temperature was [-40; +100°C], with dwell effective times of 20min and ramp of 10°C/min. The test conditions were defined according to FEM simulation results [6]. All the tests are not completed and the preliminary results show few electrical defects. After 1100 cycles, a DO213 SMT component electrically failed indicating probably a complete crack of the solder joint. A similar defect was measured on a R1206 component on repair board after 1400 cycles. These are the only failures at this time after 2000 cycles. For BGA components, different events were registered from 1000 cycles with no complete failure. Cross-sections will confirm if the balls are completely cracked.

The defects of plated through-hole components did not lead at this time to electrical failure in spite of important pad lifting. Destructive failure analyses have to be performed to complete

these first results and to inspect propagation of cracks in the solder joints.

Conclusion

The GEAMCOS project focuses on lead-free assemblies in aeronautical and military communications systems. Due to strategic stock of components, leaded components have to be soldered in a near future with a lead-free process. The first step of this project is thus to assess the forward compatibility in terms of metallurgical risks, and components risks and, to evaluate the impact on the solder joint reliability.

The first experimental tests have allowed the choice of a lead-free solder paste for convection reflow and a cleaning method for wave soldering. Then, physico-chemical analyses and optical inspection has showed that soldering process did not alter the integrity of the used components and no metallurgical issue was identified at the solder joint level. Nevertheless, some defects were observed with plated through-hole components: fillet lifting, pad lifting and fillet tearing. If some of these defects are cosmetic, the pad lifting issue has leaded to crack inside the PCB after soldering. Some intermittent events were detected on BGA components from 1000 cycles. Cross-sections have to be performed to identify the degradation of the BGA solder joints.

At this time, ageing tests are not totally completed, but the first electrical failures has occurred after 1100 cycles in the range [-40; +100°C].

Acknowledgments

The authors wish to thank the European commission for the support of this project and, A. Delye, T. Boutaric, for their participation to the tests and analyses.

References

[1] JCAA/JG-PP Lead-Free Solder Project Joint Test Report, http://acqp2.nasa.gov/JTR.htm, downloaded on January 2007

[2] C. Hunt, "Development of surface insulation resistance measurement of electronic assemblies", NPL Report MATC(A)70, 2001

[3] E. Bradley, "Lead-free Solder Assembly: Impact and Opportunity", ECTC Conference, 2003

[4] P. Biocca, "Lead-Free and Lead Contamination", Surface Mount Technology, October 2004

[5] C. Hunt, "Technology mission to assess the status of lead-free soldering in Japan", NPL Report MATC(A)12, 2001

[6] A. Chaillot, G. Massiot & al., "Finite element modelling (FEM) of green electronics in aeronautics and military communication systems (GEAMCOS)", London, Eurosime Proceeding, 2007

Eco-design workflow process

Cyril Vaško[1], Ivan Szendiuch[1], Karsten Schischke[2]

[1] Brno University of Technology, FEEC, Dept. of Microelectronics, Udolni 53, 602 00 Brno, Czech Republic

Phone: +420 5 4114-6136, Fax: +420 5 4114-6298, E-mail: cyril.vasko@phd.feec.vutbr.cz

[2] Technical University of Berlin, Gustav-Meyer-Allee 25, 13355 Berlin, Germany

Phone: +49 30 46403-156, Fax: +49 30 46403-131, E-mail: schiskche@izm.fhg.de

Abstract

Eco-design is a frequently applied concept, but mostly as a case study or based on a given product, which has to be improved. Rarely design for environment is used as part of the development of a new product, because environmental assessments usually need a sound data basis, see e.g. common life cycle analysis concepts. Hence, this paper presents an integrated approach, how to deal with environmental issues during product design, when knowledge about the later product is still fairly limited - and the possibility to implement major changes is still given. The approach presented focuses on lean and smart measures, which work without additional extensive data acquisition and scenarios. However, they impose uncertainties, but remain applicable for the designer.

Key words: Eco-design, environment, EuP, design workflow, product life cycle

Introduction

Early steps of product development determine essential features in all phases of the life cycle of a product. Thus, design offers the largest degree of freedom in influencing the characteristics of a product. To optimize the environmental performance of a product became a major task in the past, as awareness of environmental issues related with electronic products increased. Nowadays eco design is applied frequently in the electronic sector, but mostly as an "add on" activity or a separate case study - rarely as an integrated aspect of electronic design. In the near future the trend towards eco design will be pushed even more by the European Energy-using Products (EuP) directive [1], which will make eco design an essential requirement for electronic products.

Unfortunately many aspects of eco-design concepts and conventional design workflow are not compatible and sometimes even contradictory. Therefore, applicability of eco design within established general design workflows is limited.

The main task of this paper is to present an eco-design approach developed from the perspective of the product designer. Such an approach will lead to better acceptance among designers as a possibility is shown, how to make (environmental) decisions in their daily work.

There are a huge number of tools for eco-design [2], including checklists as well as calculation models, to assess environmental performance of products. ISO/TR 14062 gives a standardized guideline for eco-design. However, integration in generic design processes is weak as eco design has been seen by now mainly from the viewpoint of an environmental expert and rarely from the product designer's perspective.

On the other hand efforts have been undertaken to structure the multi-disciplinary design process of microelectronic systems design. Top-down versus bottom-up approach has to be balanced thoroughly.

By now, no efforts have been undertaken to link eco-design activities with the specific procedure of microelectronic systems design.

Integrated design approach

Overview

To make eco design feasible as an integrated procedure needs more detailed guidance. Therefore, this paper points out the design phases, where an optimization of environmental properties is possible, considering technological and economic constraints as well.

The design process consists of a sequence of decisions to be made, mainly based on technological and economic aspects, but in general not considering the environmental consequences.

The knowledge about the product changes throughout the design process from a rough project idea to a detailed bill of materials and production concept – and the data base on which environmental assessments can be based becomes more and more detailed as well. Hence, the main property of the suggested eco design tools has to be applicability according to the available data base: Additional extensive data acquisition is not feasible as it hinders a "short time-to-market strategy". Therefore, the main question is: What is already known about the later product at a specific stage and how can easy-to-use tools add significant environment related decision support at this point?

1) Specification

At the beginning of a design process, there exists solely a project idea, which is described as a rough outline in the specification. Already at this point some preliminary environmentally-relevant conclusions regarding the later product are possible: For example, if the product is aimed to be a long-living always switched on device the energy efficiency and consumption throughout the use phase of the product is of major environmental relevancy and should be described as an optimizing target within the specification. If the product is small and sold to the end-user it is likely that this product, despite all take back efforts, might end up in the municipal household waste stream. For such a product the specification should outline the target to avoid materials with a large "ecological footprint", as these materials will not be recovered from the municipal household waste stream, as well as hazardous materials, which impose an environmental risk at end-of-life.

It is important to distinguish between minimal and facultative environmental requirements. Minimal criteria are usually given by law or customer specifications. Facultative criteria are aspects for a product with an environmental friendly profile comprising more than just the minimal criteria. Mostly, such criteria are only qualitative, such as

- avoid / reduce raw materials with a certain "ecological footprint",
- reduction of energy and material consumption during production, or
- minimization of energy consumption at all working conditions

A first identification of relationships between development targets is helpful. E.g. the target of "low cost" might be inline with "avoiding materials with a large ecological footprint" (precious metals!), but there might be a contradiction with "avoiding hazardous materials". Probably even two environmental targets might be in contradiction to each other.

Clarifying development targets and identifying trade-offs between targets helps to guide the design process.

As an additional guiding document for the specification also environmental requirements published by potential customers are helpful.

The specification is the main document for the whole design process, but it is not static. In every design step the specification might be adapted and

refined by the iterative design control. This design control is an interaction of the designer, the project manager and of different experts. Hence, also the specification has to include environmental targets, which should be refined as the design process proceeds.

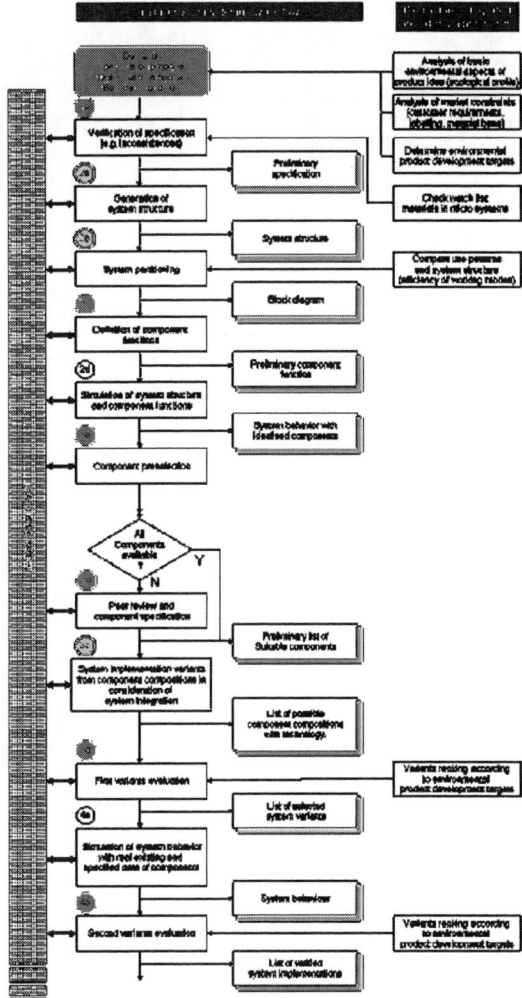

Figure 1: Microelectronic systems design work flow and integrated eco-design measures – part 1

2) System study

Following the generation of the system structure the different variants of integrating basic functions are developed (system partitioning). At this point major decisions on partitioning are made. First improvements having in mind the use phase are possible right now: For a power supply device e.g. different possibilities for transforming energy, storing energy might be alternatives. A solar charger might charge an internal battery first before charging a mobile device. Integrating an option in the system study, to directly charge the battery of a mobile device, means a significant increase in energy efficiency, although the charger might need additional components to perform both charging an integrated battery and charging an external battery directly, as different use modes. The consequence is a foreseeable trade-off between energy efficiency, material content of the product, and manufacturing costs. Having in mind the overall working conditions of a solar charger the environmental priority should be on energy efficiency instead of minimizing material contents. Regarding the trade-offs concerning energy efficiency and manufacturing costs a management decision is needed – and preferably involvement of the marketing department to communicate reduced life cycle costs versus higher product prices for the energy efficient version to the customer.

The definition of component functions can be supported by a watch list relating functions, which might be realized with a certain technology, with the materials usually used to realize this technology. For example, sensors for microelectronic systems, depending on function and technology, might need the use of specific materials.

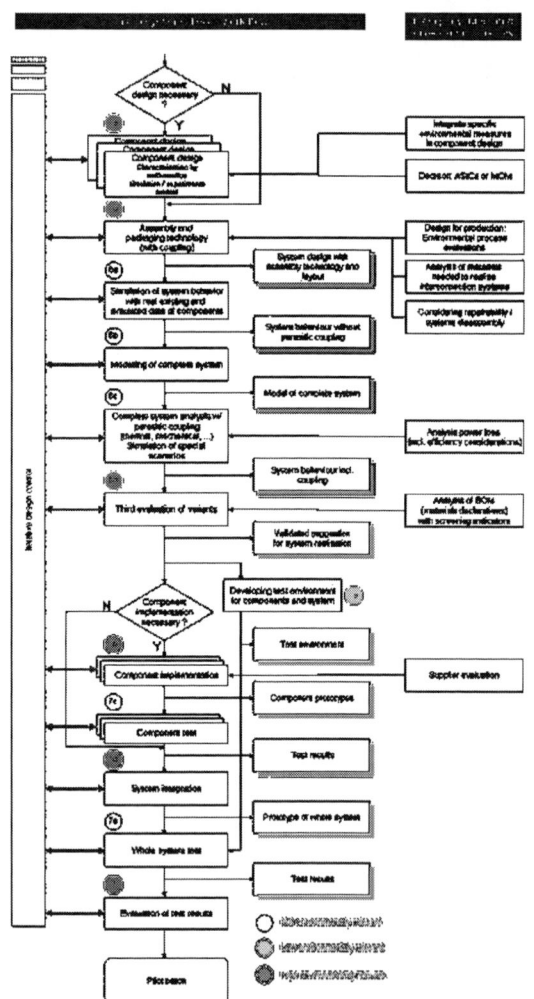

Figure 2: Microelectronic systems design work flow and integrated eco-design measures – part 2

3) Component selection and system implementation

The component functions as defined in the previous design step can be realized usually with different suitable components, and also with different system implementation variants. Hence, the result of this design step is a comparison of different variants.

4) Simulation of system behavior

Simulation of system behavior might exclude certain variants or might change priorities given to certain variants. From an environmental point of view, no new information arises at this point. The identified environmental preferences might be confronted with changed technological preferences.

5) Component design

It might be necessary to design new components, in case that not all needed components are available. Further disciplines, such as microelectronic design, require development of an ASIC (application specific integrated circuit).

Closely linked to component design is the concept for assembly and packaging. Packaging and interconnection technology affects material content of the product significantly.

6) System simulation

Based on system simulation a third variants evaluation will be done, considering the meanwhile more detailed knowledge about the product.

When availability of components is checked, and new components have been designed respectively, a preliminary bill of materials for the variants can be compiled. Depending on availability of full material declarations for these components a screening assessment of the product composition can be undertaken. Several environmental indicators are established for such assessments; one of these is the Toxic Potential Indicator, assessing the potential toxicity of the single materials and the whole product composition. For a detailed methodology description, see [3]. Fig. 3 illustrates the procedure from BOM (with material declarations) to an environmental weighting of materials and components. Thus, identification of critical components is possible and a comparison of the different variants.

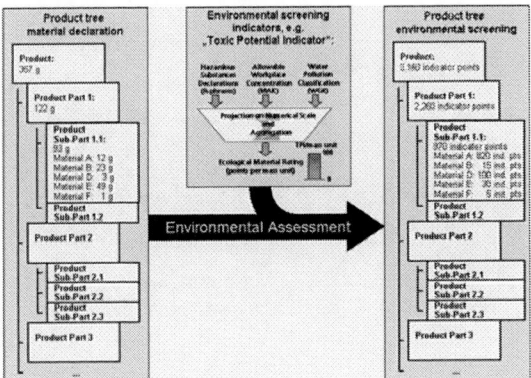

Figure 3: From Bill of Materials (BOM) to environmental assessment and ranking

7) System realization and test

Sub-task of system realization is component implementation. If the components are purchased the availability has to be assured. At this point a supplier evaluation could help to chose the appropriate supplier according to e.g. environmental aspects.

8) Pilot batch

Manufacturing the pilot batch is the first time, where real life production data for a specific product is at hand. But at this point most of the environmental impacts are fixed. A frequently cited figure states, that after product design up to 80% of environmental impacts of the whole product life cycle is fixed. Now, process optimization measures and smart recycling concepts might be developed. However, influence on environmental aspects is limited for such follow-up activities.

Conclusion

ISO/TR 14062 offers a basic guideline, how to approach eco design, but an integrated design for environment for microelectronic systems needs a more detailed understanding of the specific design process and the environmental consequences of specific design decisions. To find more environmentally benign design alternatives needs adapted lean assessment tools and guidelines as part of the integrated design control. The integrated approach presented in this paper is the basis for a common understanding of environmental aspects in microelectronic systems design. Eco design becomes an integrated part of the microelectronic systems design flow, instead of an "add on" activity. Thus, efficiency of eco design is increased and time-to-market for "green" products is improved.

Future work will focus on detailed environmental, economic, and technological assessment of exemplary design decisions within the stages of microelectronic systems design. This will enable the design team to make environmentally relevant decisions on a sound basis.

Following the presented approach will support compliance with the coming European EuP directive in an efficient way.

Acknowledgements

The paper has been prepared as a part of the research work under grant projects "Research of Microelectronics Technologies for 3D systems" GAČR 102/04/0590, and with the support of the Czech Ministry of Education in the frame of Research Plan MSM0021630503 "MIKROSYN New Trends in Microelectronic Systems and Nanotechnologies", and grant project of Ministry of Education FRVS 3217 "Introduction of Eco-design in Microelectronics Technology Education".

References

[1] Proposal for a Directive of the European Parliament and of the Council On establishing a framework for the setting of Eco-design requirements for Energy-Using Products and amending Council Directive 92/42/EEC

[2] M. Hagelueken, K. Schischke, J. Mueller, H. Griese, "Welcome to the Jungle" – Survival of the Fittest Environmental Screening Indicators? Electronics Goes Green 2004+, Berlin, September 6-8, 2004

[3] N. Nissen, Entwicklung eines ökologischen Bewertungsmodells zur Beurteilung elektronischer Systeme, Dissertation, Technical University Berlin, 2001

[4] K. Schischke, H. Griese, J. Mueller, I. Stobbe, „Micro Systems Design Work Flows – Assessment of Environmental Driven Criteria", International Asian Green Electronics Conference, March 15-18, 2005, Shanghai, China

Observations on Particle Loaded Silver Inks

Ulrike Currle, Klaus Krueger

Institute of Automation Technology, Helmut-Schmidt-University / University of the German Armed Forces,
Holstenhofweg 85, 22043 Hamburg, Germany

Ulrike Currle: Phone +49/(0)40/6541-3461, Fax +49/(0)40/6541-2004, E-Mail: currle@hsu-hh.de
Klaus Krueger: Phone +49/(0)40/6541-2722, Fax +49/(0)40/6541-2004, E-Mail: klaus.krueger@hsu-hh.de

Abstract

Many circuits in microelectronics are based on the deposition of particles. The deposition requires low cost methods with a high quality combined with low effort and high flexibility. The inkjet method fulfils these requirements perfectly for colour inks, nevertheless it is rarely used for particle loaded inks in microelectronics. We expect this method to have a high potential in microelectronics favoured by the development of new nanomaterials and discuss restrictions and possibilities of the processing. We consider silver particles which show a high density compared to the organic dispersion medium. Inkjet inks are low viscosity suspensions with a target viscosity of about 20 mPas. For low target viscosities and high particle loading promoting particle interaction sedimentation becomes a serious problem and stability issues occur as great challenge. Stability, viscosity and printability appear to be the main factors for a reliable printing process and satisfying results. Having analysed these aspects using the mentioned particles, the functionality of the inks comes to the fore. Lines are printed using a commercial piezo printhead and our self made printing equipment. The examination guides to promising inks with special focus on the evaluation of the inks.

Key words: Ink-jet, silver powder, dynamic viscosity, stability, sedimentation, light-scattering

Introduction

Digital printing attracts attention as deposition method that allows for easy handling, high flexibility and accuracy not only but especially for rapid prototyping. There is a great interest in the inkjet deposition of conductive materials and many researchers have published their successful printing experiments with nanosized silver colloids. Lee *et al.* [1] printed 50 nm silver particles on a glass substrate obtaining satisfying electrical properties. Kim *et al.* [2] deveolped a conductive ink with silver particles around 21 nm for inkjet printing. Perelaer *et al.* [3] evaluated microwave sintering of the inkjet printed silver nanoparticles on polymide substrate while Kamyshny *et al.* [4] compared an ink containing silver nanoparticles and an oil-in-water microemulsion based ink. These are only some promising examples of inkjet applications conducted during the last two years which underline the potential of inkjet printing in microelectronics.

The availability of suitable printheads which combine desired requirements concerning drop size, printing velocity etc. is essential for the relevance of inkjet printing for industrial applications. The printing experiments in **figure 1** are performed using a self made research printing station consisting of a piezo printhead from Microdrop with a nozzle diameter of 100 μm. The printing parameters are adjusted by optimizing the droplet formation in a drop-watch station and the lines are printed by fixing the printhead and moving the substrate on a planar motor. Details on the printing system and further printing samples can be found in [5, 6, 7].

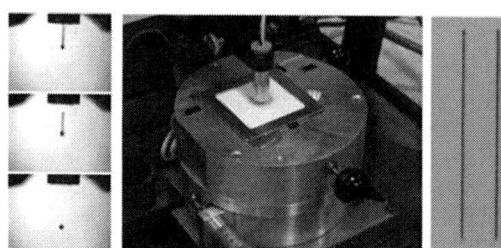

Figure 1: Droplet formation, printing process and printed lines (120 μm) on alumina substrate

The quality of the printing result depends not only on the printing system but is also strongly influenced by the properties of the ink. Viscosity and stability appear to be of particular importance especially when working with particle inks. Therefore the effect of the particle content and the particle size on viscosity and stability is evaluated.

Test Execution

The inks are build up on commercial silver particles and terpineol ($\rho = 0.934$ g/cm^3). Silver has a density of 10.5 g/cm^3 and the particles vary in size. The main properties of the silver particles are summarized in **table 1**.

Table 1: Properties of the silver particles

	S-A	S-B	S-C	S-D
D50 in μm	0.08	0.3	0.3	2.6
Surface in m^2/g	8-11.5	2.71	2.15	0.41

For the determination of the shear viscosity a Physica cone-plate rheometer (MCR301) with the cone having a diameter of 50 mm and an angle of 1° is used. The shear viscosity or dynamic viscosity η is given by

$$\eta = \tau / \dot{\gamma}$$

with shear stress τ and shear rate dγ/dt [8]. For ideal viscous fluids the ratio between shear stress and shear rate is a material's constant. If the viscosity is independent of the shear rate we also speak about a Newtonian flow or a Newtonian fluid. In case the viscosity decreases with increasing stress a so called shear thinning effect is observed while increasing viscosity with increasing stress indicates shear thickening.

Before each measurement a 10 minutes break allows the sample to recreate and to adapt to the temperature. The tests up to a shear rate of 1000 s^{-1} are performed using 30 data points and a logarithmic time schedule which gives a longer measurement interval at low shear rates. Measurements up to 10000 s^{-1} are conducted with a constant time interval for each measurement and 50 data points.

The sedimentation of the particle inks is quantified using a light-scattering method. The Turbiscan Classic from Formulaction scans in 40 μm steps the height of the sample and detects the intensity of the transmitted and backscattered light (850 nm). The measured intensity is related to a standard substance and depends on particle properties like size, concentration and adsorption. The transmission signal gives information about turbid samples and clearance effects, while the backscattering signal records changes in concentrated suspensions like the formation of a sediment. Both methods are used to compare the influence of different particles on the stability of the ink.

Discussion

Viscosity and stability are two key properties of a particle ink for inkjet applications. On the one hand a high viscosity reduces the sedimentation velocity of the particles but on the other hand the printhead requires a low viscosity and non-varying viscosities for high shear rates ensure a uniform droplet formation on constant printing conditions. Increased particle concentration allows for thicker layers in a shorter period and less liquid phase has to be dried on the substrate. The particle size highly influences the stability of the ink. Therefore the influence of the particle concentration and the effect of different particles is discussed more precisely.

Flow Properties of Particle Inks

Discussing the properties of particle inks it is of interest whether the dynamic viscosity of the inks depends on the shear stress or not. As shown in **figure 2** a change in viscosity appears for inks with different concentrations at low shear rates. In the illustrated example the ink with 10 wt% particles reveals for shear rates smaller than 1 s^{-1} a shear thinning effect at 20 °C while the other inks behave rather diffuse in this region.

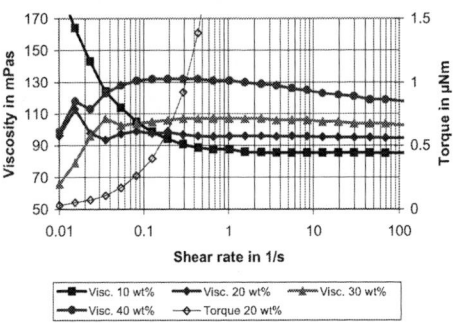

Figure 2: For small shear rates the viscosity measurement of particle inks becomes difficult.

Low shear rates correspond with very small measured torques and the experiments underline that viscosity values measured at torques smaller than 1 μNm should be handled with care. Furthermore figure 2 suggests that for shear rates bigger than 1 s^{-1} inks may show a viscosity that is independent of the shear stress. This effect is evaluated for shear rates up to 10000 s^{-1} (**figure 3**).

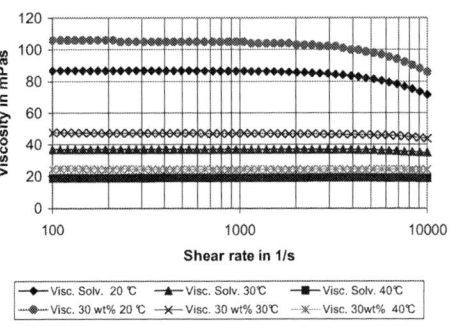

Figure 3: Particle inks show Newtonian behavior for high shear rates.

Influence of the particle concentration

In the following the effect of the particle concentration on the dynamic viscosity profile is examined. Viscosity values smaller than 1 s^{-1} are not taken into account. The considered inks vary only in the particle content. Up to 20 wt% the inks in **figure 4** show Newtonian behavior at 20 °C. Particle concentrations higher than 30 wt% feature a shear thinning effect at 20 °C that intensifies with increasing particle content.

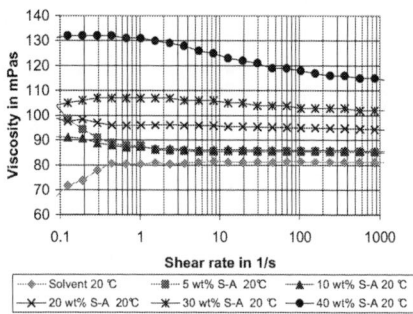

Figure 4: Inks with a particle fraction higher than 30 wt% show a shear thinning effect.

The shear thinning effect at 20 °C observed with the 40 wt% particle ink decreases with increasing temperature and decreasing viscosity as illustrated in **figure 5**. At 40 °C a Newtonian behavior is obtained at a viscosity of around 30 mPas.

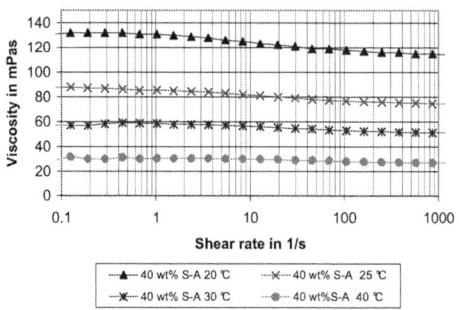

Figure 5: The shear thinning effect at high particle concentrations reduces with increasing temperature.

As the ink with 20 wt% particles in **figure 6** does not offer a shear thinning effect even not at low temperatures when the viscosity reaches values of around 300 mPas it is likely that the shear thinning effect is only related to the particle fraction of the ink.

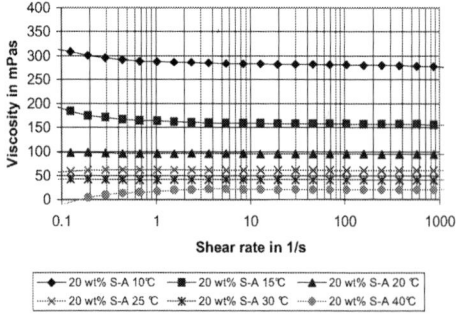

Figure 6: The shear thinning effect depends only on the concentration not on the temperature.

In order to evaluate the temperature effects, inks with different particle concentrations are analyzed at different temperatures. **Figure 7** reveals that the temperature effect on the viscosity of the ink is mainly given by the temperature dependence of the viscosity of the underlying solvent.

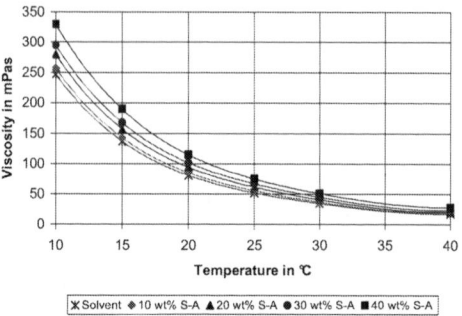

Figure 7: The effect of temperature changes on particle ink viscosities is approximated with polynomials of 4th order and is mainly influenced by the temperature behavior of the underlying fluid.

The absolute viscosity change being caused by different amounts of particles decreases with increasing temperature. In **figure 8** it looks like changes are almost negligible at 40 °C. Nevertheless the relative change increases, e.g. for the 40 wt% ink from 34% at 330 mPas to 59% at 28 mPas (40 °C).

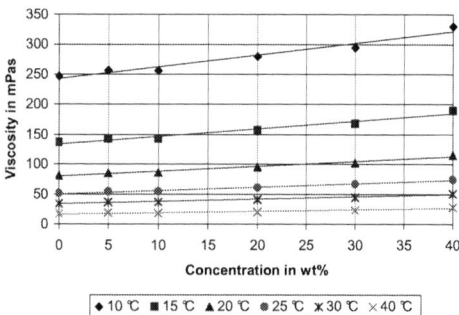

Figure 8: The effect of the particle concentration on the viscosity of the ink follows a linear approximation and decreases with increasing temperature.

Effects of different particles

The particle type influences the viscosity of an ink with 30 wt% particles in a rather small range and differences seem to be mainly caused by the size of the particles. In **figure 9** the smallest particles (D50 = 0.08 µm) show the highest viscosity and a slight shear thinning effect at 20 °C while the viscosities of inks with different particles of the same size are nearly identical with constant viscosities for different shear rates. The shear viscosity of the biggest particles is a bit unexpected as the viscosity values are bigger than the ones of the medium particles and a slight shear thickening effect seems to appear from shear rates above $100\ s^{-1}$.

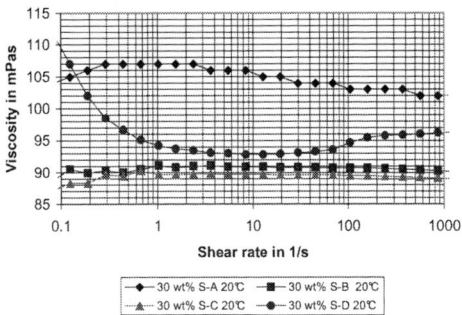

Figure 9: Different particles change the viscosity in a rather small range.

The shear thickening effect of the biggest examined particles is studied more precisely in **figure 10**. The effect disappears at higher temperatures and therefore reduced viscosities and ends up in the familiar Newtonian behavior as from 25 °C which corresponds with a dynamic viscosity of 60 mPas.

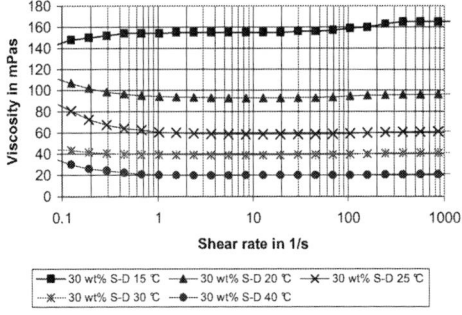

Figure 10: The shear thickening observed with particles S-D for low temperatures disappears with higher temperatures.

Comparing the viscosities at different temperatures shows a well known shape (**figure 11**). Again, the temperature effect on the dynamic viscosity of a particle ink is mainly influenced by the viscosity behavior of the underlying liquid phase. The small differences in viscosity for different kinds of silver particles decrease with increasing temperatures and at 40 °C the values coincide.

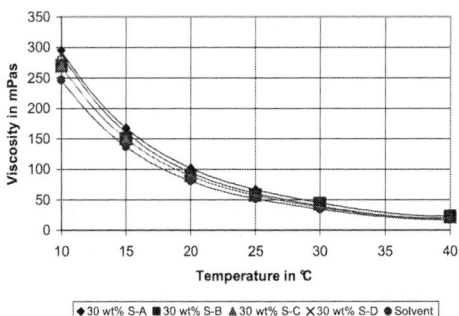

Figure 11: For different particles the temperature effect of the viscosity follows a 4th order polynomial approximation that is mainly influenced by the underlying fluid.

Stability of different particles

While the effect of different silver particles on the viscosity is rather small, the effect on the stability is significant (**figure 12**).

Figure 12: The stability of different particles depends on the particle size (from left to right: S-A, S-B, S-C, S-D, sedimentation after 2 days, 30 wt%).

In **figure 13** the sedimentation is visualized using light scattering signals of the samples. The transmission signal gives information about the clear phase on the top of the sample, the backscattering signal detects changes in the sediment at the bottom.

Only after 5 hours a sediment is detected by backscattering with the biggest particles while the other particles show no noticeable changes.

After three days the biggest particles are settled down nearly completely with an obvious clear phase at the top. The medium particles show a small sediment. The settling behavior of the 300 nm particles is similar indicating that the particle size is the key factor for stabilization. No sedimentation is detected with the smallest particles.

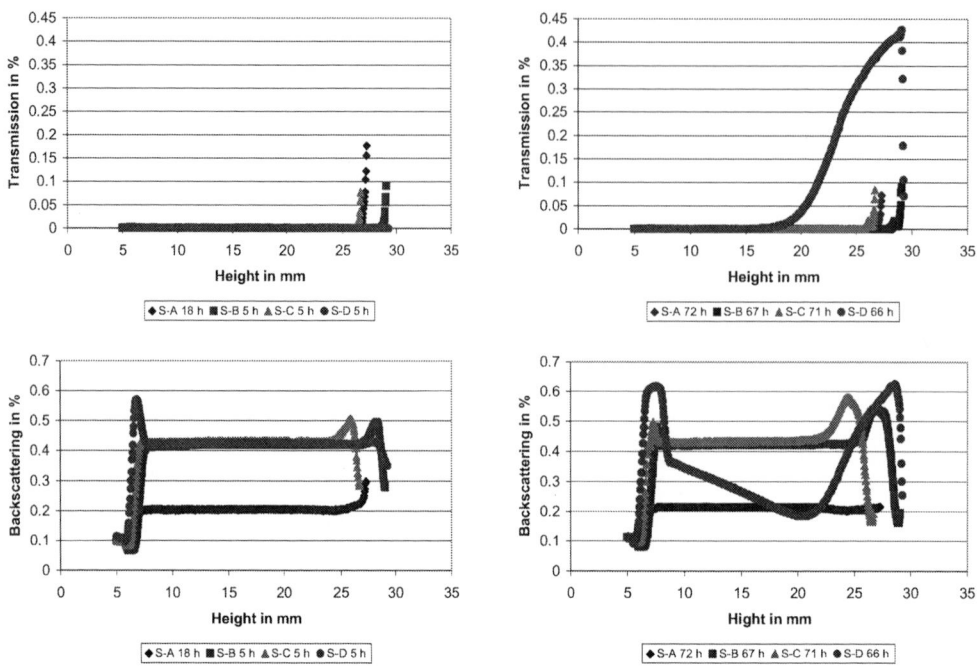

Figure 13: Evaluation of the sedimentation behavior with light scattering reveals no clearance effect in the transmission signal after five hours but the backscattering signal detects already a sediment for the ink with powder S-D. While there is no measurable sediment for the smallest particles S-A after three days particles S-D are nearly completely sedimented.

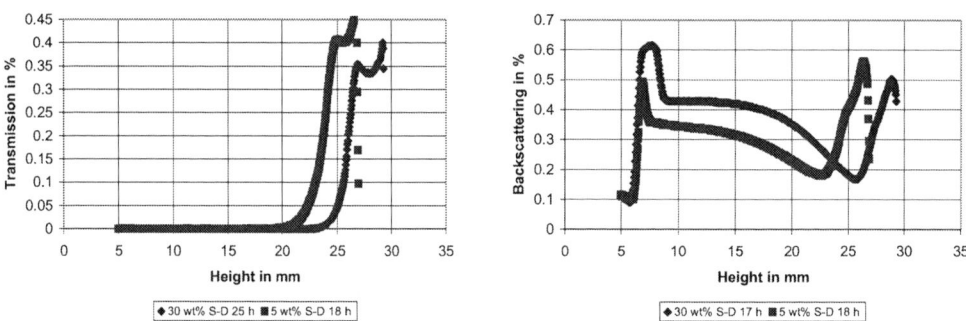

Figure 14: While reducing the particle content of S-D from 30 wt% to 5 wt% the sedimentation profile remains the same except the sediment becomes smaller with the 5 wt% ink. The position of the curves depends on the filling level.

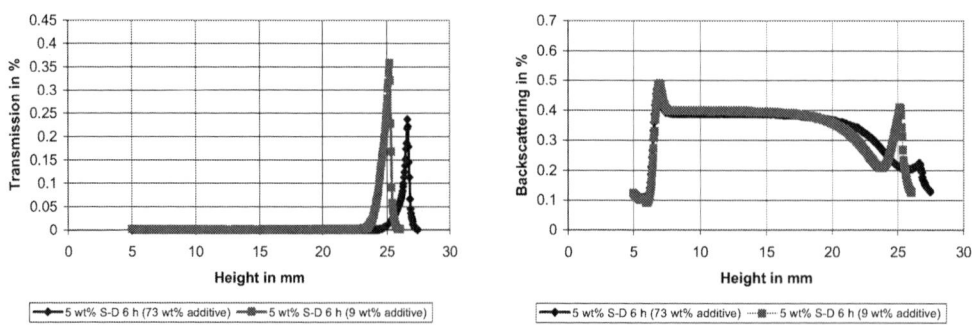

Figure 15: The sedimentation profile of the 5 wt% S-D ink is independent of the additive amount.

Factors influencing the stability of an ink

Having used the same particle concentration on identical stabilizing conditions it appears that the stabilization effect of the additive is significantly different for the varying particle sizes. In order to clarify the influence of the particle concentration on the stability of the ink the liquid phase is kept constant while the particle amount is reduced to 5 wt%.

Figure 16: The reduction of the concentration from 30 wt% to 5 wt% with particles S-D does not improve the stability.

The light scattering profiles in **figure 14** confirm the impression of **figure 16**: The reduction of the concentration has no positive effect on the stability of the ink although the amount of additive relative to the particle amount has been increased significantly as the liquid phase was not changed. The sedimentation behavior of the 30 wt% particle ink and the 5 wt% particle ink is nearly identical except there are less particles in the sediment with the 5 wt% ink.

As the reduction of the concentration does not promote stabilization of these particles the amount of additive is varied. In **figure 17** the sedimentation of two 5 wt% particle inks with different amounts of additive is shown. The first ink has the highest amount of additive namely 73 wt% related to the mass of the particles, the second has got 9 wt% which means about 100 vol% related to the particle volume. **Figure 15** verifies that changing the amount of additive has virtually no impact on the sedimentation of the biggest particles.

Figure 17: Changing the amount of additive does not improve the stability of 5 wt% S-D inks (left 73 wt% additive, right 9 wt% additive).

Conclusion

The dynamic viscosity and the stability are essential properties of an inkjet printable particle ink with high impact on the printing result. It has been shown that a shear thinning effect for shear rates smaller than 1 s^{-1} is rather random and does not give any information about the stability or quality of the particle ink. Particle inks show a Newtonian behavior which has been proved up to shear rate 10000 s^{-1}. Slight shear thinning effects occur with particle concentrations greater than 30 wt% and decrease with increasing temperature. The temperature effect of the viscosity of the particle ink is mainly affected by the temperature behavior of the underlying liquid phase. The influence of the particle concentration decreases with increasing temperature and becomes more and more negligible. Different kinds of silver particles influence the viscosity only slightly with the smallest particles showing the highest viscosities and particles with the same size have congruent shear viscosities. However, the influence of the particle size on the stability of the ink is significant.

References

[1] J. H. Lee, K.S. Chou, K.C. Huang "Ink-jet printing of nanosized silver colloids", Nanotechnology, Vol. 16, No. 10, pp. 2436-2441, October, 2005.

[2] D. Kim, S. Jeong, J. Moon, K. Kang "Ink-jet printing of silver conductive tracks on flexible substrates", Molecular Crystals and Liquid Crystals, No. 459, pp. 45-55, 2006.

[3] J. Perelaer, B.J. de Gans, U.S. Schubert "Ink-jet printing and microwave sintering of conductive silver tracks", Advanced Materials, Vol. 18, No. 16, pp. 2101-+, August, 2006.

[4] A. Kamyshny, M. Ben-Moshe, S. Aviezer, S. Magdassi "Ink-jet printing of metallic nanoparticles and microemulsions", Macromolecular rapid communications, Vol. 26, No. 4, pp. 281-288, Februar, 2005.

[5] D. Cibis, K. Krueger "DoD-printing of conductive silver tracks", Proceedings of the 1st International Conference on Ceramic Interconnect and Ceramic Microsystems Technologies (CICMT), Baltimore, April 10-13, 2005.

[6] U. Currle, K. Krueger "Factors influencing the conductivity of inkjet printed silver lines", Proceedings of the 2nd International Conference on Ceramic Interconnect and Ceramic Microsystems Technologies (CICMT), Denver, Colerado, April 25-27, 2006.

[7] U. Currle, D. Cibis, G. Steinborn, K.Krueger, "Der Inkjet-Druck – ein neues Verfahren zum Aufbringen elektrisch leitender Strukturen in der Mikroelektronik", Proceedings of the German IMAPS-Conference 2006, Munich, Germany, October 9-10, 2006.

[8] Thomas Mezger, "Das Rheologie-Handbuch", Vincentz, Hannover, 2000.

Preheating in Solderability Testing

F. Steiner, P. Harant

University of West Bohemia, Univerzitni 8, CZ 306 14 Pilsen , Czech Republic

Phone: +420 377 634 535, Fax: +420 377 634 502, steiner@ket.zcu.cz

Abstract

Paper deals with results of solderability testing. The importance of solderability testing grows up with implementing of lead free technology. This technology also requires new surface finishing of component leads or pads of printed circuit boards. On this account we need to test the solderability of new surface finishes. At solderability testing we try to simulate the soldering process. We can't forget preheating, because it is important part of machine soldering. Preheating shall activate flux and remove impurities from soldered surfaces. That means, preheating can improve solderability. For this reason the specimen for hot air preheating was prepared for solderability testing. Paper will present the comparison of results of printed circuit board solderability test with and without hot air preheating. This comparison will be also made for several surface finish types. It was tested printed circuit board coupons with surface finishing of galvanic tin, immersion tin, pure copper, OSP and nickel/gold.

Key words: solderability, solderability testing, wetting balance test, preheating

Introduction

Some metallic surfaces are more easily soldered than others. They differ in both the speed of wetting and the strength of adhesion of the solder to their surface. The speed of wetting is controlled by the combined effects of the thermal demand and the wettability of the metal surface. These combined properties are known as the solderability of the material.

The wettability of the same material may vary considerably, as it is strongly influenced by the surface condition of the metal. Thin films of oxide, grease or organic contaminants can severely affect the wettability of a metal.

For modern production soldering it is vital that components have known good solderability, before they are allowed onto the production line. The large number of solder connections that are made simultaneously by modern soldering processes mean it is vital that all the solder corrections are made correctly at the first attempt. [1]

Wetting balance test

The wetting balance test is the most useful tool for investigating the soldering properties of surface mount components, terminations and leads, and PCB pads. A wetting balance is used to measure the force involved when a molten solder wets the sample surface.

The Wetting Balance measures the vertical forces of buoyancy and surface tension as a fluxed test piece is immersed into a bath of molten solder. The wetting force is converted by a transducer into an analogue signal. This signal may be taken directly onto an X/T recorder, or may be digitised and analyzed by a computer. The digital signal is used to generate the forcetime curve (Figure 1), and is analyzed to find the required forces and times from the forcetime curve.

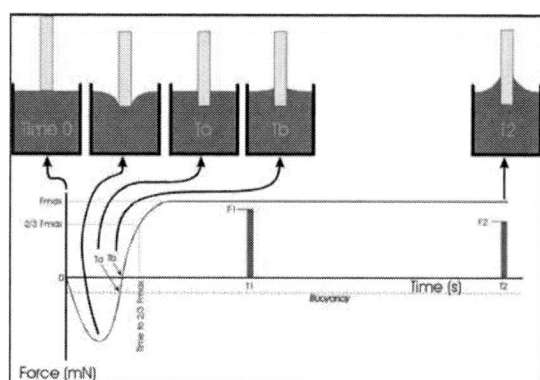

Figure 1: The Wetting Balance curve

They are a lot of methods and standards for solderability evaluation. These standards define evaluation criteria. Some examples of criteria will be presented now. Test methods use the time for the Wetting Balance curve to re-cross the buoyancy line (Ta in Figure 1) as the time for the onset of wetting. This is when the solder bath surface has returned to horizontal, and the solder contact angle has fallen to 90°. The test methods then use a minimum value for the wetting force, at a specified time of typically two seconds (F1 in Figure 1, T1 = 2 s), expressed as a percentage of the theoretical maximum or reference wetting force, as a measure of the progress of wetting.

The reference wetting force is established by finding the maximum wetting force that can be obtained on a specimen that has been pre-tinned using an active flux. The pre-tinning procedure is repeated until the maximum wetting force does not increase any further (Fmax in Figure 1).

The stability of wetting is evaluated by measuring the decline in force, if any, from the maximum wetting force, to the force at the end of the test period. This decline in force is expressed as a percentage of the maximum soldering force. Note that all forces are measured from the buoyancy line, when using this method [1].

Test System

The system MUST II was used for measurement of wetting balance curve (Figure 2). This system is able to determine wetting balance curve and evaluation parameters, of course.

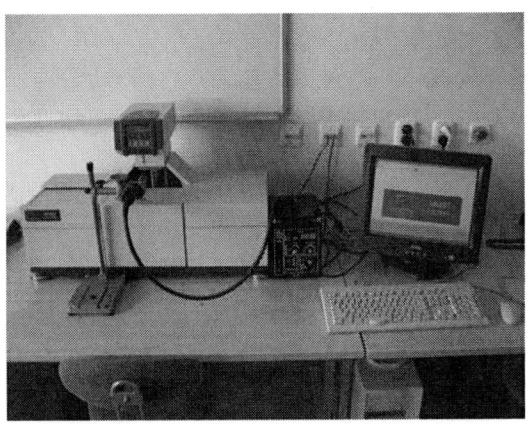

Figure 2 Solderability Test System for Surface Mount and Conventional Components–MUST II

The system MUST II offers preheating possibility by placement of a specimen above molten solder bath. But heat transmission isn't sufficient in such a case. That is why other type of preheating was chosen.

The preheating of specimens before testing was created through the use of hot-air soldering station Green Tool 21 TS (Figure 3).

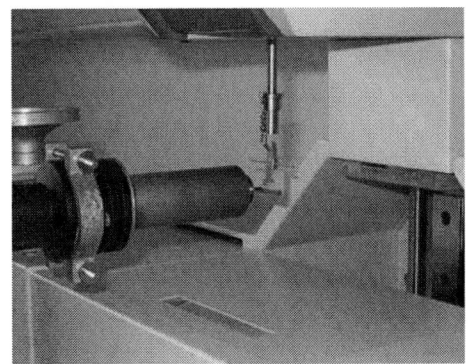

Figure 3: Hot-air preheating for MUST II

First step of hot-air preheating application was the determination of soldering station setting-up. The optimal preheating of specimens was searched. The thermocouple was soldered on the specimen and time behavior of temperature on the specimen was measured (Figure 4).

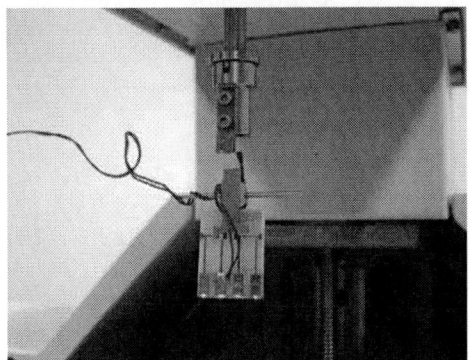

Figure 4: Measurement of specimens temperature

The optimal setting was chosen after measurement and data evaluation. This setting is 120 °C and maximal air volume (Air6).

Tested Specimens

New design required innovative solutions in production of Printed Circuit Board; ball grid arrays, wire bonding pads, press fitting, and contact switches were all outside the traditional realm of HASL. In addition, environmental concerns are focused on the elimination of lead. [3] This will change HASL from a process that uses a standard solder to a new lead-free process. That is why several types of specimens with different surface finishes were prepared for measurement. On Figure 5, you can see dimensions of the tested coupon. The coupon thickness is 1,5 mm. Coupons were made with surface finishes which are shown in Table 1.

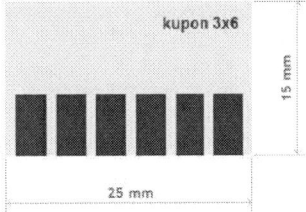

Figure 5: PCB coupon

Table 1: The labeling of PCB surface finishes

	Surface finish
Cu	Copper
OSP	Organic Solderability Preservative
SnC	Immersion tin
SnG	Galvanic tin
Au	ENIG

The international solderability specifications include a number of methods to accelerate the ageing process and provide data parallel to natural ageing, although the exact mechanism could never be the same. So, in addition to preheating the influence of ageing on wetting balance were measured too. We have chosen following ageing – *Dry Heat* (in air at 155 °C for 16 hours). In Table 2, there is shown the way of identification of specimens.

Table 2: The identification of specimens

	Ageing	Preheating
N - N	no	no
S - N	yes	no
N - P	no	yes
S - P	yes	yes

Solder SAC 305 (96,5%Sn, 3%Ag, 0,5%Cu) and pure rosin flux were used for testing. The wetting balance curves were measured at solder temperature 245 °C.

Results of Measurement

Following figures present measured values. Measured results will be resumed in *"Conclusions"*.

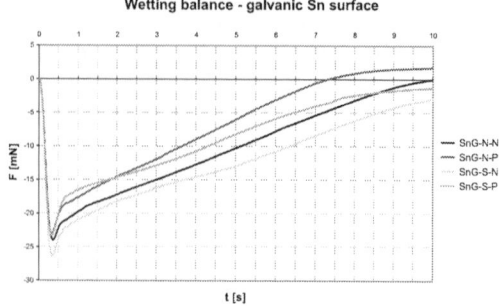

Figure 8: Wetting forces of coupons with galvanic tin surface

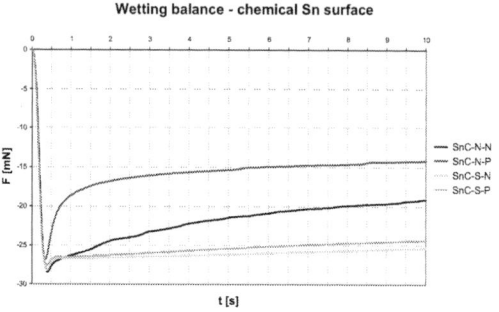

Figure 9: Wetting forces of coupons with chemical tin surface

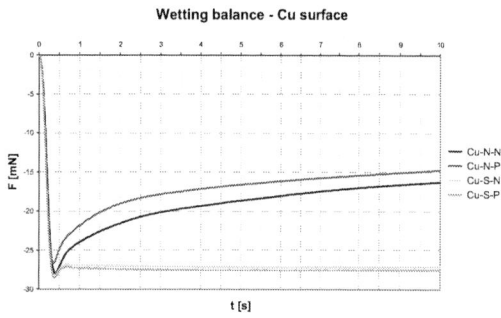

Figure 6: Wetting forces of coupons with copper surface

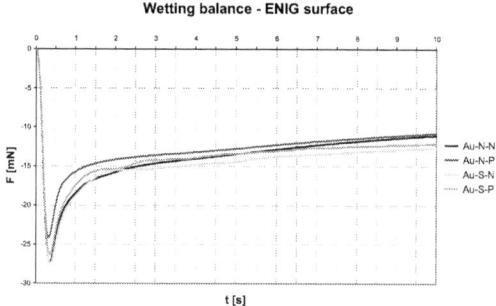

Figure 10: Wetting forces of coupons with ENIG surface

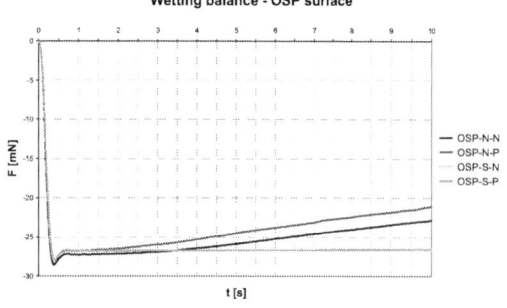

Figure 7: Wetting forces of coupons with OSP surface

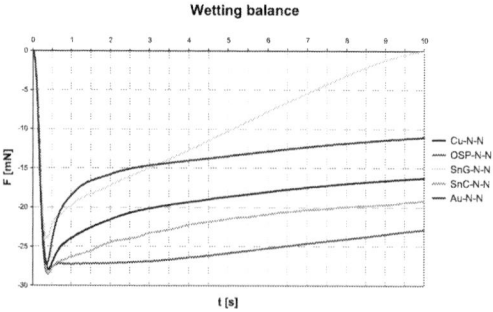

Figure 11: The comparison of coupons with different surface finishes

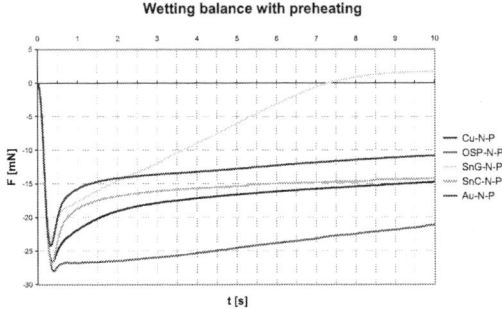

Figure 12: The comparison of coupons with different surface finishes after hot-air preheating application

Conclusions

From shown figures we can make out following conclusions. Preheating improved wetting forces almost in every case. Influence of preheating was higher at non-aged specimens. The specimens with copper and OSP surface finish (Figure 6 and 7) embody mild changes. Wetting force is higher. But the wetting forces of aged specimens are almost same with and without preheating.

The wetting forces measured on specimens with galvanic tin surface finish are shown in Figure 8. This surface finish has best wetting forces and wetting is slower. Influence of preheating is higher than in other specimens (except non-aged immersion surface specimens).

The highest improvement by preheat using is seen in Figure 9. Wetting force of non-aged specimens with immersion tin surface finish was improved, but wetting forces with preheating (after ageing) is almost same as without preheating.

Better results were also obtained at specimens with ENIG surface finish that means electroless nickel / immersion gold. In Figure 10, you can see higher speed of wetting. In contrast to Cu and OSP specimens, preheating also improve wetting force at aged coupons.

Surface finishes can be compared too. The results are shown in Figure 11 and 12. At results before ageing the specimens with galvanic tin surface have highest maximal wetting force. The specimens with ENIG surface have lower wetting forces, but they have higher wetting speed (shorter wetting time). Next sequence according maximal wetting forces is specimens with copper surface, specimens with immersion tin, and specimens with OSP surface. The amplitude of wetting forces of specimens with OSP surface was probably affected by layer quality. OSP surface wasn't high-quality.

The sequence of wetting force curves measured with preheating is almost same, but wetting force of specimens with immersion tin surface is higher than forces measured at specimens with copper surface. That is caused by the improvement of wetting forces of immersion tin finished specimens.

At the end of conclusion we have to say that determination of solderability becomes more and more important. New types of solders and surface finishes will need this evaluation in future too. And importance and influence of preheating in solderability testing were presented in this paper.

Acknowledgements

This paper is part report and has been supported by the research plan of Ministry of Education, Youth and Sports of Czech Republic No. MSM4977751310 "Diagnostic of Interactive Processes in Electrical Engineering".

References

[1] Solderability Test System for Surface Mount and Conventional Components – User manual, CONCOAT SYSTEMS 2004.

[2] L. Zou, D Lea, Ch. Hunt, "Solderability Testing of Surface Mount Components and PCB Pads" – Measurement Good Practice Guide No. 66, National Physical Laboratory, Teddington, Middlesex, United Kingdom 2004.

[3] Clyde F. Coombs Printed Circuits Handbook, McGraw-Hill, ISBN 0-07-135016-0.

[4] STEINER, F.; HARANT, P. Solderability of the lead free surface finishes. In ESTC 2006. Dresden : IEEE, 2006. s. 365-369. ISBN 1-4244-0553-X.

[5] STEINER, F.; HAMÁČEK, A. Shearing and tensile strengths of the environment-friendly interconnections. *In* Proceedings. Koszalin - Darlowko : XXIX international conference of International microelectronics and packaging society Poland chapter, 2005. s. 215-218. ISBN 83-917701-2-5.

[6] IPC/EIA J-STD-003A Joint industry standard – Solderability Tests for Printed Boards, February 2003.

Reliability Qualification of Flexible Printed Circuits

Markus Detert[1], Thomas Zerna[1] and Klaus-Jürgen Wolter[1,2]

[1]Centre of microtechnical Manufacturing

[2] Electronics Packaging Laboratory at Dresden University of Technology

Phone +49-351-463-36334, Fax +49-351-463-37069

Email Address markus.detert@tu-dresden.de

Abstract

The use of FFC and FPC constantly increases in the next years. The development of competitive products can be improved on basis of these technologies. The advantages within the range reliability and costs could be used however only with the consideration of a complete view of the entire process chain efficiently and trend-setting. The complete verification and validating of FFC and FPC require substantial time expenditures. The introduction of new and reliable constructional-technological solutions requires today too much time. For this reason we compile gradually a concept for the creation of time-efficient testing methods for FFC and FPC of products. We select substrate materials and connecting technologies with the necessary adding materials on the basis of well-known criteria. Duro plastic and thermoplastic materials can be used as substrate materials. In the starting phase we raise data from the results of the standardized accelerated aging procedures. Thus we use thermal test procedures and temperature storage with humidity. We characterize the test structures with destructive and non-destructive examinations. In the following stage we observe test superstructures under self-heating. We characterize the behaviour of the test structures again with the well-known procedures. Additionally we measure the length variations in the total structure, brought in by the self-heating. The purposeful overlay and combination of different mechanical, thermal and chemical stresses to the total structure can reduce the test phase in the future substantially. In addition we collect in the present phase data. We designed a structure of test equipment, which makes these measurements possible. Only at the end of the running investigations at present we will have data to plan the next stage with combined and selected aging procedures. Our presentation shows our motivation, investigation methods and first results of this development process.

Key words: Flexible Printed Circuits, FPC, Reliability, Solder Joints, Adhesives

Introduction

The results about the joint reliability of rigid printed boards are very often published with realizations on different connecting methods, e.g. lead containing, lead free, isotropic and/or anisotropic adhesive procedures. The goal of our work is the development of comparable reliability results on the basis of flexible wiring carriers. At this time we collect a lot of data, which are required. We evaluate by an appropriate selection of representative designs by the conception, the structure and the accelerated aging of a suitable demonstrator. The main aspect is thereby on the variation of the base material and the connecting technology. The complete qualification of building groups is currently very time-consuming and cost-intensive. In the context of this work we will determine the possibilities for the development of suitable procedures with significantly decreasing test periods. The reciprocal effects with the damage mechanisms are to be considered in a lot of details during the whole test procedures. Flexible printed circuits will be more and more popular on complex electronic systems with a high number of sub parts of very small geometrical dimensions, or such with high integration. A lot of examples of these directions are mirror reflex cameras, video cameras, mobile telephones or laptops and the great area of handheld assistance systems. A view to the patent situation showed that most innovations are promoted by the mobile telephone industry and by the automotive applications. Problems in mobile phones can particularly be found in the bending zones at the folding mechanisms. Beside the application of the flexible materials in the individual components of a product, the definition for the application in the far wiring of the building groups on flexible materials is very interesting. In the automotive electronic increased the use of flexible wiring carriers constantly during the last years. The large growth electronics in the automobile is directly combined with comfort, entertainment and the development of driving assistance units [1]. In older vehicles we already found flexible printed circuit boards for the use examples of the control units in the dash board instruments. In the focus are the very high requirements on the quality of the materials, their workability and their function in the automotive electronics fabrication. For the qualification of rigid substrate materials numerous standards (e.g. MIL, DIN, JEDEC, IPC) exist. The most standards and guidelines can also be applied to the flexible substrates without large problems. There are only few standards, which reasonably consider the flexible characteristics. In the related DIN standard some changes are already in preparation (e.g. in [2]). In the IPC standard TM-650 suitable procedures for these substrates are partly present. However special machines are often necessary for these test procedures, which are not generally available. In practice tensile strength and fatigue measurements, as well as measurements regarding the bending harness are frequently accomplished for qualification. Statements about the reliability are only available in insufficient measure. The different statements of experts do not vary from „approx. 20% more badly" over „a difference" up to „very much better". These are however only qualitative statements, concrete data are only difficult to find. Flexible base materials are generally more heat sensitive than rigid base materials, therefore the contacting of the elements predominantly takes place via the guidance sticking. Here the process temperature is clearly lower compared to soldering.

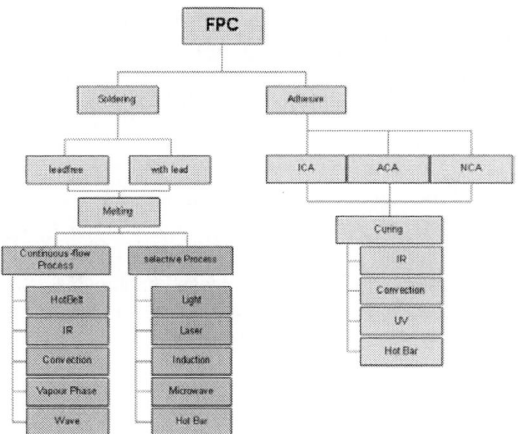

Figure 2: Flowchart of the FPC manufacturing technologies [2]

That's why we have beside the load of temperature also an additional and undefined mechanical part to the flexible printed circuits. This situation can declare, why test results of flexible printed circuits are really not safe for prognostic answers.

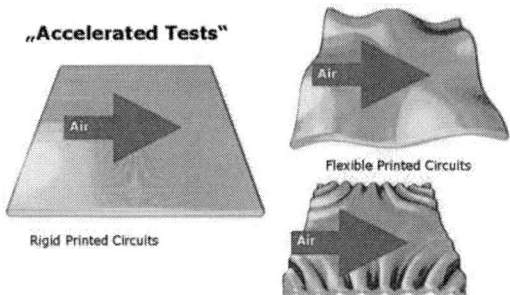

Figure 3: Behavior of rigid and flexible printed circuits during temperature cycle chambers with air stream [2]

So we can describe the initial situation at this start of our program for the investigations with flexible printed circuits in the last year. Now we can report time to time about new steps and results of this test program.

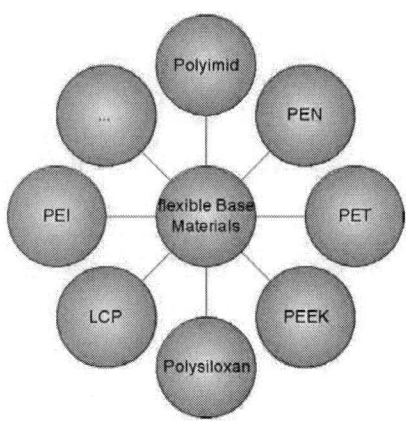

Figure 1: Examples for flexible base materials [2]

Some question will be answered again and again with the development processes of new products. What kind of material is most reliable (Figure 1)? What kind of technology is most reliable (Figure 2)?

And with these two basic questions started our investigation process with a couple of more questions for the use of accelerated temperature cycling methods on flexible printed circuits. A lot of test chambers use a strong air stream for the temperature cycling. Sophisticated by the air stream is the location of the tested flexible printed circuits not defined (Figure 3).

Common Test Methods

Figure 4: Test vehicle for common tests

For the realization of the common test methods we manufactured a test vehicle design, which contained resistor components in different sizes (1206, 0805, 0603, 0402, 0201) and a BGA (Figure 4).

Because there are some influences to the adhesion behavior during the common test loads we manufactured additional areas for peel tests to the design. As base materials we selected PI (Polyimide), PEN (Poly ethylen naphthalat) and GHE (Glass Epoxy). For joining technologies we used lead-free solder and ICA (Figure 5).

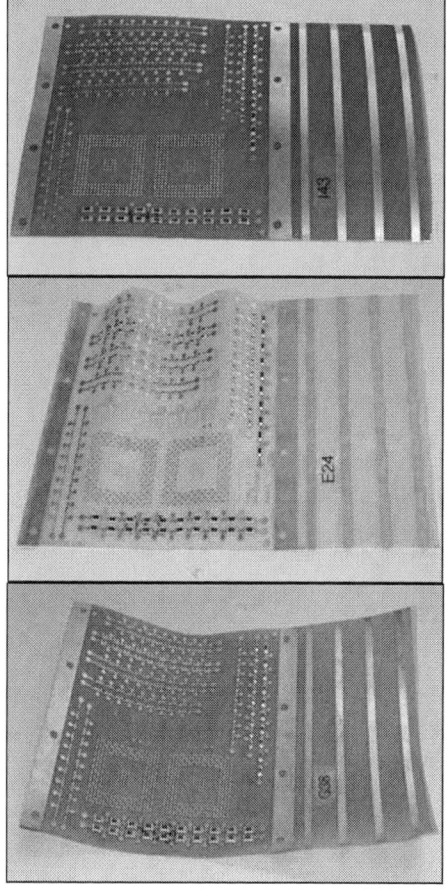

Figure 5: Test vehicle for common tests (PEN, GHE, PI)

The following accelerated test methods we selected, because there are also popular for the rigid printed boards (RPC):

- High temperature cycles with -40°C to 125°C
- 85°C, 85 % R.H.
- Pressure Cooker Test with 121°C, 100 % R.H., 205 kPa

Before we start with the test load to the test vehicles we made fully description and documentation of the characteristic parameters.

Thinking about a New Test Method

On the base of the discussion about reliability at FPC you can only think about a different way. So we developed a second test procedure for a defined sample. It will be aged by an alternative method. In order to develop a more sophisticated aging model, it is necessary to determine the aging mechanisms more exactly. Generally with the cyclic temperature change the main damage is produced over the differences of potential in the different materials. A procedure must be developed, which suitably copies these tensions. It can be realized in two ways. It is possible to put the temperature on another way on the sample test specimen. On the other hand it is possible to put the tensions over a mechanical equipment tool on the sample test specimen. In principle tensions in the contact points over two procedures can be initiated. Due to the different temperature expansion coefficients tensions are produced by change of temperature. Moreover it is possible to induce the tensions mechanically, e.g. by applying a traction power. For this attempt two procedures for the additional heat entry are possible. The warmth can be introduced by means of a heating plate directly into the sample test specimen. With an impedance matrix it is possible to place the heat entry very close at its place of destination (contacted joints) [6].

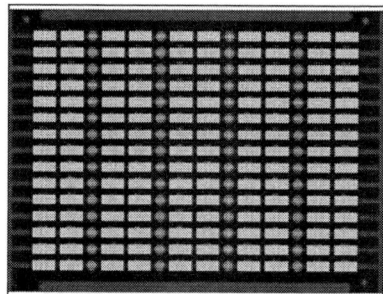

Figure 6: Impedance matrix

The impedance matrix (Figure 6) is supplied cyclically with current. An energy dissipation of approx 7 W for each resistance (size 2010) can be expect. Thereby this value goes out of preceding investigations. In order material stresses developed in such a way to measure to be able, are it possible the substrates with a defined pre-loading to be clamped. In the attempt the change of force is measured, which results from the specimen's length variation due to the heat entry.

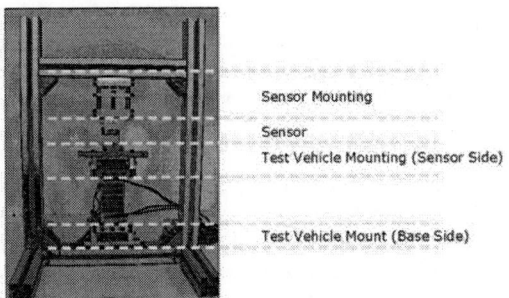

Figure 7: Test equipment for new method

Thus a possible flow process of the material is to be illustrated. The advantage of this procedure is that this kind of the damage arises also most frequently in the reality. For this attempt a special construction is necessary (Figure 7).

By course and/or torsion tests or combinations of such, tensions in contacting can be produced. Similar attempts were already accomplished with rigid substrates, e.g. producing tension by bending. It is to be noted however that this kind of tension entry and width it specific damage mechanisms occur only very rarely in the reality. It is recommended to make a FEM simulation.

Demands
* Resistance matrix as Heater (max. Achievement conversion up to 500 W)
* Clamping possibilities for the test equipment

Results
* Testlayout geometries (300 x 60) mm
* Resistance Matrix with (10 x 15) Pieces (1,8 Ω; Größe 1206; 3 mm x 1,5 mm)

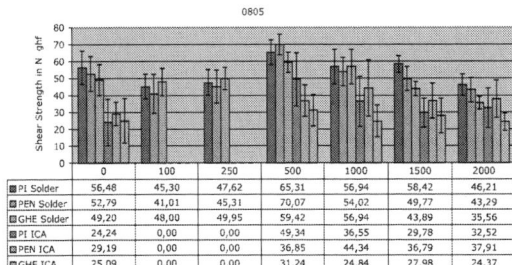

Figure 8: Test vehicle for new test procedures (PEN, GHE, PI)

Results of common Tests

	0	100	250	500	1000	1500	2000
PI Solder	56,48	45,30	47,62	65,31	56,94	58,42	46,21
PEN Solder	52,79	41,01	45,31	70,07	54,02	49,77	43,29
GHE Solder	49,20	48,00	49,95	59,42	56,94	43,89	35,56
PI ICA	24,24	0,00	0,00	49,34	36,55	29,78	32,52
PEN ICA	29,19	0,00	0,00	36,85	44,34	36,79	37,91
GHE ICA	25,09	0,00	0,00	31,24	24,84	27,98	24,37

Figure 9: Example of Shear Strain after Thermal Cycling

Only at the test vehicles with PEN we gained numerous changes in the measured values. Typical structured data after testing is shown in **Error! Reference source not found.**.

The metallurgical view of the joints under temperature cycling will show no additional effects (Figure 10). But we need a lot of data for the evaluation process of our new test method.

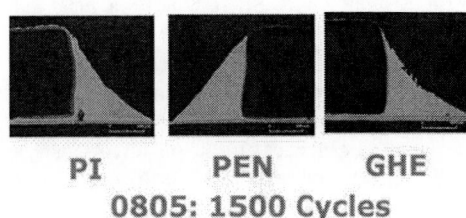

Figure 10: Metallurgical Results - 1500 Cycles

Results of new Test Methods

The greatest level of interest in the use of the new test method is the ability to make the same load like common test methods. All of the pre-tests show a good result of the temperature change during the current flow (Figure 11). At all materials we can expect the situation in the flow of current and the temperatures. The temperature cycles can be produced very well with power cycling procedure (Fig. 5). The heating-up time of the substrate amounts to approx. (30 - 40) % of the heating-up time in the conventional rapid temperature change test. The cooling rate of the GHE material corresponds to the heating-up time. The material therefore possesses a small thermal capacity. Due to this characteristic it is possible, to very simply produce a symmetrical temperature cycle in the substrate material. The testing period for a pure thermal-mechanical load can thus be reduced with this procedure to a third of the rapid temperature change tests. The temperature control with the help of the attitude of the amperage is very well possible. However the temperature gradient as a function of the amperage is to be determined before a test series for each layout and material. An on-line measurement of the temperature is not available yet. After the first test runs it is to be assumed, that the temperature distribution in the substrate is not homogeneous. Towards the edge of the sample test specimens, the heat energy can be better transferred to the environment. With variation of the individual resistance values on the inspection piece the temperature development can be produced homogeneously.

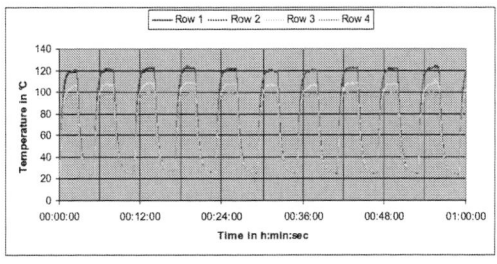

Figure 11: Temperature Cycles with new method

The behavior of the different base materials we checked for temperature distribution with the help of a thermal image camera. So we obtained, that there some small differences in the distribution of heat per time unit. So it will be a fact, that a material like polyimide will be faster in the distribution of heat in comparison to the semi-flexible GHE with glass ingredients. The final distribution of heat during the current flow has only some small differences at the different materials. The Figure 12 and Figure 13 will illustrate this situation.

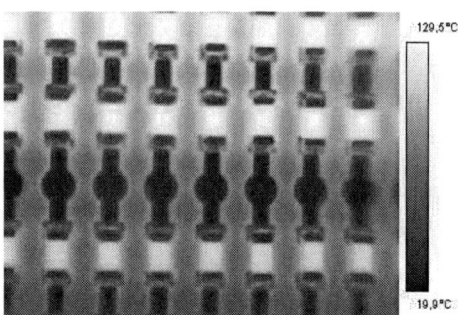

Figure 12: Thermo image at PI

Figure 13: Thermo image at GHE

After the first level of use we made the first metallurgical investigation. An typical example of the result after 4500 cycles will show Figure 14. There no failure and no great changes in the metallurgical behavior. Beside this facts it can be able, that in this case of testing we really test the temperature cycling and the combined influences.

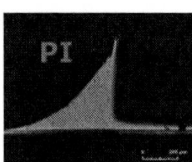 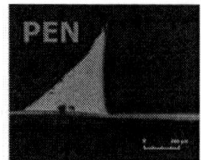

4500 Cycles, no Failure

Figure 14: Metallurgical results

Conclusions

The glass fiber fabric prepreg exhibits the highest reliability in this attempt, due to its material properties and its higher stability. The adhesive sample test specimens are altogether more critical of regarding than the soldered sample test specimens. With this technology a smaller failure rate is recognizable, than with the PI or with the glass fiber fabric prepreg after the electrical measurements with the PEN sample test specimens. Generally the handling is one of the largest problems when processing flexible wiring carriers. By the joints strong undefined tensions are exposed to breaks and bending. Improvements can be achieved with reinforcement (stiffener) under the critical places. Already during the manufacturing processes must be treated very sensitively. This should be so far automated, that negative external influences are minimized to a large extent. The first investigations with the alternative procedure could be accomplished. Here the heat entry takes place into the sample test specimen via power cycling. The substrate is clamped during the heat load, but the developing change of force due to the different coefficient of thermal expansion is measured firmly over a sensor. The whole construction process for this procedure became a very complex state. A lot of numerous details had to be considered. After the first investigations a symmetrical temperature gradient can be very well copied. The cycle duration amounts to less than 50 % in comparison to the temperature change test. The used resistances can be stressed in their needed range of stress. The temperatures thereby will reach a value, with which it is possible to loose components by melting the solder joints. The heat measurements were accomplished up to the present time. The developed procedure provides also the possibility to determine the material properties of different flexible substrate materials. Investigations from combinations of different load tests permitted a decrease in the testing time and exhibit also more room for lowering and thus should be further pursued.

Acknowledgements

We would like to thank the whole team of Mr. Löper from mechanics laboratory for the manufacturing of the parts of our new test equipment. Also we like to say "thank you" to Mr. Müller and Mrs. Schöne for the metallurgical preparations. Beside these facts we thank the colleagues of the Electronics Packaging Lab for the goal-prominent discussions in the last weeks.

References

[1] Bochtler, W.: Innovative Technologien: Flex-Applikationen zur Kosteneffizienz und Qualitätssteigerung. Fellbach: 3. DVS/GMM Tagung, 08. – 09.02.2006

[2] Detert, M.: Flexible Printed Circuits. Teaching Materials for the lessons of "Special Chapter in Electronics Packaging", January 2007

[3] Detert, M.: Reliability Qualification of Flexible Printed Circuits with Common and New Methods, ESTC 2006, $5^{th} - 7^{th}$ September 2007

[4] DIN EN 61189-2: Prüfverfahren für Elektromaterialien, Leiterplatten und andere Verbindungsstrukturen und Baugruppen, Teil 2: Prüfverfahren für Materialien für Verbindungsstrukturen. 2001

[5] http://www.taconic-add.com/technicaltopics--pim-passive-intermodulation.php

[6] Ernst, D.: Zuverlässigkeitsbewertungen von Leitklebe- und Lötverbindungen auf flexiblen Verdrahtungs-trägern, Diplomarbeit, Institut für Aufbau- und Verbindungstechnik der Elektronik, Technische Universität Dresden, 2006

Studies of Selected Inspection and Failure Analysis Techniques for LTCC Micromodules

Kari Remes, Leena Palmu and Petri Ronkainen*

University of Oulu / Meri-Lappi Institute, Technology, Tietokatu 6, FI-94600 Kemi, Finland

*Kemi-Tornio University of Applied Sciences, Technology, Kiveliönkatu 36, FI-94600 Kemi, Finland

Mobile: +358 40 724 2395, Fax: +358 16 228 552, E-mail: kari.remes@oulu.fi

Abstract

Several inspection techniques for ceramic electronic modules have been introduced, the most typical ones being ultrasonic microscopy, scanning electron microscopy (SEM) and X-ray microscopy. These methods are usually used for evaluating and quality checking of produced low temperature co-fired ceramic (LTCC) micromodules, and also as tools during research and development phases of manufacturing processes or electronics. This kind of powerful techniques can for example help to optimise a critical process step or they can detect reliability related faults in electronic devices. We have selected the following inspection and failure analysis techniques for our studies: nanofocus X-ray microscopy, SIMS (Secondary Ion Mass Spectrometry) microscopy, conventional optical microscopy and laser profilometry. A decapsulation / delayering system enabled for example SIMS based chemical analysis and visual examination of inner parts of the samples. As seen both destructive and non-destructive methods were used. All of these techniques and equipment are available at the Kemi-Tornio University of Applied Sciences. The suitability of the techniques for characterisation of multilayer LTCC micromodules with buried passive elements such as resistors was studied. Most of the techniques used are more or less imaging ones and thus give a good starting point for studies concentrating on the combination of the results obtained with different methods, and visualisation of these results. Some results of this partly EU funded research work are presented.

Key words: inspection techniques, failure analysis, LTCC, micromodules

Introduction

Ceramic substrates and modules such as low temperature co-fired ceramic (LTCC) micromodules are sometimes necessary for harsh environments because of their reliability properties: they are capable of withstanding high temperatures and thermal shocks, and are resistant to chemical attack [1, 2]. This "immunity" also brings out some challenges for the inspection and failure analysis of these ceramic electronic modules.

Several inspection techniques suitable for LTCC micromodules have been introduced. The most typical ones are ultrasonic microscopy, scanning electron microscopy (SEM) and X-ray microscopy. These methods are usually used to evaluate and quality check LTCC modules during and after production. The methods act also as valuable tools during research and development phases of manufacturing processes or electronics. Non destructive analysis is usually preferred over destructive analysis.

This kind of powerful techniques can for example help to optimise a critical process step or they can detect reliability related faults in electronic devices. Scanning acoustic microscopy (SAM) is able to detect "difficult-to-find" defects, such as

interfacial separation, solderball delamination and die attach voiding. Electron microscopy provides unique imaging with high magnification, as well as the opportunity to perform elemental analysis and phase identification. X-ray microscopy is typically used to check via registration and wire sweeping, and for void calculation. With high resolution computer tomography (microCT) the use of X-ray analysis is even enhanced.

Beside the methods mentioned above and traditional visual inspection there are some other interesting techniques available. SIMS (Secondary Ion Mass Spectrometry) microscopy and profilometry are one of them. SIMS is a very sensitive surface analytical technique which has a number of different variants. Static SIMS is used for sub-monolayer elemental analysis, dynamic SIMS for obtaining compositional information as a function of depth below the surface, and imaging SIMS for spatially resolved elemental analysis. By profilometry, the surface of the sample (for example roughness) is characterised.

One potential idea is combination of the results of these different imaging inspection and analysis methods. This may give valuable information that might not be noticed otherwise.

Experimental

We have selected the following inspection and failure analysis techniques for our studies: nanofocus X-ray microscopy (Phoenix x-ray Nanome|x X-ray system with computer tomography capabilities), SIMS microscopy (Millbrook Scientific MiniSIMS chemical microscope), conventional optical microscopy (Leica MZ 16 Plan Apo stereomicroscope) and laser profilometry (Nanofocus µscan profilometer with Confocal Point Sensor CF 2001 and Autofocus Sensor AF 2000). A decapsulation / delayering system (Ultra Tec ASAP-1 with Ultracollimator) was also used. This enabled for example SIMS based chemical analysis and visual examination of inner parts of the samples. As seen both destructive and non-destructive methods were used. All of these techniques and equipment are available at the Kemi-Tornio University of Applied Sciences.

The work had a practical approach to study the suitability of the chosen techniques for inspection and analysis of multilayer LTCC micromodules with buried passive elements. The samples contained both surface and buried resistors, as well as contact pads and lines, and also vias between layers. Mainly DuPont's sheets and pastes were used for the processing of the samples. Manufacturing process was otherwise a typical one except that all the pastes (conductor lines, resistors, glass) were sintered together with the LTCC sheets.

The samples were treated as little as possible before inspection or analysis. X-ray microscopy can image surface and buried structures without sample preparation or physical deprocessing. For SIMS analysis the samples were cut to smaller pieces so that they fit into the sample holders, also some sample preparation was nesessary. Some samples were delayered to reveal inner structures before SIMS analysis. Optical microscopy and profilometry didn't need any special sample preparation.

Image acquisition, analysis and visualisation of the results were done with the software delivered with each instrument and the combining of the results with special visualisation software.

Results and Discussion

The inspection of the samples was started with optical microscopy. Optical images of a LTCC sample are presented in the Figure 1. There the resistors on the surface of the sample and laser trimming grooves are clearly seen. Optical microscopy with a stereomicroscope gives a good general overview of the surface structures on a sample. For some purposes, like a quick inspection of a process step this might be all that is needed. With a high magnification metal microscope even the physical measures of small sample details can be determined.

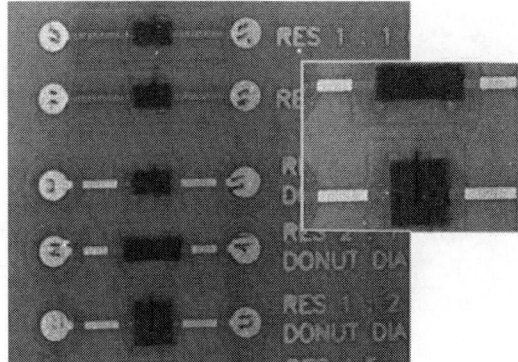

Figure 1: Optical images of a LTCC sample.

The study of the surface of the LTCC samples was continued with profilometry. Profilometric imaging of a sample reveals the fine structures - even nanoscale details - on a surface. Surface profilometry was used for example to measure film thickness and crater depths for SIMS depth profiling. Some results of this technique are presented in the Figure 2. The depth of a laser trimming groove on the LTCC module may be also measured with this technique.

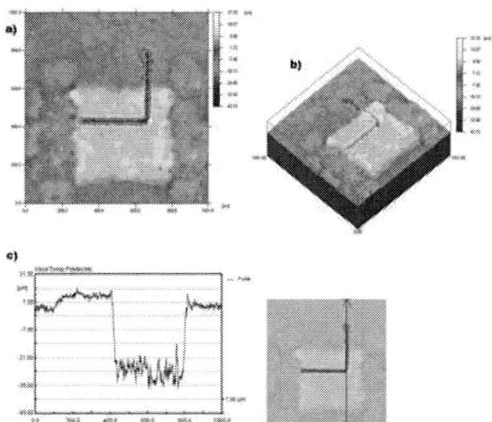

Figure 2: Profilometer images and a line profile of a LTCC sample. a) top-down view of a resistor with a trimming grove, b) 3D view of a resistor, c) line profile of a trimming groove.

To see inside the samples X-ray microscopy was utilised. High resolution X-ray microscopy is a suitable technique for a precise inspection of both surface and inner micro structures of samples. With an advanced nanofocus X-ray system detail detectability may be better than 500 nm. Some result images of this inspection are presented in the Figure 3. The X-ray system enabled the use of different views by tilting and rotating the sample to reveal the areas of interest. With computer tomography, three dimensional models of the chosen structures were reconstructed. These virtual CT models offer the possibility to examine the details of samples in an interesting way – you can even fly-thru them.

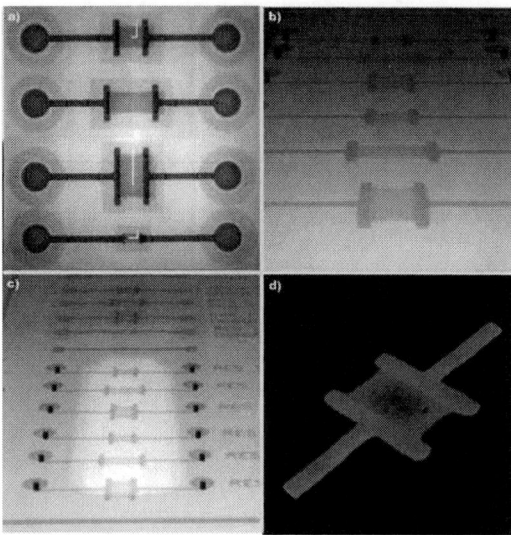

Figure 3: X-ray images of LTCC samples. a) top-down view of surface resistors, b) tilt view of embedded resistors, c) tilt view of LTCC panel with vias, d) CT view of a surface resistor.

SIMS analysis of the LTCC samples appeared to be the most challenging one of the chosen methods, because of the non-conductivity nature of ceramic LTCC substrates. To obtain acceptable results from SIMS analysis some sample preparation techniques such as the use of electrically conductive tapes and foils were needed.

Most of the techniques used during the studies are more or less imaging ones, and also supplementary to each other, which brought out the idea to combine the images and information made with the different techniques. This idea is presented in Figure 4.

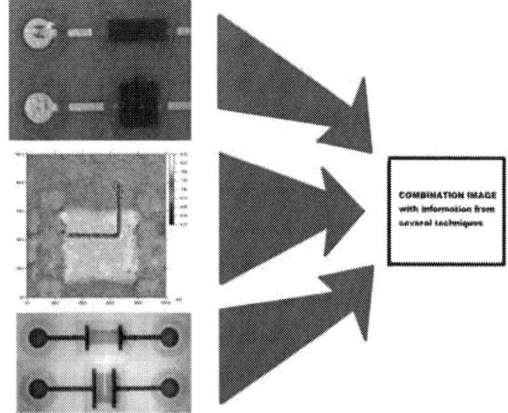

Figure 4: The idea of combining images and information of several inspection and analysis techniques.

This study actually became a starting point for further work concentrating on the combination of the results obtained with different methods and visualisation of these results.

Conclusion

The well known and much used techniques – such as X-ray and optical microscopy – for inspection and failure analysis of LTCC modules give a lot of interesting information about the samples. Other less used methods can be more challenging but with proper preparations and actions they will be useful, too. By combining the information revealed by different, complementary inspection and analysis methods something valuable may be noticed.

Acknowledgements

This work was partially funded by the European Union through Interreg III A and European Regional Development Fund (ERDF), the Finnish Funding Agency for Technology and Innovation (Tekes) and the State Provincial Office of Lapland.

The authors also want to thank Aila Petäjäjärvi and Jukka Säkkinen from the photolithography laboratory of the Kemi-Tornio University of Applied Sciences for providing the samples.

References

[1] D. Morrison, "Low-Temperature Cofired Ceramics Fuel Growth Of High-Frequency Designs", Electronic Design, Vol. 48, Iss. 20, pp. 95-102, October, 2000.

[2] T. Thelemann, H. Thust, M. Hintz, "Using LTCC for Microsystems", Microelectronics International, Vol.19, Iss. 3, pp. 19-23, 2002.

Characterization of Failure Modes and Analysis of Joint Strength Using Various Conditions for High Speed Solder Ball Shear and Cold Ball Pull Tests

Fubin Song[1], S. W. Ricky Lee[1], Stephen Clark[2], Bob Sykes[2], Keith Newman[3]

[1]EPACK Lab, Center for Advanced Microsystems Packaging, Hong Kong University of Science & Technology

[2]DAGE Group

[3]SUN Microsystems

Abstract

In the current study, solder joint integrity was investigated using high speed shear and pull tests on a newly developed bond tester. Solder ball attachment strengths and failure modes were recorded together with fracture energies. Lead-free and lead-tin solders were compared on various pad finishes. Fracture surfaces were examined to assess failure modes and fracture location. More brittle fractures were observed with increasing speed in all types of packages and in both test configurations, but differences between solder types were clear. Correlations between failure mode, strength and energy are discussed. The effect of shear tool height on high speed shear test results is also discussed.

Key words: High speed ball shear, ball pull, failure mode, fracture energy

Introduction

The rapid growth in portable electronic devices, with their increased susceptibility to mechanical shocks, has created an urgent need for package-level tests to replace or supplement the expensive and time consuming board-level drop test (BLDT) [1]. However, conventional solder ball shear and pull tests are conducted at low speeds (<1 mm/s for shear and <5 mm/s for pull), and typically yield ductile solder failures rather than the brittle solder joint failure modes observed in BLDT. Therefore, these tests can only evaluate bulk solder properties, but not the interfacial region between the solder and the package substrate pad finish.

A comprehensive evaluation was conducted to investigate the effects of solder ball alloy, package substrate surface finish, high temperature storage, and package construction on both BLDT and solder ball shear/pull tests. Due to the study complexity and large volume of test data, reporting of the results has been distributed among a series of technical papers [2, 3]. This paper, the 3rd in the series, focuses on the solder ball shear and pull test results for six unique BGA package constructions.

Experimental Procedure

Descriptions of the BGA package constructions used in the study are listed in Table 1. Reflecting the industry-wide transition to Pb-free solder, the majority of the test samples were fabricated with SnAgCu (SAC) BGA solder balls. Due to the varied fabrication sources of the test packages, however, the specific SAC alloys varied from 3-4 wt% Ag, and 0.5-0.7 wt% Cu. Similar to other studies, no clear differences in solder joint mechanical characteristics were observed among this minor range of tested SAC alloys.

Also reflecting typical industry use, the majority of the test samples used organic package substrates. The largest package (51 x 51 mm) used a glass ceramic substrate with a modified thermal coefficient of expansion.

Table 1. Description of test samples

Type	PBGA	PBGA	PBGA	CBGA	PBGA	FC-BGA
Size (mm)	27 X 27	27 X 27	27 X 27	51 X 51	27 X 27	27 X 27
Solder Ball Alloy	SnPb	SAC405	SAC405	SAC387	SAC405	SAC305
Solder Ball Size (mm)	0.76	0.76	0.76	0.635	0.63	0.6
Pad Finish	ENIG	ENIG	OSP	ENIG	Electrolytic Ni/Au	ImSn
Die Size (mm x mm)	7 X 7	7 X 7	7 X 7	22X22	6.35 X 6.35	8 X 8
I/O (Ball)	316	316	316	2386	676	544

In order to assess solder joint fracture strengths and failure modes across a wide range of solder ball shear and pull test speeds, a DAGE 4000HS tester was used. This new machine provided the most recently developed software and hardware modifications, allowing both peak force and fracture energy to be captured at all test speeds. Table 2 summarizes the solder ball shear and pull test parameters. The listed test speeds are nominal values. High-speed video indicated that test speed uniformity during the solder joint fracture generally improves with increased test speed.

Table 2. Description of mechanical tests

Test Method	HS Ball Shear Test	HS Ball Pull Test
Loading Rates	10, 100, 500, 1000 and 3000 mm/s	5, 50, 100, 250 and 500 mm/s
Shear Height	50 μm	-
Clamping Force	-	2.2 bar
Sample Status	As-received (one time reflow)	As-received (one time reflow)

Test Results – Failure Mode

Figures 1 and 2 illustrate the solder joint failure modes for the BGA packages fabricated with organic and ceramic substrates, respectively. The lower height of the top layer dielectric for the ceramic package (compared to the organic packages) resulted in less secondary interaction between the fractured solder joint and the surrounding dielectric. Further, the more robust package construction of the ceramic test samples prevented the occurrence of any pad lifts.

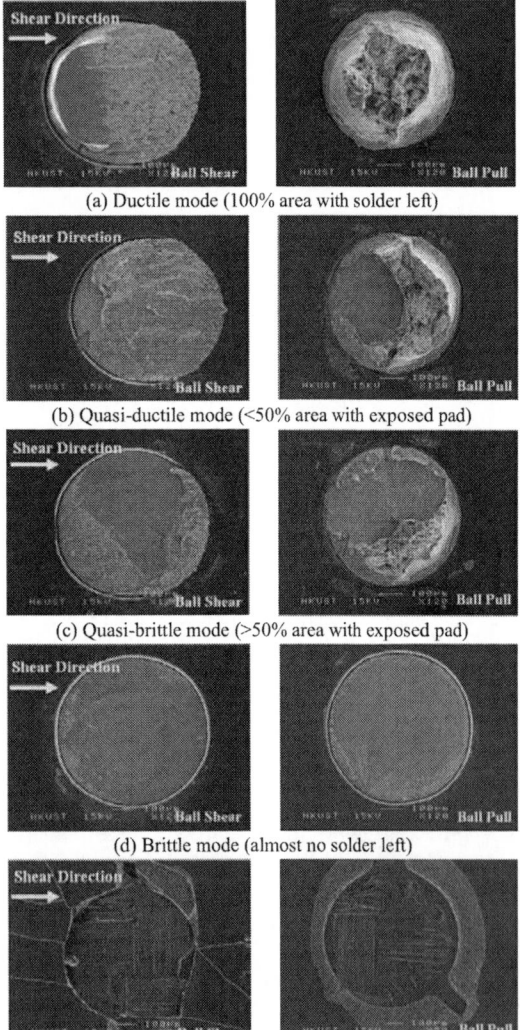

(a) Ductile mode (100% area with solder left)

(b) Quasi-ductile mode (<50% area with exposed pad)

(c) Quasi-brittle mode (>50% area with exposed pad)

(d) Brittle mode (almost no solder left)

(e) Pad lift failure

Figure 1. High-speed ball shear and pull failure modes (organic package substrates)

Figure 3 provides a comprehensive graphical summary of the solder ball shear and pull testing failure mode observations for all package samples, at all test speeds. Despite the wide range of sample constructions and materials, the failure mode trends are quite similar.

At low shear and pull test speeds, the ductile (bulk solder) failure mode predominates. Given that the typical failure mode in board-level drop tests is an interfacial failure, correlations between BLDT and low-speed solder ball shear/pull tests are necessarily problematic. As shear and pull test speeds increased, however, a parallel increase in brittle failure mode percentage was observed.

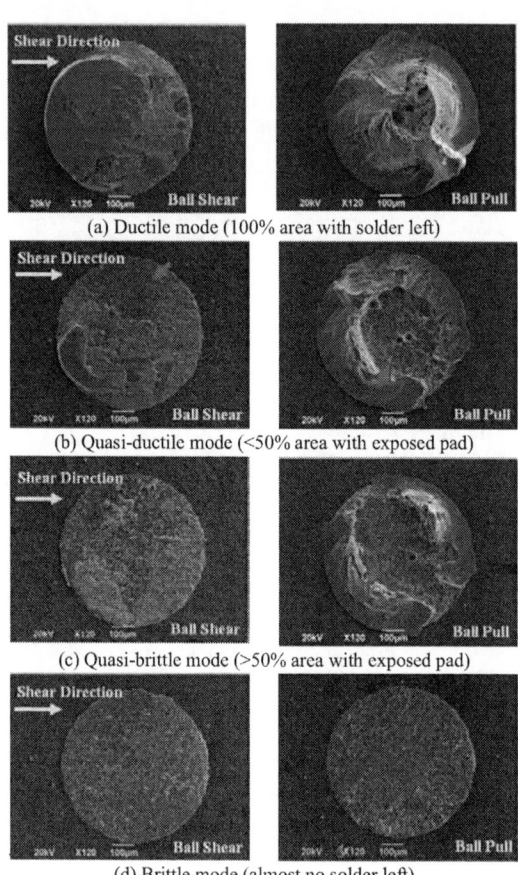

(a) Ductile mode (100% area with solder left)

(b) Quasi-ductile mode (<50% area with exposed pad)

(c) Quasi-brittle mode (>50% area with exposed pad)

(d) Brittle mode (almost no solder left)

Figure 2. High-speed ball shear and pull failure modes (ceramic package substrates)

The transition between ductile and brittle fractures occurred at higher shear and pull test speeds for the BGA sample with SnPb solder, compared to the SAC solder samples. This higher transition speed and lower brittle failure rate for the SnPb solder sample suggests that SnPb solder may prove more resistant to BLDT fracture than SAC solder; this relationship is discussed in more detail in a previous paper from this same study [3].

The failure mode trends between the solder ball shear test and pull test were remarkably similar across all test parameters and package configurations. The most noticeable difference is that the transition speed between ductile and brittle falure modes is lower for solder ball pull testing. A less obvious difference is a general reduction in mixed ductile/brittle failure modes for the solder ball pull test samples. As noted in a previous paper from the same series of study [8], secondary interactions between the fractured solder ball surface and the surrounding soldermask in solder ball shear testing tend to result in solder residue at the soldermask edge and on the fractured pad surface. Pull testing avoids these secondary interactions with the surrounding solder mask, and appears to reduce the level of mixed ductile/brittle failure modes. Other parameters such as Sn grain orientation, IMC uniformity, etc. may also contribute to mixed mode failures, and cannot strictly be avoided.

Test Results – Peak Force and Total Fracture Energy

Figures 4 and 5 graphically summarize the measured peak force and total fracture energy of the test samples for solder ball shear and pull testing, respectively. The graphs provide separate force and energy values for each observed failure mode. The error bars on the graphs identify the minimum and maximum values observed for each failure mode.

These figures illustrate that the measured peak force increases relatively steadily for all failure modes with increased test speed. Moreover, this trend is consistent for both shear and pull tests. In most cases, however, the variation in peak force between ductile solder joint failures and brittle failures falls within the min./max. range and therefore peak force does not correlate strongly with failure mode.

By contrast, solder joint fracture energy appears to correlate well with failure mode for both shear and pull tests. Indeed, mixed failure modes show intermediate fracture energy values between the extreme high values (ductile) and low values (brittle).

Interestingly, the progressive increase observed for peak force with increased test speed does not occur in solder joint fracture energy. Instead, fracture energy values appear to peak at an intermediate shear/pull speed.

Test Results – Shear Tool Height

The solder shear test results described in Figures 3 and 4 were generated using a shear tool height of 50 μm. This height was selected based upon typical industry practice and a shear height sensitivity study detailed below. For the shear height evaluation, a 316 BGA package with an ENIG surface finish and either SnPb or SAC405 solder balls was selected (see Table 3). Shear heights (distance from package surface and shear tool tip) ranging from 20 μm to 150 μm were evaluated at

test speeds of 100 mm/s and 1,000 mm/s. JESD22-B117A specifies a maximum shear tool height of 25% of the solder ball height. The solder ball height of the 316 BGA package was 600 μm; hence, the maximum shear height for this evaluation should be 150 μm,

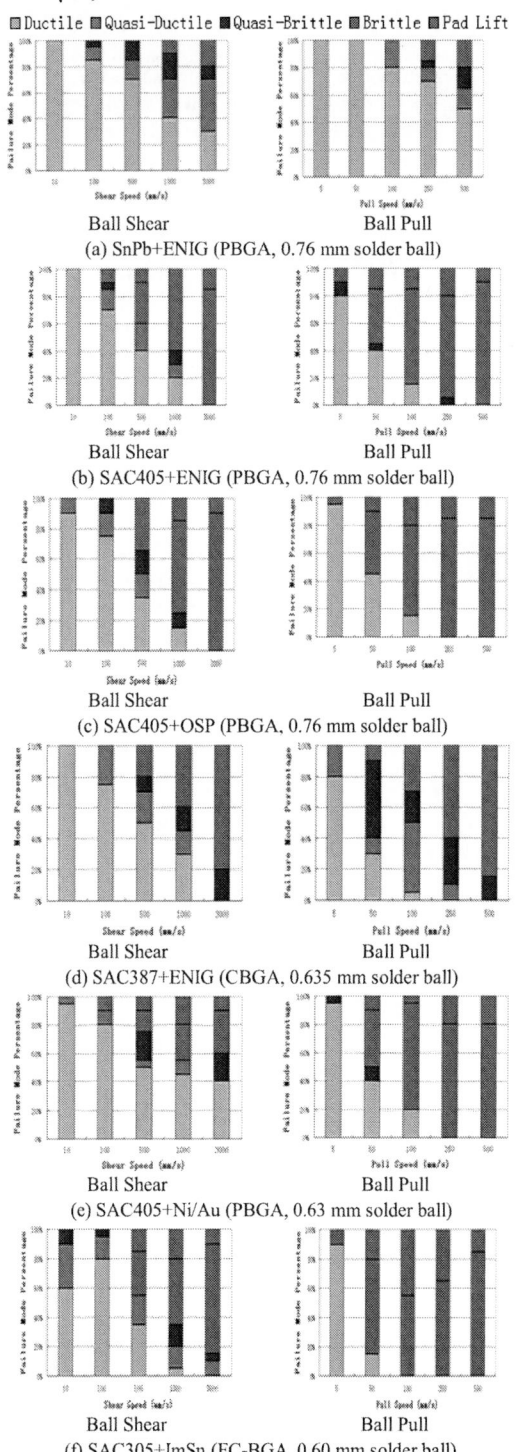

Figure 3. Failure mode distribution in ball shear and cold ball pull tests of variousn package samples with different loading speeds

heights, test speeds, and solder alloys. Clearly, the SAC405

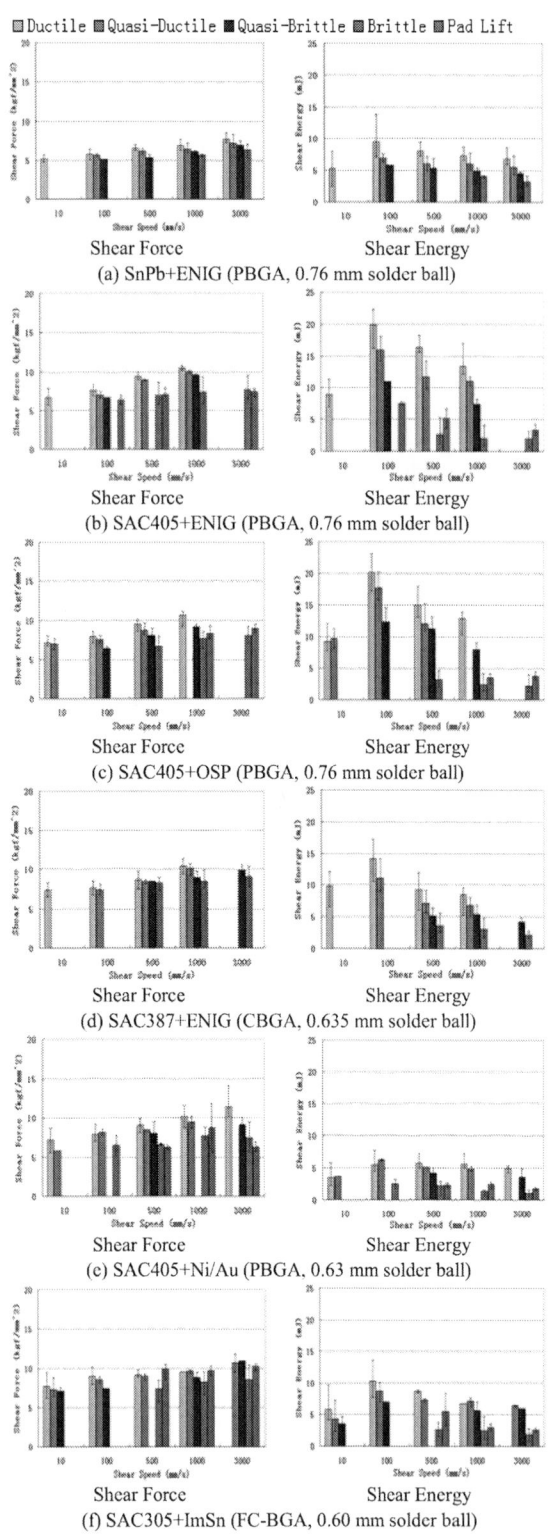

Figure 4. Ball shear strength and energy for different failure modes as a function of test speed

Figure 6 summarizes the observed solder joint failure modes across the range of shear tool

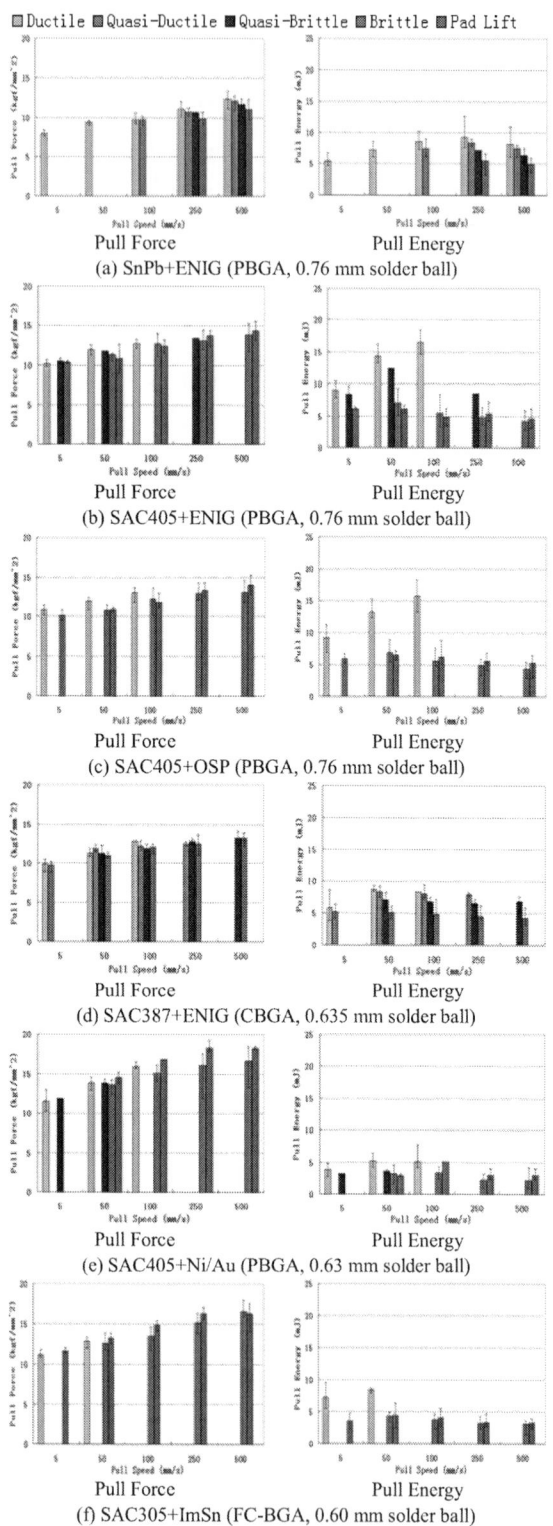

Figure 5. Ball pull strength and energy for different failure modes as a function of test speed

is more susceptible to brittle solder joint failure at the higher test speed than the otherwise identical SnPb test samples, but the failure mode distribuitions for all configurations are essentially insensitive to shear tool height.

Table 3. Description of mechanical tests (Effect of shear height)

Test Method	High speed ball shear
Loading Rates	100 and 1000 mm/s
Shear Height	20, 50, 80, 120 and 150 µm
Solder Ball Size	760 µm
Solder Composition	Sn37%Pb Sn4.0%Ag0.5%Cu
Sample Status	One time reflow
Pad Finish	ENIG

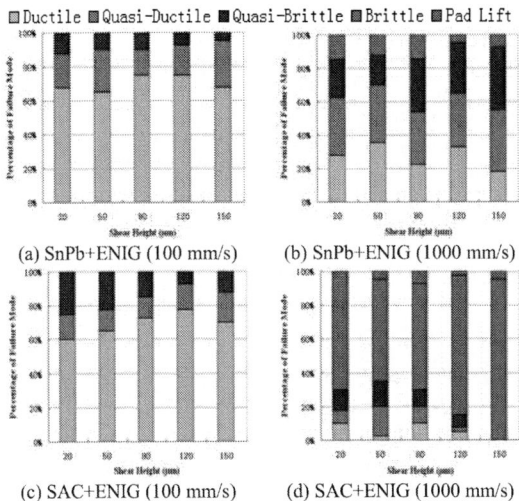

(a) SnPb+ENIG (100 mm/s)   (b) SnPb+ENIG (1000 mm/s)

(c) SAC+ENIG (100 mm/s)   (d) SAC+ENIG (1000 mm/s)

Figure 6. Failure mode distribution in ball shear tests with various shear tool heights

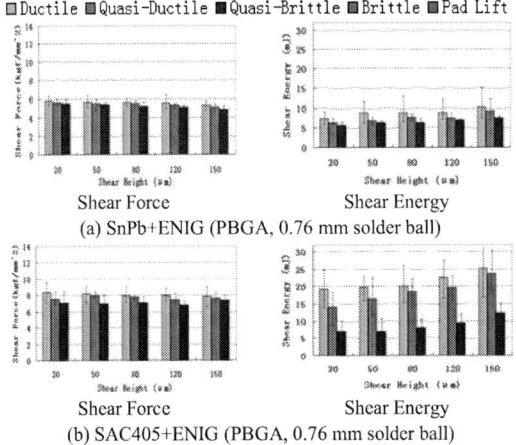

Shear Force   Shear Energy
(a) SnPb+ENIG (PBGA, 0.76 mm solder ball)

Shear Force   Shear Energy
(b) SAC405+ENIG (PBGA, 0.76 mm solder ball)

Figure 7. The effect of shear height on ball shear strength and energy for different failure modes (100 mm/s)

Peak force and total fracture energy for the various shear tool height test samples are graphically summarized in Figures 7 and 8, at shear tool test speeds of 100 mm/s and 1,000 mm/s, respectively. Generally, there is little variation in shear force across the range of tested shear tool height; however, a modest trend towards increasing fracture energy with increased shear tool height was observed. A possible explanation is increased solder ball rotation and secondary interaction with the solder mask with increased shear tool height.

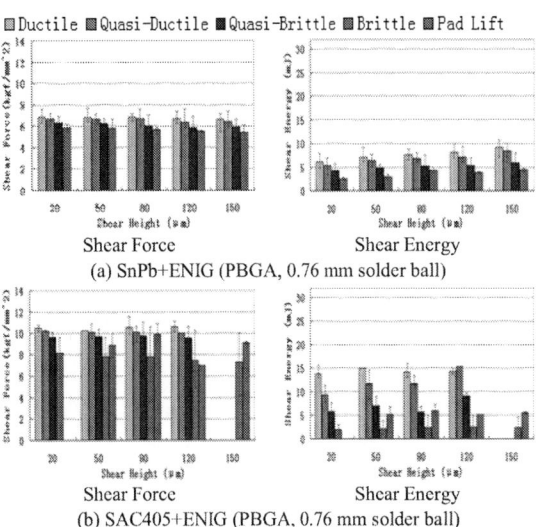

Shear Force   Shear Energy
(a) SnPb+ENIG (PBGA, 0.76 mm solder ball)

Shear Force   Shear Energy
(b) SAC405+ENIG (PBGA, 0.76 mm solder ball)

Figure 8. The effect of shear height on ball shear strength and energy for different failure modes (1000 mm/s)

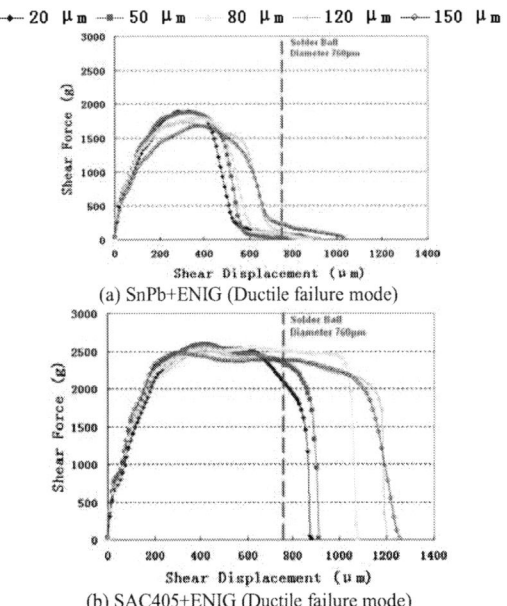

(a) SnPb+ENIG (Ductile failure mode)

(b) SAC405+ENIG (Ductile failure mode)

Figure 9. Typical shear force-displacement curve with various shear height (100 mm/s)

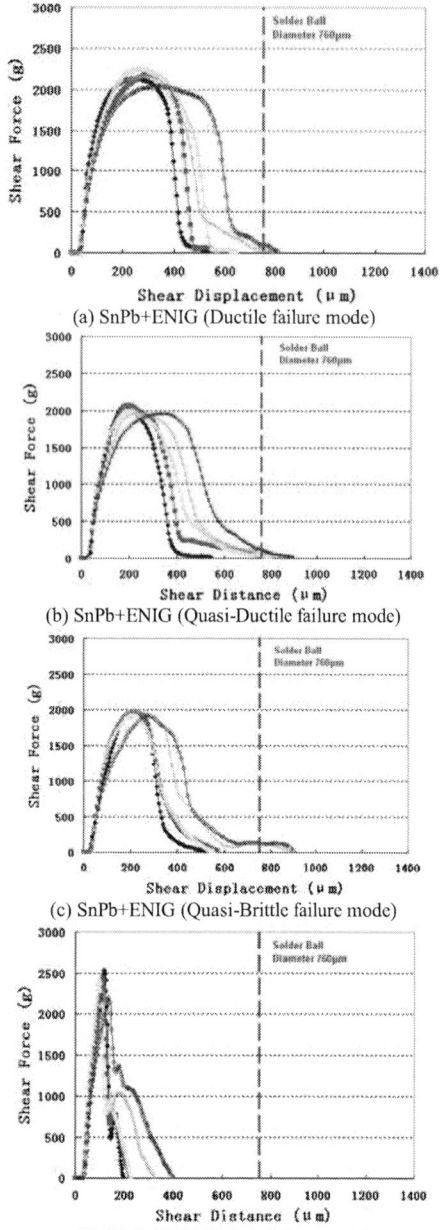

Figure 10. Typical shear force-displacement curve with various shear height (1000 mm/s)

As Figures 9 and 10 illustrate, the extended deformation and secondary interactions tend to increase total fracture energy without affecting peak force values. Although the overall significance of shear tool height is relatively minor with regard to failure mode and peak force, the fracture energy data suggests that a lower shear tool height reduces secondary interactions and extended deformation.

Conclusions

(a) As shear and pull test speeds increased, a continual increase in brittle failure mode percentage was observed, regardless of solder composition.

(b) Failure mode trends between the solder ball shear and pull tests were similar across all test parameters and package configurations.

(c) The transition between ductile and brittle fractures occurred at higher shear and pull test speeds for the BGA sample with SnPb solder, compared to the SAC solder samples.

(d) Solder ball pull tests generated a higher percentage of brittle failures than solder ball shear tests, at all test speeds.

(e) SAC solder showed more susceptibility to brittle fracture than SnPb solder alloy, both in the ball shear and pull tests.

(f) Compared to ball shear tests, pull tests showed fewer mixed ductile/brittle failure modes.

(g) Solder joint fracture energy (in contrast to peak force) appears to correlate better to failure mode for both shear and pull tests.

(h) Measured peak force increases relatively steadily for all failure modes with increased test speed, for both shear and pull tests.

(i) Fracture energy appears to peak at an intermediate shear/pull speed.

(j) Generally, there is little variation in failure mode or peak shear force across the range of tested shear tool heights; however, a modest trend toward increasing fracture energy with increased shear tool height was observed.

(k) Fracture energy data suggests that a lower shear tool height reduces secondary interactions and extended deformation.

References

[1] K. Newman, "BGA Brittle Fracture— Alternative Solder joint Integrity Test Methods," *Proc. 55th ECTC*, Orlando, FL, June (2005), pp. 1194-1200.

[2] F. B. Song and S. W. R. Lee, "Investigation of IMC Thickness Effect on the Lead-free Solder Ball Attachment Strength-Comparison between Ball Shear Test and Cold Bump Pull Test Results," *Proc. 56th ECTC*, San Diego, CA, June 2006, pp. 1196 - 1203.

[3] F. B. Song, S. W. R. Lee, K. Newman, B. Sykes and S. Clark, "High Speed Solder Ball Shear and Pull Tests vs. Board Level Mechanical Drop Tests: Correlation of Failure Mode and Loading Speed," *Proc. 57th ECTC*, Reno, NV, June 2007.

Comparison of Accelerated Life-time Test Methods of Pb-free Solder Joints

Zsolt Illyefalvi-Vitéz; <u>Pál Németh</u>; Olivér Krammer and János Pinkola

Department of Electronics Technology
Budapest University of Technology and Economics
Goldman t 3, Budapest, Hungary, H-1111

Phone: +36 1 4632740; Fax: +36 1 4634118; E-mail: illye@ett.bme.hu & nemeth@ett.bme.hu

Abstract

The electronics industry is in the transition to lead-free technology, and it is a new challenge to test and predict the reliability and the life-time of the lead-free solder joints fabricated with new materials combinations and process parameters. In the paper, the general requirements of solder joints, the theoretical considerations of the applicability of combined accelerated life-time test methods and the results of some experimental work regarding lead-free solder joints are discussed. Three types of circuit modules, manufactured in serial production, and three types of test boards were selected for the accelerated life time test experiments. The operation of the circuits and daisy chains on the test boards were monitored by electrical measurements and optical inspections during the experiments. The subsequent identification and analyzes of failure locations and types included shear force measurements, X-sectioning analysis and X-ray inspection. The failure analyses, as well as, the comparison of the concurrent life time tests showed that lead-free soldering with optimized material combinations and process parameters can provide even better reliability than SnPb, especially in the case of very small solder joints.

Key words: Quality and reliability; interconnection technologies; lead-free soldering; solder joints.

Introduction

As a consequence of increasing demands for shorter product development times, product reliability characteristics must also be understood in a shorter time. Accelerated life tests are in use as methods for understanding reliability with the minimum sample size and the shortest test time. The standards define 'accelerated tests' as 'tests carried out under conditions more severe than standard conditions for the purpose of shortening the test time'. Conducting tests under these severe conditions makes it possible to predict product failure rates in a short time using few samples, thus reducing both time and cost required confirming reliability.

Since lead-free soldering has become a big issue in terms of environmental concerns [1-2], their reliability and life-time test have also got higher emphasis. The most important issues regarding reliability testing of lead-free electronics are as follows [3-4]:
- There is no long term field reliability data for lead free solder, therefore the better understanding of failure mechanisms under different loading conditions has more importance
- The material coefficients used in many life prediction models have not been fully developed for lead free solder

- Lead-free technology implies more complex reactions resulting in more complex microstructures
- The application of concurrent testing approaches to simulate real loading conditions of components and assemblies has come into the focus
- Understanding the results of testing calls for the need of modeling and simulation tools

Considerations for Life-time Prediction

The most important motivation of the application of accelerated life tests is to predict the reliability of a product with the minimum sample size and within the possibly shortest test time. The following effects are in use for acceleration tests:
- Elevated temperature (it is almost always used in accelerated tests)
- Temperature (T) and humidity (H), including THB (TH + bias) test, pressure cooker test, highly-accelerated temperature and humidity stress test (HAST), etc
- Current, voltage or power load (bias)
- Temperature difference, based on the different thermal expansion of materials
- Concurrent methods, including thermal cycling combined with drop or vibration testing, bias cycling combined with vibration testing, vibration testing on elevated temperature and/or in corrosive environments, etc.

Newly developed concurrent accelerated test methods include the followings:

- Thermal cycling combined with drop or vibration testing
- Bias cycling combined with vibration testing
- Vibration testing on elevated temperature
- Vibration testing in corrosive environments

Lifetime models [5-8]

Lifetime estimation requires the models of each effect, which causes degradation (aging and failures) in the tested specimen. The most important models, which are applicable for soldered joints, are listed below, with a short description and formulae.

Temperature activated effects – such as based on diffusion or chemical reaction – can be described with the Arrhenius model:

$$MTTF = t = A_o \cdot \exp\left(\frac{E_a}{kT}\right) \qquad (1)$$

$$AF(T) = \frac{t_{life}}{t_{test}} = \exp\left(\frac{E_a}{k} \cdot \left(\frac{1}{T_{life}} - \frac{1}{T_{test}}\right)\right) \qquad (2)$$

where t: time, MTTF: mean time to failure; AF: acceleration factor, A_o: arbitrary scale factor, E_a: activation energy, T: temperature (K), k: Boltzman constant ($1.38 \cdot 10^{-23}$ J/K), and subscript 'life' refers to normal life while 'test' to accelerated conditions.

The activation energy (E_a) usually set to 0.7 eV, although it varies according to the specimen provided and also to the actual failure mode.

Humidity, that is, the penetration of moisture – causing mechanical stress, inelastic strain and crack growth in materials structures – has a significant effect on the degradation of solder joints and electronic assemblies. The general formula used for humidity-related life is the Coffin-Manson (inverse power law) relationship, expressed as follows:

$$MTTF = t = A_H \cdot \left(H_R\right)^{-n} \qquad (3)$$

$$AF(H_R) = \frac{t_{life}}{t_{test}} = \left(\frac{H_{Rtest}}{H_{Rlife}}\right)^{n} \qquad (4)$$

where H_R: relative humidity, A_H: arbitrary scale factor, and n: exponent. There is no standardized value for the most important model parameter, n. Each manufacturer or researcher determines own characteristic exponent with a value around 3 to evaluate the accelerated life test of the particular specimen.

Humidity is rarely applied as the sole accelerating factor to confirm moisture resistance, and instead a combination of temperature and humidity (TH) stress test is generally used. A popular standard test uses 85% relative humidity on 85 °C temperature with the duration of 1000 hours (6 weeks). Frequently, a nominal static bias is also applied to the specimen to create electrolytic cells, in order to accelerate corrosion of the metallization. This type of test is called THB.

Available combined models assume silently that during accelerated tests, in particular during TH test, the failure processes induced by temperature, by relative humidity, and maybe by other effects, are fully independent and therefore

$$AF = AF(T, H_R) = AF(T) \cdot AF(H_R). \qquad (5)$$

In the case of temperature and humidity caused effect, this leads to the following formula, called the S-H or Peck's model:

$$AF = \left(\frac{H_{Rtest}}{H_{Rlife}}\right)^{n} \exp\left(\frac{E_a}{k} \cdot \left(\frac{1}{T_{life}} - \frac{1}{T_{test}}\right)\right) \qquad (6)$$

In order to make the model more precise, we would recommend applying higher E_a value than the usual 0.7 eV, and a higher than 3, pressure dependent 'n' exponent in case of highly accelerated stress test conditions (e.g. HAST).

These humidity acceleration models generally describe the moisture penetration processes, but not the moisture condensation. Empiric extensions are in use to describe the parameters and behaviors of the surrounding materials and their effects, such as porosity, cleanliness, permeability, surface roughness, etc. Nonlinear diffusion model can be used to explain the effect of porosity of coatings or casing materials, since the humidity, deposited in pores, decreases the permeability, as well as, the diffusion coefficient of the material.

Temperature difference can be efficiently applied for accelerated life time tests. Solder joints of any electronics package are comprised of combinations of various materials and the coefficients of thermal expansion of these materials vary widely. When the package is exposed to thermal excursions and the joints experience temperature difference, the difference between the coefficients of thermal expansion of each material causes internal stress and accumulating inelastic strain, which may lead to an eventual failure as sudden breakdown. Temperature cycle and thermal shock tests using temperature differences are carried out to investigate and predict this kind of life.

Life related to temperature differences can also be modeled by the Coffin-Manson equation, and be expressed as follows:

$$MTTF = t = A_S \cdot \left(\Delta T\right)^{-m} \qquad (7)$$

$$AF = \frac{t_{life}}{t_{test}} = \left(\frac{\Delta T_{test}}{\Delta T_{life}}\right)^{m} \qquad (8)$$

where ΔT: temperature difference, A_S: arbitrary scale factor, and m: exponent, where m = 2 is in general use.

The 'strain exponent' (m) is proposed [9] to be approximately 2 for eutectic or near-eutectic solders and for long cycle dwell times, as it can be found in typical field applications. In this case the creep strain range is proportional to the cycle temperature range (ΔT), and ΔT raised to the strain exponent can then also be said to be proportional to the cycles to failure, and consequently to time to failure. For accelerated reliability testing, care must be taken to have the dwell times sufficiently long to allow complete stress relaxation in the joint and stresses are converted to creep deformation.

This equation has been further modified in an attempt to account for the conditions of field life and the accelerated test, i.e. cyclic frequencies and maximum temperatures, which are seen in the field and applied during the test:

$$AF = \left(\frac{\Delta T_{test}}{\Delta T_{life}}\right)^{1.9} \cdot \left(\frac{f_{test}}{f_{life}}\right)^{-1/3} \cdot$$
$$\cdot \exp\left(1414 \cdot \left(\frac{1}{T_{max\,life}} - \frac{1}{T_{max\,test}}\right)\right) \quad (9)$$

where T_{max}: maximum temperature during a cycle (K); f: cyclic frequency. Note that for purposes of the above equation, f_{life} should be 6 cycles per day minimum.

Such equations should be used as very rough first-order estimates and usually empirical validation of the exponents are needed to obtain reliable results from the calculations. It is also prudent to determine the actual acceleration factors from two or more different testing conditions to verify the validity of the equations, before their use in predicting field conditions to failure.

Results of Calculations for Experiments

According to the specialties of a solder joint and the surrounding structure, that are the robustness, the very simple interconnecting function to be tested, and the co-existence of materials of different coefficients of thermal expansion (CTE) in the structures, bias cycling and vibration testing were excluded and the following methods were selected for theoretical comparison and for the tests:

- TH that is acceleration by temperature (T) and humidity (H), and HAST (the highly-accelerated stress test) where – in addition to temperature and humidity – increased pressure is also applied to achieve higher temperatures than the boiling point of water (100 °C);
- TS that is acceleration by temperature difference (Thermal Shock), in particular because of the solder joint structures are the combination of various materials with different CTE.

The results of comparative calculations are summarized in Table 1, where in columns $AF_{/norm}$, $AF_{/85/85}$ and $AF_{/95/112}$, the environmental condition of the index were selected as reference, respectively.

Table 1: Calculated acceleration factors and selection of lifetime test regimes

	H_{Rlife}	T_{life}	H_{Rtest}	T_{test}	$(H_t/H_i)^{\wedge}n$	exp(...)	$AF_{/norm}$	$AF_{/85/85}$	$AF_{/95/112}$
Normal: 50/40	50	313	50	40	1,00	1,0	1,0		
TH: 85/85	50	313	85	85	4,91	26,0	127,9	1,00	
HAST1: 95/112	50	313	95	112	6,86	127,6	875,4	6,85	1,00
HAST2: 95/120	50	313	95	120	6,86	196,0	1344,6	10,51	1,54
HAST3: 100/120	50	313	100	120	8,00	196,0	1568,3	12,26	1,79
HAST4: 100/130	50	313	100	130	8,00	327,2	2618,0	20,47	2,99

	$\Delta\upsilon_{test}/\Delta\upsilon_{life}$	υ_{life-u}	υ_{life-l}	$\Delta\upsilon_{life}$	υ_{test-u}	υ_{test-l}	$\Delta\upsilon_{test}$	$AF_{/norm}$	$AF_{/85/85}$	$AF_{/95/112}$
TS1: 165/15		35	20	15	125	-40	165	121,0	1,06	7,23
TS2: 165/20		40	20	20	125	-40	165	68,1	1,88	12,86
TS3: 165/50		70	20	50	125	-40	165	10,9	11,74	80,39

On the basis of the calculated comparison of life-time test regimes, the standard 85°C-85% RH, Thermal Shock and two kind of HAST (Highly Accelerated Stress Test) tests were selected for application with the following parameters. Each test cycle duration (Δt) was determined to get similar effect of the cycles of the different regimes, as follows:

- **85/85** – 85%RH (Relative Humidity) / 85°C temperature for 160 hours, then 8 hours for circuits recovery, and for electrical and optical inspection ($AF_{/85/85} = 1$, as reference, $\Delta t = 160$ h);

- **TS** – 960 cycles at 125 °C / 5 min and -40 °C / 5 min (160 hours), then 8 hours for circuits recovery, and for electrical and optical inspection ($AF_{85/85} \cong 1$, $\Delta t = 160$ h);
- **Weak HAST** – 16 hours at 120 °C / 95 %RH / 2 bar, and 8 hours for recovery and inspection ($AF_{85/85} \cong 7$, $\Delta t = 16$ h);
- **Strong HAST** – 8 hours at 130 °C / 100 %RH / 3 bar, and 16 hours for recovery and inspection ($AF_{85/85} \cong 20$, $\Delta t = 8$ h).

Such 85/85, TS and HAST days/weeks run on the test boards until some failure rate achieved, that was expected after 5 days of strong HAST, 15 days of weak HAST, and 6 weeks (5000 cycles) of 85/85 and TS.

The following parameters were measured at the beginning for reference, and during the lifetime tests, after each test cycle, in as many points in the time scale as applicable:

- Optical inspection – photos
- Resistance measurements
- Shear force tests

Experimental Work

The degradation process of a solder joint and its surrounding (i.e. the interface between the solder and the pad, the pad and the adhesive, the adhesive and the substrate foil, the solder and the termination of the component, etc.) is generally evaluated by mounting components on a test substrate and then subjecting it to accelerated tests, with monitoring the effects with inserted inspections and measurements between the repeated cycles. The subsequent identification and analyses of failure locations and types are made to improve product performance and reliability.

Three types of circuit modules, manufactured in serial production, and three types of test boards were selected/designed for the accelerated life time test experiments. In the following, a test board, which was designed for easy qualification of the joints by resistance measurements and shear force tests, is characterized shortly.

The LTT (life time test) board was designed to measure the resistance of two joints, a zero ohm resistor and two small parts of printed wire by four-wire method, in order to determine the functionality of the solder joints during life time tests (Figure 1). On the LTT board the resistors are connected serial, in a daisy chain structure (Figure 2). Both ends of resistors are leaded out one by one to the edge connector to measure the voltage drop on each resistor separately.

Figure 1: Application of the four-wire method for measuring the resistance of two joints

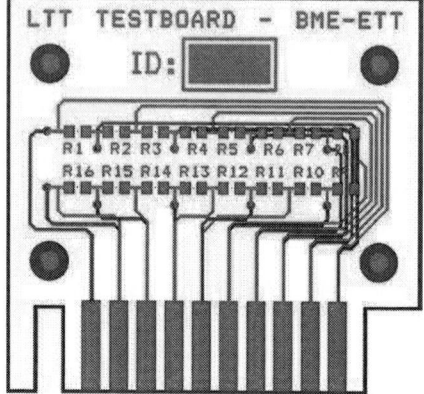

Figure 2: Daisy chain structure of the LTT board

A Measure Board was designed and applied for the easy measurement of the resistors on the LTT board. A pin array with 16+17+16 pins in three lines and two jumpers are used to select the resistor for test. By setting the jumpers on the LTT Measure Board it is possible to measure the contact resistance of the resistors one by one, and in addition it is feasible to measure the contact resistance of multiple connected resistors in any combination or the summary of the contact resistances as well. In addition, if there is a broken joint in the daisy chain, the fault can be easily found by two wire measurement, using the same jumpers to access any part of the chain or any single resistor.

The life time tests always directed to the determination of the effect of some technological parameters, like the type of board materials, the surface finish, the applied thermal profile, the solder material, the type of stencil, the shape of the aperture, etc. and always based on comparisons. The experiments were monitored by optical inspections, resistance measurements and shear force tests between the test cycles. The subsequent failure analyzes included cross-sectioning and X-ray inspections.

Results of Experiments

The following series of photos in Figure 3 – as characteristic selection of the many thousands – are presented to demonstrate the effects experienced while monitoring the assembling and the life time tests. In the figure, the assembling process of an 0603 resistor to 50 µm thick flexible polyimide foil with 18 μm thick copper, using reflow soldering is illustrated. The photos in the lower two lines show the effect caused by the 85°C/85% test and HAST, and make visual impression about the breaking mechanism of shearing.

Because of the very small resistance values of the joints as compared to the 'zero ohm' of the resistors, their change could not be seen, while the degradation was demonstrated well by the shear force. Since it was applied perpendicularly to the middle of the SM resistors, therefore the shear force values refer to the strength of two joints together.

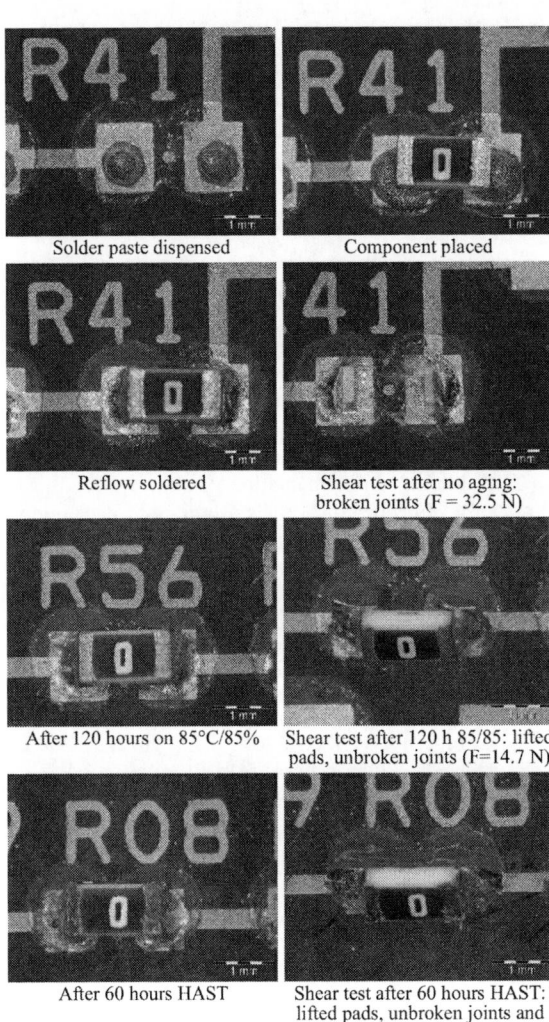

Solder paste dispensed   Component placed

Reflow soldered   Shear test after no aging: broken joints (F = 32.5 N)

After 120 hours on 85°C/85%   Shear test after 120 h 85/85: lifted pads, unbroken joints (F=14.7 N)

After 60 hours HAST   Shear test after 60 hours HAST: lifted pads, unbroken joints and torn substrate (F = 7.7 N)

Figure 3: Monitoring the assembling, as well as, the effect of 85/85 and HAST on 50-18 PI

The aging, therefore the degradation of the joints and/or its surroundings, as it is shown by the decrease of the shear force as the function of aging time in Figure 4, is fairly fast as the effect of the weak-HAST regime: within 1.5-2.5 days (36-60 hours) of aging the shear force decreases to the half of its original value. The rate of the decrease of the shear force, shown in Figure 4, was similar for all materials combinations of soldered joints. In case of flexible substrates, the shear force was always higher for thicker substrate foils of the same materials.

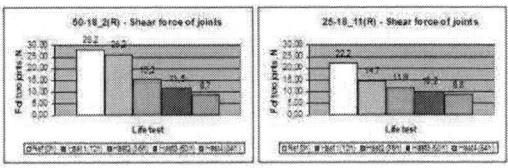

Figure 4: The effect of HAST on shear force, and comparison of PI foil thicknesses of 50 and 25 μm

The investigations included the tests of different surface finishes, i.e. lead-free HASL (Hot Air Solder Leveling), immersion tin, immersion silver and OSP (Organic Solderability Preservative) using SMT resistors with three different sizes.

The comparison of the effect of HAST as the function of dimensions of SMT resistors of 0805, 0603 and 0402 types showed similar degradation rate for all dimensions, while proportionally lower shear force for smaller resistors (see any of the rows of Figure 5).

No characteristic difference could be seen from the comparison of the surface finishes, at least when no preconditioning (reflow cycles) was applied. On the basis of the comparison of the rows of Figure 5, a slight advantage to the favor of HASL and to the drawback of OSP could be observed.

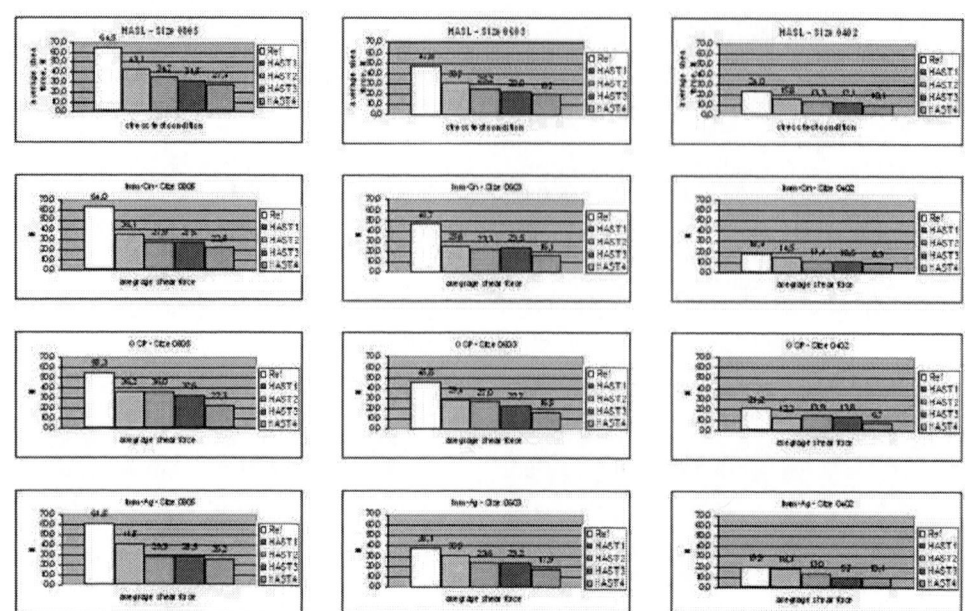

Figure 5: The effect of HAST on shear force vs surface finishes (rows) and resistor dimensions (columns)

Conclusions

The failure analyses, as well as, the comparison of the concurrent life time tests showed interesting results that can be summarized shortly as follows:

- Lead-free soldering with optimized material combinations and process parameters can provide even better reliability than SnPb, especially in the case of very small solder joints.
- Temperature-humidity, HAST and Thermal Shock life-time tests showed realistic results, even when the cycle period of TS was shorter. Reliability test time can be effectively shortened by the application of HAST, or by insertion of HAST test hours into thermal shock cycles.
- The degradation rate of the joints caused by the used 'weak' HAST test was found lower than expected from the theoretical calculations of the acceleration factors. More exact model parameters would be determined on the basis of further investigations.
- New testing approaches initiated the development of improved degradation models and simulation strategies.

Acknowledgements

The authors acknowledge the promotion of the LEADOUT Collective Research Project (COLL-CT-2004-500454) of the European Commission, whose encouragement contributed to the publication of the paper.

References

[1] Directive 2002/95/EC of the European Parliament and Council on the restriction of the use of certain hazardous substances in electrical and electronic equipment (RoHS)

[2] Directive 2002/96/EC of the European Parliament and Council on waste electrical and electronic equipment (WEEE)

[3] Zs. Illyefalvi-Vitéz; O. Krammer; J. Pinkola; B. Riegel; N. Ruzsics; G. Juhász: Lead-free Soldering Implementation Issues, Proceedings of 4th European Microelectronics and Packaging Symposium, May 21-24, 2006, Terme Catez, Slovenia, pp.231-236.

[4] H. Yu and J. K. Kivilahti: Combined Thermal, Thermodynamic and Kinetic Simulations of the Solidification of SnAgCu Interconnections. Proceedings of the 2005 IEEE/EIA CPMT Electronic Component and Technology Conference (ECTC), May 31 - June 3, 2005, Orlando, Florida, USA, p.151.

[5] P. Németh; Zs. Illyefalvi-Vitéz; N. Kánvási: Using Hast Technology for Reliability and Life Time Tests. CD Proceedings of the 14th European Microelectronics and Packaging Conference & Exhibition, Friedrichshafen, (Germany), June 23-25, 2003. pp. 506-510.

[6] Knowledge Based Reliability Evaluation of New Package Technologies Utilizing Use Conditions, INTEL white paper, No:245162-001

[7] N. Sinnadurai: The Correct Model for, and Use of, HAST, Proceedings of 33rd International Symposium on Microelectronics (IMAPS 2000), Boston, Massachusetts (USA), Sep 20-22, 2000, pp.733-736.

[8] M. R. Harrison; J. H. Vincent; H. A. H. Steen: Lead-free reflow soldering for electronics assembly, Soldering & Surface Mount Technology, Vol.13, No.3, (2001), pp.21-38.

[9] H. Solomon: Strain-Life Behavior in 60/40 Solder, GE Report #88CRD261, 1988.

On the Simulation of Flexible Circuit Boards

Luciano Arruda

Instituto Nokia de Tecnologia – Rodovia Torquato Tapajós, 7200, Manaus, Brazil

+55 92 21261034, +55 92 21261062 email: luciano.arruda@indt.org.br

Abstract

This article will compare the behavior between FR4 printed circuit boards and flexible printed circuit boards employing polyimide as a base material. Due to their significantly lower stiffness, flex polyimide boards behave differently then more commonly used FR4 rigid boards. This difference in behavior may result in different reliability characteristics for the boards as well as the components embedded or mounted to the boards. One of the most critical load conditions for rigid boards is acceleration due to impact. In this work polyimide flex boards will be submitted to similar impact loads as rigid boards are submitted during drop tests. Results will be compared and guidelines for simulation models of flex boards will be proposed.

Key words: Flex Printed Circuit Boards, FEM, drop tests, Four Point Bend Tests

Introduction

Current market is driving portable electronic devices to become smaller and to have multiple functionalities. Some manufacturers have even started to embed portable electronic devices inside clothes. The addition of multiple functionalities such as cameras and multimedia players in small packages, and the ability to be shaped inside clothes often come through the use of Flexible Printed Circuit Boards (fPCBs), as they allow the use of mechanisms (slide, flip, rotate) and are easier to be used as non flat shapes.

To be able to be applied in mass production products, fPCBs need to be reliable regarding usage conditions. Some of the main causes for failure for electronic devices are impact due to drops and bending during transportation and field use. Typically, manufacturers conduct drop tests and bend tests to quantify the durability of the components. A common tool to verify the reliability of electronic components in early stages of design is the use of the Finite Element Method (FEM).

Since fPCBs have a different behavior from traditional rigid FR4 PCBs, this work aims at proposing a test vehicle and a simulation model for fPCBs. The test vehicle will undergo experimental drop and bend tests and results will be compared with those from simulation models to validate the latter.

Different structures of fPCBs (different number of layers) will be evaluated and results compared among each other and with previously obtained results from rigid FR4 PCBs. Non linear geometric effects (stress stiffening) on bend tests and buckling on impact tests will be evaluated.

The results obtained from this work should permit proposing guidelines for the modeling of flexible printed circuit boards [1].

Bare Boards Description

There are two types of boards that will be analyzed at this work. The first is an FR4 for which there is previous simulation results correlated with previous experimental tests. The other model is Polyimide board substrate model submitted only in FEM simulation. Both have equivalent geometries and were used to compare its mechanical behavior. The materials properties considered in all analysis of copper, FR4 and polyimide can be seen in table 1 [2], [3].

Table 1 – Copper and Polyimide Mechanical Properties

Material Properties	Copper	FR4	Polyimide
E, MPa	7.50E4	19.8E3 $E_x=E_y$ 8.40E3 E_z	4.00E3
γ, g/mm^3	7.22E-3	1.9E-3	1.80E-3
υ	0.34	0.17 υ_x; 0.21 $\upsilon_y=\upsilon_z$	0.35

Where, E is Young Modulus; γ is Density, and υ is the Poisson Coefficient.

Four Point Finite Element Simulation

The first simulation is Four Point Bend model simulation as proposed in figure 1. The PCB was modeled using ANSYS with linear layered shell element SHELL99 [4].

Flex FE Models were proposed for boards with 2, 3, 5 and 7 layers, each layer containing 8 Cu traces. All these FE models were submitted to equivalent 0.255N distributed load (see figure 1). Initial simulations show that flex boards with less then 3 layers are much more flexible and show considerable deformation due to self weight (see Figure 2). In this case the Cu strips seem to have low influence on board overall stiffness. As a consequence they do not need to be modeled thus simplifying FEM and reducing simulation time and geometry complexity.

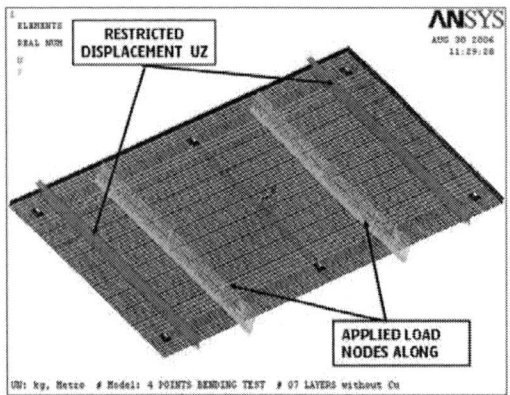

Figure 1: Four-Point Bend Model Employed in Computational Simulation with its Boundary Conditions.

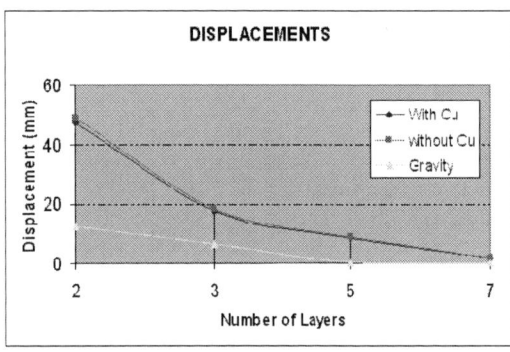

Figure 2: Maximum Displacement in Each Model Type (2, 3, 5 And 7 Layers) (Gravity Indicates Displacement Due to Self Weight of the Board – no External Load).

The strain distribution shows that flex boards behave more like shells, and less like beams (border effects). Rigid boards usually behave more like beams under bending, as illustrated on Figures 3 and 4.

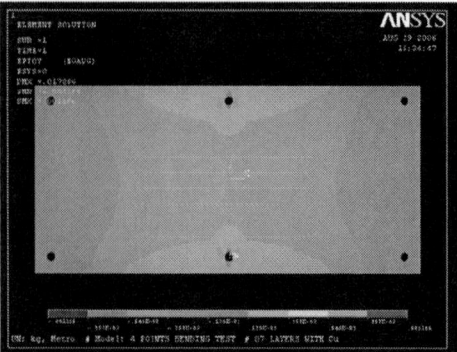

Figure 3: X Strain (Horizontal Direction) Distribution on 3 Layered Models.

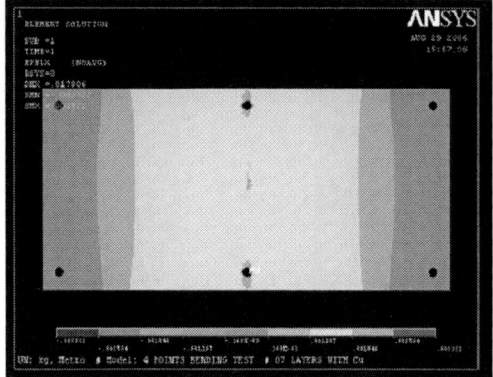

Figure 4: Y Strain (Vertical Direction) Distribution in 3 Layered Model.

Drop Test Finite Element Simulation

The drop test simulation model was done with ABAQUS FEM computational package [5]. The FR4 test vehicle was built with 7 layers and the flex polyimide with 2 and 7 layers. Figures 5, 6 and 7 exhibit the strains distribution on 7 layers FR4, 7 and 2 layer polyimide respectively. During the impact the board presents several bending modes [6]. It can see that the deformation of the FR4 is smoother and mostly in first bending mode. However, figures 6 and 7 of the FPC models show different bending deformation mode where there is local buckling due to low stiffness of these boards [7]. This is an indicator that non linear geometrical effects should be considered in these FE analyses.

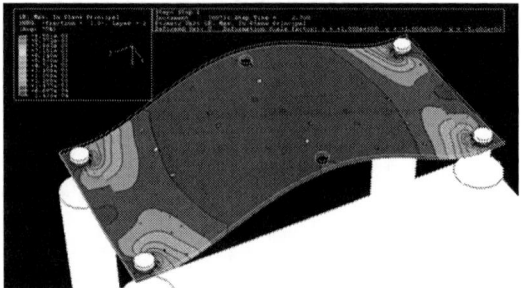

Figure 5: Maximum in plane Strain of the 7 layer FR4 Board. Approximately 9,570 microstrains.

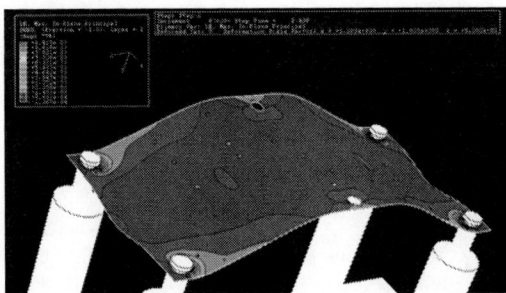

Figure 6: Maximum in plane Strain of the Polyimide Board with 5 layers. Approximately 9,410 microstrains.

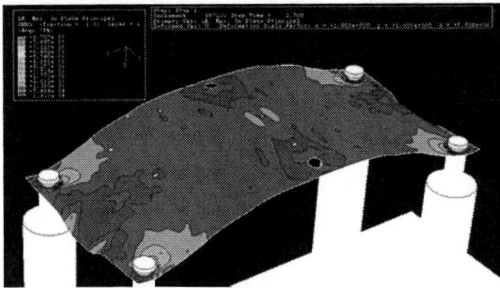

Figure 7: Maximum in plane Strain of the Polyimide Board with 2 layers. Approximately 8,710 microstrains.

Figures 8 and 9 also show that flex boards present shell behavior and rigid board behave more like a beam.

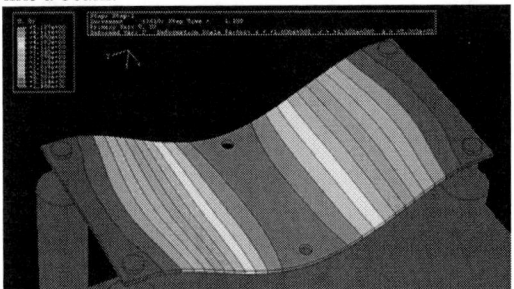

Figure 8: Displacement Field for FR4 MODEL at time: 1.2 ms with U_{3max}=5.188 m .

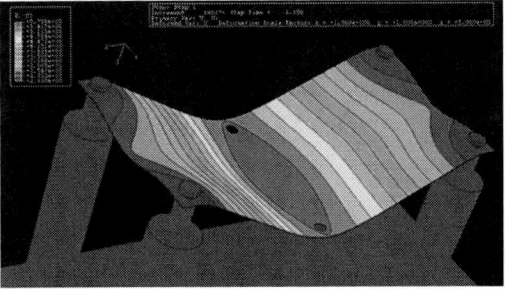

Figure 9: Displacement Field for 7 layer FPC Model at Time: 1.350ms with U3max=5.758 mm.

Due to board reduced stiffness, it is possible that less damage is induced on solder joints due to bending and more damage is induced on Cu traces.

This should be investigated on future testing and simulations.

Experimental Test Analyses

The PCB was fixed in four point vehicle base and drop height set to 1000 mm with acceleration peak limited to 1500 g [8]. The FR4 PCB (Figure 10) and drop vehicle is shown in figure 11. Detail of the tri-axial gage on center position used is shown in figure 10. The acceleration test profile applied is shown in figure 12 with impact time around 1 milisecond.

Figure 13 displays the test result measurement of FR4 PCB without active components. It could be compared with the Drop Test FEM simulation in next steps of this work.

Figure 10: FR4 Test Vehicle Board a Centered Strain Gage Position.

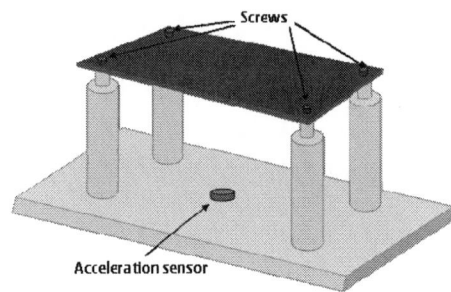

Figure 11: Assembly of the PWB on the Drop Test Level Tester Base.

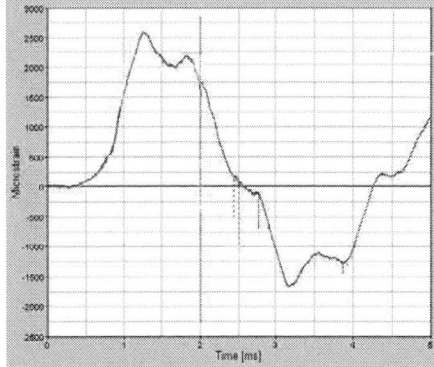

Figure 12: Strain Gage Drop Test Measurement.

Results and Discussion

Figure 13 shows displacement measurement on the center board. The displacement plot has different distribution for each board. Although maximum displacement is similar among boards as

449

well as the time history displacement on the center, the flex board presents different behavior, probably due to stress stiffness effects.

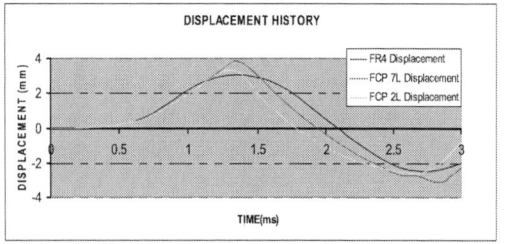

Figure 13: Displacement on Center Measurement (FE Model).

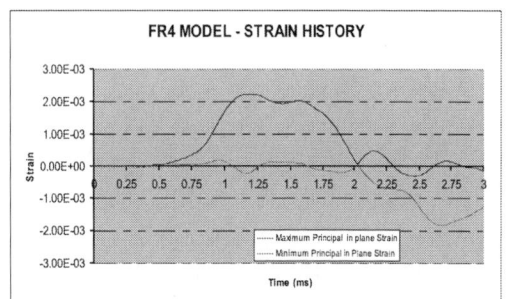

Figure 14: Strain Measurement on Center Measurement of the FR4 Model.

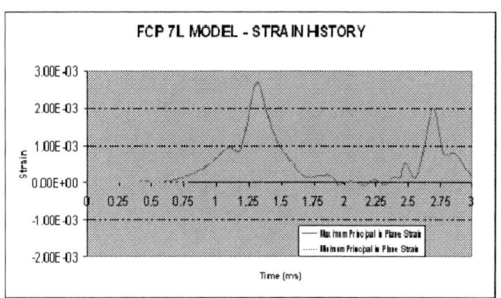

Figure 15: Strain Measurement on Center Measurement of the FCP 7L Model.

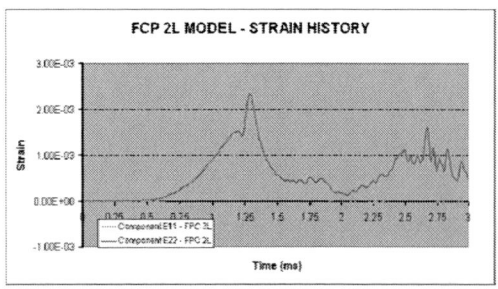

Figure 16: Strain Measurement on Center Measurement of the FCP 2L Model.

Figure 14 exhibits the maximum and minimum in plane strain for a FR4 Model. It can be seen that between 1.1 and 1.7 miliseconds the strain stays constant at about 2000 microstrains most probably due to combined effect of first and second bend modes. On the other hand, the polyimide boards (Figures 15 and 16) does not show sinusoidal

behavior, most probably due to non linear effects. Another interesting fact observed is that rigid board presents bending in opposite direction with similar peak as first peak. On flex boards bending on opposite direction is much lower. Once gain, this is probably due to non linearity [7].

Conclusion

The results obtained from the simulation show that, for the test vehicle proposed with 8 Cu traces in each layer, the effect of copper is negligible. Therefore, it is not necessary to make a complex model to insert the micro geometries of embedded capacitors, resistances, inductors and others micro copper components.

The Drop Test FEM simulation showed similar strain values but different strain history between FR4 and PI boards. The strain reached 5,000 microstrains in FPC components, area and near of the hole is 10,000 microstrains. Both FCP models (7L & 2L) have a high strain second peak response in time history.

Linear Analysis is not sufficient to simulate this physical behavior of the FPC in displacement field of Drop Test and Four Point Test. Simulation shows that geometric non linear effect (stress stiffening) is significant on PI flex boards.

Acknowledgements

The author would like to thank Gustavo Sobue to provide experimental data, Germano Freitas for supervising the program, Marco Marques for managing the SHIFT project. Thanks and appreciation also extended to Ramin Vantaparast from NRC Palo Alto as well as all the INdT members that help this work directly and indirectly.

References

[1] F. Pan, R. Vantaparast, "Reliability Analysis for the Design of a Mult-layer Flexible Board", Proceedings of the 55th Electronic Components and Technology Conference (ECTC), Lake Buena Vista, California, May 30 – June 2, pp. 1299-1304, 2005.

[2] A. Ptchelintsev, "Fatigue Analysis and Optimization of Flexible Printed Circuits", Proceedings of the 2006 Abaqus users Conference (ECTC), Lake Buena Vista, California, May 30 – June 2, pp. 1-12, 2006.

[3] Clyde F. Coombs Jr, "Printed Circuits Handbook", McGraw Hill (2001), pp. 56.3 – 56.6.

[4] ANSYS, "Electronics Manual". *Version 11.0.* 2007. USA.

[5] ABAQUS, "Electronics Manual". *Version 6.63* 2007. USA.

[6] J. Luan & T. Y. Tee, "Effects of Impact Pulse Parameters on Consistency of Board Level Drop Test and Dynamic Responses", Proceedings of the 55th Electronic Components and Technology

Conference (ECTC), Lake Buena Vista, California, May 30 – June 2, pp. 665-673, 2005.

[7] LLE, "The Properties of Polyimide Targets". University of Rochester - Laboratory for Laser Energetics. LLE Review – Quarterly Report. Vol. 92. July-September. (2002), pp. 167-180.

[8] JEDEC, "Board Level Drop Test Method of Components for Handheld Electronic Products" JESD22-B11. JEDEC Solid State Technology Association, (July 2003), USA.

Reliability of Flexible Circuits
with Different Lead-Free Technologies

Bálint Balogh[1], Péter Gordon[1], Zsolt Illyefalvi-Vitéz[1],
Graham Farmer[2], Anna Girulska[3], Tom Harvey[4], György Kotora[5], Damien Kirkpatrick[6]

[1]BME-ETT, Goldmann ter 3, 1111 Budapest,Hungary
[2]GTS Flexible Materials Ltd, Ebbw Vale, UK
[3]ELDOS Co Ltd, Wroclaw, Poland
[4]Epigem Ltd, Redcar, UK
[5]Komed Kft, Budapest, Hungary
[6]TWI Ltd, Great Abington, UK

Phone: +36-1-4634122, Fax: +36-1-4634118, E-mail: balogh@ett.bme.hu

Abstract

The demand for flexible printed circuit boards is continuously increasing as mobile phones, cameras, laptops and other portable devices become more and more widespread. The ROHS/WEEE regulations had to be applied for these types of materials as well. Some of the flexible materials can not withstand the soldering process temperatures even with tin-lead alloys, these are obviously not capable for lead-free soldering. However the influence of higher temperatures on the reliability of polyimide (PI) materials is of great interest. The other, lower temperature materials (PET, PVC, PEN), can be processed alternatively either by laser soldering or with conductive adhesives. This paper gives an overview of flexible base and joining materials, focusing on their applications and environmental specifications such as applicable temperature and humidity ranges. Environmental (moisture absorption, dimensional stability) tests have been done on the most widely used materials (PI, PEN and PET) to determine the influence of processing temperatures on material properties. Based on the results the processing temperature regimes, thus the joining materials (tin-silver-copper or low temperature solder alloy, conductive adhesive) for each base material have been chosen. Special test structures for assembly evaluation and reliability tests were built. The ageing of the samples has been accomplished traditionally by 85°C/85%RH and thermal shock tests (TS). The combination of thermal shock and HAST showed similar failure mechanism with the test duration reduced by a factor of 6 to 10. The component joints have been evaluated electronically and mechanically. The test results not only provide information on the expectable life time period of these circuits, but also help in optimizing the material choice for different application purposes.

Key words: flexible printed circuit, lead-free soldering, alternative process, reliability test

Introduction

Flexible printed circuits (FPC) started their career by interconnecting electronic devices or boards. These applications are still present and substitute the heavier and more rigid wiring in applications like ink jet print head interconnections or rigid-flex structures, where they connect rigid PCBs without the need of connectors or solder joints. Surface mount devices (SMD) can also be assembled to FPCs, which widens their application possibilities, these are among others: instrument panels, under-hood wiring, disk drives, cell phones and laptops. They have various advantages i.e. reduced weight, volume, cost and last but not least flexibility [1,2]. FPCs can be grouped according to how many times they are bent during there life. Some flexible circuits are only bent a few times during assembly (referred to as "bend and stay" or

"bend to fit"), while others are required to bend a few hundred times in their life (flexible circuit), and still others are bended thousands of times during their life (dynamic flex). Each of these has materials and design implications [3].

The RoHS/WEEE regulations also have to be applied for these products [4]. The restriction of lead from solder does not influence the application of PEN, PET and PVC films, since they could not even withstand the lower peak temperature tin-lead solder process. However the introduction also influenced the base film materials since the banned substances could have been used for additives in adhesives, so it is necessarily to investigate the reliability of the above mentioned low temperature resistance materials with conductive adhesive joining technology. Both lead and lead-free profiles can be applied to PI films. The aim of the described experiments was to reveal if the higher processing temperature has any negative effect on the reliability of the FPCs.

Properties of the tested materials

The tests of the most commonly applied flexible base materials are described in this paper. The basic properties of these films can be found in their specifications and the most common ones are summarized in the table below, however the exact data for different manufacturers and their product types can vary significantly.

Table 1. Properties of flexible base films [3]

Film	Melting Point (°C)	Glass Transition Temp. T_g (°C)	Shrinkage (%) @ 30 min/150°C	CTE (ppm/°C)	Moisture Absorption (%)
PI (polyimide)	None	360	0.07	14	2.9
PEN (polyethylene naphthalate)	265	121	0.6	19	0.8
Low shrink PEN	265	121	0.2	19	0.8
PET (polyester)	260	82	1.0	19	0.8
Low shrink PET	260	82	0.2	19	0.8
LCP (liquid crystal polymer)	350	-	0.04	18	0.1

The most important consequence from these data is that PI and LCP are the only base film materials with glass transition temperature higher than the ~270°C peak temperature of the lead-free reflow process. Although moisture absorption and shrinkage data are available for the base films, as it can be seen later, it can be useful to measure them also for the exact structures applied for our FPC.

Moisture absorption test

Moisture absorption tests have been conducted on 6 different base materials, supplied by GTS, according to IPC-TM 650 2.6.2, with a slight difference of that the samples were not immersed into water as described in paragraph 5.1.2.2, since the experience shows that the amount of water on the surfaces of the substrate influences the results. This was substituted by putting the samples for 12 hours in a humidity chamber at 85% relative humidity at 25°C. The moisture absorption or water content can be calculated in weight percent

$$\frac{W_2 - W_1}{W_1} \cdot 100$$

where W1 is the weight after 20 minutes drying at 100°C (the lowest weight was observed in this phase) and W2 is the actual weight of the samples. The chart in Figure 1 shows the water content in

weight percent after 12 hours at 85%RH, 10 minutes and 4 hours dry at 100°C. The captions below the graph show not only the base films (PEN, PET, PI, LCP) but also the adhesive (epoxy, polyurethane, acrylic) and the copper thickness in μm, except for LCP, which was only a bare film.

The highest moisture absorption was observable with the polyimide base materials. However the applied adhesive system also influences the amount of absorbed water, thus the highest absorption, more than 3%, was measured for an acrylic adhesive PI structure. While only 1.58% was measured for the polyurethane and 1.09% for the epoxy based one. These values are significantly lower than the 2.9% indicated in Table 1, even if we consider that the relative humidity was not 100% only 85% in the chamber. It is remarkable that the structure with LCP base film absorbed practically no moisture.

The significance of moisture absorption in practice is the problem of the fast desorption. This is caused when the sample is exposed to heat for example in the reflow oven. In order to avoid delamination or warpage flex circuits are always baked or pre-dried before high temperature process steps. There is no accepted industry standard for the pre-dry process. Applied temperatures are usually between 80 and 125°C and the duration can last from 30 minutes up to several hours. The experimental results showed that 10 minutes drying at 100°C is efficient to remove practically all moisture from the materials, this is indicated by that the weight differences measured after 4 hours drying are within measuring error with the ones measured after only 10 minutes. The samples used in these tests were 10x10 cm^2 single sided copper laminates, which were placed in the oven so that both of their surfaces freely contacted to air. This means on the one hand that the time needed to dry flexible circuits can be longer if the panels can not be placed so freely in the oven, while on the other hand on real patterned circuits there is no continuous copper film on neither side like in this case, thus moisture can escape easier.

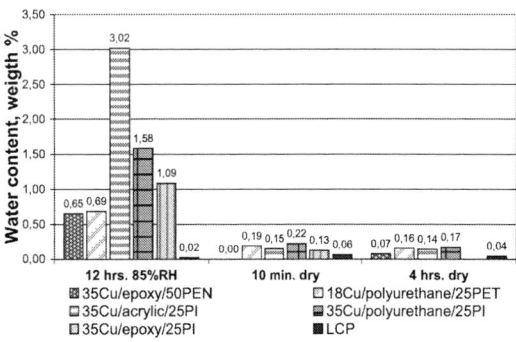

Figure 1. Moisture absorption of FPC base structures

Dimensional stability

As shown in Table 1 shrinkage of flexible materials is an important material property, but it is only given for a certain condition, (30 minutes at 150°C), while in real life FPCs are exposed to higher temperatures and steeper ramp ups. The following section describes the experiments and their results that aimed to reveal shrinkage data of real circuits. The dimensional stability of the materials was measured on test boards patterned by Eldos so that the distance of the four fiducials was measured by an X-Y translation stage combined with an optical microscope. The distance of the fiducials, as shown in Figure 2, were measured initially (unprocessed stage) and after certain thermal processes. These processes are needed to form the joints between surface mount components and copper pads of the flexible substrate. The most commonly applied peak temperature for conductive adhesive curing is 150°C, for reflow process of SnBi, SnPb or SAC, are between 175..185°C, 210..225°C, 235..255°C respectively [4].

Figure 2. Assembled test board. The dimensional stability of the boards were measured between the four fiducials A, B, C, D.

Both shrinkage (negative values) and elongation (positive values) of the materials were observed, as it can be seen in Figure 3. The investigated base materials include LCP, PEN and PI as indicated below the graph, the abbreviations after the material names refer to the Flex-No-Lead partner companiesy who patterned the test boards (FA–Flex-Ability, EP–Epigem, FT–Flexible Technology). This experiment showed that PEN tends to change its dimensions significantly when exposed to heat. One type showed 0.46% shrinkage at 180°C and averaged 0.28%, while another PEN sample tended to elongate. Such high rate of dimensional change in practice can mean several hundred μm misalignment in the range of the usual distances FPCs are applied to. Although LCP boards

were better, but still relatively high dimensional changes were measured on them as well. PI based test panels showed the smallest dimensional instability, these changes were in the range which still allows their application for fine pitch components. It is remarkable that the same types of materials showed elongation at lower temperatures, like PI-FT and PI-EP2 with tin-bismuth reflow profile while these materials shrank at higher processing temperatures. On the contrary PI-FA shrank at lower temperatures like tin-bismuth and tin-lead profiles, but elongated after the highest peak temperature SAC reflow process.

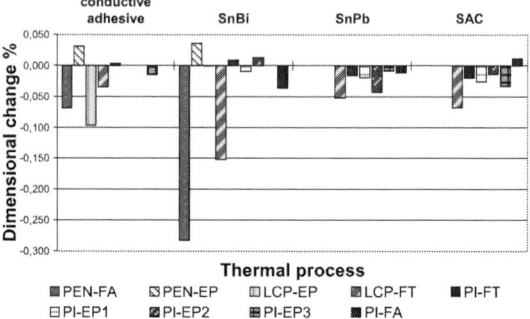

Figure 3. Dimensional stability of different flexible circuit materials at commonly applied processing temperatures.

Reliability tests

The long term reliability of all electronic assembly, thus FPCs also, can only be predicted by accelerated life-time tests. The most common causes of electronic failures are degradation of material properties and stress induced by CTE (coefficient of thermal expansion) mismatch. The corrosion and other degradation processes can be accelerated by elevated temperature and humidity, which can be achieved according to the JESD22-A101-B standard by 85°C-85%RH test, usually referred as damp-heat test. The stresses induced by CTE mismatch of materials assembled together appear in real life either by normal switch on-off processes or by ambient temperature changes. The temperature extremes even in real life can be harsh, especially in automotive electronics where -20°C is not unusual in the winter even in Europe, while close to the engine or in the sun the temperature can rise above 90°C [2]. Thus for automotive electronics an accepted thermal shock standard is +125°C/-40°C temperature extremes with 30-30 minutes soak time [5].

In order to keep the test time within reasonable range the authors decided to halve the soak times to 15 minutes. As it can be seen in Figure 4, this soak time is more than sufficient for the components to heat up or cool down to the necessary temperature extremes, thus while the complete test time is halved the number of thermal shocks can remain the same. However the shorter soak time

454

means shorter periods at temperature extremes, which reduces the material property degradations and the creep of the solder material [2, 6, 7].

Unlike Liu at. al who investigated only flip-chip joints reliability on flexible circuits [6, 7], the authors of this paper decided to analyze more types and larger size components. Several daisy-chains can be found on the test boards, thus the solder joints of 1206, 0805, 0603, 0402 SMD resistors and a BGA component can be analyzed by measuring the resistance of these chains on the contact pads as it can be seen in Figure 2.

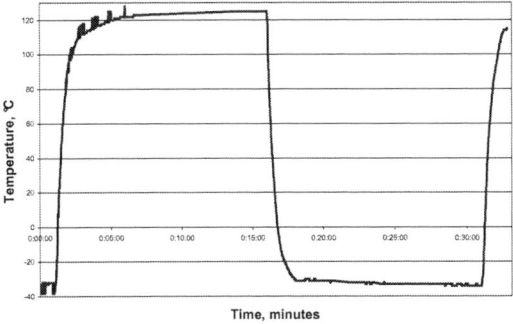

Figure 4. Temperature vs. time measured on an 0805 SMD resistor during thermal shock test. The component reaches the temperature extremes in less than 5 minutes, thus the 15 minute soak time is more than sufficient to produce thermal shocks of 165°C difference.

Ageing by damp-heat: 85°C-85RH%

The elevated temperature and humidity not only cause discoloration of the substrate and the flux residues, but the corrosion of the solder joints and conductor tracks can also be observed as shown in Figure 5.

Figure 5. Corrosion residues flowing along the substrate are marked by the red arrows and discoloration of copper tracks and solder joints can also be observed after 1000 hours 85°C-85%RH test.

As indicated by the moisture absorption data these flexible films soak up water thus it can attack the base film – adhesive – metal track interfaces. It results at least optically in more rapid degradation than observed with the traditional rigid PCBs. On those usually only slight corrosion of solder joints, copper tracks and discoloration of flux residues can be observed, like in Figure 6.

Although the optical changes are obvious due to elevated temperature and humidity, it is still the electrical reliability, which plays the most important role. Figure 7 presents the cumulative component failure percent of the BGA nets and 0805 SMD resistors with SAC solder and conductive adhesive.

Figure 6. Rigid PCB after 1000 hours 85°C-85%RH test. No corrosion residue flows and only slight discoloration of the copper tracks and the solder joints can be observed.

The conductive adhesive joints showed very poor reliability, especially in case of BGA joints, where by the end of the 1000 hours nearly 90% of all joints failed. Even with the much higher contact area SMD resistor joints the failure rate of the adhesive is significantly higher than that of lead-free solders.

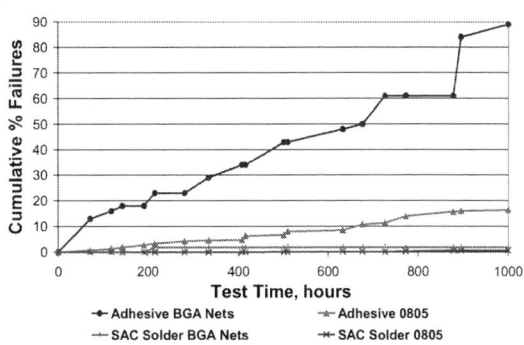

Figure 7. Cumulative component failures of SAC solder and conductive adhesive interconnection FPCs during 1000 hours 85°C-85%RH test.

The SAC solder joints showed only random failures, 1.85% of the BGA nets failed after 216 hours and their number did not increase during the remaining 784 hours. The resistor joints were intact until 500 hours and only 0.6% of them failed by the end of the test. The higher failure rate of the conductive adhesive boards, can not only be put down to the material, but also to the fact that misaligned components are not pulled back during adhesive curing like solder does when molten in reflow, thus the contact area is significantly smaller than with soldered joints. To improve the reliability of conductive adhesive joints, special attention to component placement has to be paid.

The reliability of an FPC does not only depend on the joining material, but it is also influenced by the base film. Figure 8 shows cumulative component failures grouped by different base materials. The highest failure rates are observed with LCP and PEN films. This can be explained by that these materials showed the highest dimensional change, thus the component misplacement was the most significant in their cases.

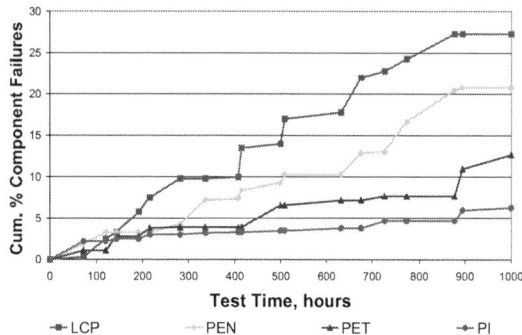

Figure 8. Cumulative % Rates of Component Failures during 85%RH/85°C Testing on Test Board Flexible Circuits

Ageing by thermal shock test

Different types of failure mechanisms can be observed by thermal shock test. TS is supposed to cause stress induced failures in solder joints due to CTE mismatch. Although it can be a case with rigid PCBs, the failures of the tested FPCs mostly occurred due to the break of the copper tracks as it can be seen in Figure 9 and 10. Cracks can be found in the substrate material as well, which indicates that it gets brittle because of the elevated temperature that is why the mechanical stresses can induce cracks in it.

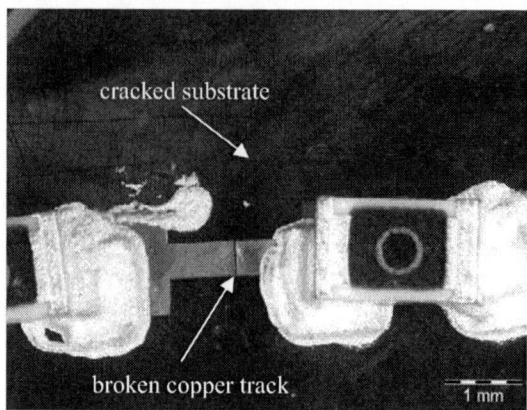

Figure 9. Broken copper track and cracked substrate on an adhesive board.

The tested FPCs were not fixed to rigid boards so that they were not impeded from free and flexible movement of their own. This fact can explain the high number of cracks since all of the circuits bent up to a certain degree and some even curled up. In real life applications the areas where the components are placed are never left without stiffening, while there are design rules i.e. symmetrical coverlay or widened tracks which can increase the durability of the copper layer to mechanical strain [1].

Figure 10. Broken copper track on a soldered board.

Figure 11 presents the contact resistance change of daisy chains consisting of 52 pieces of 0805 0 Ω SMD resistors. 1 and 2 after SAC and SnPb means different types of PI based FPCs. Based on the results it can be stated that there is no significant difference regarding contact resistance between the traditional tin-lead eutectic solder and the most commonly used SAC solder, although the lead-free solder showed slightly higher resistance in both cases. The degradation of the solder joints is not obvious, although there is approximately 2% resistance increase if the initial and the 1000 cycle data are compared, but this can implicate the degradation of the copper tracks and even the 0 Ω resistors. It is also remarkable that no change or even decrease of resistance was observed after 100 cycles.

456

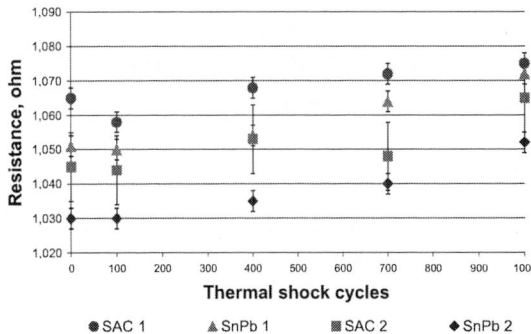

Figure 11. Resistance change of 52 pieces of 0 Ω in daisy chain soldered 0805 SMD resistors on PI test boards during thermal shock test.

Conclusions

Moisture absorption, dimensional stability, thermal shock and 85°C-85%RH ageing tests of PI, PET, PEN and LCP based flexible circuits have been conducted. Special attention was paid to the influence of the higher processing temperatures needed for lead-free soldering and to the reliability of alternative joining technology i.e. conductive adhesive. Based on the results the following conclusions can be drawn:

1. PI has the highest moisture absorption which is also influenced by the adhesive system. 10 minutes drying at 100°C was efficient to remove all absorbed water from the flexible substrates.

2. PEN and LCP showed the highest dimensional change that is in the range which impedes the application of fine pitch component on these materials. This dimensional change causes considerable problems with conductive adhesive which does not melt so self alignment of misplaced component can not be expected.

3. Elevated humidity and temperature induced more severe changes (corrosion of solder and conductor tracks, corrosion flows, discoloration) than the same conditions usually cause in case of rigid PCBs.

4. Lead-free solder joints showed very low failure rates during 85°C-85%RH test.

5. Thermal shock test induced failures mostly in copper tracks and not in solder joints. These failures were due to bending or curling up of the substrate, which can be prevented by adequate design rules and/or stiffening. Thermal shock tests of a different types of test boards are being carried out, which are stiffened to impede the bending of the substrate and the cycle number is raised to 2500 in order to see the degradation of the solder joints in more details.

6. Thermal shock test did not cause significant degradation of solder joint and copper track conductance.

The experimental results described above are complemented by a paper, to be published also at EMPC2007, about simulation results by Flex-No-Lead project partners from the University of Greenwich [8].

Acknowledgements

The authors acknowledge the promotion of the FlexNoLead Project (Flexible Circuits Processing, Performance and Reliability using Lead-Free Soldering Process, EC Contract: COOP-CT-2004-513163) of the European Commission, whose encouragement contributed to the publication of the paper. We also acknowledge the partners who put a big effort in the production of materials, test boards and in their assembly and also would like to thank the useful discussions with all of the collaborating partners on this project, namely Epigem Ltd, GTS Flexible Materials Ltd, KOMED Ltd, International Consulting Bureau, Flex-Ability Ltd, ELDOS Ltd, Flexible Technology Ltd, Freudenberg Forschungsdienste KG, TWI Ltd, University of Greenwich.

References

[1] Joseph Fjelstad, "Flexible Circuit Technology", BR Publishing, third edition, Seaside, Chapter 1-3, pp. 1-65, 2006.

[2] Rao R. Tummala, "Fundamentals of Microsystems Packaging", McGraw-Hill, New York, Chapter 16 and 22, pp. 614-918, 2001.

[3] Zs. Illyefalvi-Vitéz, B. Balogh, G. Farmer, A. Girulska, D. Kirkpatrick, "Life-time Tests of Lead-free Solder Joints on Flexible Printed Circuits", Proceedings of the 2007 Polytronic Conference, Tokyo, Japan, pp. 234-239, January 2007.

[4] Armin Rahn, "Bleifrei löten", Eugen G. Leuze Verlag, Bad Saulgau, pp. 68-79, 2004.

[5] A. Irisawa, "Töredező forrasztott kötések? ", Elektronet, pp. 43-45, December 2006.

[6] X. Liu, S. Xu, G. Lu, D. A. Dillard, " Effect of substrate flexibility on solder joint reliability", Microelectronics Reliability, Vol. 42. pp. 1883–1891, 2002.

[7] Y.C. Lin, X. Chen, X. Liu, G. Lu "Effect of substrate flexibility on solder joint reliability - Part II: finite element modeling", Microelectronics Reliability, Vol. 45. pp. 143–154, 2005.

[8] M. J Rizvi, C. Y. Yin, C Bailey, H Lu, "Performance and Reliability of Flexible Substrates when subjected to Lead-Free Processing", Proceedings of the 2007 EMPC, Oulu, Finland, in press

Prospects and Yield of Electrochemical Wafer Plating for Bumping and Signal Routing

L. Dietrich, M. Töpper, Th. Fritzsch, O. Ehrmann, and H. Reichl

Fraunhofer-Institut für Zuverlässigkeit und Mikrointegration
Gustav-Meyer-Allee 25, D-13355 Berlin

phone: +49-30-46403-605 fax: +49-30-46403-123 e-mail: lothar.dietrich@izm.fraunhofer.de

Abstract

A micro-electroplating method to fabricate smallest interconnection bumps and highly dense routing layers on semiconductor wafers is being presented and discussed. The single process steps of this wafer-level packaging (WLP) are similar to the machining widely used in front-end technology at the wafer fabs, mainly differing in the structure sizes and layer thicknesses. The process steps as sputtering, lithographical printing, electrochemical metal deposition (ECD), and selective etching are highlighted in detail. Two kinds of a plating base are used for various electrodeposits. They also act as an under-bump metallization (UBM). A sputtered Ti:W(N)/Au seed layer is used for the electroplating of Au and Au/Sn, and a Ti:W/Cu metallization is sputtered for the Cu and Ni/Au deposition as well as for the solder (SnPb37, SnPb5, SnAg3.5, SnCu0.7). Either spin-coating or spray-coating technique is used to deposit the liquid photoresist onto planar or else topography wafers. By applying highly viscous and UV sensitive resist systems, layer thicknesses from 5 µm up to 120 µm with excellent thickness homogeneity and a precise pattern resolution for all standard wafer sizes can be achieved. The commercially available electrolytes have been modified and specially adapted to the micro-electroplating requirements. Adequate wet etching solutions had to be specially formulated to minimize the affect and the undercut of the electroplated microstructures.

Key words: wafer-level packaging, bumping, redistribution, electroplating

Introduction

The challenge of a semiconductor package is the formation of the physical structure on a chip to get the electrical connection with external power, ground, and signal source and to protect the device from external environments. The interconnection must provide a geometric translation from the chip scale elements of the device to the circuit board scale of the system, complying with the electrical, mechanical, and thermal demands of the assembly. The continuous claim of highest integration density of microelectronic packages is confirmed by a high ratio of die-to-substrate foot print area as well as by the possibility to use the whole chip surface for a high number of I/O pads in area array. To relax the I/O pitch of front-end fabricated IC wafers, a rerouting of the mostly peripheral pads onto the inner chip area is often done prior to the bumping step [1].

The permanent miniaturization of electronic systems and increasing complexity of microchips lead to further reduced I/O pitches also in the future [2]. In addition, cost efficiancy is a major criterion for the successfulness of a certain technique. Here, thin-film technology can offer reliable and flexible packaging solutions on wafer level in consideration of quality, yield, and cost. The idea is to apply as much of the packaging sequence onto the wafer as possible with the approach to an efficient batch production. The current transition to 300 mm wafers forces the WLP, as the cost factor becomes more efficient with larger wafer size [3].

Circuit performance will be increasingly determined by back-end integrated passive components (IPDs). Resistors, inductors, capacitors, and filters have already been implemented into a thin-film multilayer using WLP technology [4]. WLP for micro-electromechanical systems (MEMS) is already widely used to protect fragile and moving elements by hermetic or near-hermetic cavities. Moreover, IR microbolometers and RF switches have been packaged in BCB-sealed cavities and then bumped with electroplating technique [5]. Au bumped pressure sensor chips find their way even into human eyes to controll its intraocular pressure.

Thin-Film Technology

The wafer or substrate to be bumped may be covered with different kinds of organic and inorganic thin-film layers defining the I/O pads. So, front-end fabricated IC wafers generally have Al or Cu bond pads on top. InP or GaAs semiconductors for photonic and other high-end devices are mostly covered with gold as pad metal. The wafer passivation consisting of silicon compounds or sometimes PI should protect the IC against ambient influences.

The ECD technique accommodates a variety of metals and metal alloys to support the demands of a great number of applications. Besides the conventional lead-tin solders, several lead-free alloys are being introduced to the microelectronic market, since the governmental restrictions on the use of Pb-based solders become valid for most of the electronic products. Reliability studies have indicated that the eutectic SnAg3.5 and SnCu0.7 alloys show an adequate reliability in comparison to SnPb37 solder. The eutectic Au/Sn system provides a good corrosion resistance and enables a flux-free flip-chip assembly. Hence, it is the most suitable interconnection material for optical and optoelectronic devices. Electroplated straight-wall bumps consisting of pure Au, Cu, or Ni either get the solder from the joining bond pad or else can be contacted by wire bonding (WB), thermo-compression bonding (TCB), or tape automated bonding (TAB).

In case of standard IC wafers, the bumping sequence will be applied directly onto the I/O pads. If the original I/O layout has been redistributed, the bumping sequence will take place onto the routed metal layer with a newly arranged interconnection grid [6]. Cu routing layers are electrodeposited mostly onto a previously coated and photostructured dielectric layer showing small vias to the chip I/Os. This insulating film consisting of PI or BCB should prevent short-cuts between the Cu lines and the ICs. After depositing the Cu routing layer, a second insulating film will be coated and structured on the wafer forming the bump pads and working as a solder mask. Finally, the Cu or Ni/Au bumps required for soldering or wire bonding are electrochemically deposited in new area array configuration, or – on the other hand – the solder depot is subsequently electroplated onto the Ni or Cu socket. The placement of solder bumps on top of a polymer layer offers reduced self-capacitance which is desirable for RF applications.

The fabrication sequence of ECD bumps and lines can be divided into fundamental process steps which are sputtering of the UBM, lithographical printing, electroplating of microstructures, removal of the photoresist, and differential etching of the plating base schematically shown in Fig. 1. In case of solder bumps, a final reflow is usually done to homogenize the tin alloy and to form the spherical bump shape. Each solder composition requires an adapted temperature-time profile. An inert liquid organic medium privileged with a slightly reductive potential has been chosen. In contrast to all other solder alloys, the Au/Sn bump is electroplated in a two-stage process, in fact by deposition of a pure Au socket with a Sn cap on it. The eutectic AuSn20

solder is formed during a thermal conditioning at 200 °C.

If MEMS areas are present on the wafer, one's attention must be directed to a possible damaging during thin-film processing. In some cases it will be necessary to protect the micromechanical elements by covering with photoresist prior to the sputter step, or by local etching the plating base prior to the electroplating process.

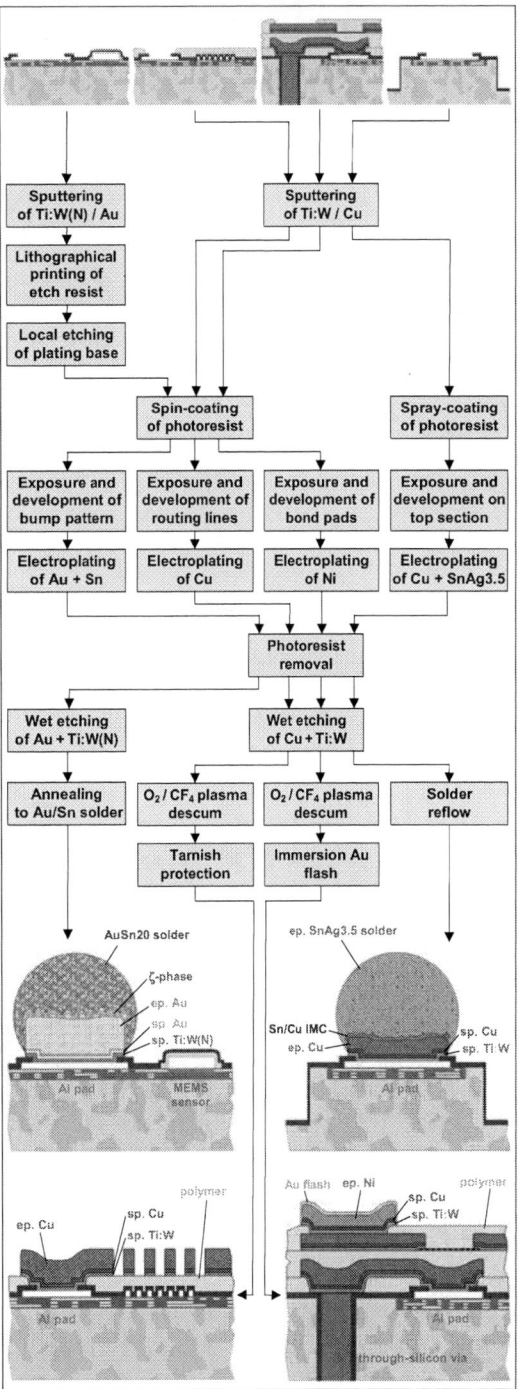

Fig. 1: Selection of process sequences for ECD wafer plating

After concluding the bumping sequence, the wafers or substrates can be electrically tested, visually inspected by automated optical inspection (AOI), and then inked for known good dies. If no further additional packaging step is intended, the wafer-level processing will be concluded by thinning and dicing.

All kinds of thin-film built-ups presented in this paper are well established on 100 mm through 200 mm wafers, going on to be qualified on 300 mm wafers. Prototyping as well as volume production of bumped wafers is possible. High-volume manufacturing equipment with fully automated handling is available for the common wafer sizes.

Deposition of Plating Base

All kinds of semiconductors such as silicon, SiGe, GaAs, and InP as well as ceramic and quartz substrates can be processed by electrochemical wafer plating. Besides, there is no restriction in passivation types such as silicon oxide, oxy-nitride, silicon nitride, and polymers like polyimide (PI), polybenzoxazole (PBO), and benzocyclobutene (BCB). After building up a WLP, yet other materials such as BCB, Epoxy, Ni, NiCr, or NiFe can be patterned on top of the additional layer sandwich.

The UBM provides the critical interface between the previous pad metallization on the wafer and the subsequent ECD bump. Therefore, the sputtered plating base must provide a good mechanical and electrical connection to the I/O pad and to the previous WLP metallization respectively. Prior to the sputter process, the native metal oxides and other contaminations have to be removed from the initial surface by back-sputtering with argon to enhance the mechanical and electrical performance of the wafer/UBM interface.

The selection of UBM and bump metal mainly depends on the melting point of the solder, the thermal and mechanical reliability of the interfaces between UBM and bump, the integrity of adjoining pad metallization, the bumping process capability, the operating conditions of the assembly, and the reliability demands on the whole package.

Thus, a 100 nm to 230 nm thick Ti:W layer is first sputtered onto the wafer acting as an adhesion promoter and diffusion barrier. The Ti:W has to protect the pad metal against the formation of intermetallics and stands out due to its high thermal stability, low contact resistance, and good adhesion properties to the adjoining layers. The diffusion barrier properties can be influenced by addition and variation of the N_2 concentration within the plasma. In case of Ti:W sputtering onto a BCB layer, the penetration of metal atoms into the polymer matrix in a depth of around 100 nm is responsible for the strong adhesion of more than 80 MPa. Without breaking the vacuum, the essential plating base consisting of 150 nm through 300 nm Cu or otherwise 200 nm Au will be sputtered on. The plating base should protect the underlying Ti:W(N)

from oxidation and also provides the electrical conductivity which is necessary for the ECD process. With optimized sputter conditions, a low-stress deposition and a good homogeneity of layer thickness are being achieved without any loss of semiconductor functionality.

Fabrication of Photoresist Pattern

Two types of a liquid and highly viscous photoresist are needed to create a temporary mask for all kinds of electrodeposits. The used positive resist system from *AZ Electronic Materials* shows a high transparency within the near UV spectrum, therefore a high depth-to-width aspect ratio and steep slopes of the generated structures are possible even in thick films. This photoresist basically is a three-component system containing a Novolak resin matrix, a photosensitive agent, and a solvent combination. The photo-active compound makes the exposed areas soluble in an alkaline solution such as diluted sodium carbonate.

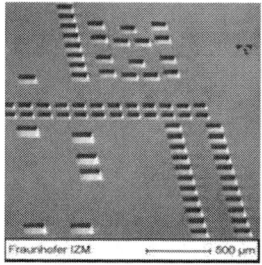

Fig. 2: Spin-coated Novolak **photo-resist** with thickness of 30 μm on planar CMOS wafer **Fig. 3: Close-up view of bump pattern with size of 80 x 80 μm²**

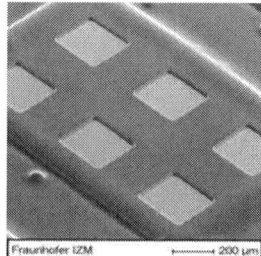

Fig. 4: Spray-coated Novolak **photo-resist** with thickness of 16 μm on Si wafer with 40 μm topography in detail deep-etched trenches **Fig. 5: Bump pattern with size of 250 x 250 μm² on top of**

As in solid state technology, spin-coating technique is used to deposit the photoresist onto the wafer in the desired layer thickness from 5 μm through 60 μm. At this, layers of more than 45 μm can be realized by multi-spin-coating with good thickness homogeneity for all standard wafer sizes up to 300 mm. On wafers with high surface topography, spray-coating technique is used to generate photoresist layers with sufficient thickness uniformity and well defined bump structures. After fixing the coated film during a pre-bake step, the wafer is aligned to a glass mask which is coated with a chromium pattern. Those pad regions are being exposed where the bumps will be located later. In order to get an accurate pattern transfer, a contact between the resist surface and the lithographic mask is necessary and can be achieved by vacuum mode operation. The exposed areas dissolve during the development in an alkaline fluid, and the resist pattern appears.

To apply liquid photoresist layers with a thickness range from around 50 µm up to 120 µm, a negative working system from *Rohm & Haas* is used. It particularly suits for the formation of large solder balls and tall pillar bumps. This UV sensitive resist system shows similar actinic properties as AZ photoresist does. It also can be processed with the same equipment for spin-coating, baking, exposing and developing.

The used photoresist systems are well compatible to a large variety of electroplating solutions working in a strongly acidic through a slightly alkaline pH range. The bleed-out of soluble resist components will not impair the bump quality, if both the lithographical patterning and the ECD process are well attuned to each other.

Especially for fine-pitch plating, a fine tuning of the exposing, developing, and baking procedures is inevitable to get a high lithographic performance. All intermediate process steps have to be carefully adapted to the desired layer thickness, otherwise pinholes, microcracks, or partial delamination can be obtained. The liquid resist systems can be patterned with aspect ratios better than 5:1. Concerning this, the desired photoresist layer should be slightly thicker than the required bump height in order to keep cost-efficiency. But if a large bump pitch exists and a large solder ball volume is required as well, a so-called mushroom bump can be deposited by overplating the original photoresist pattern.

Besides the structurability, another important aspect for the suitability of the photoresist for ECD bumping technique is an easy removal after the metal deposition. The bump metallurgy must not corrode by the remover, and absolutely no mechanical damage shall occur due to a swelling effect of the resist.

Pre-Conditioning of Plating Base

Care has to be taken to ensure complete removal of any organic residues at the bottom of the resist openings. This issue must be considered during the adjustment of exposure and development parameters. If necessary, a final post-bake at moderate temperature is made to enhance the mechanical stability and the chemical resistance to the following ECD process.

Before doing the electroplating, an oxygen plasma treatment of the lithographical printed wafer is necessary to improve the wettability and remove nano-thin organic residues from the plating area. In case of Cu plating base, a subsequent micro etch step is cleaning the plating base from metal oxides and supplies an active metal interface.

Electroplating of Bumps and Routing Lines

The formation of microstructures by electrodeposition precisely replicates the photoresist pattern in lateral dimensions. In principle, two different hardware configurations of micro-electroplating equipment are commonly used for metal deposition. In a conventional rack plater the wafer is vertically diving in an immersion tank, whereas in a fountain plater the wafer is horizontally located on top of an electrolyte fountain, which rises straight up through a plastic tube. A proper tool design ensuring uniform current density and equalized bath agitation across the wafer allows a bump height deviation of less than ± 5% at 300 mm in diameter. The incorporation of smallest particles during electroplating can induce fatal bump faults which consequently can lead to a total failure of the packaged device. To overcome this issue, a continuous and efficient filtration using high-grade 0.2 – 0.5 µm cartridges is required.

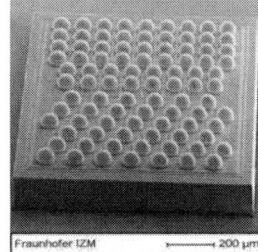 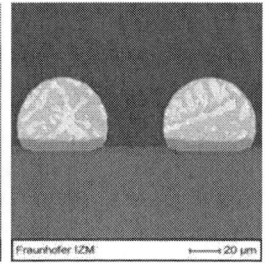

Fig. 6: Electroplated Cu/SnPb37 solder depots on CMOS diode chip

Fig. 7: Cross-section of left diode chip with bump size of 60 µm

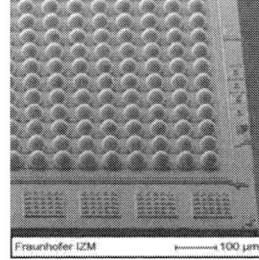 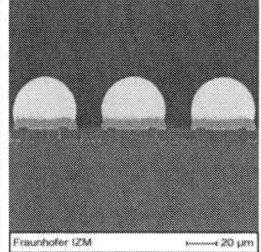

Fig. 8: ECD fabricated Cu/SnAg3.5 array on X-ray detector chip

Fig. 9: Cross-section of left detector chip with bump size of 40 µm

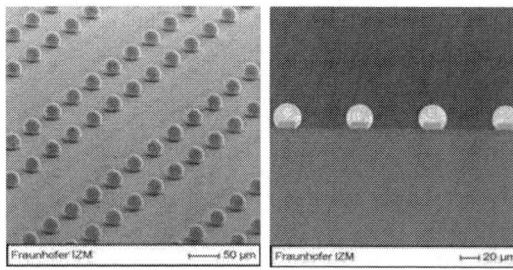

Fig. 10: Electroplated Cu/SnPb37 bumps on read-out chip for pixel detection

Fig. 11: Cross-section of left read-out chip with bump size of 26 µm

All kinds of electrodeposits presented here have been fabricated by DC electroplating. With sophisticated cell construction, proper bath chemistry, and optimized deposition parameters well defined bump structures across the full wafer area have been achieved. The pattern layer thickness is generally determined by deposition time and local current turn-over. Furthermore, the deposits have to show a well defined bump shape and grain size, a solder alloy composition as requested, low internal stress, non-porosity, and a negligible number of defects caused by dendrite formation or pittings. These attributes as a whole have to be adjusted by a suitable combination of all relevant deposition parameters such as current density, agitation strength, bath temperature, and the concentrations of electrolyte components.

For electroplating pure Au, both sulphitic and cyanidic electrolytes have been used. The advantages of the sulphitic bath type are its excellent compatibility to the applied photoresist system, its well-known non-toxicity, and the easy softening of the Au bumps during annealing. On the other hand, cyanidic Au electrolytes are easy to handle due to the more stable Au complex and offer deposits with strong adhesion even on Ni and Cu as base metal. Cyanidic solutions tend to underplating, which can become a critical fact for fine-pitch bumping. The metal content of most commercially available Au electrolytes is between 8 and 15 g/l, and the applicable current density of exemplarily 5 through 20 mA/cm² results in a deposition rate of 0,3 up to 1,2 µm/min. All kinds of Au bumping electrolytes work at higher temperatures, mostly between 50 and 70°C. In many cases of thermo-compression bonding, the ductility of as-plated Au bumps is not sufficient and has to be improved by a subsequent annealing step. During thermal aging at 200°C the initial microhardness of around 130 $HV_{0.025}$ is decreasing down to 70 through 50 $HV_{0.025}$ within a few minutes.

The used Ni electrolyte is chemically based on sulphamic acid and generally origins from electroforming industry, where very thick and low-stress layers are required. The Ni deposits look semi-bright, although a very low codeposition of organic

additives happens. To guarantee a proper solderability and wire bondability of the Ni surface, the bump has to be covered with a thin Au flash of 100 to 200 nm thickness. Therefore, different kinds of Au deposition techniques as sputtering, immersion coating, and consecutive electroplating have been proven and successfully tested at Fraunhofer IZM.

The Cu bumps as well as the Cu routing layers are deposited from a sulfuric acidic bath containing an organic suppressor and an accelerator to achieve a bright and fine-grained Cu crystallization, especially developed for semiconductor applications but primary for damascene processing. As similar as for Ni, the affinity of Cu to oxygen leads to the formation of oxide films also on the Cu deposits. Cu routing layers are usually protected by a temporary anti-tarnish coating using a weak chelating agent. Cu bumps might be covered by subsequent electroplating of a 2 µm thin Sn layer or by immersion Sn coating up to a thickness of about 1,0 - 1.2 µm. A less usual but succeeding method in WLP is the Cu surface protection with reactive organic compounds (OSP).

For solder bumping, different kinds of stannous alloy electrolytes as well as a pure Sn bath are used, all of it basing on methane sulphonic acid. Here, a non-porous crystallization and a low codeposition rate of organics are required to prevent bubble formation inside the bump during the final reflow process. Besides the mature eutectic SnPb37 and high-lead containing PbSn5 solder, electrolytes for lead-free alloy depositions such as SnAg3.5 and SnCu0.7 have been widely developed in view of long-time operation and consistent solder composition. If a sputtered Ti:W/Cu UBM is used for solder bumping, a 5 to 8 µm thick Cu or Ni pedestal must be deposited prior to the stannous alloy plating. These intermediate metals act as a solderable base and work as a sacrificial bulk for the intermetallic phase formation. Fig. 6 shows a CMOS diode chip with electroplated SnPb37 solder balls on such a Cu socket.

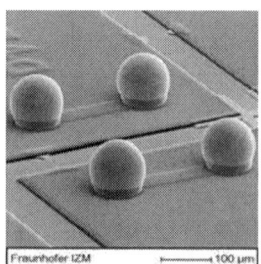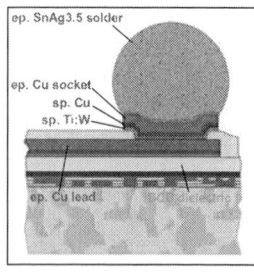

Fig. 12: Electroplated Cu/SnAg3.5 solder bumps with 110 µm in solder bumped thin-film diameter

Fig. 13: Fundamental construction of the bumped thin-film diameter on BCB/Cu built-up wiring

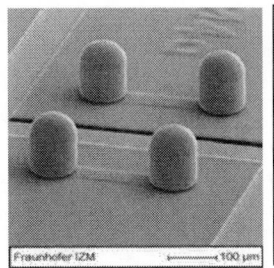 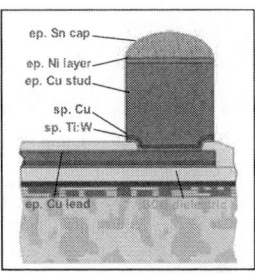

Fig. 14: Electroplated Cu/Ni/Sn pillar bumps with height of 110 µm

Fig. 15: Schematic construction of the pillar bumped thin-film on BCB/Cu package routing

Small but rapidly growing applications for very fine-pitch solder joints are radiation detector arrays. Such a high-density pixel detector consists of a silicon sensor tile for signal generation, on which the read-out IC chips are flip-chip mounted. Such a CMOS chip for pixel signal detection with ECD fabricated Cu/SnPb37 bumps is shown in Fig. 10. An X-ray detector chip with lead-free electroplated Cu/SnAg3.5 solder in highly dense area array configuration is shown in Fig. 8.

With easy changing the plating thicknesses of conventional solder bumps, comparatively high Cu or Ni pillars with a thin solder cap on top can be generated as drafted in Fig. 15 . So-called pillar bumps offer a better stand-off between chip and substrate or else between two joining dies.

Differential Etching Technique

Before the plating base is etched, the photoresist has to be stripped without any residue. Both of the applied resist systems can be easily removed by N-methyl-2-pyrrolidinone at 50 °C without needing an additional plasma ashing step. Because of low equipment and material costs, selective and accurate wet etching processes to separate the bumps and lines electrically have been developed for all the metallurgic combinations presented here.

The etchant should erode the sputtered thin-film layers uniformly and completely. Any residues could lead to transverse conductivity between two or more bumps or lines and later to a malfunction of the chip. Furthermore, the attack on the electroplated deposits should be as low as possible, and the undercut of the bumps should be very small as well. For example, the undercut caused by etching of the Ti:W layer is less than 0.5 µm on the whole wafer. For this a mixture of hydrogen peroxide with added inorganic compounds is used. In practise, the performance of an etch process depends not only on the choice of the oxidizing agent, but also on the formula of bath additives, the bath temperature, the relative motion of bath and wafer, viscosity, and surface tension. The feasibility of the process is also influenced by the thickness, composition, and crystalline structure of the sputtered layer to etch.

If the bump pattern and the UBM layer to be removed are made of the same metal, the wet etching process is not really selective but only differential. Then, a certain amount of electroplated bump will be eroded simultaneously with removal of the plating base. To set an example, during etching the Cu or Au plating base in the presence of corresponding Cu or Au bumps about 0.5 µm is lost around from the electrodeposited structure. Here, an etchant based on sulfuric acid and persulfate is used to etch the Cu layer, whereas a mixture of iodine and potassium iodide is used for Au etching.

As the electroplating baths do, the immersion etching solutions also change its chemical composition continually during operation. An accumulation of solved metal occurs along with a decomposition of the oxidizing agent. The etching rate is not only decreasing, but also the selectivity and the degree of undercut can be affected. Regenerative measures like a precipitation of metal and a replenishment of the reagents can prolong the bath life, else it has to be completely exchanged.

If an organic passivation layer covers the wafer surface, then the wet etching step is followed by an additional RIE descum. A certain amount of the upper polymer material, which is contaminated by implanted Ti and W, must be ablated in order to achieve the required insulation resistance between the electroplated microstructures.

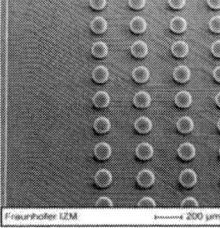 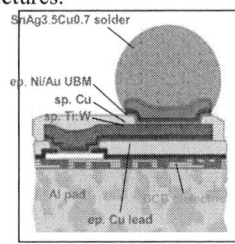

Fig. 16: ECD fabricated BCB/Cu redistribution on microprocessor, bumped with electroplated Ni/Au pads, and finally screenprinted

Fig. 17: Schematic cross-section of left SEM picture

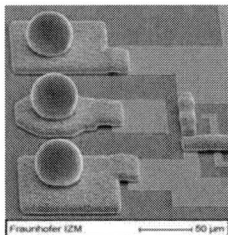 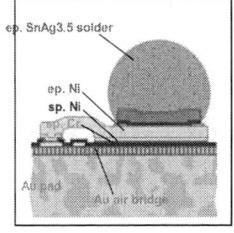

Fig. 18: ECD fabricated Au air bridges on GaAs wafer, built up with electroplated solder bumps with 40 µm in diameter

Fig. 19: Graphically displayed cut of left SEM image

Performance

The definability of photopolymer vias, the alignment and resolution of resist pattern, the integrity of electrodeposition and the accuracy of wet etching limit the minimum lead pitch and bump size. An I/O pad has to be covered by a bump hermetically in order to protect the IC against all kinds of corrosive attack. On the other hand, also the layout of the bare chip I/Os or package to bump will have direct effects on manufacturability, reliability, and cost savings from first-level interconnection through component assembly. When designing the bump pattern, the bumps should be placed as symmetrically as possible on the wafer area.

After careful consideration of a high process continuity leading to a well defined and reproducible bump quality, Sn/Pb as well as lead-free Sn/Ag and Au/Sn bumps with a lateral size of 15 x 15 μm^2 and a pitch of 25 μm can be fabricated on very small chip I/Os. The formation of straight-wall Au, Ni, and Cu bumps is mainly restricted by the achievable adjustment and resolution of photoresist pattern. 8 μm thick Cu routing layers on inter-dielectric polymers have been demonstrated to be fabricable down to a line width of 4 μm in conjunction with a pitch of 8 μm as depicted in Fig. 20 - 23.

Reliability Considerations

The ECD fabricated pillar and solder bumps offer significant advantages in connection density, mechanical properties, electrical and thermal performance, and long-term stability. Nevertheless, the metallurgical properties of the bump interconnections are changing during operation time of the device due to the grain growth of the solder and the additional formation of brittle intermetallic compounds (IMCs) within the metallic layer sandwich. Electromigration effects can result in a successive decomposition of eutectic solder alloys, and this is one of the main considerations in latest reliability studies. The fatigue life of an assembled device also depends on the influence of thermal and mechanical stress caused by the different CTEs of the joined materials.

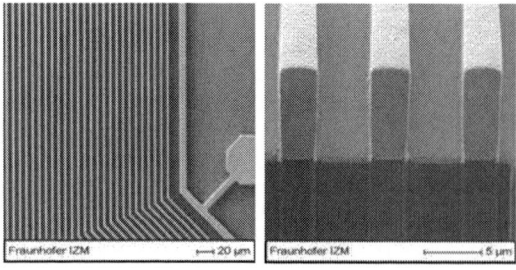

Fig. 20: Coplanar copper coil with 3,5 μm lines and 4,5 μm space on BCB

Fig. 21: FIB cut of left-hand copper coil with already etched Ti:W/Cu plating base

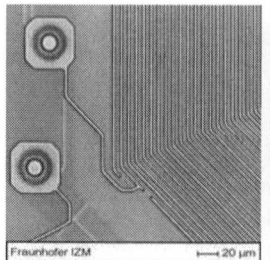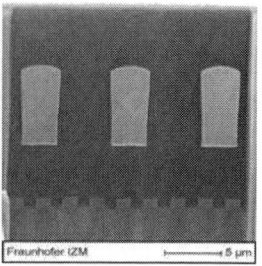

Fig. 22: IPD with 2-layer microcoil bedded into BCB dielectric layers

Fig. 23: FIB cut of left IPD with view on sputtered Al coil and electroplated Cu coil with 8 μm pitch over it

If the bump is built on to a thin-film redistribution, the reliability of the bumped package strongly depends on the adhesion between the different organic and metallic layers as well. The selection of the optimal polymer for a given application depends not only on its physical and chemical properties and processability, but also on its intrinsic interfacial characteristic. The final Ni or Cu pad metallization should create a thin intermetallic phase for a good adhesion of the attached solder. On the other hand, the IMC growth during the operation lifetime should be a minimum, so that the mechanical as well as the electrical performance of the interconnections are not significantly affected. A total diffusion of the electroplated Cu or Ni into the solder would reduce the adhesion of the bump to the redistribution. After bonding with Sn-based solders, Ni has a much slower growth rate of Ni_3Sn_4 phase compared to Cu_6Sn_5 and Cu_3Sn intermetallics on Cu bumps. Generally, lead-free solders such as SnAg3.5 or SnCu0.7 consume the pad metal faster than eutectic SnPb37 or high-lead PbSn5.

To get a statement about the reliability of the bump metallurgies under operating conditions, air-to-air thermal cycling (AATC) tests following MIL-STD-883 in the range of -55 °C and +125 °C have been carried out. During thermal cycling, the different CTEs of the joined materials are producing mechanical stresses, which can result in crack growth or even delamination of the bump. Electroplated and finally flip-chip bonded SnPb37 solder bumps on Cu as well as Ni UBM have been tested. The consumption of Cu by the eutectic solder occurred around 4 times faster than the consumption of Ni. No fracture could be detected even after 3000 cycles, indicating a very good stability of the whole layer sandwich. Pure Au as well as Au/Sn bumps which were electrodeposited onto a Ti:W(N)/Au plating base have been proven to be reliable up to 5000 thermal cycles. No significant changes in adhesion strength could be observed indicating a metallurgically intrinsic interface to the chip pad.

Yield of ECD Bumped Dies

The functional yield of electroplated devices is composed of the results on nearly each single process step. Starting from back-sputtering the wafers which came in, a following single process itself or in combination with other process steps may affect the achievable package quality [7]. In spite of cleanroom manufacturing using semiconductor equipment, one of the main problems for yield is caused by particle contamination. Particles can originate from ambient air as well as from impureness of the employed materials e.g. liquid organics, plating baths, or etching agents. Best care and attention have to be payed to bubble formation during spin-on and baking of the photoresist layer, particularly on wafers with strong surface topography. Another important influencing factor is the adequate care of wafer handling starting from incoming inspection and ending in grinding and dicing. No scratching the surface nor fracturing the substrate should happen. On the other Hand, the yield depends on the specific wafer and bump layout of each application i.e. the topography of wafer surface, and size as well as numbers and distribution of the microstructures to plate.

Exemplarily, two small volume productions of solder bumped wafers have been monitored and finally analyzed for the bumping yield. In both cases the evaluation has been done using a fully automated inspection tool which ensures an optical screening of 100 % of the bumped wafer area.

First of all, 100 mm silicon wafers with 1.4 mm² diode chips have been electrochemically bumped with SnPb37 solder. A number of 92 I/O bumps with a lateral bump size of 60 µm are array arranged on each microchip as illustrated in Fig. 24. A total of 50 wafers have been processed in a number of 8 batches. Here, the yield can be defined as the percental rate of perfect bumped dies after passing the last rinse and dry process step as graphically shown in Fig. 25. The total output of well bumped diode chips was calculated to 193581 dies, whereas 3238 chips were mainly lost by particles, defective solder bumps, faulty electrodeposition, and plating base residues. Finally, a bumping yield of 98.4 % results from this.

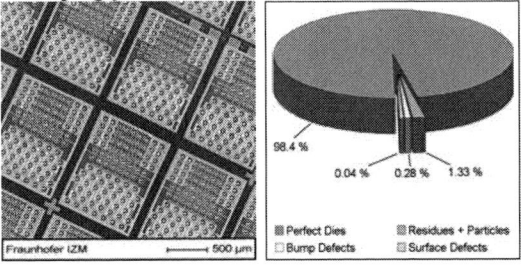

Fig. 24: 1,0 x 1,4 mm² Fig. 25: Yield of solder diode chips on 100 mm bumped diode chips CMOS wafer

Another example is the SnPb37 solder bumping of ATLAS read-out chip wafers for pixel detector modules. Here, a 200 mm CMOS wafer contains 288 microchips which show 2880 I/O bumps on its part as illustrated in Fig. 26. The bump diameter is 26 µm corresponding with a regular bump pitch of 50 µm. A number of 98 wafers have been subdivided into 13 batches and bumped using electroplating technique. A total of 20720 electrically good tested read-out chips have been bumped altogether, of which a rate of 5.4 % dies must be rejected due to defects according to Fig. 27.

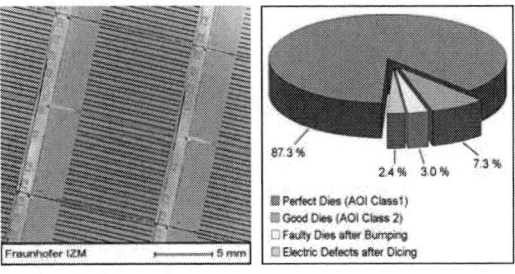

Fig. 26: 7.4 x 11,0 mm² Fig. 27: Yield of solder read-out chips on 200 bumped read-out chips mm CMOS wafer

Conclusion

WLP using electrochemical metal deposition has the largest potential for realizing highest I/O counts together with high yield output. By ECD bumping, I/O density can be increased, and high-speed devices can achieve better performance using this technique rather than wire bonding. It is particularly suited for high volume production of bumped wafers at a high quality standard and preserves high performance as well as long-term operation of the interconnections.

Pad redistribution by ECD copper routing is one of the keys for WLP over the last decade. Now, the WLP concept is engaging in integrated passives, 3-D integration, and MEMS. Basic IPD components as capacitors and inductors can be built-up on front-end wafers as a post processing technology. Along with progressive evolution of microelectronic products, WLP is moving to system-in-packaging (SiP) integrating more than one die, sensor functionality, passive components, and electrical or optical interfaces on wafer level. Here, thin-film technology can offer reliable and flexible solutions with highest performance.

As the value of wafers increases, the relatively high processing costs are less and less perceptible. The scalability to wafer sizes of 300 mm and 450 mm is given by complying lithographical printing and upgraded equipment.

Acknowledgments

The authors would like to thank the staff of HDI&WLP department at IZM for assistance in thin-film processing and sample investigation.

Furthermore, great thanks go to Mr. Oppermann, Mr. Hutter and Mr. Klein for many helpful discussions and their kind support in material characterization.

References

[1] M. Töpper, S. Fehlberg, K. Scherpinski, C. Karduck, V. Glaw, K. Heinricht, P. Coskina, O. Ehrmann, H. Reichl, "Waferlevel Chip Size Package (WL-CSP)", IEEE Trans-actions on Advanced Packaging, Vol. 23, 2000

[2] P. Garrou, "Future ICs Go Vertical", *Semiconductor International*, February 2005

[3] M. L. Hammond, "The Next Wafer Diameter Change", *Semiconductor International*, June 2004

[4] K. Zoschke, J. Wolf, M. Töpper, O. Ehrmann, Th. Fritzsch, K. Scherpinski, H. Reichl, "Fabrication of Application Specific Integrated Passive Devices Using Wafer Level Packaging Technology", 55th Electronic Components and Technology Conference (ECTC), 2005, Florida, USA

[5] L. Dietrich, M. Töpper, O. Ehrmann, H. Reichl, "Conformance of ECD Wafer Bumping to Future Demands on CSP, 3D Integration, and MEMS", 56th Electronic Components and Technology Conference (ECTC), 2006, San Diego, USA

[6] L. Dietrich, M. Töpper, V.Glaw, O. Ehrmann, H. Reichl, "Wafer Level Redistribution by ECD Copper Routing and Nickel/Gold Bumping for CSP Technology", 11th International Sym-posium for Design, Technology, and Electronics Packaging (SIITME), 2005 Sept. 22-25, Cluj-Napoca, Romania

[7] R. Roy, T. Schafer, "In-Process Bump Inspection", *Semiconductor International*, April 2006

Development of Chip to Antenna Interconnections for Contact-less Smart Card Applications

Jaakko Lenkkeri[1], Sari Kivelä[1], Eveliina Juntunen[1], Tuomo Jaakola[1], Kaj Nummila[1], Mark Allen[1],
Toni Kaskiala[2], Gerhard Hillmann[3], Alan Mathewson[4]

[1]VTT Technical Research Centre of Finland, Kaitoväylä 1, 90570 Oulu, Finland

Phone +358 20 722 2287, Fax +358 20 722 2320, E-mail: Jaakko.Lenkkeri@vtt.fi

[2]Setec/Gemalto, Finland,

[3]Datacon, Austria,

[4]CEA-LETI

Abstract

In contact-less smart cards the interconnection is usually made from packaged chip module to the antenna using soldering, welding or adhesive bonding. Because the chip itself is usually wire bonded to the module substrate, there are two levels of interconnection as well as two different substrates: antenna substrate and module substrate. Because in smart card markets there is a trend towards thinner, flexible, durable and cost efficient structures, there is a need to develop interconnection methods which allow reliable interconnection of thinned silicon chips directly onto thin flexible substrates. Various technologies of antenna manufacturing as well as methods for integration of thinned chips are discussed. Experimental results for interconnections made using anisotropically conductive adhesives (ACA) and non conductive adhesives (NCA) are reported. Some critical issues of interconnections including substrate materials, adhesive dispensing methods, curing temperatures, bonding force and surface roughness of antenna pad areas are discussed.

Key words: RFID, Smart card, antenna, flip chip, ACA, NCA

Introduction

For contact-less smart cards the card manufacturing starts from the inside out, as the card body is built up around a thin plastic sheet that contains an antenna embedded in its perimeter, a thin conductive module and a RFID chip containing microprocessor. The RFID chip is bonded to the module, which is in-turn connected to the antenna. This central layer is sandwiched between multiple plastic layers on each side, which are in turn assembled with the printed card covers. The card then undergoes the usual remaining steps of the manufacturing process, including lamination and card punching.

Most of the modules on market today have thicknesses from 320 to 430 µm. The modules are metal lead-frame based, single side moulded and contain wire bonded chips with a thickness of about 150 µm. Due to its characteristic outline and design of the lead-frame the modules are very robust during the assembly onto inlets and in the field when they are integrated into cards or tickets. The design and surface of the contact lands (leads) is compatible with almost all module to antenna interconnection techniques. The modules are delivered in a reel-to-reel format to support high throughput processes.

Highly advanced smart card controller ICs have a very complex "security" architecture and a larger memory compared to rather simple identification ICs typically used for transport or ticketing applications. Thus the design of the modules has to be adjusted to the size of the controller ICs including microprocessor.

This study reviews various technologies of antenna manufacturing as well as methods for integration of thinned chips into contact-less cards. Experimental results for interconnections made using ACA and NCA adhesives are reported.

Requirements for the Antenna

For RFID applications at 13,56 MHz the antenna should be designed so that the inductance matches the capacitance of the chip at the design frequency and the resistance of the antenna coil should be low enough in order to obtain quality factor high enough for proper operation of the system. Therefore the conductivity of the coil metallization should be good. Etched copper antennas or screen printed silver antenna metallizations are mostly used in these applications. For flip chip joining of the RFID chips the surface roughness is also a very important issue. Too rough surface may deteriorate contact between the bumps and the pad metallization. Good electrical conductivity of the antenna metallization is also advantageous in order to minimize the total thickness of the chip/antenna inlay.

In addition to etching and screen printing, ink-jet printing can be used for fabricating the antenna coils [1]. Commercially available metallic nanoparticle inks with particle sizes typically between 2 and 50 nm have such low sintering temperatures that the ink-jet printed structures can be sintered at temperatures compatible with low-cost printing substrates to form continuous, electrically conducting patterns. In order to avoid clogging the printhead during printing, piezo-driven printheads are usually utilized. Bulk conductivities of up to one fourth of the bulk conductivity of the corresponding bulk metal have been demonstrated with nanoparticle ink. Though this suggests that ink-jet printed coil antennas should easily satisfy the required series resistance values, obtaining a thick layer by printing nanoink is challenging and the typical layer thicknesses of around 500 nm for one printed layer correspond to a coil series resistance around 50 Ω.

Substrates for Contact-less Inlays

There are several issues to be taken into account when selecting substrates for contact-less inlays. First of all the substrate material should be compatible with manufacturing process, usage environment and security requirements of the product itself. The cost is also an important issue for products in mass production. Reliability of the chip to antenna interconnections as well as the yield of the interconnection technology is also a very important issue. While PET is a very common substrate material for RFID applications in general, other substrate materials like PVC, PC, PEN and paper have also been considered because of their benefits in compatibility with the processes used in the actual applications. One challenge for many polymer substrate materials is low softening temperature, which puts limits to the temperature range of the interconnection process. Conventional soldering processes are therefore not suited for these substrates. Also for interconnections made using ACA or NCA adhesive the curing temperature is often higher than 150 °C. This may be too high for some materials, because the substrate should not get too soft during the interconnection process. Otherwise there is the possibility that the substrate deforms leading to the relaxation of stresses in the joints which may deteriorate the joint quality. There is also the possibility that polymer material of the substrate has an influence on the curing process of the adhesive if the materials are not chemically compatible with each other.

For manufacturing antennas for smart cards there are three basic approaches. Wired antenna has some good benefits e.g. low resistance but it is not well suited for bare chip interconnection. Etched antennas are advantageous because the etching process allows low resistivity in the antenna coil. The subtractive nature of the process however has some drawbacks due to its environmental impacts. Various printing techniques like screen printing technology are efficient for minimal materials wastage. However the resistivity of the printed metallizations is usually much higher than the bulk resistivity of the corresponding elements and the surface roughness is also an issue for demanding applications. The challenge with ink-jet printing the coil onto a flexible substrate is that the required sintering temperature is often close to the glass transition temperature of the plastic and the obtained conductivity is directly related to the applied sintering temperature. An example of an extremely well-compatible printing substrate is polyimide, which can endure heating even above 400 °C without suffering shrinkage or deformation. With polycarbonate (PC), the corresponding temperature is around 150 °C and a coil printed onto PC should generally be sintered at max. 130 °C to preserve the original antenna dimensions. This curing temperature is on the limit of being adequate for sintering the nanoink.

In this work both etched copper metallizations, screen printed silver antenna metallizations as well as also ink-jet printed metallizations have been utilized in test structures. The screen printed metallizations were manufactured using commercial silver pastes and for ink-jet metallizations commercial nanoparticle ink was used.

Chip Interconnection Technologies

In order to make thin smart cards there are in principle two different approaches to make the interconnection from chip to antenna. One possibility is to use flip chip bonding technology for making the joints. There are several variants of flip chip bonding technology depending on the materials used in bonding and on the way of fulfilling the bonding process (see Figure 1). In flip chip bonding technology chip is joined so that bumps made onto contact area of the chip become in contact with the pads made onto the substrate metallization. In the case of contact-less inlays made onto low cost organic substrates the possible variants are limited by the temperature tolerance of the substrate material. This eliminates conventional flip chip soldering processes from the list of possible interconnection technologies. From the point of view of manufacturing process the adhesive bonding using either ACA or NCA materials are very potential methods for chip interconnection, because the technology is less critical for the accuracy in adhesive dispensing or chip assembly [2]. These technologies however need simultaneous application of pressure and heat when curing the adhesive material.

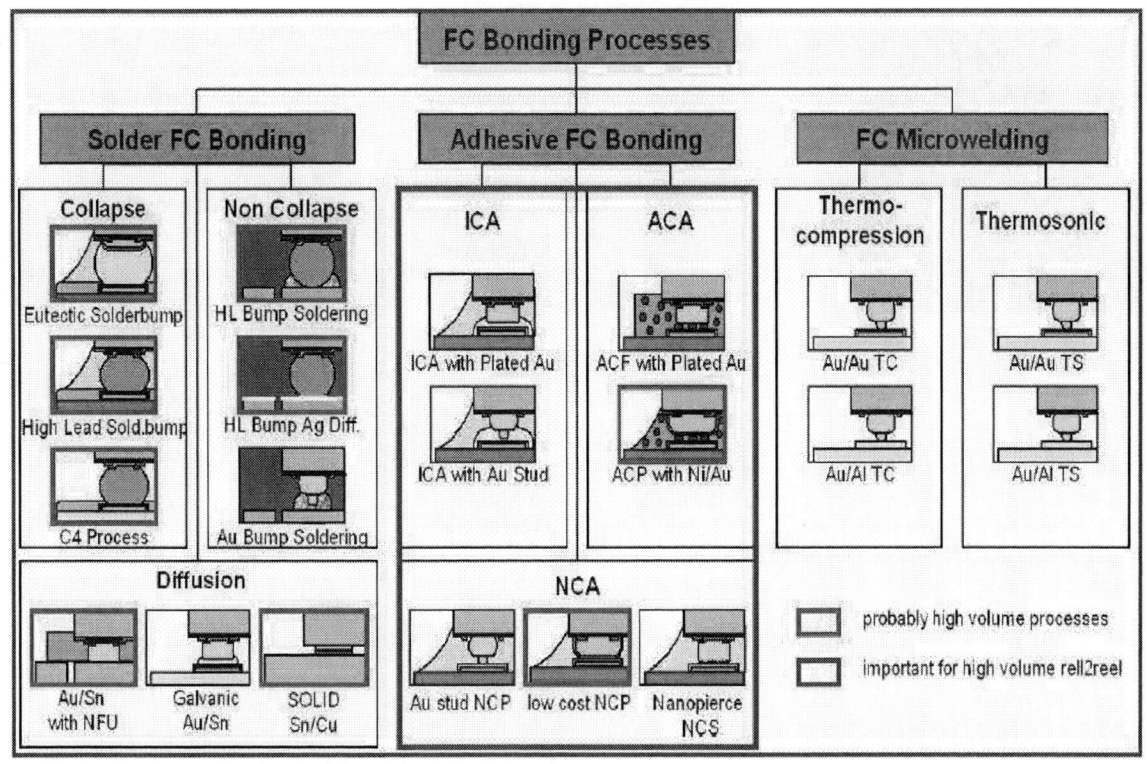

Figure 1. Classification of Flip Chip Bonding Processes.

Another possibility is to embed the chip into substrate and make the connection afterwards (see Figure 2). This helps to decrease the thickness of the assembly because in this case the thickness can be smaller than the sum of the thicknesses of the chip and substrate. This approach however is logistically very much different compared to flip chip assembly because the antenna manufacturing must be made after chip assembly whereas the order is reversed in the flip chip assembly. Chip embedding can be made by a lamination process where a chip is first die bonded onto a substrate and then a polymer layer is laminated on top of the first substrate [3]. Heat and pressure could also be used to embed the chip into the polymer substrate. The temperature must be high enough and applied force high enough to allow the material flow needed in the embedding process. There are in principle two alternative ways to put the chip into the substrate face down or back side down. In the back side down process the electrical connection to the pads of the chip must be made later in a separate plating or printing process.

This report concentrates on chip interconnection onto flexible polymer substrates using various anisotropically conducting adhesives (ACA) as well as non conducting adhesives (NCA).

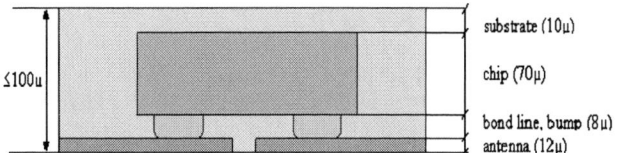

Figure 2. Principle of Chip Embedding into Substrate with a Goal of 100 μm Total Thickness.

Test Structures

Functional Test Chip

The chip used for interconnection studies was Philips Smart MX (P5CT072) secure dual interface PKI smart card controller. The size of the chip was 2.909 x 3.206 mm and the wafer was thinned to a thickness of 75 μm and it was delivered as diced and fixed to UV tape. The chip contains Au bumps (99% pure Au) with two different levels of height. Only two of the bumps (18 μm high) must be connected to the antenna. There are two other 18 μm high bumps located on the other sides of the chip giving mechanical support for the interconnection. All the other bumps (3 μm high) are only used for wafer level testing and personalization purposes and should not become into contact with the antenna metallization.

469

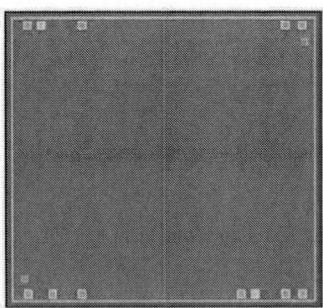

Figure 3. Layout of Philips Smart MX Functional Test Chip.

Dummy Test Chip

In order to evaluate the interconnection technology a decision was made to construct a dummy chip which has the same pin connections and die size as the Smart MX (P5CT072) controller chip. This chip was provided in two different configurations for attachment to the antenna structure. The first was with microinsert technology [4] which has been developed for making face to face interconnections in the course of 3D chip stacking technologies and the second was a more conventional gold bumped approach for mounting the components onto the antenna contact pads.

In the dummy test chip two pads corresponding to the locations of the antenna pads in the functional chip are connected together in the chip and the other two pads, which are not electrically active, are located at the alternate corners of the chip in order to provide stability during mounting. The chip can be used for direct measurement of the interconnection resistance by measuring the resistance of the chain including two interconnections.

Antenna Design and Layout

Various antenna structures have been used in the flip chip bonding tests. The antenna structure designed for the functional test chip is shown in figure 4. It includes contact pads for one flip chip component as well as for a chip module which can be assembled as a SMT component to the other pads in the layout. The resonance frequency of the system was designed to be around 15 MHz so that it still could give response with the 13.56 MHz RFID chip.

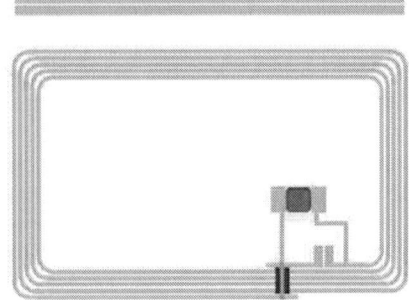

Figure 4. Antenna Layout Used for Chip Bonding Trials as well as also for Testing with Chip Modules.

Testing of the Interconnections

The aim was to study the quality of the interconnections by testing the operation of the test structure including functional chip connected to the antenna. There was a testing facility (see Figure 5) available which is able to make contact-less measurement of the operation of the system and can be used to measure the reading distance of the system. For interconnections made using dummy chips the joint resistance of the Daisy chain with two interconnections was measured using conventional resistance meter.

Figure 5. A Facility with PC and Software Used for Contact-less Functional Testing of the System.

Studies of Interconnections Using ACA and NCA Adhesives

The interconnection using ACA or NCA adhesive has the benefit that the accuracy of the adhesive dosing is less critical compared with joining using isotropically conductive adhesive. ACA and NCA however need pressure to keep the bumps of chip in contact with the pad metallization during the high temperature curing process. After curing some contact pressure must remain between the bump and pad metallization. Otherwise there are potential reliability problems in the contacts. Four different heat curable anisotropically conducting adhesives and two nonconductive adhesives were used in the bonding experiments. The bonding process includes adhesive dispensing onto the whole pad area where the chip is going to be assembled, pick up the chip from a wafer pack by the tool of the flip chip bonder, placement of the chip onto the adhesive layer on the pad area, connecting a force pressing the chip against the substrate on the working table of the flip chip bonder. While maintaining the compressive force the temperature profile is applied simultaneously to the working plate and the chip holder in order to fulfil the curing process of the adhesive.

Flip chip bonding tests were carried out with manual RD Automation M-9A bonder. In the first experiments the bonding tool was too small compared with the size of the die, which caused fracture of the die. In later experiments the size of the tool was matched to the size of the die. This avoided die fracture and minimized the camber of the polymer substrate during bonding process.

All of the adhesive materials were heat curable adhesives. The curing procedures recommended by the adhesive supplier were obeyed if possible. For some cases the softening of the polymer substrate forced the use of lower curing temperatures. In these cases the curing time used was a little bit longer. Table 1 shows typical curing conditions used for the adhesives. The pressure applied for bonding depends on the number of bumps in the die. Typical values of pressure were from 1 to 3 N.

Table 1. Typical Curing Conditions Used for the ACA and NCA Adhesive Materials.

Adhesive		Conductive particles	Curing temp.	Curing time
code	type			
A	ACA	Ag	150 °C	19 s
B	NCA	NA	150 °C	19 s
C	ACA	fusible filler	180 °C *	10 s
D	ACA	Au	180 °C *	10 s
E	ACA	Ag resin coated	150 °C	40 s
F	NCA	NA	170 °C	10 s
* 160 °C on substrate side				

One challenge in making interconnections using ACA and NCA materials is the surface roughness of the pad metallization, which is more critical for flip chip joining compared with ICA interconnection normally applied for connecting SMT components. Figure 7 shows typical surface roughness data obtained for screen printed metallization using Ag based conductor paste and for etched copper antenna. The etched antenna seems to be better suited for flip chip assembly from the point of view of surface roughness.

Results and Discussion

The interconnection tests show that flip chip joining using ACA and NCA paste material results to electrical connection from bump in the chip to antenna metallization. For etched antenna metallization on PET or PEN substrates the process conditions leading to conductive joints can be easily obtained. Thinned chips with large surface area typically show camber because of the difference in thermal expansion coefficients and temperature change after the high temperature curing

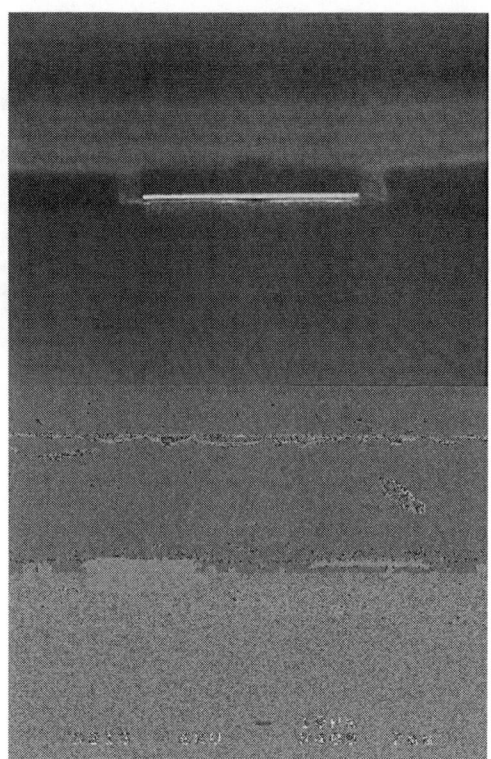

Figure 6. A Chip Bonded onto Printed Antenna Using Anisotropically Conducting Adhesive (above) and a Cross Section of a Bump Touching the Antenna Metallization (below).

step. This may be unfavourable for the stress conditions in the joints, because the stresses developed during the curing step of bonding process are allowed to relax by bending of the chip and substrate. This curvature of the assembly can be calculated using FEM simulations as shown in Figure 8. The surface roughness of screen printed silver metallization seems to be a risk for obtaining good quality flip chip interconnections. The work for making comparison between the different adhesives is still going on and final conclusions can not be made at this stage of the project. In the course of the project we found very important to have dummy chips to measure the interconnections directly because for test structures with functional chips and antennas it is sometimes very difficult to judge the effects of various contributions of the systems to the reading distance of the system.

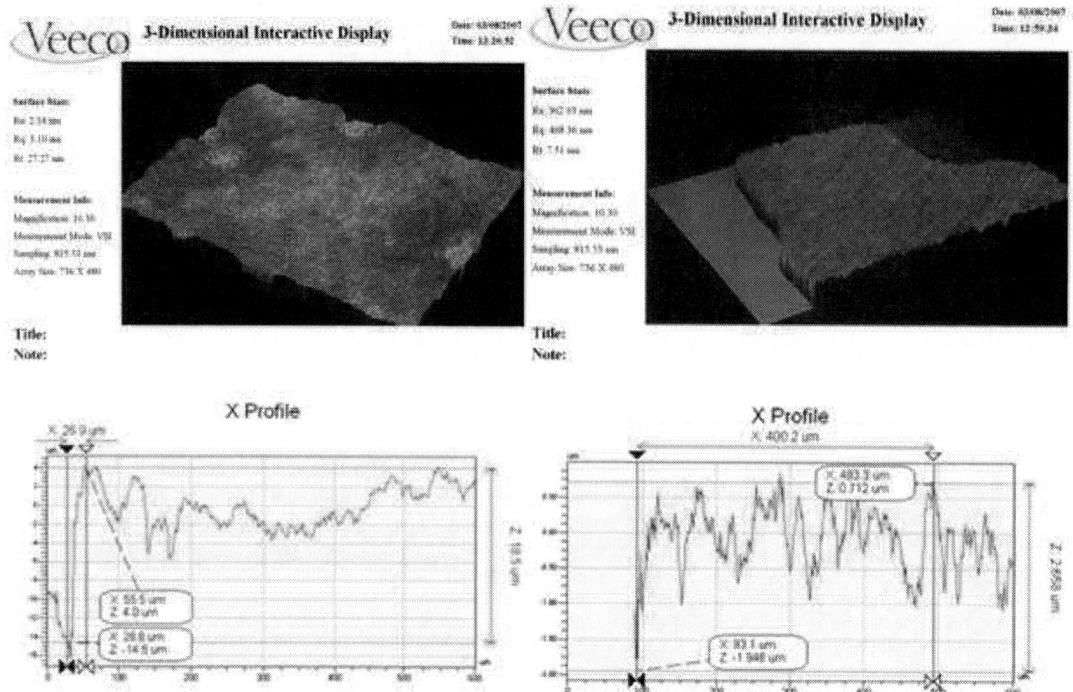

Figure 7. Surface roughness measurements of screen printed silver metallization (left) and etched copper metallization (right). The scale of roughness dimension is approximately ten times larger for the the screen printed metallization (left).

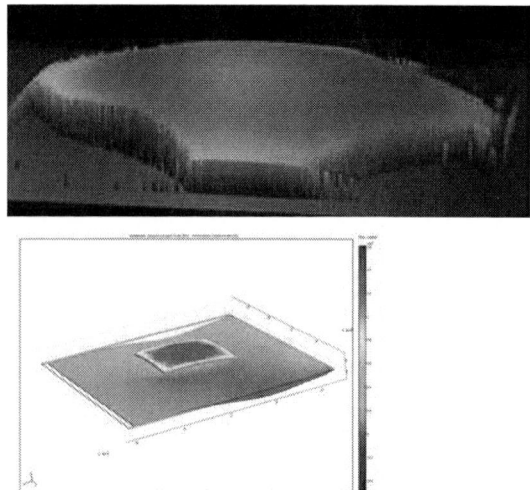

Figure 8. Typical Curvature of Chip Surface After the Bonding Process Measured by Veeco profilometer (above) and Curvature Simulated with Comsol Multiphysics software (below). The measured camber along the chip was about 25 μm.

References

[1] M. L. Allen, K. Jaakkola, K. Nummila, H. Seppä, Applicability of metallic nanoparticle inks in RFID applications, manuscript to be published.

[2] F. Ferrando, P. Stampka, Flip-chip with non-conductive adhesive adapted to high volume chip card production, 14th European Microelectronics and Packaging Conference, Germany, 23-25. June 2003.

[3] A. Ostman, J. De Baets, A. Kriechbaum, H. Kostner, A. Neumann, Technology for Embedding Active Dies. EMPC2005, June 12-15, Brugge, Belgium.

[4] N.Sillon J-C Souriau, J.Brun. Proc. "Innovative flip chip Solution For System On Wafer Concept" – First International Workshop on 3S (SOP, SIP, SOC) Electronic Technologies, Atlanta, September 23 2005.

Reliability of ACA Bonded Flip Chip Joints on LCP and FR-4 Substrates

Laura Frisk and Anne Cumini

Institute of Electronics, Tampere University of Technology, P.O.Box 692, 33101 Tampere, Finland

Phone +358 3 3115 3344, Fax +358 3 3115 3394, Email address: laura.frisk@tut.fi

Abstract

The use of anisotropic conductive adhesives (ACA) in flip chip interconnection technology has gained a great deal of popularity because of their numerous advantages. The ACA process can be used in high-density applications and with various substrates as the bonding temperature is lower than that in solder process. In this study the effect of substrate material on the reliability of ACA joints was studied. Two substrates were used in the study. One of the substrates was made of liquid crystal polymer (LCP), which is an interesting new material having excellent properties. The other substrate was made of FR-4. This is a well-known material and can meet most of the standards required for printed circuit boards. The reliability of the test samples was studied by a temperature cycling test. The test was carried out between temperatures of -40 °C and 125 °C for 20,000 cycles. To determine the exact time of a failure the resistance of each test sample was measured using continuous real-time measurement. A marked difference between the substrate materials was seen in the test. The test samples made with LCP substrates had considerably better reliability when compared to those made with FR-4 substrates.

Key words: Flip chip interconnection, Anisotropic conductive adhesive, Substrates, Reliability management

Introduction

The use of anisotropic conductive adhesives (ACA) in flip chip interconnection technology has several advantages compared to solders with underfills. These include lower process temperature [1,2] and fewer process steps, as no underfilling is required [2,3]. Furthermore, the ACA joining is a high density interconnection [3] and environmentally friendly [4]. On the other hand, the ACA joint has higher contact resistance and lower current capability than one made with solder [1,2].

The ACAs used in flip chip technology are typically epoxy-based materials to which conductive fillers, such as nickel or gold-plated polymer particles, have been added [5]. The amount of particles is below the percolation threshold [6] and the adhesive does not conduct before the interconnection is formed. During the bonding process the adhesive is pressed between mating contacts on the chip and the substrate. The ACA interconnection is established by applying pressure and heat simultaneously to the interconnection. Part of the conductive particles is trapped between the contacts and deforms forming an electrical connection. As a result, electrical conduction is restricted to the z-direction and the electrical insulation in x-y directions is maintained [6].

The reliability of the ACA joint is affected by several parameters. During the bonding process it is important that optimum bonding parameters are used to ensure the formation of good quality joints. The most important bonding parameters are time, temperature and pressure [7]. Bonding time and temperature determine the degree of cure of the polymer matrix. The bonding pressure used determines the deformation of the conductive particles. Insufficient pressure leaves the particles only slightly deformed and the resulting joints are unstable [3]. Additionally, the contact resistance of this kind of joint is high [8]. On the other hand, excessively high pressure may crush the deformable polymer particles [9]. Moreover, high bonding pressure can cause marked deformation of substrate if organic substrate materials are used.

In addition to bonding parameters, the materials used in the ACA attachment affect the reliability of the joints. Bonding of the ACA joints is done at elevated temperatures. As the coefficients of thermal expansion (CTE) of the materials differ, thermally induced stresses and strains are formed during cooling and may cause reliability problems [10]. Thus the materials chosen should have their CTEs as closely matched as possible. With organic substrates the CTE of the substrate is typically markedly greater than that of the silicon chip. As the substrate contracts much more during cooling than the silicon chip, the ACA flip chip package formed warps at low temperatures. At high temperatures this warpage evens out. If the environmental temperature fluctuates, the flip chip package warps repeatedly [11].

Several substrate materials can be used in ACA flip chip technology. In this study two substrate materials were used. One of the materials was liquid crystal polymer (LCP), which is an interesting new material having excellent properties for flexible printed circuit boards [12]. The other substrate was made of FR-4. This is a well-known material and can meet most of the standards required for printed circuit boards [13]. The effect of substrate material on the reliability of the joints was investigated using a temperature cycling test. The effect of bonding pressure was also studied by using four different bonding pressures with both substrates.

Experimental

LCP and FR-4 substrates were used in the study. The size of the LCP substrate was 110 mm × 55 mm and it contained 10 sites for 5 mm × 5 mm test chips. The thickness of the LCP film was 50 μm and the thickness of the copper tracks on it was 12 μm. The copper tracks had 1 μm thick nickel-gold plating. The LCP substrate was double-sided and had copper tracks on both sides of the substrate. It had a two-layer structure and had no adhesive layer between the copper and the LCP film. The LCP substrate had a solder resist of 20 μm on both sides of the board with openings of 6 mm × 6 mm at the bonding sites. The size of the FR-4 board was 110 mm × 55 mm and it contained 8 sites for 5 mm × 5 mm test chips. The thickness of the substrate was 100 μm and it contained only one layer of glass fibre cloth. The thickness of the nickel-gold plated copper tracks was 19-20 μm. The FR-4 substrate was single-sided. Before bonding both test boards were dried for one hour at 110 °C to remove moisture and cleaned using an organic solvent.

The test chips used were silicon chips with a daisy chain interconnection pattern. The size of the chips was 5mm × 5 mm and they had 71 square gold bumps. The size of the bumps was 100 μm × 100 μm and their thickness was 23 μm. The pitch of the peripherally situated bumps was 250 μm. The thickness of the chips was 500 μm.

Table 1: Properties of the ACF and the substrate materials.

	ACF	LCP	FR-4
T_g / °C	145	205	130-140
CTE / ppm/°C	55	12-16	12-16
Elastic modulus (at 30 C°)/ Gpa	1.4	6.7-7.0	23.4-24.8

Test samples were prepared by attaching the test chips on the test substrates using a commercially available anisotropic conductive adhesive film (ACF). The thickness of the adhesive film was 40 μm and it contained gold coated polymer particles of

3 μm in diameter. Some properties of the ACF and the substrate materials are presented in Table 1.

The chips were assembled using a Toray FC-1000 flip chip bonder. First the ACF was cut to the correct size to cover the bonding area. After this the ACF was aligned to the substrate and prebonded using light pressure and low temperature. The prebonding was performed with the flip chip bonder. After prebonding the carrier film on the ACF was removed. The final bonding was made with the flip chip bonder. First a test chip was picked by the flip chip bonder. Then the bumps on the chip and the pads on the substrate were aligned. Finally the chip was pressed onto the substrate and heat was applied to the chip and the substrate. After bonding the package was cooled down while still under pressure until its temperature was below the glass transition temperature (T_g) of the ACF.

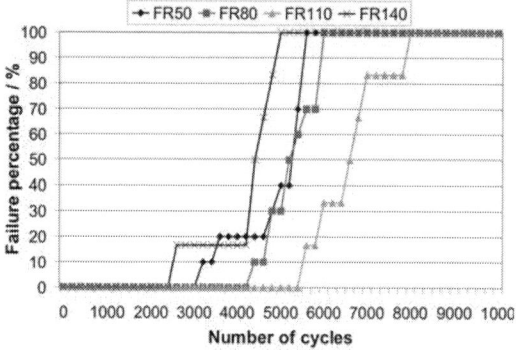

Figure 1: Results for test samples with varying pressures on FR-4 substrates during a temperature cycling test.

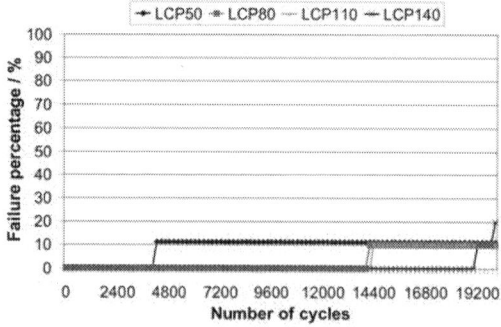

Figure 2: Results for test samples with varying pressures on LCP substrates during a temperature cycling test.

Bonding time and temperature were chosen according to the ACF manufacturer's recommendation. The bonding temperature used was 220 °C and the time was 40 s including 15 s of cooling under pressure. Four different bonding pressures were chosen from the pressure range recommended by the ACF manufacturer. The test series made are presented in Table 2.

In order to study the reliability of the test samples a temperature cycling test was used. The test was performed according to standard JESD22-A104-A between temperatures of -40 °C and 125 °C [14]. The duration of exposure at the temperature limits was 14 minutes and the transition time was 1 minute, making the total cycle time 30 minutes. The duration of the test was 20,000 cycles.

The daisy chain resistances of the test samples were measured during the temperature cycling test using continuous real-time measurements with a National Instruments data logger system. This enabled the determination of the exact time of any failures during the test. One measurement channel was used for each chip tested. During the real-time resistance measurement a constant, stable current was fed separately through shunt resistors to all the test samples to be measured. The voltage over the daisy chain structure was measured individually for every test sample. Any rise in the measured voltage indicated an increase in the resistance of the test sample measured. As open joints formed on the test samples, the measured voltage rose to the supply voltage (approx. 5V). The voltage was measured every 20 seconds.

Results and Discussion

There was a marked difference between the reliability of the substrates during testing. All test samples with FR-4 substrate failed within 8000 cycles of testing, as can be seen in Figure 1. On the other hand, only a few test samples with LCP substrate failed during the test. The results for the LCP substrate are presented in Figure 2. A test sample was considered to have failed after an open joint was observed or after a ten-fold increase in the daisy chain resistance at any time during testing.

As only a few failures occurred on the test samples with the LCP substrate, no difference was seen between the different bonding pressures. On the other hand, a clear difference was seen on the FR-4 substrate. Test series FR110 with the second highest bonding pressure had clearly the best reliability during the test. The test series with the highest pressure had the poorest reliability, as the failures

started to occur after 2500 cycles of testing for this test series. The failures in the test series with the two lowest pressures also started to occur clearly earlier than in the FR110 test series.

The daisy chain resistance of the test samples was studied in real time to see when and how the failures occurred during testing. During a temperature cycling test a failure may occur first at high or at low temperatures or simultaneously at both. It is also possible that the failures are first seen during the changes in the test temperature. This was typical in this study. All test samples excluding the one test sample from test series LCP50 failed first during the change in the test temperature. This kind of fluctuation continued typically for several hundred cycles before the test sample also showed failure at low temperature extreme. The fluctuation at low temperatures was typically very distinct. As the test progressed, the period of time during which the test samples were functioning at high temperature decreased and eventually the test samples showed open joints throughout the test. The failed test sample from test series LCP50 failed rapidly with only few indistinct fluctuations.

In order to study the relationship between the failure mechanisms and fluctuation in the contact resistance, cross sections of the test samples were prepared and studied using optical and scanning electron microscopes (SEM). Delamination has been found to cause failures first at low temperatures [8,11]. However, no delamination between the pad and the bump was found in majority of the failed test samples with the FR-4 substrate. On the other hand, severe cracking of the substrate was seen in every test sample with FR-4 substrate. An example of this cracking is presented in Figure 3. These cracks typically started from the corner of the pads and continued into the substrate until they reached the glass fibres. In some cases cracks continued through the substrate. This kind of cracking can cause delamination by providing sites for crack initiation [15]. However, only one of the test samples studied showed delamination, which was connected to the cracking of the substrate. An example of this kind of delamination is presented in Figure 4.

Table 2: Test series

	Series LCP50	Series LCP80	Series LCP110	Series LCP140	Series FR50	Series FR80	Series FR110	Series FR140
Substrate material	LCP	LCP	LCP	LCP	FR-4	FR-4	FR-4	FR-4
Bonding pressure / MPa	50	80	110	140	50	80	110	140
Number of samples tested	10	10	10	10	10	10	6	6

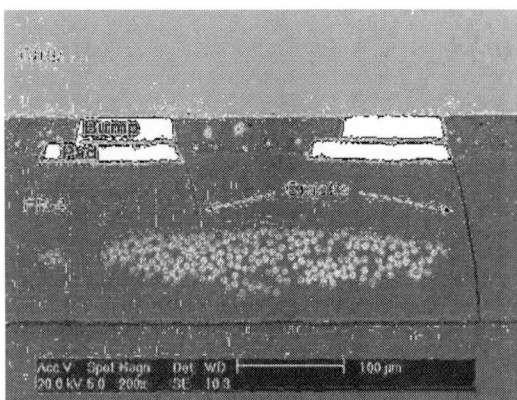

Figure 3: Micrograph showing cracking of the FR-4 substrate.

Figure 4: Micrograph showing delamination connected to a crack in the FR-4 substrate.

No cracking of substrate was seen on the LCP substrate. However, a marked deformation of the LCP film was seen. This deformation was directly proportional to the bonding pressure. With the lowest bonding pressure the deformation was only slight as can be seen in Figure 5. However, with the highest bonding pressure very marked deformation was seen (Figure 6). This kind of deformation was also seen on the FR-4 substrate. However, it was clearly less than that on the LCP substrate. Furthermore, on the FR-4 substrate the deformation was clearly related to the glass fibres. In the areas were the glass fibres were far from the surface the deformation was much more severe. When the glass fibres were near the surface, the substrate had deformed less. However, in these places, especially when the highest bonding pressure was used, clear deformation of the bumps and pads was seen. An example of this is presented in Figure 7.

Some air bubbles were seen in the space between the joints with both substrates. The number of air bubbles decreased with increasing bonding pressure. This is assumed to be caused by the increased deformation of the substrate with higher

bonding pressures. As the substrate deforms more, less adhesive is needed to fill the space between the joints and the number of air bubbles is reduced.

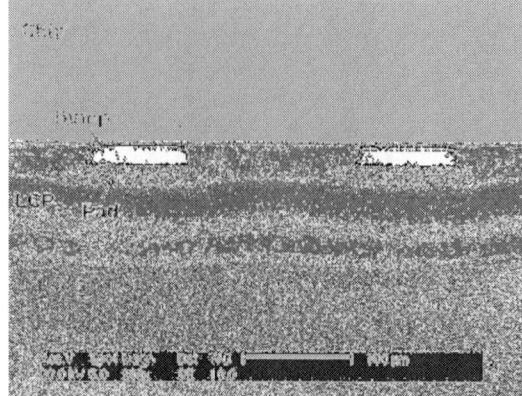

Figure 5: Micrograph presenting the deformation of LCP film for a 50 MPa bonding pressure.

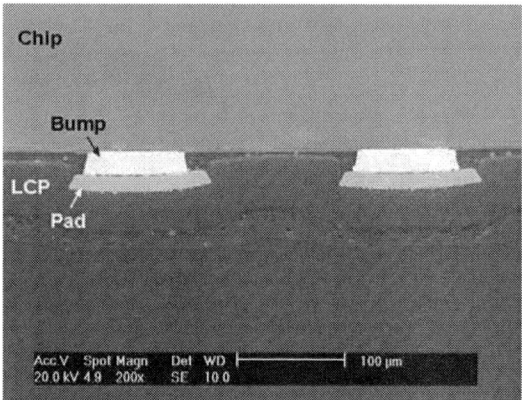

Figure 6: Micrograph presenting the deformation of LCP film for a 140 MPa bonding pressure.

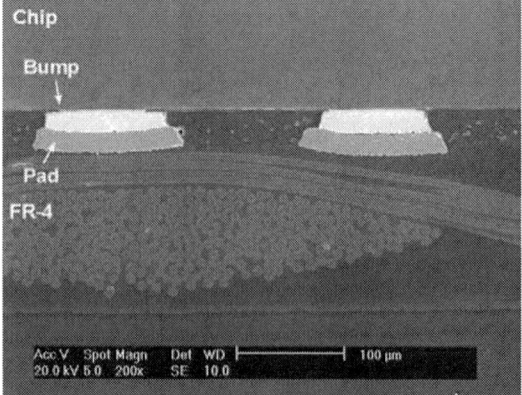

Figure 7: Micrograph presenting the deformation of the bump and the pad On FR-4 substrate for a 140 MPa bonding pressure.

Almost all the test samples failed first during the change in the test temperature before also failing at low temperature. This is assumed to be caused by different thermal conductivities of the materials used in the flip chip package. During the changes in the test temperature the silicon chip warms up and cools down faster than the organic substrate. This causes temperature gradients to form within the package and may cause warpage of the package [8]. This warping may disturb the contact between the conductive particles and the contacts and increase the resistance. During the steady temperature phase of the temperature cycle the temperature gradients even out and the resistance decreases. The SEM analysis is done at room temperature, which makes it difficult to see the effects of this kind of warping.

After failing during changes in test temperature the test samples showed failure at low temperatures. However, no reason for the failure was seen for the majority of the test samples studied. The failures at low temperature are assumed to be caused by the warpage of the flip chip package, which is caused by differences between the CTEs of the chip and the substrate. As flip chip bonding is done at high temperatures, the substrate contracts more during cooling than the chip and the package warps at low temperatures. This warpage lessens the shear strain in the joint caused by CTE mismatches [10]. Consequently, easy warpage of the package can increase its reliability. However, eventually the joints may fail, as the warpage may cause deformation of the adhesive and lead to a conduction gap between the conductive particles and the contacts. The elastic modulus of the adhesive used was relatively low (Table 1). Low elastic modulus of the adhesive increases its adhesion strength and thus reduces the probability of delamination [11]. On the other hand, low elastic modulus also reduces the stiffness of the adhesive and facilitates its deformation. It is assumed that this deformation of the adhesive matrix is the reason for the majority of the failures during testing. As the adhesive matrix deforms, the conductive particles lose contact and the joint fails. Since the warpage of the package occurs at low temperatures, the failures are first seen at low temperatures. A similar failure mechanism was seen when a thin FR-4 substrate was used with a thinned silicon chip [15].

The fibre reinforced FR-4 substrate used in this study is much more rigid than the LPC substrate even though it was only 100 µm thick and fairly pliable. This is probably the reason for the cracking of the FR-4 substrate. During the continuous warpage the stiff FR-4 substrate is not able to deform and starts to crack. From Table 1 it can be seen that the elastic modulus of the FR-4 is considerably greater than that of LCP. The greater stiffness of the FR-4 reduces its deformation during the test and causes higher shear stress between the pad and the bump. The deformation of the flexible LCP substrate occurs much more easily due to its low modulus, and some of the shear stress is absorbed by the substrate [16]. The ability of the flexible LCP substrate to absorb the stresses by deforming increases the reliability of the joints and is most likely the reason for the marked difference between the substrates.

With the FR-4 substrate the bonding pressure had a clear influence on the reliability. As the deformation of the substrate is marked, when the glass fibres are far from the surface, the deformation of the particles in these joints may be too low to ensure reliable joints when low bonding pressure is used. If the particles are only slightly deformed, less deformation of the adhesive matrix is needed for them to lose contact. As the bonding pressure is increased, the particles are adequately deformed in every joint and the reliability is increased. On the other hand, with the highest pressure the pads and bumps also started to deform. This indicates that the bonding pressure is too high and stresses are formed in the joints. These stresses may have been the reason for the early failures in the test series with highest pressure.

Conclusion

The reliability of ACA flip chip joints on LCP and FR-4 substrates was studied during a temperature cycling test. A marked difference between the substrates was seen in the test. All test samples with the FR-4 substrate failed during the 8,000 cycles of testing while only a few failures were seen with the LCP substrate during 20,000 of testing.

The failures during the test occurred first during the changes in the test temperature. The reason for this is assumed to be the differences in the thermal conductivities of the materials, which causes the warpage of the package during fast changes in test temperature. Later test samples showed failure at low temperatures before also failing at high temperatures. No reason for the failures was found in the majority of failed test samples studied using a SEM. The reason for the failures at low temperatures is assumed to be warpage of the package caused by differences in the CTEs of the chip and the substrate. As the bonding process is done at elevated temperature, this warpage occurs at low temperatures. During a temperature cycling test the package warps repeatedly. This may cause deformation of the adhesive matrix and lead to the formation of a conducting gap between the contacts and conductive particles.

The markedly lower stiffness of the LCP is assumed to be the reason for its considerably better reliability compared to that of the FR-4 substrate. During testing the deformation of the LCP substrate occurs more easily, which reduces the stresses in the joints and thus decreases the probability of failures.

Acknowledgements

We would like to thank the staff at the Institute of Material Science at Tampere University of Technology.

References

[1] M. J. Yim and K. W. Paik, "Design and understanding of anisotropic conductive films (ACF's) for LCD packaging", IEEE Transactions on Components and Packaging Technologies Part A, Vol. 21, No. 2, pp. 226 – 34, June, 1998.

[2] Z. W. Zhong, "Various Adhesives for Flip Chips", Journal of Electronic Packaging, Vol. 127, No. 1 pp. 29-32, March, 2005.

[3] Z. Lai and J. Liu, "Anisotropically conductive adhesive flip-chip bonding on rigid and flexible printed circuit substrate", IEEE Transactions on Components, Packaging, and Manufacturing Technology, Part B: Advanced Packaging, Vol. 19, No. 3, pp. 644-66, August, 1996.

[4] M.A. Uddin, Y.C. Chan, H.P. Chan, and M.O. Alam, "A Continuous Contact Resistance Monitoring during the Temperature Ramp of Anisotropic Conductive Adhesive Film Joint", Journal of Electronic Materials, Vol. 33, No. 1, pp. 14-21, January, 2004.

[5] S. C. Tan, Y. C. Chan, Y. W. Chiu, and C. W. Tan, "Thermal stability performance of anisotropic conductive film at different bonding temperatures", Microelectronics Reliability, Vol. 44, No. 3, pp. 495-503, March, 2004.

[6] J. Lau, "Flip Chip Technologies", McGraw-Hill, USA, Chapter 9, pp.301-315, 1995.

[7] P. Palm, J. Määttänen, A. Picault, and Y. Maquille, "The Evaluation of Different Base Materials for High Density flip Chip on Flex Applications", Microelectronics International, Vol. 18, No. 3, pp. 27-31, 2001.

[8] A. Seppälä and E. Ristolainen, "Study of Adhesive Flip Chip Bonding Process and Failure Mechanisms of ACA Joints", Microelectronics Reliability, Vol. 44, No. 4, pp. 639-48, April, 2004.

[9] A. Seppälä, T. Allinniemi, S. Pienimaa, and E. Ristolainen, "Effects of Bonding Parameters on Quality of Adhesive Flip Chip joints", Proceedings of the 2nd IEEE Symposium on Polymeric Electronics Packaging, Gothenburg, Sweden, October 24-27, pp. 147-52, 1999.

[10] W. S. Kwon, M. J. Yim, K. W. Paik, S. J. Ham, and S. B. Lee "Thermal Cycling Reliability and Delamination of Anisotropic Conductive Adhesives Flip Chip on Organic Substrates With Emphasis on the Thermal Deformation", Journal of Electronic Packaging, Vol. 127, No. 2, pp. 86-90, June, 2005.

[11] I. Watanabe, "ACF Flip-Chip Technology on Low-Cost Substrates", Proceedings of the 2nd IEEE Symposium on Polymeric Electronics Packaging, Goteborg, Sweden, October 24-27, pp. 153-157, 1999.

[12] L. Frisk and A. Seppälä, "Reliability of Flip Chip Joints on LCP and PI Substrate", Soldering and Surface Mount Technology Vol. 18, No. 4, Pages 12-20, 2006.

[13] T. Rapala-Virtanen, "Experiences Gained Manufacturing High Density Interconnect (HDI) Printed Wiring Boards (PWBs)," Circuit World, Vol. 19, No.1, pp. 14-18, 2002.

[14] Electronic Industries Association, "JEDEC Standard. Temperature Cycling. JESD22-A104-A", USA, pp. 4, 1989.

[15] L. Frisk and K. Kokko, "The Effects of Chip and Substrate Thickness on the Reliability of ACA Bonded Flip Chip Joints", Soldering and Surface Mount Technology, Vol. 18, No. 4, Pages 28-37, 2006.

[16] G. Connell, R. Zenner, and J. Gerber, "Conductive Adhesive Flip Chip Bonding for bumped and Unbumped die", Proceedings of 47th IEEE Electronic Components and Technology Conference (ECTC), San Jose, CA, USA, May 18-2,1 pp.274-8, 1997.

Second-Level Interconnect – "Package to PCB" – Future Challenges and Solutions

Ashok N. Kabadi

Intel Corporation, Hillsboro, Oregon, USA

Phone: (503)264-6707, Fax: (503) 264-1831, email: ashok.kabadi@intel.com

Abstract

Over the past decade, advances in packaging technologies have led to the miniaturization of features on electronic devices, resulting in denser pad-pitch, higher pin-count and faster I/O speeds. These technological advances have created new challenges in the design of the second level interconnect – "Package to Printed-Circuit-Board (PCB)." As packages became smaller and denser, the attachment of these fine-pitch and large Ball-Grid-Array (BGA) packages to PCBs became difficult, especially when the thickness of many server-type motherboards increased to 93 mils due to the need for multiple layers. Furthermore, lead-free soldering temperatures have created even more challenges for solderability of such dense pad-pitch devices. The industry trend for the next few years shows that the pad-pitch for larger BGA devices is trending to 0.5mm (19.6 mils). This reduction in the pad-pitch will involve rethinking how BGA packages are attached to the motherboard. Soldering them to PCBs may no longer be the most cost-effective and reliable solution. This paper addresses future "Package to PCB" interconnection problems by providing solutions for some solderless double-compression interconnection systems that can be scaled down to pad-pitch of less than 0.5mm (19.6 mils) for Land-Grid-Array (LGA) devices with sizes up to 49.5mm sq. The paper also demonstrates a step-by-step approach to designing a high-density interconnect system, with the focus on the force-deflection Finite-Element-Analysis (FEA) to minimize deflection of the die, package-substrate and PCB. It also addresses issues of critical mechanical tolerances by giving an example of a 0.8mm (31.4 mils) solderless double-compression system implemented at Intel Corporation.

Key words: Ball-Grid-Array, Land-Grid-Array, Printed-Circuit-Board, Solder ball, Coplanarity, Heat-sink

BGA Packages and Solderability Challenges

BGA packages are typically soldered to the PCB using leaded or lead-free reflow processes. However, in many applications these devices need to be attached to the motherboard using removal attachment methods. A large majority of BGA sizes range from 21mm square to as large as 49.5mm square. The "pitch," which is defined as the center-to-center spacing between the adjacent solder balls, plays an important role in achieving reliable solder joints. Other factors that impact the reliability of solder joints are the size of the package and the coplanarity, size and material of the solder balls. "In the electronic industry, the term 'coplanarity' means the maximum distance that the physical contact points of a surface-mount device can be from its seating plane" [1]. The seating plane is defined by the highest point of three balls that the package rests on. In general, larger packages have greater solder ball coplanarity specifications and so are more apt to warpage along the corners during the reflow soldering process, resulting in electrical opens. As the pitch becomes smaller, the ball diameter correspondingly becomes smaller to accommodate smaller pad sizes on the PCB. "The eutectic solder balls (63%Sn37%Pb) of Plastic-Ball-Grid-Array

(PBGA) packages tend to collapse at about 0.15 mm (6 mils) to 0.20 mm (8 mils) during reflow soldering process" [2]. The actual collapse depends on the package weight, package size, solder ball diameter, solder ball material, pad pitch, PCB pad diameter, etc. Furthermore, the size of the motherboard, its thickness and the number of copper layers also contribute to, to some extent, the reliability of the solder joints.

A typical BGA package consists of a plastic or a ceramic substrate with a silicon die in the center. Examples of a PBGA device with and without Integrated-Heat-Spreader (IHS) are shown in Figures 1 and 2. The need for an IHS attached to the silicon die depends on the following two factors:

Size of the device -- If the PBGA package size is large (typically > 37.5 mm square), an IHS is recommended to minimize the package warpage during the reflow soldering process. Ceramic packages have less warpage during the reflow solder process and may not require an IHS for warpage related issues.

Power dissipated by the device -- If the power dissipated by the package is too high (typically > 25 watts), a copper-based IHS is recommended for spreading the heat and cooling the die to a specified temperature. Several factors like the size of the die, power density, etc., determine the size and the material of the IHS. The IHS is often attached to the die using thermally conductive adhesive.

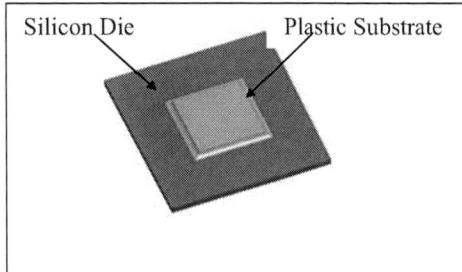

Figure 1: Package without HIS

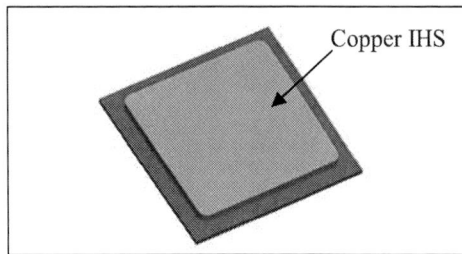

Figure 2: Package with IHS

Table 1 shows JEDEC specifications of the diameter of eutectic solder balls and package pads that are used for different solder ball pitch dimensions of PBGA packages.

Table 1: BGA Ball/Package-Pad Data [5] [6]

Pitch mm	Ball diameter mm (nominal)	Package pad diameter (min) Type (2) mm
1.27	0.75 (30 mils)	0.60 (24 mils)
1.00	0.45 (18 mils) to 0.60 (24 mils)	0.30 (12 mils to 0.50 (20 mils)
0.8	0.30 (12 mils) to 0.50 (20 mils)	0.20 (8 mils) to 0.35 (14 mils)
0.65	0.30 (12 mils) to 0.40 (16 mils)	0.20 (8 mils) to 0.30 (12 mils)
0.50	0.30 (12 mils)	0.20 (8 mils)

For smaller pitch dimensions, smaller pad diameters are used on PCBs to allow for differential pair routing of two traces between the two adjacent pads. A wide range of the packages used in the industry are PBGAs. PBGAs have larger coplanarity tolerance specifications than ceramic ones and pose more challenges to the soldering process for achieving reliable solder joints. The coplanarity tolerance specification of eutectic solder balls as defined by JEDEC for PBGA devices is shown in Table 2. These tolerances as shown are based on the three-point seating plane method.

Table 2: BGA Coplanarity Specification [5] [6]

PBGA Device Size mm square	Pitch mm	Coplanarity Tolerance mm (nominal)
<=50	1.00, 1.27, 1.50	0.20(8 mils)
< = 21	0.80	0.08 (3.2 mils) to 0.12 (4.8 mil)
<= 21	0.65	0.08 (3.2 mils) to 0.10 (4 mils)
<= 21	0.50	0.08 (3.2 mils)

Many electronic packages currently available in industry have already increased in sizes up to 37.5 mm square with pitch equal to 0.8mm (31.4 mils). As the pitch is trending to 0.5mm (19.6 mils) and the ball diameter to 12 mils, the solderability of such larger fine-pitch devices is becoming increasingly difficult due to larger solder ball coplanarity tolerance specifications and smaller pad diameters on the PCB and package. The soldering of such devices with lead-free solder alloys will be a major challenge due to higher reflow temperatures contributing to higher warpage of the package.

As the number of layers on the PCB increases, the thickness of the PCB increases. Standard motherboards used in desktop computer applications are 62 mils thick. However, for most of the server applications, the motherboard thickness is 93 mils due to a higher count of signal layers and more power and ground planes. A commonly used alloy for BGA solder balls is SAC 305 (Sn 96.5%, Ag 3% and Cu 0.5%). This alloy is liquidus at 222.81°C. Setting up the reflow profile for assembling 93 mils thick PCBs with large BGA devices and using SAC 305 alloy constantly pose a major challenge to the PCB assembly suppliers. In order to uniformly heat the board in the BGA area at the solder melting temperatures, the peak temperature often needs to be raised 30°C to 40°C higher than the melting temperature of the solder balls. The higher temperatures tend to cause more warpage problems of the plastic substrate of PBGA devices, causing electrical opens, typically at the corners. The larger sizes of BGA devices coupled with the smaller diameter of the solder balls exasperate this problem due to less forgiveness of the smaller diameter solder balls to the warpage and coplanarity issues.

Alternatives to BGA Soldering

Soldering BGA devices to the motherboard has been one of the most cost-effective methods of assembly for more than a decade. However, due to the miniaturization of geometries on electronic devices, attaching them to the PCB using a soldering process may not be the best method for a number of reasons.

- In the computer system validation environment, packages attached to the PCB often need to be replaced with new ones several times before the design of the product is finalized. If the BGA packages are soldered to the PCB, reworking such devices by desoldering the existing one and resoldering a new one often is very time-consuming and could result in damage to the expensive validation board.
- As the pitch on the BGA devices scales down to 0.5mm and the package size increases to >42.5mm, soldering such large devices could result in a very low yield due to warpage and coplanarity issues.

The alternate methods of attaching BGA devices to the PCB include the use of interconnect sockets. These sockets are either soldered to the PCB or are attached using a solderless mechanical hardware system.

Overview of BGA/LGA Interconnect Sockets

BGA/LGA interconnect sockets are primarily used for attaching electronic devices to the PCB. The sockets offer several advantages, including replacement of different packages on the same PCB without causing damage to the motherboard. The sockets used for the LGA packages typically have higher durability and reliability than those used for BGA packages due to the gold-to-gold interface between LGA gold pads and socket gold-plated contacts. The following section gives an overview of the types of LGA sockets available in the industry based on their application.

Validation: In this application, a package or several packages need to be cycled multiple times inside the PCB during the debug process of the motherboard and of the device itself. The typical insertion/removal cycle requirement for these sockets is low (typically 25). The keepout area of the socket on the PCB is critical due to critical trace routing requirements. It must be the kept as close to the space occupied by the device plus the clearance required to meet PCB Design-For-Manufacturabilty (DFM) guidelines as possible. Since these sockets are custom and used only in the validation environment, a typical quantity requirement is medium (typically less than 1,000); thus, the cost of the sockets tends to be higher than the standard production sockets described in the later section.

Test and Debug: In a test and debug application, a large number of devices are cycled in and out of the same socket, so the socket contacts must have a very high durability. The insertion/removal cycle requirement for this type of socket is typically more than 1,000. These sockets are also custom-designed and manufactured in low volumes, resulting in very high cost per socket. In most applications, the interconnect contacts for such sockets are spring-loaded pins that have a very high insertion/removal cycle life. However, the cost of such pins is also very high. As the pad-pitch is reduced to 0.5mm (19.6 mils), the pins become increasingly difficult to manufacture, resulting in even higher cost per pin.

Production: In this application, the durability requirement is much lower than the test application. Typically, these sockets are rated at 25 insertion/removal cycles. The sockets are soldered to the motherboard, and the LGA device is attached to the socket using compression-type contacts. Due to high-volume requirement, the interconnect contacts of the socket are made out of stamped-and-formed Beryllium-Copper (BeCu) material plated with Nickel (Ni) and Gold (Au). The initial tooling cost to manufacture these contacts is very high, but the cost per socket is very low due to the high volume requirement.

Table 3 shows relative comparison of the sockets used in different applications.

Table 3: LGA Interconnect Socket Comparison

Socket Type	Durability Cycle Life	Typical Volume	Cost
Validation	Low	Medium	Medium
Test	High	Low	High
Production	Low	Very high	Very Low

Validation Sockets

There are several types of validation sockets available in the industry. Their application depends on the electrical/mechanical and cost requirements. In the validation environment, it is important that the sockets designed for LGA packages have their electrical performance as close to the soldered-down part as possible especially for high-speed signals. There are two main categories of these types of sockets:

- **Socket/Interposer Two-piece Socket:** These sockets consist of the female receptacle part soldered to the PCB. The device is soldered to the interposer with pins and plugged into the female receptacle. Figure 3 shows a schematic of Socket/Interposer concept.

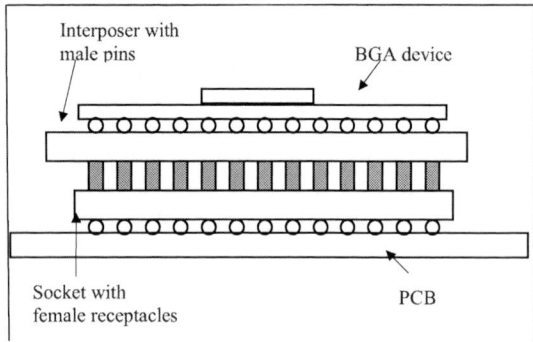

Figure 3: Socket/Interposer Schematic

- **Double-Compression Socket:** These sockets do not require soldering the device to the PCB. They are attached to the PCB using a mechanical set of hardware. The contacts are typically made out of metallized polymer or stamped-and-formed metal. An example of a socket contact manufactured by Gryphics, Inc. is shown in Figure 4.

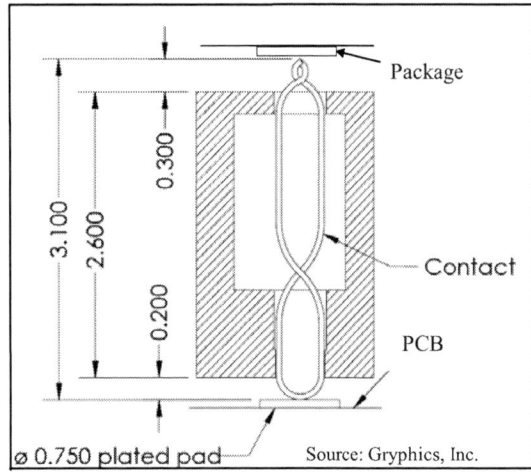

Figure 4: Gryphics® LGA contact

The Gryphics® socket contact has a large working compression-range of 10 mils. This helps accommodate noncoplanarity tolerance of the surface of the contact pads on both PCB and LGA package. The contact also provides some wiping action on the pads that helps maintain a stable contact resistance over repeated actuations.

Figures 5 show examples of metallized-polymer socket developed by Tyco Electronics.

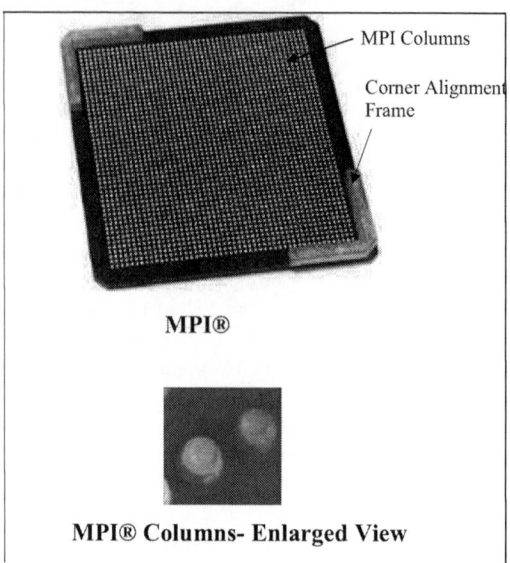

Figure 5: MPI® Socket

The MPI® socket developed by Tyco Electronics has a very short height (31 mils compressed) and is electrically equivalent to the size of a typical solder ball or half the thickness of a via in a typical 62 mils thick motherboard. However, such polymer contacts generally do not provide any horizontal wipe on the contact pads of the PCB and package. In order to implement such sockets into the product successfully, several design and assembly guidelines must be followed. Once these guidelines are followed, these sockets achieve very high reliability in many applications.

Test Sockets

Test sockets are typically of double-compression type and use spring-loaded pins that are spring operated. They are used for high-volume testing of the devices. A set of mechanical hardware is designed to minimize the insertion/removal cycle time of the sockets. Examples of a typical spring-loaded pin and socket are shown in Figures 6 and 7. These sockets occupy a large space on the PCB due to quick connect/disconnect features included in the top plate design.

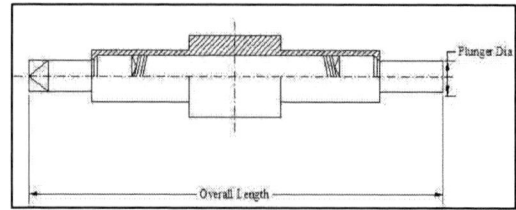

Figure 6: Typical Spring-loaded Contact

In general, the overall length of a spring-loaded pin is much more than the polymer contacts and so they are not suitable in many applications where electrical performance is critical.

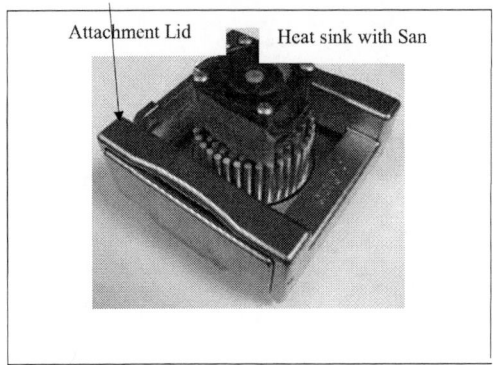

Figure 7: Typical Spring-loaded Pin Socket

Production Sockets

Production sockets are generally soldered to the PCB and use a compression-type contact design. The contacts are made out of stamped-and- formed BeCu material with solder balls attached at the bottom. The socket is provided with a feature for the pick-and-place machine to pick it from the tray and place it on the PCB. Figure 8 shows the Socket-T that Intel Corp. has used for socketing its microprocessors. As the size of the socket increases and the pitch becomes dense, the soldering of the sockets poses similar challenges as the soldering of BGA devices to the PCB. The factors that need to be considered while soldering the socket to the PCB are:

- Size of the socket – The larger the socket is, the higher the warpage;
- Ball-pitch – Smaller ball-pitch parts are more difficult to solder than those with a larger pitch;
- Weight of the socket or weight per solder ball – Heavier sockets tend to collapse the balls too much and increase the chance of shorting to adjacent solder balls;
- Diameter and material of the solder ball;
- Pad size on the motherboard; and
- Difference in the coefficient of thermal expansion of the socket substrate material and the material of the PCB.

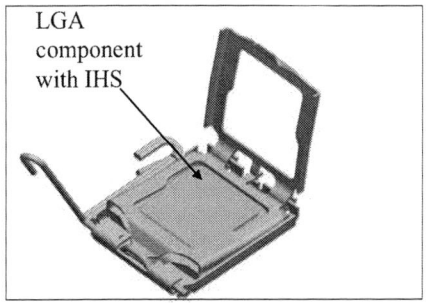

Figure 8: Production Socket-T

Future Socketing Challenges

As the packages become larger (> 42.5 mm square) to accommodate more I/Os, the sockets to hold these packages correspondingly grow in size. Larger sockets along with decreased ball-pitch make soldering the sockets to the motherboard increasingly difficult. With the pitch on LGA devices trending to 0.5mm, the double-compression type LGA sockets look more promising for attaching such devices to the PCB because they do not require a soldering process. Some of the major advantages of the double-compression type sockets are:

- Solderless design that uses a mechanical set of hardware for attachment to the PCB;
- Can be adapted to very fine pitch devices;
- Can accommodate very small pad diameters on the PCB;
- Easy to replace defective socket without losing expensive motherboard; and
- Can be adapted to thicker boards, larger than 62 mils.

However, there are many technological challenges for successful implementation of these sockets. Some of these are:

- Tight mechanical tolerances required on the motherboard. Standard PCB manufacturing tolerances may not work, especially for pad pitch smaller than 1mm.;
- Additional mounting holes required for applying compression pressure, which puts constraints on PCB trace routing on a four-layer motherboard; and
- Proper design of a mechanical hardware system to minimize the deflection of the package and PCB for a reliable interconnection.

Double-Compression Socket Design Parameters

Several design parameters need to be considered for the successful design of a double-compression socket. The most important parameter is the normal force required per contact. Traditionally, most sockets have been designed for a normal force of 45 grams per contact. However, as the pin-count per package increases, the total force required to make connection to the board increases in direct proportion. The socket with pin-count of 2,000 exerts a total force of more than 200 lbs. on the package, which can potentially cause damage to the package substrate and die if the force is not applied uniformly. For such large pin-count packages, the normal force needs to be reduced and still achieve a reliable connection. The working range or the compression range of the contact to make a reliable connection is another important design parameter. The contacts that have large working range (> 10 mils) can accommodate the larger tolerances of the mechanical hardware system. Most of these contacts are either stamped-and-formed metal type or spring-loaded pins shown in the Figures 4 and 6. Typically these metal contacts have electrical height > 2 mm or 80 mils. Other types of contacts available in the industry are polymer-based. These contacts have very short electrical height. Many polymer-based contacts available in the industry are < 1mm or 40 mils tall. These contacts, however, have a very short working range (0.1mm or 4 mils), so the design of the mechanical hardware to implement such contacts into the socket system becomes even more challenging. There are two types of interconnect socket systems for double-compression contacts, and their application is based on the working range.

• Fixed load system
• Fixed deflection system

The design of the mechanical hardware system for the fixed deflection system is much simpler than the fixed load system. The manufacturing cost for such a system is less than the fixed load system.

Manufacturing and Assembly Considerations

As the size of the contacts becomes smaller due to denser pad-pitch on the package, traditionally used stamped-and-formed metal contacts become more difficult to manufacture due to extremely fine contact geometries. In addition, they are difficult to install in the molded housing within tight mechanical tolerances. As the pitch gets down to 0.5mm (19.6 mils), the diameter of the pads is reduced to 8 mils, which is a very small target to make a reliable connection with a horizontal wipe on the surface of the pads of the PCB and package. The polymer-based contacts can hold tight tolerances as they are molded into the substrate. However, due to the inability of these contacts to provide horizontal wipe on the pads of the PCB and package, assembly considerations are becoming more important to the successful implementation of these sockets for a reliable connection. The pads on the package and PCB need to be cleaned with isopropyl alcohol before attaching such polymer-based sockets on to the motherboard.

Fixed Load System Design

This design is used for a very short working range of the contacts. Typically these are polymer-based contacts that have an electrical height of less than 1 mm or 40 mils. The interconnect system uses coil springs to accommodate the short working range of the contact and the mechanical tolerances of the hardware system. The coil springs are selected with a spring rate such that the springs compress a large amount (.250 mils in many applications) to achieve very short compression of the polymer contact (less than 5 mils). Figure 9 shows a schematic of the fixed load system design with four coil springs. Coil springs with a high spring rate allow use of mechanical hardware parts with standard mechanical tolerances, resulting in reduced cost.

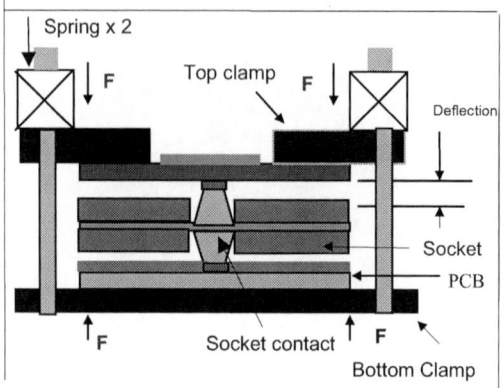

Figure 9: Fixed Load System Schematic

Fixed Deflection System

The contacts in this type of design have larger a working range (> 10 mils). This system design is more cost-effective, since no special springs are required to compress the mechanical set of hardware and the contacts. However, in general such interconnect-system designs are taller than the fixed load systems and do not have as good electrical performance as the fixed load systems. Figure 10 shows a schematic of a fixed deflection system design.

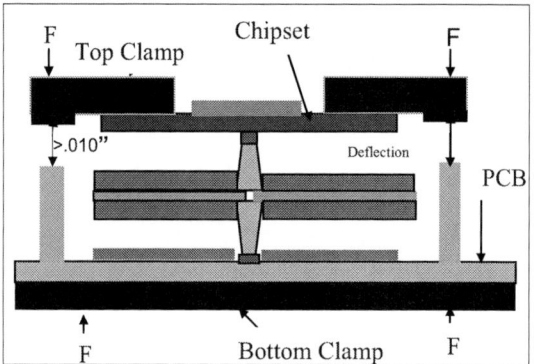

Figure 10: Fixed Deflection System

MPI® Socket System

Tyco Electronics' Metallized-Particle-Interconnect - MPI® offers many electrical benefits that commonly used metal contacts do not. Figure 11 shows the schematic of the MPI® column designed for 0.8mm (31.4 mils)-pitch. It shows the height of the column before the compression.

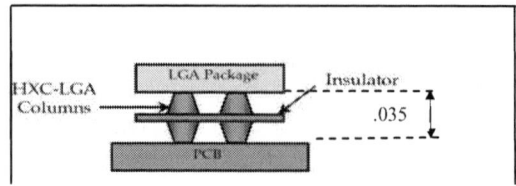

Figure 11: MPI® Column Schematic [7]

The major advantages of this contact are:

- The total electrical height after the compression is 31 mils, which is equivalent to the ½ via height in a 62 mils thick board or the height of a typical solder ball.
- The MPI® column designed for 0.8mm (31.4 mils) pitch has very low contact resistance of 11 milliohms per contact.
- Molded columns allow very tight mechanical true positional tolerance of less than 5 mils.
- The column pitch can be scaled for socketing applications for packages with 0.5mm (19.6 mils) pad-pitch and lower.

MPI® Force-Deflection

The force-deflection and force-resistance curve of 0.8mm (31.4 mils) pitch MPI® HXC-column is shown in Figure 12. Each column exerts a force of 45 grams when compressed by 4 mils. Since the compression range is very short, the interconnection system required to implement this column is of the fixed load type shown in Figure 9.

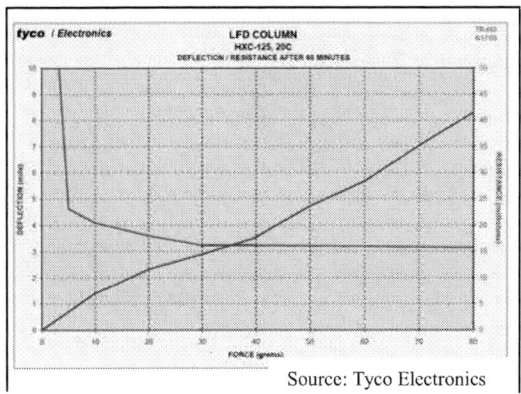

Source: Tyco Electronics

Figure 12: Force-Deflection/Resistance Curve

System Design Using MPI® Column

When MPI® columns are compressed between the package and PCB, they exert force that causes deflection of both the package and PCB. In order to minimize the deflection, a robust and a reliable mechanical hardware system is required. The mechanical hardware system is composed of a top plate assembly and a bolster plate assembly clamped together using four barrel nuts. The deflection of the PCB and the package can be estimated by assuming a uniformly distributed load exerted by the columns on a beam that is fixed at the two ends. The deflection can be approximated by the formula:

Deflection = $K *(Wx^2/ (EIL)) * (l-x)^2$ [4]

K = Constant, W= Total load exerted by all columns, L= length of the span between the attachment points, E= Young's modulus of the materials, I= Moment of inertia, x = distance from the attachment point where the deflection occurs.

Both the length of the span between the mounting points and the thickness of the plates need to be designed using Finite-Element-Analysis (FEA) to ensure that the right amount of compression of the columns is obtained with a minimum deflection of the package and PCB. Figure 13 below shows the system design for socketing a 37.5 mm square LGA package with 1226 contact pads located on a 0.8 mm (31.4 mils) pitch. The system is designed such that the package and PCB deflect less than 1 mil under the load exerted by the contact columns. The system shown in Figure 13 is a fixed load type. The fixed load is applied using custom-designed preloaded pressure-springs.

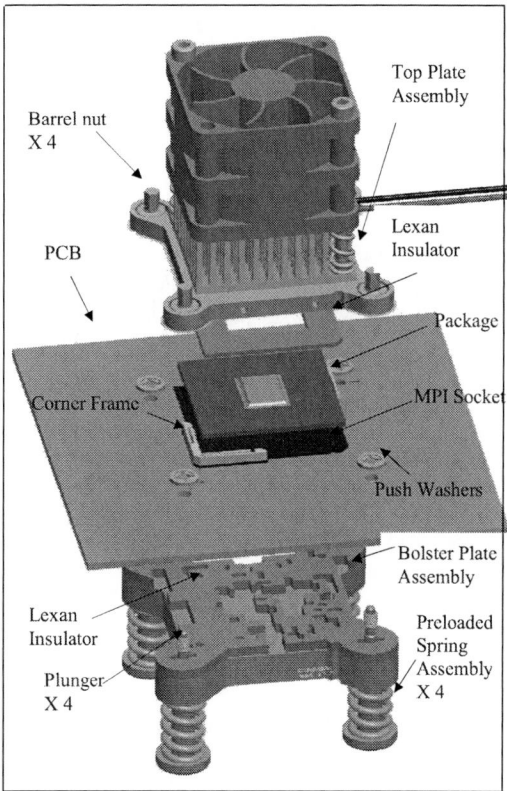

Figure 13: 1226-Pin MPI® Interconnect System

The interconnect system consists of the following main parts:

- The bolster plate assembly made out of stainless steel material. A lexan insulator is attached to the bolster plate. The insulator has cutouts to clear components on the backside of the PCB. The bolster plate is designed with plungers with preloaded pressure-springs to apply the right amount of force on the MPI® columns for a reliable electrical connection of the package pads to the PCB pads. The preloaded pressure-springs help in reducing assembly time by more than 50%. The bolster plate assembly is retained to the PCB using push washers
- The MPI® socket with two corner frames. The socket has two alignment pins, 31 mils in

485

diameter that allow accurate placement of the socket on the PCB. The plastic corner frames have spring action that allows self-centering of the package inside the socket.

- The top plate assembly with an integrated heat-sink with fan. The heat-sink applies a load of 24 lbs. on the die to prevent the deflection of the die during compression of the columns. It also provides adequate thermal connection to the die. Additionally, the top plate assembly has barrel nuts for the purpose of attaching it to the bolster plate. The depth of the threaded hole inside the barrel nut is designed such that when the plungers bottom out inside the barrel nut, the coils springs compress the right amount to provide a nominal normal force of 45 grams per column. This translates to a total force of 144 lbs. on the package substrate. The coil springs in this case compress by 250 mils to compress MPI® columns by 4 mils.

Force-Deflection Finite-Element-Analysis

Ansys Finite-Element-Analysis (FEA) tool was used to design the correct thickness of the top and bolster plate to keep the deflection of the package and PCB less than 1 mil. The total keepout of both the top and bolster plate is kept to a minimum to allow placement of components close to the package.

PCB deflection: The Figure 14 shows the FEA model of the quarter section of the bolster plate assembly. A load of 36 lbs. is applied uniformly across the quarter section of the plate. The PCB as shown in Figure 14 deflects by 0.0207 mils. This deflection is well within the MPI® column design specifications of maximum deflection of 1 mil. Table 4 shows materials and their properties used for conducting FEA to determine the PCB deflection.

Table 4: Material Properties -PCB deflection

Parts	Thickness	Material	Poisson's ratio	Young's Modulus [4] PSI
PCB	62 mils	FR4	3.49	2.55e6
Bolster Plate	250 mils	Stainless Steel 304	0.295	27.5 e6
Insulat-or	90	Polycarb-onate	0.38	3.55 e5

Condition:
Symmetry boundary condition, 36 lbs. constant force

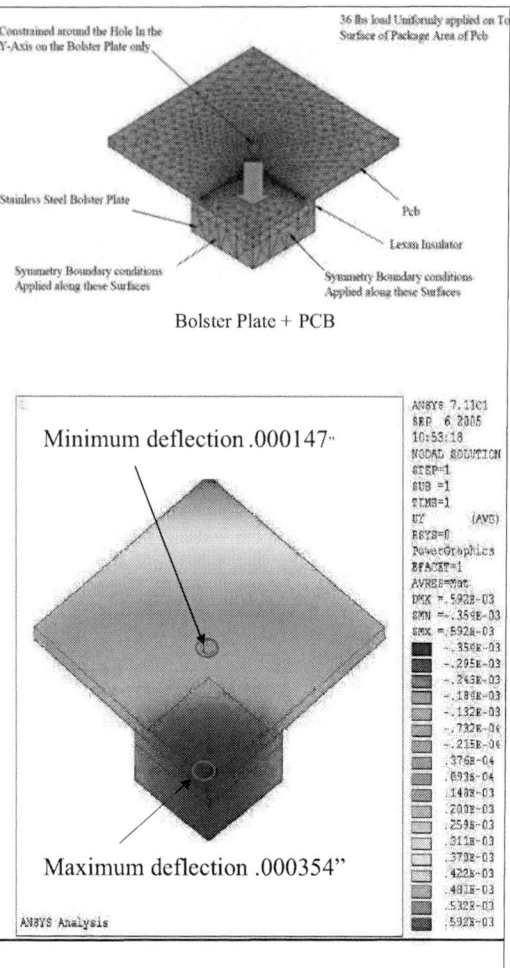

Bolster Plate + PCB

PCB Deflection = 0.0354-0.0147 =0.0207 mils

Figure 14: Bolster Plate/PCB FEA

Package Deflection: Figure 15 shows the FEA model of the quarter section of the top plate assembly. A load of 36 lbs. is applied at the bottom of the package. The heat-sink applies a load of 6 lbs. on the die. The package as shown in the Figure 15 deflects by 0.0789 mil. Table 5 shows the materials and their properties used for conducting FEA to determine the deflection of the Package.

Table 5: Material Properties –Package Deflection

Parts	Thickness	Material	Poisson's ratio	Young's Modulus [4] PSI
Package	42 mils	FR4	3.49	2.55e6
Die	31 mils	Glass	0.24	15.6 e6
Top Plate	160 mils	Stainless Steel 304	.29	27.5c6
Insulator	43 mils	Polycarb-onate	0.34	3.55e5

Condition:
Symmetry boundary condition, 36 lbs. constant force

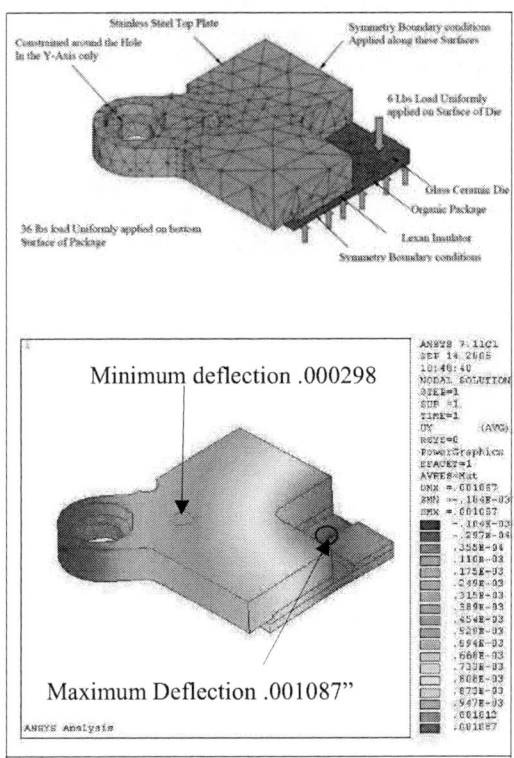

Package Deflection = 0.1087- 0. 0298 = 0.0789 mils

Figure 15: Top Plate/Package FEA

The design shown in Figure 13 can be scaled up to the larger size packages 49.5mm square and more with pad-pitch of less than 0.5mm. While using a double-compression socket with pin-counts larger than 2,000, it is recommended to reduce the normal force per contact to less than 30 grams to reduce the total load applied on the substrate or die of the package. To implement MPI® socket technology successfully for socketing 0.5mm pad-pitch LGA devices, precision mechanical tolerances need to be specified on the alignment holes of the PCB and guide-pins of the socket. This will ensure high reliability of the interconnection without any electrical opens or shorts.

Probing Plate Design

In the computer validation application, accessibility to the backside of the motherboard is required for probing critical signals especially during early debug of the system. The solid bolster plate shown in Figure 14 does not provide such accessibility. Several unique probing plate designs are shown in Figure 16. They are X, O and Q plates and are used as replacements to the solid bolster plate when backside probing is required. The mechanical design of such plates is more challenging and requires several iterations of FEAs to finalize the design that meets the MPI® design guidelines to keep the deflection of the package and PCB less than 1 mil. Since the design of these plates is more complex, the manufacturing cost is high.

The plates are used in small quantities only for probing of signals that are not accessible using the solid bolster plate.

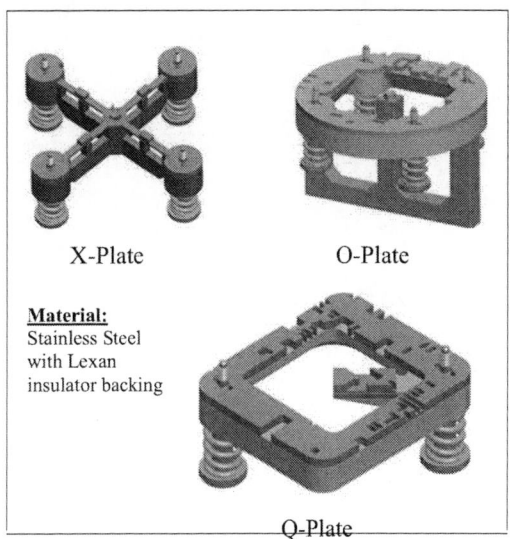

X-Plate   O-Plate

Material:
Stainless Steel
with Lexan
insulator backing

Q-Plate

Figure 16: Probing Plates

PCB Manufacturing Guidelines

There are several critical PCB manufacturing requirements for successful implementation of the MPI® technology. The connection copper pads on the board and the package must be plated with nickel and gold. The recommended surface finish is Electroless-Nickel, Immersion-Gold (ENIG). This consists of 3 to 10 micro-inches of gold over 100 micro-inches minimum of nickel. Additionally, since MPI columns are compressed by only 4 mils, the solder mask in the area of BGA pads must not exceed thickness of 1 mil. Most of the PCB manufacturers can meet this requirement without any additional cost, but it needs to be specified on the fabrication drawing. Furthermore, positional tolerance of the alignment holes to the BGA-connection pads is critical. In a standard drilling operation of a hole in a PCB, the positional tolerance of the hole to an etched feature is 8 mils. Such a large tolerance at the worst-case condition can cause electrical short of one MPI column to an adjacent BGA pad on the PCB. To fix this problem, the alignment holes need to be drilled using an optically aligned drilling method. The two alignment holes are drilled in a secondary operation using a vision system. A fiducial is placed at the center of the alignment hole to be drilled, and the hole is drilled by zeroing in at the center of the fiducial. This method of drilling gives positional tolerance of the hole to the pad on a PCB of less than 5 mils. This is shown in Figure 17.

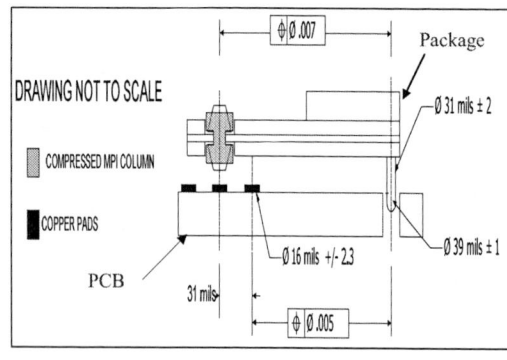

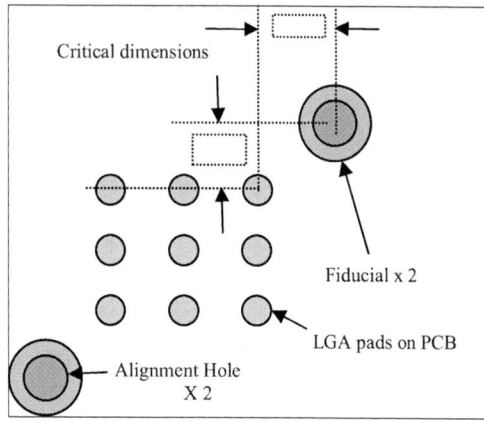

Figure 17: Critical Mechanical Tolerances

Interconnect System Reliability

The interconnect system designed with MPI® column technology has been successfully implemented at Intel Corporation on its validation platforms for internal use. Once all the design and manufacturing guidelines are followed, the interconnect system achieves very high reliability. The interconnect system was subjected to the following test conditions, and it passed all test conditions with change in the contact resistance of less than 10 milliohms:

- 85°C/85%RH – 120 hours – The purpose of this test is to ensure reliability at high humidity and temperature condition.
- 100°C – 120 hours – The purpose of this test is to ensure reliability at extreme high temperature test conditions.
- 0°C – 120 hours – The purpose of this test is to ensure reliability at extreme low temperature test condition.
- Durability of 25 insertion/removal cycles.

Interconnect System Assembly

Since MPI® columns do not provide horizontal wiping action when mated, several precautions must be taken during assembly to insure a reliable electrical connection of the socket to the PCB. The gold pads on the PCB and the package must be thoroughly cleaned with isopropyl alcohol before the start of the assembly. In order to ensure

uniform compression of the MPI® columns, the barrel nuts shown in Figure 13 must be tightened in a diagonal fashion, starting with one corner and going diagonally opposite, and then the third corner and going diagonally opposite. It is recommended that a torque driver with a torque set at a right tension to be used to tighten the screws to insure the right amount of compression force on the MPI® columns without breaking the barrel nuts.

Cross-Sectional Analysis

The primary purpose of the cross-sectional analysis is to insure the alignment of MPI® columns to the pads of the PCB and LGA package and also the right amount of compression of the MPI® columns. Figure 18A shows cross-sectional view of one MPI® column taken through the corner of the column. The compressed height of the column in this section is 31 mils, which is within the specification. Figure 18B shows the cross sectional view taken through the center of the socket. This view shows that the compressed height of the MPI® column is 32 mils, indicating that there is 1 mil deflection in the PCB and the package combined. This deflection is still within the acceptable range and complies with the finite element analysis results. The assembly shows acceptable alignment of The MPI® columns with the pads of the PCB and package.

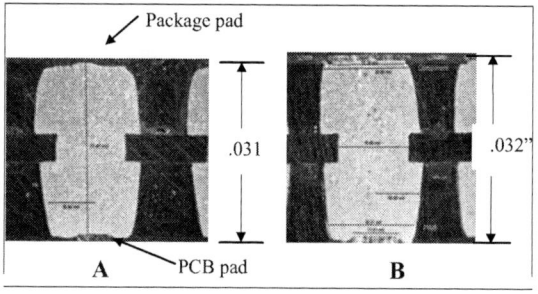

Figure 18: Cross-Sectional Analysis

Conclusion

As the pitch on the PBGA devices scales down to 0.5 mm (19.6 mils) -pitch and the package sizes increase, soldering the devices to the motherboard will be more difficult due to warpage of the substrate and large non-coplanarity tolerance of the solder balls. Reflow temperature of the lead-free alloy like SAC305 exasperates this problem. There are alternative methods of attaching the fine pad-pitch packages to the motherboard that do not require any soldering processes. The use of double-compression sockets solve many solderability issues and have shown to offer following advantages:

- Easy replacement of parts in and out of the socket;

488

- Solderless system – Easy to replace the socket if damaged, since no desoldering and resoldering required;
- The sockets can be adapted to fine-pitch devices. The polymer-based contact designs can be scaled down to 0.5mm (19.6 mils) -pitch and can hold tight tolerances since they are molded into the substrate; and
- Very short electrical height. The fixed load system typically offers much lower electrical height due to lower working range.

To implement the double-compression sockets successfully, several design and manufacturing guidelines must be followed. These include:

- Force-deflection analysis of the contact design to understand the force required per contact for a reliable interconnection;
- Proper FEA models available to characterize the load on the array of contacts and design a mechanical hardware system to minimize the deflection of the PCB and package within specifications;
- Proper design and manufacturing of the PCB to meet the design guidelines of the contact design; and
- Proper mechanical tolerances on the socket alignment pins and PCB alignment holes to ensure no electrical shorts or opens.

The MPI®-based interconnect system has proven to have very high reliability under certain test conditions. There are other double-compression socket-contacts like the one manufactured by Gryphics, Inc. Their application depends on the electrical and mechanical requirements of the system on which they are used. The reliability of such interconnect systems depends not just on the contact design, but on the supporting mechanical hardware system design. As the industry is trending to 0.5mm-pitch and the package sizes are increasing to 49.5mm square, the interconnect solutions need to focus more on the total system design rather than focusing on the individual piece part design.

Acknowledgements

Srikanth Mothukuri, Intel Corporation, for providing design models and FEA analysis data.

John-Reed Hannig, Intel Corporation, for providing mechanical tolerance data.

Jeff Mason, Tyco Electronics, for providing force-deflection and force-resistance data for the HXC MPI columns.

Jim Rathburn, Gryphics, Inc., for providing metal contact design detail.

References

Articles from conference proceedings

[1] Chris Thornton, "New power modules improve surface-mount manufacturability," Plug-in power products, Texas Instrument Incorporated, 2005.

[2] Joe Chu, "The impacts of eutectic solder ball coplanarity of PBGA package on the reliability of solder ball joints," Proceedings of the Surface Mount International Conference, August 23, 1998.

[3] Dr. Weiping Liu And Dr. Ning-Cheng Lee, "Novel SACX Solders with Drop Test Performance Outperforming Eutectic Tin-Lead," Proceedings of IPC Printed Circuit Expo APEX and Designers Summit, 2007.

Book

[4] Machinery's Handbook by Erik Oberg, Franklin D. Jones, Holbrook L. Horton and Henry H. Ryffel.

Design Standard Publication

[5] JEDEC publication 95 design guide 4.14, Ball Grid Array Package.

[6] JEDEC publication 95 design guide 4.5, Fine-pitch square Ball Grid Array Package.

[7] HXC-LGA Application Guide, Tyco Electronics, Attleboro Falls, MA 02763.

Design and Technology Considerations on LTCC Microwave Modules for Fixed Radio Link Equipment

M. Piloni, A. G. Milani

Nokia Siemens Networks S.p.A., S.S. 11 Padana Superiore km 158 - 20060 Cassina de' Pecchi (Milan), Italy

Phone: +39-02-2437 6770, Fax: +39-02-2437 6375, E-mail: marco.piloni@siemens.com

Abstract

LTCC and other similar technologies, presently used in large volume consumer products, could represent a good candidate for low cost packaging solutions, suitable for multi-chip, highly integrated (potentially 3D), fully SMT RF modules development. With the aim to check LTCC features, a Pilot project, consisting of a 23 GHz band (21.2 ÷ 23.6 GHz) fully SMT receiver front-end, was designed, developed and measured. The module comprises 8 dielectric layers in order to get adequate stiffness and to allow the integration and routing of passive elements such as baluns, filters and bias networks. Two different designs are proposed to realize I/O RF interconnection, avoiding parasitic mode propagation: a DC module using dc vias couplings and a EM module using electro-magnetic couplings for RF I/O interfaces.

Key words: LTCC, Microwave Radio Link, SMT Package

Introduction

The continuous and strong demand for reducing the cost and size of MW radio links equipment are stimulating R&D laboratories to investigate recent technology development able to offer adequate performance at low price.

Low Temperature Cofired Ceramic (LTCC) is a mature and robust technology introduced commercially in the early 1980's, mainly for hi-reliability military, aerospace and medical applications. It is presently used in large volume consumer products; these applications have considerably lowered the technology cost.

LTCC is among the best candidates for fixed radio link RF modules development, due to the availability of low loss substrates at frequencies up to 50GHz, the capability of fine-line patterning (either screen printing or thin film) and embedded components. The inherent multi-layers structure allows embedded packaging (even hermetic if necessary), RF/DC inter-layer routing, 3D integration and easy shielding of signal lines.

In order to investigate LTCC features in such applications, a fully SMT 23 GHz receiver front-end was taken into consideration as a study case. The 23 GHz band (21.2 ÷ 23.6 GHz) can be considered one of the most meaningful, being largely used and critical to implement. In particular, receivers must exploit low loss and good input matching, in order to not affect performances such as conversion gain and noise figure.

Electromagnetic simulations were performed with Ansoft HFSS™ Finite Element Method simulator, to accurately predict the impact of the 3D structures; measured performances of the assembled front-end module are reported.

Module Architecture

For the implementation of the 23 GHz receiver, an approach which uses bare die and LTCC module as a carrier has been chosen.

Chips are located on top side of the module, while metalized pads are provided on the bottom side for SMT mount on a laminated printed circuit board (motherboard).

Active devices can be mounted either by wire-bond or flip-chip technique, but in the latter case assembly cost is higher, due to bump attach process, thus wire-bonding has been used. Additionally, LTCC gold metal systems are directly suitable for wire-bond, with no need for further plating treatments.

Vertical transitions for bias and signal lines are realized thru metal filled via holes.

Table 1: Receiver's Main Performances

RF Band	IF Freq	Gain	Noise Figure	Input Match	IIP$_3$ Input
21.2÷23.6 GHz	1GHz	≥10dB	≤5dB	≥15dB	≥0dBm

The inner space of the LTCC is allocated for passive functions such as off-chip matching networks, baluns, decoupling capacitors and bias circuitry.

With the aim of keeping the complexity low and to avoid any manual tuning, a commercial GaAs monolithic integrated receiver has been used.

The MMIC (Monolithic Microwave Integrated Circuit) comprises a LNA, a cold FET variable attenuator and a image reject mixer, thus eliminating the need for any additional filtering.

De to fundamental operation of the downconverter, LO signal frequency is in the 20-25GHz range, depending on the IF value.

In order to avoid possible spurious emissions from the receiver and to simplify LO feed path, a GaAs active x2 multiplier has been integrated into the module. The receiver's expected main performances are summarized in Table 1.

Many different supplier of LTCC material and foundries are available on the market. Most of them are involved in military and hi-reliability applications, and exhibit elevated material/process performances at relatively high prices, due to little panel dimensions and un-optimized production lines.

Conversely, commercial LTCC foundries offers high yield volume lines but more severe manufacturing constraints.

Generally speaking, design rules are quite inhomogeneous, and selection of the appropriate supplier for the given application is essential.

Analysis of several different foundries has been performed, considering fixed radio link volumes and requirements; the main features of the chosen process are described in Table 2.

In particular, minimum module thickness (independently from the number of layers), represents a heavy constraint for RF feedthroughs, which must establish a proper millimeter-wave connection between top and bottom layers. Indeed, they lead the RF signal through all the dielectric layers of the LTCC module.

Table 2: Main LTCC process features

Min. Line/ Space Width	Via Diameter	Min. Via Pitch	Min. Module Thickness
75um/ 75um	100um	250um	1 mm

Two different designs are proposed to realize vertical feedthroughs, avoiding parasitic mode propagation in the inner layers: a quasi-coaxial DC module using dc vias coupling and a EM module using electro-magnetic coupling.

Module Implementation

Figure 1 illustrates the proposed module concept. The LTCC receiver consists of 8 dielectric layers having a dielectric constant $\varepsilon_r = 7.6$ and loss tangent $\tan\delta = 0.0007$; the two outer layer are 150um thick, the remaining 6 inner ones are 100um thick. Total module dimensions are 9x14mm.

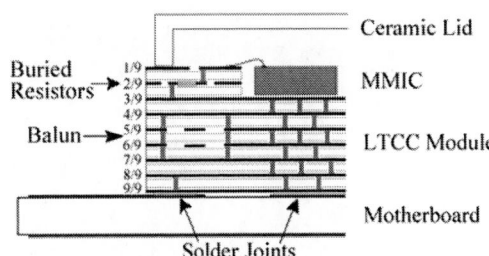

Figure 1: LTCC Receiver Module Concept

Conductors, printed with copper metal system, are 10um thick; outer metal layers are gold finished. Non-meshed inner ground planes are used.

The MMICs, along with single layer decoupling capacitor, are placed in a 250um depth slot obtained punching the two first dielectric layer, and exposing the below metallization (layer 3).

The latter acts as ground plane for chips and for the first layer interconnections; the resulting 50Ω microstrip is 330um width, allowing for higher-order mode free propagation up to 50GHz.

The use of striplines buried in the second metal layer has been considered too, but due to the proximity of ground planes and to the high dielectric constant of ceramic, the resulting line width is too narrow to be practical.

Another advantage of using microstrip over stripline is the possibility of photopatterning the conductors on the outer layers, which leads to a far better definition of geometries.

After mounting the chips, a ceramic cap is glued with epoxy on the top of the module, providing environmental and mechanical protection for chips and wire bonds. The resulting air cavity allows for proper operation of MMICs.

To keep the build-up symmetrical, in order to avoid any possible module camber problem, layer 7 is also grounded. RF lines on the bottom layer can therefore be designed as microstrips but due to the presence of the motherboard, to preserve 50Ω impedance, line width needs to be reduced down to 250um.

RF interconnection between top and bottom layers are realized using two different designs; the simplest is a quasi-coaxial structure, consisting of a center via led through holes in the ground planes, encircled by shielding vias between the ground planes, as depicted in Figure 2.

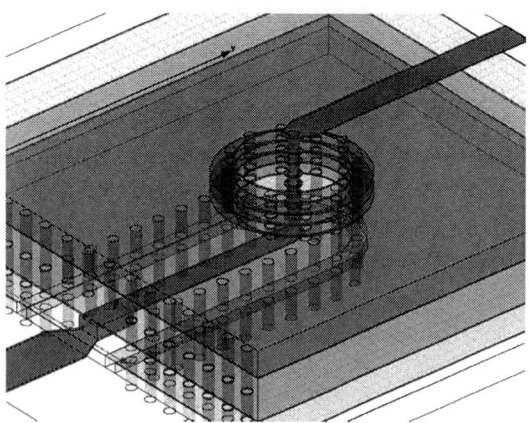

Figure 2: Via RF Feedthrough

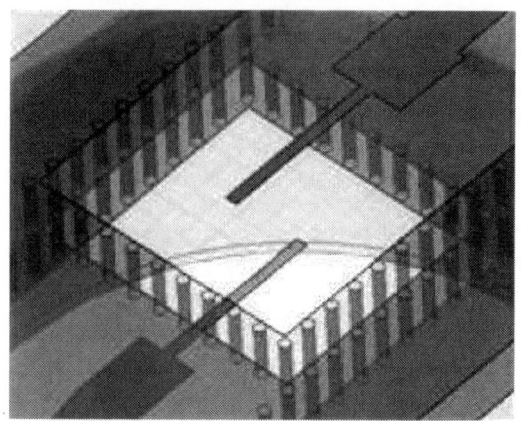

Figure 4: EM coupling RF Feedthrough

Impedance matching is obtained adjusting radius ratio between center conductor and the surrounding shielding vias ring.

Due to the fact that the two ground planes face each other, parallel plate mode can be excited, thus ground vias pitch is critical.

In order not to stress manufacturing requirements, an initial attempt using only eight shielding via has been analyzed; resulting vias pitch is 400um, corresponding to roughly λ/8 at 30GHz.

From full wave simulation, a package mode resonance has been found at approximately 25GHz. The resonance frequency is related to module dimensions and the interested region is between the two ground layers, as shown in Figure 3, meaning that shielding is not effective. Ground vias number has than been raised to twelve, leading to a via pitch of 270um, still compatible with manufacturing constraints; in this configuration, no resonances were found up to 40GHz.

The other approach used to realize RF vertical feedthroughs relies on electromagnetic coupling between transmission lines (EM coupling).

According to Kuroda's identities, two open circuit, quarter wavelength coupled lines can be arranged to behave like a capacitor, providing that $Z_{even}-Z_{odd} = 100$ Ω; this structure is usually referred to as a "DC block".

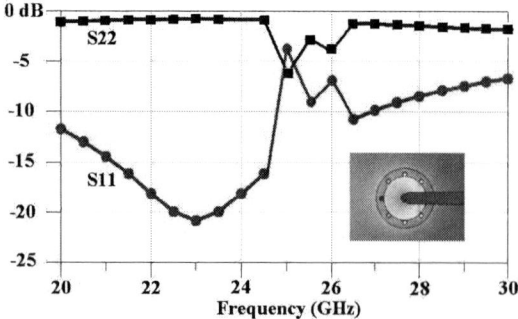

Figure 3: Package Resonance Due to Poor Shielding

In our design, a section of microstrips on top and on bottom of the module are coupled through a window opened in the ground planes, as depicted in Figure 4. Microstrips face each other for a length of 1,45 mm, which is roughly λ/4 at 23 GHz.

To obtain proper matching and coupling, microstrips impedance has been raised pulling away ground and shrinking lines width. Ground window dimensions have been numerically optimized for best coupling over the frequency band of interest.

Additional matching optimization is realized widening the feeding microstrips, thus realizing a strong impedance step. Shielding ground vias are placed around the ground window to prevent excitation of package modes.

Due to high dielectric constant of ceramic, most of the field is contained into the ceramic itself, preventing the field from radiating; nevertheless the proximity of the motherboard to the bottom coupled microstrip degrades matching.

To address this issue the motherboard has been milled in the vicinity of the coupling structure, removing the dielectric.

The motherboard has been realized using 0.26mm height RO4350 as first dielectric layer, resulting in a 0,52 mm wide microstrip for 50 Ω impedance.

Solder joints geometry must be properly designed in order to obtain high mount yield, preserving good RF matching versus SMT process parameters variation (typically solder joint thickness and shape).

Since the solder pad on bottom of the module is 250um wide, the motherboard's feeding line needs to be tapered down; assuming a solder joint thickness of 100um, from numerical simulations an optimum line width of 310um has been found; this allow also for the creation of a solder meniscus which helps to keep alignment between the module and the land pattern.

A large ground solder joint is included to better shield signal lines and to avoid any unwanted spurious mode propagation.

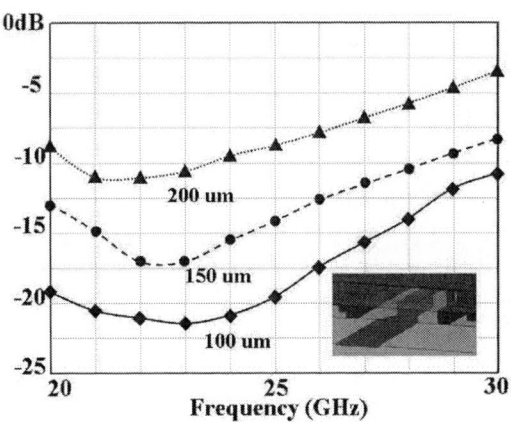

Figure 5: Via Module Input Matching versus Solder Thickness

Due to the relatively poor thermal conductivity of ceramic, thermal dissipation issues had to be addressed too; a solution was found using a thermal vias array so that heat is removed from the backside of MMICs to the bottom of the module. Heat is than transferred to the motherboard through the ground solder joint.

Tolerance sensitivity analyses have been performed, particularly versus LTCC dimensional errors, layers misalignments and solder joints parameters, as shown in Figure 5.

Modules performances are relatively insensitive to dimensional errors of typical commercial LTCC processes, particularly for EM feedthrough. Quasi-coaxial module is more affected by alignment errors.

It has also been found that solder thickness is critical, and must be kept in any case well below 200um in order not to affect input matching.

Realized Test Modules and Measurements

With the aim of experimentally evaluate the performances of the proposed module concept, several test receivers designs have been fabricated, assembled and measured.

Figure 6: Realized LTCC modules

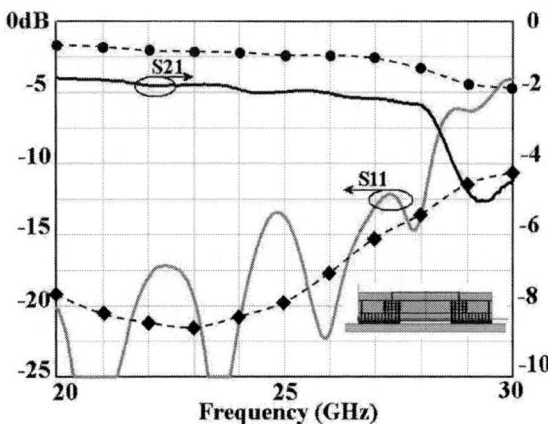

Figure 7: Simulated (Dashed) and Measured (Solid) Performances of the Via Bak-to-Back

The various implementations have been realized on the same 4 by 4 inch panel and singulated after chip mount; they include two quasi-coaxial and one EM coupling module, back-to-back transition for evaluating RF feedthroughs alone and calibration standards. Some of the realized modules are showed in Figure 6.

All the measurements have been performed with modules mounted on test motherboard with coaxial adapters, after LRL calibration.

Figures 7 and 8 show the scatter parameters of the quasi-coaxial and EM back-to-back modules respectively; the measured and simulated return loss corresponds very well, considering the multiple reflections that occur in this test structure.

It can be seen that the quasi-coaxial module shows broader band performance and is more suitable to lower band and high relative bandwidth applications.

Conversely the EM module results narrow band but less affected by manufacturing and assembly process tolerances and then applicable to high frequency bands not requiring high relative bandwidth.

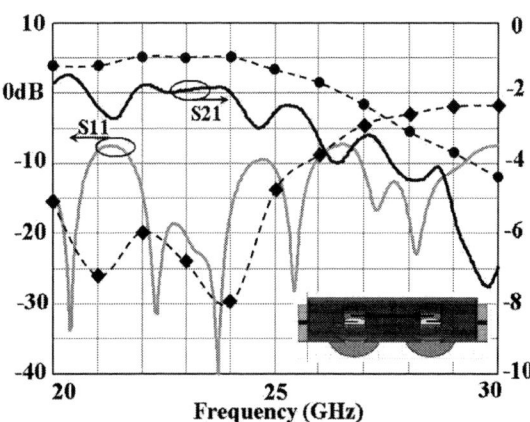

Figure 8: Simulated (Dashed) and Measured (Solid) Performances of the EM Bak-to-Back

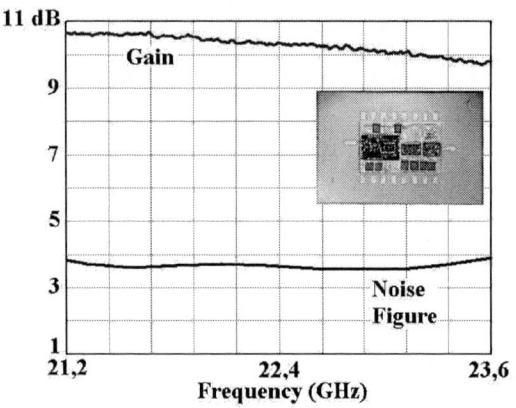

Figure 9: Measured Gain and Noise Figure of the Via Receiver

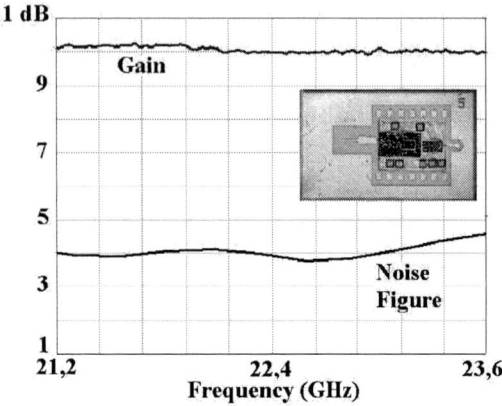

Figure 10: Measured Gain and Noise Figure of the EM Receiver

Figures 9 and 10 shows conversion gain and noise figures of quasi-coaxial and EM receivers; from a general point of view, both the designs met all the RF goal performances.

The noise figure of the EM modules is higher, and conversely the gain is lower, due to radiation losses; in fact removing the ceramic cap improves performances, meaning that the low Q ceramic of the cap absorbs part of the field.

Conclusion

A LTCC fully SMT compatible 23 GHz module receiver was designed and tested.

The module includes room for MMICs mount, bias circuitry and matching networks.

Design conditions for high frequency feedthroughs and LGA solder joint were described and discussed. Ground via locations are optimized to enhance maximum operating frequency.

The receivers have around 0.5dB and 1dB noise figure degradation for quasi-coaxial and EM implementations respectively, compared to bare die.

EM modules show narrow band behavior, which can be useful in transmitter modules to avoid spurious emissions.

Quasi-coaxial feedthroughs have better performances and occupy less space, but are not suited for applications at frequency higher than 30 GHz.

The developed modules will find useful applications for high frequency commercial radio link with remarkably low cost.

Acknowledgements

The authors wish to thanks Fabio Cortinovis for his support in modeling and EM simulations of structures and Maurizio Motta for assistance in prototypes testing.

References

[1] J. Heyen, T. von Kerssenbrock, A. Chernyakov, P. Heide, A.F. Jacob, "Novel LTCC/BGA Modules for Highly Integrated Millimeter-Wave Transceivers", IEEE Transactions on Microwave Theory and Techniques, Vol. 51, No. 8, pp. 2589-2596, December 2003

[2] W. Simon, R. Kulke, A. Wien, M. Rittweger, I. Wolff, A. Girard, J-P. Bertinet, "Interconnects and transitions in multilayer LTCC multichip modules for 24 GHz ISM-band applications", MTT-S, Int'l Microwave Symp. Dig., pp. 1047-1050, 2000

[3] A. Ziroff, M. Nalezinski, W. Menzel, "A Novel Approach for LTCC Packaging Using a PBG Structure for Shielding and Package Mode Suppression", Proceedings of the 33rd European Microwave Conference, pp. 419-422, October 2003.

[4] Ziroff, M. Nalezinski, W. Menzel "A 40 GHz LTCC Receiver Module Using a Novel SubmergedBalancing Filter Structure", Proceedings of Radio and Wireless Conference, pp. 151-154, August 2003

[5] G. Strauß, W. Menzel, "Millimeter-Wave Monolithic Integrated Circuit Interconnects using Electromagnetic Field Coupling", IEEE Trans. CPMT, Part B, vol. 19, no. 2, 5.96, pp. 278-282

LTCC Multilayer Technology Enables Very Compact 20 GHz Switch Unit for Space Applications

K.-H. Drüe[1], M. Hein[1], J. Müller[2], R. Perrone[2], S. Rentsch[2], R. Stephan[1], J. Trabert[1]

Technische Universität Ilmenau, Faculty of Electrical Engineering and Information Technology

P.O. Box 100565, D-98684 Ilmenau, Germany

[1] Institute of Micro- and Nanotechnologies, [2] Junior Research Group Functionalized Peripherics,

Phone: +49 3677 693429, Fax: +49 3677 693350, E-mail: karl-heinz.drue@tu-ilmenau.de

Abstract

A 4x4 switch matrix had to be realized within the project KERAMIS of the German space agency DLR. This project aims at the development of innovative and inexpensive components for future applications in multimedia satellite communications. The rationale of the project is to exploit the possibility of integrating passive and active components in LTCC multilayer structures. This opens possibilities to minimize the complexity of the semiconducting components. The working frequency of the device was in the 20 GHz band. A six layer construction with 7 conducting levels was chosen. The semiconductor switch dies were countersinked in cavities on both the top and the bottom of the circuit. The active chips were mounted using electrically conductive epoxy and wire-bonded afterwards. For testing purposes additional cavities are situated on the top side. The matrix was hermetically sealed by covers on both sides. These covers were also made in LTCC technology. Due to the complexity of the circuit 50 µm lines and spaces had to be applied using fine line screen printing. The minimum distance between the edges of the cavities and the wiring was 75 µm. For cavities and vias mechanical punching and laser cutting was used. The paper describes the technological processes and their particularities applied to manufacture this ambitious circuit.

Key words: Satellite application, LTCC, Microwave, Ka-Band

Introduction

The standard assembly technology for communication devices in the 20 GHz-band is to connect MMICs by microstrip lines or waveguides. Due to the high precision of structuring single layer thin film technology is mainly used to realize the connections and passive elements such as filters, couplers etc. The thick film derived LTCC technology allows multilayer designs also for RF- and microwave circuits. This leads to a substantial reduction of size and weight and makes this technology interesting for space applications. The project KERAMIS focuses on larger implementation of functionality in LTCC substrates to allow designs with „standard" MMICs. Target applications are circuits for multimedia satellite communications. This project was funded by the German Space Agency DLR.

Project KERAMIS

The KERAMIS consortium consists of a group of enterprises, institutes and universities. Different LTCC units were designed by the partners to build a modular assembly system. The units can be easily connected on PCBs by soldering like standard SMDs. A bigger part of the functionality

has moved from the MMICs to the LTCC substrate. This should reduce costs for future satellite applications. The TU Ilmenau had to design and build a 4x4 switch matrix for the Ka-Band.

Technological Preliminary Investigations

Prior to the design work technological investigations were undertaken to find adequate solutions for materials and processes that can be used in the target frequency range of 17 to 22 GHz [1], [2]. Different structuring technologies such as fine printing with special screens, photo-imageable pastes [5], etching of thick film and thin film on LTCC [2] were considered. Two material systems, the DuPont tapes 943 and 951 were examined. In addition, the via forming by mechanical punching and laser cutting was compared.

All structuring methods allow minimal lines and spaces of 50 µm and the losses at high frequencies are similar. For structures with dimensions of 30 µm and below, thin film technologies are the only choice. Of course, thin film is only possible as a postfire technology.

Finally the switching matrix was realized using fine line printing with 500 and 400 mesh screens and mechanical punching of 150 µm vias for reasons of costs and producibility.

Despite the higher dielectric losses the 951 material system was chosen. More ink materials are available for this tape and hermeticity and planarity are easier to achieve compared to commercial RF-tape systems.

For all conductors and via fillings Au inks were used (DP 5734 and DP 5738) to guarantee a high reliability.

The design with cavities at both sides of the substrate required additional tests concerning the lamination technology. Different methods using inlays and sacrificial materials (e.g. carbon) were investigated. Sacrificial layers showed good results in „pure" LTCC, but reacted with Au inks. The best results were obtained using a very soft silicone rubber mat of some mm thickness and a sheet of fibre-glass reinforced epoxy on both sides during the main lamination process.

Design and Technology for the Switch Matrix

The LTCC circuit consists of 6 layers 951 AX with 7 conductive levels (5 inside, 2 on top and on bottom respectively). Cavities are located at the top and the bottom layer for the active elements. As switches pin diode devices are used. These circuits have a Single Pole Quadruple Throw (SP4T) architecture. For the 4x4 matrix 8 of these circuits are necessary. They are mounted on both sides of the LTCC stack to get short signal paths. This construction reduces the number of signal layers and the number of transitions between layers. The chips are rotated at an angle of 45°. This leads to shorter transmission lines, decreased radiation due to lower discontinuities and reduced excitation of interfering modes.

The SP4T Pin Diode switches have the following nominal electrical specifications at 18 GHz:

Insertion loss:	1.4	dB
Isolation:	35	dB
Input return loss:	10	dB

The matrix has 4 RF inputs and 4 outputs (Figure 2). These ports are implemented as grounded coplanar lines (GCPL, GCWG) and can be contacted with coplanar measuring probes through openings in the top layers for testing. Connections inside the LTCC substrate are also manufactured as shielded coplanar lines.

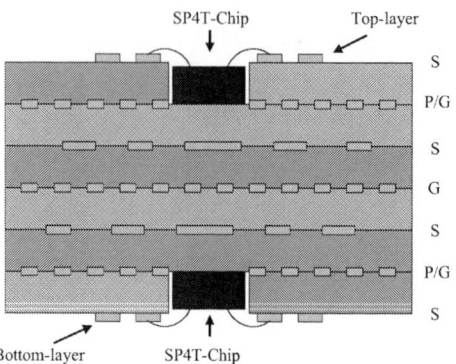

Figure 1: Principal design of the LTCC stack (S-Signal layer, P/G -Power/ground layer)

To achieve a good matching to the 50 Ω environment the signal lines had to be printed very precisely. This has be done using 500 mesh screens with a 15 µm polymer film on it. Four of the conductor layers are signal layers, the others act as ground/power planes. For these layers 400 mesh screens were sufficient.

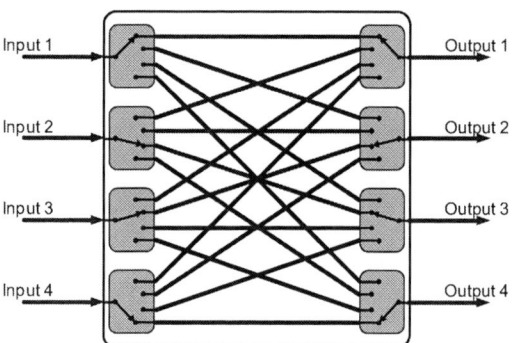

Figure 2: Principal electrical structure of the switch unit (4 incoming ports, 4 outgoing ports)

The lamination was done in two steps. After printing, drying and stacking a prelamination has been carried out. The LTCC stack was covered with mylar sheets on both sides for this process step. The mylar sheets had cut-outs corresponding to the cavities.

An uniaxial press was used for the prelamination with a temperature of 70 °C, and a pressure of about 30 Bar. The duration of this step was 2 minutes (Figure 4).

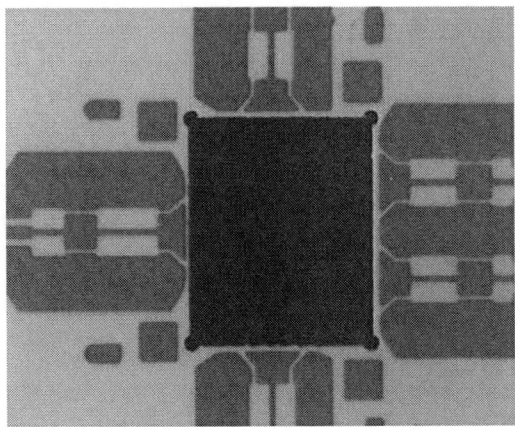

Figure 3: Cavity on top layer with signal lines and matching structures after printing

The second lamination step was in principle the standard lamination process for the 951 tape system (p=200 Bar, ϑ=70 °C, t=10 min.). For this process step an isostatic press was used and silicone mats and fibre-glass reinforced epoxy sheets (FR4) without cut-outs were added (Figure 5).

This type of lamination showed the best shape of the cavities with a high dimensional accuracy, steep edges and no cracks.

Firing was done using a muffle furnace and a standard firing profile with 875 °C peak temperature (10 minutes) and an overall duration of 443 min.

To enable brazing of the lids a solder layer was added in a postfiring step. Then the substrates were cut to size using a diamond wafer saw .

The pin diode switches were attached with conductive epoxy, cured and ribbon-bonded afterwards (Figure 7). These processes were carried out by one of our partners in the consortium, RHe Microsystems in Radeberg.

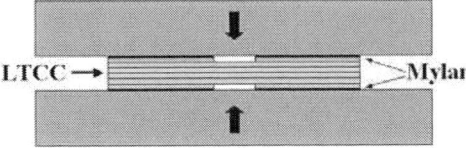

Figure 4: Uniaxial prelamination of the LTCC stack. The cut-outs in the mylar sheets were made by laser cutting.

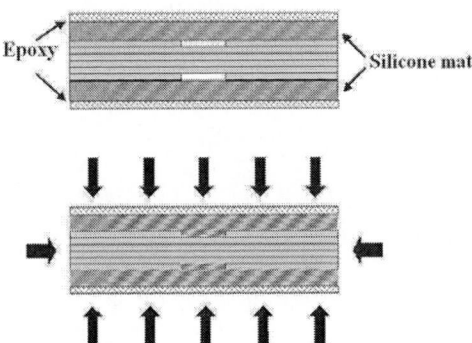

Figure 5: Isostatic lamination using silicone mats and epoxy sheets on both sides of the stack.

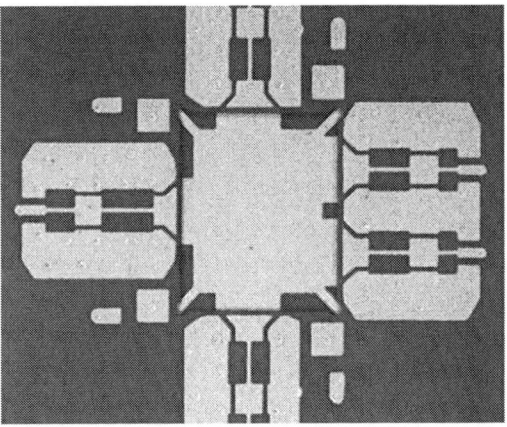

Figure 6: Close up of the top layer with empty cavity and signal lines (fired). The planar structures in the signal path near the chip enable the mode conversion between the microstrip wiring on the chip and the GCWG-environment on the LTCC substrate.

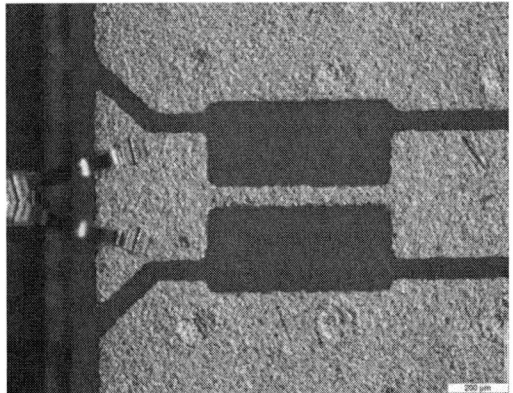

Figure 7: Signal line with matching structure and double ribbon-bond

The active elements on the bottom and the top of the circuit were covered by lids. These lids are also LTCC stacks with cavities. This design was

497

chosen to avoid thermal mismatches. Due to their shape the cavities for the lids were laser cut.

Figure 8: Four lids on one substrate after laser cutting and laminating.

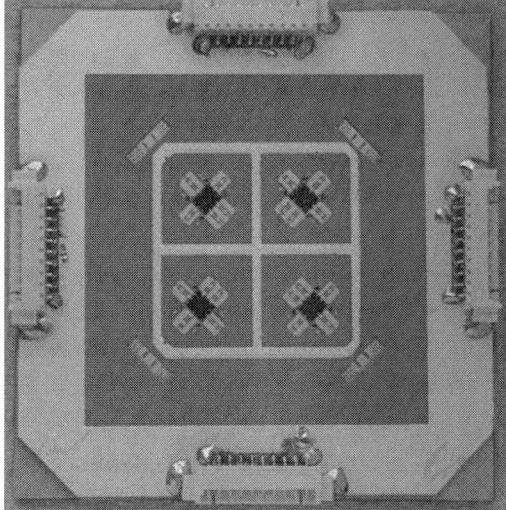

Figure 9: Complete switch unit without lid . The connectors at the borders feed the unit with DC (power supply and control voltages).

Performances of the 4x4 Switch Matrix

The following goals had to be achieved by the switch matrix:

- low insertion loss
- low input return loss
- high isolation between paths
- low weight
- low volume
- bandwidth 5 GHz

During the duration of the project the design was optimized step by step and the performance was considerably enhanced (compare [3]). Figure 10 shows some actual measuring values of insertion loss and input return loss.

The measurements were done using a network analyzer Agilent N5230A (10 MHz-40 GHz) with coplanar probes of 200 µm pitch.

The insertion loss in the target frequency range was about 5 dB, the input return loss between 15 and 20 dB (Figure 10).

The overall dimensions of the actual circuit are 56x56x4,8 mm^3 (with lids) and it weighs about 20 g. Regarding the fact, that commercial switch units have weights over 1000 g and volumes over 1000 cm^3 the savings are considerable.

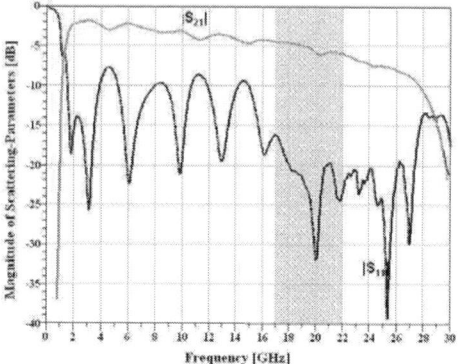

Figure 10: Insertion loss (S$_{21}$) and Input return loss (S$_{11}$) of one path of the switch unit in the frequency range up to 30 GHz. The desired working range is marked.

Conclusions and Outlook

The investigation showed that the LTCC technology allows considerable reductions of weight and volume for space qualified microwave devices.

In the second part of the KERAMIS project an on orbit verification (OOV) is planned. This means that the investigated circuits, amongst others the switch matrix, have to work on a satellite on orbit.

To make this possible another redesign is necessary. This will reduce the dimensions again due to changes in the input/output configuration. The input return losses are a field of further investigations and may be lowered for the flight exemplary. Concerning the insertion losses only minor improvements are possible due to the data of the pin diode switches and the fact that the signal always has to pass through two of these switches. The losses can only be lowered by refining the matching networks and reducing the signal path length.

The use of Kovar-lids instead of LTCC may enable additional weight savings.

Redesign and qualifying of the flight version of the switch matrix are in progress and will be done in the next months. First positive results were achieved during the vibration- and thermal-vacuum tests.

Figure 11: Two switch units (one with, the other without lid) on the test bench during vibration tests. Acceleration sensors are arranged around the substrates.

Figure 12: The DLR-Satellite BIRD was launched in 2001. A similar satellite is supposed to act as a test bed in orbit.

Acknowledgements

This work has been supported by the German Federal Ministry of Education and Research (BMBF, No. 50YB0313). We gratefully acknowledge valuable contributions from I. Koch G. Vogt, S. Humbla and D. Stöpel. The authors also wish to thank RHe microsystems for the technical support during assembly of the units.

References

[1] J. Müller et al. "Technology Benchmarking of High Resolution Structures on LTCC for Microwave Circuits", 2006 Electronics System Integration Technology Conference, Dresden, Germany.

[2] G. Reppe et al. "Development and Evaluation of Fine Line Structuring Methods for Microwave Packages in Satellite Applications", 15th European Microelectronics and Packaging Conference & Exhibition, June 12-15, 2005, Oud Sint Jan, Brugge, Belgium.

[3] J. Trabert et al. "High functional density low-temperature co-fired ceramic modules for satellite communications", 35th European Microwave Conference, October 3-7, 2005, Paris, France.

[4] R. Perrone, H. Thust, K.-H. Drüe "Progress in passive integration of planar and 3D coils by using photoimageable inks on and into LTCC", 2004 IMAPS, 37th Int. Symposium on Microelectronics, Long Beach, California USA, November 2004.

[5] R. Perrone et al. "Development and Evaluation of photodefined elements for microwave modules in LTCC for space applications", 15th European Microelectronics and Packaging Conference & Exhibition, June 12-15, 2005, Oud Sint Jan, Brugge, Belgium.

A New Integrated Waveguide Antenna using Multi-Layer Photo-imageable Thick-Film Technology

M. Henry[1], Benito Sanz Iszquierdo[2], Charles Free[1] John Batchelor[2] and Paul Young[2]

[1] Advanced Technology Institute, University of Surrey, Guildford, United Kingdom

(m.henry@surrey.ac.uk)

[2] Department of Electronics, University of Kent, Canterbury, United Kingdom

Abstract

This paper presents a new concept in antenna design, whereby a photo-imageable thick-film process is used to integrate a waveguide antenna within a multilayer structure. This has yielded a very compact, high performance antenna working at high millimetre-wave frequencies. Various demonstrators were fabricated, using up to 14 layers of photo-imageable material to form a variety of integrated antenna structures. Simulated performance data are presented with the results confirmed through practical measurement. Dimension data for the antennas are presented, together with a discussion of the critical aspects of the fabrication process and the need for a high degree of precision in the layer-to-layer registration.

Introduction

Millimetre-wave communication systems, whether for collision avoidance radar, robot control or more conventional wireless communications, require very compact, high performance antennas, giving relatively high gain. The advantage of the integrated waveguide approach discussed in this paper is that slot antennas, using the slot array principle, can be fabricated in a relatively small space, and closely integrated to the microwave circuitry [1, 2].

Various integrated antenna structures have been investigated, but they all make use of the dominant TE_{10} in rectangular waveguide. The advantage of using the dominant mode is that there is a unique frequency range over which only this mode propagates. It therefore follows that the field distribution in the waveguide is fixed and precise locations can be found for the radiating slots. A linear array can thus be formed by having a single waveguide and a series of slots, either in the broad or narrow faces of the guide. However, a single linear array has the disadvantage of producing a radiation pattern that is only narrow in the plane containing the slots, and is quite broad in a plane perpendicular to the waveguide. To overcome this problem we have refined the basic structure for form a folded waveguide, that provides a two-dimensional array of slots, and consequently gives a relatively narrow radiated beam in both planes.

Fabrication

The integrated antennas were fabricated from photoimageable thick-film material, with the layer printed onto an alumina base to gave rigidity to the structure. A schematic view of the cross-section of the basic waveguide structure is shown in figure 1.

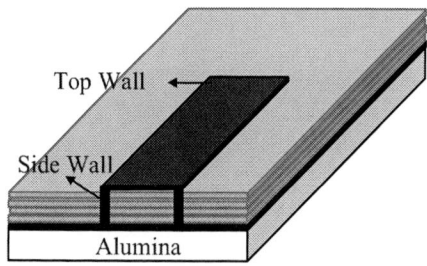

Figure 1: Schematic of cross-section of basic waveguide structure

The bottom (broad) face of the waveguide is formed from a conductor layer printed on the alumina. Successive layers dielectric layers are printed to build up the required height of the waveguide. Each dielectric layer is photo-imaged to form a vertical trench, which is subsequently filled with metal to form the side walls of the waveguide. Finally, the top layer of conductor is printed, and photo-imaged to form the radiating slots.

Typical dimensions for the cross-section of the guide are: width (a) = 1mm; height (b) = 0.12mm.

It should be noted that some care is needed in registration to ensure that each dielectric layer is photoimaged to provide uniform sidewalls to the guide. It was found necessary to use a Quintel mask aligner providing registration accuracies to within ±1 micron.

Folded waveguide antenna

The concept of the folded waveguide antenna has been established at lower frequencies, using more conventional fabrication techniques [3]. Since it is the width (a) of a waveguide that establishes the dominant mode, this parameter was unchanged as the linear guide was transformed into a folded version. In essence, folding the sides of the guide underneath the central part formed the folded structure investigated. This arrangement is shown schematically in figure 2, and it can be seen that the overall width of the structure is reduced, whilst the height is doubled. Clearly this structure will offer an advantage in terms of reduced surface area; this will be significant for highly integrated millimeter-wave circuits, where substrate area is at a premium.

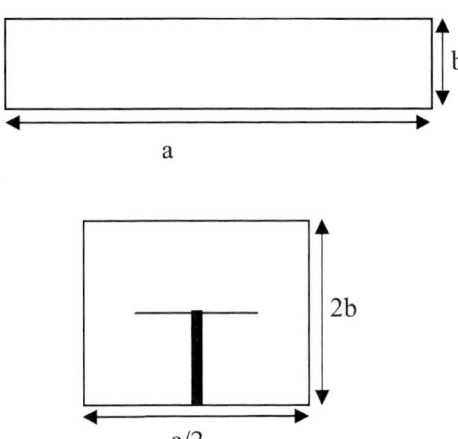

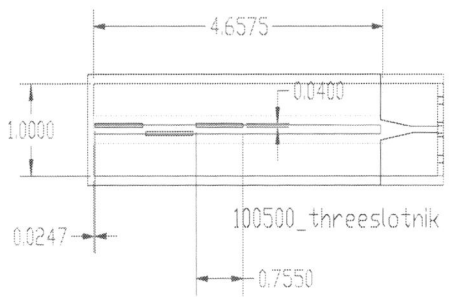

Figure 2: Folded waveguide

Test structures

The dimensions of one of the folded waveguide antennas investigated are shown in figure 3.

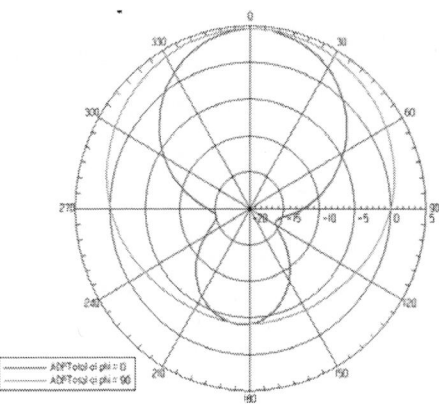

Figure 4: Simulated radiation pattern of a three-slot folded waveguide antenna.

A photograph of the fabricated four slot and three-slot folded waveguide antenna is shown in figure 5. The antenna was constructed using the techniques previously described, with silver as the conductor metal.

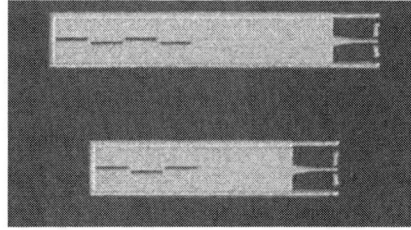

Figure 5: Photograph of the fabricated four slot and three-slot folded waveguide antenna.

The photograph of the intermediate layer in the folded waveguide is shown in Figure 6.

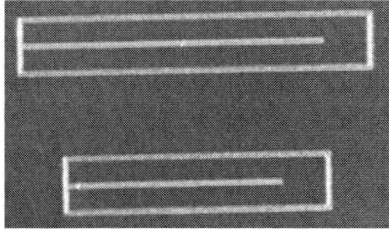

Figure 6: Intermediate Layer in the integration of folded waveguide

As an alternative to the folded waveguide approach we also considered a coupler-fed, dual waveguide antenna. Figures 7 shows the essential features of the layout, and figure 8 the dimensions of the actual test circuit.

Figure 3: Dimensions of a three-slot folded waveguide antenna.

The performance of the antenna shown in figure 3 was simulated using a 3-D electromagnetic simulator (HFSS). The resulting radiation pattern is shown in figure 4. It can be seen that the antenna has well defined radiation characteristics, with a relatively narrow main beam and a good front-to-back ratio.

501

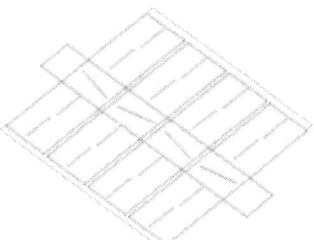

Figure 7: Layout of coupler antenna

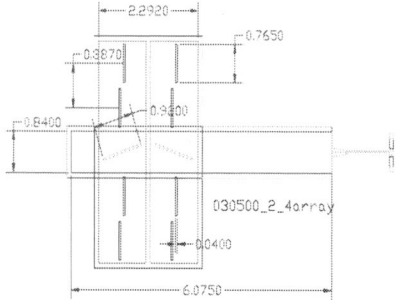

Figure 8: Dimensions of coupler antenna

The antenna is fed from a coplanar input, and thence into a coplanar-waveguide transition. Two angled (non-radiating) slots are formed in the feeding waveguide, so as to couple energy into the two parallel waveguides containing the radiating elements. The separation and angles of the coupling slots ensure that the parallel waveguides are fed in phase. Thus we have a 2 x 4 array of radiating elements, formed by the 8 axial slots in the two parallel waveguides.

The simulated radiation pattern for this coupler antenna is shown in figure 9. It can be seen that the antenna provides a well-defined radiation pattern, with a 3dB beamwidth of around 60°.

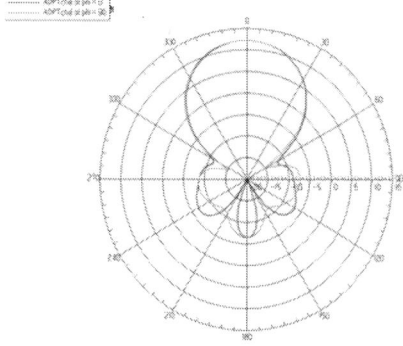

Figure 9: Simulated radiation pattern for coupler antenna.

A photograph of a fabricated 4 x 4 coupler antenna is shown in figure 10.

Figure 10: Photograph of fabricated coupler antenna.

The measured return loss of fabricated 3-slot and 4 –slot coupler antennas were measured using a vector network analyzer, and the results are shown in figures 11 and 12. It can be seen that an excellent return loss approaching 40dB is obtained close to 75 GHz.

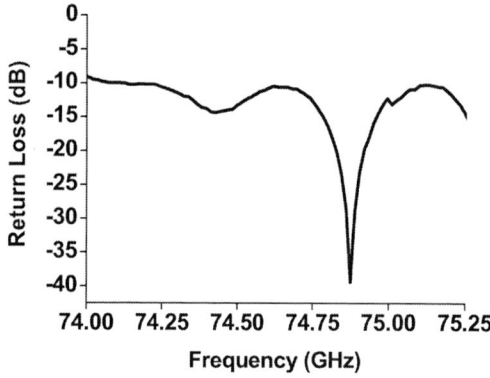

Figure 11: Return Loss plotted for a 3 slot folded waveguide antenna

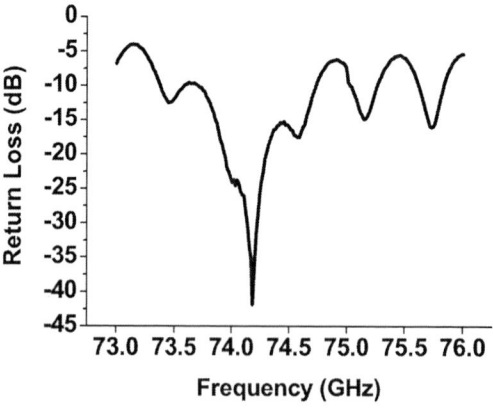

Figure 12: Return Loss plotted for a 4 slot folded waveguide antenna

Fabrication quality

Clearly with this type of antenna operating at very high frequencies, with very small wavelengths, the quality of fabrication is a key issue. In this case it is important that the radiating slots can be formed with precise dimensions and good quality edges. To demonstrate the quality of the fabrication process, an enlarged version of one of the radiating slots is shown in figure 13.

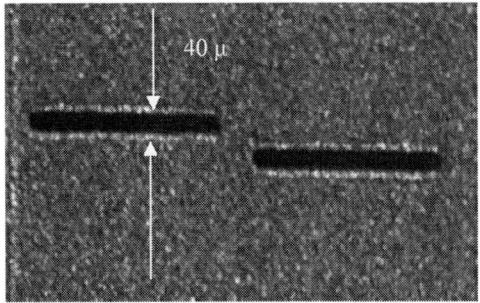

Figure 13: Enlarged view of one of the fabricated slots.

It is clear from figure 13 that the photoimageable process is capable of giving the high quality that is required for millimeter-wave devices. As a further bindication of this, figure 14 shows an enlarged view of the coplanar feed point. The circuit was fed from 100 μm GSG coplanar probes having an impedance of 50Ω. To maintain this impedance on the circuit, the signal line had to have a width of 22 μm, with a spacing between the signal line and ground pads of 30 μm. It can be seen from figure 14 that the process was able to resolve these dimensions with reasonable quality.

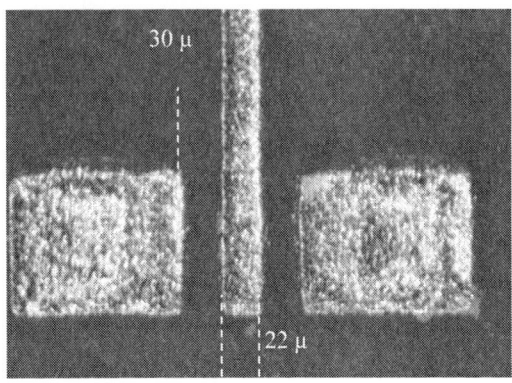

Figure 14: Enlarged view of the coplanar feed point on the fabricated circuit.

As a further indication of the quality of the fabrication process a miniature branch line coupler was fabricated, and a view of this circuit is shown in figure 15.

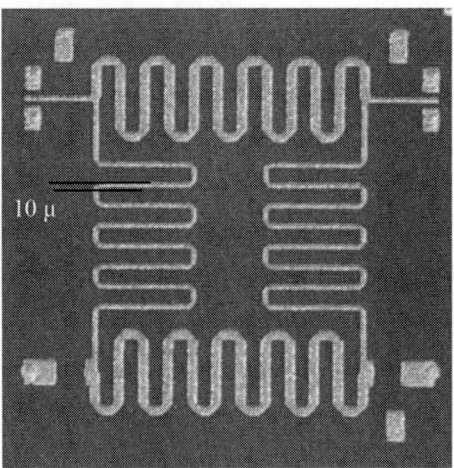

Figure 15: Miniature 3dB branch line coupler

The overall dimensions of the coupler shown in figure 15 are around 1.5mm x 1.5mm, with a minimum line width of 10μm. The fabricated lines are well defined with reasonable quality edges, this being an essential requirement for high millimeter-wave frequencies where the skin depth is less than 0.2μm.

Further measurements

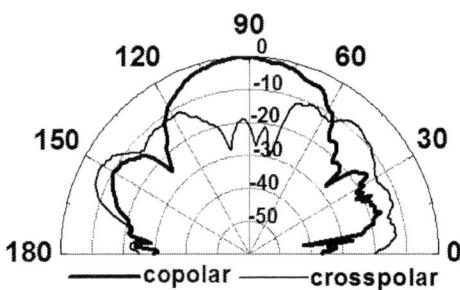

Figure 16: Typical measured radiation pattern for one of the integrated antennas.

A well defined pattern was obtained, with an gain around 5 dBi. Obviously it was important to obtain some indication of the efficiency of the antenna. Although the return loss was very good, this may have been due to excessive losses within the antenna structure, particularly in the dielectric-formed waveguide. So sections of embedded folded waveguide were fabricated to allow insertion loss measurements to be made. Figure 17 shows an array of waveguide test sections.

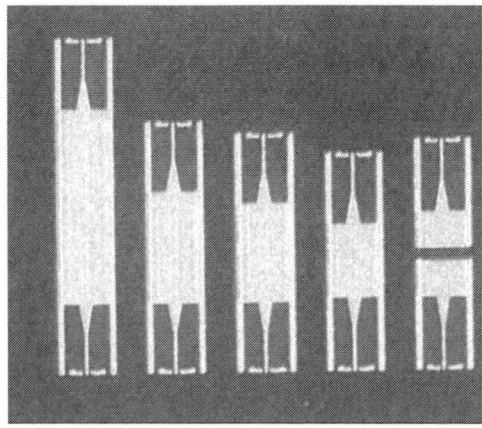

Figure 17: Array of integrated waveguide test sections.

Each section was connected to the VNA in turn, and insertion loss data obtained over a range of frequencies from 50 GHz to 80GHz. By computing the ratio of the difference in insertion loss to the difference in length, we were able to deduce the loss per unit length (dB/mm). The results are shown in figure 18.

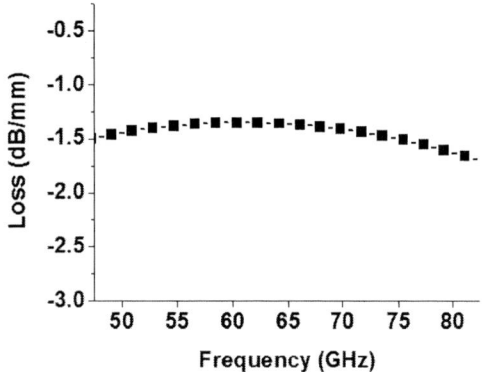

Figure 18: Measured loss in the integrated folded waveguide.

It can be seen that the loss increases slightly with frequency, with a value of around 1.5/mm at the working frequency of the antennas. This suggests a total loss of around 3-4 dB in the folded waveguide sections of the antennas under test. This is not unreasonable, given that the waveguide is filled with dielectric, and indicates that a significant portion of the input energy is being radiated, rather than being dissipated within the antenna structure.

Conclusions

(1) It has been shown that reasonably efficient antennas can be made at high millimeter-wave frequencies using slotted waveguide techniques.

(2) The fabricated antennas have demonstrated that photoimageable fabrication techniques can provide the necessary line quality for very high frequency applications.

(3) It was found to be possible to successfully integrate three-dimensional structures within multilayer package, using photoimageable thick-film, although relatively sophisticated alignment techniques were needed to provide the required registration between layers.

References

[1] Manju Henry, C.E. Free, B. S. Iszquerido, J. Batchelor, and P. Young, "Integrated Slotted Waveguide Antenna Array Using Photo-Imageable Thick-Film Techniques for E-band operation" Antennas and Propagation Symposia, APSYM, pp. 135-138, Kochi, India, 14-16 December 2006.

[2] Manju Henry, and C. E. Free, "Photo-imageable thick-film circuits up to 100 GHz, 39th International Symposium on Microelectronics IMAPS 2006, pp. 230-236, San Diego, California, USA, November 2006.

[3] N. Grigoropoulos and P.R. Young, "Compact Folded Waveguide", 34th European Microwave Conference, pp. 973-976, Amsterdam 2004.

Microstrip and Wave-guide SMT Package up to 60 GHz (MWgSP)

Carlo Buoli, Paolo Bonato, Luigi Negri, <u>Fabio Morgia</u>

Nokia Siemens Networks S.p.A .Strada Padana Superiore Km158 20060 Cassina de' Pecchi - MILANO

Phone: +39 02 24376896 Fax: +39 02 24376375 e-mail: Carlo.Buoli@siemens.com

Abstract

A new SMT package with an embedded broadband microstrip-to-waveguide transition (MWgSP) has been studied and tested for low-cost microwave applications. It employs soft lead-free substrate multilayer build-up, with an internal thick metal plate: the waveguide flange surface, the input/output and biasing pads lie on the same bottom package surface. This makes easier the manufacturing, compliant with a standard SMT process. This package allows a better performance, compared to a traditional package components line-up: it is more reliable and it doesn't need fine tuning. It enables to reach wide frequencies bands operation up to 60 GHz with good chip devices heat spreading on a wider thermal conductible area.

Key words: Microstrip and Wave-guide SMT Package (MWgSP)

Introduction

Nowadays microwave subsystems should be always smaller in size, better performing and easier manufacturing for costs saving products and low environmental impact in radio-link equipments. All these characteristics are well satisfied by new microwave SMT package [1]. The innovation is characterized to have a hermetically interfaced microstrip/waveguide transition able to be soldered like a common surface electrical connection pad in traditional SMT package (Figure1).

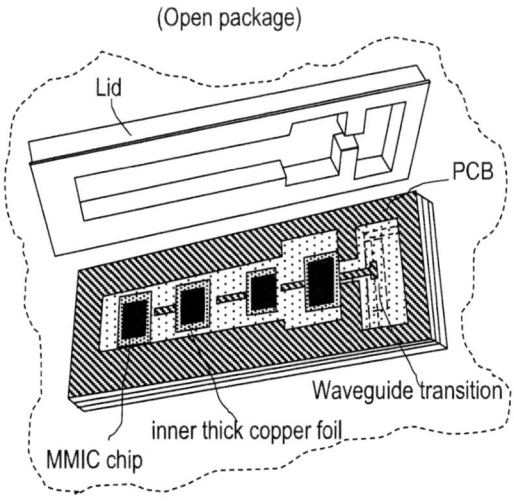

Figure1: Microstrip/Waveguide Transition Package and Lid Drawing.

The new SMT package, with an embedded broadband frequency working microstrip-to-waveguide transition (MWgSP), can house chip & wire components and matching networks in order to obtain the best mass reproducibility and microwave response [2] [4]. In this paper physical lay-out, technological approach, simulated and measured electrical results will be explained and showed.

Package Build-Up Description

The package is obtained by milling a slot in the thick metal plate and subsequently building up the layers with a standard PCB process [3], where thick metal plate is the second layer. In this way, the slot results filled with "prepreg". Figure 2 a) and b) show the worked layers before and after their pressing.

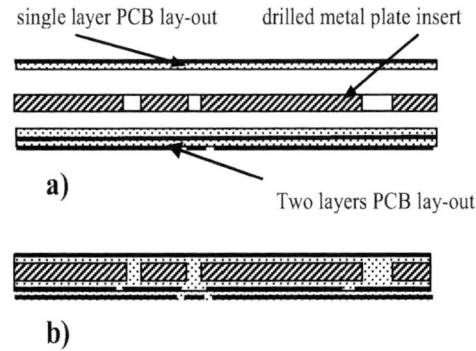

Figure 2 : Layer Build-Up

The waveguide hole milling, the metallization and final removing metal deposed excess steps are shown in Figure3 a), b) and c) respectively.

505

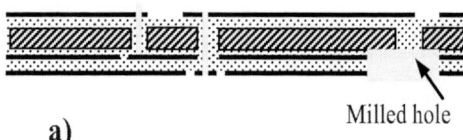

a)

Figure 3 a) : Waveguide Milling Step

In particular, in Figure3 c), the transition has been made removing metallization corresponding to the prepreg-filled slot, carving also a small portion of metal insert without any restrictive tolerance.

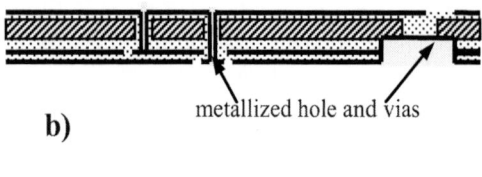

b)

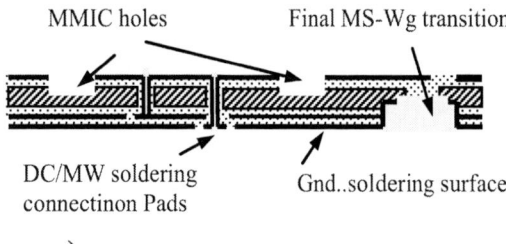

c)

Figure 3 b) c) : The Metallization and Final Removing Metal Deposed Excess Steps

Low frequency and DC connections are performed through via-holes and pads, whereas input/output microwave signals are connected to the waveguide through an embedded microstrip-to-waveguide transition. Package electrical connections pads, ground and waveguide flange are on the same bottom plane and the assembly of the package into the main circuit can be easily performed as a standard SMT component.

The thick metal plate, where the devices (i.e. MMICs) are mounted, is the ground of microstrip and at the same time, it works as a very good thermal sink, due to the heat spreading on the thick and large area.

Figure 4 shows the package soldered on the main PCB in which the waveguide aperture joins the corresponding package one.

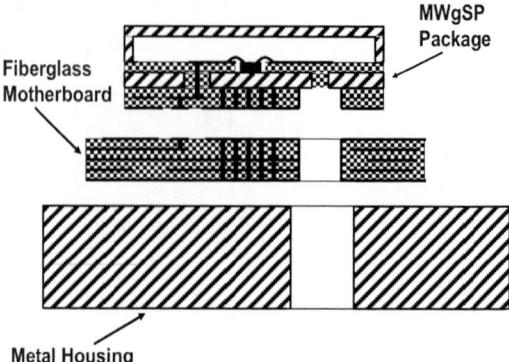

Figure 4 : Package Mounted on PC Motherboard Explosion

Chip On Board Mounting

The package can be used in TX/RX front-end in which more microwave functions can be located in very small size. The PCB process produces multiple packages array, generally 10-20 pieces on single rectangular PCB.

The array packages (Figure 5) are more suitable for automatically mounting chip-form MMIC, functional tests, quality inspection and lid-closure can be finally performed before packages separation.

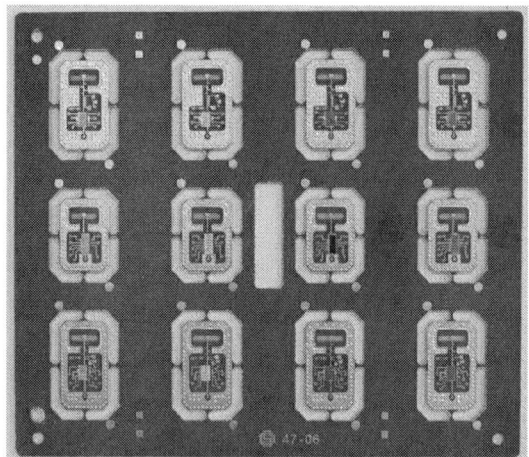

Figure 5 : 38 GHz Package Array

Figure 6 and Figure 7 show the photograph of 38 GHz transmitter and receiver packages; Figure 8 shows 18 GHz receiver top and bottom side photograph. Chips and alumina filters are automatically mounted in the packages and they are ready for testing before closure and separation from the array.

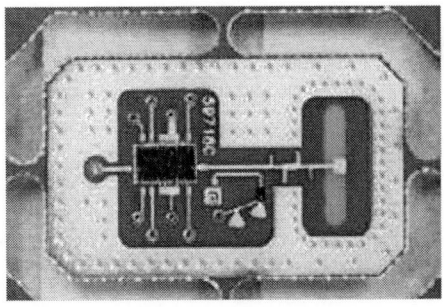

Figure 6 : 38GHz Chip Mounted TX

Figure 7 : 38GHz Chip Mounted RX

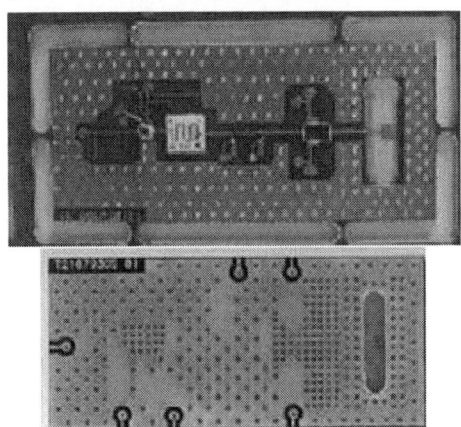

Figure 8 : Top-Bottom 18GHz RX with Chip Devices and Mounted Alumina

Advantages

The use of embedded waveguide transition and chips devices line up, instead package MMICs ones, gives a great advantage in terms of better microwave performance and repetitiveness without any tuning necessity during test process. No microwave absorber along the active devices is needed because internal transverse section size, delimited by the ground on the microstrip upper plane, can be made under cut-off frequency operation and because chip & wire technology and substrate losses minimize radiated energy. High amplification gain is therefore achievable because input-output signal are decoupled enough reaching very large and flat frequency band response. Great

importance, for avoiding internal radiation and loss, is to match without discontinuities in/out microstrip line with external waveguide. The SMT embedded waveguide transition represents the best and the cheapest way to do this. Transmitter and receiver may lie on the same package which carries the two waveguide transition and local microwave oscillator. Double front end solution doesn't limit the package manufacturability and reliability. The choice depends on many factors one of this is the external waveguides interface mechanical matching.

Simulation and Experimental Results

The microstrip-waveguide transition and the SMT joined PCB slot transformer (Figure1) has been simulated, by means of HFFS electromagnetic simulator [5], in the 18, 38 and 80 GHz frequencies band.

Despite the loss factor, the insertion loss of the microstrip-to-waveguide transition is contained below 0.5 dB; the transition is broadband and the distance to the device shall be short. The lid is about λ/4 far from the feeder of the transition; its height doesn't effect the other device performance. A low loss upper substrate could be used but poor reliability mixture with lower fiberglass substrate and metal insert has been observed. Up to now, such packages, with only fiberglass (FR4 or ISOLA 410) substrates have been well produced and mounted without any problems largely overtaking the quality standard mill tests (MIL-STD-202G – Life test 108B and humidity test 103B).

Transmitter and receiver working in 18 and 38 GHz frequencies bands have been developed to test electrical performance and SMT manufacturing process compliance. In order to verify the limit pick and place compatibility a wider than needed 18 GHz package has been manufactured. Figure 9 shows the schematic representation made for HFSS simulation.

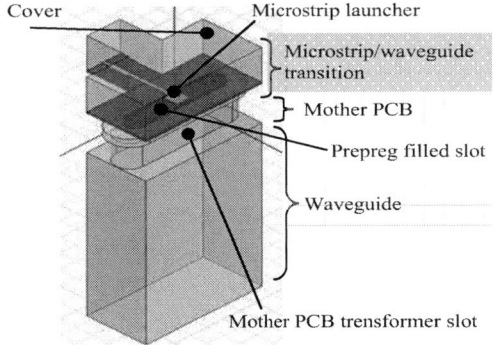

Figure 9 : HFSS Schematic Structure View

Figure 10 shows the waveguide RX input return loss measure. Figure 11 shows 18 GHz receiver noise figure and RF/IF power gain measure.
In Figure 12, 38 GHz microstrip/waveguide insertion and return losses simulation are plotted;

507

Figure 13 plots the return loss of entire RX and Figure 14 show the measures related to the insertion loss of the transition.

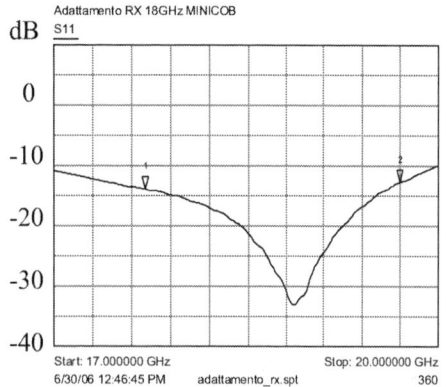

Figure 10 : 18GHz Waveguide RX Input Return Loss Measure

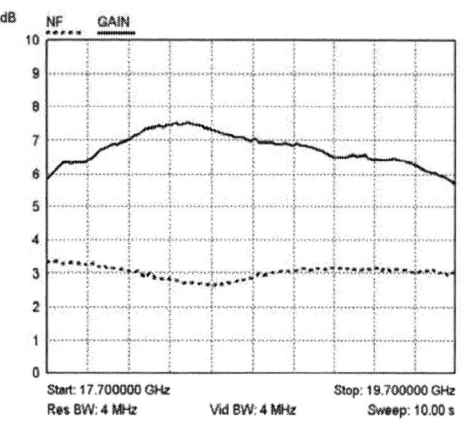

Figure 11 : 18GHz RX RF-IF Power Gain and Noise Figure

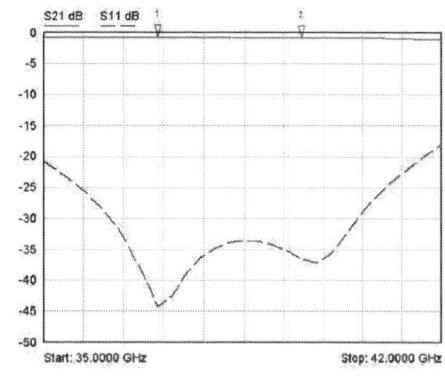

Mkr	Trace	X-Axis	Value	Notes
1 ▽	S21 dB	37.0000 GHz	-0.74	
1 ▽	S21 dB	39.5000 GHz	-0.84	

Figure 12 : 38GHz (3d HFSS) Microstrip to Waveguide Transition Simulation

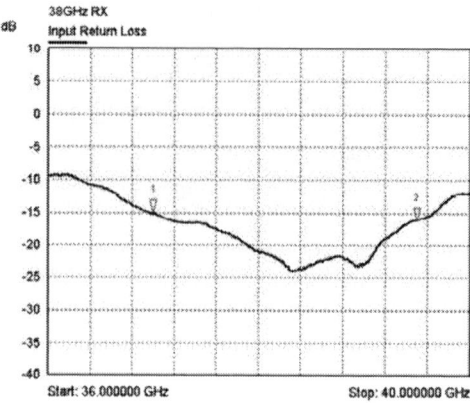

Figure 13 : 38GHz RX Waveguide Side Return Loss Measure

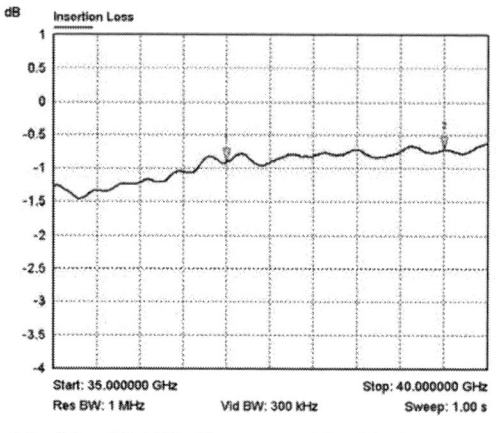

Figure 14 : 38GHz Microstrip/Waveguide Transition Insertion Loss Measure

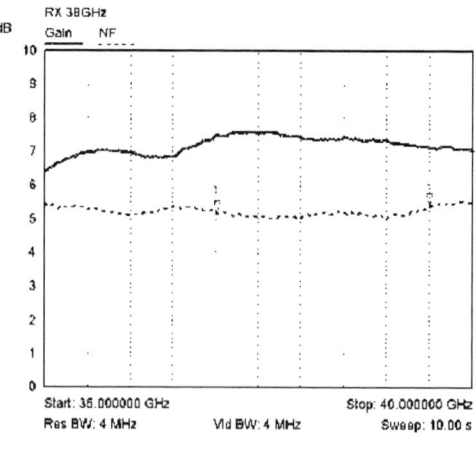

Figure 15 : 38GHz RX Noise Figure and RF-IF Power Gain Measure

In Figure 15 and Figure 16, 38 GHz RX noise figure/gain and TX saturation output Power measures are plotted.

508

Very good electrical performance without any circuit tuning or adjusting has been obtained. The mounted input-output MMIC characteristics are practically unchanged using this kind of package.

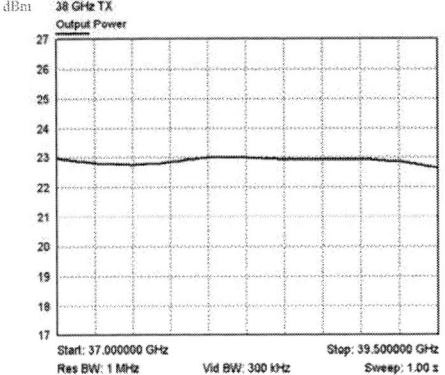

Figure 16 : 38GHz Saturation Output Power Measure

In Figure17, 77 GHz microstrip/transition S parameters simulations are plotted. Insertion losses are less than one dB in 15% frequency bandwidth.

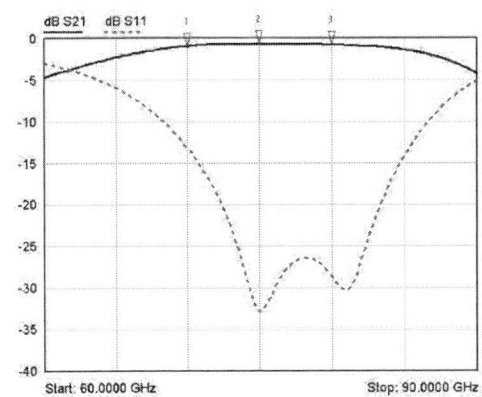

Mkr	Trace	X-Axis	Value	Notes
1	dB S21	70.0000 GHz	-0.92	
2	dB S21	75.0000 GHz	-0.67	
3	dB S21	80.0000 GHz	-0.79	

Figure17: 77GHz Microstrip/Waveguide Transition Insertion Loss and Return Loss Simulations

In the following picture (Figure18), the 38 GHz TX shape of the motherboard is shown; in particular, connections pads, slot towards waveguide and thermal vias for dissipation can be noticed.

In Figure19, the final TX and RX assembly is shown.

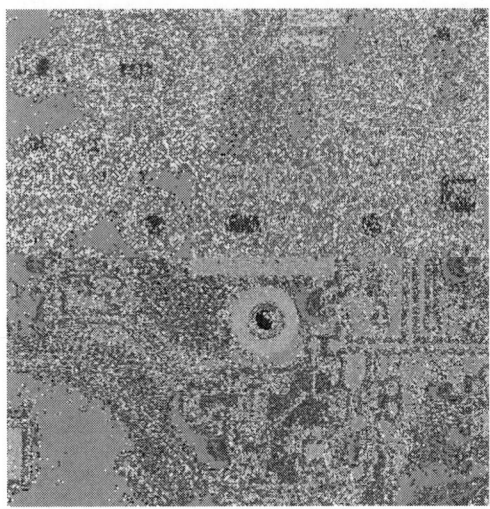

Figure 18 : 38 GHz TX Shape on Motherboard

Figure19 : Final TX (Left Side) and RX (Right Side) Assembly

Thermal Dissipation

The inner copper metal plate ensures a good heat spreading because enlarges the chip MMIC columnar heat flux. This is very important in order to decrease the overall thermal resistance greatly dues to the mother board. The package thermal resistance is about 0.16 °C x cm^2 / Watt, while the overall thermal resistance of 1 °C x cm^2 / Watt has been measured in the following conditions:

- Package is mounted (SMT) on the mother PCB
- PCB is 8 layers and has 1.6 mm thickness.
- Package metallic shape on PCB surface is reported in all the inner metal layers.
- All PCB shapes are metallized via holes connected to the ground (0.25 mm hole diameter – number 9 hole per square centimeter)
- PCB ground is mounted on a heat sink.

It's possible to increase thermal conductivity by protruding package metal insert out of fiberglass board (wings) and connecting it to a heat sink directly. In this way, thermal resistance is reduced by five times.

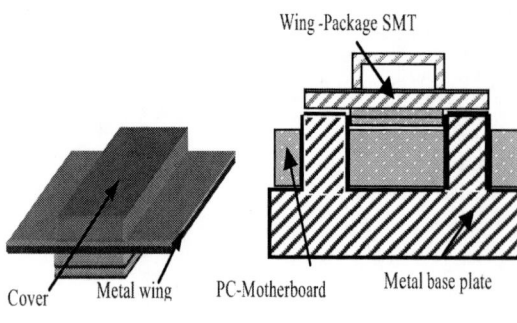

Wing-Package SMT   Complete mounting example

Figure 20 : SMT Wing Package.

In this way the package can be mounted with standard SMT process and metal wings screwed to a heat sink base plate (Figure 20). Copper wings are flexible enough to guarantee unstressed material.

Conclusion

This microwave SMT package reduces any kind of particular mother board manufacturing requirement because all the microwave functions are closed into the package. The very high frequencies up to millimeter waves can be easily managed making waveguide interface by simple mother-PCB metallized hole and avoiding any restrictive tolerance in SMT assembly process.

Many circuits can be mounted inside and this makes MWgSP package suitable for a complete microwave multifunction module. This is very useful or necessary in SDH microwave radio link equipments where linearity requires high amplification gain line up, especially in the TX chains.

A SMT single board complete radio without microwave tuning can be easily manufactured reducing costs and increasing productivity. Performance, quality and reliability are at very high level thanks to the process repetitiveness and the very good capability to increment thermal heating spread trough the mother PCB.

Acknowledgements

The authors wish to acknowledge Antonella Trombetta, Stefano Fusaroli and Antonio Cifelli of Nokia Siemens Networks S.p.A for the assistance and support during the implementation process.

References

Patent

[1] Patent pending

Articles from conference proceedings

[2] C.Buoli, V.M Gadaleta, T.Turillo, A.Zingirian, *A broadband microstrip to waveguide transition for FR4 multilayer PCBs up to 50 GHz*, 32th European Microwave Conf, Milan, Italy 23-27 September 2002

[3] C Buoli, G. Biffi, T. Turillo, A Zingirian, *Thick metal plate insertion make FR4 multilayer board a simple carrier for RF power circuits*, 31th European Microwave Conf. London, UK 23-27 September 2001.

[4] C. Buoli, S. Fusaroli, V. M Gadaleta, F. Morgia, T. Tommaso, *Microstrip to waveguide 3 dB power splitter/combiner on FR4 PCB up to 50 GHz*, 35th European Microwave Conf., Paris 2005.

Software

[5] Ansoft HFSS™ is the software used for electromagnetic simulation.

Influences of the Layout on the Lifetime of Direct Copper Bonding Substrates (DCB)

Michael Günther[1,2] and Klaus-Jürgen Wolter[2]

[1] Robert Bosch GmbH, Stuttgart, Germany

[2] Electronic Packaging Laboratory, Dresden University of Technology, Germany

Phone: +49 711 811-38511, Fax: +49 711 811-511-1704, michael.guenther3@de.bosch.com

Abstract

With the development of power modules that are stressed by increasing temperatures, heat conduction is becoming more important for the lifetime of these circuits. Typical substrates are based on ceramics. Due to their poor heat conductivity it makes sense to reduce the thickness of these materials. The probability of fracture during the desired lifetime is increased by this reduction. Therefore, strategies are necessary to diminish or control the risk of fracture. A typical substrate for a power module is manufactured as a sandwich of conductive layer, insulating ceramics, and a second conductive layer. Usually, DCB (direct copper bonding) or AMB (active metal brazing) technology is used to join these materials. By masking and etching technologies, an electrical circuit is build up from the original metal layer. During thermal cycling tests, these substrates can fail by cracking the ceramics. This kind of crack is called "conchoidal fracture", because the parts are reminiscent of a conch. Herein it is shown how a selected copper-ceramic-copper substrate can be destroyed by passive temperature cycling that is typical in automotive applications with focus on the geometry of the substrates. It is obvious that the larger the dimension of the substrate, the larger the mismatch of the absolute thermal expansion. So with increasing dimensions, the resistance against thermal cycling is decreased. Interestingly, it was also found that a decreasing dimension of the substrate causes a decrease of resistance. Hence an optimum conductive pad dimension exists for a special combination of ceramic and copper thickness. Furthermore, the length to breath ratio should be taken into account.

Key words: power module, conchoidal fracture, direct copper bonding, geometry of circuit paths

Introduction

Typical modules for automotive applications are separated in two parts; one for logic circuits and one for power circuits. Herein the focus is on the power module. The modules are stressed by temperature changes. These could be originated in temperature loads from the environment and the dissipation loss of the mounted power assemblies.

To resist this temperature level the substrates are usually made of ceramics that is covered with copper layers. If joining technologies like DCB (direct copper bonding) or AMB (active metal brazing) are used, it is necessary to cover both sides of the ceramic sheets with copper to minimise the deformation. To get the structure of an electrical circuit the substrate is etched after masking. The substrates are mounted on a heat sink that can be a block of metal or a cooler. A cross section of this typical, simplified assembly is shown in Figure 1.

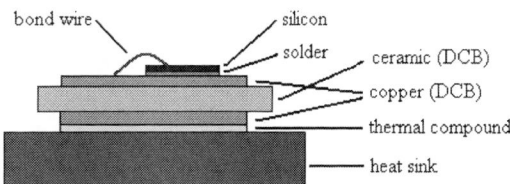

Figure 1: Cross section of a typical power module.

Real components show a more complicated structure as Figure 2 demonstrates. Here, an electrical circuit is built up after etching a copper structure by soldered components and bonding wires.

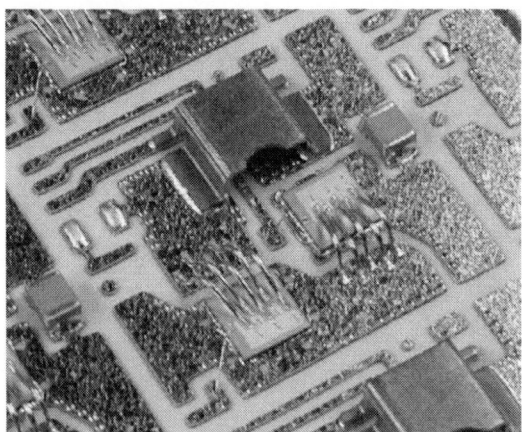

Figure 2: Typical structure of a copper on a DCB substrate after etching, soldering and bonding.

Failure of Power Modules

With the request of miniaturisation, higher temperatures and higher power density, which leads to more dissipated energy, the loads on the assembly increase. For better heat conduction, thinner ceramic sheets with higher fracture strength are desired. This does not influence the thermal mismatch of about 10 ppm/K for the combination copper and alumina but leads to a higher risk of fracture because the copper to ceramic ratio changes.

The use of lead free solder materials is preferred by law. These materials have different relaxation behaviour, so forces can be conducted better from the die to the substrate and from the substrate to the heat sink.

These two aspects lead to a higher stress in the ceramic material and a displacement of the failure locations from the joining technology to the substrate. Here, a failure of bond wires or stressed dies in packages should not be taken into account.

By the mismatch of thermal expansion, the heat conduction of power modules can fail at three locations:

- The solder connection between die and substrate.
- The ceramic component in the substrate.
- The connection between substrate and the heat sink.

The crack in ceramics can be generated by thermal cycling tests. This kind of crack is called "conchoidal fracture". If a part is broken, the two pieces are reminiscent of a conch.

Figure 3 depicts an example for this kind of fracture. The crack starts near to the copper edge at the ceramic surface and grows in the ceramic material. The crack direction changes if a certain depth in the material is reached, due to a modification of crack opening mode.

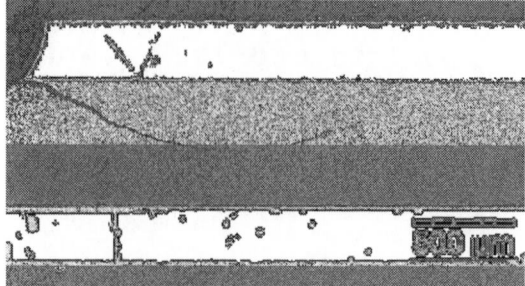

Figure 3: Cross section of a DCB substrate with a typical conchoidal fracture in the ceramics.

All further investigations describe the influence of the geometry on this failure.

Selection of Samples

It is already known that several influences determine the lifetime of power substrates. There are for example several manufacturing parameters, the metal to ceramic thickness ratio, the geometry of the etched edge, the thermal behaviour of the materials, and of course, the material properties, like the thermal expansion or the fracture tension. Usually these parameters are selected by the application, by cost and by manufacturing and/or handling reasons. Therefore, only geometric possibilities remain to influence the risk of fracture.

A material sandwich with a copper-ceramic-copper thickness of 0.3 mm – 0.63 mm – 0.3 mm was chosen for the experiments. The layers are joined with the DCB-technology as described in former patents [1]. The conductive pads are built up by etching the copper layers, whereas the geometry of the front- and backside are identical. This leads, after temperature cycling, to a conchoidal fracture that is located only at one side of the samples. This behaviour was shown before and is caused by production reasons [2].

The structures have a maximum dimension of 21 mm and a minimum dimension of 600 µm. All samples have to be separated to prevent an influence of one sample on another. To exclude an influence of the cutting edge [3], a spacing of not less than 2.2 mm between the copper layer and the cutting line is left.

Test Procedure

The samples were stressed by a temperature cycle test (150°C, -40°C, 30', 30'). To accelerate the expected damage, a temperature shock test with this large temperature amplitude is selected. This means that there is the risk of change in failure mode [2]. The aim, however, is a direct comparison of the copper geometry.

Sample storage should be executed carefully. To avoid influences on proximate samples, all specimens are separated and fixed on a metal grid. Thus, a mechanical damage by moving the samples in the two chamber furnace is avoided.

Furthermore, the testing method has an influence on the crack propagation. It is not possible to interrupt temperature cycles e.g. for non-destructive testing of a possible fracture because the strain hardening of the copper is interrupted and during this break there is enough time for the copper material to reduce the state of stress as known from similar assemblies [4]. This basic behaviour was tested on the investigated system by measurements of bending. It has to be expected that the relaxation behaviour depends on the geometry. Therefore, the result of the following temperature cycles is not the result of the temperature cycles that would be done without a break.

Samples are stored in a temperature chamber that is only opened sample removal. These samples were inspected by scanning acoustic microscopy to identify areas that are destroyed. This is illustrated by a copper pad with quadratic dimensions in Figure 4. The cracked area is coloured white, the area that still provides a good heat conduction is black.

Figure 4: Ultrasonic picture of a cracked DCB substrate. The cracked area is white, the intact area is black.

It has to be mentioned that not all identical specimens crack exactly at the same time although specimens with the same manufacturing and storage history are tested. Due to inhomogeneities in the air flow of the furnace the thermal stress on the samples is slightly different. Furthermore, the statistical behaviour of the ceramics with a Weibull's slope of $m_{corr}=11.8$ [5] leads to a distribution of the crack occurrence in reference to the number of temperature cycles.

Results

The absolute dimension has an influence, because the dimension of the pad determines the absolute thermal mismatch. In Figure 5 four sets of three different quadratic samples are compared after 200 and 300 temperature cycles. All small samples are destroyed, whereas the middle-sized samples only show small cracks at the corners. The large samples show a different behaviour due to the statistical fracture behaviour of the ceramic material.

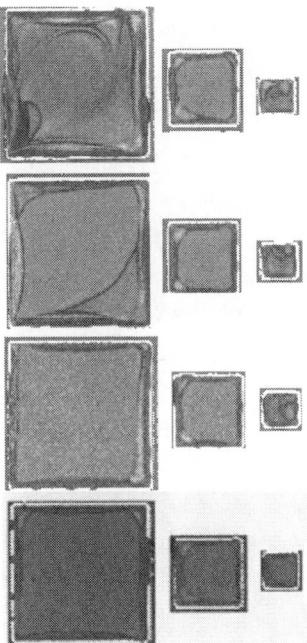

Figure 5: Ultrasonic picture of samples after 200 temperature cycles. The small squares are fully cracked. The large samples are beginning to burst.

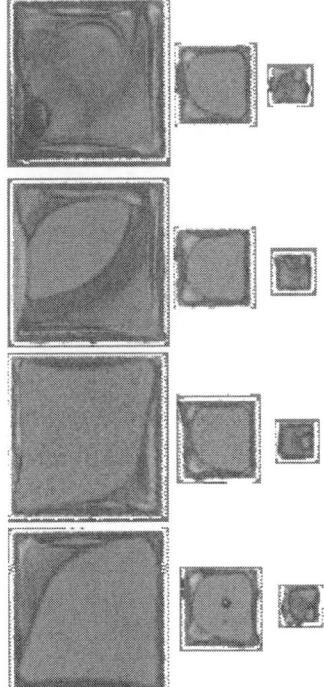

Figure 6: Ultrasonic picture of the same samples after 300 temperature cycles. The middle-sized specimens show the highest resistance against destruction.

A higher risk of fracture exists for smaller and larger parts and there seems to be an optimal dimension of the parts. This optimum probably depends on the copper to ceramic ratio, the selected materials, and the geometry of the etched edge.

Furthermore, the length to breath ratio is important. The larger the dimension of a part, the poorer is TCT behaviour. In Figure 7 and 8, samples with different dimensions are examined after 300 and 500 temperature cycles. The pictures show 28 different experiments but the ultrasonic scans are merged to only one graphic. In both pictures the typical crack starting behaviour is obvious, as shown in earlier studies [2]. The crack starts typically at the edges (and not at the corners) of the copper. The result is a thermal contacted area in form of a pillow. In a second step other cracks start at the corners and destroy the assembly quickly, if they have reached a certain dimension.

Figure 7 shows 28 samples with different size after 300 temperature cycles. Globally viewed, copper pads with a similar size of length and breadth are just slightly cracked and show the form of a "pillow". Samples with disproportionate length and breadth that are reminiscent of circuit paths are more damaged. Interestingly, horizontal and vertical orientated samples show a different picture of failure. You can find the "pillow" in the horizontal orientated circuit paths, whereas samples in vertical orientation are already completely destroyed.

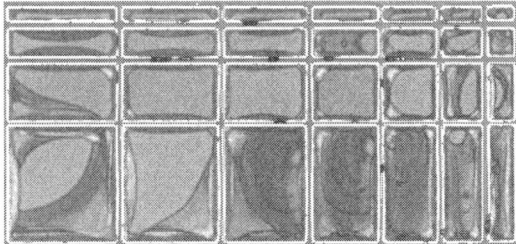

Figure 7: Ultrasonic picture of 28 samples with different size after 300 temperature cycles.

By the different failure behaviour of horizontal and vertical orientated rectangles, it seems to be suitable that the length can be slightly larger than the breadth. This behaviour has been discussed before, but with a focus on the fracture behaviour of modified ceramic sheets [5]. After the bonding process of copper and ceramic, a direction dependent fracture strength was found at one side of the ceramic sheet. Interestingly, this effect exists only at one side of the ceramic sheet and only after the modification of the surface by the DCB-process. This production caused effect in material appears here clearly in the failure behaviour of the DCB substrate.

If the samples are stored for further 200 temperature cycles (see Figure 8), nearly all samples are seriously damaged. Here these two aspects are visualised quite well:

- Samples with an "optimal size" have a good resistance against temperature cycle tests. If the copper pads become too large or too small samples are quickly destroyed.
- The length of the samples should be equal to the breadth. Otherwise the samples are also quickly destroyed.

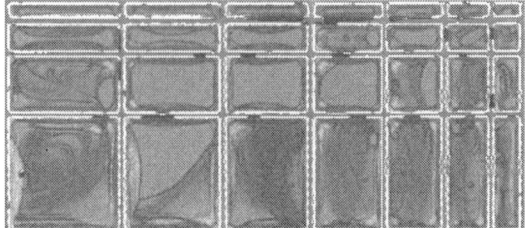

Figure 8: Ultrasonic picture of 28 samples with different size after 500 temperature cycles.

The previous investigations just dealt with structures that are larger than 2.4 mm to get information about the system-caused limitations of copper design. By having a closer look at finer structures, e.g. signal lines, it was found that the described phenomena of an optimal middle-sized dimension is completed with the result that these small copper lines also show a high resistance to temperature cycles.

This is shown in Figures 9 and 10 where different copper structures with a breadth of 0.6 mm and 2.4 mm are shown. After 200 temperature cycles, the samples were inspected. Due to the problem that small structures cannot be inspected by scanning acoustic microscopy, copper is etched and cracked ceramics is filled with a coloured agent, according to the dye penetration test [6]. The larger samples show small cracks below the copper structures whereas the smaller samples do not cause any fracture in the ceramic sheet.

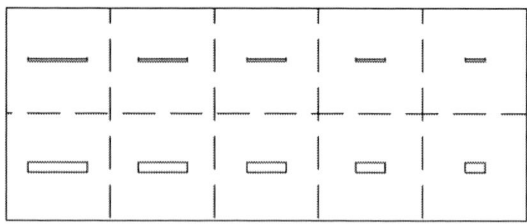

Figure 9: Geometry of ten different small sized copper lines.

It follows that signal lines are not as critical as power lines. Therefore, a copper structure on a DCB substrate should consist of quadratic pads and thin lines for signal transport and it is advantageous to avoid long power lines.

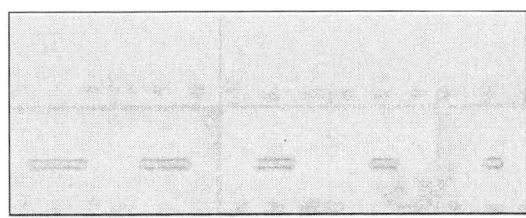

Figure 10: Visualised cracks after 200 temperature cycles (dye penetration test).

Finding this effect leads to the idea that there are advantages to a modification of the copper pad form. If thin structures can react in a way that there are no or few damage, it might be possible that a modification of the high stressed areas can delay the damage. Usually the focus is on the corners because an analytical observation suggests that a 1.4 times stronger stress should exist.

Therefore, the corners are cut off by creating a regular octagon and a circle that represents a circular state of stress. In the first column of Figure 11 three structures are shown in direct comparison after 152 temperature cycles. Astonishingly, the modified structures show a poorer resistance against testing. The quadratic (or rectangular) structure has a tension state in the material which causes a "pillow". This seems to be a kind of crash hold for a certain number of temperature cycles, whereas the circular (or nearly circular) structure has no chance to pause the crack propagation.

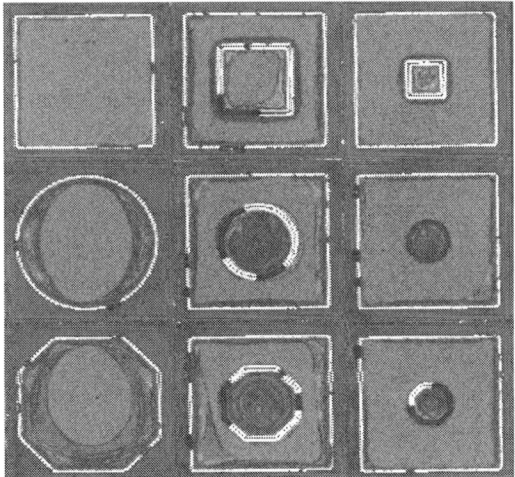

Figure 11: Different copper geometries after 152 temperature cycles. The copper pads at the front and backside are identical.

In the second and third column of Figure 11, the structures are scaled down by a factor of two and four, respectively. It is a current method to cover the surfaces around the structures with copper, as one can see in a lot of commercially available products. Astonishingly, these structures are destroyed as well

after 152 temperature cycles. In comparison with the previous test series, middle-sized samples have no advantages. The influence of the surrounding copper is evidently. A gap between the structures of about 300 μm seems to allow an influence on the small structure in the middle.

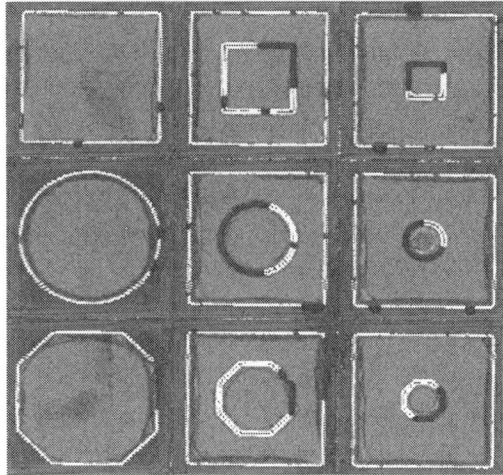

Figure 12: Different copper geometries after 152 temperature cycles. The copper pads at the backside are squares with the same dimension as the first structure.

Instead of an enlargement of the gap to avoid this accelerated failure, the backside is fully covered with copper, as shown in Figure 12. After the same testing treatment, the dimension of the cracks is reduced. The highest risk of fracture is again at the small-sized parts.

A crack starting can be observed at all circle and octagon parts. Only the largest quadratic structure shows a crack starting, at the middle of the right and left edges.

Conclusion

It is possible to design a layout of electrical circuits that is more reliable in temperature cycling tests. The selection of a system that consists of a ceramic material, a metal, and a bonding technology has several consequences. With a defined copper to ceramic thickness ratio and with consideration of the etching technology, that causes a special geometry of the copper edges, the resistance against temperature cycling is determined in dependence to the dimension of the copper structure.

Interestingly, the selection of the system by materials and the described parameters also selects the dimension of copper pads with an "ideal" dimension. That means that these pads show a higher resistance against temperature cycles than samples that are smaller or larger.

Also the length to breadth ratio of the copper structures is important. The larger the aspect ratio the higher is the risk of fracture. By taking a closer look at these rectangular structures, different crack behaviour of horizontal and vertical samples can be observed. It is known that slightly different fracture strength of this ceramic material exists after the DCB-process in both directions [2]. This is due to a correlation of resistance against temperature cycles and the texture of the ceramic crystals [7]. It is supposed that the manufacturing process of the ceramic sheets causes an anisotropic texture in the alumina, which is a reason for this different crack behaviour.

Modifications in the layout e.g. a circle or an octagon, do not show any advantages if they are tested with this accelerated temperature shock test. A copper structure of the backside benefits the crack propagation, although the layout is symmetrically. A full copper layer on the backside is more applicable.

Furthermore, there is one exception in the aspect ratio. Signal lines with a small breadth (in this example e.g. 600 µm) show a high resistance against temperature cycles, although the aspect ratio is unfavourable. This has its origins in the small dimension. These structures can not carry major forces caused by the different thermal expansion due to their stiffness.

So the advantage of these technologies, the combination of signal and power circuits on a single substrate is not available with a closer look on the lifetime. The higher and the lower area limit of the conduction pads should be taken into consideration.

References

[1] K. Bunk, G. Wahl, H. Keser, "Verfahren zum direkten Verbinden von Metallstücken mit Oxidkeramiksubstraten", patent DE 32 04 167 A1, August 11, 1983.

[2] M. Günther, K.-J. Wolter, M. Rittner, W. Nüchter, "Failure mechanisms of direct copper bonding substrates (DCB)", Proceedings of the 1st Electronics Systemintegration Technology Conference (ESTC), Sept 5-7, Dresden, Germany, pp. 714-718, 2006.

[3] J. Schulz-Harder, "Zuverlässigkeit und Integration von Leistungsmodulen für die Weitverkehrstechnik", Teilprojekt: Weiterentwicklung und Prozessoptimierung der Aluminiumnitridsubstrattechnologie für Leistungsmodule. Verbundprojekt HighModule (Förderprogramm Mikrosystemtechnik); Berichtszeitraum 1998 – 2001, Eschenbach, 2002.

[4] H. Dannheim, M. Faber, D. Brunner, E. Auerswald, R. Dudek, W. Faust, "Active brazed copper layer on AlN-ceramic" Proceedings of the 5th Internat. Conf. on Joining Ceramics, Glass and Metal, Jena, Germany, May 12-14, 1997.

[5] M. Günther, K.-J. Wolter, "Material characterization of substrates for power modules", Proceedings of the 29th International Spring Seminar on Electronics Technology (ISSE), May 10-14, St. Marienthal, Germany, pp. 20-25, 2006.

[6] Deutsches Institut für Normierung e. V.: DIN EN 571-1, Non-destructive testing, penetrant testing, part 1 "General principles", 1997.

[7] B. Waibel, W. Martin, H.-J. Bunge, patent DE 38 14 486 C2, April 26, 1990.

Solvent Resistance of Silicones when Used in Electronic Chemical Cleaning Environment

Bill Riegler, Michelle Velderrain, Scott Duffer

NuSil Technology LLC, 1050 Cindy Lane, Carpinteria, CA 93013, USA

Tel: 1(805) 684-8780, fax: 1(805) 566-9905, BillR@nusil.com

Abstract

Silicones are becoming more popular in advanced packaging of electronics because of their thermal stability above 200°C and ability to protect the electronic package from environmental factors. The electronic package may be exposed to a variety of different cleaners by fabricators in the cleaning process. Problems arise when the silicone swells with the solvent used in the cleaners. When the solvent evaporates, the silicone will become harder and put stress on the metal bonds, potentially bending and even shearing them. Fundamentals of silicone manufacturing allow silicones to have different chemical characteristics that can respond differently to various solvents. For example, some silicones are more resistant to hydrocarbon solvents, whereas others are more resistant to halogenated solvents. The purpose of this study is to evaluate the solvent resistance of silicone materials that can be used for electronic packaging. The cleaners/solvents chosen for this study are commonly used in the industry and the silicone materials chosen were based on the chemical composition. The change in thickness and specific gravity (% swell) was measured over time after silicone was exposed to various cleaners. By understanding how the electronic packaging is affected by different cleaners/solvents, the appropriate cleaner and silicone system can be chosen.

Key Words: Silicone, Swell resistant, Electronic cleaners

Introduction

Silicones are becoming more popular for their use as adhesives, encapsulants and interface materials due to their thermal stability and low modulus that can protect the components within the microelectronic package [1,2,3]. This electronic package may be exposed to a variety of different solvents in the cleaning process.

Problems arise when the volume of silicone swells with solvent. Although when the solvent evaporates the silicone may return to its original volume, this process puts stress on the metal bonds, potentially bending and even shearing them. If the solvent is unable to be removed, it can also cause out gassing (bubbles) once the equipment is in use.

This paper will discuss what can distinguish one type of silicone from another based on the chemical composition and reinforcement mode of the silicone polymers. The focus will be on how the chemical composition of the silicone can have a direct effect on the solvent resistance, as measured by % swell and change in thickness, after prolonged exposure to common solvents used in the electronics industry. By understanding how the electronic package is affected

by different solvents, the appropriate solvent can be chosen for cleaning.

Silicone

Silicones have an inherent large free volume within the amorphous structure due to the large bond lengths and angles between the repeating silicon and oxygen atoms that make up the majority of the polymeric structures [4]. This, along with unusually weak intermolecular forces, gives silicone many unique properties such as thermal stability in extreme temperatures. Other unique properties are:
• Tg typically < - 115 ° C
• Low Modulus
• Low shrinkage, < 1%
• Dielectric Strength 500V/mil (0.001 inch) or 20 kV/mm
•Versatile usage configurations:
–Optically Clear, Thermally Conductive, Electrically Conductive, Silica filled

Silicones for Microelectronic Packaging

There are several areas within, and on, a microelectronic assembly that may need an elastomeric or 'rubber' material such as an adhesive, encapsulant or interface material. Thermal Interface Materials (TIMs) are typically a polymeric material that is able to absorb stress and transfer heat away from heat sensitive components. These thermally conductive materials can have a consistency such as grease or a hard elastomer. Below are examples of where silicone can be applied within a Flip Chip assembly. They also can be used directly in contact with the die as die attach or underfill. Another common application for silicones is as an encapsulant, also known as "glob top", that protect the entire microelectronic assembly from damage during transport, and other environmental conditions that can damage the components.

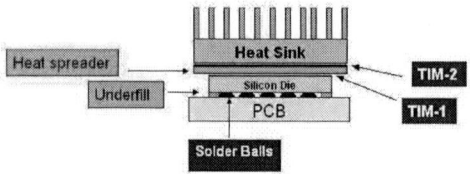

Figure 1: Example of Flip Chip assembly

Silicone Polymer Chemistry

The primary structures of many silicone products used in microelectronics are silicone polymers, which can be up to 70 percent of the total formulation. These are linear structures with alternating silicon-oxygen atoms that make up the backbone of the polymer. For a given silicone polymer formulation; the number of monomeric units dictates the chain length which affects viscosity. The R group will affect the chemical performance and the "end blocker" typically contains the functional group that partakes in the crosslinking (curing) reaction. The resulting polymer comprised of several repeating monomers is a poly (diorgano) siloxane polymer (see Figure 2)

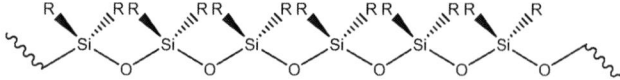

Figure 2: Diorganodisiloxane polymer. R can be methyl, trifluropropyl, phenyl, etc.

The diorganodisiloxane structures in Figure 3 shows that the R groups on the back bone can vary which allows different types of organic groups to be incorporated as pendant groups within one polymer chain (R and R').

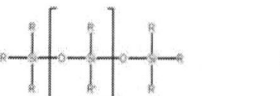 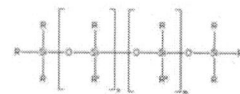

Figure 3: Generic polymer and Generic Co-Polymer

Different pendant groups can provide a variety of excellent properties that can be chosen according to the specific application. The organic pendant groups that will be evaluated for this discussion are methyl and trifluoropropylmethyl substituents [5].

Dimethylsilicones:

Dimethylpolysiloxanes (PDMS) are the most common silicone polymers used industrially. These polymers are typically the most cost effective to produce and generally yield good physical properties in silicone elastomers and gels (see Figure 4).

Figure 4: Vinyl terminated polydimethysiloxane

Fluorosilicones are based on trifluoropropyl methyl polysiloxane polymers and have historically been used for applications that require fuel or hydrocarbon resistance. The trifluoropropyl group contributes a slight polarity to the polymer, resulting in swell resistance to gasoline and jet fuels (Figure 5).

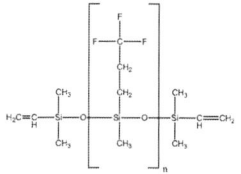

Figure   5:   Vinyl   terminated polytrifluoropropylmethylsiloxane

Silicone Polymer Reinforcement:

When silicone polymers alone are crosslinked together, the cured material is typically referred to as a "gel". Since the cured silicone has minimal tensile and tear strength properties. Gels are also soft and have very low modulus. Silicone polymers can be reinforced by adding reinforcing fillers, such as fumed silica, and/or silicone resins, that increase the elastic properties of cured silicone.

Silica reinforces the cured silicone polymers through van der Waals forces and hydrogen bonding between hydroxyl groups on silica surface and siloxane backbone of polymer. These weak interactions allow the cured silicone to absorb stresses as well as increase the viscosity of uncured silicone through the same polymer-filler interaction [6]. The silica is typically added to the polymer and treated with organosilicones to help increase the compatibility of silica in polymer, which subsequently stabilizes the viscosity to a certain degree. Silica reinforced silicones typically have non-Newtonian flow characteristics where the viscosity will decrease when shear is applied and are also translucent.

Silicone resins are highly branched polyorganosiloxanes. Silicone resins can reinforce the silicone polymer through more complex crosslinked architecture once cured, which helps distribute applied stress as well as entanglement of resin and polymer molecules. Of course, van der Waals forces and hydrogen bonding also play a role in stress response. Resin reinforced silicones are typically transparent and have Newtonian flow characteristics where the viscosity is not greatly affected by the applied shear.

Solvent Resistant Evaluation

The general rule for dissolving substances is "likes dissolve likes" [7]. The more chemically similar the solute (in this case silicone rubber) is to the solvent, the more the silicone will absorb the solvent into its cured matrix. Solvents can be readily removed by exposing the elastomer to moderate heat. Within those elastomer systems, high molecular weight, unreacted siloxane polymers are readily dissolved and removed from the cured elastomer systems when exposed to certain solvents. As a result, silicones can lose volume upon exposure to solvent systems. This volume loss can impact the package in several ways such as delamination of the silicone from the substrate surface. It is important to note the stresses associated with solvent shrinkage are considerable (roughly 3 orders of magnitude more) when compared to expansion and contraction due to thermal cycling in the solder reflow and operating phases of package. The more dissimilar they are chemically, the less solvent the cured silicone will absorb, leading to less swelling.

Polarity is the chemical property most responsible for how similar the solute and solvents are. Organic solvents may have a polar functional group such as alcohol (-OH) or a halogen (-F, -Cl, or -Br) to increase the ability to dissolve substances that have similar functional groups on them.

Materials:

The solvents chosen for this study are commonly used solvents (or solvent families) used in the microelectronics industry (Table 1). The silicone materials were chosen based on polymer chemical composition and mode of reinforcement (resin or silica). Table 2 shows the abbreviations given to the silicone systems used for the study. The polymers used for DMR and DMS are the same but one uses silica for reinforcement (DMS) and the other silicone resin (DMR). Since one mode of reinforcement for resins is additional crosslinks, the durometer for DMR is higher than DMS. The flurosilicones, FS1 and FS2, use similar polymers but have different formulations where FS1 is an adhesive and FS2 is an encapsulant.

Table 1: Solvents

Solvent	General Description and Use
Isopropyl alcohol (IPA)	Polar solvent and commonly used to wipe down a surface before applying silicone.
HFE [8]	Fluorinated ether solvents are polar compared to hydrocarbon solvent. Common family of solvents used for cleaning microelectronics.
Terpene [9]	Limonene/Ester mixture. Aromatic product of citrus fruits and replacement for CFC.
Hexane	Hydrocarbon solvent with no double bonds. Least polar of solvents evaluated.

Table 2: Silicones:

Sample Code	Polymer Chemical Composition	Reinforcement	Durometer Type A
DMR	Dimethyl	Resin	50
DMS	Dimethyl	Silica	25
FS1	Fluoro	Silica	30
FS2	Fluoro	Silica	45

Experimental

% Swell, which measures change in volume based on specific gravity, was measured to evaluate the differences before and after exposure to solvent. The method and calculations were adapted from ASTM D 471-79 where the higher % change in volume, the more solvent the silicone was absorbing.

% change in thickness was also measured before and after submersion. The thickness was measured again after a 48 hour dry out period to evaluate how much solvent would evaporate in ambient conditions after 48-hour exposure to solvent and see if there were any further dimensional changes.

Calculations

a. % Swell (% V)

$$\% \text{ Swell} = \frac{(M3 - M4) - (M1 - M2) \times 100}{M1 - M2}$$

M1 = initial weight of sample in air
M2 = initial weight of sample in water
M3 = weight of sample in air after immersion
M4 = weight of sample in water after immersion

b. Δ % Thickness (%ΔT)

$$\% \Delta \text{ Thickness} = \{(T2 - T1)/T1] \times 100$$

T1 = Original sample thickness
T2 = Final sample thickness

Sample Preparation

1. Silicones were cured per individual standard cure schedule.
2. Sample size 1" x 1" x 0.025".
3. Volume of the solvents was 50 ml in 2 oz glass container.
4. The samples were submersed for 1 hour, 6 hours and 48 hours and kept at 25° C.
5. Specific Gravity and Thickness were measured immediately after submersion (after excess solvent was allowed to drip off).

Results

% Swell

The silicone absorbed not much more solvent between 1 and 48 hours so the 6-hour results are not reported.

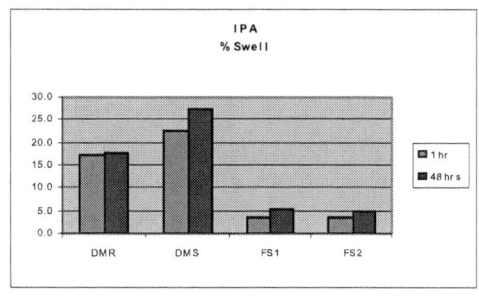

3.1 % Swell, continued

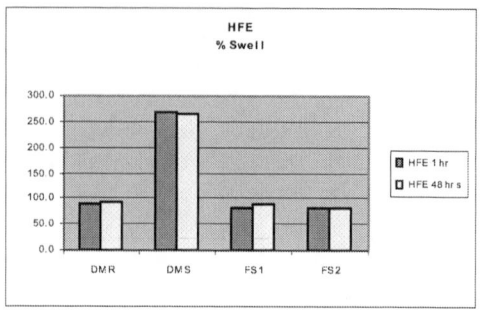

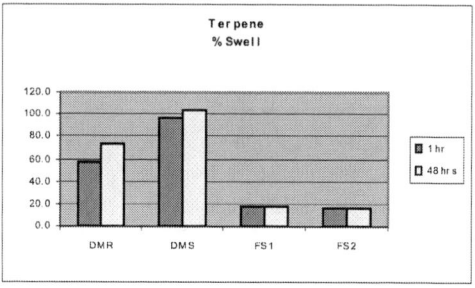

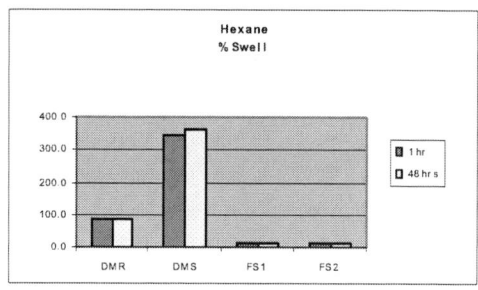

% Thickness

There were minimal changes in thickness between 1, 6 and 48 hours so only the change in thickness after 48 is reported to show "worst case scenario".

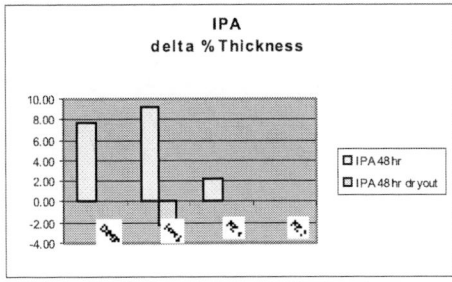

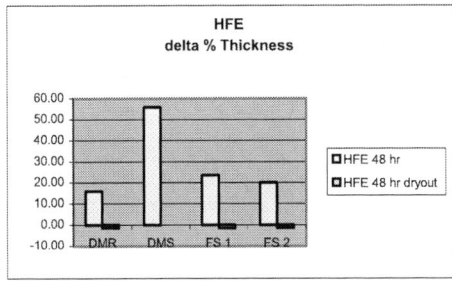

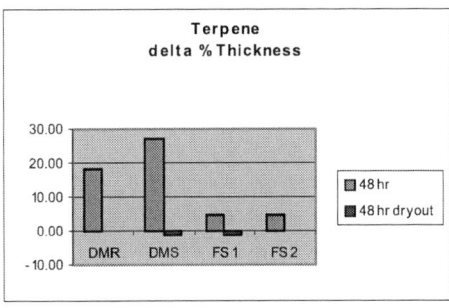

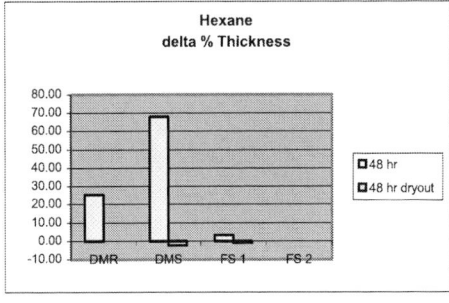

Relative Differences:

Solvent Resistance	% V	% Δ T
4 Excellent	≤5	≤1
3 Good	≤7	≤2
2 Fair	≤10	≤3
1 Poor	>10	> 3

Silicone	IPA		HFE	
	%* V	% Δ¤T	%* V	% Δ¤T
DMR	1	1	1	1
DMS	1	1	1	1
FS 1	4/3	4	1	1
FS2	4	4	1	1

Silicone	Terpene		Hexane	
	%* V	% Δ¤T	%* V	% Δ¤T
DMR	1	1	1	1
DMS	1	1	1	1
FS 1	1	1	1	1
FS2	1	1	1	4

Discussion:

There were no significant differences in results between 1 and 48 hours for % V and % Δ T. % Δ T gives a more accurate idea of effect of solvent on silicone materials based on dimensional change. The % Δ T does not have to take into account the specific gravity of solvent or silicone in the dimensional change after submersion.

All the silicones and solvents used contained some methyl (-CH3) groups so they all exhibited some solvent absorption. Fluoro silicones performed very well against IPA and hexane even though one solvent is non-polar (hexane) and the other polar (IPA). We can deduce that hexane is much less polar than the fluorosilicones and IPA must be much more polar than the fluorosilicone since it performs well with both.

The terpene is an aggressive solvent. Fluorosilicones did perform better than the standard dimethyl silicones where FS1 and FS2 had 115 Relative Percent Difference (RPD) % *lower* % Δ T than DMR.

DMR was most resistant to HFE and had 22 % RPD lower % Δ T than FS1 and FS2. Standard Silica reinforced dimethylsilicone (DMS) is least resistant to all solvents.

Conclusion:

Not all silicones are the same with respect to their response to solvent cleaning processes. Fundamentals of silicone manufacturing allow silicones to have different chemical characteristics that can respond differently to various solvents.

The fluorosilicones (FS1 and FS2) were much more resistant to IPA and hexane than the other silicones evaluated. FS2 had the best overall solvent resistance.

In general, even though IPA is a polar organic solvent, this demonstrates that solubility is a complex subject and experiments may be used to determine the effects of a particular solvent on silicone.

Solvents are designed to remove contamination that can be from chemical compounds that have a wide range of polarity. By understanding how the electronic package is affected by different solvents, the appropriate silicone/solvent can be chosen.

References

[1] H. Yu, S.G Mhaisalkar, E.H. Wong, J.F.J.M. Caers, " Evolution of Mechanical Properties and Cure Stresses for Non-Conductive Adhesives Used for Flip Chip Assemblies".

[2] R.Viswanath, V.Wakharkar, A.Watwe, V.Lebonheur, 'Thermal Performance Challenges from Silicon to Systems'.

[3] R.Mahajan, C. Chiu, R.Prasher, "Thermal Interface Materials: A Brief Review of Design Characteristics and Materials", Electronics Cooling Feb 2004.

[4] W. Noll, Chemistry and Technology of Silicones, Academic Press, New York, 1968.

[5] B.Riegler, S.Bruner, R.Thomaier, 'Low Outgassing Materials for Electro-Optic and electronic Systems', IMAPS Conference on Device Packaging March 2005.

[6] Mark, James E. and Burak Erman, Rubberlike Elasticity A Molecular Primer, John Wiley and Sons, 1988.

[7] L.G Wade J., Organic Chemistry, Prentice-Hall Inc., New Jersey, 1987.

[8] HFE-7100 3M ™ Novec ™ Engineered Fluid

[9] Citrus Burst ™ 7

[10] NuSil Technology R-2615

[11] NuSil Technology R-2186

[12] NuSil Technology FS1-3730

[13] NuSil Technology LSR-9696-30

Silicone polymer coating for piezo actuator protection

Marko Pudas*, Markus Polet and Jouko Vähäkangas

Microelectronics and Materials Physics Laboratories and EMPART Research Group of Infotech Oulu,
University of Oulu, P.O. Box 4500, 90014 Oulu, Finland

Phone +358 8 553 2719, Fax +358 8 553 2728, e-mail marko.pudas@ee.oulu.fi

Abstract

Piezoelectric bimorph components are commonly used as actuators, and are also well suited for micro tweezers/grippers, when combined with sharp metal or ceramic tips. We used them e.g. in positioning micro scale objects or studying detachment and strength properties of wood or tendon collagen fibers. The environmental factors such as humidity and temperature, can significantly effect the operation of the components. The operations may even be performed in corrosive environments. On the other hand, the object or environment under study often needs protection from high voltages (±150 V) of the piezo actuator. Large actuator movement sets limitations to the coating material, as it must be flexible and not restrict the actuation. Also the protective material must be compatible with those used in the actuator assembly. We have evaluated various coating materials and ended up using elastic silicone polymers. In this paper we describe processes for applying modified elastic coatings on piezo actuator surfaces with different methods. The scheme of the constructed bimorph piezo component assembly is also shown, and examples of its use in practical applications are expressed.

Key words: Bimorph piezo, electrical isolation, silicone polymer, micro manipulation

Introduction

The operating environment of conventional piezoelectric bending elements is usually neither liquid nor even a highly humid atmosphere. For such environments, because the danger of high voltages (±150 V), other actuators are commonly used instead. This is also because moisture and temperature have significant effect on e.g. bending piezo actuation.[1] These effects can be compensated with closed loop control in the control circuitry, using feedback of the position. High humidity or aqueous environment, however, limits the use of feedback sensors too. Also depending on system assembly, it has been found that high voltages used in conventional bending piezos can affect digital sensor electronics, if not adequately protected.

Linear piezo actuators are becoming more and more commercially available. They are based on such effects as surface waves, slip-stick or lamp/ unclamp and expand/contract cycles.[2] Such components are also sensitive to environment, the slip-stick surface can be even more affected by environmental effects as the friction is the key factor for operation. This work presents an option to use elastic polymer coating for large actuation piezo components, to enable gripping function for fibers.

Materials and methods

The work aimed to develop a light weight actuator to operate in high humidity or aqueous conditions. The target application of the piezo-actuated gripper was to grip a single paper or tendon fiber in ambient conditions of the sample. In addition, the chemical requirements created by these working environments require from the actuator withstanding of highly ionic and on the other hand high pH liquids. There is only little prior knowledge of e.g. single paper fiber adhesion, range of 200 mN has been estimated.[3]

Because the gripper mass had to be minimized, the used technical solution rules out the use of a closed loop sensor. For this purpose, non-polar poly-dimethyl siloxane (PDMS), or more widely silicone polymers, were used. They are known to be highly elastic, and unlike rubber it is highly homogenous. However the usage environment is still limited for polar liquids, as non-polar solvents, such are petroleum based, can absorb into the polymer.[4]

Bimorph piezo bending actuator (Philips BIMP30/12/0.6/PX5-N) was glued to alumina gripper tips frame, which was cut with laser (Siemens MicroBeam 3200). Ceramic gripper tips were glued to both ends of actuator, as shown in the Figure 1. Wires were soldered to contacts with SnAg solder, which enabled the wetting best. One component RTV silicone polymer adhesive sealant, were applied by dip coating.

Figure 1: Piezo-gripper assembly

The measurement assembly in the test is shown in the Figure 2. The used actuator enables motion 1 cm³ with step size of ~20 nm. Force sensor utilized was for force range of 500 mN to ~1 mN.

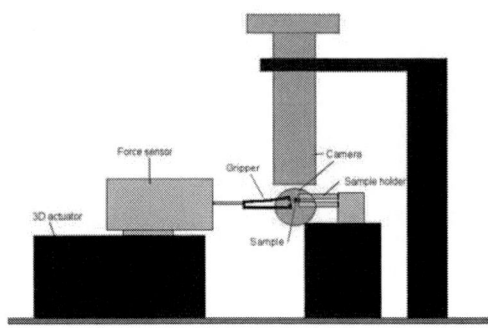

Figure 2: Measurement assembly for paper fiber adhesion.

Silicone polymers can be used also as conductors, while still working as chemical protection. This has been shown with commercial silicone polymers filled with carbon black, and lately with carbon nanotubes [5]. Such features are needed, when the gripping surface is sensing the pressure applied or the conductivity of the material in between gripper teeth.

Results and Discussion

Break through voltage of the silicone polymer was tested on metal substrates by coating it with conductive layer (conductive ink). The thickness of the layer was varied by the amount of solvent and by coating process variables. Measured coating thickness as a function of spinning speed and solvent concentration on planar substrate is shown in Figure 3. The breakthrough voltage vs. thickness behavior is shown in Figure 4 for the silicone polymer.

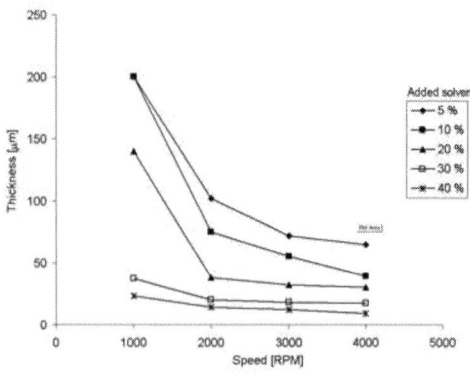

Figure 3: Thickness of silicone polymer on planar substrate vs. spinning speed, with different solvent concentrations

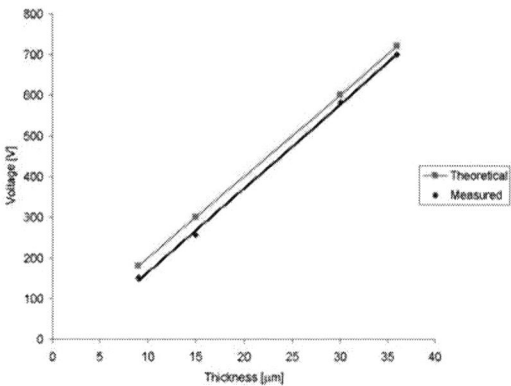

Figure 4: Break through voltage to silicone polymer vs. polymer thickness; Theoretical and measured

The coated grippes were tested in fiber adhesion test in co-operation with KCL (Oy Keskuslaboratorio - Centrallaboratorium AB). The measurements showed that drippers could be used to catch the paper fiber and pull it out, while the force was measured. The early work showed that the adhesion of a single paper fiber to a porous paper was in range of 100 mN for dry paper, and 10 mN for wet paper. Adhesion or paper bending measurement can be done with existing equipments on sheet of dry paper, but not on wet sample. Also with the new equipment, the frictional force between one dry fiber to another one was measured to be in the range of 1 mN.

3D actuator sealing from measurement system humidity was realized with silicone polymer film. It was fabricated by making a mould for the actuator assembly place, and pouring a two component RTV-type silicone polymer on it (Figure 5). Hot air flow was used to aid in manufacturing process, to create down to 200 μm thick protective film.

Figure 5: Silicone polymer film or 3D piezo actuator protection. 1x1 cm² magnetic mounting fixed on top.

Conclusions

The silicone polymer coating of bimorph piezo bender, to enable the gripping action, was demonstrated. Manufactured gripper had opening of >100 um, with high enough gripping force to pull a paper fiber out from paper.

The light weight assembly of the gripper, enabled it to be attached on force sensor. The sensor had high response speed, while measuring forces down to 1 mN with specially developed electronics.

The gripper design is planned to be used in tendon property study, where the exactly same assembly can be used. During the work, it was found essential, that the gripping and pulling action is fast, repeatable and has easy interface for user. This is also the goal of the work as actuator control is realized as 3D haptic (force feedback) controller.

Acknowledgements

Authors wish to thank Tuomas Haapa-Aho for laser processing work. The work was done in interdisciplinary research project *Tools for Nanotechnology* (TNT), funded by European union ERDF-fund, The Provincial State office of Oulu and Tekes the National Technology Agency of Finland.

References

Articles from conference proceedings

[1] P. Ronkanen, P. Kallio, Q. Zhou and H.N. Koivo, "Current control of piezoelectric actuators with environmental compensation" 2nd VDE World Microtechnologies Congress, Munich, Germany, October 13-15, VDE Verlag GMBH, pp. 323-327, 2003.

[2] "Piezo Ceramic Linear Motors & Stages" http://www.physikinstrumente.com/en/products/piezo_motor/index.php, (4.4.2007).

[3] A.V. Byvshev et. av, The effect of volumetric mass of paper moulds to the strength of connection between single paper fibers. Celljuloza 3, pp. 14-15, 1992

[4] M. Pudas, J. Hagberg and S. Leppavuori, "The absorption ink transfer mechanism of gravure offset printing for electronic circuitry" Electronics Packaging Manufacturing, 25(4), pp. 335-343, 2002.

[5] J. Engel, J. Chen, N. Chen, S. Pandya and L Chang, "Multi-Walled Carbon Nanotube Filled Conductive Elastomers: Materials and Application to Micro Transducers" 19th IEEE International Conference on Micro Electro Mechanical Systems. Istanbul, 22-26 Jan. pp.246-249, 2006.

Research partially funded by European union ERDF-fund

TEKES

The Provincial State office of Oulu

Selected Perovskite Type LSFO Thin Films for the Infrared Detectors

Andrzej Łoziński[1/.2/], Paweł Wierzba[2/]

[1/]Gdynia Maritime University, ul. Morska 81-87, 81-225 Gdynia, Poland

[2/]Gdańsk University of Technology, ul. Narutowicza 11, 80-952 Gdańsk, Poland

Phone: +4858 3471976, Fax: +4858 347 1848, E-mail: alozi@eti.pg.gda.pl

Abstract

The objectives of presented research were: development of thin-film production technology of a new class of materials for application in thermal radiation detectors and comprehensive examination of films made by using this technology. Non-stoichiometric lanthanum-strontium iron oxides (La,Sr)FeO₃ (LSFO), were the subject of research. These materials are solid solutions of LaFeO₃ and SrFeO₃, have perovskite-type structure and exhibit electrical conductivity. As we have verified, electrical and thermal properties of these materials can be varied in a broad range. In particular, some of LSFO compositions have the thermal coefficient of resistivity reaching a few %/K at room temperature. A sol-gel manufacturing process of LSFO thin films, has been developed. Films made of several $(La_xSr_{1-x})FeO_3$ compositions were manufactured, their resistance-temperature characteristics and thermal coefficients of resistivity were measured and the optimum composition $La_{0.5}Sr_{0.5}FeO_3$ was selected. Its sheet resistivity at room temperature is about $R_{20}=2\ M\Omega/\square$ (depending on number of layers), while calculated value of thermal coefficient of resistivity $\alpha_{20}=-4.2\ \%/K$. The long-term stability of resistance of these films was also investigated. It can be concluded the described LSFO films are a promising material for bolometer applications. Finally, using the described technology, thin-film bolometers were built and are presented. Their frequency responses were measured to determine thermal time constants. Noise measurements were performed for different bias current values, which allowed us to choose the operating points at which the signal-to-noise ratios reach the maximum.

Key words: LSFO, perovskite, bolometer, thermistor, thin film

Introduction

Properties of materials having a perovskite-type oxide structure ABO_3 can easily be modified by means of partial substitution of cations located at A or B sites.

One of the interesting parameters that can be changed is the electrical conductivity of some perovskite-type ABO_3 oxides.
After modification, the current conduction in non-stoichiometric, semiconductor-like ABO_3 structures has a mixed ionic and electron character.

The $LaFeO_3$ oxide is an interesting material to carry out the above-mentioned modification.

As we have verified, substitution of some Sr ions for La ions at A sites changes its electrical conductivity considerably.

The solid solution $(La,Sr)FeO_3$ system is called LSFO. The thermal coefficient of resistivity α, which for some of these compositions reaches a few %/K at room temperatures, has promoted our endeavours to produce LSFO films for infrared bolometers.

The voltage responsivity R_v of a bolometer is [1]:

$$R_v = \frac{IR\alpha\varepsilon}{G_{th}\sqrt{1+\omega^2\tau_{th}^2}}, \qquad (1)$$

where I – current polarizing the bolometer, R – electrical resistance of the bolometer, G_{th} – its thermal conductance, ε - its emissivity, ω - angular frequency at which incident radiation is modulated, and τ_{th} – thermal time constant of the bolometer.
Assumption, that

$$\omega << \frac{1}{\tau_{th}} \qquad (2)$$

allows us to simplify the equation (1) which then becomes:

$$R_v = \frac{IR\alpha\varepsilon}{G_{th}} \qquad (3)$$

In this case the low-frequency normalized detectivity D_{lf}^* of bolometer can be calculated as

$$D_{lf}^* = \frac{R_v}{V_n}\sqrt{A} = \frac{IR\alpha\varepsilon}{G_{th}V_n}, \qquad (4)$$

where V_n – noise voltage and A – absorbing radiation surface of detector.

If condition (2) is not fulfilled the equation (1) must be used. Figure 1 shows the dependence D^*/D^*_{lf}

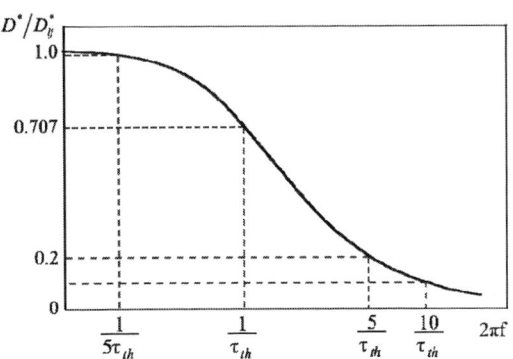

Figure 1: Normalized detectivity vs. radial frequency

Assumption that G_{th} contains only its radiative component $G_R = 4A\varepsilon\sigma T^3$, where σ - Stefan constant and T – temperature in K [2], the low-frequency normalized detectivity of bolometer can be expressed as follows [3]:

$$D^* = \frac{IR\alpha}{8\sqrt{A}\sigma T^3 V_n} \qquad (5)$$

When simple-supported bolometer is connected to the readout circuit with thin wires, this condition is nearly fulfilled. Another small error occurs when the bolometer is not isolated from the environment under vacuum.

Experiment

To produce the thin-film infrared detectors a water based sol-gel process was applied. The starting materials are aqueous solutions of nitrates of lanthanum, strontium and iron.

Basing on the former investigations [4] the $La_{0.5}Sr_{0.5}FeO_3$ composition has been selected. Starting solutions for sol-gel process were prepared according to the flow chart presented in Fig. 2. Produced solutions were used for preparing thin films by spin coating on, possessing thermal conductivity reaching 210 W/(m·K), aluminium nitride (AlN) substrates in the process presented in Fig. 3.

As resistivity of the produced films was unstable, they were aged in 200 °C temperature for 60 days. The ageing caused the stable resistivity in time. Subsequently, the bolometers were cut from the substrates covered with the films.
The bolometers of dimensions 5x5x0.3 mm³ and 2.5x2.5x0.3 mm³ were cut.

Figure 4 shows construction of the detector.

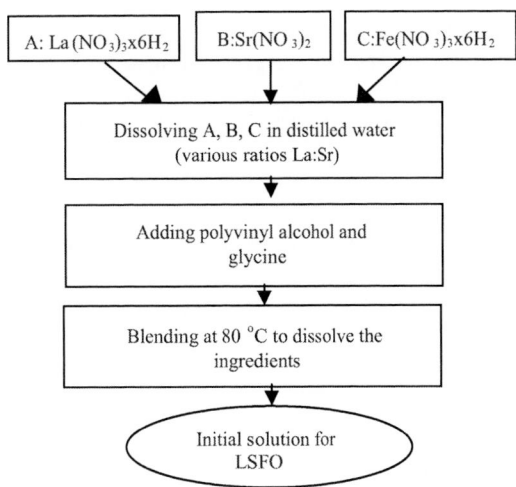

Figure 2: Flow chart of preparation of initial solution for LSFO film

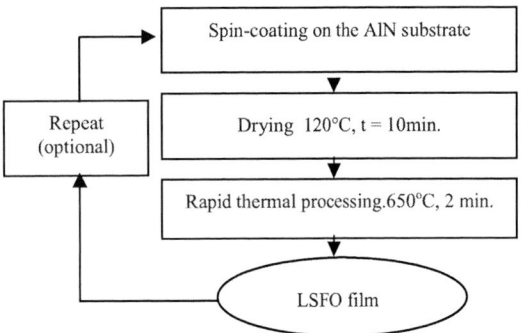

Figure 3: Flow chart of LSFO film production

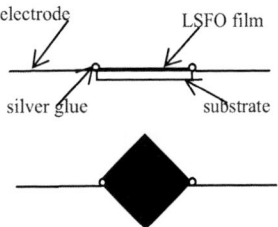

Figure 4: Construction of the detector

Thin electrode wires are connected to the LSFO film with the silver conducting glue.

Measurements, Results and Discussion

The resistance R_{20} was calculated from the current and voltage across the bolometer measured at 293 K. At the same conditions was determined the thermal coefficient of resistivity a_{20} defined as

$$\alpha = R\frac{dR}{dT}100[\%] \qquad (6)$$

The emissivity of both surfaces of bolometers was determined by using the contactless thermometer PT-34 and the temperature micro-controller measuring real temperature of bolometer. The full emissivity was calculated from the Stefan-Boltzmann's law. As the spectral range of the PT-34 thermometer is limited to $8 \div 11$ μm, it was assumed that emissivity is not worse than the calculated value. The calculated emissivity $\varepsilon = 0.95$ is comparable with highest emissivities of other materials.

The thermal time constants of the bolometers were determined by their response to the step of radiation shown in figure 5.

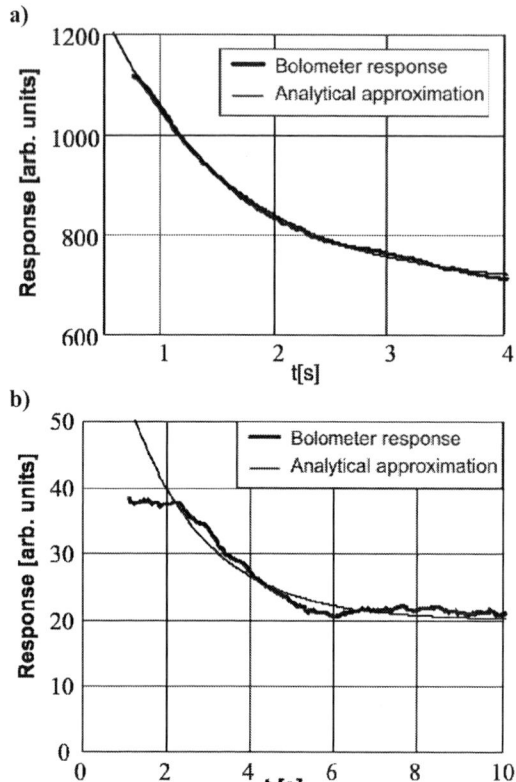

Figure 5: Thermal time constant of 6.25 mm² a) and 25 mm² b) bolometer

The smaller bolometer has 1 sec. while the bigger one has 1.8 sec. thermal time constant.
Following, the current-voltage characteristics of the bolometers were measured for a broad range of polarizing currents.

Presented in figure 6 the current-voltage characteristics show features typical of thermistors. For low polarizing currents it is linear, as the effect of self-heating in the element is negligible in this region, while for higher polarizing currents an increasing departure from linearity, due to self-heating of the bolometers, can be seen. Finally, noise measurements of the bolometers were performed in the circuit consisting of a low noise pre-amplifier shown in figure 7, connected to a SR785 Dynamic Signal Analyzer [5].

Low-frequency noise-voltages U_n were measured in the $10 + 14$ Hz band at various polarizing currents I_p. These measurements allowed

us to determine the polarizing currents at which the ratio I_p/U_n reaches the maximum. Figure 8 shows results of these measurements.

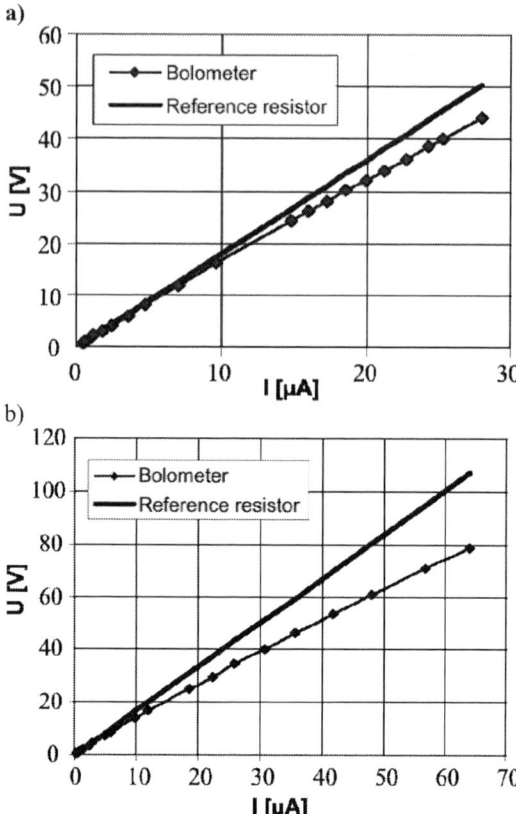

Figure 6: Current-voltage characteristics of 6.25 mm² a) and 25 mm² b) bolometer

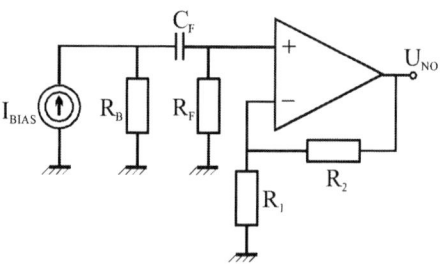

Figure 7: Low noise pre-amplifier

Figure 9 shows the frequency spectrum of the noise voltage of 2 MΩ bolometers at the optimum polarizing currents measured at the 22 °C ambient temperature. Both bolometers show similar dependences. For comparison purposes the noise voltage of a 2 MΩ metal film resistor measured in the same setup is also shown.

The noise voltage spectrum of bolometers is typical of the thermistors with significant contribution of 1/f noise at low frequencies (up to 30 Hz).

According to the equation (5) the low-frequency normalized detectivity D^*_{lf} can be estimated as $1.5 \cdot 10^6$ for the smaller bolometer and $5 \cdot 10^8$ for the bigger one.

a)

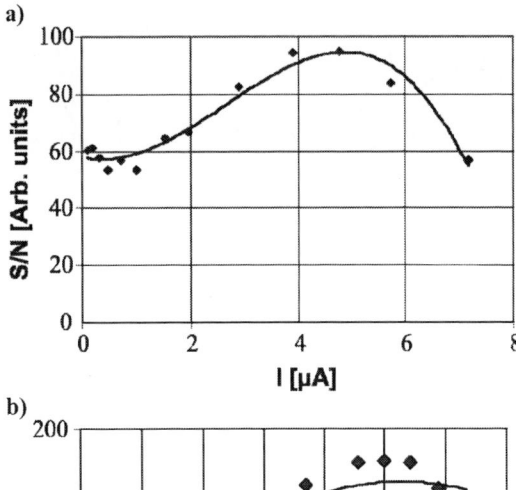

b)

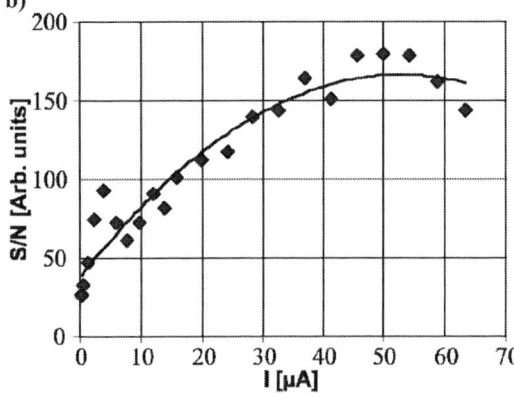

Figure 8: Sygnal to noise ratio vs polarising current of 6.25 mm² a) and 25 mm² b) bolometer

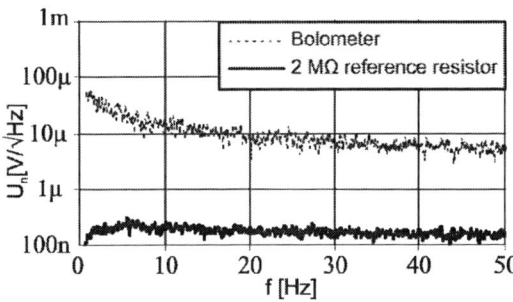

Figure 9: Frequency spectrum of the noise voltage of 2 MΩ bolometers and 2 MΩ metal film resistor

Figure 10 presents the frequency spectrum of the output of bolometer readout circuit with no optical signal falling on the bolometer a) and with bolometer stimulated with an optical signal from the black body of 311 K (38 °C) temperature b). The stimulation frequency was 2 Hz. Distance between the black body source of 70 mm in diameter and the bolometer was 300 mm. The bolometer field of view was 2π. Apparently, the additional spectral lines present above the noise background in the diagram

10 b) are odd harmonics of square shape of optical signal.

a)

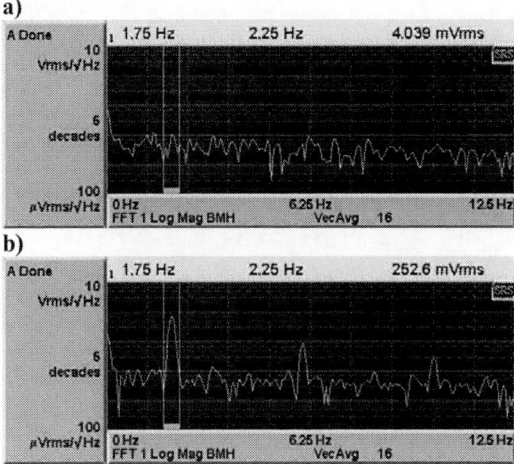

b)

Figure 10: Frequency spectrum of the output of bolometer with no optical signal a) and stimulated with the 38 °C black body b)

Conclusions

The production method of lanthanum, strontium, iron oxide (LSFO) thin films for application in thermal radiation detectors is described. The bolometers utilizing these films are realised. The results of their parameters measurements are presented. The thermal coefficient of resistivity $a_{20} = -4.2$ %/K of the produced LSFO films is typical of thermistors. High emissivity $\varepsilon = 0.95$ is advantageous. The noise voltage V_n at lower frequencies, up to about 30 Hz, shows significant influence of 1/f noise that is typical of thermistors. The optimal polarizing currents maximizing the signal-to-noise ratio of the bolometers have been found. Because of contradictory requirements, the sensitivity and response time cannot be optimised simultaneously.

References

[1] A. Rogalski, "Infrared Detectors", Gordon and Brech Science Publishers, pp. 91-92, 2000.

[2] E. H. Putley, "Optical and Infrared Detectors", Springer-Verlag Berlin second edition Chapter 3, p. 74, 1980.

[3] A. Łoziński, P. Wierzba, S. Rydzewska, "Optimisation of working conditions of thick-film LSFO bolometers", Proceedings of SPIE Vol. 6348, pp. 63480B, 2006.

[4] S. Rydzewska, A. Łoziński, "Thin LSFO Films for Bolometers", Proceedings of XXIX International Conference of IMAPS Poland Chapter, Koszalin, Poland, September 19-21, pp. 243-246, 2005.

[5] P. Wierzba, S. Rydzewska, "Design of Low Noise Preamplifiers for Characterization of LSCO Bolometers", Proceedings of SPIE Vol. 5957, pp. 286-292, 2005.

Immersion Tin wetting behaviour with Lead-free soldering

Mustafa Oezkoek and Nigel White

Atotech Deutschland GmbH, Berlin, Germany

Atotech UK Ltd West Bromwich England

00441216067173 Nigel.White@atotech.com

Abstract

In response to the trend of increasing component complexity and the need for more cost-effective processes, one common approach is to eliminate N_2 protection during Pb-free soldering. These harsh soldering conditions will pose challenges for all surface finishes, not only for copper diffusion rates but also for oxidation of the solder pastes and surface finishes. Responding to this trend the wetting behaviour of Pb-free solder pastes on immersion tin surface finishes using N_2 or air atmosphere during reflow soldering are being evaluated. To follow the progress of these next generation immersion tin surface finishes the effectiveness of solder paste reflow is examined by a "wetting / dewetting test". This paper summarizes the results of in-depth investigations of the wettability of different available immersion tin surface finishes for lead-free assembly applications. The impact of solder reflow cycling was examined with respect to wettability of the various surfaces. In addition, the effect of conducting solder reflow operations in both air and nitrogen atmospheres (with controlled residual oxygen concentration) was also investigated.

Introduction

Circuit complexity and component density, which is leading to decreasing pitch, surface finishes providing Cu/Sn IMCs like OSP, immersion tin and immersion silver are being in favor.

As follower of this trend immersion tin coatings have been widely implemented by many OEMs.

Both electrolytic and immersion plated tins have been around for several years but, both have suffered from several problems. These include whiskers, poor process control and high levels of Cu/Sn IMC formation resulting in detrimental conditions impacting Solderability and performance of the coating.

Well described in industry, the harsh soldering regime of Pb-free will challenge thin surface finishes over copper. Not only copper diffusion rates are leading into tarnishing of the surface, but also into excessive IMC formation.

Additionally, if no countermeasures are taken, oxidation of the soldering pastes and surface finishes occur. The capability of proper solder wetting of the surface will be inhibited limiting the spread of solder paste during reflow, up to an extent were de- or no wetting occurs.

As remedy typically nitrogen atmosphere is used during reflow soldering. By this, the negative impacts of oxidation and tarnishing are minimized, ensuring proper wetting and avoiding discoloration of the surface. On the other hand, nitrogen consumption during reflow is an essential part of the overall cost of the product and of a continuous target of cost saving efforts.

Experimental procedure

Table 1: The four sections of this Investigation

Section	Task
Thickness measurement	Pure tin thickness for surface conditions "as received", "Aged with one reflow cycle" and "Aged with two reflow cycles"
Surface cleanliness	Ionic contamination level prior to investigation
Visual inspection	Surface appearance for different surface conditions
Wetting behaviour	Wettability for different surface conditions

Test vehicle

The test vehicle (fig. 1) was constructed from a commercially available HTG-170 FR4 substrate, plated with 35µm DC (direct current) electrolytic copper and covered by Probimer 65 solder resist. The plated thicknesses were measured by XRF technique, using thickness standards that were verified by wet analytical method ICP-AA.

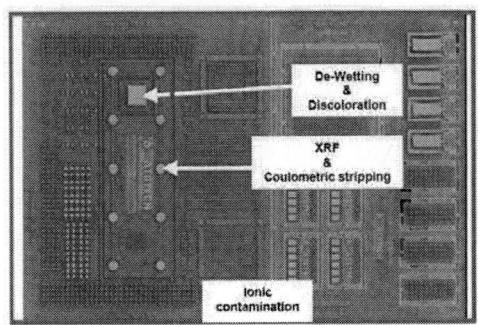

Figure 1: Test vehicle used throughout this investigation, incorporating design features for the different experiments

Surface Finishes

Two individual immersion tin surface finishes were examined, being applied in vertical and horizontal mode, representing the capability of each application. Table 2 presents a summary of the surface finish specifications for the individual deposits.

Table 2: Immersion tin surface finishes used during investigation

Surface finishes	Specifications
Immersion tin (vertical)	≥ 1.2µm, incl. anti-whisker additive
Immersion tin (horizontal)	≥ 1.2µm, incl. anti-whisker additive

Vertically applied immersion tin

Decisions towards vertical plating equipment are divers. Vertical plating equipment shows highest flexibility with regard to the production mix. Being only basket design dependent a product range from thinnest flex boards up to thickest backpanels can be produced in the same plating line.

Process times, product mix dependent, can also be easily adjusted by altering immersion times for the individual process steps, treatment times for cleaning, micro etching and immersion tin plating can be changed individually without compromising one of the other process steps.

Additionally high productivity is realized on less floor space, when compared to horizontal plating

For this, a vertical high volume installation is chosen to produce the test vehicles "Immersion tin (vertical)".

Key parameters are: 18min immersion tin plating followed by cascade rinsing, incorporating a treatment step lowering the ionic contamination level to ≤ 0.5 µg NaCl equ. / Cm². Prior to drying the rinse water quality is maintained to a conductivity level ≤ 2 µS.

Horizon applied immersion tin

Horizontal systems are engineered to meet the most demanding manufacturing schedules at highest reliability since it is often required to operate in excess of 7,000 hours per year.

Horizontal systems offer latest transport technology like panel dimensions

- Rigid panels minimum size 50 x 100 mm
- Flexible panels reel-to-reel or 25 µm single sheet

As horizontal plating systems are specialized for e.g., panel dimensions, plated thickness, surface cleanliness or final application, adjusting chemistry and plating equipment as a total system optimize process parameters.

Adding to the benefits of horizontal systems, chemical maintenance packages are available, including fully automatic dosing and solution control devices. Generally, maintenance is simplified for the key operating parameters, removing the guesswork from production, leading to manufacturing in tighter production tolerances.

For this, a horizontal 0.75m/min installation is chosen to produce the test vehicles "Immersion tin (horizontal)".

Key parameters are: 1.2µm tin at 0.75m/min followed by cascade rinsing, incorporating a treatment step lowering tin oxide formation as well as the ionic contamination level to ≤ 0.5 µg NaCl equ. / cm². Due to the unique capability of horizontal processing to operate in an automatic and controlled process environment, this approach adds to a specialized system.

Prior to drying the rinse water quality is maintained to a conductivity level ≤ 2 µS.

Surface conditions

For subsequent testing all surface finishes were systematically reflow aged (table 3), according to the reflow profile and atmosphere shown in fig. 2 (for Pb-free).

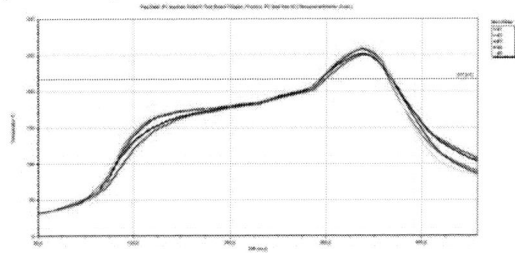

Figure 2: Used reflow profile to simulate Pb-free soldering at 260°C peak temperature and 90 s above liquidus.

Based on previous investigations, not all soldering pastes are able to show changes of surface conditions.

In order to demonstrate the phenomenon of solder spreading, wetting or dewetting more clearly, a solder paste with typical alloy composition (table 4) as used during assembly of immersion tin was used, but with a higher sensitivity of the organic composition of the solder paste towards PCB surface conditions. Additionally a reflow profile was chosen that exposes the surface to 90s above liquidus (217°C) of the solder alloy.

Table 3: Surface conditions for subsequent experiments with the Pb-free reflow profile shown in fig. 2

Surface condition	Atmosphere during ageing / reflow	
As received	Air atmosphere	Nitrogen atmosphere
Aged with one reflow cycle	Air atmosphere	Nitrogen atmosphere
Aged with two reflow cycles	Air atmosphere	Nitrogen atmosphere

Table 4: Soldering paste

Soldering paste	Alloy
Senju M705	SnAg3.0Cu0.5

Major equipment

Major equipment used in the course of the evaluation for thickness measurement, solder paste printing and reflow ageing, respectively reflow soldering:

Table 5: Major equipment used in subsequent experiments

Major equipment	Equipment / Description
XRF Thickness Measurement	CMI 931 / X-ray fluorescence
Coulometric stripping	ECI Technology, Surface Scan QC-100
Ionic contamination	SCS Instruments, Ionograph 500M STD
Semi Automatic Solder Paste Stencil Printer	DEK 248
Solder Reflow Oven	Rehm Nitro 2100 / air atmosphere or controlled residual O_2 concentration (100ppm*)
Visual Inspection	Zeiss Axioplan MRC5

*Following used as "nitrogen atmosphere" – achieved by 20m³ N_2/h

Results

The findings of the four sections of this investigation are shown separately, due to their sequence shown in table 1.

Thickness measurement

Using XRF (X-Ray Fluorescence) thickness measurement technique, all tin atoms over base material are detected and taken into account for calculating the tin thickness.

XRF technique does not differentiate if a tin atom is surrounded by other tin atoms (pure tin) or is surrounded by copper atoms (Cu_3Sn; Cu_6Sn_5 IMCs). Therefore XRF is not suitable to detect the pure tin layer.

In opposite, Coulometric stripping does detect a difference after IMC formation e.g., due to reflow cycling. By applying a fixed current density, each metallic layer requires an individual potential to be solved.

As the pure tin layer covers the Cu_6Sn_5 IMC, the tin layer will be dissolved firstly. After all pure tin is dissolved, further dissolution would require a different potential. The time needed for dissolution (at the fixed potential) and the current density is taken into account to calculate the thickness of the pure tin layer. quires an individual potential to be solved.

Table 6: Results for thickness measurements of total tin thickness and pure tin thickness comparing XRF and Coulometric stripping

Surface condition prior to testing	Immersion tin (vertical)		Immersion tin (horizontal)	
	XRF (Total tin)	Coulometric stripping (Pure tin)	XRF (Total tin)	Coulometric stripping (Pure tin)
As received	1.29 μm	1.31 μm	1.24 μm	1.24 μm
Aged with one reflow cycle	1.20 μm	0.39 μm	1.21 μm	0.40 μm
Aged with two reflow cycles	1.24 μm	0.18 μm	1.23 μm	0.16 μm

- For "as received" condition thickness readings using XRF, respectively Coulometric stripping show same values. This is due to no IMC is being formed at this stage.
- During the first reflow cycle the highest amount of tin is consumed for Cu_3Sn and Cu_6Sn_5 IMCs formation. This is independent if the tin layer is vertically or horizontally applied, and within expected range for this reflow profile.
- As the formed IMCs during the first reflow acted as diffusion barrier for copper diffusion into the tin matrix, the second

reflow showed a lower amount of pure tin being consumed for the IMC formation. This again is independent if the tin layer is vertically or horizontally applied, and within expected range for this reflow profile.

Surface cleanliness

Cleanliness of the PCB is a critical factor and especially residual ions can cause failures in electronic devices. These residues are measured quantitatively as ionic contamination.

Prior to final rinsing and drying latest immersion tin systems incorporate post treatments, to lower ionic contamination levels far below IPC's specification (J-STD-001-D $\leq$ 1.56µg NaCl equ. /cm^2) to less than 0.5 µg NaCl equ. /cm^2.

PCBs contaminated from ions such as chloride, bromide, sodium and organic acids can cause failures in electronic devices.

These conductive contaminants can be responsible for corrosion, metal migration, electrical leakage or tarnishing.

In order to quantify the degree of the potential problem the whole panel is washed in a solvent and the ionic conductivity is measured. This value relates to the ionic contamination level. The result is an average value across the whole surface. The measurement is translated into µg NaCl equivalence per square centimetre.

IPC with J-STD-001-D specify a value, which is not exceeding 1.56µg NaCl equ. /cm^2. as a pass. The instrument used for the determination has to be mentioned next to the test result, as there can be differences from device to device.

Nowadays immersion tin users introduced 0.5 µg NaCl equ. /cm^2 as limit. This approach is being adapted and implemented as post-treatment within the immersion tin plating process.

As root cause it was found out that most of the surface contaminants on PCBs are attributed to the soldermask with a clear interdependence to the type.

Better results can be accomplished by performing UV bumping prior to the immersion tin process and hot water rinsing subsequently. This improvement is still not sufficient to pass constantly the criteria of $\leq$ 0.5µg NaCl equ. /cm^2.

Both immersion tins pass the strict request of end users of a contamination criterion of $\leq$ 0.5µg NaCl equ. /cm^2.

Table 7: Results for ionic contamination measurement for "as received" boards

µg NaCl equivalent / cm² (SCS Instruments, Ionograph 500M STD)	Immersion tin (vertical)	Immersion tin (horizontal)
	0.25	0.18

Visual inspection

As tarnished surfaces interfere with the visual alignment systems for e.g., solder paste printing, task is to avoid any change in surface appearance. As previous section "surface cleanliness" proved that both finishes have a nearly ionic contaminant free surface, any discoloration should not be based on ionic residuals. Root cause then could be a higher oxidation level on the surface.

For visual inspection a 10mm square pad with four printed solder depots is chosen. The tarnishing level between the solder depots, exposed surface finish, is then recorded. This sequence, after paste printing, eliminated artefacts e.g., automatic colour adjustment, based on this automatic digital camera system.

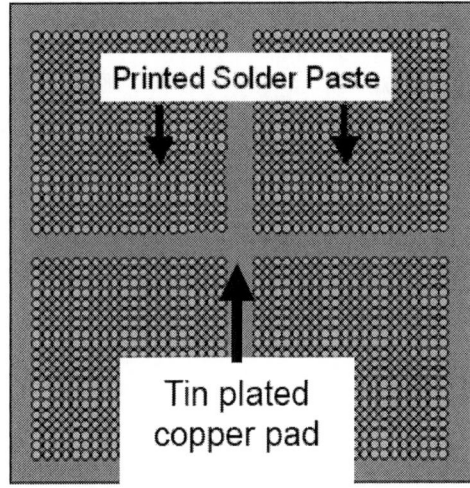

Figure 3.: immersion tin plated pad used for detecting tarnishing after reflow ageing

- For "as received" all surfaces show an identical appearance.
- As only surface finish "immersion tin (vertical)" showed tarnishing after reflow using air atmosphere. This can be seen already after the first reflow cycle.

533

- The same surface finish (immersion tin (vertical)) did not show tarnish after reflow using nitrogen atmosphere. This could be based on the low O_2 level (100ppm) in the reflow atmosphere, limiting the quantity of oxides being formed.
- "Immersion tin (horizontal)" didn't show tarnishing after 2x ageing with reflow in air atmosphere.

Wetting Behaviour

This method incorporates solder paste printing (125μm height) of four square (4.25mm x 4.25mm) solder depots on a 10mm square pad

(Fig.4). The prepared specimens are then reflow soldered and the wetted / dewetted area of the liquefied and solidified solder is inspected.

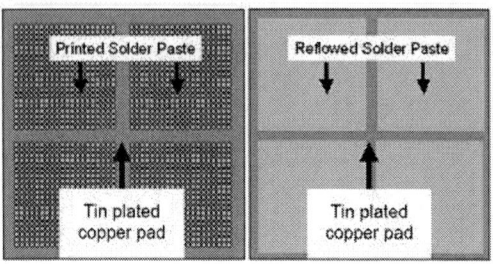

Figure 4: schematic view of the wetting behaviour test

Typically, same solder pastes are chosen, which are used for subsequent assembly.

Table 9: Results for wetting behavior comparing air vs. nitrogen atmosphere

- Again, "immersion tin (vertical)" showed tarnishing on all samples, which where reflowed in air atmosphere. Additionally this was the only surface that showed evidence of dewetting.
- "Immersion tin (vertical)" soldered in nitrogen atmosphere showed best wetting for "as received" condition of the surface with a wide spread of the molten solder. The exposed and not by solder covered areas did not show any discoloration, even

not after totally 3x reflowing (surface aging with 2x reflow + 1x reflow with solder paste).
- "Immersion tin (horizontal)", even by ageing and soldering in air atmosphere, did not show any evidence of tarnishing, discoloration or dewetting. The surface wetting for "as received" did not lead into complete wetting of the 10mm squared pad (nor did "immersion tin (vertical)" using nitrogen atmosphere) but outperformed the same tin thickness applied vertically and processed in the same soldering atmosphere.

Discussion

The investigations presented in this paper are part of a comprehensive project focused on examining immersion tin surface finishes. The objectives of this project are to identify test methods for surface wettability testing under the Pb-free soldering regime.

None of the performed tests are documented as industry standard. Joint effort, within the industry, needs to take place to clarify pass / fail criteria's for the new regime of Pb-free wettability testing, especially as described in other investigations that different solder pastes / flux systems show good wetting of immersion tin using air atmosphere during ageing and soldering.

For the wetting behavior test the governing roles is seen in the solder paste covered area (approx. 18mm²), as the surface tension of the alloy overcomes the "wetting force" of the four corners of the squared printed solder depot. If the surface tension is greater than the "wetting force" dewetting is seen here easiest.

Conclusions

As tin thickness for both (vertical and horizontal) are nearly the same with 1.24 and 1.29μm, as well as the residual pure tin thickness after each reflow cycle (after 1st reflow 0.39-0.40μm, after 2nd reflow 0.16- 0.18μm) the tarnishing of the vertically applied tin layer cannot be thickness related e.g., due to copper diffusion.

Even the ionic contamination levels do not indicate to be the root cause. And therefore an ionic contaminant related or accelerated tarnishing can't be the responsible factor.

Having a close look to the difference between vertical and horizontal applied tin, the horizontal process shows the benefit of processing in a specialized system. All process parameters are optimized in a way that chemistry and plating equipment act as a total system. In this horizontal production mode, features like post treatments to avoid tarnishing, are successfully implemented. With regards to vertical manufacturing not enough experience is present due to the lack of feasibility studies.

Based on the investigations done in this comprehensive project, the following conclusions are suggested:

1. Tin thickness is no guarantee to avoid tarnishing in air atmosphere.
2. Nitrogen atmosphere avoids the tarnishing effect, even the same
3. surface would tarnish using air atmosphere.
4. With special post treatments, time being reserved for horizontal application; air atmosphere during soldering can easily be used, eliminating the additional cost of nitrogen gas during assembly.

With respect to the here used solder paste and reflow profile only the vertically applied tin surface, aged and soldered in air atmosphere, showed dewetting. Other investigations using same solder alloy but different flux / activation systems did wet the "immersion tin (vertical)" surface during multiple reflows in air atmosphere. Nevertheless, a different paste does not overcome the tarnishing only the wetting behaviour will change.

Acknowledgements

The authors would like to acknowledge the contributions of the following individuals, which supported this study to develop the ideas and made preparation of this technical paper possible:

Iris Barz (Atotech Deutschland GmbH, Berlin, Germany),

Kuldip Johal (Atotech USA, Inc, Rock Hill, SC, USA),

Hugh Roberts (Atotech USA, Inc, Rock Hill, SC, USA),

Shozo Nishida (Atotech Japan KK, Yokohama, Japan),

Sven Lamprect (Atotech Deutschland GmbH Berlin,Germany),

Dr. Dieter Metzger (Atotech Deutschland GmbH, Berlin, Germany)

Advanced Thin Film Substrates in Cu-AlN Technology

E. Feurer, B .Holl, J. Vanselow, K. Ruess and <u>A. Kaiser</u>

Reinhardt Microtech GmbH, 89077 Ulm, Germany

Phone: +49-731-392-3275; Fax: +49-731-392-7355, Email: a.kaiser@reinhardt-microtech.de

Abstract

For applications dealing with high power, like laser diodes or RF amplifiers, thermal management is a crucial topic. Thus, for packaging of these devices, high thermal conductivity materials are essential. AlN, compared to diamond or SiC, hereby has the advantage of matching closely the thermal expansion coefficients of Si or GaAs, additionally reduced substrate costs are an important issue. The second task is to provide metallization with high thermal and electrical conductivity for interconnection of devices and coupling of devices, substrate and periphery. For this, copper should be the material of choice. Our outstanding technology, e.g. already used for laser sub mounts, combines the benefits of AlN substrates and Cu metallization by providing low resistivity interconnects and high thermal conductivity substrate material. Our paper presents recent technological approaches on the new technology as well as application examples using the Cu filled vias.

Key words: thin film, AlN, filled vias, copper

Introduction

With increasing system complexity in both, RF and optical communication systems, increasing requirements on substrate technology arises. Reduced electrical resistance for proper grounding and low impedance lines are some of the important issues in RF modules. For high power RF or laser applications also a low DC resistance is required to reduce the overall power consumption of the system and the generated heat from power loss.

For advanced systems or modules improved substrate technologies, with improved electrical and thermal properties, are needed to fulfil the above mentioned issues.

Conventional substrate technologies normally use metallised through holes for the interconnection of the front side to the backside. Insufficient electrical conductivity and problems during further assembly processing may be some of the disadvantages, as only the perimeter of the hole is used.

Lower resistivity can be reached by using the whole cross section of the drilled holes, filled with high conductivity metal (see also Figure 1).

This brings up one of the benefits of our new technological approach with the copper filled vias. Thus, improved electrical and also thermal conductivity between different layers can be realized.

Figure 1 shows the comparison between conventional metallised via holes and a solid Cu filled via. One clearly recognises the different conducting cross sections of the interconnecting vias.

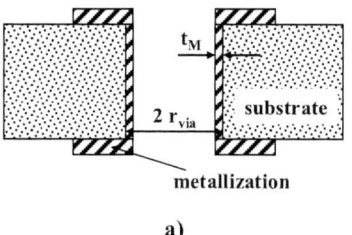

a)

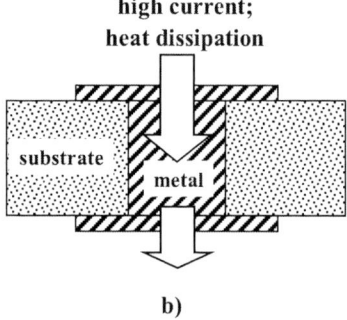

b)

Figure 1: Comparison of the two different metallization schemes for via interconnects: (a) conventional metallized through-holes; (b) Cu filled via.

With the filled vias, higher currents are possible, thus more reliable high power applications without degrading the circuits are possible. Figure 2 shows a theoretical calculation of the interconnect resistance R_{via} for Cu ($\rho_{theory} = 1.7 \ \mu\Omega cm$) and Au ($\rho_{theory} = 2.2 \ \mu\Omega cm$) metallization. t_M/r_{via} denotes the thickness of the metal in the via in relation to the via radius (see also figure 1). One can find that for increased metallised cross section, the resistance reaches closely the minimum value. But according to high currents in every case a completely metallised via is to be preferred.

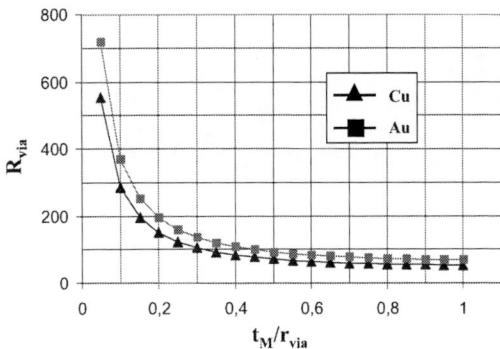

Figure 2: Via resistance as a function of used metallized cross section (theoretical calculation with via diameter: 400 µm; substrate thickness: 400 µm)

Besides the improved electrical properties, DC and RF, there are also some manufacturing issues which make the filled vias more preferable.

For many thin film applications, the use of standard photo resist is desired, which normally is applied using spin coating. With more or less large holes in the substrate, homogeneous resist spinning is not feasible. Thus, besides the holes, fine pattern (like high accuracy resistors or couplers) are nearly impossible to realize. But with filled vias, a higher lithography performance can be achieved.

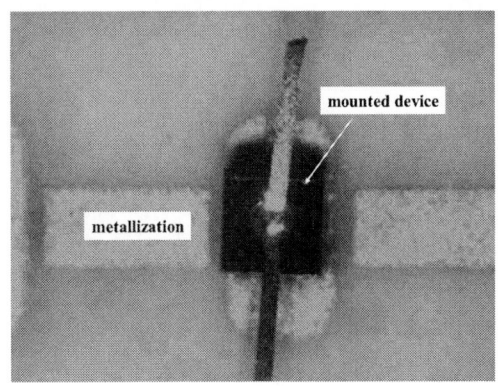

Figure 3: Direct soldering of diode on filled via.

A second thing is the mounting of devices on top of the substrate. This step, normally located at the end of the fabrication routine, is preferably done automatically. With holes in the substrate, the dispensing of adhesive or the placement of solder bumps has to be exactly controlled. Whereas with the filled vias technology, automated assembly can be performed easily. As additional side effect, the devices can directly be mounted on the via interconnect, leading to further reduction of electrical and thermal wire lengths (see figure 3).

With the mounting of conventional substrates on carriers or similar package parts, the problem of upcoming solder or adhesive through the holes during assembly arises. This also can be prevented by applying the filled vias technology.

Besides good electrical conductivity, especially in high power RF systems thermal management is a major task. Generated heat in amplifiers leads to degradation of semiconductor properties and reduced reliability of circuits. Thus, removal of the generated heat is necessary.

High thermal conductivity in this case is the important parameter in selecting the substrate material. According to thermal conductivity values, diamond is the most impressive material, but also the most expensive.

Table 1 gives an overview of important thermal properties of different substrate and device materials. Besides SiC which also has a high thermal conductivity but is rather expensive, only AlN of the technically available substrate materials has a rather high value for the thermal conductivity λ. Utilizing this sort of material as a substrate for RF or DC circuits, one gets an integrated heat spreader without additional effort.

Table 1: Thermal properties of different selected substrate materials (CTE: coefficient of thermal expansion; λ: thermal conductivity [1,2])

Material	CTE [10^{-6} K^{-1}] (at 20°C)	λ [W/mK] (at 20°C)
Si	4.2	151
GaAs	6.5	54
Diamond	1	2000
SiC	2.7	200-270
Al$_2$O$_3$	6.5	20-30
AlN	4.5	170-200

Besides the thermal conductivity, also the CTE of the substrate material is important. If one thinks about mounting GaAs or Si FETs on top of the substrate, the difference in CTE accounts for thermally induced stress in the assembly.

According the CTE, AlN matches nearly perfectly that of silicon and is near to the value of GaAs. Thus, failure originating from thermally induced stress can be minimized.

Further on, also the combination of Cu filled vias with high thermal conductivity substrate material contributes to excellent thermal behaviour.

Technology

The technological fabrication flow for the "Filled Vias" substrates consists of the following major steps:

- substrate preparation;
- drilling of holes;
- metallization of vias (front to back side interconnects);
- patterning of front side according to application (custom layout);
- mounting of surface devices;
- packaging of circuits.

For various applications different surface finishes can be required. Also for demanding lithographic structures, the surface plays an important role. Thus, in the first steps, grinding or polishing of the substrate is performed.

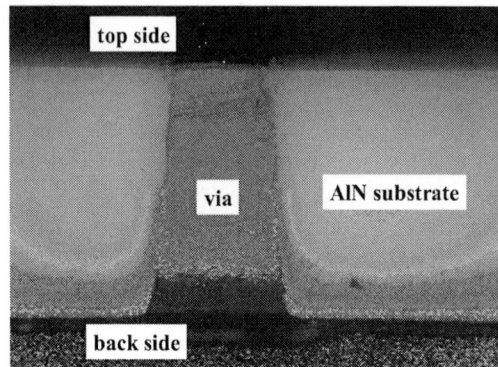

Fig. 4: Cross section of a substrate incorporating Cu filled via interconnects.

The drilling of the holes, which is normally realized by laser, with parameters adjusted to the substrate properties, is the second step. Figure 4 shows a micrograph of a filled via. One recognizes the taper originating from the laser micro machining.

For filling of the holes, Cu electroplating is used. This, compared to other techniques, provides the advantage of high conductivity and relatively cheap process costs. Also a great amount of parallelism can be achieved. Nevertheless, the specific resistance of electroplated material never reaches the ideal pure metal value.

A high density of via interconnects can be realized, only limited by the mechanical strength of the substrate material. Thus, excellent front to back side contacts for grounding or DC feeds can be achieved.

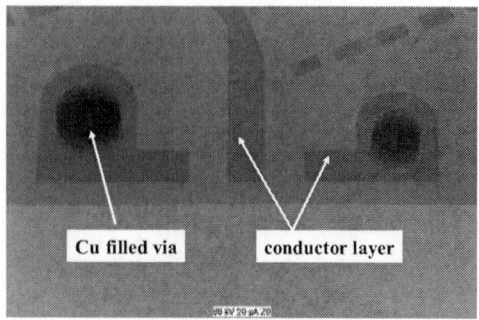

Fig. 5: X-Ray imaging of filled via substrate (parameters: V = 80 kV, I = 20 μA). No voids are visible in the metallized via interconnects.

Figure 5 shows an X-Ray image of Cu filled vias. This method can be used to examine the metallised cross section according to voids or similar defects. In our case, perfect filling by electroplating can be documented. Thus, ideal properties of the interconnections are achieved.

The next step in the fabrication cycle is the patterning of the front side of the substrate. Nearly every kind of metallization shapes can be realized, giving the designers a great amount of freedom. Especially RF applications with couplers or micro strip filters require precise manufacturing. High aspect ratio, small pitch metal lines can be realized with minimum feature sizes in the range of 10 μm to 20 μm using semiconductor-like lithography processes.

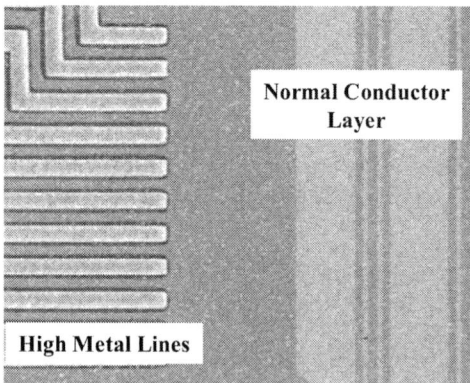

Fig. 6: High Cu Stand-Offs (in this case metal lines) on the substrate surface. Additional surface finish with Au, like shown here, is also possible.

For specific applications, different heights of metal pattern can be required. Figure 6 shows high Cu pattern (Stand-Offs) with a height of more than 50 μm which can be realized besides the normal conductor layers with thickness of several microns.

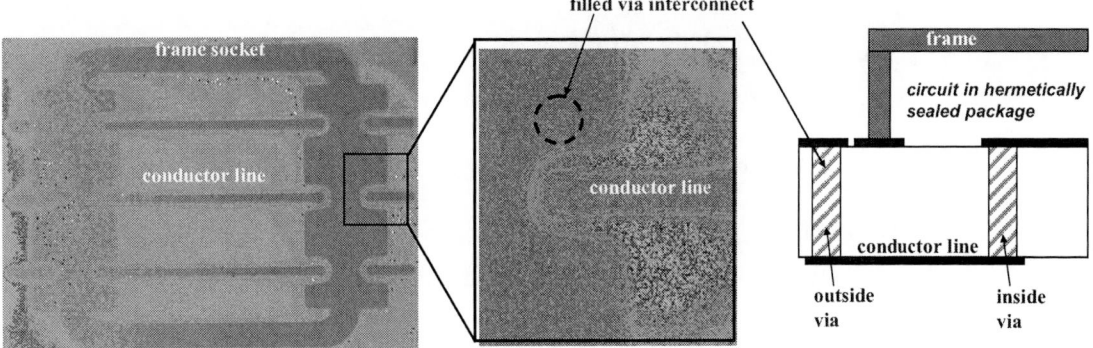

Fig. 7: Demonstrator for "Filled Vias" Technology including hermetic sealing of circuits: left and middle: micrographs of substrate with metallization pattern; right: schematic cross section for placement of the sealing frame

In a next step, the mounting of devices on the substrate, if required can be performed, whereas different metal stacks on the substrate enable the use of soldering or gluing attachments.

One of the special benefits of using Cu filled vias technology also lies in the possibility of using advanced packaging options.

To protect the assembled circuit or system from environmental influences, hermetic sealing is desired. With our technology, hermetic sealing can be achieved by only adding minor process steps. Figure 7 shows an image of a test substrate for attaching a sealing frame on top of a circuit. The area of the frame support can be bridged by several vias, which naturally are hermetic sealings (see also figure 7, far right).

Inside the frame, all kind of active or passive structures can be realized. As also the back side of the substrate can be patterned in any shape, additional connections to the outside can be done by attaching connectors on the backside of the substrate. Thus, also easy RF or DC feeds can be installed and used for integration of the complete module into more complex systems.

Conclusion

In this paper we presented our advanced thin film substrate technology using solid Cu filled vias. The high thermal conductivity AlN substrate gives the possibility of improved thermal managing capabilities, especially important for high performance, high power RF applications.

Our attempt of using solid Cu filled vias improves the electrical and also the thermal conductivity of the interconnections between front side and back side of the circuits. Thus, with our technology excellent grounding and low resistivity connections for sensitive applications can be offered.

Additional benefit brings the fact, that the substrate itself can be used as part of the package. The hermeticity allows to use the substrate as bottom of the package and thus, only the attachment of a proper top cover is necessary.

Acknowledgment

The authors would like to thank the staff at Reinhardt Microtech for support on the technological development and the management for the possibility of publishing this work.

References

[1] http://www.memsnet.org/material~
[2] http://www.matweb.com/~

Laser Soldering of LTCC Hermetic Packages with Minimal Thermal Impact

F. Seigneur, Y. Fournier, T. Maeder, J. Jacot

Ecole Polytechnique Fédérale de Lausanne, STI, IPR, LPM, 1015 Lausanne, Switzerland

+41 21 693 59 45, +41 21 693 38 91, frank.seigneur@epfl.ch

Abstract

A novel laser soldering method for hermetic packaging of temperature sensitive devices such as organic electronics, micro- and nanostructures is presented in this work. The package combines a thermally optimized LTCC (Low Temperature Co-fired Ceramic) base with a glass lid. These two parts are soldered together by the use of a laser diode. The advantages of the laser soldered joint is its hermeticity to water and air in regard to glue and plastic, as well as the possibility to heat only the soldered joint. The power of the laser diode has to be controlled during the soldering process. We propose a solution based on temperature monitoring by the mean of a pyrometer. Heat transfer from the heated solder joint to the encapsulated device can be reduced by structuring the LTCC base, which also reduces the required optical power. Several schemes such as cavities under the joint and local thinning of the base are studied.

Key words: Packaging, LTCC, laser soldering, thermal impact

Introduction

The goal of the study presented in this paper is to describe a method to reduce thermal impact during soldering of a long-term hermetic package for microsystems. First, a description of the packaging method is done in chapter 2. Chapter 3 describes the proposed solution to reduce thermal impact. The advantages and drawbacks of several models are discussed in chapter 4. The setup and the measuring method are described in chapter 5. Results are presented and discussed in chapter 6. Finally, conclusions are drawn in chapter 7.

Description of the Problem

The Laboratoire de Production Microtechnique (LPM) is working on the development of a two-part soldered hermetic packaging. One part of the package is ceramic; the other part is made of glass (Figure 1). These two parts are soldered together using a laser diode. The advantages of the laser soldered joint are its hermeticity to water and oxygen in regard to glue and plastics, as well as the possibility to heat only the soldered joint, without affecting its contents.

This allows encapsulating of thermally sensitive components, such as OLEDs, organic electronics and biological specimens. This also is a solution to ensure long term protective atmospheres for optical devices such as micro mirror arrays.

The goal of this development is to propose a robust packaging method for heat sensitive microsystems. That means that we have to understand the parameters that have an effect on the resulting solder joint, and that we can understand how to make it repetitive, and thus insure a high yield of this operation.

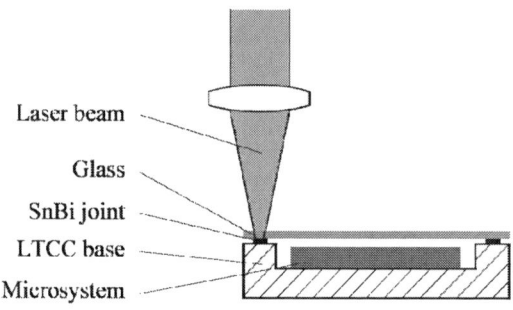

Figure 1 - Schematic view of the packaging method

Proposed Solution

When bringing energy to the joint, the heating of the microsystem occurs due to conduction inside the package (see figure 2). Several parameters can be changed to reduce the heating of the microsystem, such as active cooling, or selecting of low thermal conduction materials. Working on convection might also be a possibility. Radiation is not shown on figure 2, as the temperatures are well below 1000K, even under the laser spot.

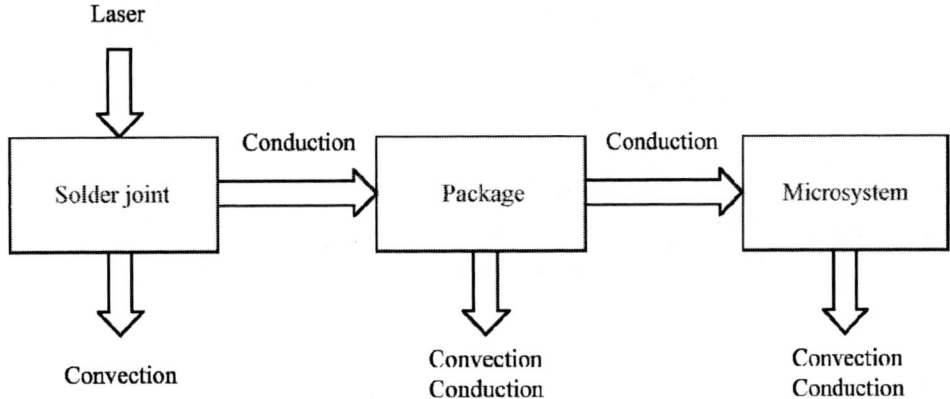

Figure 2 - Heat flux in the package

The proposed idea is to reduce the conduction in the package, between the solder joint and the microsystem, by creating cavities [2]. This solution has several advantages. The first one is to reduce the amount of heat that is transferred to the microsystem. The second advantage is to reduce the energy needed to heat the soldered joint. This decreases the global energy brought to the package during the soldering operation, thus reducing the heating of the microsystem.

These cavities are positioned between the soldered joint and the microsystem (Figure 3). They can be passive, or cooled by a liquid flowing through them. The chosen solution is not to use fluid for cooling the cavity; this is done for several reasons. The first one is that it is not convenient to create such a system, which would increase the cost and complexity of the soldering operation. A second reason is that active cooling of the part directly under the soldering zone can create stresses in the package, which is to be avoided. Finally, active cooling implies that the laser energy needed is higher than for no cooling; which is also undesirable.

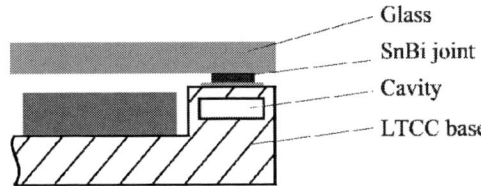

Figure 3 - Thermal resistivity increased due to reduction of the cross section

The cavities are created by structuring of the LTCC tapes. They are cut on two different layers for the vertical and horizontal walls of the package (Figures 3 &5), thus allowing an easy manipulation of the green LTCC tapes, and also offering the possibility to measure the effect of the position and the size of the cavity on the process. This also allows to easily change the height of the cavity, by choosing the number of layers that are stacked before firing of the LTCC.

The LTCC part of the package is manufactured using a standard procedure. The green tape is laser cut, then the several layers are stacked and laminated together, before firing. Note that for electrical connections, several screen-printing operations of conductive tracks might be necessary before stacking.

The green LTCC tapes are laminated between two metal plates. By doing this way, the cavities are not deformed, as it would occur by using a rubber plate. But the drawback is that the bottom of the main cavity is not laminated, which leads to possible delamination of this part. But the advantage is that the shrinkage of non-laminated zones is higher than the rest of the part, which implies that the bottom of the cavity is stretched during firing. This ensures that the bottom of the cavity remains flat, assuming that the walls of the cavity are large enough to sustain this stress.

Finally, metallization (Ag-Pd) and soldering paste (Sn-Bi) are screen-printed on the top of the LTCC part of the package, to form the soldering joint. The Sn-Bi solder is reflowed before inserting the microsystem, which allows to eliminate the majority of the solvents included in the solder paste. All these operations are done at high temperature, but they can be done before the microsystem is placed and connected.

The glass part is also screen printed with metallization to ensure a proper adhesion of Sn-Bi on the glass. The microsystem is then physically and electrically attached in the LTCC base, and then the glass is soldered to the base by the mean of the laser diode. The heating of the soldered joint occurs through the glass.

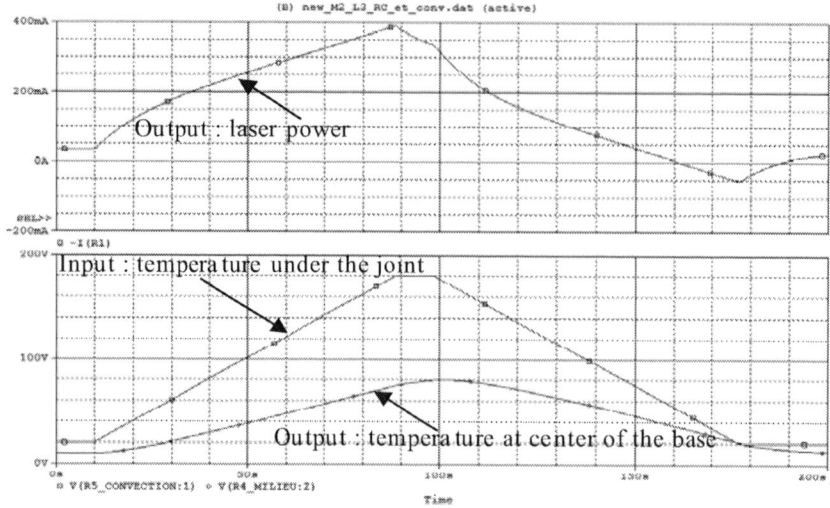

Figure 4 - Analogical model to electrical diagram. Note that cooling is needed to follow the given temperature profile

Models

Several models were used to estimate the effects of cavities on the temperature at the center of the package. Three approaches were used: analytical model, numerical simulation and electrical analogy. They allowed determining which parameters had an influence on heating of the microsystem, and which of them had to be precisely determined.

The analytical model is quite difficult to obtain, due to the complex structure of the base. Moreover, it is difficult to solve the analytical equation, when taking convection into account.

A convenient way to determine the effects of several parameters is by analogy to electrical diagrams. The idea is that thermal conduction and electrical conduction are similar and have an electrical equivalent diagram. The thermal resistance is the electrical resistance, the temperature is the voltage, and the heat flux is the electrical current. This allows to estimate the effect of the cavities when a given heat flux is applied to the system. Moreover, one can roughly determine the power needed to follow the temperature profile.

Setup and Measuring Method

The setup is composed of a 30W laser diode, combined with a galvanometer head, allowing the laser spot to move on a 50mm x 50mm surface. The scanning speed can reach up to 1m/s. The power can be adjusted between 0 and 30W, during the soldering process. The shooting mode is continuous, allowing heating of the joint for times as long as 2 minutes or more. A pyrometer is used to measure the temperature at the center of the package, as well as at the point heated by the laser spot. In order to obtain a good quality joint, a given heating profile has to be followed. The problem is to find the power needed to follow this temperature profile. The first possibility is to control the power with an a-priori curve, previously determined empirically, or with the help of a thermal model.

This is a convenient way to do, but the variability of several parameters has to be measured in order to ensure a good repeatability through a same set of packages. Moreover, this a-priori curve has to be determined for each different package configuration.

Figure 5 - LTCC piled up before laminating. Note that the top layer is not present, showing the cavities cut in two directions.

Another possibility is to use the pyrometer to control the power of the laser during the soldering process [1]. This allows to precisely follow the heating profile. We must note that the pyrometer can only measure temperatures at one point of the package, but the laser spot scans the entire joint simultaneously.

Figure 6 - LTCC before and after firing. Packages are individualized once fired by breaking them along dotted lines.

Further tests need to be undertaken, to determine if measuring temperature at one point is sufficient, or whether several measuring points are necessary to ensure a uniform heating of the joint. If this is necessary, a solution might be to use screen-printed PTC resistors, as done in a previous work [3]. This way, a continuous temperature monitoring could be done at several points of the package, ensuring a uniform heating, and control of the maximum temperature reached. Screen-printed resistors also have a second advantage over pyrometer measurement: it is not dependent on reflectivity of the measured material.

Figure 7 - Setup used for the measurements

Results

The measurements were done for several cavities configurations. The first sample measured is used as a reference; it has no cavity. Other measurements are done with the same heating profile, at the same measurement point, but with samples having a cavity.

The first observation is that when adding a cavity, the thermal resistance is higher, due to cross section reduction. This is due to the fact that the thermal capacity is smaller. Thus, for the same

energy brought to the system, heating of the soldering joint is higher than without cavity. This effect is an advantage, as less power is needed to obtain the same soldering temperature if a cavity is used.

We can see that with a cavity, the center of the package reaches a lower temperature than the version without cavity (Figure 8). The difference might seem small (4°C), but it represents 17% of the overall heating (23°C). Another observation is that for times as long as 2 minutes (which is enough for a good soldering profile), the temperature reached at the center does not exceed 50°C.

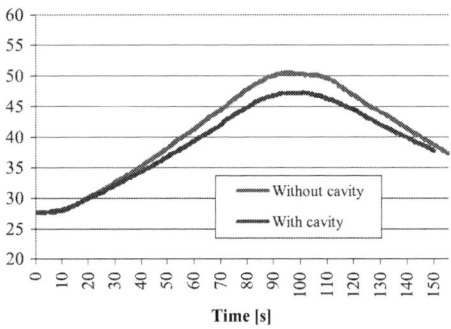

Figure 8 - Heating of the center of the package

To decrease temperature at the center of the package, a solution might be to add the cavities at a bigger distance from the soldering joint, for example creating a bridge on which the microsystem is placed. This bridge has two functions: the base would accumulate heat brought by the laser, and the bridge increases the thermal resistance. In this way, we can combine a high thermal capacity with a high thermal resistance. This configuration has not been tested yet. The drawback of this solution might be in the case where heat generated by the microsystem has to be evacuated. In this case, the use of thermal vias can increase the heat evacuation.

Further measurements have to be done to understand the influence of air convection inside the cavity. This might also have an effect on regulating the temperature of the joint. To observe this effect, we propose to use small cavities separated by walls, so that the convection is confined in a small volume.

Conclusion and Future Work

The development of a new packaging method is drawn by the need in microsystems technology. The proposed packaging method allows long-term hermeticity and low closing temperature.

A solution for reducing thermal impact on microsystems was proposed. By using cavities in

543

the LTCC body, we were able to reduce the heating of the microsystem. Further tests need to be done to measure the effect of other solutions regarding temperature reduction, such as placing the microsystem on a structure increasing thermal resistance.

The control of the temperature with the pyrometer is successful. This method allows to follow precisely a given heating profile. The uniformity of the temperature of the whole joint has not been measured yet. Nevertheless, PTC resistors can give more information regarding temperature of the whole joint; this will be tested in a future work.

References

[1] N. Boryszewski, "Développement d'un Système de Contrôle de Puissance Laser", Semester project, EPFL, February 2007.

[2] L. Brocard, "Etude et Optimisation du Comportement Thermique d'un Boîtier Hermétique", Semester project, EPFL, February 2007.

[3] F. Seigneur, "Laser Sealed Packaging for Microsystems", Proceedings of the 2006 Third International Precision Assembly Seminar (IPAS), Bad Hofgastein, Austria, February 19-22, pp. 307-314

[4] R. Bauer, V. Strickert, K.-J. Wolter, W. Sauer, D. Leonescu, P. Svasta, "Investigation on an Integrated Liquid Cooling System in LTCC-Multilayer", 24th International Spring Seminar on Electronics Technology, 5-9 mai 2001, Calimanesti-Caciulata, Romania

The Deposition of Thick Film paste by direct writing

J.Hladik, J. Vanek and I. Szendiuch

Brno University of technology, Department of Microelectronics, Udolni 53, 612 00, Brno, Czech Republic

xhladi06@phd.feec.vutbr.cz, vanek@phd.feec.vutbr.cz, szend@feec.vutbr.cz

Abstract

The main aim of this paper is to show possibility of direct writing method for thick film paste deposition. This method can replace screen printing in early develop version for the hybrid integrated circuit (HIC') on inorganic substrates as well for many other types of substrates.

Key words: thick film, paste deposition, conductive paste, direct writing, sheet resistance, reliability

Introductions

The term Direct Writing (DW) is used to describe a range of technologies, which allows the fabrication, possibly in reconfigurable short production runs, of two or three-dimensional functional structures using processes that are compatible with being carried out directly onto potentially large complex shapes.

Many diverse technologies are encompassed within DW and, together with related materials are at different stages of maturity. Key underpinning technologies include nozzle dispensing processes (notably ink jet based), moulding and transfer methods (such as soft lithography) and laser systems for modifying or adding materials. [1]

Our approach

Direct write processing uses a computer automated device for precision printing of metallic pastes. Rather than being screen printed or etched, patterns are drawn on a substrate with non-contact nozzle. The direct write approach may be utilized to print a single layer of circuitry or to sequentially stack layers of material to produce multilayer circuit.

The printing device consists of XY axis table, Z axis holder of syringe and dispensing nozzle and software capable process CAD data. The paste is extruded with compressed air of micro dosing unit through a hollow needle, working as a dispense nozzle. Constant pressure and constant movement of either the needle or the ceramics substrate enables a uniform and homogeneous paste application. (See Figure 1)

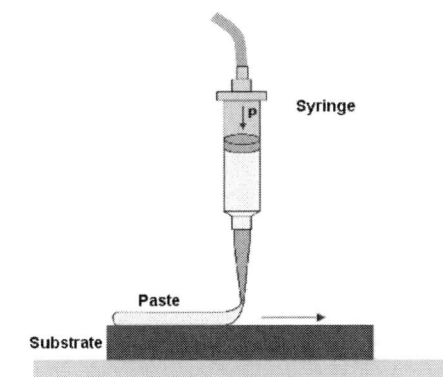

Figure 1: Principle of dispensing

Electrical properties

For determination the sheet resistance we used an Ag conductivity paste. The test shape and its dimension is shown on the picture Figure 2.

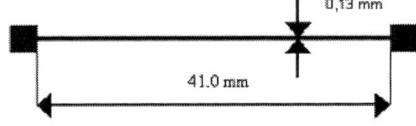

Figure 2: The test line

The paste was deposited on the Al_2O_3 substrate, fired and measured. For measuring the line resistance we used the four point probe and results are shown on the Figure 3.

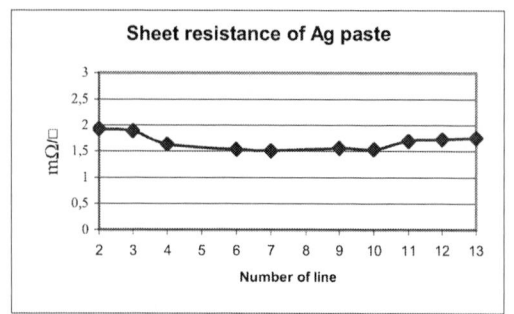

Figure 3: The sheet resistance of the test lines

These results are from the test samples, where each test line was printed alongside the next one (13 lines). It is clear visible that resistance slightly varies and it is because the deposition process is sensitive to deposition condition (in this case the distance between nozzle and substrate).

Cross-section measurement

We conducted tests of deposited line on Sloan Dektak surface profile measuring system. It is based on a small diamond stylus scanning over the surface. We prepared the test coil by direct write technique. The coil design is shown on the Figure 4 and its surface profile on Figure 5, 6.

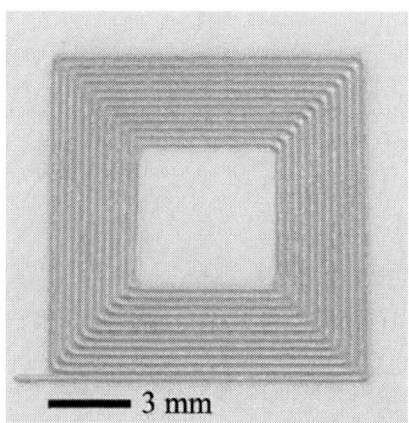

Figure 4: The DW test coil

As we noticed by the line resistance measurement - 'variation of resistance' here is demonstration what does it cause. There are clear visible variations of the cross-section (See Figure 5)

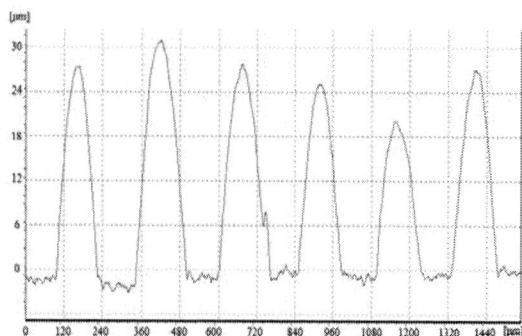

Figure 5: The DW coil surface profile

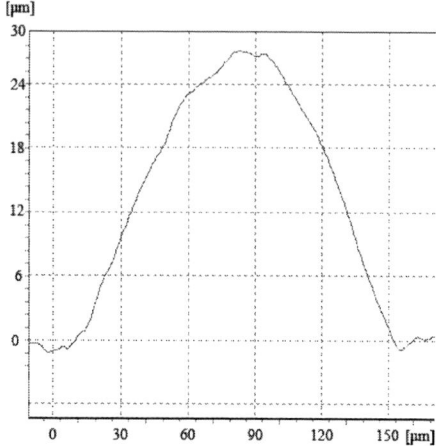

Figure 6: The surface profile of one line the DW coil

The obtained width of the line is 130 μm and width of the space is 110 μm.

Practical application

As we demonstrated before the direct write technique is capable produce circuits on the inorganic substrates and therefore it is the way to rapid prototyping. The example of circuit is on Figure 7. There is interconnection module board for TQFN-EP package.

This deposition method can be used for solder paste deposition as well for various adhesives deposition.

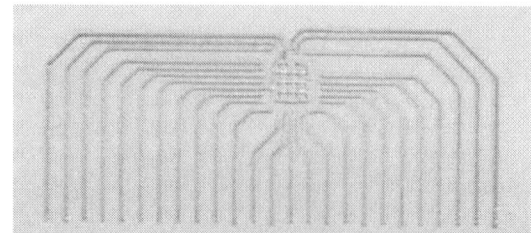

Figure 7: 28 TQFN-EP package expansion board

Future steps

All tests we made were performed on dispensing device prototype, which is in developing stage. Therefore we work on the new construction of this dispensing printer. We assume to achieve resolution below 100 μm, supposing 90 μm line and 80 μm space width.

Acknowledgement

This research has been supported by the Czech Ministry of Education in the frame of Research Plan MSM 0021630503 MIKROSYN New Trends in Microelectronic Systems and Nanotechnologies and the research work under project VaV-SN-172-05 of Ministry of Environment "Low-energetic structures for photovoltaic cells and systems".

References

[1] P. Gay, "Global status and opportunities for the UK in advanced manufacturing", 2004.

[2] A. Luque and S. Hegedus: "Handbook of Photovoltaic Science and Engineering", John Wiley & Sons, 2003.

[3] M. R. Haskard and K. Pitt , "Thick-film technology and applications", Electrochemical Publications , 2004.

Direct write technique used for solar cell fabrication

J.Hladik [1], R. Barinka[2] and I. Szendiuch[1]

[1]Brno University of technology, Department of Microelectronics, Udolni 53, 612 00, Brno, Czech Republic
[2] Solartec Ltd., Roznov pod Radhostem, Televizni 2618, Czech Republic

xhladi06@phd.feec.vutbr.cz, rbarinka@solartec.cz, szend@feec.vutbr.cz

Abstract

The main aim of this contribution is to inform about research work in application of thick film paste for buried back side solar cells, prepared by direct writing. There is described the new process for solar cell realization using direct write technique for deposition of Ag and Ag/Al thick film paste on the back side of solar cell. This approach helps to make more flexible experimental work, not only for solar cells.

Key words: direct write, solar cell, back side contacts, thick film pastes Ag , sheet resistance

Fabrication steps

Solar cells were fabricated in steps described in Figure 1. All prints of the thick film pastes were deposited by direct write approach (etching a shape of N^+ layer, print of metallization layer). This is a great advantage of shape and dimensional variability in solar cells development without need of the new screens.

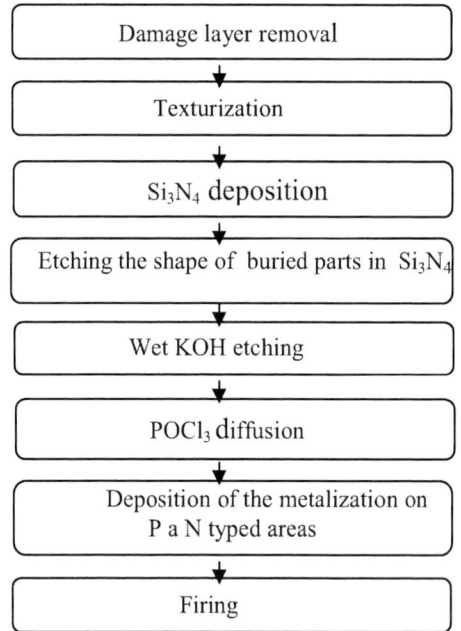

Figure 1: Solar cells fabrication steps

The final shape of metallization layer is shown in Figure 2 and 3. Passivation and antireflexion layer Si_3N_4 was used as a mask for wet etching and for diffusion.

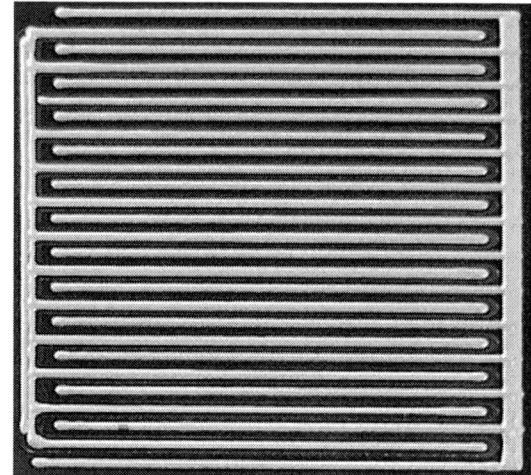

Figure 2: The hedge shape of the back side metallization

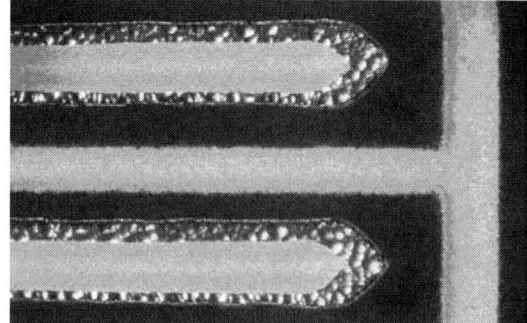

Figure 3: Detail of the contacts. Light areas are buried N typed areas

Contacts characterization

Ohmic properties of the metallic contacts were investigated by diagnostic aproach described by D.E.Reimer [1]. The contact resistance of the interface metal-semiconductor was investigated on the surface with Si_3N_4 layer and on the etched areas without SiN_3 layer. The results are shown on Figure 4. The standard value achieved by the screen printed contacts is about 9 $m\Omega cm^2$

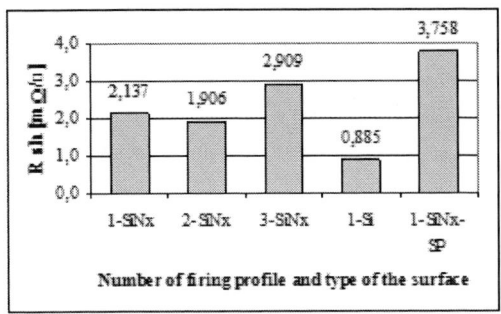

Figure 4: The influence of firing profile and type of the surface on contact resistance Rc

Next investigated parameter was line resistance of metallization. It was measured on samples fired in another firing profile. The results are shown in Figure 5.

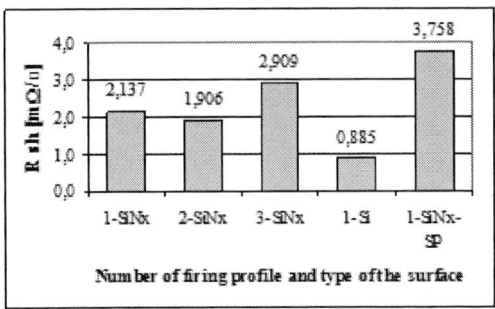

Figure 5: The influence of firing profile and type of the surface on the sheet resistance of metallic layer. SP = screen printed

From the Figure 5 is good to see that direct write contacts achieved better values of line resistance than standard screen printed contacts. It is caused bigger amount of deposited paste in case of direct write contacts. In the etched areas (N^+) are results even better and probably it is caused by smooth surface in the etched areas.

Electric characterization of tested solar cell structures

After firing process were measured VA characteristics on solar tester under AM 1,5 and 1000 Wm^{-2} conditions. The VA characteristic of two samples is shown on Figure 6.

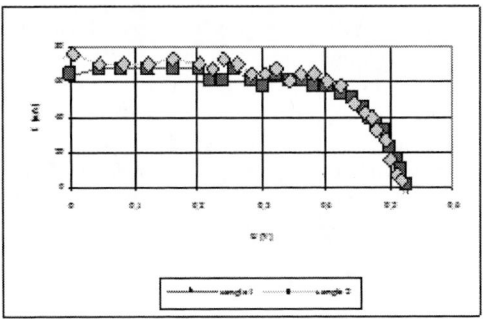

Figure 6: VA characteristics under illumination

LBIC

You can see on the Figure 7 results of diagnostic method called LBIC (Light Beam Induced Current). Light was coming from the front side (without metallization) Light areas are buried N typed areas. It is clear the generation of current is only in the buried areas and probably this is a result of poor surface passivation.

Figure 7: The distribution of generated current

Conclusion

In this work we have shown the results of back side, buried contacts solar cells. The experiment has shown the contact resistance between metal paste and semiconductor was better for conductive paste deposited by direct writing.

Acknowledgement

This research has been supported by the Czech Ministry of Education in the frame of Research Plan MSM 0021630503 MIKROSYN New Trends in Microelectronic Systems and Nanotechnologies and the research work under project VaV-SN-172-05 of Ministry of Environment "Low-energetic structures for photovoltaic cells and systems".

References

[1] D.E. Reimer, "Evaluation of thick film materials for use as solar cell contacts". Proceedings of the 13th IEEE Photovoltaic specialist conference, pp. 603-608, 1978.

[2] A. Luque and Steven Hegedus, "Handbook of Photovoltaic Science and Engineering", John Wiley & Sons, 2003.

[3] M. A. Green, "High efficiency silicon solar cells", Trans tech publications, Switzerland, 1987.

Study of the impact of high-voltage trimming on several characteristics of model TFRs and their stability

N. Johner, T. Maeder, C. Jacq and P. Ryser

EPFL, STI-IPR-LPM, Station 17, 1015 Lausanne, Switzerland

+41216935822, +41216933891, Niklaus.Johner@epfl.ch

Abstract

In this work we study in more details the impact of voltage trimming on different characteristics of model thick-film resistors (TFRs). Two series of RuO$_2$-based TFRs with different conducting grain sizes are studied. As the composition of model pastes is completely known, it allows connecting the effects of trimming with the microscopic structure of the resistors.This study focuses on the change of resistivity, thermal coefficient of resistance, piezoresistivity and their post-trim stability. Those characteristics are studied as a function of conducting filler volume fraction. We show that the relative change of conductivity and the change in pezoresistivity due to voltage trimming are both diverging as the critical volume fraction, where the conductor/insulator transition occurs, is approached. It is also shown that voltage trimming changes the critical DC exponent, and can lead to a crossover from a nonuniversal to a universal behavior. We propose a description of voltage trimming which can explain the observed changes in characteristics of the samples.

Key words: High-voltage Trimming, Thick-film resistors, Stability

Introduction

The difference between the design and the as-processed value of thick-film resistors (TFR) can be as large as 20-30% [1], therefore a trimming step is usually necessary in order to fit the requirements for the production of reliable electronic devices. This is usually done by laser trimming. Pulse voltage trimming has several advantages compared to laser trimming. It's a cheap and effective method (it was shown to allow adjustments to less than 1% for RuO$_2$ based TFRs [1]), allowing trimming of buried or very small resistors. It was shown to be reversible [2-3] and to make the trimmed resistors less sensitive to voltage pulses. Nevertheless this technique is not used in industrial applications because it has the disadvantage to make resistors more sensitive to temperature, limiting its application to fields where the resistors aren't exposed to temperature higher than 100°C [4].

We have found evidence in a previous study [5] that the trimmability (relative change of resistance) mostly depends on the topology of the underlying current carrying network. Therefore we think it is important to consider the composition of the resistors to better understand the impact of trimming on their properties and try to build up a theoretical model.

In this work we study the impact of trimming on the resistance, piezoresistance and thermal coefficient of resistance of model RuO$_2$-based TFRs and their thermal stability and try to link our results with the microscopic changes occurring in the TFRs during trimming.

Description of experiments

We study two series of low temperature RuO$_2$-based thick-film resistors. In both cases the insulating matrix is a lead borosilicate glass composition called V6, produced in our laboratory. The glassy particles have a typical size of 1-3 µm and the glass transition takes place at around 450°C. RuO$_2$ is used as the conducting phase, the first class of samples has a typical RuO$_2$ grain size of 40 nm and the other of 400nm. In each set of samples, different conducting filler volume fractions are studied. For more details about the sample preparation refer to Vionnet-Menot et al. [6] and papers cited therein. For simplicity we will designate the samples with the name of the glass, followed by the diameter of conducting particles and finally the volume fraction of RuO$_2$. For example V6-40-0.06 is composed of V6 glass and 6% volume of 40 nm RuO$_2$ particles.

The samples were prepared by screen printing on 93% alumina substrates. For the measurements of conductivity a layout for 4-wire resistive measurements was used, whereas for the piezoresistive measurements a Wheatstone bridge geometry was used.

The setup used for the high-voltage trimming of the resistors is very simple. A capacitor is charged to a given high voltage (500V in our experiments) and then discharged through the resistor to be trimmed. 2300 such shots were made for each sample and the evolution of the resistance was monitored. We used a capacity of 0.33 nF for the simple resistors and 1.33 nF for the Wheatstone bridges. Indeed in the Wheatstone geometry we trimmed the four resistors at once, so that the current flowing through each resistor was one fourth of the total current.

The piezoresistivity was measured by means of the resistance change under hydrostatic pressure. The pressure was varied from 0-1000 bars using a DH-Budenberg 540 VHX manual hydrolic press.

The post-trim stability was assessed by placing the trimmed samples in an oven for 80 hours at 100°C and 85 hours at 250°C. Their characteristics were periodically measured during this period of time.

Experimental Results and discussion

Typically the conductivity of trimmed samples changes a lot during the first pulses and then tends to an asymptotic value Σ_T. Fig. 1 shows the evolution of the conductivity of the V6-400 series during trimming and their thermal stability as a function of RuO_2 volume fraction. It shows the initial conductivity Σ_0, the conductivity after trimming Σ_T, the conductivity Σ_{100} at the end of the thermal treatment at 100°C and finally the conductivity Σ_{250} after the 85 hours at 250°C. It can be noticed that samples with lower filler content are more sensitive to trimming. Changes in conductivity of more than an order of magnitude can be attained for samples close to the percolation threshold as for V6-400-0.064 that has $\Sigma_T/\Sigma_0 = 69$.

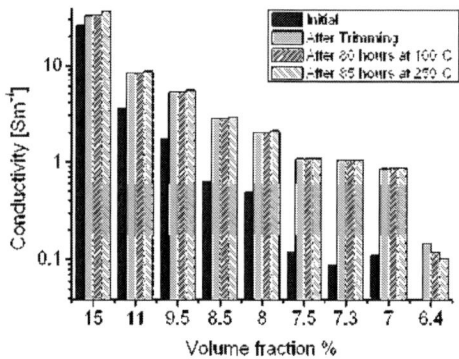

Figure 1: Change of conductivity and its thermal stability as a function of conducting filler volume fraction for the V6-400 sample set.

We notice that for all samples but one (V6-40-0.23 not shown here), the trimming lead to an increase in conductivity. More interestingly we can see that for all samples except V6-400-0.064, the thermal treatment also leads to an increase in conductivity, and not to a recovery of the initial value of the conductivity.

In fig. 2 we show the evolution of the normalized conductivity Σ/Σ_T during the thermal treatment. We can see that the change in resistivity occurs mostly during the first hours in the furnace,

followed by a slower drift. Notice that the lines are just a guide for the eye, but that the slopes of the true evolutions of the conductance are probably even larger during the first moments at 100°C and at 250°C than it seems on this graph.

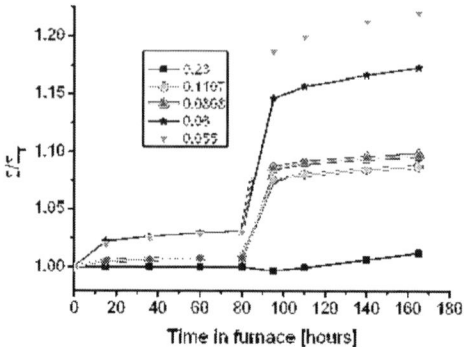

Figure 2: Evolution of the normalized conductivity of several samples from the V6-40 series (Σ_T is the conductivity of the sample after trimming). The first 80 hours were at 100°C and then 250°C.

In fig. 2 we see that samples with lower conductive filler content have a larger post trim drift for the V6-40 series, but we obtained the opposite result for the V6-400 series. Intuitively we would think that the larger the effect of trimming, the larger the post trim drift. We verify this in Fig 3 showing the relative variation of conductivity during thermal treatment as a function of the relative variation of conductivity during trimming. As we can see no clear relation can be established with our results for the V6-40 series. For the V6-400 it seems that the drift is larger for small changes of resistivity during trimming (Σ_T/Σ_0 close to 1), which goes against our intuition and is left unexplained for now.

There is one point (V6-400-0.064) that is quiet disparate from the others in fig.3. It is the closest point to the percolation threshold of this set of samples. We have shown evidence in a previous study [5] that Σ_T/Σ_0 diverges as the percolation threshold is approached, meaning that the sensitivity to voltage pulse becomes very large, which explains that we observe such a high difference for samples with very similar concentrations (the RuO2 volume fraction of the next point in the graph is only 0.5% larger).

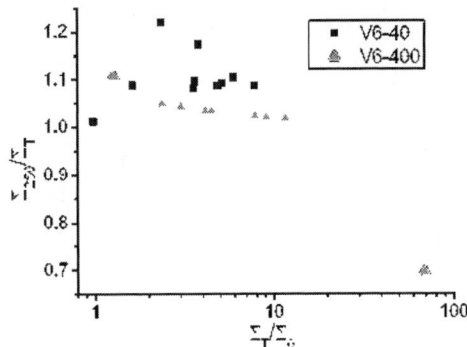

Figure 3: Post-trim drift as a function of relative change of conductivity obtained during trimming. Σ_T is the conductivity after trimming, Σ_0 the initial one and Σ_{250} after the end of the thermal treatment

For all samples the thermal coefficient of resistance (TCR) was shifted upwards by the voltage trimming as shown in fig. 4. But contrary to what is observed for the conductivity, the shift is maximal for values of the volume fraction X in the middle of the range studied here. It decreases for very small and large values of X (it almost vanishes in the V6-40 series at the extreme values of X). Moreover we see that the TCR evolves towards its original value during thermal treatment, contrary to what is observed for the conductivity (this is also the case for almost all samples of the V6-40 series).

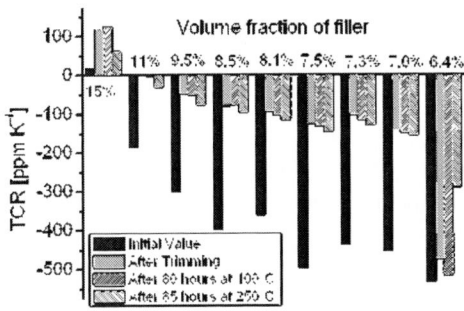

Figure 4: shift of the thermal coefficient of resistance during trimming and its thermal stability for the V6-400 series.

Let's now look at the change of piezoresistivity induced by voltage trimming. We study here the piezoresistive response coefficient G under hydrostatic pressure defined by:

$$(1) \qquad G = \frac{\partial R}{R \partial \varepsilon_z} = \left[\Gamma_T - 1 + \frac{\Gamma_L + \Gamma_T}{1 + w} \right]$$

$$(2) \qquad w = \frac{1 + v_R}{1 - v_R} \left(\frac{E_S}{E_R} \frac{1 - 2v_R}{1 - 2v_S} - 1 \right),$$

Where E and v are the elastic modulus and Poisson's coefficient and the subscripts R and S

stand respectively for the resistor and the substrate. Γ_L and Γ_T are the longitudinal and transverse piezoresistive coefficients. To calculate G from our hydrostatic measures we need to know the relation between the strain and the applied hydrostatic pressure. This is given by:

$$(3) \qquad \varepsilon_z = -\frac{1 - 2v_S}{E_S} (1 + w) P$$

Justification of those equations can be found elsewhere [7].

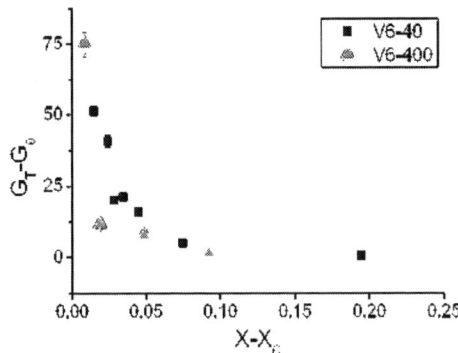

Figure 5: Change of piezoresistivity as a function of X-X_C. G_0 and G_T are respectively the initial piezoresistive response coefficient and the one after trimming

In fig. 5 we show the shift of piezoresistivity due to high-voltage trimming as a function of X-X_C, where X_C is the critical volume fraction at which the insulating/conducting phase transition occurs. For all samples G diminished after trimming and the shift seems to diverge as the percolation threshold is approached.

These results can be explained if we consider the mechanisms involved in voltage trimming and the topology of the underlying current-carrying network. We suppose that the global change of resistance and piezoresistance is due to changes in the local conductances between neighboring conducting grains. We can assume that these changes are induced by local heating, as proposed for example by Feldbaumer in [2]. The energy locally deposited in the sample is maximal in singly connected bonds with a large resistance. For samples far from the percolation threshold, the current can avoid the large local resistances, because there are several parallel routes and very few singly-connected bonds. Trimming will therefore have little influence on the characteristics of the samples with volume fractions of conductor far above the percolation threshold. As the percolation threshold is approached the probability to have singly-connected bonds with large resistances increases and trimming becomes more effective. The largest local

553

resistances will be affected by local heating, leading to their decrease which could be caused by several phenomena such as diffusion of RuO_2 into the insulating matrix, or diminution of the interparticle distance caused by the attraction due to electric potential difference and enabled through local melting of the insulating matrix.

This explains the increase of conductivity observed in our experiments and its dependence on conductive filler content. In the tunneling-percolation model the piezoresistivity is even more dominated by the large local resistances than the resistivity of the sample. Therefore it is not surprising to find a divergence of the shift in G observed in fig. 5.

Finally to support the above explanation of characteristics change due to trimming we will look at the evolution of the DC critical exponent t. It is well known that close to the percolation threshold X_C the conductivity of the sample follows a power law of the form:

$$(4) \qquad \Sigma = \Sigma_0 (X - X_C)^t$$

where Σ_0 is a prefactor. As was first shown by Kogut and Straley [8] t has a universal value of 2 in three dimensions, as long as the distribution function of the local bond conductances h(g) isn't diverging for g→0. If it does, meaning that there are very large local resistances in the sample, t can become larger than 2 and it will depend on h(g). The larger the divergence, the larger the exponent t. It is also well known (see for example [9]) that in the frame of the tunneling-percolation model, such diverging h(g) can be obtained, explaining the nonuniversal exponents observed in systems such as TFRs.

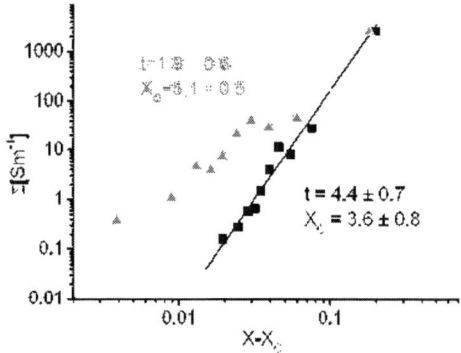

Figure 6: Conductivity of V6-40 as a function of X-X$_C$. Black squares are the values before trimming and gray triangles after trimming

In fig.6 we show the conductivity of the V6-40 series as a function of X-X$_C$ before and after trimming and the fits of equation (4) to these data (in fact the logarithm of equation (4) was fitted to ln(Σ) which gives better fits). As we can see this set of samples initially showed a nonuniversal behavior, with an exponent t=4.4±0.7. From our above considerations we expect trimming to lower the highest local resistances in the sample and therefore to diminish the strength of the divergence of the distribution function h(g). This of course should lead to a diminution of the critical exponent t and could even lead to a universal exponent provided that the divergence of h(g) is weakened enough. Indeed if we look at the gray curve in fig. 6 we see that after trimming the critical exponent has become universal, t=1.9±0.6. For the V6-400 sample the exponent changes from t=2.5±0.2 to t=1.7±0.13. In this case the diminution is still present though the change is less significant.

These results shed a new light on high-voltage trimming and support the explanations given above. High-voltage trimming changes h(g), cutting off its tail for small g values, leading to a decrease of DC critical exponent t. The fact that t diminishes explains the divergence of the change of conductivity and piezoresistivity as the percolation threshold is approached. To further verify this fact it would be interesting to perform such a study on a series of samples having a universal behavior. In that case t should remain constant and no divergence of the relative change of conductivity should be observed.

Conclusion

We have studied the changes of characteristics induced by voltage trimming on model thick-film resistors. We have shown that the relative change of conductivity increases monotically with decreasing conductive filler content. Exposure to 100°C during 80 hours and 250°C during 85 hours leads to a post-trim drift in conductance in the same direction than the trimming. On the contrary it was observed that the TCR would tend to recover its original value during thermal treatment.

We observed that the difference of piezoresistive coefficient before and after trimming diverges as the critical volume fraction X_C is approached. Most importantly we showed that high-voltage trimming changes the critical DC exponent and could lead to a crossover from nonuniversal to universal behavior. This result is important because it gives some insight on the microscopic changes occurring during high-voltage trimming and allows to explain the observed divergences of the change of conductivity and piezoresistivity.

Acknowledgements

Many thanks to S. Vionnet-Menot and M. Garcin for the fabrication of the samples and to G. Corradini for his help.

References

[1] T. Tobita and H. Takasago, "New Trimming Technology of a thick Film Resistor By the

Pulse Voltage Method", IEEE Transactions on Components, hybrids and manufacturing technology conference, Vol. 14 , pp. 613-617, September, 1991

[2] D.W. Feldbaumer, J. A. Babcock, V. M. Mercier and C. K. Y. Chun, "Pulse Current Trimming of Polysilicon Resistors", IEEE Transactions on electron Devices, Vol. 42, No. 42, pp. 689-696, April, 1995

[3] J.A. Bacock, D.K. Schroder, S.J. Prasad and K. Egan, "Precision Electrical Trimming of Very Low TCR Poly-SiGe Resistors", IEEE Electron Device Letters, Vol. 21, No. 6, pp. 283-285, June, 2000

[4] A. Dziedzic, A. Kolek, W. Ehrhardt and H. Thust, "Advanced electrical and stability characterization of untrimmed and variously trimmed thick-film and LTCC resistors", Microelectronics Reliabability, Vol. 46, pp 352-359, 2006

[5] N. Johner, T. Maeder, C. Grimaldi, A. Kambli, I. SAglini, C. Jacq and P. Ryser, "High-Voltage Sensitivity Studies of Model Thick-Film Resistors", Proceedings of the XXX International Conference of IMAPS Poland Chapter, Krakow, September 24-27, pp. 157-160, 2006

[6] S. Vionnet-Menot, T. Maeder, C. Grimaldi, C. Jacq and P. Ryser, "Properties and stability of thick-film resistors with low processing temperatures – effect of composition and processing parameters", Journal of Microelectronics and Electronic Packaging, Vol. 3, pp.37, 2006

[7] T. Maeder, B. Afra, N. Johner and P. Ryser, "Ultra high isostatic LTCC pressure sensors", Proceedings of 2007 EMPC conference, Oulu, 2007

[8] P. Kogut and J. P. Straley, "Distribution-induced non-universality of the percolation conductivity exponents", Journal of Physics C, Vol. 12, pp. 2151-2159, 1979

[9] I. Balberg, D. Azulay, D. Toker and O. Millo, "Percolation and Tunneling in Composite Materials", International Journal of Modern Pysics B, Vol. 18, No. 15, pp. 2091-2121, 2004

Analysis of Fine-pitch BGA Placement Accuracy

Johannes Hurtig and Timo Liukkonen

Nokia Corporation, P.O. Box 86, 24101 Salo, Finland
johannes.hurtig@nokia.com, timo.liukkonen@nokia.com

Abstract

Component bump pitches are becoming smaller. Lead-free soldering makes the board assembly process window tighter than ever by e.g. changing self-alignment properties. BGA type components must be placed more accurately than before. Bump recognition can be made using individual offset recognition and calculation of average offset, or using various best fit methods to maximize overlapping of bump area in expected and observed vision images. Fitting process is described in detail and new method for maximizing bump coverage is introduced. Principles are simulated with experimental study. Results and recommendations are discussed.

Key words: Placement accuracy, SMD placement, BGA components, pattern recognition

Introduction

Bump pitch in BGA (Ball Grid Array) and WL-CSP (Wafer Level Chip Scale Package) components is becoming smaller and smaller. Fine pitch technology in conjunction with lead-free process is making the assembly process window more challenging; e.g. component self-alignment properties during reflow soldering are changing. Publications indicate both larger standard deviations in placement accuracy after reflow soldering and larger average displacements before reflow, when the lead-free paste is used in board assembly process. Component placement must be better controlled within lead-free process. This is most probably due to the bigger variability in self-alignment capability during the board assembly process when using a lead-free paste [1, 2]. This means that BGA components must be placed more accurately than before on solder pads (Figures 1, 2).

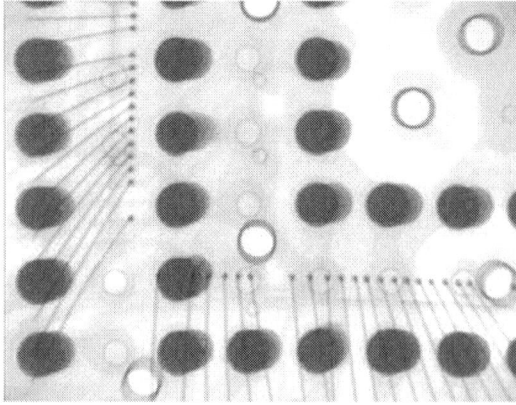

Fig 1. A misaligned BGA component after reflow, magnified in X-ray.

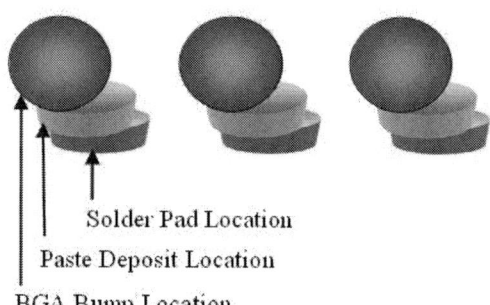

Solder Pad Location

Paste Deposit Location

BGA Bump Location

Fig 2. Some offsets in BGA placement process.

Until now many electronics manufacturers have used recognition and alignment algorithms in placement based on outline of the BGA package. This is not possible any more. Bump matrix recognition is the only way to assemble fine-pitch BGAs accurate enough in the very near future. One additional reason for this is the offset found between body centre and bump matrix on every BGA component.

In bump based offset recognition the vision system recognizes the location of each bump separately in relation to the center of the component and then calculates the average offset based on the CAD-data of expected locations and the real data of observed locations.

Placement machine manufacturers use various best fit methods for bump matrix recognition. In these methods the vision system locates the expected image (shape) of the bump matrix upon the observed image of the real bump matrix so that the total area of bump overlapping is maximized. Commercial best fit algorithms are also available together with widely used vision sensors for applications of more general nature [3].

A best fit method for bump matrix recognition is developed in this work. Method can be utilized in Vision process of placement machine but it is also applicable for placement accuracy analysis.

Development of fitting method

A. Fitting process and coverage value

Locations of bumps are measured through glass board after component placement process (see next chapter). N is the number of measured bumps. Measured coordinates for bump i are $(X_m, Y_m) = (a_i, b_i)$, $i = \{1, 2, ..., N\}$ (see Fig 3). Axes of the measurement coordinates system are denoted by (X_m, Y_m).

After measurement, the bump matrix with perfect nominal dimensions (*fitting matrix* in further text) is fitted on the measured bumps as shown in Fig 3. When the best possible fit is achieved, the centre coordinates $(X_m, Y_m) = (x_0, y_0)$ of the fitting matrix and its rotation angle θ_0 in relation to the measurement coordinate system are used as location and orientation of the measured matrix.

The bumps of the fitting matrix have coordinates $(X_m, Y_m) = (x_i, y_i)$ in the measurement coordinate system. These coordinates are determined by coordinates $(X_f, Y_f) = (c_i, d_i)$ and matrix centre coordinates (x_0, y_0) and rotation angle θ_0 as shown in equation (1). Coordinates (c_i, d_i) are the nominal x and y coordinates of bumps and these are determined in the coordinate system (X_f, Y_f). The centre of the fitting matrix is the origin of the coordinate system (X_f, Y_f). Coordinates (c_i, d_i) are easily expressed by multiples of pitch p between bumps.

$$\begin{cases} x_i(x_0,\theta_0) = x_0 + c_i \cos\theta_0 + d_i \sin\theta_0 \\ y_i(y_0,\theta_0) = y_0 - c_i \sin\theta_0 + d_i \cos\theta_0 \end{cases} \quad (1)$$

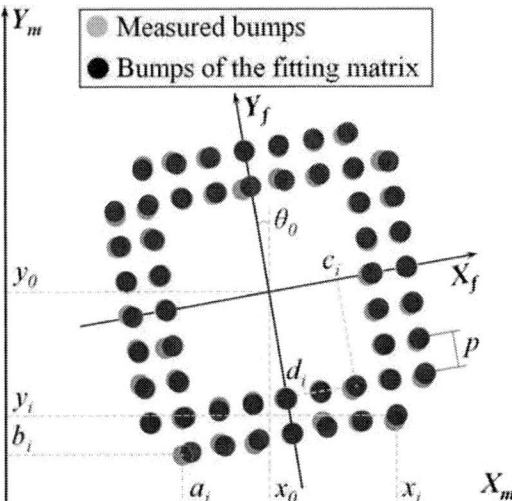

Fig 3. The fitting matrix placed on the measured bumps in the measurement coordinate system.

The perfect fit means that every bump in fitting matrix overlaps totally the corresponding bump in the measured matrix. The criterion built in this study is based on coverage value *cov*, which tells the proportion (%) of overlapping area of total bump area. Calculation method for coverage is shown in equation (2). Bump radius is denoted by r. Distance D_i, which also occurs in equation (2), is the distance between the measured bump i and the bump i of fitting matrix and it is introduced in equation (3).

$$\text{cov}(x_0, y_0, \theta_0) \\ = \frac{\sum_{i=1}^{N} \left[2r^2 \cos^{-1}\left(\frac{D_i}{2r}\right) - D_i \sqrt{r^2 - \frac{D_i^2}{4}} \right] \cdot 100\%}{N\pi r^2} \quad (2)$$

$$D_i(x_0, y_0, \theta_0) \\ = \sqrt{(a_i - x_i(x_0,\theta_0))^2 + (b_i - y_i(y_0,\theta_0))^2} \quad (3)$$

Best possible fit of the fitting matrix on the measured matrix is reached by maximizing coverage *cov*. Maximum coverage can be achieved by minimizing the sum of squared distances D_i. Calculation method for the sum $S(x_0, y_0, \theta_0)$ of squared distances is shown in equation (4).

$$S(x_0, y_0, \theta_0) = \sum_{i=1}^{N} [D_i(x_0, y_0, \theta_0)]^2 \quad (4)$$

The minimum of sum S can be found by setting partial derivatives $\partial S/\partial x_0$, $\partial S/\partial y_0$ and $\partial S/\partial\theta_0$ to zeros. Solutions for x_0, y_0 and θ_0 at the zeros of partial derivatives are given in equation (5), (6) and (7), respectively.

$$x_0 = \frac{1}{N}\sum_{i=1}^{N}(a_i - c_i) \quad (5)$$

$$y_0 = \frac{1}{N}\sum_{i=1}^{N}(b_i - d_i) \quad (6)$$

$$\theta_0 = \arctan\left[\frac{\sum_{i=1}^{N}(a_i d_i - b_i c_i)}{\sum_{i=1}^{N}(a_i c_i + b_i d_i)}\right] \quad (7)$$

If bumps are symmetrically distributed around the centre of matrix $(X_f, Y_f) = (0, 0)$, equations (5) and (6) are simplified, because averages of c_i:s and d_i:s are then zeros.

B. Classification of bump measurements

Measurement results for bump locations include noise and due to nature of optical measurement also some larger measurement errors can occur. Because the number of the measured bumps is usually quite large, it is often reasonable to leave some poor measurements out and determine location and orientation of matrix by utilizing only the remaining bumps. Algorithm for classification of bump measurements is introduced in details in the next chapter.

In the classification algorithm, bumps are compared to each other by calculating difference in distances between their nominal locations and their measured locations. If a bump is poorly measured, large deviations should occur in comparisons with neighboring bumps. Offsets e_{ij} are calculated for every possible bump pairs i and j according to equation (8). Offset e_{ij} is the absolute value of difference between nominal and measured Euclidean distances d_{ij}^0 and d_{ij}^m of bumps i and j.

$$
e_{ij} = \left| d_{ij}^m - d_{ij}^0 \right| \\
= \left| \sqrt{(a_j - a_i)^2 + (b_j - b_i)^2} - \sqrt{(c_j - c_i)^2 + (d_j - d_i)^2} \right|
\tag{8}
$$

$N(N-1)/2$ combinations have to be calculated to go through all possible pairs within N bumps. Pairs are sorted in ascending order by their e_{ij} value. Points are given for bumps based on their occurrences in the created table: bumps i and j in n^{th} row in the table get both n points. After points are given based on all $N(N-1)/2$ rows, the poorest bump measurements have the highest points and the most accurate measurements have the lowest points.

After points are determined for all bumps the poorest measurements can be rejected prior to fitting process. Coverage value can be used as a rejection criterion: one bump at a time is rejected based on its points (bump with the highest points first) and x_0, y_0 and θ_0 are calculated and coverage is determined after every rejection event. When some defined minimum number of bumps K (K could be e.g. $N/2$) is left, the best coverage so far is selected and bumps are rejected according to this result.

C. Classification algorithm

Calculating points

1) Values e_{ij} are calculated for all possible pairs i and j. Values are placed on the table with three columns: i, j and e_{ij}.

2) Table is sorted by e_{ij} value in ascending order.

3) Points are given for bumps based on their occurrences in the table, bumps i and j in the nth row of the table get both n points and n = {1, 2, ..., $N(N-1)/2$}

Rejecting bumps

4) After all points are given, the initial coverage is determined by performing fitting process with all bumps and using its results (x_0, y_0 and θ_0) for coverage calculation.

5) Bump M_1 with the highest points is left out from fitting process and it is done with bumps {1, ..., $M_1 - 1$, $M_1 + 1$, ..., N} resulting values for x_0, y_0 and θ_0. Coverage cov_1 is calculated with all bumps {1, 2, ..., N} to get idea, how good fit is

achieved by fitting process without bump M_1. Next cov_2 is calculated by leaving M_1 and M_2 out from fitting process (M_2 is the bump with the 2nd highest points) and process is continued until some reasonable limit K is reached. K could be e.g. $N/2$.

Selecting the best combination of bumps

6) Group of bumps, which produces the maximum coverage, is selected because it provides best fit. Values for x_0, y_0 and θ_0 for component are then calculated with this group.

Experimental setup

A glass board experiment was designed to analyze methods in practice. Fine-pitch BGA components were assembled on a sticky-taped glass board with a modern fine-pitch surface mounted device placement machine. After placement process each bump location was measured through glass using high-accurate 3D measuring device, the measurement capability of which was measured and analyzed beforehand (see next chapter).

Locations and orientations of bump matrices are then analyzed utilizing fitting and classification methods introduced in previous chapter. Finally the component-specific measurement results were analyzed using six sigma tools provided by Minitab software.

Because desired manipulation of algorithms inside a placement machine software is not possible, the following experiment was designed to emulate placement machine's calculation principles for BGA placement. 20 pieces of BGA component each with 28 bumps to be inspected (see Fig 5) were placed on a glass board. Then the same procedure was repeated so that one bump row (7 bumps) in the vision recognition description file (targeted for the 28 bumps) was modified and shifted 150 μm in y direction (half bump diameter), to see if the placement coordinate on the glass board shifts accordingly or not. From now on the expected image and observed image did not match perfectly any more.

It was expected that machine would not reject any bumps (in this case bump rows) and would do a simple averaging of the whole observed data; thus moving the component by 37.5 μm (7·150 μm / 28). Resulting bump measurement through glass board showed that observed movement in y was ca. 40 μm, very close to expectations (see picture from Minitab below). Mean coverage with the modified vision description file was 86.0 % and with the original non-modified file 89.5 %.

Experiment shows that e.g. in a case of a bump having lot of offset the machine will place all the bumps non-ideally because of averaging of the bump matrix. However rejection of bad bumps in calculation of the coordinate (i.e. maximizing bump coverage) would give better placement accuracy and better solder joint reliability. WL-CSP packages do not have very many bumps and one bad bump may easily produce slightly misaligned component after reflow due to poorer self-alignment properties of lead-free paste. This could result in poorer mechanical solder joint reliability of the component. Half bump misalignment of one or two individual bad bumps would still produce perfect electronic connection but much better reliability with most of the bumps being perfectly aligned in solder joints.

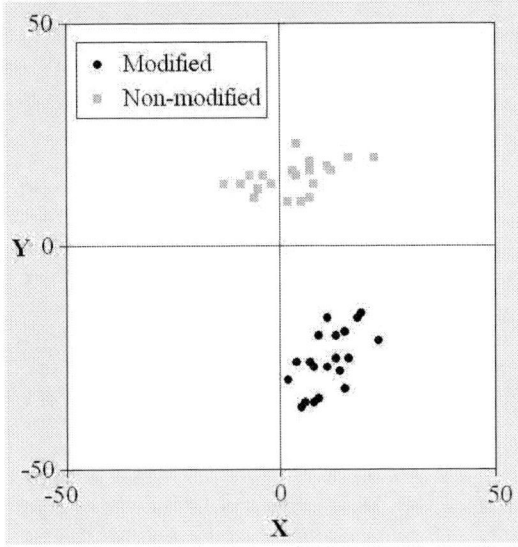

Fig 4. Scatterplot of measured components placed with modified and non-modified (i.e. original) vision description file.

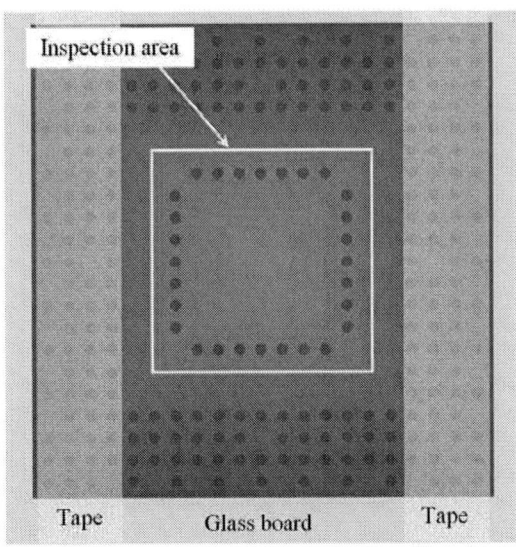

Fig 5. BGA component placed on tapes stuck on the glass board. The inner matrix (28 bumps) can be measured through the glass board.

Gage for 3D measurement device

Bumps were measured with an optical 3D granite-based measuring device having resolution of 0.1 μm and measuring accuracy 3.0 μm. For Gage R&R (Repeatability and Reproducibility) test one populated glass board was measured ten times, totally including 640 components and 23040 observations (bumps). Total GR&R result was good: % Contribution was 1.58% and Precision to Tolerance value 15.18% (comparable acceptance criteria for "good" are <4% and <20% [4]). Gage was done with a couple of randomly selected components. Result shows that 3D measuring device is fully capable as a measurement system for the case.

Discussion and conclusions

Coverage is used in this study for evaluating the goodness of fit during the introduced fitting process. Coverage is also convenient criterion for evaluating results of the whole placement process. When the proportion of contact pad area on printed wired board (PWB), which is covered by bumps of BGA component, is given as percentage value, it provides direct insight into pre-soldering situation in solder joints. Coverage value includes the effects of placement inaccuracy in x, y and rotational directions and combines these effects into one percentage value.

Future work

Most probably creating generally applicable criteria for those bump measurements that should be rejected will be a topic for future investigations.

References

[1] Nguty, T.A., Budiman, S., Rajkumar, D., Solomon, R. and Ekere, N.N., "Understanding the process window for printing lead-free solder pastes", Proceedings of the 50th IEEE Electronic Components and Technology Conference, May 2000, Las Vegas, USA, pp. 1426-143

[2] Liukkonen, T., Nummenpää, P. and Tuominen, A., "The effect of lead-free solder paste on component placement accuracy and self-alignment during reflow ", Soldering & Surface Mount Technology, Vol. 16, No. 1, 2004. Pp.44-47.

[3] "Create best fit lines and circles through a series of points", Products – Vision Tools – Gauging Tools – Best Fit Geometry. Available (February 2007) at http:// www.cognex.com.

[4] Breyfogle, F., "Implementing Six Sigma", 2nd Edition, John Wiley & Sons Inc., New Jersey, USA, 2003, pp. 306-346.

Fine Line Technology And Panel Plating
- Opposing Directions, One Solution

Stephen Kenny and Bert Reents

Atotech Deutschland GmbH, Berlin, Germany

E-mail: Stephen.Kenny@atotech.com, Bert.Reents@atotech.com

Abstract

The increasing requirements for high density interconnect can be summarised as a demand for ever finer line and spaces together with the need for improved high frequency operating characteristics. Both of these factors must be met under an ever increasing price pressure this now coming from significant increases in raw materials as well as the expectations from the OEM. The standard method to achieve fine lines and spaces is by using a variation of pattern plate copper metallisation, however this technique suffers from the well known variation in surface distribution due to varying track density and width. The resulting track profile variation is difficult to reconcile with the increasing demands for high frequency application and in particular a narrow impedance control range. In contrast the technique of panel plate copper metallisation offers the advantage of best possible surface distribution over the whole surface of a circuit together with uniformity of production giving obvious benefits for high frequency application. The critical disadvantage of the panel plating technique is that the required thickness of copper to achieve blind micro via filling, now a common requirement in HDI, cannot be etched to give the line and space tolerances which will be required. This paper presents latest results with the Uniplate InPulse 2 horizontal copper plating system which combines excellent surface distribution together with a novel panel plate metallisation which ensures filled blind micro vias with minimum surface plated copper. Blind micro vias typically seen in hand held devices with 70 µm depth and 100 µm diameter can be easily filled with only 15 µm copper deposited on the surface. This process offers the possibility to meet the requirement for 50 µm line and space. Also due to the low thickness of plated copper, savings in materials are very significant particularly in copper metal but also in solder mask and etching chemistry. The process has already reached a high acceptance in the mass production of HDI circuit boards.

Key words: Electrolytic copper, InPulse 2, Uniplate, Fine lines, Blind micro via filling.

Introduction

Production processes for blind micro via filling are currently in widespread use for both IC substrates and in the recent development for hand held applications. Reference [1] discusses the implementation of such processes in both vertical and horizontal systems. Standard processes for blind micro via filling for hand held devices can require plating of up to 30 µm to achieve a remaining dimple of less than 10 µm depending on the aspect ratio of the vias and the drilling quality. This copper thickness requirement is very dependant on the substrate material used for the micro vias. A significant quantity of production is now being made for mobile phone applications using FR4 glass fibre reinforced materials which are more demanding for the laser drilling process in comparison to RCF types of substrates. In this aspect the cost of drilling becomes critical and is very often the driving force which leads to poor final micro via shape after drill. Generally the more difficult the blind micro via for the filling process the more copper must be plated to reach the required quality of final dimple. The schematic shown in figure 1 illustrates the variation which may be seen in blind micro via shape after drilling, the easiest via for the copper plating processed is normally the most time consuming or the most expensive for mass laser drill production. Coupled to this is the inherent variation seen in the glass reinforced material due to the weave of the glass itself.

The "tapered" shape is the preferred blind micro via for copper plating but it is seldom seen except possibly in IC substrate applications where homogenous base material and low volume blind micro vias make laser drilling less demanding.

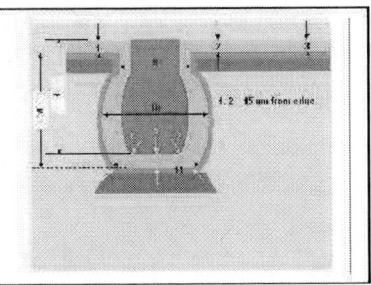

"Barrel" Shape

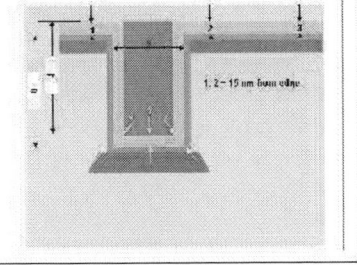

"Straight" Shape

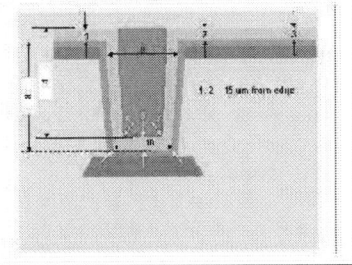

"Tapered" Shape

Figure 1: Schematic to show blind micro via variation after drilling

Figure 2 shows a typical filled blind micro via from a vertical processing system, the blind micro via form is a "barrel" shape with a slight over hang of copper foil together with protruding glass fibres more on one side of the via than the other.

Surface plate copper 30μm
Diameter 110 μm
Depth 75 μm

Figure 2 Filled blind micro via produced in vertical processing equipment.

The process used was panel plating in DC using insoluble anodes to give the best possible surface distribution. The applied current density was 1.5 A/dm² which means a plating time of approx 90 minutes was necessary to give a dimple of less than 10μm.

The demand for fine line and space is an ever present fact of life for high density interconnect. Currently 75 μm line and space is standard for mobile phone applications and requirements for 50 μm line and space are already in discussion. The panel plating process as illustrated in figure 2 above is limited in the capability for fine line imaging simply by the plated copper thickness. The problem may be overcome by use of thin copper foil but assuming a copper foil of 5 μm there will be a line and space limit of approx. 75 μm to 80 μm using the panel plating technology illustrated.

To overcome the limitations pattern plating may be considered or a combination of pattern and panel plating but then the difficulties of pattern plate surface distribution will be apparent. This aspect cannot be ignored when high frequency applications with defined impedance control range are required, pattern plating cannot meet the standard in surface distribution together with the productivity at high current densities of panel plating. The demands of blind micro via filling also pose problems for pattern plating, figure 3 illustrates this with the dimple variation depending on copper plated thickness.

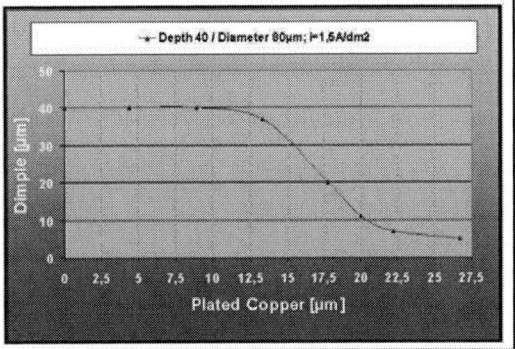

Figure 3: Variation in dimple with plating thickness for blind micro via filling.

Figure 3 shows the dimple variation in a relatively simple blind micro via diameter 80 μm and depth 40 μm with varying copper thickness plated on the surface. With 10 μm plated copper the dimple is still 40μm, no filling has occurred and the dimple is equivalent to the original micro via depth. After plating of 20 μm on the surface the dimple has reached approx. 10μm, normally the minimum requirement for a blind micro via filling process. In the surface plating range between 15 μm to 20 μm there is a dimple reduction from 30 μm dimple to 10 μm dimple. This means that an absolute surface copper thickness variation of 5 µm can give a dimple variation of 20 µm.

Current and future applications in hand held devices require blind micro via filling with low dimple and thin copper plating together with a uniform plating quality over the entire panel surface. Together these factors can enable fine line production with possible 50 µm line and space tolerance.

Fine Line Production In InPulse 2

The InPulse 2 system gives an inherently good copper plated surface distribution due to the use of insoluble segmented anodes as shown in figure 4.

Segmented insoluble anode showing the four individually controlled segments

Figure 4: Segmented anode unit to ensure optimum surface distribution.

Production panels with thin copper foil even down to 1 µm can be successfully plated using the specially designed equipment. Also standard blind micro via filling can be made as shown in figure 5.

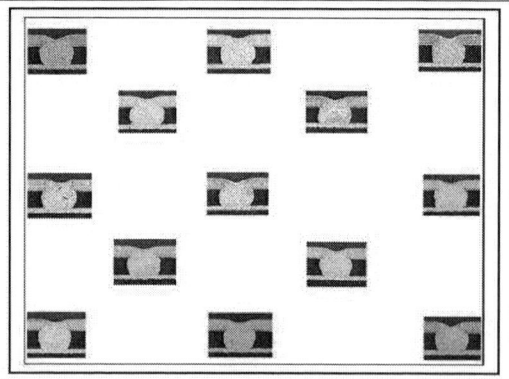

Micro-section results of filled blind micro vias over entire panel

Figure 5: Uniformity in blind micro via filling over standard panel.

Standard blind micro via filling uses standard pulsed plating parameters and will achieve results as shown in detail in figure 6.

Diameter: 110 µm
Depth 80 µm
Plating thickness 27 µm
Dimple 0 µm

Figure 6: Standard blind micro via filling result in horizontal process equipment.

The key to fine line production in panel plate with the InPulse 2 is the use of modified pulse parameters as described in [2]. Generally use of strong reverse pulse plating current density gives a significant reduction in the plating time needed to achieve the required dimple. The inorganic parameters are modified to promote blind micro via filling, this means higher copper concentration in relation to sulphuric acid and the organic additives must be monitored in the correct range. Using the "Super Filling" process the surface plated copper required is reduced whilst maintaining dimple after filling at less than 10 µm. Figure 7 shows an example of "Super Filling" with non-ideal blind micro via.

The filling quality is good in this example, there is no dimple seen in micro-section and the filling result is uniform over the panel but the plated copper thickness is in the range as seen in vertical DC plating systems and cannot reach the target of 50 µm line and space. In contrast however to vertical systems with 60 to 90 minutes plating time the higher applied current density means that the plating time will be in the region of 30 to 40 minutes. Even standard blind micro via filling in horizontal systems has production time advantages over vertical systems.

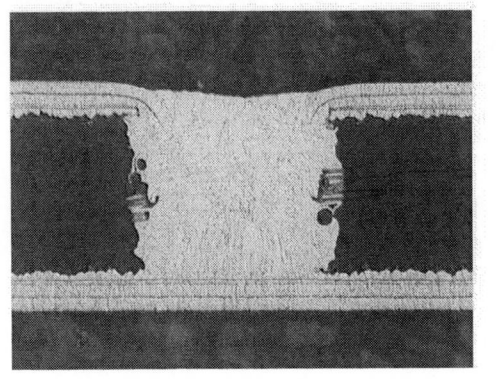

Diameter 130 µm
Depth   110 µm
Plating Thickness 20 µm
Dimple < 10 µm
Plating time 29 minutes

Figure 7: "Super BMV Filling" with protruding glass fibres in FR4 material.

The result in figure 7 is an excellent filled blind micro via with strongly protruding glass fibres but dimple less than 10 µm.

The production result in figure 8 shows a filled blind micro via with a plating thickness less than 15 µm. Assuming a copper foil of 5 µm and metallisation of 2 – 3 µm the total copper thickness to be etched is 20 –25µm, this can meet the requirements for fine line production.

Diameter 90 µm
Depth 65 µm
Plating Thickness 12 µm
Dimple < 5 µm
Plating time 25 minutes

Figure 8: "Super BMV Filling" with minimum surface thickness.

Summary

Horizontal plating systems equipped with insoluble anodes have offered advantages for production quality and cost savings for some years.

The elimination of anode maintenance and constant surface distribution are key factors in uniform production quality together with the potential to have wet to wet production of metallisation and copper plating. Higher applied current densities can be used in particular for blind micro via filling which gives advantages in processing time over conventional vertical and also vertical conveyorised equipment.

True "Super Filling" now gives further advantages for blind micro via filling from the technical side as well as offering large cost savings for the highest technology production. The low thickness of plated copper is an enabling technology which allows fine line and space to be realised with panel plate. The obvious cost savings when less copper is plated can mean in the region of US$1.5 saving per plated m² panel depending on market copper price. The excellent surface distribution is necessary for good blind micro via filling distribution but this also gives potential for savings in solder mask as illustrated in the schematic in figure 9.

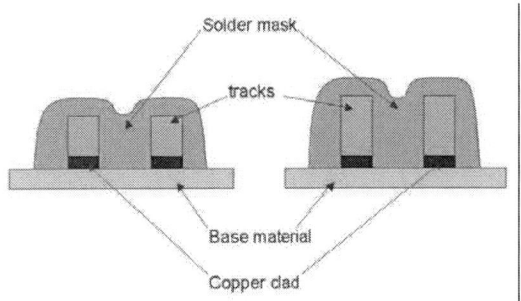

Figure 9: Schematic of savings possible in solder mask with reduced copper thickness.

The uniform plated copper over the panel and the entire production run means that less solder mask is consumed, also equipment set up is simplified due to the uniformity. Further cost savings are possible due to lower thickness of copper to be etched and also the necessary effluent treatment can be reduced.

The "Super Filling" process is already in mass production for mobile phone applications for the latest production models.

References

[1] S. Kenny and B. Reents, "Production processes in horizontal and vertical technology for blind micro via filling", EIPC 2004 Proceedings of the European PCB Convention.

[2] B. Chao, S. Chien and B. Reents, "Horizontal plating process for blind microvia and through hole filling", TPCA 2006 Proceedings of the TPCA Forum.

Electro-ultrasonic Spectroscopy of Polymer Based and Cermet Thick Film Resistors

V. Sedlakova

Physics Department, Brno University of Technology, Technicka 8, 616 00 Brno, Czech Republic

Tel/Fax: +425 7261 666, E-mail: sedlaka@feec.vutbr.cz

Abstract

We have studied the quality of polymer based and cermet thick film layers. The samples were made using different resistive and conducting pastes and dipping silvers. We have applied new principle for non-destructive testing of conducting solids which is based on the phonon interaction with conducting electrons. The ultrasonic signal changes the contact area between conducting grains and then resistance is modulated by frequency of ultrasonic excitation. Resultant intermodulation voltage depends on the value of ac current varying with frequency f_E and on the ultrasonic excited resistance change ΔR varying with frequency f_U. The intermodulation component of frequency $f_m = f_E - f_U$ varies linearly with electric excitation. The dependence of the intermodulation component on the ultrasonic excitation is influenced by the resistive layer material. The amplitude of the intermodulation component U_m is proportional to the ultrasonic voltage $U_U{}^a$, with the power $a=0.7$ to 2. For high values of ultrasonic excitation the saturation of U_m is observed. We have used the short current pulse from capacitor discharge as a stressing method for resistor degradation. The amplitude of intermodulation component increases after resistor degradation. The method can be used a diagnostic tool for thick film resistor quality and reliability assessment.

Key words: Electro-Ultrasonic Spectroscopy, Cermet, Polymer Based, Thick Film Resistors

Introduction

Electrical conductivity of cermet or C/Gr thick resistive film is smaller than that of the bulk material. For the explanation of conductivity it is supposed, that the charge carrier transport is due to: i) drift in electric field through the point contact between conducting particles, ii) thermionic emission of electrons from conducting particle and theirs transfer to the other particle by tunnelling through the thin glass or polymer layer, iii) Poole-Frenkel emission, and iv) Schottky emission. The thickness of glass or polymer layer between two conducting particles and potential barrier between the particle and glass or polymer vehicle determine the tunnelling, Poole-Frenkel and Schottky current component. Drift of charge carriers in conducting particle, g-r process on the interface between two particles, and tunnelling are sources of noise. The noise spectrum of 1/f type is observed in most of the samples [1, 4]. The noise is related to the microscopic sample structure and then it gives information on the manufacturing technology. The magnitude of noise spectral density is dependent on the resistive material composition, on the processing factors, on the size and shape of the resistor, etc. Both the noise and the resistance are affected by the thickness of glass or polymer layer between the grains, and by the effective area of the point contact between the neighbouring grains. The properties of point contacts can be changed by the ultrasonic vibrations. That leads to the modulation of resistance by the frequency of ultrasonic excitation.

Standard measuring method of resistor quality is based on the distortion of pure harmonic signal by the nonlinearity of resistance. The non-linearity of the thick film layer structure is connected with the different kinds of defects like cracks etc. High voltage pulse stressing can be used as a method of accelerated ageing for screening of unstable devices. Resistance drift is correlated to the change of non-linearity after the pulse stressing [5].

Electro-Ultrasonic Spectroscopy

We have applied new principle for the non-destructive testing of conducting solids which is based on the non-linear ultrasonic spectroscopy. There are two signals – electrical ac signal of frequency f_E and ultrasonic signal of frequency f_U. This method is based on the monochromatic phonon interaction with conducting particles. The ultrasonic signal changes the contact area between conducting grains and then resistance is modulated by frequency of ultrasonic excitation. Resultant voltage u_m on the measured sample depends on the ac current varying with frequency f_E and resistance change ΔR varying with frequency f_U. Voltage u_m is given by

$$u_m = I_M \sin \omega_E t \cdot \Delta R_M \sin \omega_U t \qquad (1)$$

Where: I_M - *electric current amplitude*

ω_E, ω_U - *angular frequency of electric and ultrasonic excitation*

ΔR_M - *amplitude of the resistance change due to the ultrasonic excitation*

Resultant signal has the frequency f_E - f_U and f_E + f_U. Our measurement was performed in the low frequency range for the intermodulation frequency $f_m = f_E$ - f_U.

The block diagram of electro-ultrasonic spectroscopy set-up is shown in Fig. 1.

Measured voltage has discrete values in the frequency domain and the amplitude of the intermodulation component follows from:

$$u_m = \frac{1}{2} U_m \sin(\omega_E - \omega_U) \cdot t \qquad (2)$$

Where: U_m - *amplitude of the intermodulation component*

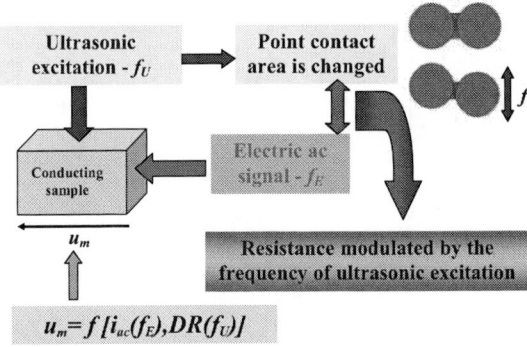

Figure 1: Block diagram of electro-ultrasonic spectroscopy set-up

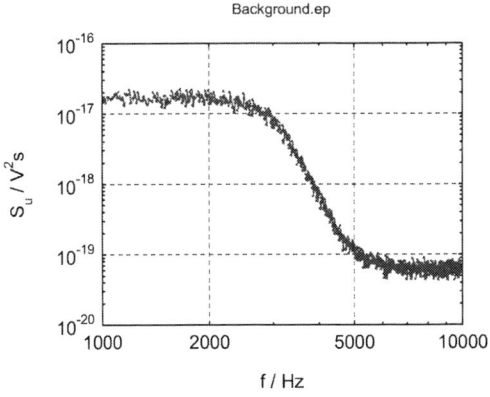

Figure 2: Measuring set-up background noise

We have used the measuring set up for the ultra low noise measurements and then we measured the resultant intermodulation signal spectral density S_U. We can calculate the amplitude of the intermodulation signal U_m from the measured noise spectral density S_u.

$$U_m = \sqrt{S_u \cdot \Delta f} \qquad (3)$$

Where S_u - *the value of the signal spectral density on frequency f_m*

Δf - *the distance between two successive lines in the noise spectra*

The background noise voltage of the measuring set-up is of the order of 2×10^{-17} V^2/Hz (see Fig. 2). The signal of the intermodulation component U_m is monochromatic as is shown in detail on Fig. 3.

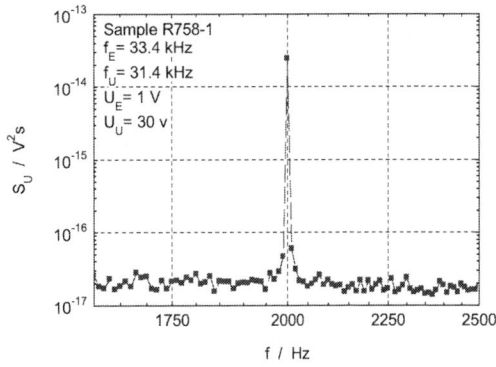

Figure 3: The signal of the intermodulation component

Measuring set-up

The block scheme of the electro-ultrasonic measurement setup is shown in Fig. 4. It consists of two parts, the electric and the ultrasonic one.

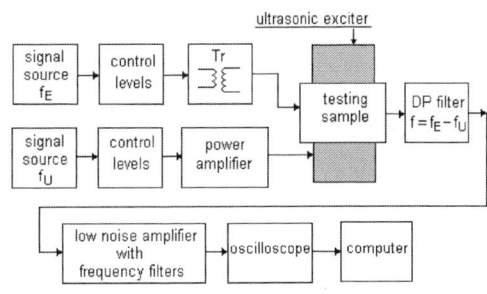

Figure 4: Electro-ultrasonic measurement set-up

The ultrasonic part consists of generator Agilent 33220A. The power amplifier consists of WPD 100 in which it is necessary to have power linear actuating harmonic signal on ultrasonic transducer. The measured sample was fixed on the power piezoceramic transmitter (HTP04) which is used for ultrasonic signal generation. Electric part consists of generator Tesla BM492 which has convenient linearity and frequency stability. Signal from the generator is transformed on higher voltage from transformer Tr. This signal is led to the measured sample over the protective resistor. Harmonic signals of frequencies higher than the differential frequency component $f_m = f_U - f_E$ are trimmed by the low pass passive filter. The passive filter has limited frequency 4200 Hz with inhibition 50 dB per decade. The amplifier (AM 22) has the adjustable input gain in the range from -20 to 50 dB by 10 dB step, the frequency band filter with lower frequency 30 mHz, 300 mHz, 0.3 Hz, 3 Hz, 30 Hz, 300 Hz, 3 kHz, 30 kHz and 300 kHz, the high frequency filter adjustable in range 3 Hz, 30 Hz, 300 Hz, 3 kHz, 30 kHz and 300 kHz, and adjustable output gain in range from 0 to 50 dB by 10 dB. All parameters are programmed over GPIB or on the front panel of the amplifier. The amplified signal is led to the A/D converter. As the A/D converter is used digital oscilloscope Agilent 54624A with sampling rate 200 Msa / s. The digitized signal is stored in the computer and signal spectral density frequency dependence is evaluated using discrete FFT. The control software was written in Borland C++ Builder and this version is based on Windows operating system.

Samples

Two sets of thick film resistors were studied.

SET 1 - The samples were made with different types of resistive pastes (C/Gr particles suspended in different polymer vehicles). Resistive paste was applied on the alumina substrate of dimensions 5 by 40 mm. Resistive layer thickness was about 20μm. The contacts were made by DiAg – one type used for all the samples. The dependence of the amplitude of intermodulation signal U_m on the electric and ultrasonic excitation was measured for these samples. Sample structure is shown in Fig. 5.

Figure 5: Sample structure – SET 1

SET 2 - Thick conducting films were made with DuPont resistive pastas DP2041 and HS8039 with sheet resistance 10 kΩ/square. Samples with nominal dimensions from 0.5×0.5 mm^2 were screen printed on the alumina substrate.

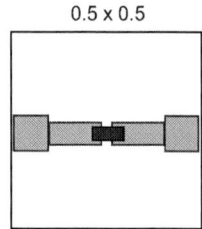

Figure 6: Resistor test pattern, resistor size 0.5 x 0.5 mm^2

Samples were terminated with pre-fired Pd/Ag thick film conductors. Resistor test pattern is on Fig. 6. We have used short current pulse from capacitor discharge as a stressing method for the resistor degradation. We have measured the amplitude of intermodulation signal A_m and non-linearity of cermet resistors before and after the pulse stressing.

Experimental results

We have tested the dependence of the intermodulation component signal on the amplitude of the electrical excitation and the results are shown in Figs. 7 and 8. Electric current is linear function of electric voltage U_E and then the intermodulation component U_m varies linearly with the electric excitation. These results are measured for both cermet and polymer based thick film resistors.

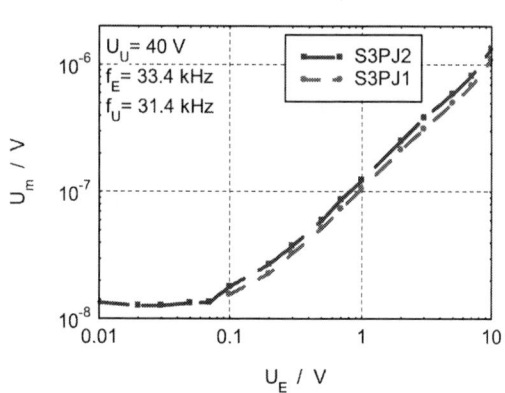

Figure 7: Amplitude of the intermodulation component U_m vs. electric excitation for $U_U = 40V$ at frequency $f_U = 31.4$ kHz and $f_E = 33.4$ kHz, for two cermet samples – resistive paste DP2041, AgPd conductors, resistor size 0.5 x 0.5 mm^2

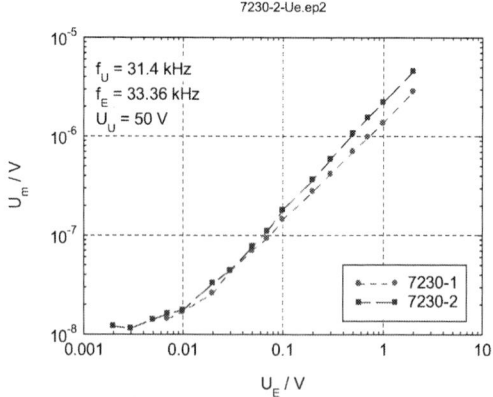

Figure 8: Amplitude of the intermodulation component U_m vs. electric excitation for $U_U = 50V$ at frequency $f_U = 31.4$ kHz and $f_E = 33.36$ kHz, for two samples of SET 1 – made by one resistive paste – No. 7230-1 and 7230-2

The dependences of the intermodulation component signal on the amplitude of the ultrasonic excitation are shown in Figs. 9 and 10.

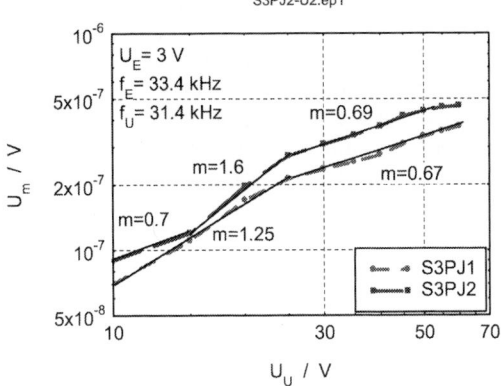

Figure 9: Amplitude of the intermodulation component U_m vs. ultrasonic excitation for $U_E = 3V$ at frequency $f_U = 31.4$ kHz and $f_E = 33.4$ kHz, for two cermet samples – resistive paste DP2041, AgPd conductors, resistor size 0.5 x 0.5 mm^2

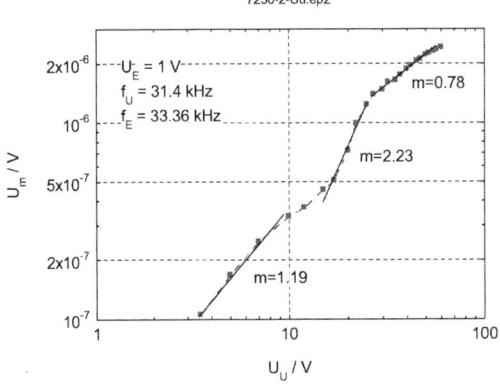

Figure 10: Amplitude of the intermodulation component U_m vs. ultrasonic excitation for $U_E = 1V$ at frequency $f_U = 31.4$ kHz and $f_E = 33.36$ kHz, for one sample of SET 1 – No.7230-2

The amplitude of the intermodulation component U_m is proportional to the ultrasonic voltage U_U^a, with the power $a = 0.7$ to 2. Higher values of a are observed for polymer based samples. For high values of ultrasonic excitation the saturation of U_m is observed. We suppose that the ultrasonic vibrations affect the effective area of conducting particles interaction in the resistive layer and hence the shape of dependence U_m vs. ultrasonic excitation is influenced by resistive layer material.

The dependences of the intermodulation signal spectral density S_U on the amplitude of the electrical excitation for two cermet samples (different resistive paste, Ag contacts) and one polymer based sample are shown in Figs. 11 and 12. Sensitivity of this method depends on the measuring set-up background noise spectral density. In Figs. 11

and 12 the background noise spectral density is about $1,6 \times 10^{-17}$ V^2/Hz, which corresponds to the equivalent noise resistance 1 kΩ. This value affects the measuring set-up resolution. For sample of resistance 1 kΩ the testing can be performed for the current 1 mA. Then the amplitude of intermodulation signal will be for more then one order above the background noise. In this case we can avoid the sample heating.

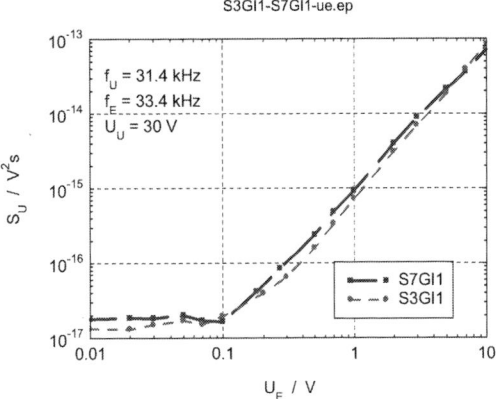

Figure 11: Intermodulation signal spectral density S_U vs. electric excitation for $U_U = 30V$ at frequency $f_U = 31.4$ kHz and $f_E = 33.4$ kHz, for two cermet samples – S3GI1 (resistive paste DP2041, Ag conductors, resistor size 0.5 x 0.5 mm^2) and S7GI1 (resistive paste HS8039, Ag conductors, resistor size 0.5 x 0.5 mm^2)

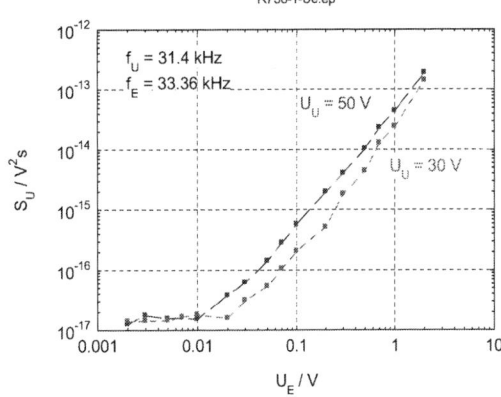

Figure 12: Intermodulation signal spectral density S_U vs. electric excitation for $U_U = 30V$ and 50V at frequency $f_U = 31.4$ kHz and $f_E = 33.36$ kHz, for one sample of SET 1 – No.R758-1

Thick Film Resistor Quality Evaluation

We have used short current pulse from capacitor discharge as a stressing method for resistor degradation [5, 6]. The high current density value leads to power density increase in the vicinity of defects and other imperfections. This creates destructive local changes in these sites. A minimization of this destructed area can be supposed with shortening of pulse time. This principle of local destruction by pulse stressing corresponds to accelerated destructive process in a resistor with low reliability.

The stressing circuit is shown in Fig. 13. The capacitor C is used as a source of constant energy pulse. The high voltage stress is applied by switching of tested resistor R_x to capacitor C. The total pulse energy is $E = \frac{1}{2} C_T . U_0^2$, where C_T is capacitance of capacitor C, U_0 is the voltage on capacitor C before it's switching on tested resistor. Current peak value is $I_{max} = U_0/R_x$. Sample stressing depends both on the total energy and current peak value.

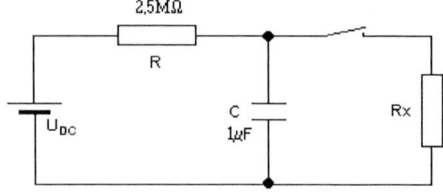

Figure 13. Testing circuit for current pulse stressing

The value of amplitude of intermodulation component U_m after the pulse stressing is shown in Fig. 14. We have applied 3 pulses, the voltage U_0 was 280 V for the first and second pulse, and 320 V for the third pulse, respectively. Resistance of measured sample was 8.6 kΩ. The highest change of amplitude of intermodulation component U_m is induced by the first testing pulse (see Fig. 14).

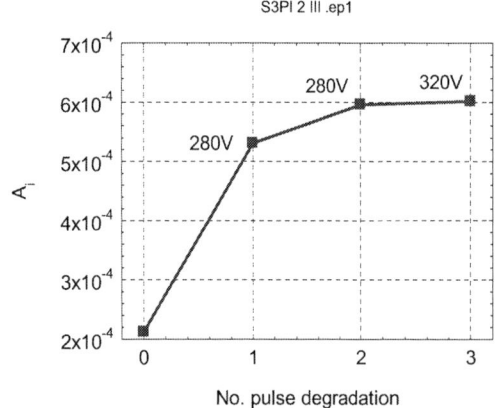

Figure 14. Amplitude of intermodulation component U_m vs. the number of testing pulses for U_U =28 V at frequency f_U = 20 kHz, f_E = 22 kHz and U_E = 10V

Standard measuring method of resistor quality is based on the distortion of pure harmonic signal by nonlinearity of resistance. We have applied the signal of the frequency 10 kHz and we have measured the response on the frequency 30 kHz. The value of the third harmonic voltage (THV) before and after the pulse stressing is shown in Fig. 15. We can see that the highest change of THV is induced by the first current pulse. Comparing results shown in Figs. 14 and 15 we can see, that the amplitude of intermodulation component U_m increases approximately 3 times, while the THV value changes for less than 5%. Current pulse stressing and subsequent nonlinearity measurement is used as a method for resistors quality screening [5, 6]. The application of new method where the amplitude of intermodulation component U_m is measured allows using lower stressing pulse energy to evaluate quality and reliability of thick film resistors.

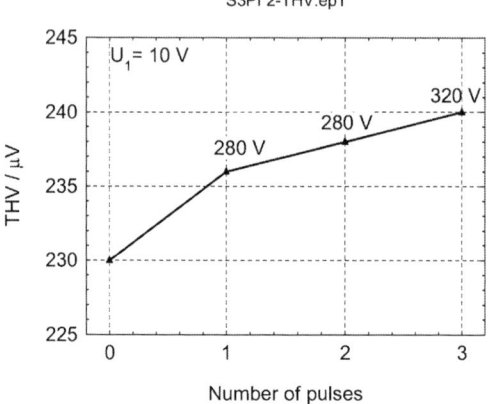

Figure 15. Third harmonic voltage vs. the number of testing pulses for applied first harmonic voltage U_1 = 10V

Conclusion

New principle for non-destructive testing of conducting solids based on the non-linear ultrasonic spectroscopy was introduced. This method is based on the phonon interaction with conducting particles. The ultrasonic signal changes the contact area between conducting grains and then resistance is modulated by frequency of ultrasonic excitation. Resultant voltage u_m on the measured sample depends on the ac current varying with frequency f_E and on the resistance change ΔR varying with frequency f_U. The intermodulation component U_m varies linearly with electric excitation and is proportional to the ultrasonic voltage U_U^a, with the power $a = 0.7$ to 2. Higher values of a are observed for polymer based samples. For high values of ultrasonic excitation the saturation of U_m is observed. We suppose that the ultrasonic vibrations affect the effective area of conducting particles interaction in the resistive layer and hence the shape of dependence U_m vs. ultrasonic excitation is influenced by resistive layer material.

The short current pulse from capacitor discharge was used as a stressing method for resistor degradation. After the sample degradation the amplitude of intermodulation component U_m

increases approximately 3times, while the THV value changes for less than 5%. The application of new method where the amplitude of intermodulation component U_m is measured allows using lower stressing pulse energy to evaluate quality and reliability of thick film resistors.

Acknowledgements

This work was supported by the Grant Agency of the Czech Republic under Grant GACR 102/07/P482, and under the project MSM 002160503.

References

[1] Dziedzic A., Kolek A. "1/f Noise in Polymer Thick-Film Resistors". Journal of Physics D: Applied Physics, **31** (1998), p. 2091-2097.

[2] Soliman L.I., Sayed W.M. "Some Physical Properties of Vinylpyridine Carbon-Black Composites". Eyptian Journal of Solids, **25** (2002), No.1, p. 103-113.

[3] Sedlakova V., Brustlova J., Sikula J., Hlavka J., Coocker J., Adams K., Greenhill D.A. "Noise of Carbon/Graphite Thick Conducting Films". Proceedings of the 17th International Conference on Noise and Fluctuations, 2003, Prague, Czech Republic, p. 201-204.

[4] Sedlakova V., Sikula J. "Charge Carrier Transport and Noise in Polymer Based Thick Films". Proccedings of the 4th Europeand Microelectronics and Packaging Symposium, 2006, Terme Catez, Slovinia, p. 15-20.

[5] Sedlakova V., Melkes F., Dobis P., Sikula J., Tacano M., Hashiguchi S. "Non-linearity changes induced by current stress in thick film resistors". Proceedings of CARTS 2004, San Antonio, Texas, (U.S.A.), 2004, p. 154 – 157.

[6] Hajek, K., Sedlakova, V., Majzner, J., Hefner, S., Sikula, J. "Non-linearity and noise characterisation of thick-film resistors after high voltage stress". In Proceedings of 3rd European Microelectronics and Packaging Symposium. Prague (Czech Republic), 2004, June 16 – 18, pp. 421 – 426

Experimental Study of Solder Joint Reliability on Injection Moulded Substrates

Minna Arra, Ilkka Härkönen and Esko J. Pääkkönen

Perlos Oyj, Äyritie 8 A, 01510 Vantaa, Finland

Tel. +358 40 7381284, Fax +358 9 2500 7210, E-mail: minna.arra@perlos.com

Abstract

3-D MID (3-Dimensional Moulded Interconnect Device) –technology enables the creation of electrical circuitries and component attachments directly onto 3-dimensional injection moulded plastic parts. 3-D MID has found popularity in the automotive sector and more recently for example in sensor packaging applications and camera modules. One of the challenges related to 3-D MIDs is the long-term reliability of the component attachments. The CTE (coefficient of thermal expansion) of injection moulded substrates is higher that that of typical PCB laminates and therefore greater stresses are induced to the component attachments. Furthermore, the CTE is anisotropic, depending on the direction and magnitude of orientation of the plastics molecules during the injection moulding process. This paper presents the results from an experimental reliability study. Two different kinds of SMD (surface mount device) components were soldered with Sn/Ag/Cu solder paste onto injection moulded, thermoplastic LCP (liquid crystal polymer) substrates patterned with MID/LDS technology. The same circuit pattern and assembly was realized on FR-4 PCBs (printed circuit boards), which served as control samples during testing. The assemblies were exposed to thermal shock and thermal cycling tests. Failures during these tests were monitored by real-time resistance measurements. As predicted, failures in solder joints occurred earlier with the LCP thermoplastic substrates than with the FR-4 PCB laminates. However, the CTE was no found to be the major factor determining the lifetime of solder joints. The most important reason was the dimensional changes that occurred in the LCP substrates during the reliability testing. Furthermore, a remarkable difference in the magnitude of these changes between the two LCP grades included in the study was observed.

Key words: LCP, MID, LDS, injection moulding, solder joint reliability

Introduction

Three-dimensional moulded interconnect devices (3-D MID) are injection moulded parts whose surface serves as a support both for conductive wiring and for electro-mechanical or electronic components. 3-D MID technology allows electrical and mechanical functions to be integrated into one single part. Electronic components can be assembled on the wiring, and at the same time the moulded part may act as a cover or a casing of a device. MID technology can reduce manufacturing and assembly costs and the weight and size of the product. Furthermore, the reduced number of components and the fact that MIDs are made of thermoplastic materials are favourable aspects for the disassembly and recycling operations.

Today, several different technologies to create conductive wiring patterns on 3-D MID thermoplastic substrates exist and more are being developed. In this study, the conductive wiring patter was created using the LPKF-LDS (LPKF-Laser Direct Structuring) technology. A non-conductive organic metal complex is blended to the thermoplastic base material. After injection moulding, the conductor traces are drawn on the

plastic surface by a laser beam. Laser radiation cuts the chemical bonds between the copper atoms and the organic ligands in the organic metal complex molecule, leaving copper atoms as seeds on the surface. Laser beam also ablates the plastic surface, which makes it microscopically irregular and rough. Therefore the copper is firmly anchored to the surface during metallization. The parts can be plated in an electroless and, if necessary, in electrolytic copper bath. Copper grows only on the laser-treated conductor areas. The process flow is illustrated in Figure 1. [1], [2], [3]

LDS –process is flexible regarding layout changes, since only a change in the CAD drawing is needed to change the pattern. Conductive wiring patterns can be easily created on 3-D shapes, however with certain limitations. The manufacturing costs are low especially for small and medium manufacturing volumes. Commercially available plastic materials for LDS process include at least PBT (polybutylene terephtalate), PA6/6T (polyamide), LCP (liquid crystal polymer) and PET/PBT (polyethylene terephtalate/ polybutylene terephtalate).

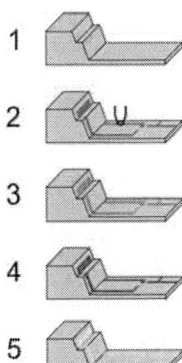

Figure 1: Schematic diagram of the LPKF-LDS process. 1: Injection moulding of plastic part, 2: Laser structuring of areas to be metallized, 3: Electroless copper plating, 4: Electrolytic copper plating (if necessary), 5: Surface finishing.

Injection moulding is a cost-effective method to manufacture 3-D thermoplastic parts in high volumes. Parts with complicated, asymmetric and high-precision 3D shapes can be realized with this technology. Therefore, besides providing mechanical functions, injection moulded structures have an interesting potential to act as low-cost current-carrying substrates over complicated 3-D shapes.

In this study, the test parts were injection moulded using LCP (liquid crystal polymer) material. LCP has been one of the first thermoplastic materials commercially available for LDS technology, i.e. containing the organic metal complex necessary for laser activation. The benefits of LCP are its suitability for parts requiring high precision and thin wall structures, and its resistance to lead-free soldering temperatures. LCP polymers are widely used today for example in connector housings, chip carriers, sockets and switches. They are becoming popular also in MEMS sensor devices providing low-cost, 3-D shaped cavity packages [4].

The most challenging aspect regarding the reliability of solder joints is the relatively high and anisotropic CTE (coefficient of thermal expansion) of the thermoplastic materials. In addition, CTE values are usually different parallel and perpendicular to the direction of orientation of the polymer chains. Such orientation of polymer chains and/or reinforcing filler materials happens during the injection moulding process due to the high shear stresses caused by the rapid injection process of the molten plastics. It has been reported elsewhere that placing rigid, two-pole electronic components, such as chip resistors, in the direction of the orientation (where the CTE value is smaller) enhances the solder joint reliability [1].

Experimental

The test parts, injection moulded from LCP material, had a simple structure as shown in Figure 2. The nominal wall thickness of the mould cavity was 1 mm. SMD components were assembled on the flat bottom area of the parts.

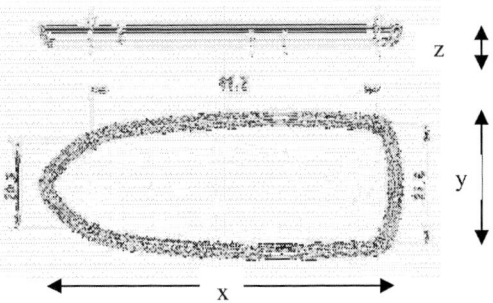

Figure 2: Mechanical drawing of the test parts.

In injection moulding the part will have a typical skin-core structure and different molecular orientation due to shear flow and the cold mould wall. The molecular orientation in different areas of the substrates was studied with MoldFlow 2.5D simulation software. The orientation is different between the skin layer (typical thickness is in the order of 50-100 μm) and the core layer as shown in Figure 3. Since the skin layer is thin, the magnitude and direction of the CTE values in different areas of the parts are mainly determined by the core layer.

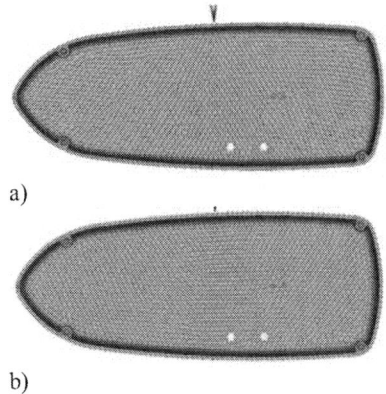

Figure 3: Orientation of plastics in the a) Skin layer and b) Core layer.

Two mineral-filled LCP grades were included in the study, a commercial one (A) and a developmental one (B). Some basic properties of these grades and a comparison with the printed circuit board's (PCB) epoxy-glass fibre laminate (FR-4) material used to manufacture the control samples are shown in Table 1. The CTE value of the LCP grade A, especially in the direction perpendicular to the plastics flow direction, is large (55 ppm/°C), which can be expected to cause earlier failures for solder joints.

Table 1: Properties of the LCP grades and the PCB laminate material

	Grade A	Grade B	PCB laminate
Glass transition temperature Tg (°C)	~120-130	~120-130	135
Mineral filler content, wt-%	40	44	N/A
CTE (ppm/°C), parallel, between 50°C and 200°C	23	12.5	N/A
CTE (ppm/°C), perpendicular, between 50°C and 200°C	55	30.3	N/A
CTE (ppm/°C), weft direction, below Tg/ above Tg	N/A	N/A	16/ 14
CTE (ppm/°C), warp direction, below Tg/ above Tg	N/A	N/A	13/ 7
CTE (ppm/°C), vertical, below Tg/ above Tg	N/A	N/A	70/ 280
Melting temperature (°C)	335	Close to that of grade A	N/A

The injection moulded parts were laser activated and then plated with Cu/Ni/Au –type of layer. Plated layer thickness was ~5 µm Cu/ 5-7 µm Ni/ flash Au. In the 1 mm thick, single-layer PCB substrates, the plating type was 18 µm Cu/ 4-5 µm Ni/ flash Au.

Two SMD (surface mount device) component types were chosen for the solder joint reliability study: BGA (ball grid array) and 1206 chip resistors. The body size of the daisy chain BGA components was 10 mm x 10 mm, solder balls were arranged in 10 x 10 full matrix and the carrier substrate was made of BT epoxy laminate. The composition of the solder balls of the BGA was 95.5Sn/4.0Ag/0.5Cu. The terminations of the ceramic 0-ohm 1206 chip resistors were coated with 100% Sn. Two BGAs and five 1206 resistors were assembled on each substrate. One BGA100 component was assembled near the sharp end of the part, and the other near the middle of the part. The 1206 resistors were arranged in two individual chains at different locations of the part.

For the component assembly, a conventional SMD assembly line was employed. The injection moulded parts were placed onto transportation jigs for the assembly. Solder paste was dispensed onto the pads, followed by component pick & place process and forced convection reflow in air atmosphere. The dispensing of the solder paste presented some challenges due to twisting and warping of the injection moulded parts. For the PCB control samples, solder paste was applied with stencil having a thickness of 127 µm and a screen printer. Solder paste was of no-clean 95.5Sn/4.0Ag/0.5Cu type.

The assemblies were exposed both to air-to-air temperature shock and temperature cycling tests to see whether the same number of test cycles in the two tests creates the same amount of cumulative failures of solder joints. The test method is described in JESD22-A104C –standard [5]. For both tests, the nominal minimum temperature was –40°C and the nominal maximum temperature was +125°C, but the temperature shock test had shorter dwell and transition times. The temperature profiles and the summary of the characteristics of the tests are shown in Figure 4, Figure 5 and Table 2. Real-time measurement was used to detect the failures of the solder joints.

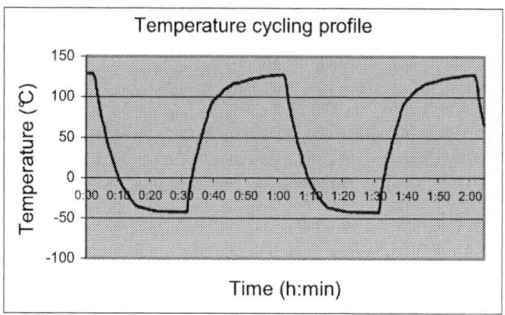

Figure 4: Temperature cycling test profile.

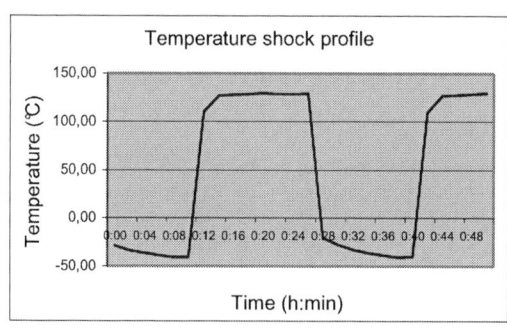

Figure 5: Temperature shock test profile.

Table 2: Temperature shock and temperature cycling test parameters.

	Temperature shock	Temperature cycling
Dwell time at maximum	12 min	13 min
Transition time max -> min	7 min	13 min
Dwell time at minimum	6 min 30 sec	16 min
Transition time min -> max	2 min 45 sec	17 min 30 sec

Results and Discussion

Temperature shock and temperature cycling tests were both maintained until 2000 shocks/ cycles. The observed failures were analyzed, and some of the results are presented in Figure 6 and Figure 7. When comparing the failures of BGA100 or 1206 chip components assembled on LCP substrates A and B, it can be concluded that no major difference exists between the two tests in terms of cumulative failures vs. number of cycles. There seemed to be a slightly bigger difference with BGA100 assemblies on FR-4 substrates between temperature shock and cycling test, but since the number of samples was small, no firm conclusions could be drawn.

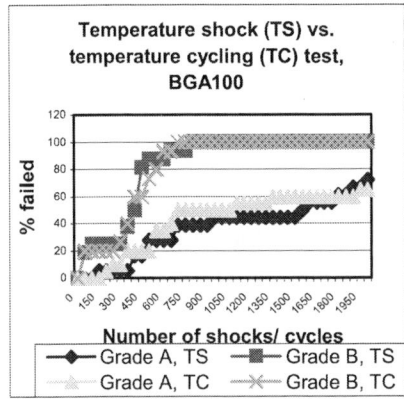

Figure 6: Comparison of BGA100 failures between temperature shock (TS) and temperature cycling (TC) tests.

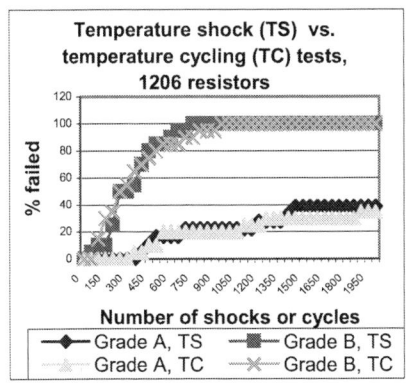

Figure 7: Comparison of 1206 resistor failures between temperature shock (TS) and temperature cycling (TC) tests.

The failure mode was verified from cross-sectioned samples and was found to be the same in both tests (solder fatigue). Two examples of cross-sectioned 1206 resistor joints after 1000 cycles of testing are shown in Figure 8 and Figure 9. After

2000 cycles, a decision to continue further with only the temperature cycling test was made.

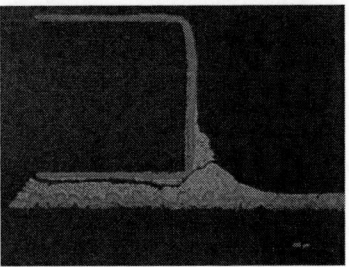

Figure 8: Failure in 1206 solder joint after 1000 temperature shocks.

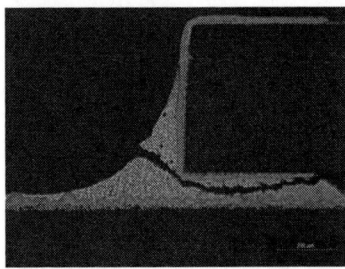

Figure 9: Failure in 1206 solder joint after 1000 temperature cycles.

Temperature cycling test was continued until 3438 cycles, until which all BGA100 components on LCP substrates A and B had failed. The collected failure data for BGA100 components is shown in Figure 10 as a two-parameter Weibull cumulative failure distribution. The data for BGA100 assembled on FR-4 is not shown because one third of those samples were still functioning after 3438 cycles. It is important to note that BGA100 assembled on LCP substrate B failed much earlier than those on substrate A, although grade B had smaller CTE values and therefore was expected to cause less stress to the solder joints.

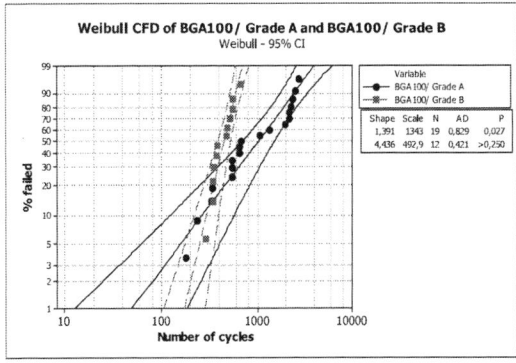

Figure 10: Weibull CFD for BGA100 assembled on LCP substrates A and B.

All 1206 resistor chains on LCP substrate B failed by the first 1000 cycles (Figure 11), whereas only 65% of the chains assembled on substrate A had failed by 3438 cycles. None of the 1206 resistors chains on FR-4 substrates showed failures by 3438 cycles.

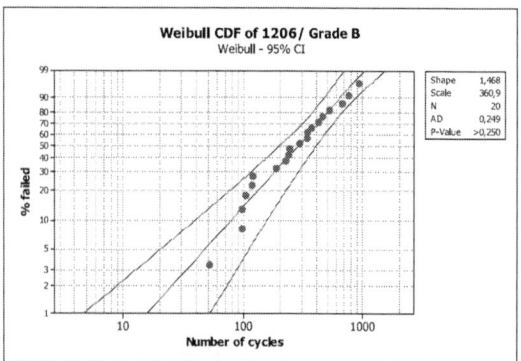

Figure 11: Weibull CFD for 1206 resistors assembled on LCP substrate B.

The reason for the unexpected results between LCP substrates A and B was found during the inspection of the test parts after reliability testing. Although LCP is known for its dimensional stability over a large temperature range, the residual (frozen-in) stresses existing in the part had caused remarkable dimensional changes in the substrate made of LCP grade B. These parts showed more severe warping after temperature cycling than parts made of grade A. This explains the earlier failures of solder joints on LCP substrates B. 3-D scanning results for the parts in the Z-direction (referring to Figure 2), are shown in Figure 12, Figure 13 and Figure 14.

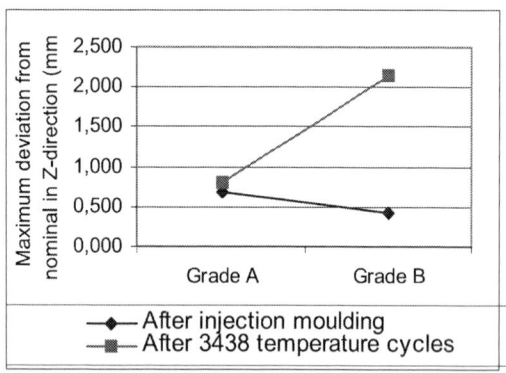

Figure 12: Maximum deviation from nominal dimension in the Z-direction.

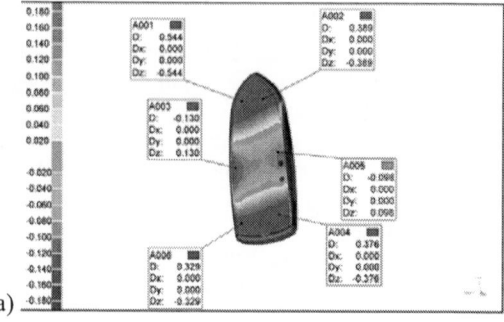

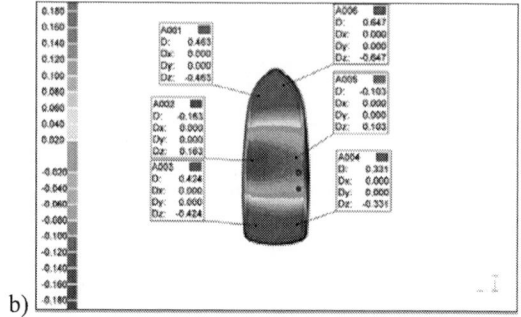

Figure 13: Deviations from nominal dimension in the Z-direction with LCP grade A, a) After injection moulding and b) After temperature cycling for 3438 cycles.

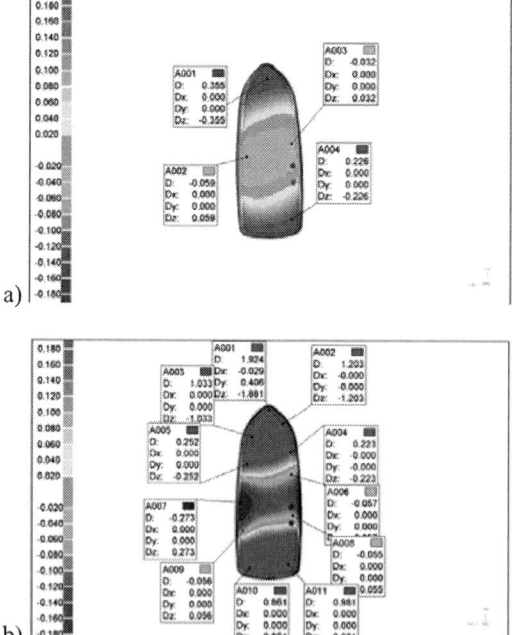

Figure 14: Deviations from nominal dimension in the Z-direction with LCP grade B, a) After injection moulding and b) After temperature cycling for 3438 cycles.

The differences observed between the two LCP grades may originate from the type of filler materials used in the two grades. Filler particles cause physical obstacles and this way generally reduce macroscopic anisotropy preventing (but not eliminating) the orientation of the polymer molecules in the x-y direction and the formation of layered structures in the z-direction during the injection moulding process. This attribute depends on the shape of fillers. Fillers may have for example a spherical or flaky shape. Also the thermal conductivity and specific heat capacity are higher in metal and mineral fillers. The efficiency of the filler to prevent the orientations determines the quantity of metastable molecular states that remain in the parts after the fast cooling of the injection moulding process. Those unstable molecular conformations may then relax when temperature is raised above the glass transition temperature and cause permanent changes in the dimensions of the parts. The intensity of these changes is also influenced by the design, shape and the thickness of the part as well as by the positioning of the injection gate point(s). For example, increasing the wall thickness or ribs in the parts makes them stiffer and thus resistant against the molecular relaxation phenomena induced by the temperature increase.

One of the original targets of the study was to evaluate the difference in the reliability of solder joints when the components are placed in different orientations in proportion to the orientation of the polymer molecules. This target was not met because of the large dimensional changes, whose influence on the lifetime of the solder joints was dominant over the effect of the CTE values. The influence of the CTE must be evaluated with test parts and constructions, which do not undergo deformations during the reliability testing.

Conclusions

The experimental test results show that both the cumulative failures of solder joint vs. the number of cycles and the failure mode of solder joints are similar in temperature shock test having a 30 min cycle time and in temperature cycling test having a 60 min cycle time at least with the component types covered by this study (BGA and 1206 chip resistor). If the same kinds of LCP plastic grades are used as substrate materials in any future assemblies, the faster reliability test can be employed to optimize time and cost.

One of the critical factors defining the reliability of solder joints is understood to be the CTE mismatch of the substrate, solder material, possible underfill and the component materials. When 3-D thermoplastic substrates are used, a new important parameter is the structure of the part. Even if the use temperature of the device or the reliability test temperatures did not violate the recommended temperature ranges for the material, if the structure and injection moulding process parameters of the part are not optimized, large dimensional deformations may occur and lead to early failures of the solder joints. Naturally in all cases, the lifetime of solder joints on a typical injection moulded thermoplastic substrate is inferior to that on typical PCB laminates (provided that the use or test conditions are the same) because of the larger CTE values of the thermoplastic materials.

When planning an electronics assembly on a 3D thermoplastic part, the first step is to reduce the frozen-in stresses caused by the orientation of the plastic molecules or increase the stiffness of the parts Stiffness can be improved for example by increasing the material thickness or by adding ribs. As a second step, samples of the parts should be manufactured and exposed to the desired temperature testing. If dimensional deformations are observed, the structure can be optimized until no deformations occur. If structural changes are not allowed, experimental testing and verification of solder joint reliability is required. Estimation of solder joint reliability by only simulation tools (for example based on finite element method) in these cases is difficult, since the complicated dimensional changes of the substrate would have to be considered.

Acknowledgements

The authors would like to thank Lotta Ekebom, Pertti Tuhkanen, Sinikka Parviainen, Jukka Ihalainen, Timo Väisänen and Juha Räsänen at Perlos Oyj for their help in the manufacturing of the samples and in carrying out the 3-D scanning of the samples.

References

[1] T. Krautheim, F. Pöhlau, W. Lorenz, S. Stampfer and R. Meier, "Production Processes, Service Requirements and Material Characteristics of Molded Interconnect Devices 3-D MID, Manual for Users and Manufacturers, 2nd ed.", Research Association Moulded Interconnect Devices 3-D MID e.v., Chapter 11, pp. 11-19 – 11-24, 2002.

[2] K. Feldmann and A. Kunze, "3D-MID Technology", Kunststoffe Plast Europe, 4/2004, pp. 17-24.

[3] Information of LDS process at www.lpkf.de

[4] K. Gilleo, "MEMS Packaging Update", Advanced Packaging, pp. 20-22, September 2005.

[5] JEDEC Standard, "Temperature Cycling, JESD22-A104C", JEDEC Solid State Technology Association, May, 2005.

ASPACT® Additive Circuit Transfer Technology

Juha Hagberg[1], Teija Kekonen[2] and Terho Kutilainen[2]

1) EMPART research group of Infotech Oulu, Microelectronics Laboratory,

P.O.Box 4500, FI-90014 University of Oulu

Email: juha.hagberg@ee.oulu.fi

2) Selmic Oy, Veistämötie 15, P.O.Box 350, FI-90501 Oulu, Finland

Email: firstname.lastname@selmic.com

Abstract

ASPACT® circuit transfer is a novel technology to make molded interconnection devices (MID) circuits on plastic parts or sheets, or to form high quality circuiting for high end modules. The circuit patterns are high quality electrodeposited metal and in the processing etching is not needed. In the ASPACT® technology, the circuit pattern is first electroplated on a flexible or rigid temporary carrier substrate. The pattern is then treated for adhesion enhancement and transferred on to a plastic component or a film by pressure and heat. The circuit is buried inside the plastic, giving a leveled surface after transfer. By using conventional photoresist materials, down to 50 μm line/spaces can be achieved. Using more advanced materials, line/spaces even below 10 μm are possible. ASPACT® transfer technology enables planar, 2.5D and limited 3D forms of plastic MID components. If pattern is first transferred to an IML film or an insert and then 3D formed, more complex forms are possible. Several thermoplastics materials have been found to be suitable in ASPACT® technology like PC, PC/ABS, PA, PS, and COC. Plastic can be in the form of a thick molded component, extrusion sheet or a very thin film e.g. in-mold labeling or decoration (IML or IMD) insert.

Key words: additive, circuits, MID, thermoplastic, IML

Introduction

In the past decades many molded interconnection devices (MID) technologies has been developed for metallising plastics, mainly for automotive and hand held electronics applications. In automotive the aim is to replace cable harness and decrease assembly steps and in hand held devices to integrate electronics as a part of the plastic cover. The main drivers have been cost efficiency together with increased functionality.

A novel ASPACT® circuiting technology has been evaluated to make high quality circuiting for MID and module use. ASPACT® circuiting method is a competitive technology to make high quality copper circuits on flat, 2.5D and limited 3D thermoplastics objects. ASPACT® is suitable for small and large products, for thin and thick conductor lines and, in addition, for very fine line demands.

ASPACT® Technology

In ASPACT® circuiting technology the circuit patterns are first electroplated on a temporary carrier substrate and then transferred to the desired object. For thermoplastic objects the transfer can be made by applying pressure and heat. The carrier substrate can be a rigid one or flexible enabling reel to reel processing. The patterning is made by resist printing or using photo imaging. Figure 1 shows ASPACT® copper circuits on a flexible carrier substrate.

In ASPACT® technology the circuits are made by electroplating and the circuit thickness can vary from very thin up to 60 μm. Even thicker lines are possible if one compromises the line definition and high resolution. After plating an adhesion enhancement treatment is applied to the circuit patterns. The optimal adhesion treatment is material dependent because the shape of the roughened

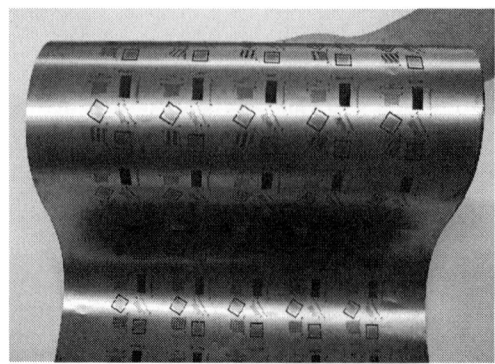

Figure 1: ASPACT® copper circuits on a flexible carrier foil.

surface and the chemical bonding characteristics vary from treatment to another.

The resolution of the resist used for patterning defines the resolution of the line/space. By using conventional resist materials, down to 50 μm line/spaces can be achieved and with more advanced materials line/spaces even below 10 μm are possible.

In Figure 2 a SEM top view of 8 μm wide ASPACT® conductor lines transferred to PC+ABS thermoplastic plastic can be seen. As a result of heat and pressure of the transfer process the circuits are embedded inside to the plastic leaving planar surface.

Figure 2: SEM top view image of 8 μm wide ASPACT® circuit lines (light grey) transferred to PC+ABS. The lines are embedded in plastic and the surface area is planar.

ASPACT® circuits can be transferred from a temporary carrier substrate to plastic components by using electrical heating, hot press tool, hot roll laminator or conventional hot press. A hot roll laminator is well suitable for reel to reel processes. A hot pressing tool can be for instance a silicone or a steel tool that can be 2.5D or 3D shaped. A pressing tool is heated up to a certain temperature and the pattern is pressed from temporary carrier substrate to the plastic object using reasonably high pressure. If movement of the plastic object during the transfer is accomplished, more complicated 3D shapes can be transferred.

To obtain deeper real 3D shapes, In-Mold Labeling or Decoration (IML/IMD) technology can be combined with ASPACT® technology. The circuits are first transferred to an IML/IMD foil (e.g. PC, PA or PET films) and then the foil is 3D formed. The insert obtained in this way is put in an injection mold and during molding the insert adheres to the plastic part as shown in Figure 3 (3D shaping and moulding made by Mosen).

Figure 3: ASPACT® copper circuits in a plastic cover. The circuits are formed by combining ASPACT® and IML technologies.

In order to achieve other metal layers on the top of the copper circuit they can be electroplated before copper in reverse order. Plating like nickel/gold or silver can be applied to prevent oxidation or to enhance hardness.

Peel Strength Measurement

Several common plastic materials shown in Table 1 were selected to evaluate their ASAPCT® MID circuits peel strength characteristic.

Table 1: Peel strength tested polymer materials. The difference in the COC A and B material is the softening point. Two qualities of HDPE was used, black and white one.

Polymer material	Abbr.	*
Polycarbonate	PC	M
PC and acrylonitrile butadiene styrene blend	PC+ABS	E, M
Polyamide 12	PA12	M
Glass reinforced PA12	PA12+GF	M
High density poly ethylene (2)	HDPE	E
Polypropylene	PP	E
Polystyrene	PS	E
Cyclic olefin copolymer	COC A	M
Cyclic olefin copolymer	COC B	M

*)   M = injection moulded
      E = extrusion sheet

The peel strength measurements were accomplished according the IPC-TM-650 standard (2.4.8). In measurements 1 mm and 3 mm wide copper strips were used with average thickness of 23 μm. The transfer was made by applying electric current trough the carrier foil strip to heat up the circuit lines on it and by pressing them in the plastic, Figure 4. Simultaneously, the voltage drop trough the carrier foil was measured and the current was controlled so that a constant power was applied during the heating period. By the voltage drop measurement the resistance change and the temperature rise of the carrier foil was possible to

577

calculate. In addition the effect of copper line thickness to the peel strength was studied for PA+GF and PC+ABS.

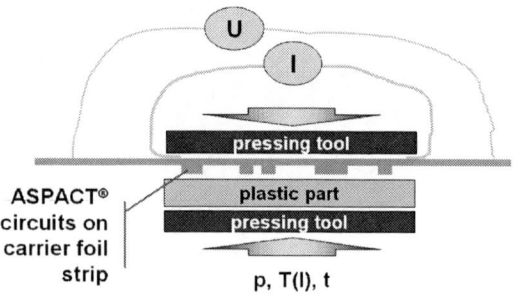

Figure 4: The ASPACT® peel strength measurement strips were transfer to the plastic substrate by electrical heating. By the voltage drop the temperature was able to calculate.

Peel Strengths

Results of the peel strength measurements are shown in Figures 5 – 8. In each Figure the measured peel strength (N/mm) is presented as a function of the end temperature of the carrier foil during transfer.

The best adhesion was measured for glass reinforced PA12, values over 1.4 N/mm were achieved, Figure 5. The glass reinforced PA12 quality behaves differently compared to the plain PA12, the glass reinforced have a much wider temperature range where adhesion is very good, over 1.1 N/mm over the transfer temperature range from 150°C up to 270°C.

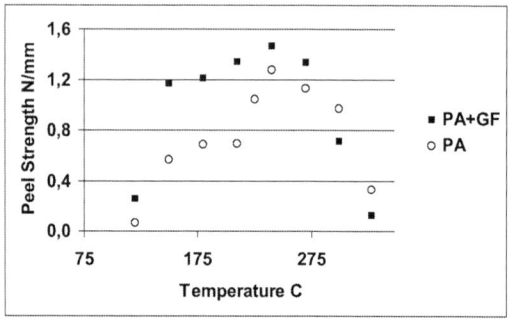

Figure 5: Peel strength as a function of transfer temperature of the ASPACT® circuit lines on glass reinforced PA12 and PA12.

For PC+ABS the best peel strength values are around 0.7 N/mm and for PC 0.6 N/mm, Figure 6. The injection molded PC+ABS has a narrow temperature range for good adhesion compared to extruded one for which the adhesion is over 0.7 N/mm over the transfer temperature range from 160°C up to 270°C. Injection molded qualities of PC+ABS and PC behaves in very similarly way as seen in Figure 6.

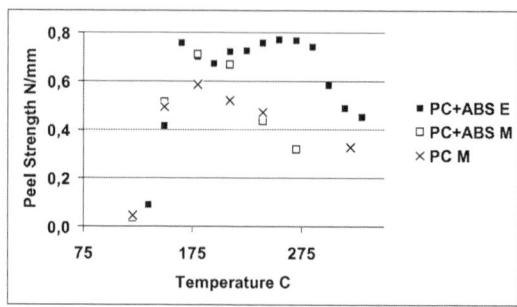

Figure 6: Peel strength as a function of transfer temperature of the ASPACT® circuit lines on extruded and injection molded PC+ABS blend and injection molded PC.

The same quality of HDPE but two different colors, black and white, gave similar peel strength dependence except at higher temperatures where the black one showed better adhesion, Figure 7. PP had higher peel strength values compared to the HDPE's, between 0.6 and 0.8 N/mm at wide temperature range.

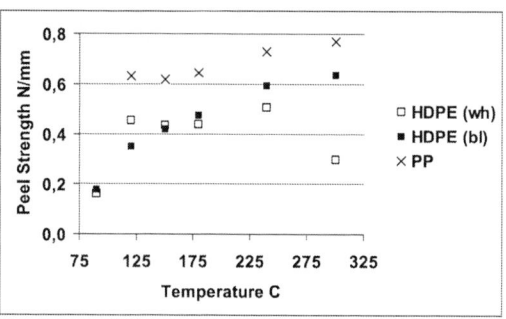

Figure 7: Peel strength as a function of transfer temperature of the ASPACT® circuit lines on PP and black and white HDPE.

The both qualities of COC behave very similarly; the adhesion is around 0.7 N/mm for relatively wide transfer temperature range, Figure 8. For PS the behavior is similar, but it had peel strength values up to 0.9 N/mm at 120°C transfer temperature.

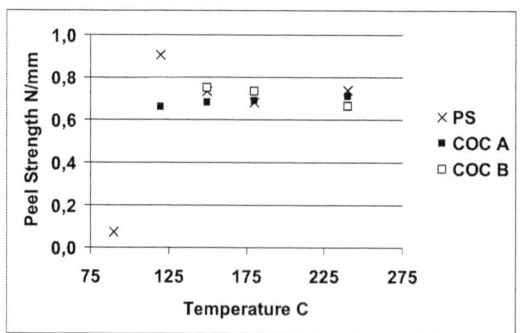

Figure 8: Peel strength as a function of transfer temperature of the ASPACT® circuit lines on PS and two different COC qualities.

Each material has an optimal temperature range where the pattern adheres to the thermoplastic material. Low peel strength can be noticed when temperature has not reached softening/melting point but also when the temperature rises too high. For some material the usable temperature range can be over 100°C but for some only few tens of degrees centigrade (°C). The samples tend to deform especially when higher temperatures for the transfer were used. To minimize residual stress it is advisable to use as low temperature as possible for the transfer taking in to account the peel strength requirements.

Circuit Thickness Peel Strength Dependence

Peel strengths between copper and PC+ABS and PA12+GF were measured using different copper thicknesses, Figure 9.

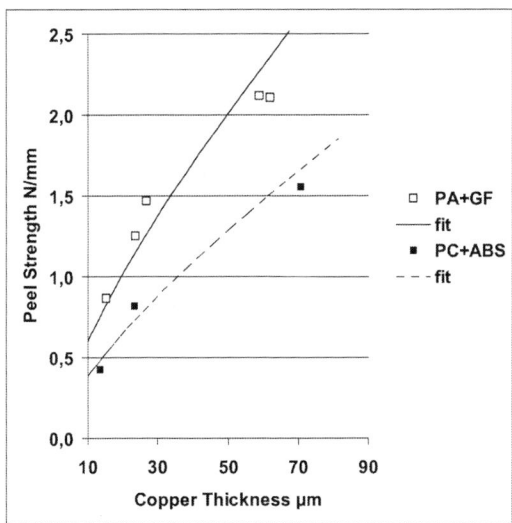

Figure 9: Peel strengths between copper line and PC+ABS and PA12+GF substrates using different copper thicknesses. Copper strips were transferred to plastic sheets by using electrical heating.

As seen the influence of the copper thickness on the measured peel strength can be quite strong. The behavior can be explained by the fact that during peeling the foil radius depends on the foil thickness and therefore for thinner foil the stress concentrates in a smaller active area. For cantilever in elastic material one can expect the peel strength to be a function of foil thickness in the power of 0.75 [1]. This seems to be close to the behaviors found for ASPACT® circuits in PA+GF and PC+ABS, fitted trend lines with power of 0,75 are shown in Figure 9.

Discussion

In printed circuit board manufacturing, peel strength is one of the main issues. Peel strength specifications for base materials for rigid and multilayer printed boards are defined in IPC-4101A standard. According to this standard peel strength values for base materials vary between 0.5 and 1.45 N/mm. The recorded peel strengths by hot embossing or electroless plating for several plastics materials are found to vary between 0.4 and 2.8 N/mm [2]. Guidelines of peel strength value requirements for automotive industry; television, hifi and video equipment; and system and control technology are 0.8 N/mm, 0.6 N/mm and 0.7-1.4 N/mm, respectively [2].

For PA12+GF peel strength value over 2 N/mm was measured for copper thickness of 50 μm and over 1 N/mm for copper thickness of 20 μm. The IPC-TM-650 standard states that copper thickness can be built up to 31.5-38.5 μm to provide strip strength. Thus all the peel strengths obtained in Figures 5 - 8 for 23 μm thick circuit lines are undervalued compared to practice accordance with the standard.

Summary

ASPACT® technology provides high quality copper patterns that can easily be transferred almost on to any thermoplastic material for several thicknesses and shapes. Technology enables fine lines, even below 10 μm, that are not possible with conventional MID techniques. Peel strength between copper and thermoplastic depends on adhesion treatment, materials and copper thickness and it can be over 2 N/mm.

With ASPACT® transfer circuits it is possible to form conductive patterns on flat, 2.5D and limited 3D thermoplastics objects. The technology can be combined with IML technology to make, for instance, antenna or connector patterns embedded in a plastic cover.

Acknowledgements

Part of injection molded samples were provided by Perlos Oyj and Mosen Co., Ltd..

References

[1] R. Wiechmann, "Experiences with peel strength", Circuit World, Vol. 32, No. 3, pp. 30-39, 2006.

[2] T. Krautheim, F. Pöhlau, W. Lorenz, S. Stampfer and R. Meier, "Manual for Users and Manufacturers: Production Processes, Service Requirements and Material Characteristics of Molded Interconnect Devices 3-D MID", Research Association Molded Interconnect Devices 3-D MID e.V., 2nd edition, June 2002.

Advanced Electronics Packaging via M³D Direct Writing

Martin Hedges[1], Bruce King[2], Mike Renn[2]

[1]Neotech Services MTP, Petzolt Str. 3, 90443, Nuremberg Germany

Tel: +49 911 274 5501, Fax: +49 911 274 5502, info@neotechservices.com

[2]Optomec Inc., 3911 Singer, NE, Albuquerque, NM 87109, USA.

Tel: +1 505 761 8250, Fax: +1 505 761 6638, info@optomec.com

Abstract

Maskless Mesoscale Materials Deposition (M³D) is a CAD driven, direct write process used for creating fine electronics features, down to 10 micron resolution, on almost any substrate material. The process uses aerodynamic focussing of fine aerosol streams to write features and components in metals, alloys, polymers, adhesives, ceramics and even bio-materials. For non-sensitive substrates the deposits are post processed using traditional sintering or curing methods. For delicate substrates, such as thin polymer films, an integrated laser sintering process is applied. The end result is a high-quality thin film deposit with excellent edge definition, smooth surface profile and near-bulk electronic properties. Current application areas span packaging and assembly, for example HDIs, direct die attach, embedded passives and flex circuits through electronic components such as resistors, capacitors and inductors. With no physical contact with the substrate by any portion of the tool other than the deposition stream, conformal writing is easily achieved. This allows for the processing of both planar and 3D substrates, opening the way for the development of novel SIP, MID and MEMS applications. The process also has the ability to add multiple layers of materials to electronic devices such as fuel cells and micro-batteries. The ability to process a wide range of materials combinations in 3D space opens the potential for new and novel device designs. This paper will introduce the fundamentals of the technology and detail the benefits of M³D technology in creating mesoscale features for electronics assembly and semiconductor packaging applications and outline some of the current application areas. The first high volume manufacturing applications using M³D will come on stream in 2007. The paper will also briefly discuss the scalability of the process to deal with the requirements for high volume manufacturing.

Key Words: HDI, high density interconnects, microelectronics packaging, direct write, 3D.

Introduction

In recent years, a new class of manufacturing techniques has become established which offer manufacturers significant cost, time and quality benefits across a broad spectrum of industries. These new techniques are collectively known as additive manufacture.

During *additive* manufacture, material is deposited layer by layer to build parts or features. This is in contrast to traditional manufacturing methods which are mainly *subtractive*, i.e. material is removed from a part to get the final form.

Features of additive manufacturing processes include direct CAD-driven, "Art-to-Part" processing which eliminates expensive tooling, masks and vertical/horizontal integration which leads to fewer overall manufacturing steps. These features combine to offer diverse benefits:

- Time Compression and Increased Manufacturing Agility. CAD driven, tool-less processes speed up the product development and manufacturing, whilst allowing greater flexibility in mass customisation.

- Lower Costs – This benefit arises because tooling and mask costs are eliminated. Process costs in terms of operator input, supplier chain complexity and work flows are reduced. Raw material is used more efficiently reducing waste levels and costs of "manufacturing" scrap. Life-cycle costs are reduced by lower design development costs, increasing product quality and the ability to repair components.
- Better Product Designs. Greater design and manufacturing flexibility offers the potential for revolutionary new end-products with improved performance based on novel size, geometries, materials and material combinations.

This paper will introduce an additive manufacturing technique designed for the electronics industry: Maskless Mesoscale Materials Deposition[TM] (M^{3}D). The process is aimed at small scale (10-100+ micron) electronic structures and devices as well as biological materials and applications.

M^{3}D - Maskless Mesoscale Materials Deposition

M^{3}D was originally developed to fill a neglected middle ground in microelectronic fabrication. Current techniques create very small electronic features, for example by vapour deposition, and relatively large ones for example by screen printing. No technology was capable for satisfactorily creating crucial mesoscale-sized (1-100μm) production of interconnects, components, and devices. As electronic devices continue to shrink, thick-film fabricators are approaching the physical limits of stencil printing. Thin-film technology can deposit mesoscale features, but it requires a highly skilled workforce and a major investment in new manufacturing capability for each new application. It is for these reasons that M^{3}D was developed by Optomec Inc.

How M^{3}D Works

M^{3}D is a CAD driven, mask less process. It uses aerodynamic focusing for the high-resolution deposition of chemical precursor solutions and/or colloidal suspensions. An aerosol stream of the deposition material is focused, deposited, and patterned onto a planar or non-planar substrate. The system consists of 3 modules, **Fig. 1**:

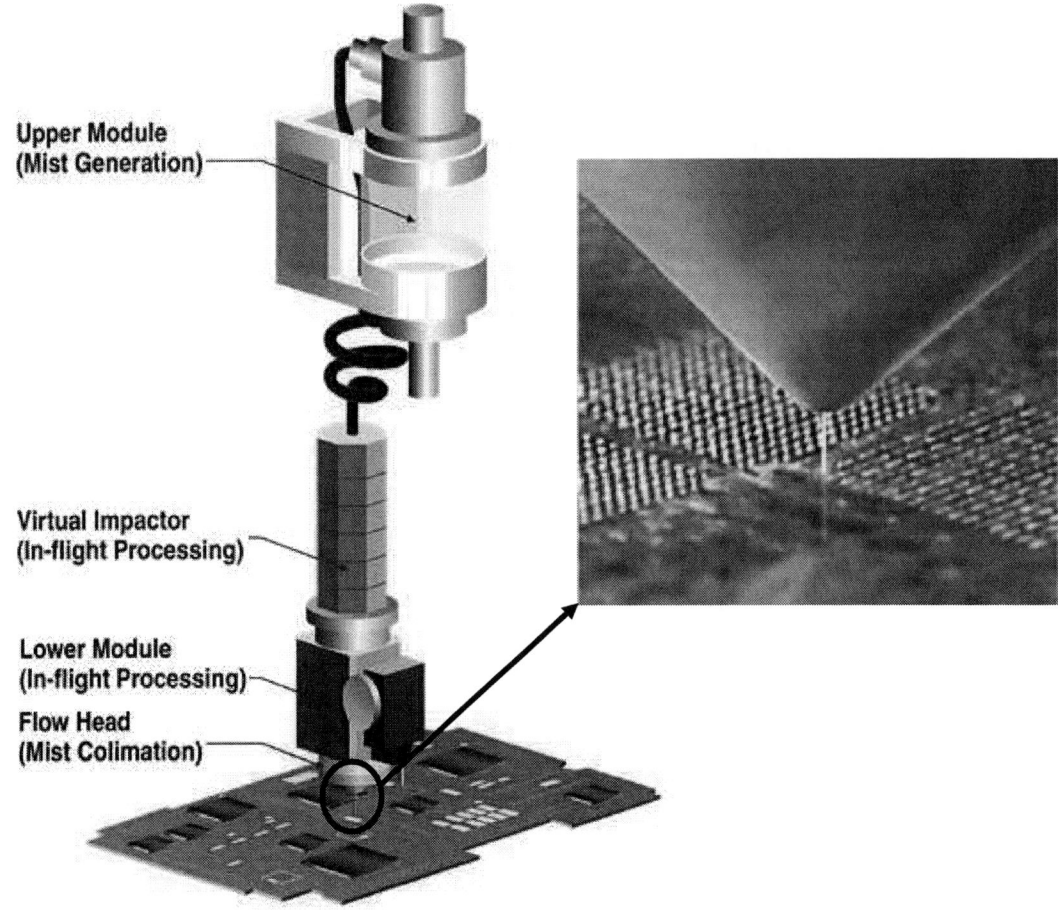

Figure 1: Schematic of the M^{3}D process and photo of the deposition head

Table 1. Application areas for M³D

Packaging and Assembly	Electronic Devices
<u>High Density Interconnects</u>	Flat Panel Displays
Flip-Chip / Direct Die Attach	Solar Cells & Fuel Cells
Embedded / Integrated Passives	Micro-Sensors
Flex Circuits	MEMS & RFID
Meso-Dispensing	<u>Hybrid Manufacture</u>
<u>Electronic Components</u>	Smart Structures
Resistors, Capacitors and Inductors	<u>BioTech</u>
Micro-Antennae	Bio-Sensors& Implantable Devices
Micro-Batteries	Micro-Arrays

1. A module for atomizing liquid and colloidal suspension raw materials (Mist Generation).
2. A second module for focusing the aerosol and depositing the droplets (In-Flight Processing).
3. The final module is a laser for post treatment sintering of the deposits. This is used where sensitive substrates which cannot be conventionally sintered are present.

Mist Generation is accomplished using an ultrasonic transducer or pneumatic nebuliser. The aerosol stream is then focused using a flow guidance deposition head, which forms an annular, co-axial flow between the aerosol stream and a sheath gas stream. The co-axial flow exits the flow guidance head through a nozzle directed at the substrate. The M³D flow guidance head is capable of focusing an aerosol stream to as small as a tenth of the size of the nozzle orifice. Patterning is accomplished by attaching the substrate to a computer-controlled platen, or by translating the flow guidance head while the substrate position remains fixed.

Thermal post processing of the deposited material is often needed to sinter or cure the material to increase properties such as electrical conductivity. Depending on the application, either conventional oven sintering is used, or for low temperature substrate materials a solid state laser is used to locally heat the deposited material without affecting the surrounding substrate. The M³D system can write electronic features at high speeds. Deposition rates are as fast as 200mm per second, allowing the system to meet the requirements of both rapid prototyping and high-volume electronic manufacturing.

M³D Features, Benefits and Applications.

Rapid Product Development. The M³D system is driven by standard .DXF CAD data, which allows designers to quickly and cost-effectively test new prototypes and products. This eliminates the delays and costs associated with mask sets and other upfront capital required by conventional electronics manufacturing techniques. This feature also makes it much easier to implement and validate design changes without the need for "re-tooling". The elimination of masks and resists also permits on-the-fly changes and rapid design iterations. The result is faster time-to-market for new products.

Materials Range. M³D can deposit a wide variety of materials, including metals, conductors, insulators, ferrites, polymers, adhesives and biological materials. Deposits can be made on virtually any surface material - silicon, glass, plastics, metals, ceramics, polyimides, and polyesters. This flexibility opens the way for many different applications using a single process, **Table 1**. Additionally, the material requirements for the process are much wider than for ink jet printing. M³D can process materials with a viscosity range of 0.7-1000cP, compared to a typical range of <20cP for Ink Jet. Thus a wider range of inks can be applied with greater ease and reduced dilution.

M³D's ability to deposit conductive, insulating, and adhesive materials layer-by-layer within a single system makes it an attractive solution for the partial or complete production of microelectronic devices. M³D uses in this area include sub-micron layers of Platinum for fuel cell applications, high density interconnect backplanes (organic and metal) for flat panel displays, and micro-sensors for avionics. Since many such markets are evolving, M³D can be a powerful product development tool, as well as a viable production solution. The process can also be used as a repair technique for devices such as electronic displays: M³D can precisely place material to fix open circuits on the backplane or coloured inks on the colour mask.

Fine Accurate Features. M³D reliably produces ultra fine feature circuitry well beyond the capabilities of thick-film and ink-jet processes, making it ideal for next generation packaging and high-density interconnect applications at both the chip and circuit board level. Depending on the material, the technology can produce electronic features as small as 10 microns at high print speeds ,

Fig. 2. This capability is being extended. Recently sub-5 micron lines have been produced on a lab scale.

Figure 2. 10micron Ag lines on glass

The small size of the interconnects allow the use of smaller devices. For example, the current size of an RFID chip is limited to the size of the interconnects created by screen-printing. With the cost of the RFID tag being dependant on the size of the chip, this acts as a barrier to reducing costs. Using M^{3}D instead of screen printing allows for the use of a smaller chip and less interconnect material, driving the cost down.

M^{3}D offers a solution for the production of smaller (down to 10μm), high performance components critical to size-sensitive applications like those in the wireless and hand-held device markets where component density is increasing dramatically. The ability of M^{3}D technology to create complex geometries from a wide range of materials makes it suitable for the production of both passive and active components, including resistors, **Fig. 3-4**, inductors, capacitors, filters, micro-batteries and micro-antennae.

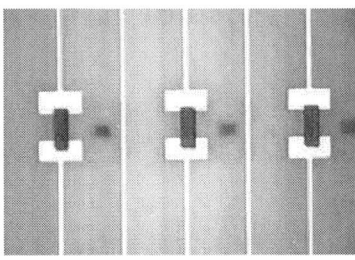

Figure. 3. Carbon PTF (Polymer Thick Film) Resistors on FR4.

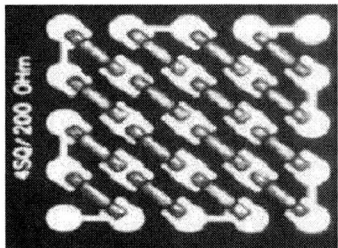

Figure 4. PTF (Asahi Chemical Co.) on a BT substrate. Resistor Area ~10^{-3} mm^2

The excellent edge definition and repeatability of the process are particularly relevant to high frequency requirements. In comparison to screen printing, embedded resistors can be made smaller and more accurately with M^{3}D such that no laser-trimming is needed to tune the resistor to the right value. Since M^{3}D can deposit both conductive and insulating materials, one layer at a time, it has the potential to directly embed passive components.

Low Temperature Laser Sintering. Once the material has been deposited, conventional approaches require high-temperature treatment. These might include re-flow or cure ovens, or a chemical decomposition process. By contrast, M^{3}D can locally process the deposition on substrates, such as polymers, that have low temperature tolerance (150°C or less) through using the laser module. The end result is a high-quality thin film with excellent edge definition and near-bulk electronic properties.

Conformal Deposition. M^{3}D can also precisely deposit materials on non-planar substrates. This is made possible by the relatively high [5mm+] stand off point of the deposition head. With no physical contact with the substrate by any portion of the tool other than the deposition stream, conformal writing is easily achieved. This allows for the ability to build 3D features on to structural component, write into trenches, **Fig 5**, or create MID devices.

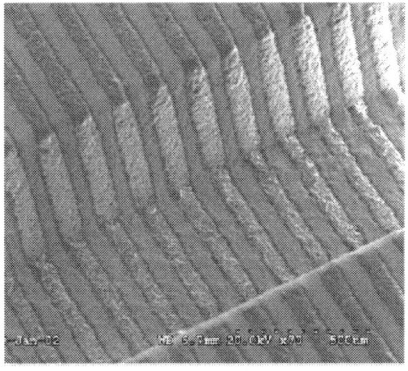

Figure. 5. Ag lines, 50micron, over a 500 micron deep trench.

This 3D capability can also be used to carry out "direct die attachment". For example **Fig. 6.** shows how M^{3}D has been used to replace wire-bonds and directly attach the die to the smart card circuit. This involves directly writing over several different kinds of materials to create the interconnect. The interconnect is written from the Cu pad, over Kapton, and epoxy adhesive (ca. 150micron height) and over the edge of the IC. This was done using 2D motion control. For height profiles greater than ca. 3mm up to 50mm an automated 3axis is applied, **Fig. 7**.

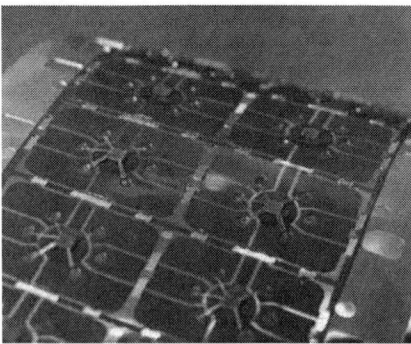

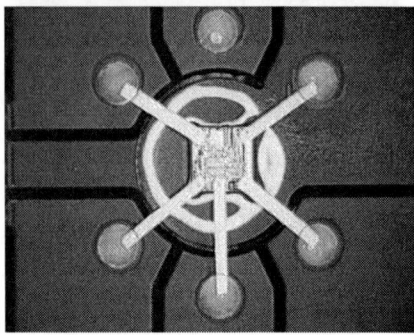

Figure 6. Direct Die Attachment. M³D deposited Ag interconnects from the smart card contact pad over epoxy adhesive and connection to the IC. Line width: ca. 150 microns

Figure 7. Ag circuit with 150 micron line width written on a ceramic substrate (25mm height).

Reduced Processing & Environmental Impact. M³D provides significant environmental benefits and reduced processing requirements because it is an additive fabrication process, which eliminates the waste associated with traditional subtractive (e.g. mask and etch) processes. Rather than coating an entire masked surface, manufacturers can deposit exact amounts of material exactly where it needs to be.

Another application for reducing processing steps and reduced chemical usage is the Catalyst Layer approach for producing interconnects or other features. The standard PCB method consists of dip coating, plating, masking, etching and secondary plating steps. M³D can cut the number of steps by directly depositing the activator solution followed by a standard plating step. In this way the mask and etch steps are removed reducing processing and environmental loading.

Materials Efficiency. The tiny droplets created by M³D allow for very thin coatings (i.e. 10's-100's of nanometers thick) which allows for good interaction between differently applied layers. These same femto-liter sized droplets allow for very careful control of dosages dispensed. Since many electronics materials are expensive to produce, the technology is a key enabler for reducing the cost of each device.

Process Scalability

Currently 25 M³D systems are in operation in product development and R&D in the Europe, Asia and the USA. 2007 will see the first volume production applications coming on stream. Consequently significant efforts are being undertaken to scale the process for high volume manufacture. These include the development of multi-nozzle systems with increased atomiser throughput as well as developments to ensure allow reliable and consistent low maintenance operation over long run times.

Summary

The emergence of CAD driven, additive manufacturing techniques, such as M³D, are set to significantly impact on many market segments including the electronics industry.

M³D can economically and rapidly deposit a wide variety of materials on many different substrate types. The process offers the potential to develop advanced packaging solutions and also to reduce the number of processing steps and environmental impact of production operations. Designers can harness the unique features of M³D to create revolutionary designs which offer a wide range of time, cost and quality benefits.

NanoCT: Visualizing of internal 3D-Structures with Submicrometer Resolution

André Egbert

Phoenix|x-ray Systems + Services GmbH, Niels-Bohr-Str. 7, D-31515 Wunstorf, Germany

AEgbert@phoenix-xray.com, Tel.: +49 5031 172-168, Fax: +49 5031 172-299, www.phoenix-xray.com

Abstract

High-resolution X-ray microscopy allows the visualisation and failure analysis of the internal microstructure of small micro devices – even if they have complicated 3D-structures where 2D X-ray microscopy would give unclear information. The recently launched nanotom® of phoenix|x-ray is a very compact laboratory system specialised for the analysis of small samples with the exceptional submicron voxel-resolution of <500 nm (0.5 microns). It is the first 180 kV nanofocus computed tomography system in the world which is tailored specifically to the highest-resolution applications in the fields of micro mechanics, electronics or material science. Therefore it is particularly suitable for nanoCT-examination of sensors, actuators, complex micro electronic components with concealing parts such as capacitors or stacked dies and material samples of any type like synthetic materials, ceramics or composites. Any internal difference in material, density or porosity within a sample can be visualised and data like distances or pore volumes can be measured. By granting the user the ability to navigate the internal structure of an object slice-by-slice in a non-destructive manner, the nanotom® creates new possibilities for sample analysis per mouse click which have thus far been unreachable. NanoCT widely expands the spectrum of detectable micro-structures and is ideal for the non-destructive inspection of compact but complex micro mechanic parts or electronic devices. The nanotom opens a new dimension of 3D-microanalysis and will partially replace destructive methods – saving costs and time per sample inspected.

Key words: Computed Tomography, nanoCT, 3D-Analysis, Failure Analysis

Introduction

For many years, the only way to determine the interior structure of a sample with resolution in the sub-micron range was to section the part. Not only was this technique time-consuming, but in this destructive process a valuable sample was lost. With advances in x-ray technology, however, this is no longer necessary. In the fields of electronics, micromechanics or engineering, nanoCT allows the researcher to explore a sample's structures into the sub-micron level as never before (see e.g. [BONSE, BARUCHEL]).

Before nanoCT could be realized, there were two major obstacles to overcome. First, computers capable of processing large amounts of data in a reasonable amount of time were necessary, and until recently they were cost prohibitive. Second, and more important, an extremely high resolution x-ray system was needed; this included both a tube with a sub-micron focal spot that was powerful enough to penetrate dense samples and a high-resolution detector. The new phoenix|x-ray nanotom® combines the power of a cluster server with a high-powered nanofocus tube and high resolution detector. An example application for nanoCT is shown in Fig. 1. The sample shows some CSP bonds, diameter ca. 400 μm.

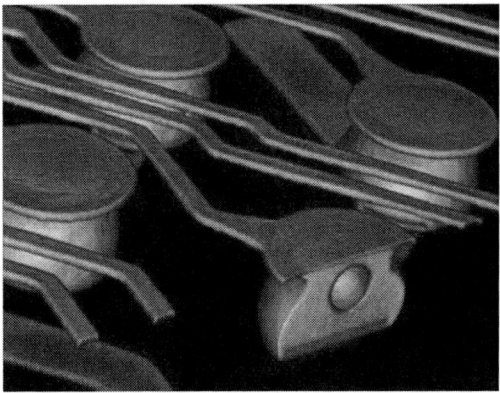

Figure 1: Example for failure analysis: nanoCT of some CSP bonds. Clearly visible is the three-dimensional structure of the package as well as the large void in the cut solder joint.

Principle of Operation

nanoCT is truly an extension of conventional 2D X-ray. A sample is fixed between an X-ray source and a detector, and the part is rotated n steps a full 360degrees, as shown in fig 2.

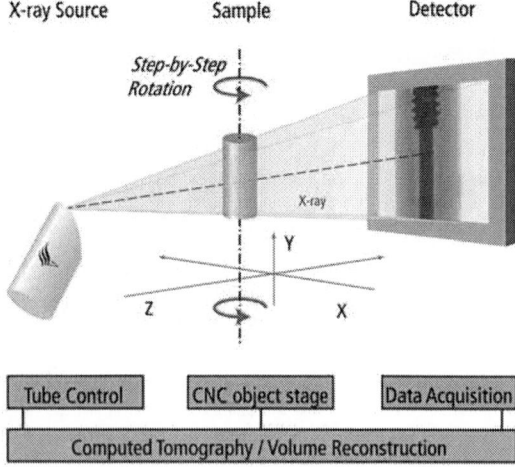

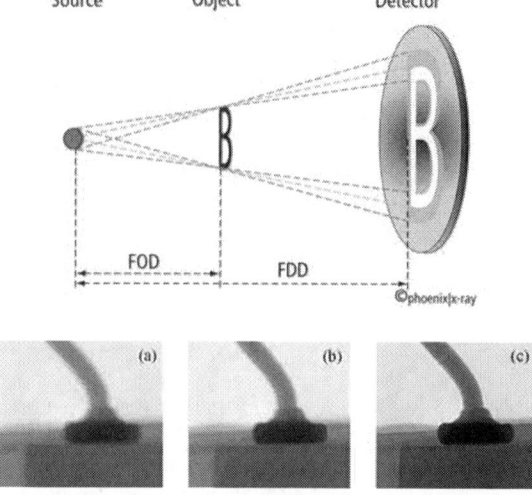

Figure 2: Schematic set up of a CT system. For the volume reconstruction, the region of interest must remain within the field of view and cone of radiation throughout the whole 360⁰ rotation.

The nanotom® uses the principles of cone beam CT (see [KAK]): The high-power nanofocus tube with a maximum acceleration voltage of 180 kV and a maximum tube power of 15 W is generating a cone-shaped X-ray beam which images the sample onto a fully digital detector with an active area of 120 x 120 mm (2300 x 2300 pixels). The image or voxel resolution results from the pixel width of 50 μm divided by the geometric magnification which is given by the ratio of source-detector distance and source-object distance. Thus the sample size determines the achievable voxel size i.e. highest magnification can be reached for smallest samples.

The sharpness (i.e. the actual image resolution) is principally determined by the focal spot size of the X-ray tube (see Figure 3). For a spot size larger than the voxel size the so called penumbra effect is limiting the image quality. A smaller X-ray spot will reduce this effect and result in a sharper image on the detector. Due to the submicron focal spot size of the high power nanofocus tube, the nanotom® can resolve image features as small as 200-300 nm.

Figure 3: nanofocus-X-ray technique: The smaller the focal spot, the higher the sharpness of the wire: a = 10 microns, b = 5 microns. A spot size <1 μm (c) allows a detail detectability of 200 to 300 nm.

Sample handling and preparation is straight forward and simple: The object is fixed between X-ray tube and detector on the rotation stage using a chuck and optionally a precise XY-table for centring. Generating tomographic images starts with the acquisition of a stack of two-dimensional X-ray images. The series is a collection of images acquired while progressively rotating the sample step by step through a full 360° within the field of view at angular increments of about 0.25-0.5° per step. These projections contain information on the position and density of absorbing object features within the sample. This accumulation of data is then used for the numerical reconstruction of the volumetric data using a filtered back projection algorithm [FELDKAMP]. The reconstruction time depends on the size of the volume: since the reconstruction is performed parallel to acquisition, the result for smaller volumes is available immediately after finishing the acquisition process. An optionally available PC cluster further reduces reconstruction time.

The resulting volume data set is visualised by slices perpendicular to the three dimensions or compiled in a three-dimensional view which can be displayed in various ways, e.g. in pseudo-colours (see section 4). It also allows slicing and sectional views in any direction of the volume i.e. this technique is capable of substituting destructive mechanical slicing and cutting in many applications.

The nanotom

In 2006, phoenix|x-ray launched the nanotom®, an ultra-precise high-resolution CT system (Fig 4). It is designed specifically for laboratory applications, scanning samples of up to 1 kg and 120 mm diameter. The maximum voxel resolutions of the system can be down to 500 nm (0.5 microns) and below. The nanotom® is the first 180kV nanofocus CT system in the world set up for the highest resolution applications in a variety of fields such as materials science, micro mechanics, electronics, geology, and biology to name a few. Therefore, it is particularly suitable for nanoCT examination of microelectronic components, sensors, complex mechatronic samples as well as for material samples such as synthetic materials, ceramics, sintered alloys, composite materials, mineral and organic samples.

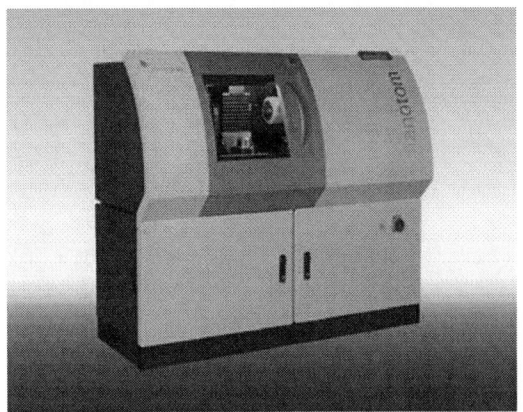

Figure 4: With its small footprint of only 160 x 74 cm, the nanotom® is uniquely suited for laboratory applications.

Computed Tomography at such exceptionally high spatial resolutions requires careful design, taking into account any features which might influence the resulting resolution. These special needs demand special manipulation systems, detectors and X-ray tubes. For example, the nanotom uses a 180 kV high power nanofocus tube which can penetrate high-absorption samples like copper or steel alloys. A 5-megapixel flat panel detector with 50 μm pixel size and a 3-position virtual detector (i.e. 360 mm detector width) give rise to a wide variety of experimental possibilities. To avoid any negative influence of vibrations or thermal expansion, tube, detector and manipulator are mounted on a granite structure. The special door construction as well as the advanced sample fixture permits simple and precise positioning of samples. Furthermore, special materials and construction details are used to minimize the variation of the focal-spot/detector distance during a scan. In addition, minimal vibrations of the system are suppressed by air bearings of the rotation unit.

Example applications

The CT results obtained with the nanotom allow analysis of the spatial microstructure of sample materials with submicrometer resolution. The figures shown here (fig. 5-7) present only a few of the possible applications for the nanotom. As it

can be seen, nearly any internal detail that corresponds to a contrast in material, density or porosity can be visualized and internal distances can be measured. Highly accurate three-dimensional measurements facilitate reverse engineering processes. It is also possible to image different phases of alloys in solder joints, with the ability to see crystalline structures that differ from those of the constituent metals. Components of a sample may be visualised individually by suppressing the contrast of all but the material of interest. In this way, the 3D-structure of the normally hidden pore network of light metal castings or rock samples may also be analysed.

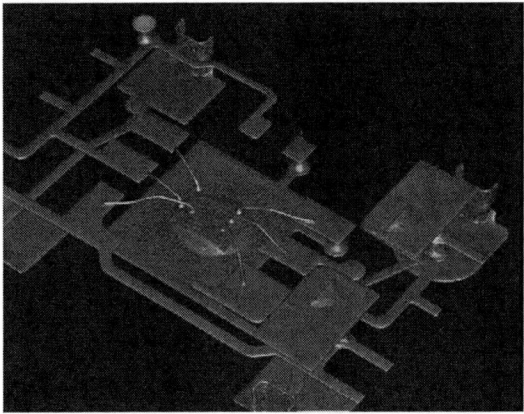

Figure 5: 3D image of a ceramic SMD IC. The cap was blinded out to visualise the wire bonds.

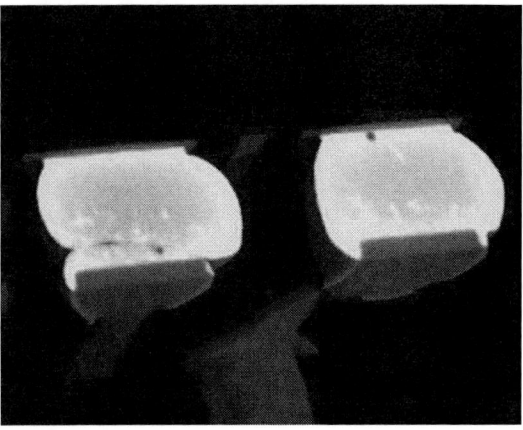

Figure 6: nanoCT of BGA solder joints diameter 500μm each. The left ball is open showing the head-in-pillow-effect: the ball did not melt with soldering paste. Bright crystal structures indicate different phases in the eutectical alloy.

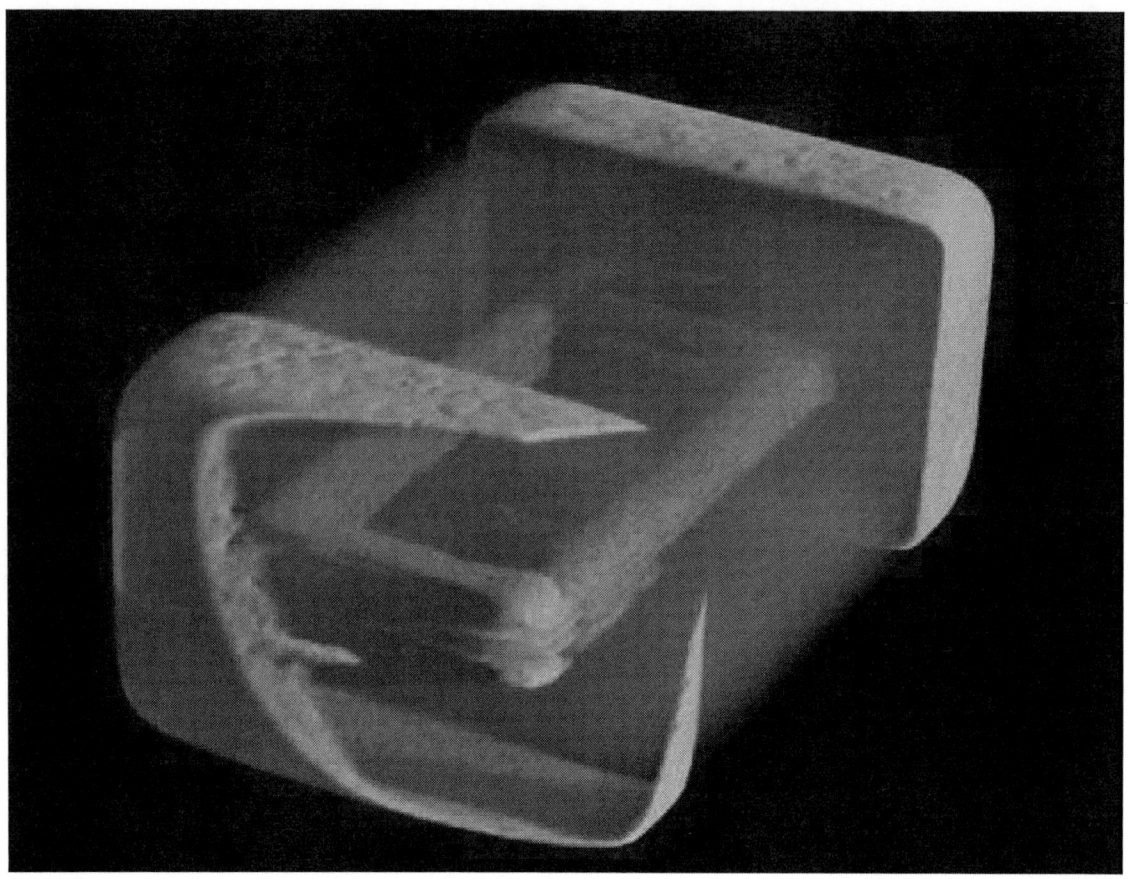

Figure 7: Example for product development: 3D-visualisation of a SMD ceramic chip inductor showing the internal coil. The sample size is about 2 mm.

The first results of high-resolution nanoCT demonstrate that this technique widely expands the spectrum of detectable micro-structures. Hence, the nanotom opens new possibilities of 3D-microanalysis and will partially substitute destructive methods – saving costs and time per sample inspected.

References

[1] BONSE, U. (Editor) (2004), SPIE Conference Proceedings: Developments in X-Ray Tomography IV, SPIE Press

[2] BARUCHEL, J., J.-Y. BUFFIERE, E. MAIRE, P. MERLE, AND P. GILLES (2000), X-Ray Tomography in Material Science, Hermes Science Publications.

[3] KAK, A.C. AND M. SLANEY (1999), Principles of Computerized Tomographic Imaging, IEEE Press

[4] FELDKAMP, L.A., L. DAVID, AND J. KRESS (1984), Practical Cone Beam Algorithm. Journal of the Optical Society of America, 1 612-619.

Performance and Reliability of Flexible Substrates when subjected to Lead-Free Processing

M. J Rizvi, C. Y. Yin, <u>C Bailey</u>, H Lu

School of Computing and Mathematical Sciences, University of Greenwich,

Greenwich, London SE10 9LS, United Kingdom

Phone +44-20-8331-8660, Fax +44-20-8331-8665, e-mail: c.bailey@gre.ac.uk

Abstract

The use of flexible substrates is growing in many applications such as computer peripherals, hand held devices, telecommunications, automotive, aerospace, etc. The drive to adopt flexible circuits is due to their ability to reduce size, weight, assembly time and cost of the final product. They also accommodate flexibility by allowing relative movement between component parts and provide a route for three dimensional packaging. This paper will describe some of the current research results from the Flex-No-Lead project, a European Commission sponsored research program. The principle aim of this project is to investigate the processing, performance, and reliability of flexible substrates when subjected to new environmentally friendly, lead-free soldering technologies. This paper will discuss the impact of specific design variables on performance and reliability. In particular the paper will focus on copper track designs, substrate material, dielectric material and solder-mask defined joints.

Key words: Flexible substrate, lead-free soldering, computer modeling, stress, reliability

Introduction

There is significant growth in the use of flexible substrates for electronic product applications. In fact current estimates put the size of the flexible substrate market at 4bn Euros with a growth rate of 15% per annum. In Europe a number of companies both supply and use flexible substrates.

Environmental legislation such as Reduction of Hazardous Substances (RoHS) means that electronics manufacturing companies have to adopt lead-free soldering practices and technologies [1-3]. This means that printed circuit boards will be subjected to higher assembly temperatures during the reflow process because the melting point of lead-free solders is in general higher than tin-lead solder. With the requirement to adopt lead-free soldering practices there are concerns that:

- higher processing temperatures will affect the behavior of the substrate during the reflow assembly process
- adoption of lead-free solders will affect the reliability of the interconnects between the substrate and the components

Flex-No-Lead, a European Commission funded project (EC Contract COOP-CT-2004-513163), is aiming to address the above concerns. The project consisting of a consortium of partners ranging from materials suppliers, product assemblers, and research organizations have been collaborating for two years.

Further information on the Flex-No-Lead project can be found at the project website (http://www.flexnoleadproject.com/).

This paper presents some of the modeling results from this project. The results from these models, together with the experimental work undertaken by our partners, is helping formulate design rules for the adoption of reliable flexible substrates with lead-free solders.

Although the focus of the paper is on lead-free solders and in particular Tin-Silver-Copper (SAC), it should be noted that the project also investigated a range of other materials including conductive adhesives. Also other assembly processes which include laser soldering.

Flexible Substrates

Flexible substrates are made from organic materials where adhesives and copper are used to establish the circuit tracks on the substrate. Typical base materials for flexible substrates include polyimide (PI), Polyethylene Terephthalate (PET), Polyethylene Naphthalate (PEN), Polyvinyl Chloride (PVC) or Liquid Crystal Polymer (LCP). All of these materials have been investigated in the Flex-No-Lead project.

The coverlay concept is also being used for flexible substrates to protect the substrate from moisture and to define the joints as solder mask defined (SMD) or non-solder mask defined (NSMD). A typical polyimide flexible substrate used in this study is shown in Figure 1.

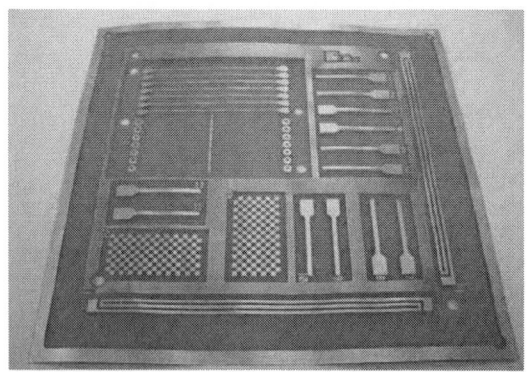

Figure 1: Typical flexible substrate (Courtesy: Budapest University of Technology and Economics).

Flexible substrates can be prone to moisture uptake and this can affect their behavior during the reflow process and during reliability testing. Together with cure (or chemical) shrinkage this can change the physical appearance of a flexible substrate after thermal processing. For example, a flexible substrate made from polyvinyl chloride (PVC) could be distorted severely after thermal processing if not handled carefully (see Figure 2).

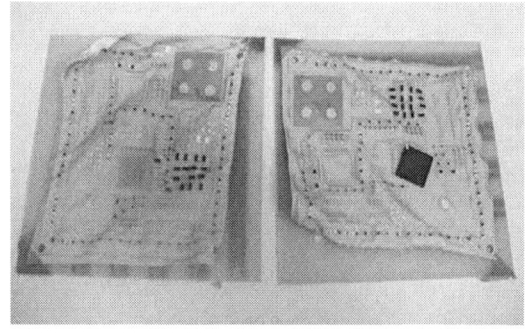

Figure 2: Distortion of PVC flexible substrate after thermal processing (Courtesy: TWI Ltd, UK).

Modeling

(a) Substrate Design

Figure 3 illustrates a portion of a flexible substrate containing the base carrier film material, dielectric and copper tracks. Both the copper tracks and the dielectric are bonded to the base material using suitable adhesives. Three types of base carrier material, namely, PI, LCP and PEN are discussed in this paper.

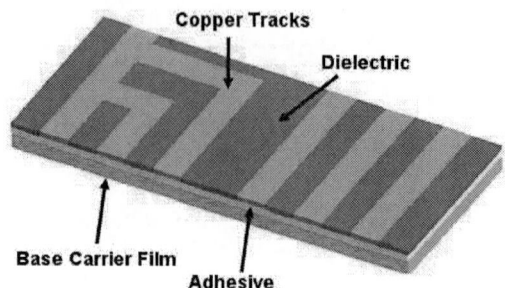

Figure 3: Illustration of a section of board with a dielectric

Finite element calculations have been undertaken to predict the behavior of a flexible substrate as it passes through a reflow process with a lead-free temperature profile. The model is based on both a lumped parameter shell model to predict the behavior of the whole substrate, and a full three-dimensional continuum model to predict the behavior locally at the material interfaces.

The lead-free (SnAgCu) reflow temperature is applied as ambient conditions to the top and bottom of the substrate with a heat transfer coefficient of 100 W/m^2-K. The mechanical material properties used in the following simulation are shown in Table-1. The computer simulations were carried out using the modeling software PHYSICA [6]. Due to a lack of any visco-elastic data for the carrier base materials, at present these are assumed to behave elastically in the following analysis.

Table 1: Material properties used in computer modeling

Materials	Properties		
	Young's Modulus (MPa)	Poisson's Ratio	CTE (10-6/K)
Insulator	4400	0.22	50
Copper	132400	0.34	16.7
Adhesive	300	0.3	150
Polyimide	2500	0.34	30
LCP	340000	0.4	17
PEN	6100	0.45	20

(b) Solder Joint Design

These simulations have investigated the behavior of BGA type solder joints connected to a flexible substrate and subjected to a temperature cycle.

Figure 4 illustrates a section of the computer model for these BGA joints. For joint design a coverlay has been added onto the flexible substrate. In this example, the coverlay is made up of another layer of the base carrier material placed over and connected to the dielectric layer with adhesive.

The addition of a coverlay allows the solder joints to be either solder mask defined (SMD) or non-solder mask defined (NSMD). In both cases the openings in the coverlay expose the copper pads for soldering onto. The diameter of this opening defines whether the pad is SMD or NSMD as shown in Figure 5.

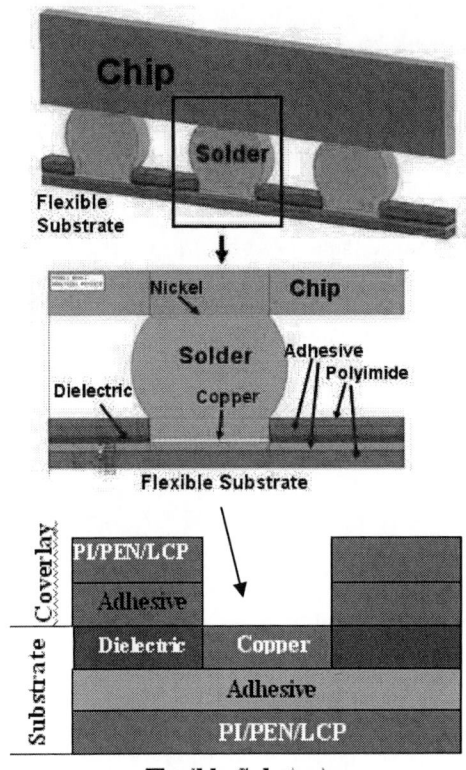

Figure 4: BGA type solder joint on flexible substrate with coverlay.

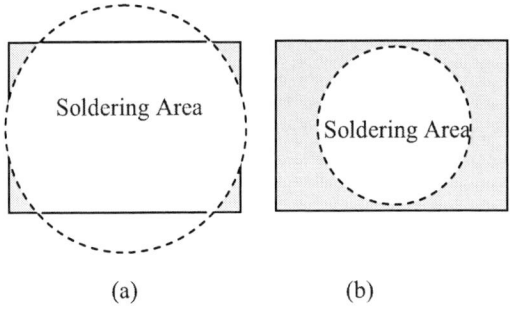

Figure 5: Types of pad defined by the opening in coverlay (a) NSMD and (b) SMD.

Simulations have been undertaken for air to air thermal shock cycling where the temperature ranges from -40 to +125 °C. Upper and Lower dwells are set for 15 minutes each and the transfer time between hot and cold dwell is 10 seconds only.

In this analysis the solder is assumed to have creep behavior defined by an elasto-viscoplastic constitutive law for SnAgCu (SAC) solder. The overall modeling procedure is described in [7]. It

should also be noted that all other materials are assumed to behave elastically.

Modeling Results

(a) Substrate Design

The copper tracks are 100 µm in width. The width of dielectric insulator is adjusted when the copper track width varies as 40, 60, 80 and 100 µm.

A number of simulations have been undertaken to see how different flexible substrate materials behave when passed through a reflow profile. Due to affects such as (i) CTE miss-match, and (ii) moisture loss (especially for substrates not pre-heated) the substrate will deform during the reflow process and possibly show shrinkage or expansion at the end of the reflow process.

The amount of deformation that can occur in a 14cm * 14cm substrate has been modeled where the substrate had different base carrier films and thickness of copper.

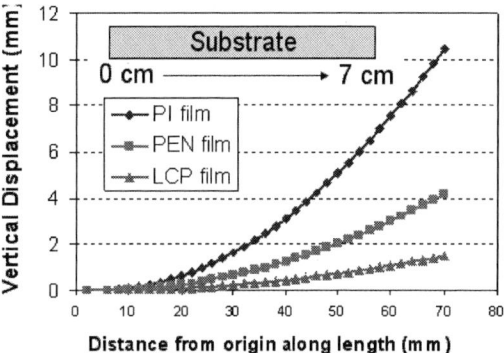

Figure 6: Upward displacement of substrate at peak temperature in a reflow process.

Figure 6 illustrates the upward displacement of a flexible substrate made of PI, PEN and LCP. In these calculations the base carrier material layer thicknesses was 50µ, 125µ and 100µ are for PI, PEN and LCP respectively.

Clearly we can see that the thinner PI material shows much greater amounts of displacement than the others at the peak reflow temperature. This is aided with the higher CTE for this material. It should be noted that no moisture affects have been included at present in the above calculations.

Visualization of substrate materials inside a prototype reflow furnace used at the site of our industrial partner, GTS Flexible Materials Ltd, seems to show similar amounts of displacement at the substrate edges for PI material. It is also interesting to note in these experiments that the substrate materials usually return to their original shape once cooled down. Although a certain amount of shrinkage can also be observed at the end of reflow which can be more pronounced for PI material. This is reported by our partners in a paper also published in the proceedings of EMPC-2007 [9].

At present it should be noted that the above calculations are only capturing the CTE effects on substrate deformation. Any moisture loss from the materials (or chemical change) is ignored in these calculations although these effects are currently under theoretical investigation.

It is obvious from the above results that having components on a substrate that is allowed to deform during the reflow process can be problematic and needs to be addressed.

Another design variable that is being investigated is the dimension of the copper tracks in the substrate and the impact this can have on the stresses at the copper/adhesive interface. Figure 7 illustrates this region of a flexible substrate.

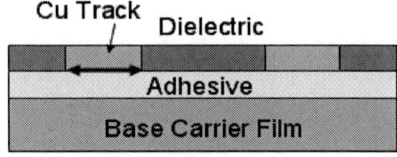

Figure 7: Arrow in the illustration show the region of interests where delamination can occur.

As seen in the material data provided in Table-1, the dielectric insulator has comparatively higher coefficient of thermal expansion (CTE) value than copper. Therefore, this insulator will try to expand more than copper when heated and push from both sides onto the copper conductors. This phenomena could cause damage to the tracks during any temperature processing (reflow or accelerated testing)

Moreover, the adhesive possesses a very high value of CTE that will cause higher expansion of the adhesive layer. In this case the possible delamination can occur either in adhesive-substrate or in copper-adhesive interfaces.

Shear stress along the copper track-adhesive interface is illustrated in Figure 8. Cleary we can see significant changes in the shear stress at the copper-adhesive interface near the dielectric material. This region could be a cause for concern as delamination or crack initiation in the copper which could occur during temperature processing due to the CTE miss-match.

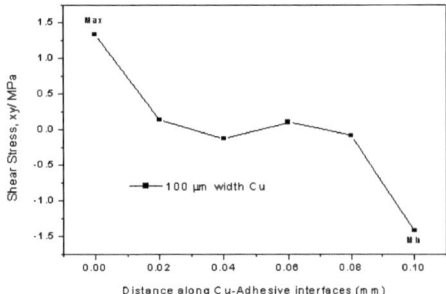

Figure 8: Shear stress (σ_{xy}) at peak reflow temperature.

The effect of copper track width has been investigated, for the layout illustrated in figure 3, and is shown in figure 9. The impact of copper track width on the stress is also shown where a wider track reduces the stress. The decreasing tendency is almost similar to all the three carrier films.

It is important to note that any increase in the copper track width eventually increases the volume percentage of copper in the circuit boards. This higher amount of copper in fact determines the behavior of flexible substrate and can of course reduce the distortion during lead-free reflow soldering.

Clearly we can see that the maximum stress at the copper-adhesive-dielectric interface is given when PI is used as the base carrier film. Therefore, it is evident that the type of carrier film also determines the failure criteria in flexible substrate.

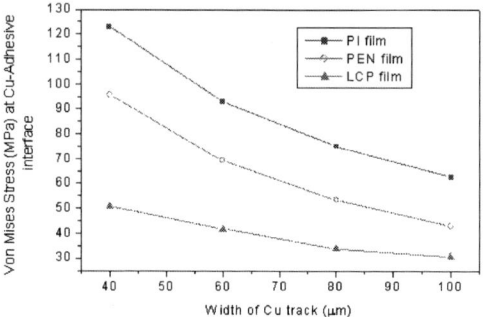

Figure 9: Stresses along copper-adhesive interface as a function of copper track width.

Figure 10 represents the comparative shear stress, σ_{xy} along the copper adhesive interfaces. The stress is plotted at peak reflow temperature of 250 °C. It is obvious from this figure that using dielectric material as an insulator in between the copper conductors can reduce the shear stress generation along adhesive-copper interfaces.

It is seen that when dielectric material is used shear stress is almost zero along the interfaces except the edges whereas a significant increase in shear stress values is noticed for the flexible substrate without dielectric material. The increased shear stress values in the adhesion interface could cause a lift of copper tracks from the substrate leaving the circuit's electrical performance at risk.

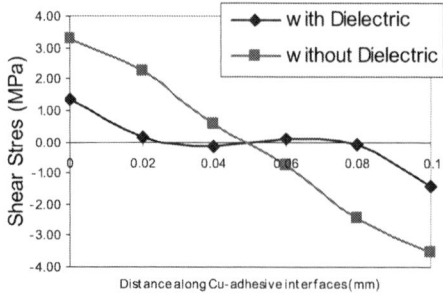

Figure 10: Shear stress distribution along the copper-adhesive interfaces.

(b) Reliability of SMD and NSMD Solder Joints on Flexible Substrate

Finite element calculations for BGA type solder joints, as illustrated in figure 4, have been undertaken for both solder masked and non-solder masked defined joints. In this analysis the following flexible substrate materials were investigated PI, PEN, LCP..

The accumulated creep energy density generated in the solder is used as the damage parameter for crack initiation and propagation which will affect reliability and performance of the joint. Figure 11 details the damage (high in red) for a solder joint subjected to a thermal cycle. Clearly we can see the location where cracks are likely to start and propagate.

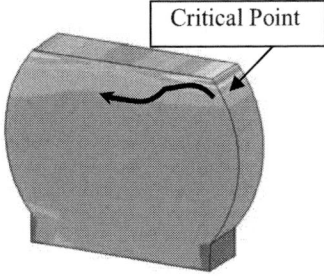

Figure 11: Damage in solder due to thermal loading.

Figure 12 details the accumulated damage in the joint at the critical point after three cycles. The most damaged joint being the one at the corner of the BGA. This trend will continue and it can clearly be see that solder mask defined joints are not degrading as fast as non-solder mask defined joints.

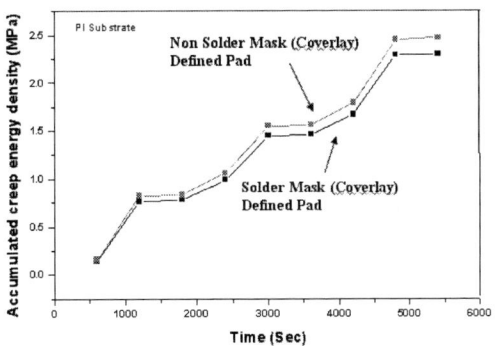

Figure 12: Damage in solder for flexible (PI) substrate.

The above is an interesting result. When using thicker FR4 material as the substrate it is widely acknowledged that the use of a solder mask defined pad will promote degradation of the joint. This is due to the mechanical interaction of the solder mask with the solder at its interface near the base of the solder. This can lead to crack initiation at this point and the solder will fail at the substrate side.

To check that the above model is consistent, simulations have been undertaken using the above joint dimensions but in this case with a standard FR4 board (1.6 mm thick) and solder mask. Figure 13 details these predictions. Clearly we can see that when using a standard FR4 board construction that the solder mask defined pad results in greater degradation of the solder over time during thermal cycling as is reported in the reported literature.

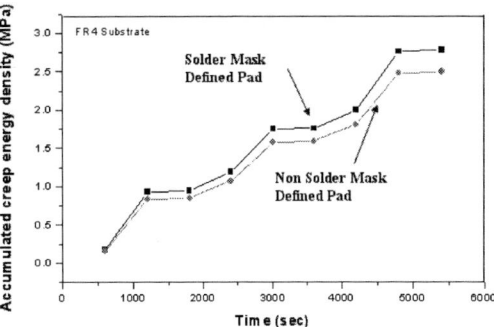

Figure 13: Damage in solder for an FR4 substrate.

Clearly the joint for the flexible substrate with coverlay is behaving differently to that observed when using FR4. When using a flexible substrate the largest amount of degradation in the solder is observed at the solder-chip interface. One reason for this could be due to the flexibility of the PI substrate. Figure 14 illustrates a magnified view of the mode of deformation that can occur in the substrate during a thermal cycle.

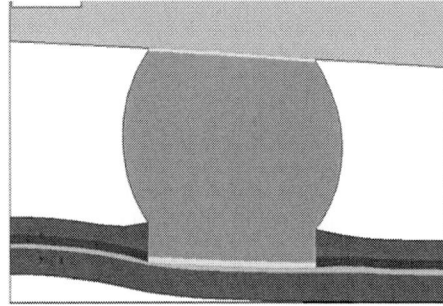

Figure 14: Mode of deformation for solder mask defined joint.

Figure 15 details the impact of different base carrier material on the degradation in the solder joint which is non-solder mask defined. Clearly we can see that a flexible substrate design that uses PI material will result in the least amount of degradation to the solder joint and LCP the most, This can be explained by the very high modulus of the LCP material, as detailed in Table 1, and this will result in this substrate being less flexible than either PI or PEN.

593

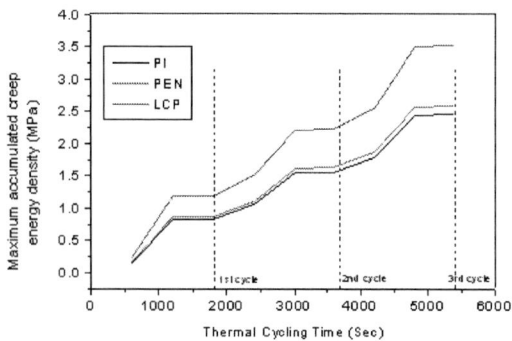

Figure 15: Damage accumulation in solder joints for three base carrier materials.

Although the work reported in [8] is for an HAST test, the results in that paper detailing some of the experimental work undertaken in the project also show that solder mask defined pads result in less solder degradation. It should also be noted that the greater amount of movement in flexible susbtrates, if allowed, can produce other modes of failure such as substrate or copper track cracking. This has been observed in some of the latest experimental results [9].

Conclusions

The impact of lead-free soldering on the thermo-mechanical performance and reliability of flexible circuit boards and their solder joints has been discussed. Based on the above results, and the assumptions used in the models, the following conclusions can be provided:

1. Flexible substrates with PI base carrier material demonstrate the most warpage during the relow profile.
2. A dielectric material in between the copper tracks can reduce the shear stresses along the copper/adhesive interface.
3. SMD pad, defined by a coverlay, can reduce the amount of damage seen in the solder joint.
4. The PI substrate results in least amount of damage to the solder joint during thermal cycling. This is probably due to its low modulus and ability to easily deform.
5. Although flexible substrates may lower the degradation on solder joints, compared to FR4, other failure modes could take place such as substrate cracking.

The modeling results described above are complimented by the experimental results, also to be published in EMPC-2007 [9],

Future Work

The results detailed above are based on the modeling assumptions and materials data used in the analysis. Some interesting trends are observed and further research is underway to confirm the current conclusions. The modeling work is being extended to investigate the effect of moisture loss and gain and its impact on warpage of the substrate during the reflow process. Models are also underway for the HAST test as reported in [8, 9].

Acknowledgements

The authors acknowledge the financial support provided by the European Union (EU) CRAFT program and the project partners in FLEXNOLEAD (Project Reference: COOP-CT-2004-513163). We also acknowledge the useful discussions with all of the project partners. These are TWI Ltd, GTS Flexible Materials Ltd, Eldos Co. Ltd, Flexible Technology Ltd, Flex-ability Ltd, Kmed Manufacturing, Epigem Ltd, International Consulting Bureau, ICI Belgium N.V., KIC International Sales Inc., Freudenberg Forschungsdienste KG, and Budapest University of Technology and Electronic.

References

[1] J. Shen, et al, "Strenthening effects of ZrO2 nanoparticles on the microstructure and microhardness of Sn-3.5Ag lead-free solder", Materials Science & Engineering A, vol. 441, no. 1-2, 2006, pp. 135-141.

[2] F. P. McCluskey et al, "Reliability of high temperature solder alternatives", Microelec. Reliab., vol. 46, no. 9-11, 2006, pp. 1910-1914.

[3] J. W. Yoon et al, "Effects of reflow and cooling conditions on interfacial reaction and IMC morphology of Sn-Cu/Ni solder joint", Journal of Alloys and Compounds, vol. 415, no. 1-2, 2006, pp. 56-61.

[4] http://www.allflexinc.com/bene.shtml

[5] http://www.lenthor.com

[6] PHYSICA, http://www.gre.ac.uk/~physica

[7] S. Ridout et al, "The effect of thermal cycle profiles on solder joint damage", 6th Int. Conf. on Electronics materials and Packaging (EMAP2004), Penang, Malayasia, 5-7 December, 2004, pp. 436-441.

[8] Z. I. Vitez et al, "Life time tests of lead free solder joints on flexible printed circuits", 6th Int. Conf. on Poly. & Adh. In Micro. Elec. & Phot., Miraikan, Odaiba-Tokyo, Japan, January 15-18, 2007, pp. 234-239.

[9] B Balogh, P Gordon, Z Illyefalvzi-Vitez, G Falmer, A Girulska, T Harvey, G Kotora, D Kirkpatrick, "Reliability of Flexible Circuits with Different Lead-Free Technologies", Proceedings of the 2007 EMPC, Oulu, Finland, in press.

Multilayer Flexible Wiring Board based on screen printed Conductors

J. Petäjä, K. Kautio, H. Funck*, M. Karppinen, P. Karioja & R. Vatanparast*

VTT Technical Research Centre of Finland, Kaitoväylä 1, PO BOX 1100, FI-90571 Oulu, Finland

Tel. +358 20 722 2441, Fax +358 20 722 2320, jarno.petaja@vtt.fi

*) Nokia Research Center, Helsinki

Abstract

For high-volume products, such as, mobile terminals, low-cost techniques for multilayer polymer-based thick film wiring board manufacture are needed. Screen-printing is a cost-efficient technology candidate to build up approximately 6 conductor layers on both sides of a flexible substrate, for example. In order to experimentally evaluate the feasibility of screen-printing technique, the printing resolution was tested on different substrate materials, such as, polycarbonate (PC), polyethylene terephtalate (PET), polyimide (PI) and liquid crystal polymer (LCP). Conventional screen printed polymer thick film pastes were characterized on polymer substrates to form multilayer fine-line patterning and through hole vias. The final demonstrator was a double sided PI substrate having two conductor layers separated by dielectric layers on both sides of the substrate and through substrate vias. The screen-printed conductor material was an Ag based nanoparticle ink and the dielectric layer was a polyimide-based material. Several challenges were identified that might hinder the applicability of the technology for mass-production. The stability of the polyimide substrate is a problem if the curing temperature of the printed materials is above 200°C. Layer-to-layer alignment tolerances are feasible if the printed area is small, 5"x5", in our case. The flatness of the substrate, however, is not very good after printing several layers on each other. The tested nanoparticle ink is a promising conductor system; however, lowering of the curing temperature from 230°C below 200°C would have a major impact on production friendliness.

Key words: Nanoparticle ink, polymer thick film paste, dielectric ink, electrical via

Introduction

Freely-bendable, light-weight, multilayer, printed wiring board is advantageous when developing high-volume hand-held equipment, such as, mobile phones. Manufacturing cost is one of the key criteria when screening technologies for high-volume production.

Common technology for multilayer flex circuit manufacturing is etched copper laminates. Printed circuits are primarily produced by selectively etching away metal that is bonded to a dielectric substrate. An alternative technology is the polymer thick film, PTF. The most contrasting characteristic of PTF compared to etched copper is that PTF is a 100% additive process where conductors are added only where required, not subtracted by etching a full sheet of conductor. The advantages such as cost reduction and plant size minimization can be achieved because there are less processing steps and waste material generated during manufacturing. In addition, the materials used can be environmentally favorable. Some limiting factors of PTF are that the most conductive PTF ink is still one order of magnitude higher in electrical resistance than pure copper or silver. In addition, the commonly used screen printing method limits the minimum PTF line width to a practical value of around 200-250µm. There is no doubt that lines down to 50 µm can be produced under the best conditions, but not consistently and not at the prints speeds needed for a cost-effective process. [1]

In this paper, we show potential substrate candidates and the screen-printing results of conductor lines on flexible substrates, describe materials and methods for through substrate vias, and finally present the printing results of multilayer boards, as well as discussions and conclusions.

Test Structures

Our target was to evaluate low cost methods to fabricate polymer based multilayer conductor structures on a flexible substrate. Screen-printing was chosen as a method to pattern conductors and dielectric layers and mechanical punching to fabricate through substrate vias.

Our target was to demonstrate fine-line screen-printing, less than 100µm line width, on a flex substrate. Two alternative test structures, shown in Figures 1a and 1b, were realized. The first design has four conductor layers and three layers of multilayer dielectric on the top side of the flex substrate. The second design has two conductor layers and one dielectric layer on each side of the flex substrate, with additional through hole metallization.

For the substrate materials, good thermal stability during the ink curing and high quality punched via holes were required. In addition to the fine-line printing of conductors, low resistivity was required for the printed conductors. During the via hole filling, in the optimal case, the via ink should completely fill the via hole and the filling should be at the surface level on both sides of the substrate. For printed dielectric layers, layer uniformity, pinhole freeness and surface smoothness were required.

In order to find compatible material systems, we tested nine polymer thick film conductor inks, three via filling inks and five dielectric inks on nine plastic substrates. After initial tests, we chose two material systems for additional adhesion, bending endurance and thermal shock tests.

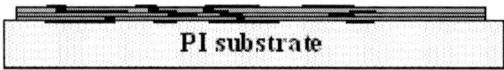

Figure 1a: Single-sided polymer multilayer board with four conductor layers and three dielectric layers.

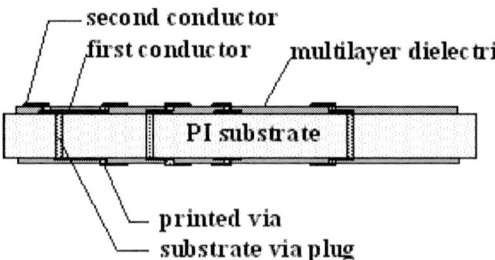

Figure 1b: Double-sided polymer multilayer board with through hole vias and two conductor layer on both sides of the substrate.

Flexible Substrates

Nine plastic substrate materials were tested. These substrates included one PI and PC, 2 types of PEN and LCP substrates and three types of PET substrates. 125-μm thick substrates were used except the PC and LCP substrates that were 175μm and 50μm, respectively.

Via holes were made with a mechanical puncher. In the initial test 200μm vias and 1mm alignment holes were punched and the quality of the via holes was inspected visually and with an optical profiler. The dimensional stability of substrates was studied by tempering at 80°C, 120°C and 150°C temperatures. Tempering was done three times at each temperature and dimensional changes were measured after each tempering cycle. We were interested in the substrate shrinkage, if it is homogenous and if the material stabilizes after the heat treatment or if it shrinks during every tempering cycle.

During the via filling experiments, it was found that it is difficult to achieve properly filled vias on the 50μm thin substrates with the used via fill ink systems. According to the tempering and via punching tests, the most suitable substrates were the PI substrate, one PET substrate and both LCP substrates. DuPont Kapton HN polyimide material was the only substrate in the tested ones that could withstand high curing temperatures required for a NPS nanoparticle conductor ink, which was tested for conductor printing.

Substrate Via filling

Altogether three substrate via inks were tested. A punched stainless steel stencil was used for the via fill tests. In the optimal case, the via hole should be completely filled with the via ink and the via should be level with the substrate surface. Spreading of the ink around the via hole may sometimes be a problem and may cause shorting to the adjacent lines. Therefore, the viscosity of the via fill ink should be optimal.

All three via inks had quite low viscosity compared to e.g. via fill inks used in low temperature co-fired ceramics, LTCC, technology. Low viscosity together with the smooth hydrophilic surface of the plastic film led to ink spreading and resulted to vias that were not filled properly or were not filled at all. The best via fill ink from the tested ones was Asahi LS-106D-1 that filled the vias relatively well but the spreading of the ink was still a problem. The other two inks were conventional conductor inks with too low viscosity for via filling.

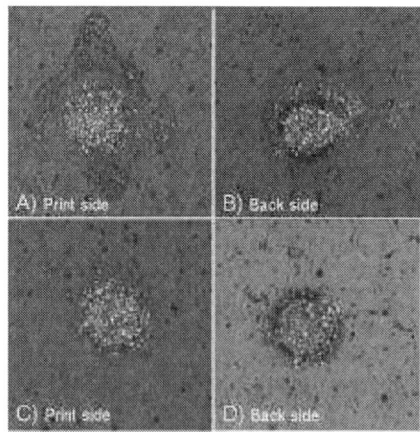

Figure 2: 80-μm vias filled with the modified Asahi LS-106D-1. Pictures a and b represent bad quality and pictures c and d good quality via filling.

Test structures included punched via holes sized 80μm, 100μm, 150μm and 200μm. Different process versions were tested for the modified version of LS-106D-1 via ink with a little higher viscosity. The viscosity, however, was not high enough and the spreading was not avoided totally, see Figures 2 a and b. The test indicated that the larger the via hole, the bigger the spreading.

With the 80μm via hole diameter, together with the notification that the bottom side of the stencil (the side that is in contact with the substrate) should be cleaned regularly, the ink spreading was reduced, as can be seen in Figures 2 c and d. Nevertheless, the spreading was not avoided completely. According to the tests, with the modified Asahi LS-106D-1 ink, the vias with collar diameter (splash diameter) about 110...150 μm are feasible.

80-μm vias were also tested using a higher viscosity LTCC via paste. The test gave good results, shown in Figure 3, and indicated that the via ink itself has the major effect for the spreading behavior.

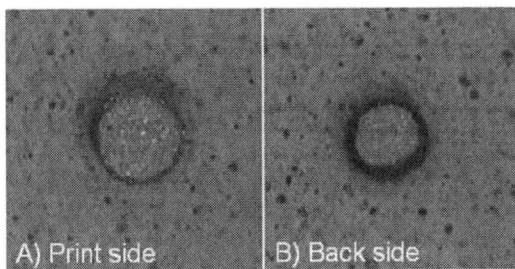

Figure 3: 80-μm via holes filled with a high viscosity LTCC via filling paste.

Conductor Inks

Printing test was made with semi-automatic Baccini screen printer. Murakami trampoline screen was used. Screen parameters were; mesh 400, wire thickness 18μm and emulsion thickness 15μm.

In order to find compatible materials, nine polymer thick film conductor inks were tested. It was assumed that only some of these materials would be suitable for fine-line screen printing (under 100μm line width) on flex substrates and, therefore, the purpose of this initial test was to screen out the poor ink candidates and to find potential candidates for the more thorough testing. The most important properties of conductor inks to be characterized were the suitability for fine-line screen printing process and the resistivity of printed conductors.

The tested inks were conventional PTF inks: one Cu and seven Ag based inks. In addition, we tested an Ag nanoparticle ink, NPS from Harima Chemicals. At that time, the ink was under product development and samples were received for test purposes. We decided to compare the best performing conventional PTF ink, Asahi SW-1100-1, and Harima NPS nanoparticle ink.

Figure 4 shows a 0.9×0.9 mm^2 area measured with an optical profiler; the conductor lines are along the printing direction. The thickness of the 75-μm line is 10±7μm and 6.5±2 μm for SW-1100-1

and NPS ink, respectively. Figures 5 shows visual appearance of conductors printed using Asahi SW-1100-1 on a PET substrate and Harima NPS on a PI substrate.

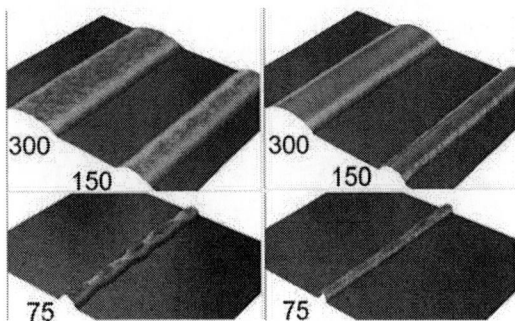

Figure 4: Conductor profiles with 300μm, 150μm and 75μm nominal widths, Asahi SW-1100-1 on the left and Harima NPS on the right.

For the NPS ink, the curing condition is 240°C / 60 minutes. Resistivity is ~3.8 μΩcm (bulk silver has 1.6 μΩcm), square resistance ~8 mΩ/□ (line width 150μm and average thickness 4.5 μm), and feasible line width/spacing ~75/100 μm. The particle size of the ink was studied with SEM. The resolution of used SEM restricted the measurement but it could be noted that particle size was less than 100nm. To prevent the agglomeration of nanoparticles there apparently is chemical shielding around the nanoparticles. To break that shielding a certain threshold temperature is needed. It was tested to cure the ink at lower temperatures but the ink did not cure (sintered). Due to the high curing temperature, polyimide was the only possible substrate material from the tested ones.

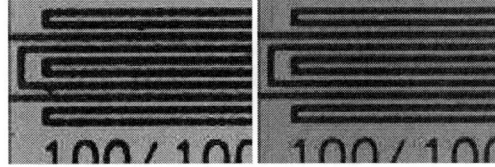

Figure 5: Microscope photos of 100μm lines with 100 μm spacing. On the left, SW-1100-1 conductor on a PET substrate, and on the right, NPS conductor on a PI substrate.

For SW1100-1, the curing condition is 150°C / 20 minutes. Resistivity is ~69 μΩcm, square resistant ~ 83 mΩ/□ (line width 150μm and average thickness 8.5 μm), and feasible line width/spacing ~100/125 μm.

The tests showed that only the nanoparticle ink fulfilled the specified fine-line and low resistivity requirements. Therefore, the nanoparticle ink, Harima NPS, was chosen as a conductor material for two-layer test circuits. The only substrate from the tested ones that can withstand the high curing temperature of Harima NPS is DuPont Kapton HN polyimide material. The best conventional Ag-based PTF ink, Asahi SW-1100-1, was chosen as a reference material for the two-layer test circuits. The substrate via ink for both systems, conventional PTF and nanoparticle based systems, was the modified, higher-viscosity version of the conventional Ag-based ink, Asahi LS-106D-1. DuPont Teijin Film Melinex ST507 PET was chosen as the substrate material for the conventional PTF system. The material systems for the fabrication of the two-layer test circuits are shown in Table 1.

	PTF-system	Nanoparticle-system
Conductor	SW-1100-1	NPS
Via ink	LS-106D-1 modified	LS-106D-1 modified
Substrate	PET, ST507	PI, Kapton HN

Table 1: Material systems for the two layer test circuit.

Bending strength, thermal shock and adhesion tests were carried out for the two-layer test circuits. Bending strength tests were using a 3.75-mm bending radius and 30 cycles per minute cycling frequency. Conductor line patterns included vias with 100, 150, 200-μm diameters and lines without vias. Every via pattern included 60 vias in a chain, see Figure 6. Samples were flexed 0…5000 times. Resistances were measured before test and after 100, 500, 1000, 3000 and 5000 cycles.

The conventional PTF system and NPS system degraded differently during the tests: the conventional system showed relatively large resistance increase and the NPS system fully broke during the tests. For SW-1100-1, the resistance increase was around 20% after 5000 cycles regardless of via diameters. A 10% resistance increase for the bare conductor without vias was measured. With the NPS system, some lines were broken after 100 cycles and all lines were broken after 5000 cycles. The electrical connection was broken or the resistance increase was more than 100%.

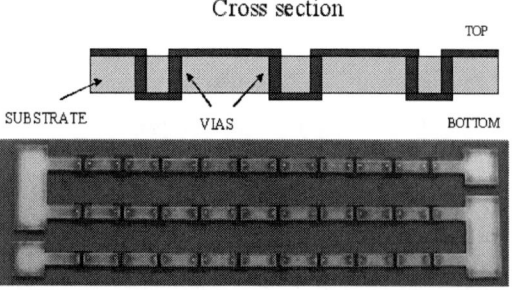

Figure 6: Cross section and photo of the bending strength test structure.

Figure 7 depicts a typical failure mechanism. One can see cracks, bends and even loss of conductor material. Bending radius 3.75mm is too tight for the NPS conductor. This does not mean that the NPS does not withstand any flexing. The question is how much it withstands and what are the requirements of the application, since larger bending radiuses were not used for the NPS system.

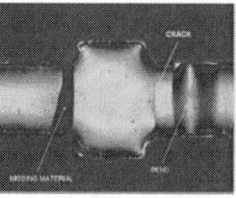

Figure 7: Microscope and SEM images from the NPS conductor line. Cracks, bends and missing material are detected.

In the thermal shock tests, the samples were similar to those used in the bending tests. The samples were packed in waterproof plastic packs that were vacuumed. Ten cycles from 1°C (ice water) to 94°C (nearly boiling water) were performed. The test cycle was ~20 seconds: 10 second at 1°C and 10 seconds at 94°C. No resistance increase was noticed with the conventional and NPS systems after 10 cycles.

The adhesion of the screen-printed conductors on the PI and PET substrates was experimented using a simple "tape test" in which a strip of adhesive tape was pressed firmly on the test pads or patterns and ripped off. Scotch 600 Crystal clear tape was used. Conductor adhesion to the substrate was tested three times: in the beginning, after the flexural endurance and after thermal shock tests. The conventional PTF system passed the test in all three cases; the NPS system did not pass the tests in any case.

Dielectric Inks

Five screen printable dielectric inks were tested to screen out less favorable candidates for the four-layer test circuit; therefore, a few prints were made per ink and minor variations in printing parameters were made, only. Stainless steel screen was used. Screen parameters were; mesh 325, wire diameter 30μm and emulsion thickness 15μm. The tests gave indication on feasible layer thicknesses, printable via sizes and layer uniformities. Two out of five inks withstand the high curing temperature of the NPS ink. The best candidate for the demonstrator proved to be a polyimide based dielectric ink, Asahi Pirica PIR-35.

Printed dielectrics structures are shown in Figure 8. A 8.5±1.5-μm dielectric layer thickness was obtained. The surface of the layer is a little wavy as can be seen from Figure 8a. The phenomenon is typical for the screen printing method. Air bubbles or other voids were not noticed. With a 400-μm nominal via diameter, the actual size of the via was about 250μm. With smaller nominal sizes, the possibility of via blocking increased. In the four-layer test circuits, we used vias with the 400-µm nominal diameter.

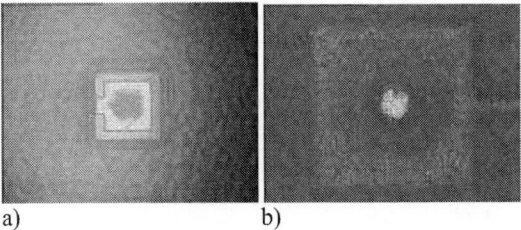

a)   b)

Figure 8: a) The printed dielectric layer of Pirica PIR-35 on the PI substrate. b) 400-µm nominal via size reduced to a 250-µm via. In both pictures, a 1.5×1.5 mm² NPS conductor pad can be seen in the via opening.

Multilayer Demonstrator

Two different four-conductor layer demonstrators were made, shown in Figure 1. Design 1a has all the layers printed on the same side of the substrate. It was assumed that this causes cambering due to shrinkage of printed dielectric layers. After a couple of printed layers, it was clear that the cambering would be quite significant, as can be seen from Figure 9. The left row includes design 1a and right row design 1b. First samples from the back contain one printed conductor layer and the front samples are ready four conductor layer circuits. The design 1b is more favorable because there is equal number of printed layers on both sides of the substrate. It did not camber much, but some waviness is seen.

Figure 9: Samples of the four-conductor layer test circuits.

Some problems were seen with certain test patterns, for example, electrical connections were broken in some samples. No broken conductors were visually seen when inspecting the samples under a microscope; therefore, it was obvious that the via connections were defective. In general, the test results of the double-sided demonstrators were promising.

Layer to layer alignment accuracy was ± 60…80µm. It could be improved by taking the substrate shrinkage into account during the layout design. In the experiments, we also noticed that the accuracy of the punching machine was not in the manufacturer's specification. If these things will be taken into account, the alignment deviation could be ± 30…50µm. Decreasing the curing temperature of the NPS ink, is the most important issue for better alignment accuracy between layers.

Discussion

From the production point of view, there are still many problems and challenges to bring these techniques into high-volume production. The stability of the polyimide substrate is a problem as long as the curing temperature of the materials is higher than 200°C. Also, the printed dielectric layer should maintain dimensions and not shrink during the curing cycles. Layer-to-layer alignment tolerances are feasible for the 5"×5" printing area. However, the flatness of the substrate was not very good after several layers were successively printed on the substrate. This might cause problems in the automated handling of the parts.

Harima nanoparticle ink is a promising conductor system. However, lowering the curing temperature from the presently used 230°C below 200°C would increase production friendliness. The poor mechanical strength and adhesion of Harima NPS conductor material observed in the tests may also cause problems in some applications.

According to the tests, 400µm was the minimum feasible nominal via size for screen printed multilayer dielectric layers. With the 400-µm nominal size, the obtained via diameter was about 250µm. This will obviously limit the wiring density of the multilayer circuits. For smaller vias, more advanced methods, such as, lasering or photoimaging could be applicable.

Via filling methods for substrate vias should be developed to assure high yield in high-volume production. Especially, the optimization of via ink viscosity should be made to enable accurate and high quality via filling with high yield. The punching of small via holes to the PI substrate may be problematic in high volume production due to tool wear. If punching would turn out to be unusable, a potential method for the via hole machining could be lasering.

Conclusions

The double-sided multilayer test board showed the following main results: 75/100-µm feasible line/space with Harima NPS nanoparticle ink printed on the PI substrate, ~8 mΩ/square sheet resistance (150µm line width and 4.5µm average thickness) and ~3,8 µΩcm resistivity. 80-µm diameter substrate via holes and 80-μm filled vias with ~110...150 μm collar diameter (splash diameter) were feasible. Printed vias with the 400-μm nominal diameter produced 250-μm vias. The double-sided multilayer design was more favourable compared to the one-sided design due to better dimensional control and smaller camber of the substrate. Layer to layer alignment accuracy was $\pm$ 60...80μm and could be reduced to about $\pm$ 30...50μm.

Acknowledgements

The authors want to acknowledge Airi Weissenfelt, Miia Aitta and Sami Karjalainen who carried out the processing experiments.

References

[1] Polymer Thick Film, Ken Gilleo, 1996.

Investigation on Printed Wiring Board Failures during Reliability Assessment for Telecommunication Products

Yujie Dong, Markku Tammenmaa, Visa Ruuhonen, and Virpi Pennanen

Nokia Networks, Kaapelitie 4, 90630, Oulu, Finland

+358(0)40 546 7475, yujie.dong@nokia.com

+358(0)40 546 7204, markku.tammenmaa@nokia.com

+358(0)40 508 8311, visa.ruuhonen@nokia.com

+358(0)40 546 0385, virpi.pennanen@nokia.com

Abstract

Telecommunication products are under continuous demand to offer more services in a cost-efficient way. High integration level, good power handling capability and long term reliability are the main drivers in product development. As an important part of the system build-up, printed wiring boards need therefore to consist of increasing number of layers, have finer geometries and more complicated structures. In addition, lead free manufacturing process poses higher thermal stress on both components and PWB during the assembly process. As a consequent, more PWB defects are expected to occur at product level. Hence, a tight quality control of the PWB appears necessary for the equipment manufacturer, as well as a wider knowledge and deeper understanding of various defects and root causes behind them. This paper reviews some of the PWB-related failures, which have recently been noticed during technology qualification and testing. Examples are tarnishing and creep corrosion of the immersion silver PWB finish, micro-voiding along the solder joint interface, micro-via opening after reflow and conductive anodic filament (CAF) growth inside the PWB. Literature review of mechanisms behind these failures is also presented.

Key words: PWB, Immersion silver finish, micro via, CAF

Introduction

Considering different requirements, printed wiring boards (PWB) have been undergoing continuous technology updates in the last decade. The updates include replacing Sn/Pb Hot Air Solder Levelling (HASL) surface finish with Electroless Nickel Immersion Gold (ENIG), Organic Solderability Preservative (OSP) or Immersion Silver (IAg) to eliminate lead; developing high density interconnection (HDI) board to include more layers, finer feature and micro / blind via for higher level of integration. Naturally, reliability concerns of the products build on these new technology PWBs have been raised through the years. PWB manufacturers have conducted various tests to address the potential risks of these new technologies. However, the equipment manufacturers still have to test and verify their products separately, due to the assembly process variation and product service environment complexity.

Findings and Discussions of the Test Results

Tarnish Resistance of IAg Finish

At the end of 1990s, immersion silver was introduced by PWB industry to achieve a low cost Cu surface finish with multiple reflow possibility and good surface conductivity. Comparing to other finishes such as ENIG, OSP and Immersion Tin (ISn), the IAg finish does not suffer from solderability degradation from multiple reflow process like ISn and OSP do. It also costs less than ENIG and yields flat soldering surface for fine circuit feature design with reliable solder joints [1, 2]. However, the drawback of IAg coating is silver finish tarnishing. The tarnish resistance of IAg finish is different from different plating process at a given thickness [3]. It also depends on the porosity and structure of the silver deposit as corrosive atmosphere could penetrate through the defects to attack the copper underneath [3].

From manufacturing and reliability point of view, the IAg finish should not only be sufficiently tarnish resistant and good solderability preservation, but also corrosion free in the product's service life. Therefore, it is necessary for the product manufacturer to evaluate the immersion silver boards from different PWB suppliers.

To test the PWB tarnish resistance, flowing mixed gas (FMG EIA364-65A) was selected. This test is considered to be a realistically accelerated environment test that can simulate the kinetics and degradation mechanism found in indoor and outdoor environments. After the test, different degree of tarnish appearance was found. Some samples showed clear discolouration, and some boards had the silver surface totally tarnished. In addition aggressive corrosion attack to the Cu track, and even creep corrosion was noticed from some samples. These results are shown in Figure 1.

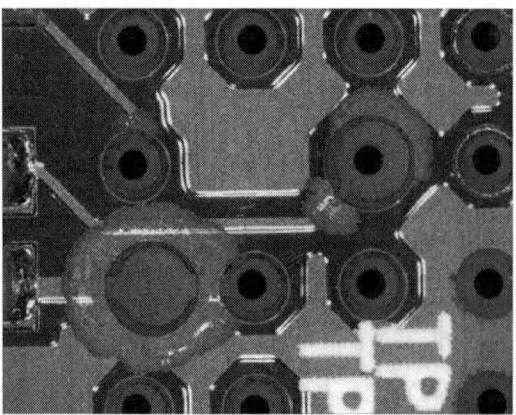

Figure 1: Immersion silver finish surface with different degree of tarnishing after testing. First photo shows obvious tarnished IAg finish. The second photo had significantly corroded IAg finished Cu pad. Creep corrosion was noticed in some boards, as shown in the last photo.

For product used in harsh environment, partly or even totally tarnished silver coating is unavoidable. But quick copper corrosion due to poor quality of silver coating will significantly shorten product life time. Moisture enhanced creep corrosion can impose high risk to the product due to shorts between conductors.

IAg Finish Planar Microvoiding

Immersion silver surface finish was introduced with the confidence that it would result in a well studied tin copper metallurgical bond between the pad and the solder, due to fast dissolving and dispersing of Ag into melt solder. Experimental study also confirmed that the silver was present in the solder - copper interface with negligible amounts [1]. This advantage overcomes ENIG finish, where the thick nickel layer prevents copper diffusion so that nickel tin is the main metallurgical bond. The solder joint reliability of ENIG finish is generally considered weaker due to its specific metallurgical bond [4]. Also 'black pad' failure has been reported to degrade solder joint mechanical strength after reflow.

Few years back, a phenomenon called planar microvoiding was observed in IAg finished board solder joints. Differing from normal processing voids, planar microvoids are small in size (less than 50 micrometer in diameter), and located in the plane between the interface of solder and intermetallic compound layer. Excessive amount of planar microvoiding inside the solder joint seriously degrades the interconnection reliability by weakening the joint strength against mechanical shock. It also accelerates solder cracking propagation during thermal cycling. Figure 2 shows cross section picture (top) of planar microvoiding found after reflow. The other picture (bottom) in Figure 2 shows a failed solder joint after vibration test. The fractured flat surface of the solder along intermetallic compound layer is obvious.

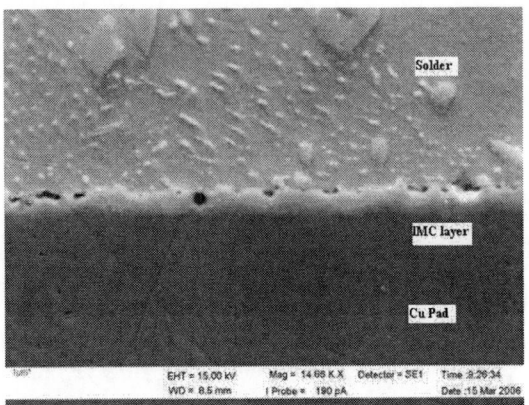

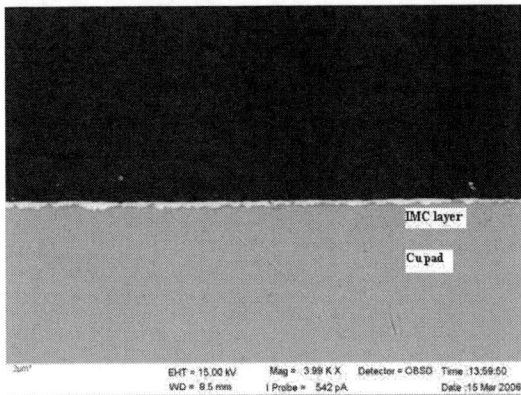

Figure 2: The upper SEM image shows a cross section view of a solder joint. Microvoiding on top of IMC layer is clearly seen. The lower image shows a failed joint after dropping test. The IMC layer stayed on Cu pad while the bulk of solder fractured along the IMC layer.

According to some chemistry supplier's study, the occurrence of microvoiding varies from immersion silver chemistry as well as PWB batch. One explanation is that the aggressive pre-etching of copper land before silver coating could be the reason of microvoiding formation. Other factors which could contribute to microvoiding include flux type, silver thickness and supplier's coating process control [5]. Our experience showed that the occurrence and amount of microvoiding do vary from supplier to supplier, and batch to batch. Microvoiding was not found from all supplier's boards, but tiny cavity between silver and copper was commonly observed.

The amount of microvoiding may or may not be related to the thickness of silver coating. However, one supplier's board with thicker Ag layer did show less amount of microvoiding comparing to another manufacturer's board with thinner silver layer. Still more studies would be needed to address the root cause of microvoiding phenomenon.

Microvia Quality

It appears that for high density interconnection (HDI) technology, laser drilling overcomes other technologies to become the main method of microvia manufacturing on PWB [6]. This technique consists of via drilling by laser beam, thoroughly cleaning after laser processing and copper plating to form the electrical connection path of the microvia. Using different types of laser sources, via of diameter as low as 25 to 30 micrometer is achievable in nowadays production [6]. Microvia technology provides an efficient way of increasing the interconnection density of PWB by reducing the feature size, easing circuit routing and lowering layer counts. It is also cost effective, because the technology can be adapted with standard fabrication processes with minimal equipment investment.

Due to its advantages, microvia PWBs are widely used in telecommunication products now. Microvia reliability has been studied by some authors. Main concerns are thermal fatigue failure of via due to material CTE (Coefficient of Thermal Expansion) mismatching. It shows that aramid reinforced epoxy has better performance compared to RCC/FR4 (Resin Coated Copper / FR4) combination due to its better CTE match [7, 8].

During our quality assessment, electrical open failures were found immediately after reflow in microvias due to poor PWB manufacturing process control. Figure 3 below is a cross section picture of one microvia that was detected electrically open after reflow. A failure analysis performed on one batch of problematic boards indicated that the root cause of the open via could be due to poor cleaning after laser drilling. The resin residue left in bottom of via hole was so slight that the boards could pass the PWB suppliers' electrical testing. However, thermal stress from lead free reflow at assembly site resulted in electrical opening at via bottom because of the copper separation. With assembler's feedback, the PWB manufacturer tightened the cleaning process after via drilling to improve the board quality.

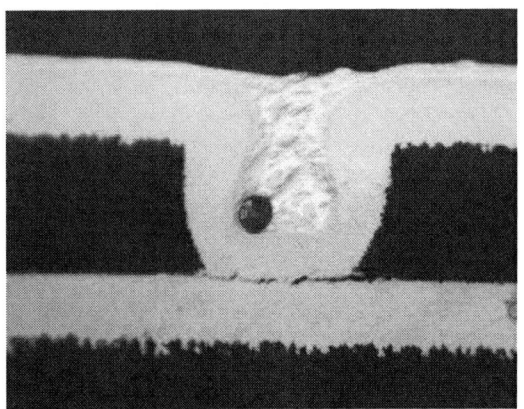

Figure 3: Cracked microvia after reflow.

Conductive Anodic Filament Growth

Conductive anodic filament formation in epoxy glass substrates was observed few decades ago in lab tests. Since then the mechanism of CAF growth has been well studied and understood [9]. It is an electrochemical reaction where copper ions are dissolved, transferred, and deposit as copper salt. The process is influenced by electrical gradient, impurities and pH of the aqueous solution. The result is conductive filament of copper salt growing from anode to cathode along the glass fibres. Based on these studies and understanding, models to assess the product reliability due to CAF formation were established for manufacturer's reference [9]. The development of conductive filament inside PWB is assumed to be a two step process: 1), delamination at the resin/glass interface to form moisture absorption path; and 2); copper electrochemical corrosion process under the voltage bias. A thorough study of CAF showed that many factors could affect the Cu filament growth: the board material, conductor clearance, hole drilling process, potential gradient, solder flux type, time above glass transition temperature during reflow, cleaning agents and the board storage/usage ambient conditions [10].

As the consequence of high integration, high density packing and high power handling demands, the PWB feature minimization and layer count increasing are pushed steadily. In addition, lead free reflow process also exposes boards to higher thermal stress, which can degrade the normal laminate material that was compatible with eutectic Sn/Pb process. CAF related field failure has been reported by some authors. Catastrophic failure of multilayer PWB build up was assessed to be due to copper conductive filament growth, which shorted the power and ground layers. The authors also believe that extra bromide inside the PWB promoted the CAF growth and the source of extra bromide was tracked to be the fluid used in PWB HASL finish [11].

HASL finish has been removed from most of our products. However, we noticed susceptible CAF caused failure in some sample under thermal stress testing. The failure site is between a plated through hole and inner Cu layer with the electrical potential difference. Conductive short between them was found. The cross section revealed some burnt out and unburned filament, propagating along the epoxy glass fibre interface as shown in Figure 4. EDS (Energy Dispersive X ray Spectrometry) analysis of the filament showed it to be composed of copper and bromine. It is not clear if the bromine is there as an impurity or part of the PWB materials.

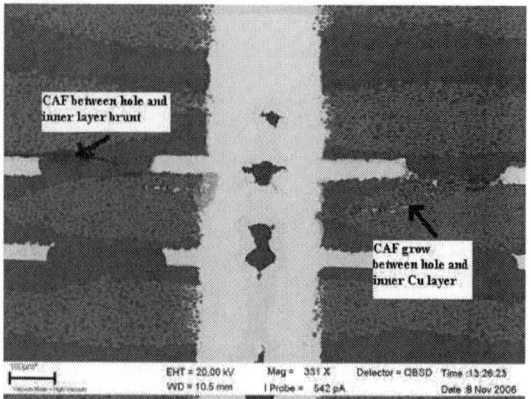

Figure 4: A suspected CAF growth between PTH and the inner Cu layer.

Conclusions

New PWB technologies related defects or failures noticed during system reliability assessment were presented in this paper. Some defects are shown to be related to the PWB manufacturing process, when new material or techniques were deployed. Examples for these defects include planar microvoiding in solder joint and laser drilled microvia opening after reflow. Product failure immediately after assembly due to these problems can significantly decrease the production yield. Other problems such as IAg finished board corrosion and conductive filament growth are PWB degrading phenomenon developed with time. These could seriously decrease the product's service life if not addressed in manufacturing phase.

Preventing of these failures from happening, thus improving the product reliability, will still require more study and understanding of the mechanisms behind them. The article also demonstrated that with the joint effort of PWB supplier and assembler, risks imposed by some of these problems can be mitigated.

Acknowledgements

The authors of the article would like to acknowledge the Failure Analysis Team of Nokia Networks in Oulu, Finland, for the excellent working experience and environments of performing the failure analysis of all works cited in this paper.

References

[1] D. Cullen, "New generation metallic solderability preservatives: immersion silver performance results", Proceedings of the 1993 Electronic Components and Technology Conference (ECTC), Atlanta, Georgia, August 12-15, pp. 21-33, 1993.

[2] F. Houghton, "Alternative metallic finishes for PWB, an ITRI/October project", IPC Expo, March, 1998.

[3] Y-H. Yau & al., "The chemistry and properties of a newly developed immersion silver coating for PWB", IPC/APEX, Anaheim, CA, 2004.

[4] L. Chase & al., "Comparison of Ag Ni/Au and solder PWB surface finishes on the second level reliability of fine pitch area array assemblies", SMTA International Conference Proceedings (Rosemont, IL), 2000.

[5] Y-H. Yau & al., "A study of planar microvoiding in lead free solder joints", IPC/JEDEC International Conference on Lead Free Electronic Components and Assemblies, Singapore, Oct, 2006.

[6] C. Dunsky, "High-speed microvia formation with UV solid-state lasers", Proc. Of the IEEE, Vol. 90, No. 10, October 2002.

[7] M. Weinhold & al., "High speed laser ablation of microvia holes in nonwoven aramid reinforced printed wiring boards to reduce cost", Circuit World, Vol. 22, Issue: 3, Dec 1996, pp. 16-22.

[8] S. Khan & al., "Effect of solder ball pitch and substrate material of printed wiring board on reliability under thermal cycling - a finite element analysis", Proc. Of Electronic Components and Technology Conference, 2004, pp. 1652-1657.

[9] B. Rudra & al., "Failure-mechanism models for conductive-filament formation", IEEE Transactions on Reliability, Vol. 43, No. 3, pp. 354-360, September, 1994.

[10] W. Ready & al., "Conductive anodic filament enhancement in the presence of a polyglycol-containing flux [PWBs]", Reliability Physics Symposium, 1996. 34th Annual Proceedings., IEEE International, 30 April - 2 May, 1996, pp. 267 - 273

[11] W. Ready & al., "Analysis of catastrophic field failures due to conductive anodic filament (CAF) formation", Mat. Res. Symp. Proc, 1998.

Global joint effort to solve microelectronics supply chain technology issues

Paul Collander[1], Marshall Andrew[2], Ruben Bergman[2]

[1]Oy Poltronic Ab, Espoo, Finland, E-mail: paul@poltronic.fi

[2]High Density Packaging User Group, Scottsdale, AZ, USA www.hdpug.org

E-mail: marsh57@hdpug.org, ruben.bergman@hdpug.se

Abstract

High Density Packaging User Group, HDPUG, is an international consortium association working with specific, most urgent, technology issues related to the changing world of electronics industry needs and complex supply chains. The organisation has for 12 years offered industrial top players a collaborative base for analysing, testing and improving emerging technologies at and before their public introduction. Suppliers and users working together at that stage improve business efficiency and performance up and down the supply chain. Concluded projects include new and improved reliability test methods for second level interconnects, early adoption of BGA and FlipChip and reliability issues in PWB vias and terminal finishes. Most important impact on today's global industry stems probably from HDPUG's long time work on Pb-free technologies, now being complemented with halogen free projects. The paper will cover the unique working method of enabling companies to collaborate on a joint, neutral ground, and adding research inputs from universities and other institutions. The presentation will focus on specific projects like the many year long General Purpose Lead-free (GPLF) project now coming to a completion.

Key words: Technology evaluation, reliability evaluation, industrial collaboration, lead free.

Introduction

The cost pressure on all electronics manufacturing is sensed by everyone involved, so does the continuous evolution of new technologies. In order to survive, some companies are gathering together and collaborating on essential but non-competitive issues related to new technologies. One such group is HDPUG, focusing on medium to high end electronics and the assessment of new design and manufacturing technologies and their reliability.

In addition to the necessary discussions on trends in needs and supply, the main activity is in Member performed project work. The small Group staff is there to support these centrally steered projects and to monitor their progress.

Starting points of the organization

It was Ericsson, Nokia and Nortel Networks that in 1992 found new micro packaging so demanding that they went together to discuss implementation and supply. Soon it was found clear rules are needed in such collaboration between competitors, especially as the Group was growing in size and complexity. Still, all activity is based on a personal reliance on each other that no rule or document can provide.

Soon the demand from Member company management put pressure on factual results, not only nice discussions. Working in projects having clear leadership, rules, tight timetables and set goals became a necessity. Next improvement was to actually secure funding, resource, technology availability and realistic goals BEFORE starting a project. Only later did IP rules and anti-trust-laws acquire a significant role in the rules.

From almost the beginning, this was a joint effort of (Northern) European and North American companies and personnel.

The global packaging infrastructure

Dramatic change has occurred in the global packaging infrastructure the last 5-7 years such as:

- Increased manufacturing capability in low labor cost countries
- The recent financial downturn forced system integrators to outsource assembly and technology development
- System integrators are depleted of recourses for packaging development

Many companies have therefore changed their corporate strategy from competition only to more focus on cooperation on issues which are pre-competitive and of common interest in resolving. This means more cooperation is needed in areas such as:

- System integrators need to cooperate aimed at giving **well-coordinated requirements** about their products to the supply chain
- Supply chain companies need to cooperate in order to **reduce cost and improve quality** of their products

Member companies roles, duties and rights

The Group consists of Member companies, paying a yearly fee (depending on company size). There is no restriction of how many employees of a Member company can participate, or in how many projects. It seems to be even more efficient the more people and the more projects each member is participating in, since learning to collaborate demands some training of both project workers and their internal management.

Each Member company has one contact person being the negotiator between the Group and the Member.

Any Member can decide to move up to Executive level, paying an additional yearly fee and getting a seat in the decision making Board of Directors. This Board decides on all Group issues like policies and project starting and closing, publicity and funding of external services.

To avoid free riders, there is a pressure on every member to participate in at least one project by providing resources. The free rider risk is, however, limited also by the fact that probably more than half the benefit members get is through the actual project teamwork with other members, learning by doing and getting aware of the external world. The wheel has been invented enough times already.

Each Member Company has full access to all project results achieved prior to and during their membership, but only for internal use. Project reports are Group confidential for about 1 year after completion. Selected parts may be published in conferences upon Board of Directors' decision.

Networking laterally and horizontally

In today's industry, manufacturing is split in short chains forming long supply chains. Therefore HDPUG has gained a lot from not only having a lateral collaboration between competing system houses, but also including their suppliers like EMS suppliers all the way down to material suppliers. Especially with the recent focus on environmentally sensitive processes this has been beneficial for all players in the supply chains. New materials can be brought in only if all steps in the manufacturing ladder are compatible.

Working in projects

The hardest negotiations are taking place in the project planning phase. In this stage we even accept non-members to participate as they may have something important to bring to the project but they have to join as members before starting to work in a project. A difficult element here is also to assure that adequate resources are committed to complete the project. Those working in a starting project are eager to get the work done and they understand the cantilever effect (your company put down x man months and as the other 7 or 9 put the same, the total power is 7 times x), but they need to convince their management even before the project has started that they can spend their resources on the project, although it is not 100 % aligned with the employer's needs.

The staff needs to learn to know the project participants and their management and need to make sure logistics are perfect. Frequent conference call, Internet based tools and a few face-to-face project meetings a year are needed. Unfortunately the cantilever effect also applies to delays; one critical player tied up for internal only duties may halt the progress for all project members.

The Staff also needs to assure no biased assessments are possible; test criteria have to be documented up front. Reliability testing methology has in fact made major steps forward in this Group, including the Rapid Temperature Cycling using Peltier heat pumps and its comparisons to different field environments.

Idea – Definition – Implementation: The Project Life Cycle

Every project has its own characteristics, but all HDP User Group projects follow a general pattern with specific rules and expectations for each step.

Idea Stage – A Problem is Identified

All new projects begin as an idea or issue that someone suggests might be appropriate for the HDP User Group to explore. That someone might be an individual from a Member Company, a Staff Member, or even someone who is not associated with the organization. A brief description of the idea or issue, together with a possible approach to resolve the issue is presented to the Membership for discussion. If two or more members are interested, and the consensus of the organization is that the idea fits within the charter of the organization, the idea moves into the second, or Definition Stage. Currently the HDP User Group has a list of over 50 ideas, topics, and issues that are under discussion.

Definition Stage – Project Plan is Prepared

Once it is determined that an idea has the support of at least two members, a project team is formed and a call for participation is sent to all HDP Members and interested parties. At this point, the project is open to everyone, even if they are not with a member company. That way, we have the project reviewed by the most knowledgeable people in the industry. This "peer review" helps us insure that the project is technically sound, meaningful, and has a good chance of success.

The project team has the task of preparing the Project Plan, including: the project description or statement of work, the project execution plan and schedule, deliverables, resources needed, and who will provide those resources. A Team Leader is appointed from the participating companies, and an HDP User Group Staff Project Facilitator is named. The final plan is then submitted to the HDP User Group Board of Directors for approval.

Implementation Stage – The Work is done

Once the Board of Directors approves the Project Plan for implementation, the project moves into the Implementation Stage. At this point, the project is open only to companies that are members of the HDP User Group and special invited guests. Non-member companies that participated in the Definition Stage must either join or drop out of the project.

The Project is managed by the Project Leader who organizes the team from member companies working on the project, gets resources, and makes reports to the board and the membership. He is assisted in this by the HDP User Group Staff Facilitator, who provides whatever services and support is necessary to keep the project on task and on schedule. The actual work of the projects, as detailed in the Project Plan, is done by volunteers and assignees from the participating companies, using their company's facilities and resources.

Periodic conference calls and occasional meetings are used to keep everyone in the project coordinated. Any issues or problems are brought up at these calls and addressed by the Project Team and the Staff Facilitator. Minutes of the calls are sent to all project members. Update presentations are given at all quarterly Member Meetings. This provides an opportunity to inform the general membership of progress, and recruit additional resources if required. All presentations, documents, and minutes of the Member Meetings are posted on the HDP User Group web site Members Section.

When the project completes the planned activities, the final report or other promised deliverables are presented to the Board of Directors for final approval. The Board determines if the deliverables are complete, as promised by the Project Plan, and meet the high technical standards of the HDP User Group. The Board also determines how the information will be handled outside of the organization, based on the recommendations of the Project Team. They decide whether the information will be kept for Members only and for what period of time, sold to the public as an HDP User Group Technical Report, or released free to the public.

At this point, the project formally ends, and the participants move on to join other new or existing projects.

Some finished projects

- Lead-free definition concept
- Lead-free marking concept
- Optimization of Lead-free assembly process
- Assembly of a GPLF Application Guideline GPLF stands for General Purpose Lead-free
- Several Lead-free reliability characterization projects
- Lead-free Acceleration Factors
- Several Halogen-free studies
- Halogen-free definition concept
- Halogen-free printed wiring board project 1
- Halogen-free printed wiring board project 2
- DfE project Environmental properties of Halogen-free printed wiring boards
- Project team for assembling A Halogen-free Product Application Guideline
- Reliability characterization of packaging concepts
- BGA
- BGA with Flip Chip
- CSP
- Wafer Scale CSP
- BMPS Application Guideline BMPS stands for Board Mounted Power Supplies
- Power/Temperature Cycling Correlation

Current HDP User Group Projects

Idea Stage Projects:

1. **Lead-free Mixed Technologies – Rework:** Understand the reliability and manufacturing techniques for reworking assemblies with different lead free and lead based solders

2. **Packaging for Portable Applications**: Evaluate the reliability and test methods for fine pitch components in portable applications

3. **PWB Embedded Passives:** Evaluate different embedded passive techniques in high speed applications

Definition Stage projects:

4. **BMPS Modules, 2nd Level Interconnection:**
   - Move to implementation late 2007
   - Finish late 2008
   Establishing design rules for Interconnecting BMPS modules to PWB

5. **Lead-free Copper Erosion:**
   - Move to implementation Mid 2007
   - Finish 2008
   - Understanding copper erosion rates for various lead free solders and establishing design rules for PWB interconnections and rework

6. **Halogen-free Product Properties:**
   Profile:
   - Move to implementation mid 2007
   - Finish late 2008
   - Establishing a Guideline for using halogen-free Materials in Electronics Manufacturing, and Establishing a distributed halogen-free Product Properties Database

Implementation Stage Projects:

7. **GPLF Reliability Characterization:**
   - Finish date Mid 2007
   - Comprehensive reliability characterization of solder joint reliability using a test vehicle with many different types of components

8. **Lead-free Board Materials Reliability**
   - Finish early 2008
   - Comprehensive reliability characterization of FR4 and Halogen-free boards using lead-free processing

9. **Lead-free Acceleration Factors**
   - Finish date Mid 2007
   - Comprehensive characterization of lead free solder joint reliability using different testing conditions

10. **SAC Microvoids**
    - Finish date late 2007
    - Establishing an understanding of the cause and reliability impact of microvoids in lead free solder joints

11. **Component Terminal Finishes Phase 2**
    - Finish date Early 2007
    - Reliability characterization of Nickel/Palladium type terminal finishes, particularly moisture level sensitivity

12. **BMPS Guideline**
    - Finish date Early 2007
    - Establishing a comprehensive guideline for using Board Mounted Power Supplies (BMPS)

13. **SAC Reliability - Mild Acceleration Project:**
    - Finish Late 2008
    - Evaluate reliability of SAC mounted devices vs tin/lead in mild temp cycle to test hypothesis: Field condition ΔT (and thus strain) is significantly $<\Delta T$ (and thus strain) of industry accelerated tests, accelerated test conditions are (may be) introducing failure modes that don't exist in the field, and therefore high strain

component types will (may) perform better in actual field conditions.

Project Focus – Halogen Free Properties

Flame-retarded plastics are commonly needed to meet strict fire safety codes for electronic equipment. Certain halogenated compounds (of which brominated flame retardants, or BFRs, are a subset) are used as flame retardants in a variety of applications including thermoplastics, insulation materials, component mold compounds, solder masks and printed circuit board laminates. In addition, polyvinyl chloride or PVC (a resin that contains chlorine, a halogen) is a commonly used base resin for certain cable jacketing. However, concerns have arisen that these materials may pose certain risks to health or the environment particularly at end-of-life. At this time, several governments are considering regulation to prohibit or restrict the use of these types of substances in electronic (and other) products. Within the marketplace, environmentally-preferable purchasing standards (such as TCO, Blue Angel, Nordic Swan, etc) also include restrictions on the use of these substances in certain products. In order for the electronics industry to continue its long-standing commitment to product stewardship, companies throughout the supply chain will need to understand which "halogen-free" alternatives are available, as well as the electrical, mechanical and environmental, health and safety properties of these alternatives.

Since 2001, the HDP User Group has been at the forefront of evaluating halogen-free materials within the electronics industry. In 2007, HDP User Group initiated the Halogen-free Properties Project aimed at assembling a comprehensive Halogen-Free Guideline and Halogen-free Materials Database. The Halogen-free Materials Database will serve as a centralized database allowing suppliers to list their halogen-free product offerings and the properties of those offerings in a uniform, concise format that is easily accessible to product designers. Increased access to this information will enhance supply chain adoption of halogen-free components.

The first part of this project involves the preparation of a Halogen Free Guideline. This document will contain best practices and experience from companies actually implementing Halogen Free products, as well as information on materials available for PWBs as well as cables, connectors, components, housings and other electronic hardware. Information gathered for writing the Guideline will be used in planning the Database. HDP User Group has prepared guidelines for Lead Free implementation, and Board Mounted Power Supplies (BMPS). These guidelines have been well received by the industry. Figure 1 is the planned schedule for the Guideline part of the project.

Project Task	When Complete
Plan Project	2/27/07
Editorial staff & core contributors identified	Q1 2007
Project moves into implementation stage	Q2 2007
Table of contents complete & chapter editors assigned	Q2 2007
First Draft complete	Q3 2007
Second Draft complete	Q4 2007
Final Draft complete	Q1 2008

Figure 1: Halogen Free Guideline Schedule

The Database part of the project will define the format of the Supplier data, the specific properties to be reported, and the test method to be used to measure the data. The Project Team will also prepare a list of potential suppliers and contact them for inclusion, and monitor and adjust the database as it begins to form. Figure 2 is a schematic of how we envision the Database.

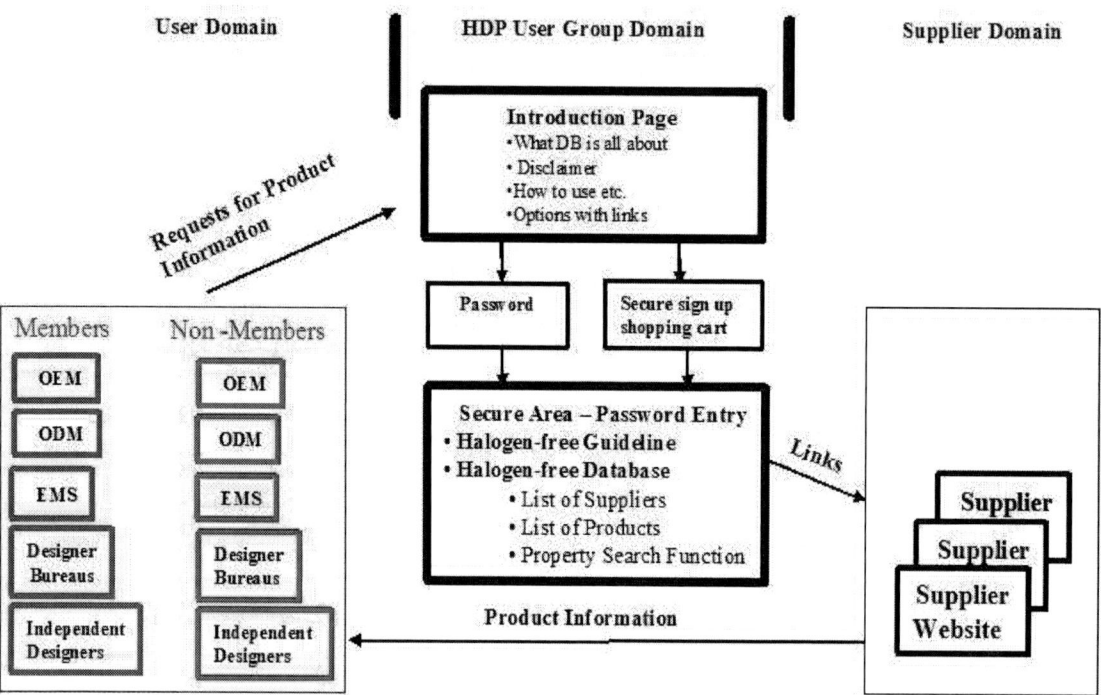

Figure 2: Halogen Free Product Properties Database Concept

It can be seen from the figure that all of the specific product data will be located on a web page created by and controlled by the supplier of that product. The HDP User Group Halogen Free Project will tell the suppliers what information the customers wants, what test method he wants used for collecting the data, and the format for presentation of that data. The Project will also provide a searchable area containing information about the suppliers and links to all of their database pages. A user can then enter the HDP User Group web page, and search for all of the information concerning a specific halogen free product or material. The results will be presented as a series of links to pages containing that data.

The Project Team believes that this is the best way to provide the industry the information they need to quickly make the decision to move to halogen free, and if that decision is made, to do so as reliably and cost effectively as possible.

Conclusions

The HDP User Group has developed a very effective methodology for allowing companies all over the world to work together to solve the many technical problems that face the electronics industry. This methodology, developed over more than 10 years of managing cooperative projects, provides economies of scale and access to experts in different companies that no single company can match.

Presently, the organization is focusing on major reliability and environmental issues through a series of projects.

As a Member-driven organization, HDP User Group is one of the many tools that companies must use to remain competitive in today's Global Economy.

Integration of Thin Flexible RF Structures into Flexible PCB

W. Christiaens[1], H. Burkard[2], J. Link[2], and J. Vanfleteren[1]

[1]IMEC, Gent, Belgium
[2]Hightec MC AG, Lenzburg, Switzerland

Tel: +32 9 2645371, Fax: +32 9 2645374, E-mail: Wim.Christiaens@elis.ugent.be

Abstract

In the framework of the European SHIFT project, both IMEC Gent as Hightec MC AG are investigating ways how to integrate RF structures in flexible printed circuit boards. This paper presents two polyimide based technologies for producing flexible multilayer RF circuits using the conventional thin film techniques; this results in very thin, highly flexible RF circuits, cables and interconnections. Narrow and well-defined lines and gaps - with pitches down to 25 μm - enabled by thin film technology ensure a perfect high frequency performance, at least up to 40 GHz. In both technologies, the multilayers are built-up on temporary rigid carrier substrates by repetitive application of polyimide layers by a spin-on process and metal layers by sputtering and if needed enforced by electroplating. The IMEC Gent technology uses three metallization layers. Sputtered NiCr is used as resistor material, a thin spincoated PI as capacitor dielectric. Inductors, capacitors and resistors are interconnected with microstrip transmission lines.The Hightec MC AG thin film on flex build-up includes 2 to 4 metallization layers; starting from PI spincoated on rigid carrier and integrating NiCr resistors, PI based capacitors and inductors. The RF passives are interconnected by striplines or microstrip. Alternative solutions are investigated by applying the thin film techniques on commercial foils, e.g. Kapton, LCP or other polymers with suitable RF properties. After processing and testing the flex multilayer is released from the substrate. The resulting highly flexible foil-like circuits have excellent mechanical and electrical properties. These small foil elements with integrated RF structures or thin film resistors can be laminated into conventional printed flexboards as local high-resolution parts and connected to the wiring of the print. These somewhat expensive high-resolution features such as RF structures in high frequency circuits can be limited to the areas where they are needed.

Key words: flexible, RF structures, embedding

Introduction

In the world of electronic circuitry, especially in devices for portable applications, there is a clear trend towards mechanically flexible circuits. In most of these applications compactness and weight are highly important factors. The compactness of the resulting circuit can be improved even more by integrating electronic components in the substrate, and not only on front- and backside. Thus the development of technologies for embedding electronic components (both passives, and RF structures as well as active components) in flexible substrates is the logical next step.

This paper presents two polyimide (PI) based technologies of integrating RF components for flexible electronic circuits: the IMEC thin film on flex technology and the HiCoFlex technology of Hightec MC AG.

IMEC Gent is developing a polyimide based technology for RF thin-film type resistors, capacitors and inductors, connected by RF microstrip lines. A suitable sequence of process steps has been established. The integrated RF structures on the thin film on flex consist of 3 metallization layers and will use microstrip transmission lines for the interconnection. The base substrates consist of spincoated PI on a rigid glass carrier. Sputtered NiCr is used as resistive material for the thin film resistors, spin-on PI as dielectric and Cu is sputtered for the metallization layers. Different test vehicles have been realized.

The HiCoFlex technology is an alternative technology for realizing ultra-thin, highly flexible multilayer foil elements with integrated thin film components like resistors, RF lines and structures. This build-up contains up includes 2-4 metallization layers, starting from spin-on polyimide on rigid carrier and integrating NiCr resistors, capacitors and inductors. The passives can be interconnected by striplines or microstrip.

An alternative solution is the application of thin film on commercial foils e.g. Kapton, Liquid Crystalline Polymer (LCP) or other polymers with suitable RF properties.

Both technologies realize PI based RF structures that can either be incorporated in standard flexible substrates or assembled on a rigid or flex board, as local high-resolution parts.

This contribution mainly focuses on the technological side of both embedding technologies. This work was carried out under the EC IST funded project IP-SHIFT (contract number 507745) [1].

IMEC's Thin Film on Flex Technology

IMEC developed a polyimide based thin film on flex technology for producing flexible multilayer RF circuits, using the conventional thin film techniques. This technology results in very thin, highly flexible RF circuits.

The multilayers are built-up on temporary rigid carrier substrates by repetitive application of polyimide layers by a spin-on process and metal layers by sputtering and if needed enforced by electroplating. An overview of the different process steps is shown in figure 1.

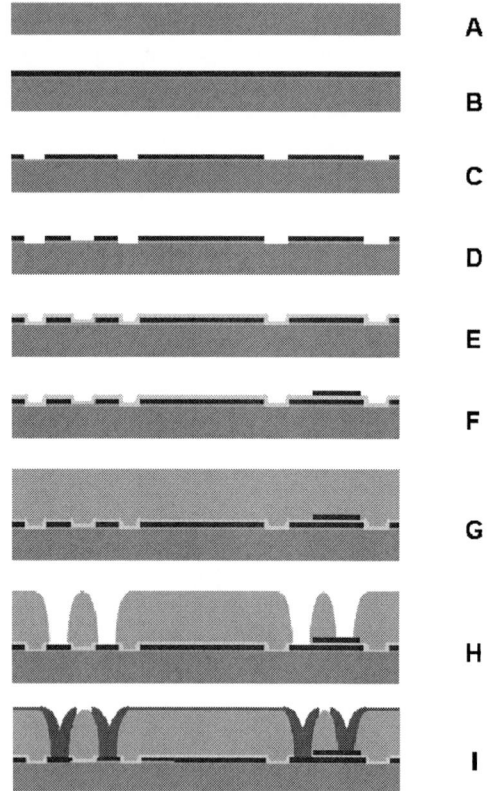

Figure 1: build-up of polyimide based RF structures

The different process steps are:

A. PI spincoated on rigid glass carrier
B. Metal 1: sputtered NiCr/Cu
C. Cu etch
D. NiCr etch
E. PI 2: thin capacitor dielectric
F. Metal 2: sputtered TiW/Cu (+ TiW and Cu etch)
G. PI 3: thick (10 µm) spincoated PI layer

H. Drilling of the vias to the contacts of the passives
I. Metal 3: sputtered TiW/Cu, top metallization + filling vias
J. Release from rigid carrier

The layer build-up of this thin film on flex technology consists of 3 polyimide layers and 3 metallization layers. The base substrates are PI 1 spincoated on rigid glass carrier. The resistors, inductors and bottom electrodes of the capacitors are defined in Metal 1. PI 2 is used as dielectric for the capacitors, which top electrodes are defined in the Metal 2 layer. Metal 3 provides the microstrip interconnections between the different RF passives.

The different process steps are discussed below.

Substrate preparation

The base substrate is a 10 0 m spincoated PI layer (PI2611) on a rigid glass carrier. After processing the RF structure has to be released from the rigid carrier. An easy release of the package from the rigid substrate is obtained in a special way: before spinning the first polyimide layer, the 4 edges of the square glass substrate are coated with an adhesion promoter. The consequence of this is that the first layer of polyimide adheres well to the edges of the substrates, and has marginal adhesion strength to the centre of the substrate. However the adhesion to the edges is sufficient to allow for the whole process cycle. After processing the multilayer structures can easily be cut out from the rigid carrier.

Metal 1

The first metal layer is a sputtered NiCr/Cu stack. The thickness of the sputtered NiCr is 20 nm, with targeted sheet resistance of 50 Ω per square. The thickness of the sputtered Cu is 1 µm. One could also consider sputtering a NiCr/Cu/TiW stack. The 50 nm TiW top layer can protect the Cu layer against oxidation and will ensure better contacting to Metal 3 layer.

Resistors, inductors and the bottom electrodes of the capacitors can be defined in this Metal 1 layer.

Before sputtering the polyimide is plasma treated in order to have better adhesion of the Metal 1 on the cured PI.

After sputtering the NiCr and Cu stack has to be patterned, so we will need two lithographic steps. First the Cu layer is patterned (etched in FeCl3 etchant). After a second lithographic step the NiCr layer is etched, in a Ce(SO4)2.2(NH4)2SO4.2H2O solution. The problem here is that this NiCr etchant also etches Cu layers. So during the NiCr etch step the remaining Cu has to be protected because the Cu is attacked by this NiCr etchant. Protection is achieved by oversizing the second mask, so that the photoresist is covering the edges of the Cu pattern. See figure 2.

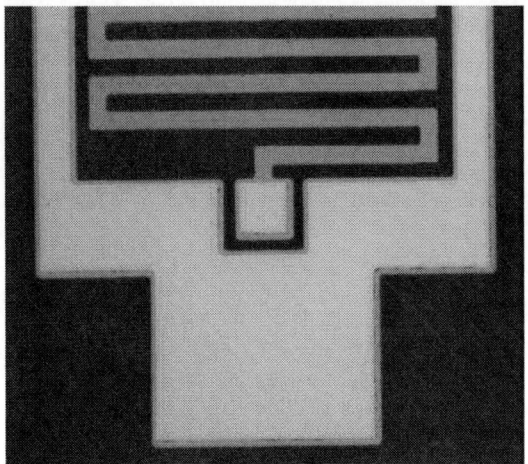

Figure 2: resistor structure after NiCr etch

Polyimide 2

After plasma treatment of PI 1 for improved adhesion, the next polyimide layer is spincoated. For the first test samples a 2 μm PI2610 layer is applied. The final goal is to apply 1 μm or thinner thin pinhole free polyimide layers. Thinned PI2610 can be used to obtain lower film thicknesses: Pyralin T9039 (HD Microsystems) is a thinner suitable for reducing the viscosity and solids contents of polyimide solutions [2]. This polyimide 2 layer will be used as dielectric for the capacitors.

Metal 2

In this second metal layer the top electrodes of the capacitors will be defined, using the thin PI 2 layer as dielectric.

Metal 2 is a sputtered 50nm TiW + 1μm Cu + 50nm TiW stack. The lower TiW ensures a good adhesion to the PI, the upper TiW protects the underlying Cu from oxidation and ensures better contacts after laser drilling.

Before metallization the PI 2 layer is plasma treated, for optimum adhesion strength of the metal on this cured PI.

Polyimide 3

The next polyimide layer is a 10 μm PI2611 layer. This is the dielectric layer for the microstrip interconnects of Metal 3.

Laser Drilling

Before the application of the top (interconnecting) metal layer, the openings for contacting the RF passives are laser drilled. In this case we use our CO2 laser (infrared, working at about 10 μm). The advantage of the use of the CO2 laser is that it doesn't ablate metal layers. So it makes it much easier to stop at the thin Metal 1 and

Metal 2 layers. The minimum spot size of the CO2 laser is about 100 μm. The smallest contacts for the RF design were 150x150 μm^2. Figures 3 and 4 show the laser drilled contacts to the resistor and microstrip structures.

Figure 3: vias to the meandered resistor structure

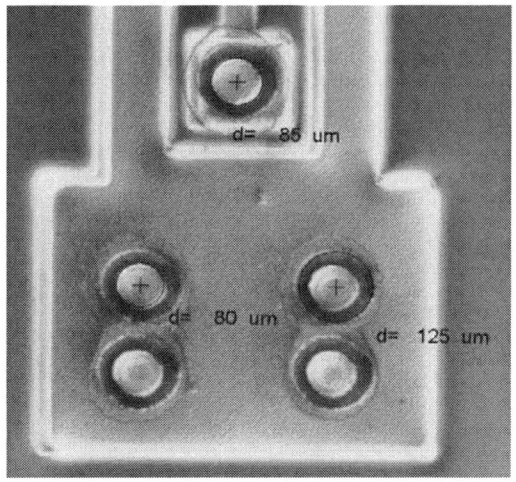

Figure 4: realized vias with bottom diameter of 80 μm and top diameter of 125 μm

Metal 3

After plasma treatment, Metal 3 is sputtered: 50 nm TiW + 1 μm Cu. RF passives can be interconnected by microstrip transmission lines, defined in this Metal 3 layer.

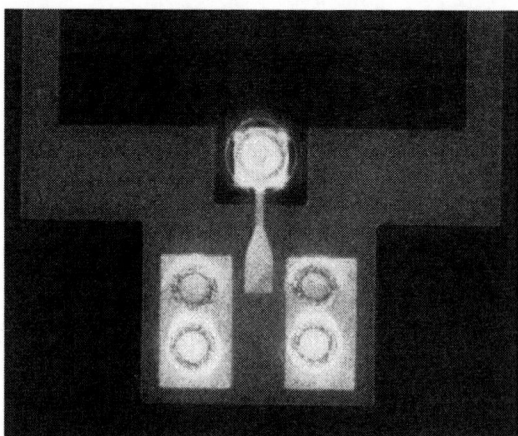

Figure 5: metallized ground-signal-ground (GSG) contact pads are foreseen for use with coplanar waveguide launching (CPW) wafer probes

The first test structures for the above described thin-film on flex technology are realized. The RF passives test structures include:

- SOLT calibration structures
- 24 µm wide microstrip transmission lines
- spiral inductors
- series and shunt capacitors
- straight and meandered resistors

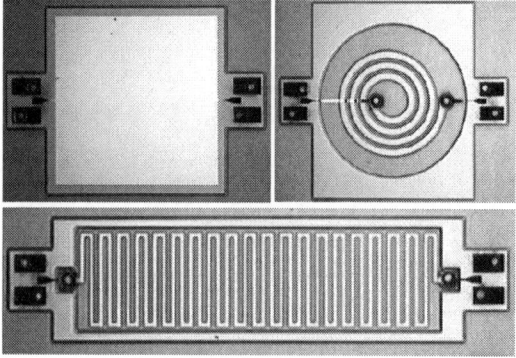

Figure 6: realized RF capacitor, inductor and resistor

HiCoFlex Technology

The HiCoFlex technology of Hightec MC AG [3] is quite similar to IMEC's thin film on flex technology. The process flow starts also with rigid substrates, which may be alumina or (cheap) glass plates, and which are used as a carrier during the multilayer build-up process and the assembly of components. The release from the carrier in this case is obtained by the application of a thin 'release layer'. The know-how about the nature and the processing of this release layer is company property of Hightec MC AG and is not disclosed. Next follows the application of a first polyimide layer by a spin-on process from the liquid solution. The layer is then dried and cured. The first metal layer is applied by a sputtering process. The metal layer is structured by photolithography and may also be enforced by galvanic deposition. A next polyimide layer is applied by spin-on, followed again by drying and curing. Vias to the underneath conductors can be opened by laser or plasma processing.

These steps (metal layer by sputtering, photolithographic and galvanic processes / polyimide layer by spin-on, drying and curing / via opening) can be repeated several times, resulting in a multilayer structure. After assembly, it's possible to release the flex multilayer from the substrate by use of the release layer. An assembled foil is shown in figure 7. The structuring technique allows line widths of 15 µm, spacing of 10 µm and vias of 30 µm. Actually circuits with up to 4 metal layers are realized. The total thickness of such a foil is around 50 µm, resulting in highly flexible foil-like circuits with excellent mechanical and electrical properties. The minimum bending radius is smaller than 0.5 mm. It is even possible to fold the material without effect to the electrical properties.

Figure 7: highly flexible multilayer assembled with SMD

Thin Film on Foils

Alternative process flows are starting from commercial foils, e.g. Kapton, Liquid Crystalline Polymer (LCP) or other polymers with suitable RF properties, and applying thin film techniques on this foils. Methods for temporary attachment of the foils on rigid carriers during the thin film coating and their detachment afterwards have been studied.

Thin Film Resistors

Integrated resistors on polyimide (PI) were produced by standard thin film methods: sputtering NiCr, photolithography, Ti/Cu/Ni contacts, annealing, laser trimming, protection by a further PI, release from the carrier substrate [4]. Examples of such resistor foil elements are shown in figure 8. Test structures were analyzed by measurement of the resistance drift during PI curing (380°C peak temp), the temperature coefficient of resistance TCR, by a temperature life test for 1000h at 125°C, a humidity test for 1000h at 85% r.H. / 85°C and a bias of 18 VDC and a simple bending test.

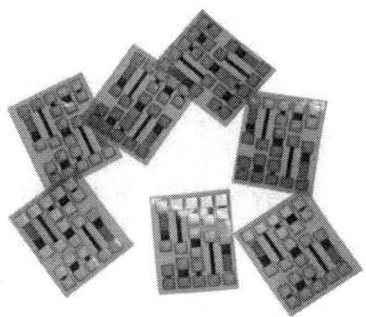

Figure 8: resistor foil elements

The evaluation showed that NiCr thin film resistors in a range of 10 Ω to 100 kΩ can be integrated, that laser trimming is possible and that the properties are nearly as on rigid alumina substrates. The test results are: TCR: -18 ± 8 ppm/°C; the temperature life test at 125°C, 1000h showed a drift < 0.1% (see figure 9); drift in the humidity tests after 1000h with 18 VDC is typically < 1%; the bending effect is typically < 1% for bending radius of 1.25 mm.

Mean abs. Drift Integrated Resistors

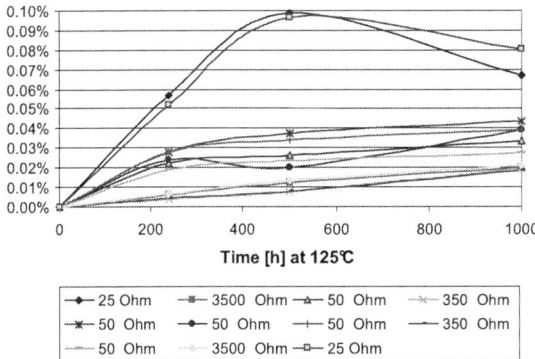

Figure 9: temperature life test 1000 h @ 125°C

RF Structures and Properties

Narrow and well-defined lines and gaps enabled by the thin film technology ensure a perfect high frequency performance. This allows the realization of very thin, highly flexible microstrips, stripline and waveguide structures for RF cables and interconnections. Line width and space tolerances are in the range of +/- 1-2 μm. Coplanar waveguide test structures are shown in figure 10.

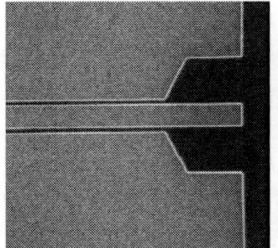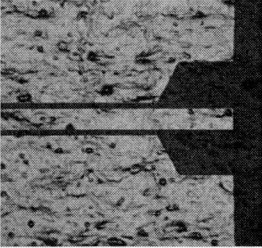

Figure 10: coplanar waveguide test structure on PI (left) and LCP (right), width of 100 μm and 20 μm spacing

RF thin-film type resistors, capacitors and inductors can be interconnected realizing RF structures that way.

Different polymers were evaluated and qualified for RF applications, e.g. spin-on polyimides, BCB, Kapton, LCP and HyRelex (fiberglass fabric with PTFE coating). The handling of the foil-type polymers is described in a previous chapter. BCB, well-known for its excellent RF properties, but due to the brittleness was not of use for detachable films. Only PI-BCB multilayers and PI-BCB-PI sandwiches make flexible films possible (a build-up of a PI-BCB-BCB-PI sandwich is shown in figure 11). LCP and BCB have the advantage of a low water uptake, an important point for high frequencies. Losses of coplanar waveguides were measured to verify the performance until 20 GHz. Polyimides and LCP showed to be acceptable, at least to 20 GHz with a bandwidth ≥ 20 dB.

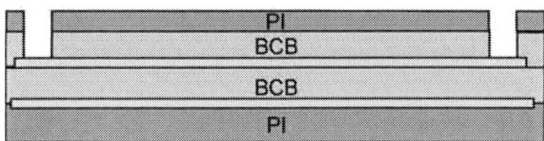

Figure 11: multilayer sandwich PI-BCB-BCB-PI (section along transmission line)

PI-BCB, PI-BCB-BCB and PI-BCB-BCB-PI multilayer sandwiches were analyzed with a microstrip test pattern up to 40 GHz. The results indicate a good RF performance for these material combinations which is not susceptible to moisture and still very flexible [5].

Lamination of Thin Flexible Foils into PCB

Small foil elements with integrated thin film components like resistors, RF lines and RF structures can be laminated into conventional printed flex-boards as local high-resolution part and connected to the wiring of the print. By this method expensive high-resolution features such as precision resistor arrays or RF structures or resistor arrays in high frequency circuits, can be limited to the areas where they are needed. The concept has been proven

615

by lamination of integrated chip foils into PCB [6]. Further test samples are under way.

Applications and Conclusions

The integration of passive components, as thin film resistors and RF structures into flexible multilayers has been demonstrated. The resulting very thin and flexible multilayer foils can be laminated into conventional rigid and flexible PCB as a local high-resolution part. These somewhat expensive high-resolution features such as RF structures in high frequency circuits can be limited to the areas where they are needed.

Applications of this multilayers are in the fields of high-density interconnect (HDI) technologies for sensors, industrial and medical micro systems, 3D packaging, and recently also for flexible RF cables and high-frequency connections, e.g. between optoelectronics submodules.

Acknowledgements

The authors would like to thank AVANEX High Frequencies Lab for the measurements of the CPW test structures, and also Steven Brebels from IMEC Leuven for the design of the RF structures and the technical discussions.

This work has been supported by the EU 6th framework program under contract 507352 SHIFT and funded by the Swiss State Secretariat for Education and Research SER, under project 03.0233.

References

[1] www.shift-project.org

[2] Information from the material datasheets.

[3] A. Fach et al., "Multilayer polyimide film substrate for interconnections in microsystems", Microsystem Technologies, Vol. 5, pp. 166-168, 1999

[4] H. Burkard, "Thin Film Resistor Integration into Flex-Boards", 5th International Workshop ‚Flexible Electronic Systems‘, Munich, November 29, 2006

[5] H. Burkard et al., "Ultra-Thin, Highly Flexible Cables and Interconnections for Low and High Frequencies", MicroTech 2006, Cambridge UK, March 7-8, 2006

[6] T. Gottwald, "How to make chip integration technology suitable for high volume production", IPC 3rd Intern. Conference on Embedded Technology, May 3-5, 2006

3D Chip Size Packaging for highly integrated memory cards for consumer-products

Reiner Götzen, Andrea Reinhardt, Dr. Helge Bohlmann

microTEC Gesellschaft für Mikrotechnologie mbH

Bismarckstr. 142 b, 47057 Duisburg, Germany

T: +49 (0)203 306 2050, F: +49 (0)203 306 2069, E-mail: info@microtec-d.com

Abstract

RMPD® Rapid Manufacturing processes based on parallel manufacture of Microsystems and photo-polymerisation methods. Parallel means that many parts are manufactured simultaneously by generative production methods. The RMPD® technologies, including 3D-CSP, are used for production of customer-specific production of polymer-based Microsystems (also called microelectronic mechanical systems – MEMS) and micro-parts. Based on microTEC's 3D-CSP method, silicon chips can be integrated into parallel structured packages and electrically connected without the need for serial processes, e.g. wire bonds. As a result of the processes working in parallel, this technology allows the cost-effective production of small memory cards (Smart cards). The batch-oriented process generates hundreds of memory cards simultaneously, as a result of which part costs are reduced. Moreover, 3D-CSP allows vertical integration of memory components (die-stacking), i.e. the storage capacity of a card can be increased simply by stacking and vertical interconnection of several memory components. The process opens up means of simply increasing the storage capacity of high-capacity memory cards. The successes in nanotechnology moreover enable scalable properties, e.g. with a view to mechanical, optical or fluidic use

Key words: System-in-a-Package, 3D-CSP, RMPD, INOS

Introduction

The use of powerful and inexpensive manufacturing processes for high density packaging has a significant share in the success of a product. The requirements are particularly high in the Microsystems engineering sector, because components of the most varied technologies have to be integrated hybrid in smallest space. Examples are microchips with digital or analogue functions or different manufacturing technologies, RF-components, different substrate materials, SMT-components, micro-lenses, micro-sensors, micro-optics, micro-mechanics etc. In microelectronics the talk is often of a "System in a Package" if, for example, the system contains a microprocessor and memory components and, as an enclosed unit, has an interface to the outside world.

The RMPD®-processes [1] together with the 3D-CSP-process allow products which integrate different components in the smallest space to be developed and manufactured quickly and inexpensively. The processes are quick and inexpensive firstly as a result of parallel manufacture but also because the process only requires 4 different types of process and is complete without expensive tools (as in injection moulding or 3D-MID).

In RMPD®-Mask, like microelectronics, the mask is the means of production tailored to the product. At microTEC 5-inch, 9-inch and 14-inch masks are currently available. The number of parallel-produced products is specified in the batch process by means of an adapted selection of the mask format in conjunction with the maximum component size. The smaller the system, the more of them can be produced in parallel. The smallness of the product becomes a cost advantage compared to injection moulding-based production processes. This makes production regardless of batch-size possible. The production time for a mask is several days, which leads to faster innovation cycles. The next innovative step must take no notice of the cost-intensive already existing tools.

3D-CSP Manufacture

Design and construction phase

Both the electrical and the mechanical layout are interlinked in the design phase for application of the 3D-CSP process to manufacture a "System in a Package". An "E-CAD" 3D model is produced, which describes the package completely both in electrical and mechanical form:

- An EDA-system is used to design the electrical system. A particular advantage in this is that circuit construction design rules are integrated into EDA-systems, to calculate the required geometry in high-frequency structural elements such as baluns, antennae or coils. The use of widespread EDA-systems also has the advantage that systems can be constantly upgraded to state-of-the-art technology.
- A CAD-system, with which all mechanical components can be specified and constructed, is used for the design of the mechanical system. Even a microelectronic chip (die) here becomes a mechanical component, which must have all its dimensions and tolerances defined.

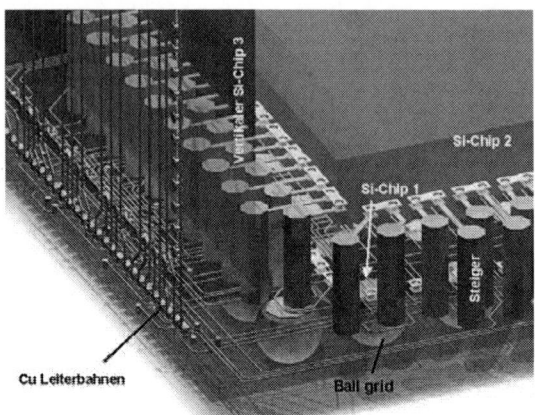

Figure 1   Complete E-CAD Model

The circuit layout is entered into the 3D-CAD system and specifies all metallic structures in the complete system. The connections can be led down to the pad of the next component. The complete system is available as a three-dimensional system and, dependent on the materials used, can be specified with its parameters of interest for the simulation. In the simulation EMC, heat dispersion, flow characteristics and the mechanical stresses can be optimised. The combining of 3D-CAD and EDA-system into an integrated design tool was developed by microTEC and called E-CAD. Computing capacities available at this point are partially exceeded, so that only the components relevant to the simulation have to be simplified.

The four basic manufacturing steps

Step 1

RMPD®-Mask processes, established 10 years ago by microTEC, are used for the production of micro-systems. As part of this, components grow in parallel in the batch process by photo-polymerisation on the substrate. Layer-thicknesses and component dimensions down to the μm-range are possible. Multi-material (industry-specific material properties in a component) has long been in use, versatile materials from soft to hard, transparent to opaque or hydrophilic to hydrophobic are used in the products.

Step 2

The second step is component integration. In this, the components are set into cavities produced using the RMPD®-Mask process, so that they bond to the surface. To do that the substrate with the parallel-built components are put in the automatic placement machine.

Step 3

The third step is metal-plating the batch after it has been coated using the RMPD®-Mask process, which only contains slots for vertical electric or fluidic connections to the lower levels. In metal-plating all components are covered in parallel with a metallic layer, which also electrically contacts the components lying under it. All contacts are manufactured in parallel.

Figure 2 Intermediate step, batch with ground layer and integrated chip prior to the second metal-plating

Step 4

The fourth step is to structure the metal-plating. A variety of processes are available for this, which can be used dependent on the precision of the circuit level required. The precision of strip-conductor geometries required for high-frequency requirements and the gaps to the grounding levels is achieved and was demonstrated at the EC-sponsored project INOS (www.INOS-ist.org).

Therefore the most varied functions (microelectronics, micro-fluidics, integrated optics) can find a place in the system through a repeated application of steps 1 to 4.

Figure 3 3D-CSP Model with stacked chips

"Smart Cards" – an example application of 3D-CSP

Smart Cards are a medium in the smallest format for storage of digital information. They are used in a variety of products, particularly mobile telephones. These cards can be manufactured particularly cheaply using 3D-CSP engineering:

- The small external dimensions of approx. 11 mm x 15 mm x 0.75 mm enable many cards to be built in parallel on a pre-defined substrate surface, which leads to low packaging costs. The only tool required is a lithographic mask, which can be manufactured at low cost (compared with the cost of an injection-moulding tool).

- The increasing demand for greater storage capacity leads to the compulsion of having to integrate storage components one on top of the other (Stacking). The stacking of storage components by using 3D-CSP offers the particular advantage that the vertical gap between 1 storage chips can be significantly smaller than in wire-bonded storage stacks. With the card having a maximum structural height of 0.75 mm, 8-12 chips can be stacked, for example, dependent on the thickness of the memory chip.

- Short innovation cycles in the case of new models of memory chips lead to frequent changes of circuit diagrams and layouts. Through the interlocking of EDA and CAD new designs can be quickly processed in the case of 3D-CSP and converted into new masks. Prototypes of new cards can be produced and mass-production started within a few weeks. As a result faster introduction to the market is possible and a competitive advantage achieved.

Time and cost analyses were carried out in the context of the EC-sponsored compound project "INOS" [2], which highlighted that the time required to manufacture a card using 3D-CSP is determined for the most part only by the pick-and-place times for handling of the discrete components. Steps 1, 3 and 4, which work in parallel, ensure that manufacturing times per package are short. As applies to the manufacture of semiconductor chips: the smaller the component, the more units can be produced simultaneously.

Outlook

Numerous applications have shown that the 3D-CSP Packaging process can be used universally and inexpensively in the miniaturisation industry. Some examples are described in [3]. The materials used can be tailor-made with additives for a range of applications, e.g. strength can be increased by the addition of nanoparticles or higher temperature stresses reached.

The RMPD®-processes will in future benefit from developments in the field of polytronics, i.e. the development of polymer components with electrical, optical and other functions. Polytronics in conjunction with RMPD® will then lead to other cost benefits.

Literature

[1] microTEC Website www.microTEC-d.com

[2] EC project „INOS", www.inos-ist.org

[3] Sensorpackaging im Batchverfahren mit RMPD® und 3D-CSP, Sensormagazin 4/2006, S. 16-18

[4] Reinhardt, A, Götzen, R: New business based on customized rapid manufacturing from polymers mems used in industry, life science and consumer products, Research, Development and Application of High Technologies in Industry Nr. 8, ISBN 5-7422-1451-0, Page 5-6

[5] Reinhardt, A.; Goetzen, R.; Goetzen, J.; Bohlmann, H.; Nanobased materials and micro systems applications in life science and sensor technology (invited presentation); Conference Proceedings FLAMN 07, St. Petersburg in print

Realization of large Area Stretchable Electronic Systems using Lamination Processes

Thomas Loeher, Dionysios Manessis, Andreas Ostmann, and Herbert Reichl

Technische Universitaet Berlin and Fraunhofer IZM

Phone: + 49 (0) 30 46403 648, E-mail: Loeher@izm.fhg.de

Abstract

Electronic systems that can be stretched to a certain extend have attracted much attention in recent years. A multitude of potential applications ranging form medical implants and health monitoring to skin like electronic surfaces for robots and application of electronics systems to complex three dimensional surfaces are conceivable. Stretchability is in short an additional feature to an electronic system that increases the freedom in design of potential products. Besides more freedom in design such systems will offer a higher degree of reliability if properly designed and fabricated. A variety of approaches to fabricate stretchable electronic systems are pursued at present. At TU Berlin a process that makes as much as possible use of conventional printed wiring board technologies was developed that offers the capability produce stretchable systems with large areas at relatively low cost. Core process is the attachment of stretchable material to full area copper sheets, the subsequent structuring of the copper layer into separated conductor lines and bond pads for component assembly. After assembly of the components a covering layer of the stretchable material is applied to fully embed and protect the components in the stretchable matrix. The attachment of a stretchable material to a copper foil in the first step yields best results in a conventional lamination process. Structuring of the copper layer is achieved by photolithography or comparable processes combined with etching. In order to impart stretchability to the copper, conductors will be meandering. The transition between more (areas free of components) and less stretchable portions (surrounding of embedded compact electronic components) deserves special attention in the design of the stretchable system. Therefore mechanical reinforcements structured out of the copper layer have been used, different kinds of stiffeners and encapsulations are conceivable. The final embedding of the electronics by a capping layer of the stretchable matrix material is realized by printing or lamination. Experiences, results and process variations for the production of large area stretchable systems in the framework of lamination, structuring, assembly and embedding will be presented.

Key words: Stretchable Electronics, Large Area Processing, Lamination, Embedding

Introduction

Electronic systems in future applications tend to have large varieties of functions comprising sensing, evaluating and regulating in different kinds of physical/chemical environments without disturbing the host environment. Novel concepts like the disappearing computer or ubiquitous computing, medical and wellness appliances close to the human body, miniaturized control and sensor networks for safety measures are examples of possible applications. The sub units of such systems may not even have complex physical and electronic structures. Only through the mutual interaction of sub-systems new functionalities of the whole system are expected to emerge.

Non-perceptible or de-materialized electronic systems spur on new physical requirements. One of which is to increase electronics versatility by allowing a certain degree of stretching of at least parts of the system.

The conventionally used carrier of electronic systems is the printed wiring board. With quite robust process highly complex three dimensional architectures of copper interconnections can be realized. The electronic system is realized by mounting and interconnecting of active and passive components onto the board.

A most obvious strategy to introduce stretchability into the electronic system is the use of stretchable substrate material instead of the conventional rigid or flexible materials. Components or functional sub systems, however, must not be stretched. The development of stretchable electroncs refers only to the complete system. That will comprise a stretchable carrier and conventional rigid components. The challenge is thus to engineer the transistion between stretchable and non-stretchable parts of the electronic systems. Local stiffening structures or the use of gradient materials may be a solution here.

In the past years concepts and first results on the fabrication of stretchable electronic substrates have been published [1-4]. S. Wagner et al. have demonstrated "skin like" metal structures that may be extended to more than 20 % of their relaxed length. The technology, however, is restricted to extremely thin films and requires high cost equipment.

In this presentation we will demonstrate the basic concepts, process flows and results on the manufacturing of stretchable substrates and simple electronic systems using conventional printed wiring board manufacturing and assembly equipment. The work was funded in the framework of the European Project STELLA (STretchable ELectronics for Large Area applications) [4].

Insulating Base Materials for the Stretchable Substrates

The potential use of stretchable electronic systems defines the requirements for the insulating base material that is used as the matrix and substrate. A maximum reversible elongation of 5 % to 20 % for electrical systems with full functionality is targeted within the framework of STELLA. Stretchability up to 50% will be a topic of research within the consortium. The material shall withstand electric fields of up to 1000 V/mm and have a resistance of in the GOhm region. Working temperatures for the systems will be 30 to 90 °C in field use. For some applications the material is additionally required to be premeable for air and (body) fluids. This requirement is subsumed as breathability of the substrate. Component assembly on breathable substrates will result only in low component and wiring density on rather large areas, thus long interconnections between the components. On the non breathable substrate a higher component and wiring density can be achieved.

According to the two distinct classed of applications on one side open structured permeable fabric, textile or non-woven materials and on the other side homogeneous stretchable materials like rubber, silicone or polyurethane can be utilized as insulating stretchable base materials. Results for substrates based on silicone and poly urethane foils will be given in the following.

Polyurethane Materials Thermoplastic polyurethane (TPU) can be tailored to exhibit a considerable number of physical properties and combinations thereof. The molecular building blocks of bulk TPU are linear segmented block copolymers composed of hard and soft segments, see fig. 1. They are synthesized by combining isocyanates with short-chain diols to form the hard segements and polyether or polyester type chains yielding the soft portions. Depending on the field conditions TPU can either be made water resistant by using polyether chains or oil resistant by using polyester chains. Molecular weight as well as the ratio and type of soft and hard segments can be largely varied and thereby make TPU materials highly versatile.

The elasticity of TPU is an entropic effect. In the stretched state the polymer chains become more and more aligned and entropy is reduced due to

higher order of the chains, the force driving the material back is the increased in entropy in the non-stretched state.

The base structure and the variability of TPU yields in high resilience, good compression set, resistance to impact, abrasion, trearing, wheather and even hydrocarbons. Even without addition of plasticizers a broad range of hardness and elasticity can be given to the material.

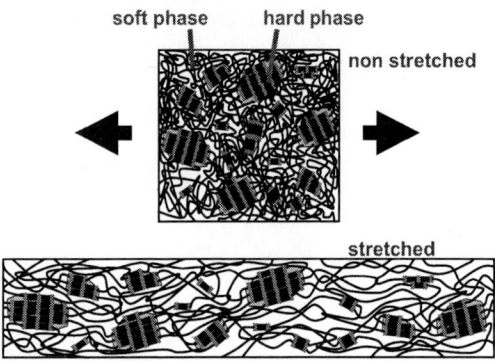

Fig. 1: Schematic representation of the molecular build up TPU.

Of the commercially available TPU materials AK-TPU 1 and Laripur 7025 were chosen to be used throughout the project. The former has a Shore A hardness of 81 and an 500 % elongation before break, tear strength is 48 N/mm and the tensile strength is 36 N/mm2. The latter material exhibits even an enlongation of 760 % before break, Shore A hardness of 70, and a tear strength of 76 N/mm respectively.

Since first experiments were also made with silicone, ELASTOSIL RT 705 from Wacker Chemicals, the respecibe mechanical parameters shall be mentioned here: Shore A hardness 45, elongation before break 200%, tear strength 3,3 N/mm, and tensile strength of 3 N/mm2. Although this material was easy to process it was discarded because of the rather low tear strength, which makes thin foils of the material prone to fracturing.

Stretchable Conductors

In order to set up the electronic system on the stretchable substrate, components have to be interconnected by stretchable electrical wires. Since the insulating substrate materials are considered for quite different applications and display very distinct microstructures here again two strategies are presued. The use of intrinsically stretchable and at the same time conducting materials is an obvious approach for wirings on non woven substrates, where the wiring density and the requirements for conductivity are not too high. On the other side metallic structures, that are almost non-stretchable themselves, but can be made stretchable by appropriate design of the wiring traces will have a high conductivity and are closer to present day electric system assembly and interconnection technologies. Only the latter type will be discussed in the following.

Materials which have intrinsically only a very low stretchablity, as e.g. most metals, can be made stretchable by forming them into an appropriate 3 or 2 dimensional shape. Metallic springs (three dimensional) have been patented as stretchable conductive interconnections between rigid electronic sub systems [6]. However, processing of three dimensional structures is highly sophisicated and needs costly equipment.

Aiming at a flat two dimensional realization of a stretchable metallic conductor a wavy shaped structure seems to be an obvious solution. However, sinusodial forms suffer from extreme stresses at the apex of the waves. Mechanical simulations led to the conclusion, that meandering horseshoe type structure meets the requirements for multiple stretchablity without fracture much better. The specific design types are classified according to bending angle of each horseshoe, as shown in figure 2. For the horseshoe type wires still most of the stress gathers in the apex. Only if the metal trace would be infinitely narrow those stresses will vanish. A design proposal to get around this problem is to split a wide single wire up into a bunch of multiple narrow wires that are randomly interconnected for stability as depicted in figure 5.

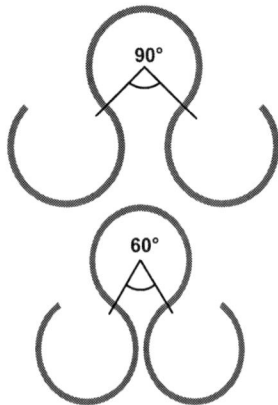

Fig. 2: Horseshoe design for stretchable metal wires. Horseshoes are classified by their bending angle.

Fabrication of Stretchable Substrates and Component Assembly

For conductors on stretchable foils the use of metal wires is most promising. For the realization of the metal wiring on the insulating base foil a couple of approaches are possible. One of these is the metal deposition onto the stretchable substrate by sputtering through a shadow mask, leading to a metal with an inherently threedimensional stretchable structure as decribed [5]. In the framwork of the project STELLA technologies are under development that aim at the development of large industrial scale fabrication technologies for stretchable systems. This will facilitate the introduction and dissemination of stretchable systems into various segments of the consumer applications. The basic fabrication technologies developed at TU Berlin will be described in the following section. The process flow is shown in figure 4.

Onto a thin sheet of Cu a layer of stretchable polymer material is deposited (a,b). The Cu foil has undergone a pre-treatment for surface roughening to promote the adhesion of the polymer layers to the Cu. The Cu sheets are commercially available and are widely used in printed circuit board manufacturing.

Various deposition methods have been investigated, screen printing, casting, and lamination of preformed stretchable sheets. Printing was quite successful when silicone was used as substrate material, see figure 3. Adhesion between the copper and the silicone was satisfactory.

Fig. 3: Stretchable electronic substrates fabricated using silicone as stretchable substrate.

Due to the superior material properties (especially the much higher tear strength) of thermo plastic polyurethane this material is actually preferred as substrate. Printing and casting of TPU onto copper foils resulted generally in poor adhesion, i.e. copper peeled off easily after a structuring. Lamination of preformed TPU sheets in a vacuum press has by contrast yielded robust mechanical anchoring between the polyurethane and the roughened copper. Cu with thicknesses von 17 μm and 32 μm was generally used throughout the experiments. The TUP foil thickness was 100 μm. The press parameters are such that the melting point of the TPU (melting regime of the used material is 155°C to 170°C) was exceeded by about 50 °C for a 10 minutes. The pistil pressure was 5 bar, while the chamber pressure during process was kept below 10 mbar. Cu and TPU sheets up to size of 30 cm by 40 cm can be processed, see fig. 5.

After lamination of the TPU the uncovered Cu surface is equipped with a mask layer which can be structured either by photo lithography or using a laser, refer to fig. 4 (c). The Cu is structured with

622

common in printed wiring board technologies, once the stretchable polymer is deposited. After structuring components are assembled and connected to the Cu either by adhesive gluing or even by soldering (d). Care has to be taken to prevent a stretching of the component and the vicinity of the interconnection of the component to the Cu wires. The latter is still a matter of investigations. Stiffening structures in the Cu layer and the use of encapsulating hard polymers around the component are used in trials, see figure 6. Finally the electronic structures are encapsulated by the printing or lamination of the stretchable polymer (e).

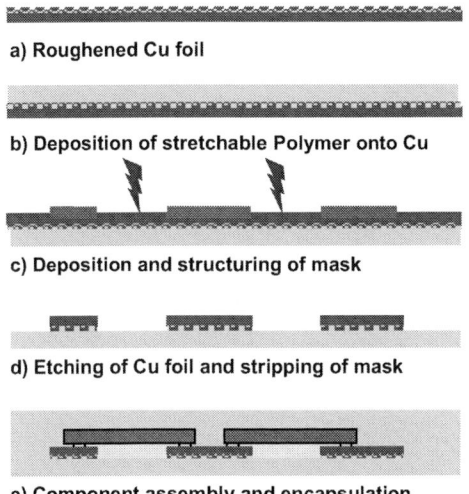

a) Roughened Cu foil

b) Deposition of stretchable Polymer onto Cu

c) Deposition and structuring of mask

d) Etching of Cu foil and stripping of mask

e) Component assembly and encapsulation

Fig. 4: Process flow for the fabrication of a stretchable system using printed wiring board technologies only.

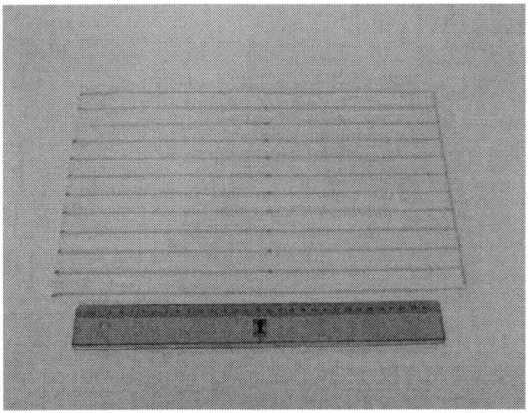

Fig. 5: Large area stretchable substrate with Cu wiring.

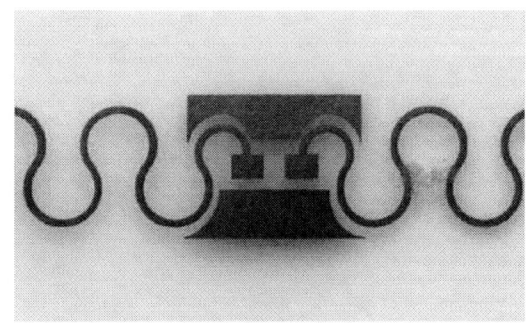

Fig. 6: Detail of the Cu structures in the vicinity of the component landings.

Assembly of the components onto the stretchable substrates can be done in principally the same way as with conventional electronics assembly. However, since the elasic substrates are likely to deform and quibble during placement and bonding processes high precision placement is not to be expected for the present modules.

Contact formation is another topic of investigation, see figure 7. Most of the stretchable substrate materials can withstand max. temperature up to 180° C for a short time. Thus lead free soldering is precluded. The only solder alloy applicable therefore is SnBi (T_m=142 °C), which has at least a small share as industrial used solder. More specialized solder alloys with lower melting temperatures (e.g. InSn or InSnBi) are discarded because of their high cost. Alternatively conductive adhesives can be used. Here materials have to be selected such that curing conditions comply with substrate robustness.

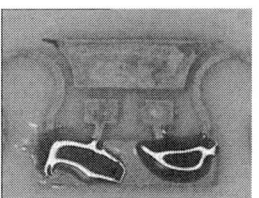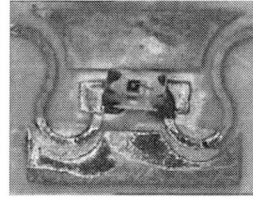

Figure 7: Wiring with locally applied solder mask (left). Component soldered to the pads with SnBi solder

In field life and especially under stretching conditions the parts of the electronic system that are exposed to severest shear stresses are the transitions between solid and stretchable material. In order to relieve and distribute those stresses away from the component stiffening structures close to the component and/or additional encapsulation of the component have been structured into the Cu layer, see figure 6. The stiffening effect of such structures has proven to be reasonable. In more elaborate processes also polymeric stiffeners can be used.

After component assembly the boards are laminated into a covering polyurethane layer. The components are thereby fully embedded into the stretchable matrix, see figure 8.

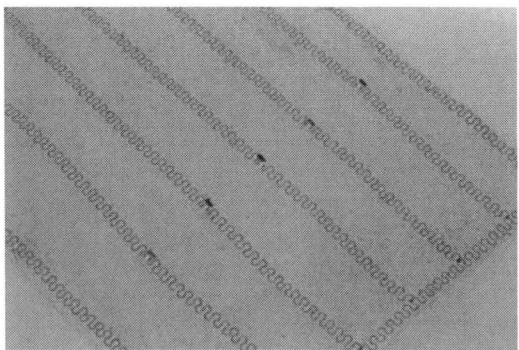

Figure 8: Embedded components after lamination of a covering layer of polyurethane.

Reliability testing for stretchability of electronic systems is not standardized today. Within the consortium of the STELLA project therefore requirements and constraints for stretchable systems will be defined. Some hundred cycles of reversible stretching up to 50 % without remanence of the substrate and without failure of the electronic systems will be investigated. However, more relaxed conditions like a 5 % or 10 % reversible stretch are more realistic to be attainable.

Conclusions

It has been shown, that stretchable electronic substrates can be manufactured making use of conventional printed circuit board technologies. By mounting electronic components onto appropriately prepared substrates and subsequent embedding of the components stretchable electronic systems are realized.

Acknowledgements

The authors express their gratitude to the European Commission for the financial support of the Project STELLA, contract no. IST-028026.

References

[1] S. P. Lacour, S. Wagner, Z. Huang and Z. Suo, Stretchable gold conductors on elastomeric substrates, Appl. Phys. Lett., vol. 82, no. 15 (2003), pp. 2404-2406

[2] T. Li, Z. Huang, Z. Suo, S. P. Lacour, S. Wagner, Stretchability of thin metal films on elastomer substrates Appl. Phys. Lett., vol. 85, no. 16 (2004) pp. 3435-3437

[3] S. P. Lacour, J. Jones, S. Wagner, T. Li, Z. Suo, Stretchable interconnects for elastic electronic surfaces, Proceedings of the IEEE on Flexible Electronics Technology vol. 93, no. 8 (2005), pp. 1459-1467

[4] www.stella-project.eu

[5] D. K. Biegelsen, D. Fork, J. Reich, Stretchable interconnects using stress gradient films, US Patent 6,743,982 June 1 (2004)

3DμTune: High-Q Micromachined Cavities for Millimetre-wave Filters and Oscillators

J B Mills[1], B Giesbers[1], M Matters-Kammerer[1], I Ocket[2], B Nauwelaers[2], A Jourdain[3], W Gautier[4], B Schönlinner[4]

[1]Philips Research, High Tech Campus 4 (WAG12), 5656AE Eindhoven, The Netherlands
[1] +31 402742669, +31 402743352, john.b.mills@philips.com, ben.giesbers@philips.com, marion.matters@philips.com
[2] K.U. Leuven, ESAT/Telemic Division, Kasteelpark Arenberg 10, 3001 Leuven (Heverlee), Belgium
[2] +32 16321875, +32 16321112, ilja.ocket@esat.kuleuven.be, bart.nauwelaers@esat.kuleuven.be
[3] Interuniversity Microelectronics Centre (IMEC), 3D Integration Division, Kapeldreef 75, 3001 Leuven (Heverlee), Belgium
[3] +32 16281909, jourdain@imec.be
[4] EADS Deutschland GmbH, EADS Innovation Works Germany, TCC4 // Dept LG SI MW, Microwave Technologies, 81663 Munich Germany
[4] +49 8960722075, +49 8960724001, william.gautier@eads.net, bernhard.schoenlinner@eads.net

Abstract

An analysis of the translation of RF frequency precision specifications into thin-film processing technology requirements for high Q-factor Silicon cavity resonators is presented. Two alternative methods of coupling RF power into- and out-of Silicon cavity resonators using existing high-Ohmic Silicon passive integration platforms from IMEC (MCM-D) and NXP (PASSI) are discussed.

Key words: micromachined Silicon, cavity resonator, filter, oscillator

Introduction

The trend towards higher carrier frequencies for mass-market applications such as Ka-band satellite television, 60GHz for future Wireless LAN and 77GHz for automotive RADAR systems brings the need for simpler, more power efficient and high performance RF front-ends. To this end the European 3DμTune project aims to develop a generic process for producing reliable, high quality-factor fixed and frequency agile millimeter-wave Silicon cavity resonators embedded in a 3D passive integration technology. Cavities will be combined with SiGe ICs to create hybrid multi-technology modules. The intention is to enable the design and construction of stable, free-running low phase noise oscillators and precise, high quality factor narrowband filters. The potential high performance of Silicon micromachined structures for microwave and millimeter-wave use has been widely reported [1-3]. This paper examines how RF frequency precision specifications translate into Silicon processing technology requirements and looks at two methods of coupling RF power into- and out-of micromachined cavity resonators.

Cavity Fabrication – Technology Requirements

Figure 1 shows the basic process for bulk micromachining of cavities in Silicon wafers. Starting with float-zone wafers, a 200nm thick Silicon Nitride (Si_xN_y) layer is grown by low pressure CVD at 800°C. This is patterned by plasma etching with CF_4-O_2 to produce a physical mask for the bulk machining of the Silicon. This is performed by wet etching with 33% Potassium Hydroxide (KOH) at 70°C. The KOH etches the <100> crystal plane preferentially to the <111> axis by around two orders of magnitude (~ 45μmhour^{-1} versus 0.45μmhour^{-1}). This anisotropic etching produces cavities with sidewalls angled inwards at a constant angle $\phi = 54.74°$.

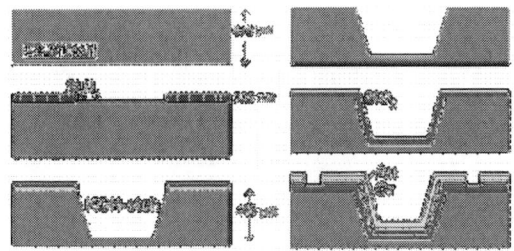

Figure 1: Silicon cavity fabrication process

The effect of this sloped sidewall upon the resonant frequency of the desired TE^z_{101} mode can be accounted for by substituting an effective, depth averaged, cavity width given by equation (1) into the solution of the wave equations (2) for the resonant frequency of an ideal straight walled cavity [4] of width and length a and c and depth b. For maximum quality factor square cavities are used hence $a = c$.

$$a_{eff} = mean[a(z)] \equiv \frac{a - b}{\tan \phi} \quad (1)$$

$$(f_r)^{TE^z}_{101} = \frac{1}{2\sqrt{\mu_o \mu_r \varepsilon_o \varepsilon_r}} \sqrt{\left(\frac{1}{a_{eff}}\right)^2 + \left(\frac{1}{a_{eff}}\right)^2} \quad (2)$$

Figure 2 shows a comparison between the resonant frequency calculated using (1) and (2) with values obtained using the commercial 3D electromagnetics finite element solver HFSS from Ansoft.

The unloaded quality factor (Q_u) of the TE^z_{101} mode in cavities after metallisation can be estimated as a function of the depth and frequency.

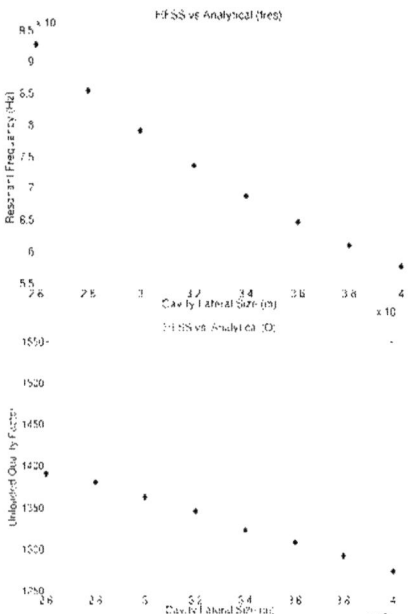

Figure 2: Comparison between analytical (solid lines) and 3D FEM (diamonds) of f_{res} (top) and Q_u (bottom) for 460µm deep KOH-etched cavities.

by combing (1) with the (3) for an ideal square cavity [4]:

$$(Q)^{TE^z}_{101} = \frac{\pi\eta}{2R_s}\left[\frac{b\left(a^2_{eff} + a^2_{eff}\right)^{3/2}}{2a^4_{eff} + 4a^3_{eff}b}\right] \quad (3)$$

where R_s is the surface resistance of the metal and η is the impedance of freespace. Figure 3 plots (3) assuming bulk conductivities of 5.76×10^7 Sm^{-1},

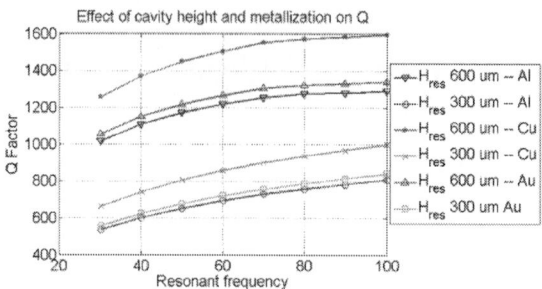

Figure 3: Effect of cavity depth and metallisation on cavity Q_u

The three target applications in the 3DµTune project, 2.5% bandwidth 20GHz bandpass filter, free-running low phase noise 60GHz VCO and a 74.3GHz low phase noise free-running reference oscillator require resonators with values of Q_u of 1000 (20GHz) and 1200 (60/74.3GHz) corresponding to cavity depths of 1mm and 460µm respectively.

All three applications require resonators within ±0.1% of their target frequencies of 20, 60 and 74.3GHz. Carrying out a partial differentiation of (2) with respect to a_{eff} enables the sensitivity of the resonant frequency to changes in width and depth to be evaluated. Assuming an equal error distribution i.e. 0.05% error in width and in depth gives (4). The lateral etch precision requirement for a 20GHz cavity (a_{eff} = 10.6mm) is 5.3µm, for 60GHz (a_{eff} = 3.53mm) it is 1.8µm and at 74.3GHz (a_{eff} = 2.85mm) it is 1.4µm.

$$\delta f_{res} = \delta f_{res1} + \delta f_{res2}$$

$$\delta f_{res} = \delta a_{eff}\frac{df_{res}}{da_{eff}} + \delta b\frac{df_{res}}{da_{eff}}$$

$$\frac{df_{res}}{da_{eff}} = \frac{-c}{\sqrt{2}a^2_{eff}} \qquad \frac{df_{res}}{da_{eff}} = \frac{df_{res}}{da_{eff}}\cdot\frac{da_{eff}}{db}$$

$$\therefore \frac{df_{res}}{da_{eff}} = \frac{-c}{\sqrt{2}a^2_{eff}}\frac{-1}{\tan \phi}$$

$$\frac{\delta f_{res1}}{df_{res}} = \frac{-\delta a_{eff}}{a_{eff}} \quad (4a)$$

$$\frac{\delta f_{res2}}{df_{res}} = \frac{\delta b}{a_{eff}\tan \phi} \quad (4b)$$

To meet the cavity width precision specifications requires that the Si_xN_y hard mask which defines the Silicon area for KOH etching to be well aligned to the <111> crystal plane. The required alignment precision can be calculated from Figure 4. For a 74.3GHz cavity with a_{eff} of 2.85mm the Si_xN_y mask needs to be within ±0.028° of the <111> crystal plane of the wafer.

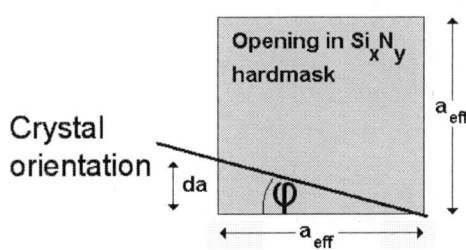

Figure 4: <111> Si crystal plane alignment for ±0.1% resonant frequency precision

Such precision is not obtainable by simple alignment to the flat on the wafer which can be oriented with errors from ±0.5 to ±1° relative to the <111> plane. A variety of procedures have been reported in the literature [5-8] which can be used to meet the tight alignment requirements for the cavity resonators.

From (4b) the cavity depths of 1000μm for the 20GHz cavities and 460μm for the 60 and 74.3GHz cavities need to be etched to better than 7.5μm, 2.5μm and 2μm respectively. Achieving such precision is currently done by frequency interruption of the KOH etching to check intermediate cavity depths to predict the final endpoint. Work is underway on a more efficient automatic endpoint detection method. To try and ensure consistency in etching times the KOH is maintained at a constant temperature and stirred both mechanically and with a flow of Nitrogen gas through the solution to give a continuous refreshment of KOH over the etch surfaces. This has been found to be particularly important in achieving repeatability in the 'notching' observed in the cavity depth. Figure 5 shows a white light interferometry measurement of the depth profile of a 13.53mm x 6.69mm cavity. There is an over-etch effect observed around the perimeter of the cavity of around 8μm. The form that this notching takes varies somewhat with cavity size.

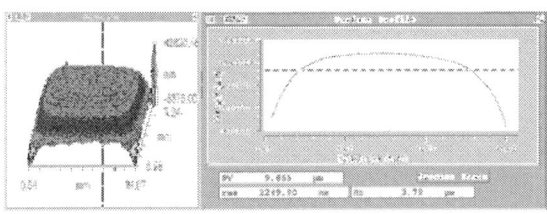

Figure 5: 'Notching' in etched Si cavity floor and diagonal floor profile

Larger cavities like the example in Figure 5 show a central 'plateau' area at the desired etch depth. Away from the plateau the depth increases as the cavity edge is approached. For cavities smaller than around 5mm, the 60 and 74.3GHz designs, there is no central plateau region as the two curved regions from the cavity walls overlap each other giving a continuous depth profile peaking in the cavity center. Regardless of cavity size the uneven etch depth only causes a small perturbation of the resonant frequency as the extra depth at the cavity walls causes only a minor increase in the effective cavity width a_{eff}. This is demonstrated in Figure 6 with results from a 3D finite-difference-time-domain simulation of a two-port 20GHz cavity resonator, carried out using the EMPIRE package from IMST, for both an ideal and 'notched' cavity.

RF Power Coupling

Two principle configurations are considered for coupling RF power into and out-of the Silicon cavity resonators. The first of these couples power directly down through the Silicon of a top wafer that is attached on top of the cavity to seal it by Gold-Gold thermocompression bonding or by transient-liquid-phase solder bonding. Such an approach has been described previously by others [9]. RF power is carried by a microstrip line, the metallisation of the cavity below acting as the microstrip line's groundplane. A slot is opened in the cavity metallisation to allow the E-field from the microstrip line to couple into the cavity.

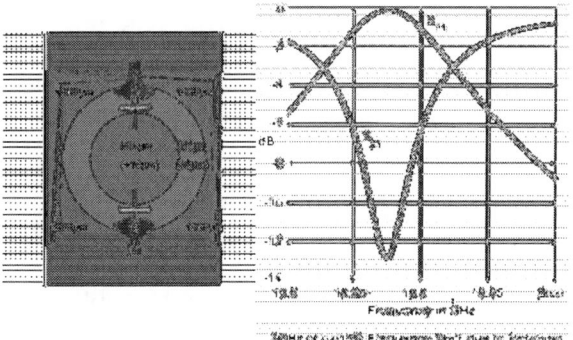

Figure 6: Shift in cavity resonance frequency due to uneven etch depth

In order to minimize the dielectric losses the top wafer uses high resistivity (> 5kOhmcm⁻¹) Silicon as well as mechanical thinning to give a 100μm thick substrate. Figure 7 shows a two-port implementation of such a structure split along its symmetry plane and scaled to a 65.7GHz resonance frequency to investigate the effect upon losses compared with [9]. Figure 8 shows the simulated performance of the 65.7GHz two-port cavity of Figure 7 when the top substrate uses NXP's PASSI high-Ohmic Silicon passive integration process.

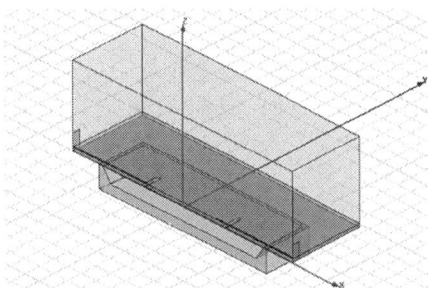

Figure 7: Microstrip / slot cavity feed structure

This has three Aluminium metal layers separated by thin (< 0.6μm) SiO$_x$ and Si$_x$N$_y$ dielectrics. The extracted Q$_u$ for a 100μm thick substrate with the microstrip line and slot fabricated in the top 5μm thick Aluminium layer is 721.

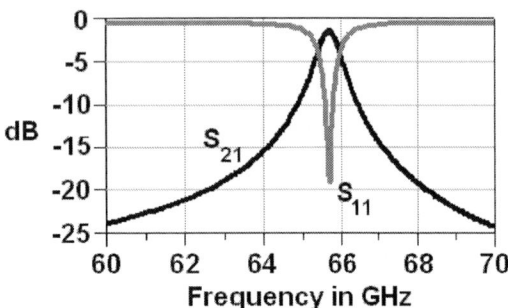

Figure 8: 65.7GHz cavity fed by microstrip / slot on PASSI substrate HFSS simulation

Switching to IMEC's MCM-D high-Ohmic (>5kOhmcm^{-1}) Silicon passive integration platform with its 5μm Copper metallisation and thick (> 5μm) BCB inter-metal dielectrics gives only a marginal improvement in Q$_u$ to 785. The principle cause of loss with this microstrip / slot structure is due to slot radiation.

In some applications it is desirable for the metallised cavity to be flipped on top of its passive integration platform carrier. In this case the microstrip / slot approach can only be used with the MCM-D process and not with PASSI. This is because the thin inter-metal dielectrics would require the use of excessively narrow microstrip lines and coupling slots leading to extremely high losses which would greatly reduce the loaded quality factor of the cavity. To avoid this problem a coplanar waveguide (CPW) slot antenna terminated in either an open- or short-circuit is fabricated in the cavity groundplane which is formed in the top metal of the PASSI / MCM-D process as shown in Figure 9. The CPW antenna is placed as close as physically possible to the outer wall of the cavity to reduce the interconnect length, hence line losses, and to reduce the likelihood of exciting spurious resonances in the Silicon substrate.

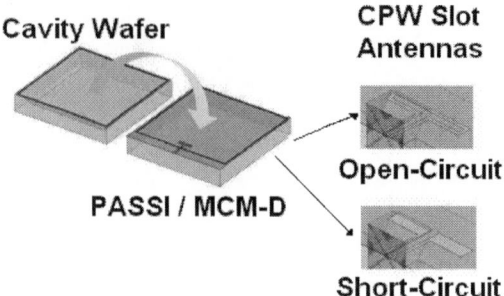

Figure 9: CPW slot antenna coupling structures

S-Parameter data from HFSS simulations for both the open- and short-circuit CPW slot antennas feeding a nominal 60GHz cavity is shown in Figure 10.

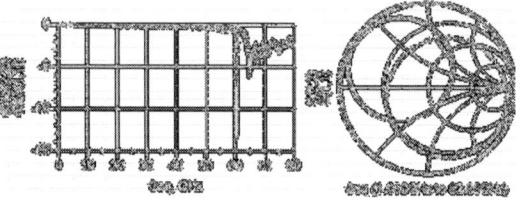

Figure 10: S-parameter data for 60 GHz designs using both coupling options: Open-ended structure (solid line) and shorted structure (dash dotted line)

For clarity, due to excitation of substrate modes, the Smith chart data has been truncated at 60GHz. The extracted Q$_u$ of the cavity for the open-circuit feed is 851 at a frequency of 60.42GHz while the short-circuited CPW antenna gives a Q$_u$ of 1044 at 60.57GHz. The difference in resonance frequency and Q$_u$ results from the different E-field configurations generated by the two antennas and their match to the field distribution of the desired TE$^z_{101}$ mode of the cavity. Figure 11 shows HFSS simulated data for a 77GHz version of the CPW excited cavities using a slightly shorter slot antenna length. The Smith chart data is once again truncated, this time to 90GHz, for clarity. As with the 60GHz versions the open-circuit antenna gives a slightly lower Q$_u$ value of 720 at 76.32GHz compared with the short-circuit design's value of 1022 at 76.66GHz.

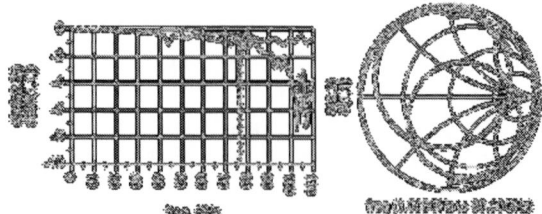

Figure 11: S-parameter data for 77 GHz designs using both coupling options : Open-ended structure (solid line) and shorted structure (dash dotted line)

628

To try and raise the cutoff frequency of some of the undesired resonant modes the CPW antennas also excite in the Silicon wafer the 77GHz simulations considered a wafer thinned from 675μm to 300μm. The effect of this in helping to suppress substrate modes is illustrated in Figure 12 with simulations for a 60GHz cavity with both a full and thinned substrate.

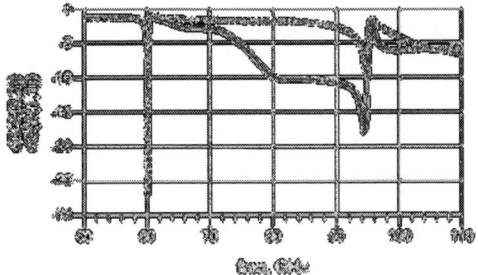

Figure 12: Effect of wafer thinning on substrate mode coupling: without substrate thinning (solid line) and with thinning to 300 μm (dash dotted line).

It is likely that further measures, such as use of a lossy backside absorber such as TiW or TiN, may be required to attenuate dielectric substrate modes. This may need to be combined with micromachining of the Silicon in the vicinity of the slot antenna to give a localized reduction in the effective substrate permittivity to raise the substrate modes' cutoff frequency [10].

Conclusions

The translation of RF frequency precision specifications into practical requirements for thin-film processing technology for the manufacture of micromachined Silicon cavity resonators operating at 20, 60 and 74.3GHz have been shown. The negligible impact of uneven cavity etch depth upon operating frequency has been demonstrate by 3D FDTD simulation. Two alternative methods of coupling RF power into and out-of cavity resonators using microstrip / slot structures and CPW slot antennas using the IMEC MCM-D and NXP PASSI high-Ohmic Silicon passive integration platforms have been explored using 3D FEM simulation data.

Acknowledgements

This work was performed within the framework of the European research project 3DμTune (IST-2005-027768). The work of Ms Guo Peng in the investigation of the 'notching' effect when etching Silicon with KOH and production of the white light interferometry images is gratefully acknowledged.

References

[1] L.P.B Katehi, J.F Harvey, E. Brown, "MEMs and Si micromachined Circuits for High-Frequency Applications", IEEE Transactions on Microwave Theory and Techniques, Vol 50,No 3, 585-866, March 2002

[2] K.J. Herrick, L.P.B. Katehi, "W-band Micromachined Circuit Combining Networks", IEEE Transactions on Microwave Theory and Techniques, Vol.50, No.6,1647-1651, June 2002

[3] K.J. Herrick, L.P.B Katehi, "Microtechnology in the development of three-dimensional circuits", IEEE Transactions on Microwave Theory and Techniques, Vol. 46, No. 11, 1832-1844, November 1989

[4] C.A. Balanis, "Advanced Engineering Electromagnetics", John Wiley & Sons, New York, Chapter8, 388-394, 1989

[5] G. Ensell, "Alignment of mask patterns to crystal orientation", Proc. Of the International Solid-State Sensors & Actuators Conference, Stockholm, Sweden, June 25-29, Vol. 1, 186-189, 1995

[6] M. Vangbo, Y. Backlund, "Precise mask alignment to the crystallographic orientation of silicon wafers using wet anisotropic etching, Journal of Micromechanics and Microengineering, Vol. 6, No. 2, 279-284, 1996

[7] J.M Lai, W.H Chieng, Y.C Huang, "Precision alignment of mask etching with respect to crystal orientation, Journal of Micromechanics and Microengineering, Vol. 8, No. 4, 327-329, 1998

[8] W.H Chang, Y.C Huang, "A new pre-etch pattern to determine <100> crystallographic orientation on both (100) and (110) silicon wafers, Microsystem Technologies, Vol. 11, No. 2-3, 117-128, 2005

[9] J. Papapolymerou, J.C Cheng, J. East, L.P.B. Katehi, "A micromachined high-Q X-band resonators", IEEE Microwave and Guided Wave Letters, Vol. 7, No. 6, 168-170, 1997

[10] I. Ocket, B. Nauwelaers, G. Carchon, A. Jourdain, W. De Raedt, "60GHz Si Micromachined Cavity Resonator on MCM-D", Proc. 6th Topical Meeting on Silicon Monolithic integrated Circuits in RF Systems, San Diego, California, January 18-20, 2006

DAVID - a Strategic Research Project for Chip-Scale MEMS / ASIC Co-integration

N. Marenco, W. Reinert, S. Warnat

Fraunhofer Institut für Siliziumtechnologie (ISIT), D-25524 Itzehoe

Phone: +49-4821-17-4620, E-mail: info@isit.fraunhofer.de

Abstract

Advanced packaging solutions offer new exciting perspectives for system integration of MEMS devices. Not only a more compact realization of smart sensor clusters, but also an enhanced quality and reduced overall costs can be expected from a consequent use of waferlevel packaging technologies.

Key words: MEMS packaging, system-in-package (SiP), inertial measurement unit, chip-scale package (CSP), waferlevel packaging

Abstract

A hybrid waferlevel packaging approach is presented that targets basically for the industrial high-volume production of inertial measurement units. In a joint effort of six renowned European partners under the FP6 project DAVID, key technologies like post-CMOS through-silicon vias, vacuum compliant chip-to-wafer bonding and waferlevel transfer molding are developed. The objective is a Chip-Scale System-in-Package (CSSiP) solution that can be implemented with a moderate effort and thus allows a realization of smart sensor systems in a short loop between design and manufacturing.

Introduction

Mobility and ambient intelligence are beneath the main innovation drivers in consumer electronics. A special field are inertial sensors that can measure linear acceleration or rotational movements. Their ability to track movements can be used for human interface devices, pedestrian navigation, image stabilization and comparable applications.

To build a complete inertial measurement unit (IMU), sensors for all spatial directions have to be combined. Although some smart multiaxial sensor systems are already available, today's technologies still require the assembly of discrete sensor devices for a full cluster, which implies additional calibration steps to compensate mounting tolerances.

The European DAVID project addresses a set of key technologies that could lead to a completely integrated smart IMU in a chip-scale package. DAVID stands for "Downscaled Assembly of Vertically Interconnected Devices" and basically investigates the following process elements:

- Through-Silicon Vias (TSV)
- Chip-to-Wafer (C2W)
- Wafer-to-Wafer (W2W) bonding
- Vacuum sealing
- Waferlevel transfer molding

Additionally, the limits of ISIT's industrialized surface micromachining process are investigated down to a 400 nm structural gap width (today: 2 µm).

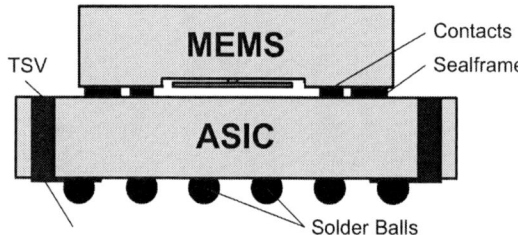

Figure 1: Chip-to-Wafer integration approach: Technology independence and optimized yield control are achieved by bonding good MEMS dies to good parts on the ASIC wafer.

The 4 x 4 mm² challenge

Taking into account the current dimensions of accelerometers and gyroscopes, a realistic and marketable size for an inertial sensor cluster lies in the order of 4 x 4 mm². This estimation assumes a MEMS shrink potential of min. 50% by reducing its structural width.

Additional design-related size reduction can be achieved by the improved signal routing that is possible with the hybrid integration approaches studied in DAVID: Due to the direct face-to-face assembly of ASIC and MEMS, the CMOS interconnect levels can replace a large part of the MEMS wiring. In this way, shortest interconnects are realized, thus minimizing the parasitic capacitances that use to deteriorate the sensor signals. In return, the sensor design can become smaller with less effective capacitance area needed.

Chip-to-Wafer (C2W) vacuum bonding

MEMS resonators like gyroscopes work under defined vacuum conditions: A well-controlled damping allows to keep driving energy very low, while a minimum of gas included in the cavity is still beneficial for an accurate signal response to fast movements.

The direct C2W bonding therefore has two functions: Ensuring a vacuum-tight hermetic encapsulation of the MEMS and providing the electrical interconnect between MEMS and ASIC. Molding stress can even add a third function to this interface: Without any contact bumps in the cavity, the transfer molding pressure could lead to considerable chip deformation.

The C2W process itself basically consists of two successive steps: A pre-fixation of each single die on the substrate wafer and a final bonding step where all dies are bonded down simultaneously.

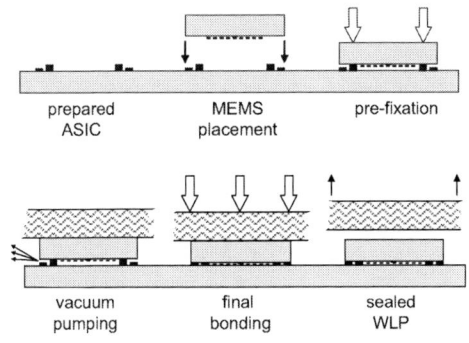

Figure 2: Chip-to-Wafer bonding process: After a die-level pre-fixation step (top), all MEMS are bonded to the ASIC wafer simultaneously and under vacuum (bottom).

The main advantages of the C2W concept are an optimized yield control (only good dies on good dies) and the independence from wafer formats: MEMS fabs are currently just beginning to shift from 6" to 8" format, while ASICs for the focused product category usually are produced on 8" or 12" wafers.

A comparable approach targeting for logic devices and flash memories has been presented by the project partner Datacon Technology in collaboration with EVG [1]. The economic evaluation of their C2W concept using the SOLID process has shown that the additional processing costs for a TSV-based C2W process is in a competitive range compared to conventional packaging methods.

To ensure a reliable vacuum level in the extremely small cavity, a getter film is used. Excellent experience in microcavity getter technology has already been gained in the preceeding VABOND project (IST-2001-34224) by Fraunhofer ISIT and SAES getters (Italy). The new challenge is to fully exploit the small available area by a fine-structured getter film on the ASIC surface.

Wafer-to-Wafer (W2W) approach

A waferbonding approach is investigated by the project partner STMicroelectronics: With ASICs and MEMS now both being produced in 8" format on the Agrate (Italy) location, this concept can benefit from a short time-to-market. However, the success depends on the availability of a Through-Silicon Via (TSV) technology as described below, while in the C2W concept alternative interconnect methods can be applied. In return, it has certain advantages regarding the handling aspects: Since the C2W approach uses die bonding, particle control will play a major role.

Through-Silicon Via (TSV)

Among a large number of TSV concepts that have been investigated in the past years, the post-CMOS via process seems to become the industrially preferred solution. Although there are some process restrictions for CMOS compliance and a certain handling risk regarding the valuable fully-processed wafers, it is the least intrusive way for the frontend facilities with their well-established and highly complex processes.

Current attempts to develop post-CMOS compliant TSV technologies use temporary carrier substrates that allow to thin down the active wafer to 10…100 µm. This allows a fast etching and reduces problems in conformal coating of the insulator and metallization layers that will finally define the electrical performance of a TSV. The largest part of these activities aims at the logic device market, where chip stacking and 3D-interconnection are the only way to further increase the functional density or storage capacity of processors and memories.

In DAVID, the focus lies on inertial MEMS packaging. The sensitivity of these devices against thermomechanical stress from packaging and field application requires a high stiffness of the assembly itself. Consequently, the wafers remain relatively thick. Two approaches are developed: For W2W, a removable carrier is used for the TSV and waferbonding step, while the C2W concept keeps a wafer thickness of 300 µm that can be handled with conventional equipment [2].

631

In both cases, the Bosch process based on deep reactive ion etching (DRIE) is used to create holes in the ASIC wafer. On ISIT side, etching trials on a silicon wafer covered with a 1 µm thin Al layer showed that it is possible to reach the frontside without perforating the thin remaining membrane. A pin-hole free SiN film was deposited by PECVD, other processes are equally under investigation. For the metallization, a conformal Cu film shall be deposited. First results using a modified sputtering process showed that a deposit down to the bottom of the 300 µm deep holes (100 µm in diameter) can be achieved within a short time. The target diameter is 30 µm according to current fine-pitch pad arrays on the chip periphery.

Final encapsulation

After bonding the MEMS and ASIC to each other, the final device could be encapsulated by traditional techniques, e.g. wire bonding and leadframe-based molding. A further cost reduction, however, is expected by wafer molding and solder balling to create a Chip-Scale System-in-Package (CSSiP). Very short cycle times can be achieved by transfer molding. Based on an extensive material screening performed by the project partner FICO B.V. (Netherlands), specific materials are being developed to fill very thin mold gaps over the long distance given for 8" wafers. Equally, liquid materials with low molding pressure are under consideration.

Design aspects in 3D ASIC/MEMS integration

The frame of the DAVID project is limited to technology and process development with the target to demonstrate a full process flow that involves all partners. Although it is not targeted to build an inertial sensor, the basic functional elements of gyroscopes and accelerometers are tested in form of a linear resonator structure with comb-structured drive and sense electrodes, suspended by thin polysilicon springs. This type of MEMS will allow to develop and evaluate the shrinked surface micromachining process and to test the vacuum level inside the waferlevel package.

Pre-studies for first design measures are performed by Wroclaw University of Technology (Poland) using FEM simulations. Investigations are related to wafer handling, processing and residual stress in the final device caused by packaging. The expected results shall be, for example, an optimized TSV distribution to avoid wafer cracking and a design rule regarding the required minimum of contact pads in the sensor cavity to prevent deformation in molding.

Reliability evaluation

A major concern in the lifetime reliability of a resonant sensor like a gyroscope is that the vacuum level inside the cavity remains in a range of 0,1 mbar for a controlled damping of the moving structures. An ultra-fine leak test has been developed at ISIT based on an initial neon bombing [3]. The test criterion is the Q-factor of the electromechanical resonator before and after submitting the device to a high pressure neon ambient.

Further reliability evaluation regarding the encapsulation and 2nd level integration (soldering and underfill on various substrates) will be performed by conventional pressure cooker tests and subsequent metallographic analyses of cross-sections.

Acknowledgements

The DAVID project is funded from beginning of 2006 until end 2008 by the European Commission within the Sixth Framework Programme (ref. IST-027240). Contractors are Fraunhofer ISIT (Germany), STMicroelectronics (Italy), Datacon Technology (Austria), FICO BV (The Netherlands), SAES Getters (Italy) and Wroclaw University of Technology (Poland).

References

[1] H. Kostner (Datacon Technology GmbH): "Prozesse und Anlagen zum Flip-Chip-Bonden auf Wafersubstrate (Processes and Equipment for FlipChip on Wafer Bonding)", IMAPS Workshop 8.2.2007, TU-Illmenau (Germany)

[2] S. Warnat, N. Marenco, D. Kähler, W. Reinert: "Design Rules for Post-CMOS Through-Silicon Vias in an Industrial Environment"; Presentation at the 8th Electronics Packaging Technology Conference, 6-8 December 2006 (Singapore)

[3] W. Reinert, D. Kähler, G. Longoni: "Assessment of vacuum lifetime in nL-packages", Proceedings of 7th Electronics Packaging Technology Conference, Singapore, 07/12/2005-09/12/2005

Project homepage

http://www.david-project.eu

3D Package-on-Package Solution for Next Generation Cameras

Vern Solberg and Giles Humpston

Tessera, Inc., 3099 Orchard Drive, San Jose, California USA

Tel: 408-894-0700 Fax: 408-894-0768, E-mail: Vsolberg123@aol.com, ghumpston@tessera.com

Abstract

Next-generation portable phones will provide expanded imaging capabilities, be made smaller and producible in high-volume at lower cost. Market research experts have forecasted that the market for image sensors within portable phones is expected to experience significant growth, increasing to 940 million units by 2009. Because the market for these enhanced products is extremely spirited, rapid deployment of sub-system components becomes a key factor in capturing a competitive advantage. Likewise, the end user attracted to the new functionality, expects that each generation of product will be smaller and lighter than its predecessor. To avoid expanding the product size, the manufacturers will need to rely more on functional integration. That is, combining a maximum number of functional features into a minimal number of sub-system components. This paper will focus on a 3D Package-on-Package solution that includes new packaging and optics technology, developed for the portable phone market. The actual imaging system (camera) utilizes an innovative wafer level silicon-glass sandwich structure to protect the image-sensing elements. The lens module is a composite of several refractive and diffractive layers manufactured using a silica based material in a wafer format. The lens units are produced in mass using lithographic processes, similar to both IC manufacture and wafer-level chip scale packaging. The end result is a finished, RoHS compliant, FBGA camera module package ready for reflow solder attachment with X/Y dimensions that are only slightly greater than the original die size, and a total product thickness that is far less than that currently in the market.

Key words: Digital Imaging, CMOS Sensors, Chip-Scale, Package-on-Package, Camera

Introduction

Combining silicon camera imaging with precision digital micro-optics technology can provide solid state imaging solutions in a small package, suitable for a variety of high-growth and high-volume consumer optics applications. Micro-optics is defined as the use of microscopic structures to shape and influence light. These digital optics' technologies have already been utilized for a number of applications including semiconductor equipment optics, communications and photonics and by providing specialty lenses and optical sub-assemblies for the semiconductor lithography and communications markets.

Solid state imaging began with the invention of the charge-coupled device (CCD) in 1969 at the Semiconductor Components Division of Bell Laboratories. Since its inception, digital imaging has progressed through improvements with the emergence of complementary metal-oxide semiconductor (CMOS) technology.

Dr. Eric Fossum conducted the research that made CMOS image sensors practical for outer space applications while at the National Aeronautics and Space Administration's (NASA) Jet Propulsion Laboratory in Pasadena, CA. Dr. Fossum's research eventually led to the development of CMOS active-pixel sensors. Although CMOS image sensor development emerged in the late 1970s, they became a really competitive digital imaging solution in the mid 1990s and now dominate the market [1].

Today, image sensor die are manufactured by many semiconductor companies with optical resolutions ranging over five orders of magnitude, to suit various digital camera applications. The examples shown in Figure 1 are a sampling of CMOS image sensors developed by a leading North American Company.

Source: Micron Semiconductor

Figure 1: Image sensor elements.

The imager portfolio shown above includes VGA and 1.3-megapixel through 5-megapixel sensors to provide both still and video images for a number of current and future mobile phone products. Active-pixel CMOS is able to furnish many on-chip functions, allowing for more portability, lower power consumption, and in the extreme a complete imaging system. This technology breakthrough has led to a huge growth in digital camera market for both professional and consumer applications. In regard to the market for miniature cameras, the researchers are projecting that nearly 80% of all mobile phones will eventually integrate one or more cameras [2].

Camera Sensor Fabrication

The CMOS manufacturing process uses reasonably standard semiconductor technology, which lowers the production cost significantly and can make integration simpler. CMOS sensors often have three or four transistors in each photo-site, which amplify and move the charge provided by incoming photons of light. They enable the pixels to be read individually.

To produce an acceptable image, the sensor array requires a number of controlling functions from color balance, pixel correction, timing, interface logic, focus control to name a few. The silicon platform of the sensor allows the integration of these and other critical functions in very close proximity to the sensor elements.

The diagram illustrated in Figure 2 is an example of a typical system-on-die image sensor package with interface to several functional elements mounted within the die elements outline.

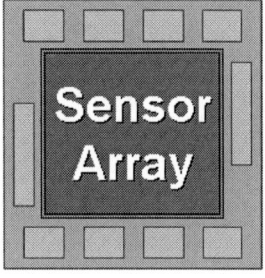

Figure 2: Image sensor die element.

With the centered sensor configuration, the controlling functions are distributed toward the die's perimeter with bond pads at the edge of the die (typically accommodating chip-n-board assembly). The lens system for capturing and projecting the image is added over the sensor area later in the camera module assembly process.

Sensor Packaging

Solid state image sensors require protection from the ambient atmosphere. The principal failure mechanisms are corrosion, mechanical damage and obscuration due to process related particle deposits.

Corrosion: Air, or more accurately the moisture contained in normal atmospheric air, is highly corrosive towards semiconductor die. Functional semiconductors are complex, multi-layer assemblies, where the outer most layers are very fine bus bars of aluminium. Aluminium possesses a very high electrode potential so that rapid corrosion ensues when placed in contact with other metals and in the

presence of moisture to complete a galvanic circuit. The external interconnects to the semiconductor are, by necessity of function, made of metals other than aluminium so it is essential to keep moisture away from image sensor die at all times.

Mechanical damage: The light-sensitive area of an imager is covered by an array of minute, hemispherical lenses. These lenses serve to focus the light that falls on each pixel in the imaging area onto the light-sensitive regions of the semiconductor. The remaining area of the imager is insensitive to light, because of the electronics and wiring lines associated with each pixel. These micro-lenses are made of soft polymers and are extremely delicate. Any contact from handling and assembly processing can result in catastrophic damage to them.

Obscuration: Solid state imagers have individual pixels that typically measure 2 microns per side (or smaller). This is tiny in comparison with dust particles in normal ambient air. The micro lenses that are aligned over each pixel are not only soft and easily damaged, but are also slightly tacky, due to a combination of electrostatic charging and their surface chemistry. This means it is near impossible to remove particulate matter once it lands on a pixel. Clearly, if the size of the particle approaches that of the pixel, it will block the incident light, resulting in a black spot in the image. The human eye is very sensitive to static defects and even a single dead pixel is an annoyance.

First-generation imagers were housed in standard semiconductor packages with glass lids (see Figure 3). These were hermetic and delivered exceptional protection for the die, but were also bulky and expensive to manufacture. Nevertheless, many high performance imagers (e.g. 20+ megapixel) still use this form of enclosure.

Source Micron Semiconductor

Figure 3: Ceramic packaged image sensor.

Wafer Level Packaging

An alternative approach to using discrete packages is wafer-level packaging. The basis is to encapsulate all die simultaneously while they are still in wafer form and finally separate the units into individually packaged parts (see Figure 4).

634

Source: Tessera/SHELLCASE

Figure 4: Wafer level sensor package model.

Formation of a wafer level cavity package:
- Left - the device wafer containing five die.
- Middle - application of a seal material to form a picture frame around the perimeter of each die.
- Right – attachment of the glass lid material to seal the cavity over each die.

Standard wafer saw singulation frees the now protected, glass sealed image sensor die from the wafer.

This wafer scale packaging process has the singular advantage that the process costs are shared among the good die on the wafer. With typically between 750 and 1,500 die on a 200 mm diameter image sensor wafer, this results in an order of magnitude decrease in the package cost per die when compared with discrete ceramic packaging.

Wafer-level packages (WLP) for image sensors are currently available in three styles, distinguished by the interconnect scheme; wire bond, flip-chip or µBGA as shown in Figure 5. They are typically an order of magnitude thinner than conventional imager packages [3].

Source: Tessera/Shellcase

Figure 5: WLP image sensor variations.

This need for these variations arise because there is no industry defined standard for the design of camera modules. Therefore, for an imaging die to achieve significant market penetration it must be made available in all three package styles.

Package manufacture at the wafer scale involves fabricating a picture frame of adhesive around the optically active area on each die. This adhesive is then used to secure a glass wafer to the semiconductor wafer, leaving a small space between them. The picture frame of adhesive and glass cover form a sealed cavity over the vulnerable area on each die, thereby protecting it against mechanical damage and corrosion. Although the glass cover does not prevent obscuration, any particles or dirt

that do land on it can be easily cleaned off without damaging the imager.

For the wire bond style of package, the part of the glass wafer over the bond pads on the die is removed and then the silicon wafer is diced. If required, this package can then be converted to flip-chip style by stud bumping each bond pad. The third alternative is to use photolithographic processes or a through hole technology, to route the bond pads on the front face of the die to an array of lands on the rear face. Deposition and reflow of solder on these lands creates a µBGA-style package.

The wafer level packaging technology commercially available for image sensors provides a low-cost, chip-size, solution with a total package thickness of less than 500 microns.

Integrated Optics

Typically the glass cover on an imager protected by a wafer level package is around 300 microns thick. It forms part of the protective enclosure for the imager but performs no other function. Clearly, it would be advantageous if this component could be utilised for additional duty. Possibilities include as a substrate for the infrared filter that is present in every (optical) camera module, for the glass to be shaped so it provides a lens action, or as a planar substrate to which other optical components can be attached, such as a lens turret assembly.

Camera Module Assembly

Traditionally, the camera lens module is a compliment of optical elements stacked in a vertical format to capture reflected light images. At a minimum, a solid state camera module consists of a lens assembly and an image sensor typically attached to a laminate substrate. The number of lens elements varies based on the requirements of the optical design, but in most camera phones, there are usually two to four lenses. A cross section through a typical solid state camera module is illustrated in Figure 6.

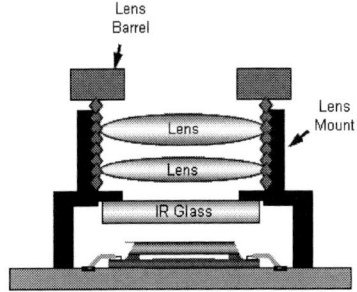

Figure 6: Camera module assembly.

One lens is often made of glass, while the others are manufactured by injection molding of plastic. An infrared filter will also be included somewhere in the optical train, either as a separate element or as a coating on one of the lenses, because silicon photo detectors are sensitive to longer wavelengths than the human eye can perceive.

Addressing Camera Lens Concerns

In developing an alternative camera lens technology, several issues need to be addressed. The traditional camera lens assembly, when compared to the small image sensor element, is rather bulky. The assembly and alignment procedure is complex as well requiring expert technicians and extensive manual handling to implement. The lens systems have other detractors as well. The plastics used to mold lenses and body elements are susceptible to damage from harsh cleaning solvents and the thermal extremes required for solder attachment of the camera assembly onto the products circuit structure [4]. To avoid exposure to solder process temperatures, many of these plastic lens products are assembled onto a dedicated substrate furnished with a connector or a flexible circuit extension for subsequent attachment to the primary substrate in the cell 'phone. Cameras developed for consumer products are extremely cost sensitive so when additional interface material and excessive labor is required for their manufacture, the price of the finished product is significantly impacted.

Advancing Camera Lens Technology

Every photonics application requires a custom optical solution to maximize its value and balance performance, unit size and cost. Tessera's Digital Optics Corporation (DOC) subsidiary in North Carolina is currently leading in the development, design and manufacture of micro-optic technology for an integrated optical sub-assembly solution for miniaturization of digital cameras for portable phones.

The DOC lens module is a unique composite of several refractive and diffractive layers manufactured using a silica based material in a wafer format. The lens units are produced using lithographic processes, similar to both IC manufacture and wafer-level chip scale packaging. The process begins with fabricating the multiple wafer size layers to furnish both refractive and diffractive optical elements. The wafers layers are coated with appropriate filters, apertures and anti-reflective films. The completed wafers are aligned, stacked and bonded together to create an array of hundreds of camera lens assembly units.

After bonding, the lens units are singulated from the wafer and made ready for inspection (see Figure 7) and forwarded onto the actual bonding to the glass surface of the sensor module. Because all of the Digital Optics wafer-to-wafer construction and back-to-front lens alignment is performed using lithographic registration, the 'block like' lens unit require no additional alignment steps [5].

Figure 7: Micro-optic lens unit.

Merging Optics and IC Package Technologies

Combining the attributes of Tessera's SHELLCASE® wafer scale technology for image sensor packaging and the Digital Optics wafer scale lens fabrication process, it is now practical to manufacture a camera module for the mobile phone market that is smaller, lighter, physically more rugged and, most significant, a lot cheaper than those used in earlier product generations. Because the micro-lens system requires no manual assembly or complex alignment process, the construction of the camera module can be a fully automated procedure [6].

The camera module assembly is quite simple because the lens site on the sensor, furnished with an optically clear adhesive, is ready to accept the micro-lens block. The lens blocks can be robotically aligned, placed onto the sensor unit and transferred to an oven for curing. The final vertically configured sensor/lens assembly (see Figure 8) is ready for attachment to a substrate interposer or, for extreme miniature product applications, directly onto the products primary substrate.

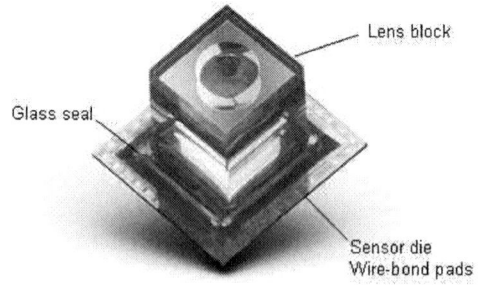

Source: Tessera

Figure 8: The µZ® camera module assembly.

Because the SHELLCASE/DOC camera units are one integral component, the assembly methodology for attachment is application dependent. The camera can be developed as a stand-alone component or attached to a substrate structure using conventional COB assembly. The substrate design may be dedicated solely to interfacing the camera to a host circuit board or include a number of passive elements to enhance performance, typical of that shown in Figure 9.

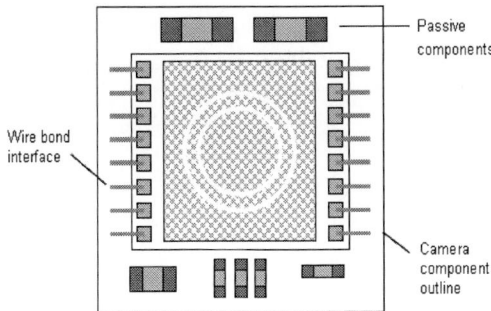

Figure 9: Wire-bond assembled camera

Addressing Future Market Demand

Hand-held communication and entertainment products continue to dominate the consumer markets worldwide and, with each generation offering more and more features and/or capability, system level integration and miniaturization becomes more of a priority. A key feature that has significantly impacted the portable phone market is the integration of imaging capability. The insatiable need to provide more functionality in wireless products while decreasing the products size posses an ongoing challenge to the IC packaging industry. To allow more space for the additional functions companies are combining number of related functions into a single package outline.

Combining logic and memory within a single package, however, poses a very real problem. The testing of these two functions is substantially different and will further compromise the level of confidence the OEM will have in its use. To ensure that the uncased bare die will meet quality and reliability criteria, some sort of electrical testing must be provided prior to board or package level assembly. The yield of ASICs, simple logic circuits, some processors and linear circuits, at some point, tends to stabilize but testing is the only way to guarantee quality and reliability.

Package-on-package (PoP) configurations are finding favor by both suppliers and users. The PoP process has proved to have less risk because the packages can be thoroughly tested before conversion to the stacked format. To facilitate the ideal PP camera module platform concept, the package assembly could include the processor, flash memory and related passive devices while maintaining a relatively short interconnect path as illustrated in Figure 10.

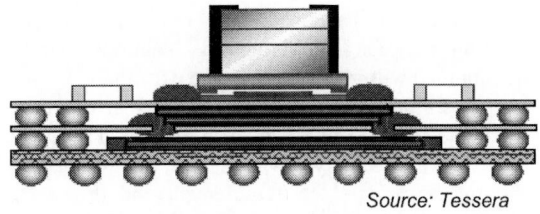

Source: Tessera

Figure 10: Vertically stacked µZ® camera module platform.

The vertically stacked package-on-package process can maintain a relatively low finished package profile, further enabling the developer to minimize functional area and reduce overall height of the module level assembly.

Conclusion

The glass seal protecting the sensor array is very economical because it is applied in mass while the sensor die remain in the wafer format. The DOC micro-optic lens system is similarly economical because it too is fabricated in mass using a fully automated process. Furthermore, it is a solid block of optic elements ready for mounting directly onto the glass sealed sensor elements. Most significant, the sensor and micro-optic lens block can withstand the elevated temperature exposure typically experienced during reflow solder processing. Traditional multiple, lens cameras, on the other hand, are bulky, heat sensitive and typically require a costly flex-circuit and connector for joining to the host module. In addition, with the micro-optic lens system, there is no manual alignment required, a procedure that is typical of conventional optical lens camera systems.

In regard to the glass sealed camera module, it will likely become the preferred image sensor package method, already tested and ready to join into the PoP configuration. The benefit gained by providing the SHELLCASE® CF glass seal is two-fold;

1. The package protects the delicate surface of the sensor throughout the die-bond and wire-bond process.

2. The package glass surface provides an ideal planer surface for attaching the miniature micro-optic lens unit.

Although a great deal of the paper focused on the new development for manufacturing miniature cameras for the next generation smaller, thinner and cheaper portable phones, vertically configured 3D packaging will play a major role in reducing the size and complexity of the substrate. This is due in part to the efficiency of the in-package circuit routing potential. Rather than relying on a complex high-density circuit board to provide the electrical interconnect between individually packaged ICs and their associated passive devices, most of the circuit routing between complimentary functions can be maintained within the multiple die package.

As far as logistical issues, whether or not to join one package to the other before or during the board level assembly process is a decision that may be influenced by the requirement for in-process configuration flexibility. In addition, the handling and testing of the base or logic portion of the package may require more dedicated fixture preparation, software and system support. Furthermore, the concern of ownership of total quality and reliability can be alleviated. The sections can be supplied as separate units and joined together at the board level assembly stage or furnished as a single package level product, tested and ready for PCB mounting.

If the decision is to join the sections at the board-level assembly, the base package can be placed onto the board and the additional package and camera module sections placed sequentially onto the mating contact matrix of the base for simultaneous reflow soldering. This alternative has two benefits. It allows the user to specify multiple variations (different camera resolution or memory functions) as well as efficiently accommodating secondary sources of supply.

References

[1] The Evolution of Digital Imaging: From CCD to CMOS *A Micron White Paper*

[2] Semiconductor and Packaging *A Prismark Report Summary, July 2006.*

[3] Humpston, G, Nystrom, N, Compromising Silicon Layout to Reduce the Cost and Form Factor of Solid State Cameras *Video/Imaging Design Line, December 2006*

[4] Chowdhury, A, Camera Module Assembly and Test Challenges, *Semiconductor International Magazine, February 2006.*

[5] Plyler, J, Morris, J, Kathman, A, Fabricating wafer-based micro-optics enables high bandwidth in a small package, *OE Magazine, January 2005.*

[6] Brady, D, Micro-Optics and Megapixels *OPN Magazine, November 2006.*

Note: Tessera's patented package assembly methodologies and fabrication processes are available only from authorized licensees of Tessera, Inc. 'SHELLCASE', 'µBGA' and 'µZ' are registered trademarks of Tessera, Inc. headquartered in San Jose, California.

Competitive Environment for 3D Semiconductor Assembly; Applications, Strategies & Cost

C. E. Bauer, H. J. Neuhaus

TechLead Corporation, Evergreen, Colorado, USA

F. Ciontu, NanoSprint, Grenoble, France

Abstract

The semiconductor industry only recently recognized the power of the third dimension in package and system assembly, yet the competitive environment already encompasses almost every aspect of electronic product manufacturing! The authors survey a wide range of application opportunities assessing the market, technology and business implications. Next a review of strategic technology options including chip stacks, origami and package on package (PoP) helps clarify the drivers for 3D semiconductor assembly development and deployment, and leads to a better understanding of potential matches between various 3D technologies and specific product applications. More in depth cost analysis provides a comparative assessment of most likely to succeed 3D technologies in real world markets. The paper then provides a well defined matrix of technical options vis-à-vis application opportunities considering technical functionality, manufacturability and cost expectations. Finally the authors review actual market trends in the development and deployment of 3D semiconductor assembly comparing existing trends with expectations derived from the selection matrix.

Introduction

The electronics industry's long history of compact assembly, originally driven by military and aerospace applications and more recently by portable consumer and business products, always employed three-dimensional (3D) assembly. However, 3D assembly of semiconductor chips only found wide spread adoption recently, yet the proliferation of technologies described since the early 1990's exceeds 3000 issued patents!

As early as 1979[1] origami type structures described by IBM indicated the potential and complex assemblies developed by the implantable medical electronics sector demonstrated the utility of 3D IC assembly for miniaturization. Prior to 1995 only the demand for higher density memory by relatively low volume workstations and main frame computers drove interest in 3D assembly, which the stacking of dual in line (DIP) packages and a variety of customized ceramic packages satisfied. But unit volumes remained low and costs high for these unique configurations and companies providing such products remained outside the mainstream of semiconductor assembly.

The advent of mobile telephony brought high volume applications for highly miniaturized packaging to the market and drove packaging technology starting in the early 1990s. Smart phones now drive demand for increased functionality and the consequent need for more, as well as more diverse, memory than conveniently available in single chip form factors. This drive for

low cost, high volume memory spurred the development of memory packaging combining SRAM and Flash memory devices into compact 3D package assemblies.

More recently the move from historical multi-chip module (MCM) and multi-chip package (MCP) configurations toward system in package (SiP) concepts facilitates the inclusion of more complex functionality into 3D chip structures where a wide range of devices combined into a single 3D package or package stack provide complex functionality adding significant value to the entire packaging food chain. Such assemblies provide functions such as global positioning systems (GPS), finger print readers, internet accessibility, etc. in very small standardized foot prints.

Applications

The authors previously analyzed the intellectual property landscape[2] for 3D semiconductor assembly technologies and discovered a strong relationship between the types of technologies under investigation at various companies and their product applications. Chart 1 displays the number of patents granted to such companies by the type of 3D semiconductor assembly technology described in the patents. The citation analysis presented in Chart 1 provides a reasonable indication of the level of industry interest in the specific technologies cited and thus their perceived applicability to real products.

More careful consideration provides insight into the perceived "best fit" applications for the various families of 3D semiconductor assembly techniques. Note that memory specialists such as Micron and Samsung appear to focus largely on stacked die approaches, including TSV technologies that provide complex memory capability and capacity in a single package, e.g. flash and SRAM. Telecommunications groups such as FreeScale and Texas Instruments appear to prefer stacked package technologies that allow integration of memory stacks with logic for miniaturized mobile phone functionality while maintaining the independence of intellectual property ownership and management. While TSV technology largely finds its home today in an academic environment (Princeton, Arkansas, etc.), implantable medical companies such as Medtronics, St. Jude and Boston Scientific also continue to pursue its development. Origami structures perhaps find the widest interest among companies, but the perfect niche remains elusive at this time.

Further refinement found through specific industrial needs analysis created Table 1 showing the market, technical function and business expectations of selected markets including the military & aerospace, implantable medical devices, mobile telephony and computer memory segments. These specific characteristics drive the compatibility, and therefore the selection, of 3D semiconductor assembly technology in each of these major market segments.

Three basic approaches to 3D assembly offer distinct benefits depending on the type of devices combined as well as the intended application. These include die stacking or stacked die packages (SDP), package on package or stacked package techniques (PoP), and Origami or folded flex packages.

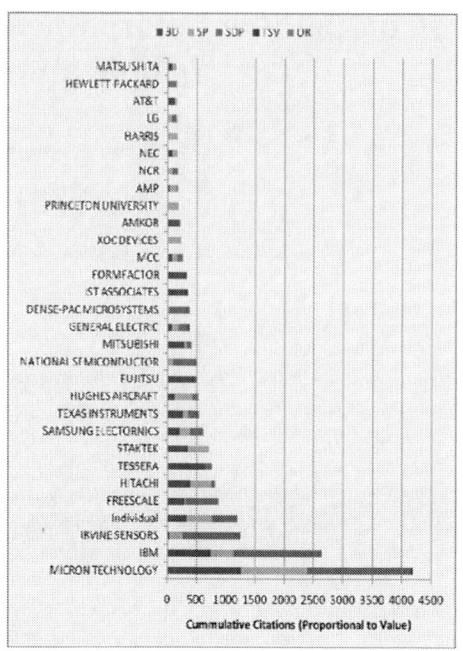

Chart 1: Citations/patent portfolio by technology type.

Technical Options

Stacked Die: Stacked-Die-Packages (SDP)[2-4] consist of bare stacked die interconnected using conventional wire-bond, flip-chip, or tape automated bonding (TAB) technologies. Spacers between die of the same size often add clearance for wire bonds. SDPs generally result in a very small footprint as well as a very low height profile. Low assembly cost results from the use of existing high volume package assembly infrastructure, but total cost can vary depending on the availability and cost of known good die, high yielding chips and/or infant failure susceptibility. Figure 1 illustrates a typical die attach/wire bonded Stacked-Die-Package.

Table 1. Market segment semiconductor packaging technology drivers.

	Market	Technical Functionality	Business
Military/Aerospace	Low/moderate volume Long product life	Low weight & volume High performance High reliability	Moderate cost sensitivity
Implantable Medical	Low/moderate volume Highly regulated	Low profile High reliability	Moderate cost sensitivity
Memory	High volume	Performance Density	High cost sensitivity
Mobile Phones	High volume Short product life	Minimal footprint Modular upgradeability	High cost sensitivity

A special subset of stacked die packaging employs through silicon via (TSV) Technology,[5,6] or holes created vertically through the die, to facilitate 3D assembly into a package. The TSV approach greatly

Fig 1: Stacked-Die-Package (STATSChipPAC).

reduces interconnect parasitic impedance (resistance, capacitance and inductance) improving overall electrical performance. TSV packaging also offers the potential for the smallest footprint and lowest height profile. Manufacturability for TSV structures remains relatively unproven at this point in time and the total cost implications continue to rely on the same KGD, die yield and infant failure considerations associated with all chip stacking approaches. A select few TSV structures may partially resolve the KGD issue by providing an intermediate testable form factor allowing both test and burn in of devices prior to stack assembly. Figure 2 illustrates a TSV package approach.

Origami: Origami or Folded-Flex packages[7,8] employ a flex substrate with multiple die sites. Following die assembly onto the flex base, typically using either die attach/wire

Fig 2: TSV Package (Samsung).

bond or TAB technology, and encapsulation at each site, the flex folds into a compact form arranging the die into a vertical configuration. The flex package substrate provides the wiring for die-to-die interconnection as well. And an adhesive between the vertically configured die holds the entire assembly together. While not strictly a stacked die packaging technology, origami packaging faces the same KGD, die yield and infant failure considerations associated with the true chip stacking techniques discussed above. Figure 3 illustrates an Origami Package.

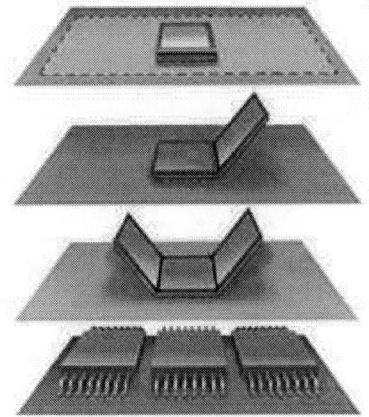

Fig 3: Origami Package (Tessera)

Stacked Packages: Stacked Packages (SP or PoP)[9,10] consist of stacked, packaged and pre-tested devices, either single chip or MCP, interconnected vertically by solder (either balls or joints), specialty lead frames or flex circuits. SPs often achieve a similar foot print to SDPs, but generally result in a higher profile. Many early SiP approaches employed similar strategies taking advantage of the larger foot print of MCM structures to accommodate vertical interconnections for the

Fig 4: PoP Package (STATSChipPAC)

modules. Since package test and burn in precedes final assembly the total yield for PoP technologies generally exceeds that for SDPs. Figure 4 illustrates a PoP Package.

Cost Analysis

Recently[11,12] the authors reported a methodology to accurately predict relative manufacturing costs among various alternative technologies. Using materials consumption and cost, cycle times and

Table 2: Technical Option Characteristics

	Technical Functionality	Manufacturability	Normalized Cost
Chip Stack	Small footprint Low height profile	Existing high volume infrastructure KGD dependant	1.00
TSV	Low parasitic impedance Small footprint Very low height profile	Existing IC fabrication infrastructure KGD dependant	2.43
Origami	Compact form factor Very low height profile	Existing high volume infrastructure KGD dependant	2.19
PoP	Modular partitioning Moderate height profile	Existing high volume infrastructure Test and burn-in precedes assembly	1.71

Table 3: Cost Analysis

	SDP	TSV	Origami	PoP
Device Cost/Pkg	13.03	13.03	13.03	13.02
Substrate Cost/Pkg	0.42	0.62	1.18	0.85
Assembly Cost/Pkg	0.20	0.06	0.20	0.23
Other Cost/Pkg	0.00	0.86	0.01	0.01
Total Cost	$13.65	$14.57	$14.42	$14.11
Packaging Cost	$0.65	$1.57	$1.42	$1.11
Normalized Packaging Cost	1.00	2.43	2.19	1.71

capital equipment requirements as inputs; capital, material, labor, and utility costs for each unit operation in a manufacturing flow result from the model calculations. Yield effects, forecast from established first-principle calculations at the unit operation level, ensure the model accurately represents the comparative total cost among the various technology options under consideration. Applied to a five chip package (one $5 control chip plus four $2 memory die) for packages employing Chip Stack, TSV, Origami and PoP technologies this methodology provides the relative cost comparison shown in Table 3. Total cost includes devices, substrates, assembly operations, special operations (e.g. TSV processing) and the value of chips and packages associated with yield losses. Packaging cost does not include the cost of the packaged devices but, like total cost, includes the die cost contribution due to any yield losses. Normalizing the packaging costs to the conventional Stacked Die technology provides a clearer picture of the relative cost structure among the four techniques; stacked die, TSV, origami and PoP.

While all device acquisition costs remain constant, PoP techniques result in a lower "effective" device costs since test and burn in precedes final assembly. Origami technology, with the highest substrate cost, endures a penalty because each device occupies a given area on the substrate in both the assembled and folded configuration. PoP results in roughly twice the substrate cost of SDP because each module (controller and memory stack) in the PoP assembly requires its own substrate. Assembly costs remain similar for all approaches except TSV where pre-assembly of the devices into a single component precedes final package assembly. Finally, significant costs associated wafer-level via formation and assembly operations drive TSV cost higher relative to the other 3D technologies.

For cases with more devices per package and/or lower individual device yield, the pre-assembly test and burn in of the PoP approach increases its attractiveness compared to the other three approaches which test only after complete package assembly.

Application Requirements versus Technology Characteristics

A quantitative comparison of market applications and technology alternatives involves scoring each technology characteristic (Table 2) for each market segment (Table 1). For purposes of consistency, +1 represents characteristics supportive of an application, -1 represents characteristics incompatible with the application, and 0 represents those characteristics with little or no perceived impact on the specific application. For each technology and application pair, the total scores listed in Table 4 help identify preferred technology to application matches. Shaded cells in Table 4 indicate the apparent best technology fit between 3D assembly technology and final product application market.

Table 4: Application Requirements to Technology Characteristics.

	Military/Aerospace	Implantable Medical	Memory	Mobile Phones
Chip Stack	+2	+2	+2	+3
TSV	+3	+1	-1	-1
Origami	+1	+2	0	+1
PoP	+0	+1	+3	+5

For example TSV satisfies many military applications due to very low parasitics, a very high level of miniaturization and the relatively low cost sensitivity in this market segment.

Market Trends vs Expectations

The military/aerospace market segment historically concentrated its efforts on stacked die packages with growing interest in PoP. Despite the apparent technical match this market segment demonstrates little interest in TSV. The implantable medical device segment splits its interest between origami and TSV, both providing the very low height profile that remains a primary driver for implantable medical devices. The memory segment, focused mostly on stacked die technologies, also demonstrates some interest in PoP. The mobile phone segment, while concentrated on stacked die packages, should definitely see a transition towards PoP as a more attractive solution to functional modularity and ease of device upgrading.

Summary & Conclusions

Potential fits between 3D packaging approaches and market segments have been identified and gaps between these fits and recent trends have been discussed.

A detailed cost analysis for 3D packaging including assembly and device yield effects has been developed and used to quantify the relative attractiveness of each approach in various market segments.

References

1. "Vertical semiconductor integrated circuit chip packaging," US Patent 4266282 filed 12 March 1979, International Business Machines Corporation.

2. "High-density electronic modules - process and product" US Pat. 4706166 - Filed 25 April 1986 - Irvine Sensors Corporation.

3. "Multichip module having a stacked chip arrangement" US Pat. 5323060 - Filed June 2, 1993 - Micron Semiconductor, Inc.

4. "Three dimensional interconnected integrated circuit" US Pat. 4801992 - Filed 1 December 1986 - Motorola Inc.

5. "Surface mount IC using silicon vias in an area array format or same size as die array" US Pat. 5973396 - Filed February 16, 1996 - Micron Technology, Inc.

6. "Three dimensional IC package module" US Pat. 6444576 - Filed 16 June 2000 - Chartered Semiconductor Manufacturing, Ltd.

7. "Forming folded-stack packaged device using vertical progression folding tool" US Pat. 7017638 - Filed 8 July 2002 - Intel Corporation

8. "Volumetrically efficient electronic circuit module" US Pat. 6862191 - Filed 19 August 2003 - Cardiac Pacemakers, Inc.

9. "Integrated circuit stackable package" US Pat. 4734825 - Filed 5 September 1986 - Motorola Inc.

10. "High-density electronic package comprising stacked sub-modules which are electrically inter-connected by solder-filled vias" US Pat. 5128831 - Filed 31 October 1991 - Micron Technology, Inc.

11. C Bauer, H Neuhaus, "Yield Based Manufacturing Cost Methodology for Advanced Packaging Technology Comparison," Proc. ICEP 2006, March 2006.

12. C Bauer, H Neuhaus, "Embedded Chip Build-Up in a Wafer-Level Packaging Environment," ECTC 2007, May 2007.

Effects of Underfill and Molding Compounds on Reliability of System in Package

Shan Gao, Jupyo Hong, Jinsu Kim, Seogmoon Choi and Sung Yi

Manufacturing Engineering R&D Institute, Samsung Electro-Mechanics
314, Maetan3-Dong, Yeongtong-Gu, Suwon, Gyunggi-Do, Korea 443743

Tel: 82-31-210-6619, Email: gao1.shan@samsung.com

Abstract

System in package (SiP) has the ability to integrate other components, such as passive component and antenna, into a single package to realize complete system functions. However, there are many electrical and mechanical reliability issues including the reliability issue for embedded structures. A mismatch of thermal coefficients of expansion among packaging materials and devices can lead to warping or delamination in the package. In this study, the effect of material properties of underfill and EMC, such as Young's modulus and CTE, are investigated through FEM simulation. In the FEM analysis, the warpage of the package, the maximum principle stress in the die and the maximum shear stress on the interface between the substrate, undefill and EMC's surface are considered. Experimental investigation on the warpage measurement of the package is carried out to verify the simulation results. In addition, some geometry parameters, such as the underfill's profile and EMC thickness, are also considered as the influencing parameters for the reliability of the package. Process optimization study, i.e., replacing underfill with EMC, is also carried out to improve the manufacturing process. The results show that the reliability of the system in package is closely related to the material properties and the geometry structures of underfill and EMC. The replacement of underfill with EMC improves the reliability performance of the package significantly. Results of this study provide a good guidance for the structure/process design and material selection when developing a SiP module.

Key words: SiP, Thermo-mechanical, reliability, warpage

Introduction

In the application product market, SiP is a good option for electronic products that feature low cost, small size, high frequency and high speed — especially the portable products, such as global positioning systems (GPS), Bluetooth, image sensors and memory cards. The power of SiP is the ability to bring together many ICs and package assembly and test technologies to create highly integrated products with optimized cost, size and performance [1].

Bluetooth is a standard technology for wireless connectivity of mobile computers, mobile phones, portable handheld devices and providing internet connectivity. It is based on a low-cost, low power, short-range radio link which cuts the cords used to tie up digital devices and therefore become the fast adopted technology in industry.

The system in package contains many components as shown in Fig.1. During the manufacturing process of the system in package, it has become susceptible to defects and internal residual stresses as dies, components, electric functionality and geometric complexity have increased. The mismatch of thermal expansion coefficient (CTE) among packaging materials and devices may lead to various failure modes during manufacturing processes, such as die broken, solder crack, substrate interface delamination. Die broken is mainly caused by the stress concentration of the principle stress in the die while the interface delamination is due to the overloading of shear stress on the interface among die, substrate, undefill and EMC. Many studies of the effect of underfill and EMC on the reliability of package have been performed [2-8]. However, limited studies have been found for the system in package.

Figure 1 A system in package for a BT module

In this study, the thermo-mechanical behaviors of a system in package for a bluetooth module have been investigated during reflow process. The finite element method (FEM) is used for the simulation analysis. The maximum principle stress in the die, the maximum shear stress on the interface between substrate and EMC and the warpage of the SiP package are considered. The effects of the material properties, process conditions and geometry parameters on the reliability issues are also investigated.

FEM Model

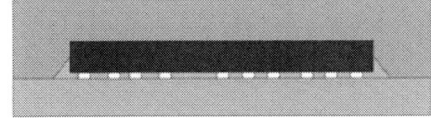

Figure 2 Geometry structure of BT package

2D plane strain finite element model of Bluetooth package are constructed as shown in Fig.2. This module is composed of a silicon die, EMC, underfill, lead free solder bumps and the FR4 organic substrate. The thermo-mechanical FEM analysis is carried out using Abaqus v6.6 [14].

Table 1 Material properties of packaging components used in the FEM analysis

Material	Young's Modulus (GPa)	CTE (ppm/°C)	Poisson's ratio	Tg (°C)
Die	130	2.6	0.27	
EMC	26(<Tg)	8	0.25	143
	0.38 (>Tg)	30		
Underfill	7.8 (25 °C)	32	0.33	137
		100		
Substrate	23.3	13.5	0.17	
Solder	56	20.04	0.35	

The elastic material properties for the die, EMC, underfill, solder and substrate materials are considered and given in Table.1. The temperature dependent Young's modulus of underfill material is considered. However, only the property at 25°C is shown in the table. CTE values for all the packaging mateirals are measured with TMA and Tg and Young's modulus of EMC, Underfill and substrate material are measured with DMA.

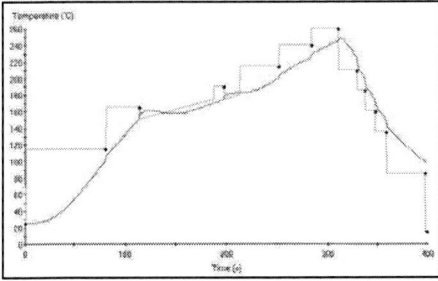

Figure 3 Temperature profile in reflow process

N_2 reflow profile is considered (shown in Fig.3). The peak temperature of reflow process is 245°C. The stress-free temperature of the package is assumed to be the peak temperature of reflow process. After the reflow, the package is cooled to room temperature. This study focuses on the thermo-mechanical behaviors of the package after cooling.

Results and discussion

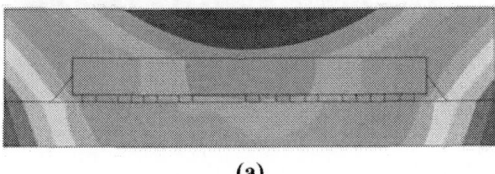

(a)

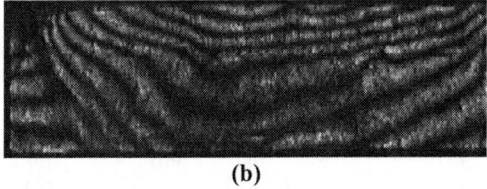

(b)

Figure 4 Warpage distribution of Bluetooth package after reflow process: (a) FEM simulation result (b) shadow moiré observation

The present FEM model needs to be verified before it can be applied for the further analysis. The warpage of the package after reflow is calculated from FEM simulation with material properties shown in Tab.1 and measured by Moire interferometer, respectively. Smile faced warpage shape is observed (shown in Fig.4). Comparing these two graphs, the simulation results are quite comparable to the experimental observations, e.g. the warpage direction, shape and the fringe pattern. The simulation results match well with the experimental ones. Therefore, the model is validated and can be used in the parametric study.

Standard model analysis

The analysis with material properties given in Tab.1, which are used in the current manufacturing production line, is considered as a benchmark. This analysis is to be used as a reference for the following material and process optimization. Since the two main failure modes of Bluetooth package are die broken and interface delamination, the stress distribution of die and substrate are considered. Fig.5 shows the principle stress distribution of the die and shear stress distribution on the substrate, respectively. The Maximum Principle Stress (MPS) lies on the bottom of the die where it contacts with the underfill and solder, especially at the corners where the three kinds of materials are contacted. The Maximum Shear Stress (MSS) on the substrate lies in the area where it contacts with underfill material.

These phenomena coincide with the observation of the failed Bluetooth package.

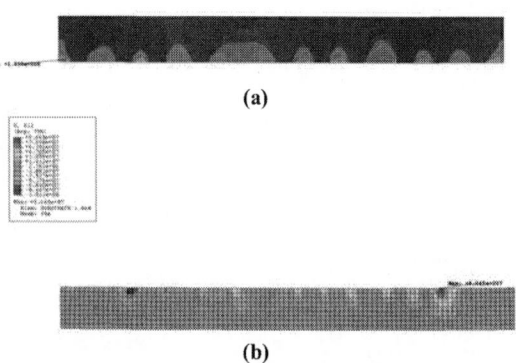

(a)

(b)

Figure 5 Stress distribution of package components after reflow: (a) Maximum principle stress of die (b) Shear stress on the substrate

Effect of Underfill

The effects of material properties, such as CTE and Young's modulus of underfill, have been investigated in this analysis. These values are artificially varied to analyze the warpage and stress trends compared with the benchmark material groups used in the current manufacturing process.

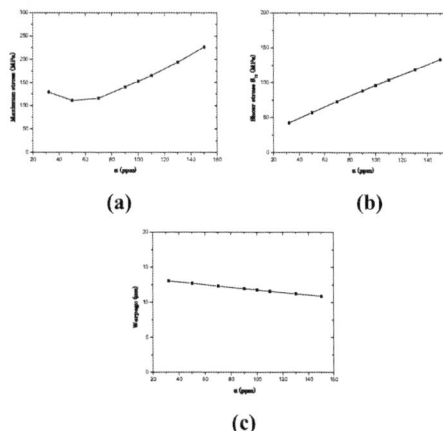

(a)   (b)

(c)

Figure 6 Relationship between the CTE of underfill and the reliability of package: (a) MPS in the die (b) MSS on the substrate (c) Maximum warpage of package

Fig.6 shows the effects of underfill CTE on the reliability of the Bluetooth package with (a) MPS in the die Vs CTE, (b) MSS on the substrate

Vs CTE and (c) Maximum warpage Vs CTE. When CTE of underfill is higher than 50ppm/°C, MPS in the die increases with increase of CTE. However, MPS in the die decreases with increase of CTE when underfill CTE is lower than 50ppm/°C. The minimum MPS is achieved when underfill CTE is 50 ppm/°C. MSS on the substrate increases monotonically while the maximum warpage of the package decreases with the CTE increase.

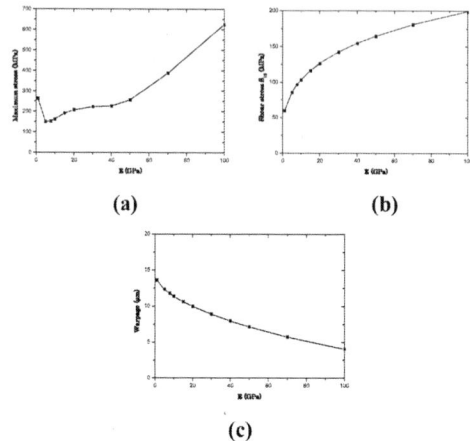

(a)   (b)

(c)

Figure 7 Relationship between the Young's modulus of underfill and the reliability of the package: (a) MPS in the die (b) MSS on the substrate (c) Maximum warpage of package

Fig.7 shows the effects of underfill Young's modulus on the reliability of the Bluetooth package with (a) MPS Vs Young's modulus, (b) MSS Vs Young's modulus and (c) Maximum warpage Vs Young's modulus. MPS increases with increase of underfill Young's modulus when Young's modulus is above 5GPa. MPS in the die decreases with increase of underfill Young's modulus when Young's modulus is below 5GPa. The minimum MPS in the die is achieved when underfill Young's modulus is 5 GPa. MSS increases while maximum warpage of package decreases monotonically with increase of Young's modulus of Underfill (see Fig.5).

Effect of EMC

The material property effect of EMC has been investigated. Fig.8 illustrates the relationship between the CTE of EMC material and MPS in the die, MSS on the substrate and maximum warpage of the package, respectively. The minimum MPS is achieved when CTE of EMC is 14 ppm/°C, which is close to the CTE of substrate. The MPS in the die increases with CTE of EMC increase when its value is above 14 ppm/°C. MPS decreases with CTE of EMC increase when it is below 14 ppm/°C. Both MSS on the substrate and the maximum warpage of the package increase with crease of EMC CTE (see Fig.7).

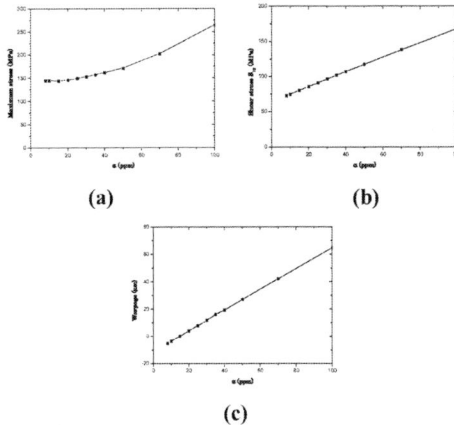

(a)   (b)

(c)

Figure 8 Relationship between the CTE of EMC and the reliability of the package: (a) MPS in the die (b) MSS on the interface (c) Maximum warpage of package

Young's modulus effect on the reliability of the Bluetooth package is considered. Fig.9 shows the relationship between Young's modulus of EMC material and MPS in the die, MSS on the substrate and the maximum warpage of the package. The MPS in the die increases with Young's modulus of EMC increase when its value is above 20MPa. MPS decreases with Young's modulus of EMC increase when it is below 20 MPa. The minimum MPS is achieved when Young's modulus of EMC is at 20 MPa. Both MSS on the substrate and the maximum warpage of the package increase with crease of EMC Young's modulus (shown in Fig.8).

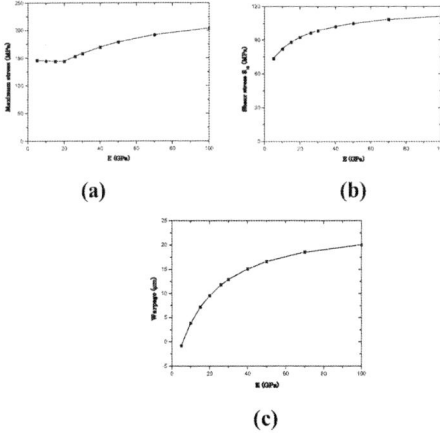

(a)   (b)

(c)

Figure 9 Relationship between the Young's Modulus of EMC and the reliability of the package: (a) MPS in the die (b) MSS on the interface (c) Maximum warpage of package

Effect of Geometry Parameter

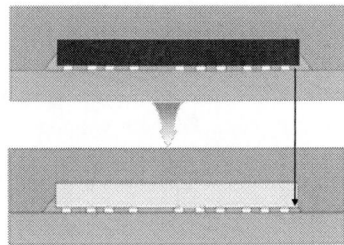

Figure 10 Fully underfilled and partly underfilled package

The effect of geometry parameters is considered. The underfilling process depends on the dispensing time, capillary performance of underfill mateiral as well as the geometry structures, such as stand off, layout of solder bumps, etc. [9-12]. During underfilling process, the gap between the die and substrate can not always be fully filled with the underfill material. Sometimes it is only partly underfilled. Fig.10 schematically illustrates the geometry variations for the fully underfilled and partly underfilled package models. It is necessary to investigate whether the reliability of the package is sensitive to the underfill's geometry. In addition, it also needs to be found out how much effect this geometry variation has on the reliability of the Bluetooth package.

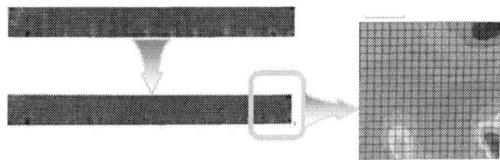

Figure 11 Maximum Principle stress of the die with partly underfilled package model

Table 2 Comparison of maximum stress and warpage between the two underfill profiles

Geometry Type	MPS in die (MPa)	MSS on the substrate (MPa)	MW of the package (um)
Fully U/F	126.1	153	11.81
Not fully U/F	151.5	153	11.658

Fig.11 shows MPS distributions of the die in the fully underfilled and partly underfilled package, respectively. It can be seen that the stress concentration area has changed from the corner, where die, underfill and solder contact in the fully underfilled package, to the area, where the die contacts with the boundary of underfill. Partly underfilled structure changes the stress distribution and makes the transition of stress concentration zone. Tab.2 summarizes the values of MPS in the die, MSS on the substrate and maximum warpage of package in these two geometry structures. It can be seen that the underfill geometry increases the MPS significantly while MSS has no change and MW only changes a little bit. This is due to the reason that underfill is in contact with die while it has no contact with substrate. The low volume of underfill material compared with EMC material results in the tiny effect of geometry changes in underfill on the warpage variation of the package.

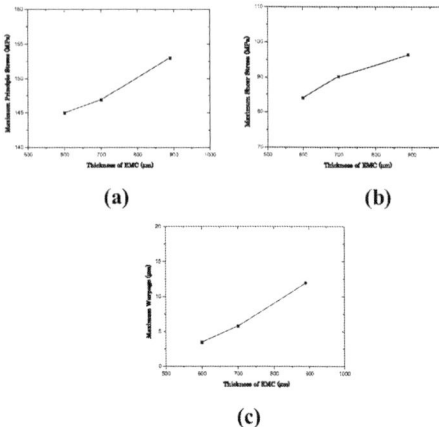

(a)   **(b)**

(c)

Figure 12 Relationship between the EMC thickness and the reliability of the package: (a) MPS in the die (b) MSS on the interface (c) Maximum warpage of package

The thickness of the EMC is considered. Epoxy molding compound is used to protect the chip and passive components from the environment, mechanical damage, and to provide structural support and electrical insulation for the package [1, 13]. The development trend of SiP is to make it slimmer and smaller. One of the slimming methods is to reduce the thickness of the EMC of the package. However, this thickness reduction may cause reliability issue. It is necessary to investigate what effect the EMC thickness has on the reliability of package. Fig.12 shows the relationship between the thickness of EMC and MPS in the die, MSS on the substrate and the maximum warpage. They all decrease with the EMC thickness decrease. Therefore, it is an effective way to improve the reliability of packaging by reducing the thickness of EMC. However, the reduction of EMC thickness may weakens the impact reliability of the package on mechanical shock. This issue is not covered in this paper.

Process Optimization

One of the objectives in package design is to minimize the thermal stresses/warpage, and to improve the reliability of the package. To this end, process optimization study is carried out. A new process, which uses EMC to replace underfill, is proposed for investigation. Together with the standard process, these two processes are numbered by

1. Fill the gap between die and substrate with Underfill (Standard)
2. Fill the gap between die and substrate with EMC

Table 3 Comparison of maximum stress and warpage between the two processes

Case #	Maximum Warpage (μm)	MPS in the die (MPa)	MSS on the substrate (MPa)
1	25.51	228.8	81.87
2	29.85	187.2	52.28

Tab.3 summarizes the results of FEM analysis. Compared with the current used process, the new process reduces the MPS in the die and MSS on the substrate significantly, MPS by 20% and MSS by 35% respectively. The warpage will increase when EMC is used to fill the gap between die and substrate. The results mean that replacing Underfill with EMC will not only save a manufacturing process (underfilling), but also improve the reliability of the package significantly. However, this replacement requires higher quality of EMC (smaller filler size) which will increase the material costs of the package.

Conclusion

Thermal stress and warpage arise in the Bluetooth package after reflow process due to the CTE mismatch between the components of the package. Failures of the package are mainly caused by the stress concentration in the die and on the substrate interface or the over values of warpage of the package. CTE and Young's modulus of underfill and EMC materials have significant effects on the maximum stress as well as the maximum warpage of the package. The geometry parameters, such as shape of underfill and EMC thickness also have great effect on the reliability of the package. The new process, in which underfill is replaced by EMC, can improve the reliability effectively.

References

[1] R.R. Tummala, 'Microsystems Package', 2001, McGraw-Hill NY.
[2] D. Maslyk, M. Privett, and B. Toleno, 'Using Underfill to Enhance Lead-free Drop Test Reliability', 2006, SMT, pp. 20~23
[3] G. Chen and X. Chen, 'Finite element analysis of flexBGA reliability', 2006, Soldering & Surface Mount Technology, vol.18, No.2, pp. 46~52

[4] Z. Cheng, B. Xu, X. Cai, W. Huang, and X. Xie, 'Underfill Delamination Analysis of Flip Chip on Low-Cost Board' 2001, EMAP

[5] H. Jiun, S. Krishnan, Jatinder, A. Aripin, G. Omar, 'Some Aspects of Materials Selection and Overhang Die Bonding for Multichip Module GFN Package', 2004, International Conference on Electronics Materials and Packaging, pp. 301~308

[6] S. Yi, P. D. Daharwal, Y.J. Lee, and D.R. Harkness, 'Study of Low-modulus die Attach Adhesives and Molding Componds on Warpage and Damage of PBGA', 2006, Electronic Components and Technology Conference, pp. 939~945

[7] J. Lu, J. Wu, Y.P. Lew, T.B. Lim, and X. Zong, 'The impact of underfill properties on the thermo-mechanical reliability of FCOB assembly', 2003, Soldering & Surface Mount Technology, pp. 37~41

[8] T.H. Wang, Y. Lai, J. Wu, 'Effect of underfill thermo-mechanical properties on fatigue life prediction of flip-chip BGA', 2003, EMAP, pp.196~202

[9] Y.C. Chia, H.S.Yam, S.H. Lim, K.S. Chian, S. Yi, W.T. Chen, 'An optimization study of underfill dispensing volume', 2003, IEEE Transactions on Electronics Packaging Manufacturing, Vol.26, No.3, pp. 205~210.

[10] A.A.O. Tay, Z.M. Huang, J.H. Wu, C.Q. Cui, 'Numerical simulation of the flip-chip underfilling process', 1997, IEEE/CPMT Electronic Components Technology Conference, pp263-269.

[11] W.H. Loeng 'Development of an underfill process for dense flip chip application', 1996, IEEE/CPMT International Electronics Manufacturing Technology Symposium, pp10-17.

[12] M. Edward, 'Some factors that affect performance of capillary flip chip underfill', 1998, 4th International Symposium on Advanced Packaging Materials, pp21-28.

[13] S.L. Liu, G. Chen, M.S. Yong, 'EMC characterization and process study for electronics packaging', 2004, Thin Solid Films, pp. 454~458

[14] Abaqus User Manual.

Via Fabrication Techniques for High Density Vertical Interconnections

Zsolt Illyefalvi-Vitéz; László Gál; Olivér Krammer; János Pinkola

Department of Electronics Technology, Budapest University of Technology and Economics,

Goldman t 3, Budapest, Hungary, 1111

Phone: +36 1 4632740, Fax: +36 1 4634118; E-mail: illye@ett.bme.hu or gal@ett.bme.hu

Abstract

The performance of electronic products is moving towards higher functionality and faster operation. Vertical interconnect technology provides smaller interconnect lengths, thus reduces parasitic effects and decreases propagation delays, therefore improves performance in high frequency range. The key technology for vertical interconnects is through-board or even through-wafer via fabrication, which consists of three process steps, namely via generation (drilling or window opening), metallization (conductive layer deposition to the wall or via filling) and contact preparation to the conductive surface pattern. The experimental work discussed in this paper deals with the latter two process steps, in particular with via filling and contacting, which processes in some cases can be carried out in a single step. Test boards were designed and samples were fabricated to compare different via filling and contacting technologies. The comparative analyses, the applicability and the criteria of selection of the examined via fabrication techniques are summarized and presented..

Key words: Interconnection technologies; 3D packaging; process optimization; manufacturing technologies; quality characterization.

Introduction

The main motivation in the research and development activities in electronics is to improve the performance of electronic products, in particular to move them towards higher functionality and faster operation. All industrial performers anticipate that improved performance can be assured by higher packaging density. Feature-rich cell phones, pocket PCs, broadband and visual info communication devices, digital cameras, and other handheld, mobile and portable consumer products require maximum functional integration, including memory, DSP, ASIC, RF, MEMs, and other devices, in a package with the smallest footprint and lowest profile available.

Application of vertical interconnects in packaging may satisfy these needs, which require more compact packaging solutions and efficient area utilization. Vertical interconnect technology provides smaller interconnect lengths resulting in less parasitic effects and smaller propagation delays, therefore better performance in high frequency range. The key technology for vertical interconnects is through-board or through-wafer via fabrication.

Outline of the Experimental Work

Via fabrication consists of three process steps, namely via generation (drilling or window opening), metallization (conductive layer deposition to the wall or via filling) and contact preparation to the conductive surface pattern. The experimental work discussed in this paper deals with the latter two process steps, in particular with via filling and contacting, which processes in some cases can be carried out in one single step.

Test boards were designed and samples were fabricated to compare different via filling and contacting technologies. The holes were always mechanically drilled and with no metallization on the walls. Via filling included the insertion of pins or pressing conductive paste, with the use of different materials and process combinations. The contacts were made either by conductive layer deposition using electroplating, electroless and immersion coating, or soldering; or by exploiting the inherent contacting properties of the pastes.

The following parameters are tested, examined and analyzed on the samples:

- the quality of the contacts (electric resistance) between the two interconnected parts, e.g. between the pads and the cured ink,
- the perfection of via filling as well as the resulted geometry of the filled vias and the caps on the ends, by X-ray inspection and cross-sectioning,
- the perfection of curing, i.e. the materials properties of the vias and the contacts (by SEM analysis),
- the corrosion resistance (reliability, life time) of the same contacts tested by TH (temperature-humidity), thermal shock and other accelerations.

The Via and the Surrounding PCB Structure

The process to design the experiments began by the development of test vehicle concept that is the definition of the simple geometric structures of the via types (micro connector pins or components) and the surrounding Printed Circuit Board (PCB) layouts. The basic structures are characterized below together with some remarks regarding fabrication and test strategy.

The major objective of the project was the fabrication of highly robust and reliable printed circuit boards using embedded micro connector or micro component injection (EMCI) to serve as vias, for the benefit of the PCB manufacturing industry, with a special consideration of environmental and cost cutting factors. The EMCI process together with PCBs of microvias using a traditional printed wire patterning process sequence is requested to establish a robust and reliable High Density Interconnect (HDI) EMCI board technology. The EMCI boards are developed to be fully compatible with SMD technology, as well. Micro component implantation using advanced injection methods would create the possibility of a 3-D circuit board space.

The steps of the insertion process are illustrated in Figure 1, in particular regarding the geometries before and after the insertion process. All modeling, simulation and experimental investigations aimed at the determination of these geometric parameters in connection with each other, taking into consideration the materials properties, as well as, the possibilities and limitations of the injection process.

Figure 1 shows that the process sequence starts with a double-sided patterned printed circuit board, and a suitable via-hole is drilled into the board before injection. The most important dimensions are also presented in the figure, as follows:

b = board thickness

t = thickness of the copper layer on the copper clad laminate

p = pitch of the pattern and via arrangement

d_1 = diameter of the pad, where hole is drilled and micro connector inserted

w = width of traces of the PCB pattern

$d_{2before}$ = diameter of the via-hole as it is drilled into the board

d_{2after} = diameter of the EMCI via after injection

To have an impression about the resulted structure, a via made by the insert & press injection process is illustrated in Figure 2.

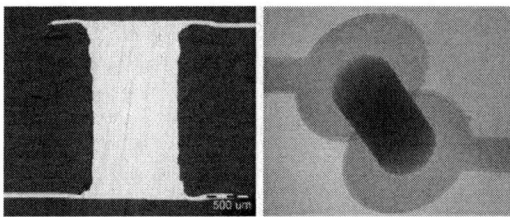

Figure 2: Cross Section and X-ray Image of a Copper Pin Injected by Insert & Press Process

Electrical Performance and the Injection Process

The final geometry and the electrical performance of an injected via are highly dependent on the drilling and the injection processes. Regarding the electrical performance, the resistance of a track containing a via with connecting pads and traces (see Figure 1) was selected for evaluation by simulation and measurement.

Drilling can be carried out by the traditional mechanical process or by the application of a sequence of laser spiraling and punching. When mechanical drilling is applied, a through hole can be drilled, which goes through both copper layers and the isolating board. The tip of the drilling tool usually smears some melted epoxy or phenol resin onto the inner cylindrical rings of the copper layer, in particular onto that of the lower layer. Oxidation can also take place on the copper surface. Both effects may result in highly resistive or isolating layers, which cause contact problems.

An important design parameter to be determined is the ratio (or difference) of the drilled diameter of the via-hole ($d_{2before}$) and its diameter (d_{2after}) after the injection process. The latter diameter is mainly determined by the diameter of the inserted pin, in particular when the applied pin

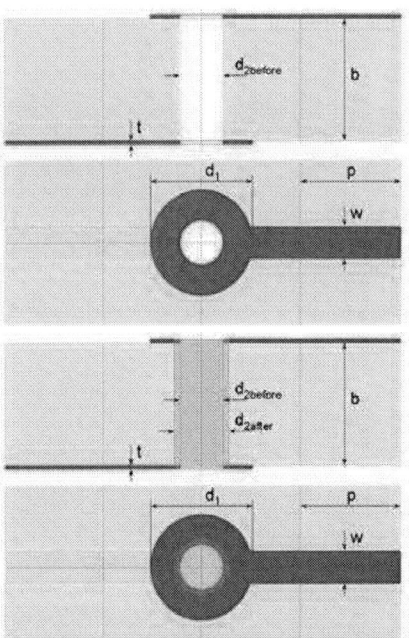

Figure 1: The Insertion Process: Drilled Hole before Insertion (upper), the Inserted Pin (lower)

material is hard. It is assumed that $d_{2before} < d_{2after}$, and a close fit type joint takes shape, however the selection of the optimum (d_{2after} / $d_{2before}$) ratio or (d_{2after} - $d_{2before}$) difference is essential. If the difference is small or if it is negative, a clearance fit type joint comes about, and the contact resistance would be high and poorly defined, increasing during the lifetime. On the other hand, if the difference is too large, then the entering pin may tuck the edge of the layer pad in the hole at the entering side of the board, or even can tear the pad off the board at the leaving side. At both sides this can result in breaking of the conductive tracks and can cause the damage of the dielectric properties of the surrounding board material.

The contact formation is highly influenced by the injection process and the pin material as well. For the insert & press injection process finally a pin, whose faces are flat at its both ends, is used. The dimensions of the pin are properly determined to have a diameter with clearance to the diameter of the predrilled hole, and a length, which is a little bit longer than the width of the board. The proper volume of the pin is essential; it should be slightly bigger than the volume of the pre-drilled hole, in a suitable measure. A soft, e.g. pure copper material of the pin is assumed. An anvil, which stops the pin at the leaving side of the board, is applied. Thus the entering pin, when it is stopped by the anvil, clenches into the hole, fills the entire volume up, presses against the wall of the pads and forms sufficient (or insufficient) compression contacts.

When there is no clearance but a slight overlapping of the hole and the pin diameters is applied, then the press & fit technology is occurred. It means that a little bit larger pin is inserted into the hole with applying a delicate pressure, and thus the pin properly fit into the hole. Using the press & fit technology, not only pins but MELF components (Figure 3), or even surface mount (SM) components with proper dimensions can also be inserted.

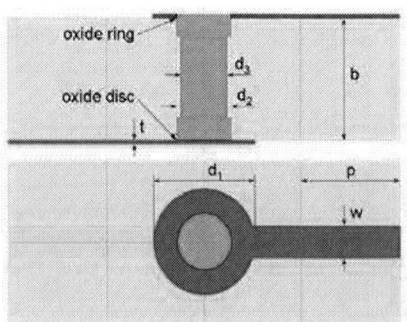

Figure 3: Injected MELF Component in a Board

According to theoretical considerations and preliminary results, the following design rules for dimensions and materials properties of pins and boards were determined:

- Apply standard board dimensions and materials
- Apply standard diameters of the via-holes with the dimensions from 0.4 to 1.0 mm, by 0.05 mm steps; 0.4 and 0.8 mm diameters are preferred. The deviation of the diameters from the nominal value should be low, max ±1% is allowable.
- Determine and use proper diameter and length of the pins. For insert & press (riveting) the diameter of the pin should be about 2 % smaller than the diameter of the hole, with a maximum deviation of ±1%. For the press & fit technology a 5 % larger diameter of the pin than the hole is proposed.
- Apply an additional contacting process (electroplating of copper, PCB surface finishing and lead-free soldering) after insertion to complete the injection process.
- Include a new type of pin insertion by the application of Transient Liquid Phase Sintering (TLPS) paste materials in the investigations.
- Exclude the application of conductive polymer pastes.

Design of Experiments

Tests and experiments were focused on the following process and materials combinations applying both copper pins, as well as, SM and MELF components. See the table below (Table 1) for an overview of the tested combinations.

Processes, which were tested (6 variants):
- Two insertion processes: insert & press or press & fit, always continued by contacting;
- Three contacting processes: copper electroplating; or surface finish & solder paste reflow; or Ormet ink applying & curing.

Components, which were tested:
- Soft copper wire D=0.80±0.02 mm, cut to L=1.75±0.05 mm;
- KEKON soft copper pins made by different technologies;
- KOA CC10 zero ohm MELF resistors;
- KOA RN41 0805 metal film MELF resistors, dimensions: L=2.0±0.1mm; D=1.25±0.05mm (D=max1.3mm);
- Standard 0603 SM (Surface Mount) resistors and capacitors, with zero ohm for test, requested values for demonstration;

Selected pin materials and insertion process combinations:
- Ormet ink: pressure filling with Power Squeegee, drying & curing;
- Ormet ink: stencil printing together with the head on both sides, drying & curing;
- Copper wire pins: insert & press, with optimum hole/wire dimensions;
- Melf (Metal Electrode Leadless Face): resistors or other components: press & fit;
- SM (Surface Mount) components (zero ohm resistors or functional resistors, capacitors or

diodes) insertion vertically by press & fit technology.

Selected contact producing processes:
- Ormet ink: stencil printing, drying & curing;
- Silver and Ormet ink: immersion Ag deposition, Ormet ink stencil printing, drying & curing;
- Lead-free solder: stencil printing & reflow;
- Silver and solder: immersion-silver deposition, solder paste stencil printing & reflow;
- Copper and silver: copper electroplating and immersion-silver deposition;
- Copper, silver and solder: copper electroplating, immersion-silver deposition, solder paste stencil printing & reflow.

Further variants
- Wire diameter: 0.4 or 0.8 mm
- R diameter: 0.9 or 1.3 mm
- Stencil aperture diameter: 0.8; 1.3 or 1.8 mm

Table 1: Tested Pin/Contact Variants at a Glance

Via type / Contact	Ormet-PS	Ormet-SP	Cu wire	Melf R	SMD R
Ormet	Test	Test	Test	Test	Test
Ag + Ormet	N/A	Test	Test	N/A	N/A
Solder	Test	N/A	Test	N/A	N/A
Ag + Solder	N/A	N/A	Test	Test	Test
Cu + Ag	N/A	N/A	Test	N/A	N/A
Cu+Ag+Solder	N/A	N/A	Test	N/A	N/A

Test Methodology: Design of Test Vehicles

The EMCI Testboards (Figure 4) were designed to measure the contact resistance of EMCI vias by four wires measuring method. The vias are connected serial in a daisy chain. Top and bottom sides of vias are leaded out one by one to the edge connector to measure the voltage drop on each via separately.

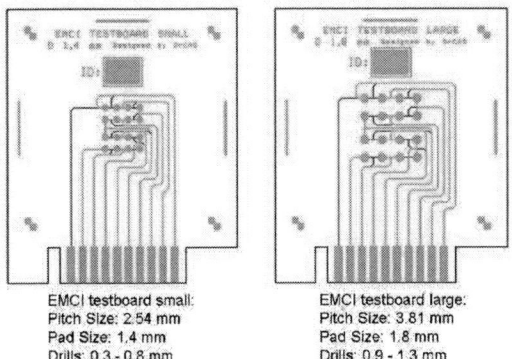

EMCI testboard small:
Pitch Size: 2.54 mm
Pad Size: 1.4 mm
Drills: 0.3 - 0.8 mm

EMCI testboard large:
Pitch Size: 3.81 mm
Pad Size: 1.8 mm
Drills: 0.9 - 1.3 mm

Figure 4: Design of the EMCI Testboards

The EMCI Measure Board (Figure 5) was designed and applied for the easy selection and measurement of the EMCI vias produced in any of the EMCI Testboards connected to the Measure Board by the edge connector. A pin array with 16+17+16 pins in three lines and two jumpers are used to select a single via for test. By setting the jumpers on the EMCI Measure Board it is possible to measure the contact resistance of the EMCI vias one by one, and in addition it is feasible to measure the contact resistance of multiple connected vias in any combination or the summary of the contact resistances as well. If a break occurred at any pin of the daisy chain, the resistance of all other pins still can be accessed and measured by the two wire method, although with lower accuracy.

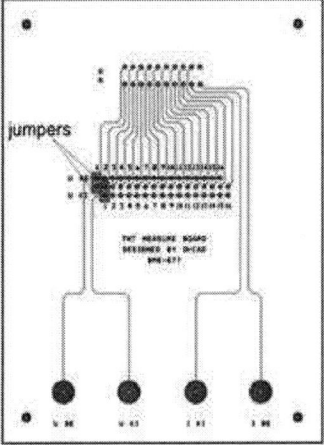

Figure 5: The EMCI Measure Board

In addition to the test of quality of contacts by measuring the electric resistance between the two interconnected parts, using the test and measure board, the following parameters are tested, examined and analyzed on the samples:
- the perfection of via filling and the resulted geometry of the filled vias as well as the caps on the ends (by X-ray inspection and X-sectioning),
- the perfection of curing or solder reflow, i.e. the materials properties of the vias and the contacts (by SEM analysis),
- the corrosion resistance (reliability, life time) of the same contacts tested by TH (temperature-humidity), thermal shock and HAST (highly accelerated stress test) acceleration.

Comparative Analysis of Different Insertion and Contacting Sequences

In the followings, Figures 6-14 are presented to characterize the results of the injection process using different insertion and contacting sequences. The injection sequences were selected according to the design of experiments, summarized in Table 1.

The figures are composed of photos, X-ray images and tables. Each figure is made on the basis of the results of an EMCI Test Board (see Figure 4), that test board was fabricated with a specific injection sequence. The applied sequence, including the type of the inserted micro connector or component and the contacting process, is described by the figure caption. In each figure, one can find the followings:

- the identity number of the board, from which the photos are taken;
- the measured total, average, maximum and minimum contact resistances of the 14 vias of the same board;
- the sum of the resistances of the other 1 to 5 test boards, which are fabricated with the same injection sequence;
- a one-word qualification of the analyzed and presented injection sequence.

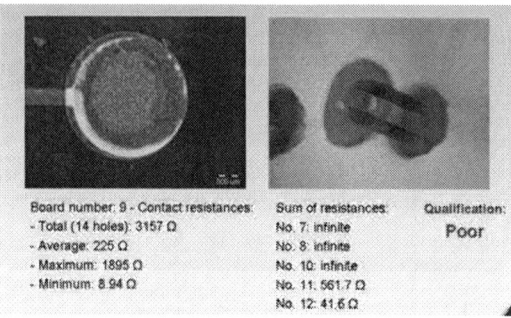

Figure 6: Inserted component: zero ohm 0603 SM Resistor to ∅ 0.9 mm hole, contacted by Stencil Printed Ormet Ink, aperture ∅ 1.8 mm

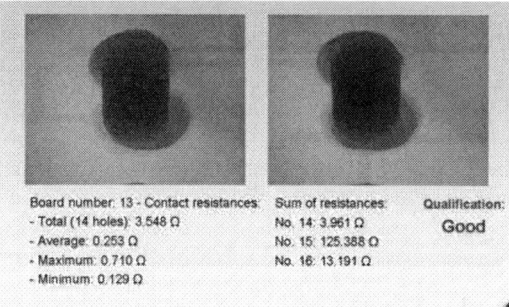

Figure 7: Inserted component: Ormet Ink filled by stencil printing to ∅ 1.3 mm hole, contacted by Stencil Printed Ormet Ink, aperture ∅ 1.8 mm

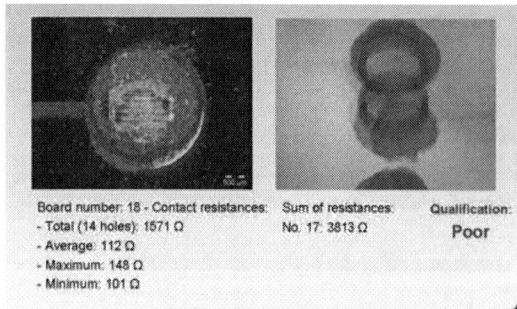

Figure 8: Inserted component: 100 ohm MELF Resistor to ∅ 1.3 mm hole, contacted by Stencil Printed Ormet Ink, aperture ∅ 1.8 mm

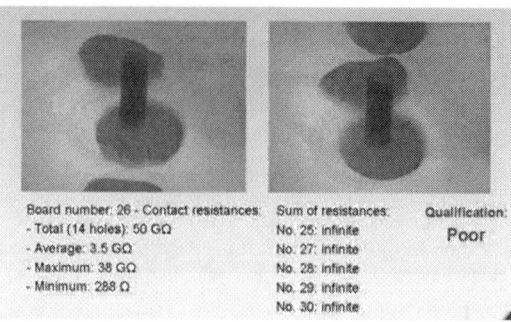

Figure 9: Inserted component: Ormet Ink filled by Power Squeegee to ∅ 0.4 mm hole, contacted by Stencil Printed Ormet Ink, aperture ∅ 1.3 mm

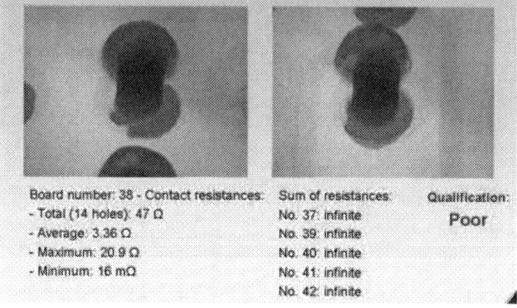

Figure 10: Inserted component: Cu Wire insert & press to ∅ 0.8 mm hole, contacted by Stencil Printed Ormet Ink, aperture ∅ 1.3 mm

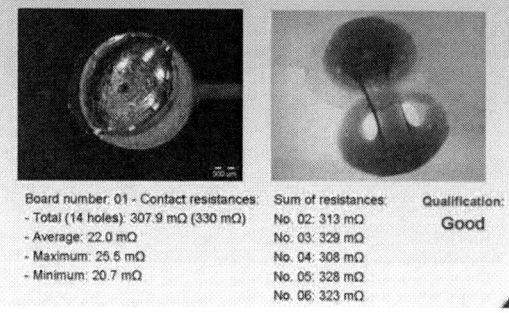

Figure 11: Inserted component: zero ohm 0603 SM Resistor press & fit to ∅ 0.9 mm, contacted by lead-free solder (SP ∅ 1.8 mm) and reflowed

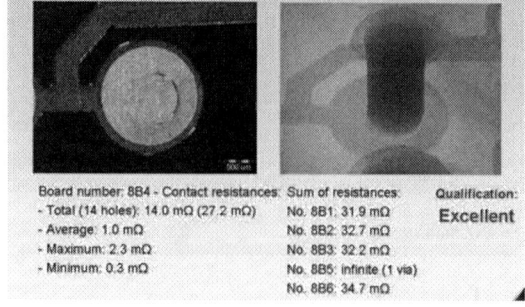

Figure 12: Inserted component: Cu Wire insert & press to ∅ 0.8 mm, contacted by Electroplated 50 μm thick Copper and imm-Ag surface finish

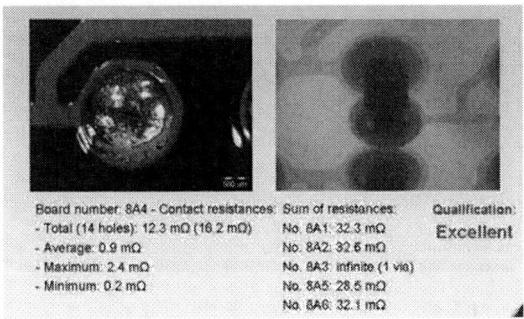

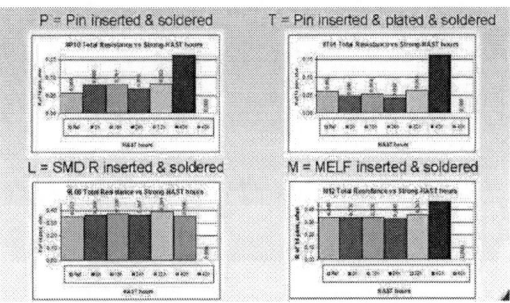

Figure 13: Inserted component: Cu Wire insert & press to ∅ 0.8 mm, contacted by Electroplated 50 µm thick Copper and imm-Ag surface finish, plus lead-free solder (SP ∅ 1.3 mm) and reflowed

The verification of the design rules was prepared by the analyses and the comparison of ca 4000 photos, 1500 X-ray images, 2000 resistance values, and lots of other inspections, which are not recorded, neither presented here.

Reliability and Accelerated Lifetime Test

The reliability of the vertical interconnect vias were also verified by accelerated life-time tests in ALT (TH – Temperature-Humidity, HAST – Highly Accelerated Stress Test, and TS – Thermal Shock) chambers.

The standard 85°C-85% RH, Thermal Shock and two kinds of HAST tests were applied to Test Boards made by the selected technologies. The test regimes were as follows:

- **85/85** – 85%RH (Relative Humidity) / 85°C temperature for 160 hours, then 8 hours for circuits recovery, and for electrical and optical inspection.
- **TS** – 960 cycles at 125 °C / 5 min and -40 °C / 5 min (160 hours), then 8 hours for circuits recovery, and for electrical and optical inspection.
- **HAST-Strong** – 8 hours at 130 °C / 100 %RH / 3 bar, and 16 hours for recovery and inspection (strong HAST).
- **HAST-Weak** – 16 hours at 120 °C / 95 %RH / 2 bar, and 8 hours for recovery and inspection (weak HAST).

Such 85/85, TS and HAST days/weeks run on the test boards until some failure rate achieved, that was expected after 5 days of strong HAST, 15 days of weak HAST, and 6 weeks (5000 cycles) of 85/85 and TS. Photos for optical inspection were taken and the resistances were measured at the beginning for reference, and during the lifetime tests in as many points in the time scale as applicable.

Figure 14 shows just one diagram from the 23, which contains the results of the life time tests. This diagram proves that the injection technologies, which were qualified as very good in Table 2, have excellent reliability as well.

Figure 14: Comparison of the life time tests on different technologies: Resistance change in case of strong-HAST test was applied to
P = silvered + insert & press + soldered
T = insert & press + Cu plated & Ag + soldered
L = silvered + 0603 SM R inserted + soldered
M = silvered + MELF resistor inserted + soldered

Conclusions

Conclusions regarding the applicability and the qualification of tested via materials and injection process sequences are summarized in Table 2.

Table 2: Qualification summary of EMCI via materials and injection processes

Via type Contact	Ormet-PS	Ormet-SP	Cu wire	Melf R	SMD R
Ormet	**Poor**	**Good**	**Poor**	**Poor**	**Poor**
Ag + Ormet	N/A	**Good**	**Poor**	N/A	N/A
Solder	**Poor**	N/A	**Poor**	N/A	N/A
Ag + Solder	N/A	N/A	**Very good**	**Very good**	**Very good**
Cu + Ag	N/A	N/A	**Very good**	N/A	N/A
Cu+Ag+Solder	N/A	N/A	**Very good**	N/A	N/A

The best vertical interconnects via materials and injection technologies are as follows:

- Copper micro connector (pin, wire): press & fit, with optimum hole/pin dimensions;
- MELF resistor: press & fit, with optimum hole matching to MELF dimensions;
- SMD resistor: press & fit, with optimum hole matching to SMD dimensions.

The best contact producing technologies are:

- Copper pin with Cu plating and imm-Ag: excellent solution, compatible with PCB;
- MELF with imm-Ag and solder: excellent solution, very simple, compatible with SMT;
- SM components with imm-Ag and solder: excellent solution, compatible with SMT.

Acknowledgements

The authors acknowledge the promotion of the EMCI Cooperative Research Project (COOP-CT-2003-508172) of the European Commission, whose encouragement contributed to the publication of the paper.

3D-Fluidic Cooling Structures in LTCC

M. Mach and J. Müller

BMBF Center for Innovation Competence MacroNano®, TU Ilmenau,

Gustav-Kirchhoff-Str. 7, 98693 Ilmenau, Germany

Phone: +49 3677 69-3385, Fax: +49 3677 69-3379, E-mail: matthias.mach@tu-ilmenau.de

Abstract

LTCC is being widely used for MMIC packaging due to its excellent microwave behaviour (low or moderate dielectric loss, high conductive inks) and its hermeticity. Unfortunately, LTCC materials have a relative low thermal conductivity between 2 and 5 W/m*K. Regarding the thermal conductivity of LTCC, there are several ways to improve it. One way is to embed material with high thermal conductivity such as metal thermal vias or head spreaders. These elements can be manufactured in silver or copper for example. Another way to improve the thermal conductivity of LTCC is to implement fluidic cooling channels at the hot spot or the heat source. But the heat has to be transported from the liquid to a heat sink through an actuator. The actuator pushes the liquid through the fluidic channels. All these parts of an integrated fluidic cooling system have to be implemented in LTCC and the liquid cooling structure should be integrable at MMIC packages. The aim of this work is to integrate a complete fluidic cooling system with all parts of a cooling system in a ceramic package. In this paper a micro heat pipe in LTCC will be presented. The heat pipe is a natural convection device and doesn't need an external engine or power. The energy for cooling the heat source comes from the heat source itself. For another active cooling system different designs of heat sinks were integrated in the first step. A part of the implemented active cooling system is a MMIC dummy to test the manufacturability of fluidic interfaces between different ceramic modules. The functionality of the manufactured cooling system parts was investigated and the results will be presented and discussed.

Key words: fluidic cooling system, heat pipe, LTCC

Introduction

Cooling of electronics is becoming the major topic for further miniaturizations and increasing power densities. For example a popular power amplifier on GaAs-Technology (TGA9083-SCC) of Triquint has 80 gates with a maximum loss of 22 Watt at these gates. New generations using other semiconductor materials will have the same losses on only a quarter of gates. These power amplifiers are used for microwave applications in MMIC packages.

LTCC is being widely used for these packages due to its excellent microwave behaviour and its good hermetic packaging capability. But the relatively low thermal conductivity between 2 and 5 W/m*K brought a limit for cooling such mounted high power devices.

The physics know three fundamental possibilities to cool a heat source.

One Way is to cool a heat source with radiation such as the sun or a radiator. But this way has a poor efficiency on such a small heat source such as a power amplifier.

Another way is to cool with conduction. If the conduction is used for cooling, the next heat sink should be close to the power device and the thermal

conductivity of the heat sink should be high, such as silver, copper or alumina which have a thermal conductivity of 410, 390 and 230 W/m*K [14]. In comparison to LTCC the thermal conductivity of silver or copper is about 100 times higher. To increase the thermal conductivity of LTCC locally, metal filled thermal vias can be used. This concept is explained in [1] and [16]. The heat source and the heat sink should be connected by thermal vias and directly under a hot spot of the die should be a thermal via [9], [4]. Large thermal vias may decrease the manufacturability and hermeticity of the package.

The third way is the usage of convection. An overview about these cooling capabilities is shown in several articles [8], [7] and [15]. In these articles three types of convection are shown: natural convection, single-phase forced convection and boiling. These types differ in the heat transfer coefficients.

Conduction is the standard cooling method for packages. However, this method has limits when used in applications with a high power density, as the thermal conductivity is restricted. The distance between heat source and sink should be kept as short as possible as well. In contrast to the standard method, modules in LTCC using convection cooling should be shown in this paper.

Heat Pipe

A heat pipe uses both the advantages of natural convection and boiling. In principle, this is a way for cooling electronics without external power or generator. The heat pipe is a natural convection device and does not need an extra power supply. The power for the convection in the heat pipe comes from the heat source which has to be cooled. The pipe consists of an evacuated sealed cavity, which is particular filled with a working fluid. The heat pipe is divided in three regions: an evaporator, a condenser and an adiabatic region, whereas the adiabatic region separates the evaporator and the condenser.

Water, ethanol, methanol, mercury and so on can be used as working fluids. The working fluid and the vapour pressure in the cavity define the working temperature. In the evaporator region the working fluid is vaporized by the heat of the source. The hot vapour travels from the evaporator through the adiabatic region to the condenser. In the adiabatic region the hot vapour does not loose thermal energy. The vapour flow ends at the condenser where the vapour is cooled down and condenses at the walls. The heat will be removed of the working fluid. The working fluid is transported back by capillary forces to the evaporator.

There are two ways to use the heat pipe principle: as a heat conducting route from one point to another or as a heat spreader to transform power densities. The former is schematically shown in Figure 1.

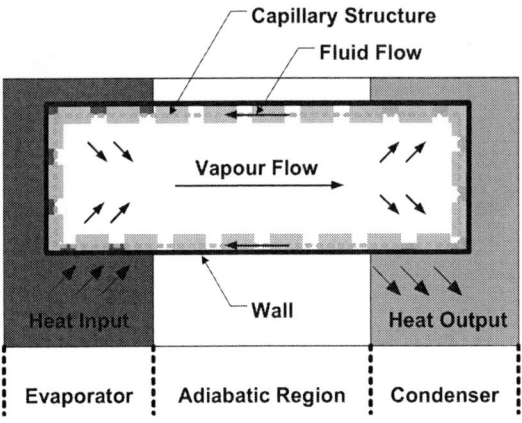

Figure 1: A typical working scheme of a heat pipe

Normally, standard heat pipes for electronics cooling are produced in copper with a capillary grove, mesh or wick. However, these are extra parts which need to be produced and mounted on the package.

Integrating the heat pipe into LTCC solves the mentioned problem and offers the advantage to cool the mounted power devices directly. Thermal vias which can cause leakage are eliminated, the whole package can act as a heat sink and costs due to the soldered metal heat sink on the package backside are eliminated.

First Design

In this project two heat pipes were designed and built. The first one was constructed of four separate parts: the evaporator, the adiabatic flow channel, the condenser and the capillary channel. These four regions are shown in Figure 1. Pressure drops and flow behaviour were calculated based on the procedure given in [13].

The heat pipe is 2 mm thick, 19 mm wide and 55 mm long and was manufactured using 10 layers green tape DP 951. Internal cavities were manufactured by punching. The channels were punched by means of a 1 mm square pin and the capillary via using a 200 µm round pin. The integrated thermal vias were punched using a 200 µm round pin and 400 µm pitch between heat source and cavity. After punching and via filling the green tapes were laminated using a special lamination procedure to prevent the separate layers from squashing and buckling in the cavity region. First, both bottom and top layers were uniaxially laminated at a pressure of 153 kN for 5 minutes. The inner layers were laminated in two blocks each at 80 kN for 5 minutes. Finally, all separate laminated blocks were stacked and uniaxially laminated at 60 kN for 5 minutes. The heat pipe was fired with the standard firing profile given by the vendor. Only the burnout time at 450°C was extended by half an hour. After co-firing the laminate, a resistor working as heat source was soldered onto the LTCC substrate (Figure 2a). The heat source is a 3 x 3 mm³ thick film resistance on Al_2O_3.

The heat pipe was filled with 15 µl of water and sealed at ambient pressure. The pipes began to work at 100 °C.

Discussion

The temperature distribution was measured with an infrared camera. Figure 2b shows an IR-picture of a running heat pipe.

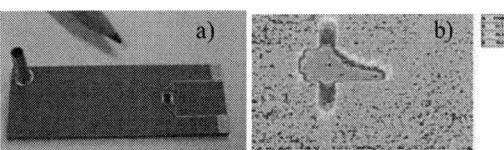

Figure 2: a) Manufactured heat pipe with mounted heat source. b) Running heat pipe.

The heat pipe dissipates 5.5 Watt at a heat source temperature of 150 °C. During the measurements the heat pipe was hung in air so that natural convection arises. An open heat pipe dissipates without a working fluid only 2.5 Watt.

657

The advantage of this heat pipe design is that the evaporator and condenser can be replaced without big redesigns; only one thing which has to be controlled is the capillary pumping limit.

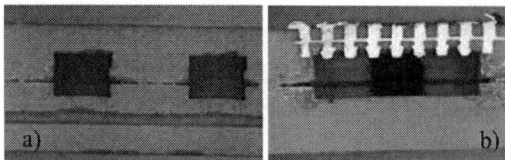

Figure 3: a) Transition between condenser and capillary channel. b) Evaporation chamber without mounted heat source but with thermal vias

In Figure 3a and 3b the delamination at the channel walls are shown. This problem can be solved by optimizing the lamination procedure and parameters. The white clouds at the cutting surface are dust from sawing.

Second Design

The second heat pipe design is an integrative design which was manufactured. The capillary channels are located at the four edges of the vapour channel (Figure 7). The channels were formed in green tape with a 125 µm thick carbon inlay. The size of the second design is 11 mm x 51 mm x 1.3 mm. The heat pipe was manufactured using 7 layers of DP 951 AX and one layer of DP 951 C2. First, the vapour channel and all vias (capillary vias, electrical vias and thermal vias) were punched and filled were needed. The vapour channel was punched using a 1 mm square pin in the layers three to five. In the second step the conductors were printed. On layer seven a flag for the carbon inlays were printed on the backside of layers three and six. The carbon inlays were put into the fresh flag to form the capillary channels.

The carbon inlays were cut off on 125 µm thick carbon tape by laser cutting. Next, the green tapes were stacked with the first layer on the bottom and the following on top. A second carbon inlay (Figure 4b) was inserted in layers three to five at the vapour channel during the stacking. The substrate was laminated isostatically at 210 bars for 10 minutes at 70 °C. The capillary channels were formed in the tape through the carbon inlay which is used as lost mould during the lamination (Figure 4a).

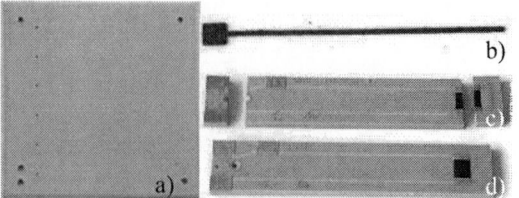

Figure 4: a) Unfired substrate of the heat pipes; b) Carbon inlay for the vapour channel; c) Heat pipe with cuttings; d) LTCC dummy

The firing process was made at a muffle furnace. The manufacturer's desired firing profile was extended during the burnout by 2 hours and the cool down by one and half an hour. Afterwards two post-fire printings were performed to build up the heat source at the substrate. The heat source consists of a layer DP 951 C2 tape and a printed resistor. The pipes were filled with purified water. All capillaries were filled with an over head of 10%.

Discussion

The evaporation chamber was changed from quadratic to cylindrical at the second design. The capillary channel width was changed from 0.5 mm to 2 mm in 0.5 mm steps to take a look at the capillary pumping limits. All these heat pipes have a quadratic chamber.

The bumpiness on the surface of these heat pipes is around 50 µm despite of the height that two carbon inlays (totally inlay thickness = 250 µm) were inserted between the tapes.

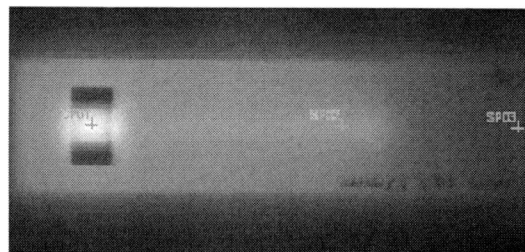

Figure 5: IR-Image of the running heat pipe with measurement points

The temperature distribution (Figure 5) was measured with an infrared camera ThermoVision A40M from Flir Systems (Figure 6).

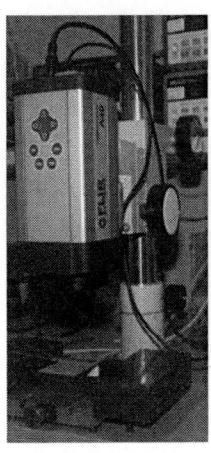
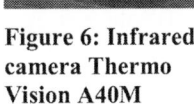
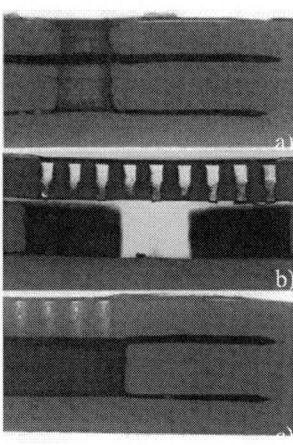

Figure 6: Infrared camera Thermo Vision A40M

Figure 7: a) Condenser with four capillaries on each corner; b) Evaporator with thermal vias and connection to condenser; c) Evaporator with two capillaries

This IR-camera has a resolution of 320x 240 pixels with 50 Hz video refresh rate and with a temperature measurement range steps from -40 to 1500°C. For thermal measurements with high resolution requests, e.g. on conductor lines smaller than 100µm, an 18 µm and 100 µm objective was bought with the IR-camera.

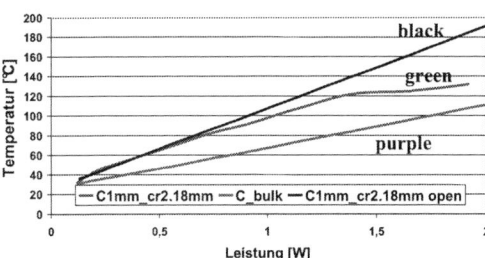

Figure 8: the heater temperature vs. power for the different modules (working heat pipe (green) using free convection, unfilled heat pipe (black) on a heat sink and a ceramic block (purple) also on a heat sink)

A problem is to put heat into the working fluid of the heat pipe. First measurements have shown that the heat pipe have the same cooling capability as a dummy on a heat sink. In Figure 8 the heater temperature of a working heat pipe (green), an unfilled heat pipe (black) and a ceramic block (purple) are shown. The black and purple charts are measured when the modules lie on an aluminium heat sink and the green one in free convection.

The isostatic lamination process is a standard lamination method which is performed at 210 bars by around 15 minutes. The second design should

investigate if the heat pipe could be integrated in a MMIC using standard lamination.

The lamination is the critical process for 3D-Channels inside a substrate. The channels were squashed without a filling or support. In the heat pipe were used a 250 µm thick carbon tape for supporting the vapour channels during lamination. The carbon burned out between 400 °C and 600 °C. Another thing was shown that a carbon inlay can be used as a lost mould in LTCC to form the capillary channels.

Liquid Cooling System

Other liquid cooling systems require an extra force for circulating the working fluid (e.g. a pump). If it is desired to integrate such a cooling system on a package a micro pump should be integrated as well. To completely implement a pump, a connection, a channel, an expansion vessel and a heat sink are needed on the package. All these parts are required to build up an integrated liquid cooling system. In several papers [2, 10, 11, 12] some parts of a fluid cooling system were shown.

Fluidic Connection

A fluidic connection is an essential part of a fluidic cooling system. It can provide an interface between LTCC-modules, between LTCC and PCB and between LTCC and a semiconductor.

First, a re-useable connection system was designed at the Center for Innovation Competence MacroNano® [5]. The universal fluid connector was designed to build up a modular chemical laboratory including several reaction modules e.g. mixers, heaters, cooler and sensor modules. The connector was used to test different parts of a cooling system.

In a next step a solder connection was designed to connect a package with another, a pump or with other separate manufactured parts. A package was build to investigate the solder connection. Six layers of green tape DP 951 AX were used. All layers were processed (punching, printing, etc). The fluid channels are two layers high and were punched using a 400 µm round pin from the universal connector to the solder connection. The cooling channels at the mounted dummy package are only one layer high and were punched using a 400 µm pin. On the top layer a solder metallization (Ag, Pd) was printed. Afterwards all layers were uniaxially laminated at 80 kN for 10 minutes. If this substrate were isostatically laminated the fluid channels would clench. The substrate was fired at 850 °C with the desired profile. In the following step a lead-free solder paste were printed and reflowed on the fired substrate. After sawing a 19 mm x 28 mm x 0.7 mm carrier plate (Figure 10a) and a 10 x 10 x 0.7 mm package dummy (Figure 10b) were singled. The package dummy was soldered on the carrier plate.

In the 3D-Model (Figure 9) the in- and outflow channels in the carrier plate (Figure 10a), the solder sealing rings around the channels and the cooling structure at the package dummy (Figure 10b) can be seen.

Figure 9: 3D-Model of a mounted liquid cooling module

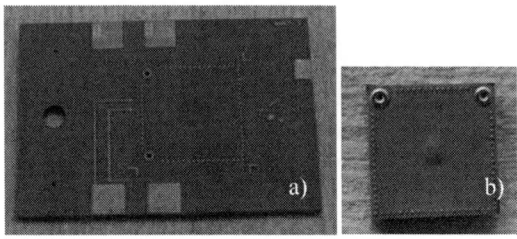

Figure 10: a) Carrier plate of the fluidic package; b) Package dummy with the fluidic connections

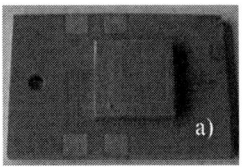

Figure 11: a) Mounted fluidic single package dummy on the carrier plate; b) Mounted three stacked package dummy

Discussion

Four modules were manufactured, three modules with one package dummy (Figure 11a) and one module with three stacked package dummies (Figure 11b) on top. The printed metallization pads on the single package dummy differ from these on the stacked one. The inner diameter of the printed pad lies at 0.6 mm and the outer diameter at 1.6 mm, 1.1 mm and 0.9 mm. All connection pads of the fourth module have an outer diameter of 1.6 mm. The third module with the smallest solder connection shows a leakage in the first series of measurements. All other modules are waterproof up to 34 bars.

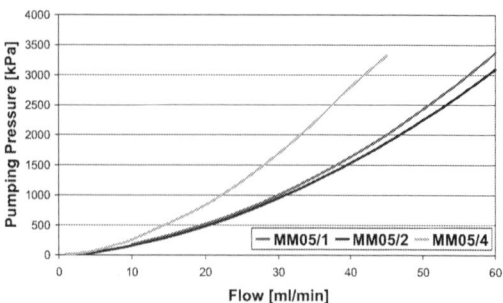

Figure 12: Pumping pressure versus flow rate

In Figure 12 the modules show a high flow resistance with a wide spreading. The three stacked module amazingly have the highest flow resistance. The roughness of the punched walls can cause those high flow resistances. Also delamination or buckles in the substrate at the channel walls can cause those high flow resistances and the spreading. Through these packages between 3 and 12 ml/min can flow with a commercially available micro pump which has a maximum pumping pressure between 0.1 to 2 bars.

Conclusion

The heat pipe is a closed cooling solution which does not need an extra device for transportation of the fluid. The manufactured heat pipes show the feasibility to make a small heat pipe in LTCC. The first constructed heat pipe is only 19 mm x 55 mm x 2 mm large. The heat pipe shows some manufacturing difficulties, which occur by filling and closing the heat pipe. The manufacturing process of this pipe is going to be optimized. And so the lamination processes is going to be controlled better. However, the uniaxial lamination process is not a standard for producing LTCC substrates. Normally, green tapes are laminated isostatically. If the first construction is laminated isostatically all cavities will be crushed. Another way to produce the cavities using isostatical lamination has to be founded. Finally, the first construction shows the cooling capabilities of 5.5 Watt at 150 °C working temperature. If this pipe is closed at lower pressures it will begin to work at lower temperatures.

The second heat pipe does not show the disadvantages and problems of the first design. This design shows that a heat pipe can be integrated in a package for a manufacturing process using an isostatical lamination process. The heat pipe was manufactured to show that a heat pipe can be integrated in the substrate. The second version shows that an integrated heat pipe can replace an alumina heat sink at the back side of the module. Both, the heat pipe and the module with Al heat sink, can dissipate 2 Watt at 130 °C.

The heat pipe shows that a heat pipe could be integrated in microwave packages.

The next step for industrial applications is to design a standard heat pipe for several power steps on different substrates and a working fluid with a working temperature from -45 °C till 125 °C. These pipes should be characterised for different field of applications.

The fluidic connection shows that a connection with 600 µm inner diameter and 1100 µm outer diameter is proof till 34 bars. The flow resistance of the dummy packages differs much. A reason can be the manufacturing process. All channels were punched with a circular pin so that the walls of the channels are rough. Also the walls of the channels can be squashed during the lamination process.

All 3D-structures were produced with higher effort. Normally, the substrates were laminated isostatically at a standard manufacturing process in industry. If unfilled 3D-structures are laminated isostatically all cavities will be crushed. Another way for producing cavities was shown at the second heat pipe design. The second heat pipe design shows the benefits of using carbon tape as a spacer during lamination and as a lost mould to form a cavity in LTCC.

Acknowledgements

The authors would like to thank the German Government for the financial support. The work was carried out in the project *MultiSysTeM* (Centre for Innovation Competence MacroNano) which is funded by the BMBF. Additional financial support was provided by the Thuringian Ministry of Culture. Special thanks to Mrs Koch for her valuable cooperation during the manufacturing processes.

References

[1] "Ceramic Circuit Thermal Desing Using Tailored Vias," published on www.dupont.com/mcm/.

[2] L. Bergstedt and K. Persson, "Printed Glass for Anodic Bonding - A Packaging Concept for MEMS and System On a Chip," Advanced Microelectronics: January/ February (2002), 29-31.

[3] H. Birol, T. Maeder, C. Jacq, and P. Ryser, "3-D Structuration of LTCC for sensor micro-fluidic applications", Proc. IMAPS European Microelectronics and Packaging Symposium, (2004), pp. 366-371.

[4] V. A. Chiriac and T.-Y. T. Lee, "Thermal Assessment of RF-Integrated LTCC Front End Module", Proc. IEEE Transactions on advanced packaging, (2004), pp. 545-557.

[5] M. Fischer, T. Thelemann, and M. Stubenrauch, "LTCC Interconnects In Different Scales," Proc. 1st MacroNano Colloquium on LTCC RF- and Microsystem Interconnect (Ilmenau, 2006).

[6] J. Kita, A. Dziedzic, L. Golonka, and T. Zawada, "Laser treatment of LTCC for 3D structures and elements fabrication" , Microelectronics Journal, 19:3 (2002), 14-18.

[7] C. J. M. Lasance and R. E. Simons, "Advances In High-Performance Cooling For Electronics", Electronics Cooling, 11:November (2005).

[8] I. Mudawar, "Assessment of High-Heat-Flux Thermal Management Schemes", Proc. IEEE Transactions on Components and Packaging Technologies, (2001), pp. 122-141.

[9] J. Müller, M. Mach, H. Thust, C. Kluge, and D. Schwanke, "Thermal Design Considerations for LTCC Microwave Packages", Proc. 4th European Microelectronics and Packaging Symposium (Terme Čatež, Slovenia, 2006), pp. 159-164.

[10] K. A. Peterson, K. D. Patel, C. K. Ho, S. B. Rohde, C. D. Nordquist, C. A. Walker, B. D. Wroblewski, and M. Okandan, "Novel Microsystem Applications with New Techniques in LTCC", Proc. CICMT (Baltimore USA, 2005), pp. 156-173.

[11] K. A. Peterson, K. D. Patel, C. K. Ho, B. R. Rohrer, C. D. Nordquist, B. D. Wroblewski, and K. B. Pfeifer, "LTCC Microsystems and Microsystem Packaging and Integration Applications", Proc. CICMT (Denver, 2006).

[12] C. Rusu, K. Persson, B. Ottosson, and D. Billger, "LTCC interconnects in microsystems", Journal of Micromechanics and Microengineering, 16 (2006), 13-18.

[13] C. C. Silverstein, Design and technology of heat pipes for cooling and heat exchange, Hemisphere Publishing Corporation, Washington, 1992.

[14] M. Wutz, Wärmeabfuhr in der Elektronik, Vieweg, 1991.

[15] M. A. Zampino, W. K. Jones, and Y. Cao, "Substrate Embedded Heat Pipes Compatible with Ceramic Cofired Processing", The International journal of microcircuits and electronic packaging, 21:1 (1998), 52-58.

[16] M. A. Zampino, R. Kandukuri, and W. K. Jones, "High performance thermal vias in LTCC substrates", ITherm 2002, Thermal and Thermomechanical Phenomena in Electronic Systems (2002), 179-185.

High-brightness RGB LED Modules Based on Alumina Substrate

Veli Heikkinen[1], Eveliina Juntunen[1], Kari Kautio[1], Antti Kemppainen[1], Pentti Korhonen[1], Jyrki Ollila[1], and Aila Sitomaniemi[1], Timo Kemppainen[2], Heikki Korkala[2], Terho Kutilainen[2], Hannu Sahavirta[2]

[1]VTT, Kaitoväylä 1, 90570 Oulu, Finland

Phone +358 20 722 2242, Fax +358 20 722 2320, E-mail: veli.heikkinen@vtt.fi

[2]Selmic Oy, Veistämötie 15, 90550 Oulu, Finland

Phone +358 10 820 2907, Fax +358 8 551 1511, E-mail: timo.kemppainen@selmic.com

Abstract

Light emitting diodes (LED) have penetrated into various lighting applications, such as backlighting of keypads and displays in mobile phones, signs, video screens, traffic signals, and exterior and interior lighting for automobiles. A new trend is to use color-tunable light to set or dynamically vary the ambience of the illuminated space. We developed high-brightness RGB LED modules that use bare chips on an alumina (Al_2O_3) substrate. The substrate has dimensions of 14.7×25.4 mm^2 and an electrical power consumption of 4.4 W. Thermal simulations were carried out to find the optimum solution among various substrate thicknesses, interface layer materials and geometry of thermal vias in the substrate. Two commercially available plug pastes were experimentally evaluated for making thermal vias. We realized three module series that use a 0.63 mm, 1 mm or 1.27 mm thick substrate. LED chips were die-bonded using electrically-conductive adhesive and wire-bonded with 25-μm Au wire. All chips and bonding wires were protected with UV-curable adhesive. The simulated thermal resistance for the 0.63-mm thick substrate was 6.3 K/W. The optimum solution using one large thermal via and heat distribution layer on top surface of the substrate decreases the resistance down to 2.8 K/W.

Key words: light emitting diodes, alumina, thermal design, thermal vias, module integration

Introduction

Light emitting diodes (LED) have strongly penetrated into mobile phones and other portable devices for backlighting keypads and full-color displays. High-power LEDs are used in signs, video screens, traffic signals, exterior and interior lighting for automobiles, and a variety of niche illumination applications. Solid-state lighting technology is emerging as a cost-competitive and energy-efficient alternative to conventional electrical lighting [1].

There are two approaches for generating white light from solid-state sources, namely phosphor LEDs and multichip LED modules. The phosphor LEDs can be considered as solid-state replacement of fluorescent tubes. The multichip LED lamps offer many advantages, such as chromaticity control, better light quality, and higher efficiency [2].

A new trend is to use color-tunable light to set or dynamically vary the ambience of the illuminated space [3]. One option is to mix the colors of red, green and blue LED chips. We developed prototype series of color-tunable high-brightness RGB LED modules that use bare red, green and blue chips on an alumina (Al_2O_3) substrate. Here one of the key issues was the thermal design of the module. Our overall objective is to realize color-tunable LED modules that can easily be customized to different applications and produced in automated production lines. In this paper we present the details of the thermal design, experiments with the thermal plug pastes, and realized modules together with measurement results.

Module specification

We selected blue and green LEDs manufactured by Cree, USA and red chips made by Epigap, Germany. The blue and green chips had dimensions of 900×900 μm^2 and a thickness of 250 μm. The red chips had a size of 1.0×1.0 mm^2 with a thickness of 170 μm. The maximum operating current of the chips were 350 mA corresponding to a total power consumption of 4.4 W. The module substrate was specified to be 14.7×25.4 mm^2 and it also contained three Zener diodes for ESD protection and a thermistor for temperature measurement.

Thermal Design

The thermal conductivity of 96% alumina, k_{Al2O3} is about 25 W/mK. This is much higher than in the epoxy laminate printed circuit boards, such as FR-4, but quite low for high power-density applications. One method to improve the thermal properties is to make holes to the substrate and fill them with high thermal conductivity material. These structures are called thermal vias. We carried out simulations using Flotherm thermal analysis software to find out the optimum solution among various substrate thicknesses, interface layer materials and geometry of thermal vias.

The thermal model of the LED module used in the simulations shows in Figure 1 and the material values and layer dimensions in Table 1. Blue and green LED chips consisted of the SiC substrate and InGaN active layer. Red chips have Si substrate and AlGaInP epilayers. The heat generated in the chip was assumed to be distributed uniformly in the active layer.

The chip was assumed to be die-bonded either by solder or electrically-conductive adhesive. The outside atmosphere was stagnant air at 25 °C. Flotherm software cannot model cylindrical layers. Therefore, the bonding wire and thermal vias were replaced with hypothetical structures that have square cross-section with equivalent volume.

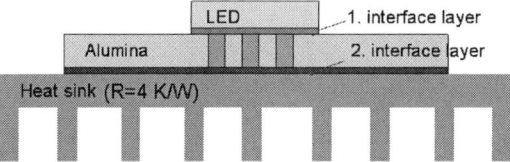

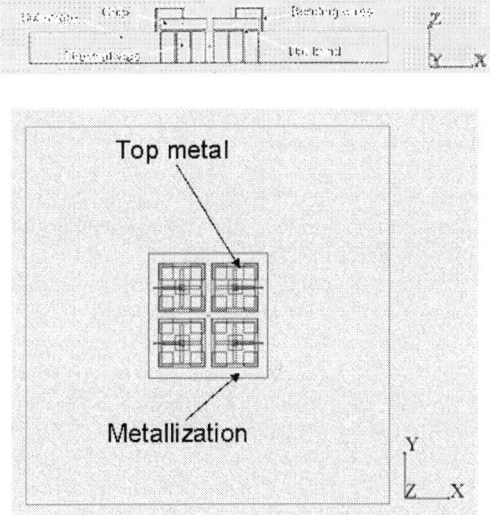

Figure 1: Thermal Model of LED Module.

The interface layers are essential for the thermal resistance of the structure. Thermal simulations were utilized to study the effect of interface material selection. The test structure used in the simulations was otherwise similar as in Figure 1, but the module was assumed to be attached on an ideal heat sink with a thermal resistance of 4 K/W. Here the assumed dissipated power was 5.44 W. The interface layers of the simulated structure are illustrated in Figure 2.

Table 1: Materials Used in Thermal Simulations.

Layer	Material	Size (x, y, z) (mm)	k (W/mK)
LED chip	SiC	$0.9 \times 0.9 \times 0.23$	270
	InGaN	$0.79 \times 0.79 \times 0.02$	100
	Si	$1.0 \times 1.0 \times 0.17$	118
	AlGaInP	$1.0 \times 1.0 \times 0.02$	100
Top metal	Au	$0.06 \times 0.7 \times 0.001$	296
Bonding wire	Au	0.022×0.022	296
Die bond	solder adhesive	$0.76 \times 0.76 \times 0.02$	57 3
Metallization	Ag/Pt1%	$2.3 \times 2.3 \times 0.008$	322
Substrate	alumina	$7 \times 7 \times 0.63$	25
Thermal vias	Ag/Pt1%	$\varnothing\, 0.3 \times 0.63$	322

The interface layer between the LED chip and alumina substrate (1. interface layer in Figure 2) was either AuSn solder or electrically conductive adhesive. The second interface between alumina substrate and heat sink was electrically isolating because the LEDs had to be isolated from each other and the heat sink. The isolation could be realized either with electrically non-conductive adhesive (k = 3.6 W/mK) or glass layer (k = 1.05 W/mK). If the glass layer is used, adhesive is needed to attach the substrate on the heat sink.

Figure 2: Interface Layers of Simulated Structure.

The simulation results are shown in Table 2. If the material on the chip-alumina interface is AuSn solder the thermal resistance, R_T, from the active junction to the bottom of the LED module is 2.8 K/W less than it would be if the interface material was conductive adhesive. The thermal resistance further increases 1.6 K/W if the electrical isolation is made with glass instead of non-conductive adhesive.

Table 2: Thermal Simulation Results with Different Interface Materials.

Interface layer 1		Interface layer 2		R_T (K/W)
Material	z (μm)	Material	z (μm)	
AuSn	20	Non-cond. adhesive	50	3.5
AuSn	20	Glass + adhesive	54	5.1
Conductive adhesive	20	Non-cond. adhesive	50	6.3
Conductive adhesive	20	Glass + adhesive	54	7.9

Thermal simulations were also used to study the effect of thermal vias and heat distribution layers. The test structure used in the simulations is illustrated in Figure 1. AuSn solder was used on the LED-alumina interface and non-conductive adhesive on the alumina-heat sink interface. The bottom of the structure was fixed on temperature 25 °C assuming that the structure sits on an ideal heat sink. The assumed dissipated power was 5.44 W. Figure 3 shows the variations made in this simulation series and Table 3 lists the main results.

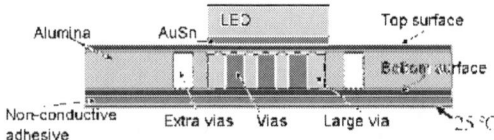

Figure 3: Simulated Structure with Variations of Thermal Vias and Heat Distribution Layers.

The simulations showed that thermal vias were the most effective heat management method for this module. Thermal vias with different locations and sizes were simulated. The simulations demonstrated that the vias were useful right under the heat source. Some extra vias were added around the LED chips, but their effect was minimal. Also the effect of one large via ($\varnothing$ = 0.763 mm) in comparison with 5 smaller ones ($\varnothing$ = 0.3 mm) was simulated. The thermal resistance decreased ~ 2% with the large via.

Another heat management option was the addition of heat distribution layers made with silver paste (k = 200 W/mK) on alumina substrate. For these layers two different thicknesses (22 µm and 55 µm) and three different sizes (large: 7×7 mm^2, mid-size: 3.5×3.5 mm^2, and small: 2.1×2.1 mm^2) were simulated. The layer can locate either on the top surface, on the bottom surface or on both surfaces. The simulations showed that if there were thermal vias the effect of the heat distribution layer was quite small. However, if the vias were not an option, the heat distribution layer enhanced the thermal management of the structure significantly.

The heat distribution layer was useful only on the top surface. The benefit of the bottom layer was close to zero, or even negative if the layer was too thin. The thermal resistance of the module with heat distribution layer on both surfaces of the substrate was the lowest, but very close to the result of one layer on the top surface. Thus, there is no use of wasting the silver paste on both sides of the alumina. Another conclusion based on the simulations is that it is beneficial to use as thick a layer as possible. The size of the layer is not as significant and increasing the size of the heat distribution layer reduces the thermal resistance only up to a certain limit. Therefore, Table 3 lists the results only for the thicker mid-size (3.5×3.5 mm^2) heat distribution layer on top surface of the alumina substrate.

Table 3: Simulation Results with Different Heat Management Methods.

Vias (mm)	Heat distribution layer	R_T (K/W)	Change (%)
-	-	5.95	0
-	55 µm, mid-size, top surface	4.51	- 24
5 × $\varnothing$ = 0.3	-	3.25	- 45
5 × $\varnothing$ = 0.3	55 µm, mid-size, top surface	2.95	- 50
1 × $\varnothing$ = 0.76	55 µm, mid-size, top surface	2.83	- 52

Final simulations were made with structure consistent with the actual prototype LED module. Now the alumina substrate size was 25.4×14.7 mm^2 and three different thicknesses, 0.63 mm, 1 mm, and 1.27 mm were tested. Heat distribution layer made with silver paste (k = 200 W/mK) was applied on top surface of the substrate. The layer thickness was 200 µm and total area under the LEDs 78 mm^2. There were four LED chips on the substrate, namely one blue, two green and one red. The power consumption used in the simulations was 4.96 W. The interface between LEDs and heat distribution layer was electrically-conductive adhesive (thickness of 20 µm), because the chips could not be soldered on the silver. The substrate-heat sink interface was non-conductive adhesive (thickness of 50 µm). The bottom of this adhesive layer was fixed to a temperature of 25 °C assuming that the structure sits on an ideal heat sink.

The simulated thermal resistance of the structure with 0.63 mm thick substrate was 6.3 K/W. The resistances with 1 mm and 1.27 mm thick substrates were 6.7 K/W and 6.8 K/W, respectively. The surface temperatures of structure with 1-mm thick alumina substrate are illustrated in Figure 4. Figure shows that heat conducts poorly along the substrate even with the thick heat distribution layer. The red LED is cooler than the other ones due to its smaller forward voltage and heat dissipation.

Figure 4: Simulated Surface Temperatures of Prototype Module with 1-mm Thick Substrate.

Thermal Via Experiments

Silver based thermal plug pastes are fairly high viscosity pastes that are used to fill lasered through holes on alumina substrates by stencil printing. Due to small amount of solvent, the deposited paste does not sag too much upon drying. Most importantly, the firing shrinkage of plug paste is very small, enabling crack-free thermal vias. Two commercially available plug pastes were evaluated to make thermal vias for high power LED chips.

Alumina substrates ($4'' \times 4''$) with lasered through hole structures were used in the tests. The diameter of holes varied between 0.25 mm and 0.9 mm. Substrate thickness was either 0.5 mm or 0.63 mm. A 50-μm thick punched stainless steel stencil was used. The printer table was equipped with a porous metal nest to supply a uniform through-hole vacuum. A sheet of paper was used between the nest and the substrate. The stencil apertures corresponded to the lasered substrate holes. The actual filling of the holes was made in double print mode with fairly slow squeeze speed and the stencil in contact with the substrate. The drying of the paste was made at 80 °C with the backing paper on a metal tray. After drying, the backing was removed and the substrate was sintered in a belt furnace using the standard 850 °C thick film profile. Figures 5 and 6 show examples of the filled and dried through-hole structures.

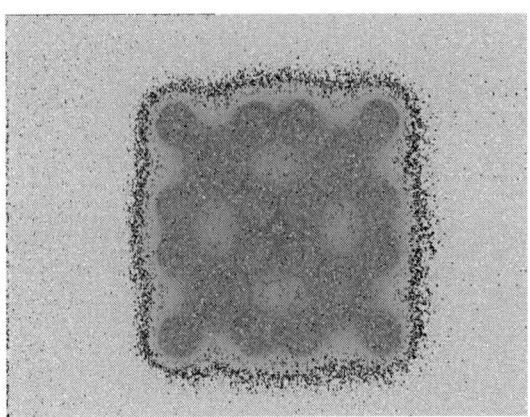

Figure 5: Thermal Via Structure for 2×2 LED Array on 0.5-mm Thick Alumina Substrate. Via diameters are 0.4 mm and 0.25 mm.

The biggest challenge with the processing of thermal plug pastes was how to fill the hole completely with a flat surface, to enable LED chip assembly. Naturally, the metallization is very flat on the bottom side of the substrate, because the paste stops to the backing paper. However, the paste adheres to the backing and when the backing is removed after drying, small cavities can be seen on the via plugs. On the top side or stencil side of the substrate, small holes up to a diameter of 0.4 mm can be filled totally in one print cycle. Larger holes,

up to a diameter of 0.9 mm probably require two print-dry cycles to obtain a smooth via surface.

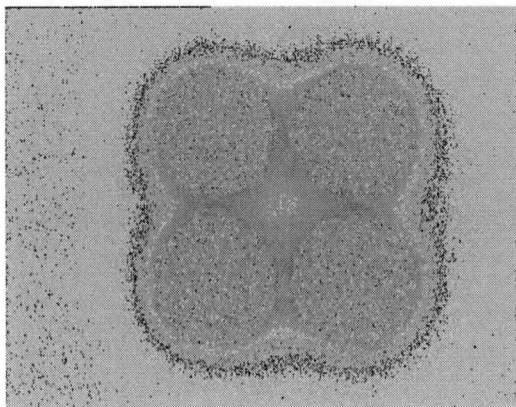

Figure 6: Thermal Via Structure for 2×2 LED Array on 0.63 mm Thick Alumina Substrate. Via Diameter is 0.9 mm.

The shrinkage of via plug paste is very small during sintering and there was no sign of via plug separation from the wall of the lasered hole, as can be seen in Figure 7.

Figure 7: Sintered and Diced Via Plugs with Diameter of 0.4 mm, 0.3 mm and 0.25 mm. Substrate thickness is 0.5 mm.

Module Realization

We realized three module series that use a 0.63 mm, 1 mm or 1.27 mm thick alumina substrate. Due to production technology reasons we did not use thermal vias. A 200-μm thick Ag paste layer with an area of 4×4 mm^2 was printed to the substrate under each LED chip. The chips were die-bonded using electrically-conductive adhesive and wire-bonded with 25-μm Au wire. A thermistor was placed on the substrate for temperature monitoring. All chips and bonding wires were protected with UV-curable adhesive. One prototype module shows in Figure 8.

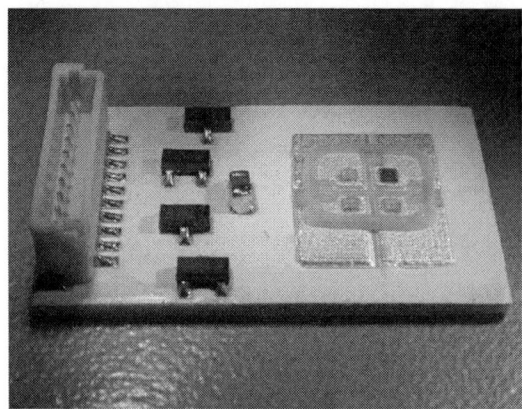

Figure 8: Prototype LED Module.

Module Measurements

The operational characteristics of the LED modules were measured using the setup shown in Figure 9. The module under test was placed on top of the thermoelectric cooler that was set to a temperature of 25 °C. The drive current of each LED was set to 350 mA. The optical power emitted from the LEDs was monitored using an optical fiber that was coupled to an optical power meter (Ando AQ-1135E with AQ-1972 detector). The forward voltage of the chips was also measured.

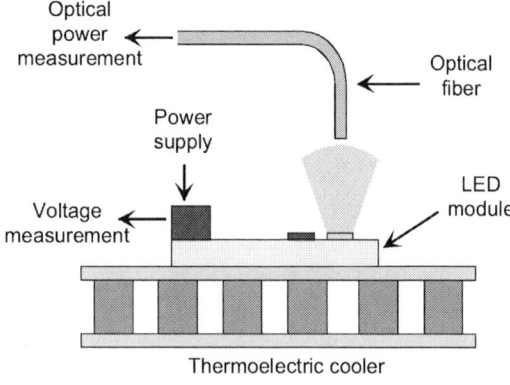

Figure 9: Measurement Setup for LED Modules.

The optical power of a blue LED as a function of time shows in Figure 10. In a warm-up time of one hour the power decreased 11% from its initial value. At the same time the forward voltage LED decreased 17 mV, Figure 11. According to manufacturer's data this corresponds to a temperature increase of 5.7 °C. This is equivalent to a thermal resistance of 4.8 K/W.

The module series will be subjected to reliability stress testing. This will include high temperature operating life (+85 °C, 1000 h), wet high temperature operating life (+70 °C/85%RH, 1000 h) and non-operating temperature cycle (−40 … +85 °C, 400 cycles) tests.

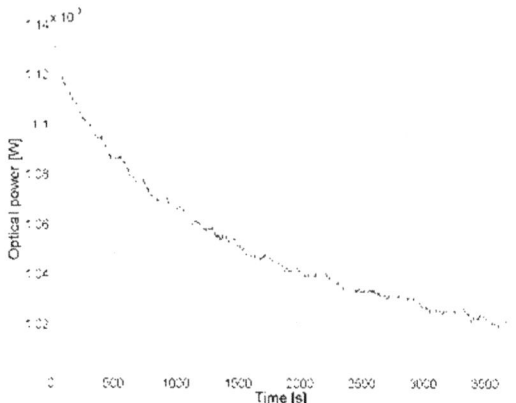

Figure 10: Optical Power of Blue LED on 1 mm Thick Alumina Substrate.

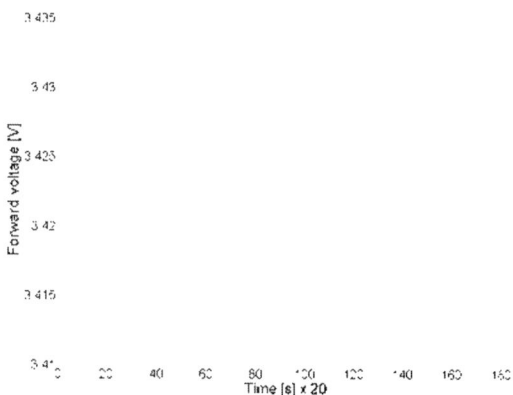

Figure 11: Forward Voltage of Blue LED on 1 mm Thick Alumina Substrate.

Conclusions

We realized three series of high-power RGB LED modules using bare chips. The modules used a 0.63 mm, 1 mm or 1.27 mm thick alumina substrate. A 200-μm thick Ag paste layer with an area of 4×4 mm^2 was printed to the substrate under each LED chip. The chips were die-bonded using electrically-conductive adhesive and wire-bonded with 25-μm Au wire. All chips and bonding wires were protected with UV-curable adhesive.

The simulated thermal resistance of the module with 0.63 mm thick substrate was 6.3 K/W. The resistances with 1 mm and 1.27 mm thick substrates were 6.7 K/W and 6.8 K/W, respectively. According to simulations, one large thermal via and heat distribution layer on top surface of the 0.63-mm thick substrate decreases the thermal resistance down to 2.8 K/W.

Two commercially available plug pastes were evaluated to make thermal vias for high power LED chips. The diameter of holes varied between 0.25 mm and 0.9 mm, and substrate thickness was either 0.5 mm or 0.63 mm. The biggest challenge with the processing of thermal plug pastes was how to fill the hole completely with a flat surface, to enable LED chip assembly.

Acknowledgements

The authors thank Eero Hietala, Jouni Kangas, Risto Karjalainen, Sami Karjalainen, Jukka-Tapani Mäkinen, Jaakko Pennanen, Kaarlo Sipola, and Airi Weissenfelt for the technical assistance in the realization and testing of the LED modules.

References

[1] R. V. Steele, "High-brightness LED market cools after three years of stellar growth", Laser Focus World, Vol. 42, No. 6, pp. 57-59, June 2006.

[2] M. S. Shur and A. Žukauskas, "Solid-state lighting: toward superior illumination", Proceedings of the IEEE, Vol. 93, No. 10, pp. 1691-1703, Oct. 2005.

[3] C. Hoelen, J. Ansems, P. Deurenberg, W. van Duijneveldt, M. Peeters, G. Steenbruggen, T. Treurniet, A. Valster, and J.W. ter Weeme, "Color tunable LED spot lighting", Proceedings of SPIE, Vol. 6337, 63370Q, 2006.

Performance of Thin and Thick Film Resistors Exposed to High Temperature and High Pressure (200°C @ 1000 Bar).

Rolf Johannessen, Frøydis Oldervoll, Frode Strisland and Per Ohlckers*

SINTEF Information and Communication Technology, Instrumentation and Microelectronics

PO Box 124 Blindern, N-0314 OSLO, Norway.

Phone: +47 22067300 Fax: +47 22067350

Email: rolf.johannessen@sintef.no, froydis.oldervoll@sintef.no, frode.strisland@sintef.no

* Vestfold University College, 3103 Toensberg, Norway

Email: Per.Ohlckers@hive.no

Abstract

Performance and failure mechanisms of directly pressurized NiCr thin film and screen printed thick film resistors on ceramic have been investigated for times ranging up to 6000 hours. Resistors submerged in silicone oil, exposed to 1000 Bar at 200°C have been compared to performances of hermetically sealed resistors at 200 and 250°C. The motivation for this work is to reach the quality and reliability required for electronic systems operating in harsh environments with high temperatures and pressure. Directly pressurizing the electronic compartment allows miniaturization, relax requirements for feed through and increase reliability. Oil well instrumentation, process monitoring and distributed engine control are typical examples of applications where reduced package size is desirable. High sheet resistance thick film resistors show stronger temperature dependence than low sheet resistance version and thin film resistors do. This applies for both directly pressurized and hermetically sealed resistors. After 6000 hours at 200°C, 200 − 400 ppm resistance increase was observed for the thin film and the low sheet resistance thick film resistors, compared to 750 - 900 ppm observed for the high sheet resistance thick film resistors. Pressure exposure does not influence the long term behavior for the thin and thick film resistors investigated. Ageing effects are clearly accelerated at 250°C compared to 200°C, and are more pronounced for high sheet resistance thick film resistors. The work shows that thin and thick film resistors have characteristics suitable for high pressure, high temperature applications.

Keywords: Thick film resistors, thin film resistors, high temperature, high pressure

Introduction

In sensor application systems, there is a drive towards placing microelectronic systems close to the ultimate point of use. This drive causes more electronics in harsh environments, i.e. exposed to elevated temperature, vibration, shock, pressure or chemically corrosive fluids. Oil well instrumentation, process industry, automotive, geothermal, avionic and military systems are examples of applications of microelectronic systems exposed to harsh environments. The motivation for placing microelectronics close to the ultimate point of use is to decrease connection complexity and length, reduce weight, improve reliability and benefit from cost saving.

SINTEF together with six partner companies are running a research project, aiming at establishing reliable packaging procedures for long term high pressure (HP) and high temperature (HT) electronics [1]. A key issue in this project is to develop electronics that can operate directly under the actual ambient pressure. Furthermore it is of crucial importance to investigate the failure mechanism and component behavior in HT, HP environment. Some work have been reporting behavior of thick film resistors exposed to combined HP, HT conditions [2,3], while only one work on behavior of thin film resistors exposed to high pressure have been reported [4]. Our study investigates pressure dependence and ageing characteristics of thin and thick film resistors subjected to combined HP and HT environment. As a reference the performances of hermetically sealed resistors at HT environment is also investigated. We have previously found excellent performance for thin film resistors exposed to pressure cycling at elevated temperature [5]. Negligible resistor change was observed for thin film resistors in this study. It is expected to find equivalent performance for combined long term HP, HT conditions as well.

Experimental

Two types of thin film chip resistors and three types of thick film resistors were selected for this study. The thin film resistors were NiCr on alumina substrate supplied by Vishay [6] and the thick film resistors were based on DuPont 2000 series of screen printed resistor materials [7]. Resistor specifications are given in Table 1.

Table 1 Specification of resistors selected for test

Table 1 A Thin film

Manufacturer	R	Size	T [°C]
Vishay	750 Ω	0805	155
Vishay	1,4 k Ω	0805	155

Table 1 B Thick film

Manufacturer	R	Size	Type
DuPont	1 k Ω/□	1 mm²	DP2031
DuPont	10 k Ω/□	1 mm²	DP2041
DuPont	1 M Ω/□	1 mm²	DP2061

Thin film resistors were subject to ambient temperature above recommended maximum operation temperature. Three types of thick film materials were selected according to Table 1 B, firing profile 850°C. No recommended maximum operation temperature is specified for thick film resistors.

The MCM test units were based on an AD-96 alumina substrate from CoorsTec [8] with the substrate metallization system shown in Figure 1.

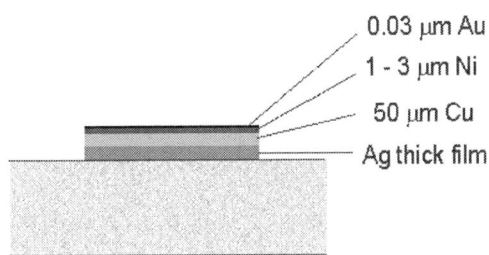

Figure 1 Ceramic substrate with CuNiAu plating on top of Ag thick film.

Each test unit consist of one 750 Ω and one 1.4 kΩ thin film resistor, four 1 MΩ, five 10 kΩ and five 1 kΩ thick film resistors. The thin film resistors were bonded to the substrate by use of electrically conductive epoxy (Epo-tek 3084 manufactured by Epoxy Technology). 56 pins Kovar flat packages were used for packaging.

Table 2 shows the ambient conditions for this study. Test units for HT testing were hermetically sealed in nitrogen atmosphere. For combined HP, HT testing, the test unit lid was punctured to allow silicone oil (Dow Corning LL 200/100 cs) to fill the package. The unit with punctured lid was placed in a silicone filled pressure tank as shown in Figure 2. Electrical connections to test unit were provided by a 21 pin electrical connector in the lid of the pressure tank. The pressure tank was placed in a laboratory oven for thermal ageing test.

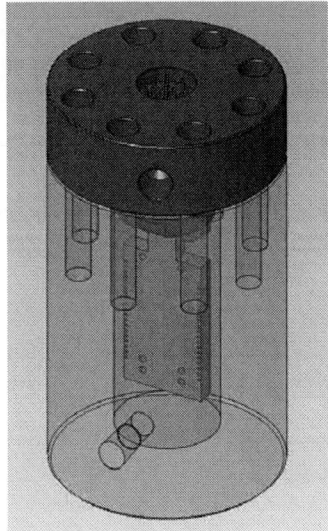

Figure 2 Pressure Tank

Thin and thick film resistors were characterized versus pressure and temperature in the range 1 – 1000 bar and 20 – 200°C. Tests were performed at temperatures 20, 50, 100, 150 and 200°C in combination with pressures 1, 250, 500, 750 and 1000 bar. Test units were subjected to thermal ageing for 6000 hours at ambient conditions according to Table 2. Resistor measurements were recorded regularly using an Agilent 34970A Data Acquisition/Switch Unit.

Table 2 Environmental stress condition matrix

Ambient condition	P [bar]	T [°C]	Ambient medium
A	1000	200	Silicone oil
B	1	200	N₂
C	1	250	N₂

Reference and thermally aged resistors were externally inspected by light microscopy. Cross sectioned resistors were inspected by light microscopy and electron probe for micro analysis (EPMA) to investigate the effects of the environmental conditions in this study. Comparative analyses were performed on reference units and test units after thermal ageing test. The EPMA analysis, performed on CAMECA model Sx100, included element scan on selected resistors.

Results

Initial measurements were carried out to characterize the thin and thick film resistors over the temperature range from 20 – 200°C and the pressure range from 1 – 1000 bar. Figure 3 shows the pressure dependence of the thin and thick film resistors. The thin film resistors have pressure changes less than 200 ppm from 1 – 1000 Bar. For the thick film resistors, there is linear pressure dependence and decrease in resistance of 1.5 - 3 % for the pressure range from 1 – 1000 Bar. The pressure dependence is stronger for high sheet resistance thick film resistors than low sheet resistance versions. The pressure dependence for thin and thick film resistors is insensitive to exposure temperature from 20 - 200°C.

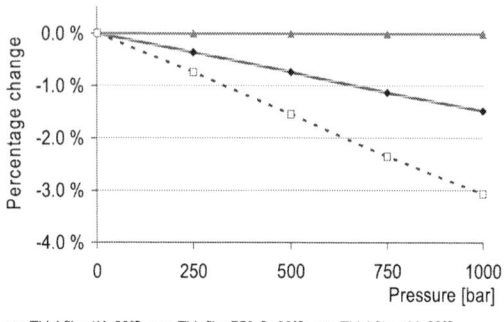

Figure 3 Change of resistance versus pressure for 750 Ω thin film and 1 k and 1 MΩ thick film resistors recorded at 20°C

Figure 4 shows resistance change after 6000 hours for resistors that have been exposed to three different environmental conditions: Test units at combined 200°C/1000 Bar, hermetically sealed at 200°C and hermetically sealed at 250°C. The hermetically sealed thin film resistors and the low sheet resistance thick film resistors (1 k and 10 kΩ) show similar ageing characteristic for long term exposure to 200°C. The relative resistance change is in the range 200 – 400 ppm. These resistors show a similar development when submerged in silicone oil at 1000 Bar; the relative change is in the range 200 – 450 ppm. The 1 M Ω thick film resistor shows higher relative resistance changes for long term exposure to 200°C, for both hermetically sealed resistor and resistor submerged in silicone oil at 1000 Bar. The relative resistance change is up to 900 ppm.

Ageing effects are clearly accelerated for resistors exposed to 250°C compared to 200°C. The effect is most pronounced for thick film resistors where relative changes in the range from 1750 to 6000 ppm are observed. The ageing effect is stronger for the high sheet resistance resistors compared to the low sheet resistance versions. For the thin film resistors relative changes up to 1100 ppm is observed.

The thin film resistors have not as pronounced increase in relative resistance change with increase in temperature as the corresponding thick film resistors.

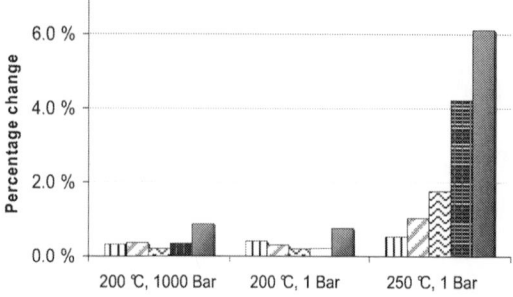

Figure 4 Resistance change versus ambient condition. Result recorded after 6000 hours (250 days) thermal ageing.

Figure 5 shows the resistance change versus time for hermetically sealed resistors exposed to 250°C. From this figure we see that resistance changes rapidly when first exposed to high temperature and then these changes are slowed down with time.

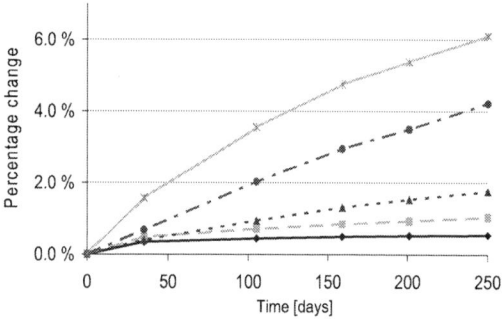

Figure 5 Resistance change vs. time for test object hermetically sealed in N_2 atmosphere, exposed to ambient temperature 250°C

Comparison of ageing characteristic for the 750 Ω thin film and the 1 kΩ thick film resistors submerged in silicone oil, exposed to long term combined HP, HT condition is shown in Figure 6. The thin and thick film resistors change rapidly when first exposed to combined HP, HT condition, changes are then slowed down with time. In Figure 6 we see that the 1 kΩ thick film resistor has stabilized at a new level after the initial rapid change. Similar development is obtained for the hermetically sealed resistors.

External visual inspection of thin and thick film resistors thermally aged at 250°C shows a color change, resulting in a grayish color shade of the passivation layer. Light microscopy and EPMA analysis on cross sectioned resistors show no change in resistor material or passivation layer for thermally

670

aged samples. This observation applies to both thin and thick film resistors.

Element composition linescans of the thick film resistors show that these are made up of a glass frit (Si-, Al- and Pb- oxides) and the conductive material (mainly Ru-oxide). High sheet resistance resistors have higher lead and lower ruthenium content than the lower sheet resistance counterpart. Various property modifying additives (Mg, Ca, Zn, Zr, Nb, Ba) in the resistive materials have been observed. The type and amount of the additives differ among the sheet resistance materials.

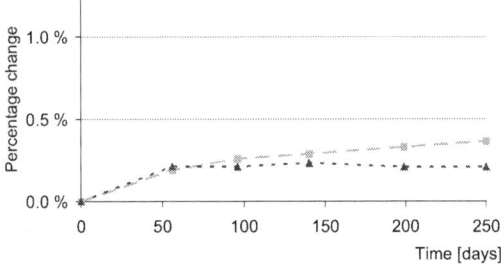

Figure 6 Resistance change vs. time for 750 Ω thin and 1 kΩ thick film resistor submerged in silicone oil, directly pressurised at 1000 Bar and exposed to ambient temperature 200°C

Figure 7 and Figure 8 shows Ruthenium distribution for the 10 k and 1 MΩ thick film resistor material. Presence of Ruthenium is indicated by the white shading in the pictures. We can see that the Ruthenium is uneven distributed for both resistor materials. The 10 kΩ resistor has higher Ruthenium concentration than the 1 MΩ resistor. The 10 kΩ resistor have bigger and more bubbles present than in the 1 MΩ resistor.

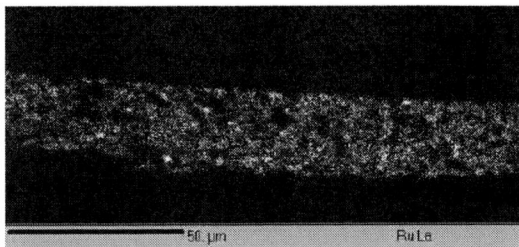

Figure 7 Ruthenium distribution for the 1 MΩ thick film resistors

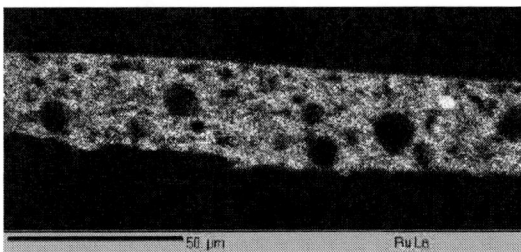

Figure 8 Ruthenium distribution for the 10 kΩ thick film resistor

Discussion

The thin film resistor consist of a thin, firm NiCr metal layer deposited on a firm ceramic substrate and covered by a mineral passivation layer. The thin conducting layer together with the constitution of firm materials will oppose external applied pressure, resulting in pressure independence for the thin film resistors. Our result, less than 200 ppm resistance changes from 1 – 1000 Bar agrees well with results obtained by Trégon et al [4]. They reported variations less than 0.1% from 1 – 1200 Bar.

The thick film resistors have a more complex structure and conduction mechanism than the thin film resistors. The conductive layer of the thick film resistors consists of a porous cermet (conductive particles inserted in a glass frit). This cermet is formed by screen printing a thick film paste onto a firm ceramic substrate and fired for curing. A passivation layer is normally covering the cermet. The linear pressure dependence of thick film resistors observed in this study is in accordance with [2,9,10], and may be explained by the difference in the mechanical property of the porous cermet versus the firm substrate material. A uniform hydrostatic pressure will result in a dominating stress perpendicular to the sheet resistance. This stress causes reduced effective distances between conducting particles resulting in a linear resistance decrease with increase in applied pressure.

The sheet resistance for the cermet is determined by the mix ratio of the conductive particles (RuO_2) and the glass frit. The relationship between conductive particle concentration in the cermet and the sheet resistance is highly non-linear in nature, as presented in [11].

Difference in pressure dependence among the different sheet resistance cermets can be explained by the combination of the porous cermet and the highly non linear relationship between content of conducting particles (RuO_2) and sheet resistances. Low sheet resistance cermets will have high concentration level of RuO_2, forming long chains of conductive particles. The resulting cermet conductivity will be dominated by the conductivity and concentration level of RuO_2. High sheet resistance cermets will have low concentration level of conductive particles; relatively short chains, with considerable gap. The resulting cermet conductivity will be dominated by the conduction of the glass matrix and distances between adjacent RuO_2 chains, and will be highly sensitive to volume changes.

The thin film resistors in this study showed excellent ageing characteristics for temperatures ranging up to 250°C. EPMA analysis shows no anomalies for aged thin film resistors. Oxidation and inter diffusion processes have been observed for NiCr thin film resistors operating at highly elevated temperatures (500 °C) showing up 100% resistance change in approximately an hour [12]. The resistance changes observed in our study during 6000 hours ageing is small indicating that oxidation and inter diffusion processes are not severe problems for temperatures ranging up to 250 °C. In some applications resistance deviations less than 1 % may be significant. Burn-in procedure can be used to stabilize the resistor values somewhat prior to operational use. Some ageing must be accounted for in the design tolerances.

The thick film resistors in this study showed ageing characteristic depending on sheet resistance. There is no observation from EPMA analysis in this study indicating that there are any oxidation or inter diffusion processes taking place during long term exposure to elevated temperatures. Ageing characteristics may be optimized somewhat by selecting proper firing profile. The difficulty in such solution will be to select a single firing profile optimizing different sheet resistance cermets.

According to [11] thick film resistor drift is due to metallurgical effects in the cermet and may be expressed as an Arrhenius type process described by Equation 1 where the value of a is approximately ½ indicating a diffusion controlled effect and an activation energy E_a close to 0.1 eV .

$$\frac{\Delta R}{R_0} = Ct^a \exp(-E_a / kT) \quad \textbf{Equation 1}$$

When applying Equation 1 to the results obtained in our study, we find that the value for a is close to ½ and the activation energy, E_a is approximately 0.85 eV for the 1 k and the 1 MΩ resistors. The high activation energy found in our study may be due to the characteristics of the thick film cermet used, the firing profile applied during the production stage and effects taking place in the thick film cermet due to the high ambient temperature.

The 10 kΩ thick film resistor has a low drift when exposed to 200° and the above equation did not fit well for this resistor.

Conclusion

The reported results show that thin and thick film resistors have characteristics suitable for HP, HT environments with temperatures up to 200 °C and pressure up to 1000 bar.

Hydrostatic pressure exposure has negligible effect on the ageing characteristics of both thin and thick film resistors.

Thin film resistors show excellent performances for combined HP, HT applications. The insignificant pressure dependence and the minor ageing effects are characteristics of main importance.

Thick film resistors are a good low cost alternative to thin film resistors for combined HP, HT applications, especially when high precision values are not required. An advantage with this technology is the substrate integration which reduce the number of interconnects and the volume occupied by the component.

Acknowledgement

The authors gratefully acknowledge the funding from Infinion Technologies Sensonor AS, Roxar Production Management AS, Presens AS, Kitron AS, Aker Kværner ASA, Read Well Services Ltd. and the Norwegian Research Council.

References

[1] "Reliable microelectronics for harsh environment applications", the Norwegian Research Council project number 159377.

[2] N. Fawcett, M. Hill, "The Electrical Response of Thick-film resistor to hydrostatic pressure and uniaxial stress between 77 and 535 K", Sensors and Actuators, Vol 78, pp 114 – 119, 1999.

[3] A. Dziedzic, A. Magiera, R. Wisniewski, "High Pressure Effect on Polymer Thick-Film Resistors", Proceedings of the 21st International Conference on Microelectronics, NIS Yugoslavia, 14. – 17. September 1997.

[4] B. Trégon, Y. Ousten, Y. Danto, L. Bechou, B. Parmentier, "Behavioral Study of Passive Components and Coating Materials under Isostatic Pressure and Temperature Stress Conditions", Microelectronics Reliability, Vol 42, pp. 1113 – 1120, 2002.

[5] R. Johannessen, F. Oldervoll, T. Fallet, P Ohlckers "Passive Component and Interconnect – Pressure Cycling at High Temperature", Proceedings from HiTec 2006, Santa Fe USA 2006.

[6] http://www.vishay.com/resistors-discrete/thin-film/surface-mount/

[7] http://www2.dupont.com/MCM/en_US/PDF/datasheets/2000.pdf

[8] http://www.coorstek.com/materials/ceramics.asp

[9] F. Oldervoll, H. Refsum, F. Strisland, "Reliable Electronics Packaging for long term 200°C Applications", Proceedings of the IMAPS Nordic Annual Conference 2005, Tønsberg, Norway 11.-13. September 2005

[10] A. Dziedzic, R. Poprawski, A. Kolarz, "Thick-film and LTCC resistors under high hydrostatic pressure", Proceedings of the 27th International Conference of IMAPS-Poland 2003, Podlesice, Poland 16. – 19. September 2003

[11] Keith Pitt, "Handbook of Thick Film Technology", Second Edition Electrochemical Publications Limited, Chapter 7, pp. 156 - 170, 2005.

[12] F. P. McCluskey, R Grzybowski, T. Podlesak, "High Temperature Electronics", CRC Press Inc Chapter 4, pp. 105 - 110, 1997

Novel Diamond Al and Diamond Cu Composites

Renaud de Langlade & Maxim Seleznev

NovaPack Technologies & ReMetAls LLC

NovaPack Technologies : Tel + 33 4 76 53 71 30; Fax + 33 4 76 53 71 31; info@novapacktech.com

Abstract

The relentless drive for miniaturization pushes operating temperatures of modern integrated circuits (ICs or chips) to their physical limits. Each additional degree of temperature spells exponentially shorter time between IC failures. Today the need to dissipate excess heat quickly becomes the priority for an electronic package designer. A well-proven approach to heat dissipation is to use high thermal conductivity materials as heat spreaders, heat sinks and base plates or substrates in IC packages. However, such metals as silver, copper and aluminium (number 1, 2 and 3 among metals ranked by thermal conductivity) are not up to the task anymore. As the metals heat up they expand much more than silicon or other semiconductor chip materials. Such expansion mismatch puts too much stress on semiconductor chip attached to a heat spreader and could ultimately destroy the chip. On the other hand, in present IC packaging technologies, the heat sinking and / or heat spreading still are largely done using copper-tungsten materials or more recently copper molybdenum. If these materials are offering acceptable TCE matching characteristics to avoid either die cracking or ceramic cracking, they have limited thermal characteristics and are quite difficult to machine, thus are quite expensive. New materials and technologies engineered to dissipate heat as good as or better than metals and control thermal expansion at the same time are urgently needed in technology marketplace. Carbon Fibre/ Al and Cu composites were initially presented. Now, several results of experience can be reported to the audience. Additionally, in the last 9 months, Diamond Particles reinforced metals have been developed and tested. Further significant advantages over the Carbon Fibre / Al and Cu composites are reported.

Key words: Heat sink, heat spreader, composites, carbon fiber, diamond particles

Basic Problem

The relentless drive for miniaturization pushes operating temperatures of modern integrated circuits (ICs or chips) to their physical limits. Each additional degree of temperature spells exponentially shorter time between IC failures. Today the need to dissipate excess heat quickly becomes the priority for an electronic package designer.

A well-proven approach to heat dissipation is to use high thermal conductivity materials as heat spreaders, heat sinks and base plates or substrates in IC packages. However, such metals as silver, copper and aluminium (number 1, 2 and 3 among metals ranked by thermal conductivity) are not up to the task anymore. As the metals heat up they expand much more than silicon or other semiconductor chip materials. Such expansion mismatch puts too much stress on semiconductor chip attached to a heat spreader and could ultimately destroy the chip. On the other hand, in present IC packaging technologies, the heat sinking and / or heat spreading still are largely done using copper-tungsten materials or more recently copper molybdenum. If these materials are offering acceptable TCE matching characteristics to avoid either die cracking or ceramic cracking, they have limited thermal characteristics and are quite difficult to machine, thus are quite expensive. New materials and technologies engineered to dissipate heat as good as or better than metals and control thermal expansion at the same time are urgently needed in technology marketplace.

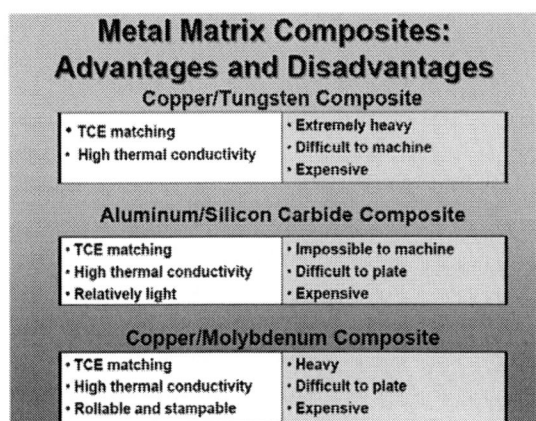

Figure 1: Metal Matrix composites advantages and disadvantages

Carbon Fiber composites as step 1

The Al and Cu carbon Fiber filled composites feature space-age technology, offer low Cost, are now widely available, light weight, with high strength and stiffness. They come with better thermal conductivity and are hermetic. They are easy to machine and plate

Figure 2 : Cu (red), Al (grey) carbon filled composites, Ni+Au plated examples

Their characteristics can be summarized in the below table. To be memorized are the better thermals with compliant CTE and low density

PHYSICAL & MECHANICAL PROPERTIES, (UNITS)	Al filled composite	Cu filled composite	NOTES
Thermal Conductivity, (W/mK)	260 - 300	300 - 340	In-plane values
Thermal Conductivity, (W/mK)	180 - 200	220 - 250	Through-thickness values
Average CTE 25 - 200C, (ppm/C)	7.0 - 8.0	7.0 - 8.0	In-plane values
Average CTE 25 - 400C, (ppm/C)	(6.5-7.5)*	6.5 - 7.5	In-plane values
Specific Heat, (J/kgK)	810	305	Calculated value
Density, (g/cm3)	2.5	6.0	Max. value
Young's Modulus, (GPa)	120	190	In-plane value

Figure 3 : Table of Characteristics

Carbon Fiber composites issues

While surpassing the good old Cu-W heat sinks by a light year, the Cu and Al carbon fiber filled composites do have issues as recognized and reported by the field of experience. It is not at all the price, very well accepted and giving a substantial competitive advantage, but rather some technical inconvenient which ever demanding applications have revealed. Their anisotropy leading users do evolve their feature into either heat sink or heat spreader but not both.

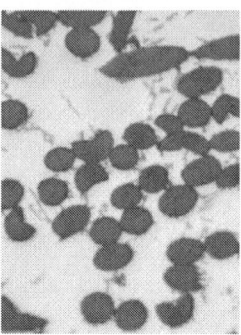 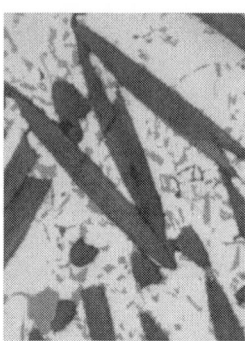

Figure 4 : anisotropy : in plane vs Z axis carbon fiber layout

Second is the brazing temperature limitation of the Al version which is not Ag-Cu compatible for ceramic attachments.

Third feedback from the market was the request for further thermal performance in the region of 500 W/m.K or in the region of 300 W/m.K but with a substantially lighter weight than Cu based existing materials.

It is therefore suggested by the market to consider new materials addressing these issues.

Diamond composites as an option

With exception to the perception of their price, diamond materials seem ideal for the challenge. Indeed, advanced Diamond Particles filled Al and Cu Matrixes can be engineered for the utmost demanding electronic thermal management applications. In fact, Diamond thermal conductivity exceeds 2000 W/mK @RT. It is now available at reasonable price as by-product and powder. Combining diamond particles with metal matrix, such as Al or Cu, results in composite materials far superior in thermal conductivity to either of the un-reinforced metal matrices.

On the other hand, introduction of diamond into metal matrix leads to difficulties in machining. Diamond metal matrix cannot be easily milled but can be wire EDM machined. Therefore at initial stage only relatively simple shapes can be offered for heat spreaders and base plates.

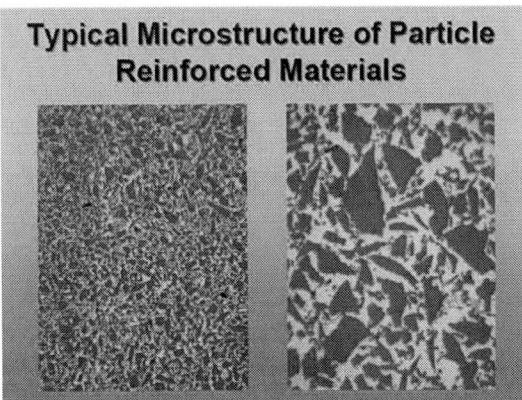

Figure 5 : isotropy is met for particles reinforced materials.

Developed Materials

With respect to the goals set by the main market drivers, both Al and Cu versions of the diamond particles composites were created. The Al version was focused on light weight while improving the thermal to its best. The Cu version was created for the utmost demanding electronic thermal management applications.

Results were met with :
- Isotropic thermal conductivity of 320 W/mK (Al) and 510 W/mK (Cu).
- Hermetic in helium leak tests performed on unplated material.
- Barely higher than Aluminium specific weight (Al Di P) or slightly highter than Titanium specific weight (Cu Di P),
- Platable with nickel or nickel/gold for reliable brazing (up to 800°C for the Cu version) or soldering (up to 450°C for the Al version).

PHYSICAL & MECHANICAL PROPERTIES, (UNITS)	AlDip	CuDip	NOTES
Thermal Conductivity, (W/mK)	320 ± 25	510 ± 30	Isotropic values
Average CTE RT-20C, (ppm/C)	N/A	6.0 ± 0	Isotropic values
Average CTE 150 - 20C, (ppm/C)	7.0 - 9.0	6.0 ± 0	Isotropic values
Specific Heat, (J/kgK)	630	410	Calculated values
Density, (g/cm³)	3.3	6.2	Max. values
Young's Modulus, (GPa)	230	360	Isotropic values
Max. soldering or brazing temperature, (degrees C)	450	800	

Figure 6 : Table of Characteristics

Shaping

Diamond metal matrix cannot be easily milled but can be wire EDM machined. Therefore at this point only relatively simple shapes have been offered for heat spreaders and base plates. With EDM developments intricate heat sink or module features can be achieved. NovaPack Technologies and its manufacturing partner AcuiTech have perfected EDM machining using plunging EDM and wire EDM to reach relatively sophisticated shapes. Additionally, the technology to produce net-shape components is under active development under the leadership of ReMetAls.

At this time a new production sequence has been validated as follows: from a block, slicing by Wire EDM to thickness of plate, make an electrode with the 3D shape, plunging EDM into the plate or block, in array or as a singulated part.

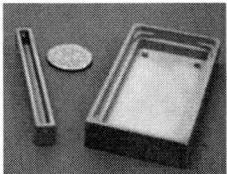

Figure 7 : example of 3D shaped objects

Plating & finishing

After the above described shaping of the newly developed materials, a plating finish is necessary prior to usage in assembly. ReMetAls has developed surface preparation of the composites so that they can be Ni or Ni+Au plated. A proprietary system has been qualified to meet industry standard tests.

Conclusion

With the careful attention and immense respect of the market material goals and assembly sequence, ReMetAls and NovaPack Technologies have reached the goal of proposing a range of new materials suitable for dramatically improved thermal management at market acceptable costs. Various proven examples now exist and confirm the validity of this technology. These are pre-molded air cavity packaging using LCP materials, ceramic assemblies and packages, and also glass-to-metal feed-through and packages or seals.

Acknowledgements

We acknowledge great support from our customer base and extend thanks to them for their outstanding contribution in setting up the goals as well as performing the tests.

References

[1] Renaud de Langlade, Novel Al and Cu filled carbon graphite heat spreader, heat sink materials for demanding die and package thermal management, ATW on Thermal Management, La Rochelle, Feb 1st, 2006.

[2] Renaud de Langlade, Novel Diamond-Al and Diamond-Cu composites, ATW on Thermal Management, La Rochelle, Feb 1st, 2007.

[3] Professor A. Kelly University of Cambridge, UK Professor C. Zweben, Composites Consultant, Pennsylvania, USA. COMPREHENSIVE COMPOSITE MATERIALS, ISBN: 0-08-042993-9, 5300 pages, publication date: 2000 Imprint: PERGAMON ; Volume 3: Metal Matrix Composites Editor: T.W. Clyne

[4] Zweben, *Advances in composite materials for thermal management in electronic packaging* JOM (USA). Vol. 50, no. 6, pp. 47-51. June 1998

[5] G. H. Ayers* and L. S. Fletcher; *Review of the Thermal Conductivity of Graphite-Reinforced Metal Matrix Composites,* JOURNAL OF THERMOPHYSICS AND HEAT TRANSFER, Vol. 12, No. 1, January– March 1998, p10-15

[6] Shen, Y. L., Needleman, A. and Suresh, S. (1994)
`*Coefficient of thermal expansion of metal-matrix composites for electronic packaging'*, Metall. Trans., 25A, 839±849.

[7] Kuniya, K., Arakawa, H., Kanai, T. and Chiba, A. (1987) `*Thermal conductivity, electrical conductivity and specific heat of copper-carbon fibre composites'*, Trans. Jap. Inst. Metals, 28, 819±826.

[8] Ellis, D. L. and McDanels, D. L. (1993) `*Thermal conductivity and thermal expansion of graphite fibre-reinforced copper matrix composite'*, Metall. Trans., A24, 43±52.

Packaging Concepts for Neuroprosthetic Implants

M. Töpper[1,2], M. Klein[1], M. Wilke[1], H. Oppermann[1], S. Kim[2], P. Tathireddy[2], F. Solzbacher[2], H. Reichl[1]

[1]Fraunhofer-Institut für Zuverlässigkeit und Mikrointegration
Gustav-Meyer-Allee 25, D-13355 Berlin
Phone: +49-30-46403-603, E-mail: toepper@izm.fraunhofer.de

[2]Dept of Electrical and Computer Engineering, University of Utah
50 S Central Campus Drive, Salt Lake City, UT 84112, USA

Abstract

A wireless neuroprosthetic implant is under development to provide a device which is able to amplify, transmit, receive and apply data for max. hundred neurons. The brain interface consists of a 100 channel amplifier, data compression, RF communication, power recovery module, two 60-turn planar coils (Au on Polyimide) on a ferrite substrate, SMD components and a 10x10 Electrode Array. The Array is the Utah Array which is a Silicon based structure with tapered Si spikes which have a length of 1.8 mm. For the mentioned components biocompatible wafer level integration and interconnect technologies have been developed based on stacking the coil/ferrite combination of the amplifier IC which is itself flip chip bonded onto the array. The SMD components were reflow-soldered beside the stack as well as special designed ceramic spacer which ensured an electrical interconnection between coil and chip. The whole assembly will finally be coated by a Parylene layer to ensure the biocompatibiliy. Test assemblies and in-vivo experiments showed already the proof of concept. An additional capping technology is under development to ensure very high reliability.

Introduction

Neuroposthetic aids require chronically implanted microelectrodes for measuring neural signals [1,2]. A fully integrated, wireless neural recording device has been developed to free the patient from the hindrance and risk of infection associated with a wired connection and to allow distribution of a network of interface nodes through the central and peripheral nervous system. Chronically implantable, wireless neural interfaces require biocompatible and long term stable high density integration of sensing, data processing, communication and power supply for operation in wet electrochemical environments. The objective of this research was to develop a biocompatible wafer level integration and interconnect technology for a stacked hybrid assembly of Silicon, Polyimide, Ceramics and SMD components for the next generation wireless Utah neural interface. The interface consists of a 100 channel amplifier, data compression, RF communication, power recovery module, two 60-turn planar coils (Au on Polyimide-Flex) on a ferrite substrate, SMD components and a modified Utah Electrode Array (UEA) including a re-routing metallization plane. The Utah Electrode Array is a Silicon based structure consisting of a 10x 10 array of tapered Si spikes with a base width of 80 micrometer and a length of 1.8 mm (see Figure 1). The tips are metallized with 200 nm Pt and 200 nm Ir, which is oxidized in an annealing process. The planar coils for powering the

electronics IC can be operated as single coils or switched in parallel or series to modify frequency range and voltage gain.

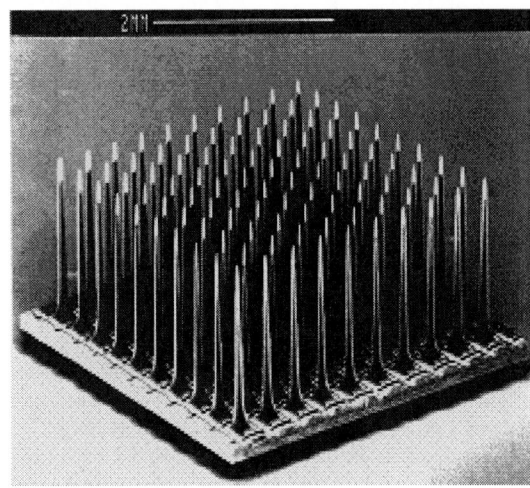

Figure 1: SEM photograph of the Utah Electrode Array.

A multi level assembly process has been developed using the modified UEA as base plate. AuSn/Au reflow flip chip bonding was used to connect the IC to the UEA and SnCu0.7 soldering for SMD, spacer and coil assembly. The ferrite substrate as well as the coil was in addition adhesive bonded. The schematic of the packaging concept is shown in Figure 2. A key challenge was to

a) limit the metallization schemes to biocompatible materials and
b) eliminate electrochemical potentials between adjacent materials wherever possible.

In following sections, individual interconnection technologies will be shown along with the components design and fabrication in which integration strategy were reflected.

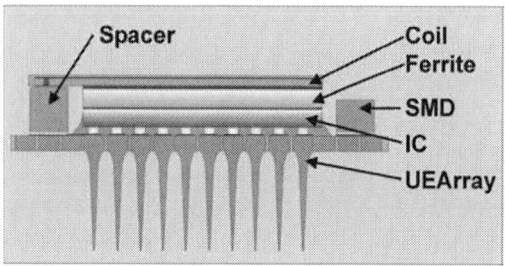

Figure 2: Schematic of the integration and packaging concept of the integrated wireless neural interface.

AuSn Reflow Flip Chip Bonding

The assembly of the electronics IC on the UEA is an essential task in the fabrication of the integrated wireless neural interface system, which can eliminate wire connections with the help of on-chip wireless data transmission. Each electrode should be connected electrically to the corresponding neural signal amplifier. The ICs are bumped by electroplating [4]. For the assembly process AuSn reflow soldering was used. AuSn reflow soldering was chosen as bumping method, since it requires lower mechanical force during assembling which minimizes stress on the UEA. Test chips were manufactured using a Si monitor wafer with 240 μm thickness. After Au/Sn electroplating the wafer was diced to get chips with the same dimensions as the functional ICs. A setup was produced with test UEAs having Au top metallization and the test ICs. initial bonding tests, functional ICs were bumped with AuSn. Figure 3 depicts images of a bumped IC after plating and after singulation. The AuSn bumped Si test chips were reflow soldered on test arrays followed by underfilling, solder paste dispensing and SMD and coil spacer mounting.

Figure 3: Light microscopy images after plating and after singulation

The biocompatible under-bump-metallization (UBM) consists of a sputtered thin film sequence of Pt/TiW/Pt/Au/TiW with respective thicknesses of 240/150/250/200/100 nm. The top TiW serves as wetting stop on leads. AuSn bumps were deposited on the Al pads of the IC chip by sputtering TiW(N)/Au and plating Au and Sn in two layers. The modules were connected using a fluxless soldering process consisting of pick & place and reflow. The Au/Sn bumps have not been reflowed prior to the reflow soldering process but were aged at elevated temperature [5].

Design and Manufacturing of Polyimide Based Coil

In order to optimize the design of coils for powering the integrated neural interface, simulation works were performed with various geometrical parameters that affect the coils' electrical characteristics [6]. Taking into account the coil design proposed based on the simulation results, the technological considerations, the UEA re-routing and the device assembly process, several optimized coil designs were manufactured based on polyimide. Single and double layer coils were manufactured with 15 μm line width and spacing and a thickness of larger than 10 μm of the Au layer. 60 turns could be realized on each layer. To avoid undesirably high inter-winding capacitance that may exist between coil layers when using double layer coils, stackable coils are used, that can be switched in parallel or series in order to tune the coil parameters and therefore the resonance frequency of the circuit. PI/Au coils were fabricated on a 4" Si wafer. The process is based on a release layer which is deposited directly on the monitor wafers. This layer based on non-filled thermoplastic polymer is necessary for the final separation of the thin film PI/Met stack. Each wafer carries 100 coils with different designs. The process for the double layer coils consists of four plating steps, starting with the polymer layer and ending with the formation of Ni/Au interconnection pads.

The plating process is done using the combination of sputtering TiW/Au and deposition of photoresist (AZ 9000 from AZ Electronic Materials). The PI is a polyamic acid type from Fuji-Film. Six different geometries of Au coils were electroplated on PI substrates. The coil stack for inductive power coupling consists of two Polyimide based electroplated Au coils with 60 turns, each. The coils are glued to the back of the IC using a 20 micrometer thick epoxy resin together with a low temperature- co-fired-ceramics (LTCC) ferrite platelet. The coils can be operated as single coils or switched in parallel or series to compensate for varying parasitic capacitances and voltage gain.

Assembly

In addition the SnCu0.7, solder paste was dispensed on the SMD pads on the array and 0402 SMDs as well as the ceramic spacer components were successfully assembled. To analyze the adhesion of SMDs and ICs, shear testing was performed.

The next assembly step was the adhesive bonding of the ferrite plate to the coil. The adhesive was dispensed on the coil which was fixed on a vacuum chuck. The ferrite plate was then aligned to and mounted on the coil applying a low bond force followed by the temperature exposure under load. In order to reduce the number of process steps, a simultaneous adhesive bonding and reflow soldering process was developed to assemble the coil/ferrite on the ICs' backside and to contact the coil to the pads on the spacer. Figure 4 depicts a light microscopy image of a fully assembled module with AuSn bonded and underfilled IC, reflow soldered SMD and spacer components and mounted coil/ferrite.

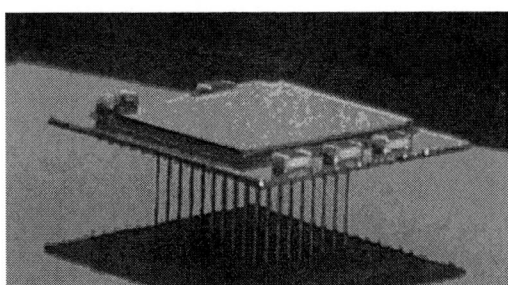

Figure 4: Light microscopy image of a fully assembled module with mounted coil/ferrite.

Capping techniques for a SiP Approach

In order to protect the system from humidity, ingression of ions and mechanical impact, a solid shell is developed that covers the most sensible components on the backside of the electrode array. To keep the assembly feasible it is split up into two parts [7]. The IC is flip chip bonded and underfilled as done before while the shell is loaded with all passive components (Fig 5).

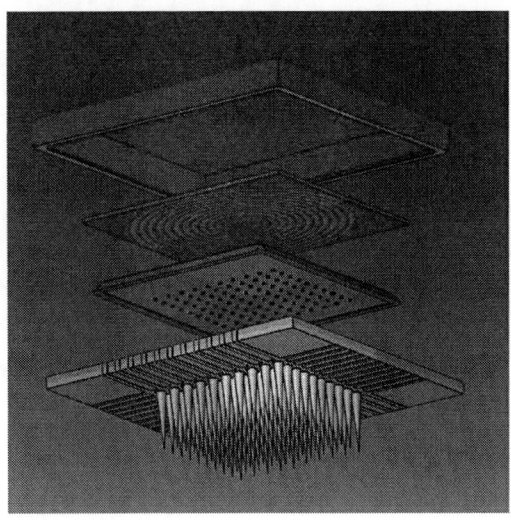

Fig. 5: Concept of capping technology for neuroprothetics

In one solder step both parts are bonded together, which ensures a hermetic sealing as well as an electrical interconnection.

Two different SIP technologies using Silicon respectively Low Temperature Cofired Ceramics as encapsulation material are presented to realize the shell. Both have to be covered with a SiC and Parylene Layer to guarantee the biocompatibility.

1. Silicon is KOH etched in order to receive tapered side walls on which an electric rewiring can be done by common photolithography and plating steps. In this case, a main issue is to deposit and expose the photo resist in an approximately 600μm deep cavity. An electro deposited photo resist is therefore compared with a spray coating process. The adaption of the mask design is evaluated in order to guarantee the resolution of the deeper located structures.
2. The LTCC shell is fabricated by laminating, stencil printing and cofiring of green tape ceramics. The advantage of this solution is that the hull serves as encapsulation as well as ferrite core for the coil and therefore saves space. Furthermore it is easy to realize the electrical rewiring. It is investigated how the thermal mismatch between silicon and LTCC influences the use of this material.

The advantage of the Silicon approach is the existing infrastructure of the MEMS industry. In addition the power transmission is not much restricted by the Silicon. In the case of LTCC the coil has to be assembled outside the lid.

Conclusions

The subject of this work was to develop new packaging concepts for the existing implant in order to improve its reliability and facilitate the assembly looking forward to establish a feasible production process. Because the system should be powered wirelessly, additional requirements arise regarding the electrical and magnetical properties of the used packaging material as well as the partitioning of the components. The purpose of the implant demands furthermore to keep the height of the package smaller than 1mm, measured from the bottom of the electrodes to the upper assembly edge.

In order to protect the system from humidity, ingression of ions and mechanical impact, different concepts have been developed and have been discussed.

Specifically, the Au/Sn electroplating process for the single chip bumping was investigated and optimized. Assembly tests were successfully performed. The process makes use of the most biocompatible combination of materials available for interconnects, takes into account process compatibility, electrochemical effects and stability in electrolytic environments, and is a general technology base for high density interconnects for implantable microdevices.

Acknowledgments

This work was supported by NIH contract No. HHSN265200423621C.

References

[1] Santhanam, G., Shenoy, K.V., "Methods for estimating neural step sequences in neural prosthetic applications," Proceedings IEEE EMBS. 1st International Conference on Neural Engineering, 2003, pp. 344-347.

[2] Shenoy, K.V. et al., "Neural prosthetic control signals from plan activity," NeuroReport 14 (2003), pp. 591-596.

[3] M. Töpper M. Klein, K. Buschick, V. Glaw, K. Orth, O. Ehrmann, M. Hutter, H. Oppermann, K.-F. Becker, T. Braun, F.Ebling, H. Reichl, Biocompatible hybrid flip chip microsystem integration for next generation wireless neural interfaces, Proceedings ECTC 2006, San Diego

[4] Töpper, M., Tönnies, D., "Microelectronic Packaging" in Semiconductor Fabrication Handbook, Ed. Geng, M.H., McGraw-Hill (2005), pp. 21.1 - 21.54.

[5] Hutter, M., Hohnke, H. Oppermann, M. Klein, G. Engelmann., "Assembly and Reliability of Flip Chip Solder Joints Using Miniaturized Au/Sn Bumps," Proc 54th Electronic Components and Technology Conf, Las Vegas, 2004.

[6] Rieth, L. et al., "Switchable LTCC/ Polyimide Based Thin Film Coils," presented in 2005 Neural Interfaces Workshop, Bethesda, MD, Sep 2005

[7] M. Wilke, M. Töpper, S. Kim, M. Klein, K.-H. Drüe, J. Müller, M. Wiemer, V. Glaw, H. Oppermann, F. Solzbacher, H. Reichl, Development of a caping process based on Si and LTCC for a wireless neuroprosthetic implant, S3 Workshop, Atlanta, October 2006

Low Cost, Biocompatible Elastic and Conformable Electronic Circuits and Assemblies Using MID in Stretchable Polymer

F. Axisa[1], D. Brosteaux[1], E. De Leersnyder[1], F. Bossuyt[1], M. Gonzalez[2], M. Vanden Bulcke[2], N. DeSmet[3] J. Vanfleteren[1]

1: IMEC/TFCG Microsystems, Ghent, Belgium

2: IMEC/MCP, Leuven, Belgium

3: University of Ghent, Belgium

Phone +32 9 264 53 54, fax + 32 9 264 53 74, E-mail: Fabrice.Axisa@elis.UGent.be, url: http://TFCG.elis.ugent.be

Abstract

For user comfort reasons, electronic circuits for implantation in the human body or for use as smart clothes should ideally be soft, stretchable and elastic. In this contribution the results of an MID (Molded Interconnect Device) technology will be presented, showing the feasibility of functional stretchable electronic circuits. In the developed technology rigid or flexible standard components are interconnected by meander shaped metallic wires and embedded by molding in a stretchable substrate polymer. Several technologies have been developed to this purpose, which combine low cost and good reliability under mechanical strain. In this way reliable stretchability of the circuits above 100% has been demonstrated. Enhanced reliability has been reached using an additional conductive polymer layer. Different demonstrators have been built and demonstrated, like a simple stretchable thermometer circuit with 4 components embedded in Dow Corning Silastic® PDMS silicone material, or a stretchable coil, used as inductive link for signal and power transfer.

Keyword: Stretchable electronics, self-healing structure, conductive polymer, MID, meander, electroplating, biomedical system

Introduction

Our daily life is more and more accompanied by electronic systems. Nowadays these systems show an increasing degree of complexity; however, this enhanced functionality may not lead to a decrease in comfort for the user. Therefore the increased complexity must be combined with advanced electronics packaging and interconnection solutions, so that compact and lightweight systems become available and do not hamper the comfort of the user. Ideally the electronics should be almost non-noticeable to the user. It is clear that in this perspective, the electronics should preferably take the shape of the object in which they are integrated. Therefore, there is currently a strong tendency to replace common rigid electronic interconnection substrates and assemblies by mechanically flexible equivalents. A further step is to not only ascertain the flexibility of the substrate, but also to make it stretchable. This emerging technology will be applied for biomedical electronics and for wearable electronics incorporated in clothes. Biomedical applications include implantable devices and electronics on skin. By making the electronic circuits stretchable, maximal comfort can be reached if the electronics behave like the tissue itself.

The main challenge in making stretchable electronic circuits consists in the development of elastic conductors, interconnecting rigid or flexible islands which hold the components. Today metals are the best option to realize this type of interconnection with high performance and low cost. Few research groups [1-4] have recently reported work on the development of stretchable metallic interconnections in elastic substrates. In this contribution we propose an approach based on the one of Gray et al. [1]. The interconnection between two points is not a straight line but a two-dimensional spring-shaped metal track.

Experiments/Method

Stretchable metallic patterns

The general process to embed tortuous metallic interconnections into a stretchable material is shown in Figure 1. A photoresist is spin coated on a copper foil and patterned with the desired conductor shape by illuminating it (with UV) through a mask (Figure 2(a)). In step 3 of Figure 1 a nickel seed layer followed by a 4 μm thick gold layer are electroplated. At last a Ni-Au finish is electroplated in order to be able to solder components on pad areas. As stretchable substrate material polysiloxanes (silicones) are considered. They are used as IC device encapsulants and protection against moisture [6]. The most common silicones used in the electronic industry are polydimethylsiloxanes (PDMS). Some types have a soft and stretchable nature, conformable to tissues. Silastic MDX4-4210 from Dow Corning has been used, which is a biomedical grade silicone elastomer. It has a low Young's modulus and a high elongation (470%). In step 5 of Figure 1 the sample is overmoulded with the viscous (≥ 60,000 cP) silicone. In step 7 of figure 1 the interconnection is embedded completely by applying another silicone layer. The total film thickness is 1mm. A stretchable electronic circuit can be made if thicker silicone areas at the soldered components can protect them against excessive strain.

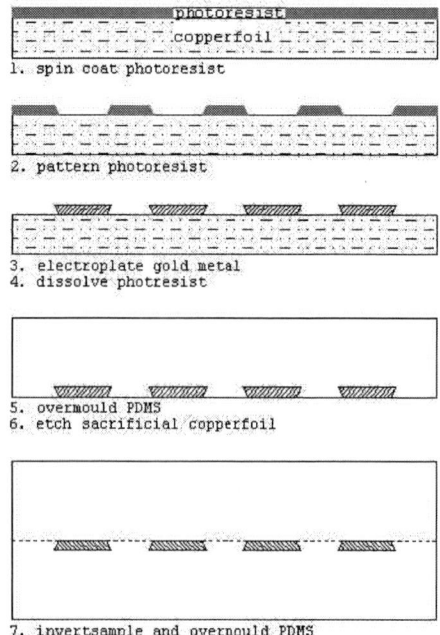

1. spin coat photoresist

2. pattern photoresist

3. electroplate gold metal
4. dissolve photresist

5. overmould PDMS
6. etch sacrificial copperfoil

7. invertsample and overnould PDMS

Figure 1: Process sequence for metallic stretchable interconnections embedded in PDMS.

Finite Element Analysis (FEA) has been used to characterise the shape of the conductors in order to allow high deformations without permanent damage. Even if simple conductor shapes like triangular or sinusoidal allow higher deformations compared to a straight line, they present a high concentration of stresses in the crest and trough, giving rise to early failures at fairly small deformations. Based on these results, a horseshoes shape was designed. Stresses are distributed in a wider region instead of being concentrated in a small zone. The horseshoe pattern is created by joining a series of circular arcs. Three different patterns were included in the mask design with a radius R of 450 μm and an angle H of 0°, 30° and 45° respectively.

The induced stresses in the metal under deformation increase drastically if a wide metal track is used. In order to increase even more the stretchability of the conductors without sacrifying electrical performance, each conductor track has been split in four parallel lines of 15 μm each. The single tracks have been made as narrow as the photolithography process allowed it with sufficient reliability, resulting in a width of 15μm. The spacing between two single tracks is also 15μm, making the whole "multitrack" 105 μm wide. On regular points, where the deformation will be minimal according to the simulations, neighboring single tracks are connected to each other in order to compensate single track interruptions caused by process faults or mechanical failure.

ID encapsulation in a biocompatible stretchable matrix

To provide a reliable way to embed components in a polymer, MID (Moulded Interconnect Device) strategy had been chosen. This technology consists in the encapsulation of the electronic system using a mould. If the electronic system is flexible, the result can be shaped in many ways. Using a mould enables to control thickness, to provide a 3D shape, and to shorten the processing time.

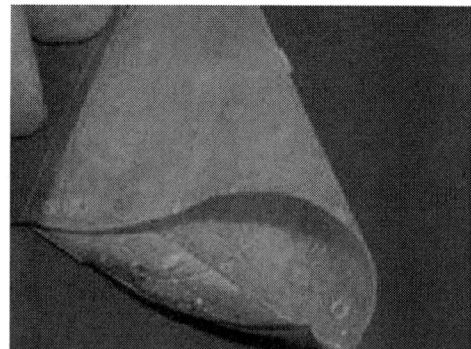

Figure 2: A moulded PDMS membrane of 500μm thick, with gold meanders embedded 250μm deep.

The MID has 4 steps (Figure 3.a). Firstly (Figure 3.a.1), the substrate with electroplated meanders and soldered components is maintained in place using a vacuum aspiration. The upper layer of PDMS is then injected and cured. Secondly (Figure 3.a.2), the vacuum plate is removed and the substrate is etched chemically. Then another injection plate is mounted on the mould to inject the final layer of

PDMS (Figure 3.a.3). After curing this layer at 150°C during 30 minutes the system is unmoulded to obtain the stretchable system (Figure 3.a.4). The material chosen for the plates was the polyetherimide ULTEM® 1000, based on its broad chemical resistance, its high heat resistance and its high mechanical strength and rigidity. A result of MID is shown on Figure 3. The moulded device is a PDMS membrane of 7cm x 7cm x 500 µm, in which gold meanders are embedded at precisely 250 µm.

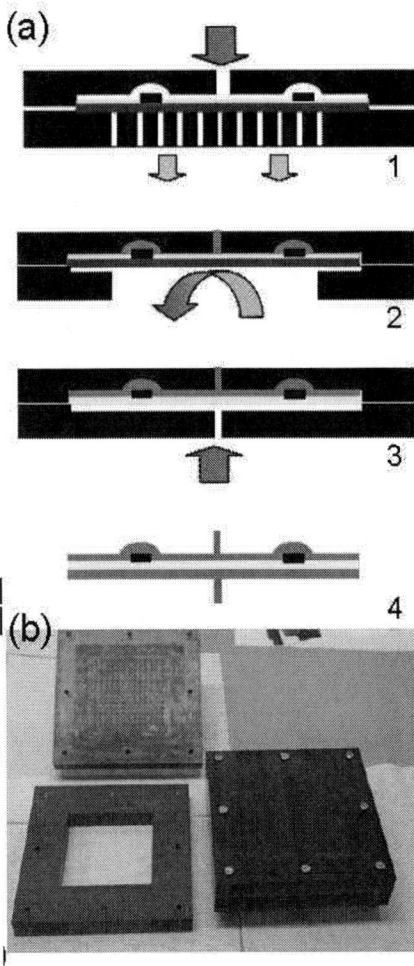

Figure 3: (a) MID sequence for stretchable circuits. (b) Mould in 4 parts for the three steps of MID.

Biocompatibility enhancement.

Silastic MDX4-4210 from Dow Corning is a biocompatible PDMS which can be used for implants. However PDMS is hydrophobic polymer. For applications which need biocompatible hydrophilic surfaces, an additional monolayer of vinyl monomer is chemically stitched to the surface. The treated surface is biocompatible and hydrophilic.

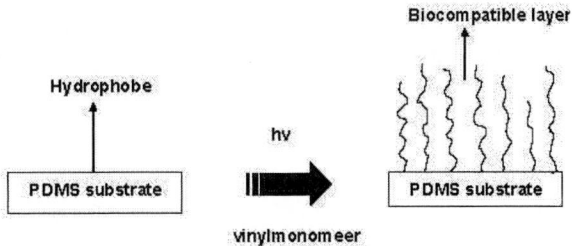

Figure 2: Treatment surface of the PDMS.

Results and Discussion

Influence of the meander shape

The goal is to create stretchable interconnections which can support at least 50% and up to 100% elongation. These stretchable interconnections links functional non-stretchable islands together in order to create stretchable electronic systems for biomedical applications, in which common SMD components, platinum plated electrodes, antennas and flexible batteries are incorporated. The first step is to create interconnections with uniaxial stretchability or 1D stretchable interconnection.

The length before elongation of an interconnection λ is 3cm (length of the straight line between the two interconnection points). L is the length of the metal track (distance when following the meander shaped track). Characterisation of an interconnection is done using two parameters:

- The resistance per running length unit of the interconnection with 0% of elongation r_0, (Ω/cm) r_0 is proportional to L/λ. The total resistance R (Ω) of a track doesn't significantly increase under stretch (see Table I), where $R = r_0 * \lambda$.

- The maximum elongation of a conductive stretchable interconnection $E_{max}=100*(\lambda F-\lambda 0)/\lambda 0$ where λF is the length at which the interconnection fails and $\lambda 0$ is the length without elongation. The theoretical maximum elongation of an interconnection is expected to be proportional to the ratio L/λ. We define the quality of an interconnection as $Q=E_{max}/r_0$. Q represents the efficiency of a stretchable pattern.

Table I: Characterization result of different patterns for stretchable interconnections.

Shape	Resistance increase at mid-elongation (%)	Mean r0 resistivity (W/cm)	Mean Emax Elongation (%)	Mean Quality Q (cm/W)
H0	1,76	0,94	57,53	61,57
H30	1,19	1,86	74,82	40,73
H45	1,28	2,23	77,68	34,77

H0, H30 and H45 have a very low increase of resistivity during elongation (below 2% of increase after mid elongation). Mean Emax of H0 is 57%, of H30 75% and of H45 77%. H0 has the higher quality (mean value 61 cm/Ω) and H45 has a significantly lower quality (see table I). When the elongation is above 50%, fabrication defaults affect definitively the reliability of stretchable interconnections and are responsible for 66% of the breaks.

In order to design reliable stretchable electronic interconnections, two solutions are available. One is using stretchable conductive polymer, the other one is to use metal meanders, which are like springs. However this first solution has high resistivity, electrical properties are too much affected by deformation and it is not solderable. The second solution is solderable, has excellent electrical properties, is not affected by deformation, but is fragile. To combinate all the advantages, the idea of shunted meander is to deposit on the top of the metal meander a layer of stretchable conductive polymer. If the metallic meander is broken somewhere, conductive polymer acts like a bridge or a shunt and ensure global connectivity. Even if the conductivity of the polymer is low, it shunts the meanders only over a short distance, and therefore the global resistivity remains high. When the global structure is stretched, the structure is still highly conductive and reliable at 100% of elongation. Conductive polymer can be deposited on a large band, or a narrow band, or multi narrow band, or in a meander shape. The polymer can be silver-filled polymer, organic conduction polymer (polypyrrole, Pedot, polyannilline, etc..), but should be stretchable. The combination of metallic meanders and a conductive polymer is a self healing structure, which enables us to have metallic stretchable conductor with high reliability.

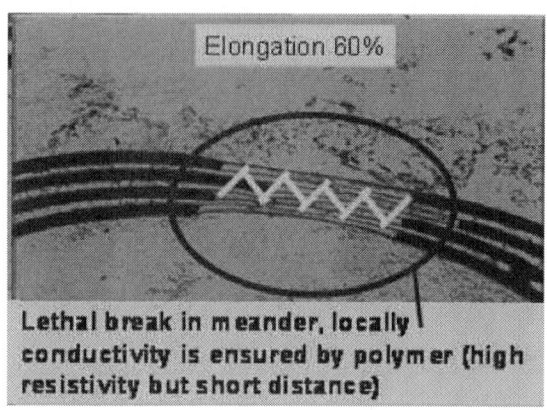

Figure 4: Shunting effect of conductive polymer on the metallic meanders.

Figure 7 represents the evolution of linear resistivity of shunted meander (meanders : H45, with a diameter of 450μm, Ni-Au-Ni-Au ; conductive polymer : XCS80091-1 from Emerson and cumming, 300μm width, 300μm height). Without deformation, the resistivity is 0.7 Ω/cm. Until a deformation of 60% the resistivity is less or equal to 0.8Ω/cm. Then the linear resistivity increase linearly to 1.2Ω/cm under a deformation of 100%. If the tension is released, the resistivity decreases linearly to 0.8 Ω/cm at 60% of deformation, and then goes back to 0.6Ω/cm. Until 60%, the meander is intact, and then all the structure behaves exactly like a simple meander. Beyond 60%, cracks in the meander appear and the conductive polymer repairs it, which induces a larger increase of the resistivity.

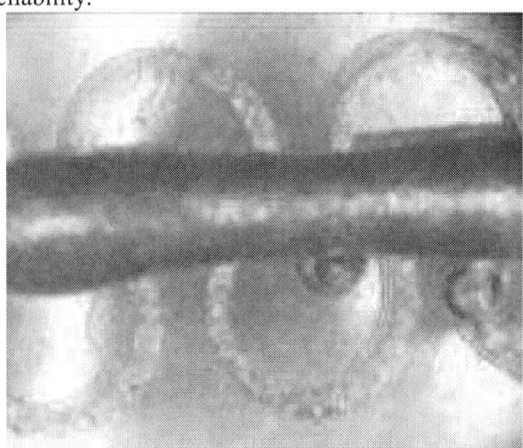

Figure 3: Conductive silver filled silicone (XCS80091-1 Emerson and Cuming) dispensed of conductive metallic meander.

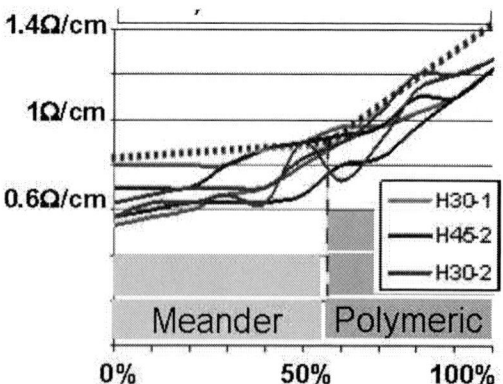

Figure 5: Shunted meander's evolution of the resistivity under deformation of 100%.

Multilayer and electrodes

The need for connecting different layers in a multilayer electronic circuit embedded in PDMS has driven us to make vias. In a layer of 25 um PDMS, the possibility of making vias with reasonable dimensions is tested. Laser ablating using a CO_2- and YAG-laser is the chosen technology. We are able to make vias with a diameter of 50 um by using a YAG-laser and 85 um by using a CO_2-laser. Copper electroplating is technique to fill the vias . If using platinum instead of copper electrodes, vias can be used as electrodes as well.

Figure 8: Profilometer view of the vias through PDMS Layer.

Integration in breathable substrate

PDMS is biocompatible, and has a very low permeability to water. However this last feature is a problem for the conception of wearable stratechable electronic system, as evaporation of sweat is not possible. For wearable system, there is a need of stretchable and wearable system.

The way to embed stretchable and breathable electronic system is to embed only the components and their junctions in PDMS and to use another material (e.g. a non-woven) to hold the separate functional islands of the stretchable material together. This last technique is used to produce electronic and breathable system. The presence of a non-woven also leads to a better stretchability of the sample on the parts that are not covered with silicone.

Biomedical stretchable electronic

The first operating electronic circuit (figure 9) using this type of stretchable interconnection is a standard SMD mounted LED HSMC-S690, soldered on interconnection pads at one end of a 3 cm long interconnection, and linked with two other pads to an external circuit. The LED and the interconnection are embedded in 500 μm thick Silastic MDX4-4210 from Dow Corning. This circuit is very flexible, twistable and has a maximum stretchability above 60%. Pads and SMD component are protected from excessive strain with thicker silicone pad.

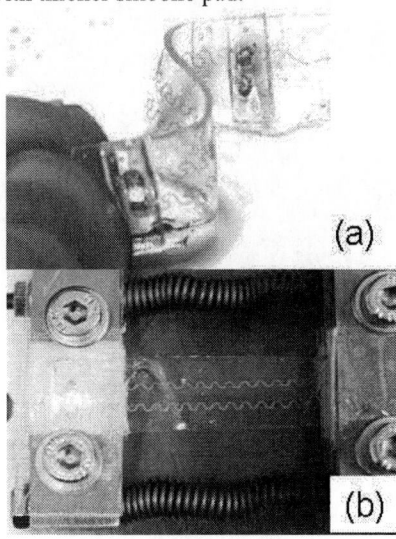

Figure 9: (a) Twistable blue LED circuit embedded in silicone and (b) operating under 35 % elongation.

The second step is to produce larger, functional, multi-component systems. The chosen system is a thermometer to monitor fever. This thermometer is composed of a head band to set it on the front and of a stretchable circuit with a temperature sensor to measure the skin temperature. The stretchable sensor circuit is attached to the head band. The whole system is a wearable healthcare monitoring system. This thermometer (Figure 10) is composed of a SMD embedded thermometer LM92 from National Semiconductor, a 0.33°C accurate 12bit temperature sensor with an I2C serial interface, with a current supply of 625 μA, and 8 pads. A decoupling capacitor, a resistor and a On/Off LED are also implemented in the stretchable circuit.

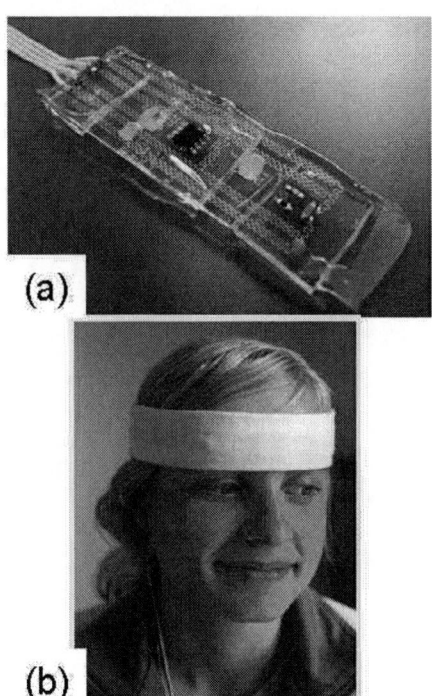

Figure 10: (a) Stretchable thermometer for fever monitoring. (b)This sensor can be embedded in a wearable textile head band.

One of the key point of implantable system is energy. We have developed induction power transmission system in PDMS matrix. This system is not stretchable. However it is biocompatible, very flexible and very soft. It is a subcutaneous 10kHz induction antenna which can feed implant in deep body using a stretchable power transmission line.

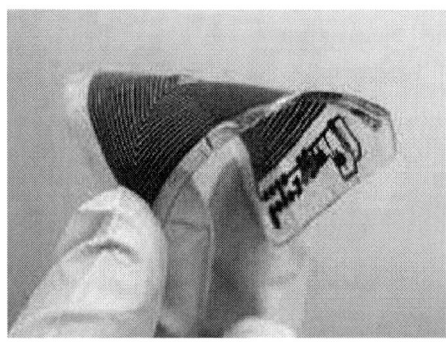

Figure 11 Induction power coil in PDMS matrix.

Conclusion

We have developed a simple, low-cost process, using standard PCB manufacturing technologies, for stretchable interconnections with an elongation capability of up to 100%. Those stretchable interconnections can be used to produce stretchable electronic systems, using standard SMD components and soldering technologies. Embedded in biocompatible silicone like Silastic MDX4-4210 from Dow Corning, these systems can be implantable or be a part of a biomedical system. Moulding technologies applied to stretchable interconnections enable us to produce 3D stretchable systems, with any type of shape. Moreover different polymer types can be used for other types of applications like implantable biomedical systems, smart textiles, 3D shaped flexible or stretchable systems, strain absorbing systems, sensors, actuators, robotic skins, etc.

Moreover functional design had been processed and tested for wearable biomedical systems. It is possible now to design more complex systems, including stretchable antennas and batteries, for wireless systems.

Acknowledgement

The work is supported by the Institute for the Promotion of Innovation by Science and Technology in Flanders (IWT) through the SBO-Bioflex project (contract number 04101). This work is also supported by European Commission Research programme STELLA (contract number 028026) www.stella-project.eu

References

[1] D. S. Gray, J. Tien, and C. S. Chen, "High-conductivity elastomeric electronics," Adv. Mater., vol. 16, no. 5, pp. 393-397, Mar. 2004.

[2] S.P. Lacour, J. Jones, S. Wagner, T. Li, S. Zhigang, "Stretchable Interconnects for Elastic Electronic Surfaces", Proceedings of IEEE, Vol. 93, N°. 8, august 2005.

[3] S.P. Lacour, J. Jones, S. Wagner, "Design and Performance of Thin Metal Film Interconnects for Skin-Like Electronic Circuits", IEEE Electron Device Letters, Vol.25, N°4, April 2004

[4] M.N. Maghribi, PhD Thesis LLNL, Preprint UCRL-LR-153347, 2003

[5] T.A. Green, M.J. Liew, S. Roy, "Electrodeposition of Gold from a Thiosulfate-Sulfite Bath for Microelectronic Applications", Journal of the electrochemical society, 150(3),C104-C110,2003

[6] C.P Wong, "High Performance Screen Printable Silicone as Selective Hybrid IC Encapsulant", IEEE Transactions on components, hybrids and manufacturing technology, Vol. 13, N°4, December 1990

[7] D. Brosteaux, F. Axisa, J. Vanfleteren, N. Carchon, M. Gonzalez, "Elastic Interconnects for Stretchable Electronic Circuits using MID (Moulded Interconnect Device) Technology", MRS spring 2006.

Packaging of an Implantable Accelerometer for Measurements of Heart Motion

Kristin Imenes[1], Knut Aasmundtveit[1], Ellen M. Husa[1], Jan Olav Høgetveit[2], Steinar Halvorsen[2], Ole Jakob Elle[2], Erik Fosse[2], and Lars Hoff[1]

[1]Vestfold University College, Horten, Norway

[2]Rikshospitalet-Radiumhospitalet Medical centre, University of Oslo, Oslo, Norway

Phone: +47 33037714, Fax: +47 33031103, E-mail: kristin.imenes@hive.no

Abstract

In heart bypass surgery there is always a risk for blood vessel occlusion which may result in heart infarction. Previously we have shown in animal experiments that a three-axis accelerometer attached to the heart surface can monitor the heart movements and give valuable information about the heart condition. We have now developed an assembly and packaging method for a new generation sensors for clinical trials. The sensor is a commercially available three-axis capacitive micro machined accelerometer. This has been soldered onto a ceramic substrate and connected to a cable for power supply and signal output. Medical silicone was chosen as encapsulation material to meet the strict requirements for implantable sensors. It is electrically insulating, prevents body fluid from penetrating, withstands a sterilization process and is approved for implantation up to 29 days. The assembled sensor was press moulded and heat cured with good results; no entrapment of air, no sticking to the mould and always good vulcanization. Tests of the encapsulation showed leakage currents well below the 10μA limit for direct heart applications. The hydrogen peroxide gas plasma sterilization method was selected mainly because its low temperature does not harm the electronics. In animal experiments the sensor was sutured to the heart surface and performed very well, the oscillating forces from the beating heart did not cause fatigue failures in the silicone. The sensor has now been qualified for clinical trials. Our ongoing work heads for miniaturization of the sensor and the packaging to also allow for postoperative use.

Key words: micro accelerometer, silicone, moulding, heart infarction, assembly, packaging

Introduction

In heart bypass surgery there is always a risk for blood vessel occlusion during and after surgery [1]. A blood vessel occlusion may result in heart infarction and therefore "real time" monitoring of heart function during and after surgery is vital. We have previously demonstrated that a three-axis accelerometer can be used to detect heart infarction in animal experiments [2]. This study used a prototype accelerometer. To perform further studies on humans, a second implantable prototype sensor for clinical trials has been developed.

A challenge in developing a new implantable microsystem in medical applications is the lack of standards and suitable packaging. The assembly and packaging methods diverge greatly depending on the application and its purpose [3, 4]. Furthermore implantable devices must also be biocompatible and able to survive a sterilization procedure. Another challenge arising when sensors are miniaturized is that the packaging and connections tend to enlarge the size of the final product. This paper presents the assembly and packaging methods of an implantable micro accelerometer, for use in clinical trials to detect heart infarction. During open-heart surgery the sensor is to be attached to the heart surface and wire-connected to an analysis system.

Materials and Methods

I. Sensor and Assembly

The sensor to be encapsulated was a three-axis accelerometer from Kionix Inc (KXM52-1050, Ithaca, NY, USA). This is a capacitive micro machined accelerometer with integrated signal conditioning front end circuitry. The casing was a "Dual Flat No-lead" package with the outer dimensions 5.0 mm x 5.0 mm x 1.8 mm and 14 I/O terminals underneath. The performance specifications are given in table 1.

External capacitors were used to set the -3 dB low-pass filter frequency for each sensor output. In this case a bandwidth harmonizing with the frequency spectrum of the heart motion (0 – 100 Hz) was desired. For that reason we chose 47 nF capacitors, giving a 105 Hz bandwidth.

A medical cable from New England Wire Technologies (Lisbon, NH, USA) was used for power supply and signal output. This cable is round with a silicone rubber jacket and an outer diameter of 2.06 mm. The silver plated copper conductors are 0.23 mm in diameter and are twisted in pairs. The cable overall is very flexible which is essential to ensure that the cable and the sensor do not influence the heart motion.

Table 1. Performance specifications of the KXM52-1050 from Kionix Inc.

Property	Value
Range	±2 g[1]
Sensitivity[2]	660 mV/g[1]
Supply voltage	2.5 - 5.5 V
Current draw[2]	1.5 mA
3 dB bandwidth	0 – 3000 Hz (x and y direction)
	0 – 1500 Hz (z direction)

[1] $g = 9.8 \text{ m/s}^2$
[2] at 3.3 V

Accelerometer, capacitors (surface mounted, size "0402") and cable leads were soldered on to an alumina substrate with thick film printed conductors (MicroComponent A/S, Horten, Norway). The size of the substrate is 5.0 mm x 11.5 mm x 0.625 mm. By means of a wire, the cable was also anchored to the substrate. Figure 1 shows the assembled sensor.

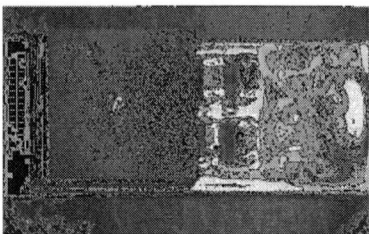

Figure 1. Assembled sensor (A) with capacitors (B) and cable termination (C).

II. Encapsulation

The choice of encapsulation material depends on the physical requirements of the implant. All categories of materials are used in vivo; metals, polymers, ceramics and composites [5]. In our case the main requirements were a material approved for at least 7 days of implantation, it had to be electrically insulating and prevent body fluid from penetrating. Silicone was chosen because it is a well established material for use in invasive devices and it is suitable for prototyping and moulding. The material also satisfied our encapsulation requirements.

To minimize the injury to heart tissue and the surrounding body tissue, an encapsulation with a round shape was preferred. Combined with a smooth surface, this shape also ensured easy and proper cleaning in case of the need to reuse the sensor. A reliable fixation to the heart was important to minimize the self-motion of the sensor, which could interfere with the sensor signals and cause misinterpretations. It was decided to fix the sensor to the heart surface by sutures, thus the encapsulation was designed with suture holes in each corner.

As a result of these conditions, an encapsulation as shown in figure 2 was designed. The outer dimensions of the encapsulation are 11.0 mm x 14.5 mm x 5.2 mm. A two part negative mould was milled out of aluminium according to this drawing.

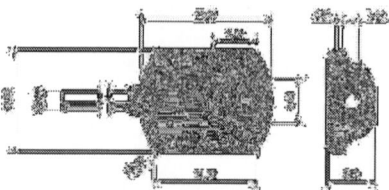

Figure 2. Dimensional sketch (mm) of encapsulation design

The platinum-catalyzed addition curing silicone, Elastosil R Plus 4001/40, from Wacker Chemie AG (Munich, Germany) was selected as encapsulation material. This silicone is a one-component high viscosity material which is transparent and possesses very good mechanical properties. Some of the material characteristics are presented in table 2.

Table 2. Elastosil R plus 4001/40 specifications (from data sheet)

Property	Value
Appearance	Transparent
Press cure	15 min / 165 °C
Post cure	4 h / 200 °C
Cure system	Platinum
Tensile strength	11.0 N/mm^2
Elongation at break	940 %

The silicone was formed around the sensor by hand and then placed in the aluminium mould. To get the right amount of silicone each time, the silicone was weighed. The two parts of the mould were pressed together, by use of a clamp, and placed in a heating chamber. The assembly was pre-cured at 145 °C for 2 hours and then the sensor was removed from the mould and post cured for 7 days at 145 °C. In order to obtain good sealing between the silicone and the cable, a primer (G790, Wacker Chemie AG, München, Germany) was used. The primer was applied to the cable, substrate and components and was allowed to dry for 15 minutes at 100 °C prior to moulding. Figure 3 shows the moulded sensor.

Figure 3. Encapsulated sensor.

III. Cleaning and Sterilization

The assembly of the sensor was done in a non-sterile environment. Thus, the sensor and part of the cable were cleaned with 100% isopropanol to remove surface contaminations prior to moulding. Isopropanol is also an effective bactericide [6]. The preparations and the moulding were carried out in a clean room class 100 000.

The sensor with encapsulation and cable had to undergo a sterilization process, and a low-temperature hydrogen peroxide gas plasma sterilization method was selected. This method benefits from the short cycle time (less than 1 hour), low process temperature and humidity; no aeration is required and no toxic residue is left. It is also compatible with a wide range of materials, among them silicone [7]. A Sterrad 100S system (Advanced Sterilization Products, Irvine, CA, USA) was used to sterilize the device.

Steam sterilization was also assessed but this method is not recommended for materials and devices that cannot tolerate high temperatures and humidity. Another method which also operates in the lower range of temperature and relative humidity is ethylene oxide (EtO) sterilization. It is convenient for silicone and electronics, but the drawback with this method is the toxicity of the EtO gas [6].

IV. Approval for Clinical Use

Use of medical devices is regulated by the national authorities in Europe through the Medical Device Directive (Council Directive 93/42/EEC of 14 June 1993 concerning medical devices), and is based on the IEC 60601 standard (CEI/IEC 60601-1 Medical electrical equipment, General requirements for safety) in Norway. At Rikshospitalet-Radiumhospitalet Medical centre a local procedure has been established in agreement with the national authorities in order to allow non-approved equipment to be used on patients in a legal and secure way. For the accelerometer, a major concern was to keep the current leakage below the limits specified in the IEC 60601. The current should not exceed 10 µA for cardiac floating devices under normal conditions or 50 µA under "single fault condition".

Experiments

I. Leakage Current

The encapsulation must seal the electronic components, allowing no moisture to diffuse through the silicone or penetrate along the cable causing leakage currents. To preliminary inspect the silicone sealing we immersed the sensor into a saline solution and measured the leakage current over a period of 4 days.

Current leakage tests were done on devices with the sensors replaced by platinum electrodes to avoid any unwanted chemical reactions at the electrodes. The actual electrode was a rectangular platinum piece with a platinum wire on one end. By removing the leads from the medical cable (New England Wire Technologies, Lisbon, NH, USA), the left-over silicone rubber jacket was used as an insulation for the platinum lead. The electrode was then moulded in silicone.

One moulded platinum electrode and one bare platinum electrode were immersed in a 0.9 % NaCl water which was heated to 37 °C to simulate physiological conditions. The electrodes were connected to a source meter (2602 System SourceMeter, Keithley Instruments Inc., Cleveland, OH, USA) and a 5 V DC signal with 10 mA current limitation was applied. Figure 4 shows the test set-up. The leakage current was measured over a period of 4 days on 10 devices with voltage applied only when measuring.

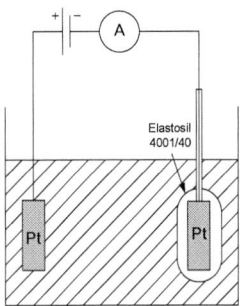

Figure 4. Leakage current measurement set-up. Electrodes immersed in saline solution. The left electrode is bare and the electrode to the right are moulded in silicone.

In addition one operating encapsulated accelerometer was subjected to two different high voltage tests. The sensor was immersed in a tub with 0.9% NaCl water and all sensor leads were first connected to 230 V AC. Furthermore the sensor leads were directly connected to an isolation measurer (Metriso 5000, Metrawatt GmbH, Nürnberg, Germany) which generated 5000 V DC.

II. Tensile Test

When the sensor is sutured to the heart wall it is exposed to cycling strain forces. Strain values up to 20 % have been measured in the heart [8]. Consequently large forces apply to the suture holes, threads and the heart tissue. The suture thread is thin and has a high tensile strength and depending on how tight the suturing is, it can, over time, start cutting the silicone and/or the epicardium. In the data sheets for Elastosil R Plus 4001/40 the tensile strength is stated to be 11.0 N/mm^2 and the elaongation at break is 940 %. These numbers are given with reference to test procedures not necessarily valid for our application. Therefore a static tensile test was performed on the actual silicone encapsulation.

A dummy device without sensor inside was moulded. The device was fastened in two of its suture holes on one side by suture thread (Prolene 5-0, Ethicon Inc., Somerville, NJ, USA) in an axial stretch loader (LR50K, Lloyd Instruments Ltd, Fareham, UK) with a 500 N loadcell (NLC 500N, Lloyd Instruments Ltd, Fareham, UK). The initial distance between the two holes was 7.5 mm.

III. Animal Experiments

The encapsulated three-axis accelerometer was tested in a porcine animal model according to a protocol approved by the Institutional Animal Care and Use Committee. The sensor was sutured on the left ventricle free wall in the region of blood supply of the left anterior descending coronary artery. The picture in figure 5 is from one of the experiments.

Figure 5. Accelerometer sutured on the left ventricle free wall.

Data from the sensor was acquired by a NI-USB 6009 data acquisition board and software written in Lab View (both from National Instruments Inc, Austin, TX, USA). As a reference, the ECG signal was sampled synchronously. The results were stored in a computer, and Matlab (The MathWorks Inc, MA, USA) was used for signal processing.

Results and Discussion

I. Sensor and Assembling

Both the three-axis accelerometer and the assembling satisfied our requirements for this prototype sensor. It has given us valuable experience in the further work on miniaturizing the sensor and the packaging to also allow for postoperative use.

The size of the assembled prototype sensor was limited by the given size of the three axis accelerometer. Most of the assembly was done by hand resulting in practical limits as to how small the dimensions could be. Soldering of the leads onto the substrate together with a mechanical anchoring of the cable also added to the total size.

In the design of the ceramic substrate there where not drawn separate test points. In addition, all the traces on the substrate were covered with an insulation layer. This made trouble shooting in faulty circuits more difficult as only the component leads were accessible for probing.

II. Moulding

Press moulding with Elastosil 4001/40 was a simple and straight forward method which gave very good results; no entrapment of air, no sticking to the mould (unnecessary with mould release agents) and always good vulcanization.

The lack of a fixation of the sensor inside the mould made it difficult to orientate the sensor exactly. Only the cable was to some extent guiding the sensor during moulding. However the thick silicone walls allowed for small disorientations so that the entire sensor was always covered with silicone. If the wall thickness is to be reduced, in order to gain a thinner and smaller encapsulation, a fixation of the sensor must be implemented.

According to the data sheet, the Elastosil 4001/40 silicone can be pre cured at 165 °C for 15 min and post cured at 200 °C for 4 h. But the specified max storage temperature for the accelerometer (150 °C) limited the upper curing temperature, and both the pre and post curing time was prolonged.

III. Leakage Current

Equipment intended for direct cardiac application shall, under normal operation, not exceed 10 μA in patient leakage current (IEC 60601). Our tests showed effective zero leakage current since the measurements were within the noise floor. Nor did we observe any change or drift in the measurements over the period of 4 days. During moulding the alignment of the electrodes was difficult and the wall thicknesses varied from 0.5 mm to 1.5 mm. However, the variation did not affect the test results. The leakage current results are presented in table 3.

Table 3. Leakage current measurements[1] of 10 Elastosil 4001/40 encapsulated electrodes immersed in saline solution.

Immerse time	Mean	Standard deviation	% of max patient leakage current (10μA)
4 min	0.014 nA	0.008 nA	$1.4 \cdot 10^{-6}$ %
4 days	0.002 nA	0.006 nA	$0.2 \cdot 10^{-6}$ %

[1]The current measurement range of the source meter was set to 100.00 nA which according to the instruments specification gives a measurement accuracy of ± (.0005 CMR + 0.100 nA) = ± 0.15 nA. The average noise floor was measured to 0.005 nA.

The operating encapsulated accelerometer that was subjected to high voltage trials did not generate a current leakage higher than 5 μA. This was also true after enduring stress testing of the accelerometer.

IV. Tensile Test

The results from the static tensile test are presented in table 4 and they are within expected values compared to the given values in the material data sheet. If we assume a moderate strain value of 10 % at the heart surface, the sensor would experience an elongation of 0.7 mm, and worse, if the strain value is 20%, the elongation would be 1.5 mm. The test showed that Elastosil 4001/40 has a good margin in relation to the assumed strain values to which the encapsulation would be exposed. Another observation when performing the tensile test was the sutures' cutting the silicone. To some extent the Elastosil 4001/40 showed cutting resistance.

Table 4. Results from tensile test of Elastosil R plus 4001/40 encapsulation.

Property	Value
Load at break	10 N
Elongation at break	41.5 mm
% elongation at break	553 %

V. Animal Experiments

During animal experiments the encapsulated accelerometer sensor performed very well. The sensor replicated our previously reported results [2] and gave additional support to the claim that accelerometer data can be used to detect regional cardiac ischemia. There was no biological response to the encapsulation and no damaging of the tissue caused by the shape of the sensor. Suturing the sensor to the heart resulted in good fixation and made the sensor follow the heart motion closely.

When the sensor is subjected to the cyclic strain force from the heart for a long period of time it may eventually fail because of fatigue. The heart frequency of an adult at rest is 70 beats/min or 1.2 Hz. During the pig experiments with duration up till 4 hours, no cutting in the Elastosil 4001/40 silicone was observed. Instead the suture started cutting the heart tissue. This problem was reduced by using thicker sutures. Alternative suturing techniques was also tried without changing the encapsulation design. 3 fixation points were acheived by two sutures at the upper part of the sensor and one around the cable. This suturing technique also performed well. Because the sensor is relatively large, less fixation points would increase the self-motion of the sensor.

Further miniaturization

Our further work heads for a miniaturization of the sensor and the packaging to also allow for postoperative use. After use the device is to be pulled out, like a temporary pacing lead. As a comparison the size of a pacing lead tip from Medtronic (Minneapolis, MN, USA) is about 0.7 mm in diameter and 4 mm in length. This device is pulled out after 4-5 days of surveillance without medical complications. Therefore it is naturally to let the dimensions of the prototype accelerometer be in this range.

The size of the actual sensing element is the first challenge to overcome. Hitachi Metals America, Ltd. (Arlington Heights, IL, USA) claims to have the smallest 3D accelerometer in the world; 2.9 mm x 2.9 x 0.92 mm. But even this is too big when the dimension of the device is to be about the size of a temporary pacing lead tip. One part of our ongoing work is to design and fabricate a silicon three axis accelerometer with a strict specification regarding size. The design is based on seismic masses supported by cantilever beams. Acceleration is transduced by piezoresistors in the top of the beams [9].

With the first prototype sensor we experienced that the termination of a round cable occupied almost 1/3 of the total size. This can be considerable reduced by using a micro ribbon cable. Today Temp-Flex, Inc (South Grafton, MA, USA) and W. L. Gore & Associates GmbH (Pleinfeld, Germany) produces micro ribbon cables with wire diameter down to 25 µm and a pitch of 100 µm or less. One drawback with ribbon cables could be higher stiffness compared to a round cable. A ribbon cable may also allow for avoiding the use of a substrate. This requires the bonding pads on the silicon die to be placed in a pattern corresponding to the wire pitch. It also requires that no off-chip components are needed.

When the device is to be pulled out through the chest it is essential that the interconnection of cable and die has high strength. A fracture, leaving the sensor inside the body when the cable is being pulled out, is not tolerated. One possible way to encapsulate the sensor and the cable, and at the same time strengthen the interconnection, is by use of an additional tubing. A very thin and flexible tubing also encapsulates the components in a uniform manner. As an example, Dow Corning (Midland, MI, USA) offers silicone tubing for use in medical devices with wall thickness/outer diameter down to 0.17 mm / 0.64 mm.

Conclusion

A silicone encapsulated three-axis accelerometer with a signal output cable has been successfully assembled and packaged. Press moulding with silicone was found to be a simple and straight forward method which gave very good results: there was no entrapment of air, no sticking to the mould and always good vulcanization. During implantation no moisture may diffuse through the silicone or penetrate along the cable causing leakage currents. Our test showed leakage currents well below the 10 µA limit for direct cardiac applications. The sensor with encapsulation and cable were sterilized with the low-temperature hydrogen peroxide gas plasma method. In animal experiments the sensor was sutured to the heart and performed very well, no fatigue failures ensued due to the alternating forces from the heart. The sensor is now approved for use in clinical trials. Our ongoing work heads for miniaturization of the sensor and the packaging to also allow for postoperative use.

Acknowledgements

This work was supported by the Research Council of Norway.

References

[1] P. K. Hol, P. S. Lingaas, R. Lundblad, K. A. Rein, K. Vatne, H.-J. Smith, S. Nitter-Hauge and E. Fosse, "Intraoperative angiography leads to graft revision in coronary artery bypass surgery", The Annals of Thoracic Surgery, Vol. 78, No. 2, pp. 502-505, 2004.

[2] O. J. Elle, S. Halvorsen, M. G. Gulbrandsen, L. Aurdal, A. Bakken, E. Samset, H. Dugstad and E. Fosse, "Early recognition of regional cardiac ischemia using a 3-axis accelerometer sensor", Physiological Measurement, No. 4, pp. 429-440, 2005.

[3] Y. Haga, T. Matsunaga, W. Makishi, K. Totsu, T. Mineta and M. Esashi, "Minimally invasive diagnostics and treatment using micro/nano machining", Minimally Invasive Therapy & Allied Technologies, Vol. 15, No. 4, pp. 218-225, 2006.

[4] W. Mokwa, "Medical implants based on microsystems", Measurement Science and Technology, Vol. 18, No. 5, pp. R47-R57, 2007.

[5] J. Black, "Biological performance of materials. Fundamentals of biocompatibility", Dekker, 1999.

[6] A. D. Russell, W. B. Hugo, G. A. J. Ayliffe, A. P. Fraise and P. A. Lambert, "Russel, Hugo & Ayliffe's principles and practice of disinfection, preservation & sterilization", Blackwell Publ., 2004.

[7] L. A. Feldman and H. K. Hui, Medical Device & Diagnostic Industry Magazine, p. 57, 1997.

[8] H. Skulstad, S. Urheim, T. Edvardsen, K. Andersen, E. Lyseggen, T. Vartdal, H. Ihlen and O. A. Smiseth, "Grading of Myocardial Dysfunction by Tissue Doppler Echocardiography: A Comparison Between Velocity, Displacement, and Strain Imaging in Acute Ischemia", Journal of the American College of Cardiology, Vol. 47, No. 8, pp. 1672-1682, 2006.

[9] C. Lowrie, C. Grinde, L. Hoff and M. Desmulliez, Piezoresistive three-axis accelerometer for monitoring heart wall motion, Symposium on Design, Test, Integration and Packaging of MEMS/MOEMS, Montreux, Switzerland, 2005.

Development of a Reliable LTCC-BGA Module Platform for RF/Microwave Telecommunication Applications

Tero Kangasvieri[1], Olli Nousiainen[2], Jouko Vähäkangas[1], Kari Kautio[3] and Markku Lahti[3]

[1]Microelectronics and Materials Physics Laboratories and EMPART Research Group of Infotech Oulu, University of Oulu, P.O. Box 4500, 90014 Oulu, Finland

[2]Materials Engineering Laboratory and EMPART Research Group of Infotech Oulu, University of Oulu, P.O. Box 4200, 90014 Oulu, Finland

[3]VTT - Kaitoväylä 1, P.O. Box 1100, FIN-90571 Oulu, Finland

Abstract

This paper describes the development of a LTCC-BGA module platform for use in telecom applications up to the K_a-band (25 GHz) frequencies. The presented module platform can serve as a physical building block for various wireless System-in-Package (SiP) products and it is designed to comply with adequate board-level reliability and RF performance requirements. The module with an area of 15x15 mm^2 is mounted to an organic motherboard (PCB) encompassing 38 DC/low-frequency inputs/outputs (I/Os) and two broadband RF I/Os. In order to fulfill the board-level reliability requirements, BGA solder joints composed of plastic-core solder balls (PCSBs) in a diameter of 1100 μm and a select lead-free solder on the LTCC side are applied. The optimized RF I/O includes high-performance BGA-via transition path, all the way from the motherboard up to the top surface of the LTCC module. The RF performance of the developed transition path was validated with on-wafer microwave scattering parameter measurements. The measurement results correlated well with electromagnetic simulations, exhibiting insertion and return loss values better than 0.8 dB and 15 dB, respectively, up to 25 GHz. The preliminary results of the thermal cycling test in the -40 to 125 °C temperature range showed that the failure free-time of the module assemblies was over 900 thermal cycles, which is considered to be adequate for many telecom applications.

Key words: BGA, LTCC, package, reliability, transition, wideband.

Introduction

Low temperature co-fired ceramic (LTCC) based system-in-package (SiP) is an emerging module technology for wireless communication applications. The SiP concept aims at integrating ICs along with passive components for achieving a single 3-D sub-system/system module package with optimized size, cost and performance. The multilayer LTCC substrate technology is a viable candidate for realization the 3-D integrated high-frequency RF/analog/digital or mixed-signal module packages, since it provides excellent microwave material properties (low loss and dispersion) and enables embedding of high-quality passive components, such as filters, baluns, power dividers, and antennas. To interconnect the module packages to other system, they are typically mounted to a common organic motherboard (PCB) using surface-mount compatible land-grid-array (LGA) or ball-grid-array (BGA) interconnection techniques. Therefore, there is a distinct need to route the broadband high-frequency digital and/or RF signal content to the motherboard via the board-level solder joints. At high frequencies, the implementation of wideband vertical package transitions is challenging, since they can easily deteriorate the signal integrity, e.g., due to reflections, excessive radiation and spurious resonances, if not properly accounted for in the design stage [1].

The LGA interconnection method enables smaller solder joints compared with BGA technique. The smaller joints results in smaller interconnect parasitics and thus improved RF properties. Moreover, the LGA joints are easier and cheaper to fabricate. However, the LGA joints are well known to be more susceptible for thermal fatigue failures due to the reduced standoff height between the module package and PCB. This is believed to restrict the maximum LTCC-LGA package size to ~ 8x7 mm^2 [2], when a coefficient of thermal expansion (CTE) difference between the LTCC and organic motherboard materials is in the order of 10 ppm/°C.

Considering the design of a non-collapsible LTCC-BGA joint structure from a reliability perspective, the stand-off height, BGA ball material, creep/fatigue resistance of applied solders and the composition of the LTCC solder pad metallization together with the solder pad structure (i.e. non-solder or solder mask defined pad) are the main factors that determine the fatigue life of the composite BGA solder joints under the thermal loading conditions. Moreover, the size, density and configuration (peripheral, full or partial area-array etc.) of the BGA joint array have an effect on the overall stress levels and distributions of solder joints and, hence on their dominant failure mechanism.

Previously conducted thermal cycling tests (TCTs) have shown that lead-free LTCC-BGA modules with a size of 15x15 mm^2, including full area-array joint configuration and elastic plastic-core solder balls (PCSBs) in diameter of 800 µm, result in the failure-free time of 500 cycles in the -40 to 125°C temperature range [3]. Moreover, there are many cases in practice, which require placing components (discrete or ICs) on the module backside along with the BGA balls in order to, e.g., obtain compact module size or leave sufficient space for antenna locating on the topside of the module. The shift from full to partial area-area array or peripheral joint configuration leads likely to reduced board-level reliability. Hence, an improvement in thermal-fatigue performance of composite BGA solder joints in LTCC/PCB assemblies is clearly needed.

This paper presents the development of a LTCC-BGA module platform, which can serve as a physical building block for various LTCC-SiP applications up to the K_a-band (25 GHz). The module platform is designed to comply with adequate board-level reliability and RF performance requirements. The module includes 38 low-frequency I/Os and two wideband RF I/Os from DC up to 25 GHz. In order to fulfill the reliability requirements, two significant changes have been made to the joint structure compared with our earlier work [3]. These changes are: 1) applying larger 1100 µm PCSBs and 2) selection and use of commercially available lead-free solder material (named here as 'SOL1'), which has better endurance against thermal fatigue than widely used ternary Sn4Ag0.5Cu solder, on the LTCC side of the joint.

Design of a Wideband BGA-via Transition

Fig. 1 shows a cross-section of the developed wideband LTCC-BGA package transition, all the way from the organic motherboard up to the top surface of the module package. The motherboard is composed of a hybrid Rogers Ro4003 / FR-4 construction. In this way the more expensive RF laminate (ε_r=3.55, tanδ=0.0027, thickness=0.41 mm) is only employed on top layer where the high-frequency signal content is routed. The feeding transmission lines on the motherboard and module side are 50 Ohm grounded coplanar waveguides (GCPWs). The BGA transition consists of 8 solder joints with a height of 1100 µm. The BGA joint pitch is 1.9 mm. One BGA joint is for a signal connection while the remaining seven are used for proper RF grounding and shielding. The design of a

wideband via transition through the entire LTCC substrate (ε_r=7.8, tanδ=0.0055) with thickness of 1.2 mm was adopted from [4].

Dimensions of the vertical BGA and via transitions are optimized with Ansoft-HFSS field simulator to provide nearly constant 50 Ohm transition path. In order to prevent excitation of detrimental package/substrate modes at the vertical transitions, it is critical to ensure sufficient electromagnetic shielding along the whole transition path. To maximize the lumped-element BGA transition bandwidth, it is necessary to minimize its associated parasitic reactances. Hence, to reduce capacitive loading in the BGA transition, an air-cavity with a size of 2x1.9x0.6 mm^3 was made on the top of the BGA signal pad, as depicted in Fig.1. For the same reason, the BGA signal pad sizes were reduced by 25 percent compared with the other joints, although being well aware that this may jeopardize the mechanical integrity of the RF signal joint under thermal loading. The EM-simulated S-parameters of the optimized BGA package transition is presented in Fig. 2. The simulated results predict excellent RF performance over a wide frequency range from DC up to 25 GHz. The simulated insertion and return loss values are better than 0.5 dB and 20 dB, respectively, up to 25 GHz.

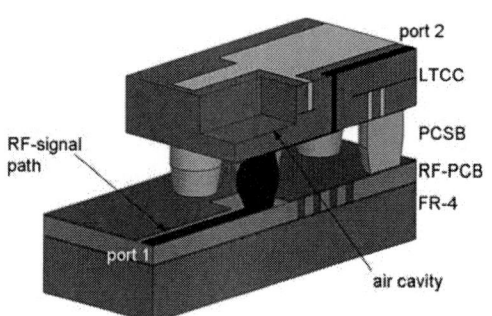

Figure 1: Designed wideband RF transition structure.

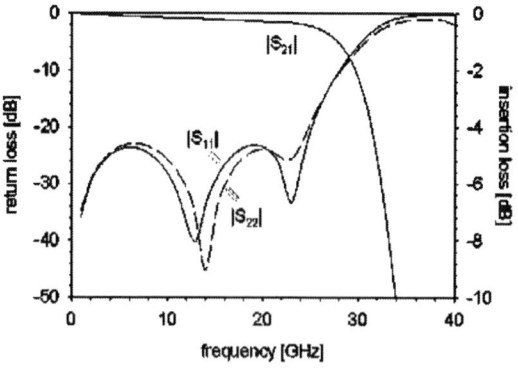

Figure 2: EM-simulated S-parameters of the optimized RF transition structure.

According to simulations, the RF performance of the BGA package transition is deteriorated soon after 25 GHz. In order to study the cause of this performance degradation, the electric-field distribution of the propagating wave mode in the transition structure is examined before and after the cut-off frequency (25 GHz). The electric-field magnitude distributions at 23 GHz and 37 GHz are presented in Fig. 3. The signal power is fed from the port 1 (i.e. from the motherboard side). It can be noticed that at 23 GHz (Fig. 3a) the power is transferred through the entire transition path, where as at 37 GHz (Fig. 3b) it is strongly reflected back from the BGA transition. This is explained that the BGA transition, approximated as a lumped element LC network, presents constant 50 Ohm impedance only over a finite bandwidth. Near the cut-off frequency the transition input impedance becomes increasingly inductive, and thus more power is reflected back. Hence, the results indicate that the BGA transition section limits the bandwidth of the complete transition structure.

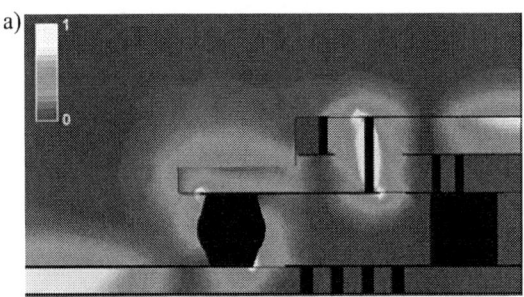

Fig. 3. Side view of the simulated electric-field magnitude distribution (a) at 23 GHz (b) and 37 GHz for the RF transition structure.

Description of the LTCC-BGA Module Platform

The developed module platform and an illustration of its partial area-array BGA joint configuration are presented in Fig. 4. The size of the LTCC module is 15 x 15 mm^2, while the module thickness is 1.2 mm (excluding the BGA joints). The module encompasses 38 DC/low-frequency I/Os and two broadband RF I/Os. For reliability testing purposes, the low-frequency BGA joints are interconnected with daisy-chain test pattern. The used BGA balls are PCSBs with a diameter of 1100 μm. The minimum BGA joint pitch is 1.9 mm.

a)

b)
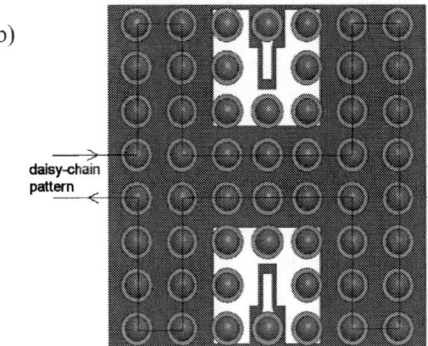

Fig. 4. (a) A photograph of the developed LTCC-BGA module platform, (b) and a schematic illustration of its BGA joint distribution with daisy-chain test pattern.

Experimental Procedures

LTCC/PCB test assembly composed of 16 LTCC-BGA modules was fabricated for validating the reliability and RF performance of the module platform. In order to investigate the effect of the cavities on the board-level reliability, 8 of 16 test modules were made without the cavities. The LTCC test panels were manufactured in a standard LTCC process with screen-printed conductor lines. A widely used and mature DuPont 951 LTCC material system with Ag metallization was employed. The fired thickness of the test panels was 1.2 mm.

The PCSBs were first attached on the LTCC modules using lead-free 'SOL1' solder paste. After that, the modules were mounted to the PCBs with OSP surface finish using standard surface-mount assembly processes and ternary Sn4Ag0.5Cu solder paste. In order to control solder flow on the PCB side, circular solder masks were printed onto the substrate. A photograph of the fabricated LTCC-BGA module assembly is shown in Fig. 4a.

Microwave measurements were carried out using an Agilent 8510C Network Analyzer, a Cascade Microtech probe station and air-coplanar probes with a pitch of 600 μm. The measurement system calibration was performed from 45 MHz up to 35 GHz using the ISS substrate and the LRRM calibration technique. To enhance the measurement accuracy, the parasitic capacitance and inductance effects of the measurement probe pads were de-embedded out using Cascade Microtech's "Wincal" calibration software. This necessitated fabricating additional open and shorted pad elements on both LTCC and PCB substrates.

In the TCT, the test assemblies were exposed to temperature cycles from -40°C to +125 °C, using one-hour cycle duration. A fifteen minutes dwell time at the temperature extremes was applied in the tests to allow creep/stress relaxation. DC resistance changes in the daisy-chained test assemblies were monitored using a Fluke 2635A data logger. Since a continuous resistance measurement cannot be done by the data logger, a test module was defined to be damaged when its initial resistance was doubled. To investigate the effect of thermal cycling on the performance of the RF transition structure, several transitions were measured after 250, 500, 1000 and 1400 cycles. It should be noted that the microwave measurements were performed only at the room temperature.

Scanning acoustic microscopy (SAM) was used to detect the initiation and propagation of cracks in the test modules. Test boards were taken out of the test chambers for SAM imaging after 100, 250, 500 and 1400 cycles. C-images were produced by using the 'interface scan' technique [5] and the determination of the crack paths and the failure modes related to them have been described in the previous study [6]. The equipment consisted of a Sonoscan D-9000 C-mode scanning acoustic microscope and 50 MHz and 100 MHz focused transducers (F# = 2, pulse-echo mode). Since the TCT was going on during the writing of this paper, no metallographic examination was done yet.

Results

Microwave Measurements

Fig. 5 shows the measured and EM simulated S-parameters of the optimized wideband RF transition structure. The measurements correlated well with the simulations. The measured insertion and return loss values were better than 0.8 dB and 15 dB, respectively, up to 25 GHz.

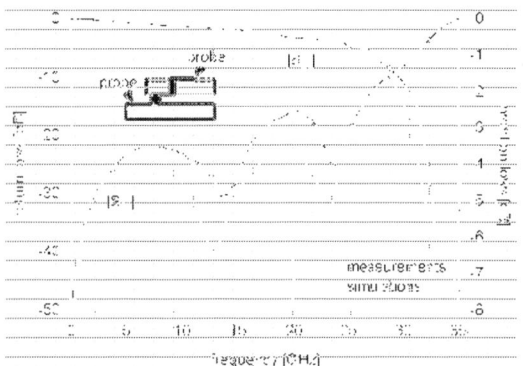

Fig. 5. Measured and EM simulated S-parameters of the RF transition structure.

Thermal Cycling Test

The lifetime determination of the test assemblies in the TCT was solely based on the DC resistance measurements. Since the TCT is still continuing at the time of writing this manuscript, the parameters of the Weibull distribution cannot be given. However, the failure-free time of the test modules with and without the cavities was 922 and 1241 cycles, respectively.

Microwave measurement results of the thermally cycled RF transition structure are presented in Fig. 6. Since the measured S-parameters varied very little between 0-1000 cycles, the results at 250 and 500 cycles were omitted from the figure for the sake of clarity. All the measured RF transitions worked properly up to 1000 cycles, showing no distinct indication of performance degradation. However, after 1400 cycles the measurements exhibited a strong detrimental resonance at ~3 GHz, as shown in Fig. 6. This phenomenon was consistent with several other RF transition structures.

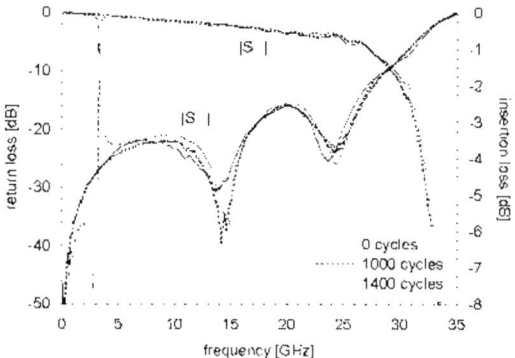

Fig. 6. Measured S-parameters of the RF transition structure after 0, 1000 and 1400 cycles.

In order to study this resonance phenomenon in more detail, additional EM simulations were performed on the RF transition structure with an attempt to capture the same resonance effect. This would allow determining the cause for the resonance. Since the RF signal and its two adjacent ground joints carry the main current flow (i.e. signal and its return current) through the BGA transition, they are then the most critical joints for the proper transition operation. Therefore, the RF transition structure was simulated with an open joint failure located in either the signal joint or its adjacent ground joint. The results are presented in Fig. 7. It can be noticed that the simulation results with the open ground joint match nicely with the measurements. Similarly, the simulation results with the open signal joint resemble to the response of a high-pass filter with 1-dB cut-off frequency at ~ 2 GHz. Hence, it can be concluded that the observed resonance were caused by the open failure in either of, or both, the ground joints, and not by the open RF signal joint.

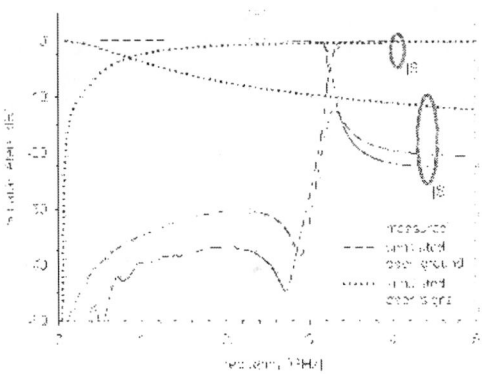

Fig. 7. EM simulated results of the RF transition structure with an open joint failure located in either RF signal or its adjacent ground joint.

SAM Analysis

The SAM imaging after 100 cycles showed that cracks located at the inner edge of the daisy-chained solder joints and the growth direction of a crack was away from the neutral point of a module, as shown in Fig. 8. Other cracks propagating to the opposite direction can be seen after 500 cycles, but their propagation rate seems to be notably slower compared with the cracks located at the inner edges of the joints (Fig. 8c). The large damaged area can be seen in joints after 1400, as shown in Fig. 8d.

It was also noticed that the cracks propagated similarly in the RF signal joints although the joint geometry was different compared with the other joints (Fig. 9). The condition of the joints used in RF grounding was impossible to monitor, since multiple grounding vias in the middle of the substrate obstructed the propagation of the acoustic signal.

According to the SAM investigation, only a few ceramic cracks existed in the joints of the test modules. Moreover, the ceramic cracks seemed to be randomly located in the joint matrix. These observations suggest that the primary failures detected by DC resistance measurements are located in the solder joints and the occasional ceramic cracks do not have significant effect on the lifetime of the test modules.

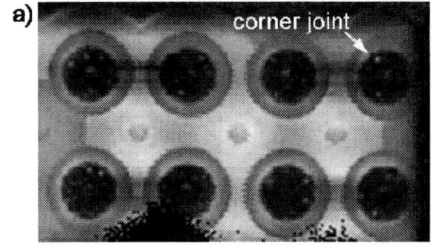

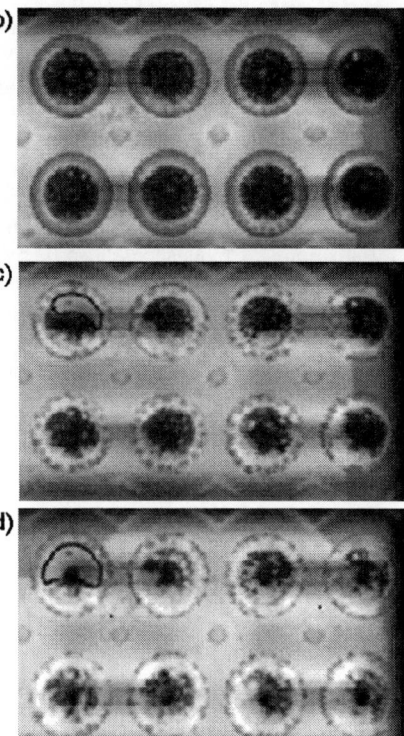

Fig. 8. SAM imaging showing (a) the initial condition, the position and growth direction of the cracks in daisy-chained solder joints after (b) 100 cycles, (c) 500 cycles, and (d) 1400 cycles.

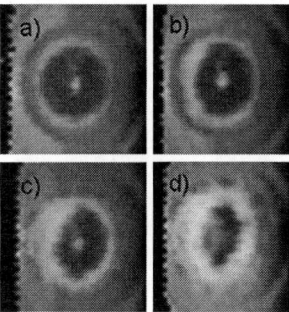

Fig. 9. SAM imaging showing (a) the initial condition and (b) the position and growth direction of the cracks in a RF signal joint after (b) 100 cycles, (c) 500 cycles, and (d) 1400 cycles.

Discussions

The microwave measurements verified that the developed RF transition structure provided excellent electrical signal transmission properties up to 25 GHz. Also, the microwave measurements performed on the thermally cycled RF transition structures showed that their performance remained practically the same up to at least 1000 cycles. After 1400 cycles, the measurements exhibited the strong detrimental resonance at ~3 GHz. According to the EM simulations the observed resonance was caused by the open failure in the RF ground joint(s). The validity of this will also be checked by SEM investigations after the TCT. Although all the measured RF transition structures were fully functional at least up to 1000 cycles, it must be noted that the measurements were made at the room temperature, which may result in overly optimistic estimations of the RF performance. Hence in future, the RF transition structures should also be measured during the TCTs in order to clarify the effect of joint cracks on the RF performance. However, the results indicated that the thermal-fatigue performance of the RF signal joints with the reduced solder pad size were comparable with the other joints in thermal loading. Thus at this point, it seems that there is no need to enlarge the solder pad size in the RF signal joint.

The results of DC resistance measurements showed that the failure-free time of both test module assemblies (i.e. with and without the cavities) was over 900 cycles, which is at acceptable level for many telecom applications [7]. On the basis of above mentioned results, the use of larger 1100 µm PCSBs and 'SOL1' solder paste on the LTCC side, allowed to significantly improve the board-level reliability performance when compared with [3], although the number of composite solder joints were reduced by 30%. The results suggested also that the LTCC module assembly without the cavities has better life-time duration in the TCT than the modules with the cavities. Thus, the cavity structure may have an adverse effect on the stress distribution in the present LTCC-BGA module assemblies. However, the effect of the cavity on the board-level reliability of the LTCC module cannot be adequately determined from the present data, but more research work is needed on this issue.

The SAM results indicated that the two separate cracks were formed at the interface of the composite solder joints and the LTCC substrate. The crack that propagated from the inner edge to the center of a joint was related to the failure induced at the high (125 °C) temperature extreme. Such failure is typically located at the IMC/solder interface of LTCC modules with Ag-based metallization due to the inadequate adhesion [3]. Since this failure mode seemed to be dominant according to the SAM results, it is assumed that it also governed the reliability performance of the present assemblies. This assumption will be verified in metallographic examinations later on. Further work is planned to test more advantageous solder material/pad metallization pairs in order to further enhance the board-level reliability of the developed LTCC-BGA module platform.

Summary

This paper reported the development of a reliable LTCC-BGA module platform, which can be used in various broadband RF/mixed-signal LTCC-SIP packaging applications.

The wideband RF transition structure included in the module platform demonstrated excellent RF performance up to the K_a-band frequencies. The measured results exhibited return and insertion loss values better than 15 dB and 0.8 dB, respectively, up to 25 GHz.

The board-level reliability of the LTCC-BGA module platform was managed to raise to at adequate level regardless of the large global thermal mismatch between the used LTCC and organic substrate materials. This was achieved by using the PCSBs as large as 1100 µm and applying proper solder material on the LTCC side.

Acknowledgements

The authors acknowledge the financial support of Tekes (the Finnish Funding Agency for Technology and Innovation) under the ACERMI project (40114/06) and the Graduate School of Infotech Oulu. Technical support from Selmic Oy, Finland is appreciated. Also, Sekisui Chemical Co., Ltd., Japan, is gratefully acknowledged for providing the PCSBs in this experiment.

References

[1] T. Kangasvieri, et al., "An Ultra-Wideband BGA-Via Transition for High-Speed Digital and Millimeter-Wave Packaging Applications", Accepted to IEEE MTT-S Int. Microw. Symp. Dig., 2007.

[2] A. Ziroff,, et al.: 'A novel approach for LTCC packaging using a PBG structure for shielding and package mode suppression', IEEE EuMC Conf., pp. 419-422, 2003.

[3] O. Nousiainen, et al., "Failure mechanisms of thermomechanically loaded SAC/PCSB composite joints in LTCC/PWB assemblies" Accepted to Journal of Electronic Materials.

[4] T. Kangasvieri, et al., "Ultra-Wideband Shielded Vertical Via Transitions from DC up to the V-band", IEEE EuMIC Conf., pp.476-479, 2006.

[5] J.E. Semmens and L.W. Kessler, "Characterization of Flip Chip Interconnect Failure Modes using High Frequency Acoustic Micro Imaging with Correlative Analysis", in Proc. of IEEE IRPS Conf., pp. 142 - 148, 1997.

[6] O. Nousiainen, et al., "Acoustic Micro Imaging of Fatigue Cracking in Solder Joints of LTCC Modules", in Proc. of the IMAPS EMPC Conf., Strasbourg, pp. 172-177, 2001.

[7] R. Darveaux and A. Syed, "A. Reliability of Area Array Solder Joints in Bending", In Proc. of the Surface Mount Technology Association (SMTA), p. 313-324, 2000.

2D and 3D X-ray Inspection for Nano-technology

Keith Bryant, David Bernard

Dage X-ray Systems, Rabans Lane, Aylesbury, HP19 8RG, England

E-mail k.bryant@dage-group.com, d.bernard@dage-group,com

Abstract

I should start by pointing out that the wording "nano-technology" has become very popular and is used to describe many types of research where the characteristic dimensions are less than about 1,000 nano-meters. For example, continued improvements in lithography have resulted in line widths that are less than one micron: this work is often called "nano-technology." This is the definition I shall be using; I will not be addressing atomic level structure growth for biomechanics or similar technologies that are not linked to electronics.

Introduction

The arrival of nano-technology

The size of a single transistor has been reducing in an exponential manner for several decades, leading to integrated circuits containing tens of millions of transistors. But as the size of the transistor decreases a physical limit is encountered where the transistor becomes too small and quantum effects become significant. When this limit is reached the exponential growth in computing power that has been characteristic of the 1980s and 1990s will come to an end. This event is expected to occur somewhere between 2010 and 2020. This will be the end of the road for pure silicon technology. At this point completely new technologies will be needed. So the question is, what is coming next?

Micro-Electro-Mechanical Systems (MEMS) is the integration of mechanical elements, sensors, actuators, and electronics on a common silicon substrate through microfabrication technology. While the electronics are fabricated using integrated circuit (IC) process sequences (e.g., CMOS, Bipolar, or BICMOS processes), the micromechanical components are fabricated using compatible " micromachining" processes that selectively etch away parts of the silicon wafer or add new structural Layers to form the mechanical and electromechanical devices.

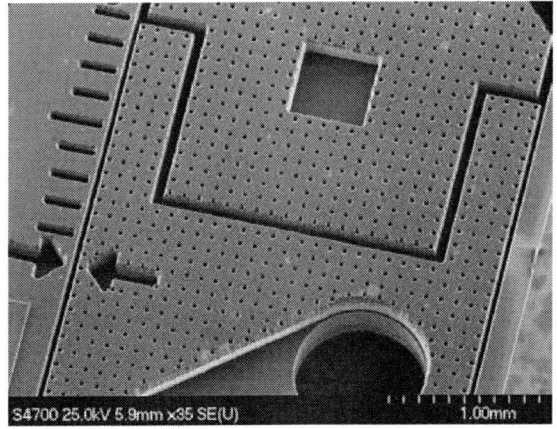

MEMS promises to revolutionise nearly every product category by bringing together silicon-based microelectronics with micromachining technology, making possible the realisation of complete **systems-on-a-chip**. MEMS is an enabling technology allowing the development of smart products, augmenting the computational ability of microelectronics with the perception and control capabilities of microsensors and microactuators and expanding the space of possible designs and applications.

Microelectronic integrated circuits can be thought of as the "brains" of a system and MEMS augments this decision-making capability with "eyes" and "arms", to allow microsystems to sense and control the environment. Sensors gather information from the environment through measuring mechanical, thermal, biological, chemical, optical, and magnetic phenomena. The electronics then process the information derived from the sensors and through some decision making capability direct the actuators to respond by moving, positioning, regulating, pumping, and filtering, thereby controlling the environment for some desired outcome or purpose. Because MEMS devices are manufactured using batch fabrication techniques similar to those used for integrated circuits, unprecedented levels of functionality, reliability, and sophistication can be placed on a small silicon chip at a relatively low cost.

There are numerous possible applications for MEMS and Nano-technology. As a breakthrough technology, allowing unparalleled synergy between previously unrelated fields such as biology and microelectronics, many new MEMS and Nanotechnology applications will emerge, expanding beyond that which is currently identified or known. Here are a few applications of current interest:

Challenges for X-ray Technology

In a similar way to the silicon technology coming to a technology plateau, the current open tube, analogue x-ray system technology has a "real life" minimum feature recognition of around 3,0000 nano-meters or 3 microns.

Standard imaging technology and existing greyscale resolution is also limited in this area, with 8 bit technology, limited pixel counts and poor greyscale sensitivity mean that differentiating between very small and similar density materials is not possible.

In summary the established equipment will not be suitable for nano-technology inspection.

The way forward

Digital technology brings with it a quantum leap in x-ray ability overcoming the issues of grey scale sensitivity and pixel count, these systems can have up to 65,000 greyscale levels and run 12 or 16 bit technology

.

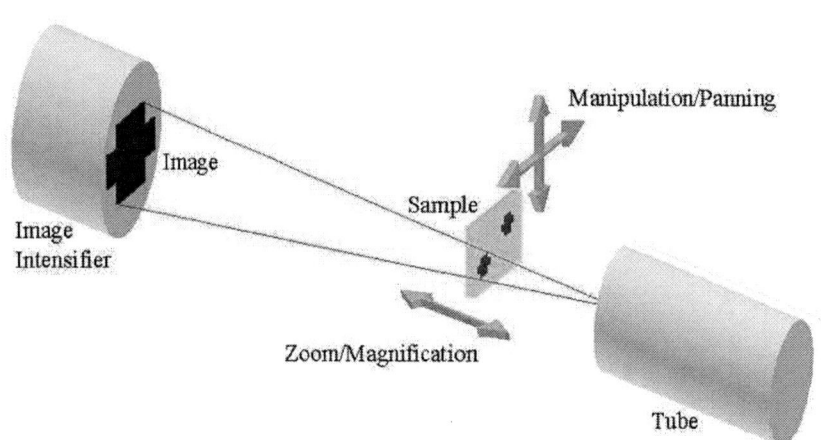

The pathway from the Image Intensifier to the monitor is digital, whereas the earlier systems processed the analogue signal within the PC. This also allows smaller features to become visible and the minimum feature recognition moves down to around 600 nanometers or 0.6 microns. This feature recognition is a major step forward but not really good enough for nano-technology inspection, unfortunately this level is at the limit of existing open tube technology.

The image can be significantly improved by removing vibration from the sample tray. Vibration can be a major contributing factor to degradation of image quality at high magnification, this is because the image on the screen is an average of many "real time" pictures and the more stable the sample, the clearer the image. Mounting the sample table, x-ray source and digital Image Intensifier on active air

mounts dramatically reduces vibration, both from external sources and within the equipment.

This system of vibration damping improved the quality of the image, allowing features which were not previously visible to be seen easily, the picture below shows a clearance area and via within a wafer.

The future of X-ray technology

Clearly if open tube technology has reached the limit of the envelope at about 600 nano-meter feature recognition, we cannot move further without a new method of producing and controlling x-rays.

This technology has been developed, unfortunately I cannot go into too much detail on the technology, suffice to say that it is a filament free system with the x-ray source contained in a sealed for life vacuum chamber. However it is not similar to the older "closed tube" systems that were replaced by "open tubes" due to their lack of magnification and resolution.

This exciting new technology can give feature recognition below 250 nano-meters together with superior resolution and improved grey scale sensitivity. The resultant images are a major improvement on current x-ray technology allowing inspection of MEMS and other nano-technologies together with other small features including die cracks.

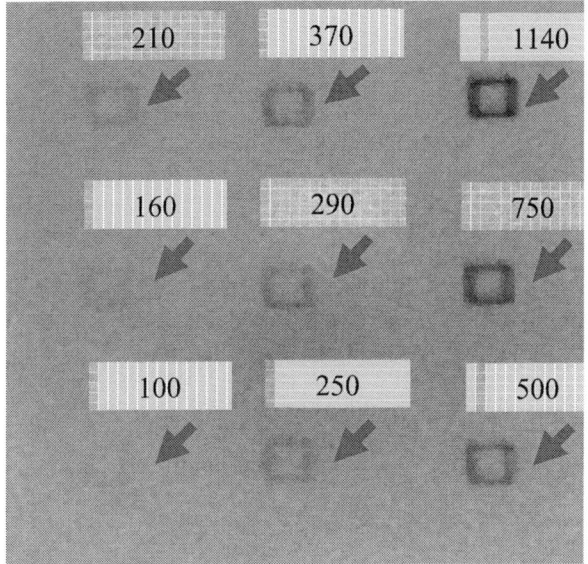

This special sample, whose dimensions have been independently verified by SEM, has 3 rows of three 10-micron squares. The first column (right hand side) from the top down has wall thickness of 1140nm, 750nm and 500nm. Second column (middle) is 370nm, 290nm and 250nm. Third column is 210nm, 160nm and 100nm.

At 250 nm wall size, the walls of this square are only 900,000 atoms wide!

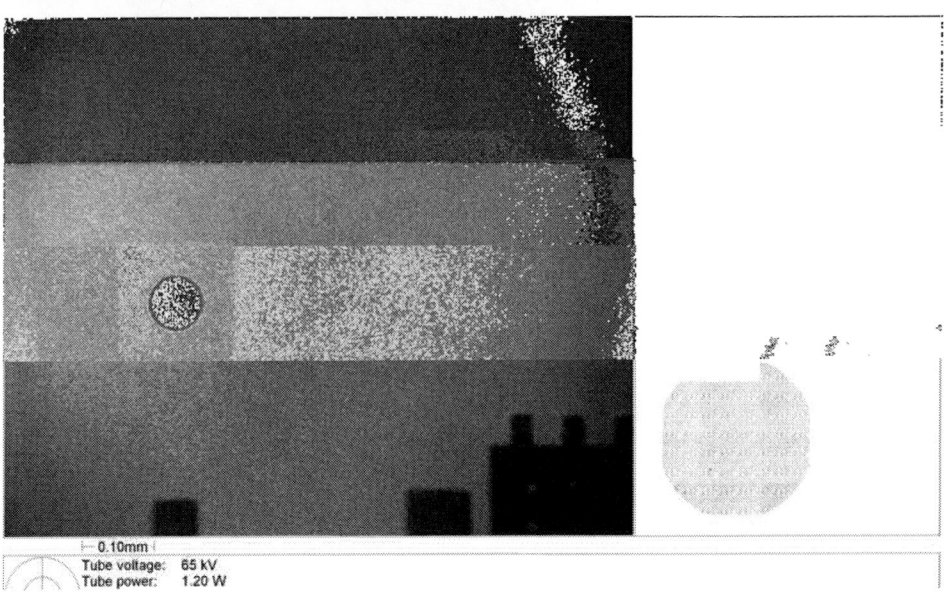

Computerized Tomography

Computerized tomography is an imaging method whereby computational geometric processing is used to generate a three-dimensional (3D) model of an object from a large series of individual two-dimensional (2D) x-ray images, taken about a single axis of rotation. Since its invention in 1972, CT has emerged as a transformational technology that has aided medical diagnosticians around the world to predict, diagnose and treat disease - through the 'CAT scan'. This is the same technique that is being applied to inspecting these advanced 3D packages.

Until recently, the use of CT for semiconductor and 3D package inspection has been limited by its slow computer processing times, low resolution and relatively high expense. While some CT systems have been employed in the past for semiconductor inspection, these conventional systems typically have not delivered the analytical performance required for such critical, high-density applications.

However, recent advancements in CT have taken place that greatly improve the imaging speed, increase the resolution to a suitable level to allow complete analysis of the most detailed package features, and all at a reasonable price! Therefore, computerized tomography is now an ideal inspection methodology for complex 3D packages since it generates a 3D model of the entire electronic package.

The resulting 3D model can be viewed with real-time manipulation so that interconnections normally obscured by other joints or components within the package can be diagnosed, assuring complete package inspection A CT model is developed computationally from a series of 2D x-ray images taken as the sample, or semiconductor device, is rotated in an x-ray beam. The density variations within those images, and how their relative locations change as the sample rotates, are evaluated in a computer to reconstruct a 3D model of the sample. This 3D model can then be viewed and manipulated, so as to provide analytical images, or slices, through any 2D plane in the object CT model. The final, and vital, element of the CT process is the advanced processing software necessary to manipulate the reconstructed 3D model and facilitate visualization of the necessary slices for the correct analytical view. It is this critical step that allows the user to see the density contours within the sample CT model and change the viewed slice.

The latest CT systems provide hardware accelerator graphics cards in the PC to specifically handle the manipulation and rendering of the CT model in real time. All of this can be viewed off-line of the image acquisition, so allowing analysis of one model while the 2D images of the next sample are being acquired. In this way individual slices can be viewed through any plane in the model

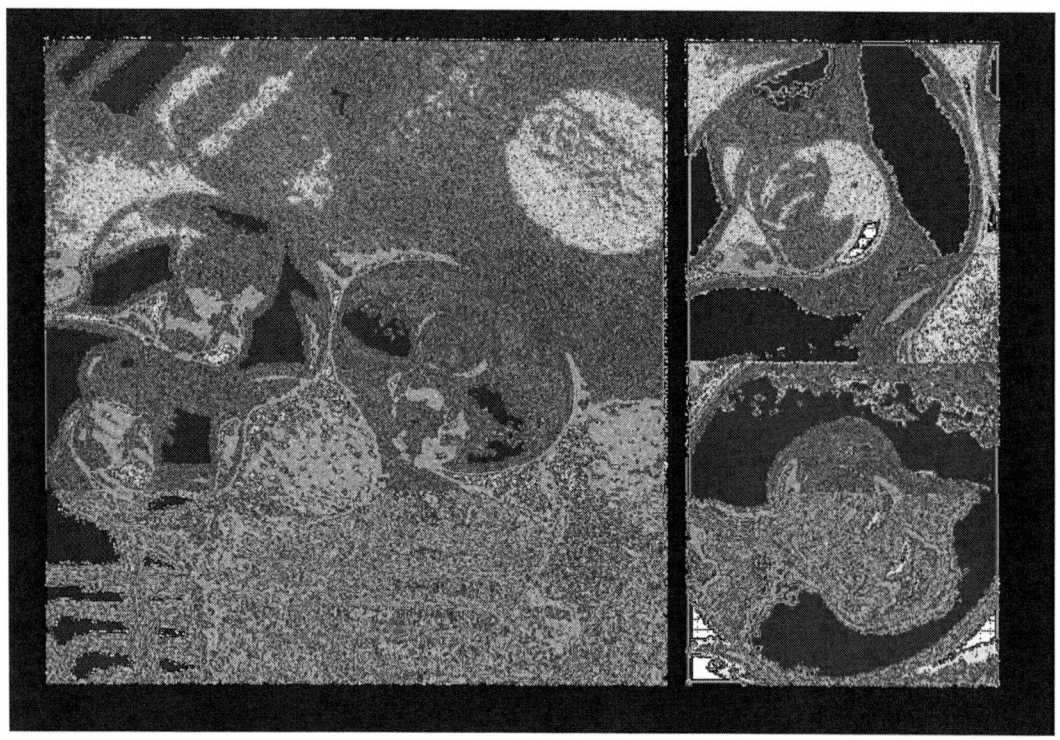

AUTHOR INDEX

Aasmundtveit, Knut..............687
Abraham, Eitan..............140
Afra, B...........370
Alajoki, T...........239, 250
Alastalo, Ari..............1
Allen, Mark 467
Andrew, Marshall 606
Arndt, D..........376
Arra, Minna..........570
Arruda, Luciano..........447
Aschenbrenner, R..........324
Axisa, F..........681
Bailey, C..........589
Ball, Ken 3
Ballandras, S. 152
Balogh, Balint 452
Barinka, R. 85, 548
Barnwell, P. 35
Bartek, M..........254, 299
Batchelor, John 45, 500
Bauer, C..........125, 639
Beh, Jiun Kai..........103
Belavic, Darko 304
Beleran, John D..........346
Bellenger, S..........265
Bergman, Ruben 606
Bernard, David..........699
Beyne, Eric..........340
Biermans, P. 265
Blanc, D..........239
Bohlmann, Helge..........617
Bonato, Paolo..........505
Bonfert, D..........165
Bos, Arnold..........163
Boschman, Frank 163
Bosman, E..........146
Bossuyt, F..........681
Bousek, Jaroslav..........85
Bousquet, S. 406
Brebels, S..........25
Brosteaux, D..........681
Bryant, Keith 699
Buoli, Carlo 505
Burghoorn, M..........352
Burkard, H. 611
Buttiglione, Roberta..........364
Caccioli, Danilo 130
Canali, Arturo..........130
Carr, John 140
Catoni, Simone 364
Cauwe, M..........309, 330
Chaillot, A..........406
Chastanet, C. 406
Chatras, M..........205
Chen, Chang-Sheng 19

Chen, L. C. 25
Chen, Wei-Ting 19
Chin, Kuo-Chiang 19
Choi, Seogmoon 644
Chong, C. 152
Christiaens, W. 25, 146, 611
Chung, C. L. 171
Cianci, Elena 364
Cilensek, Jena 304
Clark, Stephen 435
Codreanu, Norocel Dragos 98
Collander, Paul 280, 606
Conway, Paul P. 50
Coppa, Andrea 364
Corradini, G. 179
Cortese, Mario 194
Couderc, Pascal 31
Cumini, Anne 473
Currle, Ulrike 416
De Baets, J. 330, 336
De Langlade, Renaud 673
De Leersnyder, E. 681
De Raedt, Walter 340
Debaes, Christof 73, 79
Desmet, N. 681
Desmulliez, Marc P. Y..........140, 153
Destouches, N. 239
Detert, Markus 426
Dietrich, L. 458
Dispenza, Massimiliano..........364
Dong, Yujie 601
Drnovsek, Silvo 304
Drue, K.-H. 495
Duffer, Scott 517
Dumonteil, R. 406
Dvorak, T. 95
Effenberger, Erwin..........260
Egbert, Andre 585
Ehlert, M. 35
Ehrmann, O. 113, 458
Elle, Ole Jakob..........687
Erho, Tomi 1
Farmer, Graham 452
Feiertag, G. 125
Feller, Claudia 200
Ferhanoglu, O. 289
Feurer, E. 536
Fiorello, Anna Maria..........364
Fleischer, D. 376
Foglietti, Vittorio 364
Fontanelli, Anna 13
Fosse, Erik..........687
Fournier, Y. 293, 370, 540
Franc, J. 239
Free, Charles..........45, 500

AUTHOR INDEX

Frisk, Laura 473
Fritzsch, T. 113, 458
Fu, S. L. 171
Fujita, H. 152
Funck, H. 595
Gal, Laszlo 650
Gao, Shan 644
Gautier, W. 625
Gebhardt, S. 189, 376
Geerinck, P. 146
Geißler, U. 119
Georgi, H. 376
Giesbers, B. 625
Gieser, H. 165
Girulska, Anna 452
Golonka, Leszek J. 159
Golumbeanu, Virgil 98
Gonzalez, M. 681
Gordon, Peter 452
Gotzen, Reiner 617
Grech, Anne Marie 194
Grob, T. 352
Groger, Barbara 90
Gruchow, M. 376
Gunde, Marta Klanjsek 384
Günther, Michael 511
Hagberg, Juha 576
Halvorsen, Steinar 687
Hapenciuc, Iaroslav-Andrei 175
Harant, P. 422
Harkonen, Ilkka 570
Harvey, Tom 452
Hast, Jukka 1
Hauck, Karin 107
Hauptman, Nina 384
Hedges, Martin 580
Hegr, Ondrej 85
Heikkinen, Veli 662
Hein, M. 285, 495
Hendrickx, N. 68, 73, 146, 239, 250
Henry, M. 45, 500
Henttinen, K. 152
Hetschel, Thomas 395
Hillmann, Gerhard 467
Hin, Tze Yang 50
Hladik, J. 545, 548
Hlavka, J. 389
Hoff, Lars 687
Hogetveit, Jan Olav 687
Holc, Janez 304
Holl, B 536
Hong, Jupyo 644
Hong, Tan H. 346
Hoskio, P. 275
Hrdy, Radim 207

Hrovat, Marko 304
Hu, Tao 285
Hubalek, Jaromir 207, 211
Humpston, Giles 633
Hurme, Eero 1
Hurtig, Johannes 556
Husa, Ellen M. 687
Ikarashi, Sen-Ichi 270
Illyefalvi-Vitez, Zsolt 441, 452, 650
Imenes, Kristin 687
Ionescu, Ciprian 98
Isamoto, K. 152
Iszquierdo, Benito Sanz 45, 500
Jaakola, Tuomo 467
Jacot, J. 540
Jacq, C. 179, 551
Jantunen, H. 285
Jerlah, Mitja 304
Johannessen, Rolf 668
Johner, N. 179, 370, 551
Johnson, M. 136
Jones, Alun 3
Jordan, Rafael 113
Jourdain, A. 625
Jourlin, Y. 136
Juntunen, Eveliina 467, 662
Jurkow, Dominik 159
Kabadi, Ashok N. 479
Kaija, Kimmo 56
Kaiser, A. 536
Kangasvieri, Tero 693
Kansakoski, Markku 1
Karila, Tanja 354
Karioja, P. 136, 239, 250, 275, 289, 595
Karppinen, M. 136, 250, 595
Kaskiala, Toni 467
Kattelus, H. 152
Kautio, K. 364, 595, 662, 693
Kekonen, Teija 576
Kemppainen, Antti 1, 662
Kemppainen, Timo 662
Kenny, Stephen 560
Keranen, K. 136, 275
Kidd, Matt 140
Kim, Jinsu 644
Kim, S. 677
King, Bruce 580
Kirkpatrick, Damien 452
Kita, Jaroslaw 359
Kivela, Sari 467
Klein, M. 677
Klink, G. 165
Klosova, Katerina 207, 211
Koh, Y. C. 346
Kolan, Ravi 346

AUTHOR INDEX

Kololuoma, Terho 1
Komagata, Michinori 270
Kopola, Harri 1
Korhonen, Pentti 364, 662
Korkala, Heikki 662
Kosec, Marija 304
Kostner, H. 336
Kotora, Gyorgy 452
Krammer, Oliver 441, 650
Kretzschmar, Christel 200
Kriechbaum, A. 315, 336
Krueger, Klaus 416
Kruger, H. 125
Kuchiki, Mikiharu 56
Kuhner, Thomas 245
Kulawik, J. 183
Kulke, R. 40
Kutilainen, Terho 576, 662
Kuusisto, Jani-Mikael 1
Lahdes, Manu 364
Lahti, Markku 693
Laine, Eric 107
Lang, K.-D. 119
Langton, Conrad 140
Lee, S. W. Ricky 435
Lehto, A. 275
Lenkkeri, Jaakko 467
Link, J. 611
Liu, Changqing 50
Liukkonen, Timo 556
Loeher, Thomas 620
Lombaert-Valot, I. 406
Lorenzotti, Stefano 130
Low, Richard C. 221
Lozinski, Andrzej 526
Lu, H. 589
Luetzelschwab, Markus 153
Maaninen, Arto 1
Macek, Marijan 384
Mach, M. 656
Mackrodt, Wolfgang 3
Maeder, T. 179, 293, 370, 540, 551
Maggi, Luca 130
Maire, O. 406
Makinen, J. T. 136, 275
Malek, C. Khan 152
Manessis, D. 107, 309, 324, 620
Mansikkamaki, Pauliina 56, 227
Mantysalo, Matti 56, 227
Marcelli, Romolo 364
Marenco, N. 630
Maron, D. 406
Mathewson, Alan 467
Matters-Kammerer, M. 625
Maulwurf, K. 40

McFarland, Geoff 140
McKee, Andrew 140
McKendrick, David 140
Mehta, Gaurav 346
Meredith, Wyn 140
Meyer-Berg, Georg 3
Midl, Manuela 62
Miessner, Ralf 221
Miettinen, Jani 56, 227
Milani, A. G. 490
Mills, J. B. 625
Miyashiro, Fumio 7
Mizumura, Noritsuka 270
Mollenbeck, G. 40
Moos, Ralf 359
Moreno, J.-O. 265
Morgia, Fabio 505
Morosawa, A. 152
Mukerjee, Indro 264
Muller, J. 285, 495, 656
Munier, C. 406
Munier, E. 406
Nauwelaers, B. 625
Navarova, H. 389
Negri, Luigi 505
Nemeth, Pal 441
Neubert, H. 376
Neuhaus, H. J. 639
Neuilly, F. 265
Newman, Keith 435
Nigg, P. 205
Nousiainen, Olli 693
Novotny, M. 95
Nummila, Kaj 467
Obi, Samuel 250
Ocket, I. 625
Oezkoek, Mustafa 530
Ohlckers, Per 668
Ojapalo, A. 275
Okoro, C. 25
Oldervoll, Froydis 668
Ollila, J. 275, 289, 364, 662
Onda, N. 205
Ong, Wilson 346
Oppermann, H. 677
Oppermann, Martin 233
Oprins, H. 25
Ostmann, A. 107, 309, 324, 336, 620
Paakkonen, Esko J. 570
Palmu, Leena 432
Parriaux, O. 136
Partsch, U. 189, 376
Pavelka, J. 389
Peels, W. 320
Pekkanen, Ville 227

AUTHOR INDEX

Pennanen, Virpi 601
Perrone, R. 495
Petaja, J. 595
Pierce, B. 35
Piloni, M. 490
Pinkola, Janos 441, 650
Plouseau, D. 406
Pochesci, Daniele 364
Polet, Markus 523
Pommier, M. 265
Poruba, Ales 85
Pressel, Klaus 3
Preve, Gianni 130
Pudas, Marko 523
Quinones, Horatio 216
Qvintus-Leino, Pia 1
Raynal, P. 406
Reents, Bert 560
Reichl, H. 113, 119, 324, 458, 620, 677
Reinert, W. 630
Reinhardt, Andrea 617
Reinhardt, Kathrin 200
Reitbauer, Roland 62
Remes, Kari 432
Renn, Mike 580
Rentsch, S. 285, 495
Riegler, Bill 517
Rittweger, M. 40
Rizvi, M. J. 589
Rodig, Thomas 189
Roggen, Jean 3
Roguszczak, Henryk 159
Ronka, Kari 364
Ronkainen, Petri 432
Ronkka, Risto 56
Roozeboom, F. 352
Ruess, K. 536
Ruhmer, Klaus 107
Ruuhonen, Visa 601
Ryser, P. 179, 293, 370, 551
Sahavirta, Hannu 662
Salmi, T. 275
Sanders, F. 352
Sauer, Wilfried 233
Scheel, W. 119
Schischke, Karsten 412
Schneider, Marc 245
Schneider-Ramelow, M. 119
Schonecker, Andreas J. 189
Schonlinner, B. 625
Schorpp, M. 275
Schuetz, Reiner 3
Sedlakova, V. 389, 564
Seigneur, F. 540
Seleznev, Maxim 673

Serrano, Christophe 31
Shaw, Mark 194
Shirai, Yukio 270
Shrier, Karen 280
Shyu, Chin-Sun 19
Sikula, J. 389
Silvennoinen, M. 275
Sinaga, S. 254
Sita, Z. 389
Sitomaniemi, Aila 662
Skwarek, A. 401
Sobota, Jaroslav 85
Sodergard, Caj 1
Solberg, Vern 633
Solzbacher, F. 677
Sommer, Johann-Peter 309
Song, Fubin 435
Sosin, S. 299
Steiner, F. 422
Stephan, R. 285, 495
Stocksreiter, Wolfgang 62
Stoukatch, Serguei 340
Strisland, Frode 668
Suni, T. 152
Svasta, P. 98, 165, 175
Sykes, Bob 435
Szendiuch, I. 95, 412, 545, 548
Szwagierczak, D. 183
Tammenmaa, Markku 601
Tathireddy, P. 677
Thienpont, H. 68, 73, 79, 239, 250
Tian, J. 254, 299
Töpper, M. 107, 458, 677
Torfs, Tom 340
Toshiyoshi, H. 152
Toy, M. F. 289
Trabert, J. 495
Tsai, Cheng-Hua 19
Tschanun, Wolfgang 205
Tsubouchi, Mikihiko 56
Tuominen, Jarkko 364
Uhlig, P. 40
Urey, H. 289
Ursic, Hana 304
Vahakangas, Jouko 523, 693
Val, Christian 31
Van Daele, P. 68, 73, 146, 239, 250
Van Den Heuvel, F. 352
Van Der Lugt, A. 320
Van Erps, J. 68, 73, 239, 250
Van Grunsven, E. 352
Van Hoof, Chris 340
Van Steenberge, G. 146
Van Weelden, Ton 163
Vanden Bulcke, M. 681

AUTHOR INDEX

Vandevelde, B. 25
Vanek, J. 545
Vanfleteren, J. 25, 146, 611, 681
Vanselow, J. 536
Vasko, Cyril 412
Vatanparast, R. 595
Velderrain, Michelle 517
Vervaeke, Michael 79
Villard, S. 406
Vincent, T. 35
Warnat, S. 630
Wei, Chang-Lin 19
Weiland, Dominik 153
Weston, Nick 140
White, Nigel 530
Wiedmer, S. 293
Wierzba, Pawe- 526
Wilde, Jurgen 221
Wilke, M. 677
Willems, G. 336
Winters, Christophe 340
Witek, K. 401
Wolf, H. 165
Wolter, Klaus-Jurgen 233, 395, 426, 511
Yannou, J.-M. 265
Yen, S-F. 324
Yi, Sung 644
Yin, C. Y. 589
Young, Paul 45, 500
Zaraska, Krzysztof 90
Zarnik, Marina Santo 304
Zerna, Thomas 233, 426
Ziglioli, Federico 194

International Microelectronics and Packaging Society, Europe
PO Box 277
SE-431 24 Molndal, Sweden

ISBN 978-1-62276-466-2